GATE PREP SERIES

GATE

Graduate Aptitude Test in Engineering

2022

Topic-wise

Previous Solved Papers

35 Years'

Solved Papers

Mechanical Engineering

G. K. PUBLICATIONS (P) LTD.

Title : GATE 2022 : Mechanical Engineering 35 Years Topic-wise Previous Solved Papers

Language : English

Editor's Name : Vinit Garg

Typeset & Published by :

Career Launcher Infrastructure (P) Ltd.

A-45, Mohan Cooperative Industrial Area, Near Mohan Estate Metro Station, New Delhi - 110044

Marketed by :

G.K. Publications (P) Ltd.

Plot No. 9A, Sector-27A, Mathura Road, Faridabad, Haryana-121003

ISBN : **978-93-90820-91-7**

Printer's Details : Made in India, New Delhi.

For product information :
Visit ***www.gkpublications.com*** or email to ***gkp@gkpublications.com***

Contents

Unit V Machine Design

Unit VI Fluid Mechanics and Hydraulic Machinery

Unit VII Heat-Transfer

Unit VIII Thermodynamics, Power Plant Engineering and Internal Combustion Engines

Unit IX Refrigeration and Air-conditioning

Unit X Materials Science and Manufacturing Engineering

Preface

Whether you want to work with esteemed PSUs including the Maharatna and Navratna companies or do your Masters from IITs and IISc, GATE is your entry to career and academic avenues post your graduation. Thousands of students write the GATE paper annually. The level of competition is fierce, owing to the increasing competition every year for a limited number of seats. If you are a serious aspirant, it is advisable to start preparing for GATE from the second year of your Graduation with the right books.

GK Publications has been the guiding force for GATE aspirants for the last 3 decades. Better known as the **"Publisher of Choice"** to students preparing for GATE and other technical test prep examinations in the country, we published the first set of books in 1994 when GATE exam, both objective and conventional, was conducted in the paper and pencil environment and used as a checkpoint for entry to postgraduate courses in IITs and IISCs. At that time, students had little access to technology and relied mainly on instructor-led learning followed by practice with books available for these examinations.

A lot has changed since then!

Today, GATE is conducted in an online-only mode with multiple choice and numerical based questions.The exam pattern too has evolved over the years. Instead of theory, the questions today test a student's basic understanding and his capacity to put fundamentals to practical applications.The exam only rewards those with a concrete foundational clarity of basics.

We, at GKP, have also embraced change. Today, our preparatory manuals are not mere books but torchbearers in your GATE preparation journey. We understand that tackling the syllabus one chapter at a time is not only easy but also allows students to cope with the looming exam pressure. A major game changer is the habit to practice and solve previous year questions and this is why, our **GATE 2022 Topic-wise Solved Papers of Mechanical Engineering** is your best bet to be GATE READY by 2022!

Each chapter carries detailed analysis of 5 previous years' solved papers (all sets) so that you are well-versed with the latest exam pattern and how it has evolved over time. Both MCQs and Numerical have been explained in a lucid and detailed fashion, so that you take your preparation forward - one question at a time.

Moreover, the Manufacturing Engineering Unit is supplemented with solved questions from Production Engineering. This will ensure that you can prepare for the toughest portion of your syllabus in a comprehensive, well-planned manner.

With each question, its marks' weightage has also been given so that you can gauge the level of question and make judicious use of your time.

Today's students have easy access to technology and the concept of a monologue within the classroom has changed to dialogue where students come prepared with concepts and then discuss topics. We have launched a web portal and mobile app to help you take your preparation on the go - so that you learn, revise and practice mock tests online.

The mobile app provides access to video lectures, short tests and provides regular updates about the exam. The web portal in addition to what is available on the app provides full-length mock tests to mimic the actual exam to help you gauge your level of preparedness. The combination of practice content in print, video lectures, and short and full-length tests on mobile and web make this product complete courseware for GATE preparation.

We also know that improvement is a never-ending process and hence we welcome your suggestions and feedback or spelling and technical errors if any. Please write to us at gkp@gkpublications.com

We hope that our small effort will help you prepare well for the examination.

We wish you all the best!

Team GKP

GATE

Graduate Aptitude Test in Engineering

IIT Institutes

GATE 2021 will be conducted by
Indian Institute of Technology, Mumbai

Indian Institute of Technology, Kharagpur

GATE 2020 conducted by
Indian Institute of Technology, Delhi

Indian Institute of Technology, Chennai

Indian Institute of Technology, Guwahati

Indian Institute of Technology, Roorkee

Indian Institute of Technology, Kanpur

About GATE

The Graduate Aptitude Test in Engineering (GATE) conducted by IISc and IITs has emerged as one of the bench mark tests for engineering and science aptitude in facilitating admissions for higher education (M.Tech./Ph.D.) in IITs, IISc and various other Institutes/Universities/Laboratories in India. With the standard and high quality of the GATE examination in 27 disciplines of engineering and science, Humanities and Social Sciences subjects, it identifies the candidate's understanding of a subject and aptitude and eligibility for higher studies. During the last few years, GATE score is also being used as one of the criteria for recruitment in Government Organizations such as Cabinet Secretariat, and National/State Public Sector Undertakings in India. Because of the importance of the GATE examination, the number of candidates taking up GATE exams has increased tremendously. GATE exams are conducted by the IITs and IISc as a computer based test having multiple choice questions and numerical answer type questions. The questions are mostly fundamental, concept based and thought provoking. From 2017 onwards GATE Exam is being held in Bangladesh, Ethiopia, Nepal, Singapore, Sri Lanka and United Arab Emirates. An Institute with various nationalities in its campus widens the horizons of an academic environment. A foreign student brings with him/her a great diversity, culture and wisdom to share. Many GATE qualified candidates are paid scholarships/assistantship, especially funded by Ministry of Human Resources Development, Government of India and by other Ministries. Indian Institute of Technology Bombay is the Organizing Institute for GATE 2021.

Why GATE?

Admission to Post Graduate and Doctoral Programmes

Admission to postgraduate programmes with MHRD and some other government scholarships/ assistantships in engineering colleges/institutes is open to those who qualify through GATE. GATE qualified candidates with Bachelor's degree in Engineering/Technology/Architecture or Master's degree in any branch of Science/Mathematics/Statistics/Computer Applications are eligible for admission to Master/Doctoral programmes in Engineering/Technology/Architecture as well as for Doctoral programmes in relevant branches of Science with MHRD or other government scholarships/ assistantships. Candidates with Master's degree in Engineering/Technology/Architecture may seek admission to relevant Ph.D programmes with scholarship/assistantship without appearing in the GATE examination.

Financial Assistance

A valid GATE score is essential for obtaining financial assistance during Master's programs and direct Doctoral programs in Engineering/Technology/Architecture, and Doctoral programs in relevant branches of Science in Institutes supported by the MHRD or other Government agencies. As per the directives of the MHRD, the following procedure is to be adopted for admission to the post-graduate programs (Master's and Doctoral) with MHRD scholarship/assistantship. Depending upon the norms adopted by a specific institute or department of the Institute, a candidate may be admitted directly into a course based on his/her performance in GATE only **or** based on his/her performance in GATE **and** an admission test/interview conducted by the department to which he/she has applied **and/or** the candidate's academic record. If the candidate is to be selected through test/interview for post-graduate programs, a minimum of 70% weightage will be given to the performance in GATE and the remaining 30% weightage will be given to the candidate's performance in test/interview and/or academic record, as per MHRD guidelines. The admitting institutes could however prescribe a minimum passing percentage of marks in the test/interview. Some colleges/institutes specify GATE qualification as the mandatory requirement even for admission without MHRD scholarship/assistantship.

To avail of the financial assistance (scholarship), the candidate must first secure admission to a program in these Institutes, by a procedure that could vary from institute to institute. Qualification in GATE is also a minimum requirement to apply for various fellowships awarded by many Government organizations. Candidates are advised to seek complete details of admission procedures and availability of MHRD scholarship/assistantship from the concerned admitting institution. The criteria for postgraduate admission with scholarship/assistantship could be different for different institutions. The management of the post-graduate scholarship/assistantship is also the responsibility of the admitting institution. Similarly, reservation of seats under different categories is as per the policies and norms prevailing at the admitting institution and Government of India rules. *GATE offices will usually not entertain any enquiry about admission, reservation of seats and / or award of scholarship / assistantship.*

PSU Recruitments

As many as 37 PSUs are using GATE score for recruitment. It is likely that more number of PSUs may start doing so by next year. Below is the list of PSUs:

MDL, BPCL, GAIL, NLC LTD, CEL, Indian Oil, HPCL, NBPC, NECC, BHEL, WBSEDCL, NTPC, ONGC, Oil India, Power Grid, Cabinet Secretariat, Govt. of India, BAARC, NFL, IPR, PSPCL, PSTCL, DRDO, OPGC Ltd., THDC India Ltd., BBNL, RITES, IRCON, GHECL, NHAI, KRIBHCO, Mumbai Railway Vikas Corporation Ltd. (MRVC Ltd.), National Textile Corporation, Coal India Ltd., BNPM, AAI, NALCO, EdCIL India.

Important :

1. Admissions in IITs/IISc or other Institutes for M.Tech./Ph.D. through GATE scores shall be advertised separately by the Institutes and GATE does not take the responsibility of admissions.
2. Cabinet Secretariat has decided to recruit officers for the post of Senior Field Officer (Tele) (From GATE papers of EC, CS, PH), Senior Research Officer (Crypto) (From GATE papers of EC, CS, MA), Senior Research Officer (S&T) (From GATE papers EC, CS, CY, PH, AE, BT) in the Telecommunication Cadre, Cryptographic Cadre and Science & Technology Unit respectively of Cabinet Secretariat. The details of the scheme of recruitment shall be published in National Newspaper/Employment News by the concerned authority.
3. Some PSUs in India have expressed their interest to utilize GATE scores for their recruitment purpose. The Organizations who intend to utilize GATE scores shall make separate advertisement for this purpose in Newspapers etc.

Who Can Appear for GATE?

Eligibility for GATE

Before starting the application process, the candidate must ensure that he/she meets the eligibility criteria of GATE given in Table.

Eligibility Criteria for GATE 2021

Degree/Program	Qualifying Degree/Examination	Description of Eligible Candidates	Expected Year of Completion
B.E. / B.Tech. / B.Pharm.	Bachelor's degree in Engineering / Technology (4 years after 10+2 or 3 years after B.Sc. / Diploma in Engineering / Technology)	Currently in the 3rd year or higher grade or already completed	2022
B. Arch.	Bachelor's degree of Architecture (5- year course) / Naval Architecture (4- year course) / Planning (4- year course)	Currently in the 3rd year or higher grade or already completed	2023 (for 5-year program), 2022 (for 4-year program)
B.Sc. (Research) / B.S.	Bachelor's degree in Science (Post-Diploma/4 years after 10+2)	Currently in the 3rd year or higher grade or already completed	2022
Pharm. D. (after 10+2)	6 years degree program, consisting of internship or residency training, during third year onwards	Currently in the 3rd/4th/5th/6th year or already completed	2024
M.B.B.S.	Degree holders of M.B.B.S. and those who are in the 5th/6th/7th semester or higher semester of such programme.	5th, 6th, 7th or higher semester or already completed	2022
M. Sc. / M.A. / MCA or equivalent	Master's degree in any branch of Arts/Science/Mathematics/Statistics/ Computer Applications or equivalent	Currently in the first year or higher or already Completed	2022
Int. M.E./ M.Tech. (Post-B.Sc.)	Post-B.Sc Integrated Master's degree programs in Engineering/ Technology (4-year program)	Currently in the 1st / 2nd/3rd/4th year or already completed	2024
Int. M.E./ M.Tech. or Dual Degree (after Diploma or 10+2)	Integrated Master's degree program or Dual Degree program in Engineering/Technology (5-year program)	Currently in the 3rd /4th/5th year or alreadycompleted	2023
B.Sc. / B.A. / B.Com.	Bachelor degree in any branch of Science / Arts / Commerce (3 years program)	Currently in the 3rd year or already completed	2021
Int. M.Sc. / Int. B.S. / M.S.	Integrated M.Sc. or 5-year integrated B.S.-M.S. program	Currently in the 3rd year or higher or already completed	2022
Professional Society Examinations (equivalent to B.E. / B.Tech. / B.Arch.)	B.E./B.Tech./B.Arch. equivalent examinations of Professional Societies, recognized by MoE/UPSC/AICTE (e.g. AMIE by Institution of Engineers-India, AMICE by the Institute of Civil Engineers-India and so on)	Completed Section A or equivalent of such professional courses	NA

In case a candidate has passed one of the qualifying examinations as mentioned above in 2020 or earlier, the candidate has to submit the degree certificate / provisional certificate / course completion certificate / professional certificate / membership certificate issued by the society or institute. In case, the candidate is expected to complete one of the qualifying criteria in 2021 or later as mentioned above, he/she has to submit a certificate from Principal or a copy of marks card for section A of AMIE.

Certificate From Principal

Candidates who have to submit a certificate from their college Principal have to obtain one from his/her institution beforehand and upload the same during the online submission of the application form.

Candidates With Backlogs

Candidates, who have appeared in the final semester/year exam in 2021, but with a backlog (arrears/failed subjects) in any of the papers in their qualifying degree should upload a copy of any one of the mark sheets of the final year,

OR

obtain a declaration from their Principal along with the signature and seal beforehand and upload the same during the online submission of the application form.

GATE Structure

Structure of GATE

GATE 2021 will be conducted on 27 subjects (papers). Table below shows the list of papers and paper codes for GATE 2021. A candidate is allowed to appear in ANY ONE or UP TO TWO papers of the GATE examination. However, note that the combination of TWO papers in which a candidate can appear MUST be from the pre-defined list as given in Table. Also note that for a paper running in multiple sessions, a candidate will be mapped to appear for the examination in one of the sessions ONLY.

List of GATE Papers and Corresponding Codes

GATE Paper	Code	GATE Paper	Code
Aerospace Engineering	AE	Instrumentation Engineering	IN
Agricultural Engineering	AG	Mathematics	MA
Architecture and Planning	AR	Mechanical Engineering	ME
Bio-medical Engineering	BM	Mining Engineering	MN
Biotechnology	BT	Metallurgical Engineering	MT
Civil Engineering	CE	Petroleum Engineering	PE
Chemical Engineering	CH	Physics	PH
Computer Science and Information Technology	CS	Production and Industrial Engineering	PI
Chemistry	CY	Statistics	ST
Electronics and Communication Engineering	EC	Textile Engineering and Fibre Science	TF
Electrical Engineering	EE	Engineering Sciences	XE*
Environmental Science & Engineering	ES	Humanities & Social Sciences	XH**
Ecology and Evolution	EY	Life Sciences	XL***
Geology and Geophysics	GG		

Note: *Environmental Science and Engineering (ES) and Humanities and Social Sciences (XH) are two new papers introduced in GATE 2021.*

*XE Paper Sections	Code	**XH Paper Sections	Code	***XL Paper Sections	Code
Engineering Mathematics (Compulsory) (15 marks)	A	Reasoning and Comprehension (Compulsory) (25 marks)	B1	Chemistry (Compulsory) (25 marks)	P
Any TWO optional Sections (2x35 = 70 marks)		Any ONE optional Section (60 marks)		Any TWO optional Sections (2x30 = 60 marks)	
Fluid Mechanics	B	Economics	C1	Biochemistry	Q
Materials Science	C	English	C2	Botany	R
Solid Mechanics	D	Linguistics	C3	Microbiology	S
Thermodynamics	E	Philosophy	C4	Zoology	T
Polymer Science and Engineering	F	Psychology	C5	Food Technology	U
Food Technology	G	Sociology	C6		
Atmospheric and Oceanic Sciences	H				

*XE (Engineering Sciences), **XH (Humanities & Social Sciences), ***XL (Life Sciences), papers are of general nature and will be comprised of Sections listed in the above table

Note: *Each subject/paper is of total 100 marks. General Aptitude (GA) section of 15 marks is common for all papers. Hence remaining 85 marks are for the respective subject/paper code.*

Combination of Two Papers Allowed to Appear in GATE 2021 (subject to availability of infrastructure and schedule)

Code of The First (Primary) Paper	Codes of Papers Allowed as The Second Paper
AE	XE
AG	ES
AR	CE
BM	BT / XL
BT	BM / XL
CE	AR / ES
CH	CY / PE / XE
CS	MA
CY	CH / XL
EC	IN / PH
EE	IN
ES	AG / CE
EY	XL
GG	MN / PE / PH
IN	EC / EE / PH
MA	CS / PH / ST
ME	XE
MN	GG / XE
MT	PH / XE
PE	CH / GG / XE
PH	EC / GG / IN / MA / MT / ST
PI	XE
ST	MA / PH
TF	XE
XE	AE / CH / ME / MN / MT / PE / PI / TF
XH	----
XL	BM / BT / CY / EY

General Aptitude Questions

All the papers will have a few questions that test the General Aptitude (Language and Analytical Skills), apart from the core subject of the paper.

Duration and Examination Type

All the papers of the GATE 2021 examination will be for 3 hours duration and they consist of 65 questions for a total of 100 marks. Since the examination is an ONLINE computer based test (CBT), at the end of the stipulated time (3-hours), the computer screen will automatically close the examination inhibiting any further action.

Candidates will be permitted to occupy their allotted seats 40 minutes before the scheduled start of the examination. Candidates can login and start reading the instructions 20 minutes before the start of examination. The late login time (if any) recorded by the computer system MUST NOT be beyond 30 minutes from the actual starting time of the examination. Under NO circumstances, will a candidate be permitted to login after 30 minutes from the actual examination starting time. Candidates will NOT be permitted to leave the examination hall before the end of the examination.

Pattern of Question Papers

GATE 2021 may contain questions of THREE different types in all the papers:

(i) **Multiple Choice Questions (MCQ)** carrying 1 or 2 marks each, in all the papers and sections. These questions are objective in nature, and each will have choice of four answers, out of which ONLY ONE choice is correct.

Negative Marking for Wrong Answers: For a wrong answer chosen in a MCQ, there will be negative marking. For 1-mark MCQ, 1/3 mark will be deducted for a wrong answer. Likewise, for 2-mark MCQ, 2/3 mark will be deducted for a wrong answer.

(ii) **Multiple Select Questions (MSQ)** carrying 1 or 2 marks each in all the papers and sections. These questions are objective in nature, and each will have choice of four answers, out of which ONE or MORE than ONE choice(s) is / are correct.

Note: There is NO negative marking for a wrong answer in MSQ questions. However, there is NO partial credit for choosing partially correct combinations of choices or any single wrong choice.

(iii) **Numerical Answer Type (NAT)** Questions carrying 1 or 2 marks each in most of the papers and sections. For these questions, the answer is a signed real number, which needs to be entered by the candidate using the virtual numeric keypad on the monitor (keyboard of the computer will be disabled). No choices will be shown for these types of questions. The answer can be a number such as 10 or –10 (an integer only). The answer may be in decimals as well, for example, 10.1 (one decimal) or 10.01 (two decimals) or –10.001 (three decimals). These questions will be mentioned with, up to which decimal places, the candidates need to present the answer. Also, for some NAT type problems an appropriate range will be considered while evaluating these questions so that the candidate is not unduly penalized due to the usual round-off errors. Candidates are advised to do the rounding off at the end of the calculation (not in between steps). Wherever required and possible, it is better to give NAT answer up to a maximum of three decimal places.

Note: *There is NO negative marking for a wrong answer in NAT questions.*
Also, there is NO partial credit in NAT questions.

Marking Scheme – Distribution of Marks and Questions

General Aptitude (GA) Questions

In all papers, GA questions carry a total of 15 marks. The GA section includes 5 questions carrying 1-mark each (sub-total 5 marks) and 5 questions carrying 2-marks each (sub-total 10 marks).

Question Papers other than GG, XE, XH and XL

These papers would contain 25 questions carrying 1-mark each (sub-total 25 marks) and 30 questions carrying 2-marks each (sub-total 60 marks) consisting of some MCQ type questions, while the remaining may be MSQ and / or NAT questions.

GG (Geology and Geophysics) Paper

Apart from the General Aptitude (GA) section, the GG question paper consists of two parts: Part A and Part B. Part A is compulsory for all the candidates. Part B contains two sections: Section 1 (Geology) and Section 2 (Geophysics). Candidates will have to attempt questions in Part A and questions in either Section 1 or Section 2 of Part B. The choice of Section 1 OR Section 2 of Part B has to be made at the time of online registration in GOAPS. At the examination hall, candidate cannot request for change of section.

Part A consists of 25 questions carrying 1-mark each (sub-total 25 marks and some of these will be MSQ and/or numerical answer type questions while remaining questions will be MCQ type). Either section of Part B (Section 1 and Section 2) consists of 30 questions carrying 2-marks each (sub-total 60 marks and some of these will be MSQ and/or numerical answer type questions while remaining questions will be MCQ type).

XE Paper (Engineering Sciences)

A candidate appearing in the XE paper has to answer the following:

- GA – General Aptitude carrying a total of 15 marks.
- Section A – Engineering Mathematics (Compulsory): This section contains 11 questions carrying a total of 15 marks: 7 questions carrying 1-mark each (sub-total 7 marks), and 4 questions carrying 2-marks each (sub-total 8 marks). Some questions will be of numerical answer type while remaining questions will be MCQ type.
- Any two of XE Sections B to H: The choice of two sections from B to H can be made during the examination after viewing the questions. Only TWO optional sections can be answered at a time. A candidate wishing to change midway of the examination to another optional section must first choose to deselect one of the previously chosen optional sections (B to H). Each of the optional sections of the XE paper (Sections B through H) contains 22 questions carrying a total of 35 marks: 9 questions carrying 1-mark each (sub-total 9 marks) and 13 questions carrying 2-marks each (sub-total 26 marks). Some questions will be of MSQ and/or numerical answer type while remaining questions will be MCQ type.

XH Paper (Humanities and Social Sciences)

A candidate appearing in the XH paper has to answer the following:

- **GA - General Aptitude** carrying a total of 15 marks.
- **Section B1 - Reasoning and Comprehension (Compulsory):** This section contains 15 questions carrying a total of 25 marks: 5 questions carrying 1-mark each (sub-total 5 marks) and 10 questions carrying 2-marks each (sub-total 20 marks). Some questions will be of MSQ and/or numerical answer type while remaining questions will be MCQ type.
- **Any ONE of XH Sections C1 to C6:** The ONE choice of section from C1 to C6 has to be made at the time of online registration in GOAPS. At the examination hall, candidate cannot request for change of section. Each of the optional sections of the XH paper (Sections C1 through C6) contains 40 questions carrying a total of 60 marks: 20 questions carrying 1-mark each (sub-total 20 marks) and 20 questions carrying 2-marks each (sub-total 40 marks). Some questions will be of MSQ and/or numerical answer type while remaining questions will be MCQ type.

XL Paper (Life Sciences)

A candidate appearing in the XL paper has to answer the following:

- **GA – General Aptitude** carrying a total of 15 marks.
- **Section P–Chemistry (Compulsory):** This section contains 15 questions carrying a total of 25 marks: 5 questions carrying 1-mark each (sub-total 5 marks) and 10 questions carrying 2-marks each (sub-total 20 marks). Some questions will be of MSQ and/or numerical answer type while remaining questions will be MCQ type.
- **Any two of XL Sections Q to U:** The choice of two sections from Q to U can be made during the examination after viewing the questions. Only TWO optional sections can be answered at a time. A candidate wishing to change midway of the examination to another optional section must first choose to deselect one of the previously chosen optional sections (Q to U). Each of the optional sections of the XL paper (Sections Q through U) contains 20 questions carrying a total of 30 marks: 10 questions carrying 1-mark each (sub-total 10 marks) and 10 questions carrying 2-marks each (sub-total 20 marks). Some questions will be of MSQ and/or numerical answer type while remaining questions will be MCQ type.

GATE Score

After the evaluation of the answers, the actual (raw) marks obtained by a candidate will be considered for computing the GATE Score. For multi-session papers (subjects), raw marks obtained by the candidates in different sessions will be converted to Normalized marks for that particular subject. Thus, raw marks (for single session papers) or normalized marks (for multi-session papers) will be used for computing the GATE Score, based on the qualifying marks.

Calculation of Normalized Marks for Multi-Session Papers

In GATE 2021 examination, some papers may be conducted in multi-sessions. Hence, for these papers, a suitable normalization is applied to take into account any variation in the difficulty levels of the question papers across different sessions. The normalization is done based on the fundamental assumption that "in all multi-session GATE papers, the distribution of abilities of candidates is the same across all the sessions". This assumption is justified since the number of candidates appearing in multi-session papers in GATE 2021 is large and the procedure for allocation of session to candidates is random. Further, it is also ensured that for the same multi-session paper, the number of candidates allotted in each session is of the same order of magnitude.

Based on the above, and considering various normalization methods, the committee arrived at the following formula for calculating the normalized marks for the multi-session papers.

Nsormalization mark of jth candidate in the ith session $\widehat{M_{ij}}$ is given by

$$\widehat{M_{ij}} = \frac{\bar{M}_t^g - M_q^g}{\bar{M}_{ti} - M_{iq}}(M_{ij} - M_{iq}) + M_q^g$$

where

M_{ij} : is the actual marks obtained by the jth candidate in i^{th} session

$\bar{M}_t^g$: is the average marks of the top 0.1% of the candidates considering all sessions

M_q^g : is the sum of mean and standard deviation marks of the candidates in the paper considering all sessions

$\bar{M}_{ti}$: is the average marks of the top 0.1% of the candidates in the ith session

M_{iq} : is the sum of the mean marks and standard deviation of the ith session

Calculation of GATE Score for All Papers

For all papers for which there is only one session, actual marks obtained by the candidates will be used for calculating the GATE 2021 Score. For papers in multi-sessions, normalized marks will be calculated corresponding to the raw marks obtained by a candidate and the GATE 2021 Score will be calculated based on the normalized marks.

The GATE 2021 score will be computed using the formula given below.

$$\text{GATE Score} = S_q + (S_t - S_q)\frac{(M - M_q)}{(\bar{M}_t - M_q)}$$

where

- M : marks obtained by the candidate (actual marks for single session papers and normalized marks for multi-session papers)
- M_q : is the qualifying marks for general category candidate in the paper
- $\bar{M}_t$: is the mean of marks of top 0.1% or top 10 (whichever is larger) of the candidates who appeared in the paper (in case of multi-session papers including all sessions)
- S_q : 350, is the score assigned to M_q
- S_t : 900, is the score assigned to $\bar{M}_t$

In the GATE 2021 the qualifying marks (M_q) for general category student in each subject will be 25 marks (out of 100) or $\mu + \sigma$, whichever is larger. Here μ is the mean and σ is the standard deviation of marks of all the candidates who appeared in the paper.

After the declaration of results, GATE Scorecards can be downloaded by the GATE qualified candidates ONLY.

The GATE 2021 Committee has the authority to decide the qualifying mark/score for each GATE paper. In case of any claim or dispute with respect to GATE 2021 examination or score, the Courts and Tribunals in Mumbai alone will have the exclusive jurisdiction to entertain and settle them.

GATE Syllabus

SECTION 1: ENGINEERING MATHEMATICS

Linear Algebra: Matrix algebra, systems of linear equations, eigenvalues and eigenvectors.

Calculus: Functions of single variable, limit, continuity and differentiability, mean value theorems, indeterminate forms; evaluation of definite and improper integrals; double and triple integrals; partial derivatives, total derivative, Taylor series (in one and two variables), maxima and minima, Fourier series; gradient, divergence and curl, vector identities, directional derivatives, line, surface and volume integrals, applications of Gauss, Stokes and Green's theorems.

Differential Equations: First order equations (linear and nonlinear); higher order linear differential equations with constant coefficients; Euler-Cauchy equation; initial and boundary value problems; Laplace transforms; solutions of heat, wave and Laplace's equations.

Complex Variables: Analytic functions; Cauchy-Riemann equations; Cauchy's integral theorem and integral formula; Taylor and Laurent series.

Probability and Statistics: Definitions of probability, sampling theorems, conditional probability; mean, median, mode and standard deviation; random variables, binomial, Poisson and normal distributions.

Numerical Methods: Numerical solutions of linear and non-linear algebraic equations; integration by trapezoidal and Simpson's rules; single and multi-step methods for differential equations.

SECTION 2: APPLIED MECHANICS AND DESIGN

Engineering Mechanics: Free-body diagrams and equilibrium; friction and its applications including rolling friction, belt-pulley, brakes, clutches, screw jack, wedge, vehicles, etc.; trusses and frames; virtual work; kinematics and dynamics of rigid bodies in plane motion; impulse and momentum (linear and angular) and energy formulations; Lagrange's equation.

Mechanics of Materials: Stress and strain, elastic constants, Poisson's ratio; Mohr's circle for plane stress and plane strain; thin cylinders; shear force and bending moment diagrams; bending and shear stresses; concept of shear centre; deflection of beams; torsion of circular shafts; Euler's theory of columns; energy methods; thermal stresses; strain gauges and rosettes; testing of materials with universal testing machine; testing of hardness and impact strength.

Theory of Machines: Displacement, velocity and acceleration analysis of plane mechanisms; dynamic analysis of linkages; cams; gears and gear trains; flywheels and governors; balancing of reciprocating and rotating masses; gyroscope.

Vibrations: Free and forced vibration of single degree of freedom systems, effect of damping; vibration isolation; resonance; critical speeds of shafts.

Machine Design: Design for static and dynamic loading; failure theories; fatigue strength and the S-N diagram; principles of the design of machine elements such as bolted, riveted and welded joints; shafts, gears, rolling and sliding contact bearings, brakes and clutches, springs.

SECTION 3: FLUID MECHANICS AND THERMAL SCIENCES

Fluid Mechanics: Fluid properties; fluid statics, forces on submerged bodies, stability of floating bodies; control-volume analysis of mass, momentum and energy; fluid acceleration; differential equations of continuity and momentum; Bernoulli's equation; dimensional analysis; viscous flow of incompressible fluids, boundary layer, elementary turbulent flow, flow through pipes, head losses in pipes, bends and fittings; basics of compressible fluid flow.

Heat-Transfer: Modes of heat transfer; one dimensional heat conduction, resistance concept and electrical analogy, heat transfer through fins; unsteady heat conduction, lumped parameter system, Heisler's charts; thermal boundary layer, dimensionless parameters in free and forced convective heat transfer, heat transfer correlations for flow over flat plates and through pipes, effect of turbulence; heat exchanger performance, LMTD and NTU methods; radiative heat transfer, Stefan- Boltzmann law, Wien's displacement law, black and grey surfaces, view factors, radiation network analysis.

Thermodynamics: Thermodynamic systems and processes; properties of pure substances, behavior of ideal and real gases; zeroth and first laws of thermodynamics, calculation of work and heat in various processes; second law of thermodynamics; thermodynamic property charts and tables, availability and irreversibility; thermodynamic relations.

Applications: Power Engineering: Air and gas compressors; vapour and gas power cycles, concepts of regeneration and reheat. I.C. Engines: Air-standard Otto, Diesel and dual cycles. Refrigeration and air-conditioning: Vapour and gas refrigeration and heat pump cycles; properties of moist air, psychrometric chart, basic psychrometric processes. Turbomachinery: Impulse and reaction principles, velocity diagrams, Pelton-wheel, Francis and Kaplan turbines; steam and gas turbines.

SECTION 4: MATERIALS, MANUFACTURING AND INDUSTRIAL ENGINEERING

Engineering Materials: Structure and properties of engineering materials, phase diagrams, heat treatment, stress-strain diagrams for engineering materials.

Casting, Forming and Joining Processes: Different types of castings, design of patterns, moulds and cores; solidification and cooling; riser and gating design. Plastic deformation and yield criteria; fundamentals of hot and cold working processes; load estimation for bulk (forging, rolling, extrusion, drawing) and sheet (shearing, deep drawing, bending) metal forming processes; principles of powder metallurgy. Principles of welding, brazing, soldering and adhesive bonding.

Machining and Machine Tool Operations: Mechanics of machining; basic machine tools; single and multi-point cutting tools, tool geometry and materials, tool life and wear; economics of machining; principles of non-traditional machining processes; principles of work holding, jigs and fixtures; abrasive machining processes; NC/CNC machines and CNC programming.

Metrology and Inspection: Limits, fits and tolerances; linear and angular measurements; comparators; interferometry; form and finish measurement; alignment and testing methods; tolerance analysis in manufacturing and assembly; concepts of coordinate-measuring machine (CMM).

Computer Integrated Manufacturing: Basic concepts of CAD/CAM and their integration tools; additive manufacturing.

Production Planning and Control: Forecasting models, aggregate production planning, scheduling, materials requirement planning; lean manufacturing.

Inventory Control: Deterministic models; safety stock inventory control systems.

Operations Research: Linear programming, simplex method, transportation, assignment, network flow models, simple queuing models, PERT and CPM.

Chapter-Wise Analysis

GATE PAPERS (Mechanical Engineering)

Subject	2011	2012	2013	2014	2015	2016	2017	2018	2019	2020	2021
1. Engineering Maths											
1 mark Questions	6	5	5	5	6	5	5	5	4	6	5
2 marks Questions	4	5	5	7	3	4	4	5	4	4	3
Total Marks	**14**	**15**	**15**	**19**	**12**	**13**	**13**	**15**	**12**	**14**	**11**
2. Engineering Mechanics											
1 mark Questions	2	1	0	0	0	2	1	1	2	1	0
2 marks Questions	2	3	1	1	4	2	1	3	2	0	2
Total Marks	**6**	**7**	**2**	**2**	**8**	**6**	**3**	**7**	**6**	**1**	**4**
3. Strength of Materials											
1 mark Questions	4	3	4	2	4	2	4	3	5	1	4
2 marks Questions	5	4	1	3	4	4	3	4	5	3	2
Total Marks	**14**	**11**	**6**	**8**	**12**	**10**	**10**	**11**	**15**	**7**	**8**
4. Theory of Machines											
1 mark Questions	0	3	2	4	2	2	2	3	3	4	3
2 marks Questions	4	1	3	3	4	3	3	4	4	3	3
Total Marks	**8**	**5**	**8**	**10**	**10**	**8**	**8**	**11**	**11**	**10**	**9**
5. Design of Machine Elements											
1 mark Questions	0	0	2	2	1	1	1	1	0	1	0
2 marks Questions	0	0	3	2	2	1	1	1	0	2	3
Total Marks	**0**	**0**	**8**	**6**	**5**	**3**	**3**	**3**	**0**	**5**	**6**
6. Fluid Mechanics											
1 mark Questions	1	1	2	2	3	3	4	3	3	2	2
2 marks Questions	2	2	2	2	2	4	3	4	2	3	2
Total Marks	**5**	**5**	**6**	**6**	**7**	**11**	**10**	**11**	**7**	**8**	**6**
7. Heat Transfer											
1 mark Questions	2	2	2	1	2	1	1	0	0	2	2
2 marks Questions	1	2	4	2	1	3	3	1	2	1	2
Total Marks	**4**	**6**	**10**	**5**	**4**	**7**	**7**	**2**	**4**	**4**	**6**

Subject	2011	2012	2013	2014	2015	2016	2017	2018	2019	2020	2021
8. Thermodynamics, IC Engines, Refrigeration & Power Plant Engineering Air-Conditioning											
1 mark Questions	2	3	2	3	2	3	2	2	3	3	4
2 marks Questions	9	5	4	5	5	3	4	3	4	5	4
Total Marks	**20**	**13**	**10**	**13**	**12**	**9**	**10**	**8**	**11**	**13**	**12**
9. Materials, Manufacturing & Industrial Engg.											
1 mark Questions	7	7	6	6	5	6	5	7	5	5	5
2 marks Questions	8	8	7	5	5	6	8	5	7	9	10
Total Marks	**23**	**23**	**20**	**16**	**15**	**18**	**21**	**17**	**19**	**23**	**25**
10. General Aptitude											
1 mark Questions	5	5	5	5	5	5	5	5	5	5	5
2 marks Questions	5	5	5	5	5	5	5	5	5	5	5
Total Marks	**15**	**15**	**15**	**15**	**15**	**15**	**15**	**15**	**15**	**15**	**15**

SOLVED PAPER - 2021
SET 1

INSTRUCTIONS

1. Total of 65 questions carrying 100 marks, out of which 10 questions carrying a total of 15 marks are in General Aptitude (GA)
2. The Engineering Mathematics will carry around **15% of the total marks,** the General Aptitude section will carry **15% of the total marks** and the **remaining 70% of the total marks.**
3. **Types of Questions**
   (*a*) Multiple Choice Questions (MCQ) carrying 1 or 2 marks each in all papers and sections. These questions are objective in nature, and each will have a choice of four options, out of which the candidate has to mark the correct answer(s).
   (*b*) Numerical Answer Questions of 1 or 2 marks each in all papers and sections. For these questions the answer is a real number, to be entered by the candidate using the virtual keypad. No choices will be shown for these type of questions.
   (*c*) Multiple Select Questions (MSQ) carrying 1 or 2 marks each in all the papers and sections. These questions are objective in nature, and each will have choice of four answers, out of which ONE or MORE than ONE choice(s) is / are correct.
4. For **1-mark** multiple-choice questions, **1/3 marks** will be deducted for a wrong answer. Likewise, for **2-marks** multiple-choice questions, **2/3** marks will be deducted for a wrong answer. There is no negative marking for numerical answer type questions & multiple selection questions.

Chapter-Wise Analysis

#	Subject	Marks		
		1	2	Total
1	General Aptitude	5	5	15
2	Engineering Mathematics	5	3	11
3	Engineering Mechanics	-	2	4
4	Strength of Materials	4	2	8
5	Theory of Machines	3	3	9

#	Subject	Marks		
		1	2	Total
6	Design of Machine Elements	-	3	6
7	Fluid Mechanics	2	2	6
8	Heat Transfer	2	2	6
9	Thermodynamics, IC Engines, Refrigeration & Power Plant Engineering Air-Conditioning	4	4	12
10	Materials, Manufacturing & Industrial Engineering	5	10	25

GENERAL APTITUDE (GA)

Q. 1 to Q. 5 : Carry One Mark Each.

1. Consider the following sentences:

(i) After his surgery, Raja hardly could walk.

(ii) After his surgery, Raja could barely walk.

(iii) After his surgery, Raja barely could walk.

(iv) After his surgery, Raja could hardly walk.

Which of the above sentences are grammatically CORRECT?

(a) (i) and (ii) (b) (i) and (iii)

(c) (iii) and (iv) (d) (ii) and (iv)

2. Ms. X came out of a building through its front door to find her shadow due to the morning sun falling to her right side with the building to her back. From this, it can be inferred that building is facing ________

(a) North (b) East

(c) West (d) South

3. In the above figure, O is the center of the circle and, M and N lie on the circle.

The area of the right triangle MON is 50 cm^2.

What is the area of the circle in cm^2 ?

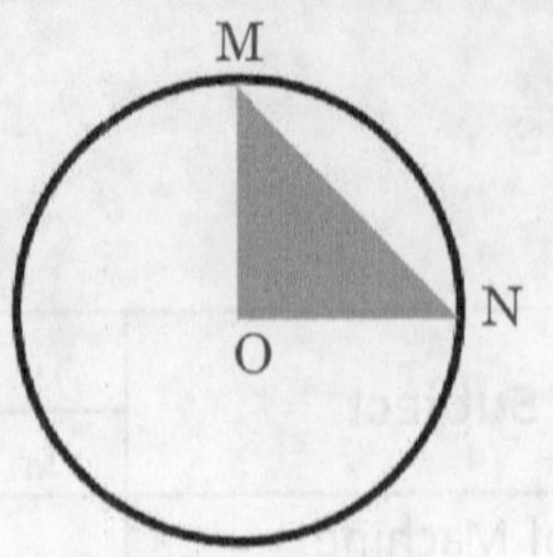

(a) 2π (b) 50π

(c) 75π (d) 100π

4. If $\begin{cases} "\oplus" \text{ means } "-", \\ "\otimes" \text{ means } "\div", \\ "\Delta" \text{ means } "+", \\ "\nabla" \text{ means } "\times", \end{cases}$

then, the value of the expression $\Delta\, 2 \oplus 3\, \Delta\, ((4 \otimes 2) \nabla 4) =$

(a) –1 (b) –0.5

(c) 6 (d) 7

5. "The increased consumption of leafy vegetables in the recent months is a clear indication that the people in the state have begun to lead a healthy lifestyle"

Which of the following can be logically inferred from the information presented in the above statement?

(a) The people in the state did not consume leafy vegetables earlier.

(b) Consumption of leafy vegetables may not be the only indicator of healthy lifestyle.

(c) Leading a healthy lifestyle is related to a diet with leafy vegetables.

(d) The people in the state have increased awareness of health hazards causing by consumption of junk foods.

Q. 6 to Q. 10: Multiple Choice Question (MCQ), Carry Two Marks Each

6. Oxpeckers and rhinos manifest a symbiotic relationship in the wild. The oxpeckers warn the rhinos about approaching poachers, thus possibly saving the lives of the rhinos. Oxpeckers also feed on the parasitic ticks found on rhinos.

In the symbiotic relationship described above, the primary benefits for oxpeckers and rhinos respectively are,

(a) Oxpeckers get a food source, rhinos have no benefit.

(b) Oxpeckers save their habitat from poachers while the rhinos have no benefit.

(c) Oxpeckers get a food source, rhinos may be saved from the poachers.

(d) Oxpeckers save the lives of poachers, rhinos save their own lives.

7.

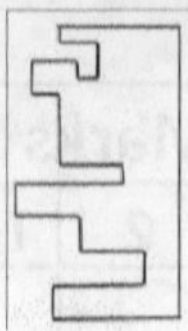

A jigsaw puzzle has 2 pieces. One of the pieces is shown above. Which one of the given options for the missing piece when assembled will form a rectangle? The piece can be moved, rotated or flipped to assemble with the above piece.

(a) 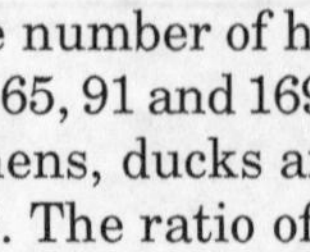

(b) 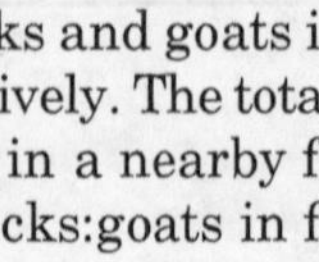

(c) 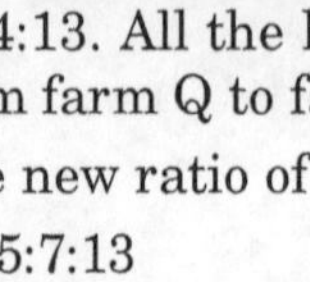

(d) 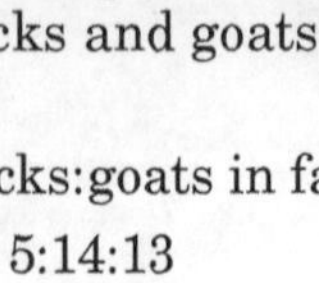

8. The number of hens, ducks and goats in farm P are 65, 91 and 169, respectively. The total number of hens, ducks and goats in a nearby farm Q is 416. The ratio of hens:ducks:goats in farm Q is 5:14:13. All the hens, ducks and goats are sent from farm Q to farm P.

The new ratio of hens:ducks:goats in farm P is

(a) 5:7:13 (b) 5:14:13

(c) 10:21:26 (d) 21:10:26

9.

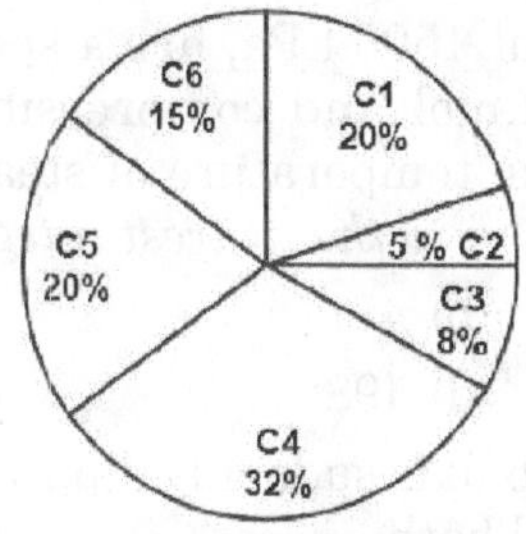

Company	Ratio
C1	3:2
C2	1:4
C3	5:3
C4	2:3
C5	9:1
C6	3:4

The distribution of employees at the rank of executives, across different companies C1, C2, ..., C6 is presented in the chart given above. The ratio of executives with a management degree to those without a management degree in each of these companies is provided in the table above. The total number of executives across all companies is 10,000.

The total number of management degree holders among the executives in companies C2 and C5 together is ________.

(a) 225 (b) 600

(c) 1900 (d) 2500

10. Five persons P, Q, R, S and T are sitting in a row not necessarily in the same order. Q and R are separated by one person, and S should not be seated adjacent to Q.

The number of distinct seating arrangements possible is:

(a) 4 (b) 8

(c) 10 (d) 16

MECHANICAL ENGINEERING

Q.1 to Q.19: Multiple Choice Question (MCQ), Carry One Mark Each

1. If $y(x)$ satisfies the differential equation

$$(\sin x)\frac{dy}{dx} + y \cos x = 1,$$

subject to the condition $y(\pi/2) = \pi/2$, then $y(\pi/6)$ is

(a) 0 (b) $\frac{\pi}{6}$

(c) $\frac{\pi}{3}$ (d) $\frac{\pi}{2}$

2. The value of $\lim_{x\to 0}\left(\frac{1-\cos x}{x^2}\right)$ is

(a) $\frac{1}{4}$ (b) $\frac{1}{3}$

(c) $\frac{1}{2}$ (d) 1

3. The Dirac-delta function ($\delta(t - t_0)$) for $t, t_0 \in \mathbb{R}$, has the following property

$$\int_a^b \phi(t)\delta(t-t_0)dt = \int_0^{\phi(t_0)} \begin{matrix} a < t_0 < b \\ \text{otherwise} \end{matrix}$$

The Laplace transform of the Dirac-delta function $\delta(t - a)$ for $a > 0$;

$L(\delta(t - a)) = F(s)$ is

(a) 0 (b) ∞

(c) e^{sa} (d) e^{-sa}

4. The ordinary differential equation $\frac{dy}{dt} = -\pi y$ subject to an initial condition $y(0) = 1$ is solved numerically using the following scheme:

$$\frac{y(t_{n+1}) - y(t_n)}{h} = -xy(t_n)$$

where h is the time step, $t_n = nh$, and $n = 0,1,2,....$ This numerical scheme is stable for all values of h in the interval________.

(a) $0 < h < \frac{2}{\pi}$ (b) $0 < h < 1$

(c) $0 < h < \frac{\pi}{2}$ (d) for all $h > 0$

5. Consider a binomial random variable X. If X_1, X_2, ..., X_n are independent and identically distributed samples from the distribution of X with sum $Y = \sum_{i=1}^{n} X_i$, then the distribution of Y as $n \to \infty$ can be approximated as

(a) Exponential (b) Bernoulli

(c) Binomial (d) Normal

6. The loading and unloading response of a metal is shown in the figure. The elastic and plastic strains corresponding to 200 MPa stress, respectively, are

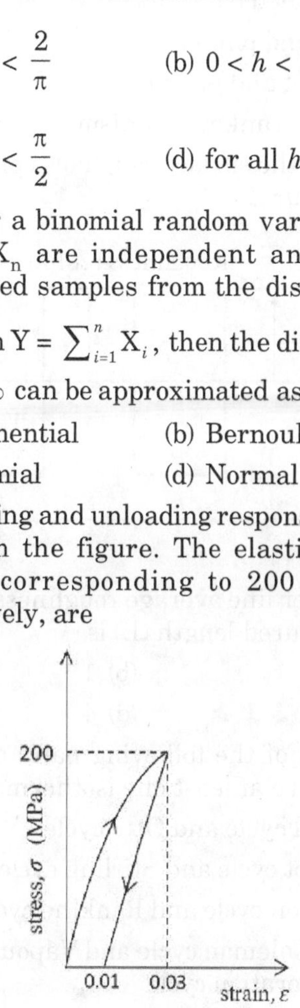

(a) 0.01 and 0.01 (b) 0.02 and 0.01

(c) 0.01 and 0.02 (d) 0.02 and 0.02

7. In a machining operation, if a cutting tool traces the workpiece such that the directrix is perpendicular to the plane of the generatrix as shown in figure, the surface generated is

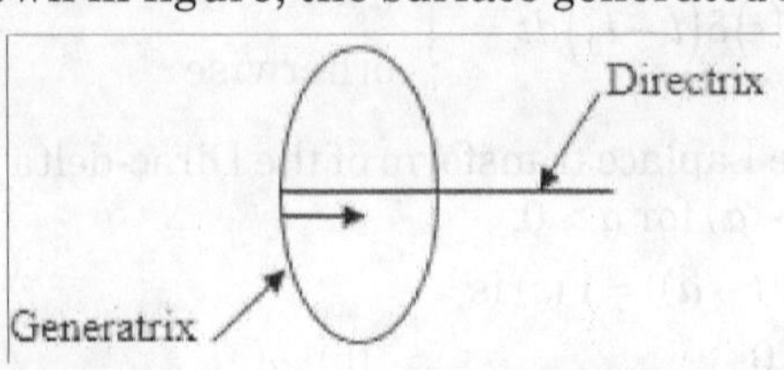

(a) plane (b) cylindrical
(c) spherical (d) a surface of revolution

8. The correct sequence of machining operations to be performed to finish a large diameter through hole is
(a) drilling, boring, reaming
(b) boring, drilling, reaming
(c) drilling, reaming, boring
(d) boring, reaming, drilling

9. In modern CNC machine tools, the backlash has been eliminated by
(a) preloaded ballscrews
(b) rack and pinion
(c) ratchet and pinion
(d) slider crank mechanism

10. Consider the surface roughness profile as shown in the figure.

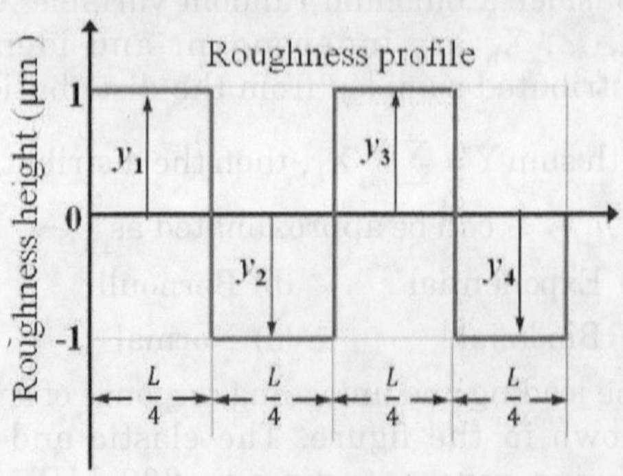

The center line average roughness (R_a, in µm) of the measured length (L) is
(a) 0 (b) 1
(c) 2 (d) 4

11. In which of the following pairs of cycles, both cycles have at least one isothermal process?
(a) Diesel cycle and Otto cycle
(b) Carnot cycle and Stirling cycle
(c) Brayton cycle and Rankine cycle
(d) Bell-Coleman cycle and Vapour compression refrigeration cycle

12. Superheated steam at 1500 kPa, has a specific volume of 2.75 m^3/kmol and compressibility factor (Z) of 0.95. The temperature of steam is ________ °C (*round off to the nearest integer*).
(a) 522 (b) 471
(c) 249 (d) 198

13. A hot steel spherical ball is suddenly dipped into a low temperature oil bath.
Which of the following dimensionless parameters are required to determine instantaneous center temperature of the ball using a Heisler chart?
(a) Biot number and Fourier number
(b) Reynolds number and Prandtl number
(c) Biot number and Froude number
(d) Nusselt number and Grashoff number

14. An infinitely long pin fin, attached to an isothermal hot surface, transfers heat at a steady rate of Q_1 to the ambient air. If the thermal conductivity of the fin material is doubled, while keeping everything else constant, the rate of steady- state heat transfer from the fin becomes Q_2. The ratio Q_2/Q_1 is
(a) $\sqrt{2}$ (b) 2
(c) $\frac{1}{\sqrt{2}}$ (d) $\frac{1}{2}$

15. The relative humidity of ambient air at 300 K is 50% with a partial pressure of water vapour equal to p_v. The saturation pressure of water at 300 K is p_{sat}. The correct relation for the air-water mixture is
(a) $p_v = 0.5\, p_{sat}$ (b) $p_v = p_{sat}$
(c) $p_v = 0.622\, p_{sat}$ (d) $p_v = 2\, p_{sat}$

16. Consider a reciprocating engine with crank radius R and connecting rod of length L. The secondary unbalance force for this case is equivalent to primary unbalance force due to a virtual crank of ________
(a) radius $\frac{L^2}{4R}$ rotating at half the engine speed
(b) radius $\frac{R}{4}$ rotating at half the engine speed
(c) radius $\frac{R^2}{4L}$ rotating at twice the engine speed
(d) radius $\frac{L}{2}$ rotating at twice the engine speed

17. A cantilever beam of length, L, and flexural rigidity, EI, is subjected to an end moment, M as shown in the figure. The deflection of the beam at Let $x = \frac{L}{2}$ is

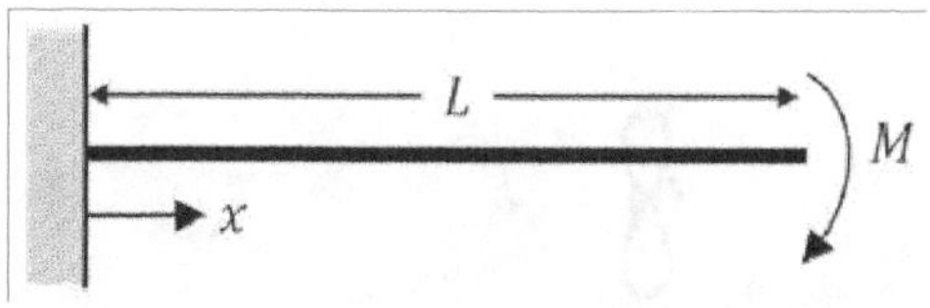

(a) $\frac{ML^2}{2EI}$ (b) $\frac{ML^2}{4EI}$

(c) $\frac{ML^2}{8EI}$ (d) $\frac{ML^2}{16EI}$

18. A prismatic bar PQRST is subjected to axial loads as shown in the figure. The segments having maximum and minimum axial stresses, respectively, are

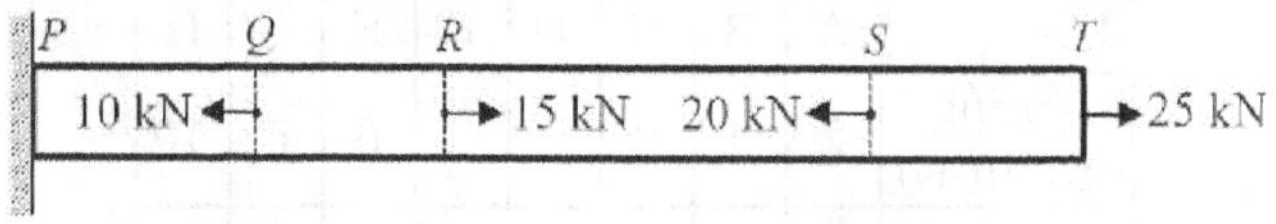

(a) QR and PQ (b) ST and PQ

(c) QR and RS (d) ST and RS

19. Shear stress distribution on the cross-section of the coil wire in a helical compression spring is shown in the figure. This shear stress distribution represents

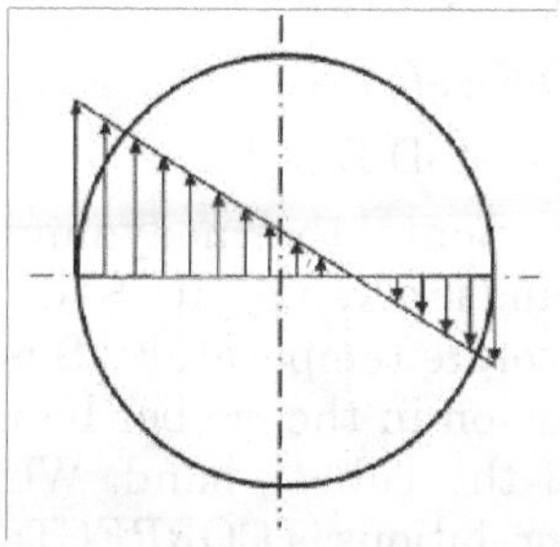

(a) direct shear stress in the coil wire cross-section

(b) torsional shear stress in the coil wire cross-section

(c) combined direct shear and torsional shear stress in the coil wire cross-section

(d) combined direct shear and torsional shear stress along with the effect of stress concentration at inside edge of the coil wire cross-section

Q.20 to Q.25: Numerical Answer Type (NAT), carry ONE mark each

20. Robot Ltd. wishes to maintain enough safety stock during the lead time period between starting a new production run and its completion such that the probability of satisfying the customer demand during the lead time period is 95%. The lead time period is 5 days and daily customer demand can be assumed to follow the Gaussian (normal) distribution with mean 50 units and a standard deviation of 10 units. Using $\phi^{-1}(0.95) = 1.64$, where ϕ represents the cumulative distribution function of the standard normal random variable, the amount of safety stock that must be maintained by Robot Ltd. to achieve this demand fulfillment probability for the lead time period is ______________ units (*round off to two decimal places*).

21. A pressure measurement device fitted on the surface of a submarine, located at a depth H below the surface of an ocean, reads an absolute pressure of 4.2 MPa. The density of sea water is 1050 kg/m^3, the atmospheric pressure is 101 kPa, and the acceleration due to gravity is 9.8 m/s^2. The depth H is ______________ m

(*round off to the nearest integer*).

22. Consider fully developed, steady state incompressible laminar flow of a viscous fluid between two large parallel horizontal plates. The bottom plate is fixed and the top plate moves with a constant velocity of U = 4 m/s. Separation between the plates is 5 mm. There is no pressure gradient in the direction of flow. The density of fluid is 800 kg/m^3, and the kinematic viscosity is 1.25×10^{-4} m^2/s. The average shear stress in the fluid is ______________ Pa (*round off to the nearest integer*).

23. A rigid insulated tank is initially evacuated. It is connected through a valve to a supply line that carries air at a constant pressure and temperature of 250 kPa and 400 K respectively. Now the valve is opened and air is allowed to flow into the tank until the pressure inside the tank reaches to 250 kPa at which point the valve is closed. Assume that the air behaves as a perfect gas with constant properties

(c_p = 1.005 kJ/kg.K, c_v = 0.718 kJ/kg.K, R = 0.287 kJ/kg.K). Final temperature of the air inside the tank is ______________ K (*round off to one decimal place*).

24. The figure shows an arrangement of a heavy propeller shaft in a ship. The combined polar mass moment of inertia of the propeller and the shaft is 100 kg.m^2. The propeller rotates at $\omega = 12$ rad/s. The waves acting on the ship hull induces a rolling motion as shown in the figure with an angular velocity of 5 rad/s. The gyroscopic moment generated on the shaft due to the motion described is ______________ N.m (*round off to the nearest integer*).

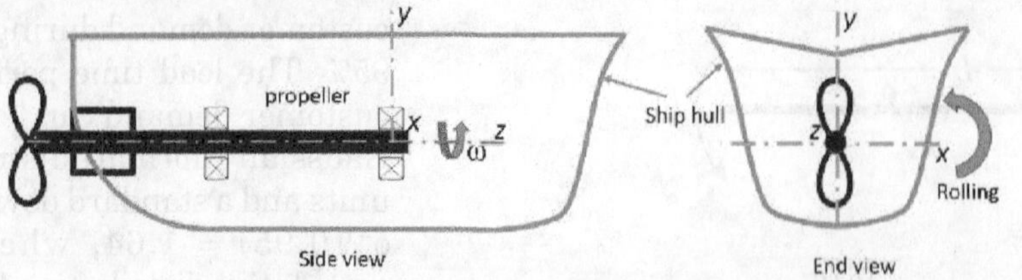

25. Consider a single degree of freedom system comprising a mass M, supported on a spring and a dashpot as shown in the figure.

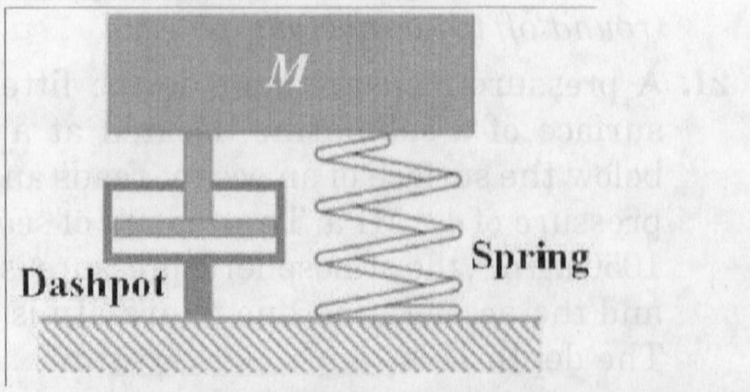

If the amplitude of the free vibration response reduces from 8 mm to 1.5 mm in 3 cycles, the damping ratio of the system is ____________ (*round off to three decimal places*).

Q. 26 - Q. 34 Carry Two Mark Each

26. Consider a vector p in 2-dimensional space. Let its direction (counter-clockwise angle with the positive x-axis) be θ. Let p be an eigenvector of a 2×2 matrix A with corresponding eigenvalue λ, $\lambda > 0$. If we denote the magnitude of a vector v by $||v||$, identify the VALID statement regarding p', where $p' = Ap$.

(a) Direction of $p' = \lambda\theta$, $||p'|| = ||p||$

(b) Direction of $p' = \theta$, $||p'|| = \lambda ||p||$

(c) Direction of $p' = \lambda\theta$, $||p'|| = \lambda ||p||$

(d) Direction of $p' = \theta$, $||p'|| = ||p||$

27. Let C represent the unit circle centered at origin in the complex plane, and complex variable, $z = x + iy$. The value of the contour integral $\int_C \frac{\cosh 3z}{2z} dz$ (where integration is taken counter clockwise) is

(a) 0 (b) 2

(c) πi (d) $2\pi i$

28. A set of jobs A, B, C, D, E, F, G, H arrive at time t = 0 for processing on turning and grinding machines. Each job needs to be processed in sequence - first on the turning machine and second on the grinding machine, and the grinding must occur immediately after turning. The processing times of the jobs are given below.

Job	A	B	C	D	E	F	G	H
Turning (minutes)	2	4	8	9	7	6	5	10
Grinding (minutes)	6	1	3	7	9	5	2	4

If the makespan is to be minimized, then the optimal sequence in which these jobs must be processed on the turning and grinding machines is

(a) A-E-D-F-H-C-G-B

(b) A-D-E-F-H-C-G-B

(c) G-E-D-F-H-C-A-B

(d) B-G-C-H-F-D-E-A

29. The fundamental thermodynamic relation for a rubber band is given by $dU = TdS + \tau dL$, where T is the absolute temperature, S is the entropy, τ is the tension in the rubber band, and L is the length of the rubber band. Which one of the following relations is CORRECT:

(a) $\tau = \left(\frac{\partial U}{\partial S}\right)_L$

(b) $\left(\frac{\partial T}{\partial L}\right)_S = \left(\frac{\partial \tau}{\partial S}\right)_L$

(c) $\left(\frac{\partial T}{\partial S}\right)_L = \left(\frac{\partial \tau}{\partial L}\right)_S$

(d) $T = \left(\frac{\partial U}{\partial S}\right)_\tau$

30. Consider a two degree of freedom system as shown in the figure, where PQ is a rigid uniform rod of length, b and mass, m.

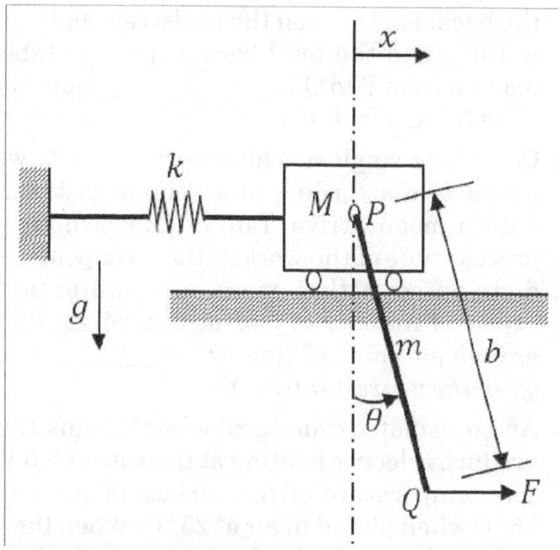

Assume that the spring deflects only horizontally and force F is applied horizontally at Q. For this system, the Lagrangian, L is

(a) $\frac{1}{2}(M+m)\dot{x}^2 + \frac{1}{6}mb^2\dot{\theta}^2 - \frac{1}{2}kx^2 + mg\frac{b}{2}\cos\theta$

(b) $\frac{1}{2}(M+m)\dot{x}^2 + \frac{1}{2}mb\dot{\theta}\dot{x}\cos\theta$

$\frac{1}{6}mb^2\dot{\theta}^2 - \frac{1}{2}kx^2 + mg\frac{b}{2}\cos\theta$

(c) $\frac{1}{2}M\dot{x}^2 + \frac{1}{2}mb\dot{\theta}\dot{x}\cos\theta + \frac{1}{6}mb^2\dot{\theta}^2 - \frac{1}{2}kx^2$

(d) $\frac{1}{2}M\dot{x}^2 + \frac{1}{2}mb\dot{\theta}\dot{x}\cos\theta + \frac{1}{6}mb^2\dot{\theta}^2 - \frac{1}{2}kx^2$

$+mg\frac{b}{2}\cos\theta + Fb\sin\theta$

31. A right solid circular cone standing on its base on a horizontal surface is of height H and base radius R. The cone is made of a material with specific weight w and elastic modulus E. The vertical deflection at the mid-height of the cone due to self-weight is given by

(a) $\frac{wH^2}{8E}$ (b) $\frac{wH^2}{6E}$

(c) $\frac{wRH}{8E}$ (d) $\frac{wRH}{6E}$

32. A tappet valve mechanism in an IC engine comprises a rocker arm ABC that is hinged at B as shown in the figure. The rocker is assumed rigid and it oscillates about the hinge B. The mass moment of inertia of the rocker about B is 10^{-4} kg.m^2. The rocker arm dimensions are a = 3.5 cm and b = 2.5 cm. A pushrod pushes the rocker at location A, when moved vertically by a cam that rotates at N rpm. The pushrod is assumed massless and has a stiffness of 15 N/mm. At the other end C, the rocker pushes a valve against a spring of stiffness 10 N/mm. The valve is assumed massless and rigid.

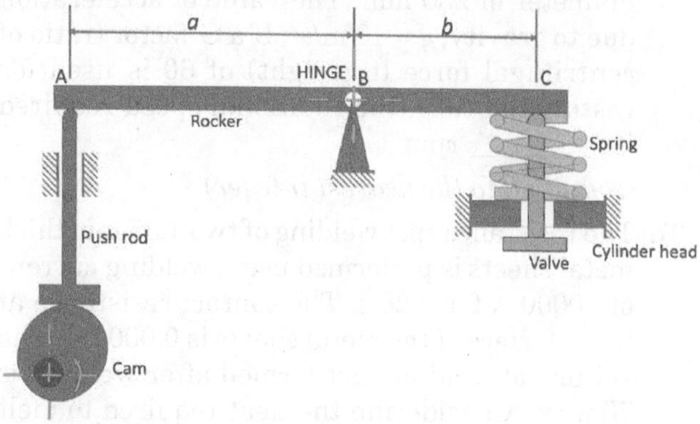

Resonance in the rocker system occurs when the cam shaft runs at a speed of ________ rpm (*round off to the nearest integer*).

(a) 496 (b) 4739

(c) 790 (d) 2369

33. Customers arrive at a shop according to the Poisson distribution with a mean of 10 customers/hour. The manager notes that no customer arrives for the first 3 minutes after the shop opens. The probability that a customer arrives within the next 3 minutes is

(a) 0.39 (b) 0.86

(c) 0.50 (d) 0.61

34. Let $f(x) = x^2 - 2x + 2$ be a continuous function defined on $x \in [1,3]$. The point x at which the tangent of $f(x)$ becomes parallel to the straight line joining f(1) and f(3) is

(a) 0 (b) 1

(c) 2 (d) 3

Q.35 to Q.55: Numerical Answer Type (NAT), carry TWO marks each

35. Activities A, B, C and D form the critical path for a project with a PERT network. The means and variances of the activity duration for each activity are given below. All activity durations follow the Gaussian (normal) distribution, and are independent of each other.

Activity	A	B	C	D
Mean (days)	6	11	8	15
Variance (days^2)	4	9	4	9

The probability that the project will be completed within 40 days is________ *(round off to two decimal places).*

(Note: Probability is a number between 0 and 1).

36. A true centrifugal casting operation needs to be performed horizontally to make copper tube sections with outer diameter of 250 mm and inner diameter of 230 mm. The value of acceleration due to gravity, $g = 10$ m/s^2. If a G-factor (ratio of centrifugal force to weight) of 60 is used for casting the tube, the rotational speed required is ________ rpm

(round off to the nearest integer).

37. The resistance spot welding of two 1.55 mm thick metal sheets is performed using welding current of 10000 A for 0.25 s. The contact resistance at the interface of the metal sheets is 0.0001 Q. The volume of weld nugget formed after welding is 70 mm^3. Considering the heat required to melt unit volume of metal is 12 J/mm^3, the thermal efficiency of the welding process is ________ % *(round off to one decimal place).*

38. An orthogonal cutting operation is performed using a single point cutting tool with a rake angle of 12° on a lathe. During turning, the cutting force and the friction force are 1000 N and 600 N, respectively. If the chip thickness and the uncut chip thickness during turning are 1.5 mm and 0.75 mm, respectively, then the shear force is ________ N *(round off to two decimal places).*

39. In a grinding operation of a metal, specific energy consumption is 15 J/mm^3. If a grinding wheel with a diameter of 200 mm is rotating at 3000 rpm to obtain a material removal rate of 6000 mm^3/min, then the tangential force on the wheel is ________ N *(round off to two decimal places).*

40. A 200 mm wide plate having a thickness of 20 mm is fed through a rolling mill with two rolls. The radius of each roll is 300 mm. The plate thickness is to be reduced to 18 mm in one pass using a roll speed of 50 rpm. The strength coefficient (K) of the work material flow curve is 300 MPa and the strain hardening exponent, n is 0.2. The coefficient of friction between the rolls and the plate is 0.1. If the friction is sufficient to permit the rolling operation then the roll force will be ________ kN *(round off to the nearest integer).*

41. The XY table of a NC machine tool is to move from P(1,1) to Q(51,1); all coordinates are in mm. The pitch of the NC drive leadscrew is 1 mm. If the backlash between the leadscrew and the nut is 1.8°, then the total backlash of the table on moving from P to Q is ________ mm *(round off to two decimal places).*

42. Consider a single machine workstation to which jobs arrive according to a Poisson distribution with a mean arrival rate of 12 jobs/hour. The process time of the workstation is exponentially distributed with a mean of 4 minutes. The expected number of jobs at the workstation at any given point of time is ________ *(round off to the nearest integer).*

43. An uninsulated cylindrical wire of radius 1.0 mm produces electric heating at the rate of 5.0 W/m. The temperature of the surface of the wire is 75 °C when placed in air at 25 °C. When the wire is coated with PVC of thickness 1.0 mm, the temperature of the surface of the wire reduces to 55 °C.

Assume that the heat generation rate from the wire and the convective heat transfer coefficient are same for both uninsulated wire and the coated wire. The thermal conductivity of PVC is ________ W/m.K *(round off to two decimal places).*

44. A solid sphere of radius 10 mm is placed at the centroid of a hollow cubical enclosure of side length 30 mm. The outer surface of the sphere is denoted by 1 and the inner surface of the cube is denoted by 2. The view factor F^{22} for radiation heat transfer is ________ *(rounded off to two decimal places).*

45. Consider a steam power plant operating on an ideal reheat Rankine cycle. The work input to the pump is 20 kJ/kg. The work output from the high pressure turbine is 750 kJ/kg. The work output from the low pressure turbine is 1500 kJ/kg. The thermal efficiency of the cycle is 50 %. The enthalpy of saturated liquid and saturated vapour at condenser pressure are 200 kJ/kg and 2600 kJ/kg, respectively. The quality of steam at the exit of the low pressure turbine is ________ % *(round off to the nearest integer).*

46. In the vicinity of the triple point, the equation of liquid-vapour boundary in the P – T phase diagram for ammonia is In P = 24.38 - 3063/T, where P is pressure (in Pa) and T is temperature (in K). Similarly, the solid-vapour boundary is given by In P = 27.92 – 3754/T. The temperature at the triple point is ________ K *(round off to one decimal place).*

47. A cylindrical jet of water (density = 1000 kg/m^3) impinges at the center of a flat, circular plate and spreads radially outwards, as shown in the figure. The plate is resting on a linear spring with a spring constant k = 1 kN/m. The incoming jet diameter is D = 1 cm.

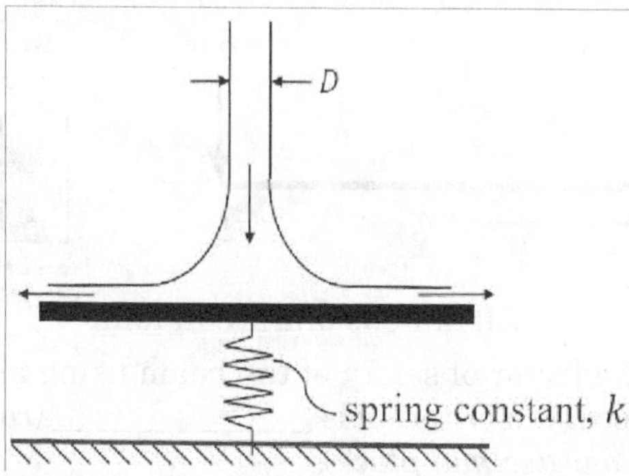

If the spring shows a steady deflection of 1 cm upon impingement of jet, then the velocity of the incoming jet is m/s (*round off to one decimal place*).

48. A single jet Pelton wheel operates at 300 rpm. The mean diameter of the wheel is 2 m. Operating head and dimensions of jet are such that water comes out of the jet with a velocity of 40 m/s and flow rate of 5 m^3/s. The jet is deflected by the bucket at an angle of 165°. Neglecting all losses, the power developed by the Pelton wheel is__________MW (*round off to two decimal places*).

49. An air-conditioning system provides a continuous flow of air to a room using an intake duct and an exit duct, as shown in the figure. To maintain the quality of the indoor air, the intake duct supplies a mixture of fresh air with a cold air stream. The two streams are mixed in an insulated mixing chamber located upstream of the intake duct. Cold air enters the mixing chamber at 5 °C, 105 kPa with a volume flow rate of 1.25 m^3/s during steady state operation. Fresh air enters the mixing chamber at 34 °C and 105 kPa. The mass flow rate of the fresh air is 1.6 times of the cold air stream. Air leaves the room through the exit duct at 24 °C.

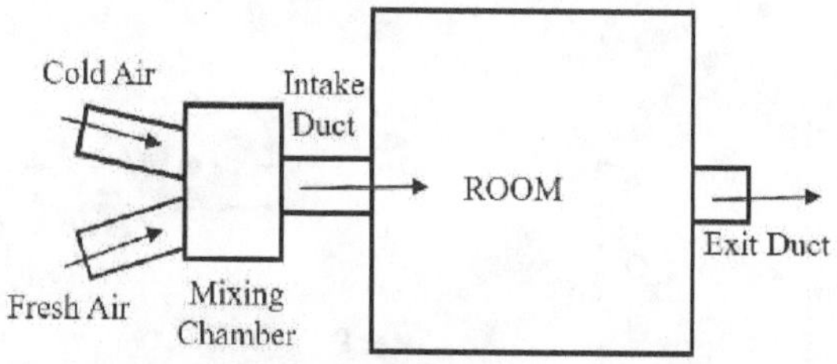

Assuming the air behaves as an ideal gas with c_p = 1.005 kJ/kg.K and R = 0.287 kJ/kg.K, the rate of heat gain by the air from the room is __________kW (*round off to two decimal places*).

50. Two smooth identical spheres each of radius 125 mm and weight 100 N rest in a horizontal channel having vertical walls. The distance between vertical walls of the channel is 400 mm.

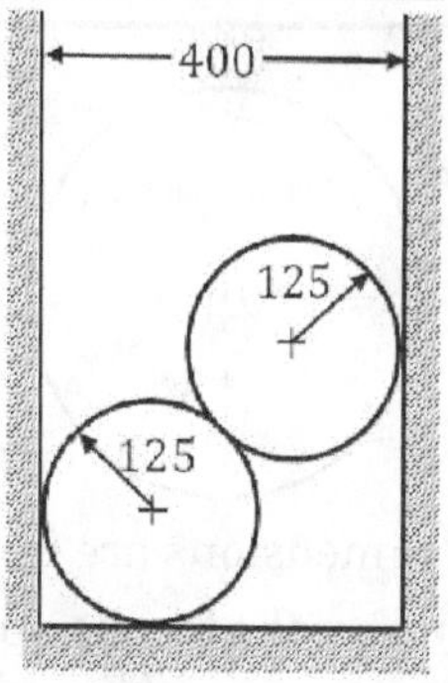

All dimensions are in mm

The reaction at the point of contact between two spheres is __________ N (*round off to one decimal place*).

51. An overhanging beam PQR is subjected to uniformly distributed load 20 kN/m as shown in the figure.

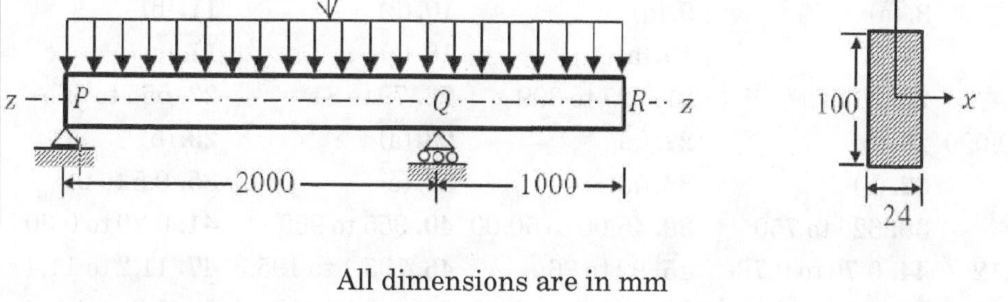

All dimensions are in mm

The maximum bending stress developed in the beam is __________ MPa (*round off to one decimal place*).

52. The Whitworth quick return mechanism is shown in the figure with link lengths as follows: OP = 300 mm, OA = 150 mm, AR = 160 mm, RS = 450 mm.

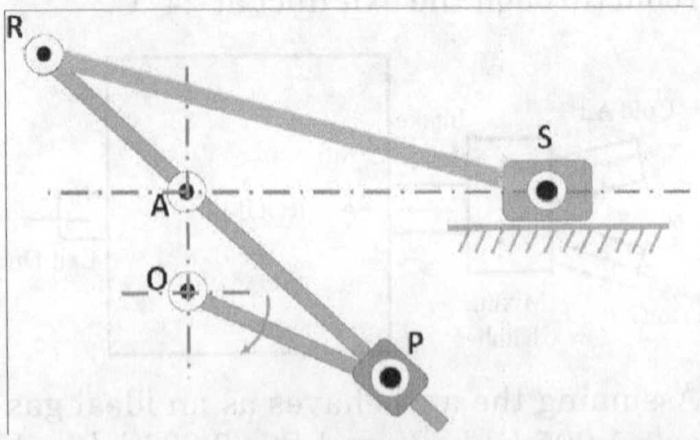

The quick return ratio for the mechanism is ________ (*round off to one decimal place*).

53. A short shoe drum (radius 260 mm) brake is shown in the figure. A force of 1 kN is applied to the lever. The coefficient of friction is 0.4.

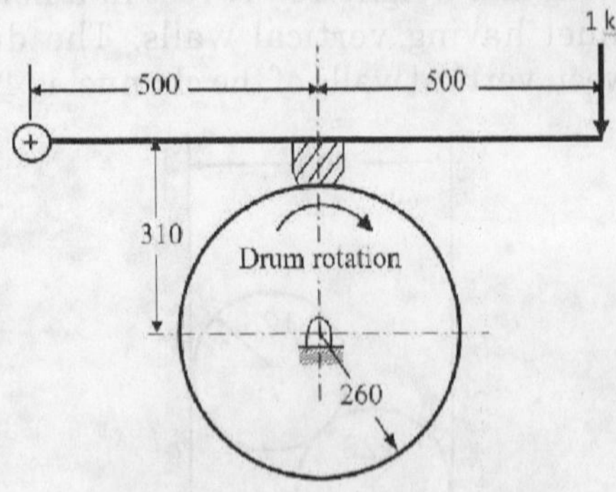

All dimensions are in mm

The magnitude of the torque applied by the brake is ________ N.m (*round off to one decimal place*).

54. A machine part in the form of cantilever beam is subjected to fluctuating load as shown in the figure. The load varies from 800 N to 1600 N. The modified endurance, yield and ultimate strengths of the material are 200 MPa, 500 MPa and 600 MPa, respectively.

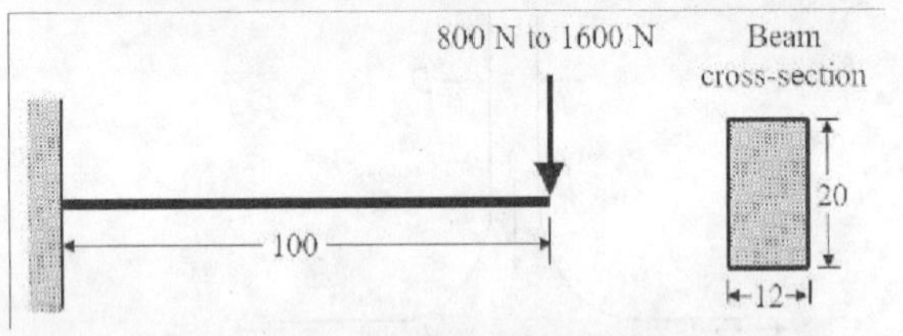

All dimensions are in mm

The factor of safety of the beam using modified Goodman criterion is ________ (*round off to one decimal place*).

55. A cantilever beam of rectangular cross-section is welded to a support by means of two fillet welds as shown in figure. A vertical load of 2 kN acts at free end of the beam.

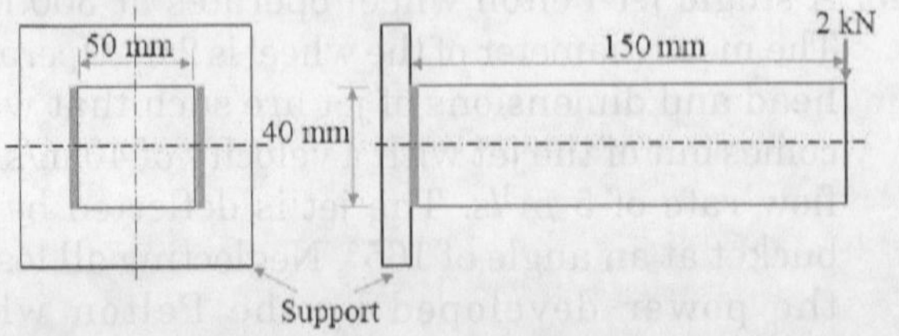

Considering that the allowable shear stress in weld is 60 N/mm^2, the minimum size (leg) of the weld required is ________ mm (*round off to one decimal place*).

ANSWERS

General Aptitude (GA)

1. (d)	**2.** (d)	**3.** (d)	**4.** (d)	**5.** (c)	**6.** (c)
7. (a)	**8.** (c)	**9.** (c)	**10.** (d)		

Mechanical Engineering

1. (c)	**2.** (c)	**3.** (d)	**4.** (a)	**5.** (d)	**6.** (b)
7. (b)	**8.** (a)	**9.** (a)	**10.** (b)	**11.** (b)	**12.** (c)
13. (a)	**14.** (a)	**15.** (a)	**16.** (c)	**17.** (c)	**18.** (d)
19. (c)	**20.** 36 to 37	**21.** 397 to 399	**22.** 79 to 81	**23.** 555 to 565	**24.** 0 to 0
25. 0.085 to 0.090	**26.** (b)	**27.** (c)	**28.** (a)	**29.** (b)	**30.** (b)
31. (a)	**32.** (b)	**33.** (a)	**34.** (c)	**35.** 0.5 to 0.5	**36.** 660 to 664
37. 30 to 36	**38.** 625 to 750	**39.** 45.00 to 50.00	**40.** 955 to 999	**41.** 0.20 to 0.30	**42.** 4 to 4
43. 0.10 to 0.12	**44.** 0.76 to 0.78	**45.** 92 to 96	**46.** 195.1 to 195.3	**47.** 11.2 to 11.4	
48. 2.50 to 2.80	**49.** 4.90 to 5.10	**50.** 124.0 to 126.0	**51.** 249.0 to 251.0	**52.** 1.9 to 2.1	
53. 199.0 to 201.0	**54.** 1.9 to 2.1	**55.** 6.4 to 6.9			

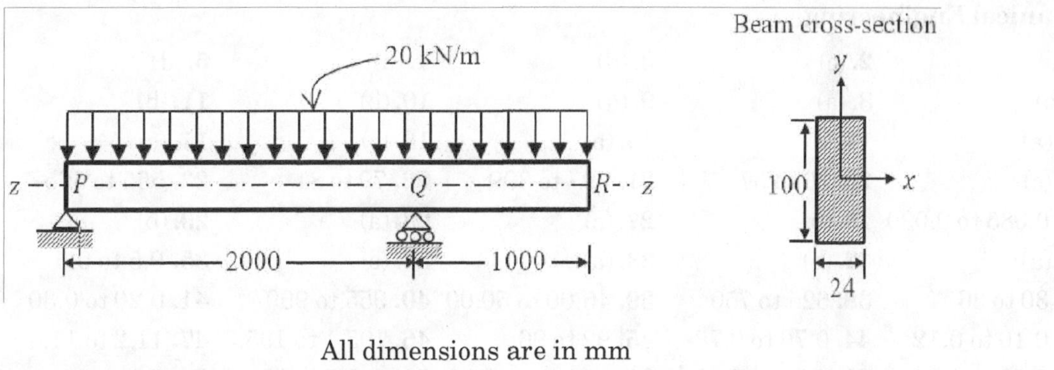

EXPLANATIONS

GENERAL APTITUDE

1. When you say you could hardly do some thing you are emphasing that it is very difficult for you to do it.

$\because$ Could + Adverb + Verb (I).

Hence, option (d) is the correct answer.

2. According to the questions:

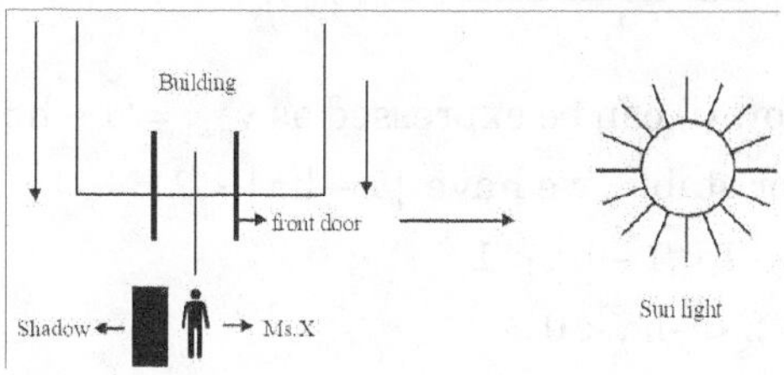

$\therefore$ From the above figure, it is clear that building is facing towards $\boxed{\text{South}}$.

Hence, option (d) is the correct answer.

3. Given that,

Area of right triangle MON = 50

According to given figure:

$$\Rightarrow \frac{1}{2} \times ON \times OM = 50$$

$$[\because ON = OM = \text{radius of circle} = r(\text{say})]$$

$$\Rightarrow \frac{1}{2} \times r \times r = 50$$

$$\Rightarrow r^2 = 100$$

$\therefore$ Area of circle with radius $r = \pi r^2$

$= \pi(100) = 100\pi$.

Hence option (d) is the correct answer.

4. As given expression =

$\Delta 2 \oplus 3\Delta((4 \otimes 2) \nabla 4)$

$= +2 - 3 + ((4 \div 2) \times 4)$

$= 2 - 3 + (2 \times 4) = 2 - 3 + 8 = 7$

Hence option (d) is the correct answer.

5. One can clearly infer that eating leafy vegetables undoubtely lead a healthy style.

7. From the options, the figure in option (b) is the missing piece. If the figure is rotated clockwise then it will form a rectangular with the given piece in the picture.

8. For farm P:

Number of hens = 65

Number of ducks = 91

Number of goats = 169

For farm Q:

Number of hens $= 416 \times \frac{5}{32} = 65$

Number of ducks $= 416 \times \frac{14}{32} = 182$

Number of goats $= 416 \times \frac{13}{32} = 169$

$\therefore$ Total number of hens = 65 + 65 = 130

Total number of ducks = 91 + 182 = 273

Total number of goats = 169 + 169 = 338

$\therefore$ The new ratio of hens: ducks: goats = 130: 273: 338 = 10: 21:26.

9. As given that,

Total number of executives, across all companies = 10,000

$\therefore$ Number of executives in $C_2 = 10{,}000 \times \frac{5}{100} = 500$

Number of executives in $C_5 = 10{,}000 \times \frac{20}{100} = 2000$

$\therefore$ Number of management degree holders among the executives in $C_2 = 500 \times \frac{1}{5} = 100$

And, Number of management degree holders among the executives in $C_5 = 2000 \times \frac{9}{10} = 1800$

So, the total number of management degree holders among the executives in companies C_5 and C_5 together = 100 + 1800 = 1900.

Hence, option (c) is the correct answer.

10. As given that

Total block is (Q_R)

Total number of ways,

$$\frac{1}{Q}\frac{2}{-} \times \frac{1}{R} \times \frac{2}{S} \times \frac{1}{-} \rightarrow 4 \text{ choices}$$

(OR)

$$\frac{2}{-} \times \frac{}{Q} \times \frac{1}{R} \times \frac{S}{-} \rightarrow 2 \text{ choices}$$

$$\frac{S}{-}\frac{2}{-} \times \frac{}{Q} \times \frac{1}{R} \rightarrow 2 \text{ choices}$$

(OR)

$$\frac{2}{R} \times \frac{}{Q} \times \frac{1}{S} \rightarrow 2 \text{ choices}$$

(OR)

$\frac{S}{} \times \frac{R}{} \times \frac{2}{} \times \frac{Q}{} \times \frac{1}{} \rightarrow 2$ choices

(OR)

$\frac{2}{S} \times \frac{2}{} \times \frac{}{R} \times \frac{1}{} \times \frac{}{Q} \rightarrow 4$ choices

Hence, total number of distinct seating arrangement possible = 4 + 2 + 2 + 2 + 2 + 4 = 16.

MECHANICAL ENGINEERING

1. As given differential equation,

$$\sin(x)\frac{dy}{dx} + y\cos x = 1$$

Differential equation can be written as

$$\frac{dy}{dx} + (\cot x)\, y = \text{cosec } x \quad .... (i)$$

$$\therefore \text{ I.F} = e^{\int \cot x\, dx} = e^{\log(\sin x)} = \sin x$$

$$\therefore \text{ solution is } y \times (\sin x) = \int \text{cosec } x \times \sin x\, dx + c$$

$$= \int 1 dx + c$$

$$\Rightarrow y \sin x = x + c \quad ...(ii)$$

Using $y\left(\frac{\pi}{2}\right) = \frac{\pi}{2}$, (ii) gives

$$\frac{\pi}{2} \times 1 = \frac{\pi}{2} + c \Rightarrow c = 0$$

$$\therefore y \sin x = x \quad ....(iii)$$

at $x = \frac{\pi}{6}$, (iii) gives $y \times \frac{1}{2} = \frac{\pi}{6}$

$\Rightarrow y = \frac{\pi}{3}$ is the value of $y\left(\frac{\pi}{6}\right)$

Hence, option (c) is the correct answer.

2. Let, $\ell = \lim_{x \to 0}\left(\frac{1 - \cos x}{x^2}\right)$ is $\frac{0}{0}$ form

Then, $\ell = \lim_{x \to 0}\frac{\sin x}{2x}$ (using L – Hospital's rule)

$$\therefore \ell = \frac{1}{2}\lim_{x \to 0}\frac{\sin x}{x} = \frac{1}{2} \times 1 = \frac{1}{2}$$

Hence, option (c) is correct answer.

3. As we know that, Dirac Delta function

$$\delta(t-a) = \lim_{\epsilon \to 0}\begin{bmatrix} 0, & -\infty < t < a \\ 1/\epsilon, & -a \le t \le a+\epsilon \\ 0, & a+\epsilon < t < \epsilon \end{bmatrix}$$

$L\{\delta(t-a)\} = e^{-as} \rightarrow$ direct result

$L\delta(t) = 1$

4. As we know that,

The Euler's iterative formula (forward) is

$$y_{n+1} = y_n + hf(t_n, y_n)$$

$$= y_n + h \times (-\pi y_n) \quad ...(i)$$

$$\Rightarrow \frac{y(t_{n+1}) - y(t_n)}{h} = -\pi y(t_n)$$

Now (i) can be expressed as $y_{n+1} = (1 - h\pi)\, y_n$

For stable, we have $|1 - h\pi| < 1$

$\Rightarrow -1 < 1 - h\pi < 1$

$\Rightarrow 2 < -h\pi < 0$

$\Rightarrow 0 < h < \frac{2}{\pi}$ $(\because |x| < \ell \Leftrightarrow -\ell < x < \ell)$

Hence, option (a) is the correct answer.

5. Since sum of independent Binomial random variables is also a Binomial random variable and Normal distribution is limiting case of Binomial distribution, if n is indefinitely large i.e., $n \rightarrow \infty$

$$\therefore y = \sum_{i=1}^{n} X_i = X_1 + X_2 + ... + X_n$$

And distribution of y as $n \rightarrow \infty$ can be approximated by Normal.

Hence option (d) is the correct answer.

6. Let, Loading path is 'OA'

Unloading path is 'AB'

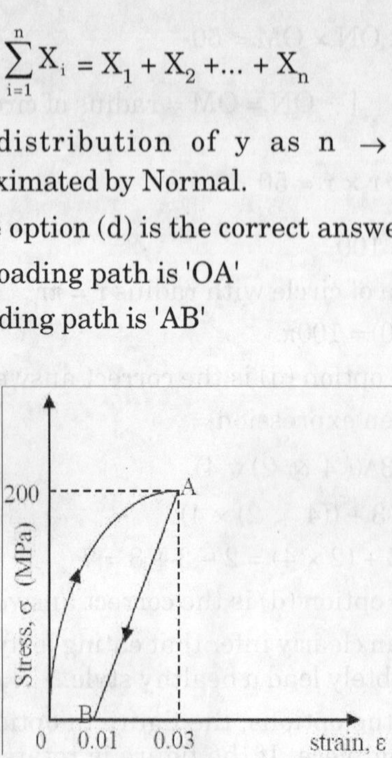

Elastic strain is nothing but the strain which is recovered = 0.03 – 0.01 = 0.02

Plastic strain is nothing but the strain which is retained = 0.01

Hence the correct option is (b)

7. As given figure

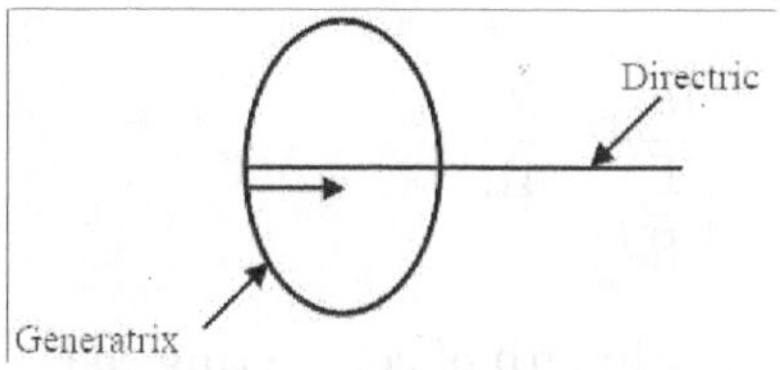

Directrix is perpendicular to generatrix means the surface obtained is cylindrical.

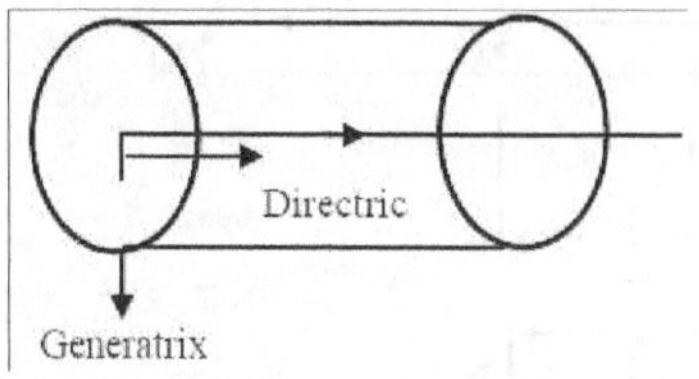

Hence, option (b) is the correct answer.

8. For large diameter through hole machining, the sequence of operation will be drilling, boring and reaming.

Hence option (a) is the correct answer.

9. Total axial displacement is eliminated by using pre-loaded ball screws, which consequently also reduces the backlash. Hence, answer is option (a).

10. The centre line average roughness, (R_a)

$$= \frac{\Sigma A}{L} = \frac{(4\times1)L/4}{L} = 1 \text{ mm.}$$

Hence, option (b) is the correct answer.

11. As we know that, Carnot cycle and striling cycle will have one isothermal process.

Carnot cycle

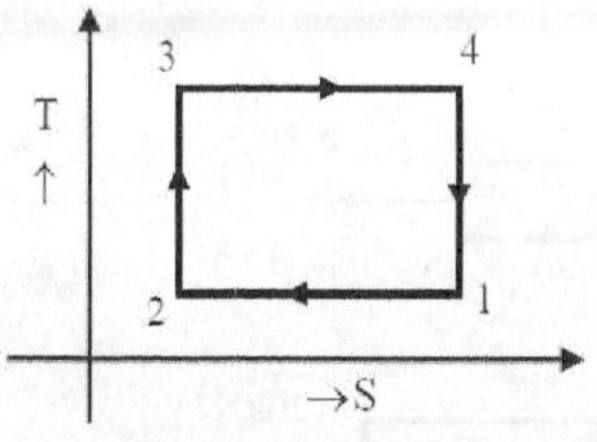

1 – 2 = Isothermal heat rejection process

2 – 3 = Isentropic compression

3 – 4 = Isothermal heat addition process

4 – 1 = Isentropic expansion

Stirling cycle

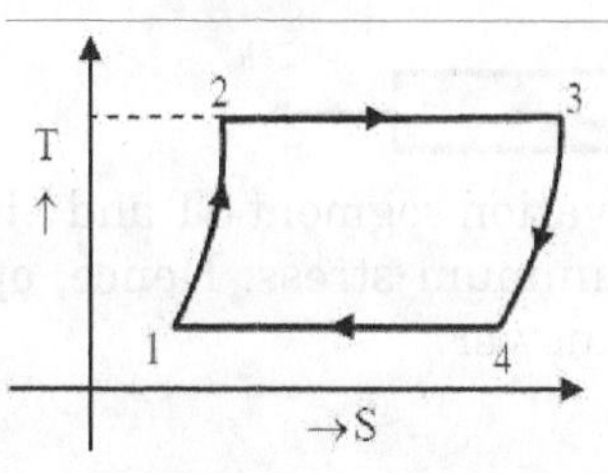

1 – 2 = Constant volume heat addition

2 – 3 = Isothermal compression

3 – 4 = Constant volume heat rejection

4 – 1 = Isothermal expansion

Hence, answer is option (b).

12. As given that,

Specific volume $(\bar{v})$ = 2.75m³/kmol

Compressibility factor (z) = 0.95

Pressure of super heated steam (P) = 1500 kPa

Temperature of steam (T) = ?

As we know that,

$$\text{Compressibility factor (z)} = \frac{P\bar{v}}{RT}$$

$$095 = \frac{500\times10^3\times2.75}{8.3143\times T\times10^3} (\because \bar{R} = 8.3143\text{kJ/kmolK})$$

$$T = 522.24K = 249.24°C$$

Hence, so answer is option (c).

13. According to the question:

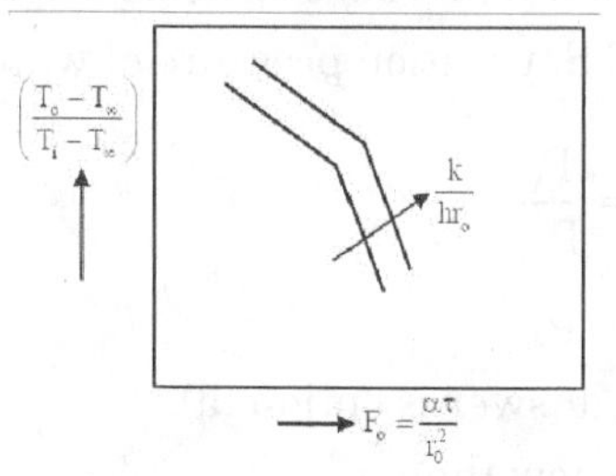

where,

T_o = Centre line temperature of sphere

T_∞ = Temperature of oil bath

T_i = Initial temperature of sphere

α = Thermal diffusivity of sphere

τ = Time

r_0 = Radius of sphere

k = Thermal conductivity of sphere

h = Convective heat transfer coefficient between Sphere and oil bath

F_o = Fourier number

$$\text{Bi = Biot number} = \frac{hr_o}{3k} \therefore \frac{k}{hr_o} \propto \frac{1}{Bi}$$

Therefore dimensionless parameters required to determine instantaneous center temperature of ball using Heisler chart are Biot number and Fourier number. Hence option (a) is the correct answer.

14. As we know that,

Heat transfer through a long fin, $(\dot{Q})$ =

$\sqrt{hPKA_C}\left(T_S - T_\infty\right)$

From the above we can say that $Q \propto \sqrt{K}$ (keeping everything else is constant)

Given that $K_2 = 2K_1$

$$\frac{\dot{Q}_2}{\dot{Q}_1} = \sqrt{\frac{K_2}{K_1}} = \sqrt{\frac{2K_1}{K_1}} = \sqrt{2}$$

$$\frac{\dot{Q}_2}{\dot{Q}_1} = \sqrt{2}$$

Hence, answer is option (a) is the correct answer.

15. As given that, relative humidity of air at 300 K is $(\phi) = 0.50$

We know that, $\phi = \dfrac{P_V}{P_{sat}}$

where,

P_v = Partial pressure of water vapour

P_{sat} = 6 Saturation pressure of water vapour

$$\Rightarrow 0.5 = \frac{P_V}{P_{sat}}$$

$P_V = 0.5\ P_{sat}$

Hence, answer is option (a).

16. As we know that,

Primary unbalance force in reciprocating engine $= mR\omega^2 \cos\theta$

$\therefore$ Secondary unbalance force in reciprocating engine $= mR\omega^2 \dfrac{\cos 2\theta}{n}$

$\therefore$ Expression for secondary unbalance force can be written as $= mR(2\omega)^2 \dfrac{\cos 2\theta}{4n}$

$$= m\left(\frac{R}{4n}\right)(2\omega)^2 \cos 2\theta$$

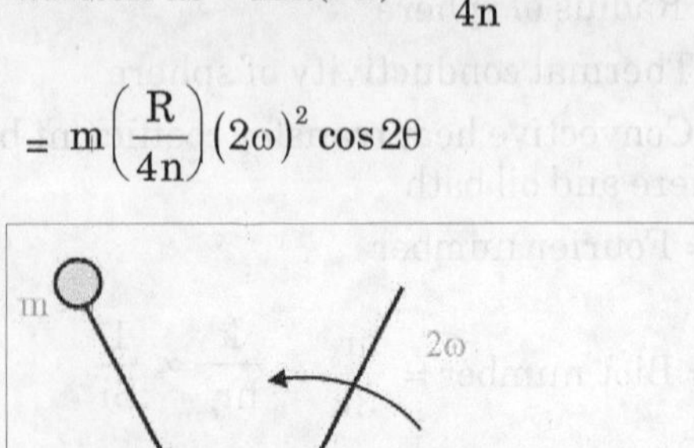

So, the effect of secondary force is equivalent to an imaginary crank of length $\dfrac{R}{4n}$ rotating at a double angular velocity, i.e., twice the engine speed.

$$\frac{R}{4n} = \frac{R}{4\left(\frac{L}{R}\right)} = \frac{R^2}{4L},$$

where L = length of connecting rod.

Hence, answer is option (a).

17.

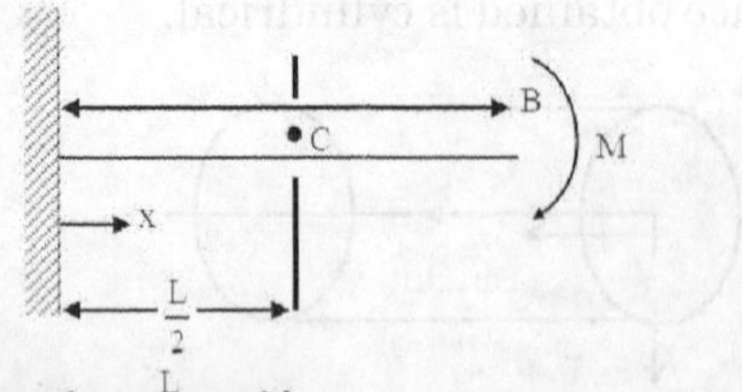

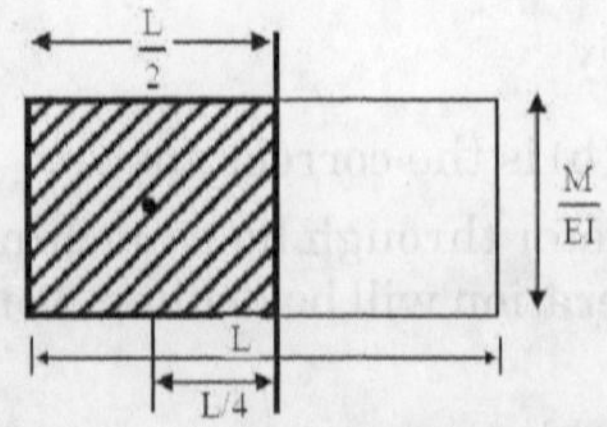

Deflection of the beam at

$x = \dfrac{L}{2}$ (at point 'C')

$= \text{Moment of} \left(\dfrac{M}{EI}\right)$ diagram about 'C' $\left(x = \dfrac{L}{2}\right)$

$$\therefore \delta_C = \left(\frac{M}{EI}\right)\left(\frac{L}{2}\right)\left(\frac{L}{4}\right) = \frac{ML^2}{8EI}\ (\downarrow)$$

Hence answer is option (c).

18. From the given figure,

$R_p = 25 + 15 - 20 - 10 \Rightarrow R_p = 10kN$

For segment PQ

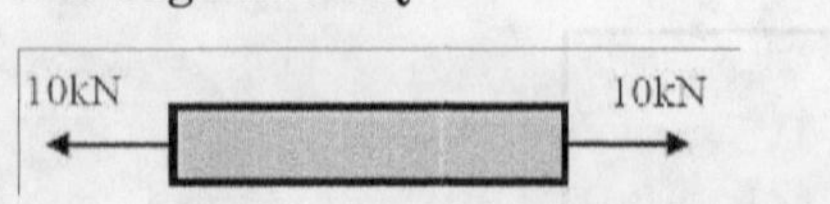

For segment QR

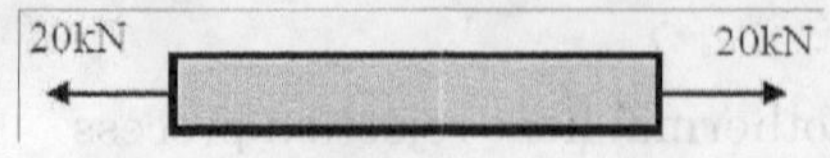

For segment SR

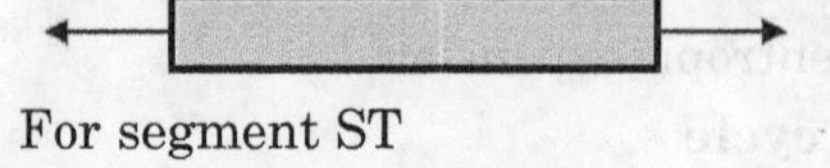

For segment ST

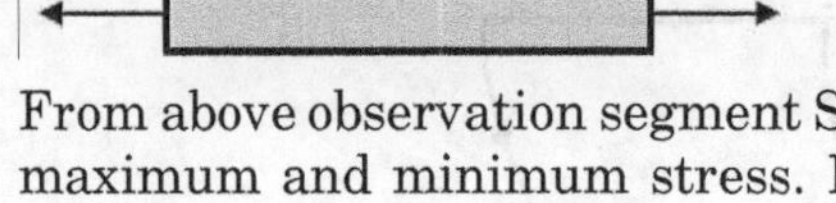

From above observation segment ST and SR will maximum and minimum stress. Hence, option (d) is the correct answer.

19. According to the question, in helical compression spring (or) helical extension spring the spring wire will be subjected to both direct shear stress (τ_1) and torsional shear stress (τ_2).

d = diameter of spring wire

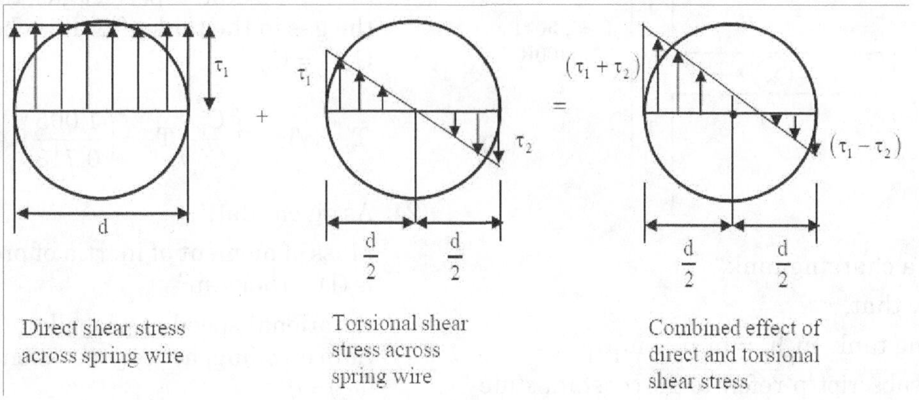

Hence, answer is option (c) is the correct answer.

20. As given that,

Probability of satisfaction = 0.95

Lead time (LT) = 5 days

Mean (μ) = 50 units

Standard deviation (σ) = 10 units

Z for 95% satisfaction = 1.64

As we know that,

$$\text{Safety stock} = Z\sqrt{LT}\,\sigma = (1.64)\sqrt{5}\times 10$$

$$= 36.67\text{units}$$

21. As given,

Density of sea water, (ρ) = 1050kg/m^3

P_{atm} = 101kPa

$= 101 \times 10^3\text{Pa} = 101 \times 10^3 \text{ N/m}^2$

$g = 9.8 \text{ m/sec}^2$

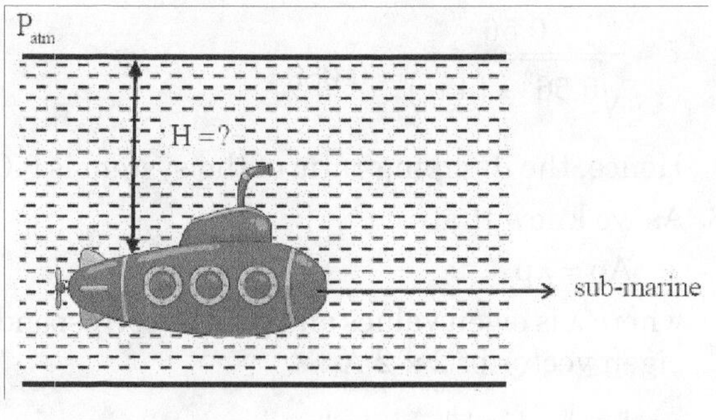

As we know that,

Pressure on the submarine,

$(P_{abs}) = P_{atm} + \rho gH$

$4.2 \times 10^6 = 101 \times 10^3 + (1050 \times 9.8 \times H)$

$H = 398.35\text{m}$

22. As given that

$u = 4\text{m/sec}$

$du = (u - 0) = (4 - 0) = 4\text{m/sec}$

$$dy = 5\text{mm} = \left(\frac{5}{1000}\text{m}\right)$$

$\rho = 800\text{kg/m}^3$

$v = 1.25 \times 10^{-4}\text{m}^2/\text{sec}$

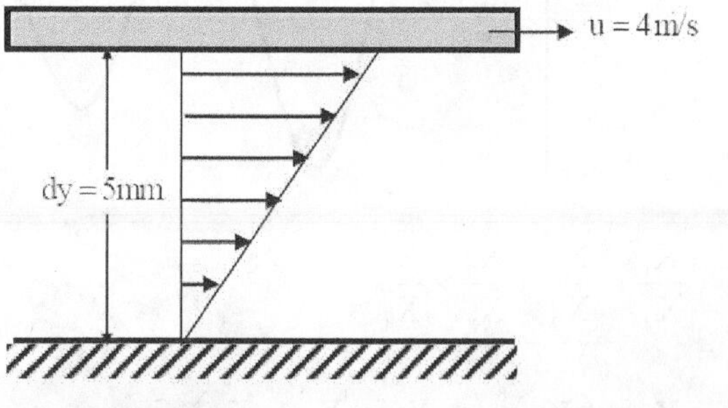

As we know that,

$$\text{Average shear stress in the fluid is } (\tau) = \mu\left(\frac{du}{dy}\right)$$

where,

$$v = \frac{\mu}{\rho} \Rightarrow \mu = (v)(\rho)$$

$\mu = (1.25 \times 10^{-4})(800) = 0.1 \text{ Nsec/m}^2$

$$\Rightarrow \tau = (0.1)\left(\frac{4}{\left(\frac{5}{1000}\right)}\right) = 80\text{N/m}^2$$

23. According to the question:

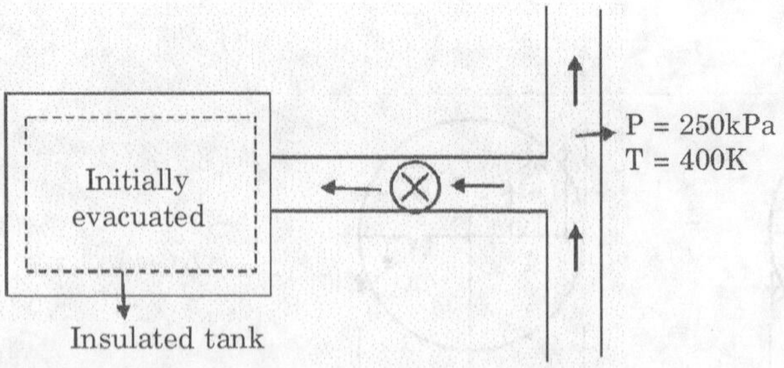

It look like a charging tank.

As we know that,

For charging tank, $m_p h_p = m_2 u_2 - m_1 u_1$

where, the subscript 'p' refers to the constant state of fluid in the pipeline. Since the tank is initially empty, $m_1 = 0$

Since, $m_p h_p = m_2 u_2$

$\therefore m_p = m_2$

$\therefore h_p = u_2$

Since, the fluid is perfect gas, the temperature of the gas in the tank after it is charged is given by $C_p T_p = C_V T_2$

$$T_2 = \gamma T_p = \left(\frac{C_P}{C_V}\right) T_p = \left(\frac{1.005}{0.718}\right)(400) = 559.89K$$

24. As given that,

Mass of moment of inertia of propeller and shaft is $(I) = 100 kg.m^2$

Rotational speed of propeller, $(\omega_S) = 12 rad / sec$

Ship is rolling, hence precession angular velocity, $(\omega_p) = 0$

$\therefore$ Gyroscopic couple (moment) generated on the shaft $(C) = I \cdot (\bar{\omega}_s \times \bar{\omega}_p) = 0$ N.m

25. As given that,

$X_1 = 8$ mm; $X_4 = 1.5$ mm

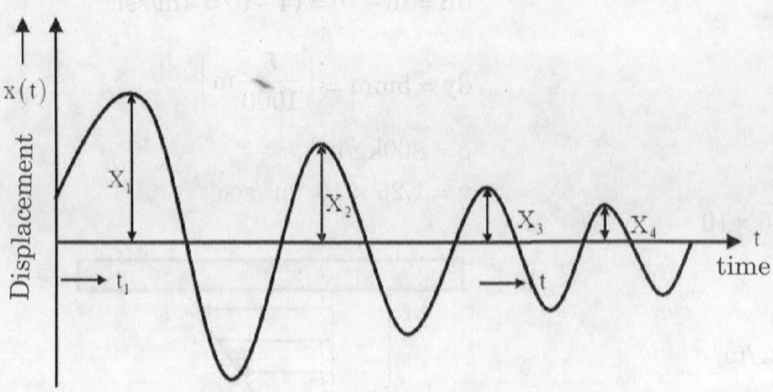

$$\frac{X_1}{X_4} = \frac{X_1}{X_2} \times \frac{X_2}{X_3} \times \frac{X_3}{X_4}$$

$$\frac{X_1}{X_4} = \left(\frac{X_1}{X_2}\right)^3 \left(\because \frac{X_1}{X_2} = \frac{X_2}{X_3} = \frac{X_3}{X_4}\right)$$

$$\frac{X_1}{X_2} = \left(\frac{X_1}{X_4}\right)^{\frac{1}{3}}$$

$$\ell n\left(\frac{X_1}{X_2}\right) = \frac{1}{3}\ell n\left(\frac{X_1}{X_4}\right)$$

Logarithmic decrement, (δ)

$$= \ell n\left(\frac{X_1}{X_2}\right) = \frac{1}{3}\ell n\left(\frac{8}{1.5}\right) = 0.56$$

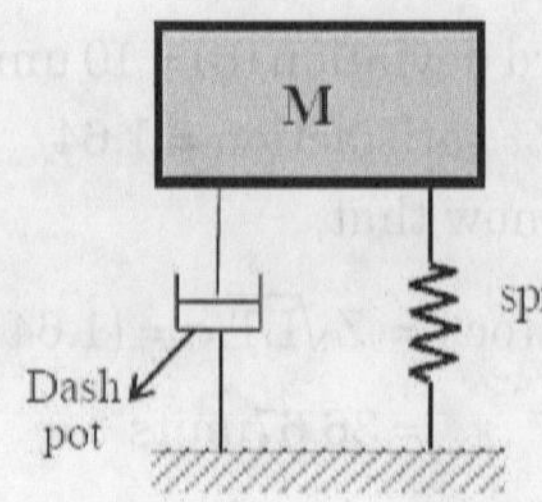

Damping factor, $(\xi) = \dfrac{\delta}{\sqrt{\delta^2 + 4\pi^2}}$

$$\xi = \frac{0.56}{\sqrt{0.56^2 + (4\pi^2)}} = 0.089$$

Hence, the damping ratio of the system is 0.089.

26. As we know that

$Ap = \lambda p$,

where λ is eigen values and p is the corresponding eigen vector of 2 × 2 A

$\Rightarrow \|p'\| = |\lambda p'\| = \lambda\ |p\|$

Since p is 2-dimensional vector

$\Rightarrow \lambda$ p and p' are also 2-dimensional vectors. Also p and λp has same direction vectors of p' = θ

Hence, option (b) is the correct answer.

27. Let, $I = \oint \frac{\cosh 3z}{2z}$, where

C : |z| = 1 and 2z = 0 gives z= 0 is a singular point lies inside C.

$$I = \int_c \frac{\cosh 3z}{2z} dz = \int_C \frac{\left(\frac{\cosh 3z}{2}\right)}{z-0} dz$$

$$I = 2\pi i \left(\frac{\cosh 3z}{2}\right)_{at\, z=0}$$

(By using Cauchys integral formula)

$$I = 2\pi i \frac{1}{2} = \pi i$$

28. As given table,

Job	M_1 Turing (Minutes)	M_2 Grinding (Minutes)
A	(2)	6
B	4	(1)
C	8	(3)
D	9	(7)
E	(7)	9
F	6	(5)
G	5	(2)
H	10	(4)

M_1								M_2
	A	E	D	F	H	C	H	B

If least time taken job is to perform on machine (M_1) should be performed first and place at near to M_1.

If least time taken job is to perform on machine 2 (M_2) should be performed at last and placed near to M_2.

If tie occurs first place near to machine 1 (M_1)

So the optimal sequence is A E D F H C G B.

Hence, the correct answer is option (a).

29. As given that,

dU = Tds + τdL

Since all the thermodynamic properties and 'dU' is exact differential like

dz = Mdx + Ndy

The condition for exact differential is

$$\left(\frac{\partial M}{dy}\right)_x = \left(\frac{\partial N}{dx}\right)_y$$

Based on above condition $\left(\frac{\partial T}{dL}\right)_S = \left(\frac{\partial \tau}{\partial S}\right)_L$

Hence, option (b) is the correct answer.

30. As we now that,

Lagrangian (L) = T – V

where,

T = kinetic energy of system

V = Potential energy of systems.

T = Linear kinetic energy of 'M' + Linear kinetic energy of 'm' in 'x' and 'y' directions

Linear kinetic energy of 'M = $\frac{1}{2}Mx^2$

Linear kinetic energy of 'm' = Sum of kinetic energy as of small strips of mass 'dm' with length 'd*l*'

dm = mass of small strip of rod

d*l* = length of strip of rod

Because force 'F' mass 'M' moves a distance 'x' and rod of mass 'm' moves an angle 'θ'.

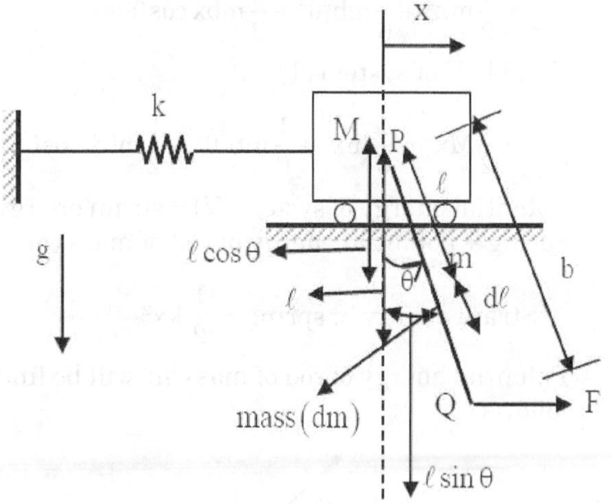

Mass 'm' is over 'b' length,

'dm' is over 'd*l*' length

$$\therefore \frac{m}{b} = \frac{dm}{dl}$$

$$dm = \left(\frac{m}{b}\right)(dl)$$

Linear displacement of strip 'dm' which is at a distance of '*l*' from P' is in x-Direction = (x + *l* sin θ)

Linear velocity of strip (dm) in x-direction = $(x + l\cos\theta\dot{\theta})$

Linear displacement of strip 'dm' in y-direction = ($\ell - \ell$ cos0)

Linear velocity of strip 'dm' in y-direction = $+\ell \sin\theta\dot{\theta}$

Resultant velocity of 'dm'

$$= \sqrt{(\dot{x} + \ell\cos\theta\dot{\theta})^2 + (\ell\sin\theta\dot{\theta})}$$

∴ Resultant K.E of small strip

$$= \frac{1}{2}(dm)\left((\dot{x} + \ell\cos\theta\dot{\theta})^2 + (\ell\sin\theta\theta)^2\right)$$

$$= \frac{1}{2}\left(\frac{m}{b}\right)(d\ell)\left(\dot{x}^2 + \ell^2\cos\theta\dot{\theta}^2 + 2\dot{x}\ell\cos\theta\dot{\theta} + \ell^2\sin^2\theta\dot{\theta}^2\right)$$

$$= \frac{1}{2}\left(\frac{m}{b}\right)(d\ell)\left(\dot{x}^2 + \ell^2\dot{\theta}^2 + 2\dot{x}\ell\cos\theta\dot{\theta}\right)$$

So, Total K.E of Rod:

$$\int_0^b \left(\frac{1}{2}\right)\left(\frac{m}{b}\right)(d\ell)\left(\dot{x}^2 + \ell^2\dot{\theta}^2 + 2\dot{x}\ell\cos\theta\dot{\theta}\right)$$

$$= \left(\frac{1}{2}\right)\left(\frac{m}{b}\right)\left(\dot{x}^2(b) + \left(\frac{b^3}{3}\right)\dot{\theta}^2(b) + 2\dot{x}\frac{b^2}{2}\cos\theta.\dot{\theta}\right)$$

$$= \frac{1}{2}m\dot{x}^2 + \frac{1}{6}mb^2\dot{\theta}^2 + \frac{1}{2}mb\dot{x}\cos\theta\dot{\theta}$$

Total K.E of system (T)

$$= \frac{1}{2}M\dot{x}^2 + \frac{1}{6}m\dot{x}^2 + \frac{1}{6}mb^2\dot{\theta}^2 + \frac{1}{2}mb\dot{x}\cos\theta\dot{\theta}$$

Potential energy of system (V) = strain energy in spring + potential energy of rod of mass (m)

∴ Strain energy in spring $= \frac{1}{2}kx^2$

Potential energy of rod of mass 'm' will be find as follows

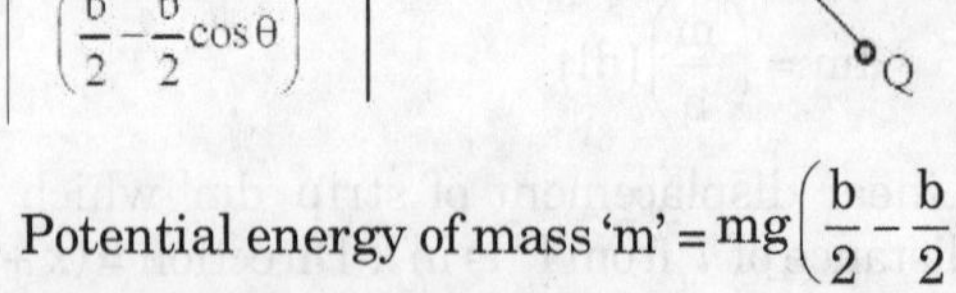

∴ Potential energy of mass 'm' $= mg\left(\frac{b}{2} - \frac{b}{2}\cos\theta\right)$

But in calculation of langrangian, will take the terms which are function 'θ' hence P.E of mass

'm' $= \frac{-mgb}{2}\cos\theta$

As we know that,

∴ Total potential energy of system (V)

$$= \frac{1}{2}kx^2 - \frac{mgb}{2}\cos\theta$$

$$L = T + V = \frac{1}{2}(m + M)\dot{x}^2 + \frac{1}{6}mb^2\dot{\theta}^2 + \frac{1}{2}mb\dot{x}\cos\theta\dot{\theta} - \frac{1}{2}kx^2 + \frac{mgb}{2}(\cos\theta)$$

$$L = \frac{1}{2}(m + M)\dot{x}^2 + \frac{1}{6}mb^2\dot{\theta}^2 + \frac{1}{2}mb\dot{x}\cos\theta\dot{\theta} - \frac{1}{2}kx^2 + \frac{mgb}{2}(\cos\theta)$$

Hence, option (b) is the correct answer.

31. As to the question:

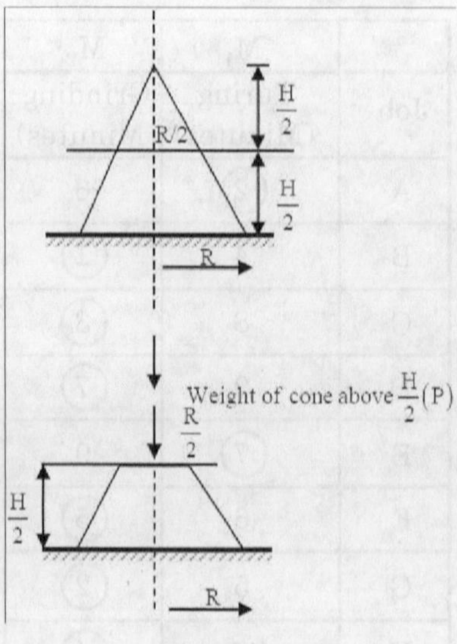

Weight of cone (P), above $\frac{H}{2}$

= volume of Cone above $\frac{H}{2}$ × specific weight of cone

$$P = \pi\left(\frac{R}{2}\right)^2\left(\frac{H}{2}\right)(W)$$

$$\text{Deflection of cone}(\delta) = \frac{4P\left(\frac{H}{2}\right)}{\pi E(4R)\left(\frac{R}{2}\right)} \quad \left(\because \delta = \frac{4PL}{\pi E d_1 d_2}\right)$$

$$\delta = \frac{(4)(\pi)\left(\frac{R}{2}\right)^2\left(\frac{H}{2}\right)(W)\left(\frac{H}{2}\right)}{(\pi E)(4R)\left(\frac{R}{2}\right)} = \frac{WH^2}{8E}$$

Hence, option (a) is the correct answer.

32. As given that

Mass moment of inertia of rocker about 'B' is $(I_B) = 10^{-4}kg.m^2$

$a = 3.5cm = 0.035m$, $b = 2.5cm = 0.025m$

Stiffness of push rod $(K_P) = 15N/mm = 15,000\ N/m$

Stiffness of valve spring $(K_V) = 10N/mm = 10,000\ N/m$

As given figure:

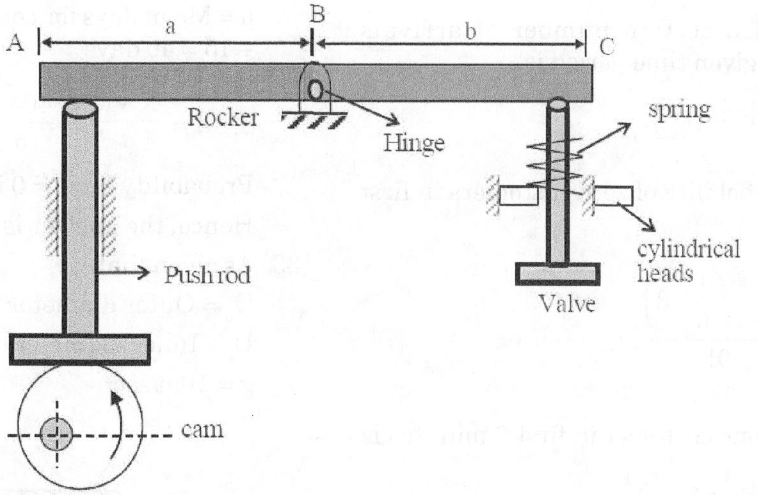

According to the question the mass of push rod and valve is negligible, then the given system will be looking like this

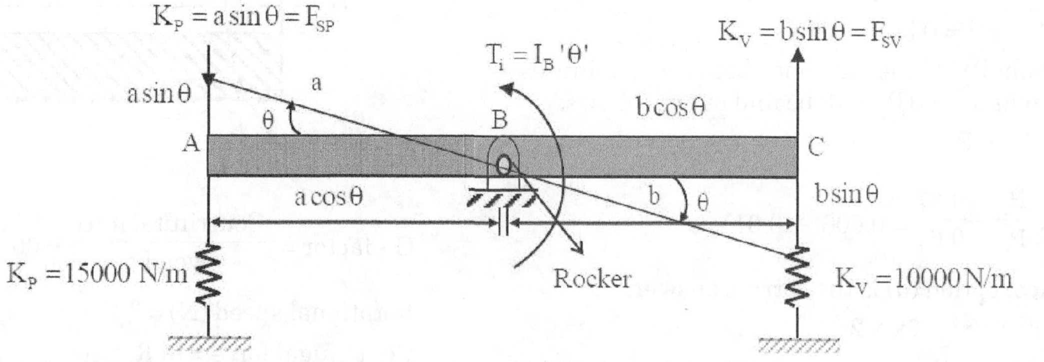

Here, that give some small angular displacement to Rocker in clock wise,

F_{SP} = Spring force of push rod = $K_p a \sin\theta$

F_{SV} = Spring force of valve spring = $K_v b \sin\theta$

$\Sigma M_B = 0$

$I_B\ddot{\theta} + (K_v b \sin\theta)(b\cos\theta) + (K_p a \sin\theta)(a\cos\theta) = 0$

For small angles

$\sin\theta \cong \theta$, $\cos\theta \cong 1$

$I_B\ddot{\theta} + (K_V b^2 + K_p a^2)\theta = 0$

$10^{-4}\ddot{\theta} + (10000 \times 0.025^2 + 15000 \times 0.035^2)\theta = 0$

$10^{-4}\ddot{\theta} + 24.6250 = 0$

$$\therefore \omega_n = \sqrt{\frac{24.625}{10^{-4}}} = 496.23\ \text{rad/sec}$$

$$\frac{2\pi N}{60} = 496.23 \Rightarrow N = 4738.65 \cong 4739\ \text{rpm}$$

Hence, option 'b' is the correct answer.

33. As given that,

Mean arrival rate, $(\lambda) = \dfrac{10 \text{ customers}}{\text{hr}}$

$= \dfrac{10}{60} = \dfrac{1}{6}$ customers/min

Probability of a certain number of arrivals occurring in a given time period is

$\therefore P(X = x) = \dfrac{e^{-(\lambda t)} (\lambda t)^x}{x!}$

Given that probability of zero customers in first 3 minutes is

$P(X = 0) = \dfrac{e^{-\left(\frac{1}{6}\times 3\right)} \left(\frac{1}{6}\times 3\right)^0}{0!}$

$\Rightarrow P_1 = 0.606 \cong 0.61$

Probability of one customer in first 6 minutes is

$P(X = 1) = \dfrac{e^{-\left(\frac{1}{6}\times 6\right)} \left(\frac{1}{6}\times 6\right)^1}{1!}$

$\Rightarrow P_3 = e^{-1} = 0.367 \cong 0.37$

Probability of one customer between 3 minutes and 6 minutes (P_2) will be find out as follows:

$P_1 \times P_2 = P_3$

$P_2 = \dfrac{P_3}{P_1} = \dfrac{0.37}{0.61} = 0.606 \cong 0.61$

Hence, option (d) is the correct answer.

34. Let $f(x) = x^2 - 2x + 2$

and $[a, b] = [1, 3]$

than $f'((x) = (2x - 2)$

Since the tangent of f(x) is parallel to the straight line joining f(1) and f(3), which is geometrical interpretation of Lagrange's mean value theorem.

$\therefore f'(x) = \dfrac{f(b) - f(a)}{b - a}$, where a = 1, b = 3

$\Rightarrow 2x - 2 = \dfrac{5-1}{2} = 2$

$\Rightarrow x = 2 \in [1, 3]$

$\therefore x = 2$

Hence, option (c) is the correct answer.

35. As given table;

Activity	A	B	C	D
Mean (days)	6	11	8	15
Variance (days^2)	4	9	4	9

Sum of variances of all activities in critical path $= 4 + 9 + 4 + 9 = 26$ days

Standard deviation of all activities in critical path $= \sqrt{26} = 5.099 \cong 5.1$

X = No. of days project to be completed = 40

μ = Mean days for completing project = 6 + 11 + 8 + 15 = 40 days

$\therefore Z = \dfrac{X - \mu}{\sigma} = \dfrac{40 - 40}{5.1} = 0$

Probability for Z = 0 is 50% i.e., 0.5

Hence, the answer is (0.5)

36. As given that

D_o = Outer diameter of casting = 250mm

D_i = Inner diameter of casting = 230mm

$g = 10\text{m/sec}^2$

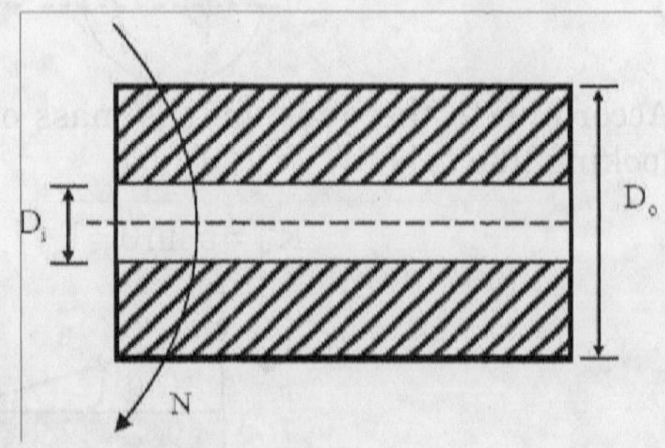

$\text{G - factor} = \dfrac{\text{Centrifutal force}}{\text{weight}} = 60$

Rotational speed (N) = ?

Centrifugal force $= m R_{mean} \omega^2$

As we know that

$R_{mean} = \left(\dfrac{D_o + D_i}{4}\right) = \left(\dfrac{250 + 230}{4}\right) = 120$ mm

$= 0.12$ m

Centrifugal force $= (m)(0.12)\omega^2$

$\text{G-factor} = \dfrac{mR\omega^2}{mg} = \dfrac{(m)(0.12)\omega^2}{(mg)} = 60$

$= \dfrac{(0.12)\omega^2}{(10)} = 60$

$\Rightarrow \omega = 70.71$ rad/sec

$\dfrac{2\pi N}{60} = 70.71 \Rightarrow N = 675.23 \cong 675$ rpm

37. As given that

Welding current (I) = 10,000A

Time taken for welding, (t) = 0.25 seconds

Contact resistance of metal sheets, (R) = 0.0001Ω

Volume of weld nugget, (V) = $70mm^3$

Heat required to melt unit volume, (Q_R)

$= 12\ J/mm^3$

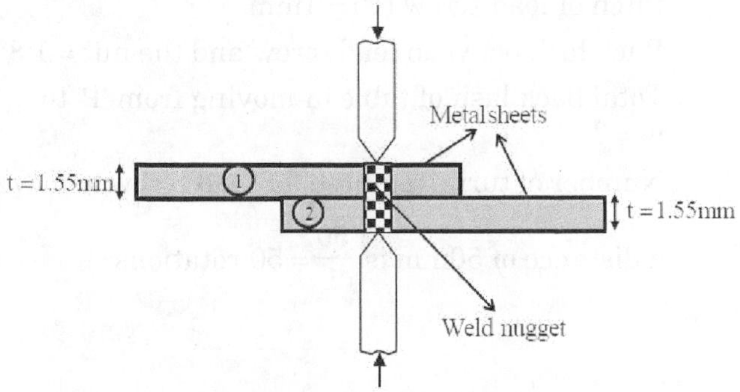

As we know that,

Heat supplied for welding, $Q_s = I^2RT$

$= (10{,}000)^2\ (0.0001)(0.25) = 2500$Joules

Total heat required, $(Q_R) = 12 \times 70 = 840J$

We also know that,

Thermal efficiency of welding process (μ)

$$= \frac{Q_R}{Q_S} = \frac{840}{2500} = 33\ 6\%$$

38. As given that

Cutting force $(F_C) = 1000N$

Friction force $(F_f) = 600N$

Uncut chip thickness $(t_1) = 0.35mm$

Chip thickness $(t_2) = 1.5mm$

Rake angle (α) = 120°

Shear force $(F_S) = ?$

Merchant's circle is as follows

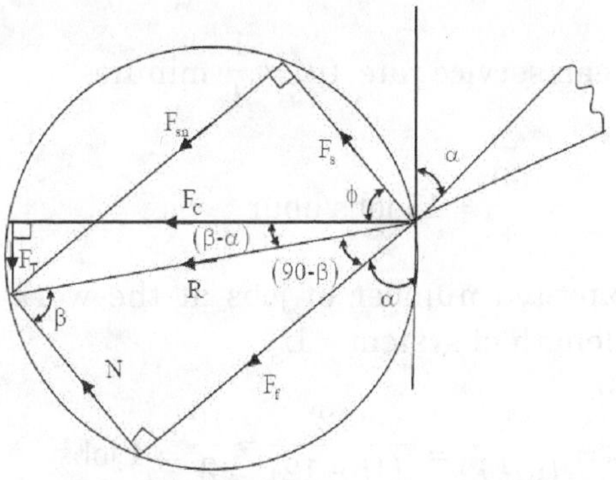

where,

φ = shear angle

$$r = \text{Chip thickness ratio} = \frac{t_1}{t_2} = \frac{0.75}{1.5} = 0.5$$

$$\therefore \tan\phi = \frac{r\cos\alpha}{1 - r\sin\alpha}$$

$$= \frac{r\cos(12)}{1 - r\sin(12)} = \frac{(0.5)\cos(12)}{1-(0.5)\sin 12}$$

φ = 28.63°

From the above circle

$R\cos(90 - \beta) = F_f = 600N$

$R\sin\beta = 600N$ ...(i)

$R\cos(\beta - \alpha) = 1000N$ ...(ii)

$$\frac{\sin\beta}{\cos(\beta-\alpha)} = \frac{600}{1000N}$$

$$\frac{\sin\beta}{\cos(\beta-12)} = \frac{600}{1000N}$$

$1000\sin\beta = 600\ (\cos\beta\cos 12 + \sin\beta\sin 12)$

$1000\sin\beta = 586.89\cos\beta + 124.75\sin\beta$

$= 875.25\sin\beta = 586.89\cos\beta$

$\tan\beta = 0.67 \Rightarrow \beta = 33.82°$

$R\sin\beta = 600$

$$R = \frac{600}{\sin(33.82)} = 1078$$

$R\cos(\phi + \beta - \alpha) = F_s$

$(1078)\cos(28.63 + 33.82 - 12) = F_s$

$F_S = 686.42N$

Hence, the shear force is 686.42N

39. As given that, grinding operations;

Specific energy consumption = $15J/mm^3$

Diameter of grinding wheel (D) = 200mm

Rotating speed of grinding wheel (N) = 3000 rpm

Material removal rate (MRR) = $6000\ mm^3/min$

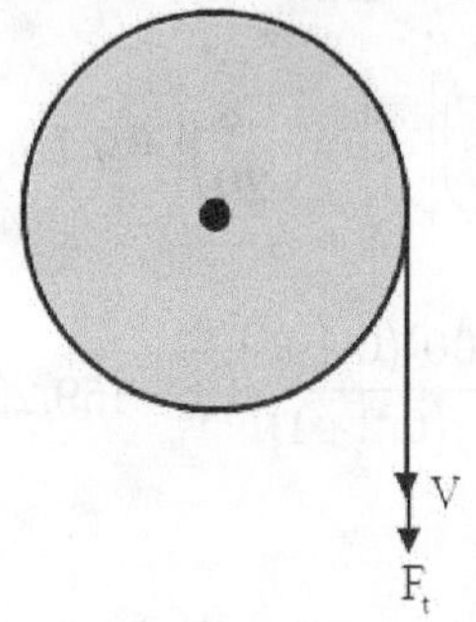

Peripheral speed of grinding wheel, (V)

$$= \frac{\pi DN}{1000}\ (m/min)$$

$$= \frac{\pi \times 200 \times 3000}{1000} = 600\ \pi\text{m/min}$$

Let F_t = Tangential force

We know that, Specific energy consumption

$$= \frac{(F_t \times V)}{MRR} = 15 = \frac{F_t \times 600\pi}{6000} = \frac{150}{\pi} = 47.77\text{N}$$

Hence, the tangential force on the wheel is 47.77N

40. As given that

Width of plate (W) = 200mm

Initial thickness of plate, (h_o) = 20mm

Radius of roller, (R) = 300mm

Final thickness of plate, (h_f) = 18mm

Speed of roller (N) = 50 rpm

Strength coefficient (K) of the work material flow curve is 300MPa

Strain hardening coefficient (n) = 0.2

Coefficient of friction between rolls and plate, (μ) = 0.1

Roll force or roll separating force = ?

As we know that

Roll force per each roller,

$$F|_{Roller} = 1.15\, LW\bar{y}_{flow}\left(1 + \frac{\mu L}{2h_{avg}}\right)$$

where,

L = Length of contact roller with plate = $\sqrt{R\Delta h}$

R = 300mm, $\Delta h = h_o - h_f = 20 - 18 = 2$mm

$L = \sqrt{300 \times 2} = 24.49$mm

W = width of plate = 200mm

$\bar{y}_{flow}$ = Flow stress = $\frac{K\varepsilon_f^n}{n+1}$

K = 300Mpa, n = 0.2

We also know that,

$$\varepsilon_f = \left|\ell n\left(\frac{h_f}{h_o}\right)\right| = \left|\ell n\left(\frac{18}{20}\right)\right| = 0.15$$

$$\therefore \bar{y}_{flow} = \frac{(300)(0.105)^{0.2}}{(0.2+1)} = 159.29\text{MPa}$$

$\mu = 0.1$

$$h_{avg} = \frac{h_o + h_f}{2} = \frac{20+18}{2} = 19\text{ mm}$$

$$F|_{Roller} = (1.15)(24.49)(200)(159.29)\left(1 + \frac{(0.1 \times 24.49)}{2 \times 19}\right)$$

= 955057.1N = 955.05kN

Hence, the roll force is 955.05 kN

41. As given that, distance moved by tool = 51 – 1 = 50 mm

Pitch of lead screw (P) = 1mm

Back lash between lead screw and the nut = 1.8°

Total back lash of table in moving from 'P' to 'Q' is = ?

Number of turns required for lead screw to move a distance of 50mm is $\frac{50}{1}$ = 50 rotations

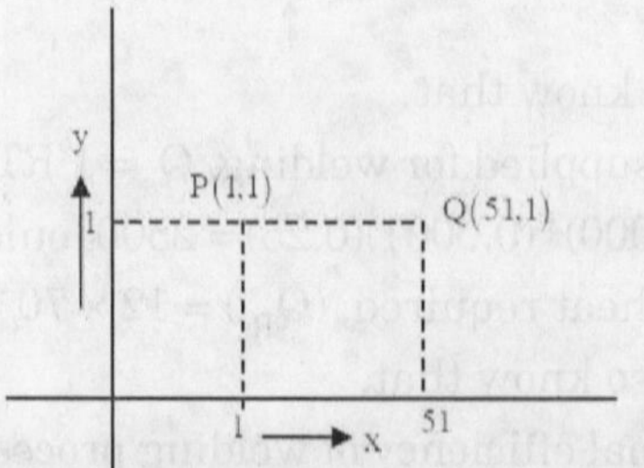

For each turn, the back lash = 1.8°

Total back lash = 50 × 1.8 = 90°

For every 360°, lead screw will move a distance of 1 mm

Back lash in mm is = $\left(\frac{90}{360}\right)$ = 0.25mm

Hence, the total backlash of the table on moving P to Q is 0.25 mm.

42. As given that,

Mean arrival rate, (λ) = 12 jobs/hour

Mean service rate, (μ) = $\frac{1}{4}$ minutes

$$= \frac{60}{4} = 15\text{ jobs/hour}$$

Expected number of jobs at the work station = length of system = L_S

$$L_S = \frac{\lambda}{(\mu - \lambda)} = \frac{12}{(15-12)} = \frac{12}{3} = 4\text{ jobs}$$

Hence, the expected number of jobs is 4.

43. As given that,

Heat rate in wire = 5 W/m

Assume unit length Heat transfer from wire,

Q = 5 watts

h = convective heat air transfer coefficient between wire and air.

As we know that,

$$\dot{Q} = \frac{T_S - T_\infty}{\left(\frac{1}{hA}\right)} \Rightarrow 5 = \frac{(348-298)}{\dfrac{1}{(h)\left(2\pi\left(\frac{1}{1000}\right)(1)\right)}}$$

So, h = 15.91 W/m²K

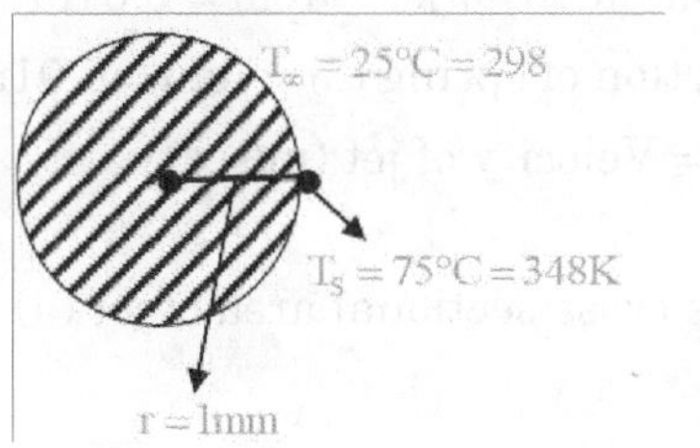

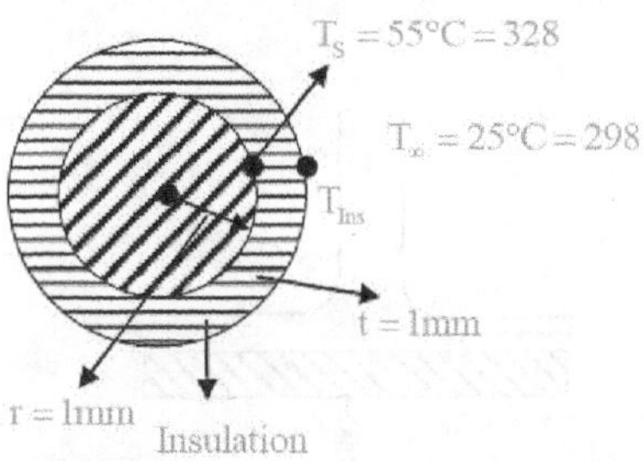

Where, T_{ins} = Temperature at the outer surface of Insulation

$$\therefore \dot{Q} = 5 = \frac{T_{Ins} - 298}{\dfrac{1}{(15.91)(2\pi)\left(\frac{2}{1000}\right)(1)}}$$

$$5 = \frac{(15.91)(2\pi)(2)(T_{Ins} - 298)}{(1000)}$$

T_{Ins} = 323K

K_{Ins} = Thermal conductivity of PVC insulation

$$\therefore \dot{Q} = \frac{\dfrac{T_S - T_{Ins}}{\ell n\left(\frac{2}{1}\right)}}{(2\pi)\times K_{Ins} L}$$

$$\Rightarrow 5 = \frac{(328-323)\left(2\pi(K_{Ins})(1)\right)}{\ell n(2)}$$

K_{Ins} = 0.11 W/mK

Hence, the terminal conductivity of PVC is 0.11 with K.

44. As given that,

Radius of sphere (r) = 10 mm

Side of cube (a) = 30mm

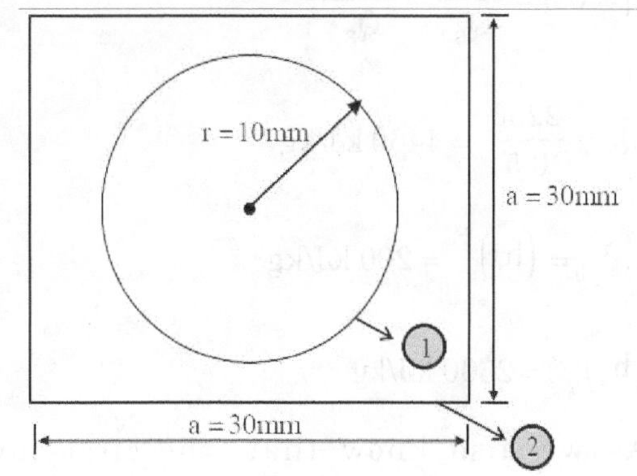

F_{22} = ?

$F_{11} + F_{12} = 1$

$F_{12} = 1$

$F_{21} + F_{22} = 1$

$A_1F_{12} = A_2F_{21}$, $A_1 = 4\pi r^2$, $A_2 = 6a^2$

$$F_{21} = \frac{A_1 \times F_{12}}{A_2} = \frac{4\pi r^2}{6a^2}\times 1 = \frac{4\pi \times 10^2}{6\times 30^2} = 0.23$$

$F_{22} = 1 - F_{21} = 1 - 0.23 = 0.77$

45. As given that, work input to the compressor (W_P) = 20 kJ/kg

Work output of high pressure turbine (W_T)
= ($h_3 - h_4$) = 750 kJ/kg

Work output of low pressure turbine $(W_T)_L$
= ($h_5 - h_6$) = 1500 kJ/kg

As we know that,

$$\text{Thermal efficiency if cycle } (\eta) = \frac{W_{net}}{Q_S} = 0.5$$

Where,

W_{net} = Network output of cycle
$= ((W_T)_H + (W_T)_L - W_P)$

So, W_{net} = (1500 + 750 – 20) = 2230 kJ/kg

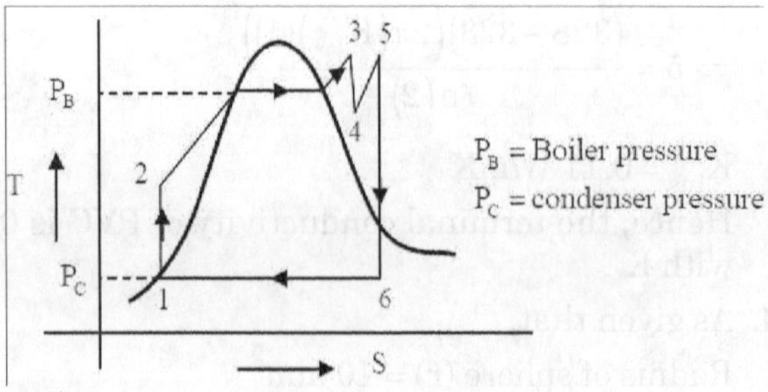

$$\eta = 0.5 = \frac{W_{net}}{Q_S} = \frac{2230}{Q_S}$$

$$Q_S = \frac{2230}{0.5} = 4460 \text{ kJ/kg}$$

$$\therefore h_1 = (h_f)_{P_c} = 200 \text{ kJ/kg}$$

$$(h_g)_{P_C} = 2600 \text{ kJ/kg}$$

As we also know that, the first law of thermodynamics:

$$\oint \delta Q = \oint dw$$

Heat supplied (Q_S) – Heat rejected (Q_R)

= Network done

$4460 - Q_R = 2230$

$Q_R = 4460 - 2230 = 2230$ kJ/kg

$$\therefore Q_R = (h_6 - h_1),\ h_6 = \left((h_f)\right)_{P_C} + x\left((h_g)_{P_C} - (h_f)_{P_C}\right)$$

$$h_f = (h_f)_{P_C}$$

$\therefore 2230 = (200 + x(2600 - 200) - (200))$

$x = 0.929 \cong 0.93 = 93\%$

Hence, answer is '93'

46. As we know that, liquid-vapour boundary in the P-T phase diagram for ammonia is

$$\ell nP = 24.38 - \frac{3063}{T}$$

Solid-vapour boundary in the P-T phase diagram for ammonia is

$$\ell nP = 27.92 - \frac{3754}{T}$$

For a triple point in P-T curve same pressure & Temperature will be there for solid-vapour, liquid – vapour phases.

$$\therefore 24.38 - \frac{3063}{T} = 27.92 - \frac{3754}{T}$$

$$\Rightarrow \frac{3754 - 3063}{T} = (27.92 - 24.38)$$

$$T = \frac{3754 - 3063}{27.92 - 24.38} = 195.2\text{K}$$

Hence, the temperature at the triple point is 195.2 K.

47. As given that

$\rho = 1000$ kg/m^2

K = 1 kN/m=1000 N/m

D = Diameter of jet = 1cm = 0.01m

Deflection of spring (δ) = 1cm = 0.01m

Let V= Velocity of jet (m/sec)

$\therefore$ The cross-sectional area of jet (a) = $\frac{\pi}{4} D^2$

$$= \frac{\pi}{4}(0.01^2) = 7.85 \times 10^{-5} \text{ m}^2$$

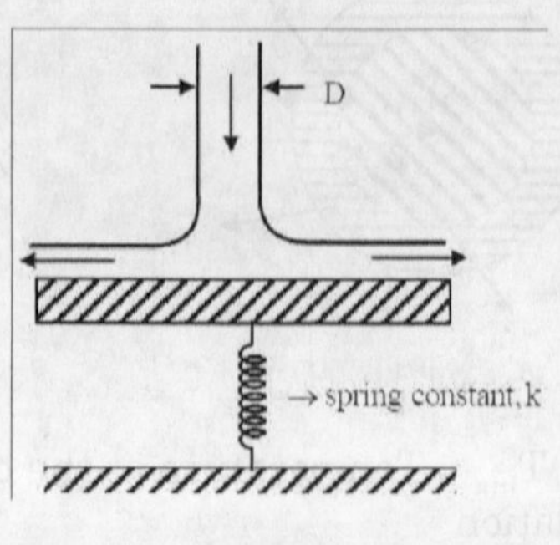

Force exerted by jet on circular plate in the direction of jet, (F) = m(V – 0)

m = mass flow rate of jet of water = ρaV

V = Velocity of jet of water striking the circular plate in the downward direction.

As we know that,

$F = \rho aV^2 = (1000)(7.85 \times 10^{-5})(V^2)$

As we also know that,

$$\text{Stiffness of spring, (K)} = \frac{F}{\delta}$$

$$1000 = \frac{(1000 \times 7.85 \times 10^{-5} \times V^2)}{(0.01)} = 11.29\text{m/sec}$$

Hence the velocity of the incoming jet is 11.29 m/sec.

48. As given that,

N = 300rpm, D = 2m

V = 40m/sec, Q = 5m³/sec

As we know that,

Power developed by

Pelton wheel (P) = $\rho av(v - u) \times u(1 + \cos\phi)$

$\phi = 180 - 165 = 15°$

$$u = \frac{\pi DN}{60} = \frac{\pi \times 2 \times 300}{60} = 10\pi \text{ m/sec}$$

$Q = \rho av = 5m^3/sec$

$\therefore P = (1000)(5)(40 - 10\pi)(10\pi)(1 + \cos 15) = 2.65MW$

Hence, the power developed by the pelton wheel is 2.65 MW.

49. As given figure:

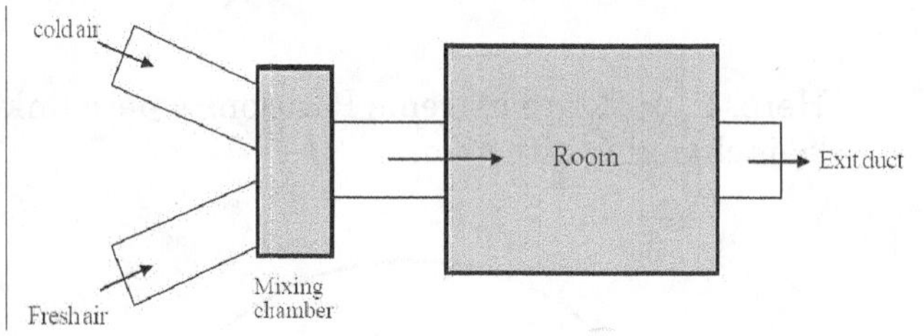

As given that,

Temperature of cold air $(T_c) = 5°C = 273 + 5 = 278K$

Pressure of cold air $(P_C) = 105kPa = 105 \times 10^3 pa$

Volume flow rate of cold air $(V_c) = 1.25 \ m^3/sec$

$(C_P)_{air} = 1.005kJ/kgK$; R = 0.287 kJ/kgK

As we know that,

ideal gas mass flow rate of cold air $(\dot{m}_c)$

$$P_C V_C = m_C RT_C$$

$(105 \times 10^3)(1.25) = \dot{m}_c (0.287 \times 10^3)(278)$

$\dot{m}_c = 1.64$ kg/sec

Mass flow rate of fresh air $(\dot{m}_F) = 1.6 \times \dot{m}_c$

$= 1.6 \times 1.64 = 2.624kg/sec$

Temperature of fresh air $(T_F) = 34°C$

$= 34 + 2.73 = 307K$

In mixing chambers,

Heat lost of fresh air = Heat gained by cold air

$$\dot{m}_F C_P (T_F - T_f) = \dot{m}_C C_P (T_f - T_C)$$

$(2.624)(1.005)(307 - T_f) = (1.64)(1.005)(T_f - 278)$

$(307 - T_f) = (0.625)(T_f - 278)$

$307 - T_f = 0.625T_f - 173.75$

$1.625T_f = 480.75$

$T_f = 298.85K$

T_f = Final temperature of mixed air = 295.85K

Final mass flow rate of air from the mixed chamber

$$\dot{m}_f = (\dot{m}_C + \dot{m}_F) = 4.264 \text{ kg / sec}$$

Exit temperature of air, $(T_E) = 24°C = 297K$

Energy balance in room:

Heat gain by air from the room $(\dot{Q}) = \dot{m}_f C_p (T_E - T_f)$

$= (4.264)(1.005)(297 - 295.85) = 4.93kW$

50. As given figure:

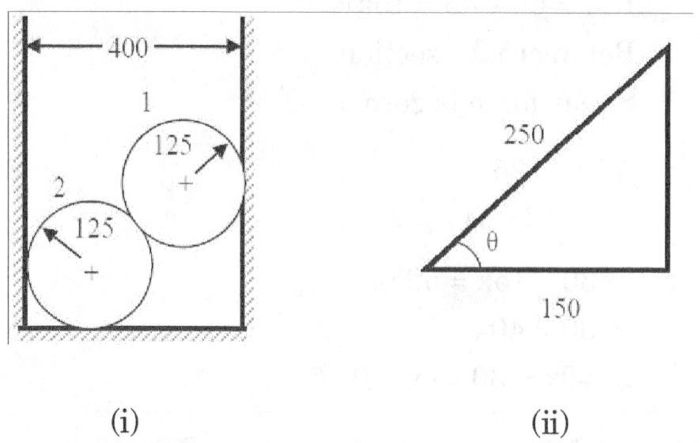

From (ii):

$$\theta = \cos^{-1}\left(\frac{150}{250}\right)$$

$\theta = 53.13°$

Considering sphere 1

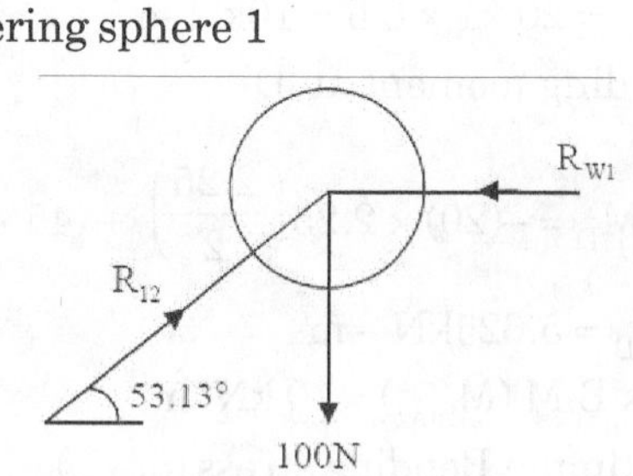

$\Sigma F_y = 0$

$R_{12} \sin 53.13° - 100 = 0$

$$R_{12} = \frac{100}{\sin 53.13} = 125N$$

Hence, the reaction at the point of constant between two sphere is 125 N.

51. As given figure:

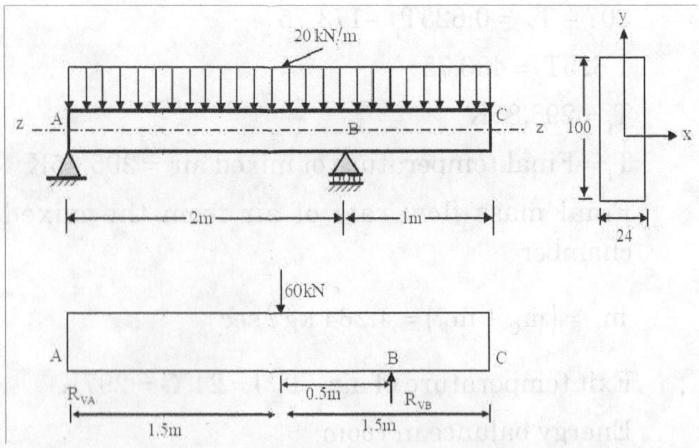

By taking the bending moment at

$\sum M_A = 0$

$R_{VB} \times 2 = 60 \times 1.5 = 45kN$

$R_{VA} + R_{VB} = 60$

$R_{VA} = 60 - 45 = 15kN$

Beam cross - section

Shear force is zero at 'D'

$$\frac{15}{x} = \frac{25}{(2-x)}$$

$\Rightarrow 30 - 15x = 525x$

$\Rightarrow 30 = 40x$

$\Rightarrow 40x = 30 \Rightarrow x = 0.75m$

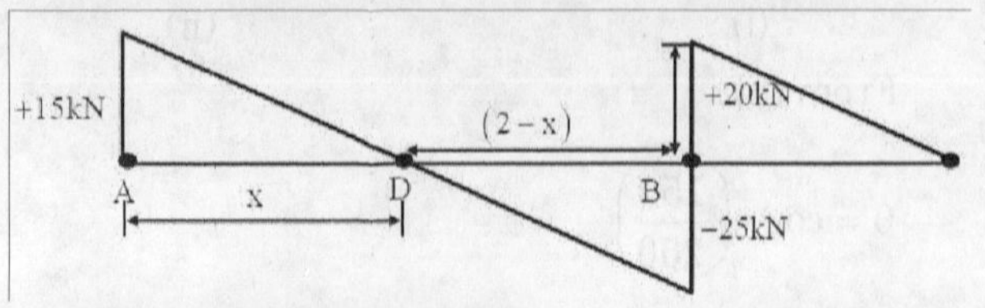

Bending moment at 'B'

$BM_B = 20 \times 1 \times 0.5 = 10kN\text{–}m$

Bending moment at 'D'

$$= BM_D = -(20) \times 2.25\left(\frac{2.25}{2}\right) + (45 \times 1.25)$$

$BM_D = 5.625kN - m$

Max B.M (M_{max}) = 10 kN–m

Maximum Bending stress (σ_{bmax})

$$= \frac{M_{max} \times y_{max}}{I}$$

$$= \frac{10 \times 10^3 \times 10^3 \times \left(\frac{100}{2}\right)}{\left(\frac{24 \times 100^3}{12}\right)} = 250\text{MPa (N/mm}^2\text{)}$$

Hence, the maximum bending stress developed in the beam is 250 mPa.

52. As given that

OA = 150mm, OP = 300mm,

AR = 160mm, RS = 450mm

The link 'S' will change from left to right or from right to left when 'P' crosses the horizontal line.

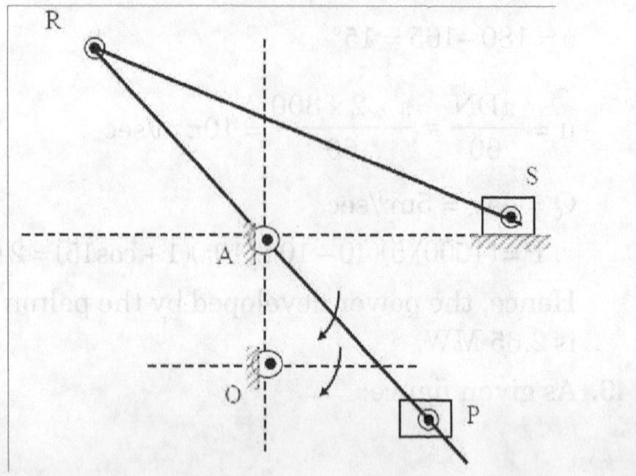

Here, P_1 & P_2 are extreme Positions where link 'S' is changing directions

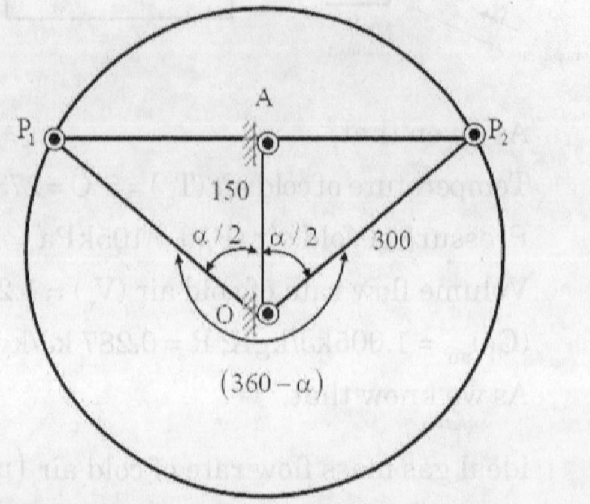

From figure above:

$$\therefore \cos\frac{\alpha}{2} = \frac{150}{300}$$

$$\frac{\alpha}{2} = \cos^{-1}\left(\frac{1}{2}\right) \Rightarrow \frac{\alpha}{2} = 60°$$

$\alpha = 120°$

As we know that,

$$\text{Quick return ratio} = \frac{360 - \alpha}{\alpha}$$

$$= \frac{(360 - 120)}{120} = \frac{240}{120} = 2.0$$

Hence, the quick return ratio for the mechanism is 2.0.

SOLVED PAPER - 2021
SET 2

INSTRUCTIONS

1. Total of 65 questions carrying 100 marks, out of which 10 questions carrying a total of 15 marks are in General Aptitude (GA)
2. The Engineering Mathematics will carry around **15% of the total marks,** the General Aptitude section will carry **15% of the total marks** and the **remaining 70% of the total marks.**
3. **Types of Questions**
   (a) Multiple Choice Questions (MCQ) carrying 1 or 2 marks each in all papers and sections. These questions are objective in nature, and each will have a choice of four options, out of which the candidate has to mark the correct answer(s).
   (b) Numerical Answer Questions of 1 or 2 marks each in all papers and sections. For these questions the answer is a real number, to be entered by the candidate using the virtual keypad. No choices will be shown for these type of questions.
   (c) Multiple Select Questions (MSQ) carrying 1 or 2 marks each in all the papers and sections. These questions are objective in nature, and each will have choice of four answers, out of which ONE or MORE than ONE choice(s) is / are correct.
4. For **1-mark** multiple-choice questions, **1/3 marks** will be deducted for a wrong answer. Likewise, for **2-marks** multiple-choice questions, **2/3** marks will be deducted for a wrong answer. There is no negative marking for numerical answer type questions & multiple selection questions.

Chapter-Wise Analysis

#	Subject	Marks		
		1	2	Total
1	General Aptitude	5	5	15
2	Engineering Mathematics	5	4	13
3	Engineering Mechanics	1	2	5
4	Strength of Materials	3	2	7
5	Theory of Machines	2	4	10

#	Subject	Marks		
		1	2	Total
6	Design of Machine Elements	-	1	2
7	Fluid Mechanics	5	3	11
8	Heat Transfer	1	2	5
9	Thermodynamics, IC Engines, Refrigeration & Power Plant Engineering Air-Conditioning	2	3	8
10	Materials, Manufacturing & Industrial Engineering	6	9	24

GENERAL APTITUDE (GA)

Q. 1 to Q. 5 : Carry One Mark Each.

1. Five persons P, Q, R, S and T are to be seated in a row, all facing the same direction, but not necessarily in the same order. P and T cannot be seated at either end of the row. P should not be seated adjacent to S. R is to be seated at the second position from the left end of the row. The number of distinct seating arrangements possible is:

(*a*) 2 (*b*) 3

(*c*) 4 (*d*) 5

2. Consider the following sentences:

(i) The number of candidates who appear for the GATE examination is staggering.

(ii) A number of candidates from my class are appearing for the GATE examination.

(iii) The number of candidates who appear for the GATE examination are staggering.

(iv) A number of candidates from my class is appearing for the GATE examination.

Which of the above sentences are grammatically CORRECT?

(*a*) (i) and (ii) (*b*) (i) and (iii)

(*c*) (ii) and (iii) (*d*) (ii) and (iv)

3. A digital watch X beeps every 30 seconds while watch Y beeps every 32 seconds. They beeped together at 10 AM.

The immediate next time that they will beep together is ________

(*a*) 10.08 AM (*b*) 10.42 AM

(*c*) 11.00 AM (*d*) 10.00 PM

4. If $\oplus \div \odot = 2; \oplus \div \Delta = 3; \odot + \Delta = 5; \Delta \times \otimes = 10,$

Then, the value of $(\otimes - \oplus)^2$, is:

(*a*) 0 (*b*) 1

(*c*) 4 (*d*) 16

5. The front door of Mr. X's house faces East. Mr. X leaves the house, walking 50 m straight from the back door that is situated directly opposite to the front door. He then turns to his right, walks for another 50 m and stops. The direction of the point Mr. X is now located at with respect to the starting point is ________

(*a*) South-East (*b*) North-East

(*c*) West (*d*) North-West

Q. 6 to Q. 10: Multiple Choice Questions (MCQ), Carry Two Marks Each

6. Given below are two statements 1 and 2, and two conclusions I and II.

Statement 1: All entrepreneurs are wealthy.

Statement 2: All wealthy are risk seekers.

Conclusion I: All risk seekers are wealthy.

Conclusion II: Only some entrepreneurs are risk seekers.

Based on the above statements and conclusions, which one of the following options is CORRECT?

(*a*) Only conclusion I is correct

(*b*) Only conclusion II is correct

(*c*) Neither conclusion I nor II is correct

(*d*) Both conclusions I and II are correct

7. A box contains 15 blue balls and 45 black balls. If 2 balls are selected randomly, without replacement, the probability of an outcome in which the first selected is a blue ball and the second selected is a black ball, is ________

(*a*) $\frac{3}{16}$ (*b*) $\frac{45}{236}$

(*c*) $\frac{1}{4}$ (*d*) $\frac{3}{4}$

8.

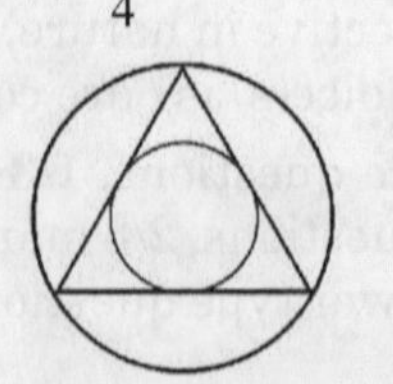

The ratio of the area of the inscribed circle to the area of the circumscribed circle of an equilateral triangle is ________

(*a*) $\frac{1}{8}$ (*b*) $\frac{1}{6}$

(*c*) $\frac{1}{4}$ (*d*) $\frac{1}{2}$

9. Consider a square sheet of side 1 unit. The sheet is first folded along the main diagonal. This is followed by a fold along its line of symmetry. The resulting folded shape is again folded along its line of symmetry. The area of each face of the final folded shape, in square units, equal to ____

(*a*) $\frac{1}{4}$ (*b*) $\frac{1}{8}$

(*c*) $\frac{1}{16}$ (*d*) $\frac{1}{32}$

10. The world is going through the worst pandemic in the past hundred years. The air travel industry is facing a crisis, as the resulting quarantine requirement for travelers led to weak demand.

In relation to the first sentence above, what does the second sentence do?

(*a*) Restates an idea from the first sentence.

(*b*) Second sentence entirely contradicts the first sentence.

(*c*) The two statements are unrelated.

(*d*) States an effect of the first sentence.

MECHANICAL ENGINEERING

Q.1 to Q.19: Multiple Choice Question (MCQ), Carry One Mark Each

1. Consider an $n \times n$ matrix A and a non-zero $n \times l$ vector p. Their product $Ap = \alpha^2 p$, where $a \in R$ and $\alpha \notin \{-1,0,1\}$. Based on the given information, the eigen value of A^2 is:

(a) α (b) α^2

(c) $\sqrt{\alpha}$ (d) α^4

2. If the Laplace transform of a function f(t) is given by $\frac{s+3}{(s+1)(s+2)}$, then f(0) is

(a) 0 (b) $\frac{1}{2}$

(c) 1 (d) $\frac{3}{2}$

3. The mean and variance, respectively, of a binomial distribution for n independent trials with the probability of success as p, are

(a) $\sqrt{np}$, $np(1-2p)$ (b) $\sqrt{np}, \sqrt{np(1-p)}$

(c) np, np (d) $np, np(1-p)$

4. The Cast Iron which possesses all the carbon in the combined form as cementite is known as

(a) Grey Cast Iron

(b) Spheroidal Cast Iron

(c) Malleable Cast Iron

(d) White Cast Iron

5. The size distribution of the powder particles used in Powder Metallurgy process can be determined by

(a) Laser scattering (b) Laser reflection

(c) Laser absorption (d) Laser penetration

6. In a CNC machine tool, the function of an interpolator is to generate

(a) signal for the lubrication pump during machining

(b) error signal for tool radius compensation during machining

(c) NC code from the part drawing during post processing

(d) reference signal prescribing the shape of the part to be machined

7. The machining process that involves ablation is

(a) Abrasive Jet Machining

(b) Chemical Machining

(c) Electrochemical Machining

(d) Laser Beam Machining

8. A PERT network has 9 activities on its critical path. The standard deviation of each activity on the critical path is 3. The standard deviation of the critical path is

(a) 3 (b) 9

(c) 27 (d) 81

9. The allowance provided in between a hole and a shaft is calculated from the difference between

(a) lower limit of the shaft and the upper limit of the hole

(b) upper limit of the shaft and the upper limit of the hole

(c) upper limit of the shaft and the lower limit of the hole

(d) lower limit of the shaft and the lower limit of the hole

10. In forced convective heat transfer, Stanton number (St), Nusselt number (Nu), Reynolds number (Re) and Prandtl number (Pr) are related as

(a) $St = \frac{Nu}{Re\,Pr}$ (b) $St = \frac{Nu\,Pr}{Re}$

(c) $St = Nu\ Pr\ Re$ (d) $St = \frac{Nu\,Re}{Pr}$

11. For a two-dimensional, incompressible flow having velocity components u and v in the x and y directions, respectively, the expression

$$\frac{\partial(u^2)}{\partial x} + \frac{\partial(uv)}{\partial y}$$

can be simplified to

(a) $u\frac{\partial u}{\partial x} + u\frac{\partial v}{\partial y}$ (b) $2u\frac{\partial u}{\partial x} + u\frac{\partial v}{\partial y}$

(c) $2u\frac{\partial u}{\partial x} + v\frac{\partial u}{\partial y}$ (d) $u\frac{\partial u}{\partial x} + v\frac{\partial u}{\partial y}$

12. Which of the following is responsible for eddy viscosity (or turbulent viscosity) in a turbulent boundary layer on a flat plate?

(a) Nikuradse stresses

(b) Reynolds stresses

(c) Boussinesq stresses

(d) Prandtl stresses

13. A two dimensional flow has velocities in x and y directions given by $u = 2xyt$ and $v = -y^2t$, where t denotes time. The equation for streamline passing through $x = 1, y = 1$ is

(a) $x^2y = 1$

(b) $xy^2 = 1$

(c) $x^2y^2 = 1$

(d) $x/y^2 = 1$

14. A plane truss PQRS (PQ = RS, and ∠PQR = 90°) is shown in the figure.

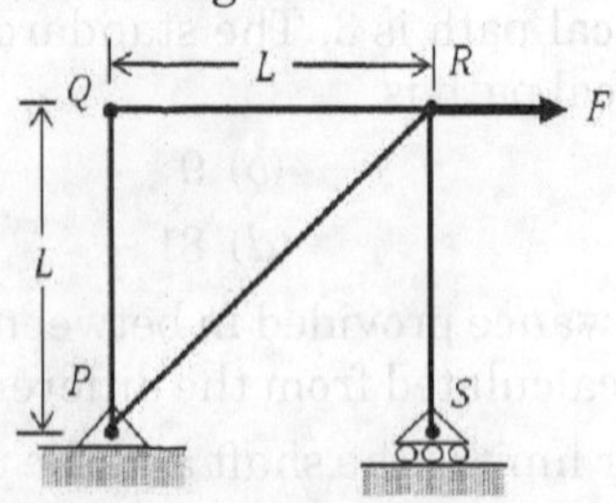

The forces in the members PR and RS, respectively, are

(*a*) F $\sqrt{2}$ (tensile) and F (tensile)
(*b*) F $\sqrt{2}$ (tensile) and F (compressive)
(*c*) F (compressive) and F $\sqrt{2}$ (compressive)
(*d*) F (tensile) and F $\sqrt{2}$ (tensile)

15. Consider the mechanism shown in the figure. There is rolling contact without slip between the disc and ground.

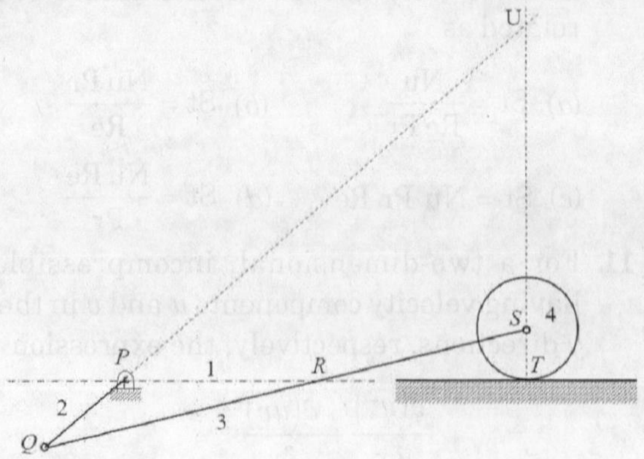

Select the correct statement about instantaneous centers in the mechanism.

(*a*) Only points P, Q, and S are instantaneous centers of mechanism
(*b*) Only points P, Q, S and T are instantaneous centers of mechanism
(*c*) Only points P, Q, R, S, and U are instantaneous centers of mechanism
(*d*) All points P, Q, R, S, T and U are instantaneous centers of mechanism

16. The controlling force curves P, Q and R for a spring controlled governor are shown in the figure, where r_1 and r_2 are any two radii of rotation.

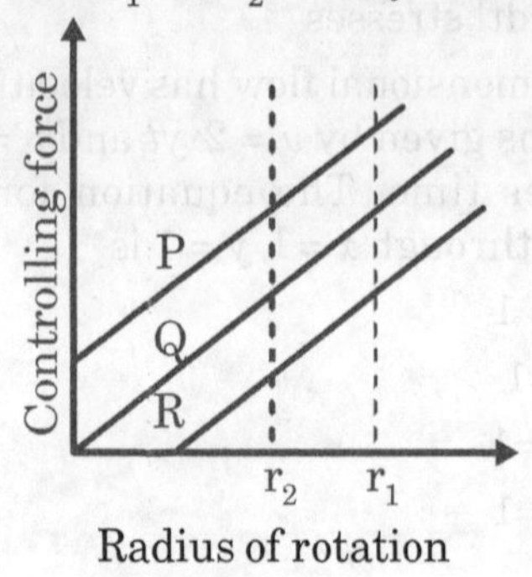

The characteristics shown by the curves are

(*a*) P - Unstable; Q - Stable; R - Isochronous
(*b*) P - Unstable; Q - Isochronous; R - Stable
(*c*) P- Stable; Q - Isochronous; R - Unstable
(*d*) P - Stable; Q - Unstable; R - Isochronous

17. The von Mises stress at a point in a body subjected to forces is proportional to the square root of the

(*a*) total strain energy per unit volume
(*b*) plastic strain energy per unit volume
(*c*) dilatational strain energy per unit volume
(*d*) distortional strain energy per unit volume

18. Value of $\int_4^{5.2} \ln x \, dx$ using Simpson's one-third rule with interval size 0.3 is

(*a*) 1.83 (*b*) 1.60
(*c*) 1.51 (*d*) 1.06

19. Value of $(1 + i)^8$, where $i = \sqrt{-1}$, is equal to

(*a*) 4 (*b*) 16
(*c*) 4*i* (*d*) 16*i*

Q.20 to Q.25: Numerical Answer Type (NAT), carry ONE mark each

20. Consider adiabatic flow of air through a duct. At a given point in the duct, velocity of air is 300 m/s, temperature is 330 K and pressure is 180 kPa. Assume that the air behaves as a perfect gas with constant c_p =1.005 kJ/kg.K. The stagnation temperature at this point is______K (*round off to two decimal places*).

21. Consider an ideal vapour compression refrigeration cycle working on R-134a refrigerant. The COP of the cycle is 10 and the refrigeration capacity is 150 kJ/kg. The heat rejected by the refrigerant in the condenser is ______kJ/kg (*round off to the nearest integer*).

22. A rigid tank of volume 50 m^3 contains a pure substance as a saturated liquid vapour mixture at 400 kPa. Of the total mass of the mixture, 20% mass is liquid and 80% mass is vapour. Properties at 400 kPa are: Tsat = 143.61 °C, V_f = 0.001084 m^3/kg, v_g = 0.46242 m^3/kg. The total mass of liquid vapour mixture in the tank is ______kg (*round off to the nearest integer*).

23. An object is moving with a Mach number of 0.6 in an ideal gas environment, which is at a temperature of 350 K. The gas constant is 320 J/kg.K and ratio of specific heats is 1.3. The speed of object is __________m/s (*round off to the nearest integer*).

24. A column with one end fixed and one end free has a critical buckling load of 100 N. For the same column, if the free end is replaced with a pinned end then the critical buckling load will be________N (*round off to the nearest integer*).

25. A steel cubic block of side 200 mm is subjected to hydrostatic pressure of 250 N/mm^2. The elastic modulus is 2×10^5 N/mm^2 and Poisson ratio is 0.3 for steel. The side of the block is reduced by_______mm (*round off to two decimal places*).

Q. 26 to Q.34: Multiple Choice Questions (MCQ), Carry Two Mark Each

26. The value of $\int_0^{\frac{\pi}{2}} \int_0^{\cos\theta} r \sin\theta \, dr \, d\theta$ is

(*a*) 0 (*b*) $\frac{1}{6}$

(*c*) $\frac{4}{3}$ (*d*) π

27. Let the superscript T represent the transpose operation. Consider the function $f(X) = \frac{1}{2}x^T Qx - r^T x$, where x and r are $n \times 1$ vectors and Q is a symmetric $n \times n$ matrix. The stationary point of $f(x)$ is

(*a*) $Q^T r$ (*b*) $Q^{-1} r$

(*c*) $\frac{r}{r^T r}$ (*d*) r

28. Consider the following differential equation

$$(1+y)\frac{dy}{dx} = y$$

The solution of the equation that satisfies the condition y(1) = 1 is

(*a*) $2ye^y = e^x + e$ (*b*) $y^2 e^y = e^x$

(*c*) $ye^y = e^x$ (*d*) $(1+y)e^y = 2e^x$

29. A factory produces m(i = 1, 2, ..., m) products, each of which requires processing on n(j = 1,2,..,n) workstations. Let a_{ij} be the amount of processing time that one unit of the ith product requires on the jth workstation. Let the revenue from selling one unit of the ith product be r_i and h_i be the holding cost per unit per time period for the ith product. The planning horizon consists of T (t = 1,2,..., T) time periods. The minimum demand that must be satisfied in time period t is dit, and the capacity of the jth workstation in time period t is c_{jt}. Consider the aggregate planning formulation below, with decision variables S_{it} (amount of product i sold in time period t), X_{it} (amount of product i manufactured in time period t) and I_{it} (amount of product i held in inventory at the end of time period t.

$$\max \sum_{t=1}^{T} \sum_{i=1}^{m} \left(r_i S_{it} - h_i I_{it}\right)$$

subject to

$S_{it} \geq d_{it} \ \forall \ i,t$

<capacity constraint>

<inventory balance constraint>

$X_{it}, S_{it}, I_{it} \geq 0; I_{i0} = 0$

The capacity constraints and inventory balance constraints for this formulation are

(*a*) $\sum_i^m a_{ij} X_{it} \leq C_{jt} \ \forall \ j, t$ and $I_{it} = I_{i,t-1} + X_{it} - S_{it} \ \forall \ i, t$

(*b*) $\sum_i^m a_{ij} X_{it} \leq C_{jt} \ \forall \ i, t$ and $I_{it} = I_{i,t-1} + X_{it} - d_{it} \ \forall \ i, t$

(*c*) $\sum_i^m a_{ij} X_{it} \leq d_{it} \ \forall \ i, t$ and $I_{it} = I_{i,t-1} + X_{it} - S_{it} \ \forall \ i, t$

(*d*) $\sum_i^m a_{ij} X_{it} \leq d_{it} \ \forall \ i, t$ and $I_{it} = I_{i,t-1} + S_{it} - X_{it} \ \forall \ i, t$

30. Ambient pressure, temperature, and relative humidity at a location are 101 kPa, 300 K, and 60%, respectively. The saturation pressure of water at 300 K is 3.6 kPa. The specific humidity of ambient air is_________g/kg of dry air.

(*a*) 21.4 (*b*) 35.1

(*c*) 21.9 (*d*) 13.6

31. A plane frame PQR (fixed at P and free at R) is shown in the figure. Both members (PQ and QR) have length, L, and flexural rigidity, EI. Neglecting the effect of axial stress and transverse shear, the horizontal deflection at free end, R, is

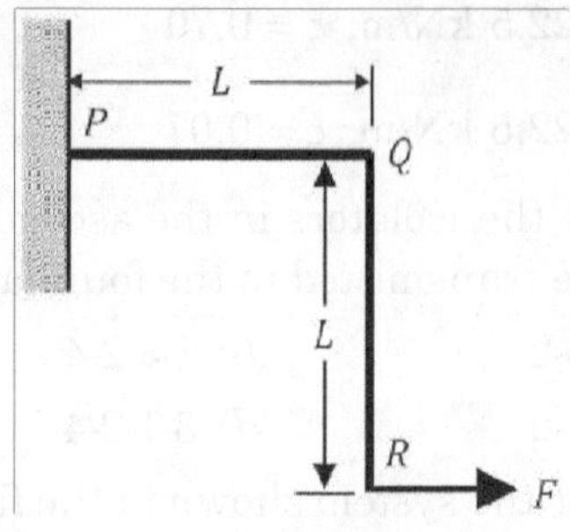

(*a*) $\frac{5FL^3}{3EI}$ (*b*) $\frac{4FL^3}{3EI}$

(*c*) $\frac{2FL^3}{3EI}$ (*d*) $\frac{FL^3}{3EI}$

32. A power transmission mechanism consists of a belt drive and a gear train as shown in the figure.

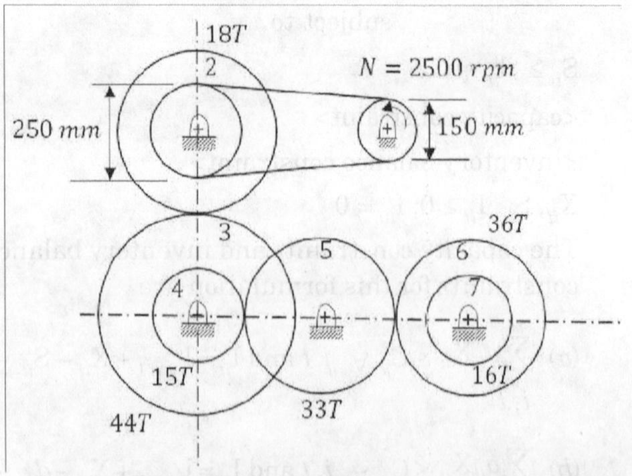

Diameters of pulleys of belt drive and number of teeth (T) on the gears 2 to 7 are indicated in the figure. The speed and direction of rotation of gear 7, respectively, are

(*a*) 255.68 rpm; clockwise

(*b*) 255.68 rpm; anticlockwise

(*c*) 575.28 rpm; clockwise

(*d*) 575.28 rpm; anticlockwise

33. A machine of mass 100 kg is subjected to an external harmonic force with a frequency of 40 rad/s. The designer decides to mount the machine on an isolator to reduce the force transmitted to the foundation. The isolator can be considered as a combination of stiffness (K) and damper (damping factor, ξ) in parallel. The designer has the following four isolators:

1) K = 640 kN/m, $\xi = 0.70$

2) K = 640 kN/m, $\xi = 0.07$

3) K = 22.5 kN/m, $\xi = 0.70$

4) K = 22.5 kN/m, $\xi = 0.07$

Arrange the isolators in the ascending order of the force transmitted to the foundation.

(*a*) 1-3-4-2 (*b*) 1-3-2-4

(*c*) 4-3-1-2 (*d*) 3-1-2-4

34. Consider the system shown in the figure. A rope goes over a pulley. A mass, *m*, is hanging from the rope. A spring of stiffness, *k*, is attached at one end of the rope. Assume rope is inextensible, massless and there is no slip between pulley and rope.

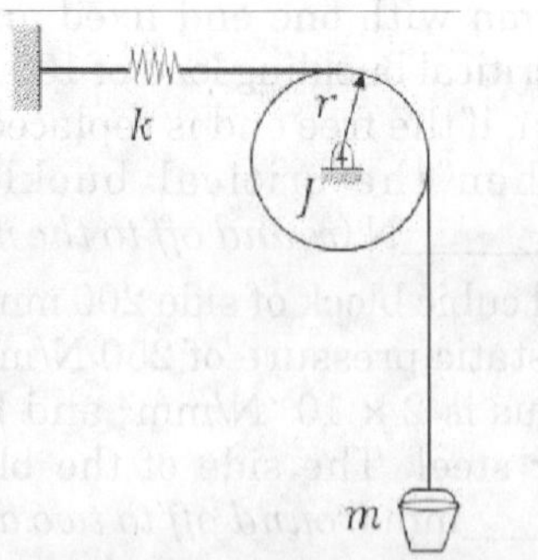

The pulley radius is *r* and its mass moment of inertia is *J*. Assume that the mass is vibrating harmonically about its static equilibrium position. The natural frequency of the system is

(*a*) $\sqrt{\dfrac{kr^2}{J - mr^2}}$ (*b*) $\sqrt{\dfrac{kr^2}{J + mr^2}}$

(*c*) $\sqrt{k/m}$ (*d*) $\sqrt{\dfrac{kr^2}{J}}$

Q.35 to Q.55: Numerical Answer Type (NAT), Carry Two Mark Each

35. Find the positive real root of $x^3 - x - 3 = 0$ using Newton-Raphson method. If the starting guess (x_0) is 2, the numerical value of the root after two iterations (x_2) is__________(*round off to two decimal places*).

36. Daily production capacity of a bearing manufacturing company is 30000 bearings. The daily demand of the bearing is 15000. The holding cost per year of keeping a bearing in the inventory is ₹20. The setup cost for the production of a batch is ₹1800. Assuming 300 working days in a year, the economic batch quantity in number of bearings is__________(*in integer*).

37. A cast product of a particular material has dimensions 75 mm × 125 mm × 20 mm. The total solidification time for the cast product is found to be 2.0 minutes as calculated using Chvorinov's rule having the index, $n = 2$. If under the identical casting conditions, the cast product shape is changed to a cylinder having diameter = 50 mm and height = 50 mm, the total solidification time will be ________ minutes (*round off to two decimal places*).

38. A spot welding operation performed on two pieces of steel yielded a nugget with a diameter of 5 mm and a thickness of 1 mm. The welding time was 0.1 s. The melting energy for the steel is 20 J/mm^3. Assuming the heat conversion efficiency as 10%, the power required for performing the spot welding operation is_______kW (*round off to two decimal places*).

39. A surface grinding operation has been performed on a Cast Iron plate having dimensions 300 mm (*length*) × 10 mm (*width*) × 50 mm (*height*). The grinding was performed using an alumina wheel having a wheel diameter of 150 mm and wheel width of 12 mm. The grinding velocity used is 40 m/s, table speed is 5 m/min, depth of cut per pass is 50 μm and the number of grinding passes is 20. The average tangential and average normal force for each pass is found to be 40 N and 60 N respectively. The value of the specific grinding energy under the aforesaid grinding conditions is ________ J/mm^3 (*round off to one decimal place*).

40. In a pure orthogonal turning by a zero rake angle single point carbide cutting tool, the shear force has been computed to be 400 N. The cutting velocity, V_c = 100 m/min, depth of cut, t = 2.0 mm, feed, s_0 = 0.1 mm/revolution and chip velocity, V_f = 20 m/min, the shear strength, τ_s of the material will be________ MPa (*round off to two decimal places*).

41. The thickness, width and length of a metal slab are 50 mm, 250 mm and 3600 mm, respectively. A rolling operation on this slab reduces the thickness by 10% and increases the width by 3%. The length of the rolled slab is ________mm (*round off to one decimal place*).

42. A 76.2 mm gauge block is used under one end of a 254 mm sine bar with roll diameter of 25.4 mm. The height of gauge blocks required at the other end of the sine bar to measure an angle of 30° is________mm (*round off to two decimal places*).

43. The demand and forecast of an item for five months are given in the table.

Month	Demand	Forecast
April	225	200
May	220	240
June	285	300
July	290	270
August	250	230

The Mean Absolute Percent Error (MAPE) in the forecast is________% (*round off to two decimal places*)

44. A shell and tube heat exchanger is used as a steam condenser. Coolant water enters the tube at 300 K at a rate of 100 kg/s. The overall heat transfer coefficient is 1500 $W/m^2.K$, and total heat transfer area is 400 m^2. Steam condenses at a saturation temperature of 350 K. Assume that the specific heat of coolant water is 4000 J/kg.K. The temperature of the coolant water coming out of the condenser is________K (*round off to the nearest integer*).

45. Ambient air flows over a heated slab having flat, top surface at y = 0. The local temperature (in Kelvin) profile within the thermal boundary layer is given by T(y) = 300 + 200 exp(–5y), where y is the distance measured from the slab surface in meter. If the thermal conductivity of air is 1.0 W/m.K and that of the slab is 100 W/m.K, then the magnitude of temperature gradient $|dT/dy|$ within the slab at y = 0 is________K/m (*round off to the nearest integer*).

46. Water flows out from a large tank of cross-sectional area A_t = 1 m^2 through a small rounded orifice of cross-sectional area A_0 = 1 cm^2, located at y = 0. Initially the water level, measured from y = 0, is H = 1 m. The acceleration due to gravity is 9.8 m/s^2.

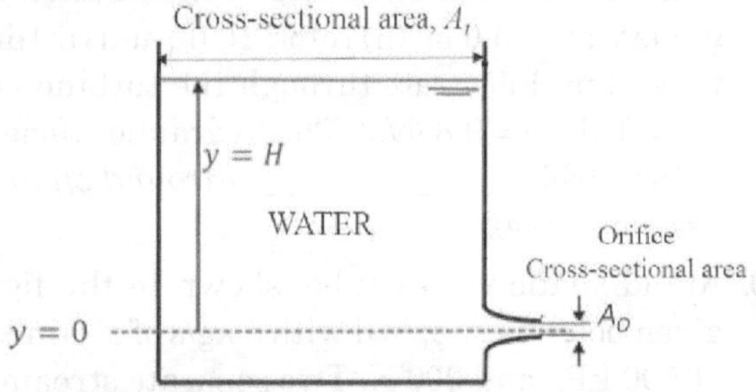

Neglecting any losses, the time taken by water in the tank to reach a level of y = H / 4 is ________seconds (*round off to one decimal place*).

47. Consider the open feed water heater (FWH) shown in the figure given below:

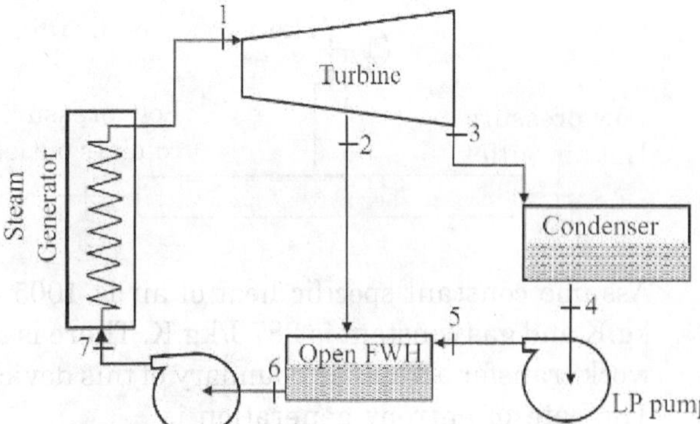

Specific enthalpy of steam at location 2 is 2624 kJ/kg, specific enthalpy of water at location 5 is 226.7 kJ/kg and specific enthalpy of saturated water at location 6 is 708.6 kJ/kg. If the mass flow rate of water entering the open feed water heater (at location 5) is 100 kg/s then the mass flow rate of steam at location 2 will be ________ kg/s (*round off to one decimal place*).

48. A high velocity water jet of cross section area = 0.01 m^2 and velocity = 35 m/s enters a pipe filled with stagnant water. The diameter of the pipe is 0.32 m. This high velocity water jet entrains additional water from the pipe and the total water leaves the pipe with a velocity 6m/s as shown in the figure.

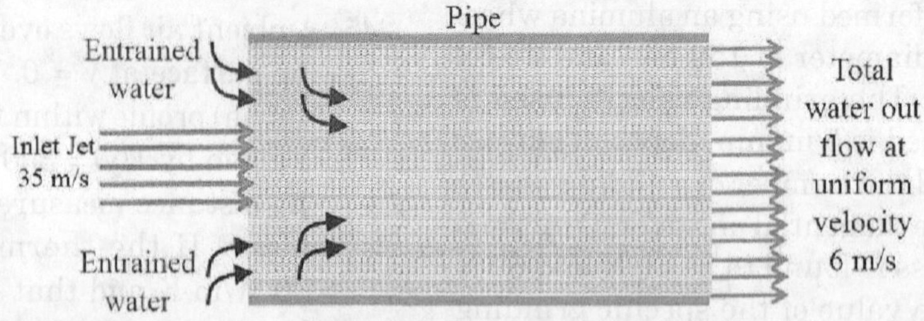

The flow rate of entrained water is ________ litres/s (*round off to two decimal places*).

49. A vertical shaft Francis turbine rotates at 300 rpm. The available head at the inlet to the turbine is 200 m. The tip speed of the rotor is 40 m/s. Water leaves the runner of the turbine without whirl. Velocity at the exit of the draft tube is 3.5 m/s. The head losses in different components of the turbine are: (i) stator and guide vanes: 5.0 m, (ii) rotor: 10 m, and (iii) draft tube: 2 m. Flow rate through the turbine is 20 m^3/s. Take g = 9.8 m/s^2. The hydraulic efficiency of the turbine is ________ % (*round off to one decimal place*).

50. An adiabatic vortex tube, shown in the figure given below is supplied with 5 kg/s of air (inlet 1) at 500 kPa and 300 K. Two separate streams of air are leaving the device from outlets 2 and 3. Hot air leaves the device at a rate of 3 kg/s from outlet 2 at 100 kPa and 340 K, and 2 kg/s of cold air stream is leaving the device from outlet 3 at 100 kPa and 240 K.

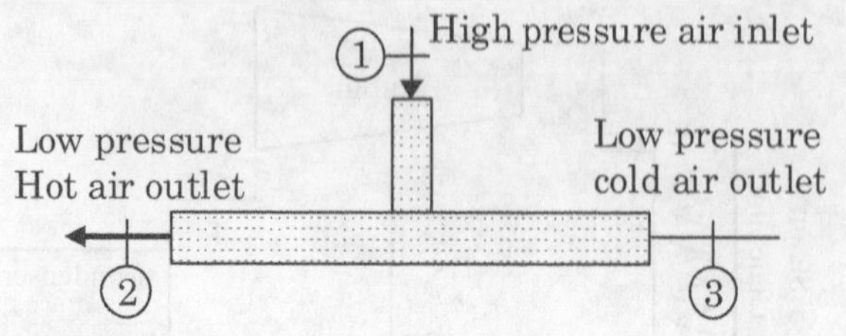

Assume constant specific heat of air is 1005 J/kg.K and gas constant is 287 J/kg.K. There is no work transfer across the boundary of this device. The rate of entropy generation is ________ kW/K (*round off to one decimal place*).

51. A block of negligible mass rests on a surface that is inclined at 30° to the horizontal plane as shown in the figure. When a vertical force of 900 N and a horizontal force of 750 N are applied, the block is just about to slide.

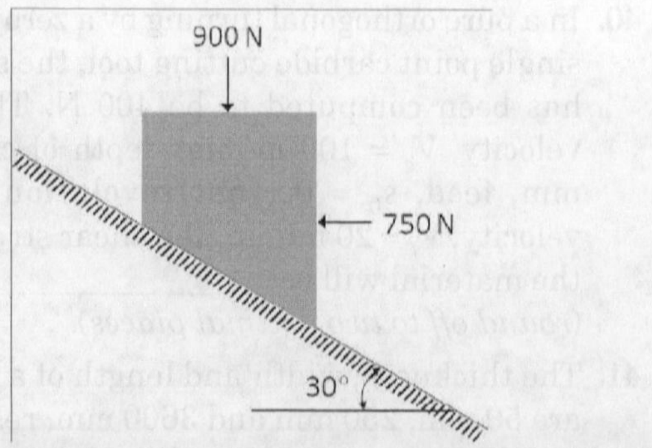

The coefficient of static friction between the block and surface is ________ (*round off to two decimal places*).

52. The wheels and axle system lying on a rough surface is shown in the figure.

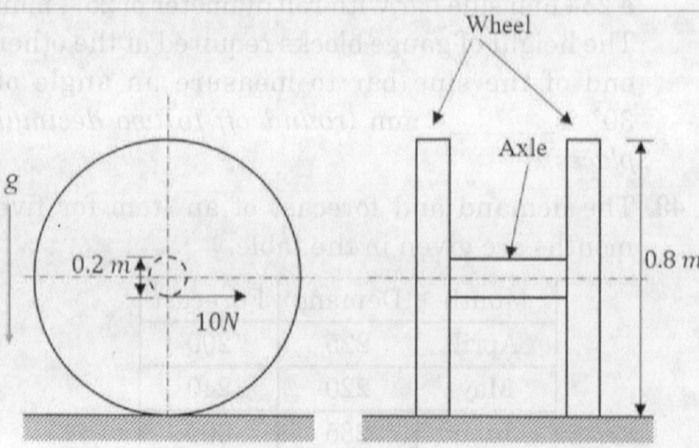

Each wheel has diameter 0.8 m and mass 1 kg. Assume that the mass of the wheel is concentrated at rim and neglect the mass of the spokes. The diameter of axle is 0.2 m and its mass is 1.5 kg. Neglect the moment of inertia of the axle and assume g = 9.8 m/s^2. An effort of 10 N is applied on the axle in the horizontal direction shown at mid span of the axle. Assume that the wheels move on a horizontal surface without slip. The acceleration of the wheel axle system in horizontal direction is________ m/s^2 (*round off to one decimal place*).

53. A cantilever beam with a uniform flexural rigidity (EI = 200×10^6 N.m^2) is loaded with a concentrated force at its free end. The area of the bending moment diagram corresponding to the full length of the beam is 10000 N.m^2. The magnitude of the slope of the beam at its free end is ________ micro radian (*round off to the nearest integer*).

54. The torque provided by an engine is given by $T(\theta) = 12000 + 2500 \sin(2\theta)$ N.m, where θ is the angle turned by the crank from inner dead center. The mean speed of the engine is 200 rpm and it drives a machine that provides a constant resisting torque. If variation of the speed from the mean speed is not to exceed ±0.5%, the minimum mass moment of inertia of the flywheel should be ____________kg.m^2 (*round off to the nearest integer*).

55. The figure shows the relationship between fatigue strength (S) and fatigue life (N) of a material. The fatigue strength of the material for a life of 1000 cycles is 450 MPa, while its fatigue strength for a life of 10^6 cycles is 150 MPa.

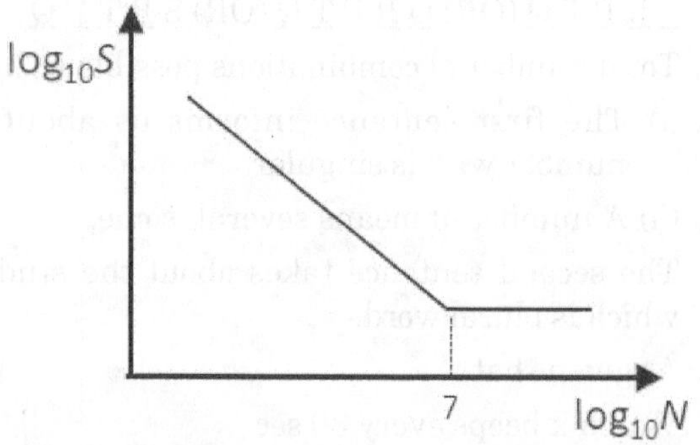

The life of a cylindrical shaft made of this material subjected to an alternating stress of 200 MPa will then be____________cycles (*round off to the nearest integer*).

ANSWERS

General Aptitude (GA)

1. (*b*)	**2.** (*a*)	**3.** (*a*)	**4.** (*b*)	**5.** (*d*)	**6.** (*c*)
7. (*b*)	**8.** (*c*)	**9.** (*b*)	**10.** (*d*)		

Mechanical Engineering

1. (*d*)	**2.** (*c*)	**3.** (*d*)	**4.** (*d*)	**5.** (*a*)	**6.** (*d*)
7. (*d*)	**8.** (*b*)	**9.** (*c*)	**10.** (*a*)	**11.** (*d*)	**12.** (*b*)
13. (*b*)	**14.** (*b*)	**15.** (*d*)	**16.** (*b*)	**17.** (*d*)	**18.** (*a*)
19. (*b*)	**20.** 373.00 to 377.00		**21.** 165 to 165	**22.** 134 to 136	**23.** 228 to 230
24. 800 to 840	**25.** 0.08 to 0.12	**26.** (*b*)	**27.** (*b*)	**28.** (*c*)	**29.** (*a*)
30. (*d*)	**31.** (*b*)	**32.** (*a*)	**33.** (*c*)	**34.** (*b*)	
35. 1.66 to 1.68	**36.** 40200 to 40300		**37.** 2.60 to 3.00	**38.** 39.00 to 40.00	
39. 57 to 58	**40.** 388.00 to 400.00		**41.** 3880.0 to 3886.0		
42. 200.00 to 206.00		**43.** 6.00 to 10.00	**44.** 337 to 341	**45.** 10 to 10	
46. 2257.0 to 2259.0		**47.** 25.0 to 25.4	**48.** 130.00 to 134.00		
49. 90.0 to 92.0	**50.** 2.1 to 2.3	**51.** 0.16 to 0.19	**52.** 1.0 to 1.5	**53.** 48 to 52	**54.** 560 to 580
55. 152000 to 165000					

EXPLANATIONS

GENERAL APTITUDE

1. Positive arrangements are

S R P T Q (OR) Q R P T Q (OR) S R T P Q

Total number of combinations possible is 3.

2. (i) The first sentence informs us about the number with is singular.

(ii) A number of means several, some.

The second sentence takes about the students which is plural word.

3. As given that,

Watch x beeps every 30 sec

Watch y beeps every 32 sec

L.C.M of 30, 32 = 480 seconds = 8 minutes

They beeps together every 8 min

$\therefore$ i.e., 10 AM, 10.08 AM; 10.16 AM, 10.24 AM, ...

Hence option (a) is the correct answer.

4. (i) $\oplus \div \odot \Rightarrow \frac{\oplus}{\odot} = 2$

(ii) $\oplus \div \Delta = 3 \Rightarrow \frac{\oplus}{\Delta} = 3$

(iii) $\odot + \Delta = 5 \Rightarrow \frac{\oplus}{2} + \frac{\oplus}{3} = 5,\ \oplus = 6$ and $\Delta = \frac{6}{3} = 2.$

(iv) $\Delta \times \otimes = 10 \Rightarrow 2 \times \otimes = 10\ \otimes = 5$

So, $(\otimes - \oplus)^2 = (5-6)^2 = 1$

5. From the figure, it is clear that, the direction of X is North West

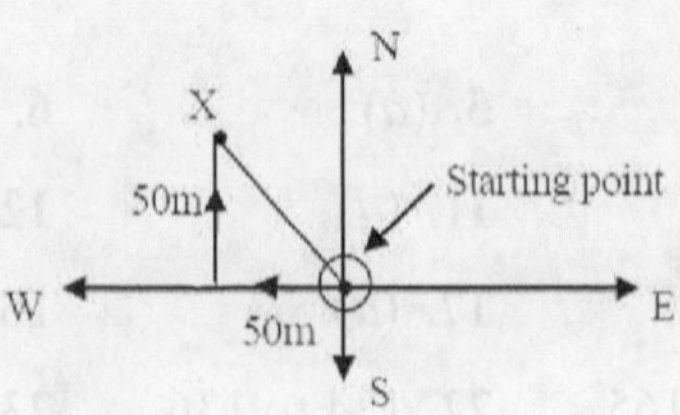

6. E $\rightarrow$ Entrepreneurs

W $\rightarrow$ Wealthy

R $\rightarrow$ Risk seekers

Statement-1

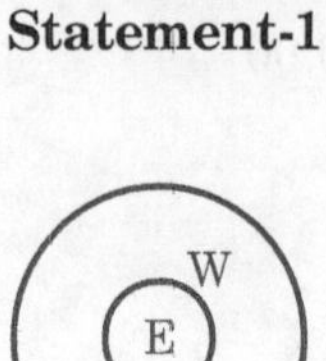

All entrepreneurs are wealthy

Statement-2

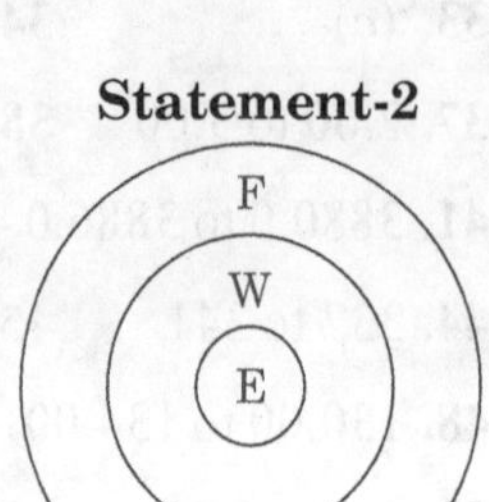

All wealthy are risk seekers

$\therefore$ All risk sectors are not wealthy $\Rightarrow$ Conclusion I is incorrect.

All entrepreneurs are risk seekers $\Rightarrow$ Conclusion II is incorrect

Hence, neither conclusion I nor II is correct.

7. Box

Total balls = | 15 Blue | 45 Black |

If the two balls are drawn without replacement the required probability of getting first ball blue and second ball black will be

$\therefore$ Required probability = $\frac{15}{60} \times \frac{45}{59}$

$= \frac{45}{236}.$

8. $\therefore$ From the figure,

$\sin 30° = \frac{r}{R} \Rightarrow \frac{1}{2} = \frac{r}{R} \Rightarrow R = 2r$

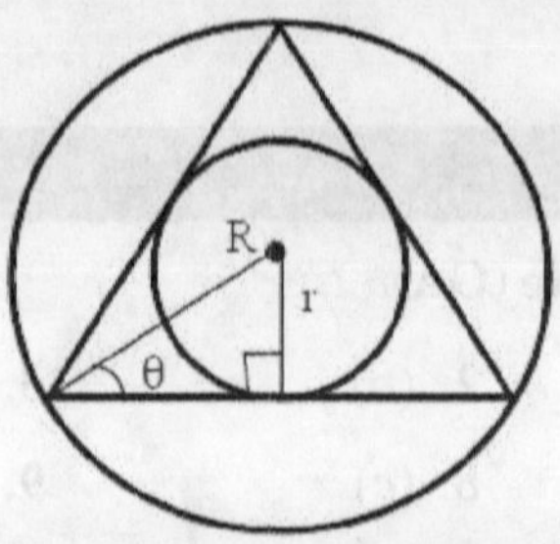

r $\rightarrow$ radius of inscribed circle

R $\rightarrow$ radius of circumscribed circle

$\therefore$ Area of inscribed circle = πr^2

Area of circumscribed circle = πR^2

$= \pi(4r^2)$

$\therefore$ Required ratio = $\pi r^2 : 4\pi r^2$

$= 1 : 4$

$= \frac{1}{4}$

9. $\therefore$ Area of each face of the final folded shape. (Right angle triangle)

$= \frac{1}{2} \times \frac{1}{2} \times \frac{1}{2}\left(\frac{1}{2} \times \text{Box} \times \text{weight}\right)$

$= \frac{1}{8}$ square units

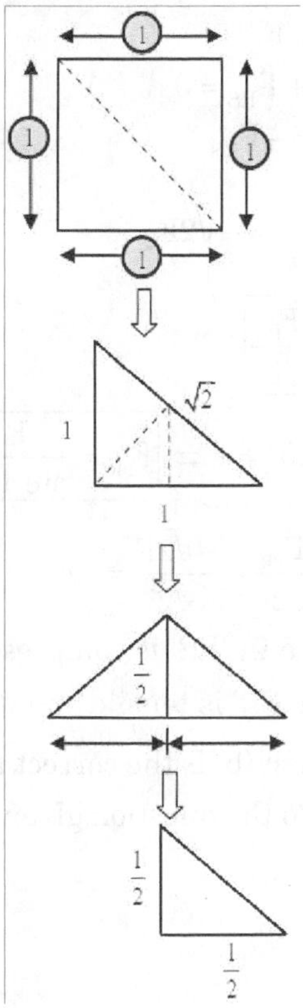

Hence, option (b) is the correct answer.

10. Due to worst pandemic the air travel industry is facing a crisis is nothing but restarting the idea from first sentence.

MECHANICAL ENGINEERING

1. As given that,

$Ap = \alpha^2 p$ ...(i)

Pre-multiply both sides by A, we get

$A(AP) = A(\alpha^2 p)$

$\Rightarrow (AA)p = \alpha^2 (Ap)$

$\Rightarrow A^2 p = \alpha^2(\alpha^2 p) = \alpha^4 p$ ($\because$ From (i))

Hence, α^4 is the eigenvalue of A^2

2. As given that,

$$F(s) = L\{f(t)\} = \frac{s+3}{(s+1)(s+2)}$$

$\therefore f(0) = \lim_{s\to\infty} sF(s)$ (Using initial value theorem)

$$= \lim_{s\to\infty} \frac{s(s+3)}{s(s+1)(s+2)}$$

$$= \lim_{s\to\infty} \frac{\left(1+\frac{3}{s}\right)}{\left(1+\frac{1}{s}\right)\left(1+\frac{2}{s}\right)} = \frac{1}{1\times 1} = 1$$

Hence, option (c) is correct answer.

3. Mean and variance of a binomial distribution are np and npq respectively, where q = 1 – p i.e.., np and np(1 – p)

Hence, option (d) is correct answer.

4. While cast iron has carbon in the form of cementite hence the answer is option (d).

5. Laser diffraction (or) laser scattering used in size distribution of powder particles used in powder metallurgy process. Hence option 'a' is the correct answer.

6. In CNC machine tool, the function of an interpolator is to generate reference signal prescribing the shape of the part to be machined. Hence option (d) is correct.

7. Ablation means removal of material by melting or evaporation.

From the given options, the only laser beam machining method will remove the material by melting and vaporization.

Hence the answer is option (d).

8. Number of activities on the critical path = 9

Standard deviation of each activity in the critical path $(\sigma) = 3$

Variance of each activity in the critical path $= \sigma^2 = 9$

Sum of variances of all activities in the critical path $= 9\sigma^2 = 9 \times 9 = 81$

Standard deviation of critical path $= \sqrt{81} = 9$

Hence answer is option (b).

9. As we know that, Allowance

= Lower limit of hole – Upper limit of shaft

Hence option (c) is the correct answer.

10. As we know that,

$$\text{Stanton Number (st)} \frac{j}{ec_p\mu_m} = \frac{h\Delta T}{ec_p\mu_m\Delta T}$$

$$= \frac{\dfrac{h\Delta T}{k\left(\dfrac{\Delta L}{L}\right)}}{\dfrac{ec_p\left(\mu_m\Delta_T(\mu)\right)}{k\left(\dfrac{\Delta T}{L}\right)\mu}} = \frac{Nu}{(Pr)\left(\dfrac{e\mu_m L}{\mu}\right)} = \frac{Nu}{Pr.Re}$$

Hence option (a) is the correct answer.

11. As we know that,

For a two-dimensional compressible flow, continuity equation is

$$\frac{\partial u}{\partial^2 x} + \frac{\partial v}{\partial y} = 0$$

$$\Rightarrow \frac{\partial u}{\partial x} = \frac{-\partial v}{\partial y} \Rightarrow \frac{\partial v}{\partial y} = \frac{-\partial u}{\partial x}$$

As given that,

$$\frac{\partial(u^2)}{\partial x}+\frac{\partial(uv)}{\partial y}=2u\frac{\partial u}{\partial x}+\left[\frac{\partial u}{\partial y}(v)+(u)\frac{\partial v}{\partial y}\right]$$

$$=2u\frac{\partial u}{\partial x}+(v)\left(\frac{\partial u}{\partial y}\right)+(u)\left(\frac{\partial v}{\partial y}\right)$$

$$=2u\frac{\partial u}{\partial x}+(V)\left(\frac{\partial u}{\partial y}\right)+(u)\left(\frac{-\partial u}{\partial x}\right)$$

$$=(2u-u)\left(\frac{\partial u}{\partial x}\right)+v\frac{\partial u}{\partial y}+u\frac{\partial u}{\partial x}+v\frac{\partial u}{\partial y}$$

Hence option (d) is the correct answer.

12. Boussinesq stress is responsible for eddy viscosity in turbulent boundary layer. Hence, option (c) is the correct answer.

13. $u = 2xyt, \qquad v = -y^2t$

Equation for steam line is

$$\frac{dx}{u}=\frac{dy}{v}$$

$$\frac{dx}{2xyt}=\frac{dy}{-y^2t}$$

$$\frac{dx}{2xy}=\frac{-dy}{y^2}\Rightarrow\frac{dx}{2x}=\frac{-dy}{y}$$

Integrating on both sides

$$\frac{1}{2}\log x=-\log y+c$$

at x = 1, y = 1

$$\frac{1}{2}(\log 1)=-\log(1)+c$$

$$c = 0$$

∴ equation of stream line is

$$\frac{1}{2}\log x=-\log y$$

$$\log^{\sqrt{x}}=-\log y\Rightarrow\log^{\sqrt{x}}+\log y=0$$

$$\sqrt{xy}=1$$

$$xy^2=1$$

Hence, option (b) is correct answer.

14. PQ = RS

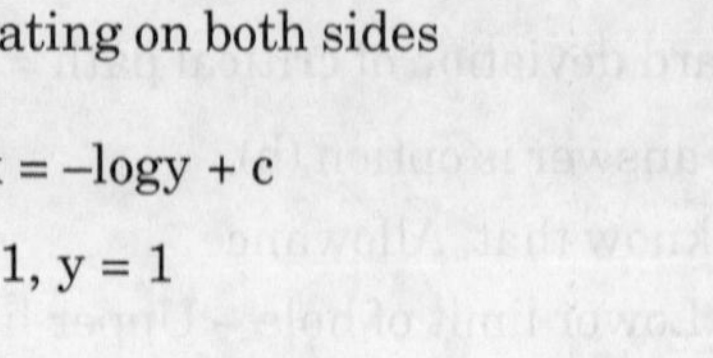

$\angle PQR = 90°$

At junction 'F'

$$F_{PR}\cos 45° + F_{RS} = 0$$

$$F_{PR}\cos 45° = -F_{RS}$$

$$F_{PR}=\frac{-F_{Rs}}{\cos 45°}=-\sqrt{2}F_{RS}$$

So, $\boxed{F_{Rs}=\frac{-F_{PR}}{\sqrt{2}}}$

$$\therefore\quad F_{Pr}\sin 45° = F \Rightarrow \boxed{F_{PR}=\frac{F}{\sin 45°}=\sqrt{2}F}$$

$$F_{RS}=\frac{-F_{PR}}{\sqrt{2}}=\frac{-\sqrt{2}F}{\sqrt{2}}=-F\Rightarrow\boxed{F_{RS}=-F}$$

So, the force in 'RS' is compressive and it is 'F' the force in 'PR' is tensile and it is $\sqrt{2}$ F

Hence, option (b) is the correct answer.

15. According to the question given figure is :

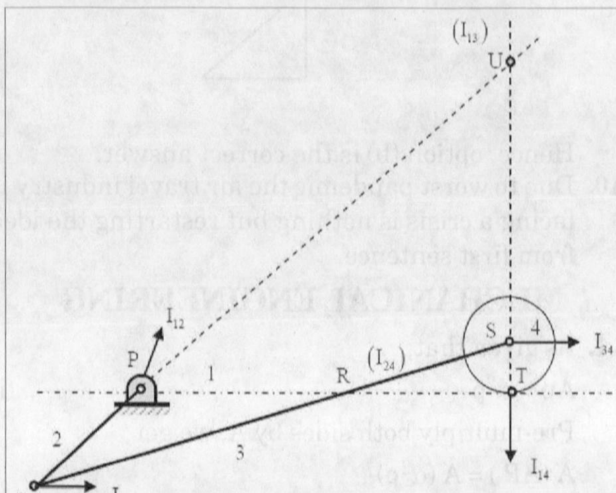

Number of instantaneous centresare

$$\frac{n(n-1)}{2}=\frac{4(4-1)}{2}$$

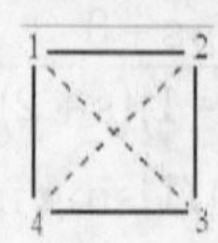

So, P, Q, R, S, T and U are instantaneous centres of mechanism

Hence option (d) is the correct answer.

16. The controlling force (F) (Vs) radius of rotation (r) in a string controlled governor is as follows.

F = ar + b

If b < 0 i.e., F = ar – b → It is a stable governor

'R' is stable governor

If b = 0 i.e., F = ar → It is isochronous governor

'Q' is isochronous governor

If b > 0 i.e., F = ar + b → It is a unstable governor

'P' is unstable governor

Hence, option (b) is the correct answer.

17. According to distortion energy theorem,

For 2-D dimensional case,

$$\sigma_1^2 + \sigma_2^2 - \sigma_1\sigma_2 \le s_{yt}^2$$

$$\sqrt{\sigma_1^2 + \sigma_2^2 - \sigma_1\sigma_2} \le s_{yt}$$

Von mises stress $(\sigma_{von}) = \sqrt{\sigma_1^2 + \sigma_2^2 - \sigma_1\sigma_2}$

Where, $\sigma_1^2 + \sigma_2^2 - \sigma_1\sigma_2$ represent distortion strain energy per unit volume Hence von mises stress is proportional to distortion strain energy per unit volume

Hence answer is option (d).

18. As given that,

f(x) = fnx, h = 0.3; a = 4, b = 5.2

$y_0 = f(4) = \ln 4 = 1.386$

$y_0 = f(a + h) = \ln(4.3) = 1.458$

$y_2 = f(a + 2h) = l.n(4.6) = 1.526$

$y_3 = f(a + 3h) = \ln(4.9) = 1.589$

$y_4 = f(a + 4h) = \ln(5.2) = 1.648$

∴ By Simpson's one-third rule,

$$\int_4^{5.2} \ln x dx = \frac{h}{3}\left[y_0 + y_4 + 4\times(y_1 + y_3) + 2(y_2)\right]$$

$$= \frac{0.3}{3}[(1.386 + 1.648) + 4(1.458 + 1.589) + 2\pi(1.526)]$$

$$= 1.827 \approx 1.83$$

Hence, option (a) is the correct answer.

19. As we know that,

$(1 + i)^2 = 1^2 + i^2 + 2i = 2i$

$(1 + i)^4 = 2^2i^2 = -4$

∴ $(1 + i)^8 = (-4)^2 = 16.$

Hence, option (b) is the correct answer.

20. As given that,

Velocity of air (V) = 300 msec

Temperature of air (T) = 330K

Pressure of air (P) = 180 kPa

$C_P = 1.005$ kJ/kg.K

As we know that,

Stagnation temperature $(T_{stag}) = T + \frac{V^2}{2C_p}$

$$T_{stag} = 330 + \left(\frac{300^2}{2\times1.005\times10^3}\right) = 374.78 \text{ K}$$

21. As given that, COP = 10

Refrigeration capacity = 150 kJ/kg

Work input to refrigeration (W) = ?

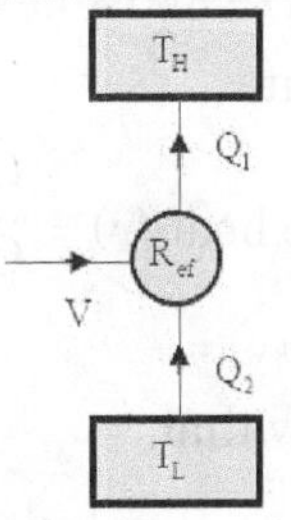

As we know that,

$$COP = \frac{Q_2}{W}$$

$$10 = \frac{150}{W}$$

$$W = = \frac{150}{10} = 15 \text{ kJ/kg}$$

Q_1 = Heat rejected by refrigeration

$W = (Q_1 - Q_2)$

$1.5 = Q_1 - 150$

$Q_1 = 150 + 15 = 165$ kJ/kg

22. As given that,

∴ Volume of tank (V) = 50m³

Saturation temperature at 400 KPa, (T_{sat}) = 143.61° C

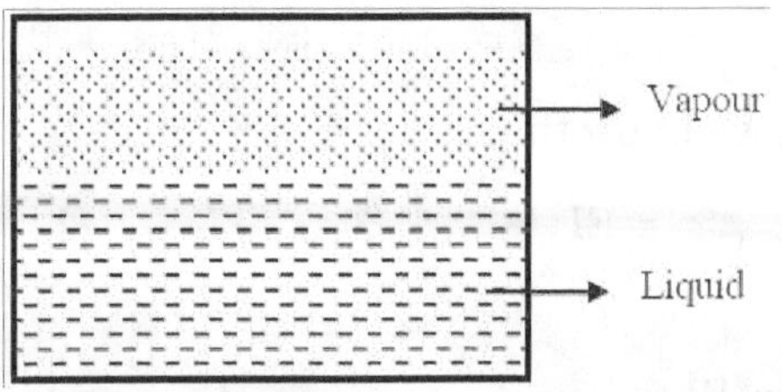

$V_f = 0.001084$ m³/kg

$V_g = 0.46242$ m³/kg

Total mass of liquid-vapour-mixture (m) = ?

$m = m_L + m_v$

Specific volume of vapour $(V_V) = 0.46242$ m³/kg

Mass of vapour $(m_v) = \frac{V}{V_g} = \left(\frac{50}{0.46242}\right)$

= 108.13kg

According to the question.

Mass of vapour (m_v) + mass of liquid (m_L) = Total mass (m) of liquid vapour mixture.

So, $108.13 + m_L = m$

$1081.13 + (0.2m) = m$

$108.13 = 0.8m$

m = 135.16 kg

23. As given that,

Mach number (M) = 0.6

Temperature of air (T) = 350K´

Gas constant (R) = 320 J/kg.K

As we know that

Ratio of specific heat (r) = $\dfrac{C_p}{C_v}$ = 1.3

Speed of object (v) = ?

As we also know that

$$\text{Mach number (M)} = \frac{\text{velocity of object(V)}}{\text{sonic velocity(C)}} = 0.6$$

$$\text{And } M = \frac{V}{\sqrt{rRT}}$$

$$\text{So, } V = M\sqrt{rRt} = (0.6)\sqrt{1.3 \times 320 \times 350}$$

$$= 228.9 \cong 229 \text{ m/sec}$$

24. According to the question,

Case (i)

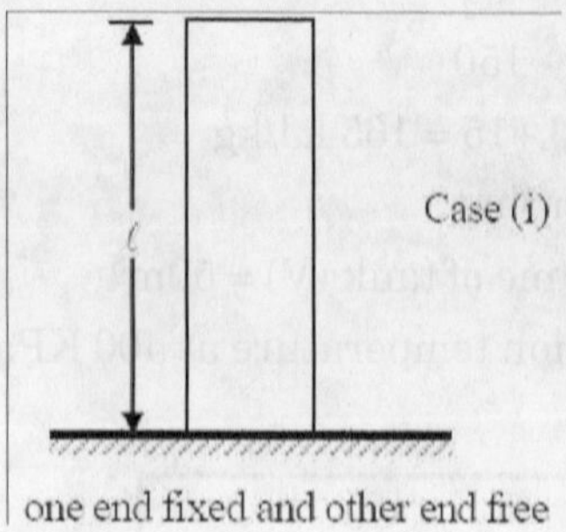

one end fixed and other end free

As we know that

$$(P_{cr})_1 = \frac{\pi^2 EI}{L_e^2}$$

$$= \frac{\pi^2 EI}{(2\ell)^2} = 100\text{N}$$

$$\Rightarrow \frac{\pi^2 EI}{\ell^2} = 400\text{N}$$

Case (ii)

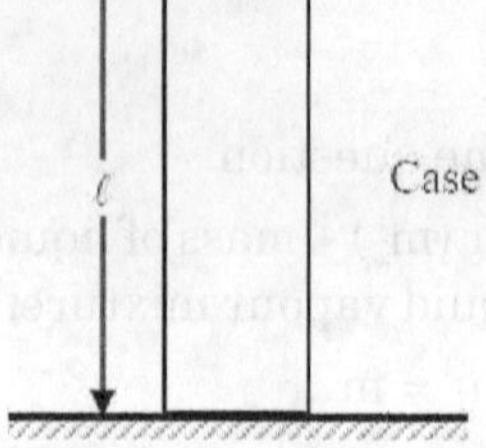

one end fixed and other end pinned

$$(P_{cr})_2 = \frac{\pi^2 EI}{L_e^2}$$

$$= \frac{\pi^2 EI}{\left(\frac{\ell}{\sqrt{2}}\right)^2} = \frac{2\pi^2 EI}{\ell^2}$$

$$= 2 \times 400 = 800\text{N}$$

Hence, answer is 800N

25. As given that,

$E = 2 \times 10^5 \text{N/mm}^2$, $\mu = 0.3$

$E = 3k(1 - 2\mu)$

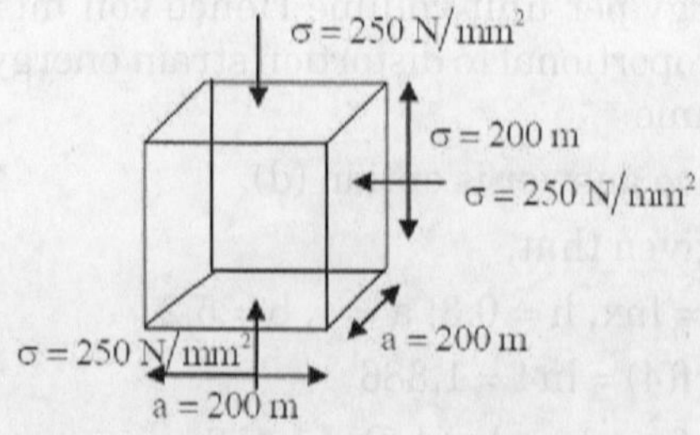

As we know that,

$$\text{Bulk modulus (k)} = \frac{E}{3(1-2\mu)} = \frac{(2 \times 10^5)}{3(1 - 2 \times 0.3)}$$

$k = 166666.67 \text{ N/mm}^2$

Bulk modulus (k)

$$= \frac{\text{volumetric stress(or) hydrostatic stress}(\sigma)}{\text{volumetric strain}(e_v)}$$

$$166666.67 = \frac{250}{e_v}$$

$$e_v = \frac{250}{166666.67} = 1.499 \times 10^{-3} \cong 1.5 \times 10^{-3}$$

$$e_v = \frac{\text{change in volume}(\Delta V)}{\text{original volume(or) initial }(a^3)}$$

$$\Delta V = e_v(a^3) = (1.5 \times 10^{-3})(200)^3 = 12{,}000$$

ΔV = Initial volume – Final volume

$12{,}000 = V_i - V_f$

$V_f = (200)^3 - (12{,}000) = 7988000$

Let a' = New size of cube

$(a')^3 = 7988000$

a' = 199.89

a – a' – = 200 –199.89 = 0.11mm

Hence, the side of the block is reduced by 0.11 mm.

26. As given double integral

$$\Rightarrow \int_{\theta=0}^{\pi/2}\left[\int_{r=0}^{\cos\theta} r\sin\theta dr\right]d\theta$$

$$= \int_{\theta=0}^{\pi/2}\sin\theta\left(\frac{r^2}{2}\right)_0^{\cos\theta} d\theta = \frac{1}{2}\int_0^{\pi/2}\sin\theta\cos^2\theta d\theta$$

$$= \frac{1}{2}\int_0^1 t^2 dt$$

[∵ By putting $\cos\theta = t, \sin\theta\, d\theta = dt$ & $\theta = \frac{\pi}{2} \Rightarrow t = 1$]

$$= \frac{1}{2}\left(\frac{t^3}{3}\right)_0^1 = \frac{1}{6}$$

Hence, option (b) is the correct answer.

27. As we know that

If x is $(n \times 1)$ vector $= \begin{bmatrix} x_1 \\ x_2 \\ \vdots \\ x_n \end{bmatrix}$

So, x^T is $(1 \times n)$ vector $= [x_1, x_2, \ldots x_n]$

As given that,

Take n = 2 and Let $Q = \begin{bmatrix} a & c \\ c & b \end{bmatrix}$ be a symmetric (2 × 2) matrix, $x = \begin{bmatrix} x_1 \\ x_2 \end{bmatrix}; r = \begin{bmatrix} r_1 \\ r_2 \end{bmatrix}$

Then $f(x) = \frac{1}{2}\{x, x_2\}\begin{bmatrix} a & c \\ c & b \end{bmatrix}\begin{bmatrix} x_1 \\ x_2 \end{bmatrix} - \begin{bmatrix} r_1 & r_2 \end{bmatrix}\begin{bmatrix} x_1 \\ x_2 \end{bmatrix}$

$$= \frac{1}{2}\left[ax_1^2 + bx_2^2 + 2cx_1x_2\right] - r_1x_1 - r_2x_2$$

Let $f(x) = \phi(x_1, x_2) = \frac{ax_1^2}{2} + \frac{bx_2^2}{2} + cx_1x_2 - r_1x_1 - r_2x_2$

For stationary points, we have $= \frac{\partial\phi}{\partial x_1} = 0$ and $\frac{\partial\phi}{\partial x_2} = 0$

$\Rightarrow ax_1 + cx_2 - r_1 = 0$ ....(i)

And $bx_2 + cx_1 - r_2 = 0$ ....(ii)

Equation (i) and (ii) in matrix form is

$$\begin{bmatrix} a & c \\ c & b \end{bmatrix}\begin{bmatrix} x_1 \\ x_2 \end{bmatrix} = \begin{bmatrix} r_1 \\ r_2 \end{bmatrix}$$

$\Rightarrow Q.x = r$

Multiply by Q^{-1}, we get

Hence, $x = Q^{-1}r$, option (b) is the stationary point.

28. As given that,

The differential equation

$\Rightarrow \left(\frac{1+y}{y}\right)dy = dx$, which is VSF

Integrating, we get

$$\int\left(\frac{1}{y}+1\right)dy = \int dx + C$$

$\Rightarrow \ell ny + y = x + C$ ...(1)

Using y(1) = 1, (1) gives 0 + 1 = 1+ C

$\Rightarrow C = 0$

$\therefore \ell ny + y = x$

$\Rightarrow y = e^{x-y}$ or $ye^y = e^x$

Hence option (c) is the correct answer.

29. As given that,

j → work station

n → number of work station

i → product

m → number of products.

Capacity constraint is :

$a_{ij}x_{it} \leq c_{jt}$

Where, a_{ij} → processing time that one unit of ith product requires on j^{th} work station

x_{it} → Amount of product 'r' manufacture in time period 't'.

C_{ij} → Capacity work station 'j' in time 't'

Inventory balance constraint is :

$$I_{it} = I_{i,t-1} + x_{it} - s_{it}$$

Where, I_{it} → Amount of product 'i' held in inventory at the end of time period 't'

$I_{i,t-1}$ → Amount of product 'i' held in inventory at the time period 't–1'

X_{it} → Amount product 'I' manufacture in time period 't'

$I_{i,t-1}$ → Amount of product 'I' held C_n Inventory at the time period 't-x'

X_{it} → Amount product is manufactured in time period 't'

S_{it} → Amount of product 'i' sold in time period 't'.

Hence option (a) is the correct answer.

30. As given that,

Ambient pressure (P_{atm}) = 101kPa = P

Temperature of air (T) = 300K

Relative humidity (ϕ) = 0.6

Saturation pressure of water at 300K is

(P_{sat}) = 3.6 kPa

Specific humidity of air (W) = ?

As we know that,

$$\phi = \frac{P_V}{P_{sat}} \Rightarrow P_V = (0.6)(3.6)$$

$P_V = 2.16$ kPa

We also know that,

$$W = \frac{0.622P_V}{P - P_V} = \frac{(0.622)(2.16)}{(101 - 2.16)}$$

W = 0.01359 kg of water vapour/kg of dry air

= 13.59 gms of water vapour/kg of dry air

Hence option (d) is the correct answer.

31. According to the question.

Flexural rigidity of members PQ and QR having length 'L' are 'EI'

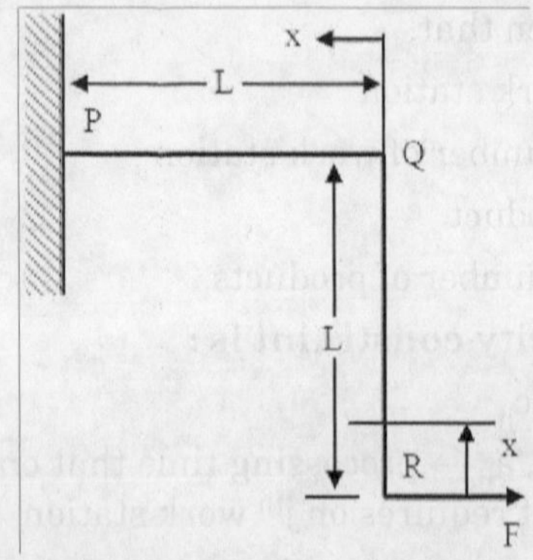

Bending moment at any distance 'x ' from 'R' is $(m_x)_{QR} = Fx$

Bending moment in 'PQ' portion is constant and is equal to $M_{PQ} = FL$

Strain energy stored in frame ($U = U_{QR} + U_{PQ}$)

$$\int_0^L \frac{(m_x)_{QR}^2\, dx}{2EI} + \int_0^L \frac{(m_{PQ})^2\, dx}{2EI}$$

$$U = \int_0^L \frac{(Fx)^2\, dx}{2EI} + \int_0^L \frac{(FL)^2\, dx}{2EI}$$

$$U = \frac{F^2}{2EI}\left[\frac{x^3}{3}\right]_0^L + \frac{F^2L^2}{2EI}[dx]_0^L$$

$$= \frac{F^2}{2EI}\left[\frac{L^3}{3}\right]_0^L + \frac{F^2L^2}{2EI}[L]$$

$$= \frac{F^2L^3}{6EI} + \frac{F^2L^3}{2EI}$$

$$= \frac{F^2L^3 + 3F^2L^3}{6EI} = \frac{2}{3}\frac{F^2L^3}{EI}$$

Horizontal deflection at 'R' is $(\delta_H)_R$

$$U = \frac{\partial U}{\partial F} = \left(\frac{2}{3}\right)\left(\frac{2FL^3}{EI}\right) = \frac{4FL^3}{3EI}$$

Hence option (b) is the correct answer.

32. As given that,

$d_1 = 150$ mm, $d_2 = 250$ mm

$$\frac{N_2}{N_1} = (2500)\left(\frac{150}{250}\right) = 1500 \text{ rpm (ACW)}$$

$$\frac{N_3}{N_2} = \frac{d_2}{d_3} = \frac{z_1}{z_3}$$

$$\frac{N_3}{1500} = \left(\frac{18}{44}\right) \Rightarrow N_3 = 613.64 \text{ rpm (CW)}$$

$N_3 = N_4$, since they are mounted on same shaft

$$\frac{N_5}{N_4} = \frac{d_4}{d_5} = \frac{z_4}{z_5}$$

$$\frac{N_5}{613.64} = \frac{15}{33}$$

$N_5 = 278.93$ rpm (ACW)

$$\frac{N_6}{N_5} = \frac{d_5}{d_6} = \frac{z_5}{z_6}$$

$$\frac{N_6}{278.93} = \frac{33}{36}$$

$N_6 = 255.68$ rpm (CW)

$N_7 = N_6$, since they are mounted on same shaft

Hence option (a) is the correct answer.

33. As given that,

Mass of machine (m) = 100 kg

Applied frequency (ω) = 40rad/sec

As we know that,

$$\text{Transmissibility (TR)} = \frac{\sqrt{\left(1+(2\xi r)^2\right)}}{\sqrt{\left(1-r^2\right)^2+(2\xi r)^2}} = \left|\frac{F_T}{F_0}\right|$$

Where, F_T = Force transmitted

F_0 = Applied force

For first Isolators

$\therefore$ K = 640 kN/m, ξ = 0.70

$$\therefore \omega_n = \sqrt{\frac{k}{m}} = \sqrt{\frac{640 \times 10^3}{100}} = 80 \text{ rad/sec}$$

$$\therefore r = \frac{\omega}{\omega_n} = \frac{40}{80} = 0.5$$

$$\text{So, } \left|\frac{F_T}{F_0}\right| = \frac{\sqrt{\left(1+(2\times0.70\times0.5)^2\right)}}{\sqrt{\left(1-0.5^2\right)^2+(2\times0.7\times0.5)^2}}$$

$$\left|\frac{F_T}{F_0}\right| = \frac{1.22}{1.026} = 1.189$$

For second Isolators

$\therefore K = 640$ kN/m, $\xi = 0.07$

$$\therefore \omega_n = \sqrt{\frac{k}{m}} = = 80 \text{ rad/sec}, r = 0.5$$

$$\left|\frac{F_T}{F_0}\right| = \frac{\sqrt{\left(1+(2\times0.07\times0.5)^2\right)}}{\sqrt{(1-0.5^2)^2+(2\times0.07\times0.5)^2}}$$

$$\left|\frac{F_T}{F_0}\right| = \frac{1}{0.753} = 1.32$$

For third Isolators

$\therefore K = 22.5$ kN/m, $\xi = 0.70$

$$\therefore \omega_n = \sqrt{\frac{k}{m}}\sqrt{\frac{225\times10^3}{100}} = 15 \text{ rad/sec}$$

$$\therefore r = \frac{\omega}{\omega_n} = \frac{40}{15} = 2.67$$

$$\left|\frac{F_T}{F_0}\right| = \frac{\sqrt{\left(1+(2\times0.7\times2.67)^2\right)}}{\sqrt{(1-2.67^2)^2+(2\times0.7\times2.67)^2}}$$

$$\left|\frac{T_T}{F_0}\right|\frac{3.87}{4.18} = 0.54$$

For forth Isolators

$\therefore K = 22.5$ kN/m, $\xi = 0.07$

$\therefore \omega_n = 15$ rad/sec, $r = 2.67$

$$\left|\frac{F_T}{F_0}\right| = \frac{\sqrt{\left(1+(2\times0.07\times2.67)^2\right)}}{\sqrt{(1-2.67^2)^2+(2\times0.07\times2.67)^2}}$$

$$\left|\frac{T_T}{F_0}\right| = \frac{1.067}{6.14} = 0.17$$

So ascending order is 4 – 3 – 1 – 2.

Hence, option (c) is the correct answer.

34. As we know that,

A small angular displacement to pulley

F_s = spring force = $k r\sin\theta$

$$F_i = m\frac{d^2}{dt^2}(r\sin\theta)$$

For small angles $r\sin\theta \cong r\theta = (mr)\frac{d^2}{dt^2}(\theta) = mr\theta$

$T_i = J\,\theta$

Where,

J = Mass moment of inertia of purely about '0'

$\Sigma M_0 = \theta$

$J\ddot{\theta} + (F_i)(r\cos\theta) + (F_s)(r\cos\theta) = 0$

$J\ddot{\theta} + (mr\ddot{\theta})(r\cos\theta) + (kr\sin\theta)(r\cos\theta) = 0$

For small angels $\cos\theta \cong 1$, $\sin\theta = \theta$

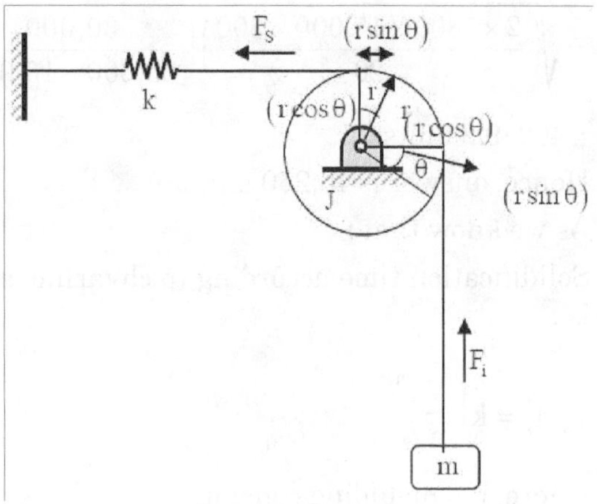

$J\ddot{\theta} + mr^2)\ddot{\theta} + kr^2\theta = 0$

$(J + mr^2)\ddot{\theta} + kr^2\,\theta = 0$

$$\omega_n = \sqrt{\frac{kr^2}{J+mr^2}}$$

Hence option (b) is the correct answer.

35. As given that,

$f(x) = x^3 - x - 3$

$\Rightarrow f'(x) = 3x^2 - 1$ and $x_0 = 2$

As we know that,

Newton-Raphson method :

$$x_1 = x_0 - \frac{f(x_0)}{f'(x_0)} = 2 - \frac{(3)}{11} = 1.73$$

$$x_2 = x_1 - \frac{f(x_1)}{f'(x_1)} = 1.73 - \frac{(0.48)}{7.98} \approx 1.67$$

Hence the numerical value of the root after two iterations (x_2) is 1.67.

36. As given that,

Daily Demand = 15,000 bearings

Annual Demand (R) = 15,000 × 300

C_3 = ordering cost (or) set up cost = 1800 Rs

C_1 = carrying cost = 20 Rs | Bearing | year

K = Production rate = 30, 000, Bearings per day

As we know that,

$$\text{Economic batch quantity } (E_BQ) = \sqrt{\frac{2C_3R}{C_1\left(1-\frac{R}{K}\right)}}$$

$$= \sqrt{\frac{2C_3R}{C_1}\left(\frac{k}{k-R}\right)}$$

$$= \sqrt{\left(\frac{2\times1800\times15{,}000\times300}{20}\right)\left(\frac{30{,}000}{30{,}000-15{,}000}\right)}$$

= 40249.2 units

Hence, answer is 40,250 units.

37. As we know that :

Solidification time according to chvarinovs rule is :

$$t_s = k\left(\frac{V}{a}\right)^{\eta}$$

where, k = moulding constant

n = 2

For cube :

Volume of the cube (V) = 75 × 125 × 20

= 18750 mm^3

Surface are of cube (A) = 2(75 × 125 + 125 × 20 + 20 × 75) = 26750 mm^2

Solidification time (t_s) for cube is = 2 minutes

$$2 = (k)\left(\frac{187500}{26750}\right)^2$$

k = 0.041

For cylinder :

Volume of cylinder (V)

$$= \frac{\pi}{4}(d^2)(h) = \frac{\pi}{4}(50^2)(50) = 31250 \text{ imm}^3$$

Surface area of cylinder (A)

$$= \left(\pi dl + 2\times\frac{\pi}{4}d^2\right) = \left(\pi(50)(50) + \frac{\pi}{2}(50)^2\right)$$

= 3750 πmm^2

Solidification of time (t_s) for cylindrical is

$$t_s = (k)\left(\frac{V}{A}\right)^n = (0.041)\left(\frac{31250\pi}{3750\pi}\right)^2 = 2.85 \text{ mins}$$

Hence total solidification time will be 2.85 mm.

38. As given that,

Diameter of Nugget (d) = 5mm

Thickness of Nugget (t) = 1mm

Welding time (t) = 0.1 seconds

Melting energy of steel or Heat required for melting = +20 J/mm^3

Heat conversion efficiency (η) = 0.10

Heat supplied (Qs) = ?

As we know that

$$\text{Volume of Nugget (V)} = \left(\frac{\pi}{4}d\right)(t)$$

$$V = \left(\frac{\pi}{5}(5^2)(1)\right) = 6.25\ \pi\ \text{mm}^3$$

Total heat required (Q_R) = (20 × 6.25m) = 125πJ

we also know that,

$$\eta = \frac{Q_R}{Q_S}$$

$$0.10 = \frac{125\pi}{Q_S} \Rightarrow Q_S = \frac{125\pi}{0.1}$$

= 3926.99 ≅ 3927 joules

So, heat supplied per seconds.

$$= \frac{Q_S}{t} = \frac{3927}{0.1} = 39270 \text{ watts} = 39.27 \text{ kW}$$

39. As given that,

Grinding wheel diameter (D) = 150 mm

Width of wheel (b)= 12 mm

Grinding velocity (V) = 40 m/sec

Table speed (V_{Table}) = 5m/min

Depth of cut per pass = 50μm = 50 × 10^{-6}m

Number of grinding passes (n) = 20

Tangential force = 40N = F_t

Normal force = 60N = F_n

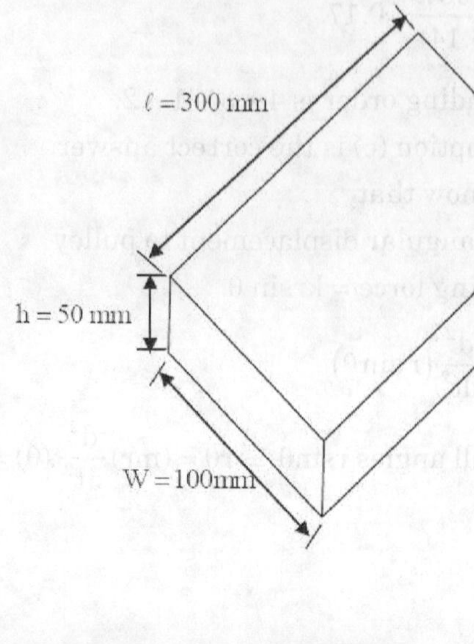

As we know that

Resultant force acting on grinding wheel

$$(F)_{grinding} = \sqrt{F_t^2 + F_n^2}$$

$$F_{grinding} = \sqrt{40^2 + 60^2} = 72.11N$$

In grinding : We also know that,

Total grinding force $(F_{grinding}) = u \times \frac{\partial db}{V}$

Where,

u = specific grinding (J/mm^3)

∂ = work piece velocity = 5 mmin = 500 mm/min

V = grinding wheel velocity = 40 m/sec

d = depth of cut = 50 × 10^{-6}m = 50 × 10^{-3}mm

b = width of cut = 12 mm

By putting respective values:

So, $72.11 = \frac{(u)\times(5000mm/min)(50\times10^{-3})mm(12)mm}{(40\times60)(m/mm)}$

u = 57.7 J/mm^3

40. As given that,

Rake angle (α) = 0 9

Shear force (F_S) = 400N

Cutting velocity (V_C) = 100 min/min

Depth of cut (t) = 2mm

Feed (S_0) = 0.1mm/rev

Chip flow velocity (V_f) = 20m/min

Shear strength of material (τ_s) = ?

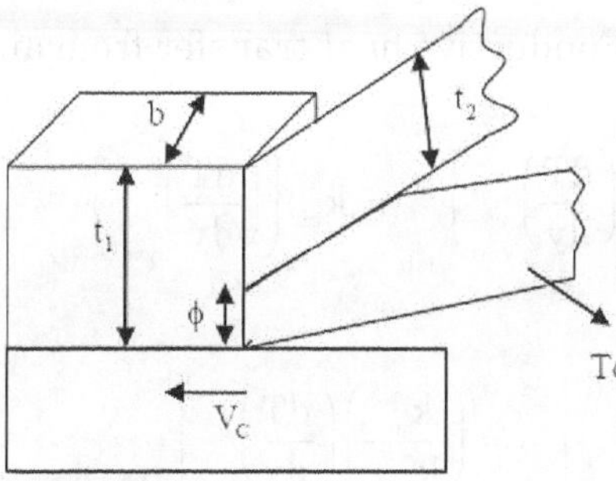

As we know that,

$$\text{Chip thickness ratio } (r) = \frac{t_1}{t_2} = \frac{V_f}{V_c} = \frac{20}{100} = 0.2$$

Shear angle 'ϕ' will be find out as follows,

$$\tan\phi = \frac{r\cos\alpha}{1 - r\sin\alpha} = \frac{(0.2)\cos 0°}{1-(0.2)\sin 0°}$$

$\phi = \tan^{-1}(0.2) = 11.31°$

We also know that,

Shear strength (or) shear stress in the work piece is

$$(\tau_s) = \left(\frac{F_s}{t_1 b}\right)(\sin\phi)$$

$$= \left(\frac{400}{2\times0.1}\right)\sin(1.31)$$

$$= 392.23 \text{ MPa}$$

t_1 = uncut chip thickness = depth of cut = 2mm

b = width of chip = feed = 0.1 mm

Hence, the shear strength, τ_s of the material is 392.23 MPa.

41. As given that,

t_1 = 50 mm

W_1 = 250 mm

ℓ_1 = 3600 min

t_2 = 0.9 × 50 = 45 mm

W_2 = 1.03 × 250 = 257.5 mm

ℓ_2 = ?

According to the question

Volume of slab before rolling (V_1) = Volume of slab after (V_2) Rolling

So, $t_1 w_1 \ell_1 = t_2 w_2 \ell_2$

$(50)(25)(3600) = (45)(257.5)(\ell_2)$

$$\ell_2 = \left[\frac{50\times25\times3600}{45\times257.5}\right]$$

Hence, Length of slab after rolling (ℓ_2) = 3883.5 mm

42. As given that,

Length of sine bar (l) = 254 mm

Diameter of roll (d) = 25.4 mm

Angle measured (θ) = 30°

Height of gauge blocks

On one side (h_2) = 76.2 mm

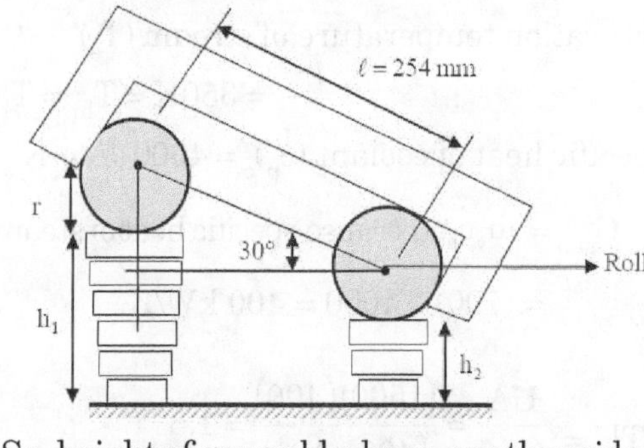

So, height of gauge blocks on another side (h_1) = ?

$$\sin 30° = \frac{(h_1 + r) - (h_2 + r)}{1} \quad (\because \text{ from figure})$$

$$\frac{1}{2} = \frac{(h_1 - h_2)}{254} = \frac{(h_1 - 76.2)}{254}$$

$h_1 = 203.2$ mm

43. According to given table :

Month	Demand(D)	Forecast(F)	Percentage absolute error $\left(\left\|\frac{D-F}{D}\right\|\right)$
April	225	200	$\frac{25}{225} = 11.11\%$
May	220	240	$\frac{20}{220} = 9.09\%$
June	285	300	$\frac{15}{285} = 5.26\%$
July	290	270	$\frac{20}{290} = 6.89\%$
August	250	230	$\frac{20}{250} = 8\%$

Mean absolute percentage :

$$\text{error (MAPE)} = \frac{11.11 + 9.09 + 5.26 + 6.89 + 8}{5}$$

$= 8.07\%$

Hence (MAPE) in the forecast is 8%.

44. As given that,

Mass flow rate of coolant (m_c) = 100 kg/sec

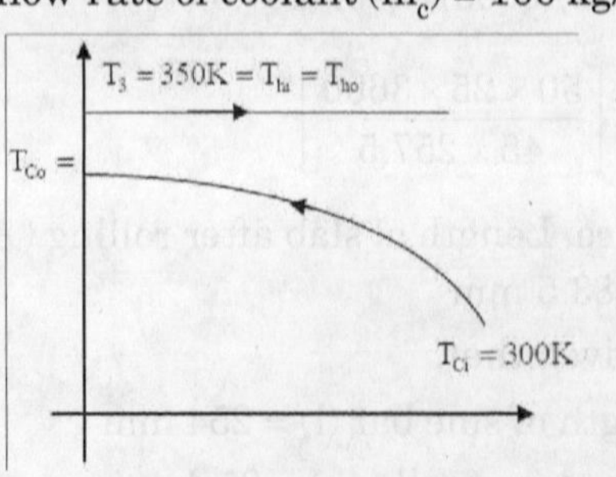

Inlet temperature of coolant water overall heat transfer T_{ci} = 300K

Coefficient (U) = 1500 W/m^{2}K

Total heat transfer area (A) = 400 m^2

Saturation temperature of stream (T_s)

$= 350K = T_{hi} = T_{ho}$

Specific heat of coolant $(c_p)_c$ = 4000 J/kg.K

$\therefore \; C_{min} = \dot{m}_c c_{pc}$ (because specific heat of steam is ∞)

$= (100) * 4000 = 400$ kW/K,

$$NTU = \frac{UA}{C_{min}} = \frac{(1500)(400)}{(400 \times 10^3)} = 1.5$$

For condensing of boiling effectiveness of heat exchanger is

$\varepsilon = 1 - e^{-NTU} = 1 - e^{-15} = 0.78$

As we know that

$$\varepsilon = \frac{m_c C_{pc}(T_{co} - T_{ci})}{C_{min}(T_{hi} - T_{ci})} \quad (\because C_{min} = \dot{m}C_{pc})$$

$$\text{So, } 0.78 = \frac{(T_{co} - 300)}{(350 - 300)}$$

$T_{co} = 339K$

Hence, the temperature of the coolant water coming out of the condenser is 339 K.

45. As given that,

Thermal conductivity of air (k_{air}) = 1 W/mK

Thermal conductivity of slab (k_{slab}) = 100 W/mK

According to the question :

At y = 0

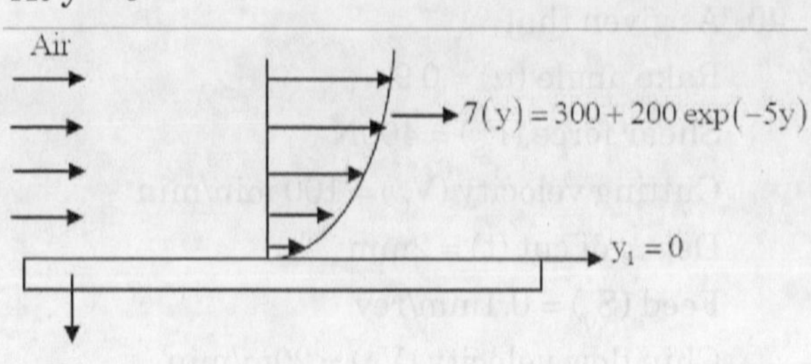

So, conduction heat transfer into slab

= Conductive heat transfer from air to slab

$$k_{slab}\left(\left(\frac{dT}{dy}\right)_{y=0}\right)_{slab} = k_{air}\left(\left(\frac{dT}{dy}\right)_{y=0}\right)_{Air}$$

$$\left(\frac{dT}{dy}\right)_{y=0}\Bigg|_{slab} = \left(\frac{k_{air}}{k_{slab}}\right)\left(\frac{dT}{dy}\right)_{y=0}\Bigg|_{air}$$

$$= \left(\frac{1}{100}\right)\left(-(5)(200)e^{-5y}\right)_{y=0}$$

$$-\left(\frac{1}{100}\right)\left(-(5)(200)\right) = -10 \text{ k/m}$$

Hence, magnitude of temperature gradient within the slab at y = 0 is '10'

46. As given that,

Cross sectional area of tank $(A_t) = 1m^2$

Cross sectional of area orifice $(A_0) = 1cm^2$

$= 1 \times 10^{-4}\ m^2$

Initially Height of water, $(H) = 1m$

$g = 9.8\ m/sec^2$

According to question.

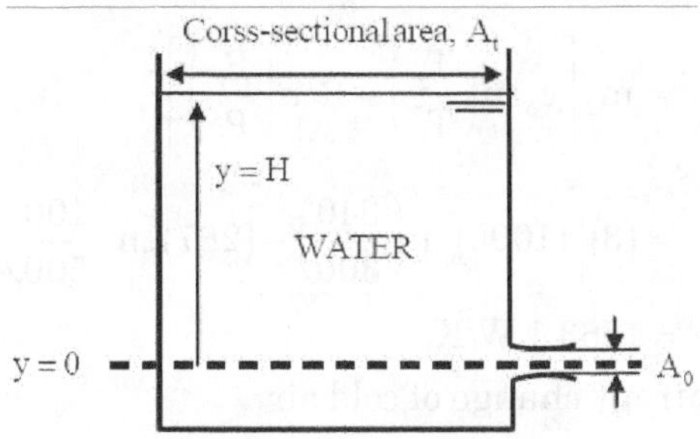

Now, time taken to reach a level of water $y = \frac{H}{4}$ will be find out as follows,

$m_{in} - m_{out} = m_{cu}$

$m_{in} = 0$

$$-\rho A_0\sqrt{2gH} = \frac{d}{dt}(\rho A_t dH)$$

$$-A_0\sqrt{2gH} = \frac{d}{dt}(A_t dH)$$

$$-A_0\sqrt{2gH} = d(A_t dH)$$

By integration on both sides :

$$-A_0\sqrt{2gH}(t-0) = A_t(dH)$$

$$t = \frac{-A}{A_0\sqrt{2g}}\left(\frac{dH}{\sqrt{H}}\right)$$

$$t = \frac{-A_t}{A_0\sqrt{2g}}\int_1^{0.25}\frac{dH}{\sqrt{H}} = \frac{A}{A_0(\sqrt{29})}\int_{0.25}^{1}\frac{dH}{\sqrt{H}}$$

$$= \frac{(1)}{(1\times10^{-4}\times\sqrt{2\times9.8})}\left[2\sqrt{H}\right]_{0.25}^{1}$$

t = 2258 seconds

47. As given figure :

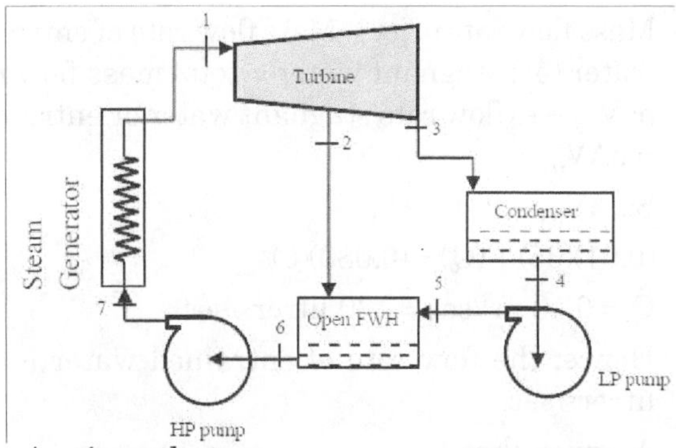

As given that,

$h_2 = 2624$ kJ/kg

$h_5 = 226.7$ kJ/kg

$h_6 = 708.6$ kJ/kg

$m_5 = 100$ kg/sec

$m_2 = ?$

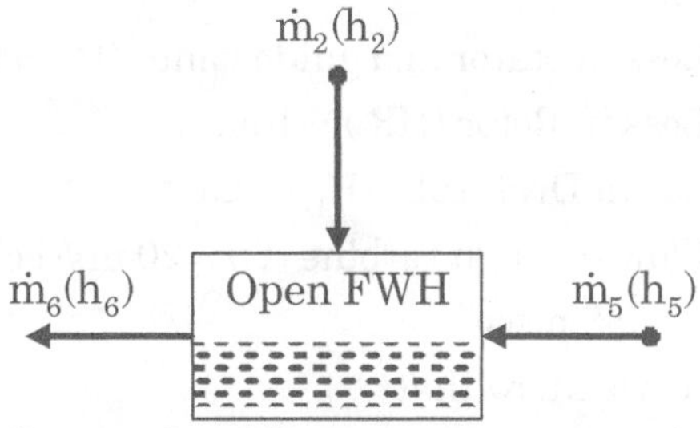

According to above figure

$m_2h_2 + m_5h_5 = m_6h_6$

$m_2(2624) + (100)(226.7) = m_6(708.6)$

Let 'm_5' be the mass of flow rate steam taken at location '2'

Let m_1 be the total mass flow rate of steam entering at location '1'

$m_1 = m_2 + m_5$ and $m_1 = m_6$

$m_6 = m_2 + m_5 = m_2 + 100$

$(m_2)(2624) + (100)(226.7) = (m_2 + 100)(708.6)$

$2624m_2 + 22670 = 708.6h + 70860$

$1915.4m_2 = 48190$

$m_2 = 25.16$ kg/sec

Hence, the mass flow rate of stream at location 2 is 25.16 kg/sec.

48. As given figure :

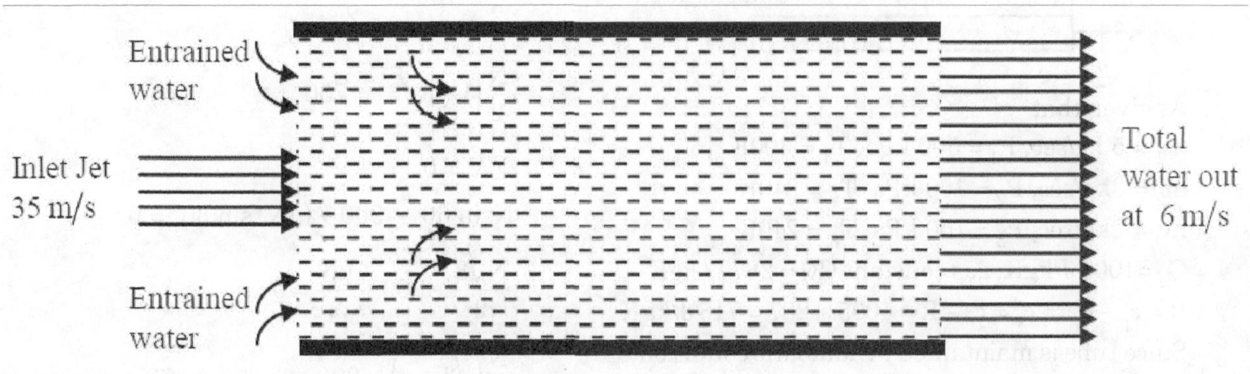

Now, applying conservation of mass of principle,
Mass flow rate of jet + Mass flow rate of entrained water (or) stagnant water = total mass flow rate
$\rho a V_{jet}$ + e (flow rate stagnant water or entrained) $= \rho A V_0$

So, $aV_{jet} + (Q) = AV_0$

$(0.01)(35) + (Q) = (0.080)(6)$

$Q = 0.13\ m^3/sec$ = 130 litrers/sec

Hence, the flow rate of entrained water is 130 litrers/sec.

49. As given that,

N = 300 rpm

H = 200 mm

u = 40 msec

Velocity of water at exit of draft tube $(V_2) = 3.5$ m/sec

Loss in stator and guide vanes $(H_s) = 5m$

Loss in Rotor (HR) = 10m

Los in Draft tube $(H_D) = 2m$

Flow through turbine (G) = 20 m^3/sec

g = 9.8 $msec^2$

As we know that

Net available head for turbine

$$(H_{net}) = H - H_S - H_R - H_D - \frac{V_0^2}{2g}$$

$$H_{Net} = 200 - 5 - 10 - 2 - \left(\frac{3.5^2}{2\times 9.8}\right) = 182.375m$$

$\therefore$ Hydraulic efficiency $\eta_H = \frac{H_{net}}{H} = \frac{182.375}{200}$

$\eta_H = 91.187\%$

Hence, answer is 91.2

50. As given figure

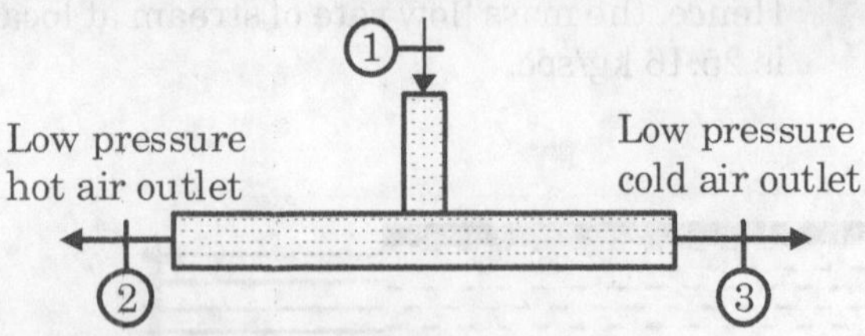

As given that

$m_1 = 5$ kg/sec, $P_1 = 500$ kPa, $T_1 = 300K$

$m_2 = 3$kg/sec, $P_2 = 100$ kPa, $T_2 = 340K$

$m_3 = 2$kg/sec, $P_3 = 100$ kPa, $T_3 = 240K$

C_p =1005 J/kgK; gas constant (R) = 287J / kgK

$R = c_p - c_v \Rightarrow c_v = c_p - R = 1005 - 287 = 718$ J/kgK

Since tube is maintained in adiabatic condition

No entropy transfer from the vortex tube

$\therefore$ Entropy generation = Entropy change – Entropy transfer

$\therefore$ Entropy generation = Entropy change ($\because$ entropy transfer = 0)

As we know that

Entropy change of hot air

$$= m_2\left[c_p \ell n\left(\frac{T_2}{T_1}\right) - R\ell n\left(\frac{P_2}{P_1}\right)\right]$$

$$= (3)\left[(1005)\ell n\left(\frac{340}{300}\right) - (287)\ell n\left(\frac{100}{500}\right)\right]$$

= 1763.1 W/K

Entropy change of cold air

$$= m_3\left[c_p \ell n\left(\frac{T_3}{T_1}\right) - R\ell n\left(\frac{P_3}{P_1}\right)\right]$$

$$= (2)\left[(1005)\ell n\left(\frac{240}{300}\right) - (287)\ell n\left(\frac{100}{500}\right)\right]$$

= 475.30 W/K

$\therefore$ Entropy generation (s_{gen})

= Total entropy change

= 1763.1 + 475.30

= 2238.4 W/K = 2.24 kW/K

51. As given that, free body diagram of block is

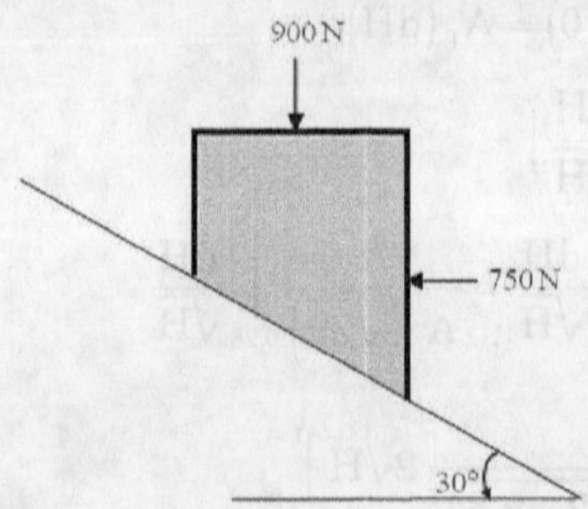

$\Sigma H = 0$

$N \cos 60° - 750 - \mu N \cos 30° = 0$

$$\frac{N}{2} - 750 - \frac{\mu N\sqrt{3}}{2} = 0$$

$$\frac{N}{2}\left(1 - \mu\sqrt{3}\right) = 750 \quad ....(i)$$

$\Sigma V = 0$

$N\sin 60° - 900 + (\mu N) \sin 30° = 0$

$$\frac{N\sqrt{3}}{2} - 900 + \frac{\mu N}{2} = 0$$

$$\frac{N}{2}\left(\sqrt{3} + \mu\right) = 900 \quad ....(ii)$$

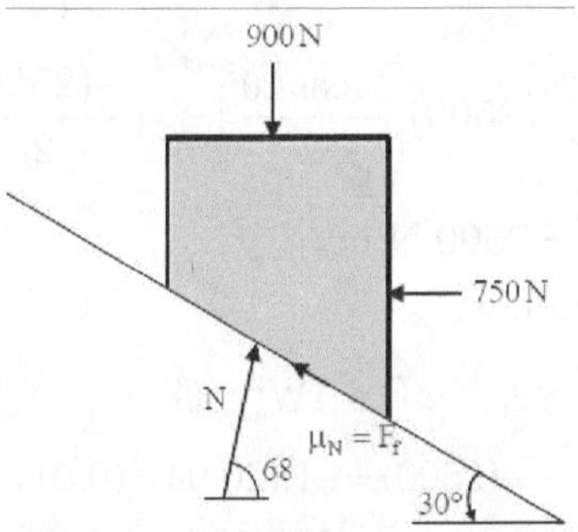

$$\text{Equation (i)/(ii)} = \frac{(1-\mu\sqrt{3})}{(\sqrt{3}+\mu)} = \frac{750}{900}$$

$$(1-\mu\sqrt{3}) = \left(\frac{5}{6}\right)(\sqrt{3}+\mu)$$

$$6 - 6\sqrt{3}\mu = 5\sqrt{3} + 5\mu$$

$$6 - 5\sqrt{3} = (5 + 6 + \sqrt{3})\mu$$

$$\mu = \frac{(6-5\sqrt{3})}{(5+6\sqrt{3})} = -0.17$$

'μ' is negative means our assumed direction of frictional force is not correct.

Hence, answer is μ = 0.17

52. As given figure :

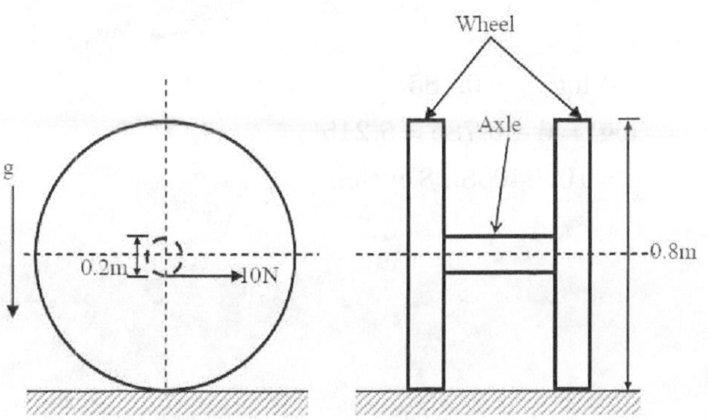

Force applied (F) = 10N

As given that,

Diameter of wheel (D) = 0.8m, Radius of wheel

$(R) = \frac{D}{2} = 0.4\text{m}$

Mess of wheel (m_w) = 1 kg

Diameter of axle (α) = 0.2m,

Radius of axle $(r) = \frac{d}{2} = \frac{0.2}{2} = 0.1\text{m}$

Mass of axle (m_a) = 1.5kg

g = 9.8 m/sec^2

Mass moment inertia of wheels and axle is (I) = $2m_wR^2$ (Neglecting mass moment of inertia of axle)

$I = (2)(1)(0.4)^2 = 0.32\text{kg.m}^2$

Torque applied on wheel axle (T) = (F)(r)

$= (10)(0.1) = 1$ N.m

T = Iα

$$\alpha = \frac{T}{I} = \frac{1}{0.32} \simeq 3.125 \text{ rad/sec}^2$$

Tangential acceleration of wheel axle system (a_t) = Rα

$a_t = (0.4)(3.125) = 1.25\text{m/sec}^2$

53. As given that,

Area of bending moment diagram between A and B is = 10,000 N.m^2

$E_I = 200 \times 10^6\text{N.m}^2$

According to the question:

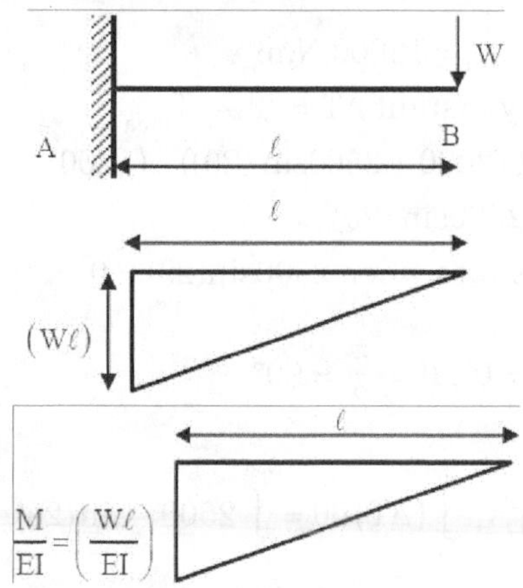

$$\therefore 10{,}000 = \frac{1}{2}(W\ell)(\ell)$$

$$10{,}000 = \frac{W\ell^2}{2} \Rightarrow W\ell^2 = 20{,}000 \text{ N.m}^2$$

∴ Slope at 'B' = Area of $\left(\frac{M}{EI}\right)$ diagram between A and B.

$$= \frac{1}{2} \times \left(\frac{W\ell}{EI}\right)(\ell) = \frac{1}{2} \times \frac{W\ell^2}{EI}$$

$$= \frac{1}{2} \times \frac{20{,}000}{200 \times 10^6} = 5 \times 10^{-5} \text{ radians}$$

= 50 micro radians

54. As given that,

$T(\theta) = 1200 + 2500\sin(2\theta)$

$N_{mean} = 200\text{rpm}$

Coefficient of fluctuation of speed $(C_S) = 2 \times \dfrac{0.5}{100}$

$C_s = 0.01$

$$W_{mean} = \frac{2\pi N_{mean}}{60} = \frac{2\pi \times 200}{60} = 20.94 \text{ rad/sec}$$

ΔE = maximum fluctuation in energy

Since T (θ) is a function of 'θ'

$$\text{So, } T_{mean} = \frac{1}{\pi}\int_0^{\pi} T d\theta = \frac{1}{\pi}\int_0^{\pi}(12000 + 2500\sin(2\theta))\,d\theta$$

$$= \frac{1}{\pi}\left[12000(\pi - 0) + \frac{2500}{2}(-\cos 2\theta)_0^{\pi}\right]$$

$$= \frac{1}{\pi}\left[1200\pi + 1250(-\cos 360° + \cos 0°)\right]$$

$= 12000$ N.m

At any instant $\Delta T = T - T_{mean}$

$\Delta T = (12000 + 2500\sin(2\theta)) - (1200)$

$\Delta T = 2500\sin(2\theta)$

'ΔT' is zero when $2500\sin(2\theta) = 0$

$$\Rightarrow \theta = 0°, \theta = \frac{\pi}{2} = 90°$$

$$\therefore \Delta E = \int_0^{90°}(\Delta T)\,d\theta = \int_0^{90°} 2500 - \sin(2\theta)\,d\theta$$

$$= (2500)\left[\frac{-\cos 2\theta}{2}\right]_0^{90°} = \frac{-(2500)}{2}[-1-1]$$

$= 2500$ N.m

As we know that,

$$\Delta E = I W_{mean}^2 C_S$$

$$(2500) = (I)(20.94)^2(0.01)$$

$$I = 570.15 \text{ kg.m}^2$$

Hence, the minimum mass moment of inertia of flywheel should be 570.15 kg.m^2.

55. According to question,

⇒ Comparing similar triangle ΔABC and ΔADE

$$\frac{AB}{BC} = \frac{AD}{ED}$$

$$\frac{\log_{10}^{450} - \log_{10}^{150}}{(6-3)} = \frac{\log_{10}^{200} - \log_{10}^{150}}{(6 - \log_{10}^{x})}$$

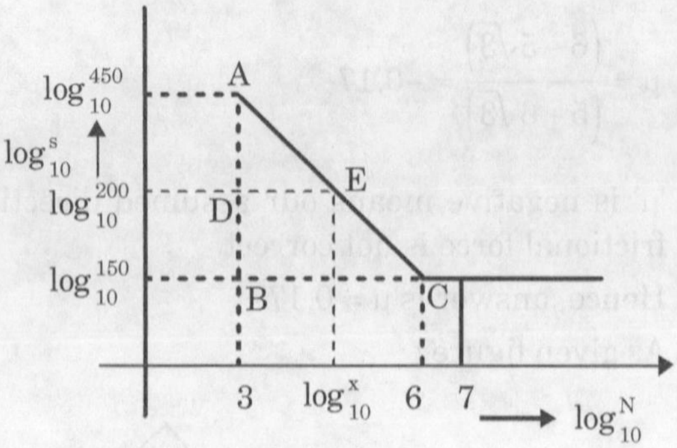

$6 - \log^{x}{}_{10} = 0.785$

$\log^x = 6 - 0.785 = 5.215$

x = 10164058.98 cycles

SOLVED PAPER - 2020

SET 1

INSTRUCTIONS

1. Total of 65 questions carrying 100 marks, out of which 10 questions carrying a total of 15 marks are in General Aptitude (GA)
2. The Engineering Mathematics will carry around **15% of the total marks,** the General Aptitude section will carry **15% of the total marks** and the **remaining 70% of the total marks.**
3. **Types of Questions**
   (*a*) **Multiple Choice Questions (MCQ)** carrying 1 or 2 marks each in all papers and sections. These questions are objective in nature, and each will have a choice of four options, out of which the candidate has to mark the correct answer(s).
   (*b*) **Numerical Answer Questions** of 1 or 2 marks each in all papers and sections. For these questions the answer is a real number, to be entered by the candidate using the virtual keypad. No choices will be shown for these type of questions.
4. For **1-mark** multiple-choice questions, **1/3 marks** will be deducted for a wrong answer. Likewise, for **2-marks** multiple-choice questions, **2/3** marks will be deducted for a wrong answer. There is no negative marking for numerical answer type questions.

Chapter-Wise Analysis

#	Subject	Marks		
		1	2	Total
1	General Aptitude	5	5	15
2	Engineering Mathematics	6	4	14
3	Engineering Mechanics	1	0	1
4	Strength of Materials	1	3	7
5	Theory of Machines	4	3	10
6	Design of Machine Elements	1	2	5
7	Fluid Mechanics	2	3	8
8	Heat Transfer	2	1	4
9	Thermodynamics, IC Engines, Refrigeration & Power Plant Engineering Air-Conditioning	3	5	13
10	Material Manufacturing & Industrial Engineering	5	5	15

GENERAL APTITUDE (GA)

Q. 1 to Q. 5 : Carry One Mark Each.

1. He is known for his unscrupulous ways. He always sheds ________ tears to deceive people.

(a) fox's (b) crocodile's

(c) crocodile (d) fox

2. Jofra Archer, the England fast bowler, is _______ than accurate.

(a) more fast (b) faster

(c) less fast (d) more faster

3. Select the word that fits the analogy:

Build : Building : : Grow :

(a) Grown (b) Grew

(c) Growth (d) Growed

4. I do not think you know the case well enough to have opinions. Having said that, I agree with your other point.

What does the phrase "having said that" mean in the given text?

(a) as opposed to what I have said

(b) despite what I have said

(c) in addition to what I have said

(d) contrary to what I have said

5. Define [x] as the greatest integer less than or equal to x, for each $x \in (-\infty, \infty)$. If y = [x], then area under y for $x \in [1, 4]$ is _______ .

(a) 1 (b) 3

(c) 4 (d) 6

Q. 6 to Q. 10 : Carry Two Marks Each.

6. Crowd funding deals with mobilisation of funds for a project from a large number of people, who would be willing to invest smaller amounts through web-based platforms in the project.

Based on the above paragraph, which of the following is correct about crowd funding?

(a) Funds raised through unwilling contributions on web-based platforms.

(b) Funds raised through large contributions on web-based platforms.

(c) Funds raised through coerced contributions on web-based platforms.

(d) Funds raised through voluntary contributions on web-based platforms.

7. P, Q, R and S are to be uniquely coded using α and β. If P is coded as $\alpha\alpha$ and Q as $\alpha\beta$, then R and S, respectively, can be coded as

(a) $\beta\alpha$ and $\alpha\beta$ (b) $\beta\beta$ and $\alpha\alpha$

(c) $\alpha\beta$ and $\beta\beta$ (d) $\beta\alpha$ and $\beta\beta$

8. The sum of the first n terms in the sequence 8, 88, 888, 8888, ... is ________ .

(a) $\frac{81}{80}(10^n - 1) + \frac{9}{8}n$

(b) $\frac{81}{80}(10^n - 1) - \frac{9}{8}n$

(c) $\frac{80}{81}(10^n - 1) + \frac{8}{9}n$

(d) $\frac{80}{81}(10^n - 1) - \frac{8}{9}n$

9. Select the graph that schematically represents BOTH $y = x^m$ and $y = x^{1/m}$ properly in the interval $0 \le x \le 1$, for integer values of m, where m > 1.

(a)

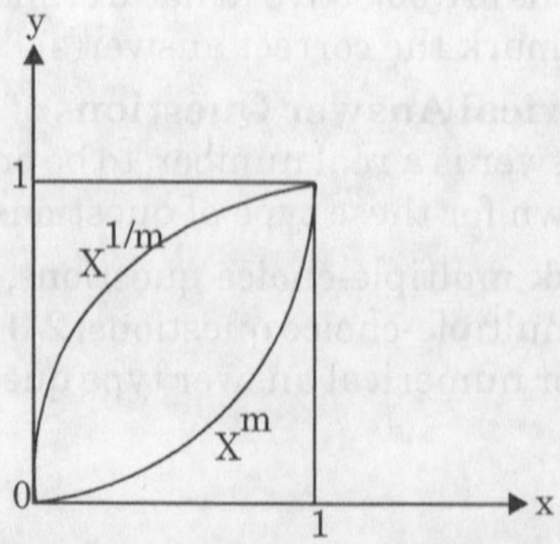

(b)

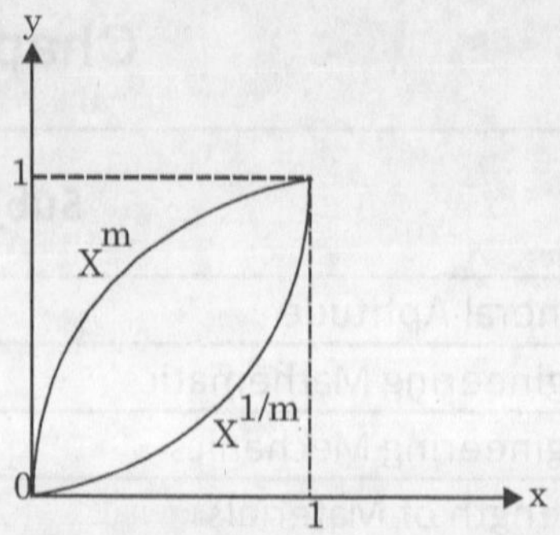

(c)

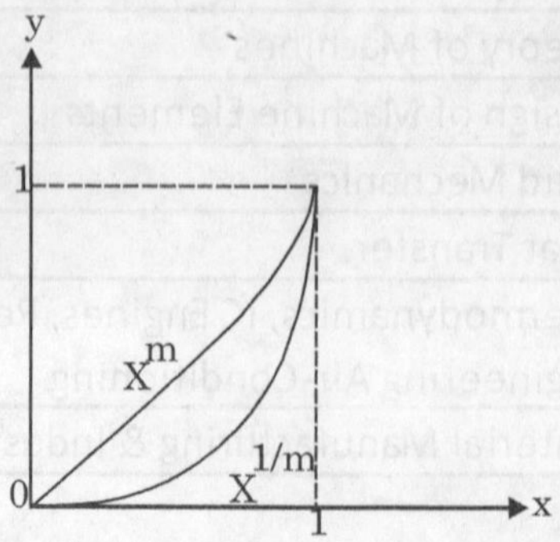

(d)

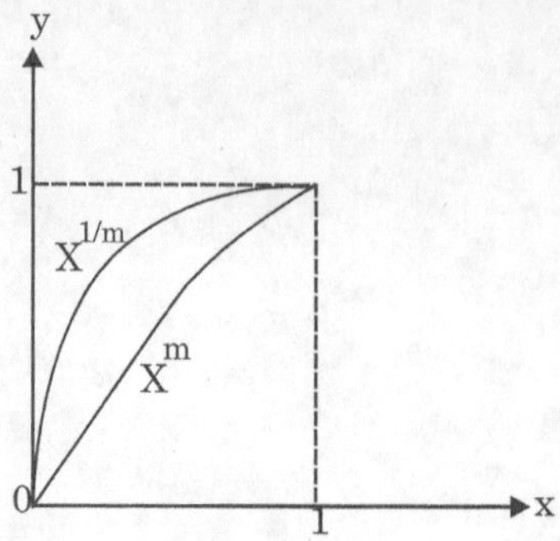

10. The bar graph shows the data of the students who appeared and passed in an examination for four schools P, Q, R and S. The average of success rates (in percentage) of these four schools is _____.

Performance of schools P, Q, R and S

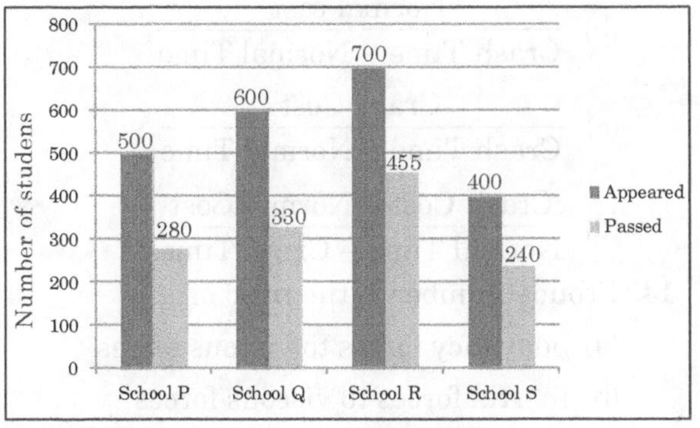

(a) 58.5% (b) 58.8%

(c) 59.0% (d) 59.3%

MECHANICAL ENGINEERING

Q. 1 to Q. 25 : Carry One Mark Each.

1. Multiplication of real valued square matrices of same dimension is

(a) associative

(b) commutative

(c) always positive definite

(d) not always possible to compute

2. The value of $\lim_{x\to 1}\left(\frac{1-e^{-c(1-x)}}{1-xe^{-c(1-x)}}\right)$ is

(a) c (b) c + 1

(c) $\frac{c}{c+1}$ (d) $\frac{c+1}{c}$

3. The Laplace transform of a function f(t) is $L(f) = \frac{1}{(s^2+\omega^2)}$. Then, f(t) is

(a) $f(t) = \frac{1}{\omega^2}(1 - \cos \omega t)$

(b) $f(t) = \frac{1}{\omega}\cos \omega t$

(c) $f(t) = \frac{1}{\omega}\sin \omega t$

(d) $f(t) = \frac{1}{\omega^2}(1 - \sin \omega t)$

4. Which of the following function f(z), of the complex variable z, is NOT analytic at all the points of the complex plane?

(a) $f(z) = z^2$ (b) $f(z) = e^z$

(c) $f(z) = \sin z$ (d) $f(z) = \log z$

5. The members carrying zero force (i.e zero-force members) in the truss shown in the figure, for any load P > 0 with no appreciable deformation of the truss (i.e with no appreciable change in angles between the members) are

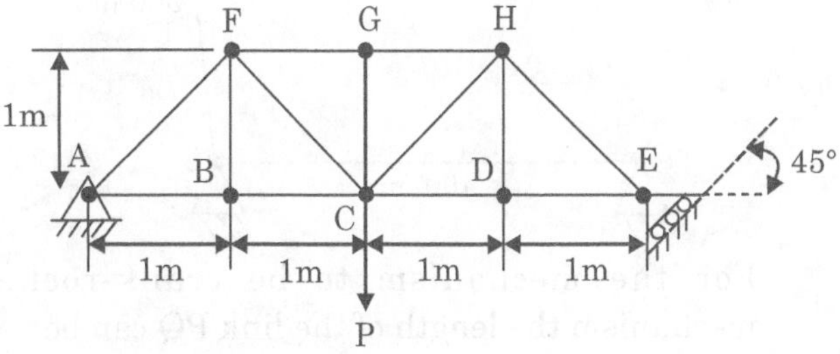

(a) BF and DH only

(b) BF, DH and GC only

(c) BF, DH, GC, CD and DE only

(d) BF, DH, GC, FG and GH only

6. A single-degree-of-freedom oscillator is subjected to harmonic excitation $F(t) = F_0 \cos(\omega t)$ as shown in the figure.

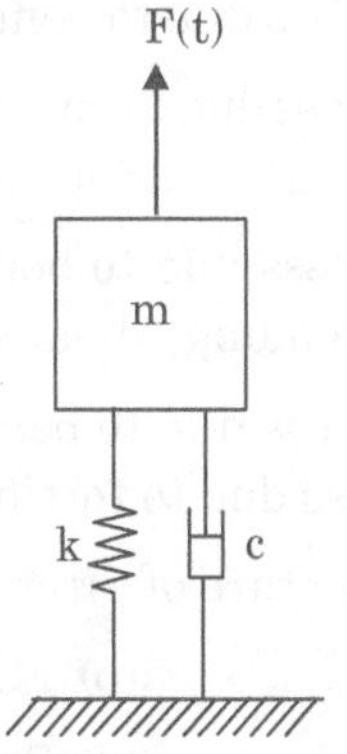

The non-zero value of ω, for which the amplitude of the force transmitted to the ground will be F_0, is

(a) $\sqrt{\frac{k}{2m}}$ (b) $\sqrt{\frac{k}{m}}$

(c) $\sqrt{\frac{2k}{m}}$ (d) $2\sqrt{\frac{k}{m}}$

7. The stress state at a point in a material under plane stress condition is equi-biaxial tension with a magnitude of 10 MPa. If one unit on the σ – τ plane is 1 MPa, the Mohr's circle representation of the state of stress is given by

(a) a circle with a radius equal to principal stress and its center at the origin of the σ – τ plane

(b) a point on the σ axis at a distance of 10 units from the origin

(c) a circle with a radius of 10 units on the σ – τ plane

(d) a point on the τ axis at a distance of 10 units from the origin

8. A four bar mechanism is shown below.

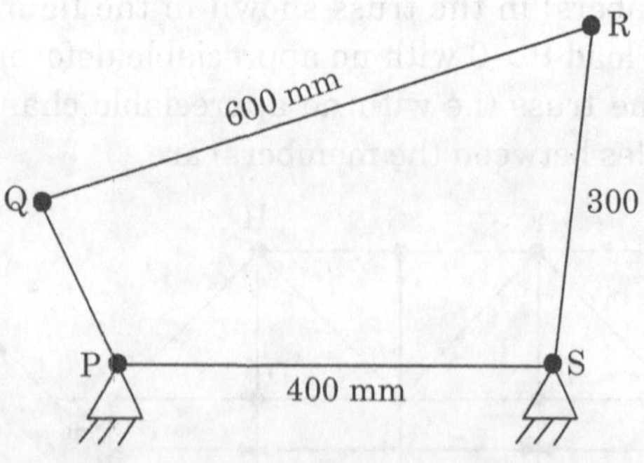

For the mechanism to be crank-rocker mechanism the length of the link PQ can be

(a) 80 mm (b) 200 mm

(c) 300 mm (d) 350 mm

9. A helical gear with 20° pressure angle and 30° helix angle mounted at the midspan of a shaft that is supported between two bearings at the ends. The nature of the stresses induced in the shaft is

(a) normal stress due to bending only

(b) normal stress due to bending in one plane and axial loading; shear stress due to torsion

(c) normal stress due to bending in two planes and axial loading; shear stress due to torsion

(d) normal stress due to bending in two planes; shear stress due to torsion

10. The crstal structure of γ iron (austenite phase) is

(a) BCC (b) FCC

(c) HCP (d) BCT

11. Match the following

Heat treatment process	Effect
P : Tempering	1. Strengthening
Q : Quenching	2. Toughening
R : Annealing	3. Hardening
S : Normalizing	4. Softening

(a) P-2, Q-3, R-4, S-1

(b) P-1, Q-1, R-3, S-2

(c) P-3, Q-3, R-1, S-3

(d) P-4, Q-3, R-2, S-1

12. The base of a brass bracket needs rough grinding. For this purpose,t he most suitable grinding wheel grade specification is

(a) C30Q12V (b) A50G8V

(c) C90J4B (d) A30D12V

13. In the Critical Path Method (CPM), the cost-time slope of an activity is given by

(a) $\dfrac{\text{Crash cost} - \text{Normal cost}}{\text{Crash Time}}$

(b) $\dfrac{\text{Normal cost}}{\text{Crash Time} - \text{Normal Time}}$

(c) $\dfrac{\text{Crash Cost}}{\text{Crash Time} - \text{Normal Time}}$

(d) $\dfrac{\text{Crash Cost} - \text{Normal Cost}}{\text{Normal Time} - \text{Crash Time}}$

14. Froude number is the ratio of

(a) buoyancy forces to viscous forces

(b) inertia forces to viscous forces

(c) buoyancy forces to inertia forces

(d) inertia forces to gravity forces

15. Match the following non-dimensional numbers with the corresponding definitions:

Non-dimensional number	Definition
P : Reynolds number	1. $\dfrac{\text{Buoyancy force}}{\text{Viscous force}}$
Q : Grashof number	2. $\dfrac{\text{Momentum diffusivity}}{\text{Thermal diffusivity}}$
R : Nusselt number	3. $\dfrac{\text{Inertia force}}{\text{Viscous force}}$
S : Prandtl number	4. $\dfrac{\text{Convective heat transfer}}{\text{Conduction heat transfer}}$

(a) P-1, Q-3, R-2, S-4

(b) P-3, Q-1, R-2, S-4

(c) P-4, Q-3, R-1, S-2

(d) P-3, Q-1, R-4, S-2

16. The velocity field of an incompressible flow in a Cartesian system is represented by

$$\vec{V} = 2(x^2 - y^2)\hat{i} + v\hat{j} + 3\hat{k}$$

Which one of the following expressions for v is valid?

(a) $-4xz + 6xy$ (b) $-4xy - 4xz$

(c) $4xz - 6xy$ (d) $4xy + 4xz$

17. For an ideal gas, the value of the Joule-Thomson coefficient is

(a) Positive

(b) Negative

(c) Zero

(d) Indeterminate

18. For an ideal gas, a constant pressure line and a constant volume line intersect at a point, in the Temperature (T) versus specific entropy (s) diagram. Cp is the specific heat at constant pressure and Cv is the specific heat at constant volume. The ratio of the slopes of the constant pressure and constant volume lines at the point of intersection is

(a) $\frac{C_P - C_V}{C_P}$ (b) $\frac{C_P}{C_V}$

(c) $\frac{C_P - C_V}{C_V}$ (d) $\frac{C_V}{C_P}$

19. For three vectors $\vec{A} = 2\hat{j} - 3\hat{k}, \vec{B} = -2\hat{i} + \hat{k}$, and $\vec{C} = 3\hat{i} - \hat{j}$ where $\hat{i}, \hat{j}$ and $\hat{k}$ are unit vectors along the axes of a right-handed rectangular Cartesian coordinate system, the value of $(\vec{A} \cdot (\vec{B} \times \vec{C}) + 6)$ is ________ .

20. A flywheel is attached to an engine to keep its rotational speed between 100 rad/s and 110 rad/s. If the energy fluctuation in the flywheel between these two speeds is 1.05 kJ then the moment of inertia of the flywheel is ________ kg.m^2 (round off to 2 decimal places).

21. A balanced rigid disc mounted on a rigid rotor has four identical point masses, each of 10 grams, attached to four points on the 100mm radius circle shown in the figure.

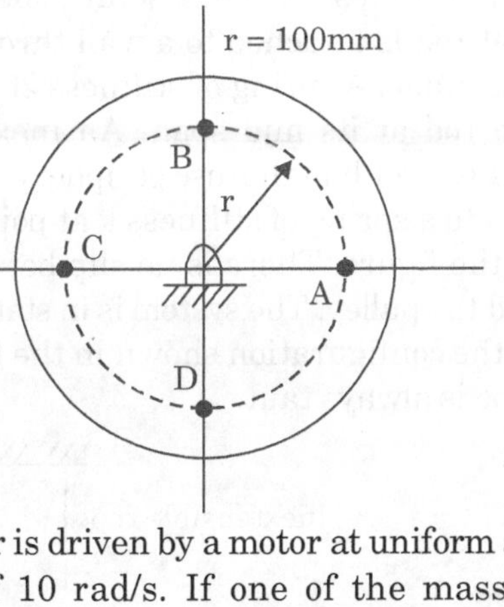

The rotor is driven by a motor at uniform angular speed of 10 rad/s. If one of the masses gets detached then the magnitude of the resultant unbalance force on the rotor is ______ N (round off to 2 decimal places).

22. A sheet metal with a stock hardness of 250 HRC has to be sheared using a punch and a die having a clearance of 1mm between them. If the stock hardness of the sheet metal increases to 400 HRC, the clearance between the punch and the die should be ________ mm.

23. A company is hiring to fill four managerial vacancies. The candidates are five men and three women. If every candidate is equally likely to be chosen then the probability that at least one woman will be selected is ________ (round off to 2 decimal places).

24. The compressor of a gas turbine plant, operating on an ideal intercooled Brayton cycle, accomplishes an overall compression ratio of 6 in a two-stage compression process. Intercooling is used to cool the air coming out from the first stage to the inlet temperature of the first stage, before its entry to the second stage. Air enters the compressor at 300 K and 100 KPa. If the properties of gas are constant, the intercooling pressure for minimum compressor work is ______ kPa (round off to 2 decimal places).

25. In a concentric tube counter flow heat exchanger, hot oil enters at 102°C and leaves at 65°C. Cold water enters at 25 °C and leaves at 42°C. The log mean temperature difference (LMTD) is ______ °C (round off to one decimal place).

Q. 26 to Q. 55 : Carry Two Marks Each.

26. The evaluation of the definite integral $\int_{-1}^{1.4} x|x|\,dx$ by using Simpson's 1/3rd (one third) rule with step size h = 0.6 yields.

(a) 0.914 (b) 1.248

(c) 0.581 (d) 0.592

27. A vector field is defined as

$$\vec{f}(x, y, z) = \frac{x}{\left[x^2 + y^2 + z^2\right]^{\frac{3}{2}}}\hat{i} + \frac{y}{\left[x^2 + y^2 + z^2\right]^{\frac{3}{2}}}\hat{j} + \frac{z}{\left[x^2 + y^2 + z^2\right]^{\frac{3}{2}}}\hat{k}$$

where, $\hat{i}, \hat{j}, \hat{k}$ are unit vectors along the axes of a right handed rectangular– Cartesian coordinate system. The surface integral $\iint \vec{f} \cdot d\vec{S}$ (where $d\vec{S}$ is an elemental surface area vector) evaluated over the inner and outer surfaces of a spherical shell formed by two concentric spheres with origin as the center, and internal and external radii of 1 and 2, respectively is

(a) 0 (b) 2π

(c) 4π (d) 8π

28. Bars of square and circular cross-section with 0.5 m length are made of a material with shear strength of 20 MPa. The square bar cross-section dimension is 4 cm × 4 cm and the cylindrical bar cross-section diameter is 4 cm. The specimens are loaded as shown in the figure.

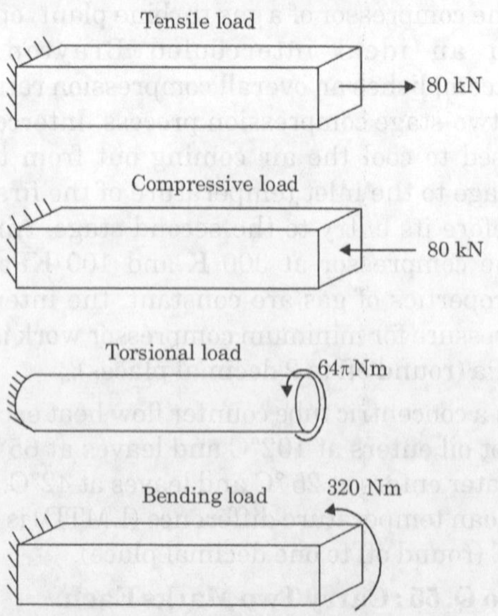

Which specimen(s) will fail due to the applied load as per maximum shear stress theory?

(a) Tensile and compressive load specimens
(b) Torsional load specimen
(c) Bending load specimen
(d) None of the specimens

29. The 2 kg block shown in figure (top view) rests on a smooth horizontal surface and is attached to a massless elastic cord that has a stiffness 5 N/m.

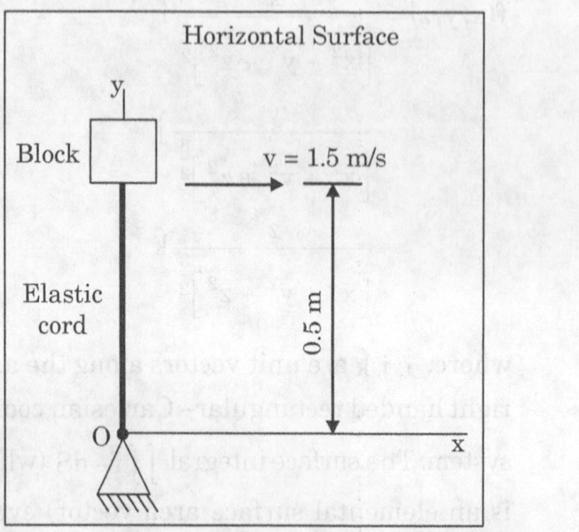

The cord hinged at O is initially unstretched and always remains elastic. The block is given a velocity v of 1.5 m/s perpendicular to the cord. The magnitude of velocity in m/s of the block at the instant the cord is stretched by 0.4 m is

(a) 0.83 (b) 1.07
(c) 1.36 (d) 1.50

30. The truss shown in the figure has four members of length l and flexural rigidity EI, and one member of length $l\sqrt{2}$ and flexural rigidity 4EI. The truss is loaded by a pair of forces of magnitude P, as shown in the figure.

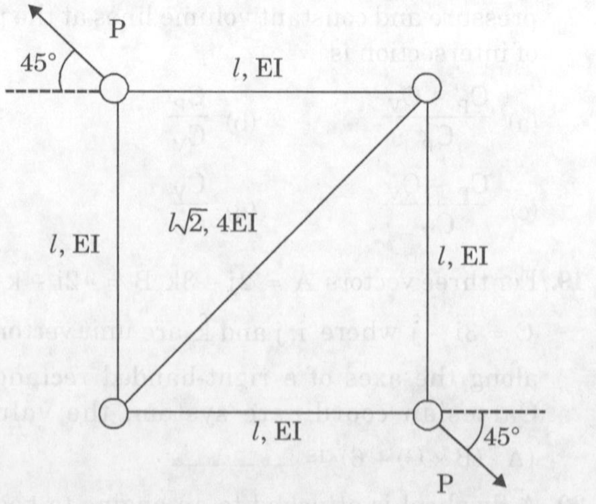

The smallest value of P, at which any of the truss members will buckle is

(a) $\frac{\sqrt{2}\pi^2 EI}{l^2}$ (b) $\frac{\pi^2 EI}{l^2}$
(c) $\frac{2\pi^2 EI}{l^2}$ (d) $\frac{\pi^2 EI}{2l^2}$

31. A rigid mass-less rod of length L is connected to a disc (pulley) of mass m and radius r = L/4 through a friction less revolute joint. The other end of that rod is attached to a wall through a friction less hinge. A spring of stiffness 2k is attached to the rod at its mid span. An inextensible rope passes over half the disc periphery and is securely tied to a spring of stiffness k at point C as shown in the figure. There is no slip between the rope and the pulley. The system is in static equilibrium in the configuration shown in the figure and the rope is always taut.

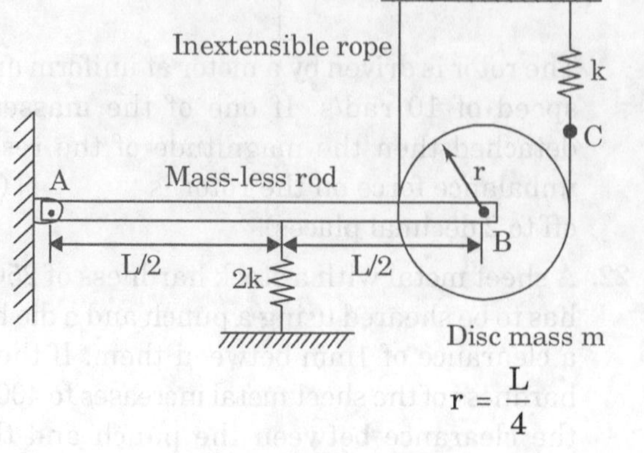

Neglecting the influence of gravity, the natural frequency of the system for small amplitude vibration is

(a) $\sqrt{\frac{3}{2}}\sqrt{\frac{k}{m}}$ (b) $\frac{3}{\sqrt{2}}\sqrt{\frac{k}{m}}$

(c) $\sqrt{3}\sqrt{\frac{k}{m}}$ (d) $\sqrt{\frac{k}{m}}$

32. A strip of thickness 40 mm is to be rolled to a thickness of 20 mm using a two-high mill having rolls of diameter 200 mm. Coefficient of friction and arc length in mm, respectively are

(a) 0.45 and 38.84 (b) 0.39 and 38.84

(c) 0.39 and 44.72 (d) 0.45 and 44.72

33. For an assembly line, the production rate was 4 pieces per hour and the average processing time was 60 minutes. The WIP inventory was calculated. Now, the production rate is kept the same, and the average processing time is brought down by 30 percent. As a result of this change in the processing time, the WIP inventory.

(a) decreases by 25% (b) increases by 25%

(c) decreases by 30% (d) increases by 30%

34. A small metal bead (radius 0.5 mm) initially at 100°C. When placed in a stream of fluid at 20°C, attains a temperature of 28°C in 4.35 seconds. The density and specific heat of the metal are 8500 kg/m^3 and 400 J/kg.K, respectively. If the bead is considered as lumped system, the convective heat transfer coefficient (in W/m^2.K) between the metal bead and the fluid stream is

(a) 283.3 (b) 299.8

(c) 149.9 (d) 449.7

35. Consider two exponentially distributed random variables X and Y, both having a mean of 0.50. Let Z = X + Y and r be the corrrelation coefficient between X and Y. If the variance of Z equals 0, then the value of r is ________ (round off to 2 decimal places)

36. An analytic function of a complex variable

$z = x + iy\left(i = \sqrt{-1}\right)$ is defined as

$f(z) = x^2 - y^2 + i\psi(x, y)$

where $\psi(x, y)$ is a real function. The value of the imaginary part of f(z) at z = (1 + i) is ________ (round off to 2 decimal places)

37. In a disc-type axial clutch, the frictional contact takes place within an annular region with outer and inner diameters 250 mm and 50 mm, respectively. An axial force F_1 is needed to transmit a torque by a new clutch. However, to transmit the same torque, one needs an axial force F_2 when the clutch wears out. If contact pressure remains uniform during operation of a new clutch while the wear is assumed to be uniform for an old clutch, and the coefficient of friction does not change, then the ratio F_1/F_2 is ________ (round off to 2 decimal places)

38. A cam with a translating flat-face follower is desired to have the follower motion

$$y(\theta) = 4\left[2\pi\theta - \theta^2\right], 0 \le \theta \le 2\pi$$

Contact stress considerations dictate that the radius of curvature of the cam profile should not be less than 40 mm anywhere. The minimum permissible base circle radius is ________ mm (round off to one decimal place)

39. A rectangular steel bar of length 500 mm, width 100 mm and thickness 15 mm is cantilevered to a 200 mm steel channel using 4 bolts as shown. For an external load of 10 kN applied at the tip of the steel bar, the resultant shear load on the bolt at B, is ____ kN (round off to one decimal place).

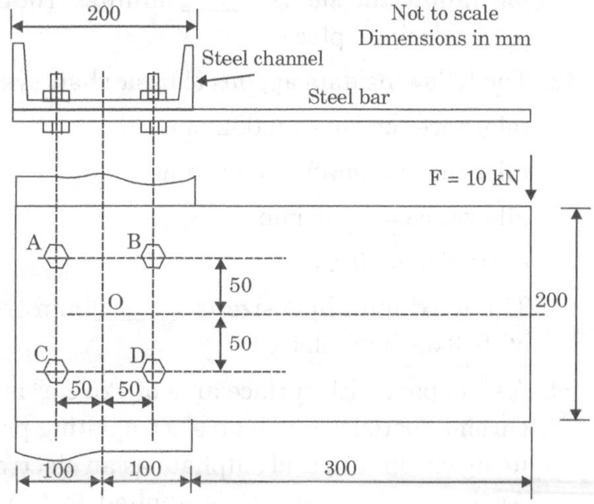

40. The barrier shown between two water tanks of unit width (1m) into the plane of the screen is modeled as a cantilever.

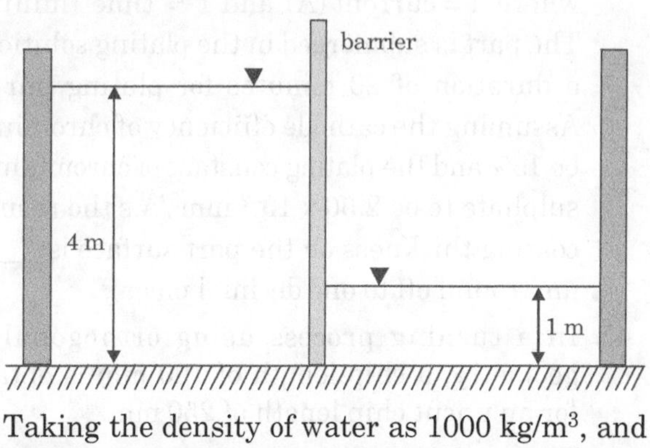

Taking the density of water as 1000 kg/m^3, and the acceleration due to gravity as 10 m/s^2, the maximum absolute bending moment developed in the cantilever is ________ kN-m (round off to the nearest integer)

41. The magnitude of reaction force at joint C of the hinge-beam shown in the figure is _____ kN (round off to 2 decimal places)

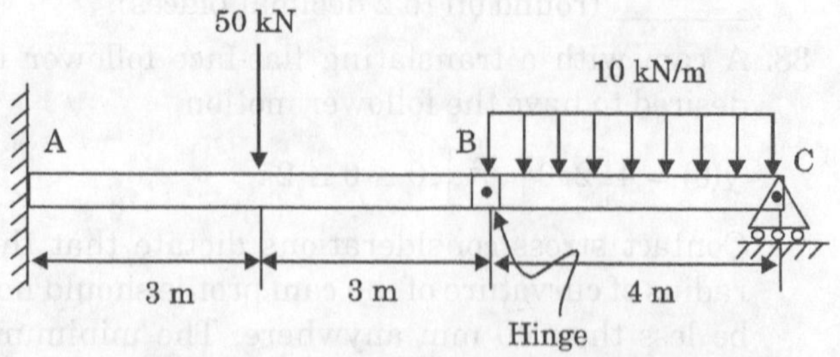

42. A slot of 25 mm × 25 mm is to be milled in a workpiece of 300 mm length using a side and face milling cutter of diameter 100 mm, width 25 mm and having 20 teeth.

For a depth of cut 5 mm, feed per tooth 0.1 mm, cutting speed 35 m/min and approach and over travel distance of 5 mm each, the time required for milling the slot is _____ minutes (round off to one decimal place).

43. The following data applies to basic shaft system :

tolerance for hole = 0.002 mm

tolerance for shaft = 0.001 mm

allowance = 0.003 mm

basic size = 50 mm

The maximum hole size is _____ mm (round off to 3 decimal places)

44. A steel part with surface area of 125 cm^2 is to be chrome coated through an electroplating process using chromium acid sulphate as an electrolyte. An increasing current is applied to the part according to the following current time relation:

$$I = 12 + 0.2t$$

where, I = current (A) and t = time (minutes). The part is submerged in the plating solution for a duration of 20 minutes for plating purpose. Assuming the cathode efficiency of chromium to be 15% and the plating constant of chromium acid sulphate to be 2.50×10^{-2} mm^3/A.s the resulting coating thickness on the part surface is _____ μm (round off to one decimal place).

45. In a turning process using orthogonal tool geometry, a chip length of 100 mm is obtained for an uncut chip length of 250 mm.

The cutting conditions are :

cutting speed = 30 m/min, rake angle = 20°

The shear plane angle is _____ degrees (round off to one decimal place).

46. The thickness of a steel plate with material strength coefficient of 210 MPa has to be reduced from 20 mm to 15 mm in a single pass in a two high rolling mill with a roll radius of 450 mm and rolling velocity of 28 m/min. If the plate has a width of 200 mm and its strain hardening exponent, n is 0.25, the rolling force required for the operation is _____ kN (round off to 2 decimal places).

Note: Average flow stress = Material strength coefficient × $\frac{(\text{True strain})^n}{(1+n)}$

47. Two business owners Shweta and Ashok run their businesses in two different states. Each of them, independent of the other, produces two products A and B, sells them at Rs. 2,000 per kg and Rs. 3,000 per kg, respectively, and uses Linear programming to determine the optimal quantity of A and B to maximize their respective daily revenue. Their constraints are as follows: (i) For each business owner, the production process is such that the daily production of A has to be at least as much as B, and the upper limit for production of B is 10 kg per day, and (ii) the respective state regulations restrict Shveta's production of A to less than 20 kg per day, and Ashok's production of A to less than 15 kg per day. The demand of both A and B in both the states is very high and everything produced is sold.

The absolute value of the difference in daily (optimal) revenue of Shveta and Ashok is _____ thousand Rupees (round off to 2 decimal places).

48. Consider two cases as below.

Case 1: A company buys 1000 pieces per year of a certain part from vendor 'X'. The changeover time is 2 hours and the price is Rs. 10 per piece. The holding cost rate per part is 10% per year.

Case 2: For the same part, another vendor 'Y' offers a design where the changeover time is 6 minutes, with a price of Rs. 5 per piece and a holding cost rate per part of 100% per year. The order size is 800 pieces per year from 'X' and 200 pieces per year from 'Y'.

Assume the cost of downtime as Rs. 200 per hour. The percentage reduction in the annual cost for case 2, as compared to case 1 is _____ (round off to 2 decimal places).

49. Consider steady, viscous, fully developed flow of a fluid through a circular pipe of internal diameter D. We know that the velocity profile forms a paraboloid about the pipe centre line, given by :

$$V = -C\left(r^2 - \frac{D^2}{4}\right) \text{ m/s. Where C is a constant.}$$

The rate of kinetic energy (in J/s) at the control surface A-B, as shown in the figure is proportional to D^n. The value of n is ________

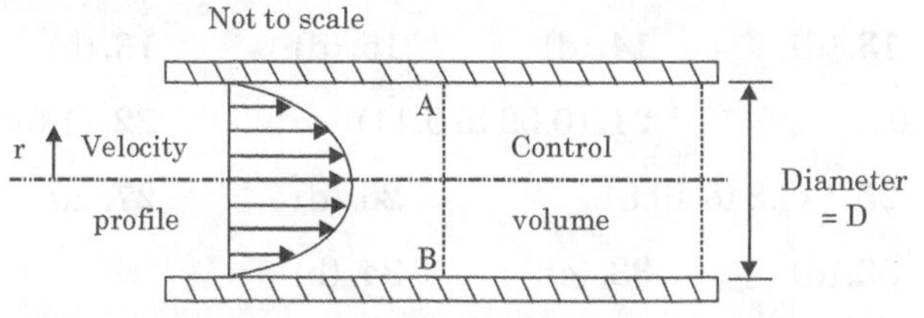

50. Air discharges steadily through a horizontal nozzle and impinges on a stationary vertical plate as shown in figure.

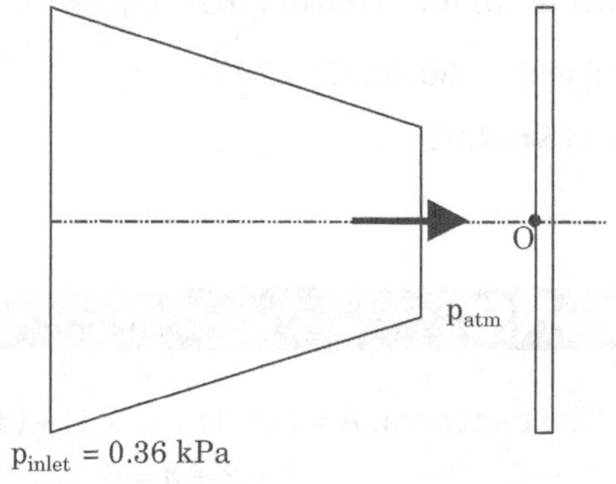

The inlet and outlet areas of the nozzle are 0.1 m^2 and 0.02 m^2, respectively. Take air density as constant and equal to 1.2 kg/m^3. If the inlet gauge pressure of air is 0.36 kPa. the gauge pressure at point O on the plate is ________ kPa (round off to two decimal places).

51. Air (ideal gas) enters a perfectly insulated compressor at a temperature of 310 K. The pressure ratio of the compressor is 6. Specific heat at constant pressure for air is 1005 J/kg.K and ratio of specific heats at constant pressure and constant volume is 1.4. Assume that specific heats of air are constant. If the isentropic efficiency of the compressor is 85 percent, the difference in enthalpies of air between the exit and the inlet of the compressor is_______ kJ/ kg (round off to nearest integer)

52. One kg of air, initially at a temperature of 127°C, expands reversibly at a contant pressure until the volume is doubled. If the gas constant of air is 287 J/kg.K, the magnitude of work transfer is __________ kJ (round off to 2 decimal places).

53. For an ideal Rankine cycle operating between pressures of 30 bar and 0.04 bar, the work output from the turbine is 903 kJ/kg and the work input to the feed pump is 3 kJ/kg. The specific steam consumption is ________ kg/kW.h. (round off to 2 decimal place)

54. For a Kaplan (axial flow) turbine, the outlet blade velocity diagram at a section is shown in figure.

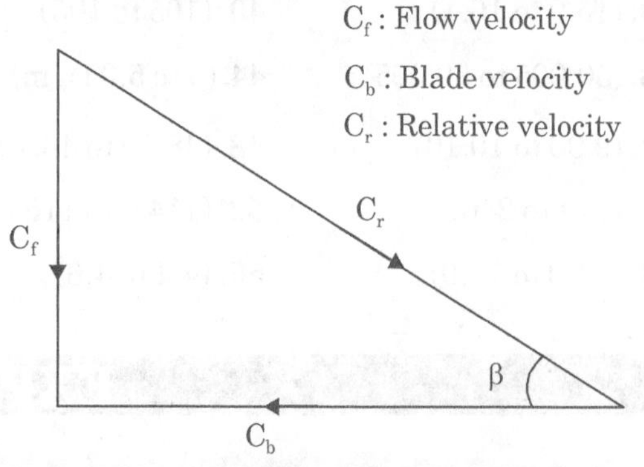

The diameter at this section is 3m. The hub and tip diameters of the blade are 2m and 4m, respectively. The water volume flow rate is 100 m^3/s. The rotational speed of the turbine is 300 rpm. The blade outlet angle β is _______ degrees (round off to one decimal place)

55. The indicated power developed by an engine with compression ratio of 8, is calculated using an air-standard Otto cycle (constant properties). The rate of heat addition is 10 kW. The ratio of specific heats at constant pressure and constant volume is 1.4. The mechanical efficiency of the engine is 80 percent.

The brake power output of the engine is _____ kW (round off to one decimal place)

ANSWERS

General Aptitude (GA)

1. (c) **2.** (a) **3.** (c) **4.** (b) **5.** (d) **6.** (d) **7.** (d) **8.** (d)

9. (a) **10.** (c)

Mechanical Engineering

1. (a) **2.** (c) **3.** (c) **4.** (d) **5.** (b) **6.** (c) **7.** (b) **8.** (a)

9. (c) **10.** (b) **11.** (a) **12.** (a) **13.** (d) **14.** (d) **15.** (d) **16.** (b)

17. (c) **18.** (d) **19.** (6 to 6) **20.** (0.98 to 1.02) **21.** (0.09 to 0.11) **22.** (1 to 1)

23. (0.90 to 0.95) **24.** (243 to 247) **25.** (48.8 to 49.8) **26.** (d) **27.** (a)

28. (a) **29.** (c) **30.** (c) **31.** (c) **32.** (d) **33.** (c) **34.** (b)

35. (–1.00 to –0.98) **36.** (1.99 to 2.01) **37.** (0.85 to 0.89) **38.** (47.9 to 48.1)

39. (15.9 to 16.1) **40.** (105 to 105) **41.** (19.95 to 20.05) **42.** (1.5 to 1.7)

43. (50.005 to 50.005) **44.** (5 to 5.04 μ m) **45.** (21 to 25) **46.** (1166 to 1168)

47. (9.90 to 10.10) **48.** (48.00 to 48.70) **49.** (8 to 8) **50.** (0.37 to 0.45)

51. (244 to 246) **52.** (114.6 to 115.0) **53.** (3.98 to 4.02)

54. (11.0 to 14.0) **55.** (4.4 to 4.6)

EXPLANATIONS

GENERAL APTITUDE

1. Crocodile tears: A false, insincere display of emotion.

2. Jofra Archar, the England fast bowler, is more fast than accurate.

3. Build : Building (Noun)

so, Grow : Growth (Noun)

4. having said that : despite what one just said.

5. Graph of $y = [x]$ for $x \in [1, 4]$

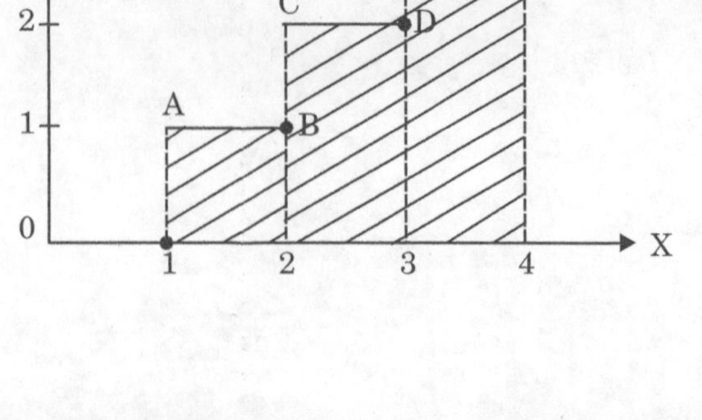

The total area, $A = (1 \times 1) + (2 \times 1) + (3 \times 1)$

$$= 1 + 2 + 3 = 6$$

6. The correct statement about crowd funding is to fund raised through voluntary contributions on web-based platforms.

7.

P,	Q,	R,	S
αα	αβ	$\boxed{\beta\alpha}$	$\boxed{\beta\beta}$

8. $S = 8 + 88 + 888 + ... n$

$$= \frac{8}{9}(9 + 99 + 999 + ...n)$$

$$= \frac{8}{9}\left[(10^1 - 1) + (100^2 - 1) + ... + (10^n - 1)\right]$$

$$= \frac{8}{9}\left[(10^1 + 100^2 + ...100^n) - n\right]$$

$$= \frac{8}{9}\left[\frac{10 \times (10^n - 1)}{10 - 1}\right] - \frac{8}{9}n$$

$$= \frac{80}{81}(10^n - 1) - \frac{8}{9}n$$

9. Graph of $y = x^m$ and $y = x^{1/m}$ for $m > 1$

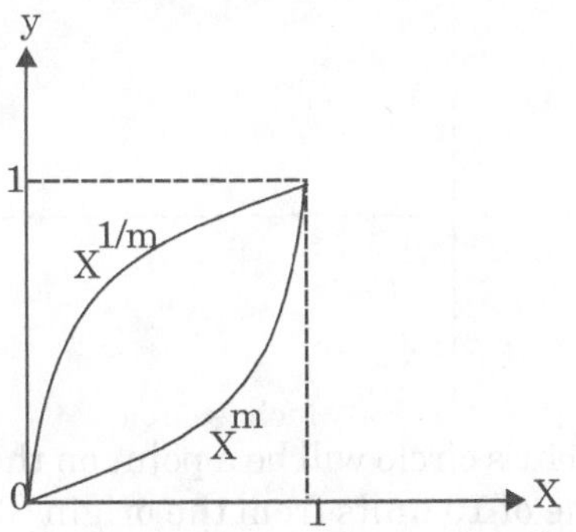

Correct option is (a).

10. $$\text{Success rate} = \frac{\text{Number of candidtes who passed the exam}}{\text{Number of appeared candidates}}$$

So, Average success rate,

$$= \frac{\frac{280}{500} + \frac{330}{600} + \frac{455}{700} + \frac{240}{400}}{4} = 0.59 \text{ or } 59\%$$

Correct option is (c).

MECHANICAL ENGINEERING

1. If A, B and C are real valued square matrices of same dimension,

(1) $AB \neq BA$

i.e. multiplication of matrices are not commutative.

(2) $A(BC) = (AB)C$

i.e. multiplication of matrices are associative.

so, the correct option is (a)

2. $$\lim_{x\to 1}\left(\frac{1 - e^{-c(1-x)}}{1 - xe^{-c(1-x)}}\right)$$

At x = 1, $$\frac{1 - e^{-c(1-x)}}{1 - xe^{-c(1-x)}} = \frac{0}{0}$$

Apply L-Hospital rule,

$$\lim_{x\to 1}\left[\frac{-ce^{-c(1-x)}}{0 - e^{-c(1-x)} - xe^{-c(1-x)} \times c}\right]$$

At x = 1,

$$\Rightarrow \lim_{x\to 1}\left[\frac{-ce^{-c(1-x)}}{0 - e^{-c(1-x)} - xce^{-c(1-x)}}\right] = \frac{-c \times 1}{-1 - c}$$

$$= \frac{c}{1 + c}$$

3. $$L(\sin \omega t) = \frac{\omega}{s^2 + \omega^2} \quad \text{(Formula to be used)}$$

Given, $$L(f) = \frac{1}{s^2 + \omega^2} = \frac{1}{\omega} \times \frac{\omega}{s^2 + \omega^2}$$

$$L(f) = \frac{1}{\omega} L(\sin \omega t)$$

$$\boxed{f(t) = \frac{1}{\omega} \sin \omega t}$$

4. Analytic function: A function f(z) is said to be analytic in a given region R, if it is continuous, single-valued and has definite derivartive at every point inside the region R.

f(z) = log z is differentiable and analytic everywhere except at origin.

5. To determine zero force members:

(1) If three members are connected at a joint and there is no external force applied to the joint and two of the members are colinear.

i.e, Joint B, D and G.

(i)

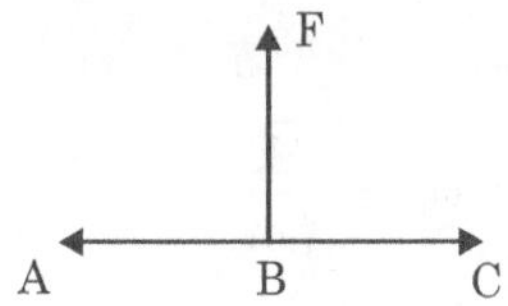

here, $F_{BF} = 0$

(ii)

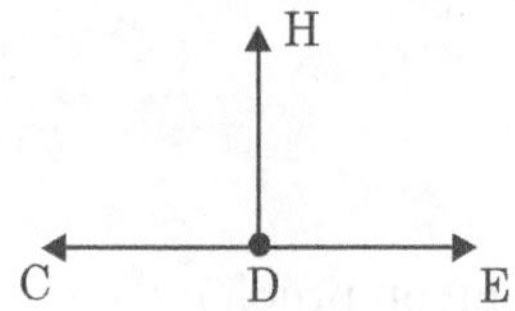

here, $F_{DH} = 0$

(iii)

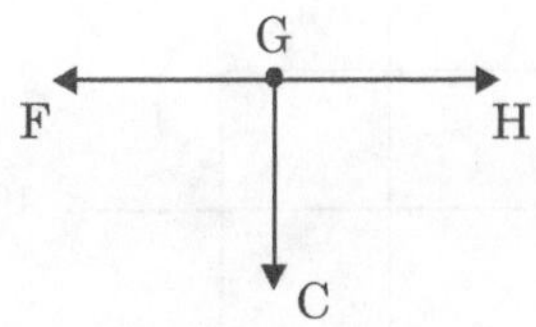

here, $F_{GC} = 0$

so, the members carrying zero force in the truss are BF, DH and GC.

6. Transmissibility ratio,

$$\frac{F_t}{F_o} = \sqrt{\frac{1 + \left(2\xi \frac{\omega}{\omega_n}\right)^2}{\left[1 - \left(\frac{\omega}{\omega_n}\right)^2\right]^2 + \left[2\xi \frac{\omega}{\omega_n}\right]^2}}$$

$\because F_t = F_o$ (Given)

$$\text{so, } 1 = \frac{\sqrt{1+\left(2\xi\frac{\omega}{\omega_n}\right)^2}}{\sqrt{\left[1-\left(\frac{\omega}{\omega_n}\right)^2\right]^2+\left[2\xi\frac{\omega}{\omega_n}\right]^2}}$$

$$\Rightarrow \left[1-\left(\frac{\omega}{\omega_n}\right)^2\right]^2+\left(2\xi\frac{\omega}{\omega_n}\right)^2 = \left(2\xi\frac{\omega}{\omega_n}\right)^2+1$$

It can be written as, $\left[\left(\frac{\omega}{\omega_n}\right)^2-1\right]^2 = 1$

$$\Rightarrow \left(\frac{\omega}{\omega_n}\right)^2 = 2$$

$$\Rightarrow \frac{\omega}{\omega_n} = \sqrt{2}$$

$$\Rightarrow \omega = \sqrt{2}\,\omega_n \qquad \because \omega_n = \sqrt{\frac{k}{m}}$$

$$\text{so, } \omega = \sqrt{2}\times\sqrt{\frac{k}{m}}$$

$$\boxed{\omega = \sqrt{\frac{2k}{m}}}$$

7. According to the given problem:

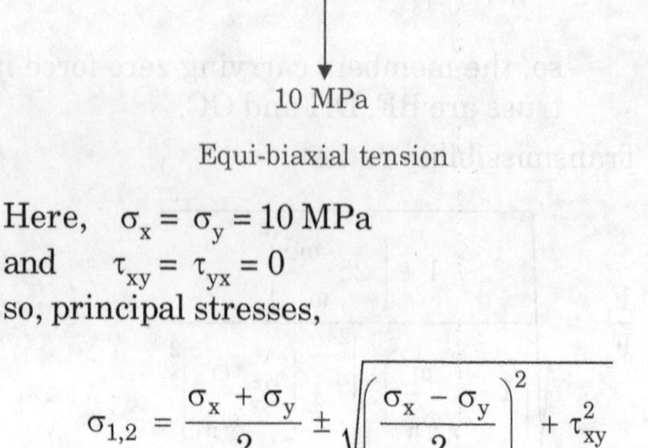

Equi-biaxial tension

Here, $\sigma_x = \sigma_y = 10$ MPa

and $\tau_{xy} = \tau_{yx} = 0$

so, principal stresses,

$$\sigma_{1,2} = \frac{\sigma_x+\sigma_y}{2} \pm \sqrt{\left(\frac{\sigma_x-\sigma_y}{2}\right)^2+\tau_{xy}^2}$$

$$\sigma_1 = 10 \text{ MPa} \quad \text{and} \quad \sigma_2 = 10 \text{ MPa}$$

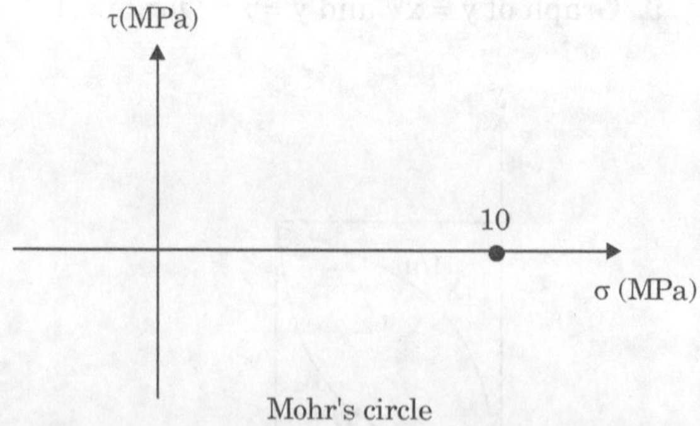

Mohr's circle

i.e. Mohr's circle will be a point on the σ-axis at a distane of 10 units from the origin (0, 0).

8. According to Grashof's Law, (for the continuous relative motion

$$l + S \le P + Q$$

Here, l = length of largest side

S = length of shortest side

P and Q = length of other two sides for, S = 80 mm, the above condition is satisfied.

i.e. 600 + 80 < 300 + 400 to obtain crankrocker mechanism, the adjacent side of shortest link (s = 80 mm) should be fixed.

So, for the mechanism to be a crank-rocker mechanism, the length of the link PQ should be 80 mm.

9.

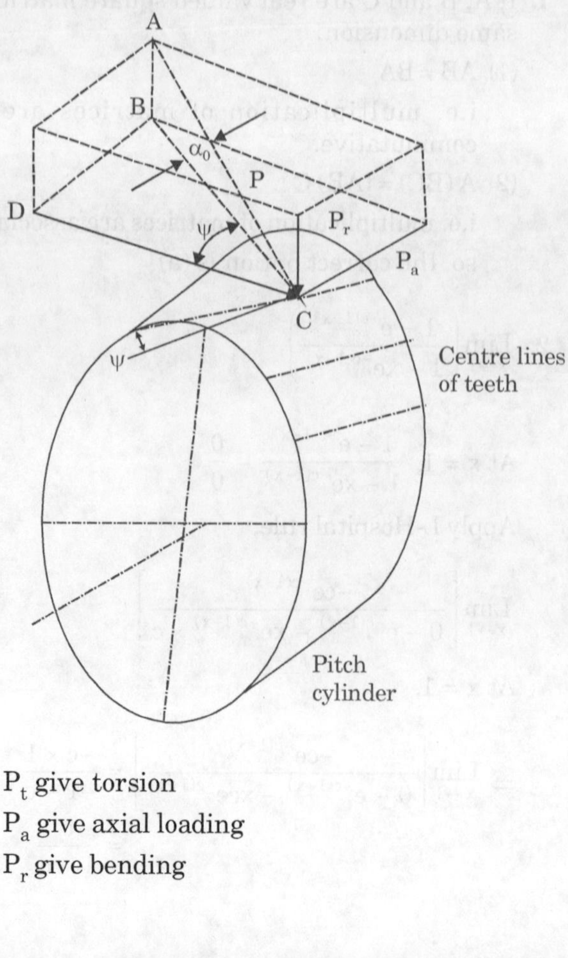

P_t give torsion

P_a give axial loading

P_r give bending

10. The crystal structure of γ-iron (Austenite phase) is FCC i.e. Face Centered Cubic.

11. (i) Tempering: The main effect of tempering is to impart toughness in the metal.

i.e $P \rightarrow 2$

(ii) Quenching : It increases the hardness of the material.

i.e $Q \rightarrow 3$

(iii) Annealing: It is a heat treatment process which 'soften' a metal

i.e $R \rightarrow 4$

(iv) Normalizing: This heat treatment process is used for strengthening and grain refinement.

i.e $S \rightarrow 1$

12. For rough grinding of brass material:

(1) Silicon carbide (SiC) is preferred for grinding of brass material.

(2) For rough cut, open structure is preferred.

Range of structure from dense to open varies from 0 to 15 i.e. more the number; more the open structure is.

So, according to that the suitable grinding wheel is,

$\boxed{C}\ 30\ Q\ \boxed{12}\ V$

13. In critical parth method (CPM), the cost-time slope of an activity.

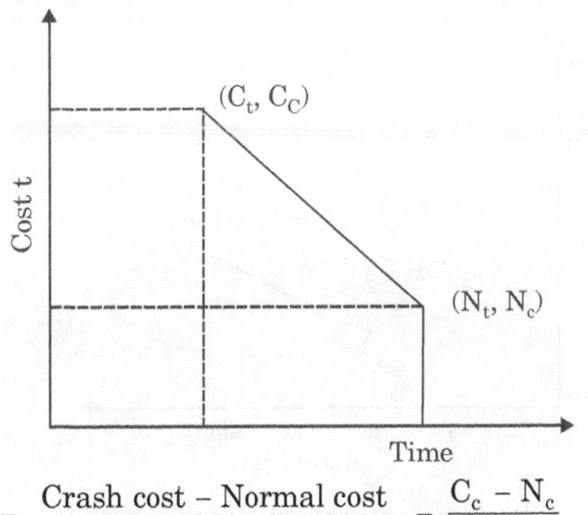

$$= \frac{\text{Crash cost} - \text{Normal cost}}{\text{Normal time} - \text{Crash time}} = \frac{C_c - N_c}{N_t - C_t}$$

14. Froude number, $F_r = \dfrac{\text{Inertia force } (F_i)}{\text{Gravity force } (F_g)}$

15. (i) Reynolds number, $Re = \dfrac{\text{Inertia force}}{\text{Viscous force}}$

i.e. $P \rightarrow 3$

(ii) Grashof Number, $Gr = \dfrac{\text{Buoyancy force}}{\text{Viscous force}}$

i.e. $Q \rightarrow 1$

(iii) Nusselt Number, $Nu = \dfrac{\text{Convective heat transfer}}{\text{Conduction heat transfer}}$

i.e. $R \rightarrow 4$

(iv) Prandtl number, $Pr = \dfrac{\text{Momentum diffusivity}}{\text{Thermal diffusivity}}$

i.e. $S \rightarrow 2$

so, the correct option is (d)

16. Given : $\vec{V} = 2(x^2 - y^2)\hat{i} + v\hat{j} + 3\hat{k}$
$= u\hat{i} + v\hat{j} + w\hat{k}$

For incompressible flow,

$$\Delta \cdot \vec{V} = 0$$

$$\frac{\partial}{\partial x}(u) + \frac{\partial}{\partial y}(v) + \frac{\partial}{\partial z}(w) = 0$$

$$\Rightarrow \frac{\partial}{\partial x}[2(x^2 - y^2)] + \frac{\partial}{\partial y}(v) + \frac{\partial}{\partial z}(3) = 0$$

$$\Rightarrow 4x + \frac{\partial v}{\partial y} + 0 = 0$$

$$\Rightarrow \frac{\partial v}{\partial y} = -4x$$

$$\Rightarrow v = \int -4x\,dy + \text{cons tan t}$$

$$\Rightarrow v = -4xy + \text{constant}$$

17. Joule-Thomson coefficient is defined as,

$$\mu_j = \left(\frac{\partial T}{\partial P}\right)_h$$

Also, $\mu_j = \dfrac{1}{C_P}\left[T\left(\dfrac{\partial V}{\partial T}\right)_P - v\right]$

For ideal gas, $Pv = RT$

$$\Rightarrow \left(\frac{\partial v}{\partial T}\right)_P = \frac{R}{P}$$

$$\Rightarrow \mu_j = \frac{1}{C_P}\left[\frac{T \times R}{P} - v\right]$$

$$\Rightarrow \mu_j = \frac{1}{C_P}\left[\frac{Pv}{P} - v\right]$$

$$\boxed{\mu_j = 0}$$

i.e., for ideal gas, the value of the Joule-Thomson coefficient is zero.

18. For ideal gas,

$Tds = du + Pdv = C_v dT + Pdv$

At constant volume, $(dv = 0)$

$$Tds = C_v dT$$

$$\Rightarrow \left(\frac{dT}{ds}\right)_v = \frac{T}{C_v} \quad \text{...(i)}$$

Again, for ideal gas.

$$Tds = dh - vdP = C_p dT - vdP$$

For constant pressure,

$$Tds = C_p dT$$

$$\left(\frac{dT}{ds}\right)_P = \frac{T}{C_P} \quad \text{...(ii)}$$

so, the ratio,

$$\frac{\left(\frac{\partial T}{ds}\right)_P}{\left(\frac{dT}{ds}\right)_v} = \frac{T}{C_P} \times \frac{C_v}{T} = \frac{C_v}{C_P}$$

19. Given: $\vec{A} - = 2\hat{j} - 3\hat{k}$

$$\vec{B} = -2\hat{i} + \hat{k}$$

and $\vec{C} = 3\hat{i} - \hat{j}$

$$(\vec{B} \times \vec{C}) = (-2\hat{i} + \hat{k}) \times (3\hat{i} - \hat{j})$$

$$= -2\hat{i} + 3\hat{i} + 2\hat{i} \times \hat{j} + \hat{k} \times 3\hat{i} - \hat{k} \times \hat{j}$$

$[\because \hat{i} \times \hat{i} = 0, \hat{i} \times \hat{j} = \hat{k}, \hat{k} \times \hat{i} = \hat{j} \text{ and } \hat{k} \times \hat{j} = -\hat{i}]$

so, $(\vec{B} \times \vec{C}) = 0 + 2\hat{k} + 3\hat{j} + \hat{i}$

$$= (2\hat{k} + 3\hat{j} + \hat{i})$$

Now, $\vec{A} \cdot (\vec{B} \times \vec{C}) = (2\hat{j} - 3\hat{k}) \cdot (2\hat{k} + 3\hat{j} + \hat{i})$

$$= 4\hat{j} \cdot \hat{k} + 6\hat{j} \cdot \hat{j} + 2\hat{j} \cdot \hat{i} - 6\hat{k} \cdot \hat{k} - 9\hat{k} \cdot \hat{j} - 3\hat{k} \cdot \hat{i}$$

$\because$ $[\hat{j} \cdot \hat{k} = 0, \hat{k} \cdot \hat{k} = 1, \hat{j} \cdot \hat{j} = 1, \hat{j} \cdot \hat{i} = 0, \hat{k} \cdot \hat{j} = 0,$

and $\hat{k} \cdot \hat{i} = 0]$

So $(\vec{A} \cdot (\vec{B} \times \vec{C}) = 0 + 6 + 0 - 6 - 0 - 0$

The value of $(\vec{A} \cdot (\vec{B} \times \vec{C}) + 6)$ is $0 + 6 = 6$

20. Given : Max rotational speed, $\omega_1 = 110$ rad/s

Min rotational speed, $\omega_2 = 100$ rad/s

Energy fluctuation, $\Delta E = 1.05$ kJ

$$\because \Delta E = \frac{1}{2} I\left(\omega_1^2 - \omega^2\right)$$

where, I = Moment of inertia of fly wheel

$$1.05 \times 1000 = \frac{1}{2} \times 1 \times (110^2 - 100^2)$$

$$\boxed{I = 1 \text{ kg} - \text{m}^2}$$

21. Mass, $m = 10$ grams $= 10 \times 10^{-3}$ kg

Radius, $r = 100$ mm $= 100 \times 10^{-3}$ m

Angular speed, $\omega = 10$ rad/s

Force on each point masses,

$$F = mr\omega^2 = 10 \times 10^{-3} \times 100 \times 10^{-3} \times (10)^2$$

$$= 0.1 \text{ N} = 0.1 \text{ N}$$

22. Clearance in the punch, $C = 0.0032t\sqrt{\tau})$

where, τ is the shear strength of sheet metal.

$\because \tau \propto (\text{BHN}) \cdot (\text{HRC})$

We can write, $C \propto \sqrt{\text{HRC}}$

$\because$ At, $(\text{HRC})_1 = 250$, $C_1 = 1$ mm

then at $(\text{HRC})_2 = 400$,

$$\text{Clearance, } C_2 = C_1 \times \sqrt{\frac{(\text{HRC})_2}{(\text{HRC})_1}} = 1 \times \sqrt{\frac{400}{250}}$$

$$\boxed{C_2 = 1.26 \text{ mm}}$$

23. Number of Men, M = 5

Number of Women N = 3

We have to hire four candidates.

The required probability (i.e. the Probability that at least one woman will be selected).

$$P = \frac{{}^5C_3 \times {}^3C_1}{{}^8C_4} + \frac{{}^5C_2 + {}^3C_2}{{}^8C_4} + \frac{{}^5C_1 \times {}^3C_3}{{}^8C_4}$$

$$= \frac{3}{7} + \frac{3}{7} + \frac{1}{14} = \frac{13}{14} = 0.93$$

24.

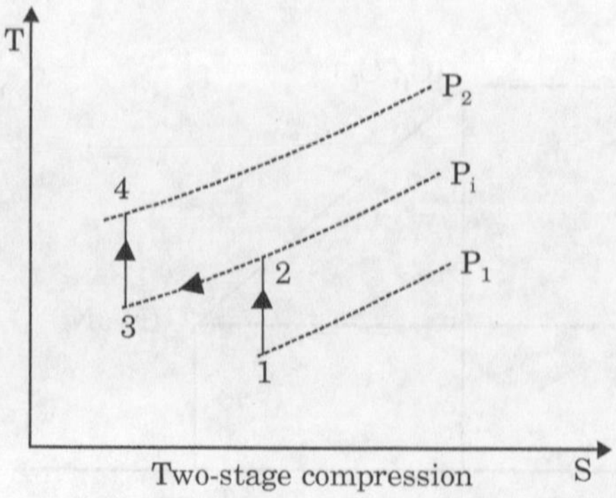

Two-stage compression process

$\because$ There are two (n = 2) at stages.

For minimum work,

$$\text{Overall pressure ratio, } r_o = \left(\frac{P_i}{P_1}\right)^{\frac{1}{2}}$$

$$\Rightarrow \quad 6 = \left(\frac{P_i}{100}\right)^{\frac{1}{2}}$$

Inter cooling pressure, $P_1 = 244.95$ kPa

25. Given :

Temperature of hot oil at the entry, $T_{h_i} = 102°C$

Temperature of hot oil at the exit, $T_{h_e} = 65°C$.

Temperature of cold water at entry, $T_{c_i} = 25°C$

and Temperature of cold water at exit, $T_{c_e} = 42°C$

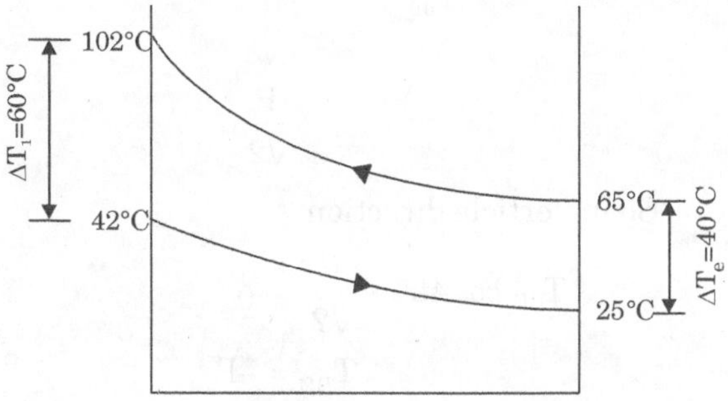

Counter-flow heat exchanger

∵ LMTD is defined as,

$$(\Delta T)_{lm} = \frac{\Delta T_i - \Delta T_e}{\ln\left(\frac{\Delta T_i}{\Delta T_e}\right)} = \frac{60-40}{\ln\left(\frac{60}{40}\right)}$$

$$(\Delta T)_{ln} = 49.3°C$$

26. $\int_{-1}^{1.4} x|x|\,dx$

$f(x) = -x^2 \quad -1 \le x \le 0$

$f(x) = x^2 \quad 0 \le x < 14$

x	–1	–0.4	0.2	0.8	1.4
f(x)	–1	–0.16	0.04	0.64	1.96
	y_0	y_1	y_2	y_3	y_4

By simpson's 1/3rd rule

$$\int_{-1}^{1} x|x|\,dx = \frac{h}{3}[(y_0 + y_4) + 2(y_2) + 4(y_1 + y_2)]$$

$$= \frac{0.6}{3}[(-1 + 1.96) + 2 \times 0.04 + 4(-0.16 + 0.64)]$$

$$= 0.592$$

27. $\vec{f}(x, y, z) = \frac{x}{(x^2 + y^2 + z^2)^{3/2}}\hat{i} + \frac{y}{(x^2 + y^2 + z^2)^{3/2}}\hat{j} + \frac{z}{[x^2 + y^2 + z^2]^{3/2}}\hat{k}$

By gauss-divergence theorem

$$\iint_s \vec{F}\hat{n}ds = \iiint_v \vec{\nabla}\cdot\vec{F}\,dV \quad ...(i)$$

$$\text{So, } \vec{\nabla}\cdot\vec{F} = \frac{\partial}{\partial x}\left(\frac{x}{x^2 + y^2 + z^2)^{\frac{3}{2}}}\right) + \frac{\partial}{\partial y}\left(\frac{y}{(x^2 + y^2 + z^2)^{\frac{3}{2}}}\right) + \frac{\partial}{\partial z}\left(\frac{z}{(x^2 + y^2 + z^2)^{\frac{3}{2}}}\right)$$

$$\vec{\nabla}\cdot\vec{F} = \frac{1}{(x^2 + y^2 + z^2)^{\frac{3}{2}}} - \frac{\frac{3}{2} \times x \times 2x}{(x^2 + y^2 + z^2)^{\frac{5}{2}}} + \frac{1}{(x^2 + y^2 + z^2)^{\frac{3}{2}}} - \frac{\frac{3}{2} \times y \times 2y}{(x^2 + y^2 + z^2)^{\frac{5}{2}}} + \frac{1}{(x^2 + y^2 + z^2)^{\frac{3}{2}}} - \frac{\frac{3}{2} \times z \times 2z}{(x^2 + y^2 + z^2)^{\frac{5}{2}}} = 0$$

So by equation (i),

$$\iint_S \vec{F}\hat{n}6dS = \iiint_V \vec{\nabla}\cdot\vec{F}\,dV = 0$$

28.

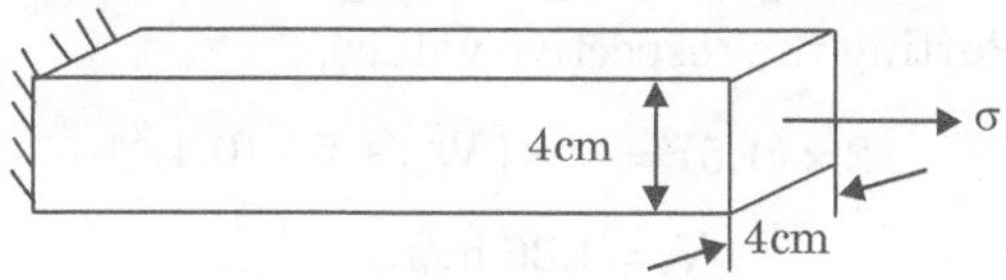

$$\sigma = \frac{P}{A} = \frac{80 \times 10^3}{40 \times 40} = 50 \text{ MPa}$$

According to maximum shear stress theory

$$\tau_{max} = \frac{\sigma_1 - \sigma_2}{2} = \frac{50 - 0}{2}$$

$$= 25\text{MPa}$$

$\because \tau_{per} < \tau_{max}$

∴ This component will fail

Similarly,

The component, loaded with same load in compression, will also fails.

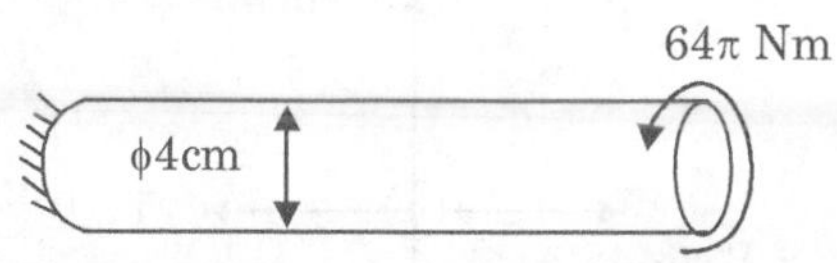

$$\tau_{max} = \frac{16T}{\pi d^3} = \frac{16 \times 64\pi \times 10^3}{\pi \times 40^3} = 16 \text{ MPa}$$

$\because \tau_{max} < \tau_{per}$

∴ Component will not fail.

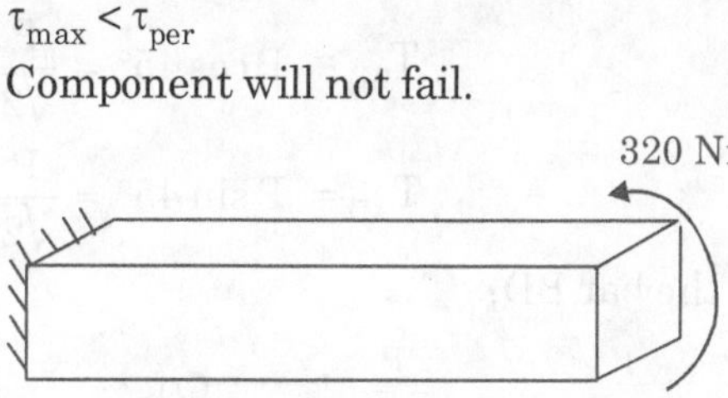

$$\sigma_{max} = \frac{M}{I} \times y = \frac{320 \times 10^3}{\frac{40 \times 40^3}{12}} \times 20$$

$$= 30 \text{ MPa}$$

According to maximum

$$\tau_{max} = \frac{\sigma_{max} - \sigma_{min}}{2}$$

$$= \frac{30-0}{2} = 15 \text{ MPa}$$

$\because \tau_{max} < \tau_{per}$

$\therefore$ component will not fail

29. Given: Mass, m = 2 kg

Stiffness of rod, k = 5 N/m

Initial velocity when the block is unstretched, $V_i = 1.5$ m/s

Now, the cord is stretched by x = 0.4 m, then velocity of the block is V_f.

Using work-energy theorem,

$$\frac{1}{2}mv_i^2 = \frac{1}{2}mv_f^2 + \frac{1}{2}kx^2$$

Putting the respective values,

$$\Rightarrow \quad 2 \times (1.5)^2 = 2 \times \left(V_f^2\right) + 5 \times (0.4)^2$$

$$\Rightarrow \quad V_f = 1.36 \text{ m/s}$$

30.

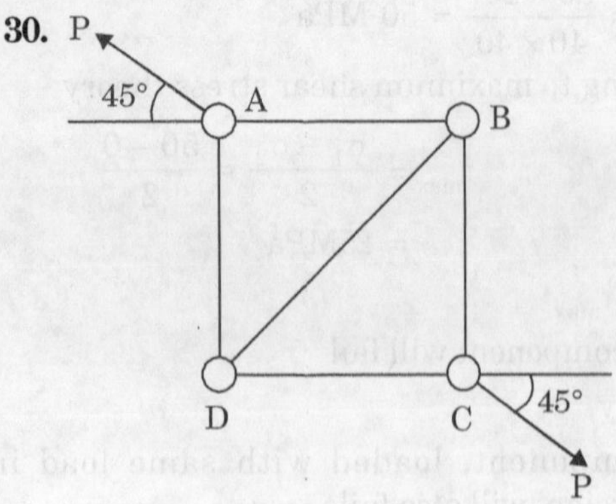

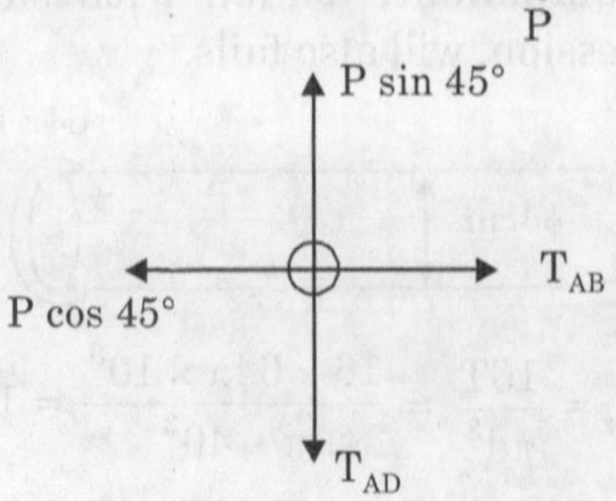

$$T_{AB} = P\cos 45° = \frac{P}{\sqrt{2}}$$

$$T_{AD} = P\sin 45° = \frac{P}{\sqrt{2}}$$

For the bar BD,

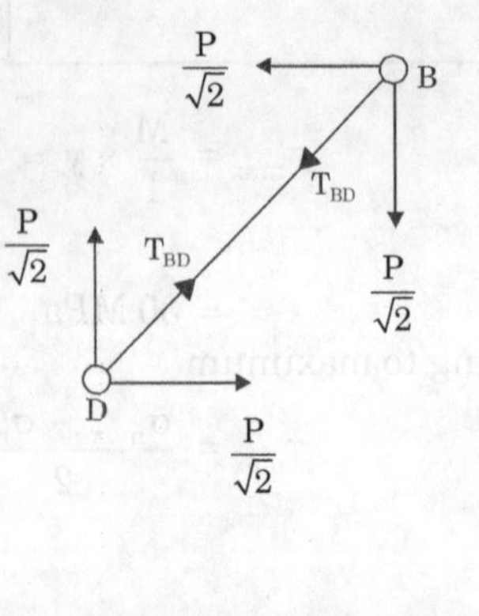

For the joint B,

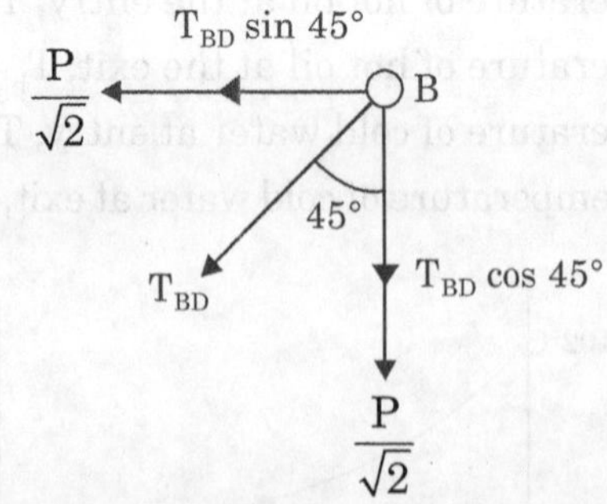

So in verticle direction

$$T_{BD}\cos 45° + \frac{P}{\sqrt{2}} = 0$$

$$T_{BD} = -P$$

so maximum load acting on the bar BD.

so bar BD buckle first

Buckling load,

$$P = \frac{\pi^2 \times 4EI}{\left(\sqrt{2}l\right)^2}$$

[both ends are hinge]

$$P = \frac{2\pi^2 EI}{l^2}$$

31.

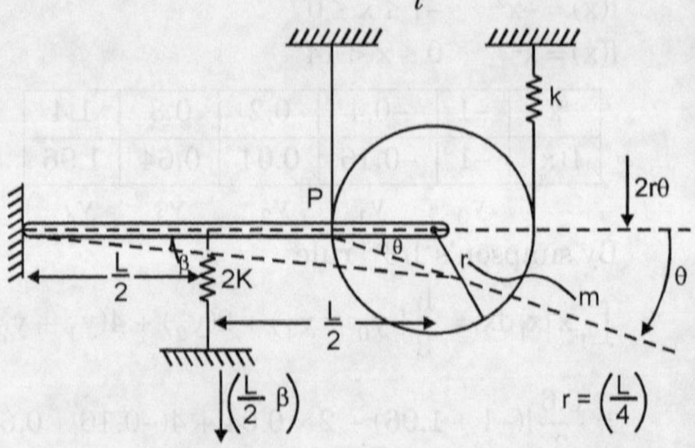

Since the disc is in pull rolling on the string:

$$L\beta = R\theta$$

$$\beta = \frac{\theta}{4} \qquad \because R = \frac{L}{4}$$

Applying Energy method

$$E_{Total} = \frac{1}{2}I_p\dot{\theta}^2 + \frac{1}{2}k(2R\theta)^2 + \frac{1}{2}2k\left(\frac{L\theta}{8}\right)^2$$

$$\frac{dE}{dt} = 0$$

$$\Rightarrow \frac{3}{2}MR^2.\ddot{\theta}\dot{\theta} + 4R^2K\theta\dot{\theta} + \frac{2KL^2\theta\dot{\theta}}{64} = 0$$

$$\Rightarrow \left(\frac{3}{2}\frac{NL^2}{16}\right)\ddot{\theta}+\frac{4L^2}{16}k\theta+\frac{2}{64}kL^2\theta=0$$

$$\Rightarrow \left(\frac{3N}{32}\right)\ddot{\theta}+\left(\frac{9k}{32}\right)\theta=0$$

$$\Rightarrow 3N\ddot{\theta}+9k\theta=0$$

$$\Rightarrow \ddot{\theta}+\left(\frac{3k}{m}\right)\theta=0 \Rightarrow \left[\omega n_1=\sqrt{3}\sqrt{\frac{k}{m}}\right]$$

32.
$H_1 = 40$ mm
$H_2 = 20$ mm
$R = 100$ mm
$\Delta H = 40 - 20 = 20$ mm

Coefficient of friction (μ)

$$\mu=\sqrt{\frac{(\Delta H)_{max}}{R}} \qquad [\Delta H=(\Delta H)_{max}]$$

$$=\sqrt{\frac{20}{100}}=0.45$$

Arc length (L) $=\sqrt{R(\Delta H)}$

$$=\sqrt{100\times 20}=44.72 \text{ mm}$$

33. WIP means work in process

In first case $(WIP)_1 = 4\times\frac{60}{60} = 4$ pieces

In second case $(WIP)_2 = 4\times\frac{(60-60\times 0.3)}{60}$

$= 2.8$ piece

So, Increase in (WIP) $=\frac{(WIP)_1-(WIP)_2}{(WIP)_1}\times 100$

$= 30\%$

34. $T_o = 100°C$ $T_\infty = 20°C$
$T = 28°C$ $t = 4.35$ s
$r = 8500$ kg/m^3 $C = 400$ J/kgK
$r = 0.5$ mm $h = ?$

Metal bead is assumed as spherical shape

By lumped system analysis

$$\frac{T-T_\infty}{T_o-T_\infty}=e^{-\frac{hA\tau}{\rho VC}}$$

$$\frac{T-T_\infty}{T_o-T_\infty}=e^{-\frac{h\times 4\pi r^2\tau}{\rho\times V\times\frac{4}{3}\pi r^3}}$$

$$\frac{T-T_\infty}{T_o-T_\infty}=e^{-\frac{3h\tau}{\rho\times V\times r}}=e^{-\frac{3\times 4\times 4.35}{8500\times 400\times 0.5\times 10^{-3}}}$$

$h = 299.95$ W/m^{2}K

35. Mean of exponential distribution $=\frac{1}{\lambda}$

Standard deviation of exponential distribution $=\frac{1}{\lambda}$

For x, $\lambda_1=\frac{1}{0.5}=2$

Standard deviation, $\sigma_x = 0.5$

Variance, $V(X) = (\sigma_x)^2 = 0.25$

For Y, $\lambda_2=\frac{1}{0.5}=2$

Standard deviation, $\sigma_Y = 0.5$

Variance $V(Y) = (\sigma_Y)^2 = 0.25$

$\because V(X+Y) = V(X) + V(Y) + 2COV(X, Y)$

Given : $V(X + Y) = 0$

$\Rightarrow$ $0 = 0.25 + 0.25 + 2\,COV(X, Y)$

$\Rightarrow COV(X, Y) = -0.25$

Corelation factor,

$$r=\frac{COV(X,Y)}{\sigma_X\sigma_Y}$$

$$=\frac{-0.25}{0.5\times 0.5}=-1$$

36.
$f(z) = u + iv$
$u = x^2 - y^2$
$v = \psi(x, y)$ ...(i)

f (z) is an analytic function,

$$\frac{\partial u}{\partial x}=\frac{\partial v}{\partial y}$$

$$\frac{\partial v}{\partial x}=\frac{\partial u}{\partial y}$$

so, $dv=\frac{\partial v}{\partial x}dx+\frac{\partial v}{\partial y}dy$

$$dv=-\frac{\partial u}{\partial y}dx+\frac{\partial u}{\partial x}dy$$

$$dv = 2y\,dx + 2x\,dy$$

$$\int dv = 2\int d(xy)$$

$v = 2xy$ ...(ii)

Given, $z = 1 + i$

So, $x = 1$

$y = 1$

From equation (1) and (2)

So, $v=\psi(x,y)_{(1,1)}=2$

37. Given:

Inner diameter, $D_i = 50$ mm

or Inner radius, $R_i = \dfrac{50}{2} = 25$ mm

and outer diameter, $D_o = 250$ mm

or outer radius, $R_o = \dfrac{250}{2} = 125$ mm

(1) For uniform pressure theory (for new clutch)

Torque transmitted, $\tau_1 = \mu_1 F_1 R'_m$

where, $R'_m = \dfrac{2\left(R_o^3 - R_i^3\right)}{3\left(R_o^2 - R_i^2\right)}$

i.e., $\tau_1 = \mu_1 \times F_1 \times \dfrac{2}{3}\left(\dfrac{R_o^3 - R_i^3}{R_o^2 - R_i^2}\right)$

(2) For uniform wear theory (for old clutch),

Torque transmitted, $\tau_2 = \mu_2 F_2 R''_m$

where, $R''_m = \dfrac{R_o + R_i}{2}$

i.e., $\tau_2 = \mu_2 \times F_2 \times \dfrac{(R_o + R_i)}{2}$

$\because$ Torque transmitted are same and coeficient of friction is also same.

so, $F_1 \times \dfrac{2}{3}\left(\dfrac{R_o^3 - R_i^3}{R_o^2 - R_i^2}\right) = F_2 \times \dfrac{(R_o + R_i)}{2}$

$\Rightarrow \dfrac{F_1}{F_2} = \dfrac{3}{4}\left(\dfrac{R_o^2 - R_i^2}{R_o^3 - R_1^3}\right) \times (R_o + R_i)$

Putting the respective values,

$$\frac{F_1}{F_2} = \frac{3}{4}\left[\frac{125^2 - 25^2}{125^3 - 25^3}\right] \times (125 + 25)$$

$$\boxed{\frac{F_1}{F_2} = 0.87}$$

38. Given: $y(\theta) = 4[2p\theta - \theta^2], 0 \le \theta \le 2\pi$

$y'(\theta) = 8\pi - 8\theta$

$y''(\theta) = -8$

The minimum permissible base circle radius,

$(r_0)_{min} = \rho_{min} - y(\theta) + y''(\theta)]_{min}$

Here, $\rho_{min} = 40$ mm

$y(\theta) + y''(\theta) = 4[2\pi\theta - \theta^2] - 8, 0 \le \theta \le 2\pi$

$[y(\theta) | y''(\theta)]_{min} = 4[2\pi \times 0 - 0^2] - 8$

$= -8$

so, $(r_o)_{min} = 40 - (-8) = 48$ mm

39.

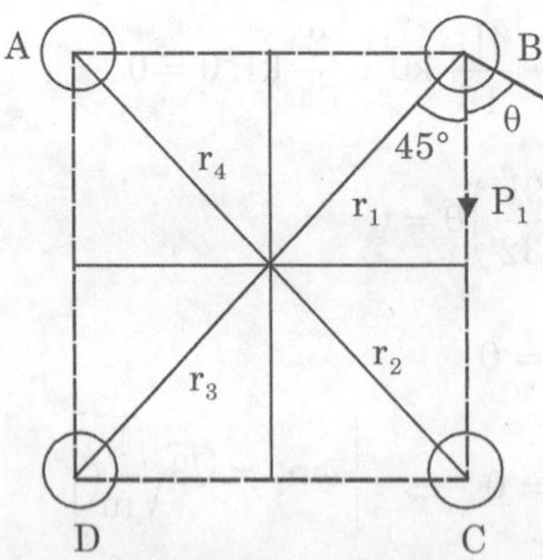

$P = 10$ kN

$r_1 = r_2 = r_3 = r_4$

$= \sqrt{(50)^2 + (50)^2} = 70.71$ mm

Eccentricity, e = 400 mm

Primary fore, $P_1 = \dfrac{P}{4} = \dfrac{10}{4} = 2.5$ kN

Secondary force, $P_2 = \dfrac{P \cdot e \cdot r_1}{r_1^2 + r_2^2 + r_3^2 + r_4^2}$

$= \dfrac{10 \times 400 \times 70.71}{4 \times (70.71)^2}$

$= 14.14$ kn

$\theta = 90° - 45° = 45°$

Resultant shear force on bolt B,

$P_B = \sqrt{P_1^2 + P_2^2 + 2P_1P_2 \cos 45°}$

$= \sqrt{(2.5)^2 + (14.14)^2 + 2 \times 2.5 \times 14.14 \cos 45}$

$= 16$ kN

40.

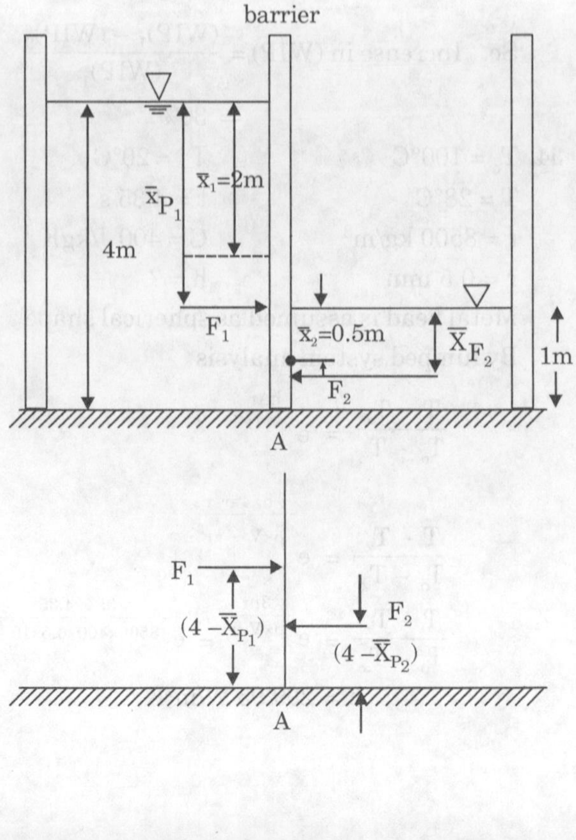

$$F_1 = wA\bar{x}_1 = 10 \times (4 \times 1) \times 2 = 80 \text{ kN}$$

$$F_2 = wA\bar{x}_2 = 10 \times (1 \times 1) \times 0.50 = 5 \text{ kN}$$

$$\bar{x}_{P_1} = \bar{x}_1 + \frac{1 \times 4^3}{12 \times 1 \times 4 \times 2} = 2 + \frac{2}{3} = \frac{8}{3} \text{ m}$$

$$\bar{x}_{P_2} = \bar{x}_2 + \frac{1 \times 1^3}{12 \times 1 \times 1 \times 0.5} = 0.5 + \frac{1}{6} = \frac{2}{3} \text{ m}$$

The maximum absolute bending moment will be at fixed support i.e. at point A.

$$(BM)_A = F_1 \times (4 - \bar{x}_{P_1}) - F_2 \times (1 - \bar{x}_{P_2})$$

$$= 80 \times \left(4 - \frac{8}{3}\right) - 5 \times \left(1 - \frac{2}{3}\right) = 105 \text{ kN-m}$$

41. At thing point moment is ero

From bar BC

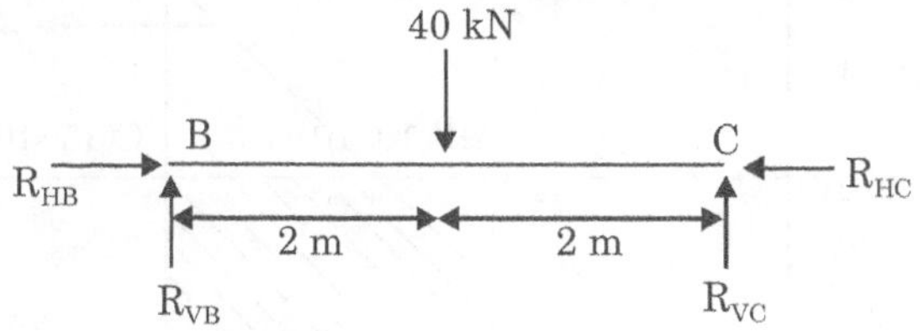

Take moment about B

$$R_{VC} \times 4 = 40 \times 2$$

$$R_{VC} = 20 \text{ kN}$$

There is no force applyinhorizontal direction so

$$R_{HB} = R_{HC} = 0\text{N}$$

Net force on C,

$$F_C = \sqrt{(R_{HC})^2 + (R_{VC})^2}$$

$$= \sqrt{(0)^2 + (20)^2}$$

$$F_C = 20 \text{ kN}$$

42. Length of workpiece, L_w = 300 mm

Diameter of cutter, D_c = 100 mm

Number of teeth, Z = 20

Depth of cut = 5 mm

Feed per tooth, f_z = 0.10 mm

Cutting speed, v = 35 m/min.

Approach (A) = over travel (O) = 5 mm

To find, Machining time (t_m)

$$t_m = \frac{\text{length of cut (L)}}{\text{feed rate } (f_r)}$$

$$L = L_w + O + A$$

$$= 300 + 5 + 5 = 310 \text{ mm}$$

$$f_r = f_z \times Z \times N = f_z \times Z \times \frac{v}{\pi D_c}$$

$$= 0.10 \times 20 \times \frac{35 \times 10^3}{\pi \times 100}$$

$$= 222.82 \text{ mm/min.}$$

$$t_m = \frac{310}{222.82} = 1.39 \text{ min.}$$

$$\boxed{t_m \simeq 1.4 \text{ min}}$$

43. For basic shaft system,

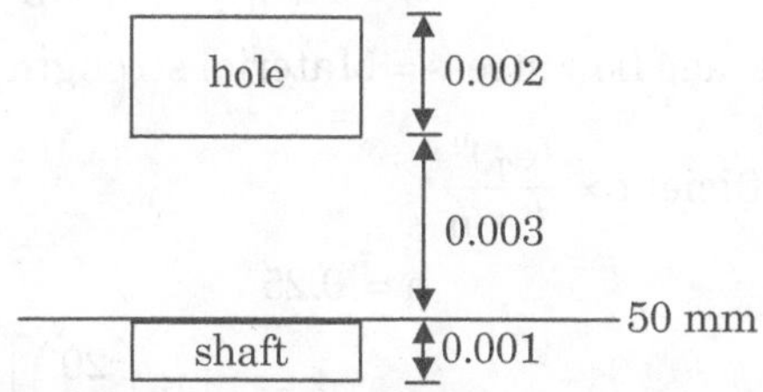

Maximum hole size = 50 + 0.003 +0.002

= 50.005 mm

44. Given: Surface area, A = 125 cm^2

Current-time corelation,

I = 12 + 0.2t, t is in minutes

Duration = 20 minutes

Plating constant = 2.50×10^{-2} mm^3/A.s

Cathode efficiency, $\eta = 0.15$

Let, the coating thickness is dx.

According to the given question,

$$A \times dx = \eta \times C \times I\,dt$$

$$\Rightarrow 125 \times 10^2 \times dx = 0.15 \times 2.50 \times 10^{-2} \times (12 + 0.2t)\,dt$$

Integrating both sides,

$$125 \times 10^2 \times \int_0^x dx = (0.15 \times 2.50 \times 10^{-2}) \times \int_0^{20} (12 + 0.2t) \times 60\,dt$$

$$\Rightarrow 125 \times 10^2 \times x = (0.15 \times 2.50 \times 10^{-2}) \times 60 \times \left[12t + \frac{0.2t^2}{2}\right]_0^{20}$$

$$\Rightarrow \quad x = 5.04 \times 10^{-3} \text{ mm}$$

$$= 5.04\ \mu\text{m}$$

45. l_c = 100 mm

l = 250 mm

V = 30 m/min.

$\alpha = 20°$

Chip thickness ratio (r),

$$r = \frac{l_c}{l} = \frac{100}{250} = 0.4$$

Shear plane angle (ϕ),

$$\tan\phi = \frac{r\cos\alpha}{1 - r\sin\alpha}$$

$$\phi = \tan^{-1}\left(\frac{0.4\cos 20°}{1 - 0.4\sin 20°}\right) = 23.53°$$

46. (Final volume) = (Initial volume)

$$l_f b_f h_f = l_i b_i h_i$$

$$l_f \times 200 \times 15 = l_i \times 200 \times 20$$

[∵ width remains same in rolling]

$$\frac{l_f}{l_i} = \frac{20}{15}$$

Now, true strain, $\in_T = \ln\left(\frac{l_f}{l_i}\right) = \ln\left(\frac{20}{15}\right)$

Average flow stress = Material strength coefficient × $\frac{(\in_T)^n}{1+n}$

$n = 0.25$ (given)

$$\sigma_f = 210 \times \frac{\left[\ln\left(\frac{20}{15}\right)\right]^{0.25}}{1+0.25}$$

$$= 123.04 \text{ MPa}$$

→ Roll contact length,

$$L = \sqrt{R\Delta h}$$

$$= \sqrt{450 \times 5}$$

$$= 47.434 \text{ mm}$$

→ Required force, $F = Lb\sigma_f$

$$= 47.43 \times 200 \times 123.04$$

$$= 1167255.87 \text{ N}$$

$$= 1167.255 \text{ kN}$$

47. Let x_1 = daily production of A

x_2 = daily production of B

Objective function

Maximise Z = 2000 x1 + 3000 x2

For shweta

Constraints $x_1 \geq x_2$

$x_2 \leq 10$

$x_1 \leq 20$

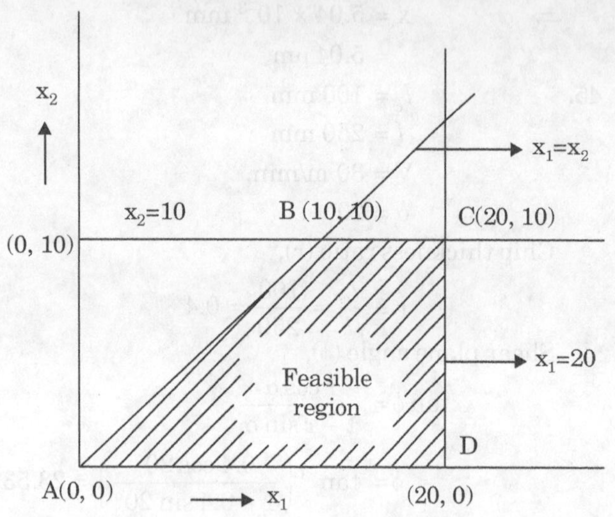

At A Z = 0

At B Z = 2000 × 10 + 3000 × 10 = 50000

At C Z = 2000 × 20 + 3000 × 10 = 70000

At D Z = 2000 × 20 + 3000 × 0 = 40000

Z_{max} = 70000

For Ashok,

Constraints, $x_1 \geq x_2$

$x_2 \leq 10$

$x_1 \leq 15$

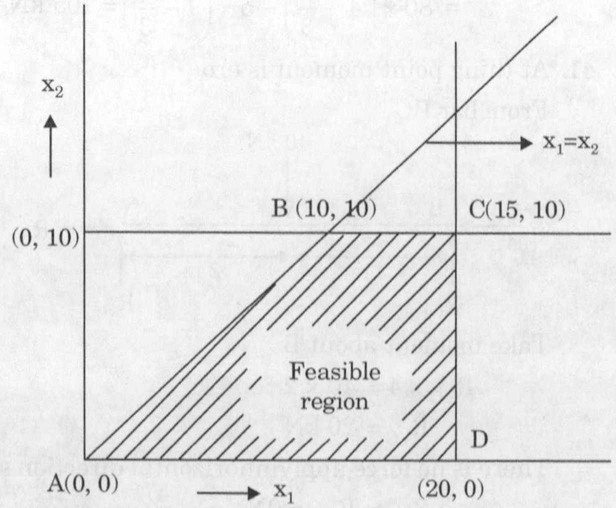

At C,

$$Z_{max} = 2000 \times 15 + 3000 \times 10$$

$$= 60000$$

Difference = 70000 – 60000 = Rs. 10000

48. For vendor x:

Setup cost $C_0 = 2 \times 200 = 400$ Rs.

Carrying cost $C_c = 0.1 \times p = 0.1 \times 10 = 1$ Rs.

p = Rs. 10/price

Assumed demand D = 1000

$$\therefore \text{Total cost} = p \times D + \frac{Q}{2} \times C_c + \frac{D}{Q} \times C_0$$

$$= 10 \times 1000 + \frac{800}{2} \times 1 + \frac{1000}{800} \times 400$$

$$= 10000 + 400 + 500 = 10900 \text{ Rs.}$$

For vendor y

Setup cost $C_0 = \frac{6}{60} \times 200 = 20$

Carrying cost $C_c = 1 \times p = 5$

p = Rs. 5/price

$$\text{Total cost} = p \times D + \frac{Q}{2} + C_c + \frac{D}{Q} \times C_0$$

$$= 5 \times 1000 \frac{200}{2} \times 5 + \frac{1000}{200} \times 20$$

$$= 5000 + 500 + 100 = \text{Rs. } 5600$$

$$\% \text{ reduction in cost} = \frac{10900 - 5600}{10900} \times 100$$

$$= 48.63$$

49.

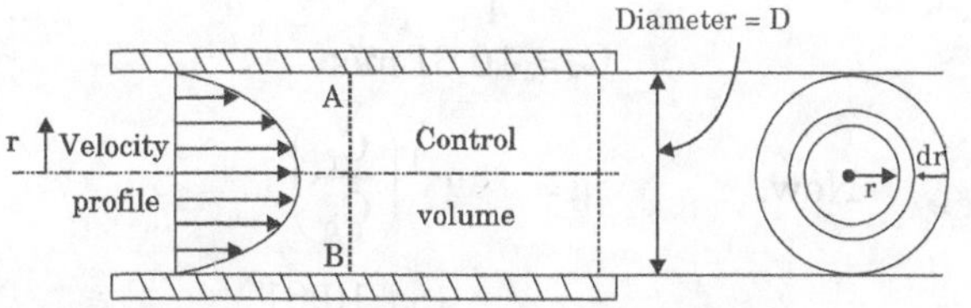

Consider a fluid element of dr thickness at a radius of r from the centre.

Now, kinetic energy rate of this element.

$$d(K\dot{}E) = \frac{1}{2} d\dot{m}v^2$$

$$= \frac{1}{2}(\rho dA\, v) \cdot v^2$$

$$d(K\dot{}E) = \frac{1}{2}\rho 2\pi r\, dr\, v^3$$

$$\int d(K\dot{}E) = \int_0^{D/2} \pi\rho \cdot r dr \left[-C\left(r^2 - \frac{D^2}{4}\right)\right]^3$$

$$K\dot{}E = -\pi\rho C^3 \int_0^{D/2} r\left(r^2 - \frac{D^2}{4}\right)^3 .dr$$

$$K\dot{}E = -\pi\rho C^3 \int_0^{D/2} r\left[r^6 - \frac{D^6}{4^3} - 3 \cdot \frac{r^2 D^2}{4}\left(r^2 - \frac{D^2}{4}\right)\right] \cdot dr$$

After integration we get,

$$= \text{fn}(D^8)$$

So, $n = 8$

50. $A_1 = 0.1 m^2$

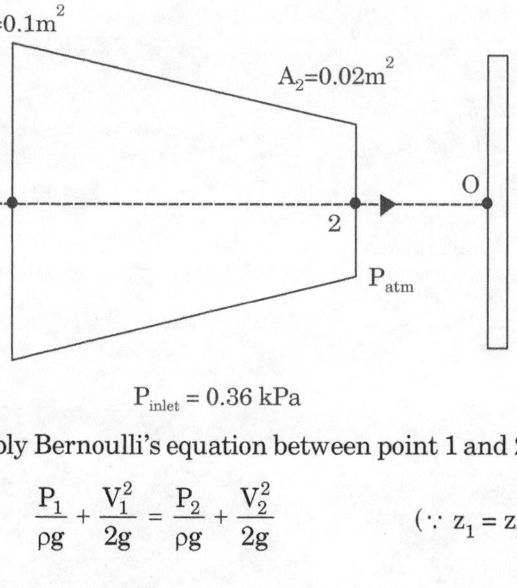

$P_{inlet} = 0.36$ kPa

Apply Bernoulli's equation between point 1 and 2,

$$\frac{P_1}{\rho g} + \frac{V_1^2}{2g} = \frac{P_2}{\rho g} + \frac{V_2^2}{2g} \quad (\because z_1 = z_2)$$

Here, $P_1 = 0.36$ kPa (gauge)

$P_{atm} = 0$ (gauge)

We get,

$$V_2^2 - V_1^2 = \frac{2P_1}{\rho} = \frac{2 \times 0.36 \times 1000}{1.2}$$

$$= 600 \quad \text{...(i)}$$

From continuity equation,

$$A_1V_1 = A_2V_2$$

$$\Rightarrow 0.1 \times V_1 = 0.02 \times V_2$$

$$\Rightarrow \frac{V_2}{V_1} = 5 \quad \text{...(ii)}$$

Solving both equations (i) and (ii),

We get, $V_2 = 25$ m/s

∵ The pressure betwen 2 and 0 is atmospheric. So, the only pressure at point O will be dynamic pressure due to air which is equal to $\frac{1}{2}\rho V^2$.

so, Gauge pressure at point O on the plate,

$$P_o = \frac{1}{2}\rho V^2 = \frac{1}{2} \times \frac{1.2 \times (25)^2}{1000}$$

$$P_o = 0.375 \text{ kPa} \approx 0.38 \text{ kPa (Gauge)}$$

51.

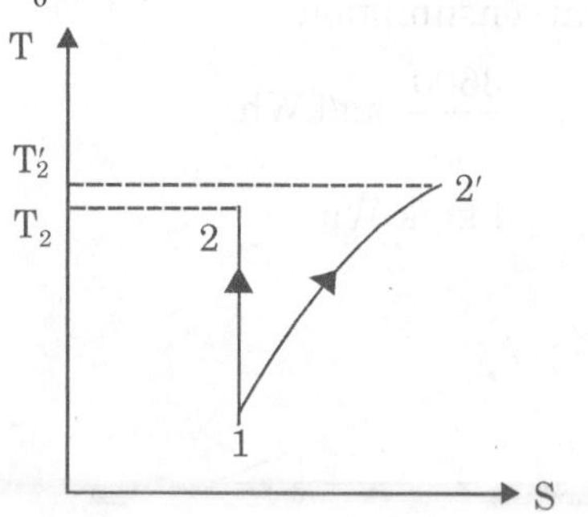

$T_1 = 310$ K

$\frac{P_2}{P_1} = 6$

$C_p = 1005$ J/kgK

$\eta_c = 85\%$

$\gamma = 1.4$

Process (1-2) (Isentropic compression)

$$T_2 = T_1\left(\frac{P_2}{P_1}\right)^{\frac{\gamma-1}{\gamma}}$$

$$T_2 = 310 \times (6)^{\frac{1.4-1}{1.4}} = 517.24 \text{ K}$$

Isentropic efficiency

$$\eta_C = \frac{T_2 - T_1}{T_2' - T_1}$$

$$0.85 = \frac{517.24 - 310}{T_2' - 310}$$

$$T'_2 = 553.81 \text{ K}$$

Inrease in enthalpy

$$\Delta h = h_2 - h_1 = C_P(T'_2 - T_1)$$
$$= 1.005\,(553.81 - 310)$$
$$= 245.03 \text{ kJ/kg} = 245 \text{ kJ/kg}$$

52.
$$m = 1 \text{ kg}$$
$$R = 287 \text{ J/kgK}$$
$$T_1 = 127°C = 127 + 273 = 400 \text{ K}$$
$$P_1 = P_2 = P$$
$$V_1 = V$$
$$V_2 = 2V$$

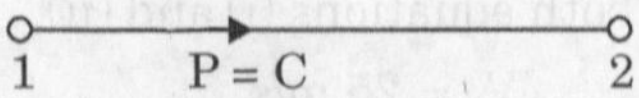

$$\text{Work done, } w = P(V_2 - V_1)$$
$$= P(2V - V)$$
$$= PV = mRT$$
$$= 1 \times 0.287 \times 400 = 114.80 \text{ kJ}$$

53. Turbine work, $w_T = 903$ kJ/kg

Pump work, $w_P = 3$ kJ/kg

$$w_{net} = w_T - w_P$$
$$= 903 - 3 = 900 \text{ kJ/kg}$$

Specific steam consumption

$$= \frac{3600}{w_{net}} \text{ kg/kWh}$$
$$= 4 \text{ kg/k Wh}$$

54.

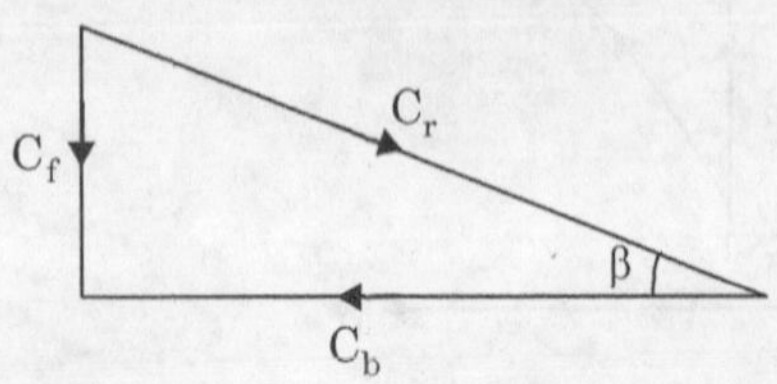

Blade velocity (C_b) at the given section

$$C_b = \frac{\pi DN}{60} = \frac{\pi \times 3 \times 300}{60} = 47.12 \text{ m/s}$$

Discharge (Q) = flow area (A_f) × flow velocity (C_f)

$$100 = \frac{\pi}{4} \times \left(D_t^2 - D_h^2\right) \times C_f$$
$$C_f = 10.61 \text{ m/s}$$

Now,
$$\beta = \tan^{-1}\left(\frac{C_f}{C_b}\right)$$
$$= \tan^{-1}\left(\frac{10.61}{47.12}\right) = 12.69°$$

55. Compression ratio, r = 8

Het supplied, $\dot{Q}_s = 10$ kW

Mechanical efficiency, $\eta_m = 80\%$

$\because$ Efficiency of otto cycle, $\eta = 1 - \frac{1}{r^{\gamma-1}}$

We also know that, $\eta = \frac{\text{work output}}{\text{heat supplied}}$

Here, work output is indicated power, I.P. (given in the question)

so,
$$\frac{\text{I.P.}}{\dot{Q}_s} = 1 - \frac{1}{r^{\gamma-1}}$$
$$\Rightarrow \quad \frac{\text{I.P.}}{10} = 1 - \frac{1}{(8)^{1.4-1}}$$
$$\Rightarrow \quad \text{IP.} = 5.65 \text{ kW}$$

Now, mechanial efficiency,

$$\Rightarrow \quad 0.8 = \frac{\text{B.P.}}{5.65}$$
$$\text{B.P.} = 4.5 \text{ kW}$$

SOLVED PAPER - 2020
SET 2

INSTRUCTIONS

1. Total of 65 questions carrying 100 marks, out of which 10 questions carrying a total of 15 marks are in General Aptitude (GA)
2. The Engineering Mathematics will carry around **15% of the total marks,** the General Aptitude section will carry **15% of the total marks** and the **remaining 70% of the total marks.**
3. **Types of Questions**
   (*a*) Multiple Choice Questions (MCQ) carrying 1 or 2 marks each in all papers and sections. These questions are objective in nature, and each will have a choice of four options, out of which the candidate has to mark the correct answer(s).
   (*b*) Numerical Answer Questions of 1 or 2 marks each in all papers and sections. For these questions the answer is a real number, to be entered by the candidate using the virtual keypad. No choices will be shown for these type of questions.
4. For **1-mark** multiple-choice questions, **1/3 marks** will be deducted for a wrong answer. Likewise, for **2-marks** multiple-choice questions, **2/3** marks will be deducted for a wrong answer. There is no negative marking for numerical answer type questions.

Chapter-Wise Analysis

#	Subject	Marks		
		1	2	Total
1	General Aptitude	5	5	15
2	Engineering Mathematics	5	4	13
3	Engineering Mechanics	1	1	3
4	Strength of Materials	1	3	7
5	Theory of Machines	3	2	7
6	Design of Machine Elements	1	2	5
7	Fluid Mechanics	2	3	8
8	Heat Transfer	1	2	5
9	Thermodynamics, IC Engines, Refrigeration & Power Plant Engineering Air-Conditioning	4	5	14
10	Material Manufacturing & Industrial Engineering	7	8	23

GENERAL APTITUDE

(Q1 – Q5) carry one mark each.

1. While I agree_____ his proposal this time, I do _____ not often agree_____ him.
   (a) to, with (b) with, to
   (c) with, with (d) to, to

2. The recent measures to improve the output would _____ the level of production to our satisfaction.
   (a) increase (b) decrease
   (c) speed (d) equalise

3. Select the word that fits the analogy :
   White : Whitening :: Light : _____
   (a) Lightning (b) Lightening
   (c) Lighting (d) Enlightening

4. In one of the greatest innings ever seen in 142 years of Test history, Ben Stokes upped the tempo in a five-and-a-half hour long stay of 219 balls including 11 fours and 8 sixes that saw him finish on a 135 not out as England squared the five-match series.
   Based on their connotations in the given passage, which one of the following meanings DOES NOT match?
   (a) upped = increased
   (b) squared = lost
   (c) tempo = enthusiasm
   (d) saw = resulted in

5. There are five levels (P, Q, R, S, T), in a linear supply chain before a product reaches customers, as shown in figure.

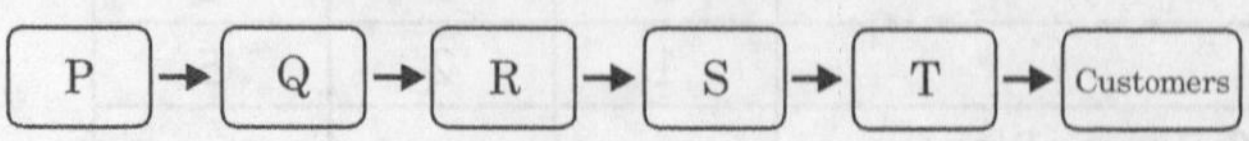

   At each of the five levels, the price of the product is increased by 25%. If the product is produced at level P at the cost of Rs. 120 per unit, what is the price paid (in rupees) by the customers?
   (a) 187.50 (b) 234.38
   (c) 292.96 (d) 366.21

(Q6 – Q10) carry two marks each.

6. Climate change and resilience deal with two aspects - reduction of sources of non-renewable energy resources and reducing vulnerability of climate change aspects. The terms 'mitigation' and 'adaptation' are used to refer to these aspects, respectively.
   Which of the following assertions is best supported by the above information?
   (a) Mitigation deals with consequences of climate change.
   (b) Adaptation deals with causes of climate change.
   (c) Mitigation deals with actions taken to reduce the use of fossil fuels.
   (d) Adaptation deals with actions taken to combat green-house gas emissions.

7. Find the missing element in the following figure.

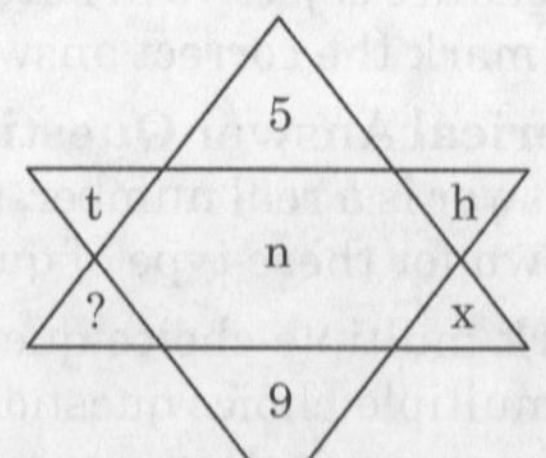

   (a) d (b) e
   (c) w (d) y

8. It was estimated that 52 men can complete a strip in a newly constructed highway connecting cities P and Q in 10 days. Due to an emergency, 12 men were sent to another project. How many number of days, more than the original estimate, will be required to complete the strip?
   (a) 3 days (b) 5 days
   (c) 10 days (d) 13 days

9. An engineer measures THREE quantities X, Y and Z in an experiment. She finds that they follow a relationship that is represented in the figure below: (the product of X and Y linearly varies with Z)

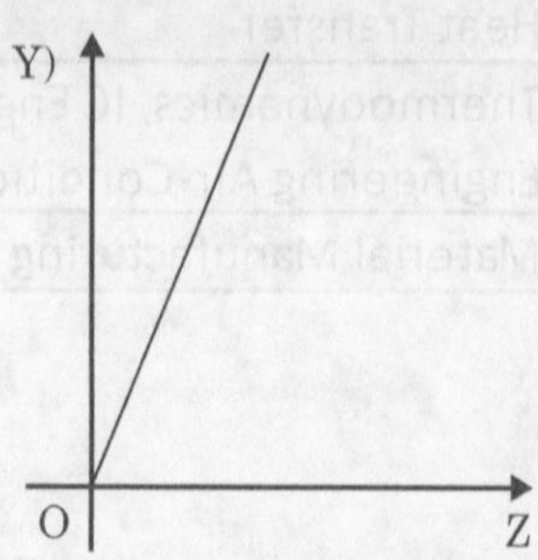

   Then, which of the following statements is FALSE?
   (a) For fixed Z; X is proportional to Y
   (b) For fixed Y; X is proportional to Z
   (c) For fixed X; Z is proportional to Y
   (d) XY/Z is constant

10. The two pie-charts given below show the data of total students and only girls registered in different streams in a university. If the total number of students registered in the university is 5000, and the total number of the registered girls is 1500; then, the ratio of boys enrolled in Arts to the girls enrolled in Management is ______ .

Percentage of students enrolled in different streams in a University

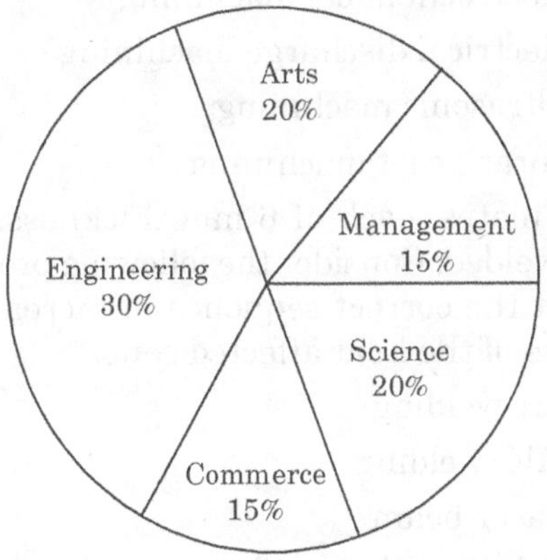

Percentage of girls enrolled in different streams

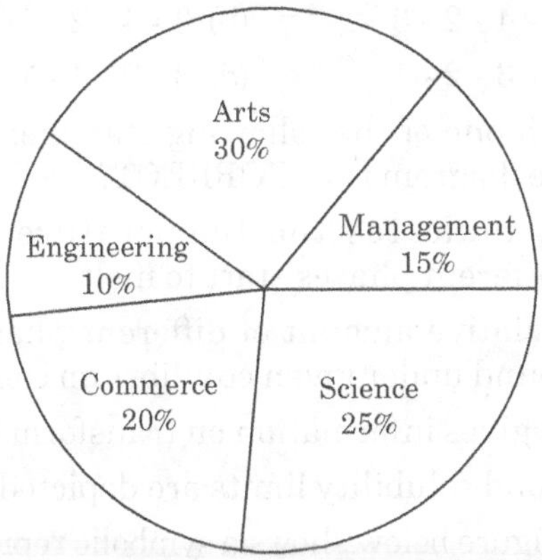

(a) 2 : 1 (b) 9 : 22

(c) 11 : 9 (d) 22 : 9

MECHANICAL ENGINEERING

(Q1 – Q25) carry one mark each.

1. The sum of two normally distributed random variables X and Y is

(a) always normally distributed

(b) normally distributed, only if X and Y are independent

(c) normally distributed, only if X and Y have the same standard deviation

(d) normally distributed, only if X and Y have the same mean

2. A matrix P is decomposed into its symmetric part S and skew symmetric part V. If

$$S = \begin{pmatrix} -4 & 4 & 2 \\ 4 & 3 & 7/2 \\ 2 & 7/2 & 2 \end{pmatrix}, \; V = \begin{pmatrix} 0 & -2 & 3 \\ 2 & 0 & 7/2 \\ -3 & -7/2 & 0 \end{pmatrix}$$

then matrix P is

(a) $\begin{pmatrix} -4 & 6 & -1 \\ 2 & 3 & 0 \\ 5 & 7 & 2 \end{pmatrix}$ (b) $\begin{pmatrix} -4 & 2 & 5 \\ 6 & 3 & 7 \\ -1 & 0 & 2 \end{pmatrix}$

(c) $\begin{pmatrix} 4 & -6 & 1 \\ -2 & -3 & 0 \\ -5 & -7 & -2 \end{pmatrix}$ (d) $\begin{pmatrix} -2 & 9/2 & -1 \\ -1 & 81/4 & 11 \\ -2 & 45/2 & 73/4 \end{pmatrix}$

3. Let $I = \int_{x=0}^{1} \int_{y=0}^{x^2} xy^2 dy\, dx$. Then, I may also be expressed as

(a) $\int_{y=0}^{1} \int_{x=0}^{\sqrt{y}} xy^2 dx\, dy$ (b) $\int_{y=0}^{1} \int_{x=\sqrt{y}}^{1} yx^2 dx\, dy$

(c) $\int_{y=0}^{1} \int_{x=\sqrt{y}}^{1} xy^2 dx\, dy$ (d) $\int_{y=0}^{1} \int_{x=0}^{\sqrt{y}} yx^2 dy\, dx$

4. The solution of

$$\frac{d^2y}{dt^2} - y = 1,$$

which additionally satisfies $y\Big|_{t=0} = \frac{dy}{dt}\Big|_{t=0} = 0$ in the Laplace s-domain is

(a) $\frac{1}{s(s+1)(s-1)}$ (b) $\frac{1}{s(s+1)}$

(c) $\frac{1}{s(s-1)}$ (d) $\frac{1}{s-1}$

5. An attempt is made to pull a roller of weight W over a curb (step) by applying a horizontal force F as shown in the figure.

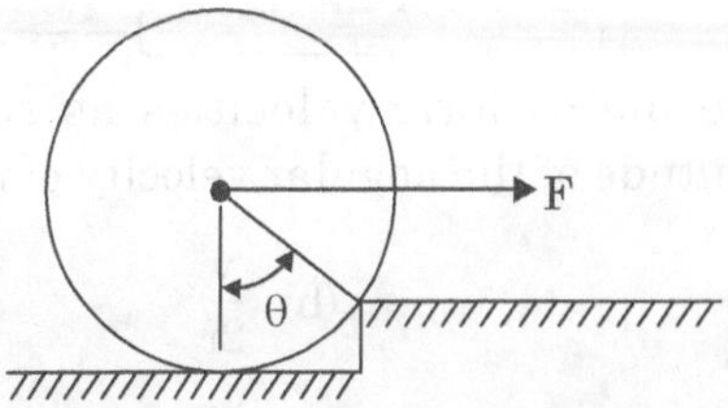

The coefficient of static friction between the roller and the ground (including the edge of the step) is μ. Identify the correct free body diagram (FBD) of the roller when the roller is just about to climb over the step.

(a)

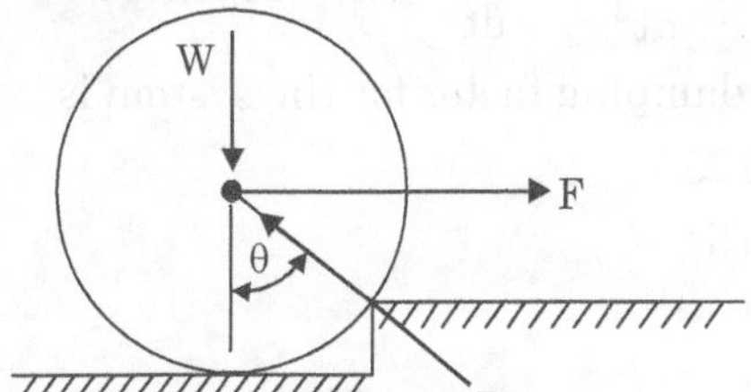

(b) 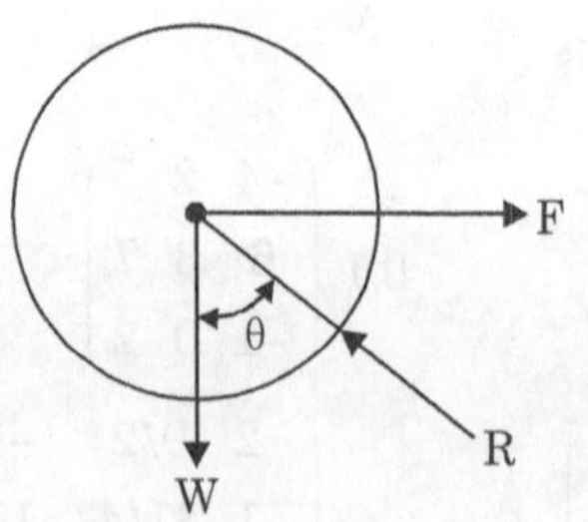

(c) 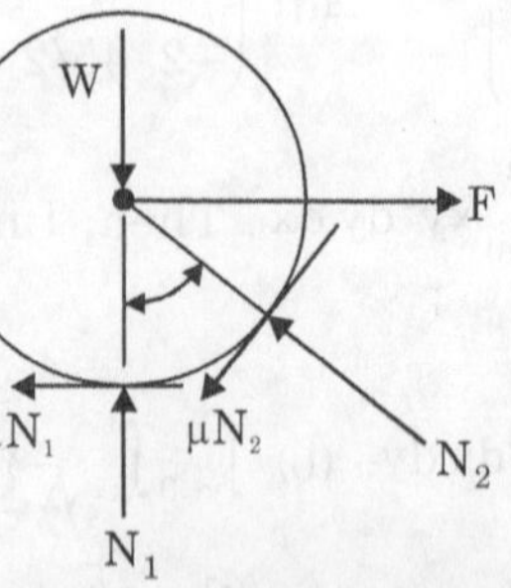

(d) 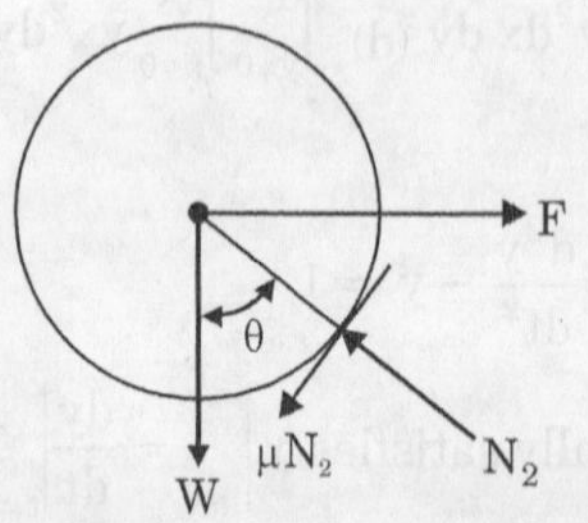

6. A circular disk of radius r is confined to roll without slipping at P and Q as shown in the figure.

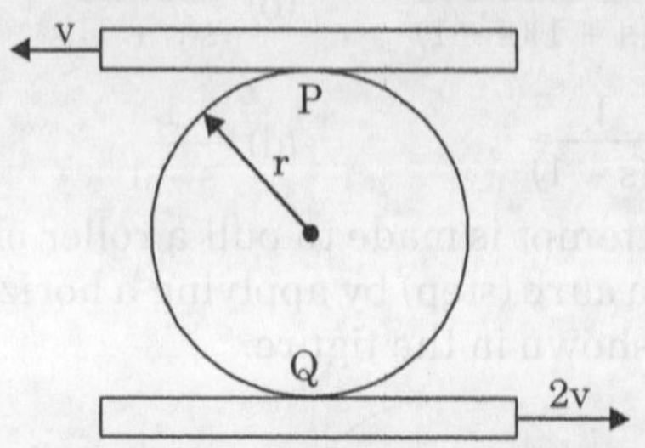

If the plates have velocities as shown, the magnitude of the angular velocity of the disk is

(a) $\frac{v}{r}$ (b) $\frac{v}{2r}$

(c) $\frac{2v}{3r}$ (d) $\frac{3v}{2r}$

7. The equation of motion of a spring-mass-damper system is given by

$$\frac{d^2x}{dt^2} + 3\frac{dx}{dt} + 9x = 10\sin(5t)$$

The damping factor for the system is

(a) 0.25

(b) 0.5

(c) 2

(d) 3

8. The number of qualitatively distinct kinematic inversions possible for a Grashof chain with four revolute pairs is

(a) 1 (b) 2

(c) 3 (d) 4

9. The process, that uses a tapered horn to amplify and focus the mechanical energy for machining of glass, is

(a) electrochemical machining

(b) electrical discharge machining

(c) ultrasonic machining

(d) abrasive jet machining

10. Two plates, each of 6 mm thickness, are to be but-welded. Consider the following processes and select the correct sequence in increasing order of size of the heat affected zone.

1. Arc welding
2. MIG welding
3. Laser beam
4. Submerged arc welding

(a) 1 - 4 - 2 - 3 (b) 3 - 4 - 2 - 1

(c) 4 - 3 - 2 - 1 (d) 3 - 2 - 4 - 1

11. Which one of the following statements about a phase diagram is INCORRECT?

(a) It indicates the temperature at which different phases start to melt

(b) Relative amount of different phases can be found under given equilibrium conditions

(c) It gives information on transformation rates

(d) Solid solubility limits are depicted by it

12. The figure below shows a symbolic representation of the surface texture in a perpendicular lay orientation with indicative values (I through VI) marking the various specifications whose definitions are listed below.

P : Maximum Waviness Height (mm);

Q : Maximum Roughness Height (mm);

R : Minimum Roughness Height (mm);

S : Maximum Waviness Width (mm);

T : Maximum Roughness Width (mm);

U : Roughness Width Cutoff (mm),

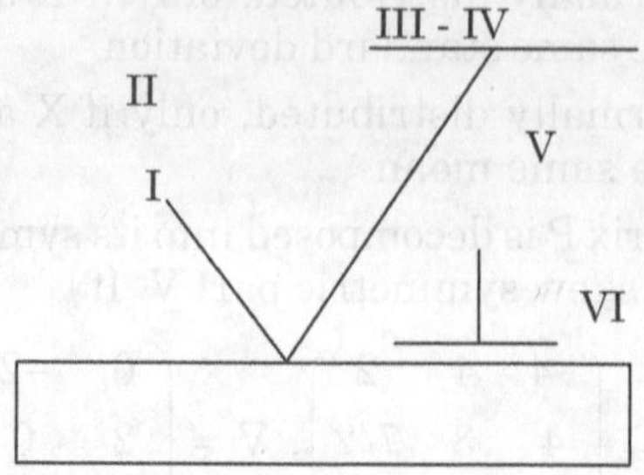

The correct match between the specifications and the symbols (I to VI) s

(a) I-R, II-Q, III-P, IV-S, V-U, VI-T
(b) I-R, II-P, III-U, IV-S, V-T, VI-Q
(c) I-U, II-S, III-Q, IV-T, V-R, VI-P
(d) I-Q, II-U, III-R, IV-T, V-S, VI-P

13. In Materials Requirement Planning, if the inventory holding cost is very high and the setup cost is zero, which one of the following lot sizing approaches should be used?

(a) Economic Order Quantity
(b) Lot-for-Lot
(c) Base Stock Level
(d) Fixed Period Quantity, for 2 periods

14. Which of the following conditions is used to determine the stable equilibrium of all partially submerged floating bodies?

(a) Centre of buoyancy must be above the centre of gravity
(b) Centre of buoyancy must be below the centre of gravity
(c) Metacentre must be at a higher level than the centre of gravity
(d) Metacentre must be at a lower level than the centre of gravity

15. In the space above the mercury column in a barometer tube, the gauge pressure of the vapour is

(a) positive, but more than one atmosphere
(b) negative
(c) zero
(d) positive, but less than one atmosphere

16. A closed vessel contains pure water, in thermal equilibrium with its vapour at 25°C (Stage # 1), as shown.

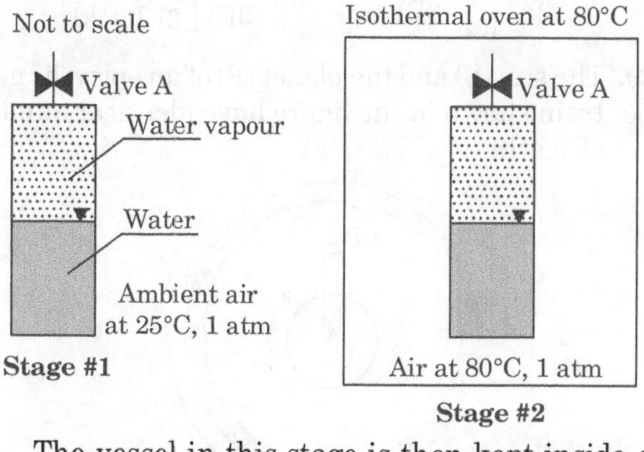

The vessel in this stage is then kept inside an isothermal oven which is having an atmosphere of hot air maintained at 80°C. The vessel exchanges heat with the oven atmosphere and attains a new thermal equilibrium (Stage #2). If the Valve A is now opened inside the oven, what will happen immediately after opening the valve?

(a) Water vapor inside the vessel will come out of the Valve A
(b) Hot air will go inside the vessel through Valve A
(c) Nothing will happen - the vessel will continue to remain in equilibrium
(d) All the vapor inside the vessel will immediately condense

17. For an air-standard Diesel cycle,

(a) heat addition is at constant volume and heat rejection is at constant pressure
(b) heat addition is at constant pressure and heat rejection is at constant pressure
(c) heat addition is at constant pressure and heat rejection is at constant volume
(d) heat addition is at constant volume and heat rejection is at constant volume

18. The values of enthalpies at the stator inlet and rotor outlet of a hydraulic turbomachine stage are h_1 and h_3 respectively. The enthalpy at the stator outlet (or, rotor inlet) is h_2. The condition $(h_2 - h_1) = (h_3 - h_2)$ indicates that the degree of reaction of this stage is

(a) zero (b) 50%
(c) 75% (d) 10%

19. Let I be a 100 dimensional identity matrix and E be the set of its distinct (no value appears more than once in E)real eigenvalues. The number of elements in E is ________.

20. A beam of negligible mass is hinged at support P and has a roller support Q as shown in the figure.

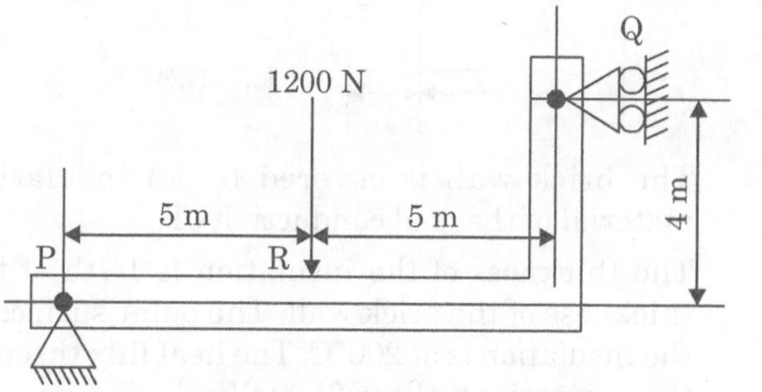

A point load of 1200 N is applied at point R. The magnitude of the reaction force at support Q is ________ N.

21. A machine member is subjected to fluctuating stress $\sigma = \sigma_0 \cos(8\pi t)$. The endurance limit of the material is 350 MPa. If the factor of safety used in the design is 3.5 then the maximum allowable value of σ_0 is ______ MPa (round off to 2 decimal places).

22. A bolt head has to be made at the end of a rod of diameter d = 12 mm by localized forging (upsetting) operation. The length of the unsupported portion of the rod is 40 mm. To avoid buckling of the rod, a closed forging operation has to be performed with a maximum die diameter of ________ mm.

23. Consider the following network of activities, with each activity named A-L, illustrated in the nodes of the network.

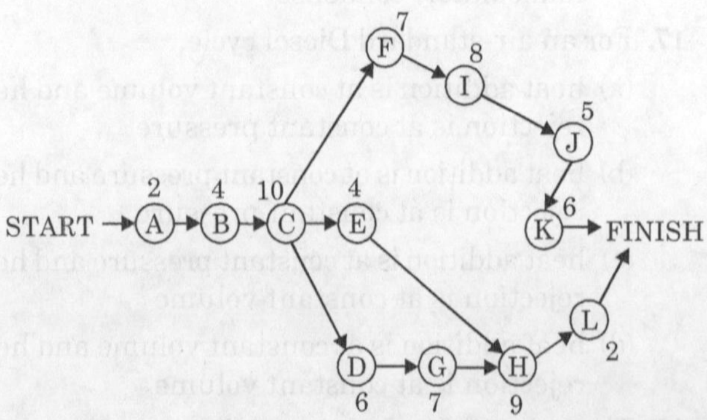

The number of hours required for each activity is shown alongside the nodes. The slack on the activity L, is ________ hours.

24. In a furnace, the inner and outer sides of the brick wall (k_1 = 2.5 W/m.K) are maintained at 1100°C and 700°C, respectively as shown in figure.

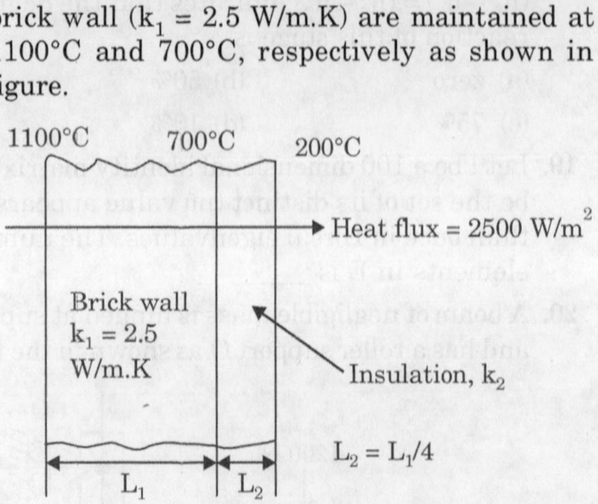

The brick wall is covered by an insulating material of thermal conductivity k_2.

The thickness of the insulation is 1/4th of the thickness of the brick wall. The outer surface of the insulation is at 200°C. The heat flux through the composite walls is 2500 W/m^2.

The value of k_2 is ________ W/m.K (round off to one decimal place).

25. If a reversed Carnot cycle operates between the temperature limits of 27°C and −3°C, then the ratio of the COP of a refrigerator to that of a heat pump (COP of refrigerator / COP of heat pump)based on the cycle is ________ (round off to 2 decimal places).

(Q26 – Q55) carry two marks each.

26. The directional derivative of f (x, y, z) = xyz at point (−1, 1, 3) in the direction of vector $\hat{i} - 2\hat{j} + 2\hat{k}$ is

(a) $3\hat{i} - 3\hat{j} - \hat{k}$ (b) $-\dfrac{7}{3}$

(c) $\dfrac{7}{3}$ (d) 7

27. The function f(z) of complex variable z = x + iy, where i = $\sqrt{-1}$, is given as f(z) = $(x^3 - 3xy^2)$ + i v(x, y). For this function to be analytic, v(x, y) should be

(a) $(3xy^2 - y^3)$ + constant

(b) $(3x^2y^2 - y^3)$+ constant

(c) $(x^3 - 3x^2y)$ + constant

(d) $(3x^2y - y^3)$ + constant

28. A cantilever of length l, and flexural rigidity EI, stiffened by a spring of stiffness k, is loaded by a transverse force P, as shown.

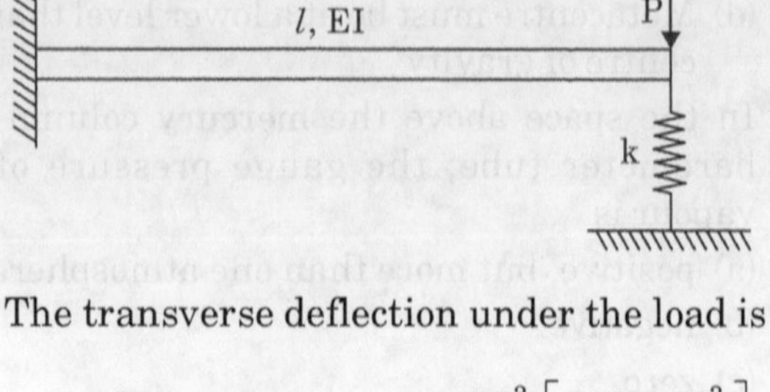

The transverse deflection under the load is

(a) $\dfrac{Pl^3}{3EI}\left[\dfrac{3EI}{3EI + 2kl^3}\right]$ (b) $\dfrac{Pl^3}{3EI}\left[\dfrac{6EI - kl^3}{6EI}\right]$

(c) $\dfrac{Pl^3}{3EI}\left[\dfrac{3EI - kl^3}{3EI}\right]$ (d) $\dfrac{Pl^3}{3EI}\left[\dfrac{3EI}{3EI + kl^3}\right]$

29. The sun (S) and the planet (P) of an epicyclic gear train shown in the figure have identical number of teeth.

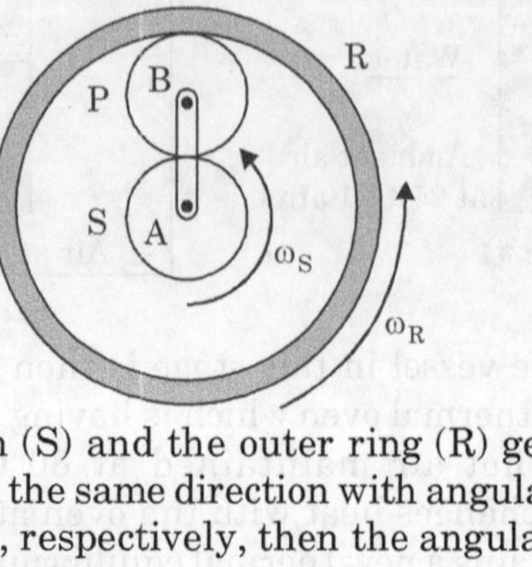

If the sun (S) and the outer ring (R) gears are rotated in the same direction with angular speed ω_s and ω_R, respectively, then the angular speed of the arm AB is

(a) $\frac{3}{4}\omega_R + \frac{1}{4}\omega_S$ (b) $\frac{1}{4}\omega_R + \frac{3}{4}\omega_S$

(c) $\frac{1}{2}\omega_R - \frac{1}{2}\omega_S$ (d) $\frac{3}{4}\omega_R - \frac{4}{4}\omega_S$

30. A thin-walled cylinder of radius r and thickness t is open at both ends, and fits snugly between two rigid walls under ambient conditions, as shown in the figure.

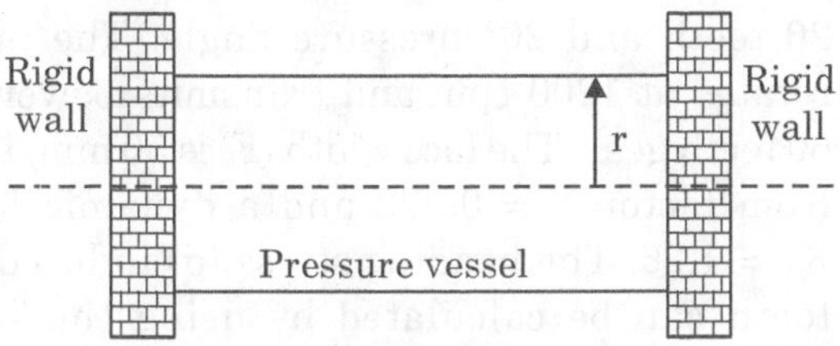

The material of the cylinder has Young's modulus E, Poisson's ratio ν, and coefficient of thermal expansion α. What is the minimum rise in temperature ΔT of the cylinder (assume uniform cylinder temperature with no buckling of the cylinder) required to prevent gas leakage if the cylinder has to store the gas at an internal pressure of p above the atmosphere?

(a) $\Delta T = \frac{3\nu pr}{2\alpha tE}$ (b) $\Delta T = \left(\nu - \frac{1}{4}\right)\frac{pr}{\alpha tE}$

(c) $\Delta T = \frac{\nu pr}{\alpha tE}$ (d) $\Delta T = \left(\nu + \frac{1}{2}\right)\frac{pr}{\alpha tE}$

31. A helical spring has spring constant k. If the wire diameter, spring diameter and the number of coils are all doubled then the spring constant of the new spring becomes

(a) k/2 (b) k

(c) 8 k (d) 16 k

32. Two rollers of diameters D_1 (in mm) and D_2 (in mm) are used to measure the internal taper angle in the V-groove of a machined component. The heights H_1 (in mm) and H_2 (in mm) are measured by using a height gauge after inserting the rollers into the same V-groove as shown in the figure.

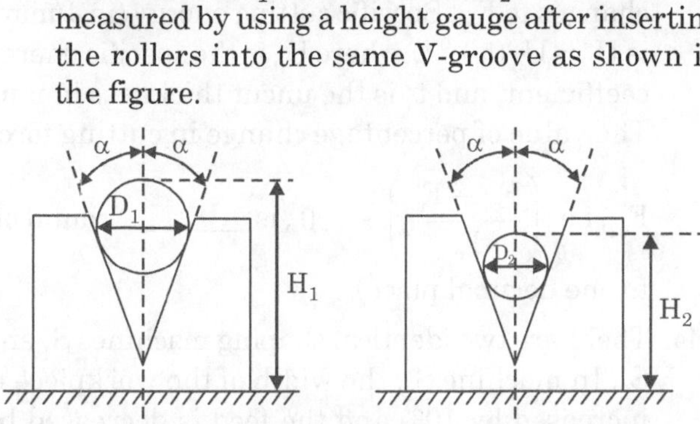

$D_1 > D_2$

Which one of the following is the correct relationship to evaluate the angle α as shown in the figure?

(a) $\sin\alpha = \frac{(D_1 - D_2)}{2(H_1 - H_2) - (D_1 - D_2)}$

(b) $\cos\alpha = \frac{(D_1 - D_2)}{2(H_1 - H_2) - 2(D_1 - D_2)}$

(c) $\operatorname{cosec}\alpha = \frac{(H_1 - H_2) - (D_1 - D_2)}{2(D_1 - D_2)}$

(d) $\sin\alpha = \frac{(H_1 - H_2)}{(D_1 - D_2)}$

33. The forecast for the monthly demand of a product is given in the table below.

Month	Forecast	Actual Sales
1	32.00	30.00
2	31.80	32.00
3	31.82	30.00

The forecast is made by using the exponential smoothing method. The exponential smoothing coefficient used in forecasting the demand is

(a) 0.10 (b) 0.40

(c) 0.50 (d) 1.00

34. One kg of air in a closed system undergoes an irreversible process from an initial state of p_1 = 1 bar (absolute) and T_1 = 27°C, to a final state of p_2 = 3 bar (absolute) and T_2 = 127°C. If the gas constant of air is 287 J/kg.K and the ratio of the specific heats g = 1.4, then the change in the specific entropy (in J/kg.K) of the air in the process is

(a) – 26.3

(b) 28.4

(c) 172.0

(d) indeterminate, as the process is irreversible

35. For the integral $\int_0^{\pi/2}(8 + 4\cos x)\,dx$, the absolute percentage error in numerical evaluation with the Trapezoidal rule, using only the end points, is ________ (round off to one decimal place).

36. A fair coin is tossed 20 times. The probability that 'head' will appear exactly 4 times in the first ten tosses, and 'tail' will appear exactly 4 times in the next ten tosses is ________ (round off to 3 decimal places).

37. A hollow spherical ball of radius 20 cm floats in still water, with half of its volume submerged. Taking the density of water as 1000 kg/m^3, and the acceleration due to gravity as 10 m/s^2, the natural frequency of small oscillations of the ball, normal to the water surface is ______ radians/s (round off to 2 decimal places).

38. Uniaxial compression test data for a solid metal bar of length 1 m is shown in the figure.

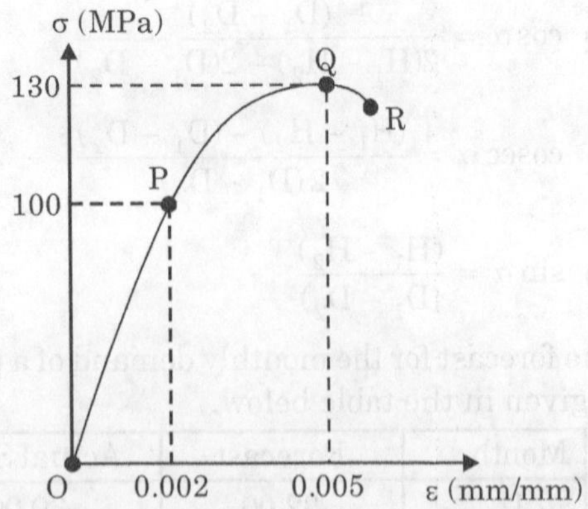

The bar material has a linear elastic response from O to P followed by a nonlinear response. The point P represents the yield point of the material. The rod is pinned at both the ends. The minimum diameter of the bar so that it does not buckle under axial loading before reaching the yield point is ______ mm (round off to one decimal place).

39. The turning moment diagram of a flywheel fitted to a fictitious engine is shown in the figure.

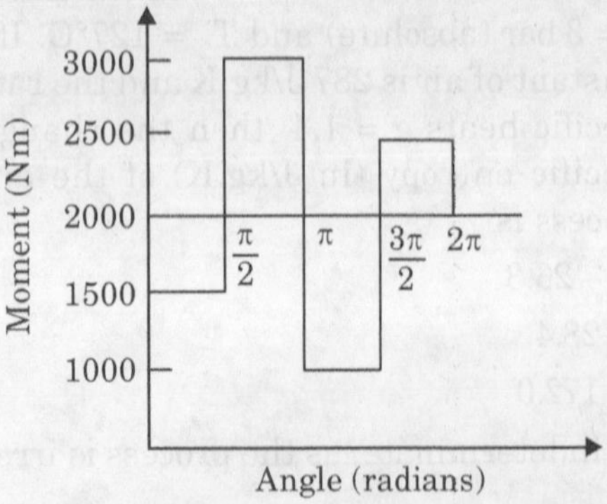

The mean turning moment is 2000 Nm. The average engine speed is 1000 rpm. For fluctuation in the speed to be within ± 2% of the average speed, the mass moment of inertia of the flywheel is ______ kg.m^2.

40. A rigid block of mass m_1 = 10 kg having velocity v_0 = 2 m/s strikes a stationary block of mass m_2 = 30 kg after traveling 1 m along a frictionless horizontal surface as shown in the figure.

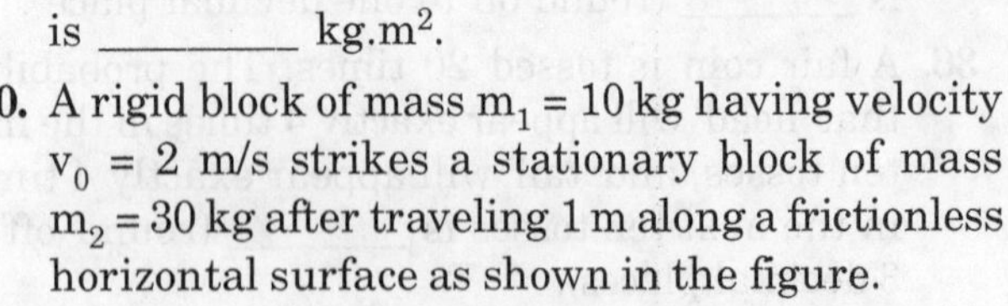

The two masses stick together and jointly move by a distance of 0.25 m further along the same frictionless surface, before they touch the massless buffer that is connected to the rigid vertical wall by means of a linear spring having a spring constant k = 10^5 N/m. The maximum deflection of the spring is ______ cm (round off to 2 decimal places).

41. A steel spur pinion has a module (m) of 1.25 mm, 20 teeth and 20° pressure angle. The pinion rotates at 1200 rpm and transmits power to a 60 teeth gear. The face width (F) is 50 mm, Lewis from factor Y = 0.322 and a dynamic factor Kv = 1.26. The bending stress (σ) induced in a tooth can be calculated by using the Lewis formula given below.

If the maximum bending stress experienced by the pinion is 400 MPa, the power transmitted is ______ kW (round off to one decimal place).

Lewis formula : $\sigma = \dfrac{K_v W^t}{FmY}$, where W^t is the tangential load acting on the pinion.

42. A mould cavity of 1200 cm^3 volume has to be filled through a sprue of 10 cm length feeding a horizontal runner. Cross-sectional area at the base of the sprue is 2 cm^2. Consider acceleration due to gravity as 9.81 m/s^2. Neglecting frictional losses due to molten metal flow, the time taken to fill the mould cavity is ______ seconds (round off to 2 decimal places).

43. A cylindrical bar with 200 mm diameter is being turned with a tool having geometry 0° - 9° - 7° - 8° - 15° - 30° - 0.05 inch (Coordinate system, ASA) resulting in a cutting force F_{c1}. If the tool geometry is changed to 0° - 9° - 7° - 8° - 15° - 0° - 0.05 inch (Coordinate system, ASA) and all other parameters remain unchanged, the cutting force changes to F_{c2}. Specific cutting energy (in J/mm^3) is $U_C = U_0 (t_1)^{-0.4}$, where U_0 is the specific energy coefficient, and t_1 is the uncut thickness in mm. The value of percentage change in cutting force F_{c2}, i.e. $\left(\dfrac{F_{c_2} - F_{c_1}}{F_{c_1}}\right) \times 100$, is ______ (round off to one decimal place).

44. There are two identical shaping machines S_1 and S_2. In machine S_2, the width of the workpiece is increased by 10% and the feed is decreased by 10% with respect to that of S_1. If all other conditions remain the same then the ratio of total time per pass in S_1 and S_2 will be ______ (round off to one decimal place).

45. Bars of 250 mm length and 25 mm diameter are to be turned on a lathe with a feed of 0.2 mm/rev. Each regrinding of the tool costs Rs. 20. The time required for each tool change is 1 min. Tool life equation is given as $VT^{0.2} = 24$ (where cutting speed V is in m/min and tool life T is in min). The optimum tool cost per piece for maximum production rate is Rs. ______ (round off to 2 decimal places).

46. A point 'P' on a CNC controlled XY-stage is moved to another point 'Q' using the coordinate system shown in the figure below and rapid positioning command (G00).

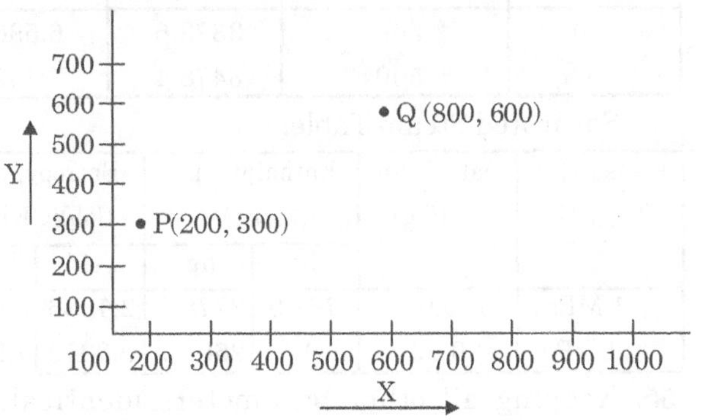

A pair of stepping motors with maximum speed of 800 rpm, controlling both the X and Y motion of the stage, are directly coupled to a pair of lead screws, each with a uniform pitch of 0.5 mm. The time needed to position the point 'P' to the point 'Q' is ______ minutes (round off to 2 decimal places).

47. For a single item inventory system, the demand is continuous, which is 10000 per year. The replenishment is instantaneous and backorders (S units) per cycle are allowed as shown in the figure.

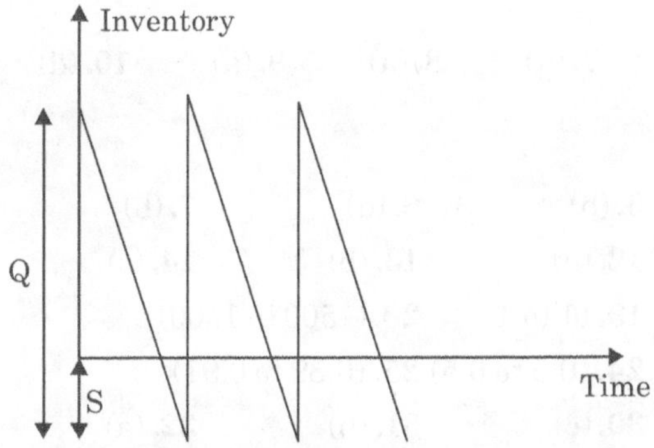

As soon as the quantity (Q units) ordered from the supplier is received, the backordered quantity is issued to the customers. The ordering cost is Rs. 300 per order. The carrying cost is Rs. 4 per unit per year. The cost of backordering is Rs. 25 per unit per year. Based on the total cost minimization criteria, the maximum inventory reached in the system is ______ (round off to nearest integer).

48. Consider a flow through a nozzle, as shown in the figure below.

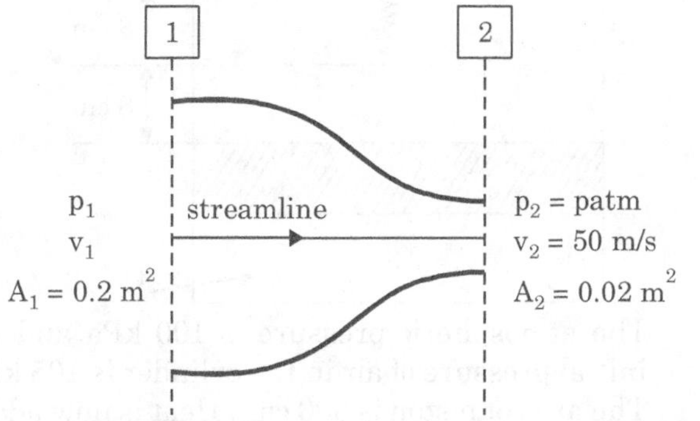

The air flow is steady, incompressible and inviscid. The density of air is 1.23 kg/m^3.

The pressure difference, $(p_1 - p_{atm})$ is ______ kPa (round off to 2 decimal places).

49. Water (density 1000 kg/m^3) flows through an inclined pipe of uniform diameter. The velocity, pressure and elevation at section A are V_A = 3.2 m/s, p_A = 186 kPa and z_A = 24.5 m, respectively, and those at section B are V_B = 3.2 m/s, p_B = 260 kPa and Z_B = 9.1 m, respectively. If acceleration due to gravity is 10 m/s^2 then the head lost due to friction is ______ m (round off to one decimal place).

50. The spectral distribution of radiation from a black body at T_1 = 3000 K has a maximum at wavelength λ_{max}. The body cools down to a temperature T_2. If the wavelength corresponding to the maximum of the spectral distribution at T_2 is 1.2 times of the original wavelength λ_{max}, then the temperature T_2 is ______ K (round off to the nearest integer).

51. Water flows through a tube of 3 cm internal diameter and length 20 m. The outside surface of the tube is heated electrically so that it is subjected to uniform heat flux circumferentially and axially. The mean inlet and exit temperatures of the water are 10°C and 70°C, respectively. The mass flow rate of the water is 720 kg/h. Disregard the thermal resistance of the tube wall. The internal heat transfer coefficient is 1697 W/m^{2}K. Take specific heat C_p of water as 4.179 kJ/kg.K. The inner surface temperature at the exit section of the tube is ______ °C (round off to one decimal place).

52. Air is contained in a frictionless piston-cylinder arrangement as shown in the figure

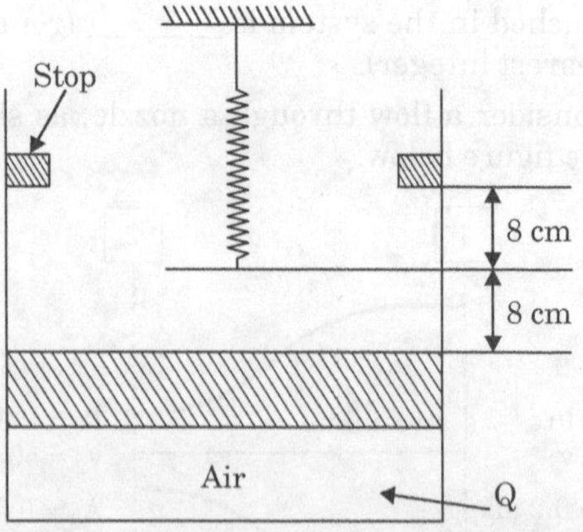

The atmospheric pressure is 100 kPa and the initial pressure of air in the cylinder is 105 kPa. The area of piston is 300 cm^2. Heat is now added and the piston moves slowly from its initial position until it reaches the stops. The spring constant of the linear spring is 12.5 N/mm. Considering the air inside the cylinder as the system, the work interaction is ______ J (round off to the nearest integer).

53. Moist air at 105 kPa, 30°C and 80% relative humidity flows over a cooling coil in an insulated air-conditioning duct. Saturated air exits the duct at 100 kPa and 15°C. The saturation pressures of water at 30°C and 15°C are 4.24 kPa and 1.7 kPa respectively. Molecular weight of water is 18 g/mol and that of air is 28.94 g/ mol. The mass of water condensing out from the duct is g/kg of dry air (round off to the nearest integer).

54. In a steam power plant, superheated steam at 10 MPa and 500°C, is expanded isentropically in a turbine until it becomes a saturated vapour. It is then reheated at constant pressure to 500°C. The steam is next expanded isentropically in another turbine until it reaches the condenser pressure of 20 kPa. Relevant properties of steam are given in the following two tables. The work done by both the turbines together is ______ kJ/kg (round off to the nearest integer).

Superheated Steam Table:

Pressure, p (MPa)	Temperature, T (°C)	Enthapy, h (kJ/kg)	Entropy, s (kJ/kg.K)
10	500	3373.6	6.5965
1	500	3478.4	7.7621

Saturated Steam Table:

Pressure, p	Sat. Temp.	Enthalpy, h		Entropy, s	
T_{sat} (°C)	(kJ/kg)	(kJ/kg)		(kJ/kg.K)	
		hr	hg	sf	sg
1 MPa	179.91	762.9	2778.1	2.1386	6.5965
20 kPa	60.06	251.38	2609.7	0.8319	7.9085

55. Keeping all other parameters identical, the Compression Ratio (CR) of an air standard diesel cycle is increased from 15 to 21. Take ratio of specific heats = 1.3 and cut-off ratio of the cycle $r_c = 2$.

The difference between the new and the old efficiency values, in percentage,

$(\eta_{new}\,|_{CR=21}) - (\eta_{old}\,|_{Cr=15}) = \%$ (round off to one decimal place).

ANSWERS

General Aptitude (GA)

1. (a) **2.** (a) **3.** (b) **4.** (b) **5.** (d) **6.** (c) **7.** (a) **8.** (a) **9.** (a) **10.** (d)

Mechanical Engineering

1. (a) **2.** (b) **3.** (c) **4.** (a) **5.** (b) **6.** (d) **7.** (b)
8. (c) **9.** (c) **10.** (d) **11.** (c) **12.** (a) **13.** (b) **14.** (c)
15. (b) **16.** (b) **17.** (c) **18.** (b) **19.** (1 to 1) **20.** (1500 to 1500)
21. (99.98 to 100.02) **22.** (18 to 18) **23.** (2 to 2) **24.** (0.5 to 0.5) **25.** (0.89 to 0.91)
26. (c) **27.** (d) **28.** (d) **29.** (a) **30.** (c) **31.** (b) **32.** (a)
33. (a) **34.** (a) **35.** (5.1 to 5.3) **36.** (0.041 to 0.043) **37.** (8.60 to 8.70)
38. (55.0 to 58.0) **39.** (745 to 755) **40.** (0.98 to 1.02) **41.** (9.8 to 10.8)
42. (4.20 to 4.30) **43.** (– 5.8 to – 5.5) **44.** (0.7 to 0.9) **45.** (26.00 to 28.00)
46. (1.40 to 1.60) **47.** (1130 to 1140) **48.** (1.50 to 1.55) **49.** (7.9 to 8.1)
50. (2499 to 2501) **51.** (85.1 to 86.1) **52.** (543 to 545) **53.** (10 to 10)
54. (1500 to 1525) **55.** (4.6 to 4.9)

EXPLANATIONS

GENERAL APTITUDE

1. "Agree to" is typically used for consent to a request.

   "Agree with" is most often used when two people have the same opinion.

2. The recent measures to improve the output would increase the level of production to our satisfaction.

3. White : Whitening

   Light : Lightening

4. Square means level or draw (in cricket game).

   So, here squared does not mean lost, it means level or draw.

5. Let, product cost at P = x = Rs. 120

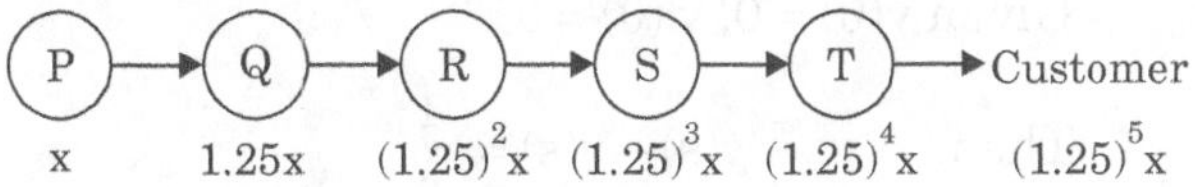

   Price paid by the customer = $(1.25)^5x$

   $= (1.25)^5 \times 120 =$ Rs. 366.21

6. • Mitigation is an intervention to reduce the emissions sources or enhance the sinks of green house gases.

   So, the correct option is (c).

7.

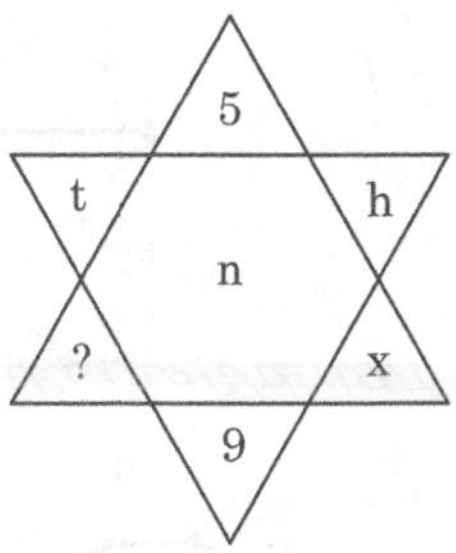

(1)	(2)	(3)	(4)	(5)	(6)	(7)	(8)	(9)
a	b	c	d	e	f	g	h	i
(10)	(11)	(12)	(13)	(14)	(15)	(16)	(17)	(18)
j	k	l	m	n	o	p	q	r
(19)	(20)	(21)	(22)	(23)	(24)	(25)	(26)	
s	t	u	v	w	x	y	z	

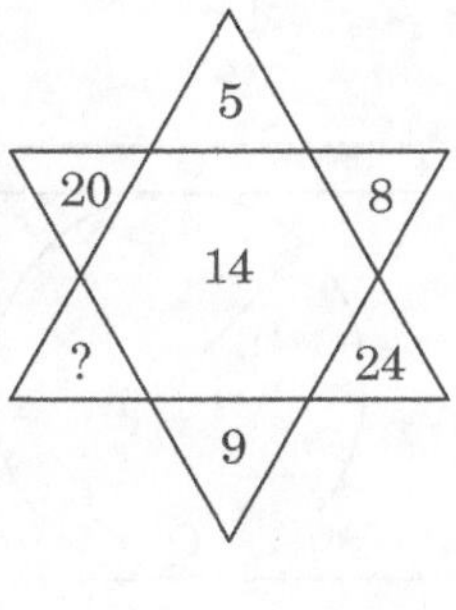

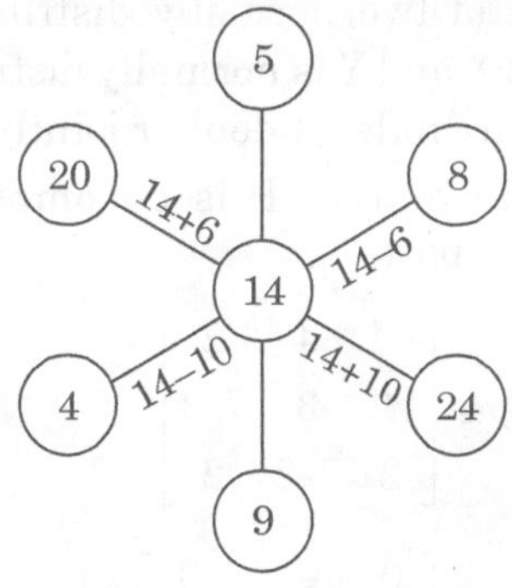

   4 represents by d so, the correct option is (a).

8. 52 men can complete a strip in 10 days

   12 men were sent to another project,

   Total men available for work on construction of strip = 52 – 12 = 40

   $52 \times 10 = 40 \times x$

   $x = 13$ days

   Number of days, more than the original is,

   = 13 – 10 = 3 days

9.

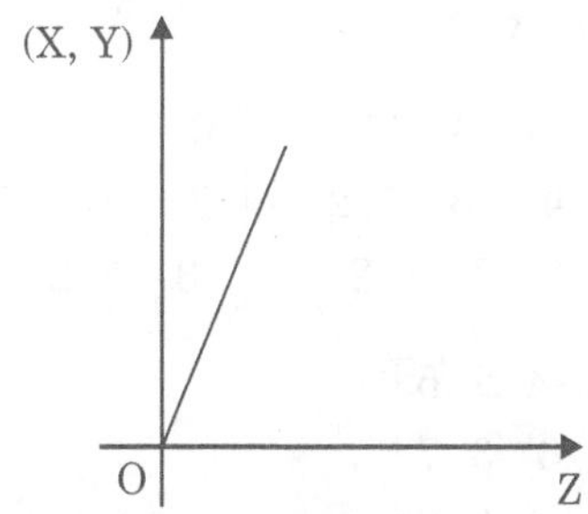

   The product of X and Y linearly varies with Z.

   XY = aZ + b, where a and b are constants.

   For fixed X, Z is proportional to Y.

   $\frac{XY}{Z}$ is constant,

   For fixed Z, X is inversely proportional to Y.

   For fixed, Y, X is proportional to Z.

10. Ratio of boys enrolled in Arts to the girls enrolled in management = $\frac{\text{Boys (Arts)}}{\text{Girls (Management)}}$

    Let, Number of boys in Arts = y

    Number of girls in Management

    $= \frac{15}{100} \times 1500 = 225$

    Number of girls in Arts = $\frac{20}{100} \times 5000 = 1000$

    Number of boys in Arts = 1000 – 450 = 550

    $\frac{\text{Boys (A)}}{\text{Girls (M)}} = \frac{550}{225} = \frac{110}{45} = \frac{22}{9}$

MECHANICAL ENGINEERING

1. The sum of two normally distributed random variables X and Y is normally distributed only if X and Y are independent or jointly normal.

2. When the matrix P is decomposed into its symmetric part.

$$S = \begin{bmatrix} -4 & 4 & 2 \\ 4 & 3 & 7/2 \\ 2 & 7/2 & 2 \end{bmatrix}$$

$$V = \begin{bmatrix} 0 & -2 & 3 \\ 2 & 0 & 7/2 \\ -3 & -7/2 & 0 \end{bmatrix}$$

For matrix P its symmetric part is

$$S = \frac{P + P^T}{2}$$

For matrix P its skew symmetric part is

$$V = \frac{P - P^T}{2}$$

$$P = S + V$$

$$= \begin{bmatrix} -4 & 4 & 2 \\ 4 & 3 & 7/2 \\ 2 & 7/2 & 2 \end{bmatrix} + \begin{bmatrix} 0 & -2 & 3 \\ 2 & 0 & 7/2 \\ -3 & -7/2 & 0 \end{bmatrix}$$

$$= \begin{bmatrix} -4 & 2 & 5 \\ 6 & 3 & 7 \\ -1 & 0 & 2 \end{bmatrix}$$

3. Given that

Then by changing the order of integration

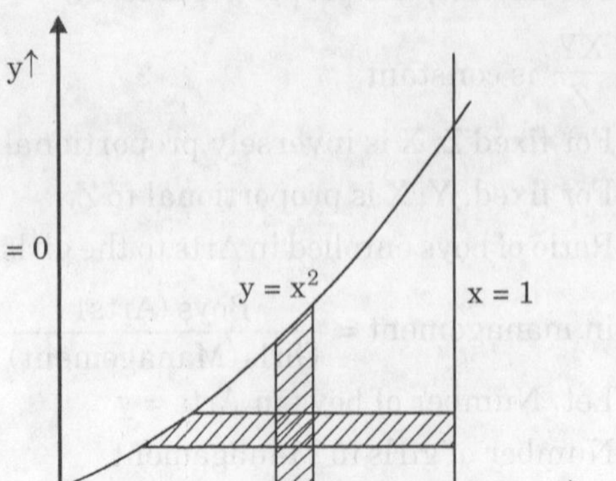

Limit $\quad x = 0 \text{ to } x = 1$

$$y = 0 \text{ to } y = x^2$$

By changing the order of the integration the strip change from vertical to horizontal.

The new limit, $y = 0$ to $y = 1$

$$x = \sqrt{y} \text{ to } x = 1$$

So the integration

$$I = \int_{y=0}^{1} \int_{x=\sqrt{y}}^{1} xy^2 dx\, dy$$

4. Given $\quad \frac{d^2y}{dt^2} - y = 1$

$$y\Big|_{t=0} = 0 \quad \frac{dy}{dt}\Big|_{t=0} = 0$$

$$\frac{d^2y}{dt^2} - y = 1$$

$$y''(t) - y(t) = 1$$

Taking Laplace both side

$$Ly''(t) - Ly(t) = L(1)$$

$$S^2y(s) - sy(0) - y'(0) - y(s) = \frac{1}{s}$$

Given $y(0) = 0, y'(0) = 0$

Then $\quad s^2y(s) - y(s) = \frac{1}{s}$

$$y(s)(s^2 - 1) = \frac{1}{s}$$

$$y(s) = \frac{1}{s(s-1)(s+1)}$$

5.

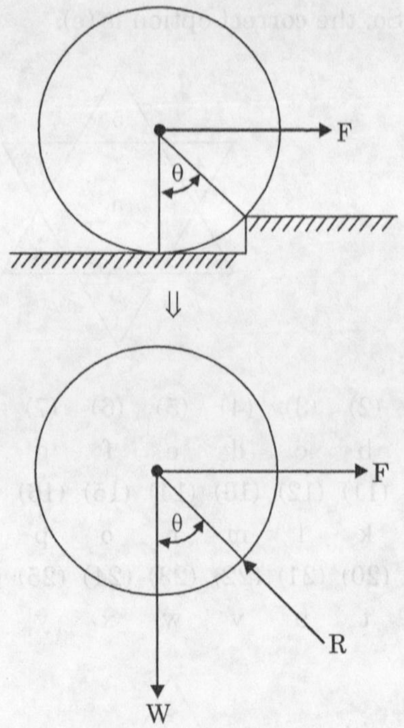

6.

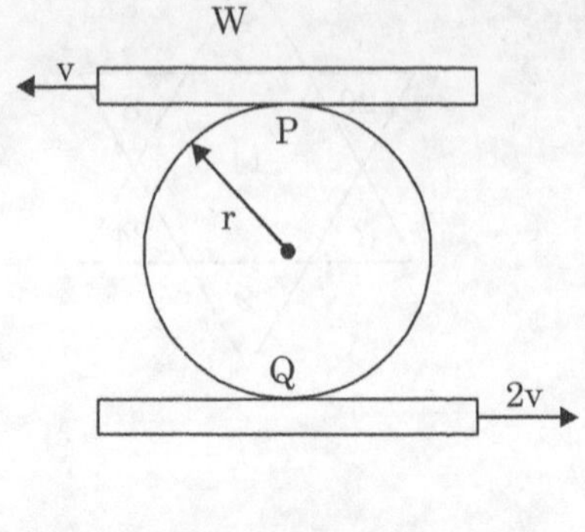

Radius r and roll without slipping at P and Q.

For pure rolling condition P and Q velocity should be zero

Assume, the roller moves with linear velocity with V_0 and rotate with ω angular velocity

So for P $\quad V_0 + V = r\omega$

$$V_0 - r\omega = -V \quad ...(i)$$

For $\quad Q, V_0 + r\omega = 2V \quad ...(ii)$

By solving equation (i) and (ii)

We get $\quad \omega = \frac{3V}{2r}$

7. The equation of motion of spring-mass damper system is given by

$$\frac{d^2x}{dt^2} + \frac{3dx}{dt} + 9x = 10\sin(5t)$$

The general equation of forced vibration system is given by

$$m\ddot{x} + C\dot{x} + Sx = F_0 \sin(\omega t)$$

$$\ddot{x} + \frac{C}{m}\dot{x} + \frac{S}{m}x = \frac{F_0}{m}\sin(\omega t)$$

$$\frac{d^2x}{dt^2} + \frac{C}{m}\frac{dx}{dt} + \frac{S}{m}x = \frac{F_0}{m}\sin(\omega t)$$

By comparing from the general equation, we get

$$\frac{C}{m} = 3, \quad \frac{S}{M} = 9$$

$$C = 3m \quad S = 9m$$

From relation, $\quad C = 2\xi\sqrt{Sm}$

$$3m = 2\xi\sqrt{9m \times m}$$

$$\xi = \frac{1}{2} = 0.5$$

8. Total qualitative inversion when Grashof's law satisfied.

(i) Crank-Crank mechanism when shortest link at the fixed position.

(ii) Rocker-Rocker mechanism when shortest link opposite to the fixed link that is coupler position.

(iii) Crank-rocker mechanism when shortest link at adjacent to the fixed position.

9. Ultrasonic machining is used for machining of glass.

10. Correct sequence is

Laser beam < MIG < SAW < Arc welding

11. • Phase diagram is a plot on the temperature composition space showing the stability of various phases.

- Phase diagram tells us what will be the melting point of alloy.
- Relative amount of different phases can be found under given equilibrium condition by lever rule.
- Phase diagram tells about solid solubility it does not gives the information on transformation rate.

12. R = minimum Roughness height

Q = maximum Roughness height

P = maximum waviness height

S = maximum waviness width

u = Roughness width cutoff

T = maximum Roughness width

13. In MRP if the inventory holding cost is very high and setup cost is zero then the lot sizing approaches should be used lot-for-lot.

14. For partially submerged floating bodies:

- For stable equilibrium metacentre (M) must be at higher level than the centre of gravity (g)
- For unstable equilibrium metacentre (M) must be at lower level than the centre of gravity (g).
- For neutral the motacentre (M) and the centre of gravity (g) coincide with each other.

15. In the space above the mercury column in a barometer tube the gauge pressure of the vapour is negative.

The absolute pressure is positive but less than one atmosphere.

17. Air standard diesel cycle:

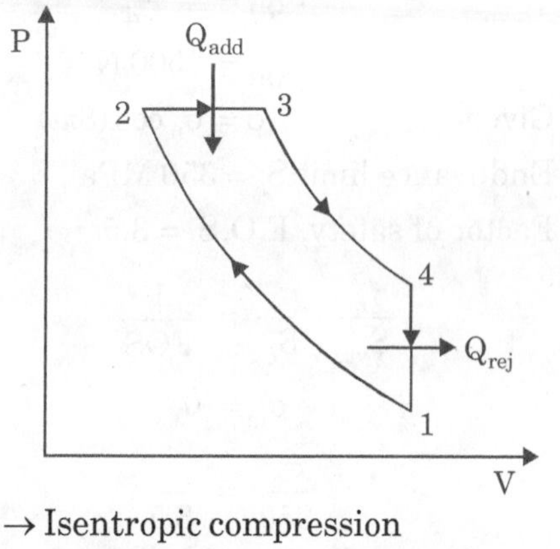

1–2 → Isentropic compression

2–3 → Constant pressure heat addition

3–4 → Reversible adiabatic (Isentropic) expansion

4–1 → Constant volume heat addition

18. Given, $\quad (h_2 - h_1) = (h_3 - h_2)$

$$\text{Degree of reaction} = \frac{\Delta h_{Rotor}}{\Delta h_{Rotor} + \Delta h_{stator}}$$

$$= \frac{(h_3 - h_2)}{(h_3 - h_2) + (h_2 - h_1)}$$

$$= \frac{1}{2} \times 100 = 50\%$$

19. $$I = \begin{bmatrix} 1 & 0 & 0 & - & - & - & 0 \\ 0 & 1 & 0 & - & - & - & 0 \\ 0 & 0 & 1 & - & - & - & 0 \\ - & - & - & - & - & - & 0 \\ - & - & - & - & - & - & 0 \\ - & - & - & - & - & - & 1 \\ - & - & - & - & - & - & 1 \end{bmatrix}_{100 \times 100}$$

Since, diagonal elements are eigen values of identify matrix.

So, E has only 1 element i.e., 1.

20.

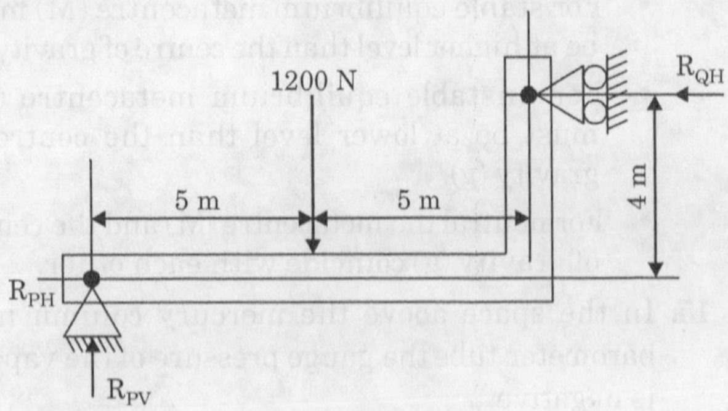

Taking moment about P, we get

$$R_{QH} \times 4 - 1200 \times 5 = 0$$

$$R_{QH} = \frac{1200 \times 5}{4}$$

$$R_{QH} = 1500 \text{ N}$$

21. Given $\sigma = \sigma_0 \cos(8\pi t)$

Endurance limit $S_e = 350$ MPa

Factor of safety, F.O.S. = 3.5

$$\frac{\sigma_m}{S_{yt}} + \frac{\sigma_a}{S_e} = \frac{1}{FOS}$$

$$\sigma_m = 0$$

$$\frac{\sigma_a}{350} = \frac{1}{3.5}$$

$$\sigma_a = \sigma_o$$

$$\sigma_o = 100 \text{ MPa}$$

22. In upsetting forging operation, if the stock is larger than 3 times the diameter then die cavity should not be more than 1.5 times the stock diameter.

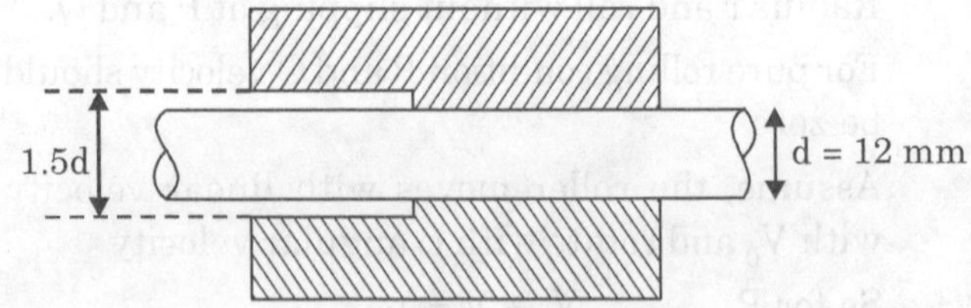

Hence, diameter of die = 1.5 d

$$= 1.5 \times 12$$

$$= 18 \text{ mm}$$

23.

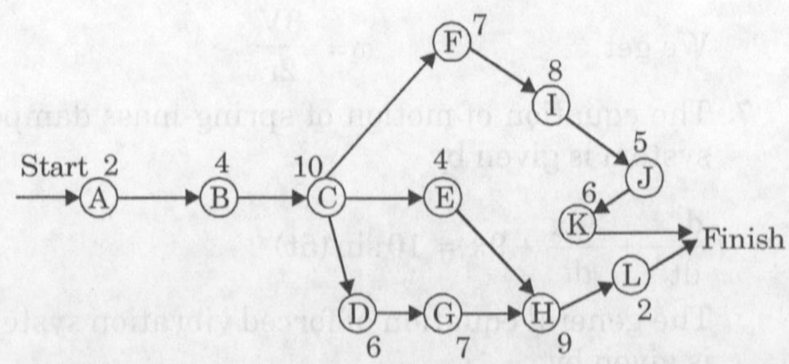

LFT = 42

EFT = 40

ST = LFT – EFT = 42 – 40 = 2

24.

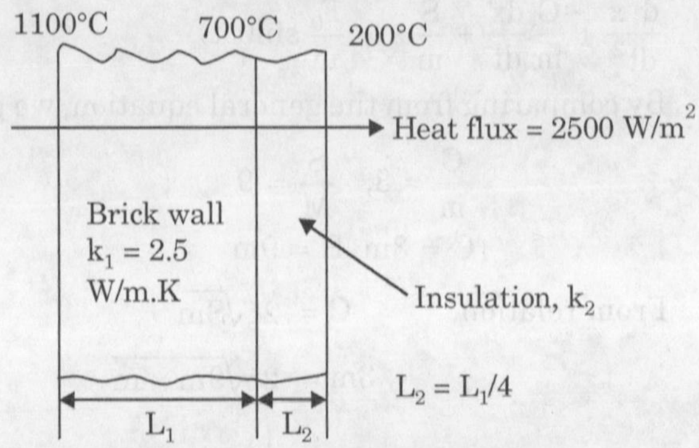

The heat flux $q = \frac{\delta T}{R_{eq.}} = \frac{\Delta T}{R_1}$ both resistance in series so

$$R = R_1 + R_2$$

$$R = \frac{L_1}{K_1 A_1} + \frac{L_2}{K_2 A_2}$$

Then $$q = \frac{\Delta T}{R_{eq.}} = \frac{\Delta T}{R_1}$$

$$2500 = \frac{1100 - 700}{\frac{L_1}{2.5}}$$

$$\Rightarrow \quad L_1 = 0.4$$

$$L_2 = L_1/4 = 0.1$$

$$2500 = q = \frac{1100 - 200}{\frac{0.4}{2.5} + \frac{0.1}{K_2}}$$

$$K_2 = 0.5 \text{ W/mK}$$

25. Reversed Carnot cycle operate between temperature limit 27°C and – 3°C

COP of the refrigerator, $(COP)_R = \frac{T_L}{T_H - T_L}$

$$T_L = -3 + 273 = 270 \text{ K}$$

$$T_H = 27 + 273\ \ 300 \text{ K}$$

$$(COP)_R = \frac{270}{300 - 270} = 9$$

COP of heat pump

$$(COP)_{H,P} = \frac{T_H}{T_H - T_L}$$

$$= \frac{300}{300 - 270} = 10$$

$$\frac{COP_R}{COP_{HP}} = \frac{9}{10} = 0.9$$

26. Directional Derivative = $(\Delta f)_\phi \cdot \hat{a}$

$$f(x) = xyz$$

$$\vec{a} = \hat{i} - 2\hat{j} + 2\hat{k}$$

$$\hat{a} = \frac{\vec{a}}{|a|} = \frac{\hat{i} - 2\hat{j} + 2\hat{k}}{\sqrt{1^2 + (2)^2 + 2^2}}$$

$$= \frac{1}{3}(\hat{i} - 2\hat{j} + 2\hat{k})$$

Gradient of (f) $= (\nabla f)_\phi = \left(\frac{\partial}{\partial x}\hat{i} + \frac{\partial}{\partial y}\hat{j} + \frac{\partial}{\partial z}\hat{k}\right)f$

$$= yz\hat{i} + xz\hat{j} + xy\hat{k}$$

Directional derivative

$$= \left(yz\hat{i} + xz\hat{j} + xy\hat{k}\right) \cdot \frac{1}{3}\left(\hat{i} - 2\hat{j} + 2\hat{k}\right)$$

$$DD = \frac{1}{3}[yz - 2xz + 2xy]$$

DD at (–1, 1, 3) is,

$$DD = \frac{1}{3}[1 \times 3 - 2 \times (-1) \times 3 + 2 \times (-1) \times 1] = \frac{7}{3}$$

27. Given: $u = x^3 - 3xy^2$

For function to be analytic it should follow Cauchy-Riemann's equations

$$u_x = v_y \text{ and } u_y = -v_x \quad \dots (2)$$

Now, $\because$ $u_x = v_y$

$$u_x = \frac{\partial u}{\partial x} = 3x^2 - 3y^2$$

$$= v_y = \frac{\partial v}{\partial y}$$

$$\Rightarrow \quad v = \int (3x^2 - 3y^2)\,dy + f(x) + C_1$$

$$\Rightarrow \quad v = 3x^2y - \frac{3y^3}{3} + f(x) + C_1 \quad \dots(3)$$

$$\Rightarrow \quad \frac{\partial v}{\partial x} = 6xy - 0 + f'(x)$$

$$\Rightarrow \quad v_x = 6xy + f'(x) \quad \dots(4)$$

Now, $u_y = \frac{\partial u}{\partial y} = 0 - 6xy \quad \dots(5)$

From (2)

$$u_y = -v_x$$

$$-6xy = -v_x \text{ [From (5)]}$$

$$-6xy = -(6xy + f'(x)) \quad \text{[From (4)]}$$

$$\Rightarrow \quad 6xy = 6xy + f'(x)$$

$$\Rightarrow \quad f'(x) = 0$$

$$\Rightarrow \quad f(x) = \text{Constant} \quad \dots(6)$$

So, from (3)

$$v = 3x^2y - y^3 + \text{Constant} + C_1$$

[From (6)]

$$v = 3x^2y - y^3 + C$$

[Let C = Constant + C_1]

28.

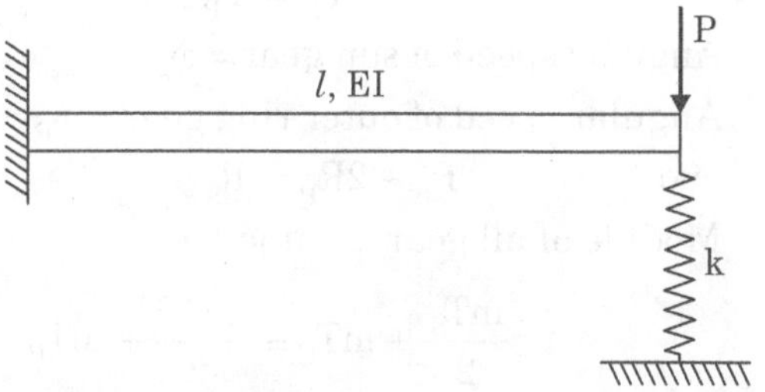

The deflection in spring and cantilever beam is same.

Therefore, stiffness of both (beam and spring) are in parallel combination.

For parallel system,

Equivalent stiffness $(K_{eq}) = K + K_{beam}$ (Stiffness of beam)

$$K_{beam} = \frac{3EI}{l^3}$$

[For point load at cantilever beam]

$$K_{eq} = K + \frac{3EI}{l^3}$$

$$= K + \frac{3EI}{l^3}$$

$$\text{Deflection} = \frac{P}{K_{eq}}$$

$$= \frac{P}{\left(\frac{Kl^3 + 3EI}{l^3}\right)}$$

$$= \frac{Pl^3}{Kl^3 + 3EI}$$

$$= \frac{Pl^3}{3EI}\left(\frac{1}{\frac{Kl^3}{3EI}+1}\right)$$

$$= \frac{Pl^3}{3EI}\left(\frac{3EI}{Kl^3 + 3EI}\right)$$

29.

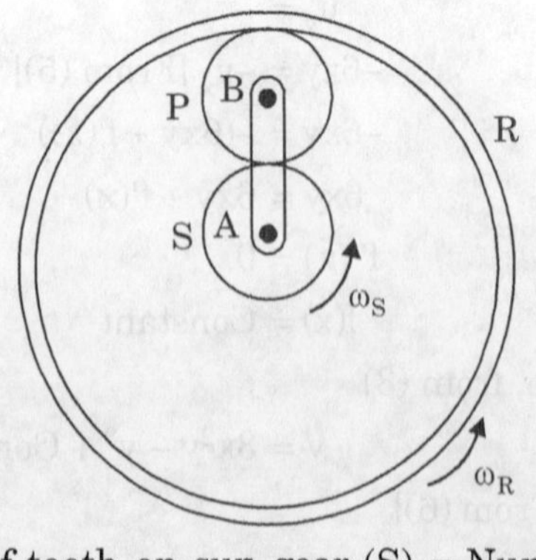

Number of teeth on sun gear (S) = Number of teeth on planet gear (P)

$$T_S = T_P$$

Angular speed of sun gear = ω_s

Angular speed of outer ring gear = ω_R

and $R_S + 2R_P = R_R$

Module of all gear is same,

$$\frac{mT_S}{2} + mT_P = \frac{mT_S}{2} + mT_P$$

$$T_S + 2T_P = T_R \quad (\because T_S = T_P)$$

$$T_R = 3\,T_P$$

Let, revolution made by (S) = x

Revolution made by arm (a) = y

Line	Action	Rev. of arm	Rev. of S	Rev. of P	Rev. of R
1.	Arm, fixed, S+1 rev.	0	1	$-\frac{T_S}{T_P}$	$\frac{T_S}{T_P} \times \frac{T_P}{T_R}$
2.	Arm fixed,	x	x	$\frac{T_S}{T_P}x$	$-\frac{T_S}{T_R}x$
3.	Add y	y	x + y	$y - \frac{T_S}{T_P}x$	$y - \frac{T_S}{T_R}x$

Angular speed of (S) = $\omega_S = x + y$

Angular speed of (R) = $\omega_R = y - \frac{T_S}{T_R}x = y - \frac{x}{3}$

$$x + y = \omega_S \quad \text{...(i)}$$

$$-\frac{x}{3} + y = \omega_R \quad \text{...(ii)}$$

By subtracting equation (ii) from equation (i)

$$x + \frac{x}{3} = \omega_S - \omega_R$$

$$x = \frac{3}{4}(\omega_S - \omega_R)$$

Putting value of x in equation (i)

$$y = \omega_S - \frac{3}{4}(\omega_S - \omega_R)$$

$\left(y = \frac{\omega_S}{4} + \frac{3}{4}\omega_R\right)$, angular speed of the arm AB.

30. $\alpha\Delta T = \frac{\nu pr}{tE}$

$$\boxed{\Delta T = \frac{\nu pr}{\alpha tE}}$$

31. Since spring constant (K) = $\frac{Gd^4}{8D^3n}$ ...(1)

where,

G = modulus of rigidity of spring matrial

D = Diameter of spring

d = Diameter of wire

n = No. of coils

After doubling the wire diameter, spring diameter and the number of coils,

$$D' = 2D$$

$$d' = 2d$$

$$n' = 2n$$

Putting these values in (1),

$$K' = \frac{G(2d)^4}{8 \times (2D)^3 \times (2n)}$$

$$= \frac{Gd^4 \times 2^4}{8 \times D^3 \times n \times 2^3 \times 2}$$

$$= \frac{Gd^4}{8D^3n} = K$$

32.

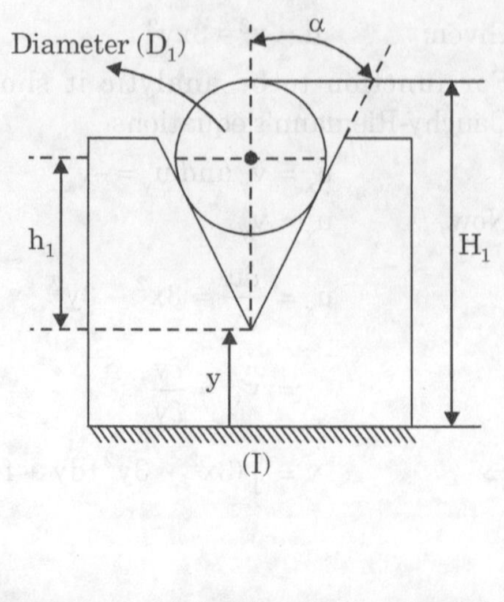

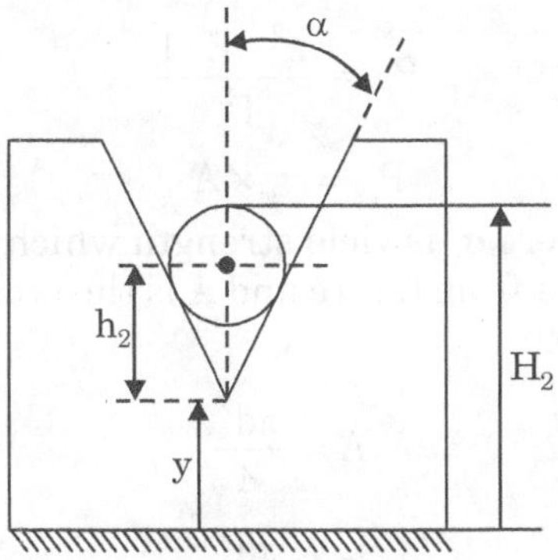

(II)

From Figure (1)

$$\sin\alpha = \frac{\frac{D_1}{2}}{h_1}$$

where $h_1 = H_1 - y - \frac{D_1}{2}$

$$\Rightarrow \left(H_1 - y - \frac{D_1}{2}\right)\cdot \sin\alpha = \frac{D_1}{2}$$

Let $\sin\alpha = K$

$$\Rightarrow \left(H_1 - y - \frac{D_1}{2}\right) = \frac{D_1}{2K}$$

$$\Rightarrow \quad -y = \frac{D_1}{2K} + \frac{D_1}{2} - H_1 \quad ...(1)$$

From Fig. (II)

$$\sin\alpha = \frac{\frac{D_2}{2}}{h_2} = \frac{\frac{D_2}{2}}{H_2 - y - \frac{D_2}{2}}$$

$$\Rightarrow K = \frac{\frac{D_2}{2}}{H_2 + \frac{D_1}{2K} + \frac{D_1}{2} - H_1 - \frac{D_2}{2}}$$

$$\Rightarrow K\left(H_2 + \frac{D_1}{2K} + \frac{D_1}{2} - H_1 - \frac{D_2}{2}\right) = \frac{D_2}{2}$$

$$\Rightarrow K\left(H_2 - \frac{D_2}{2} - H_1 + \frac{D_1}{2}\right) = \frac{D_2}{2} - \frac{D_1}{2}$$

$$\Rightarrow K\left(H_2 - \frac{D_2}{2} - H_1 + \frac{D_1}{2}\right) = \frac{D_2}{2} - \frac{D_1}{2}$$

$$\Rightarrow K = \frac{\frac{(D_2 - D_1)}{2}}{H_2 - H_1 - \frac{(D_2 - D_1)}{2}}$$

$$\Rightarrow K = \frac{D_1 - D_2}{2(H_1 - H_2) - (D_1 - D_2)}$$

$$\Rightarrow \sin\alpha = \frac{D_1 - D_2}{2(H_1 - H_2) - (D_1 - D_2)}$$

33.

$$F_t = F_{t-1} + \alpha(\Delta_{t-1} - F_{t-1})$$

$$F_2 = F_1 + \alpha(D_1 - F_1)$$

$$31.8 = 32 + \alpha\,(30 - 32)$$

$$\alpha = 0.1$$

34. Closed system undergoing irreversible process,

Air, m = 1 kg, R = 0.287 kJ/kgk, $\gamma = 1.4$

Initial state	Final state
P_1 = 1 bar (absolute)	P_2 = 3 bar (absolute)
T_1 = 27°C = 300K	T_2 = 127°C = 400 K

Change in the specific entropy of the air in the process,

$$\Delta S = S_2 - S_1$$

$$= C_p \ln\left(\frac{T_2}{T_1}\right) - R\ln\left(\frac{P_2}{P_1}\right)$$

$$= 1.005\ln\left(\frac{400}{300}\right) - 0.287\ln\left(\frac{3}{1}\right)$$

$$= -0.0262\frac{\text{kJ}}{\text{kgk}} = -26.2\frac{\text{J}}{\text{kgk}}$$

35. Conventional Method (I)

$$I = \int_0^{\pi/2} (8 + 4\cos x)\cdot dx$$

$$= (8x + 4\sin x)_0^{\pi/2}$$

$$= 4\pi + 4 = 16.566$$

Numerical Method using trapezoidal rule,

$$II = \frac{h}{2}(y_0 + y_n)$$

where, $h = \frac{\pi}{2} - 0 = \frac{\pi}{2}$

$$y_0 = 8 + 4 \times \cos(0) = 12$$

$$y_n = 8 + 4 \times \cos\left(\frac{\pi}{2}\right) = 8$$

$$\Rightarrow \quad II = \frac{\pi}{4}(12 + 8) = 15.708$$

$$\therefore \quad \%\ \text{error} = \frac{II - I}{I} \times 100$$

$$= \left(\frac{II}{I} - 1\right)\times 100$$

$$= \left(\frac{15.708}{16.566} - 1\right)\times 100$$

$$= -5.18$$

$$= 5.18\ (\text{absolute }\%)$$

36. Fair coin is tossed 20 times,

Probability (Head exactly appear 4 times in the first 10 tosses)

$$P(X) = {}^{10}C_4 \left(\frac{1}{2}\right)^4 \left(\frac{1}{2}\right)^6 = 0.205$$

Probability (Tail exactly appears 4 times in the next 10 tosses)

$$P(Y) = {}^{10}C_4 \left(\frac{1}{2}\right)^4 \left(\frac{1}{2}\right)^6 = 0.205$$

Total probability, P(T) = P(X) × P(Y)

$$= 0.205 \times 0.205 = 0.042$$

37.

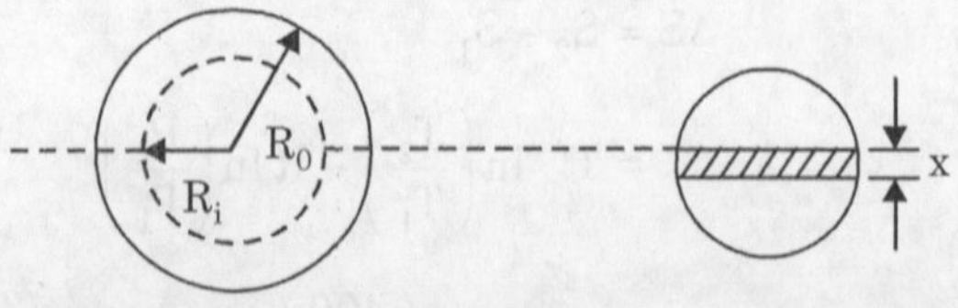

(a) (b)

Assuming G = Specific gravity of sphere

In case (a): When hollowsphere is in equilibrium (when half submerged) weight of sphere = upward buoyant force

$$\frac{4}{3} \times \pi \left(R_0^3 - R_i^3\right) \times G\rho_w g = \frac{1}{2} \times \frac{4}{3} \pi R_0^3 \times \rho_{wg}$$

$$\Rightarrow \left(R_0^3 - R_i^3\right) \times G = \frac{1}{2} R_0^3 \quad \text{...(i)}$$

In case (b): When a small downward displaement of 'x' is given

Net upward force = Additional buoyant force due to increase in submergence

$$\Rightarrow F_{net} = m.a_{net} = \left(\pi R_0^2 x\right) \times \rho_w g \quad \text{...(ii)}$$

From (ii)

$$\text{in}.a_{net} = \left(\pi R_0^2 x\right) \times \rho_w g$$

$$\frac{4}{3}\pi\left(R_0^3 - R_i^3\right) \times G\rho_w \cdot a_{net} = \left(\pi R_0^2 x\right) \times \rho_w g$$

From (i), putting $\left(R_0^3 - R_i^3\right)G = \frac{R_0^3}{2}$, we get

$$\frac{4}{3}\pi \frac{R_0^3}{2} \times \rho_w \cdot a_{net} = \pi R_0^2 x \times \rho_w g$$

$$\Rightarrow \frac{2}{3} R_o \cdot a_{net} = x.g \quad \text{...(iii)}$$

Comparing with SHM equation

$$a_{net} = \omega^2 x$$

$$\Rightarrow \quad \omega = \sqrt{\frac{3g}{2R_o}} = \sqrt{\frac{3 \times 10}{2 \times 0.2}}$$

$$\Rightarrow \quad \omega = 8.66 \text{ rad/sec}$$

38.

$$P_{cr} = \frac{\pi^2 \cdot E \cdot I}{L_e^2} \quad \text{...(1)}$$

$$P_{cr} = \sigma_y \times A$$

Where σ_y is yield strength which is equal to 100 MPa from figure and A is the cross-section area of bar,

$$A = \frac{\pi d^2}{4}$$

$$\because \quad I = \frac{\pi d^4}{64}$$

$$E = \frac{\sigma_y}{\varepsilon} = \frac{100}{0.002} = 50000 \text{ MPa}$$

From (1), we get

$$100 \times \frac{\pi d^2}{4} = \frac{\pi^2 \times 50000 \times \frac{\pi d^4}{64}}{(1)^2}$$

$$\Rightarrow \frac{100 \times \pi d^2 \times 64}{4 \times \pi^2 \times 50000 \times \pi d^4} = 1$$

$$\Rightarrow \frac{100 \times 16}{\pi^2 \times 50000 \times d^2} = 1$$

$$\Rightarrow d^2 = \frac{16}{\pi^2 \times 500} = 3.2422 \times 10^{-3}$$

$$\Rightarrow \quad d = 0.05694 = 56.94 \text{ mm}$$

39.

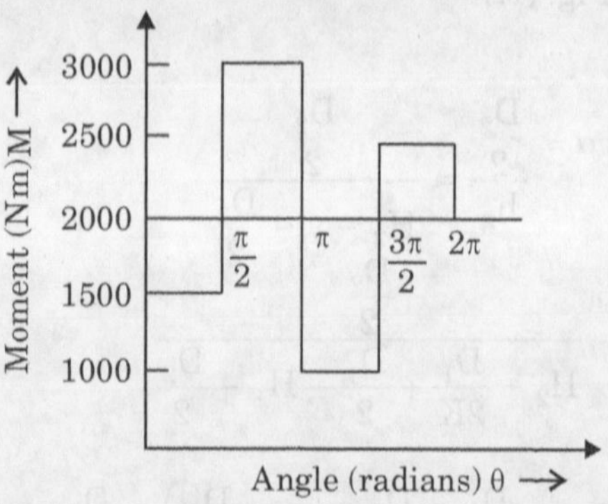

Mean turning moment, T_{mean} = 2000 Nm

Average engine speed, $\bar{N}$ = 1000 rpm

Fluctuation in speed, C_S = 0.04

Let, I = mass moment of inertia of the flywheel (kg-m²)

Maximum fluctuation of energy, ΔE,

$$\Delta E = (T - T_{mean}) \times (\text{Time period})$$

$$= (3000 - 2000) \times \left(\pi - \frac{\pi}{2}\right) = 1570.8 \text{ J}$$

$$\Delta E = I\omega^2 C_s$$

$$1570.8 = I \times \left(\frac{2\pi \times 1000}{60}\right)^2 \times 0.04$$

$$I = 3.58 \text{ kg-m}^2$$

40.

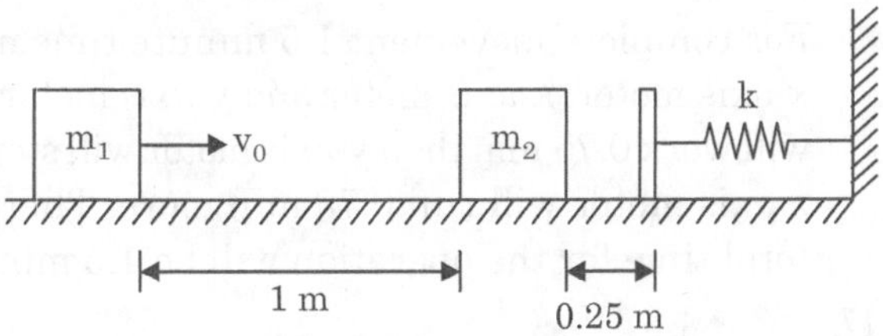

Let velocity of combined masses $(m_1 + m_2)$ just after collision is V m/s.

Since, masses are moving together after collision. So, it is a case of inelastic collision.

$\Rightarrow m_1 v_0 + m_2 \times 0 = (m_1 + m_2) \times V$

$\Rightarrow 10v_0 = (10 + 30) \times V$

$$\Rightarrow V = \frac{10V_0}{40} = \frac{V_0}{4} = \frac{2}{4} = 0.5 \text{ m/s}$$

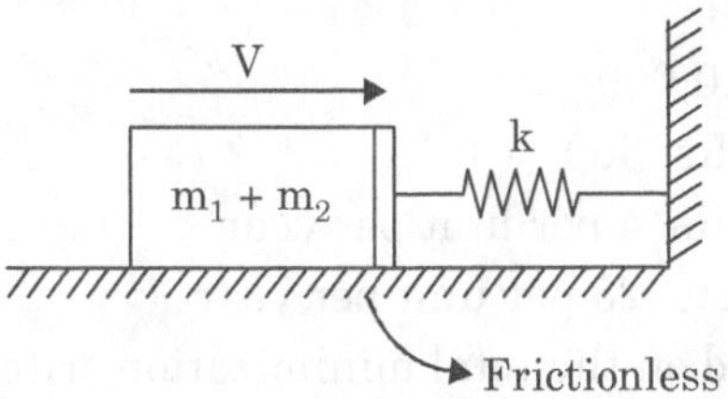

Note : Velocity V_0 will be constant before collision, i.e. during travelling of 1 m (before collision) because the horizontal surface is frictionless.

Similarly, velocity V will also be constant for 0.25 m before collision of combined masses with the spring.

Now, According to conservation of mechanical energy.

Initial K.E. + Initial P.E.

= Final K.E. + Final P.E.

when final velocity of the system (combined masses and spring) is zero. Final K.E will be zero and there will be maximum deflection in spring i.e. A.

$$\text{So, } \frac{1}{2}(m_1 + m_2)V^2 + 0 = 0 + \frac{1}{2}kA^2$$

$$\Rightarrow \frac{1}{2}(10 + 30) \times 0.5^2 + 0 = 0 + \frac{1}{2} + 10^5 \times A^2$$

$\Rightarrow 40 \times 0.5^2 = 10^5 \times A^2$

$\Rightarrow A = 0.01 \text{ m} = 1 \text{ cm}$

41. Let, W^t = Tangential load acting on the pinion

P = Power transmitted

$$\text{Bending stress } (\sigma) = \frac{k_v W^t}{FmY}$$

$$400 = \frac{1.26 \times W^t}{50 \times 1.25 \times 0.322}$$

$W^t = 6388.89$ N

Power transmitted (P) = Torque × Speed

$= (T\omega)_{\text{pinion}}$

Torque applied on pinion gear

$$= W^t \times r_p = W^t \times \frac{mZ_p}{2}$$

$$T = 6388.89 \times \frac{1.25 \times 20}{2}$$

Torque, (T) = 79861.125 N.mm

= 79.86 N.m

P = Tω = 79.86 × 125.66

= 10035.5 watt = 10 kW

42. Volume = Discharge rate × time

$$\text{time (t)} = \frac{\text{Volume}}{\text{Discharge rate}}$$

$$= \frac{\text{Volume}}{\text{Area} \times \text{Velocity}}$$

$$= \frac{1200}{2 \times \sqrt{2 \times 981 \times 10}}$$

= 4.28 seconds

43. Given diameter, d = 200 mm

Initial geometry (ASA)

0 – 9° – 7° – 15° – 30° – 0.05 inch

After some time

0 – 9° – 7° – 8° – 15° – 0° – 0.05 inch

We know that specific energy consumption

$$u_c = (e) = \frac{F_c}{1000fd}$$

$$f = \left(\frac{t}{\sin\lambda}\right)$$

$\lambda_1 = 90° - 30° = 60°$

$\lambda_2 = 90 - 0 = 90°$

Specific cutting energy,

$$u_c = u_o(t_1)^{-0.4} = \frac{F_c}{1000fd}$$

$$F_c = 40(f\sin\lambda)^{-0.4} \times 1000 \times fd$$

$$F_c \propto (\sin\lambda)^{-0.4}$$

$$\left(\frac{F_{c2} - F_{cl}}{F_{cl}}\right) \times 100 = \left[\left(\frac{\sin\lambda_2}{\sin\lambda_1}\right)^{-0.4} - 1\right] \times 100$$

$$= \left[\left(\frac{\sin 90}{\sin 60}\right)^{-0.4} - 1\right] \times 100$$

= – 5.59%

44. Two identical shaping machine, S_1 and S_2.

Let width of workpiece for machine $S_1 = b_1$

Feed of workpiece for machine $S_1 = f_1$

For machine S_2, width of workpiece, $b_2 = 1.1\ b_1$

Feed of workpiece, $f_2 = 0.9\ f_1$

Shaper total time of machining,

$$t_m \propto \left(\frac{b(1+R)}{f}\right); t_m \alpha\ b\ \&\ \tan \alpha\ \frac{1}{f}$$

$$\frac{t_{m_1}}{t_{m_2}} = \frac{b_1}{b_2} \times \frac{f_2}{f_1} = \frac{0.9}{1.1}$$

$$= 0.818 \approx 0.8$$

45. Length of bar (L) = 250 mm

Diameter (D) = 25 mm

Feed (f) = 0.2 mm/rev

Time required for tool change (T_c) = 1 min.

Tool life equation, $VT^{0.2} = 24$

For maximum production rate,

$$T = T_C\left(\frac{1-n}{n}\right) = 1\left(\frac{1-0.2}{0.2}\right)$$

$$= 4 \text{ min}$$

$$V = \frac{24}{(4)^{0.2}} = 18.18 \text{ m/min}$$

Total machining time

$$= \frac{L}{fN} = \frac{L}{f\left(\frac{V}{\pi D}\right)} = \frac{250}{0.2 \times \left(\frac{18.18 \times 1000}{\pi \times 25}\right)}$$

$$= 5.4 \text{ min}$$

Number of tool changes per component.

46.

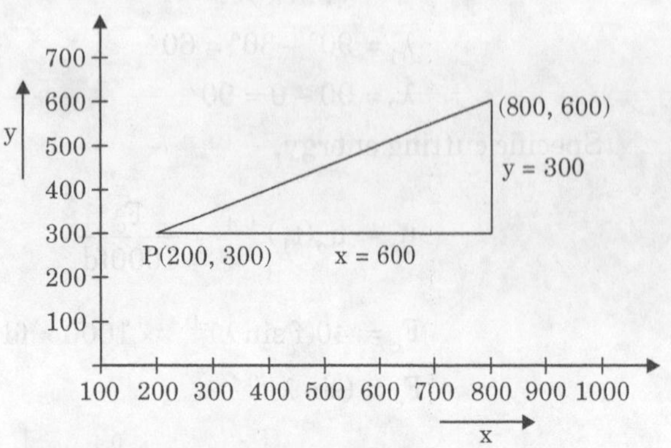

Maximum speed of N = 800 rpm,

Pitch = 0.5 mm

time taken in x direction,

$$t_x = \frac{\Delta x}{P \times N} = \frac{600}{0.5 \times 800} = 1.5 \text{ min}$$

time taken in y direction

$$t_y = \frac{\Delta y}{P \times N} = \frac{300}{0.5 \times 800} = 0.75 \text{ min}$$

For complete movement 1.5 minute time neded x-axis motor. x-axis motor and y-axis motor both will work 0.75 min then y-axis motor will stop and x axis motor will run 0.75 min more. Therefore total time for the operation will be 1.5 min.

47.

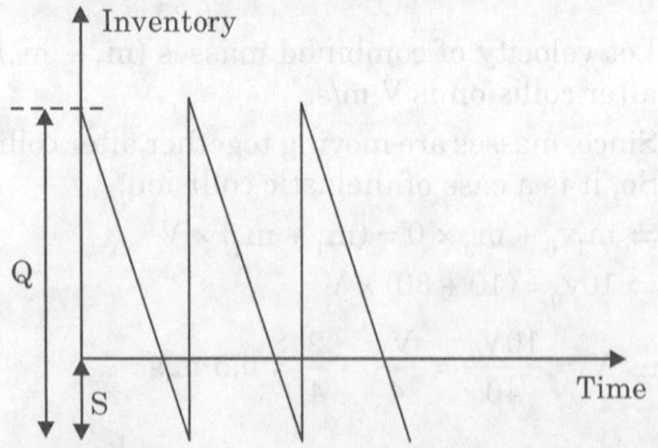

D = 10,000

C_o = Rs. 300

C_h = Rs. 4 per unit per year

C_b = Rs. 25 per unit per year

Based on the total minimization criteria,

Economic order quantity, Q* is given by

$$Q^* = \sqrt{\frac{2DC_o}{C_h}\left(\frac{C_b + C_h}{C_b}\right)}$$

$$Q^* = \sqrt{\frac{2 \times 10000 \times 300}{4}\left(\frac{25+4}{25}\right)} = 1319.09$$

Back order quantity,

$$\frac{Q^*}{S^*} = \frac{C_b + C_h}{C_h}$$

$$S^* = \frac{Q^* C_h}{C_b + C_h} = \frac{1319.09 \times 4}{25 + 4}$$

$$= 181.94$$

Maximum inventory reached in the system

$$Q_m^* = Q^* - S^*$$

$$Q_m^* = 1319.09 - 181.94$$

$$= 1137.15 \approx 1137$$

48. Flow through a nozzle,

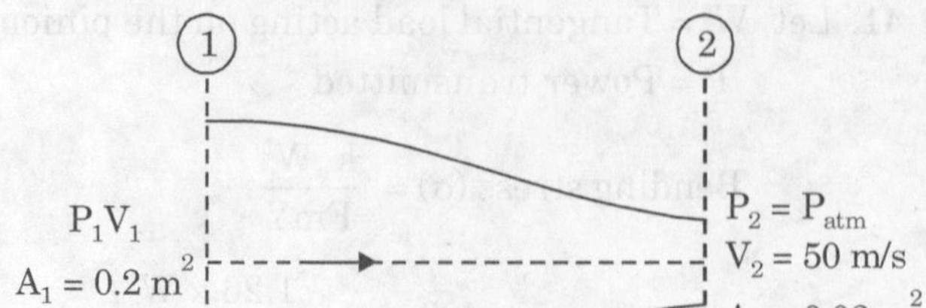

Air flow is steady, incompressible and inviscous.

Density of air (ρ) = 1.23 kg/m³

Apply Bernoulli equation between (1) and (2)

$$\frac{P_1}{\rho g} + \frac{V_1^2}{2g} + z_1 = \frac{P_2}{\rho g} + \frac{V_2^2}{2g} + z_2$$

$$\frac{P_1 - P_2}{\rho g} = (z_2 - z_1) + \frac{V_2^2 - V_1^2}{2g}$$

$$P_1 - P_{atm} = \rho g (z_2 - z_1) + \frac{\rho}{2}\left(V_2^2 - V_1^2\right)$$

$$= \frac{\rho}{2} \times V_2^2 \left[1 - \left(\frac{V_1}{V_2}\right)^2\right]$$

$$= \frac{\rho}{2} \times V_2^2 \left[1 - \left(\frac{A_2}{A_1}\right)^2\right]$$

$$= \frac{1.23}{2} \times 50^2 \times \left[1 - \left(\frac{0.02}{0.2}\right)^2\right]$$

$= 1522$ Pa $= 1.52$ kPa

49. Density of water $(\rho) = 1000$ kg/m^3

$g = 10$ m/s^2

Inclined pipe of uniform diameter,

Section A	Section B
$V_A = 3.2$ m/s	$V_B = 3.2$ m/s
$P_A = 186$ kPa	$P_B = 260$ kPa
$Z_A = 24.5$ m	$Z_B = 9.1$ m

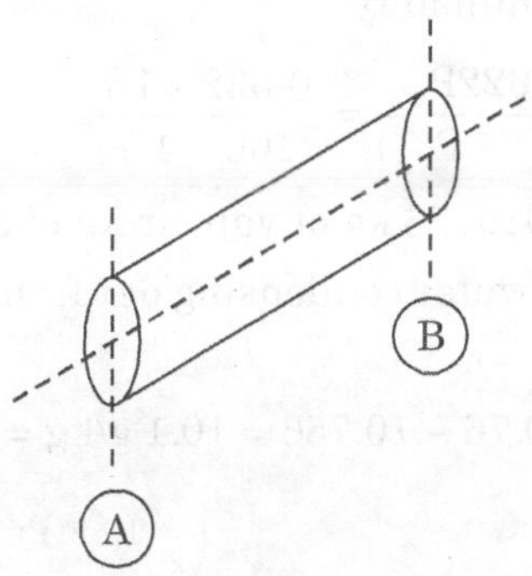

Let, Head lost due to friction $= h_L$

By applying Bernoulli principle between section A and B, we get

Total head at section A,

$$H_A = \frac{P_A}{\rho g} + \frac{V_A^2}{2g} + Z_A$$

$$= \frac{186 \times 1000}{1000 \times 10} + \frac{3.2^2}{2 \times 10} + 24.5$$

$= 43.612$ m

Total head at section B,

$$H_B = \frac{P_B}{\rho g} + \frac{V_B^2}{2g} + z_B$$

$$= \frac{260 \times 1000}{1000} + \frac{3.2^2}{2 \times 10} + 9.1 = 35.612 \text{ m}$$

Head lost due to friction $h_L = H_A - H_B$

$= 43.612 - 35.612 = 8$ m

50. From Wein's displacement Law

$$\lambda_{max} T = b$$

So, $(\lambda_{max})_1 T_1 = b$

and $(\lambda_{max})_2 T_2 = b$ ...(2)

From eqn. (1) divided by eqn. (2), we get

$$\Rightarrow \frac{(\lambda_{max})_1 T_1}{(\lambda_{max})_2 . T_2} = 1$$

Given, $T_1 = 300$ K

and $(\lambda_{max})_2 = 1.2 \times (\lambda_{max})_1$

$$\text{So, } \frac{(\lambda_{max})_1 \times 3000}{1.2 \times (\lambda_{max})_1 \times T_2} = 1$$

$$\Rightarrow T_2 = \frac{3000}{1.2} = 2500 \text{ K}$$

51.

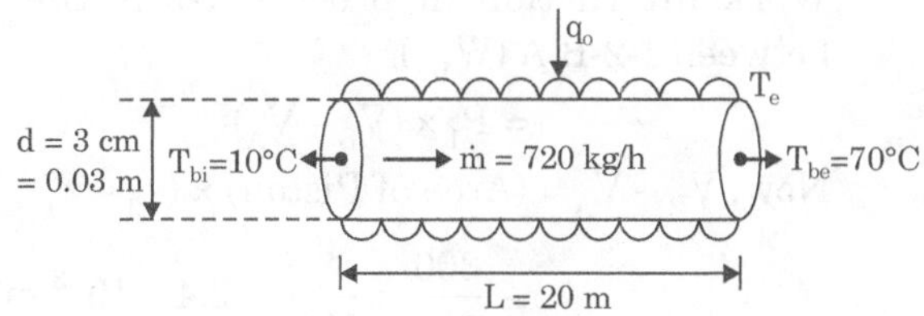

$C_p = 4.179$ kJ/kgK

$h = 1697$ W/m^{2}K

$$m = 720 \times \frac{1}{3600} = 0.2 \text{ Kg/s}$$

∵ Heat given to the tube + Heat gain by water in tube.

$$\Rightarrow q \times \pi d \times L = \dot{m} \times C_P \times (T_{be} - T_{bi})$$

$$q = \frac{0.2 \times 4.179 \times (70 - 10)}{\pi \times 0.03 \times 20}$$

$= 26.604$ kW/m^2

$= 26604.34$ W/m^2

At the end of tube,

$$q = h \times (T_e - T_{be})$$

$$\Rightarrow 26604.34 = 1697 \times (T_e - 70)$$

$$\Rightarrow T_e = 85.67 \approx 85.7°C$$

52.

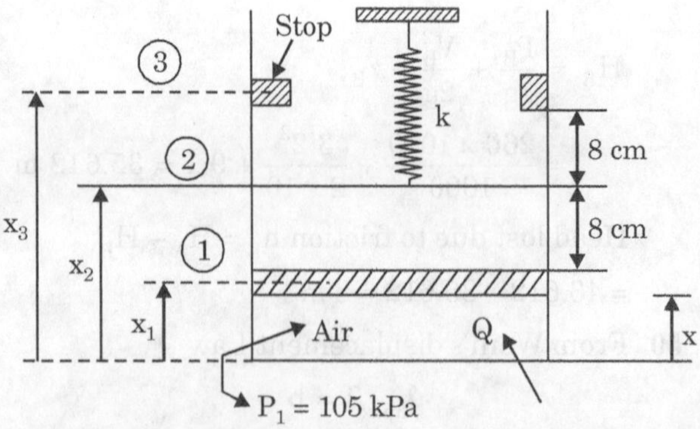

Process (1) – (2) is isobaric (constant pressure)

$\Rightarrow \quad P_1 = P_2$

Process (2) – (3) is a line as shown in P-V diagram

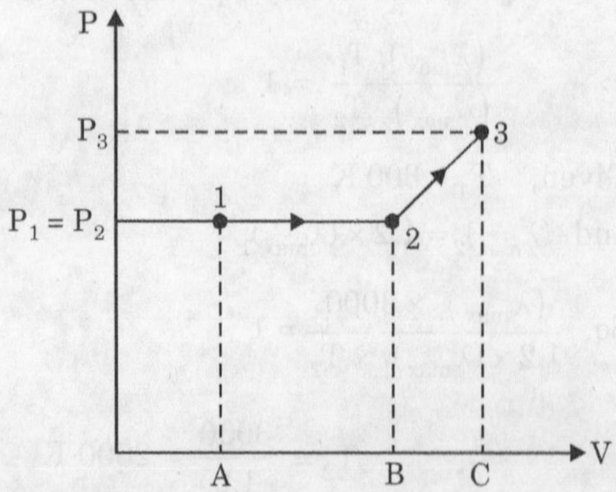

Work interaction in process 1-2 is the area between 1-2-B-A ($W_{1\text{-}2}$).

$$= P_1 \times (V_B - V_A)$$

Now, $V_B - V_A$ = (Area of Piston) × $(x_2 - x_1)$

$$= \frac{300}{10^4} \times \frac{8}{100} = 2.4 \times 10^{-3} \text{ m}^3$$

$W_{1-2} = 105 \times (2.4 \times 10^{-3}) \times 10^3 = 252\text{J}$

Now, W_{2-3} is the area between 2-3-C-B

$$W_{2\text{-}3} = \frac{1}{2}(P_2 + P_3) \times (V_C - V_B)$$

$$= \frac{1}{2}(P_2 + P_3) \times A(x_3 - x_2)$$

where A is the area of piston.

$$\therefore \quad P_3 = P_2 + \frac{Kx}{A}$$

$$= 105 \times 10^3 + \frac{12.5 \times 10^3 \times 8 \times 10^{-2}}{300 \times 10^{-4}}$$

$$= \frac{415}{3} \times 10^3 \text{Pa}$$

$$\therefore \quad W_{2-3} = \frac{1}{2}\left(105 \times 10^3 + \frac{415}{3} \times 10^3\right) \times 300 \times 10^{-4} (8 \times 10^{-2})$$

$$= 292 \text{ J}$$

$$\therefore \quad W_{Total} = W_{1-2} + W_{2-3}$$

$$= 252 + 292$$

$$= 544 \text{ J}$$

53. Moist air,

State (1), P_1 = 105 kPa, T_1 = 30°C, $(R.H)_1$ = 80%

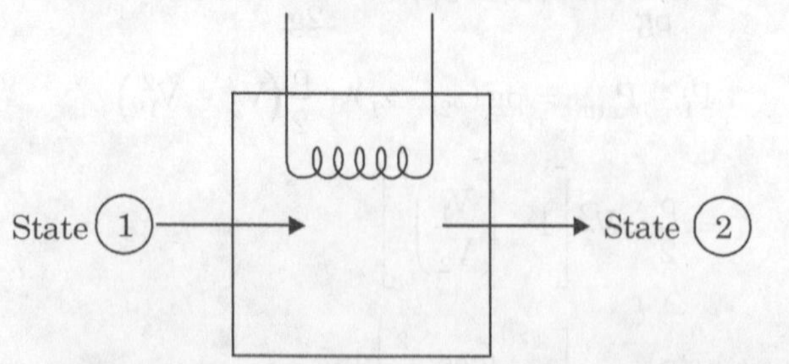

State (2), P_2 = 100 kPa, T_2 = 15°C, (R.H) = 100%

$(P_s)_{@30°C}$ = 4.24 kPa

$(P_s)_{@15°C}$ = 1.7 kPa

$(M.W.)_{water}$ = 18 g/mol, $(M.W)_{air}$ = 28.94 g/mol

Vapour pressure at state (1),

$P_{V1} = (R.H.)_1 \times (P_V)_{S1} = 0.8 \times 4.24 = 3.392$ kPa

Specific humidity,

$$\omega_1 = \frac{0.622(P_{V1})}{(P_1 - P_{V1})} = \frac{0.622 \times 3.392}{105 - 3.392}$$

= 0.02076 k of vapour/kg.d.a.

Vapour pressure at state (2),

$P_{V2} = (P_{VS})_2$ = 1.7 kPa

Specific humidity

$$\omega_2 = \frac{0.622P_{V2}}{(P_2 - P_{V2})} = \frac{0.622 \times 1.7}{100 - 1.7}$$

= 0.010756 kg of vapour/kg of d.a.

Mass of water condensing out from the duct.

$M_\omega = (\omega_1 - \omega_2)$

= 20.76 – 10.756 = 10.1 g/kg = 10 g/kg

54.

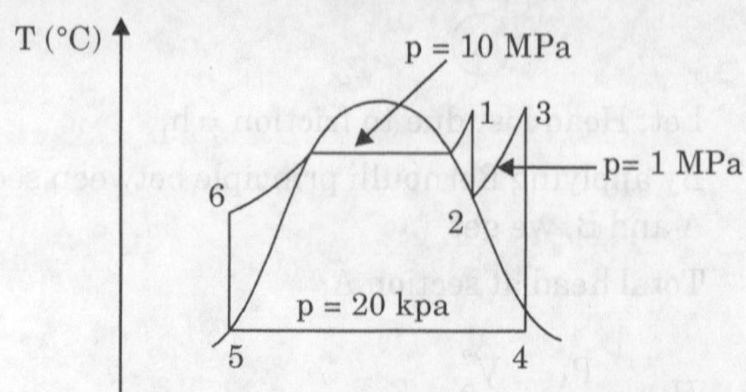

From state (1) to (2), Isentropic expansion

At state (1),

h_1 = 3373.6 kJ/kg, S_1 = 6.5965 kJ/kg-k

At state (2),

$h_2 = h_g = 2778.1$ kJ/kg, $S_2 = 6.5965$ kJ/kg-k

At state (3),

$h_3 = 3478.4$ kJ/kg, $S_4 = 7.7621$ kJ/-kg-k

At state (4),

$h_4 = h_f + xh_{fg}$, $S_3 = S_4$ (Isentropic expansion)

$S_3 = 7.7621 = 0.8319 + x(7.9085 - 0.8319)$

$x = 0.979$

$h_4 = 251.38 + 0.979\,(2609.7 - 251.38)$

$= 2560.91$ kJ/kg

Work done by H.P & L.P. Turbines

$= W_T = W_{1-2} + W_{3-4}$

$= (h_1 - h_2) + (h_3 - h_4)$

$= (3373.6 - 2778.1) + (3478.4 - 2560.91)$

$= 1513$ kJ/kg

55.

$$\eta = 1 - \frac{1}{(r)^{\gamma-1}}\left(\frac{r_c^{y} - 1}{r_c - 1)\gamma}\right)$$

$$\eta_{old}\big|_{CR=r=15} = 1 - \frac{1}{(15)^{1.3-1}}\left(\frac{2^{1.3} - 1}{(2 - 1)1.3}\right)$$

$$= 0.5 \text{ or } 50\%$$

$$\eta_{New}\big|_{CR=R=21} = 1 - \frac{1}{(21)^{1.3-1}}\left(\frac{2^{1.3} - 1}{(2 - 1)1.3}\right)$$

$$= 0.5487 \text{ or } 54.87\%$$

$$\left(\eta_{new}\big|_{CR=21}\right) - \left(\eta_{old}\big|_{CR=15}\right) = 54.87 - 50$$

$$= 4.87 \approx 4.9\%$$

Unit - I

Engineering Mathematics

Syllabus

Linear Algebra: Matrix algebra, systems of linear equations, eigen values and eigen vectors.

Calculus: Functions of single variable, limit, continuity and differentiability, mean value theorems, indeterminate forms; evaluation of definite and improper integrals; double and triple integrals; partial derivatives, total derivative, Taylor series (in one and two variables), maxima and minima, Fourier series; gradient, divergence and curl, vector identities, directional derivatives, line, surface and volume integrals, applications of Gauss, Stokes and Green's theorems.

Differential equations: First order equations (linear and nonlinear); higher order linear differential equations with constant coefficients; Euler-Cauchy equation; initial and boundary value problems; Laplace transforms; solutions of heat, wave and Laplace's equations.

Complex variables: Analytic functions; Cauchy-Riemann equations; Cauchy's integral theorem and integral formula; Taylor and Laurent series.

Probability and Statistics: Definitions of probability, sampling theorems, conditional probability; mean, median, mode and standard deviation; random variables, binomial, Poisson and normal distributions.

Numerical Methods: Numerical solutions of linear and non-linear algebraic equations; integration by trapezoidal and Simpson's rules; single and multi-step methods for differential equations.

Contents

1 CHAPTER

Linear Algebra

Analysis of Previous GATE Papers

Topic	Year/Set →Marks Distribution ↓		2019 Set 1	2019 Set 2	2018 Set 1	2018 Set 2	2017 Set 1	2017 Set 2	2016 Set 1	2016 Set 2	2016 Set 3	2015 Set 1	2015 Set 2	2015 Set 3
Algebra of Matrices	1 Mark	MCQ Type	1								1			
		Numerical Type												
	2 Marks	MCQ Type					1							
		Numerical Type												
	Total		1				2				1			
Determinant and Inverse	1 Mark	MCQ Type										1		1
		Numerical Type				1								
	2 Marks	MCQ Type												
		Numerical Type					1							
	Total					1	2					1		1
Rank and System of Linear Equations	1 Mark	MCQ Type			1				1					
		Numerical Type												
	2 Marks	MCQ Type	1											
		Numerical Type												
	Total		2		1				1					
Eigen Values and Eigen Vectors	1 Mark	MCQ Type		1			1			1		1		
		Numerical Type												
	2 Marks	MCQ Type												1
		Numerical Type						1			1			
	Total			1			1	2		1	2	1		2

Algebra of Matrices

1. Match the items in columns I and II

Column I

P. Singular matrix
Q. Non-square matrix
R. Real symmetric
S. Orthogonal matrix

Column II

1. Determinant is not defined
2. Determinant is always one
3. Determinant is zero
4. Eigenvalues are always real
5. Eigenvalues are not defined

(a) P–3, Q–1, R–4, S–2
(b) P–2, Q–3, R–4, S–1
(c) P–3, Q–2, R–5, S–4
(d) P–3, Q–4, R–2, S–1 **[2006 : 2 Marks]**

2. Multiplication of matrices E and F is G. Matrices E and G are

$$E = \begin{bmatrix} \cos\theta & -\sin\theta & 0 \\ \sin\theta & \cos\theta & 0 \\ 0 & 0 & 1 \end{bmatrix} \text{ and } G = \begin{bmatrix} 1 & 0 & 0 \\ 0 & 1 & 0 \\ 0 & 0 & 1 \end{bmatrix}$$

What is the matrix F?

(a) $\begin{bmatrix} \cos\theta & -\sin\theta & 0 \\ \sin\theta & \cos\theta & 0 \\ 0 & 0 & 1 \end{bmatrix}$

(b) $\begin{bmatrix} \cos\theta & \cos\theta & 0 \\ -\cos\theta & \sin\theta & 0 \\ 0 & 0 & 1 \end{bmatrix}$

(c) $\begin{bmatrix} \cos\theta & \sin\theta & 0 \\ -\sin\theta & \cos\theta & 0 \\ 0 & 0 & 1 \end{bmatrix}$

(d) $\begin{bmatrix} \sin\theta & -\cos\theta & 0 \\ \cos\theta & \sin\theta & 0 \\ 0 & 0 & 1 \end{bmatrix}$ **[2006 : 2 Marks]**

3. For a matrix $[M] = \begin{bmatrix} \frac{3}{5} & \frac{4}{5} \\ x & \frac{3}{5} \end{bmatrix}$, the transpose of the matrix is equal to the inverse of the matrix, $[M]^T = [M]^{-1}$. The value of x is given by

(a) $-\frac{4}{5}$ (b) $-\frac{3}{5}$

(c) $\frac{3}{5}$ (d) $\frac{4}{5}$ **[2009 : 1 Mark]**

4. Which one of the following equations is a correct identity for arbitrary 3 × 3 real matrices P, Q and R?

(a) $P(Q + R) = PQ + RP$
(b) $(P - Q)^2 = P^2 - 2PQ + Q^2$
(c) $\det(P + Q) = \det P + \det Q$
(d) $(P + Q)^2 = P^2 + PQ + QP + Q^2$

[2014 : 1 Mark, Set-4]

5. A real square matrix A is called skew-symmetric if

(a) $A^T = A$ (b) $A^T = A^{-1}$
(c) $A^T = -A$ (d) $A^T = A + A^{-1}$

[2016 : 1 Mark, Set-3]

6. Consider the matrix $P = \begin{bmatrix} \frac{1}{\sqrt{2}} & 0 & \frac{1}{\sqrt{2}} \\ 0 & 1 & 0 \\ \frac{-1}{\sqrt{2}} & 0 & \frac{1}{\sqrt{2}} \end{bmatrix}$

Which one of the following statements about P is INCORRECT?

(a) Determinant of P is equal to 1.
(b) P is orthogonal.
(c) Inverse of P is equal to its transpose.
(d) All eigenvalues of P are real numbers.

[2017 : 2 Marks, Set-1]

7. Consider the matrix :

$$P = \begin{bmatrix} 1 & 1 & 0 \\ 0 & 1 & 1 \\ 0 & 0 & 1 \end{bmatrix}$$

The number of distinct eigenvalues of P is :

(a) 3 (b) 0
(c) 1 (d) 2

[2019 : 1 Mark, Set 1]

Determinant and Inverse

8. For which value of x will the matrix given below become singular?

$$\begin{bmatrix} 8 & x & 0 \\ 4 & 0 & 2 \\ 12 & 6 & 0 \end{bmatrix}$$

(a) 4 (b) 6
(c) 8 (d) 12

[1997 : 2 Marks]

9. Given that the determinant of the matrix $\begin{bmatrix} 1 & 3 & 0 \\ 2 & 6 & 4 \\ -1 & 0 & 2 \end{bmatrix}$ is –12, the determinant of the matrix $\begin{bmatrix} 2 & 6 & 0 \\ 4 & 12 & 8 \\ -2 & 0 & 4 \end{bmatrix}$ is

(a) –96 (b) –24

(c) 24 (d) 96

[2014 : 1 Mark, Set-1]

10. The matrix form of the linear system $\frac{dx}{dt} = 3x - 5y$ and $\frac{dy}{dt} = 4x + 8y$ is

(a) $\frac{d}{dt}\begin{Bmatrix} x \\ y \end{Bmatrix} = \begin{bmatrix} 3 & -5 \\ 4 & 8 \end{bmatrix}\begin{Bmatrix} x \\ y \end{Bmatrix}$

(b) $\frac{d}{dt}\begin{Bmatrix} x \\ y \end{Bmatrix} = \begin{bmatrix} 3 & 8 \\ 4 & -5 \end{bmatrix}\begin{Bmatrix} x \\ y \end{Bmatrix}$

(c) $\frac{d}{dt}\begin{Bmatrix} x \\ y \end{Bmatrix} = \begin{bmatrix} 4 & -5 \\ 3 & 8 \end{bmatrix}\begin{Bmatrix} x \\ y \end{Bmatrix}$

(d) $\frac{d}{dt}\begin{Bmatrix} x \\ y \end{Bmatrix} = \begin{bmatrix} 4 & 8 \\ 3 & -5 \end{bmatrix}\begin{Bmatrix} x \\ y \end{Bmatrix}$

[2014 : 1 Mark, Set-1]

11. If any two columns of a determinant $P = \begin{vmatrix} 4 & 7 & 8 \\ 3 & 1 & 5 \\ 9 & 6 & 2 \end{vmatrix}$ are interchanged, which one of the following statements regarding the value of the determinant is CORRECT?

(a) Absolute value remains unchanged but sign will change

(b) Both absolute value and sign will change

(c) Absolute value will change but sign will not

(d) Both absolute value and sign will remain unchanged

[2015 : 1 Mark, Set-1]

12. For given matrix $P = \begin{bmatrix} 4+3i & -i \\ i & 4-3i \end{bmatrix}$ where $i = \sqrt{-1}$, the inverse of matrix P is

(a) $\frac{1}{24}\begin{bmatrix} 4-3i & i \\ -i & 4+3i \end{bmatrix}$ (b) $\frac{1}{25}\begin{bmatrix} i & 4-i \\ 4+3i & -i \end{bmatrix}$

(c) $\frac{1}{24}\begin{bmatrix} 4+3i & -i \\ i & 4-3i \end{bmatrix}$ (d) $\frac{1}{25}\begin{bmatrix} 4+3i & -i \\ i & 4-3i \end{bmatrix}$

[2015 : 2 Marks, Set-3]

13. The determinant of a 2 × 2 matrix is 50. If one eigenvalue of the matrix is 10, the other eigen value is ________.

[2017 : 2 Marks, Set-1]

14. If $A = \begin{bmatrix} 1 & 2 & 3 \\ 0 & 4 & 5 \\ 0 & 0 & 1 \end{bmatrix}$ then $\det(A^{-1})$ is ________.

(correct to two decimal places).

[2018 : 1 Mark, Set-2]

Rank and System of Linear Equations

15. Consider the system of simultaneous equations

$x + 2y + z = 6$

$2x + y + 2z = 6$

$x + y + z = 5$

This system has

(a) unique solution

(b) infinite number of solutions

(c) no solution

(d) exactly two solutions **[2003 : 2 Marks]**

16. A is a 3 × 4 real matrix and Ax = b is an inconsistent system of equations. The highest possible rank of A is

(a) 1 (b) 2

(c) 3 (d) 4 **[2005 : 1 Mark]**

17. With a 1 unit change in b, what is the change in x in the solution of the system of equations $x + y = 2$, $1.01x + 0.99y = b$?

(a) zero (b) 2 units

(c) 50 units (d) 100 units

[2005 : 2 Marks]

18. For what value of a, if any, will the following system of equation in x, y and z have a solution?

$2x + 3y = 4$

$x + y + z = 4$

$x + 2y - z = a$

(a) any real number (b) 0

(c) 1 (d) There is no such value

[2008 : 2 Marks]

19. Consider the following system of equations

$2x_1 + x_2 + x_3 = 0$

$x_2 - x_3 = 0$

$x_1 + x_2 = 0$

this system has

(a) a unique solution

(b) no solution

(c) infinite number of solutions

(d) five solutions **[2011 : 2 Marks]**

20. $x + 2y + z = 4$

$2x + y + 2z = 5$

$x - y + z = 1$

The system of algebraic given below has

(a) A unique solution of $x = 1, y = 1$ and $z = 1$

(b) only the two solutions of $(x = 1, y = 1, z = 1)$ and $(x = 2, y = 1, z = 0)$

(c) infinite number of solutions

(d) No feasible solution

[2012 : 2 Marks]

21. The solution to the system of equations is $\begin{bmatrix} 2 & 5 \\ -4 & 3 \end{bmatrix}\begin{bmatrix} x \\ y \end{bmatrix} = \begin{bmatrix} 2 \\ -30 \end{bmatrix}$

(a) 6, 2 (b) –6, 2

(c) –6, –2 (d) 6, –2

[2016 : 1 Mark, Set-1]

22. The rank of the matrix $\begin{bmatrix} -4 & 1 & -1 \\ -1 & -1 & -1 \\ 7 & -3 & 1 \end{bmatrix}$ is

(a) 1 (b) 2

(c) 3 (d) 4

[2018 : 1 Mark, Set-1]

23. The set of equations

$x + y + z = 1$

$a - ay + 3z = 5$

$5x - 3y + az = 6$

has infinite solution, if a =

(a) –4 (b) 3

(c) 4 (d) –3

[2019 : 2 Mark, Set-I]

Eigen Values and Eigen Vectors

24. The sum of the eigen values of the matrix given below is

$$\begin{bmatrix} 1 & 2 & 3 \\ 1 & 5 & 1 \\ 3 & 1 & 1 \end{bmatrix}$$

(a) 5 (b) 7

(c) 9 (d) 18

[1989 : 2 Marks]

25. For the matrix $\begin{bmatrix} 4 & 1 \\ 1 & 4 \end{bmatrix}$ the eigen values are

(a) 3 and – 3 (b) –3 and –5

(c) 3 and 5 (d) 5 and 0

[2003 : 1 Mark]

26. Which one of the following is an eigen vector of the matrix

$$\begin{bmatrix} 5 & 0 & 0 & 0 \\ 0 & 5 & 5 & 0 \\ 0 & 0 & 2 & 1 \\ 0 & 0 & 3 & 1 \end{bmatrix}$$

(a) $\begin{bmatrix} 1 \\ -2 \\ 0 \\ 0 \end{bmatrix}$ (b) $\begin{bmatrix} 0 \\ 0 \\ 1 \\ 0 \end{bmatrix}$

(c) $\begin{bmatrix} 1 \\ 0 \\ 0 \\ -2 \end{bmatrix}$ (d) $\begin{bmatrix} 1 \\ -1 \\ 2 \\ 1 \end{bmatrix}$

[2005 : 2 Marks]

27. Eigenvalues of a matrix $S = \begin{bmatrix} 3 & 2 \\ 2 & 3 \end{bmatrix}$ are 5 and 1. What are the eigenvalues of the matrix $S^2 = SS$?

(a) 1 and 25 (b) 6 and 4

(c) 5 and 1 (d) 2 and 10

[2006 : 2 Marks]

28. If a square matrix A is real and symmetric, then the eigenvalues

(a) are always real

(b) are always real and positive

(c) are always real and non-negative

(d) occur in complex conjugate pairs

[2007 : 1 Mark]

29. The number of linearly independent eigenvectors or $\begin{bmatrix} 2 & 1 \\ 0 & 2 \end{bmatrix}$ is

(a) 0 (b) 1

(c) 2 (d) infinite

[2007 : 2 Marks]

30. The matrix $\begin{bmatrix} 1 & 2 & 4 \\ 3 & 0 & 6 \\ 1 & 1 & p \end{bmatrix}$ has one eigenvalue equal to 3. The sum of the other two eigenvalues is

(a) p (b) p – 1

(c) p – 2 (d) p – 3

[2008 : 1 Mark]

31. The eigenvectors of the matrix $\begin{bmatrix} 1 & 2 \\ 0 & 2 \end{bmatrix}$ are written in the form $\begin{bmatrix} 1 \\ a \end{bmatrix}$ and $\begin{bmatrix} 1 \\ b \end{bmatrix}$. What is a + b?

(a) 0 (b) $\frac{1}{2}$

(c) 1 (d) 2

[2008 : 2 Marks]

32. One of the eigen vectors of the matrix A $= \begin{bmatrix} 2 & 2 \\ 1 & 3 \end{bmatrix}$ is

(a) $\begin{Bmatrix} 2 \\ -1 \end{Bmatrix}$ (b) $\begin{Bmatrix} 2 \\ 1 \end{Bmatrix}$

(c) $\begin{Bmatrix} 4 \\ 1 \end{Bmatrix}$ (d) $\begin{Bmatrix} 1 \\ -1 \end{Bmatrix}$

[2010 : 1 Mark]

33. Eigenvalues of a real symmetric matrix are always

(a) positive (b) negative

(c) real (d) complex

[2011 : 1 Mark]

34. For the matrix A = $\begin{bmatrix} 5 & 3 \\ 1 & 3 \end{bmatrix}$, ONE of the normalized eigen vectors is given as

(a) $\begin{pmatrix} \frac{1}{2} \\ \frac{\sqrt{3}}{2} \end{pmatrix}$ (b) $\begin{pmatrix} \frac{1}{\sqrt{2}} \\ \frac{-1}{\sqrt{2}} \end{pmatrix}$

(c) $\begin{pmatrix} \frac{3}{\sqrt{10}} \\ \frac{-1}{\sqrt{10}} \end{pmatrix}$ (d) $\begin{pmatrix} \frac{1}{\sqrt{5}} \\ \frac{2}{\sqrt{5}} \end{pmatrix}$

[2012 : 2 Marks]

35. The eigen values of a symmetric matrix are all

(a) complex with non-zero positive imaginary part

(b) complex with non-zero negative imaginary part

(c) real

(d) pure imaginary **[2013 : 1 Mark]**

36. Choose the CORRECT set of functions, which are linearly dependent.

(a) sin x, $\sin^2$ x and $\cos^2$ x

(b) cos x, sin x and tan x

(c) cos 2x, $\sin^2$ x and $\cos^2$ x

(d) cos 2x, sin x and cos x

[2013 : 1 Mark]

37. One of the eigenvectors of matrix $\begin{bmatrix} -5 & 2 \\ -9 & 6 \end{bmatrix}$ is

(a) $\begin{Bmatrix} -1 \\ 1 \end{Bmatrix}$ (b) $\begin{Bmatrix} -2 \\ 9 \end{Bmatrix}$

(c) $\begin{Bmatrix} 2 \\ -1 \end{Bmatrix}$ (d) $\begin{Bmatrix} 1 \\ 1 \end{Bmatrix}$

[2014 : 1 Mark, Set-2]

38. Consider a 3 × 3 real symmetric matrix S such that two of its eigenvalues are a $\neq 0$, b $\neq 0$ with respective eigenvectors $\begin{bmatrix} x_1 \\ x_2 \\ x_3 \end{bmatrix}, \begin{bmatrix} y_1 \\ y_2 \\ y_3 \end{bmatrix}$ if a $\neq$ b then $x_1y_1 + x_2y_2 + x_3y_3$ equals

(a) a (b) b

(c) ab (d) 0

[2014 : 1 Mark, Set-3]

39. At least one eigenvalue of a singular matrix is

(a) Positive (b) zero

(c) negative (d) imaginary

[2015 : 1 Mark, Set-1]

40. The lowest eigenvalue of the 2 × 2 matrix $\begin{bmatrix} 4 & 2 \\ 1 & 3 \end{bmatrix}$ is ______.

[2015 : 1 Mark, Set-3]

41. The condition for which the eigenvalues of the matrix A = $\begin{bmatrix} 2 & 1 \\ 1 & k \end{bmatrix}$ are positive, is

(a) k > $\frac{1}{2}$ (b) k > – 2

(c) k > 0 (d) k < – $\frac{1}{2}$

[2016 : 1 Mark, Set-2]

42. The number of linearly independent eigenvectors of matrix A = $\begin{bmatrix} 2 & 1 & 0 \\ 0 & 2 & 0 \\ 0 & 0 & 3 \end{bmatrix}$ is ____

[2016 : 2 Marks, Set-3]

43. The product of eigenvalues of the matrix P is

$$P = \begin{bmatrix} 2 & 0 & 1 \\ 4 & -3 & 3 \\ 0 & 2 & -1 \end{bmatrix}$$

(a) –6 (b) 2

(c) 6 (d) –2

[2017 : 1 Mark, Set-1]

44. Consider the matrix $A = \begin{bmatrix} 50 & 70 \\ 70 & 80 \end{bmatrix}$ whose eigenvectors corresponding to eigenvalues λ_1 and λ_2 are $x_1 = \begin{bmatrix} 70 \\ \lambda_1 - 50 \end{bmatrix}$ and $x_2 = \begin{bmatrix} \lambda_2 - 80 \\ 70 \end{bmatrix}$, respectively. The value of $x_1^T x_2$ is ________.

[2017 : 2 Marks, Set-2]

45. In matrix equation $[A]\{X\} = \{R\}$,

$$[A] = \begin{bmatrix} 4 & 8 & 4 \\ 8 & 16 & -4 \\ 4 & -4 & 15 \end{bmatrix} \text{ and } \{R\} = \begin{Bmatrix} 32 \\ 16 \\ 64 \end{Bmatrix}$$

One of the eigenvalues of matrix [A] is

(a) 16 (b) 8

(c) 15 (d) 4

[2019 : 1 Mark, Set-2]

ANSWERS

1. (a)	**2.** (c)	**3.** (a)	**4.** (d)	**5.** (c)	**6.** (d)	**7.** (c)	**8.** (a)	**9.** (a)	**10.** (a)
11. (a)	**12.** (a)	**15.** (c)	**16.** (b)	**17.** (c)	**18.** (b)	**19.** (c)	**20.** (c)	**21.** (d)	**22.** (b)
23. (c)	**24.** (b)	**25.** (c)	**26.** (a)	**27.** (a)	**28.** (a)	**29.** (c)	**30.** (c)	**31.** (b)	**32.** (a)
33. (c)	**34.** (b)	**35.** (c)	**36.** (c)	**37.** (d)	**38.** (d)	**39.** (b)	**41.** (a)	**43.** (b)	**45.** (a)

EXPLANATIONS

2. Given, $E = \begin{bmatrix} \cos\theta & -\sin\theta & 0 \\ \sin\theta & \cos\theta & 0 \\ 0 & 0 & 1 \end{bmatrix}$ and

$$G = \begin{bmatrix} 1 & 0 & 0 \\ 0 & 1 & 0 \\ 0 & 0 & 1 \end{bmatrix}$$

We know that product of (E × F) is unit matrix 50 F will be the inverse of E.

Now E × F = G

$\therefore$ $F = E^{-1}$

3. $A^T = A^{-1}$

so, orthogonal matrix

Now $AA^{-1} = I = AA^T$

$$\therefore \begin{bmatrix} \frac{3}{5} & \frac{4}{5} \\ x & \frac{3}{5} \end{bmatrix} \begin{bmatrix} \frac{3}{5} & x \\ \frac{4}{5} & \frac{3}{5} \end{bmatrix} = \begin{bmatrix} 1 & 0 \\ 0 & 1 \end{bmatrix}$$

or $\frac{3}{5}x + \frac{3}{5} \times \frac{4}{5} = 0$

or $\frac{3}{5}x = -\frac{12}{25}$

or $x = -\frac{4}{5}$

4. This is a standard relation.

$(P + Q)^2 = P^2 + PQ + QP + Q^2$

5. A is skew-symmetric

$\therefore A^T = -A$

6. The eigen values of an orthogonal matrix are of magnitude 1 and are real (s) complex conjugate pairs.

7. $$|P - \lambda I| = \begin{vmatrix} 1-\lambda & 1 & 0 \\ 0 & 1-\lambda & 1 \\ 0 & 0 & 1-\lambda \end{vmatrix} = 0$$

$$(1-\lambda)\left[(1-\lambda)(1-\lambda) - 0\right] = 0$$

Eigen values are,

$\lambda = 1,1,1$

So, No. of distinct eigen value = 1

8. For singular matrix, $|A| = 0$

$$\begin{vmatrix} 8 & x & 0 \\ 4 & 0 & 2 \\ 12 & 6 & 0 \end{vmatrix} = 0$$

$8(-2) - x(-24) = 0$

$-96 + 24x = 0$

$x = 4$

9. $\begin{vmatrix} 2 & 6 & 0 \\ 4 & 12 & 8 \\ -2 & 0 & 4 \end{vmatrix} = (2)^3 \begin{vmatrix} 1 & 3 & 0 \\ 2 & 6 & 4 \\ -1 & 0 & 2 \end{vmatrix} = 8 \times (-12) = -96$

10. Given that $\dfrac{dx}{dt} = 3x - 5y$

$\dfrac{dy}{dt} = 4x + 8y$

Matrix term $\dfrac{d}{dt}\begin{Bmatrix} x \\ y \end{Bmatrix} = \begin{bmatrix} 3 & -5 \\ 4 & 8 \end{bmatrix}\begin{bmatrix} x \\ y \end{bmatrix}$

11. This is a property of the determinant.

12. $P = \begin{bmatrix} 4+3i & -i \\ i & 4-3i \end{bmatrix}$

$P^{-1} = \dfrac{\begin{bmatrix} 4-3i & -(-i) \\ -i & 4+3i \end{bmatrix}}{|A|} = \dfrac{1}{24}\begin{bmatrix} 4-3i & i \\ -i & 4+3i \end{bmatrix}$

13. The product of eigen value of any matrix is equal to the determinant value of the matrix

$\lambda_1 = 10$ λ_2 = unknown

$|A| = 50$

$\lambda_1 . \lambda_2 = 50$

$10 (\lambda_2) = 50$

$\therefore \lambda_2 = 5$

14. $A = \begin{bmatrix} 1 & 2 & 3 \\ 0 & 4 & 5 \\ 0 & 0 & 1 \end{bmatrix}$

$|A| = 4$

$|A^{-1}| = \dfrac{1}{|A|} = \dfrac{1}{4} = 0.25$

15. Given equations are

$x + 2y + z = 6$

$2x + y + 2z = 6$

$x + y + z = 5$

Given system can be written as

$\begin{bmatrix} 1 & 2 & 1 \\ 2 & 1 & 2 \\ 1 & 1 & 1 \end{bmatrix} \cdot \begin{bmatrix} x \\ y \\ z \end{bmatrix} = \begin{bmatrix} 6 \\ 6 \\ 5 \end{bmatrix}$

Applying row operation

$R_2 \to R_2 - 2R_1, R_3 \to R_3 - R_1$

$\begin{bmatrix} 1 & 2 & 1 \\ 0 & -3 & 0 \\ 0 & -1 & 0 \end{bmatrix} \cdot \begin{bmatrix} x \\ y \\ z \end{bmatrix} = \begin{bmatrix} 6 \\ -6 \\ -1 \end{bmatrix}$

and applying $R_3 \to 3R_3 - R_1$

$\begin{bmatrix} 1 & 2 & 1 \\ 0 & -3 & 0 \\ 0 & 0 & 0 \end{bmatrix} \cdot \begin{bmatrix} x \\ y \\ z \end{bmatrix} = \begin{bmatrix} 6 \\ -6 \\ 3 \end{bmatrix}$

Since the rank of co-efficient matrix is 2 and rank of argument matrix is 3, which is not equal. Hence system has no solution.

16. $C = [A : B]_{3 \times 5}$

$\therefore \rho\, [C_{3 \times 5}] \le \min\, [3, 5\}$

Since the system is inconsistent

$\rho\, (A) < \rho\, (C)$

$\therefore \rho\, (A) < 3$

Hence, the maximum possible rank of A = 2

17. Given $x + y = 2$ ...(i)

$1.01\, x + 0.99\, y = b$... (ii)

Multiply 0.99 with equation (i), and substract from equation (ii); we get

$(1.01 - 0.99)\, x = 1.98$

$\therefore 0.02\, \Delta b$

$\therefore \Delta x = \dfrac{1}{0.02} = 50$ units

18. Augmented matrix is

$\left[\begin{array}{ccc|c} 2 & 3 & 0 & 4 \\ 1 & 1 & 1 & 4 \\ 1 & 2 & -1 & a \end{array}\right]$

$R_1 \to R_1 - 2R_2$

$R_3 \to R_3 - R_2$

$\left[\begin{array}{ccc|c} 0 & 1 & -2 & -4 \\ 1 & 1 & 1 & 4 \\ 0 & 1 & -2 & a-4 \end{array}\right]$

$R_1 \to R_1 - R_3$

$\left[\begin{array}{ccc|c} 0 & 0 & 0 & -a \\ 1 & 1 & 1 & 4 \\ 0 & 1 & -2 & a-4 \end{array}\right]$

will have solution if a = 0

as Rank of [A] = Rank of Augmented $[\hat{A}]$

19. $[A] = \begin{bmatrix} 2 & 1 & 1 \\ 0 & 1 & -1 \\ 1 & 1 & 0 \end{bmatrix}$

$[A/B] = \left[\begin{array}{ccc|c} 2 & 1 & 1 & 0 \\ 0 & 1 & -1 & 0 \\ 1 & 1 & 0 & 0 \end{array}\right]$

Rank of [A] = 2

Rank of $[A/B] = 2$

Since Rank of $[A]$ = Rank of $[A/B] < 3$

So infinite number of solutions are obtained.

20. $\rho[A] = [A/B] = 2$

But $\rho[A] = \rho[A/B] <$ Number of variables.

$\rho < n$

So we have infinite number of solutions.

21. Equations from this matrix are:

$2x + 5y = 2$ and $-4x + 3y = -30$

By solving these two equations, we get

$x = 6; y = -2$

22. Let $A = \begin{bmatrix} -4 & 1 & -1 \\ -1 & -1 & -1 \\ 7 & -3 & 1 \end{bmatrix}$

Applying $R_1 \longleftrightarrow R_2$ $\begin{bmatrix} -1 & -1 & -1 \\ -4 & 1 & -1 \\ 7 & -3 & 1 \end{bmatrix}$

Applying $R_2 - 4R_1$, $R_3 + 7R_1$ $\begin{bmatrix} -1 & -1 & -1 \\ 0 & 5 & 3 \\ 0 & -10 & -6 \end{bmatrix}$

Now applying $R_3 + 2R_2$ $\begin{bmatrix} -1 & -1 & -1 \\ 0 & 5 & 3 \\ 0 & 0 & 0 \end{bmatrix}$

$\therefore$ No. of non zero rows = 2

rank = 2

23. Given set of equations are,

$x + y + z = 1$, $a - ay + 32 = 5$ and $5x - 3y + az = 6$

Now $\rho(A) = \rho(AB) <$ no. of variables

$|A| = 0$

$$\begin{vmatrix} 1 & 1 & 1 \\ a & -a & 3 \\ 5 & -3 & a \end{vmatrix} = 0$$

$(-a^2 + 9) - (a^2 - 15) + (-3a + 5a) = 0$

$9 - a^2 + 15 - a^2 + 2a = 0$

$2a^2 - 2a - 24 = 0$

$a^2 - a - 12 = 0$

$a^2 - 4a + 3a - 12 = 0$

$(a = 4, -3)$

$\rho(AB) <$ (No. of variable)

$$\begin{vmatrix} 1 & 1 & 1 \\ -a & 3 & 5 \\ -3 & a & 6 \end{vmatrix} = 0$$

$1(18 - 5a) - 1(-6a + 15) + 1(-a^2 + 9) = 0$

$18 - 5a - 15 + 6a + 9 - a^2 = 0$

$a^2 - a - 12 = 0$

$a = 3, 4$ or $a = 4$

$$\begin{vmatrix} 1 & 1 & 1 \\ a & -a & 5 \\ 5 & -3 & 6 \end{vmatrix} = 0$$

$(-6a + 15) - (6a - 25) + (-3a + 5a) = 0$

$-6a + 15 - 6a + 25 + 2a = 0$

$10a = 40$

$a = 4$

if $a = 4$

$|AB| = 0$

$$\begin{vmatrix} 1 & 1 & 1 \\ a & 3 & 5 \\ 5 & a & 6 \end{vmatrix} = 0$$

$(18 - 5a) - (6a - 25) + (a^2 - 15) = 0$

$a^2 - 11a + 28 = 0$

$a = 7, 4$ or $\boxed{a = 4}$

24. Sum of eigen values is equal to the sum of diagonal elements of the matirix.

$\therefore$ Sum of eigen values = 1 + 5 + 1 = 7

25. $A = \begin{bmatrix} 4 & 1 \\ 1 & 4 \end{bmatrix}$

Now $A - \lambda I = 0$, where λ = eigen value

$\therefore \begin{bmatrix} 4-\lambda & 1 \\ 1 & 4-\lambda \end{bmatrix} = 0$

$\Rightarrow (4 - \lambda)^2 - 1 = 0$

$\Rightarrow (4 - \lambda)^2 - (1)^2 = 0$

$\Rightarrow (4 - \lambda - 1)(4 - \lambda + 1) = 0$

$\therefore \lambda = 3, \quad \lambda = 5$

Also "Summation of Eigen values is equal to the sum of diagonal elements of the MATRIX"

26. Eigen vector is independent vector. Since last two terms are zero, hence only option (a) is correct.

27. For S matrix, if eigen values are $\lambda_1, \lambda_2, \lambda_3, \ldots$ then for S^2 matrix, the eigen values will be $\lambda_1^2, \lambda_2^2, \lambda_3^2, \ldots$

For S matrix, if eigen values are 1 and 5 then for S^2 matrix, the eigen values are 1 and 25

28. $(A - \lambda I) = 0$

$A = \begin{bmatrix} a & b \\ b & a \end{bmatrix}$

Now $\begin{vmatrix} a-\lambda & b \\ b & a-\lambda \end{vmatrix} = 0$

which gives $(a-\lambda)^2 - b^2 = 0$

$\Rightarrow a - \lambda = \pm b$

$\Rightarrow \lambda = a \pm b$

Hence eigen values of the matrix are real.

29. $A = \begin{bmatrix} 2 & 1 \\ 0 & 2 \end{bmatrix}$

Now, $|A - \lambda I| = 0$

$\Rightarrow \begin{vmatrix} 2-\lambda & 1 \\ 0 & 2-\lambda \end{vmatrix} = 0$

$\Rightarrow \lambda = 2$

Hence, only 1 linearly independent eigen vector exists.

30. Sum of the diagonal vector = sum of the Eigen values,

$\Rightarrow 1 + 0 + p = 3 + S$

$\Rightarrow S = p - 2$

31. Let us find the Eigen values of the matrix,

$(1-\lambda)(2-\lambda) = 0$

$\Rightarrow \lambda = 1, 2$

For $\lambda = 1$,

$\begin{bmatrix} 0 & 2 \\ 0 & 1 \end{bmatrix}_{2\times 2} \begin{bmatrix} 1 \\ a \end{bmatrix}_{2\times 1} = \begin{bmatrix} 0 \\ 0 \end{bmatrix}$

$\Rightarrow a = 0$

$\Rightarrow$ For $\lambda = 2$, $\begin{bmatrix} -1 & 2 \\ 0 & 0 \end{bmatrix}\begin{bmatrix} 1 \\ b \end{bmatrix} = \begin{bmatrix} 0 \\ 0 \end{bmatrix}$

$\therefore\ a + b = 0 + \frac{1}{2} = \frac{1}{2}$

32. Given : $A = \begin{bmatrix} 2 & 2 \\ 1 & 3 \end{bmatrix}$

Characteristic equation is

$|A - \lambda I| = 0$

$\Rightarrow \begin{bmatrix} 2-\lambda & 2 \\ 1 & 3-\lambda \end{bmatrix} = 0$

$\Rightarrow (2-\lambda)(3-\lambda) - 2 = 0$

$\Rightarrow \lambda^2 - 5 + 4 = 0$

$\Rightarrow \lambda = 1, 4$

Eigen vector for $\lambda = 1$ is

$\begin{bmatrix} 2-\lambda_1 & 2 \\ 1 & 3-\lambda_1 \end{bmatrix}\begin{bmatrix} x_1 \\ x_2 \end{bmatrix} = 0$

$\Rightarrow \begin{bmatrix} 1 & 2 \\ 1 & 2 \end{bmatrix}\begin{bmatrix} x_1 \\ x_2 \end{bmatrix} = 0$

$\Rightarrow x_1 + 2x_2 = 0$

$\Rightarrow x_2 = -\frac{1}{2}x$

i.e. $x_1 : x_2 = -2 : 1$

Since choice (a)

$\begin{bmatrix} 2 \\ -1 \end{bmatrix}$ is in same ratio of x_1 to x_2.

$\therefore$ choice (a) is an eigen vector.

33. Eigen values of a real symmetric matrix are always real.

34. $|A - \lambda I| = 0$

$\Rightarrow \begin{vmatrix} 5-\lambda & 3 \\ 1 & 3-\lambda \end{vmatrix} = 0$

$\Rightarrow (5-\lambda)(3-\lambda) = 3 = 0$

$\Rightarrow \lambda^2 - 8\lambda + 15 - 3 = 0$

$\Rightarrow \lambda^2 - 8\lambda + 12 = 0$

$\Rightarrow \lambda = 2, \lambda = 6$

Now, at $\lambda = 2$, eign vector:

$\begin{pmatrix} 3 & 3 \\ 1 & 1 \end{pmatrix}\begin{pmatrix} x_1 \\ x_2 \end{pmatrix} = \begin{pmatrix} 0 \\ 0 \end{pmatrix}$; $x_1 + x_2 = 0$; $x_1 = -x_2$

Hence required vector is $\begin{bmatrix} \frac{1}{\sqrt{2}} \\ \frac{-1}{\sqrt{2}} \end{bmatrix}$

35. Suppose the eigen value of matrix A is n

$(= \lambda + i B_B)$ say.

and the eigen vector is n whereas the conjugate pair of eigen value and eigen vector is λ and $\bar{n}$

So, $An = \lambda n$ ...(1)

And $A\bar{n} = \bar{\lambda}\bar{x}$ ...(2)

Taking transpose of equation

$x^{-T}A^T = \bar{x}^T\bar{\lambda}$ ...(3)

$\Rightarrow x^{-T}A^T x = x^{-T}\bar{\lambda}n$

$\Rightarrow x^{-T}An = x^{-T}\bar{\lambda}n$

$\Rightarrow \lambda = \bar{\lambda}$

$\Rightarrow \alpha + i\beta = \alpha - i\beta$

$\Rightarrow 2i\beta = 0$

$\Rightarrow \beta = 0$

Hence eigen value of a symmetric matrix are real

36. Going by options,

$\cos 2x = 2\cos^2 x - 1 = 2\cos^2 x - (\sin^2 x + \cos^2 x)$

$\cos 2x = \cos^2 x - \sin^2 x$

37. Eigen values of the matrix $\begin{bmatrix} -5 & 2 \\ -9 & 6 \end{bmatrix}$ are 4, – 3

$\therefore$ the eigen vector corresponding to eigen vector λ is $Ax = \lambda x$ (verify the options)

$\begin{pmatrix} 1 \\ 1 \end{pmatrix}$ is eigen vector corresponding to eigen value $\lambda = 3$

38. We know that the Eigen vectors corresponding to distinct Eigen values of real symmetric matrix are orthogonal.

$$\begin{bmatrix} x_1 \\ x_2 \\ x_3 \end{bmatrix} \begin{bmatrix} y_1 \\ y_2 \\ y_3 \end{bmatrix} = x_1 y_1 + x_2 y_2 + x_3 y_3 = 0$$

39. Atleast one eigen value of a singular matrix is zero.

40. $\begin{vmatrix} 4-\lambda & 2 \\ 1 & 3-\lambda \end{vmatrix} = 0$

$(4-\lambda)(3-\lambda) - 2 = 0$

$\lambda^2 - 7\lambda + 12 - 2 = 0$

$\lambda^2 - 7\lambda + 10 = 0$

$\lambda^2 - 5\lambda - 2\lambda + 10 = 0$

$\lambda(\lambda - 5) - 2(\lambda - 5) = 0$

$\lambda_{12} = 2{,}5$

The lowest eigenvalve of the matrix is 2.

41. All eigen values of $A = \begin{bmatrix} 2 & 1 \\ 1 & k \end{bmatrix}$ are position

$2 > 0$

$\therefore$ 2 × 2 leading minor will be greater then zero.

$\begin{bmatrix} 2 & 1 \\ 1 & k \end{bmatrix} > 0$

$2k - 1 > 0$

$2K > 1$

$K > \frac{1}{2}$.

42. Consider $A - \lambda I = \begin{bmatrix} 2-\lambda & 1 & 0 \\ 0 & 2-\lambda & 0 \\ 0 & 0 & 3-\lambda \end{bmatrix}$

$\therefore$ characteristics equation is $|A - \lambda I| = 0$

$\lambda = 2, 2, 3$

There are only two linearly independent eigen vectors.

43. Product of eigen value = |p|

$$= \begin{vmatrix} 2 & 0 & 1 \\ 4 & -3 & 3 \\ 0 & 2 & -1 \end{vmatrix} = 2$$

44. $A = \begin{bmatrix} 50 & 70 \\ 70 & 80 \end{bmatrix}$

Eigen values of A are λ_1, λ_2

$\lambda_1 + \lambda_2 = 130$

$\lambda_1 \lambda_2 = -900$

Given that $X_1 = \begin{bmatrix} 70 \\ \lambda_1 - 50 \end{bmatrix}$

$X2 = \begin{bmatrix} \lambda_2 - 80 \\ 70 \end{bmatrix}$

$X_1^T X_2 = [70 \;\; \lambda_1 - 50] \begin{bmatrix} \lambda_2 - 80 \\ 70 \end{bmatrix}$

$= 70\lambda_2 - 5600 + 70\lambda_1 - 3500$

$= 70(\lambda_1 + \lambda_2) - 9100$

$= 70(130) - 9100$

$= 0$

45. In matrix equation A[X] = {R}

A[X] can also be substituted by λ[X]

Choosing from option '16' first

$$16[X] = 16 \begin{bmatrix} 2 \\ 1 \\ 4 \end{bmatrix} = \begin{bmatrix} 32 \\ 16 \\ 64 \end{bmatrix}$$

As 16 satisfies the equation, the correct option is (a).

2 CHAPTER

Calculus

Analysis of Previous GATE Papers

Topic	Year/Set →Marks Distribution ↓		2019 Set 1	2019 Set 2	2018 Set 1	2018 Set 2	2017 Set 1	2017 Set 2	2016 Set 1	2016 Set 2	2016 Set 3	2015 Set 1	2015 Set 2	2015 Set 3
Limit & Continuity	1 Mark	MCQ Type								1	1	1	1	
		Numerical Type												1
	2 Marks	MCQ Type						1			1			
		Numerical Type								1				
	Total							**2**		**3**	**3**	**1**	**1**	**1**
Differentiability and applications of derivative	1 Mark	MCQ Type												
		Numerical Type												
	2 Marks	MCQ Type												
		Numerical Type										1		
	Total											**2**		
Integral Calculus and Application	1 Mark	MCQ Type	1	1	1									
		Numerical Type												
	2 Marks	MCQ Type		1			1		1					
		Numerical Type	1									1		1
	Total		**3**	**3**	**1**		**2**		**2**			**2**		**2**
Vector Calculus and Application	1 Mark	MCQ Type				1								1
		Numerical Type						1						
	2 Marks	MCQ Type				1								
		Numerical Type					1			1	1		1	
	Total					**3**	**2**	**1**		**2**	**2**		**2**	**1**

Limit and Continuity

1. $\underset{x\to 0}{\text{Lt}} \dfrac{\sin^2 x}{x}$ is equal to

(a) 0 (b) ∞
(c) 1 (d) –1

[2003 : 1 Mark]

2. If $f(x) = \dfrac{2x^2 - 7x + 3}{5x^2 - 12x - 9}$ then $\lim_{x\to 3} f(x)$ will be

(a) $-\dfrac{1}{3}$ (b) $\dfrac{5}{18}$

(c) 0 (d) $\dfrac{2}{5}$

[2006 : 2 Marks]

3. The minimum value of function $y = x^2$ in the interval [1, 5] is

(a) 0 (b) 1
(c) 25 (d) undefined

[2007 : 1 Mark]

4. $\lim_{x\to 0} \dfrac{e^x - \left(1 + x + \dfrac{x^2}{2}\right)}{x^3} =$

(a) 0 (b) $\dfrac{1}{6}$

(c) $\dfrac{1}{3}$ (d) 1

[2007 : 2 Marks]

5. The value of $\lim_{x\to 8} \dfrac{x^{1/3} - 2}{(x - 8)}$

(a) $\dfrac{1}{16}$ (b) $\dfrac{1}{12}$

(c) $\dfrac{1}{8}$ (d) $\dfrac{1}{4}$

[2008 : 1 Mark]

6. The function $y = |2 - 3x|$

(a) is continuous $\forall x \in R$ and differentiable $\forall x \in R$

(b) is continuous $\forall x \in R$ and differentiable $\forall x \in R$ except at $x = 3/2$

(c) is continuous $\forall x \in R$ and differentiable $\forall x \in R$ except at $x = 2/3$

(d) is continuous $\forall x \in R$ except $x = 3$ and differentiable $\forall x \in R$

[2010 : 1 Mark]

7. What is $\lim_{\theta\to 0} \dfrac{\sin\theta}{\theta}$ equal to?

(a) θ (b) $\sin\theta$
(c) 0 (d) 1

[2011 : 1 Mark]

8. Consider the function $f(x) = |x|$ in the interval $-1 < x \le 1$. At the point $x = 0$, $f(x)$ is

(a) continuous and differentiable
(b) non-continuous and differentiable
(c) continuous and non-differentiable
(d) neither continuous nor differentiable

[2012 : 1 Mark]

9. $\lim_{x\to 0}\left(\dfrac{1 - \cos x}{x^2}\right)$ is

(a) 1/4 (b) 1/2
(c) 1 (d) 2

[2012 : 1 Mark]

10. $\lim_{x\to 0} \dfrac{x - \sin x}{1 - \cos x}$

(a) 0 (b) 1
(c) 3 (d) not defined

[2014 : 1 Mark, Set-1]

11. $\lim_{x\to 0}\left(\dfrac{e^{2x} - 1}{\sin(4x)}\right)$ is equal to

(a) 0 (b) 0.5
(c) 1 (d) 2

[2014 : 1 Mark, Set-2]

12. If a function is continuous at a point,

(a) the limit of the function may not exist at the point.
(b) the function must be derivable at the point.
(c) the limit of the function at the point tends to infinity.
(d) the limit must exist at the point and the value of limit should be same as the value of the function at that point.

[2014 : 1 Mark, Set-3]

13. The value of $\lim_{x\to 0} \dfrac{1 - \cos(x^2)}{2x^4}$ is

(a) 0 (b) $\dfrac{1}{2}$

(c) $\dfrac{1}{4}$ (d) undefined

[2015 : 1 Mark, Set-1]

14. At x = 0, the function $f(x) = |x|$ has

(a) a minimum
(b) a maximum
(c) a point of inflection
(d) neither a maximum nor minimum

[2015 : 1 Mark, Set-2]

15. The value of $\lim_{x\to 0}\left(\dfrac{-\sin x}{2\sin x+\cos x}\right)$ is ______.

[2015 : 1 Mark, Set-3]

16. Consider the function $f(x) = 2x^3 - 3x^2$ in the domain [–1, 2]. The global minimum of f(x) is _____.

[2016 : 2 Marks, Set-1]

17. The values of x for which the function

$f(x) = \dfrac{x^2-3x-4}{x^2+3x-4}$ is NOT continuous are

(a) 4 and –1
(b) 4 and 1
(c) –4 and 1
(d) –4 and –1

[2016 : 1 Mark, Set-2]

18. $\underset{x\to 0}{\text{Lt}}\ \dfrac{\log_e(1+4x)}{e^{3x}-1}$ is equal to

(a) 0
(b) $\dfrac{1}{12}$
(c) $\dfrac{4}{3}$
(d) 1

[2016 : 1 Mark, Set-3]

19. $\lim_{x\to\infty}\sqrt{x_2+x-1}-x$ is

(a) 0
(b) ∞
(c) 1/2
(d) –∞

[2016 : 2 Marks, Set-3]

20. The value of $\lim_{x\to 0}\dfrac{x^2-\sin(x)}{x}$ is

(a) 0
(b) 3
(c) 1
(d) –1

[2017 : 2 Marks, Set-1]

Differentiability and Applications of Derivative

21. If $x = a(\theta + \sin\theta)$ and $y = a(1 - \cos\theta)$, then dy/dx will be equal to

(a) $\sin\left(\dfrac{\theta}{2}\right)$
(b) $\cos\left(\dfrac{\theta}{2}\right)$
(c) $\tan\left(\dfrac{\theta}{2}\right)$
(d) $\cot\left(\dfrac{\theta}{2}\right)$

[2004 : 1 Mark]

22. Equation of the line normal to function $f(x) = (x-8)^{2/3} + 1$ at P(0, 5) is

(a) y = 3x – 5
(b) y = 3x + 5
(c) 3y = x + 15
(d) 3y = x – 15

[2006 : 2 Marks]

23. If, $y = x + \sqrt{x+\sqrt{x+\sqrt{x+\ldots\ldots\infty}}}$, then y(2) =

(a) 4 or 1
(b) 4 only
(c) 1 only
(d) undefined

[2007 : 2 Marks]

24. The length of the curve $y = \dfrac{2}{3}x^{3/2}$ between x = 0 and x = 1 is

(a) 0.27
(b) 0.67
(c) 1
(d) 1.22

[2008 : 2 Marks]

25. Let $f = y^x$. What is $\dfrac{\partial^2 f}{\partial x\partial y}$ at x = 2, y = 1?

(a) 0
(b) ln 2
(c) 1
(d) $\dfrac{1}{\ln 2}$

[2008 : 2 Marks]

26. In the Taylor series expansion of e^x about x = 2 the coefficient of $(x-2)^4$ is

(a) 1/4!
(b) $2^4/4!$
(c) $e^2/4!$
(d) $e^4/4!$

[2008 : 1 Mark]

27. A series expansion for the function sin θ is

(a) $1-\dfrac{\theta^2}{2!}+\dfrac{\theta^4}{4!}-\ldots\ldots$

(b) $\theta-\dfrac{\theta^3}{3!}+\dfrac{\theta^5}{5}-\ldots\ldots$

(c) $1+\theta+\dfrac{\theta^2}{2!}+\dfrac{\theta^3}{3!}-\ldots\ldots$

(d) $\theta+\dfrac{\theta^3}{3!}+\dfrac{\theta^5}{5!}+\ldots\ldots$

[2011 : 1 Mark]

28. At x = 0, the function $f(x) = x^3 + 1$ has

(a) a maximum value
(b) a minium value
(c) a singularity
(d) a point of inflection

[2012 : 1 Mark]

29. The best approximation of the minimum value attained by $e^{-x}\sin(100x)$ for $x \geq 0$ is ______.

[2014 : 1 Mark, Set-2]

30. Consider a spatial curve in three-dimensional space given in parametric form by $x(t) = \cos t$, $y(t) = \sin t$, $z(t) = \frac{2}{\pi}t$, $0 \leq t \leq \frac{\pi}{2}$ The length of the curve is ______.

[2015 : 2 Marks, Set-1]

Integral Calculus and Application

31. The area enclosed between the parabola $y = x^2$ and the straight line $y = x$ is

(a) $\frac{1}{8}$ (b) $\frac{1}{6}$

(c) $\frac{1}{3}$ (d) $\frac{1}{2}$

[2003 : 2 Marks]

32. The volume of an object expressed in spherical coordinates is given by

$$V = \int_0^{2\pi}\int_0^{\pi/3}\int_0^1 r^2 \sin\phi\, dr\, d\phi\, d\theta$$

The value of the integral is

(a) $\frac{\pi}{3}$ (b) $\frac{\pi}{6}$

(c) $\frac{2\pi}{3}$ (d) $\frac{\pi}{4}$

[2004 : 2 Marks]

33. $\int_{-a}^{a}(\sin^6 x + \sin^7 x)\,dx$ is equal to

(a) $2\int_0^a \sin^6 x\,dx$

(b) $2\int_0^a \sin^7 x\,dx$

(c) $2\int_0^a (\sin^6 x + \sin^7 x)\,dx$

(d) Zero

[2005 : 1 Mark]

34. Changing the order of the integration in the double integral $I = \int_0^8\int_{x/4}^2 f(x,y)\,dy\,dx$ leads to $I = \int_r^s\int_p^q f(x,y)\,dx\,dy$. What is q?

(a) 4y (b) $16y^2$

(c) x (d) 8

[2005 : 1 Mark]

35. By a change of variable $x(u, y) = uv$, $y(u, v) = v/u$ is double integral, the integrand $f(x, y)$ changes to $f(uv, v/u)\,\phi(u, v)$. Then, $\phi(u, v)$ is

(a) $2v/u$ (b) $2uv$

(c) v^2 (d) 1

[2005 : 1 Mark]

36. The right circular cone of largest volume that can be enclosed by a sphere of 1 m radius has a height of

(a) $\frac{1}{3}$m (b) $\frac{2}{3}$m

(c) $\frac{2\sqrt{2}}{3}$m (d) $\frac{4}{3}$m

[2005 : 2 Marks]

37. Consider the shaded triangular region P shown in the figure. What is $\iint_P xy\,dx\,dy$?

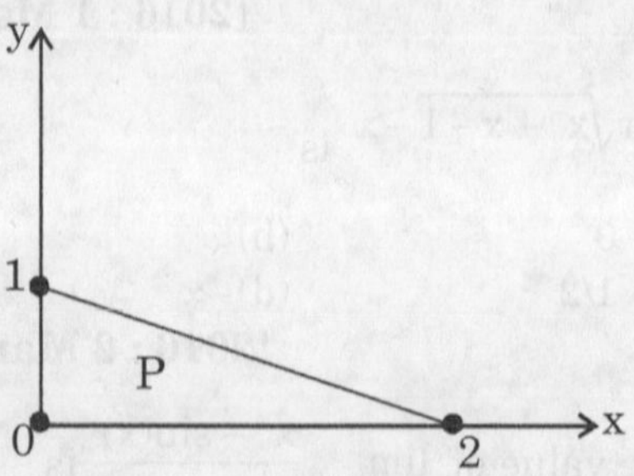

(a) $\frac{1}{6}$ (b) $\frac{2}{9}$

(c) $\frac{7}{16}$ (d) 1

[2008 : 2 Marks]

38. Which of the following intergrals is unbounded?

(a) $\int_0^{\pi/4}\tan x\,dx$ (b) $\int_0^{\infty}\frac{1}{x^2+1}dx$

(c) $\int_0^{\infty}xe^{-x}dx$ (d) $\int_0^1\frac{1}{1-x}dx$

[2008 : 1 Mark]

39. A path AB in the form of one quarter of a circle of unit radius is shown in the figure. Integration of $(x + y)^2$ on path AB traversed in a counter clockwise sense is

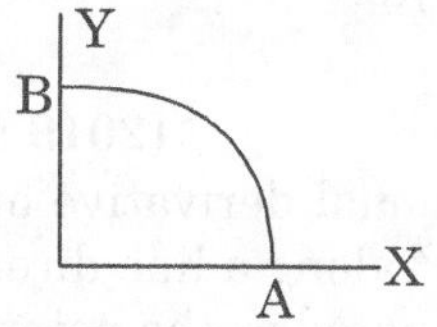

(a) $\frac{\pi}{2}-1$ (b) $\frac{\pi}{2}+1$

(c) $\frac{\pi}{2}$ (d) 1

[2009 : 2 Mark]

40. The distance between the origin and the point nearest it on the surface $z^2 = 1 + xy$ is

(a) 1 (b) $\frac{\sqrt{3}}{2}$

(c) $\sqrt{3}$ (d) 2

[2009 : 2 Marks]

41. The area enclosed between the curves $y^2 = 4x$ and $x^2 = 4y$ is

(a) $\frac{16}{3}$ (b) 8

(c) $\frac{32}{3}$ (d) 16

[2009 : 2 Marks]

42. The parabolic arc $y = \sqrt{x}$, $1 \le x \le 2$ is revolved around the x-axis. The volume of the solid of revolution is

(a) $\pi/4$ (b) $\pi/2$

(c) $3\pi/4$ (d) $3\pi/2$

[2010 : 1 Mark]

43. The value of the integral $\int_{-\infty}^{\infty}\frac{dx}{1+x^2}$ is

(a) $-\pi$ (b) $-\pi/2$

(c) $\pi/2$ (d) π

[2010 : 1 Mark]

44. If f(x) is an even function and a is a positive real number, then $\int_{-a}^{a} f(x)\,dx$

(a) 0 (b) a

(b) 2 a (d) $2\int_{0}^{a} f(x)\,dx$

[2011 : 1 Mark]

45. The area enclosed between the straight line y = x and the parabola $y = x^2$ in the x–y plane is

(a) 1/6 (b) 1/4

(c) 1/3 (d) 1/2

[2012 : 1 Mark]

46. The value of the definite integral $\int_1^e \sqrt{x}\,\text{In}(x)\,dx$ is

(a) $\frac{4}{9}\sqrt{e^3}+\frac{2}{9}$ (b) $\frac{2}{9}\sqrt{e^3}-\frac{4}{9}$

(c) $\frac{2}{9}\sqrt{e^3}+\frac{4}{9}$ (d) $\frac{4}{9}\sqrt{e^3}-\frac{2}{9}$

[2013 : 2 Marks]

47. The integral $\oint_C (ydx - xdy)$ is evaluated along the circle $x^2 + y^2 = \frac{1}{4}$ traversed in counter clockwise direction. The integral is equal to

(a) 0 (b) $-\frac{\pi}{4}$

(c) $-\frac{\pi}{2}$ (d) $\frac{\pi}{4}$

[2014 : 2 Marks, Set-1]

48. The value of the integral

$$\int_0^2 \frac{(x-1)^2\sin(x-1)}{(x-1)^2+\cos(x-1)}dx \text{ is}$$

(a) 3 (b) 0

(c) –1 (d) –2

[2014 : 1 Mark, Set-4]

49. The value of the integral $\int_0^2\int_0^x e^{x+y}\,dydx$

(a) $\frac{1}{2}(e-1)$ (b) $\frac{1}{2}(e^2-1)^2$

(c) $\frac{1}{2}(e^2-e)$ (d) $\frac{1}{2}\left(e-\frac{1}{e}\right)^2$

[2014 : 2 Marks, Set-4]

50. Consider an ant crawling along the curve $(x - 2)^2 + y^2 = 4$, where x and y are in meters. The ant starts at the point (4, 0) and moves counter-clockwise with a speed of 1.57 meters per second. The time taken by the ant to reach the point (2, 2) is (in seconds) ________.

[2015 : 2 Marks, Set-1]

51. The value of $\oint[(3x-8y^2)dx+(4y-6xy)dy]$ (where C is the boundary of the region boundary by x = 0, y = 0 and x + y = 1) is ______.

[2015 : 2 Marks, Set-3]

52. The value of the integral

$$\int_{-\infty}^{\infty}\frac{\sin x}{x^2+2x+2}dx$$

evaluated using contour integration and the residue theorem is

(a) $\frac{-\pi\sin(1)}{e}$ (b) $\frac{-\pi\cos(1)}{e}$

(c) $\frac{\sin(1)}{e}$ (d) $\frac{\cos(1)}{e}$

[2016 : 2 Marks, Set-1]

53. A parametric curve defined by $x = \cos\left(\frac{\pi u}{2}\right), y = \sin\left(\frac{\pi u}{2}\right)$ in the range $0 \le u \le 1$ is rotated about the X-axis by 360 degrees. Area of the surface generated is

(a) $\frac{\pi}{2}$ (b) π

(c) 2π (d) 4π

[2017 : 2 Marks, Set-1]

54. According to Mean Value Theorem, for a continuous function f(x) in the interval [a, b], there exists a value in this interval ξ such that

$$\int_a^b f(x)dx$$

(a) $f(\xi)(b-a)$ (b) $f(b)(\xi-a)$

(c) $f(a)(b-\xi)$ (d) 0

[2018 : 1 Mark, Set-1]

55. A parabola $x = y^2$ with $0 \le x \le 1$ is shown in the figure. The volume of the solid of rotation obtained by rotating the shaded area by 360° around the x-axis is

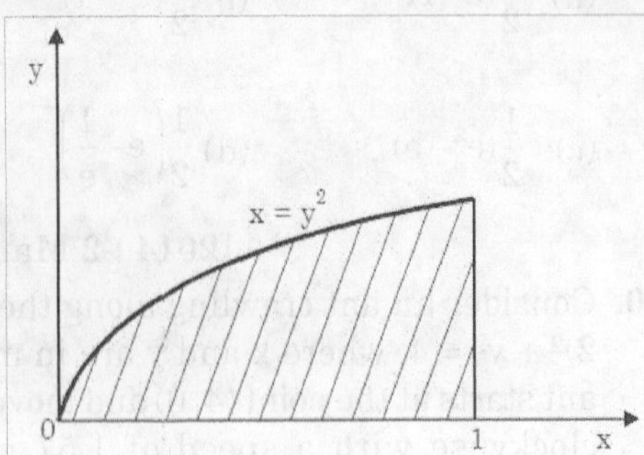

(a) 2π (b) $\frac{\pi}{2}$

(c) $\frac{\pi}{4}$ (d) π

[2019 : 1 Mark, Set-1]

56. The value of the following definite integral is ______ (round off to there decimal places).

$$\int_1^e (x\ln x)dx$$

[2019 : 2 Marks, Set-1]

57. The directional derivative of the function $f(x, y) = x^2 + y^2$ along a line directed from (0, 0) to (1, 1), evaluated at the point x = 1, y = 1 is

(a) 2 (b) $4\sqrt{2}$

(c) $2\sqrt{2}$ (d) $\sqrt{2}$

[2019 : 1 Mark, Set-2]

58. Given a vector $\vec{u} = \frac{1}{3}\left(-y^3\hat{i}+x^3\hat{j}+z^3\hat{k}\right)$ and $\hat{n}$ as the unit normal vector to the surface of the hemisphere ($x^2 + y^2 + z^2 = 1; z \ge 0$), the value of integral $\int(\nabla\times\vec{u})\cdot\hat{n}\ dS$ evaluated on the curved surface of the hemisphere S is

(a) $-\frac{\pi}{2}$ (b) π

(c) $\frac{\pi}{2}$ (d) $\frac{\pi}{3}$

[2019 : 2 Marks, Set-2]

Vector Calculus and Application

59. The vector field $\vec{F} = x\hat{i} - y\hat{j}$ (where $\hat{i}$ and $\hat{j}$ are unit vector) is

(a) divergence free, but not irrotational

(b) irrotational, but not divergence free

(c) divergence free and irrotational

(d) neither divergence free nor irrotational

[2003 : 2 Marks]

60. The angle between two unit-magnitude coplanar vectors P(0.866, 0.500, 0) and Q(0.259, 0.966, 0) will be

(a) 0° (b) 30°

(c) 45° (d) 60°

[2004 : 1 Mark]

61. Stokes theorem connects

(a) a line integral and a surface integral

(b) a surface integral and a volume integral

(c) a line integral and a volume integral

(d) gradient of a function and its surface integral

[2005 : 1 Mark]

62. The line integral $\int \vec{V}.\vec{dr}$ of the vector $\vec{V}(\vec{r}) = 2xyz\hat{i} + x^2z\hat{i} + x^2z\hat{j} + xy\hat{k}$ from the origin to the point P(1, 1, 1)

(a) is 1
(b) is zero
(c) is –1
(d) cannot be determined without specifying the path

[2005 : 2 Mark]

63. The area of a triangle formed by the tips of vectors $\bar{a}, \bar{b}$ and $\bar{c}$ is

(a) $\frac{1}{2}(\bar{a}-\bar{b}).(\bar{a}-\bar{c})$ (b) $\frac{1}{2}|(\bar{a}-\bar{b})\times(\bar{a}-\bar{c})|$

(c) $\frac{1}{2}|\bar{a}\times\bar{b}\times\bar{c}|$ (d) $\frac{1}{2}(\bar{a}\times\bar{b}).\,\bar{c}$

[2007 : 2 Marks]

64. The divergence of the vector field

$(x-y)\hat{i}+(y-x)\hat{j}+(x+y+z)\hat{k}$ is

(a) 0 (b) 1
(c) 2 (d) 3

[2008 : 1 Mark]

65. The directional derivative of the scalar function $f(x, y, z) = x^2 + 2y^2 + z$ at the point P = (1, 1, 2) in the direction of the vector $a = 3\hat{i} - 4\hat{j}$ is

(a) –4 (b) –2
(c) –1 (d) 1

[2008 : 2 Marks]

66. The divergence of the vector field $3xz\hat{i} + 2xy\hat{j} - yz^2\hat{k}$ at a point (1, 1, 1) is equal to

(a) 7 (b) 4
(c) 3 (d) 0

[2009 : 1 Mark]

67. For the spherical surface $x^2 + y^2 + z^2 = 1$, the unit outward normal vector at the point $\left(\frac{1}{\sqrt{2}}, \frac{1}{\sqrt{2}}, 0\right)$ is given by

(a) $\frac{1}{\sqrt{2}}\hat{i}+\frac{1}{\sqrt{2}}\hat{j}$ (b) $\frac{1}{\sqrt{2}}\hat{i}-\frac{1}{\sqrt{2}}\hat{j}$

(c) $\hat{k}$ (d) $\frac{1}{\sqrt{3}}\hat{i}+\frac{1}{\sqrt{3}}\hat{j}+\frac{1}{\sqrt{3}}\hat{k}$

[2012 : 1 Mark]

68. The following surface integral is to be evaluated over a sphere for the given steady velocity vector field $\vec{F} = x\hat{i} + y\hat{j} + z\hat{k}$ defined with respect to a Cartesian coordinate system having i, j and k as unit base vectors.

$$\iint_S (\vec{F}.n)\,dA$$

where S is the sphere, $x^2 + y^2 + z^3 = 1$ and n is the outward unit normal vector to the sphere. The value of the surface integral is

(a) π (b) 2π
(c) 3π/4 (d) 4π

[2013 : 2 Marks]

69. Which one of the following describes the relationship among the three vectors, $\hat{i}+\hat{j}+\hat{k}$, $2\hat{i}+3\hat{j}+\hat{k}$ and $5\hat{i}+6\hat{j}+4\hat{k}$?

(a) The vectors are mutually perpendiculaar
(b) The vectors are linearly dependent
(c) The vectors are linearly independent
(d) The vectors are unit vectors

[2014 : 1 Mark, Set-1]

70. Curl of vector $\vec{F} = x^2z^2\hat{i} - 2xy^2z\hat{j} + 2y^2z^3\hat{k}$ is

(a) $(4yz^3 + 2xy^2)\hat{i} + 2x^2z\hat{j} - 2y^2z^3\hat{k}$

(b) $(4yz^3 + 2xy^2)\hat{i} - 2x^2z\hat{j} - 2y^2z\hat{k}$

(c) $2xz^2\hat{i} - 4xyz\hat{j} + 6y^2z^2\hat{k}$

(d) $2xz^2\hat{i} + 4xyz\hat{j} + 6y^2z^2\hat{k}$

[2014 : 1 Mark, Set-2]

71. Divergence of the vector field

$x^2z\hat{i} + xy\hat{j} - yz^2\hat{k}$ at (1, –1, 1) is

(a) 0 (b) 3
(c) 5 (d) 6

[2014 : 1 Mark, Set-3]

72. Curl of vector $V(x, y, z) = 2x^2i + 3z^2j + y^3k$ at x = y = z = 1 is

(a) – 3i (b) 3i
(c) 3i – 4j (d) 3i – 6k

[2015 : 1 Mark, Set-2]

73. The surface integral

$$\iint_S \frac{1}{\pi}(9xi - 3yj).ndS$$

$x^2 + y^2 + z^2 = 9$ is _______.

[2015 : 2 Marks, Set-2]

74. Let φ be an arbitrary smooth real valued scalar function and V be an arbitrary smooth vector valued function in a three-dimensional space. Which one of the following is an identify?

(a) $\text{Curl}(\phi\vec{V}) = \nabla(\phi\text{Div}\vec{V})$

(b) $\text{Div}\vec{V} = 0$

(c) $\text{Div Curl}\vec{V} = 0$

(d) $\text{Div}(\phi\vec{V}) = \phi\text{Div}\vec{V}$

[2015 : 1 Mark, Set-3]

75. A scalar potential φ has the following gradient: $\nabla\varphi = yz\hat{i} + xz\hat{j} + xy\hat{k}$. Consider the integral $\int_c \nabla\varphi.d\vec{r}$ on the curve $\vec{r} = x\hat{i} + y\hat{j} + z\hat{k}$

The curve C is parameterized as follows :

$$\begin{cases} x = t \\ y = t \quad \text{and } 1 \le t \le 3 \\ z = 3t^2 \end{cases}$$

The value of the integral is ________

[2016 : 2 Marks, Set-2]

76. The value of the line intergral $\oint_c \vec{F}.\vec{r}ds$, where C is a circle of radius $\frac{4}{\sqrt{\pi}}$ units is ________

Here, $\vec{F}(x, y) = y\hat{i} + 2x\hat{j}$ and $\hat{r}$ is the UNIT tangent vector on the curve C at an arc length s from a reference point on the curve $\hat{i}$ and $\hat{j}$ are the basis vectors in the x – y Cartesian reference. In evaluating the line integral, the curve has to be traversed in the counter-clockwise direction.

[2016 : 2 Marks, Set-3]

77. For the vector $\vec{V} = 2yz\hat{i} + 3xz\hat{j} + 4xy\hat{k}$, the value of $\nabla(\nabla \times \vec{V})$ is ________.

[2017 : 2 Marks, Set-1]

78. The divergence of the vector $-yi + xj$ is ____

[2017 : 1 Mark, Set-2]

79. The divergence of the vector field $\vec{u} = e^x(\cos y\hat{i} + \sin y\hat{j})$ is

(a) 0 (b) $e^x \cos y + e^x \sin y$

(c) $2e^x \cos y$ (d) $2e^x \sin y$

[2018 : 1 Mark, Set-2]

80. For a position vector $r = x\hat{i} + y\hat{j} + z\hat{k}$ the norm of the vector can be defined as $|\vec{r}| = \sqrt{x^2 + y^2 + z^2}$. Given a function $\phi = \ln|\vec{r}|$, its gradient $\nabla\phi$ is

(a) $\vec{r}$ (b) $\frac{\vec{r}}{|\vec{r}|}$

(c) $\frac{\vec{r}}{\vec{r}.\vec{r}}$ (d) $\frac{\vec{r}}{|\vec{r}|^3}$

[2018 : 2 Marks, Set-2]

ANSWERS

1. (a)	**2.** (b)	**3.** (b)	**4.** (b)	**5.** (b)	**6.** (c)	**7.** (d)	**8.** (c)	**9.** (b)	**10.** (a)
11. (b)	**12.** (d)	**13.** (c)	**14.** (a)	**17.** (c)	**18.** (c)	**19.** (c)	**20.** (d)	**21.** (c)	**22.** (b)
23. (b)	**24.** (d)	**25.** (c)	**26.** (c)	**27.** (b)	**28.** (d)	**31.** (b)	**32.** (a)	**33.** (a)	**34.** (a)
35. (a)	**36.** (d)	**37.** (a)	**38.** (d)	**39.** (b)	**40.** (a)	**41.** (a)	**42.** (d)	**43.** (d)	**44.** (d)
45. (a)	**46.** (c)	**47.** (c)	**48.** (b)	**49.** (b)	**52.** (a)	**53.** (c)	**54.** (a)	**55.** (b)	**57.** (c)
58. (c)	**59.** (c)	**60.** (c)	**61.** (a)	**62.** (a)	**63.** (b)	**64.** (d)	**65.** (b)	**66.** (c)	**67.** (a)
68. (a)	**69.** (b)	**70.** (a)	**71.** (c)	**72.** (a)	**74.** (c)	**79.** (c)	**80.** (c)		

EXPLANATIONS

1. $\underset{x\to 0}{\text{Lt}} \dfrac{\sin^2 x}{x} = \underset{x\to 0}{\text{Lt}} \left(\dfrac{\sin x}{x}\right)^2 \times x$ (Applying L'Hospital's rule) $= 1 \times 0 = 0$

2. $I = \lim\limits_{x\to 3} \dfrac{2x^2 - 7x + 3}{5x^2 - 12x - 9}$

Applying L' Hospital rule, we get

$I = \lim\limits_{x\to 3} \dfrac{4x - 7}{10x - 12}$

$\therefore\ I = \dfrac{4 \times 3 - 7}{10 \times 3 - 12} = \dfrac{5}{18}$

3. Since interval given is bounded, so minimum value of function is 1.

4. Given: $\lim\limits_{x\to 0} \dfrac{e^x = \left[1 + x + \dfrac{x^2}{2}\right]}{x^3}$

We known, $e^x = 1 + \dfrac{x}{1!} + \dfrac{x^2}{2!} + \dfrac{x^3}{3!} + \dfrac{x^4}{4!} + \ldots$

$\Rightarrow \lim\limits_{x\to 0} \dfrac{\dfrac{x^3}{3!} + \dfrac{x^4}{4!} + \ldots}{x^3}$

$\Rightarrow\ \lim = \dfrac{1}{3!} = \dfrac{1}{6}$

5. $L = \underset{x\ \ 8}{\lim} \dfrac{x^{1/3} - 2}{(x - 8)}$

Let $x = a^3$

$\Rightarrow a = 2$

$\therefore\ \underset{a\ \ 2}{\lim} \dfrac{(a - 2)}{a^3 - 8}$

Now by partial fractions,

$(a^3 - 8) = (a - 2)(a^2 + 2a + 4)$

$\Rightarrow L = \underset{a\ \ 2}{\lim} \dfrac{1}{a^2 + 2a\ \ 4} = \dfrac{1}{12}$

6. Given : $y = |2 - 3x|$

At $x = \dfrac{2}{3}$, sharp point will be there.

Hence, function is continuous to all $\forall\ x \in R$ and also differentiable at all points, $\forall\ x \in R$ except at $x = \dfrac{2}{3}$.

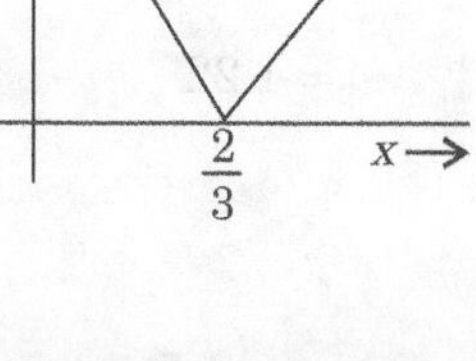

7. Applying L' Hospitals rule, we have

$\lim\limits_{\theta\to 0} \dfrac{\cos\theta}{1} = \cos 0 = 1$

8. Given function is, $f(x) = |x|$

$|x|$ is continuous at $x = 0$ but not differentiable.

9. $\underset{x\to 0}{\text{Lt}} \dfrac{1 - \cos x}{x^2} = \underset{x\to 0}{\text{Lt}} \dfrac{2\sin^2\left(\dfrac{x}{2}\right)}{x^2}$

$= 2\ \underset{x\to 0}{\text{Lt}} \dfrac{\sin^2\left(\dfrac{x}{2}\right)}{\left(\dfrac{x^2}{2}\right) \times 4} = 2 \times \dfrac{1}{4} = \dfrac{1}{2}$

10. $\lim\limits_{x\to 0} \dfrac{x - \sin x}{1 - \cos x} = \left(\dfrac{0}{0}\right)$

Applying L'. Hospital Rule,

$\lim\limits_{x\to 0} \dfrac{1 - \cos x}{\sin x} = \left(\dfrac{0}{0}\right)$

Once again, L'. Hospital rule

$\lim\limits_{x\to 0} \dfrac{\sin x}{\cos x} = \left(\dfrac{0}{1}\right) = 0$

11. $\lim\limits_{x\to 0} \dfrac{e^{2x} - 1}{\sin 4x}$ is $\left(\dfrac{0}{0}\right)$

So, Applying L' Hospital Rule,

$\lim\limits_{x\to 0} \dfrac{2e^{2x}}{4\cos 4x} = \dfrac{2}{4} = 0.5$

12. We know that $f(x)$ is continuous at $x = a$, if $\lim\limits_{x\to a} f(x)$ exists and equal to $f(a)$.

13. $\lim\limits_{x\to 0} \dfrac{1 - \cos(x^2)}{2x^4} = \dfrac{0}{0}$

Using L' Hospital Rule

$\lim\limits_{x\to 0} \dfrac{(\sin x^2)2x}{-8x^3} = \dfrac{0}{0}$

$\lim\limits_{x\to 0} \dfrac{(\cos x^2)2x2x + (\sin x^2)2}{24x^2}$

$\lim\limits_{x\to 0} \dfrac{(\cos x^2)4x^2 + 2\sin x^2}{24x^2}$

$= \lim\limits_{x\to 0} \dfrac{(-\sin x^2)4x^2 + \cos x^2(8x) + 2(\cos x^2)2x}{48x}$

$= \lim\limits_{x\to 0} \dfrac{(-\cos x^2)2x(4x^2) + (-\sin x^2)(8x) + (-\sin x^2)(2x)(8x) + 12\cos x^2 + (-\sin x^2)(2x)4x}{48}$

$= \dfrac{12}{48} = \dfrac{1}{4}$

14. At x = 0, the function f(x) = |x| has a minimum.

15. $\lim\limits_{x\to 0}\left(\dfrac{-\sin x}{2\sin x+\cos x}\right)=\lim\limits_{x\to 0}\left(\dfrac{-\sin 0}{2\sin 0+\cos 0}\right)=\dfrac{0}{1}=0$

16. $f(x) = 2x^3 - 3x^2$

$f'(x) = 0$

$\Rightarrow 6x^2 - 6x = 0$

$6x(x-1) = 0$

$x = 0$ and 1

$f(-1) = -5,\ f(1) = -1,\ f(2) = -4$

Global minimum = –5

17. $f(x)=\dfrac{x^2-3x-4}{x^2+3x-4}=\dfrac{x^2-3x-4}{(x+4)(x-1)}$

The function is not defined at $x = -4, 1$

so function f(x) is discontinuous

at $x = -4, 1$

18. $\lim\limits_{x\to 0}\dfrac{\log_e(1+4x)}{e^{3x}-1}$

Apply in L-hospital rule = $\lim\limits_{x\to 0}\dfrac{\frac{1}{1+4x}\times 4}{e^{3x}\times 3}=\dfrac{4}{3}$

19. $=\lim\limits_{x\to\infty}\left(\sqrt{2^2+x-1}-x\right)$

$=\lim\limits_{x\to\infty}\left(\sqrt{x^2+x-1}-x\right)\times\dfrac{\left(\sqrt{x^2+x-1}+x\right)}{\left(\sqrt{x^2+x-1}+x\right)}$

$=\lim\limits_{x\to\infty}\dfrac{x^2+x-1-x^2}{\sqrt{x^2+x-1}+x}$

$=\lim\limits_{x\to 0}\dfrac{x\left(1-\frac{1}{x}\right)}{x\left[\sqrt{1+\frac{1}{x}-\frac{1}{x}}+1\right]}=\dfrac{1}{2}$

20. From L′ – Hospital rule

$\lim\limits_{x\to 0}\dfrac{x^3-\sin(x)}{x}=\dfrac{3x^2-\cos x}{1}$

$=\dfrac{3\times 0^2-\cos 0}{1}=\dfrac{0-1}{1}=-1$

21. Given: $x = a(\theta + \sin\theta)$, $y = a(1 - \cos\theta)$

$\therefore\ \dfrac{dx}{d\theta} = a(1+\cos\theta)$, and

$\dfrac{dx}{d\theta} = a(\sin\theta) = a\sin\theta$

$\dfrac{dy}{dx}=\dfrac{\frac{dy}{d\theta}}{\frac{dx}{d\theta}}=\dfrac{a\sin\theta}{a(1+\cos\theta)}$

$=\dfrac{2a\sin\left(\frac{\theta}{2}\right)\cos\left(\frac{\theta}{2}\right)}{a\times 2\cos^2\left(\frac{\theta}{2}\right)}=\tan\dfrac{\theta}{2}$

22. Given, $f(x) = (x-8)^{2/3} + 1$

$m_{normal} \times m_{tangent} = -1$

where, $m_{tangent} = \dfrac{dy}{dx}$ and m_{normal}

$=\dfrac{-1}{(dy/dx)}$

$\dfrac{dy}{dx}=\dfrac{df(x)}{dx}=\dfrac{2}{3}(x-8)^{-1/3}$

$m_{normal}=\dfrac{-3}{2}(x-8)^{1/3}$ or $[m_{normal}]_{(0,5)}=3$

$\therefore$ Equation of normal is,

$(y - y_1) = \text{mnormal}\ (x - x_1)$

or $(y - 5) = 3(x - 0)$ or $y = 3x + 5$

23. $y = x+\sqrt{x+\sqrt{x+\sqrt{x+\ldots\infty}}}$

$y-x=\sqrt{x+\sqrt{x+\sqrt{x+\ldots\infty}}}$

Square both sides, we get

$(y-x)^2 = y$

Put x = 2, we get

$(y-4)(y-1) = 0$

We get

y = 4, y = 1

But y > x

$\therefore$ y = 4 only

24. The length is given by $\int\sqrt{(dy)^2+(dx)^2}$

$dx = \sqrt{x}\ .\ dx$

$\therefore\quad L=\int_0^1\left(\sqrt{x+1}\right).\,dx$

$=\dfrac{2}{3}(x+1)^{3/2}\Big|_0^1$

$= 1.22$

25. Given $f = y^x$

$$\Rightarrow \ln f = x \ln y$$

$$\Rightarrow \frac{1}{f}\frac{\partial f}{\partial y} = \frac{x}{y}$$

$$\Rightarrow \frac{\partial f}{\partial y} = \frac{x}{y}.\, y^x$$

Now $\ln\frac{\partial f}{\partial y} = \ln\frac{x}{y} + x \ln y$

$$\Rightarrow \frac{1}{\frac{\partial f}{\partial y}}\cdot\frac{\partial^2 f}{\partial x\, \partial y} = \left[\frac{y}{x}\cdot\frac{1}{y} \quad \ln y\right]$$

$$\Rightarrow \frac{\partial^2 f}{\partial x\, \partial y} = \left[\frac{1}{x} + \ln y\right]\frac{x}{y}.\, y^x = \left[\frac{1}{2} \quad 0\right] \times 2.1 = 1$$

26. Taylor series expansion of f(x) about a is given by

$$f(x) = f(a) + \frac{(x-a)}{1!} f'(a) + \frac{(x-a)^2}{2!} f''(a) + \text{- - - - -}$$

Coefficient of $(x - a)^4$ is $\frac{f''''(a)}{4!}$

Now $f(x) = e^x$

$\Rightarrow f''''(x) = e^x$

$\Rightarrow f''''(a) = e^a$

Hence for a = 2,

$$\frac{f''''(a)}{4!} = \frac{e^2}{4!}$$

27. $\sin\theta = \theta - \frac{\theta^3}{3!} + \frac{\theta^5}{5!} - \ldots$

28. Function $f(x) = x^3 + 1$ has a point of inflection at $x = 0$, since in the graph sign of the curvature (i.e., concavity) is changed.

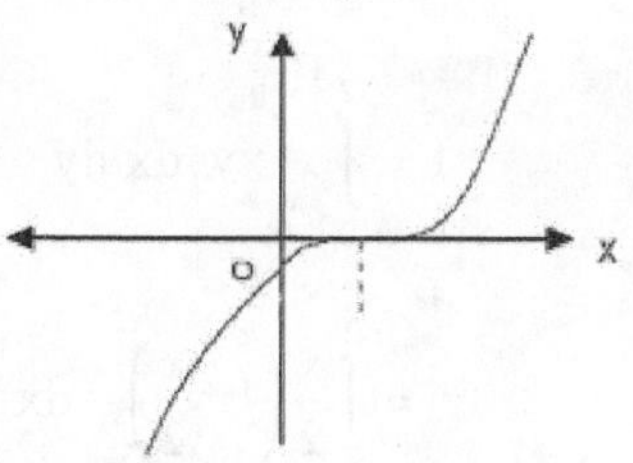

29. $f(x) = e^{-x} \sin(100x)$

$f'(x) = -e^{-x}\sin(100x) + e^{-x}\cos(100x) \times 100$

for minima $f'(x) = 0$

$\tan(100x) = 100$

$$x = \frac{1}{100}\tan^{-1}(100) = .0156$$

$f(x) = e^{-.0156} \sin(100 \times .0156) = 0.9844$

30. The length of the curve

$$= \int_0^{\pi/2} \sqrt{\left(\frac{dx}{dt}\right)^2 + \left(\frac{dy}{dt}\right)^2 + \left(\frac{dz}{dt}\right)^2}\, dt$$

$$= \int_0^{\pi/2} \sqrt{(-\sin t)^2 + (\cos t)^2 + \left(\frac{2}{\pi}\right)^2}\, dt$$

$$= \int_0^{\pi/2} \sqrt{\sin^2 t + \cos^2 t + \frac{4}{\pi^2}}\, dt$$

$$= \int_0^{\pi/2} \sqrt{1 + \frac{4}{\pi^2}}\, dt$$

$$= \sqrt{1 + \frac{4}{\pi^2}} \cdot t\Big|_0^{\pi/2} = \sqrt{1 + \frac{4}{\pi^2}} \times \frac{\pi}{2} = 1.8622$$

31. Given equations are

$$y = x^2 \ldots(i)$$

$$y = x \ldots(ii)$$

From equations (*i*) and (*ii*)

$$x^2 - x = 0$$

$$\Rightarrow \quad x(x - 1) = 0$$

$$\Rightarrow \quad x = 0, 1$$

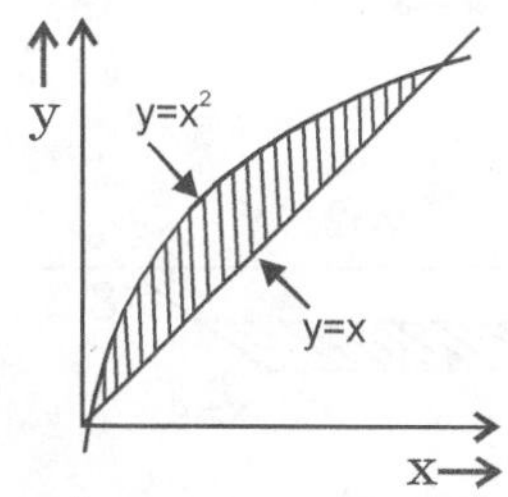

Area enclosed

$$= \int_{x=0}^{1}\int_{y=x}^{y=x^3} dy\, dx = \int_{x=0}^{1} dx \int_{y=x}^{y=x^2} dy$$

$$= \int_{x=0}^{1} [y]_x^{x^2}\, dx = \int_{x=0}^{1} (x^2 - x)\, dx$$

$$= \left[\frac{x^3}{3} - \frac{x^2}{2}\right]_0^1 = \frac{1}{3} - \frac{1}{2} = \frac{2-3}{6} = -\frac{1}{6}$$

$$\therefore \text{ Enclosed area} = \left|-\frac{1}{6}\right| = \frac{1}{6}$$

32. $$V = \int_0^{2\pi}\int_0^{\pi/3}\int_0^{1} r^2 \sin\phi\, dr\, d\phi\, d\theta$$

$$= \int_0^{2\pi}\int_0^{\pi/3}\left[\frac{r^3}{3}\right]_0^1 \sin\phi\, d\phi\, d\theta$$

$$= \int_0^{2\pi}\int_0^{\pi/3}\frac{1}{3}\sin\phi\, d\phi\, d\theta$$

$$= \frac{1}{3}\int_0^{2\pi}[-\cos\phi]_0^{\pi/3}\, d\theta = \frac{1}{3}\int_0^{2\pi}\frac{1}{2}\, d\theta = \frac{1}{6} \times 2\pi = \frac{\pi}{3}$$

33. $I = \int_{-a}^{a} \left(\sin^6 x + \sin^7 x\right) dx$

$=2\int_0^a \sin^6 x dx + 0$

$=2\int_0^a \sin^6 x dx$,

34. When $I = \int_0^8 \int_{\frac{x}{4}}^{2} f(x,y) dy dx$

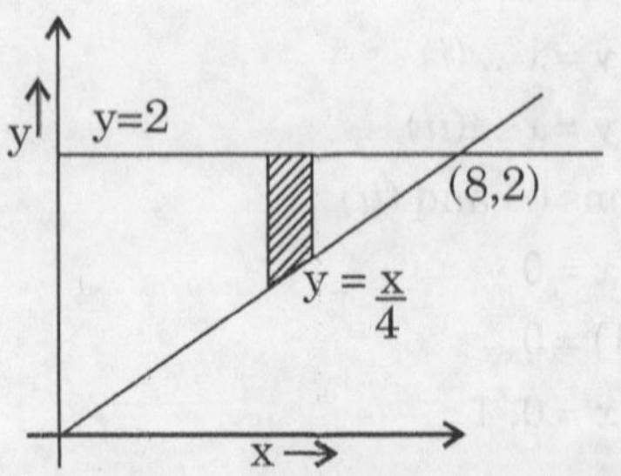

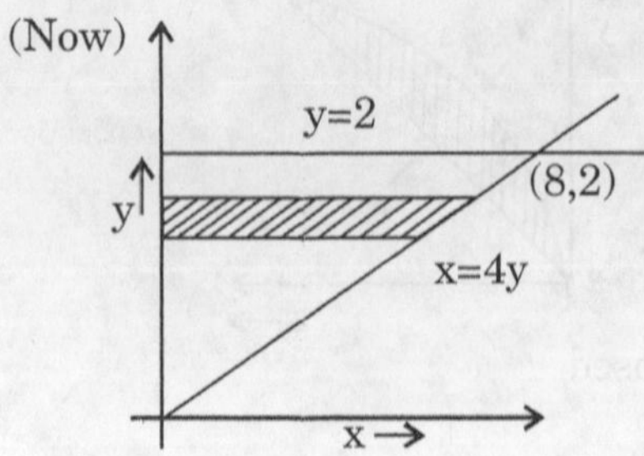

$I = \int_0^2 \int_0^{4y} f(x,y) dx\, dy$

35. $\phi(u, v) = \begin{vmatrix} \frac{\partial x}{\partial u} & \frac{\partial x}{\partial v} \\ \frac{\partial y}{\partial u} & \frac{\partial y}{\partial v} \end{vmatrix} = \frac{2v}{u}$

36.

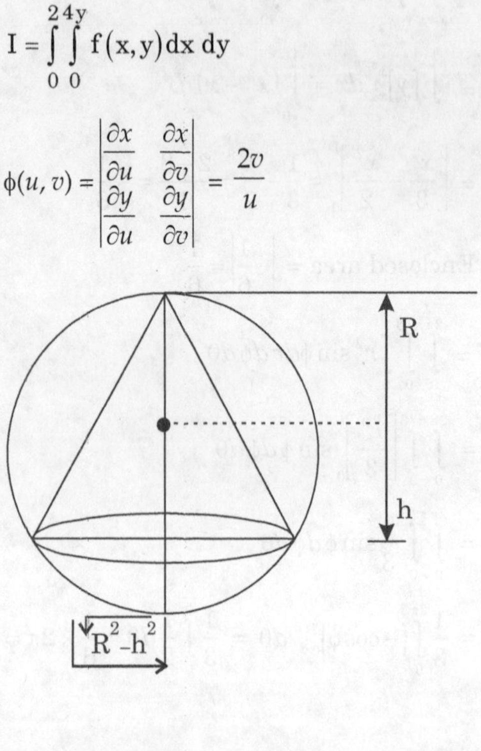

Given R = 1, radins of sphere.

Let height of cone is H = h + R

Volume, $V = \frac{1}{3}\pi \times \left(\sqrt{R^2 - h^2}\right)^2 (R + h)$ for maximum valume

$\Rightarrow \frac{dh}{dh} = 0$

$\Rightarrow \frac{d}{dh}\left[\frac{1}{3}(R^2 - h^2)(R + h)\right]$

$\Rightarrow -2h(R + h) + (R^2 - h^2) = 0$

$\Rightarrow (R + h)(R - 3h) = 0$

$h = -R, \frac{R}{3}$

Height of the come $= R + \frac{R}{3} = \frac{4R}{3}$

$= \frac{h \times 1}{3} m = \frac{4}{3} m$

37. $I = \iint xy.\ dx\, dy$

The limit of y is from 0 to $\frac{1 - \frac{x}{2}}{2}$ and limit of x from 0 to 2.

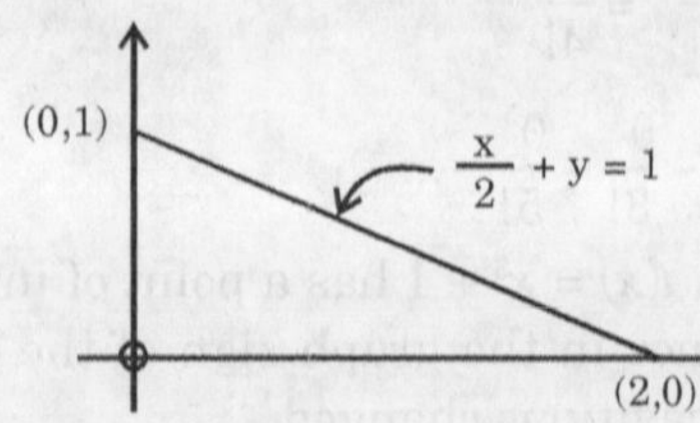

$I = \int_0^2 \int_0^{1-\frac{x}{2}} xy\ dx\, dy$

$= \int_0^2 \frac{x}{2}\left(1 \quad \frac{x}{2}\right)^2 . dx$

$= \frac{1}{8}\int_0^2 x\ (x^2 + 4 - 4x).\ dx$

$= \frac{1}{8}\int_0^2 (\ x^3 + 4x - 4x^2).\ dx$

$= \frac{1}{8}\left(\frac{x^4}{4} + 2x^2 - \frac{4x^2}{3}\right)_0^2 = \frac{1}{6}$

38. To see whether the integrals are bounded or unbounded, we need to see that the integrand doesn't reach infinity in the interval of integration. Let us write down the range of the integrands in the 4 options,

$a \to (0, 1)$; $b \to (0, 1)$; $c \to (0, \frac{1}{e})$; $d \to (0, \infty)$

Thus, (d), i.e. $\int_0^1 \frac{1}{1-x}\,dx$ is unbounded.

39. Given : $x^2 + y^2 = 1$

Put $x = \cos\theta$, and $y = \sin\theta$

$$(x+y)^2 = \cos^2\theta + \sin^2\theta + 2\sin\theta\cos\theta$$

$$= 1 + \sin 2\theta$$

$$= \int_0^{\pi/2} (1+\sin 2\theta)\, d\theta$$

$$= \left[\theta - \frac{\cos 2\theta}{2}\right]_0^{\pi/2} = \left[\frac{\pi}{2} + \frac{1}{2} + \frac{1}{2}\right]$$

$$= \left(\frac{\pi}{2} + 1\right)$$

40. Take a point on the curve $z = 1, x = 0, y = 0$

Length between origin and this point

$$l = \sqrt{(1-0)+(0-0)+(0-0)} = 1$$

This is minimum length because all options have length greater than l.

41. Given :

$$y^2 = 4x$$

$$x^2 = 4y$$

$$\Rightarrow \quad \frac{x^4}{16} = 4x$$

or $\quad x^4 = 64x$

or $\quad x(x^3 - 64) = 0$

or $\quad x^3 = 64$

or $\quad x = 4$

$\therefore \quad y = 4$

$$\therefore \text{ Required area} = \int_0^4 \left(\sqrt{4x} - \frac{x^2}{y}\right) dx$$

$$= \left[2 \times x^{3/2} \times \frac{2}{3} - \frac{x^3}{12}\right]_0^4$$

$$= \frac{4}{3} \times (4)^{3/2} - \frac{64}{12}$$

$$= \frac{32}{3} - \frac{16}{3} = \frac{16}{3}$$

Alternately

For point where both parabolas cut each other

$$y^2 = 4x, \; x^2 = 4y$$

$\therefore \quad x^2 = 4\sqrt{4x}$

or $\quad x^2 = 8\sqrt{x}$

or $\quad x^4 = 16x$

or $\quad x^3 = 16$

$\therefore$ x =(4, 0), (–4, 0)

$\therefore$ Required area

$$= \int_0^4 \sqrt{4x} - \int_0^4 \frac{x^2}{4}\, dx = \left[2 \times \frac{2}{3} x^{3/2} - \frac{x^3}{12}\right]_0^4 = \frac{16}{3}$$

42. Given:

$y = \sqrt{x}$

$\Rightarrow y^2 = x,\ 1 \le x \le 2$

$$\text{Required volume} = \pi\int_a^b y^2\, dx$$

$$= \pi\int_1^2 x\,dx$$

$$= \pi\left(\frac{x^2}{2}\right)^2 = \pi\left(2 - \frac{1}{2}\right) = \frac{3\pi}{2}$$

43. $I = \int_{-\infty}^{\infty} \frac{dx}{1+x^2}$

$$= \left[\tan^{-1} x\right]_{-\infty}^{\infty} = \tan^{-1}(\infty) - \tan^{-1}(-\infty)$$

$$= \frac{\pi}{2} - \left(-\frac{\pi}{2}\right) = \pi$$

44. $$\int_{-a}^{a} f(x)\,dx = \begin{cases} 2\int_0^a f(x)\,dx; & f(x) \text{ is even} \\ 0; & f(x) \text{ is odd} \end{cases}$$

45. Given equations are

$$y = x^2 \ldots(i)$$

$$y = x \ldots(ii)$$

From equations (i) and (ii)

$$x^2 - x = 0$$

$\Rightarrow \quad x(x-1) = 0$

$\Rightarrow \quad x = 0, 1$

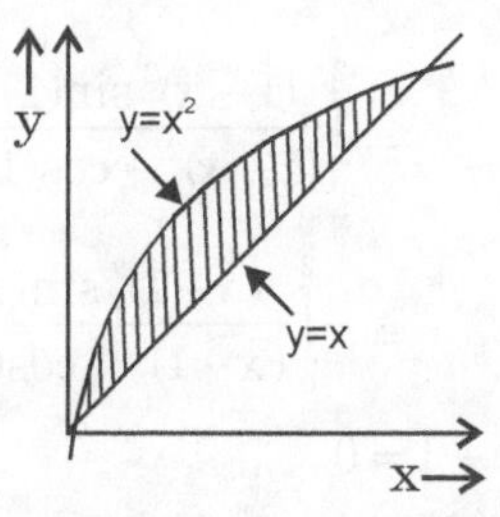

Area enclosed

$$= \int_{x=0}^{1}\int_{y=x}^{y=x^3} dy\,dx = \int_{x=0}^{1} dx \int_{y=x}^{y=x^2} dy$$

$$= \int_{x=0}^{1} [y]_x^{x^2}\, dx = \int_{x=0}^{1} (x^2 - x)\,dx$$

$$= \left[\frac{x^3}{3} - \frac{x^2}{2}\right]_0^1 = \frac{1}{3} - \frac{1}{2} = \frac{2-3}{6} = \frac{1}{6}$$

$\therefore$ Enclosed area $= \left|-\frac{1}{6}\right| = \frac{1}{6}$

46. $\int_1^e \sqrt{x}\ln x dx = \left[\angle nx \int \sqrt{x}dx - \int\left(\frac{d}{dx} < nx\right)\left(\sqrt{x}dx\right)dx\right]_1^e$

(Integrate by parts)

47. Given integral $\oint_c (ydx - xdy)$

where C is $x^2 + y^2 = \frac{1}{4}$

Applying Green's theorem

$$\oint_c \text{M}dx - \text{N}dy = \iint_R \left(\frac{\partial \text{N}}{\partial x} - \frac{\partial \text{M}}{\partial y}\right) dxdy$$

where R is region included in c

$$\oint_c ydx - xdy = \iint_R (-1-1)\,dxdy$$

$$= -2 \iint_R dxdy = -2 \times \text{Region R}$$

$= -2 \times$ area of circle with radius $\frac{1}{2}$

$$= -2 \times \pi\left(\frac{1}{2}\right)^2 = \frac{-\pi}{2}$$

48. Let $\text{I} = \dfrac{\int_0^2 (x-1)^2 \sin(x-1)}{(x-1)^2 + \cos(x-1)} dx$

We know that, $\int_0^2 \dfrac{(2-x-1)^2 \sin(2-x-1)}{(2-x-1)^2 + \cos(2-x-1)} dx$

$$= \int_0^2 \frac{(1-x)^2 \sin(1-x)}{(1-x)^2 + \cos(1-x)} dx$$

$$= -\int_0^2 \frac{(x-1)^2 \sin(x-1)}{(x-1)^2 + \cos(x-1)} dx = -1$$

$\Rightarrow \quad \text{I} + \text{I} = 0$

$\Rightarrow \quad 2\text{I} = 0$

$\Rightarrow \quad \text{I} = 0$

49. $\int_0^2\int_0^x e^{x+y}\, dy\, dx = \int_0^2 e^x\left(\int_0^x e^y dy\right) dx$

$$= \int_0^2 e^x\left(e^y\right)_0^x dx = \int_0^2 e^x\left(e^x - 1\right) dx$$

$$= \int_0^2 \left(e^{2x} - e^x\right) dx = \left(\frac{e^{2x}}{2} - e^x\right)_0^2$$

$$= \frac{e^4}{2} - e^2 - \frac{1}{2} + 1 = \frac{e^4}{2} - e^2 + \frac{1}{2}$$

$$= \frac{1}{2}(e^4 - 2e^2 + 1) = \frac{1}{2}(e^2 - 1)^2$$

50.

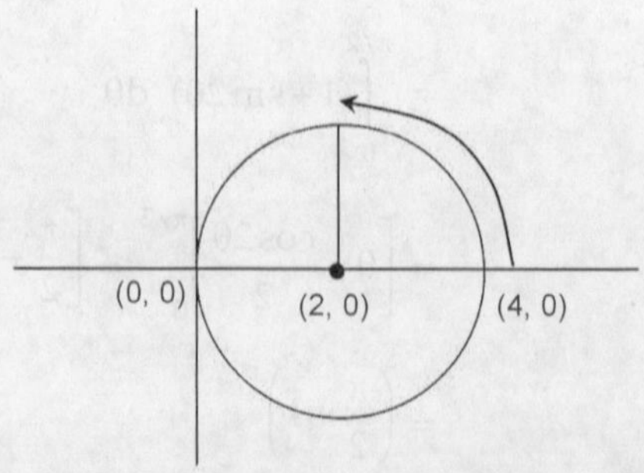

$\frac{1}{4}$ ×circumference

$\Rightarrow \frac{1}{4} \times \pi \times 4$

$$\text{time} = \frac{\pi}{1.5 \rightarrow} = 2\,\text{sec}$$

51. $\int_c \left[(3\text{x} - 8\text{y}^2)\text{dx} + (4\text{y} - 6\text{xy})\text{dy}\right],$

C is boundary of region bounded by x = 0, y =1, and z + y = 1

By using Green's theoren, we get

$$\text{I} = \oint_c (\text{Pdx} + \text{Qdy}) = \oint\oint_R \frac{\partial \text{Q}}{\partial \text{v}}\left(\frac{-\partial \text{P}}{\partial \text{y}}\right) \text{dxdy}$$

Her $\text{P} = 3\text{x} - 8\text{y}^2$

$\text{Q} = 4\text{y} - 6\text{xy}$

$$\frac{\partial \text{Q}}{\partial \text{x}} = -6\text{y}$$

$$\frac{\partial \text{P}}{\partial \text{y}} = -16\text{y}$$

$$\text{I} = \iint(-6\text{y} - (16\text{y})\text{dxdy}$$

$$\text{I} = \iint 10\text{y dx dy}$$

$$I = 10\int_0^1 dx \int_0^{1-x} \frac{y^2}{2}$$

$$I = 5\int_0^1 dx(1-x)^2$$

$$I = 1.666$$

52. $I = \int_{-\infty}^{\infty} \frac{\sin(x)}{x^2+2x+2}$

let $f(x) = \frac{I_m(e^{iz})}{z^2+2z+2}$

Then poles of $f(z)$ are given by $z^2 + 2z + 2 = 0$

$\therefore\ z = -1 \pm i$

$$R_1 = \text{Res}(f(z) : z = -1+i)$$

$$= \underset{z\to -1\pm i}{\text{Lt}}[z-(-1+i)]\frac{e^{iz}}{[z-(-1+i)][z-(-1-i)]}$$

$$= \frac{e^{i(-1+i)}}{-1+i+1+i} = \frac{e^{-i-1}}{2i}$$

$$\int_c f(z)\,dz = \int_c \frac{I_m(e^{iz})}{z^2+2z+2}dz = I_m[2\pi i(R_1)]$$

$$= I_m\left[2\pi i\left(\frac{e^{-i-1}}{2i}\right)\right]$$

$$= I_m[\pi e^{-1}(\cos(1) - i\sin(1))] = -\frac{\pi\sin(1)}{e}$$

53. $x = \cos\frac{\pi u}{2}; y = \sin\left(\frac{\pi u}{2}\right)$

$x^2 + y^2 = 1$

It represents a circle in x-y plane

$0 \le 4 \le 1$

$\therefore\ 0 \le 4 \le 1, 0 \le y \le 1$

So, $0 \le \theta < \frac{\pi}{2}$

Area of hemisphere $= 2\pi(r)^2$

$= 2\pi(1)^2$

$= 2\pi$

54. $\int_a^b f(x)\,dx = f(\xi)(b-a)$

55.

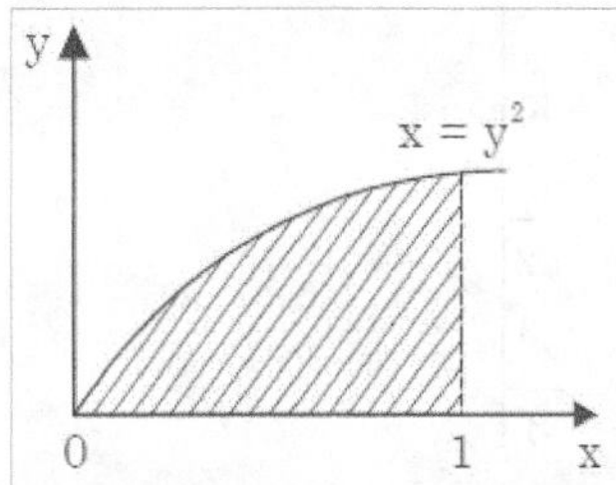

Here, volume of solid of rotation by rotating shaded area by 360° around x-axis is,

$$V = \int_0^1 \pi y^2 dx$$

$$V = \int_0^1 \pi x\,dx = \left.\frac{\pi x^2}{2}\right|_0^1 = \frac{\pi}{2}$$

56.

$$I = \int_1^e x\,ln\,x dx$$

$$= \frac{ln\,x^2}{2} - \int\left[\frac{1}{x}\times\frac{x^2}{2}\right]dx$$

$$= \left.\frac{x^2 ln\,x}{2} - \frac{x^2}{4}\right|_1^e$$

$$= \left(\frac{e^2}{2} - \frac{e^2}{4}\right) + \frac{1}{4}$$

$$= \frac{e^2+1}{4} = 2.097$$

57. Directional derivative $= \nabla f.\frac{\vec{a}}{|\vec{a}|}$

$$\nabla f = \frac{\partial x^2}{\partial x}\hat{i} + \frac{\partial y^2}{\partial y}\hat{j}$$

$= 2x\hat{i} + 2y\hat{j}$

$\vec{a}$ = line joining (0,0) and (1,1)

$= 1\hat{i} + 1\hat{j}$

$\vec{a} = \hat{i} + \hat{j}$

Directional derivative $= \frac{(2x\hat{i}+2y\hat{j})(\hat{i}+\hat{j})}{\sqrt{2}}$

$$= \frac{2x+2y}{\sqrt{2}}$$

Directional derivative at (1,1) $= \dfrac{2+2}{\sqrt{2}} = 2\sqrt{2}$

58. $\nabla \times u = \begin{vmatrix} \hat{i} & \hat{j} & \hat{k} \\ \dfrac{\partial}{\partial x} & \dfrac{\partial}{\partial y} & \dfrac{\partial}{\partial z} \\ \dfrac{-y^3}{3} & \dfrac{x^3}{3} & \dfrac{z^3}{3} \end{vmatrix}$

$= 0 + 0 + \left(x^2 + y^2\right)\hat{k}$

$\hat{n} = \hat{k}$

$(\nabla \times u) \cdot \hat{n} = (\nabla \times v) \cdot \hat{k} = x^2 + y^2$

$\int \left(x^2 + y^2\right)\hat{k} \cdot \hat{n}\, dS = \int \left(x^2 + y^2\right) dS$

$= \iint \left(x^2 + y^2\right) dx\, dy$

$= \int_0^{2\pi}\int_0^1 r^2 (r\, dr\, d\theta) = \dfrac{\pi}{2}$

59. $\bar{F} = x\bar{i} - y\bar{j}$

For divergence:-

grad $\bar{F} = \nabla F = \left(\bar{i}\dfrac{\partial}{\partial x} + \bar{j}\dfrac{\partial}{\partial y} + \bar{k}\dfrac{\partial}{\partial z}\right) \cdot (x\bar{i} - y\bar{i})$

$= 1 - 1 = 0$

Hence vector field is divergence free

For irrotational,

Curl F = $\nabla \times F$

$= \left(\bar{i}\dfrac{\partial}{\partial x} + \bar{j}\dfrac{\partial}{\partial y} + \bar{k}\dfrac{\partial}{\partial z}\right) \times \left(x\bar{i} - y\bar{j}\right)$

$= \begin{vmatrix} \bar{i} & \bar{j} & \bar{k} \\ \dfrac{\partial}{\partial x} & \dfrac{\partial}{\partial y} & \dfrac{\partial}{\partial z} \\ x & -y & 0 \end{vmatrix}$

$= \bar{i}\left[\dfrac{\partial(0)}{\partial y} - \dfrac{\partial(-y)}{\partial z}\right] - \bar{j}\left[\dfrac{\partial(0)}{\partial x} - \dfrac{\partial(-x)}{\partial z}\right]$

$+ \bar{k}\left[\dfrac{\partial(-y)}{\partial x} - \dfrac{\partial(x)}{\partial y}\right]$

$= \bar{i} \times 0 - \bar{j} \times 0 + \bar{k} \times 0 = 0$

Hence vector field is irrotational.

60. $\overline{P} = 0.866\hat{i} + 0.500\hat{j} + 0\hat{k}$

$\overrightarrow{Q} = 0.259\hat{i} + 0.966\hat{j} + 0\hat{k}$

$\therefore \vec{P}.\vec{Q} = |\vec{P}|\,|\vec{Q}|\cos\theta$

$0.866 \times 0.259 + 0.5 \times 0.966$

$= \sqrt{0.866^2 + 0.5^2} \times \sqrt{0.259^2 + 0.966^2}\,.\cos\theta$

$\therefore \cos\theta = 0.707$

$\theta = 45°$

61. Stokes theorem connects a line integral and a surface integral.

62. Since, potential function of $\vec{V}$ is $x^2 yz$

$\therefore (x^2 yz)$ at $(1, 1, 1)$ — $x^2 yz$ at $(0, 0, 0) = 1$

63. Let the vectors be

$i, j, i + j$

$(\vec{c})\,(\vec{b})\,(\vec{a})$

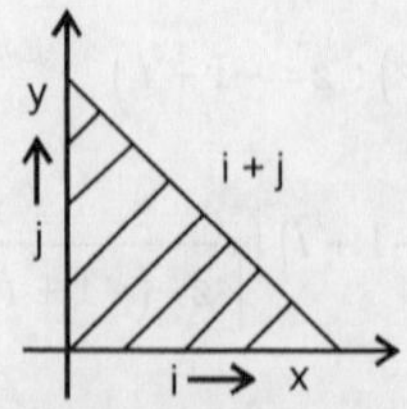

Now Area vector will be perpendicular to place of x, y i.e. z

k will be the required unit vector

Now, (a), (d) cannot give vector product

$\therefore \dfrac{1}{2}\left|(\vec{a} - \vec{b}) \times (\vec{a} - \vec{c})\right| = \dfrac{1}{2}|(i) \times (j)| = \dfrac{1}{2}k$

64. $\nabla.A = \dfrac{\partial(x-y)}{\partial x} + \dfrac{\partial(y-x)}{\partial y} + \dfrac{\partial}{\partial z}(x + y + z)$

$= 1 + 1 + 1 = 3$

65. The directional derivative of a function ϕ along a vector $\vec{A}$ is gevin by,

$(\text{grad } \phi).\dfrac{\vec{A}}{|A|}$

Hence directional derivative is

$(\text{grad } (x^2 + 2y^2 + z)).\dfrac{(3\hat{i} - 4\hat{j})}{\sqrt{3^2 + 4^2}}$

$= (2x\,\hat{i} + 4y\,\hat{j} + \hat{k}),\dfrac{(3\hat{i} - 4\hat{j})}{5}$

$= \dfrac{6x - 16y}{5}$

Hence (1, 1, 2), directional derivative

$= \dfrac{6 \times 1 - 16 \times 1}{5} = -2$

66. $\vec{F} = 3xz\vec{i} + 2xy\vec{j} - yz^2\vec{k}$

$\therefore$ div $= \vec{\nabla}.\vec{F} = \frac{\partial}{\partial x}(3xz) + \frac{\partial}{\partial y} - \frac{\partial}{\partial z}(yz^2)$

$= 3z + 2x - 2yz$

At point (1, 1, 1),

divergence $= 3 + 2 - 2 = 3$

67. Given spherical surface is $x^2 + y^2 + z^2 = 1$

and point is $\left(\frac{1}{\sqrt{2}}, \frac{1}{\sqrt{2}}, 0\right)$

Normal vector outward to $x^2 + y^2 + z^2 - 1 = 0$ at $\left(\frac{1}{\sqrt{2}}, \frac{1}{\sqrt{2}}, 0\right)$ is

$$\frac{1}{\sqrt{2}}i + \frac{1}{\sqrt{2}}j + 0.k = \frac{1}{\sqrt{2}}i + \frac{1}{\sqrt{2}}j$$

Hence outward unit normal vector

$$= \frac{\frac{1}{\sqrt{2}}i + \frac{1}{\sqrt{2}}j}{\left(\frac{1}{\sqrt{2}}\right)^2 + \left(\frac{1}{\sqrt{2}}\right)^2} = \frac{1}{\sqrt{2}}i + \frac{1}{\sqrt{2}}j$$

68. Use Gauss-Divergence theorem,

$$I = \iint_s \frac{1}{4}\vec{F}.\hat{n}dA = \iiint_V \vec{\nabla}.\vec{F}dV$$

$\vec{\nabla}.\vec{F} = 1 + 1 + 1 = 3$

$\therefore\ I = \frac{1}{4} \times 3V = \frac{3}{4} \times \frac{4}{3}\pi(1)^3 = \pi$

69. Given vectors are

$i + j + k$, $2i + 3j + k$ and $5i + 6j + k$

$$\begin{vmatrix} 1 & 1 & 1 \\ 2 & 3 & 1 \\ 5 & 6 & 1 \end{vmatrix} = 0$$

$\therefore$ Vectors are linearly dependent.

70. Given $\vec{F} = x^2z^2 i\, 2xy^2zJ + 2y^2z^3k$

$$\text{curl } \vec{F} = \nabla \times \vec{F} = \begin{vmatrix} i & j & k \\ \frac{\partial}{2x} & \frac{\partial}{2y} & \frac{\partial}{2z} \\ x^2z^2 & -2xy^2z & 2y^2z^3 \end{vmatrix}$$

$$= i\left(4yz^3 + 2xy^2\right) - j\left(0 - 2x^2z\right) + k\left(-2y^2z - 0\right)$$

$$= \left(4yz^2 + 2xy^2 + 2x^2zj\right) - 2y^2zk$$

71. Given $\vec{F} = x^2\, a\hat{i} + xy\hat{j} - yz^2\hat{k}$

div $\vec{F} = \nabla\cdot\vec{F}$

$$= \frac{\partial}{\partial x}(x^2z) + \frac{\partial}{\partial y}(xy) + \frac{\partial}{\partial z}(-yz^2)$$

$$= 2xz + x - 2yz$$

div $\vec{F}$ at $(1, -1, 1) = 2 + 1 + 2 = 5$

72. Curl of vector $= \begin{vmatrix} i & j & k \\ \frac{\partial}{\partial x} & \frac{\partial}{\partial y} & \frac{\partial}{\partial z} \\ 2x^2 & 3z^2 & y^3 \end{vmatrix}$

$$= i\left[\frac{\partial}{\partial y}\left(y^3\right) - \frac{\partial}{\partial z}\left(3z^2\right)\right] - j\left[\frac{\partial}{\partial x}\left(y^3\right) - \frac{\partial}{\partial z}\left(2x^2\right)\right] + k\left[\frac{\partial}{\partial x}\left(3z^2\right) - \frac{\partial}{\partial y}\left(2x^2\right)\right]$$

$= i[3y^2 - 6z] - j[0] + k[0 + 0]$

At $x = 1, y = 1, z = 1$

Curl $= -3i$

73. By Gauss divergence theorem,

$$\int_S \overline{F}.\overline{n}\,ds = \int_V \text{div}\,\overline{F}dV$$

Here $\overline{F} = 9x\overline{i} - 3y\hat{j}$

div $\overline{F} = 9 - 3 = 6$

$$\therefore \int_s \frac{1}{\pi}(9x\overline{i} - 3y\overline{j}).\overline{n}ds = \frac{1}{\pi}\int_V 6dV = \frac{1}{\pi}6V$$

$$= \frac{1}{\pi}6\left(\frac{4}{3}\pi r^3\right) = 8(3)^3 = 216$$

75. $\int_c \nabla\phi.\overline{dr} = \int_c (yz\hat{i} + xz\hat{j} + xy\hat{k})(dx\hat{i} + dy\hat{j} + dz\hat{k})$

$$= \int_c (yzdx + xzdy + xydz)$$

$= xyz$

Given that $x = t$, $y = t^2$, $z = 3t^2$

$= t.t^2.3t^2\Big|_1^3$

$= 3\left(t^5\right)\Big|_1^3$

$= 3(3^5 - 1)$

$= 726$

76. $\int_c \vec{F}.\vec{r}ds = \int_c \vec{F}.\vec{dr} = \int_c F_1 dx + F_2 dy$

$\iint_R \left(\frac{\partial F_2}{\partial x} \frac{-\partial F_1}{\partial y} \right) dxdy$

$F_1 = y, F_2 = 2x$

$\iint_R (2-1) dxdy$

$\frac{\partial F_1}{\partial y} = 1, \frac{\partial F_2}{\partial x} = 2 \; = \iint dxdy \; = \pi \frac{16}{\pi} = 16$

77. Given vector $\vec{V} = 2yz\hat{i} + 3xz\hat{j} + 4xy\hat{k}$

Now $\nabla \times \vec{V} = \begin{vmatrix} i & j & k \\ \frac{\partial}{\partial x} & \frac{\partial}{\partial y} & \frac{\partial}{\partial z} \\ 2yz & 3xz & 4xy \end{vmatrix} = xi - 2yj + zk$

$\nabla.(\nabla \times \vec{V}) = 1 - 2 + 1 = 0$

78. $\vec{F} = -y\hat{i} + x\hat{j}$

$\nabla.\vec{F} = \frac{\partial}{\partial x}(-y) + \frac{\partial}{\partial y}(x)$

$= 0 + 0 = 0$

79. $\vec{u} = e^x \cos y\hat{i} + e^x.\sin y\hat{j}$

$\nabla.\vec{u} = \frac{\partial}{\partial x}(u_1) + \frac{\partial}{\partial y}(u_2)$

$= \frac{\partial}{\partial x}(e^x.\cos y) + \frac{\partial}{\partial y}(e^x.\sin y)$

$= e^x \cos y + e^x \cos y$

$= 2e^x \cos y$

3

CHAPTER

Differential Equation

Analysis of Previous GATE Papers														
Topic	**Year/Set → Marks Distribution ↓**		**2019 Set 1**	**2019 Set 2**	**2018 Set 1**	**2018 Set 2**	**2017 Set 1**	**2017 Set 2**	**2016 Set 1**	**2016 Set 2**	**2016 Set 3**	**2015 Set 1**	**2015 Set 2**	**2015 Set 3**
1st order differential equation	**1 Mark**	MCQ Type	1	1		1								
		Numerical Type												
	2 Marks	MCQ Type											1	
		Numerical Type												
	Total		**1**	**1**		**1**							**2**	
nth order differential equation	**1 Mark**	MCQ Type					1							
		Numerical Type												
	2 Marks	MCQ Type		1								1		
		Numerical Type				1		1	1					
	Total			**2**		**2**	**1**	**2**	**2**			**2**		
Partial differentiation and its applications	**1 Mark**	MCQ Type				1	1							
		Numerical Type												
	2 Marks	MCQ Type												
		Numerical Type												
	Total					**1**	**1**							

1st order Differential equation

1. The solution of the differential equation

$\frac{dy}{dx} + y = 0$ is

(a) $y = \frac{1}{x+c}$

(b) $y = \frac{-x^3}{3} + c$

(c) ce^x

(d) unsolvable as equation is non-linear

[2003 : 2 Marks]

2. If $x^2\frac{dy}{dx} + 2xy = \frac{2\ln(x)}{x}$, and $y(1) = 0$, then what is $y(e)$?

(a) e (b) 1

(c) 1/e (d) $1/e^2$

[2005 : 2 Marks]

3. The solution of the differential equation

$\frac{dy}{dx} + 2xy = e^{-x^2}$ with $y(0) = 1$ is

(a) $(1+x)e^{+x^2}$ (b) $(1+x)e^{-x^2}$

(c) $(1-x)e^{+x^2}$ (d) $(1-x)e^{-x^2}$

[2006 : 1 Mark]

4. The solution of $dy/dx = y^2$ with initial value $y(0) = 1$ bounded in the interval

(a) $-\infty \le x \le \infty$ (b) $-\infty \le x \le 1$

(c) $x < 1, x > 1$ (d) $-2 \le x \le 2$

[2007 : 2 Marks]

5. The solution of $x\frac{dy}{dx} + y = x^4$ with the condition $y(1) = \frac{6}{5}$ is

(a) $y = \frac{x^4}{5} + \frac{1}{x}$ (b) $y = \frac{x^4}{5} + \frac{4}{5x}$

(c) $y = \frac{x^4}{5} + 1$ (d) $y = \frac{x^5}{5} + 1$

[2009 : 2 Marks]

6. Consider the differential equation $\frac{dy}{dx} = (1 + y^2)x$. The general solution with constant c is

(a) $y = \tan\frac{x^2}{2} + \tan c$ (b) $y = \tan^2\left(\frac{x}{2} + c\right)$

(c) $y = \tan^2\left(\frac{x}{2}\right) + c$ (d) $y = \tan\left(\frac{x^2}{2} + c\right)$

[2011 : 2 Mark]

7. The solution of the initial value problem

$\frac{dy}{dx} = -2xy$; $y(0) = 2$ is

(a) $1 + e^{-x^2}$ (b) $2e^{-x^2}$

(c) $1 + e^{x^2}$ (d) $2e^{x^2}$

[2014 : 1 Mark, Set-4]

8. The general solution of the differential equation $\frac{dy}{dx} = \cos(x + y)$, with c as a constant, is

(a) $y + \sin(x + y) = x + c$

(b) $\tan\left(\frac{x+y}{2}\right) = y + c$

(c) $\cos\left(\frac{x+y}{2}\right) = x + c$

(d) $\tan\left(\frac{x+y}{2}\right) = x + c$

[2014 : 2 Marks, Set-2]

9. Consider the following differential equation:

$\frac{dy}{dt} = -5y$; initial condition; y = 2 at t = 0

The value of y at t = 3 is

(a) $-5e^{-10}$ (b) $2e^{-10}$

(c) $2e^{-15}$ (d) $-15e^2$

[2015 : 2 Marks, Set-2]

10. If y is the solution of the differential equation

$y^3\frac{dy}{dx} + x^3 = 0$,

$y(0) = 1$

the value of $y(-1)$ is

(a) −2 (b) −1

(c) 0 (d) 1

[2018 : 1 Mark, Set-2]

11. For the equation $\frac{dy}{dx}+7x^2y=0$, if y(0) = 3/7, then the value of y(1) is

(a) $\frac{3}{7}e^{-7/3}$ (b) $\frac{3}{7}e^{-3/7}$

(c) $\frac{7}{3}e^{-3/7}$ (d) $\frac{7}{3}e^{-7/3}$

[2019 : 1 Mark, Set-1]

12. The differential equation $\frac{dy}{dx}+4y=5$ is valid in the domain $0 \le x \le 1$ with y(0) = 2.25. The solution of the differential equation is

(a) $y = e^{4x} + 5$ (b) $y = e^{-4x} + 1.25$

(c) $y = e^{-4x} + 5$ (d) $y = e^{4x} + 1.25$

[2019 : 1 Mark, Set-2]

nth order Differential equation

Linked Answer Questions 13 and 14

The complete solution of the ordinary differential equation $\frac{d^2y}{dx^2}+p\frac{dy}{dx}+qy=0$ is $y = c_1e^{-x} + c^2e^{-3x}$.

13. Then, p and q are

(a) p = 3, q = 3 (b) p = 3, q = 4

(c) p = 4, q = 3 (d) p = 4, q = 4

[2005 : 2 Marks]

14. Which of the following is a solution of the differential equation

$$\frac{d^2y}{dx^2}+p\frac{dy}{dx}+(q+1)y=0?$$

(a) e^{-3x} (b) xe^{-x}

(c) xe^{-2x} (d) x^2e^{-2x}

[2005 : 2 Marks]

15. For $\frac{d^2y}{dx^2}+4\frac{dy}{dx}+3y=3e^{2x}$, the particular integral is

(a) $\frac{1}{15}e^{2x}$ (b) $\frac{1}{5}e^{2x}$

(c) $3e^{2x}$ (d) $C_1e^{-x} + C_2e^{-3x}$

[2006 : 1 Mark]

16. Given that $\ddot{x}+3x=0$, and x(0) = 1, $\dot{x}$(0) what is x(1)?

(a) –0.99 (b) –0.16

(c) 0.16 (d) 0.99

[2008 : 1 Mark]

17. It is given that y + 2 y' = 0, y = 0, y(0) = 0, y(1)= 0. What is y(0.5)?

(a) 0 (b) 0.37

(c) 0.62 (d) 1.13

[2008 : 2 Marks]

18. Consider the differential equation $x^2\frac{d^2y}{dx^2}+x\frac{dy}{dx}-4y=0$ with the boundary conditions of y(0) with the boundary conditions of y(0) = 0 and y(1) = 1. The complete solution of the differential equation is

(a) x^2 (b) $\sin\left(\frac{\pi x}{2}\right)$

(c) $e^x \sin\left(\frac{\pi x}{2}\right)$ (d) $e^{-x} \sin\left(\frac{\pi x}{2}\right)$

[2012 : 2 Marks]

19. The solution to the differential equation $\frac{d^2u}{dx^2}-k\frac{du}{dx}=0$ is where k is constant, subjected to the boundary conditions u(0) = 0 and u(L) = U, is

(a) $u = U\frac{x}{L}$ (b) $u = U\left(\frac{1-e^{kx}}{1-e^{kL}}\right)$

(c) $u = U\left(\frac{1-e^{-kx}}{1-e^{-kL}}\right)$ (d) $u = U\left(\frac{1+e^{kx}}{1+e^{kL}}\right)$

[2013 : 2 Marks]

20. If y = f(x) is the solution of $\frac{d^2y}{dx^2}$ = 0, with the boundary conditions y = 5 at x = 0, and $\frac{dy}{dx}$ = 2 at x = 10, f(15) = ________.

[2014 : 2 Marks, Set-1]

21. Consider two solutions x(t) = x_1(t) and x(t) = x_2(t) of the differential equation $\frac{d^2x(t)}{dt^2} + x(t) = 0$, t > 0, such that $x_2 = 0, \frac{dx_2(t)}{dt}\Big|_{t=0} = 1$.

The Wronskian $W(t) = \begin{vmatrix} x_1(t) & x_2(t) \\ \frac{dx_1(t)}{dt} & \frac{dx_2(t)}{dt} \end{vmatrix}$ at t = π/2 is

(a) 1 (b) –1

(c) 0 (d) π/2

[2014 : 1 Mark, Set-2]

22. Finding the solution of $\frac{d^2y}{dx^2} = y$ which passes through the origin and the point $\left(\ln 2, \frac{3}{4}\right)$

(a) $y = \frac{1}{2}e^x - e^{-x}$ (b) $y = \frac{1}{2}(e^x + e^{-x})$

(c) $y = \frac{1}{2}(e^x - e^{-x})$ (d) $y = \frac{1}{2}e^x + e^{-x}$

[2015 : 2 Marks, Set-1]

23. If y = f(x) satisfies the boundary value problem $y'' + 9y = 0, y(0), y\left(\frac{\pi}{2}\right) = \sqrt{2}$ then $y\left(\frac{\pi}{4}\right)$ is ___

[2016 : 2 Marks, Set-1]

24. The differential equation $\frac{d^2y}{dx^2} + 16y = 0$ for y(x) with the two boundary conditions $\frac{dy}{dx}\Big|_{x=0} = 1$ and $\frac{dy}{dx}\Big|_{x=\frac{\pi}{2}} = -1$ has

(a) no solution

(b) exactly two solutions

(c) exactly one solution

(d) infinitely many solutions

[2017 : 1 Mark, Set-1]

25. Consider the differential equation 3y''(x) + 27y(x) = 0 with initial conditions y(0) and y'(0) = 2000. The value of y at x = 1 is ________.

[2017 : 2 Marks, Set-2]

26. Given the ordinary differential equation

$$\frac{d^2y}{dx^2} + \frac{dy}{dx} - 6y = 0$$

with y(0) = 0 and $\frac{dy}{dx}(0) = 1$, the value of y(1) is ________ (correct to two decimal places).

[2018 : 2 Marks, Set-2]

27. A differential equation is given as

$$x^2\frac{d^2y}{dx^2} - 2x\frac{dy}{dx} + 2y = 4$$

The solution of the differential equation in terms of arbitrary constants C_1 and C_2 is

(a) $y = C_1x^2 + C_2x + 4$ (b) $y = \frac{C_1}{x^2} + C_2x + 2$

(c) $y = C_1x^2 + C_2x + 2$ (d) $y = \frac{C_1}{x^2} + C_2x + 4$

[2019 : 2 Marks, Set-2]

Partial differentiation and its application

28. The partial differential equation

$$\frac{\partial^2 \varphi}{\partial x^2} + \frac{\partial^2 \varphi}{\partial y^2} + \left(\frac{\partial \varphi}{\partial x}\right) + \left(\frac{\partial \varphi}{\partial y}\right) = 0 \text{ has}$$

(a) degree 1, order 2

(b) degree, 1 order 1

(c) degree 2, order 1

(d) degree 2, order 2

[2007 : 1 Mark]

29. The Blasius equation, $\frac{d^3f}{d\eta^3} + \frac{f}{2}\frac{d^2f}{d\eta^2} = 0$, is a

(a) second order nonlinear ordinary differential equation

(b) third order nonlinear ordinary diferential equation

(c) third order linear ordinary differential equation

(d) mixed order nonlinear ordinary differential equation

[2010 : 1 Mark]

30. The partial differential equation

$$\frac{\partial u}{\partial t}+u\frac{\partial u}{\partial x}=\frac{\partial^2 u}{\partial x^2} \text{ is}$$

(a) linear equation of order 2
(b) non-linear equation of order 1
(c) linear equation of order 1
(d) non-linear equation of order 2

[2013 : 1 Mark]

31. Consider the following partial differential equation u(x, y) with the constant c > 1:

$$\frac{\partial u}{\partial y}+c\frac{\partial u}{\partial x}=0$$

Solution of this equation is
(a) u(x, y) = f(x + cy)
(b) u(x, y) = f(x – cy)
(c) u(x, y) = f(cx + y)
(d) u(x, y) = f(cx – y)

[2017 : 1 Mark, Set-1]

32. Consider a function u which depends on position x and time t. The partial differential equation

$$\frac{\partial u}{\partial t}=\frac{\partial^2 u}{\partial x^2} \text{ is known as the}$$

(a) Wave equation
(b) Heat equation
(c) Laplace's equation
(d) Elasticity equation

[2018 : 1 Mark, Set-2]

ANSWERS

1. (a)	**2.** (d)	**3.** (b)	**4.** (c)	**5.** (a)	**6.** (d)	**7.** (b)	**8.** (d)	**9.** (c)	**10.** (c)
11. (a)	**12.** (b)	**13.** (c)	**14.** (c)	**15.** (b)	**16.** (b)	**17.** (a)	**18.** (a)	**19.** (b)	**21.** (a)
22. (c)	**24.** (a)	**27.** (c)	**28.** (a)	**29.** (b)	**30.** (d)	**31.** (b)	**32.** (b)		

EXPLANATIONS

1. Given differential equation is

$$\frac{dy}{dx^2}+y^2=0$$

$$\Rightarrow \frac{dy}{y^2}=-dx$$

On integration, we get

$$\int\frac{dy}{y^2}=-\int dx$$

$$-\frac{1}{y}=x-c$$

where c is a constant

$$\Rightarrow -\frac{1}{y}=-(x+c)$$

$$\therefore y=\frac{1}{x+c}$$

2. $x^2y'+2xy=\dfrac{2\ln x}{x}$

$$\Rightarrow y'+\frac{2}{x}y=\frac{2\ln x}{x^3}$$

Comparing, we get $P=\dfrac{2}{x}, Q=\dfrac{2\ln x}{x^3}$

Now, $\text{I.F.}=e^{\int Pdx}=e^{\int\frac{2}{x}dx}=e^{2\ln x}=x^2$

$$y(\text{I.F.})=\int Q\ \text{I.F.}dx+C$$

$$x^2y=\int\frac{2\ln x}{x^3}.x^2dx+C$$

Solving above, find the value of C.
Putting $x = 1$, then find the value of y at $x = e$.

which is $y=\dfrac{1}{e^2}$

3. First order equation, $\dfrac{dy}{dx}+Py=Q$, where $P = 2x$ and $Q=e^{-x^2}$

Since P and Q are functions of x, then
∴ Integrating factor,

$$\text{I.F.}=e^{+\int Pdx}=e^{+x^2}$$

$$e^{x^2}\frac{dy}{dx}+(2xe^{x^2})y=e^{-x^2}\cdot e^{+x^2}$$

$$\frac{d}{dx}(ye^{x^2})=1$$

$$ye^{x^2}=x+C$$

Since, $y(0) = 1, \Rightarrow C = 1$

$\therefore \quad y = (1 + x)\, e^{-x^2}$

4. Given : $\dfrac{dy}{dx} = y^2$

$\Rightarrow \quad \dfrac{dy}{dy^2} = dx$

Integrating $-\dfrac{1}{y} = x + c$

At $\quad y = 0, c = -1$

$\therefore \quad x = 1 - \dfrac{1}{y} \Rightarrow y = \dfrac{-1}{x-1}$

Only (c) $\quad x \neq 0$

5. Given: $\quad x\dfrac{dy}{dx} + y = x^4$

or $\quad \dfrac{dy}{dx} + \dfrac{y}{x} = x^3$

Its a linear differential equation of the form

$$\frac{dy}{dx} + Py = Q$$

where P, Q are only functions of x

$$P = \frac{1}{x}, \quad Q = x^3$$

Solution is given by

$$ye^{\int Pdx} = \int\left(Qe^{\int Pdx}\right)dx + C$$

or $\quad y.e^{\int \frac{1}{x}dx} = \int\left(Q.e^{\int \frac{1}{x}dx}\right)dx + C$

or $\quad ye^{lnx} = \int x^3.e^{lnx} + C$

or $\quad yx = \int x^3.xdx + C$

or $\quad yx = \int x^4 dx + C$

or $\quad yx = \dfrac{x^5}{5} + C$

or $\quad y = \dfrac{x^4}{5} + \dfrac{C}{x}$

Given condition : $y(1) = \dfrac{6}{5}$

$\therefore \quad \dfrac{6}{5} = \dfrac{(1)^4}{5} + \dfrac{C}{1}$

or $\quad C = \dfrac{6}{5} - \dfrac{1}{5} = 1$

$\therefore \quad y = \dfrac{x^4}{5} + \dfrac{1}{x}$

6. Given differential equation is

$$\frac{dy}{dx} = (1 + y^2)\,x$$

$\Rightarrow \quad \dfrac{dy}{1+y^2} = xdx$

Integrating, we have

$\Rightarrow \quad \tan^{-1} y = \dfrac{x^2}{2} + c$

$\Rightarrow \quad y = \tan\left(\dfrac{x^2}{2} + c\right)$

7. Given $\dfrac{dy}{dx} = -2y$, $y(0) = 2$ is

$\Rightarrow \quad \dfrac{dy}{dx} + 2xy = 0$

comparing with $\dfrac{dy}{dx} + py = Q$

$$P = 2x;\ Q = 0$$

I.F. $= e^{\int pdx} = e^{\int 2xdx} = e^{x^2}$

Solution is $y(IF) = \int Q(IP)dx + C$

$$ye^{x^2} = 0 + C$$

$\therefore \quad y = Ce^{-x^2}$

Given $y(0) = 2$

$\Rightarrow \quad 2 = C$

$\therefore \quad y = 2e^{-x^2}$

8. $\quad \dfrac{dy}{dx} = \cos(x + y)$

Put $\quad x + y = v \Rightarrow 1 + \dfrac{dy}{dx}$

$$= \frac{dv}{dx} \Rightarrow \frac{dy}{dx} = \frac{dv}{dx} - 1$$

$\Rightarrow \quad \dfrac{dv}{dx} - 1 = \cos v$

$$\Rightarrow \quad \frac{1}{1+\cos v}dv = dx \text{ (variable separable)}$$

$$\int \frac{1}{2\cos^2\frac{v}{2}}dv = \int dx + c \quad \begin{bmatrix} 1+\cos 2\theta = 2\cos^2\theta \\ 1+\cos\theta = 2\cos^2\frac{\theta}{2} \end{bmatrix}$$

$$\frac{1}{2}\int \sec^2\frac{v}{2}\,dv = x + c$$

$$\tan\frac{v}{2} = x + c$$

$$\Rightarrow \quad \tan\left(\frac{x+y}{2}\right) = x + c$$

9. $\frac{dy}{dt} = -5y$

$$\frac{+1}{y}\frac{dy}{} = -dt$$

By Integrating above equation, we get

$$\frac{+1}{0}\ln y = 5t + c$$

at t = 0

y = 2

ln 2 = c

So, ln y = – 5t + ln 2

$$\ln\frac{y}{2} = -5t$$

$$\frac{y}{2} = e^{-5t}$$

$$y = 2e^{-5t}$$

at t = 3

$y = 2\,e^{-15}$

10.

$y^3\frac{dy}{dx} = -x^3$	$\frac{x^4+y^4}{4} = c$	$x^4+y^4 = 1$
$y^3dy = -x^3dx$	$y(0) = 1,$	$y^4 = 1-x^4$
$\int y^3dy = -\int x^3dx$	$\frac{0+1}{4} = c$	$y = 4\sqrt{1-x^4}$
$\frac{y^4}{4} = \frac{-x^4}{4} + c$	$c = \frac{1}{4}$	when, x = −1 y = 0

11. From question, we have

$$\frac{dy}{dx} + 7x^2y = 0,\ y(0) = \frac{3}{7},\ y(1) = ?$$

$$\frac{dy}{dx} = -7x^2y$$

$$\frac{dy}{y} = -7x^2\,dx$$

$$\ln y = -\frac{7x^3}{3} + C$$

$$y = e^{-\frac{7x^3}{3}+C}$$

$$y(0) = e^C = \frac{3}{7}$$

$$C = \ln\left(\frac{3}{7}\right)$$

$$\ln y = -\frac{7x^3}{3} + \ln\left(\frac{3}{7}\right)$$

$$\ln y - \ln\left(\frac{3}{7}\right) = -\frac{7x^3}{3}$$

$$\ln\left(\frac{7y}{3}\right) = -\frac{7x^3}{3}$$

$$y = \frac{3}{7}e^{-\frac{7x^3}{3}}$$

$$\therefore \quad y(1) = \frac{3}{7}e^{-\frac{7}{3}}$$

12. $\frac{dy}{dx} + 4y = 5$

$$\text{I.F} = e^{\int 4dx} = e^{4x}$$

$$\Rightarrow y \times \text{I.F} = \int 5 \times (\text{I.F})dx + C$$

$$\Rightarrow y.e^{4x} = 5.\frac{e^{4x}}{4} + C$$

$$\Rightarrow 2.25 \times 1 = \frac{5.e^0}{4} + C$$

$$\Rightarrow C = 1$$

$$y.e^{4x} = \frac{5}{4}e^{4x} + 1$$

$$\Rightarrow y = \frac{5}{4} + 1e^{-4x}$$

$$\Rightarrow \boxed{y = e^{-4x} + 1.25}$$

13. $\frac{d^2y}{dx^2} + p\frac{dy}{dx} + qy = 0$

$$D^2 + pD + q = 0$$

It's solutions is $y = c_1e^{-x} + c_2e^{-3x}$

$\Rightarrow \quad m = -1, n = -3$

If m and n are two roots of the above equation, then

$$m + n = -p,$$

$$\Rightarrow \quad -1 - 3 = -p,$$

$$\Rightarrow \quad p = 4$$

and $mn = q,$

$\Rightarrow$ $(-1)(-3) = q,$

$\Rightarrow$ $q = 3$

14. $\dfrac{d^2y}{dx^2} + p\dfrac{dy}{dx} + (q+1)y = 0$

$\therefore$ $D^2 + pD + (q+1) = 0$

$\Rightarrow$ $D^2 + 4D + 4 = 0$

Since, $p = 4, q = 3$, therefore

$\Rightarrow$ $D = \dfrac{-4 \pm \sqrt{16-16}}{2} = -2$

Hence, it's solution is, $= xe^{-Dx} = xe^{-2x}$

15. The given differential equaiton may be written as

$$D^2y + 4Dy + 3y = 3e^{2x}$$

$\therefore$ $\text{P.I.} = \dfrac{3e^{2x}}{(D^2 + 4D + 3)}$

Substituting D = 2, we get

$$\text{P.I.} = 3e^{2x}\left(\frac{1}{15}\right) = \frac{e^{2x}}{5}$$

16. $\ddot{x} + 3x = 0$

$\Rightarrow$ $(D^2 + 3)x = 0$

$$\text{P.I.} = \left(\frac{1}{D^2+3}\right)(0) = 0$$

Now C.F. is given by,

$$C_1 e^{m_1 t} + C_2 e^{m_2 t}$$

$\therefore$ $m^2 + 3 = 0$

$\Rightarrow$ $m = \pm i\sqrt{3}$

Hence the solution is C.F. + P.I.

i.e. $C_1 e^{i\sqrt{3}t} + C_2 e^{-i\sqrt{3}t}$

But $x(0) = 1$

$\Rightarrow$ $C_1 + C_2 = 1$

and $\dot{x}(0) = 0,\ \dot{x} = i\sqrt{3}\,C_1 e^{i\sqrt{3}t} - i\sqrt{3}\,C_2 e^{-i\sqrt{3}t}$

$\Rightarrow$ $\dot{x}(0) = 0$

$\Rightarrow$ $C_1 = C_2 = \dfrac{1}{2}$

$\therefore$ $x = \dfrac{1}{2}\left[e^{i\sqrt{3}t} + e^{-i\sqrt{3}t}\right]$

$\Rightarrow$ $x(1) = \dfrac{1}{2}\left[e^{i\sqrt{3}} + e^{\frac{1}{i\sqrt{3}}}\right]$

$= \cos\sqrt{3} = -0.16$

17. $y'' + 2y' + y = 0$

A.E is, $D^2 + 2D + 1 = 0$

$\Rightarrow$ $(D+1)^2 = 0$

$\Rightarrow$ $m = -1$

The C.F. is $(C_1 + C_2 x)e^{-x}$ and P.I. = 0

Now $y(0) = 0$

$\Rightarrow$ $C_1 = 0$

and $y(1) = 0$

$\Rightarrow$ $C_2 e^{-1} = 0,$

$\Rightarrow$ $C_2 = 0$

$\Rightarrow$ C.F. = 0

$\Rightarrow$ $y(0.5) = 0$

18. $x^2\dfrac{d^2y}{dx^2} + x\dfrac{dy}{dx} - 4y = 0$

$\Rightarrow$ $x^2y'' + xy' - xy = 0$

From Euler-Cauchy's Equation

$$a_0 x^2y'' + a_1 xy' + a_2 y = 0$$

Given: $a_0 = 1, a_1 = 1, a_2 = -4$

$\therefore$ $a_0 m(m-1) + a_1 m + a_2 = 0$

$\Rightarrow$ $m(m-1) + m - 4 = 0$

$\Rightarrow$ $m^2 = 4$

$\Rightarrow$ $m = \pm 2$

$$y(x) = Ax^2 + Bx^{-2} = Ax^2 + \frac{B}{x^2}$$

Applying Boundary condition

$$y(0) = A0 + \frac{B}{0} = 0$$

$\therefore$ $B = 0$

and $y(1) = A(1) = 1$

$\therefore$ $A = 1$

$y(x) = x^2$

20. $\dfrac{d^2y}{dx^2} = 0$

$\dfrac{dy}{dx} = C_1$

$\Rightarrow C_1 = 2$

$y = C_1 x + C_2$

at x = 0; $y = 5 = C_2$

$\therefore$ $y = 2x + 5$

at $y(15) = 2 \times 15 + 5 = 35$

21. Given Differential equation is

$$\frac{d^2x(t)}{dt^2} + x(t) = 0$$

Auxiliary equation is $m^2 + 1 = 0$

$m = 0 \pm i$

Complementary y solution is

$$x_c = c_1 \cos t + c^2 \sin t$$

Particular solution $x_p = 0$

∴ General solution $x = c_1 \cos t + c_2 \sin t$

Let $x_1(t) = \cos t\ x_2(t) = \sin t$

Clearly $x_1(0) = 1 \frac{dx_1}{dt} = 0$ and $x_2(0) = 0, \frac{dx_2}{dt} = 1$

$$W = \begin{vmatrix} x_1(t) & x_2(t) \\ \frac{dx_1}{dt} & \frac{dx_2}{dt} \end{vmatrix}_{x=\pi/2}$$

$$= \begin{vmatrix} \cos t & \sin t \\ -\sin t & \cos t \end{vmatrix}_{t=\pi/2} = \cos^2 t + \sin^2 t = 1$$

22. $\frac{d^2y}{dx^2} = y \Rightarrow (D^2 - 1)\,y = 0$

$$D^2 - 1 = 0$$

$$\Rightarrow \quad D = \pm 1$$

$$y = c_1 e^x + c_2 e^{-x}$$

Passes through (0, 0) and $\left(142, \frac{3}{4}\right)$ (0, 0)

$$\Rightarrow \quad 0 = C_1 + C_2 \qquad ...(1)$$

$$\left(142, \frac{3}{4}\right)$$

$$\frac{3}{4} = C_1 e^{142} + C_2 e^{-142} = C_1 2 + \frac{C_2}{2}$$

$$\Rightarrow 2C_1 + \frac{1}{2}C_2 = \frac{3}{4} \qquad ...(2)$$

solving (1) and (2)

$$\Rightarrow \quad C_1 = \frac{1}{2}$$

$$C_2 = -\frac{1}{2}$$

$$\therefore \quad y = \frac{1}{2}e^x - \frac{1}{2}e^{-x} = \frac{1}{2}(e^x - e^{-x})$$

23. $y'' + 9y = 0$

A.E. is $m^2 + 9 = 0$

$$m = \pm 3i$$

$y = y_c + y_p$

$$y = C_1 \cos 3x + C_2 \sin 3x \qquad (\because y_p = 0)$$

If $x = 0$, $y = 0$

$$0 = C_1(1) + C_2(0) \Rightarrow C_1 = 0$$

If $x = \pi/2$, $y = \sqrt{2}$

$$\sqrt{2} = C_1(0) + C_2 \sin\left(\tfrac{3\pi}{2}\right) \Rightarrow C_2 = -\sqrt{2}$$

$$\therefore \quad y = -\sqrt{2}\sin 3x$$

If $x = \pi/4$

$$y\left(\tfrac{\pi}{4}\right) = -\sqrt{2}\sin\left(\tfrac{3\pi}{4}\right) = -\sqrt{2}\left(\frac{1}{\sqrt{2}}\right) = -1$$

24. $(D^2 + 16)y = 0$

A.E. → $D^2 + 16 = 0$

$\Rightarrow \quad D = \pm 4i$

$$y = C_1 \cos 4x + C_2 \sin 4x$$

$$\frac{dy}{dx} = -4C_1 \sin 4x + 4C_2 \cos 4x$$

$$\left.\frac{dy}{dx}\right|_{x=0} = -4C_1 \times 0 + 4C_2 \times 1 = 4C_2$$

$$4C_2 = 1$$

$$C_2 = \frac{1}{4}$$

Now,

$$\left.\frac{dy}{dx}\right|_{x=\pi/2} = -4C_1 \times \sin 4/2 + 4C_2 \times \cos 4/2$$

$$\left.\frac{dy}{dx}\right|_{x=\pi/2} = 0 + 4C_2 = 4\left(\frac{1}{4}\right) = 1$$

But $\left.\frac{dy}{dx}\right|_{x=\pi/2} = -1$ (given)

Hence no solution.

25. The D.E. is 3y (x) + 27 y (x) = 0

The auxiliary equation is

$3m^2 + 27 = 0$

$m^2 + 9 = 0$

$m = \pm 3i$

Solution is $y = C_1 \cos 3x + C_2 \sin 3x$

Given that y (0) = 0

∴ $0 = C_1$

$y' = -3C_1 \cdot \sin 3x + 3c_2 \cos 3x$

$y'(0) = 2000$

$2000 = 0 + 3C_2$

$$C_2 = \frac{2000}{3}$$

∴ Solution is

$$y = \frac{2000}{3}\sin 3x$$

when x = 1, $y = \frac{2000}{3}\sin 3 = 94.08$

26. $(D^2 + D - 6)\, y = 0$

$y(0) = 0,$

$y'(0) = 1$

$(D+3)(D-2)\, y = 0$

$D = 2, -3$

$C.F. = C_1 e^{2x} + C_2 e^{-3x}$

$y = c_1 e^{2x} + C_2 e^{-3x}$

$y(0) = 0$

So, $0 = C_1 + C_2$ ___(i)

$$y = (1) = \frac{e^2 - e^{-3}}{5} = 1.4678$$

$y(0) = 1$

$1 = 2C_1 - 3C_2$ ___(ii)

From equations (i) & (ii), we get

$$C_1 = \frac{1}{5}, C_2 = \frac{-1}{5}$$

$$y = \frac{e^{2x}}{5} - \frac{e^{-3x}}{5}$$

when, x = 1

$$y = (1) = \frac{e^2 - e^{-3}}{5} = 1.4678$$

27. $$x^2\frac{d^2y}{dx^2} - 2x\frac{dy}{dx} + 2y = 4$$

Let $x = e^z$

$$CF : \frac{d^2y}{dx^2} = \theta(\theta - 1)$$

$$\Rightarrow \theta(\theta-1) - 2\theta + 2 = 0$$

$$\Rightarrow \theta^2 - 3\theta + 2 = 0$$

$$(\theta-1)(\theta-2) = 0$$

$$\theta = 1, 2$$

$$\theta = C_1e^z + C_1e^{2z}$$

$$y = C_1x^2 + C_2x$$

$$PI : -\frac{4}{(\theta-1)} + \frac{4}{(\theta-2)}$$

$$= +4(1-\theta)^{-1} - \frac{4}{2}\left(1 - \frac{\theta}{2}\right)^{-1}$$

$$= +4(1+\theta+...) - 2\left(1 + \frac{\theta}{2} + ...\right)$$

$= +4 - 2 = 2$ (Neglecting higher order term)

$y = CF + PI$

$$\boxed{y = C_1x^2 + C_2x + 2}$$

28. A order is 2 and degree is 1.

29. According to Blasius equation, $f''' + \frac{f}{2}.f'' = 0$, is a third order linear ordinary differential equation.

30. Order is index of a derivative present in a PDE (i.e.) maximum there.

Hence, $\frac{\partial^2 u}{\partial n^2} \Rightarrow$ highest index 2 $\Rightarrow$ order = 2

if y, y_2 solution of a PDE, then PDE is linearly independent.

$$\frac{\partial y_1}{\partial t} + \frac{y_1 \partial y_1}{\partial tx} = \frac{\partial y_1^2}{\partial x^2} \quad ...(1)$$

$$\frac{\partial y_2}{\partial t} + \frac{y_2 \partial y_2}{\partial x} = \frac{\partial^2 y_2}{\partial x^2} \quad ...(2)$$

If $y = ay_1 + a_1y_2$ is a sum of the PDE

$$\frac{\partial}{\partial t}(ay_1 + a_1y_1) + (ay_1 + a_2y_2)\frac{\partial}{d_2}(ay_1 + ay_2) = \frac{\partial^2}{d2^2}(ay_1 + ay_2)$$

Hence PDE is non linear of order 2

31. $u = f(x - cy)$

$$\frac{\partial u}{\partial x} = f'(x - cy)(1)$$

$$\frac{\partial u}{\partial y} = f'(x - cy)(-c)$$

$= -c.\, f^1(x-cy)$

$$= -c.\frac{\partial u}{\partial x}$$

$$\therefore \frac{\partial u}{\partial y} + c\frac{\partial u}{\partial x} = 0$$

32. $$\frac{\partial u}{\partial t} = \frac{\partial^2 u}{\partial x^2}$$

is called as heat equation.

4 CHAPTER

Complex Variables

Analysis of Previous GATE Papers

Topic	Year/Set → Marks Distribution →		2019 Set 1	2019 Set 2	2018 Set 1	2018 Set 2	2017 Set 1	2017 Set 2	2016 Set 1	2016 Set 2	2016 Set 3	2015 Set 1	2015 Set 2	2015 Set 3
Complex function and derivatives	1 Mark	MCQ Type										1		
		Numerical Type												
	2 Marks	MCQ Type												
		Numerical Type												
	Total											1		
Complex Integral and Analytic function	1 Mark	MCQ Type		1										
		Numerical Type								1				
	2 Marks	MCQ Type	1											
		Numerical Type												
	Total		2	1						1				
Cauchy's Integral Theorem and Residues	1 Mark	MCQ Type			1									
		Numerical Type												
	2 Marks	MCQ Type				1		1						
		Numerical Type												
	Total				1	2		2						

Complex function and derivatives

1. If $\varphi(x, y)$ and $\psi(x, y)$ are functions with continuous second derivatives, then $\varphi(x, y) + i\psi(x, y)$

(a) $\frac{\partial \varphi}{\partial x} = -\frac{\partial \psi}{\partial x}, \frac{\partial \varphi}{\partial y} = \frac{\partial \psi}{\partial y}$

(b) $\frac{\partial \varphi}{\partial y} = -\frac{\partial \psi}{\partial x}, \frac{\partial \varphi}{\partial x} = \frac{\partial \psi}{\partial y}$

(c) $\frac{\partial^2 \varphi}{\partial x^2} + \frac{\partial^2 \varphi}{\partial y^2} = \frac{\partial^2 \psi}{\partial x^2} + \frac{\partial^2 \psi}{\partial y^2} = 1$

(d) $\frac{\partial \varphi}{\partial x} + \frac{\partial \varphi}{\partial y} = \frac{\partial \psi}{\partial x} + \frac{\partial \psi}{\partial y} = 0$

[2003 : 2 Marks]

2. The argument of the complex number $\frac{1+i}{1-i}$, where $i = \sqrt{-1}$ is

(a) $-\pi$ (b) $-\frac{\pi}{2}$

(c) $\frac{\pi}{2}$ (d) π

[2006 : 1 Mark]

3. The modulus of the complex number $\left(\frac{3+4i}{1-2i}\right)$ is

(a) 5 (b) $\sqrt{5}$

(c) $1/\sqrt{5}$ (d) 1/5

[2010 : 1 Mark]

4. The product of two complex numbers (1+ i) and 2 – 5i is

(a) 7 – 3i (b) 3 – 4i

(c) –3 – 4i (d) 7 + 3i

[2011 : 1 Mark]

5. Given two complex numbers

$z_1 = 5 + \left(5\sqrt{3}\right)i$ and $z_2 = \frac{2}{\sqrt{3}} + 2i$

the argument of $\frac{z_1}{z_2}$ in degree is

(a) 0 (b) 30

(c) 60 (d) 90

[2015 : 1 Mark, Set-1]

Complex Integral and Analytic function

6. Assuming $i = \sqrt{-1}$ and t is a real number $\int_0^{\pi/3} e^{it} dt$ is

(a) $\frac{\sqrt{3}}{2} + i\frac{1}{2}$ (b) $\frac{\sqrt{3}}{2} - i\frac{1}{2}$

(c) $\frac{1}{2} + i\left(\frac{\sqrt{3}}{2}\right)$ (d) $\frac{1}{2} + i\left(1 - \frac{\sqrt{3}}{2}\right)$

[2006 : 2 Marks]

7. An analytic function of a complex variable z = x+ iy is expressed as f(z) = u(x, y) + i v(x, y) where $i = \sqrt{-1}$. If u = xy, the expression for v should be

(a) $\frac{(x+y)^2}{2} + k$ (b) $\frac{x^2 - y^2}{2} + k$

(c) $\frac{y^2 - x^2}{2} + k$ (d) $\frac{(x-y)^2}{2} + k$

[2009 : 2 Marks]

8. If(z) = u(x, y) + iv(x, y) is an analytic function or complex variable z = x + iy where $i = \sqrt{-1}$ u (x, y) = 2 xy, then v(x, y) may be expressed as

(a) $-x^2 + y^2$ + constant

(b) $x^2 - y^2$ + constant

(c) $x^2 + y^2$ + constant

(d) $-(x^2 + y^2)$ + constant

[2008 : 1 Mark]

9. An analytic function of a complex variable z = x + iy is expressed as f(z) = u(x, y)+ iv(x, y), where $i = \sqrt{-1}$. If u(x, y) = 2xy, then v(x, y) must be

(a) $x^2 + y^2$ + constant

(b) $x^2 - y^2$ + constant

(c) $-x^2 + y^2$ + constant

(d) $-x^2 - y^2$ + constant

[2014 : 2 Marks, Set-2]

10. An analytic function of a complex variable z = x + iy is expressed as f(z) = u(x, y)+ iv(x, y), where $i = \sqrt{-1}$. If $u(x, y) = x^2 - y^2$, then expression for v(x, y) in terms of x, y and a general constant c would be

(a) xy + c (b) $\frac{x^2 + y^2}{2} + c$

(c) 2xy + c (d) $\frac{(x-y)^2}{2} + C$

[2014 : 1 Mark, Set-3]

11. If z is a complex variable, the value of $\int_{5}^{3i} \frac{dz}{z}$

(a) 0.511 – 1.57i (b) –0.511 + 1.57i

(c) 0.511 – 1.57i (d) 0.511 + 1.57i

[2014 : 2 Marks, Set-4]

12. A function f of the complex variable z = x + iy, is given as f(x, y) = u(x, y) + iv(x, y), where u(x, y) = 2 kxy and v(x, y) = $x^2 - y^2$. The value of k, for which the function is analytic, is_____

[2016 : 1 Mark, Set-2]

13. A harmonic function is analytic if it satisfies the Laplace equation. If u(x, y) = $2x^2 - 2y^2 + 4xy$ is a harmonic function, then its conjugate harmonic function v (x, y) is

(a) $4y^2 - 4xy$ + constant

(b) $4xy - 2x^2 + 2y^2$ + constant

(c) $2x^2 - 2y^2 + xy$ + constant

(d) $-4xy + 2y^2 - 2x^2$ + constant

[2019 : 2 Marks, Set-1]

14. An analytic function f(z) of complex variable z = x + iy may be written as f(z) = u(x, y) + iv(x, y). Then u(x, y) and v(x, y) must satisfy

(a) $\frac{\partial u}{\partial x} = -\frac{\partial v}{\partial y}$ and $\frac{\partial u}{\partial y} = \frac{\partial v}{\partial x}$

(b) $\frac{\partial u}{\partial x} = \frac{\partial v}{\partial y}$ and $\frac{\partial u}{\partial y} = -\frac{\partial v}{\partial x}$

(c) $\frac{\partial u}{\partial x} = -\frac{\partial v}{\partial y}$ and $\frac{\partial u}{\partial y} = -\frac{\partial v}{\partial x}$

(d) $\frac{\partial u}{\partial x} = \frac{\partial v}{\partial y}$ and $\frac{\partial u}{\partial y} = \frac{\partial v}{\partial x}$

[2019 : 1 Mark, Set-2]

Cauchy's Integral theorem and residues

15. The integral $\oint f(z)dz$ evaluated around the unit circle on the complex plane for f(z) = $\frac{\cos z}{z}$ is

(a) $2\pi i$ (b) $4\pi i$

(c) $-2\pi i$ (d) 0

[2008 : 2 Marks]

16. The value of $\oint_r \frac{3z-5}{(z-1)(z-2)} dz$ along a closed path R is is equal to ($4\pi i$), where z = x + iy and i = $\sqrt{-1}$. The correct path r is

(a)

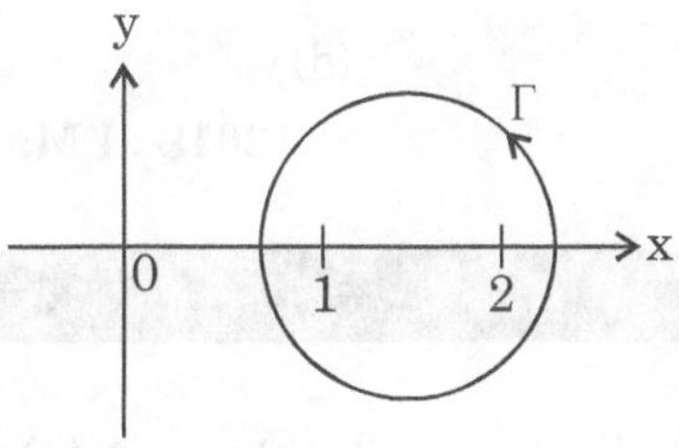

(b)

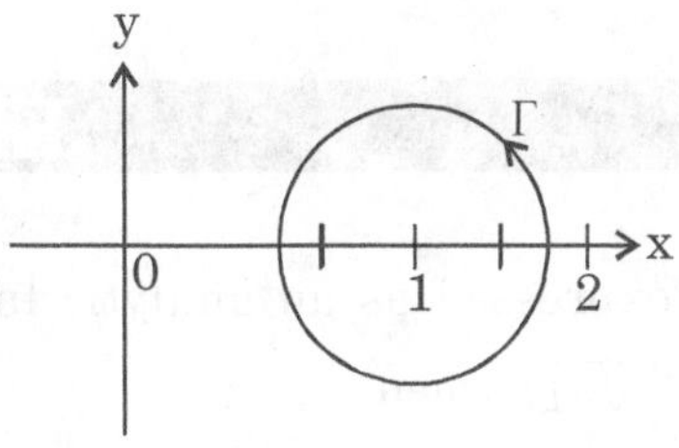

(c)

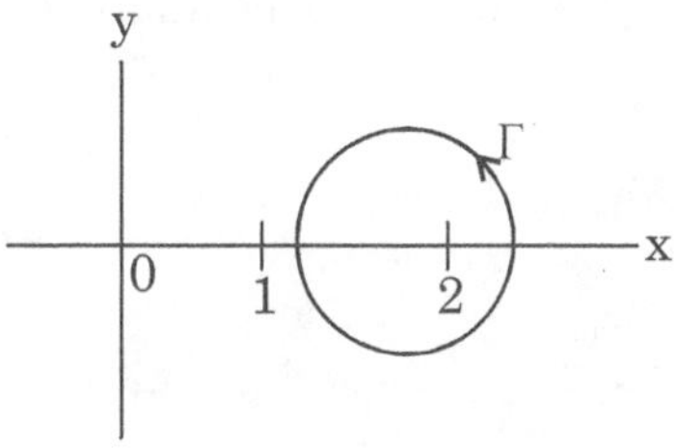

(d)

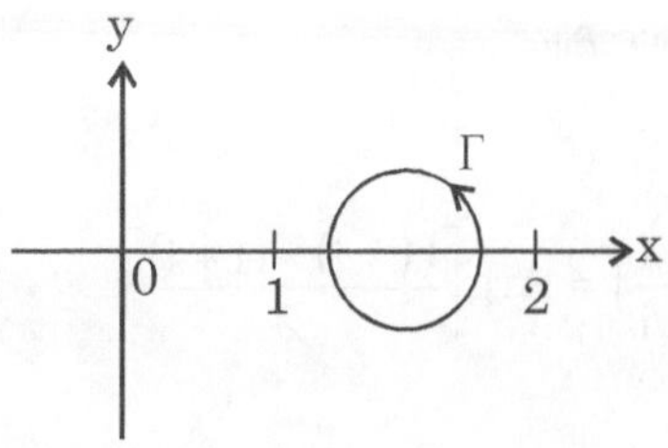

[2009 :1 Mark]

17. If f(z) = $(x^2 + ay^2)$ + ibxy is a complex analytic function of z = x + iy, where i = $\sqrt{-1}$, then

(a) a = –1, b = –1

(b) a= –1, b = 2

(c) a = 1, b = 2

(d) a = 2, b = 2

[2017 : 2 Marks, Set-2]

18. F(z) is a function of the complex variable z = x + iy given by

F(z) = iz + k Re(z) + i Im(z)

For what value of k will F(z) satisfy the Cauchy Riemann equations?

(a) 0 (b) 1

(c) –1 (d) y

[2018 : 1 Mark, Set-1]

19. Let z be a complex variable. For a counterclockwise integration around a unit circle C, centered at origin.

$\oint_c \frac{1}{5z-4} dz = A\pi i$ the value of A is

(a) 2/5 (b) 1/2

(c) 2 (d) 4/5

[2018 : 2 Marks, Set-2]

ANSWERS

1. (b)	**2.** (c)	**3.** (b)	**4.** (a)	**5.** (a)	**6.** (a)	**7.** (c)	**8.** (a)	**9.** (c)	**10.** (c)
11. (b)	**13.** (b)	**14.** (*b*)	**15.** (a)	**16.** (b)	**17.** (b)	**18.** (b)	**19.** (a)		

EXPLANATIONS

1. Can be expressed as an analytic function of $\phi + i\psi\left(i = \sqrt{-1}\right)$, when

The necessary condition for a function

f (z) = ϕ (x,y) + i ψ (x, y) to be analytic

(i) $\frac{\partial\phi}{\partial x} = \frac{\partial\psi}{\partial y}$

(ii) $\frac{\partial\phi}{\partial y} = \frac{-\partial\psi}{\partial x}$

are known as Cauchy Rieman equations, provided

$\frac{\partial\phi}{\partial x}, \frac{\partial\phi}{\partial y}, \frac{\partial\psi}{\partial x}, \frac{\partial\psi}{\partial y}$ exist.

2. $\text{Arg}\left(\frac{1+i}{1-i}\right) = \text{Arg}\left[\frac{(1+i)\times(1+i)}{2}\right]$

$\text{Arg}\left(\frac{1+i^2+2i}{2}\right)$

= Arg [i]

$\text{Arg}(x + iy) = \tan^{-1}\left(\frac{y}{x}\right)$

$\text{Arg}(0 + i) = \tan^{-1}\left(\frac{1}{0}\right)$

= tan–1 (∞)

$= \frac{\pi}{2}$

3. Given : $z = \frac{3+4i}{1-2i}$

Rationalizing the denominator, we get

$$z = \frac{3+4i}{1-2i} \times \frac{1+2i}{1+2i}$$

$$= \frac{3+6i+4i-8}{5}$$

$$= \frac{10i-5}{5} = 2i-1$$

$\therefore$ Modulus of $|z| = |2i-1| = \sqrt{2^2+(-1)^2} = \sqrt{5}$

4. (1 + *i*) (2 – 5*i*) = 2 – 5*i* + 2*i* + 5

= 7 – 3*i*

5. $z_1 = 5 + \left(5\sqrt{3}\right)i$

$$\arg z_1 = \tan^{-1}\left(\frac{5\sqrt{3}}{5}\right) = \tan^{-1}\left(\sqrt{3}\right) = 60^\circ$$

$$z_2 = \frac{2}{\sqrt{3}} + 2i$$

$$\arg z_2 = \tan^{-1}\left(\frac{2}{\frac{2}{\sqrt{3}}}\right) = \tan^{-1}\left(\sqrt{3}\right) = 60^\circ$$

$$\arg\left(\frac{z_1}{z_2}\right) = \arg(z_1) - \arg(z_2)$$

$$= 60^\circ - 60^\circ = 0$$

6. $\int_0^{\frac{\pi}{3}} e^{it}.dt = \left[\frac{e^{it}}{i}\right]_0^{\frac{\pi}{3}} = \left[\frac{\cos t + i \sin t}{i}\right]_0^{\frac{\pi}{3}}$

$= \frac{1}{i}\left[\frac{1}{2}+\frac{i\sqrt{3}}{2}-1\right]$

$= \left[\frac{-1}{2i}+\frac{\sqrt{3}}{2}\right] = \left[\frac{\sqrt{3}}{2}+i\frac{1}{2}\right]$

7. Given u = x,y

For analytic function

$\frac{\partial u}{\partial x} = \frac{\partial v}{\partial y}$

and $\frac{\partial u}{\partial y} = -\frac{\partial v}{\partial x}$

By Milne Thomson method

Let $w = u + zv$

$\therefore \frac{dw}{dz} = \frac{\partial y}{\partial x} + i\frac{\partial v}{\partial x}$

$= \frac{\partial u}{\partial x} - i\frac{\partial u}{\partial y}$

or $\frac{dw}{dz} = y - ix$

Replacing x by z and y by 0, we get

$\frac{dw}{dz} = 0 - iz$

where, $z = x + iy$

$\therefore dw = -iz\,dz$

Integrating, $w = -i\frac{z^2}{2} + C$

where C is a constant,

$\therefore V = \text{Im}\left[-i\frac{z^2}{2}+C\right]$

$= \text{Im}\left[-i\frac{(x^2-y^2+2ixy)}{2}\right]$

or $v = \frac{y^2-x^2}{2}$

8. Given $f(z) = u + iv$ is analytic and $u = 2xy$

We know that if $f(z)$ is analytic

then CR equations will be satisfied.

i.e., $\frac{\partial u}{\partial x} = \frac{\partial v}{\partial y}$ and $\frac{\partial u}{\partial y} = -\frac{\partial v}{\partial x}$

Verify the options which satisfy the above C-R equations.

$v = -x^2 + y^2$ + constant, satisfies the C-R equation.

9. Given $f(z) = u + iv$ is analytic and $u = 2xy$

We know that if $f(z)$ is analytic

then CR equations will be satisfied.

i.e., $\frac{\partial u}{\partial x} = \frac{\partial v}{\partial y}$ and $\frac{\partial u}{\partial y} = -\frac{\partial v}{\partial x}$

Verify the options which satisfy the above C-R equations.

$v = -x^2 + y^2$ + constant, satisfies the C-R equation.

10. Given $f(z) = \mu x(x, y) + iv(x, y)$ is analytic and $x = x^2 - y^2$

We know that if $f(z) = \mu + iv$ is analytic then C-R equations will be satisfied.

i.e., $\frac{\partial \mu}{\partial x} = \frac{\partial v}{\partial xy}$ and $\frac{\partial \mu}{\partial y} = \frac{-\partial v}{\partial x}$

$\therefore\ v = 2xy + c$ is correct answer

11. $\int_5^{3i} \frac{dz}{z} = \ln(z)\Big|_5^{3i} = \ln 3i - \ln 5$

$= \ln 3 + i\frac{\pi}{2} - (\ell n 5 + i0)$

$\left[\therefore \ln z = \ln r = i\theta \text{ where } r = \sqrt{x^2+y^2}, \theta = \arg z\right]$

$= \ln 3 - \ln 5 + i\frac{22}{14} = -0.511 + 157i$

12. Given that f (z) = u + i v is analytic

$u(x,y) = 2kxy$	$v = x^2 - y^2$
$u_x = 2ky$	$v_y = -2y$
$ux = v_y$	
$k = -1$	
$u_y = 2kx$	$v_x = 2x$
$u_y = -vx$	
$2kx = -2x$	
$k = -1$	

13. $u(x, y) = 2x^2 - 2y^2 + 4xy$

As harmonic function is analytic therefore,

$u_x = V_y$

$u_y = -V_x$

$u_x = 4x + 4y$

$u_y = -4y + 4x$

$\frac{\partial v}{\partial y} = V_y = 4x + 4y$

$V = 4xy + 2y^2 + f(x)$

Now $\frac{\partial v}{\partial y} = 4y + f'(x) = 4y - 4x$

$f'(x) = -4x$

$f(x) = -2x^2 + C$

So, conjugate harmonic function v(x, y) is,

$V = 2y^2 - 2x^2 + 4xy + C$

14. $f(z) = u(x, y) + iv(x, y)$

$$\frac{\partial u}{\partial x} = \frac{\partial v}{\partial y} \quad \dots (1)$$

$$\frac{\partial u}{\partial y} = -\frac{\partial v}{\partial x}$$

15. $f(z) = \frac{\cos z}{z}$ has simple pole at $z = 0$

$\therefore$ Residue of $f(z)$ at $z = 0$

$\underset{z \to 0}{\text{Lt}} f(z) = \underset{z \to 0}{\text{Lt}} \cos z = 1$

$= 2\pi i$ (Residue at $z = 0$)

16. $$\oint_r \frac{3z-5}{(z-1)(z-2)} dz = 4\pi i = 2\pi i\,(2)$$

Sum of residues must be equal to 2

$$\underset{z=1}{\text{Res}} f(z) = \underset{z\to1}{u}\,(z-1)\frac{(3z-5)}{(z-1)(z-2)} = 2$$

$$\underset{z=2}{\text{Res}} f(z) = \underset{z\to2}{u}\,(z-2)\frac{(3z-5)}{(z-1)(z-2)} = \frac{6-5}{2-1} = 1$$

There force $z = 1$ must lies inside C

Z = 2 lies outside C

17. For analytic function,

$f(2) = (x^2 + a y^2) + ibxy$

$u + iv = (x^2 + ay^2) + i\,(bxy)$

$u = x^2 + ay^2;\ v = bxy$

$u_x = 2x;\ u_y = 2\,ay$

$V_x = b_y\ ;\ V_y = bx$

$u_x = V_y;\ u_y = -V_x$

$2\,x = bx\ ;\ 2ay = -by$

By solving , we get

$a = -1\ b = 2$

18. By Cauchy -Riemann equations

$u_x = v_y$

$u + iv = i\,(x + iy) + kx + iy$

$u + iv = kx - y + i\,(x + y)$

$u = kx - y,\ v = x + y$

$u_x = k,\ u_y = -1$

$v = x + y$

$v_x = 1;\ v_y = 1$

$u_x = v_y$

$k = 1$

19. $$\oint_c \frac{1}{5z-4} dz = A\pi i$$

$$\int \frac{\left(\frac{1}{5}\right)}{\left(z - \frac{4}{5}\right)} dz = z\pi i.f\left(\frac{4}{5}\right)$$

$$= \frac{2}{5}\pi i$$

$$A = \frac{2}{5} = 0.4$$

5 CHAPTER Probability and Statistics

Analysis of Previous GATE Papers

Topic	Year/Set → Marks Distribution ↓		2019 Set 1	2019 Set 2	2018 Set 1	2018 Set 2	2017 Set 1	2017 Set 2	2016 Set 1	2016 Set 2	2016 Set 3	2015 Set 1	2015 Set 2	2015 Set 3
Fundamentals of Probability	1 Mark	MCQ Type			2									1
		Numerical Type					1	1						
	2 Marks	MCQ Type									1		1	1
		Numerical Type		1						1				
	Total			2	2		1	1		2	2		2	3
Probablity Distribution	1 Mark	MCQ Type												
		Numerical Type											1	
	2 Marks	MCQ Type			1									
		Numerical Type												
	Total				2								1	
Statistics	1 Mark	MCQ Type							1					
		Numerical Type		1										
	2 Marks	MCQ Type			1	1								
		Numerical Type												
	Total			1	2	2			1					

Fundamentals of Probability

1. A box contains 5 black and 5 red balls. Two balls are randomly picked one after another from the box, without replacement. The probability for both balls being red is

(a) $\frac{1}{90}$ (b) $\frac{1}{2}$

(c) $\frac{19}{90}$ (d) $\frac{2}{9}$ **[2003 : 2 Marks]**

2. From a pack of regular playing cards, two cards are drawn at random. What is the probability that both cards will be Kings, if first card in NOT replaced?

(a) $\frac{1}{26}$ (b) $\frac{1}{52}$

(c) $\frac{1}{169}$ (d) $\frac{1}{221}$ **[2004 : 2 Marks]**

3. A lot has 10% defective items. Ten items are chosen randomly from this lot. The probability that exactly 2 of the chosen items are defective is

(a) 0.0036 (b) 0.1937

(c) 0.2234 (d) 0.3874

[2005 : 1 Mark]

4. A single die is thrown twice. What is the probability that the sum is neither 8 nor 9?

(a) 1/9 (b) 5/36

(c) 1/4 (d) 3/4

[2005 : 2 Marks]

5. A box contains 20 defective items and 80 non-defective items. If two items are selected at random without replacement, what will be the probability that both items are defective?

(a) $\frac{1}{5}$ (b) $\frac{1}{25}$

(c) $\frac{20}{99}$ (d) $\frac{19}{495}$ **[2006 : 1 Mark]**

6. A coin is tossed 4 times. What is the probability of getting heads exactly 3 times?

(a) $\frac{1}{4}$ (b) $\frac{3}{8}$

(c) $\frac{1}{2}$ (d) $\frac{3}{4}$ **[2008 : 1 Mark]**

7. If three coins are tossed simultaneously, the probability of getting at least one head is

(a) $\frac{1}{8}$ (b) $\frac{3}{8}$

(c) $\frac{1}{2}$ (d) $\frac{7}{8}$ **[2009 : 1 Mark]**

8. A box contains 2 washers, 3 nuts and 4 bolts. Items are drawn from the box at random one at a time without replacement. The probability of drawing 2 washers first followed by 3 nuts and subsequently the 4 bolts is

(a) $\frac{2}{315}$ (b) $\frac{1}{630}$

(c) $\frac{1}{1260}$ (d) $\frac{1}{2520}$

[2010 : 2 Marks]

9. A box contains 4 red balls and 6 black balls. Three balls are selected randomly from the box one after another, without replacement. The probability that the selected set contains one, red ball and two black balls is

(a) $\frac{1}{20}$ (b) $\frac{1}{12}$

(c) $\frac{3}{10}$ (d) $\frac{1}{2}$

[2012 : 2 Marks]

10. The probability that a student knows the correct answer to a multiple choice question is 2/3. If the student does not known the answer, then the student guesses the answer. The probability of the guessed answer being correct is 1/4. Given that the student has answered the question correctly, the conditional probability that the student known the correct answer is

(a) $\frac{2}{3}$ (b) $\frac{3}{4}$

(c) $\frac{5}{6}$ (d) $\frac{8}{9}$

[2013 : 2 Marks]

11. A box contains 25 parts of which 10 are defective. Two parts are being drawn simultaneously in a random manner from the box. The probability of both the parts being good is

(a) $\frac{7}{20}$ (b) $\frac{45}{125}$

(c) $\frac{25}{29}$ (d) $\frac{5}{9}$

[2014 : 1 Mark, Set-2]

12. A group consists of equal number of men and women. Of this group 20% of the men and 50% of the women are unemployed. If a person is selected at random from this group, the probability of the selected person being employed is ______ .

[2014 : 1 Mark, Set-3]

13. Consider an unbiased cubic dice with opposite faces coloured identically and each face coloured red, blue or green such that each colour appears only two times on the dice. If the dice is thrown thrice, the probability of obtaining red colour on top face of the dice at least twice is _____.

[2014 : 2 Marks, Set-2]

14. If $P(X) = \frac{1}{4}$, $P(Y) = \frac{1}{3}$, and $P(X \cap Y) = \frac{1}{12}$, the value of P(Y/X) is

(a) $\frac{1}{4}$ (b) $\frac{4}{25}$

(c) $\frac{1}{3}$ (d) $\frac{29}{50}$

[2015 : 1 Mark, Set-3]

15. The chance of a student passing an exam is 20%. The chance of a student passing the exam and getting above 90% marks in it is 5%, Given that a student passes the examination, the probability that the student gets above 90% marks is

(a) $\frac{1}{18}$ (b) $\frac{1}{4}$

(c) $\frac{2}{9}$ (d) $\frac{5}{18}$

[2015 : 2 Marks, Set-2]

16. The probability of obtaining at least two "SIX" in throwing a fair dice 4 times is

(a) $\frac{425}{432}$ (b) $\frac{19}{144}$

(c) $\frac{13}{144}$ (d) $\frac{125}{432}$

[2015 : 2 Marks, Set-3]

17. The probability that a screw manufactured by a company is defective is 0.1. The company sells screws in packets containing 5 screws and gives a guarantee of replacement if one or more screws in the packet are found to be defective. The probability that a packet would have to be replaced is ___. **[2016 : 2 Marks, Set-2]**

18. Three cards were drawn from a pack of 52 cards. The probability that they are a king, a queen, and a jack is

(a) $\frac{16}{5525}$ (b) $\frac{64}{2197}$

(c) $\frac{3}{13}$ (d) $\frac{8}{16575}$

[2016 : 2 Marks, Set-3]

19. A six-face fair dice is rolled a large number of times. The mean value of the outcomes is ___.

[2017 : 1 Mark, Set-1]

20. Two coins are tossed simultaneously. The probability (upto two decimal points accuracy) of getting at least one head is ___.

[2017 : 1 Mark, Set-2]

21. A six-faced fair dice is rolled five times. The probability (in%) of obtaining "ONE" at least four times is

(a) 33.3 (b) 3.33

(c) 0.33 (d) 0.0033

[2018 : 1 Mark, Set-1]

22. Four red balls, four green balls and four blue balls are put in a box. Three balls are pulled out of the box at random one after another without replacement. The probability that all the three balls are red is

(a) $\frac{1}{72}$ (b) $\frac{1}{55}$

(c) $\frac{1}{36}$ (d) $\frac{1}{27}$

[2018 : 1 Mark, Set-1]

23. The probability that a part manufactured by a company will be defective is 0.05. If such parts are selected randomly and inspected, then the probability that at least two parts will be defective is ____ (round off to two decimal places).

[2019 : 2 Marks, Set-2]

Probability distribution

24. Consider the continuous random variable with probability density function

$f(t) = 1 + t$ for $-1 \le t \le 0$

$= 1 - t$ for $0 \le t \le 5\ 1$

The standard deviation of the random variable is

(a) $\frac{1}{\sqrt{3}}$ (b) $\frac{1}{\sqrt{6}}$

(c) $\frac{1}{3}$ (d) $\frac{1}{6}$ **[2006 : 2 Marks]**

25. Let X be a normal random variable with mean 1 and variance 4. The probability P{X < 0} is

(a) 0.5

(b) greater then a zero and less than 0.5

(c) greater than 0.5 and less than 1.0

(d) 1.0

[2013 : 1 Mark]

26. A nationalized bank has found that the daily balance available in its savings accounts follows a normal distribution with a mean of Rs. 500 and a standard deviation of Rs. 50. The percentage of savings account holders, who maintain an average daily balance more than Rs. 500 is ___.

[2014 : 1 Mark, Set-4]

27. The number of accidents occurring in a plant in a month follows Poisson distribution with mean as 5.2. The probability of occurrence of less than 2 accidents in the plant during a randomly selected month is

(a) 0.029 (b) 0.034
(c) 0.039 (d) 0.044

[2014 : 2 Marks, Set-4]

28. Three vendors were asked to supply a very high precision component. The respective probabilities of their meeting the strict design specifications are 0.8, 0.7 and 0.5. Each vendor supplies one component. The probability that out of total three components supplied by the vendors, at least one will meet the design specification is ______.

[2015 : 1 Mark, Set-2]

29. The variable x takes a value between 0 and 10 with uniform probability distribution. The variable y takes a value between 0 and 20 with uniform probability distribution. The probability of the sum of variables (x + y) being greater than 20 is

(a) 0.33 (b) 0.25
(c) 0 (d) 0.50

[2019 : 2 Marks, Set-1]

Statistics

30. The standard deviation of a uniformly distributed random variable between 0 and 1 is

(a) $\frac{1}{\sqrt{12}}$ (b) $\frac{1}{\sqrt{3}}$

(c) $\frac{5}{\sqrt{12}}$ (d) $\frac{7}{\sqrt{12}}$

[2009 : 2 Marks]

31. In the following table, x is a discrete random variable and p(x) is the probability density. The standard deviation of x is

x	1	2	3
p(x)	0.3	0.6	0.1

(a) 0.18 (b) 0.36
(c) 0.54 (d) 0.6

[2014 : 2 Marks, Set-1]

32. A machine produces 0, 1 or 2 defective pieces in a day with associated probability of 1/6, 2/3 and 1/6, respectively. The mean value and the variance of the number of defective pieces produced by the machine in a day, respectively, are

(a) 1 and 1/3 (b) 1/3 and 1
(c) 1 and 4/3 (d) 1/3 and 4/3

[2014 : 2 Marks, Set-3]

33. Consider a Poisson distribution for the tossing of a biased coin. The mean for this distribution is μ. The standard deviation for this distribution is given by

(a) $\sqrt{\mu}$ (b) μ^2

(c) μ (d) $\frac{1}{\mu}$

[2016 : 1 Mark, Set-1]

34. Let X_1, X_2 be two independent normal random variables with means μ_1, μ_2 and standard deviations σ_1, σ_2 respectively. Consider $Y = X_1 - X_2$; $\mu_1 = \mu_2 = 1$, $\sigma_1 = 1$, $\mu_2 = 2$, Then,

(a) Y is normal distributed with mean 0 and variance 1

(b) Y is normally distributed with mean 0 and variance 5

(c) Y has mean 0 and variance 5, but is NOT normally distributed

(d) Y has mean 0 and variance 1, but is NOT normally distributed

[2018 : 2 Marks, Set-1]

35. Let X_1 and X_2 be two independent exponentially distributed random variables with means 0.5 and 0.25 respectively. Then $Y = \min(X_1, X_2)$ is

(a) exponentially distributed with mean $\frac{1}{6}$

(b) exponentially distributed with mean 2

(c) normally distributed with mean $\frac{3}{4}$

(d) normally distributed with mean $\frac{1}{6}$

[2018 : 2 Marks, Set-2]

36. If x is the mean of data 3, x, 2 and 4, then the mode is

[2019 : 1 Mark, Set-2]

ANSWERS

1. (d)	**2.** (d)	**3.** (b)	**4.** (d)	**5.** (d)	**6.** (a)	**7.** (d)	**8.** (c)	**9.** (d)	**10.** (d)
11. (a)	**14.** (c)	**15.** (b)	**16.** (b)	**18.** (a)	**21.** (c)	**22.** (b)	**24.** (b)	**25.** (b)	**27.** (b)
29. (b)	**30.** (a)	**31.** (d)	**32.** (a)	**33.** (a)	**34.** (b)	**35.** (a)			

EXPLANATIONS

1. The probability of drawing a red ball $= \frac{5}{10}$

If the ball is not replaced, the box will have a ball, so probability of drawing the red ball in next chance $= \frac{4}{9}$

Hence probability of drawing 2 balls

$$= \frac{5}{10} \times \frac{4}{9} = \frac{2}{9}$$

2. Probability of both cards being king

$$= \frac{4}{52} \times \frac{3}{51} = \frac{1}{221}$$

3. Probability of defective item $= \frac{10}{100} = 0.1$

Probability of not defective item = 1 – 0.1 = 0.9

So, probability that exactly 2 of the chosen items are defective

$$= {}^{10}C_2 \, (0.1)^2 \, (0.9)^8 = 0.1937.$$

4. The number of ways of coming 8 and 9 are

(2, 6), (3, 5), (4, 4), (5, 3), (6, 2), (3, 6), (4, 5), (5, 4), (6, 3)

Total ways = 9

So probability of coming 8 and 9 are

$$= \frac{9}{6 \times 6} = \frac{9}{36}$$

So probability of not coming these

$$= 1 - \frac{9}{36} = \frac{3}{4}$$

5. $P\text{ (Both defective)} = \frac{N\text{(Both defective)}}{\text{Sample space}}$

$$= \frac{{}^{20}C_2}{{}^{100}C_2} = \frac{19}{495}$$

6. Number of favourable cases are given by

H H H T
H H T H
H T H H
T H H H

Total number of cases $= {}^2C_1 \times {}^2C_1 \times {}^2C_1 \times {}^2C_1 = 16$

$\therefore$ Probability $= \frac{4}{16} = \frac{1}{4}$

7. Let probability of getting atleast one head = P(H)

then P(at least one head) = 1 – P (no head)

$\Rightarrow$ P(H) = 1 – P(all tails)

But in all cases, $2^3 = 8$

$$\therefore \quad P(H) = 1 - \frac{1}{8} = \frac{7}{8}$$

Alternately:

Probability of getting at least one head

$$= 1 - {}^3C_0 \left(\frac{1}{2}\right)^0 \left(\frac{1}{2}\right)^3 = 1 - \frac{1}{8} = \frac{7}{8}$$

8. Given : Washers – 2, Nuts – 3, Bolts – 4,

Total–9 objects

Hence probability of drawing two washers from 9 objects

$$= \frac{{}^2C_2}{{}^9C_2} = \frac{1}{36}$$

Probability of drawing, 3 nuts out of remaining 7 objects is

$$= \frac{{}^3C_3}{{}^7C_3} \frac{1}{{}^7C_3} = \frac{1}{35}$$

Probability of drawing 4 bolts out of remaining 4 bolts

$$= \frac{{}^4C_4}{{}^4C_4} = 1$$

$\therefore$ Required probability $= \frac{1}{36} \times \frac{1}{35} \times 1 = \frac{1}{1260}$

9. *Given :* $\frac{\text{Red}}{4} \ \frac{\text{Black}}{6}$

Selection will be RBB or BBR of BRB

Probability of selecting RBB $= \frac{4}{10} \times \frac{6}{9} \times \frac{5}{8}$

Probability of selecting BBR $= \frac{6}{10} \times \frac{5}{9} \times \frac{4}{8}$

Probability of selecting BRB $= \frac{6}{10} \times \frac{4}{9} \times \frac{5}{8}$

P (Red = 1) = sum of above three probabilities = 0.5

10. Probability = $\dfrac{2/3\times1}{(2/3\times1)+\left(\dfrac{1}{3}\times\dfrac{1}{4}\right)}=8/9$

11. Reqired Probability $=\dfrac{15C_2}{25C_2}=\dfrac{14\times15}{25\times24}=\dfrac{7}{20}$

12. Total % of employed person

$=100-\left(\dfrac{20+50}{2}\right)=65\%$

13. $p=\dfrac{2}{6}=\dfrac{1}{3}$

$q=1-\dfrac{1}{3}=\dfrac{2}{3}$

using Binomial distribution

$p(x\geq2)=3_{C_2}\left(\dfrac{1}{3}\right)^2\left(\dfrac{2}{3}\right)^1+3_{C_3}\left(\dfrac{1}{3}\right)^3\left(\dfrac{2}{3}\right)^0$

$=\dfrac{6}{27}+\dfrac{1}{27}$

$=\dfrac{7}{27}$

14. $P(Y/X)=\dfrac{P(X\cap Y)}{P(X)}=\dfrac{\frac{1}{12}}{\frac{1}{4}}=\dfrac{1}{3}$

15. Given that the student is passing exam, i.e., if only 20 students (out of 100) are considered, of the 5 students get more than 90%

∴ 5 out of 20 is the probability

or 1/4 is the answer.

Alternative Method :

Let the student pass the examination be A and student pass the examination and got about 90% marks be B

Now P(A) = 20% and P(A∩B) = 5%

$\therefore\quad P\left(\dfrac{B}{A}\right)=\dfrac{P(A\cap B)}{P(A)}=\dfrac{5\%}{20\%}=\dfrac{1}{4}$

16. Let P be the probability that six happens in a fair dice,

$\therefore p=\dfrac{1}{6}$

$\therefore q=\dfrac{5}{6}$

Let X, be the number of times 'six' happens Probability of obtaining atleast two 'six' in throwing a fair dice 4 times is

$=1-\{P(X=O)+P(X=1)\}$

$=1-\{{}^4C_0P^{o}q^4+{}^4C_1P'q^3\}$

$=1-\left\{\left(\dfrac{5}{6}\right)^4+\left[4\times\dfrac{1}{6}\times\left(\dfrac{5}{6}\right)^3\right]\right\}=1-\left\{\dfrac{125}{144}\right\}=\dfrac{19}{144}$

17. we require $P(x\geq1)=1-P(x=0)$

$=1-5C_0(0.1)^0(0.9)^5$

$=0.4095\approx0.41$

18. $\rho=\dfrac{{}^4C_1\times{}^4C_1\times{}^4C_1}{{}^{52}C_3}$

$\rho=\dfrac{16}{5525}$

19.

x	1	2	3	4	5	6
p(x)	1/6	1/6	1/6	1/6	1/6	1/6

We know that

$\mu=\sum x.p(x)$

$=(1+2+3+4+5+6)\times\dfrac{1}{6}$

$=21\times\dfrac{1}{6}=3.5$

20. Considering all Sample Space, we gets {HH, HT, TH, TT}

For at least one head, probability is $\dfrac{3}{4}$

P (at least one head) = 0.75

21. A dice is rolled 5 times

$n=5$

$P=$ (Probability of getting 1) $=\dfrac{1}{6}$

0% $q=1-p=1-\dfrac{1}{6}=\dfrac{5}{6}$

Probability of getting 1 at least 4 times is

$P(x\geq4)=P(x=4)+P(x=5)$

$={}^nC_4p^4q^{n-4}+{}^nC_5p^5q^{n-5}\qquad\left[\because{}^nC_rp^xq^{n-x}\right]$

$={}^5C_4p^4q^1+{}^5C_5p^5q^0$

$=5\times\left(\dfrac{1}{6}\right)^4\left(\dfrac{5}{6}\right)^1+1\times\left(\dfrac{1}{6}\right)^5\times\left(\dfrac{5}{6}\right)^0$

$=\dfrac{25}{(6)^5}+\dfrac{1}{(6)^5}=\dfrac{26}{(6)^5}=0.0033$

Required probability = 0.33%

22. (4R, 4G, 4B)

P(RRR) (without replacement)

$$= \frac{{}^4C_1}{{}^{12}C_1} \times \frac{{}^3C_1}{{}^{11}C_1} \times \frac{{}^2C_1}{{}^{10}C_1} = \frac{1}{55}$$

23. $(P)_{defective} = 0.05$

$(P)_{non\text{-}defective} = 1 - 0.05 = 0.95$

Probability that atleast two parts will be defective,

$$P = \left[{}^{15}C_0 \cdot (0.05)^0 \cdot (0.95)^{15} + {}^{15}C_1 (0.05) \cdot (0.95)^{14} \right]$$

$= 1 - 0.829$

Required probability

$\boxed{P = 0.1709} \approx 0.17$

24. $\text{Mean}(\bar{t}) = \int_{-1}^{\infty} t.f(t).dt$

$$= \int_{-1}^{0} (1 + t)dt + \int_{0}^{1} t(1 - t)dt$$

$$= \left[\frac{t^2}{2} + \frac{t^3}{3}\right]_{-1}^{0} + \left[\frac{t^2}{2} - \frac{t^3}{3}\right]_{0}^{1}$$

$$= -\left[\frac{1}{2} - \frac{1}{3}\right] + \left[\frac{1}{2} - \frac{1}{3}\right] = -\left[\frac{1}{6} + \frac{1}{6}\right] = 0$$

$$\text{Variance} = \int_{-\infty}^{\infty} (t - \bar{t})^2 f(t).dt$$

$$= \int_{-1}^{0} t^2(1 + t)dt + \int_{0}^{1} t^2(1 - t)dt = \frac{1}{6}$$

$$\text{Standard deviation} = \sqrt{\text{Variance}} = \frac{1}{\sqrt{6}}$$

25. $P(x < 0) = P\left\{\frac{(X - \bar{X})}{\sigma} < \frac{0 - \bar{X}}{\sigma}\right\}$

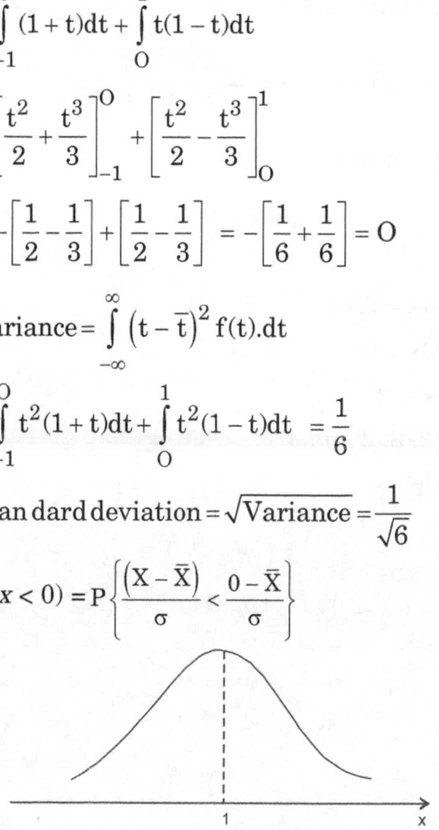

$\because \quad \bar{x} = 1, \sqrt{V} = \sigma \text{ or } \sigma = \sqrt{4} = 2$

$\therefore \quad P\left\{\frac{x - 1}{2} < -\frac{1}{2}\right\} \text{ or } P\left\{z < -\frac{1}{2}\right\}$

$$P\left(z < -\frac{1}{2}\right) = P\left(z > \tfrac{1}{2}\right)$$

$\because$ $P(x > a)$ represents area under curve for $x > a$

26. Given $M = 500$

$\sigma = 50$

$P(X > 500) = ?$

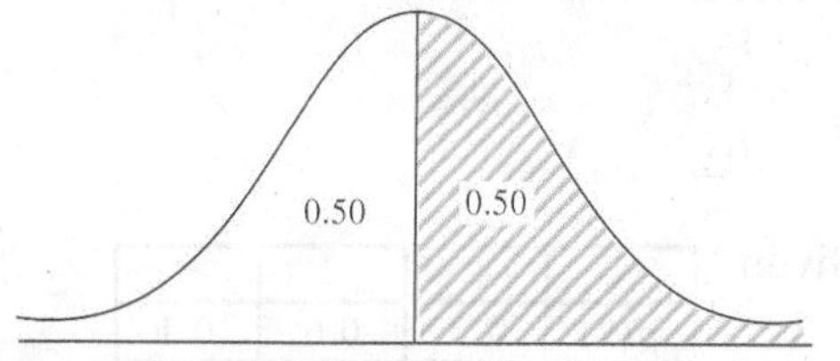

Where X follows normal distribution

We know that standard normal variable

$$z = \frac{x - M}{\sigma}$$

$$z = \frac{500 - 500}{50} = 0$$

$\therefore \quad (X > 500) = P(z > 0)$

$= 0.50$ (see figure)

27. Given $\lambda = 5.2$

Let x be random variable which follows Poisson's distribution

$$P(x < 2) = P(x = 0) + P(x = 1)$$

$$= \frac{e^{-\lambda}\lambda^0}{0!} + \frac{e^{-\lambda}}{1!}\lambda^1 = e^{-5.2}(6.2)$$

$= 0.0055 \times 6.2 = 0.034$

28. Probability (at least one will meet specification)

$= 1 -$ probability (none will meet specification)

$= 1 - (1 - 0.8) \times (1 - 0.7) \times (1 - 0.5)$

$= 1 - 0.2 \times 0.3 \times 0.5$

$= 1 - 0.03 = 0.97$

29.

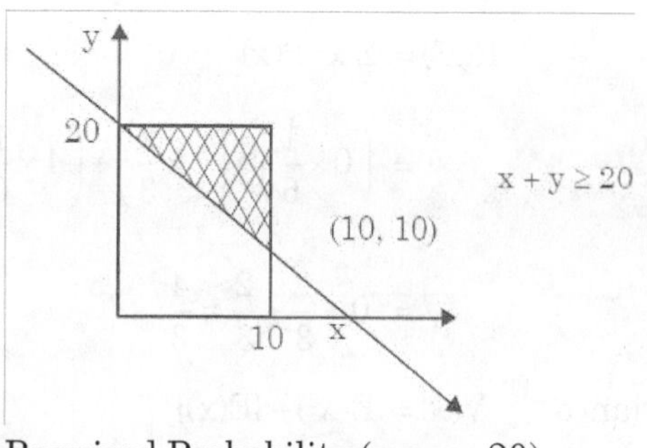

Required Probability $(x + y > 20)$

$$= \frac{\text{Shaded Area}}{\text{Total Area}}$$

$$= \frac{\frac{1}{2} \times 10 \times 10}{20 \times 10}$$

$$= \frac{50}{200} = \frac{1}{4} = 0.25$$

30. Variance, $V(x) = \frac{(b-a)^2}{12} = \frac{(1-0)^2}{12}$

∴ Standard deviation,

$$\sigma = \frac{1}{\sqrt{12}}$$

31. Given

x	1	2	3
p(x)	0.3	0.6	0.1

mean (μ) = exp(x) = 1 × 0.3 + 2 × 0.6 + 3 × 0.1

= 0.3 + 1.2 + 0.3 = 1.8

$E(x^2) = \sum x^2 P(x)$

= 1 × 0.3 + 4 × 0.6 + 9 × 0.1

= 0.3 + 2.4 + 0.9 = 3.6

Variance $v(x) = E(x^2) - \mu^2 = 3.6 - (1.8)^2$

S. D $(\sigma) = +\sqrt{v(x)}$

$= +\sqrt{3.6-(1.8)^2} = \sqrt{0.36} = 0.6$

32. Let 'x' be no. of defective pieces.

x	0	1	2
P(x)	$1/6$	$2/3$	$1/6$

Mean (μ) = E(x) = Σ x P(x)

$$= \left(0\times\frac{1}{6}\right)+\left(1\times\frac{1}{3}\right)+\left(2\times\frac{1}{6}\right)$$

$$= 0+\frac{2}{3}+\frac{1}{3}=1$$

$E(x^2) = \Sigma x^2 P(x)$

$$= \left(0\times\frac{1}{6}\right)+\left(1\times\frac{2}{3}\right)+\left(4\times\frac{1}{6}\right)$$

$$= 0+\frac{2}{3}+\frac{2}{3}=\frac{4}{3}$$

Variance, $V(x) = E(x^2) - \{E(x)\}^2$

$$= \frac{4}{3}-1 = 1/3$$

33. Standard deviation $= \sqrt{\text{Variance}} = \sqrt{\mu}$

34.

$$Y = X_1 - X_2 \begin{cases} E_{(y)} = E(x_1 - x_2) = E(x_1) - E(x_2) = 1 - 1 = 0 \\ V_{(y)} = V(x_1 - x_2) = V(x_1) + V(x_2) = 1 + 4 = 5 \end{cases}$$

Y is normally distributed with mean 0 and variance 5.

35. Mean $(X_1) = 0.5$

$$\frac{1}{\lambda_1} = 0.5$$

$$\lambda_1 = \frac{1}{0.5} = 2$$

Mean $(X_2) = 0.25$

$$\frac{1}{\lambda_2} = 0.25 \Rightarrow \lambda_2 = 4$$

y= mean (X_1, X_2)

$$\text{Mean}(y) = \frac{1}{y_1+y_2} = \frac{1}{2+4} = \frac{1}{6}$$

36. 3, x, 2 and 4

$$\text{Mean, } x = \frac{3+x+2+4}{4}$$

⇒ 4x = 9 + x

⇒ 3x = 9

x = 3

So, 3, 3, 2 and 4

Mode = 3

6 CHAPTER

Numerical Methods

Analysis of Previous GATE Papers

Topic	Year/Set → Marks Distribution ↓		2019 Set 1	2019 Set 2	2018 Set 1	2018 Set 2	2017 Set 1	2017 Set 2	2016 Set 1	2016 Set 2	2016 Set 3	2015 Set 1	2015 Set 2	2015 Set 3
Numerical Solutions of Non-linear equations	1 Mark	MCQ Type									1			
		Numerical Type							1					
	2 Marks	MCQ Type		1										
		Numerical Type												1
	Total			**2**					**1**		**1**			**2**
Numerical Solutions of ordinary differential equations	1 Mark	MCQ Type												
		Numerical Type												
	2 Marks	MCQ Type												
		Numerical Type			1				1					
	Total				**2**				**2**					
Numerical Integration	1 Mark	MCQ Type								1				
		Numerical Type	1									1		1
	2 Marks	MCQ Type					1							
		Numerical Type								1			1	
	Total		**1**				**2**			**3**		**1**	**2**	**1**

Numerical solutions of Non-linear equations

1. The values of a function f(x) are tabulated below:

x	f(x)
0	1
1	2
2	1
3	10

Using Newton's forward difference formula, the cubic polynomial that can be fitted to the above data, is

(a) $2x^3 + 7x^2 - 6x + 2$ (b) $2x^3 - 7x^2 + 6x - 2$

(c) $x^3 - 7x^2 - 6x2 + 1$ (d) $2x^3 - 7x^2 + 6x + 1$

[2004 : 2 Marks]

2. Starting from $x_0 = 1$, one step of Newton-Raphson method in solving the equation $x^3 + 3x - 7 = 0$ gives the next value (x_1) as

(a) $x_1 = 0.5$ (b) $x_1 = 1.406$

(c) $x_1 = 1.5$ (d) $x_1 = 2$

[2005 : 2 Marks]

3. Let X and Y be two independent random variables. Which one of the relations between expectation (n), variance (Var) and covariance (Cov) given below is FALSE?

(a) E (XY) = E(X) E(Y)

(b) Coy (X, Y) = 0

(c) Var (X+ Y) = Var (X) + Var (Y)

(d) $E(X^2 Y^2) = (E (X))^2 (E (Y))^2$

[2007 : 2 Marks]

4. The real root of the equation 5x – 2 cosx – 1 = 0 (up to two decimal accuracy) is _______.

[2014 : 2 Marks, Set-3]

5. Newton-Raphson method is used to find the roots of the equation, $x^3 + 2x^2 + 3x - 1 = 0$. If the initial guess is $x_0 = 1$, then the value of x after 2nd iteration is ________.

[2015 : 2 Marks, Set-3]

6. Solve the equation x = 10 cos(x) using the Newton-Raphson method. The initial guess is $x = \frac{\pi}{4}$. The value of the predicted root after the first iteration, up to second decimal, is _____.

[2016 : 1 Mark, Set-1]

7. The root of the function $f(x) = x^3 + x - 1$ obtained after first iteration on application of Newton Raphson scheme using an initial guess of $x_0 = 1$ is

(a) 0.682 (b) 0.686

(c) 0.750 (d) 1.000

[2016 : 1 Mark, Set-3]

8. The derivative of f(x) = cos(x) can be estimated using the approximation

$$f'(x) = \frac{f(x+h) - f(x-h)}{2h}$$

The percentage error is calculated as

$$\left(\frac{\text{Exact value} - \text{Approximate value}}{\text{Exact value}}\right) \times 100$$

The percentage error in the derivative of f(x) at x = π/ 6 radian, choosing h = 0.1 radian, is

(a) < 0.1% (b)> 1% and < 5%

(c) > 0.1% and < 1% (d)> 5%

[2019 : 2 Marks, Set-2]

Numerical solution of ordinary differential equations

9. Match the items in columns I and II.

Column I

P. Gauss-Seidel method

Q. Forward Newton-Gauss method

R. Runge-Kutta method

S. Trapezoidal Rule

Column II

1. Interpolation
2. Non-linear differential equations
3. Numerical integration
4. Linear algebraic equations

(a) P-1, Q-4, R-3, S-2 (b) P-1, Q-4, R-2, S-3

(c) P-1, Q-3, R-2, S-4 (d) P-4, Q-1, R-2, S,3

[2006 : 1 Mark]

10. Consider an ordinary differential equation $\frac{dx}{dt} = 4t + 4$. If $x = x_0$ at t = 0, the increment in x calculated using Runge-Kutta fourth order multi-step method with a step size of Δt = 0.2 is

(a) 0.22 (b) 0.44

(c) 0.66 (d) 0.88

[2014 : 2 Marks, Set-4]

11. Gauss seidel method is used to solve the following equations (as per the given order):
$x_1 + 2x_2 + 3x_3 = 1$
$2x_1 + 3x_2 + x_3 = 1$
$3x_1 + 2x_2 + x_3 = 1$
Assuming initial guess as $x_1 = x_2 = x_3 = 0$, the value of x_3 after the first iteration is ______.

[2016: 2 Marks, Set-1]

12. An explicit forward Euler method is used to numerically integrate the differential equation

$$\frac{dy}{dt} = y$$

using a time step of 0.1. With the initial condition y(0) = 1, the value of y(1) computed by this method is ______ (correct to two decimal places).

[2018 : 2 Marks, Set-1]

Numerical Integration

13. The accuracy of Simpson's rule quadrature for a step size h is

(a) $O(h^2)$ (b) $O(h^3)$
(c) $O(h^4)$ (d) $O(h^5)$

[2003 : 1 Mark]

14. A calculator has accuracy up to 8 digits after decimal place. The value of $\int_0^{2\pi} \sin x dx$ when evaluated using this calculator by trapezoidal method with 8 equal intervals, to 5 significant digits is

(a) 0.00000 (b) 1.0000
(b) 0.00500 (d) 0.00025

[2007 : 2 Marks]

15. Torque exerted on a flywheel over a cycle is listed in the table. Flywheel energy (in J per unit cycle) using Simpson's rule is

Angle (degree)	0	60	120	180	240	300	360
Torque (Nm)	0	1066	-323	0	323	-355	0

(a) 542 (b) 993
(c) 1444 (d) 1986

[2010 : 2 Marks]

16. The integral $\int_1^3 \frac{1}{x} dx$, when evaluated by using Simpson's 1/3 rule on two equal subintervals each of length 1, equals

(a) 1.000 (b) 1.098
(c) 1.111 (d) 1.120

[2011 : 2 Marks]

17. Match the correct pairs

Numerical Integration Scheme	Order of Fitting Polynomial
P. Simpson's 3/8 Rule	1. First
Q. Trapezoidal Rule	2. Second
R. Simpson's 1/3 Rule	3. Third

(a) P-2, Q-1, R-3 (b) P-3, Q-2, R-1
(c) P-1, Q-2, R-3 (d) P-3, Q-1, R-2

[2013 : 1 Mark]

18. The definite integral $\int_1^3 \frac{1}{x} dx$ is evaluated using trapezoidal rule with a step size of 1. The correct answer is ________.

[2014 : 1 Mark, Set-3]

19. Using the trapezoidal rule, and dividing the interval of integration into three equal subintervals, the definite integral $\int_{-1}^{+1} |x| dx$ is_____ .

[2014: 2 Marks, Set-1]

20. The value of $\int_{2.5}^{4} \ln(x) dx$ calculated using the Trapezoidal rule with five subintervals is ___.

[2014 : 2 Marks, Set-2]

21. Simpson's $\frac{1}{3}$ rule is used to integrate the function $f(x) = \frac{3}{5}x^2 + \frac{9}{5}$ between x = 0 and x = 1 using the least number of equal sub-intervals. The value of the integral is ______.

[2015 : 1 Mark, Set-1]

22. Using a unit step size, the volume of integral $\int_1^2 x \ln x dx$ by trapezoidal rule is ______

[2015 : 1 Mark, Set-3]

23. The values of function f(x) at 5 discrete points are given below:

x	0	0.1	0.2	0.3	0.4
f(x)	0	10	40	90	160

Using Trapezoidal rule step size of 0.1, the value of $\int_0^{0.4} f(x) dx$ is

[2015 : 2 Marks, Set-2]

24. Numerical integration using trapezoidal rule gives the best result for a single variable function, which is

(a) linear (b) parabolic
(c) logarithmic (d) hyperbolic

[2016 : 1 Mark, Set-2]

25. The error in numerically computing the integral $\int_0^{\pi}(\sin x+\cos x)\,dx$ using the trapezoidal rule with three intervals of equal length between 0 and π is ______.

[2016 : 2 Marks, Set-2]

26. P(0, 3), Q(0.5, 4) and R(1, 5) are three points on the curve defined by f(x). Numerical integration is carried out using both Trapezoidal rule and Simpson's rule within limits x = 0 and x = 1 for the curve. The difference between the two results will be

(a) 0 (b) 0.25
(c) 0.5 (d) 1

[2017 : 2 Marks, Set-1]

27. Evaluation of $\int_2^4 x^3dx$ using a 2-equal-segment trapezoidal rule gives value of ______.

[2019 : 1 Mark, Set-1]

ANSWERS

1. (d) **2.** (c) **3.** (d) **7.** (c) **8.** (c) **9.** (d) **10.** (d) **13.** (c) **14.** (a) **15.** (b)
16. (c) **17.** (d) **24.** (a) **26.** (a)

EXPLANATIONS

1.

x	0	1	2	3
f(x)	1	2	1	10

Difference table as

x	f(x)	Δf(x)	Δ²f(x)	Δ³f(x)
0	1			
		1		
1	2		−2	
		−1		12
2	1		10	
		9		
3	10			

We take $x_0 = 0, p = \dfrac{x-0}{k} = x \quad [\because k = 1]$

using Newton's forward interpolation formulae, we get

$$f(x) = f(0) + \frac{x}{1}\Delta f(0) + \frac{x(x-1)}{1.2}\Delta^2 f(0) + \frac{x(x-1)(x-2)}{1\cdot 2\cdot 3}\Delta^3 f(0)$$

$$= 1 + x(1) + \frac{x(x-1)}{2}(-2) + \frac{x(x-1)(x-2)}{6}(12)$$

$$= 1 + x + (x - x^2) + 2x(x^2 - 3x + 2)$$

$$= 1 + x + x - x^2 + 2x^3 - 6x + 4x$$

$$= 2x^3 - 7x^2 + 6x + 1$$

2. By N–R method, $x_{n+1} = x_n - \dfrac{f(x)}{f'(x)}, x_0 = 1$

$$f(x) = x^3 + 3x - 7,$$

$$\Rightarrow f(1) = -3,$$

$$\Rightarrow x_1 = x_0 - \frac{f(x_0)}{f'(x_0)}$$

$$f'(x) = 3x^2 + 3,$$

$$\Rightarrow f'(1) = 6,$$

$$\Rightarrow x_4 = 1 - \frac{-3}{6} = 1 - (-0.5) = 1.5$$

3. X and Y are independent

$\therefore$ (a), (b), (c) are true

Only (d) is odd one

4. Let $f(x) = 5x - 2\cos x - 1$

$$\Rightarrow f'(x) = 5 + 2\sin x$$

$$f(0) = -3;\ f(1) = 2.9$$

By intermediate value theorem roots lie between 0 and 1.

Let $x_0 = 1$ rad $= 57.32°$

By Newton Raphson method,

$$x_{n+1} = x_n - \frac{f(x_n)}{f'(x_n)}$$

$\Rightarrow \quad x_{n+1} = \dfrac{2x_n \sin x_n + 2\cos x_n + 1}{5 + 2\sin x_n}$

$\Rightarrow \quad x_1 = 0.5632$

$\Rightarrow \quad x_2 = 0.5425$

$\Rightarrow \quad x_3 = 0.5424$

5. By Newton-Raphson Method,

1st iteration, $x_1 = x_0 - \dfrac{f(x_0)}{f'(x_0)}$

$= 1 - \dfrac{f(1)}{f'(1)} = 1 - \dfrac{5}{10} = \dfrac{1}{2}$

Where $f(x) = x^3 + 2x^2 + 3x - 1 \Rightarrow f(1) = 5$

$f'(x) = 3x^2 + 4x + 3 \Rightarrow f'(1) = 10$

2nd iteration, $x_2 = x_1 \dfrac{f(x_1)}{f'(x_1)} = 0.5 - \dfrac{f(0.5)}{f'(0.5)} = 0.3043$

6. $f(x) = x - 10 \cos x;$

$f\left(\dfrac{\pi}{4}\right) = \dfrac{\pi}{4} - \dfrac{10}{\sqrt{2}} = -6.286$

$f'(x) = 1+1 \sin x$

$f\left(\dfrac{\pi}{4}\right) = 1 + \dfrac{10}{\sqrt{2}} = 8.07$

$x_1 = x_0 - \dfrac{f(x_0)}{f'(x_0)}$

$= \dfrac{\pi}{4} - \left(\dfrac{-6.286}{8.07}\right) = 1.5639$

7. $f(x) = x^3 + x-1$

$f(1) = 1+1 - 1 = 1$

$f'(x) = 3x^2 + 1$

$f'(1) = 3+1=4$

$x_1 = x_0 - \dfrac{f(x_0)}{f'(x_0)} = 1 - \dfrac{1}{4}$

$= 1 - 0.25$

$x_1 = 0.75$

8. $f(x) = \cos(x)$

at $x = \dfrac{\pi}{6}\text{rad} = \dfrac{\pi}{6} \times \dfrac{180}{\pi} = 30°$

$[f'(x)]_2 = -\sin 30° = \dfrac{-1}{2}$ ...(1)

$f'(x) = \dfrac{\cos(x+h) - \cos(x-h)}{2h}$

$\because \quad x = \dfrac{\pi}{6} = 30°$

$h = 0.1 = \dfrac{28°}{\pi}$

$[f'(x)]_2 = \dfrac{\cos x.\cosh - \sin x.\sinh \quad -\cos x.\cosh - \sin x.\sinh}{2h}$

$= \dfrac{-2\sin x.\sinh}{2h} = \dfrac{-2\sin 30° \times \sin\left(\dfrac{18}{\pi}\right)}{2 \times 0.1}$

$= -0.499$

$\%\text{error} = \dfrac{[f'(x)]_1 - [f'(x)]_2}{[f'(x)]_1}$

$= 0.166\%$ (> 0.1% and < 1%)

10. Given $\dfrac{dx}{dt} = 4t + 4$

$x = x_0$ at $t = 0$

$n = 0.2$

Calculate $x(0.2)$ value

$K_1 = f(t_0, x_0) = f(0, x_0) = 4$

$K_2 = f\left(t_0 + \dfrac{h}{2}, x_0 + \dfrac{1}{2}K_2 h\right)$

$= f(0 + 0.1, x_0 + 0.4)$

$= f(001, x_0 + 0.4) = 4(0.1) + 4 = 4.4$

$K_3 = f\left(x_0 + \dfrac{h}{2}, x_0 + \dfrac{K_2 h}{2}\right)$

$= f(t_0 + 0.1, x_0 + (2.2)(0.2))$

$= f(0.1, x_0 + 0.44) = 4(0.1) + 4 = 4.4$

$K_4 = f(t_0 + h, x_0 + K_3 h)$

$= f(0 + 0.2, x_0 + 0.88)$

$= f(0.2, x_0 + 0.88)$

$= 4(0.2) + 4 = 4.8$

$x(0.2) = x_1 = x_0 + \dfrac{h}{6}(K_1 + 2K_2 + 2K_3 + K_4)$

$= x_0 + \dfrac{0.2}{6}(4 + 2(4.4) + 2(4.4) + (4.8))$

$= x_0 + \dfrac{0.2}{6}(4 + 8.8 + 8.8 + 4.8)$

$= x_0 + 0.88$

Increment as $x = x_1 - x_0 = x_0 + 0.88 - x_0 = 0.88$

11. The equations are

$x_1 + 2x_2 + 3x_3 = 5$

$2x_1 + 3x_2 + x_3 = 1$

$3x_1 + 2x_2 + x_3 = 3$

By pivoting we get

$$x_1 = \frac{3 - 2x_2 - x_3}{3}$$

$$x_2 = \frac{1 - 2x_1 - x_3}{3} \quad \text{... (2)}$$

$$x_3 = \frac{5 - x_1 - 2x_2}{3} \quad \text{... (3)}$$

When we

put $x_2 = 0$, $x_3 = 0$ in equaiton (1), $x_1 = 1$

put $x_1 = 1$, $x_3 = 0$ in equation (3), $x_2 = -.333$

put $x_1 = 1$, $x_2 = -.333$ in equation (3), $x_3 = 1.555$

12. General formula

$$y_{n+1} = y_n + h_f(t_n, y_n)$$

For $n = 0$, $y_1 = y_0 + h_f(t_0, y_0)$

$= y_0 + hy_0$

$= 1 + 0.1\,(1)$

$y_1 = 1.1$

For $n = 1$, $y_2 = y_1 + h_f(t_1, y_1)$

$= y_1 + h.y_1$

$= 1.1 + 0.1\,(1.1)$

$y_2 = 1.21$

For $n = 2$, $y_3 = y_2 + hf(t_2, y_2)$

$= y_2 + h.y_2$

$= 1.21 + 0.1 \times 1.21$

$y_3 = 1.331$

For $n = 3$, $y_4 = y_3 + h.f(t_3, y_3)$

$= y_3 + h.y_3$

$= 1.331 + 0.1 \times 1.331$

$y_4 = 1.4641$

For $n = 4$, $y_5 = y_4 + h.f(t_4, y_4)$

$= y_4 + h.y_4$

$= 1.4641 + 0.1 \times (1.4641)$

$= 1.61051$

For $n = 5$, $y_6 = y_5 + h.f(t_5, y_5)$

$= y_5 + h.y_5$

$= 1.61051 + 0.1 \times 1.61051$

$y_6 = 1.771561$

For $n = 6$, $y_7 = y_6 + h.f(t_6, y_6)$

$= y_6 + h \times y_6$

$= 1.771561 + 0.1 \times 1.771561$

$= 1.9487$

For $n = 7$, $y_8 = y_7 + h.f(t_7, y_7)$

$= y_7 + h.y_7$

$= 1.9487 + 0.1 \times (1.9487)$

$y_8 = 2.14357$

For $n = 8$, $y_9 = y_8 + h.f(t_8, y_8)$

$= y_8 + h.y_8$

$= 2.14357 + 0.1 \times 2.14357$

$y_9 = 2.3579$

For $n = 9$, $y_{10} = y_9 + h.f(t_9, y_9)$

$= y_9 + h.y_9$

$= 2.3579 + 0.1 \times (2.3579)$

$y_{10} = 2.5937$

14. $h = \dfrac{2\pi - 0}{8} = \dfrac{\pi}{4}$

$y_0 = \sin(0) = 0$

$y_1 = \sin\left(\frac{\pi}{4}\right) = 0.7071$

$y_2 = \sin\left(\frac{\pi}{2}\right) = 1$

$y_3 = \sin\left(\frac{3\pi}{4}\right) = .7071$

$y_4 = \sin(\pi) = 0$

$y_5 = \sin\left(\frac{5\pi}{4}\right) = -.7071$

$y_6 = \sin\left(\frac{6\pi}{4}\right) = -1$

$y_7 = \sin\left(\frac{7\pi}{4}\right) = -0.7071$

$y_8 = \sin\left(\frac{8\pi}{4}\right) = 0$

Trapezoidal rule

$$\int_{x_0}^{x_0+nh} f(x).dx = \frac{h}{2}\left[(y_0 + y_n + 2(y_1 + y_2 + .. + y_{n-1})\right]$$

$$\int_0^{2\pi} \sin x.dx = \frac{h}{8}$$

[(0+0)+2 (0.7071+1+.7071+0-.7071-.7071)]=0

15. Given : $h = 60° - 0° = 60° = 60 \times \frac{\pi}{180} = 1.047$

$y_0 = 0, y_1 = 1066, y_2 = -323, y_4 = 323, y_5 = -355, y_6 = 0$

By Simpson's rule, flywheel energy

$$= \frac{h}{3}[(y_0 + y_6) + 4(y_1 + y_3 + y_5) + 2(y_2 + y_4)]$$

$$= \frac{1.047}{3}[(0+0) + 4(1066 + 0 - 355) + 2(-323 + 323)]$$

= 993 Nm rad

= 993 Joules/cycle

16. Given: $\int_1^3 \frac{1}{x} dx$

Here, $a = 1$, $b = 3$, $n = 2$

$$h = \frac{b-a}{n} = 1$$

$x_0 = 1$	$x_1 = 2$	$x_2 = 3$
$y_0 = 1$	$y_2 = \frac{1}{2} = 0.5$	$y_3 = \frac{1}{3} = 0.33$

By Simpson's rule

$$\int_1^3 \frac{1}{x} dx = \frac{1}{3} h[(y_1 + y_3) + 4(y_2)]$$

$$= \frac{1}{3}(1)[(1 + 0.33) + 4(0.51)]$$

$$= 1.11$$

17. Trapezoidal rule fits linear polynomial (1st order) simplex 1/3 rule fits a parabolic curve (2nd order polynomial slip on 3/5 rule fit a able curve (3rd order polynomial

∴ Correct Answer ⇒ D

18. Given, $\int_1^3 \frac{1}{x} dx$

$h =$ step size 1

$n =$ no. of sub intervals

$$= \frac{3-1}{1} = 2$$

Let $y =$

x	1	2	3
y	1	$\frac{1}{2}$	$\frac{1}{3}$

By trapezoidal rule

$$\int_{x_0}^{x_n} f(x)dx = \frac{h}{2}[(y_0 + y_n) + 2(y_1 + ... + y_{n-1})]$$

$$\int_1^3 \frac{1}{x} dx = \frac{1}{2}\left[\left(1 + \frac{1}{3}\right) + 2\left(\frac{1}{2}\right)\right]$$

$$= \frac{1}{2}\left[\frac{4}{3} + 1\right] = \frac{1}{2} \times \frac{7}{3} = 1.1667$$

19. $\int_{-1}^{+1} |x| \, dx$

Let $y = |x|$

$n =$ no. of subintervals = 3

$$h = \frac{x_n - n_o}{n} = \frac{1-(-1)}{3} = \frac{2}{3}$$

Values of x are,

$$-1, -1 + \frac{2}{3}, -1 + 2\left(\frac{2}{3}\right), -1 + 3\left(\frac{2}{3}\right)$$

x	-1	$-\frac{1}{3}$	$\frac{1}{3}$	1
$y = \lvert x \rvert$	1	$\frac{1}{3}$	$\frac{1}{3}$	1

Trapezodial rule

$$= \int_{x_o}^{x^n} f(x)dx = \frac{h}{2}[(y_o + y_n) + 2(y_1 + ...+ y_{n-1})]$$

$$= \int_{-1}^{1} |x| dx = \frac{1}{3}\left[(1+1) + 2\left(\frac{1}{3} + \frac{1}{3}\right)\right]$$

$$= \frac{1}{3}\left[2 + \frac{4}{3}\right] = \frac{1}{3} \times \frac{10}{3} = \frac{10}{9} = 1.1111$$

20. Given $\int_{2.5}^{4} \ln x \, dx \quad n = 5$

Let $y = \ln x \quad n = \frac{4-2.5}{5} = \frac{1.5}{5} = 0.30$

x	2.5	2.8	3.1	3.4	3.7	4.0
$y = \ln x$	0.9163	1.0296	1.1314	1.2238	1.3083	1.3863

Trapezoidal Rule is $\int_{x_0}^{x_n} f(x)dx$

$$= \frac{h}{2}[(y_0 + y_n) + 2y_1 + ...y_{n-1}]$$

$$\int_{2.5}^{4} \ln x \, dx$$

$$= \frac{0.3}{2}[(0.9163+1.3863)+(1.0296+1.314+1.2238+1.3083)]$$

$$= 0.15 (2.3026 + 9.3862) = 1.7533$$

21.

x	0	$\frac{1}{2}$	1
$y=f(x)=\frac{3}{5}x^2+\frac{9}{5}$	$\frac{9}{5}$	$\frac{39}{20}$	$\frac{12}{5}$

$$\int_0^1 y\,dx=\frac{\left(\frac{1}{2}\right)}{2}\left[\left(\frac{9}{5}+\frac{12}{5}\right)+4\left(\frac{39}{20}\right)\right]=2$$

22.

x	1	2
f(x)	0	2ln(2)

$$\therefore I=\frac{h}{2}[y_0+y_n]$$

$$I=\frac{1}{2}[0+2\ln 2]$$

$$=\ln 2=0.693$$

23.

x	0	0.1	0.2	0.3	0.4
$y=f(x)$	0	10	40	90	160
	y_0	y_1	y_2	y_3	y_4

$$\int_0^{0.4} f(x)\,dx=\int_0^{0.4} y\,dx$$

$$=\frac{h}{2}[(y_0+y_4)+2(y_1+y_2+y_3)]$$

$$=\frac{0.1}{2}[(0+160)+2(10+40+90)]=22$$

24. Trapezoidal rule gives best result when function is linear in nature.

25.

	y_0	y_1	y_2	y_3
x	0	$\frac{\pi}{3}$	$\frac{2\pi}{3}$	π
f(x)	1	1.366	0.366	–1

by trapezoidal

$$\int_0^{\pi}(\sin x+\cos x)dx=\frac{\frac{\pi}{3}}{2}(1+(-1)+2(1.366+0.366))$$

$$=1.8137$$

$$\int_0^{\pi}(\sin x+\cos x)dx=\int_0^{\frac{\pi}{2}}(\sin x+\cos x)dx+\int_{\frac{\pi}{2}}^{\pi}(\sin x+\cos x)$$

$$=2$$

Error = Exact value – Approx Value

$$=2-1.8137=0.1863$$

26.

	y_0	y_1	y_2
x	0	0.5	1
y	3	4	5

By Trapezoidal rule

$$\int_0^1 f(x)dx=\frac{h}{2}\left[(y_0+y_2)+2y\right]$$

$$=\frac{0.5}{2}\left[(3+5)+2(4)\right]=4$$

By Simpson rule

$$\int_0^1 f(x)dx=\frac{h}{3}\left[(y_0+y_2)+4y_1\right]$$

$$=\frac{0.5}{3}\left[(3+5)+4(4)\right]$$

$$=4$$

Difference between these two results will be 0.

27. Trapezoidal Rule, h = 1

$$\int y\,dx=\frac{h}{2}\left[(y_0+y_1)+2(y_1+y_2+\ldots)\right]$$

$$=\frac{1}{2}\left[(2^3+4^3)+2(3^3)\right]$$

$$=\frac{1}{2}[8+64+54]=63$$

7

CHAPTER

Transform Theory

Analysis of Previous GATE Papers

Topic	Year/Set → Marks Distribution ↓		2019 Set 1	2019 Set 2	2018 Set 1	2018 Set 2	2017 Set 1	2017 Set 2	2016 Set 1	2016 Set 2	2016 Set 3	2015 Set 1	2015 Set 2	2015 Set 3
Laplace Transform	1 Mark	MCQ Type						1	1		1		1	
		Numerical Type												
	2 Marks	MCQ Type			1									1
		Numerical Type												
	Total				2			1	1		1		1	2
Inverse of Laplace Transform	1 Mark	MCQ Type												
		Numerical Type												
	2 Marks	MCQ Type												
		Numerical Type												
	Total													
Fourier Transform	1 Mark	MCQ Type				1								
		Numerical Type												
	2 Marks	MCQ Type												
		Numerical Type												
	Total					1								

Laplace Transform

1. Laplace transform of the function sin ωt is

(a) $\frac{s}{s^2+\omega^2}$ (b) $\frac{\omega}{s^2+\omega^2}$

(c) $\frac{s}{s^2-\omega^2}$ (d) $\frac{\omega}{s^2-\omega^2}$

[2003 : 2 Marks]

2. A delayed unit step function is defined as $u(t-a) = \begin{cases} 0, \text{for } t < a \\ 1, \text{for } t \geq a \end{cases}$ Its Laplace transform is

(a) $a.e^{-as}$ (b) $\frac{e^{-as}}{s}$

(c) $\frac{e^{as}}{s}$ (d) $\frac{e^{as}}{a}$

[2004 : 2 Marks]

3. If F(s) is the Laplace transform of function f(t), then Laplace transform of $\int_0^t f(\tau)d\tau$ is

(a) $\frac{1}{s}F(s)$ (b) $\frac{1}{s}F(s)-f(0)$

(c) $sF(s)-f(0)$ (d) $\int F(s)ds$

[2007 : 1 Mark]

4. The function f(t) satisfies the differential equation $\frac{d^2f}{dt^2}+f=0$ and the auxilliary conditions, f(0) = 0, $\frac{df}{dt}(0)=4$. The Laplace transform of f(t) is given by

(a) $\frac{2}{s+1}$ (b) $\frac{4}{s+1}$

(c) $\frac{4}{s^2+1}$ (d) $\frac{2}{s^2+1}$

[2013 : 2 Marks]

5. Laplace transform of cos (ωt) is

(a) $\frac{s}{s^2+\omega^2}$ (b) $\frac{\omega}{s^2+\omega^2}$

(c) $\frac{s}{s^2-\omega^2}$ (d) $\frac{\omega}{s^2-\omega^2}$

[2014 : 1 Mark, Set-1]

6. Laplace transform of cos(ωt) is $\frac{s}{s^2+\omega^2}$. The laplace transform of $e^{-2t}\cos(4t)$ is

(a) $\frac{s-2}{(s-2)^2+16}$ (b) $\frac{s+2}{(s-2)^2+16}$

(c) $\frac{s-2}{(s+2)^2+16}$ (d) $\frac{s+2}{(s+2)^2+16}$

[2014 : 1 Mark, Set-4]

7. The Laplace transform of e^{i5t} where $i=\sqrt{-1}$, is

(a) $\frac{s-5i}{s^2-25}$ (b) $\frac{s+5i}{s^2+25}$

(c) $\frac{s+5i}{s^2-25}$ (d) $\frac{s-5i}{s^2+25}$

[2015 : 1 Mark, Set-2]

8. Laplace transform of the function f(t) is given by $F(s) = L\{f(t)\} = \int_0^\infty f(t)e^{-st}dt$. Laplace transform of the function shown below is given by

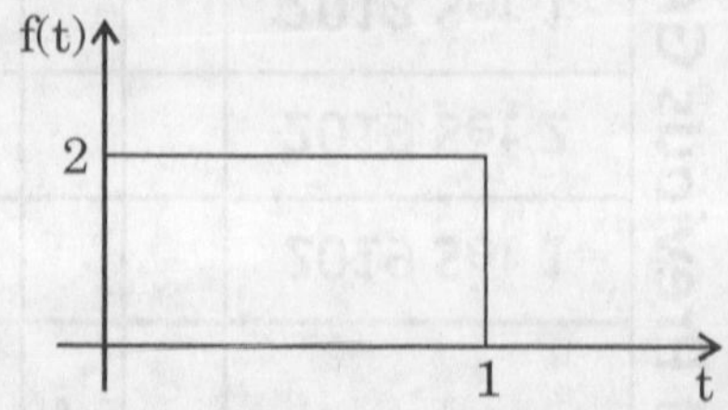

(a) $\frac{1-e^{-2s}}{s}$ (b) $\frac{1-e^{-s}}{2s}$

(c) $\frac{2-2e^{-s}}{s}$ (d) $\frac{1-2e^{-s}}{s}$

[2015 : 2 Marks, Set-3]

9. If f(t) is a function defined for all t > 0, its Laplace transform F(s) is defined as

(a) $\int_0^\infty e^{st}f(t)dt$ (b) $\int_0^\infty e^{-st}f(t)dt$

(c) $\int_0^\infty e^{ist}f(t)dt$ (d) $\int_0^\infty e^{-ist}f(t)dt$

[2016 : 1 Mark, Set-1]

10. Solutions of Laplace's equation having continuous second-order partial derivatives are called

(a) biharmonic functions

(b) harmonic functions

(c) conjugate harmonic functions

(d) error functions

[2016 : 1 Mark, Set-3]

11. The Laplace transform of te^t is

(a) $\frac{s}{(s+1)^2}$ (b) $\frac{1}{(s-1)^2}$

(c) $\frac{1}{(s+1)^2}$ (d) $\frac{s}{s-1}$

[2017 : 1 Mark, Set-2]

12. F(s) is the Laplace transform of the function $f(t) = 2t^2e^{-t}$

F(1) is ______(correct to two decimal places.)

[2018 : 2 Marks, Set-1]

Inverse of Laplace Transform

13. The inverse Laplace transform of $\frac{1}{(s^2+s)}$ is

(a) $1 + e^t$ (b) $1 - e^t$

(c) $1 - e^{-t}$ (d) $1 + e^{-t}$

[2009 : 1 Mark]

14. The inverse Laplace transform of the function

$F(s) = \frac{1}{s(s+1)}$ is given by

(a) f(t) = sin t (b) $f(t) = e^{-t}$ sint

(c) $f(t) = e^{-t}$ (d) $f(t) = 1 - e^{-t}$

[2012 : 2 Marks]

Fourier Transform

15. The Fourier cosine series for an even function f(x) is given by

$$f(x) = a_0 + \sum_{n=1}^{\infty} a_n \cos(nx)$$

The value of the coefficient a_2 for the function $f(x) = \cos^2(x)$ in $[0, \pi]$ is

(a) –0.5 (b) 0.0

(c) 0.5 (d) 1.0

[2018 : 1 Mark, Set-2]

ANSWERS

1. (b)	**2.** (b)	**3.** (b)	**4.** (c)	**5.** (a)	**6.** (d)	**7.** (b)	**8.** (c)	**9.** (b)	**10.** (b)
11. (b)	**12.**	**13.** (c)	**14.** (d)	**15.** (c)					

EXPLANATIONS

1. Laplace transforms,

$$L[\sin(\omega t)] = \frac{\omega}{S^2 + \omega^2}$$

$$L[\cos(\omega t)] = \frac{S}{S^2 + \omega^2}$$

2. $L[u(t-a)] = \int_0^{\infty} e^{-st} u(t-a)\, dt$

$$= \int_0^{\infty} e^{-st} \cdot 0 \cdot dt + \int_a^{\infty} e^{-st} \cdot 1 \cdot dt$$

$$= 0 + \left| \frac{e^{-st}}{-s} \right|_a^{\infty} = \frac{e^{-as}}{s}$$

3. From definition,

we know $\int_0^t f(t)\, dt = \frac{1}{S} F(S) - f(0)$

4. Check by options

$$L^{-1}\left\{\frac{4}{s^2+1}\right\} = 4 \text{ sint satisfies } \frac{d^2f}{dt^2} + f = 0$$

5. Laplace transform of cos (ωt) is

$$L[\cos(wt)] = \frac{S}{s^2 + \omega^2}$$

6. We know that if L{f(t)} = F(s)

Then $L\{e^{at} f(t)\} = F(s-a)$

$$\therefore \quad L\{e^{-2t}\cos 4t\} = \frac{s+2}{(s+2)^2+4^2}$$

$$= \frac{s+2}{(s+2)^2+16}$$

7. $L\left[e^{i5t}\right] = \frac{1}{s+5i}$

$$= \frac{1}{s-5i} \times \frac{(s+5i)}{(s+5i)}$$

$$= \frac{s+5i}{s^2+25}$$

8. $f(t) = 2;\ 0 < t < 1 > 0$; otherwise

$$\therefore \quad L[f(t)] = \int_0^1 2e^{-st}dt = 2\left.\frac{e^{-st}}{-s}\right|_0^1 = \frac{2-2e^{-s}}{s}$$

9. According to the Laplace transformation of

$f(t)\ \forall \in t \geq 0$ is as $F(s) = L\{f(t)\} = \int_0^\infty e^{-st} f(t)\,dt$

10. The solution of Laplace's equation having continuous 2nd order partial derivatives is called a harmonic function.

11. Laplace transform of $t = \frac{1}{S^2}$

By applying shift property we get

$$L\left[te^t\right] = \frac{1}{(S-1)^2}$$

12. Applying Laplace transform

$$L\left[t^n\right] = \frac{n!}{S^{n+1}}$$

$$L\left(t^2\right) = \frac{2}{S^3}$$

$$L\left(2t^2\right) = \frac{4}{S^3}$$

$$F(s) = L\left[e^{-t}\cdot 2t^2\right] = \frac{4}{(S+1)^3}$$

Putting s = 1 we get

$$F(1) = \frac{4}{(1+1)^3} = \frac{4}{8} = 0.5$$

13. $\frac{1}{(s^2+s)} = \frac{1}{s(s+1)} = \frac{1}{s} - \frac{1}{(s+1)}$

$\therefore$ Inverse Laplace transform = $(1-e^{-t})$

14. $L^{-1}\left(\frac{1}{s(s+1)}\right) = \frac{1}{s} - \frac{1}{s+1}$

$$L^{-1}\left(\frac{1}{s} - \frac{1}{s+1}\right) = 1 - e^{-t}$$

15. $\cos^2 x = \frac{1+\cos^2 x}{2}$

$$f(x) = \frac{1}{2} + \frac{\cos^2 x}{2}$$

$$f(x) = \frac{a_0}{2} + \sum_{n=1}^{\infty} a_n.\cos nx$$

$a_0 = 1$

$a_1 = 0$

$a_2 = \frac{1}{2}$

Unit - II

Engineering Mechanics

Syllabus

Free-body diagrams and equilibrium; trusses and frames; virtual work; kinematics and dynamics of particles and of rigid bodies in plane motion; impulse and momentum (linear and angular) and energy formulations, collisions.

Contents

1 CHAPTER

FBD, Equilibrium, Plane Trusses and Virtual Work

Analysis of Previous GATE Papers

Topic	Year/Set → Marks Distribution ↓		2019 Set 1	2019 Set 2	2018 Set 1	2018 Set 2	2017 Set 1	2017 Set 2	2016 Set 1	2016 Set 2	2016 Set 3	2015 Set 1	2015 Set 2	2015 Set 3
Free Body Diagram	1 Mark	MCQ Type			1				1		1			
		Numerical Type												
	2 Marks	MCQ Type												
		Numerical Type												
	Total				1				1		1			
Equilibrium	1 Mark	MCQ Type												1
		Numerical Type												
	2 Marks	MCQ Type	1											
		Numerical Type												
	Total		2											1
Truss	1 Mark	MCQ Type												
		Numerical Type		1								1		
	2 Marks	MCQ Type							1			1	1	
		Numerical Type	1							1				
	Total		2	1					2	2		3	2	
Virtual Work	1 Mark	MCQ Type												
		Numerical Type												
	2 Marks	MCQ Type												
		Numerical Type												
	Total													

Free body diagram

1. A spring scale indicates a tension T in the right hand cable of the pulley system shown in Neglecting the mass of the pulleys and ignoring friction between the cable and pulley the mass m is

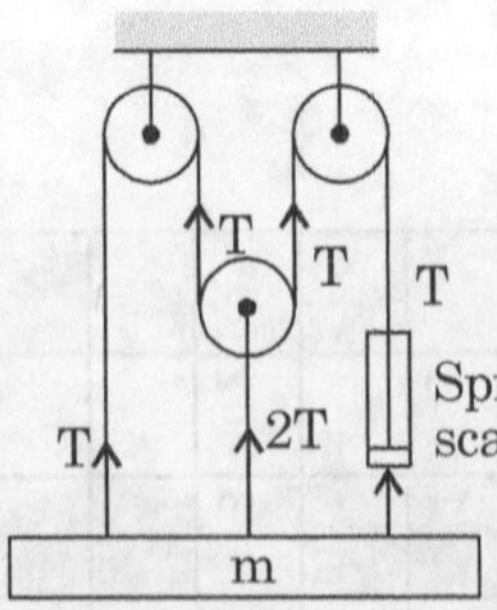

(a) 2T/g (b) $T(1 + e^{4\pi})/g$

(c) 4T/g (d) None of these

[1995 : 2 Marks]

2. A rigid ball of weight 100 N is suspended with the help of a string. The ball is pulled by a horizontal force F such that the string makes an angle of 30° with the vertical. The magnitude of force F(in N) is ______.

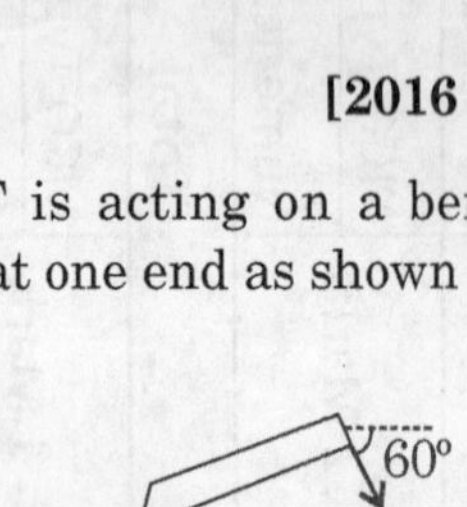

[2016 : 1 Mark, Set-1]

3. A force F is acting on a bent bar which is clamped at one end as shown in the figure.

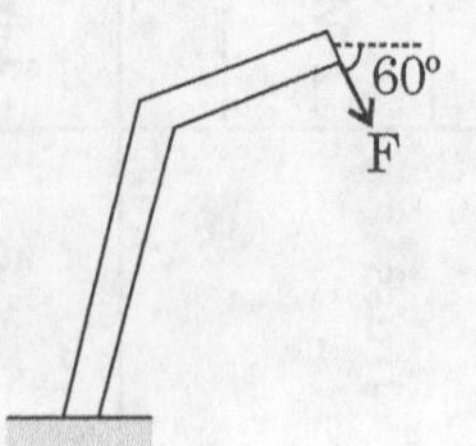

The CORRECT free body diagram is

(a)

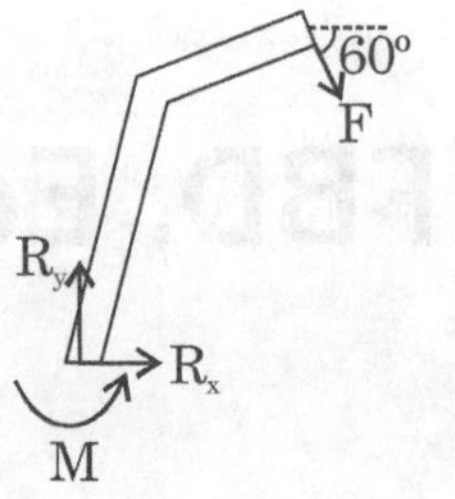

(b)

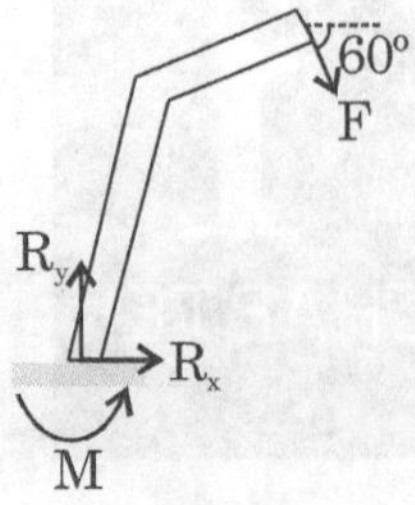

(c)

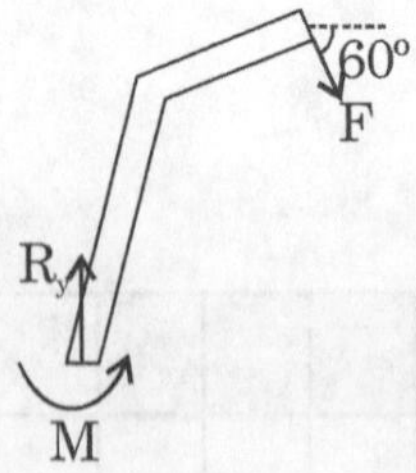

(d)

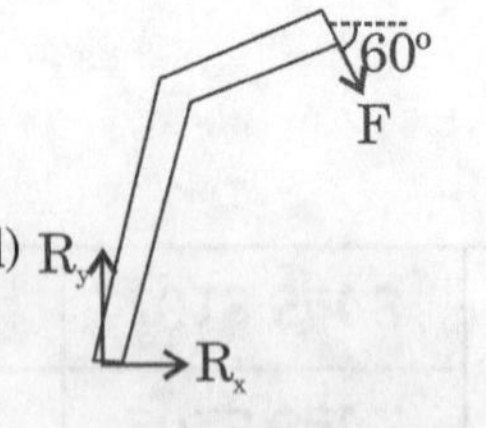

[2016 : 1 Mark, Set-3]

4. A bar of uniform cross section and weighing 100 N is held horizontally using two massless and inextensible strings S_1 and S_2 as shown in the figure.

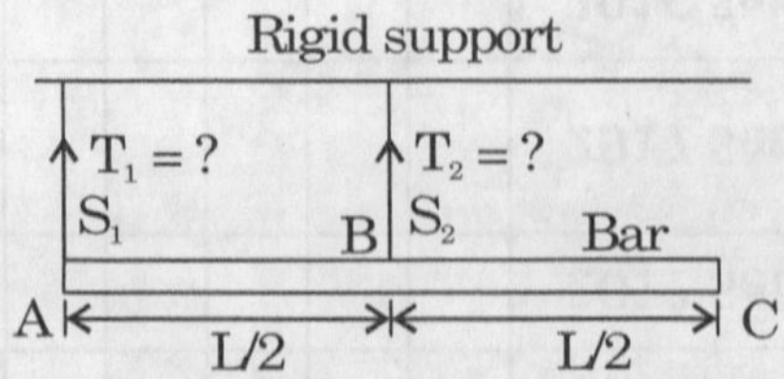

(a) T_1 = 100 N and T_2 = 0 N

(b) T_1 = 0 N and T_2 = 100 N

(c) T_1 = 75 N and T_2 = 25 N

(d) T_1 = 25 N and T_2 = 75 N

[2018 : 1 Mark, Set-1]

Equilibrium

5. If point A is in equilibrium under the action of the applied forces, the values of tensions T_{AB} and T_{AC} are respectively.

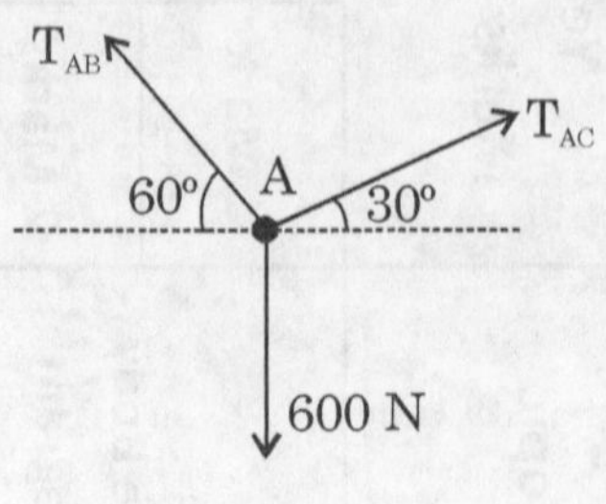

(a) 520 N and 300 N (b) 300 N and 520 N

(c) 450 N and 150 N (d) 150 N and 450 N

[2006 : 2 Marks]

6. A ladder AB of length 5 m and weight (W) 600 N is resting against a wall. Assuming frictionless contact at the floor (B) and the wall (A), the magnitude of the force P (in newton) required to maintain equilibrium of the ladder is_____.

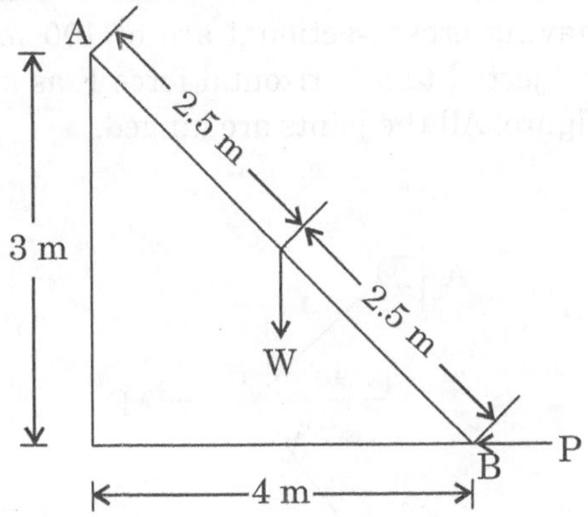

[2014 : 2 Marks, Set-4]

7. A weight of 500 N is supported by two metallic ropes as shown in the figure. The values of tensions T_1 and T_2 are respectively

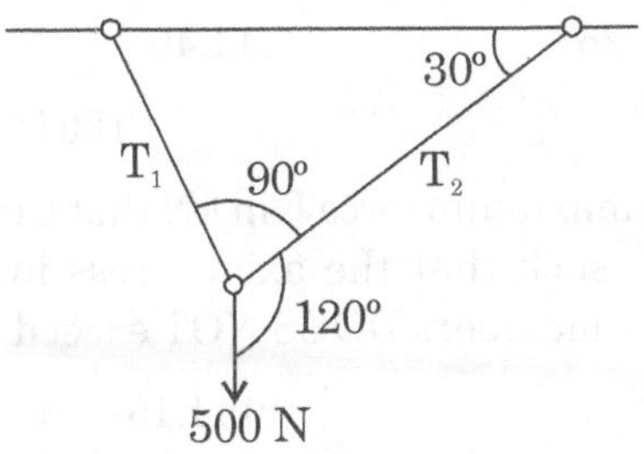

(a) 433 N and 250 N

(b) 250 N and 433 N

(c) 353.5Nand250 N

(d) 250 N and 353.5 N

[2015 : 1 Mark, Set-3]

8. A car having weight W is moving in the direction as shown in the figure. The center of gravity (CG) of the car is located at height h from the ground, midway between the front and rear wheels. The distance between the front and rear wheels is *l*. The acceleration of the car is a, and acceleration due to gravity is g. The reaction on the front wheels(R_f) and rear wheels (R_r) are given by:

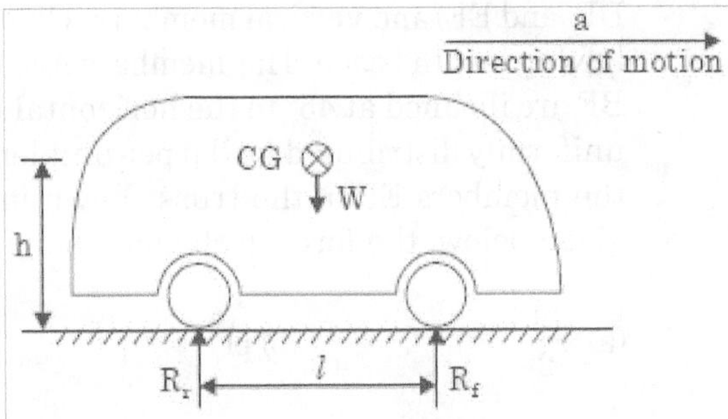

(a) $R_f = \frac{W}{2} - \frac{W}{g}\left(\frac{h}{l}\right)a : R_r = \frac{W}{2} + \frac{W}{g}\left(\frac{h}{l}\right)a$

(b) $R_f = R_r = \frac{W}{2} + \frac{W}{g}\left(\frac{h}{l}\right)a$

(c) $R_f = R_r = \frac{W}{2} - \frac{W}{g}\left(\frac{h}{l}\right)a$

(d) $R_f = \frac{W}{2} + \frac{W}{g}\left(\frac{h}{l}\right)a\ ;\ R_r = \frac{W}{2} - \frac{W}{g}\left(\frac{h}{l}\right)a$

[2019 : 2 Marks, Set-1]

Truss

9. A mass 35 kg is suspended from a weightless bar AB which is supported by a cable CB and a pin at A as shown in figure. The pin reactions at A on the bar AB are

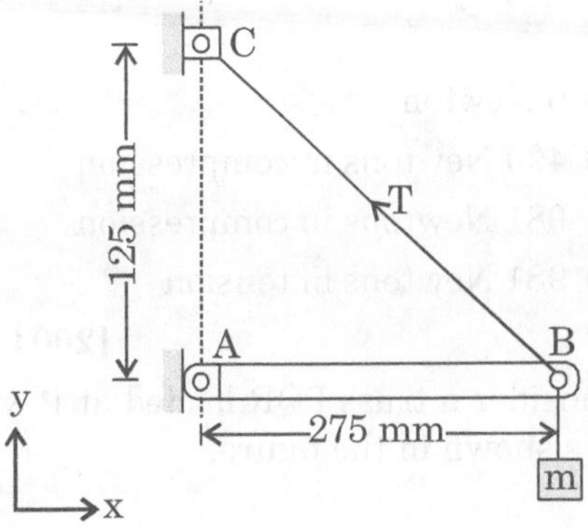

(a) $R_x = 343.4$ N, $R_y = 755.4$ N

(b) $R_x = 343.4$ N, $R_y = 0$

(c) $R_x = 755.4$ N, $R_y = 343.4$ N

(d) $R_x = 755.4$ N, $R_y = 0$

[1997 : 2 Marks]

10. A truss consists of horizontal members (AC, CD, DB and EF) and vertical members (CE and DF) having length l each. The-members AE, DE and BF are inclined at 45° to the horizontal For the uniformly distributed load p per unit length on the members EF of the truss shown in figure given below, the force in the member CD is

(a) $\frac{pl}{2}$ (b) pl

(c) 0 (d) $\frac{2pl}{3}$

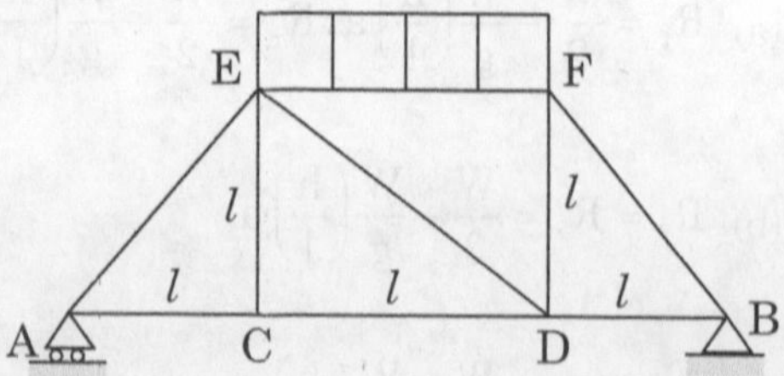

[2003 : 1 Mark]

11. The figure shows a pin-jointed plane truss loaded at the point M by hanging a mass of 100 kg. The member LN of the truss is subjected to a load of

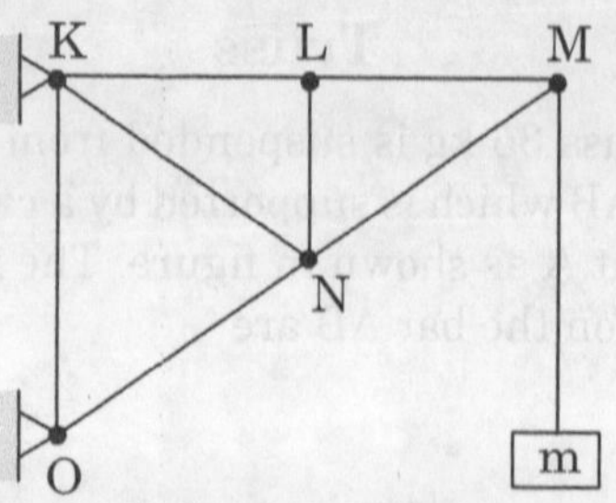

(a) 0 Newton

(b) 490 Newtons in compression

(c) 981 Newtons in compression

(d) 981 Newtons in tension

[2004 : 1 Mark]

12. Consider a truss PQR loaded at P with aforce Fas shown in the figure.

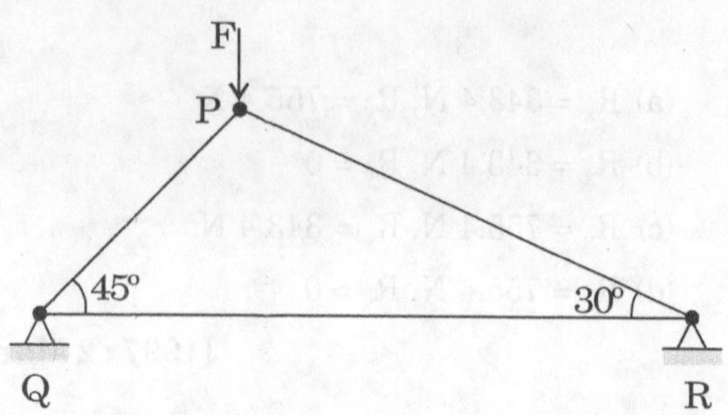

The tension in the number QR is

(a) 0.5 F (b) 0.63 F

(c) 0.73 F (d) 0.87 F

[2008 : 2 Marks]

Common Data Questions 13 and 14:

Two steel truss members, AC md BC, each having cross sectional are of 100 mm², are subjected to a horizontal force F as shown in figure. All the joints are hinged.

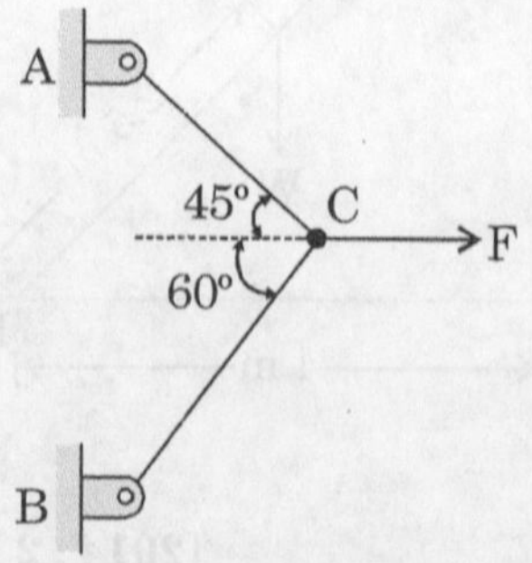

13. If F = 1 kN, the magnitude of the vertical reaction force developed at the point B in kN is

(a) 0.63 (b) 0.32

(c) 1.26 (d) 1.46

[2012 : 2 Marks]

14. The maximum force F in kN that can be applied at C such that the axial stress in any of the truss members DOES NOT exceed 100 MPa is

(a) 8.17 (b) 11.15

(c) 14.14 (d) 22.30

[2012 : 2 Marks]

15. A two member truss ABC is shown in the figure. The force (in kN) transmitted in member AB is ______

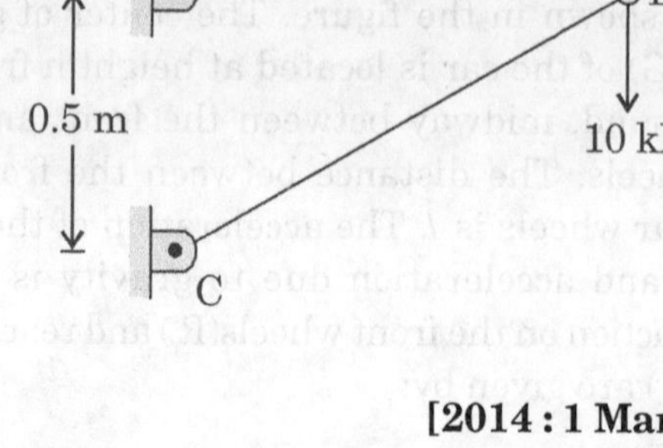

[2014 : 1 Mark, Set-1]

16. For the truss shown in the figure, the forces F_1 and F_2 are 9 kN and 3 kN, respectively. The force (in kN) in the member QS is (All dimensions are in m)

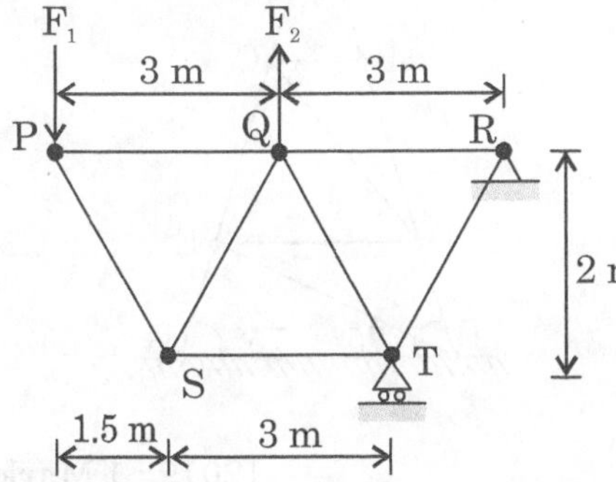

(a) 11.25 tension (b) 11.25compression
(c) 13.5 tension (d) 13.5compression

[2014 : 2 Marks, Set-4]

17. Two identical trusses support a load of 100 N as shown in the figure. The length of each truss is 1.0 m, cross-sectional area is 200 mm²; Young's modulus E= 200 GPa. The force in the truss AB (in N) is

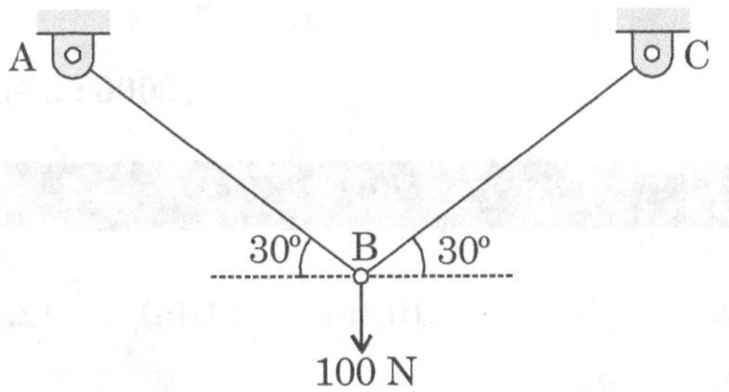

[2015 : 1 Mark, Set-1]

18. For the truss shown in figure, the magnitude of the force in member PR and the support reaction at R are respectively

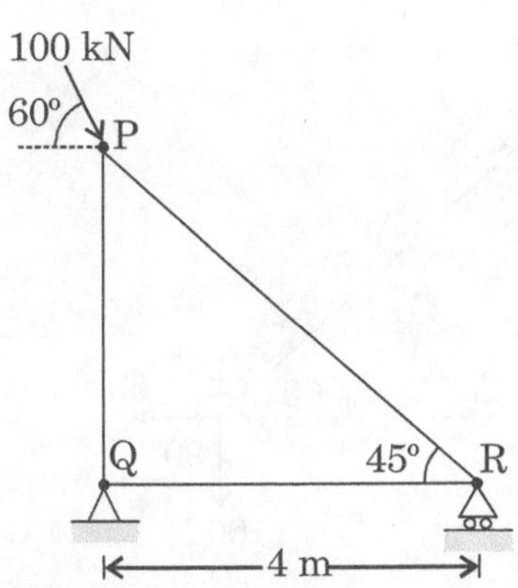

(a) 122.47 kN and 50 kN
(b) 70.71 kNand100kN
(c) 70.71 kN and 50 kN
(d) 81.65kNand100kN

[2015 : 2 Marks, Set-1]

19. For the truss shown in the figure, the magnitude of the force (in kN) in the member SR is

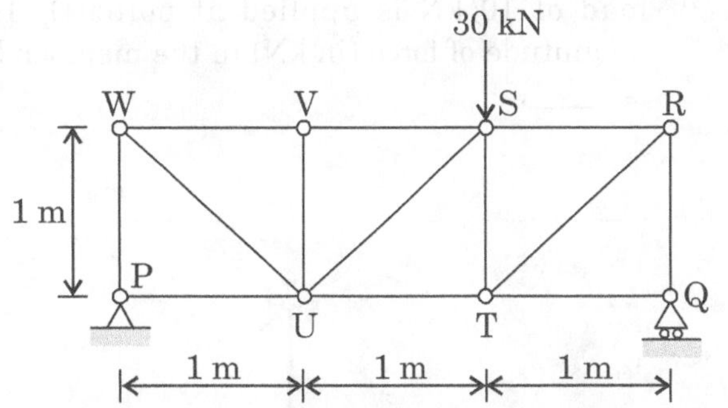

(a) 10 (b) 14.14
(c) 20 (d) 28.28

[2015 : 2 Marks, Set-2]

20. A two member truss PQR is supporting a load W. The axial forces in members PQ and QR are respectively

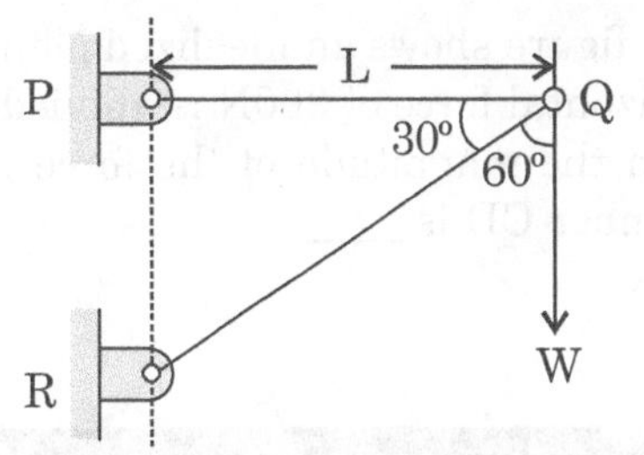

(a) 2W tensile and $\sqrt{3}W$ compressive
(b) $\sqrt{3}W$ tensile and 2W compressive
(c) $\sqrt{3}W$ compressive and 2W tensile
(d) 2W compressive and $\sqrt{3}W$ tensile

[2016 : 2 Marks, Set-1]

21. In the figure, the load P= 1 N, length L = 1 m, Young's modulus E = 70 GPa, and the cross-section of the links is a square with dimension 10 mm × 10 mm. All joints are pin joints.

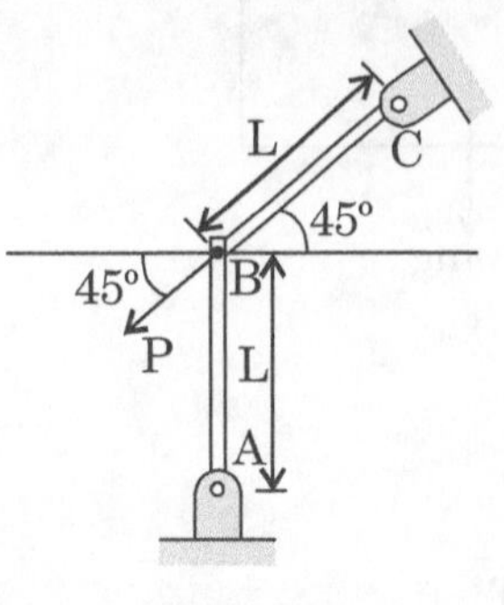

The stress (in Pa) in the AB is ____
(Indicate compressive stress by a negative sign and tensile stress by a positive sign.)

[2016 : 2 Marks, Set-2]

22. A truss is composed of members AB, BC, CD, and AD and BD as shown in the figure. A vertical load of 10 kN is applied at point D. The magnitude of force (in kN) in the member BC is _____.

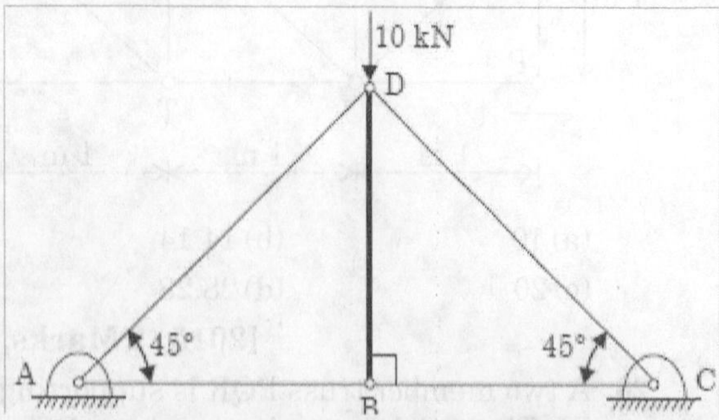

[2019 : 2 Marks, Set-1]

23. The figure shows an idealized plane truss. If a horizontal force of 300N is applied at point A, then the magnitude of the force produced in member CD is _____N.

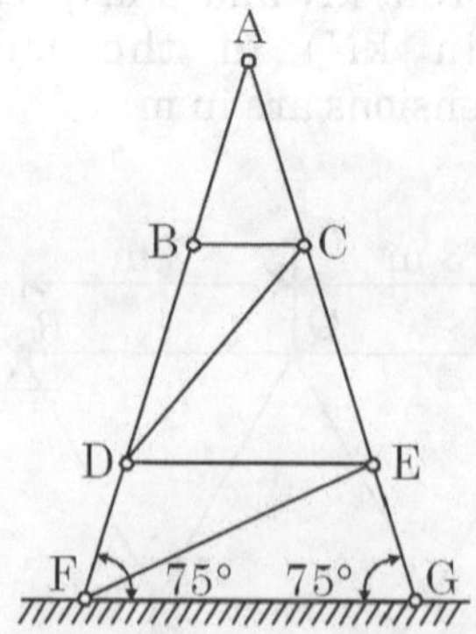

[2019 : 1 Mark, Set-2]

Virtual work

24. If a system is in equilibrium and the position of the system depends upon many independent variables, the principle of virtual- work states that the partial derivatives of its total potential energy with respect to each of the independent variable must be

(a) -1.0 (b) 0

(c) 1.0 (d) ∞

[2006 : 2 Marks]

ANSWERS

1. (c) **3.** (a) **4.** (b) **5.** (a) **7.** (a) 8. (a) **9.** (d) **10.** (a) **11.** (a) **12.** (b)

13. (a) **14.** (b) **16.** (a) **18.** (c) **19.** (c) **20.** (b) **24.** (b)

EXPLANATIONS

1.

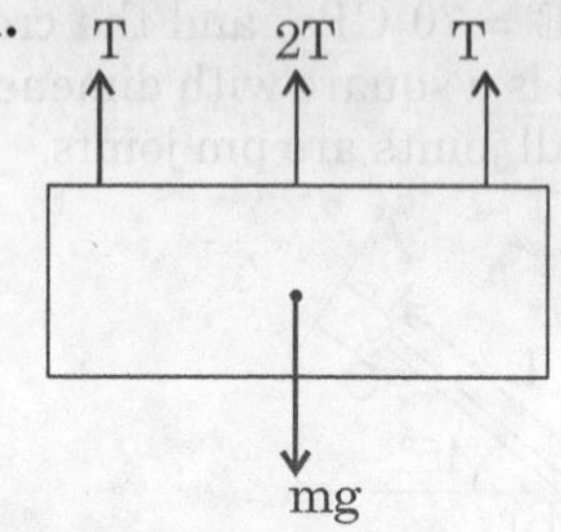

$$\sum F_V = 0$$

$$4T = mg$$

$$\text{or } m = \frac{4T}{g}$$

2.

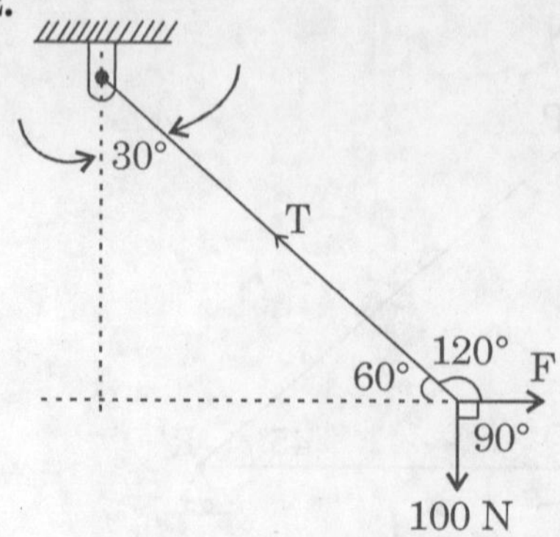

Applying Lami's theorem

$$\frac{100}{\sin 120°} = \frac{F}{\sin(60° + 90°)} = \frac{T}{\sin 90°}$$

$$F = 57.74 \text{ N}$$

3. Fixed support should be replaced by vertical reaction, horizontal reaction and a moment reaction in free body diagram.

4.

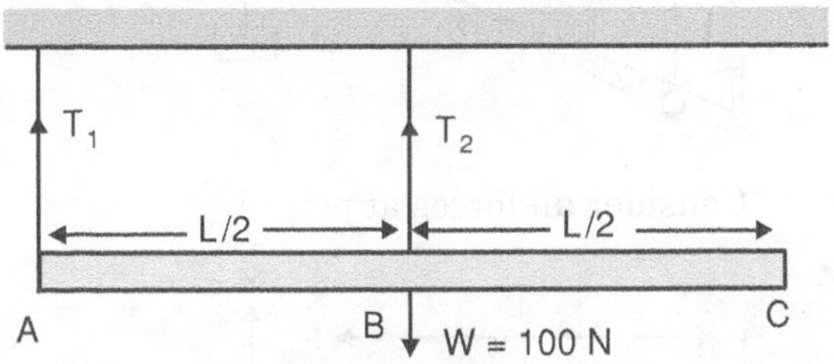

$$T_1 + T_2 = 100\text{N}$$

$$\Sigma M_A = 0$$

$$T_2 \cdot \frac{L}{2} = 100 \times \frac{L}{2}$$

$$T_2 = 100 \text{ N}$$

$$T_1 = 0 \text{ N}$$

5.

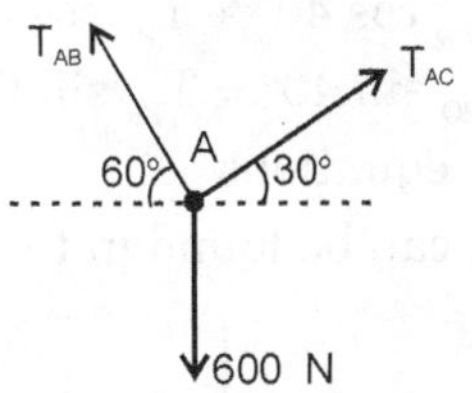

$T_{AB} \cos 60° = T_{AC} \cos 30°$ ...(*i*)

$T_{AB} \sin 60° + T_{AC} \sin 30° = 600$ ...(*ii*)

Solving equations (*i*) and (*ii*), we get

$T_{AB} = 520\text{N}$ and $T_{AC} = 300\text{N}$

6. Taking moments about B

$$\Sigma M_B = 0$$

$\Rightarrow$ $W \times 2 = R_A \times 3$

$\Rightarrow$ $R_A = \dfrac{600 \times 2}{3} = 400$

$$\sum H = 0$$

$\Rightarrow$ $R_A = P = 400 \text{ N}$

7. Using sine rule

$$\frac{T_1}{\sin 120°} = \frac{T_2}{\sin 150°} = \frac{500}{\sin 90°}$$

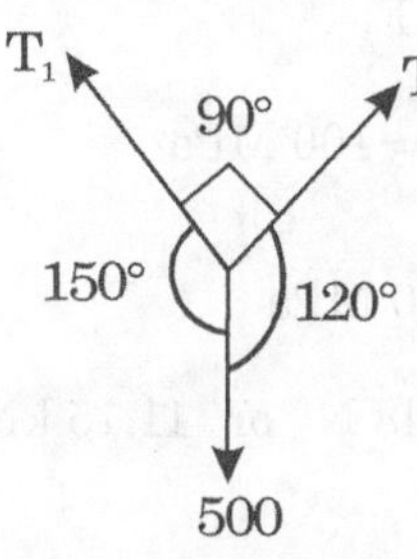

$$T_1 = 500 \times \sin 120°$$

$$T_2 = 500 \sin 150°$$

$$T_1 = 433 \text{ N}, T_2 = 250\text{N}$$

8. Here, Reaction on front wheels = R_f

Reaction on rear wheels = R_r

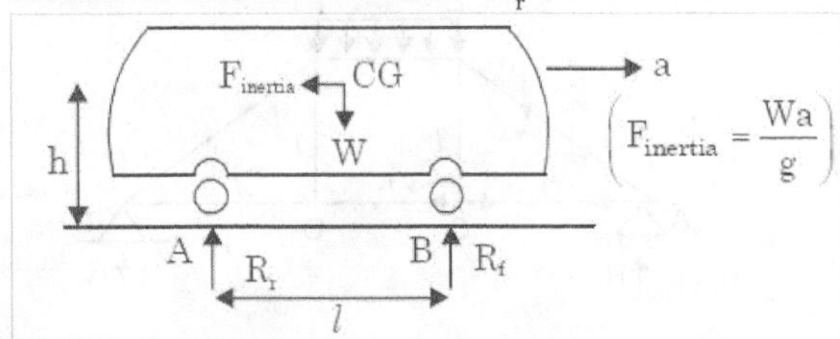

$$\Sigma M_A = 0,\ F_{inertia} \times h + R_f \times l - W \times \frac{l}{2} = 0$$

$$R_f = \frac{\frac{Wl}{2} - h \times \frac{W_a}{g}}{l}$$

$$R_f = \frac{W}{2} - \frac{W}{g}\left(\frac{h}{l}\right)a$$

$$\Sigma f_y = 0 \quad W = R_f + R_r$$

$$R_r = \frac{W}{2} + \frac{W}{g}\left(\frac{h}{l}\right)a$$

9.

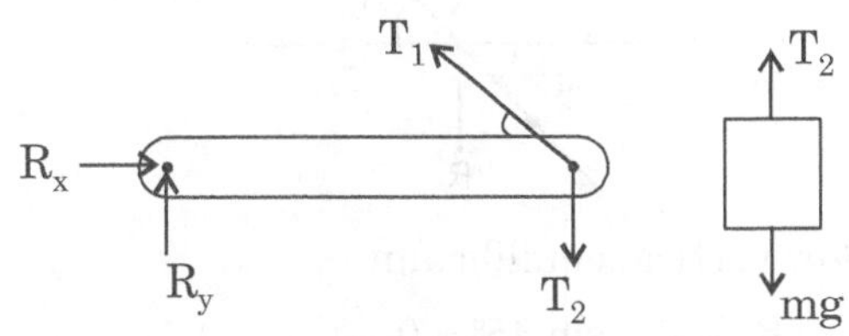

For block, $T_2 = mg = 343.35$ N ...(i)

$$\tan\theta = \frac{125}{275} = 27.44°$$

For bar, $\sum M_A = O$

$T, \sin\theta \times 1 = T_2 \times 1$

$$\Rightarrow T_1 = \frac{T_2}{\sin\theta}$$

$\therefore T_1 = 829.74$ N

$$\sum F_y = O$$

$$R_y - T_2 + T_1 \sin\theta = O$$

$$\therefore R_y = O$$

$$\sum Fx = O$$

$$R_x - T_1 \cos\theta = O$$

$$R_x = 755.4 \text{ N}$$

10. Total load on EF member = pl

where l = length of EF

For horizontal equilibrium

$$R_A + R_B = Pl$$

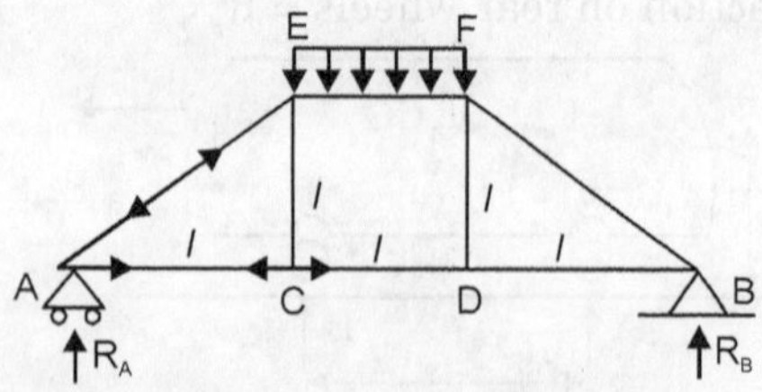

Taking moment about A, we have

$$Pl \times \frac{3l}{2} = R_B \times 3l$$

$$\Rightarrow \quad R_B = \frac{Pl}{2}$$

Considering a point A.

For horizontal equilibrium

$$F_{AC} + F_{AE} \cos 45° = 0$$

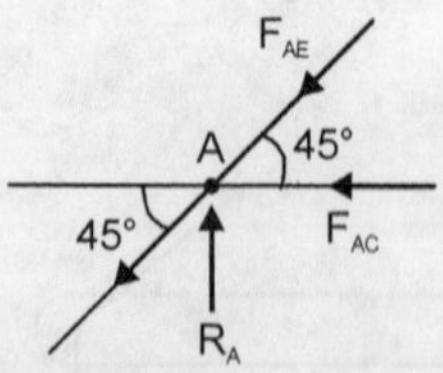

For vertical equilibrium

$$R_A - F_{AE} \sin 45° = 0$$

$$\Rightarrow F_{AE} = \frac{R_A}{\sin 45°} = \frac{Pl}{2 \times \frac{1}{\sqrt{2}}} = \frac{Pl}{\sqrt{2}}$$

$$\therefore \; F_{AC} = -F_{AE} \cos 45°$$

$$= \frac{-Pl}{\sqrt{2}} \cdot \frac{1}{\sqrt{2}} = \frac{-Pl}{2}$$

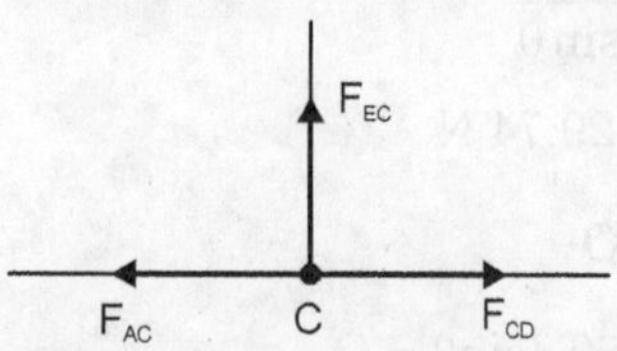

–ve sign shows that the force on member AC is opposite of assumed.

Now considering a point C

For horizontal equilibrium

$$F_{AC} = F_{CD} = \frac{Pl}{2}$$

11.

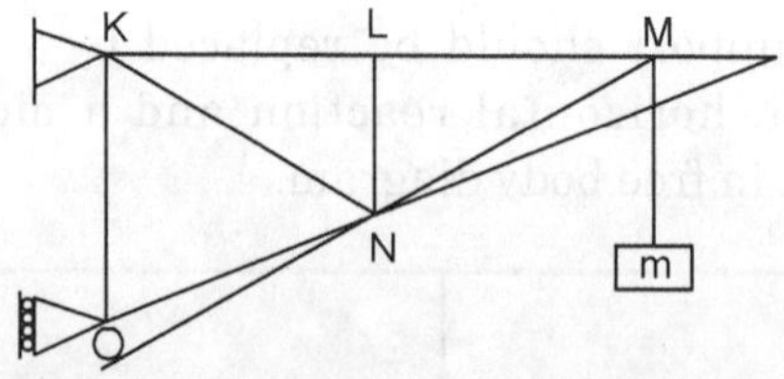

Consider all forces at point L

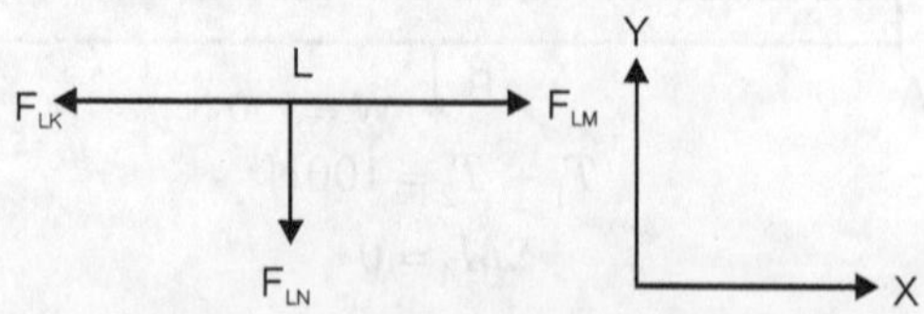

Summation of all forces along x-direction,

$$F_{LK} = F_{LM}$$

Summation of all forces along y-direction,

$$F_{LN} = 0$$

12. $T_{PQ} \cos 40° + T_{PR} \cos 60° + F = 0$

and $T_{PQ} \sin 40° = T_{PR} \sin 60°$

From these equations

T_{PQ} and T_{PR} can be found in terms of F

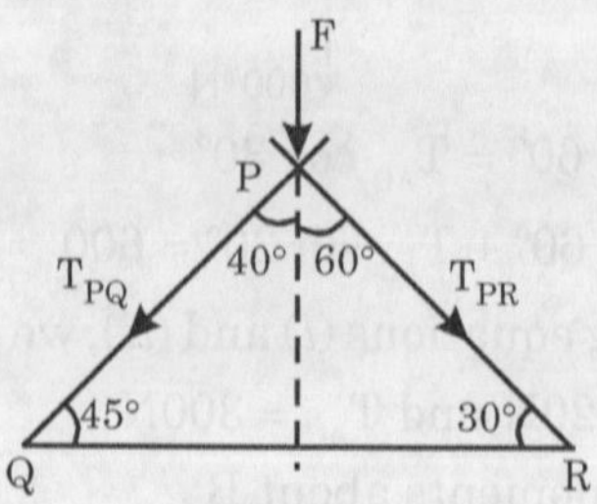

Now at Q, T $T_{QR} = T_{PQ} \cos 45°$

Solving these equations, we get

$$T_{QR} = 0.63 \text{ F}$$

13. Vertical reaction at point B,

$$R_B = F_B \cos 30 = 0.633 \text{ F}$$

$$F = 1\text{kN}$$

$$\therefore \quad R_B = 0.633 \text{ kN}$$

14. From Lami's theorem,

$$\frac{F}{\sin 105} = \frac{F_A}{\sin 120} = \frac{F_B}{\sin 135}$$

$$\Rightarrow \quad F_A = 0.8965 \text{ F}; \; F_B = 0.732 \text{ F}$$

Hence maximum force is F_A

and $\text{stress} = \frac{F_A}{\text{Area}} = 100 \text{ MPa}$

$$\therefore \quad \frac{0.8965 \text{ F}}{100} = 100 \text{ MPa}$$

$$\Rightarrow \quad F = 11154.48 \text{ N or } 11.15 \text{ kN}$$

15. 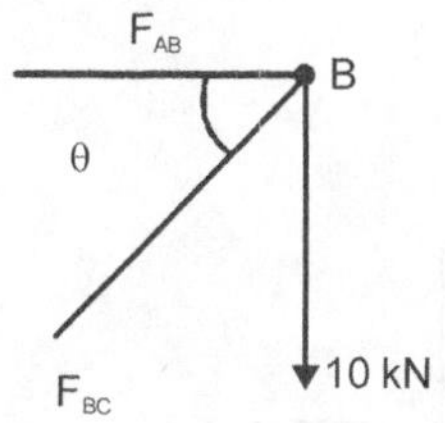

FBD of joint B

$$\theta = \tan^{-1}\left(\frac{0.5}{1}\right) = 26.5$$

$$\Sigma V = 0$$

$$\frac{F_{BC}\sin\theta}{F_{BC}\cos\theta} = \frac{10\text{ kN}}{F_{AB}}$$

$$F_{AB} = \frac{10}{\tan(26.2)}$$

$$= 20\text{ kN.}$$

16. By method of sections

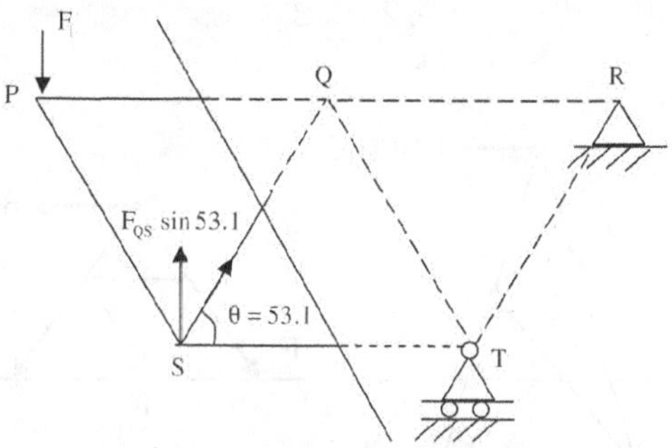

$$\sum V = 0$$

$$F_1 = F_{QS}\sin 53.1$$

$$\Rightarrow \quad F_{QS} = 11.25(T)$$

17. $\Rightarrow$ $2F\sin 30 = 100$

$\Rightarrow$ $F = 100$ N

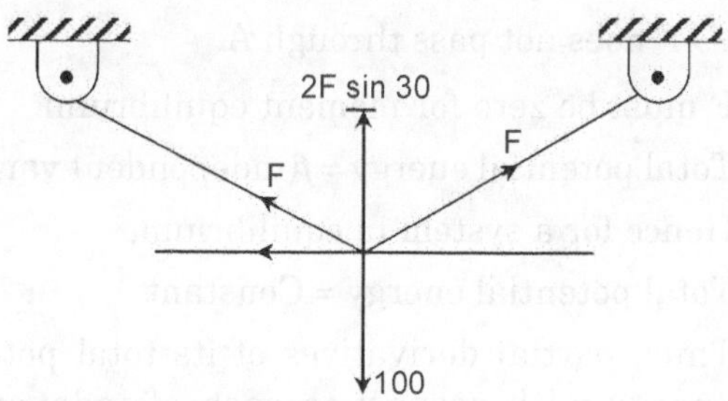

18. $\Sigma M_Q = 0$

$$100 \times \cos 60 \times 4 = R_a \times 4$$

$$\Rightarrow \quad R_a = 50\text{ kN}$$

$$F_{PR} \times \cos 45 = 100\cos 60$$

$$\Rightarrow \quad F_{PR} = 70.71\text{ kN}$$

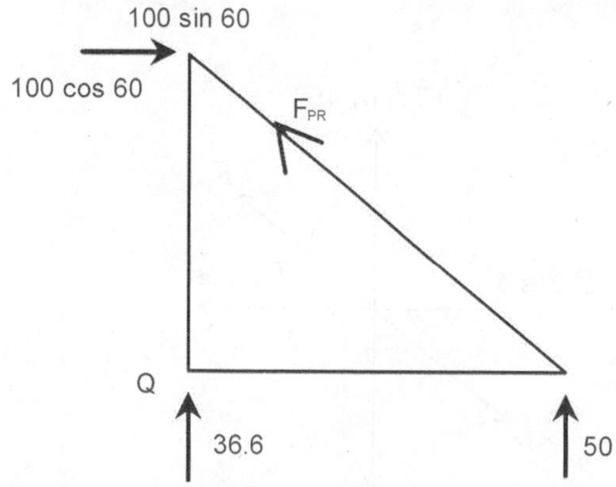

19. 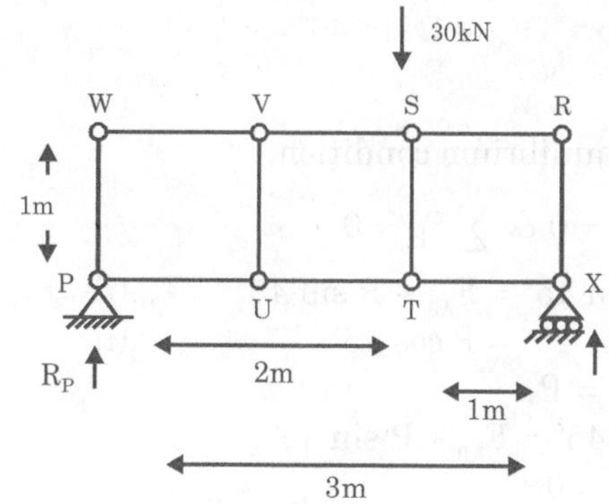

$$R_p + R_x = 30000$$

$$\Sigma M_P = 0$$

$$R_x \times 3 = 2 \times 3000$$

$$\Rightarrow \quad R_x = 20\text{ kN}$$

and $\quad R_p = 10$ kN

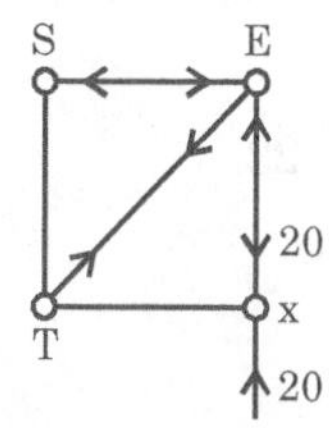

For balance at 'x' $\rightarrow F_{Rx} = 20$ kN

at 'RT' $\rightarrow F_{RT}\cos 45 = 20$

$$\Rightarrow \quad F_{RT} = \frac{20}{\cos 45} \qquad ...(i)$$

Also $\quad F_{SR} = F_{RT}\cos 45 \qquad ... (ii)$

From equations (*i*) and (*ii*),

$$F_{SR} = 20\text{ kN}$$

20. 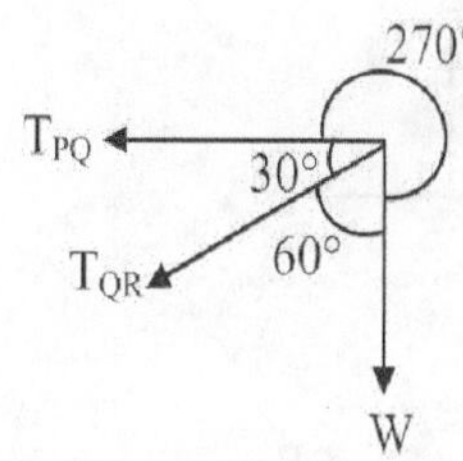

Lami's theorem is used to find the F

$$\frac{W}{\sin 30°} = \frac{T_{PQ}}{\sin 60°} = \frac{T_{QR}}{\sin 270°}$$

$$\Rightarrow \quad T_{PQ} = \sqrt{3}W(T)$$

$$\Rightarrow \quad T_{QR} = -2W = 2W(C)$$

21.

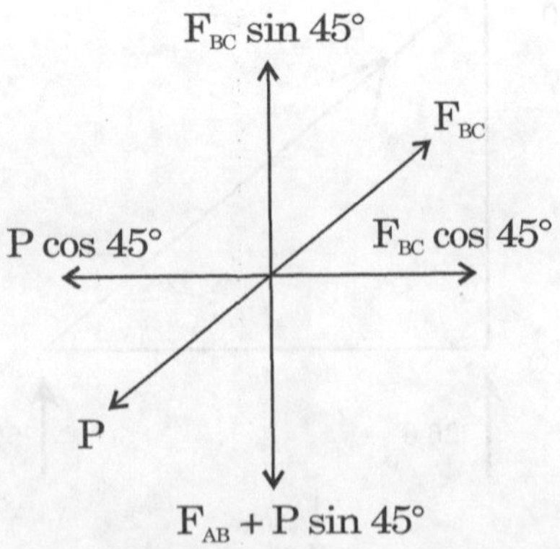

For equilibrium condition,

$\sum F_y = 0$ & $\sum F_H = 0$

F_{BC} sin 45° = F_{AB} + P sin 45° ...(i)

F_{BC} cos 45° = P cos 45° ...(ii)

$\therefore F_{BC} = P$

P sin 45° = F_{AB} + P sin 45°

$\therefore F_{AB} = 0$

$$\sigma_{AB} = \frac{F_{AB}}{A} = \frac{0}{A} = 0 \text{ Pascal}$$

22.

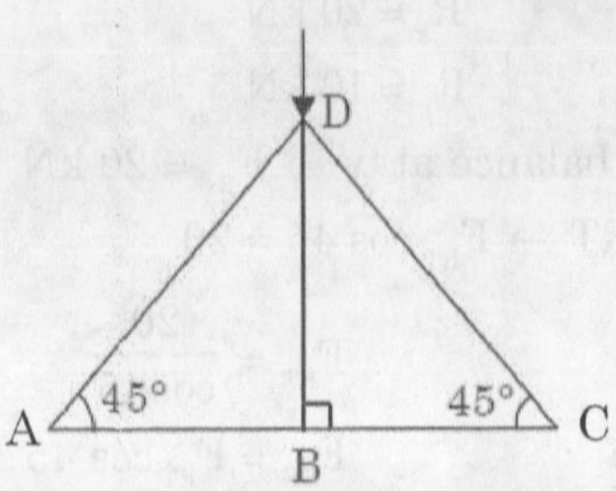

$F_{BD} = 0$

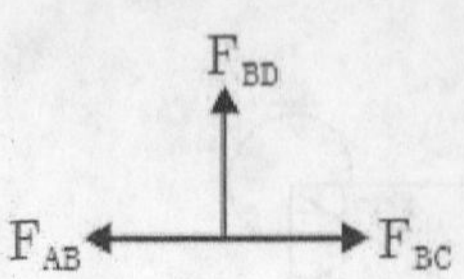

At joint D,

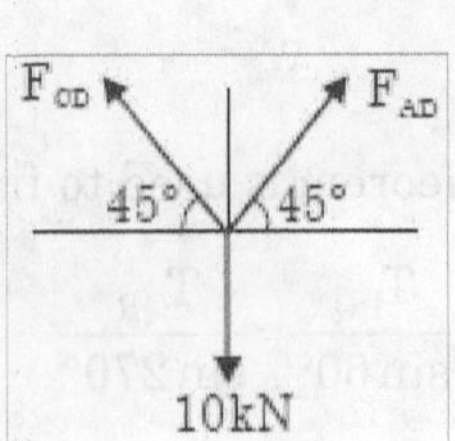

At joint C,

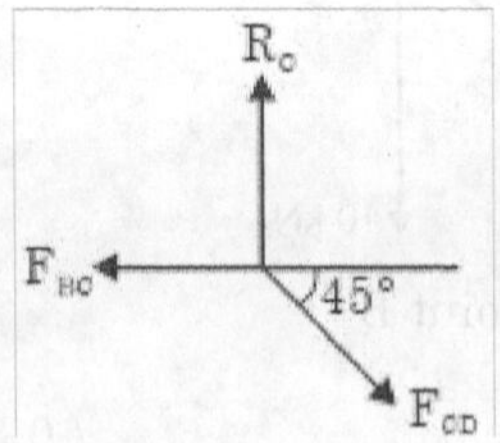

$$\Sigma F_x = 0$$

$$F_{CD} \cos 45 = F_{AD} \cos 45$$

$$F_{CD} = F_{AD}$$

$$\Sigma F_y = 0,\ 10 = 2F_{CD} \sin 45$$

$$F_{CD} = \frac{10}{\sqrt{2}} \text{kN}$$

$$F_{CD} \cos 45 = F_{BC}$$

$$F_{BC} = \frac{10}{\sqrt{2}} \times \frac{1}{\sqrt{2}} = 5 \text{ kN}$$

23.

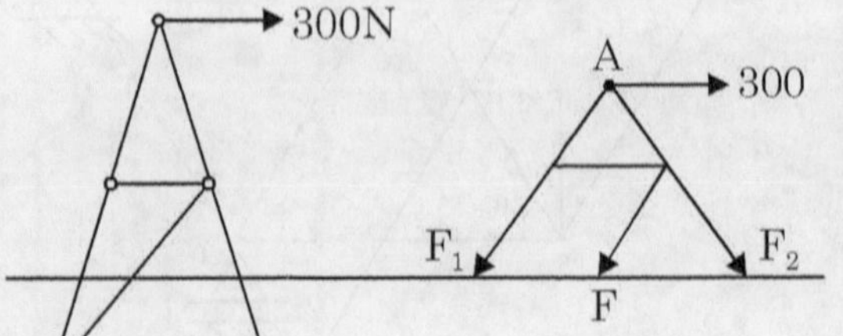

By taking moment of all forces about A, we get F = 0

Because, F_1, F_2, 300N, all passes through A

$\Rightarrow$ F does not pass through A.

F must be zero for moment equilibrium.

24. Total potential energy = *f*(independent variable)

Hence for a system in equilibrium,

Total potential energy = Constant

Thus, partial derivatives of its total potential energy with respect to each of independent variable must be zero.

2

CHAPTER

Translational and Rotational motion

Analysis of Previous GATE Papers

Topic	Year/Set → Marks Distribution ↓		2019 Set 1	2019 Set 2	2018 Set 1	2018 Set 2	2017 Set 1	2017 Set 2	2016 Set 1	2016 Set 2	2016 Set 3	2015 Set 1	2015 Set 2	2015 Set 3
Translational Motion	1 Mark	MCQ Type												
		Numerical Type										1		
	2 Marks	MCQ Type												
		Numerical Type									1		1	
	Total										2	1	2	
Rotational Motion	1 Mark	MCQ Type	1											1
		Numerical Type												
	2 Marks	MCQ Type										2		
		Numerical Type						1						1
	Total							2				4		3

Translational motion

1. AB and CD are two uniform and identical bars of mass 10 kg each, as shown. The hinges at/A and B are frictionless. The assembly is released from rest and motion occurs in the vertical plane. At the instant that the hinge B passes the point B, the angle between the two bars will be

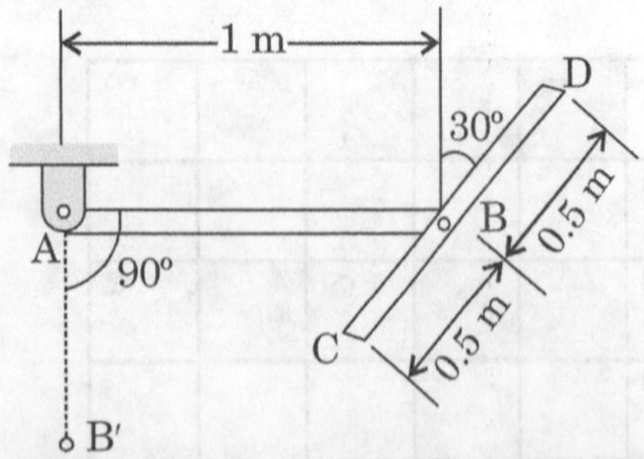

(a) 60 degrees (b) 37.4 degrees
(c) 30 degrees (d) 45 degrees

[1996 : 2 Marks]

2. A car moving with uniform acceleration covers 450 m in a 5 second interval, and covers 700 m in the next 5 second interval. The acceleration of the car is

(a) 7 m/s^2 (b) 50 m/s^2
(c) 25 m/s^2 (d) 10 m/s^2

[1998 : 1 Mark]

3. The time variation of the position of a particle in rectilinear motion is given by $x = 2t^3 + t^2 + 2t$. If v is the velocity and a is the acceleration of the particle in consistent units, the motion started with

(a) v = 0, a = 0 (b) v = 0, a = 2
(c) v = 2, a = 0 (d) v = 2, a = 2

[2005 : 1 Mark]

4. A simple pendulum of length of 5 m, with a bob of mass 1 kg, is in simple harmonic motion. As it passes through its mean position, the bob has a speed of 5 m/s. The net force on the bob at the mean position is

(a) zero (b) 2.5 N
(c) 5 N (d) 25 N

[2005 : 1 Mark]

5. A uniform rigid rod of mass M and length L is hinged at one end as shown in the adjacent figure. A force P is applied at a distance of 2L/3 from the hinge so that the rod swings to the right. The horizontal reaction at the hinge is

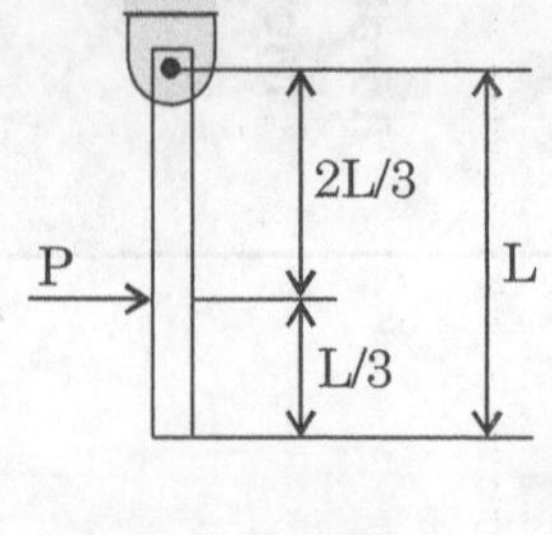

(a) -P (b) 0
(b) P/3 (d) 2P/3

[2009 : 2 Marks]

6. A swimmer can swim 10 km in 2 hours when swimming along the flow of a river. While swimming against the flow, she takes 5 hours for the same distance. Her speed in still water (in km/h) is ________.

[2015 : 1 Mark, Set-1]

7. The initial velocity of an object is 40 m/s. The acceleration a of the object is given by the following expression.

$a = -0.1v$

where v is the instantaneous velocity of the object. The velocity of the object after 3 seconds will be ______.

[2015 : 2 Marks, Set-2]

8. An inextensible massless string goes over a frictionless pulley. Two weights of 100 N and 200 N are attached to the two ends of the string. The weights are released from rest, and start moving due to gravity. The tension in the string (in N) is ________.

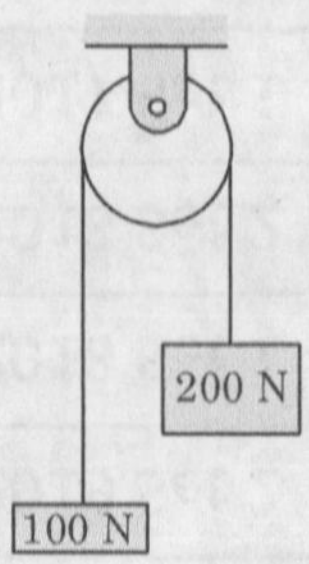

[2016 : 2 Marks, Set-3]

Rotational motion

9. A and B are the end points of a diameter of a disc rolling along a straight line with a counter clockwise angular velocity as shown in figure. Referring to the velocity vectors $\vec{V}_A$ and $\vec{V}_B$ shown in the figure

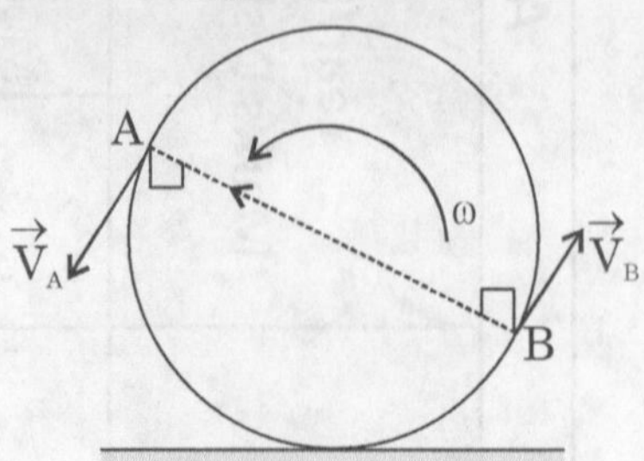

(a) $\vec{V}_A$ and $\vec{V}_B$ are both correct

(b) $\vec{V}_A$ is incorrect but $\vec{V}_B$ is correct

(c) $\vec{V}_A$ and $\vec{V}_B$ are both incorrect

(d) $\vec{V}_A$ is correct but $\vec{V}_B$ s incorrect

[1990 : 1 Mark]

10. The cylinder shown below rolls without slipping. Toward which of the following points is the acceleration of the point of contact on the cylinder directed?

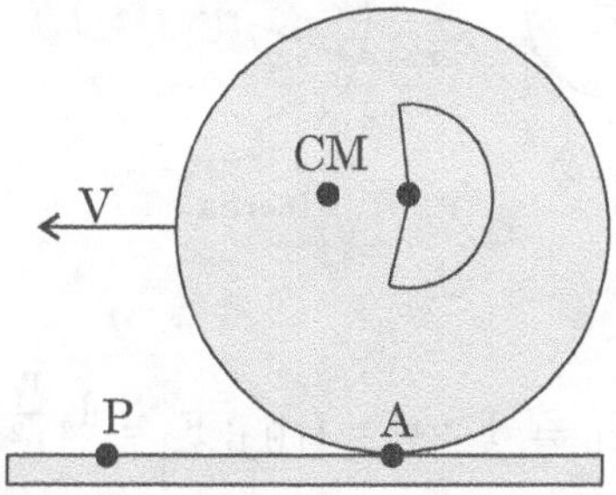

(a) The mass centre
(b) The geometric centre
(c) The point P as marked
(d) None of the above

[1993 : 1 Mark]

11. A stone of mass m at the end of a string of length l is whirled in a vertical circle at a constant speed. The tension is the string will be maximum when the stone is

(a) at the top of the circle
(b) half way down from the top
(c) quarter way down from the top
(d) at the bottom of the circle

[1994 : 1 Mark]

12. A rod of length 1 m is sliding in a corner as shown in figure. At an instant when the rod makes an angle of 60 degrees with the horizontal plane, the velocity of point A on the rod is 1 m/s. The angular velocity of the rod at the instant is

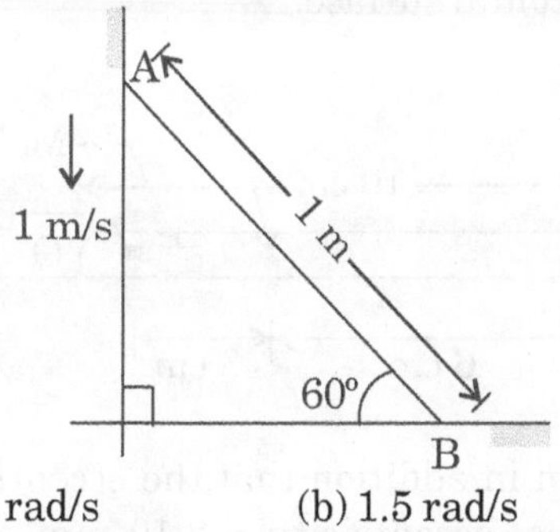

(a) 2 rad/s (b) 1.5 rad/s
(c) 0.5 rad/s (d) 0.75 rad/s

[1996 : 2 Marks]

13. As shown in figure, a person A is standing at the centre of a rotating platform facing person B who is riding a bicycle, heading East. The relevant speeds and distances are shown in given figure person, a bicycle, heading East. At the instant under consideration, what is the apparent velocity of B as seen by A?

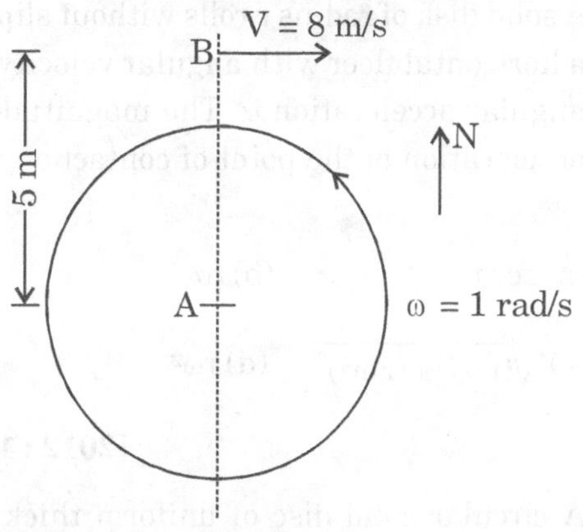

(a) 3 m/s heading east
(b) 3 m/s heading west
(c) 8 m/s heading east
(d) 13 m/s heading east

[1999 : 2 Marks]

14. A rigid body shown in the fig. (a) has a mass of 10 kg. It rotates with a uniform angular velocity ω. A balancing mass of 20 kg is attached as shown in fig. (b). The percentage increase in mass moment of inertia as a result of this addition is

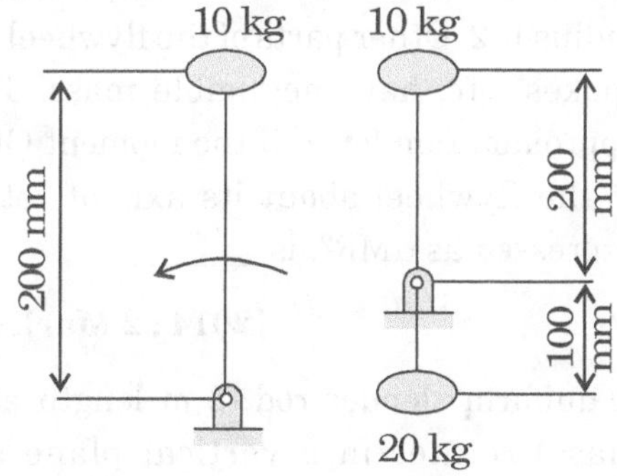

(a) 25% (b) 50%
(c) 100% (d) 200%

[2004 : 2 Marks]

15. A circular disk of a radius R rolls without slipping at a velocity V. The magnitude of the velocity at point P (see figure) is:

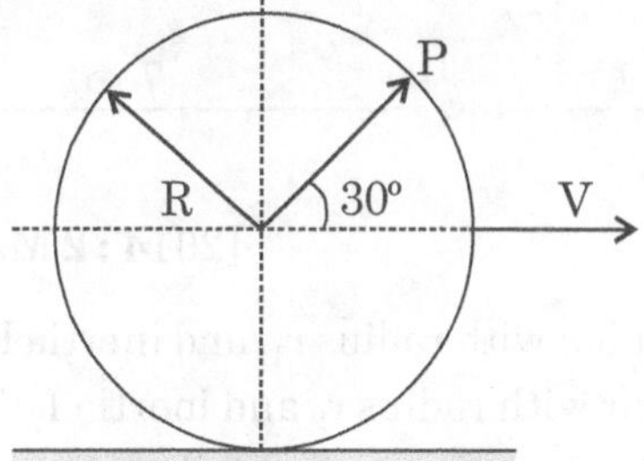

(a) $\sqrt{3}$ V (b) $\sqrt{3}$ V/2
(c) V/2 (d) 2V/ $\sqrt{3}$

[2008 : 2 Marks]

16. A solid disk of radius r rolls without slipping on a horizontal floor with angular velocity ω and angular acceleration α. The magnitude of the acceleration of the point of contact on the disc is

(a) zero (b) $r\alpha$

(c) $\sqrt{(r\alpha)^2 + (r\omega^2)^2}$ (d) $r\omega^2$

[2012 : 1 Mark]

17. A circular solid disc of uniform thickness 20 mm, radius 200 mm and mass 20 kg, is used as a flywheel. If it rotates at 600 rpm, the kinetic energy of the flywheel, in Joules is

(a) 395 (b) 790

(c) 1580 (d) 3160

[2012 : 1 Mark]

18. Consider a fly wheel whose mass M is distributed almost equally between a heavy, ring-like rim of radius P and a concentric disk-like feature of radius R/2. Other parts of the flywheel, such as spokes, etc, have negligible mass. The best approximation for α, if the moment Of inertia of the flywheel about its axis of rotation is expressed as αMR^2, is ______.

[2014 : 2 Marks, Set-2]

19. A uniform slender rod (8 m length and 3 kg mass) rotates in a vertical plane about a horizontal axis 1 m from its end as shown in the figure. The magnitude of the angular acceleration (in rad/s²) of the rod at the position shown is ______

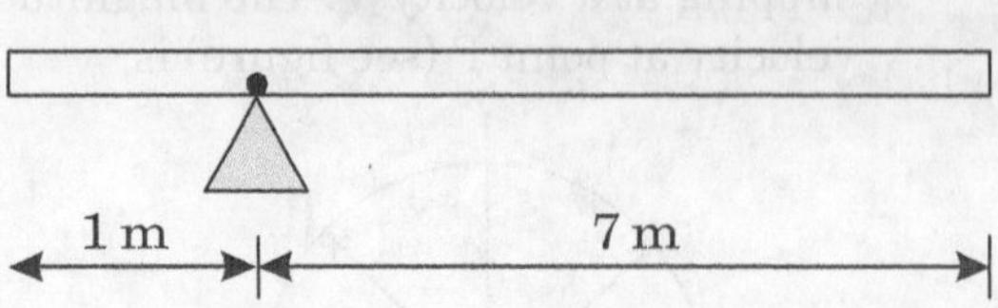

[2014 : 2 Mark, Set-4]

20. A pinion with radius r_1, and inertia I_1, is driving a gear with radius r_2 and inertia I_2. Torque τ_1 is applied on pinion. The following are free body diagrams of pinion and gear showing important forces (F_1 and F_2) of interaction. Which of the following relations hold true?

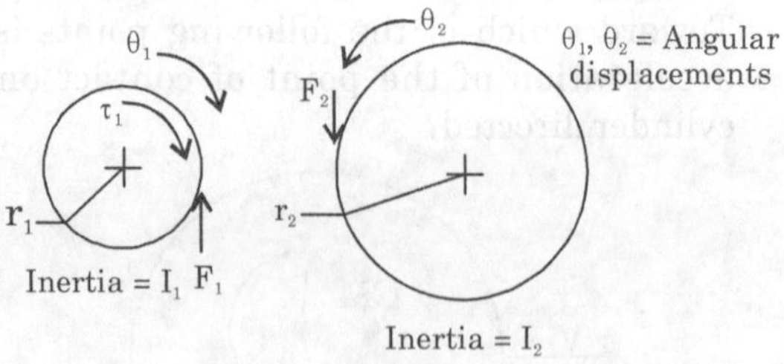

(a) $F_1 \neq F_2;\ \tau_1 = I_1\ddot{\theta}_1;\ F_2 = I_2\frac{r_1}{r_2^2}\ddot{\theta}_1$

(b) $F_1 = F_2;\ \tau_1 = \left[I_1 + I_2\left(\frac{r_1}{r_2}\right)^2\right]\ddot{\theta}_1;\ F_2 = I_2\frac{r_1}{r_2^2}\ddot{\theta}_1$

(c) $F_1 = F_2;\ \tau_1 = I_1\ddot{\theta}_1;\ F_2 = I_2\frac{r_1}{r_2}\ddot{\theta}_2$

(d) $F_1 \neq F_2;\ \tau_1 = \left[I_1 + I_2\left(\frac{r_1}{r_2}\right)^2\right]\ddot{\theta}_1;\ F_2 = I_2\frac{1}{r_2}\ddot{\theta}_2$

[2015 : 2 Marks, Set-1]

21. A mobile phone has a small motor with an eccentric mass used for vibrator mode. The location of the eccentric mass on motor with respect to center of gravity (CG) of the mobile and the rest of the dimension of the mobile phone are shown. The mobile is kept on a flat horizontal surface.

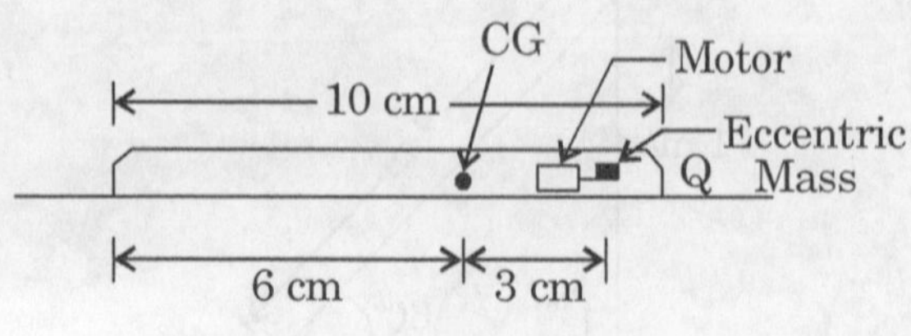

Given in addition that the eccentric mass = 2 grams, eccentricity = 2.19 mm, mass of the mobile = 90 grams, g = 9.81 m/s². Uniform speed of the motor in RPM for which the mobile will get just lifted off the ground at the end Q is approximately

(a) 3000 (b) 3500

(c) 4000 (d) 4500

[2015 : 2 Marks, Set-1]

22. For the same material and the mass, which of the following configurations of flywheel will have maximum mass moment of inertia about the axis of rotation OO' passing through the center of gravity.

(a)

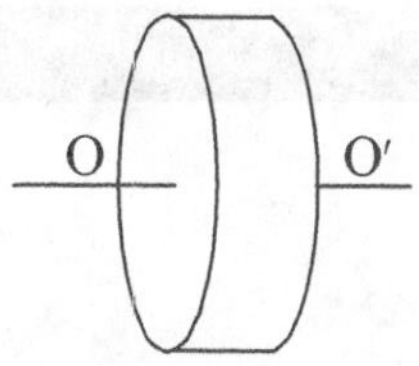

(b)

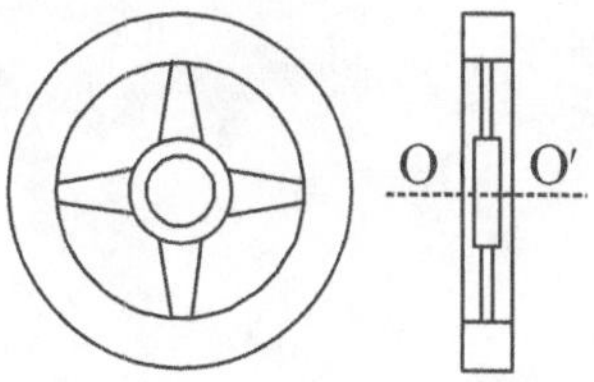

(c)

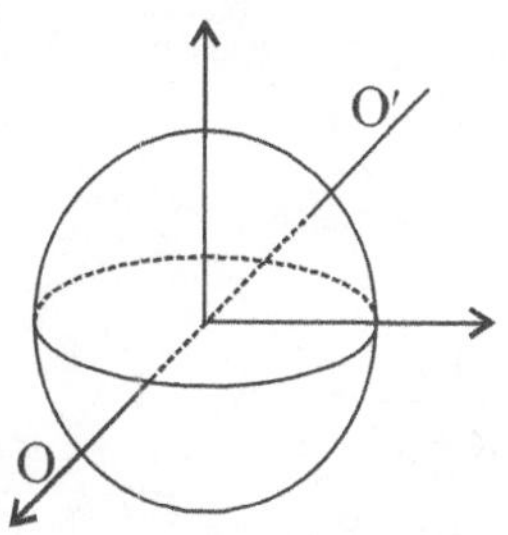

(d)

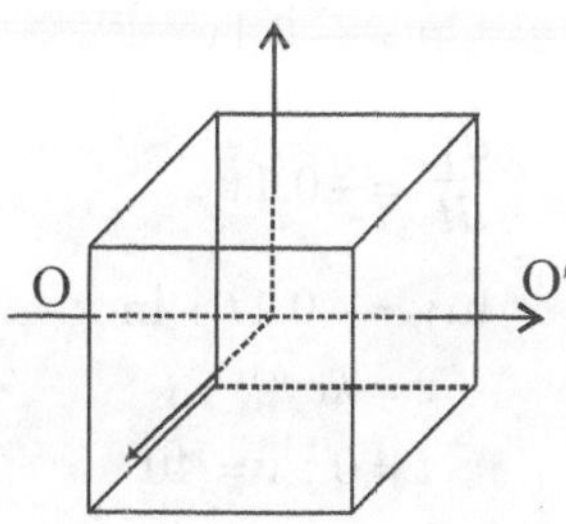

[2015 : 1 Mark, Set-3]

23. Figure shows a wheel rotating about Q_2. Two points A and B located along the radius of wheel have speeds of 80 m/s and 140 m/s respectively. The distance between the points A and B is 300 mm.

The diameter of the wheel (in mm) is ______.

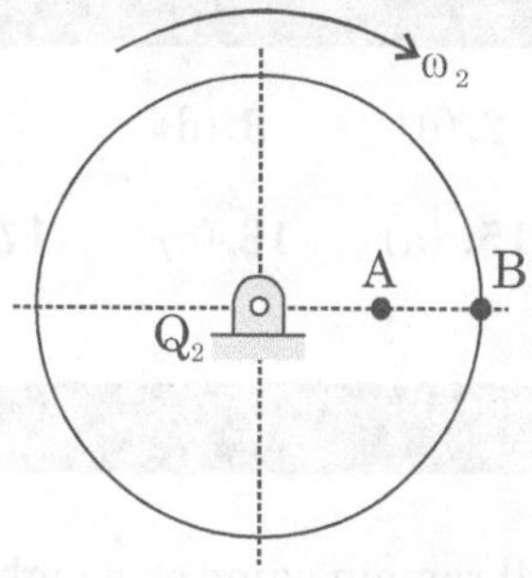

[2015 : 2 Marks, Set-3]

24. The rod PQ of length $L = \sqrt{2}$ m and uniformly distributed mass of M = 10 kg, is released from rest at the position shown in the figure. The ends slide along the frictionless faces OP and OQ. Assume acceleration due to gravity, g = 10 m/s². The mass moment of inertia of the rod about its centre of mass and an axis perpendicular to the plane of the figure is ($ML^2/12$). At this instant, the magnitude of angular acceleration (in radian/s²) of the rod is ______.'

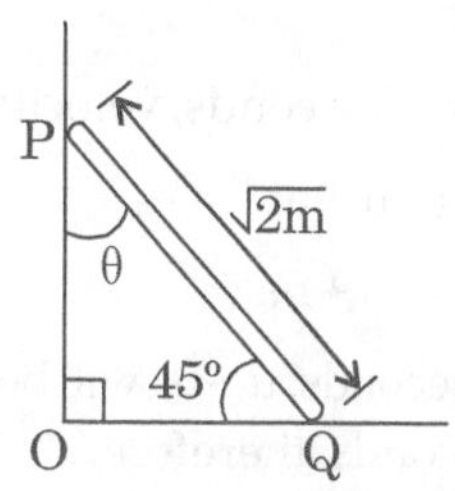

[2017 : 2 Marks, Set-2]

25. The position vector $\overline{OP}$ of point P(20, 10) is rotated anti-clockwise in X-Y plane by an angle θ = 30° a such that point P occupies position Q, as shown in the figure. The coordinates (x, y) of Q are

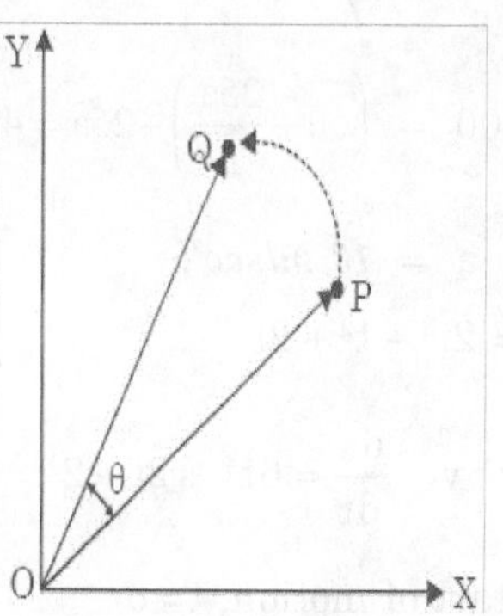

(a) (12.32, 18.66) (b) (18.66, 12.32)
(c) (22.32 ,8.26) (d) (13.40, 22.32)

[2019 : 1 Mark, Set-1]

ANSWERS

1. (a) **2.** (d) **3.** (d) **4.** (a) **5.** (b) **9.** (c) **10.** (b) **11.** (d) **12.** (a) **13.** (c)
14. (b) **15.** (a) **16.** (d) **17.** (b) **20.** (b) **21.** (b) **22.** (b) **25.** (a)

EXPLANATIONS

1. $\theta = 60°$ will remain same as weight mg of the rod CD passes through hinge point B so that there will be no torque in rod CD with respect to B.

2. For $t_1 = 5$ secs, $S_1 = 450$ m
and for $t_2 = 5$ secs $S_2 = 700$ m

From the relation $S = ut + \frac{1}{2}at^2$

$$450 = 5u + \frac{1}{2} \times a \times 25 \text{ [since initial velocity = u]}$$

or $$450 = 5u + \frac{25a}{2} \qquad \text{......}(i)$$

Also after first 5 seconds, velocity of car,

$$v = u + at$$
$$= u + 5a.$$

After first 5 seconds, u + 5a will be the initial velocity for next 5 seconds, therefore

$$700 = (4+5a)5 + \frac{1}{2}a \times 25$$
$$= 20 + 25a + \frac{25a}{2}$$

From equation (i), we have

$$700 = \left(5u + \frac{25a}{2}\right) + 25a = 450 + 25a,$$

$$a = \mathbf{\mathit{10\ m/sec^2}}.$$

3. Given $x = 2t^3 + t^2 + 2t$

Velocity : $v = \frac{dx}{dt} = 6t^2 + 2t + 2$

At the start of motion, t= o

V = 0+0+2

V = 2 m/s

Acceleration; $a = \frac{dv}{dt} = 12t + 2$

at t = 0

a = 0 + 2

$\therefore$ a = 2 m/s^2

4. The net force at the mean position will be zero.

5.

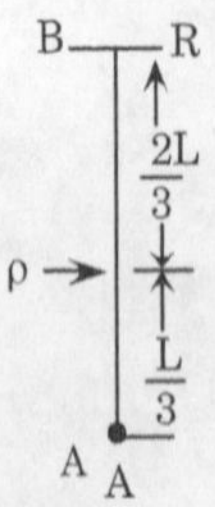

Moment about A is zero

$$\therefore\ R \times L = P \times \frac{L}{3}$$

6. Let Swimmer = x
River = y

$$2 = \frac{10}{x+y}$$

$$5 = \frac{10}{x-y}$$

On solving we get $x = 3.5$ km/h

7. $$a = -0.1v$$

$$\frac{dv}{dt} = -0.1v$$

$$\ln v = -0.1t + \ln k$$

$$v = ke^{-0.1t}$$

At $t = 0$; $v = 40$

$\therefore$ $k = 40$

$$v = 40e^{-0.1t}$$

At t = 3 seconds

$$v = 40e^{-0.1\times3} = 29.6327 \text{ m/s}$$

8.

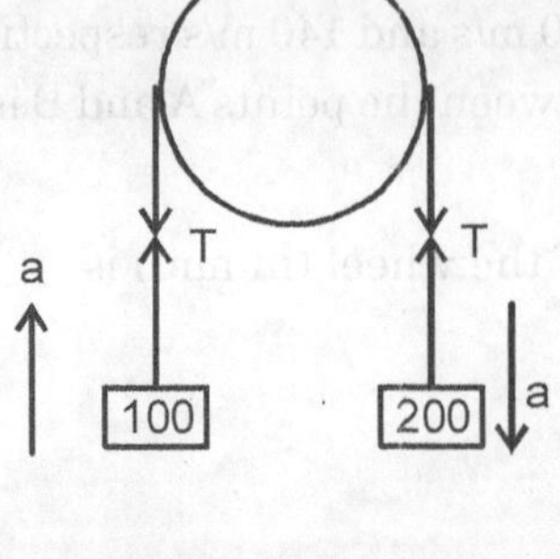

From FBD

$$200 - T = \frac{200}{g}a$$

$$T - 100 = \frac{100}{g}a$$

On adding

$$100 = \frac{300}{g}a \Rightarrow a = \frac{g}{3}$$

$$T = 133.3N$$

10. Acceleration of the point of contact A acts towards geometric centre.

11.

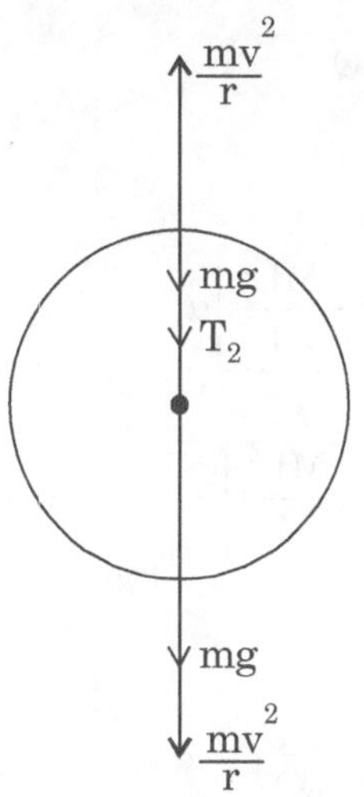

$$T_1 = mg + \frac{mv^2}{r}$$

$$T_2 = \frac{mv^2}{r} - mg$$

$T_1 > T_2$

Maximum tension will be maximum at the bottom

12.

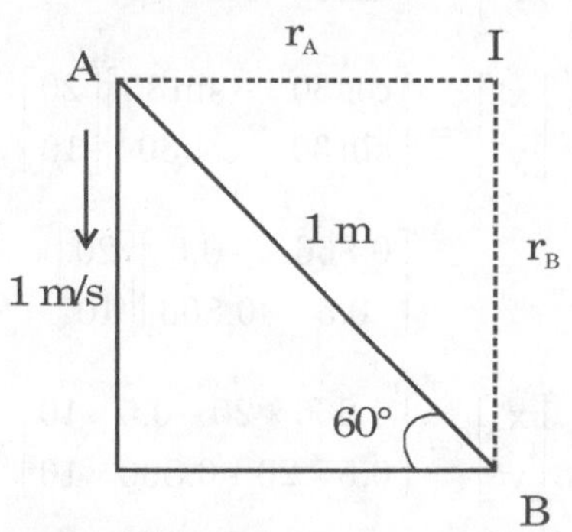

$V_A = r_A\ \omega$

$1 = 1 \cos 60° \times \omega$

$\omega = 2\text{rad/sec}$

13. Relative velocity = 8

14.

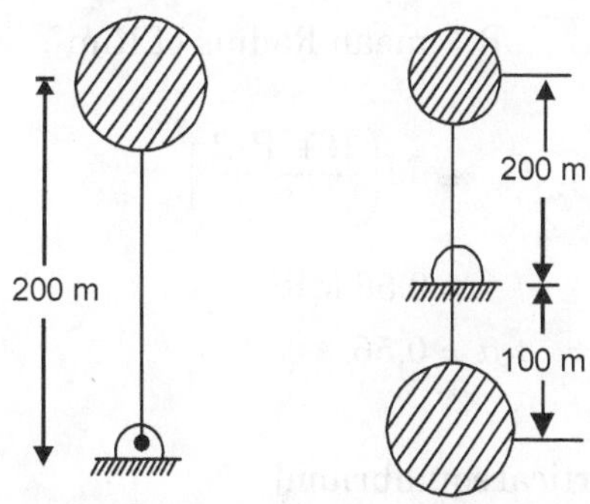

$I_1 = 10 \times (200)^2 = 4 \times 10^5$

$I_2 = 10 \times (200)^2 + 20 \times (100)^2 = 6 \times 10^5$

$\therefore$ % increase in moment of inertia

$$= \frac{I_2 - I_1}{I_1} = \frac{6 \times 10^5 - 4 \times 10^5}{4 \times 10^5} = 50\%$$

15. The instantaneous centre of rotation of the disk is 0, i.e. its point of contact with ground every other point on the disk is rotating with respect to this center with an angular velocity.

$$\omega = \frac{v}{R}$$

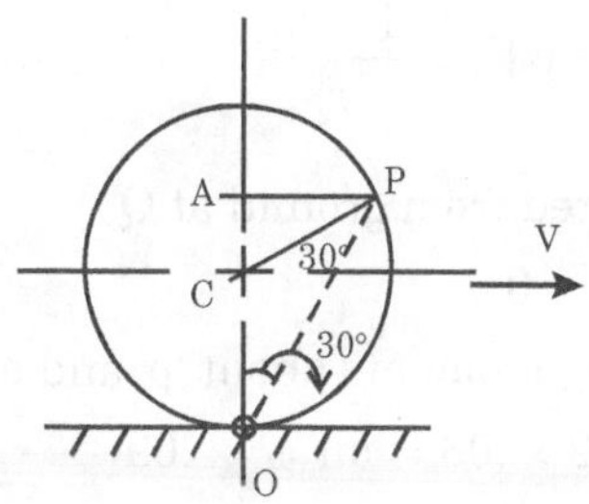

$\Rightarrow$ Velocity of P $= \omega.\ OP = \frac{v}{R}.\sqrt{3}R = \sqrt{3}v$

17. K.E. of fly wheel $= \frac{1}{2}I\omega^2$

$$\text{where, } I = \frac{mR^2}{2} = \frac{20 \times 0.2^2}{2} = 0.4 \text{ kg} - \text{m}^2$$

$$\omega = \frac{2\pi N}{60} = \frac{2 \times \pi \times 600}{60}$$

$$= 62.83 \text{ rad/s}$$

$$\therefore \quad K.E. = \frac{1}{2} \times 0.4 \times 62.83^2 = 790 \text{ J}$$

18. Moment of Inertia of fly wheel

$$I = m_r R^2$$

R = mean Radius of Rim

$$= M\left(\frac{R + R/2}{2}\right)^2$$

$$= 0.56\ MR^2$$

$$\Rightarrow \quad \alpha = 0.56.$$

19. 2.053

20. $F_1 = F_2$[Vertical equilibrium]

For Gear 1, For Gear 2,

$$\tau_1 - F_1 r_1 = I_1\ddot{\theta}_1 \quad F_2 r_2 = I_2\ddot{\theta}_2$$

$$\tau_1 - I_2\ddot{\theta}_2\frac{r_1}{r_2} = I_1\ \ddot{\theta}_1 \text{ ... using } F_2 = \frac{I_2\ddot{\theta}_2}{r_2} \text{ and}$$

$$F_1 = \tau_1 = I_1\ddot{\theta}_1 + I = \ddot{\theta}_2\frac{r_1}{r_2} = \left[I_1 + I_2\frac{\ddot{\theta}_2}{\ddot{\theta}_1}\frac{r_1}{r_2}\right]\ddot{\theta}_1$$

$$\tau_1 = \left[I_1 + I_2\left(\frac{r_1}{r_2}\right)^2\right]\ddot{\theta}_1 ......\left[\frac{\ddot{\theta}_1}{\ddot{\theta}_2} = \frac{r_2}{r_1}\right]$$

and clearly $F_2 = \dfrac{I_2\, r_1\, \ddot{\theta}_1}{r_2}$

21. When lifted from ground at Q

Reaction = 0

∴ taking moments about 'p' and equating to 0

$$.09 \times .06 = mr\ \omega^2 \times .09$$

$$9.01 \times .09 \times .06 = .002 \times 2.19 \times 10^{-3} \times \omega^2 \times .09$$

$$\Rightarrow \quad \omega = 366.50 \text{ rad/s}$$

$$= 3500 \text{ rpm}$$

22. Rim med wheel has maximum mass located away from the axis of rotation. Thus will have maximum moment of inertia.

23. 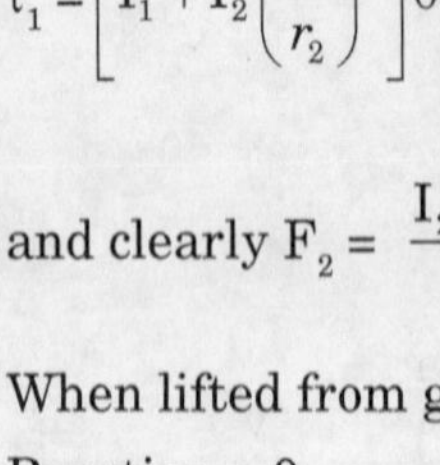

$$V_A = 80 \text{ m/s}, V_B = 140 \text{ m/s}$$

$$r_B - r_A = 300 \qquad ...(i)$$

$$\omega \times r_A = 80$$

$$\omega \times r_B = 140$$

$$\therefore \quad \frac{r_B}{r_A} = 1.75 \qquad ...(ii)$$

Solving (*i*) and (*ii*), r_B = 700 mm.

∴ Diameter of wheel is 1400 mm.

24. 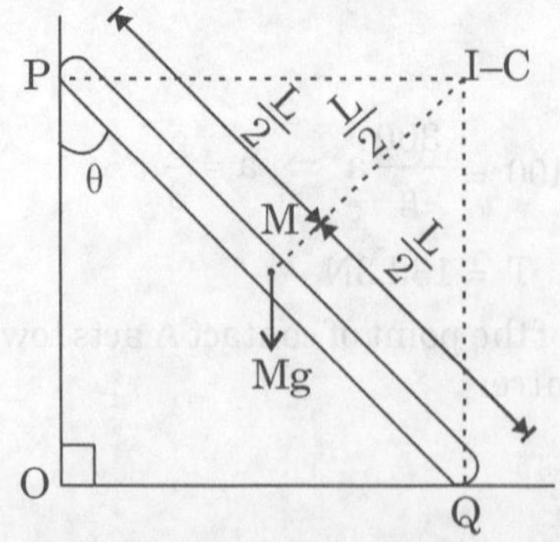

$$T_{I\text{-}C} = T_{I\text{-}C}\alpha$$

$$\Rightarrow Mg x\frac{L}{2}\cos 45° = \left[I_{CM} + \frac{ML^2}{4}\right]\alpha$$

$$\Rightarrow Mg\frac{L}{2}\cos 45° = \left[\frac{ML^2}{12} + \frac{ML^2}{4}\right]\alpha$$

$$\alpha = \frac{3g}{2L}\cos 45°$$

$$\alpha = \frac{3o}{2 \times V_2} + \frac{1}{V_2}\left(\because g = 10\frac{m}{s^2}\right)$$

$$\alpha = 7.5\frac{rad}{s^2}$$

25. Let Co-ordinates of Q are (x', y') and rotation about x-y plane anti clockwise.

$$\begin{bmatrix} x' \\ y' \end{bmatrix} = \begin{bmatrix} \cos\theta & -\sin\theta \\ \sin\theta & \cos\theta \end{bmatrix}\begin{bmatrix} x \\ y \end{bmatrix}$$

$$\begin{bmatrix} x' \\ y' \end{bmatrix} = \begin{bmatrix} \cos 30 & -\sin 30 \\ \sin 30 & \cos 30 \end{bmatrix}\begin{bmatrix} 20 \\ 10 \end{bmatrix}$$

$$= \begin{bmatrix} 0.866 & -0.5 \\ 0.5 & 0.866 \end{bmatrix}\begin{bmatrix} 20 \\ 10 \end{bmatrix}$$

$$\begin{bmatrix} x' \\ y' \end{bmatrix} = \begin{bmatrix} 0.866 \times 20 - 0.5 \times 10 \\ 0.5 \times 20 + 0.866 \times 10 \end{bmatrix}$$

$$= \begin{bmatrix} 12.32 \\ 18.66 \end{bmatrix}$$

$$(x', y') = (12.32, 18.66)$$

3 CHAPTER Friction

Analysis of Previous GATE Papers

Topic	Year/Set → Marks Distribution ↓		2019 Set 1	2019 Set 2	2018 Set 1	2018 Set 2	2017 Set 1	2017 Set 2	2016 Set 1	2016 Set 2	2016 Set 3	2015 Set 1	2015 Set 2	2015 Set 3
Friction	1 Mark	MCQ Type												
		Numerical Type	1											
	2 Marks	MCQ Type												
		Numerical Type				1			1					
	Total		1			2			2					

Friction

1. A wheel of mass m and radius r is in accelerated rolling motion without slip under a steady axle torque T. If the coefficient of kinetic friction is μ, the friction force from the ground on the wheel is

(a) μ mg (b) T/r
(c) zero (d) None of these

[1996 : 1 Mark]

2. A steel wheel of 600 mm diameter on a horizontal steel rail. It carries a load of 500 N. The coefficient of rolling resistance is 0.3 The force in Newton, necessary to roll the wheel along the rail is

(a) 0.5 (b) 5
(c) 1.5 (d) 150

[2000 : 1 Mark]

3. The ratio of tension on the tight side to that of the slack ride in a flat belt drive is

(a) proportional to the product of coefficient of friction and lap angle
(b) an exponential function of the product of coefficient of friction and lap angle
(c) proportional to lap angle
(d) proportional to the coefficient of friction

[2000 : 1 Mark]

4. The figure shows a pair of pin-jointed gripper-tongs holding an object weighing 2000 N. The coefficient of friction (μ) at the gripping surface is 0.1. XX is the line of action of the input force and YY is the line of application of gripping force. If the pin- joint is assumed to be frictionless, the magnitude of force F required to hold the weight is

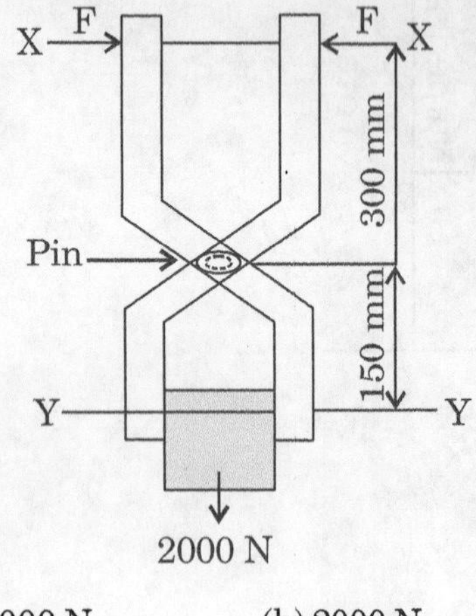

(a) 1000 N (b) 2000 N
(c) 2500 N (d) 5000 N

[2004 : 2 Marks]

5. Two books of mass 1 kg each are kept on a table, one over the other. The coefficient of friction on every pair of contacting surfaces is 0.3. The lower book is pulled with a horizontal force F. The minimum value of F for which slip occurs between the two books is

(a) zero (b) 11.77 N
(c) 5.88 N (d) 8.83 N

[2005 : 2 Marks]

6. A block of mass M is released from point P on a rough inclined plane with inclination angle θ, shown in the figure below. The coefficient of friction is μ. If $\mu < \tan\theta$, then the time taken by the block to reach another point Q on the inclined plane, where PQ = s, is

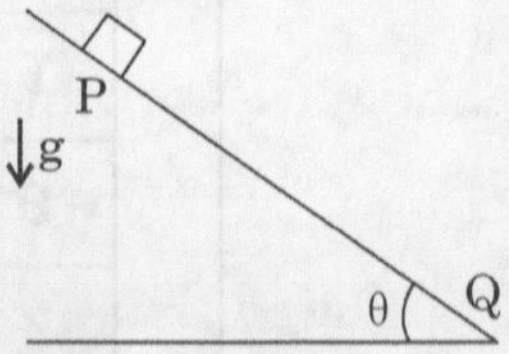

(a) $\sqrt{\dfrac{2s}{g\cos\theta(\tan\theta - \mu)}}$ (b) $\sqrt{\dfrac{2s}{g\cos\theta(\tan\theta + \mu)}}$

(c) $\sqrt{\dfrac{2s}{g\sin\theta(\tan\theta - \mu)}}$ (d) $\sqrt{\dfrac{2s}{g\sin\theta(\tan\theta + \mu)}}$

[2007 : 2 Marks]

7. A block weighing 981 N is resting on a horizontal surface. The coefficient of friction between the block and the horizontal surface is μ = 0.2. A vertical cable attached to the block provides partial support as shown. A man can pull horizontally with a force of 100 N. What will be the tension, T (in N) in the cable if the man is just able to move the block to the right?

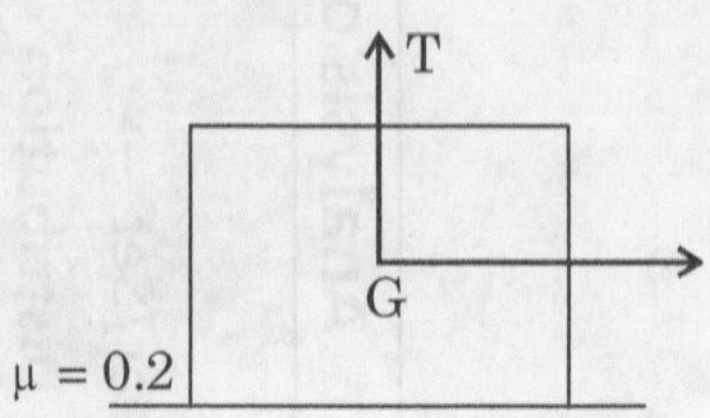

(a) 176.2 (b) 196.0
(c) 481.0 (d) 981.0

[2009 : 1 Mark]

8. A1 kg block is resting on a surface with coefficient of friction μ = 0.1. A force of 0.8 N is applied to the block as shown in the figure. The friction force is

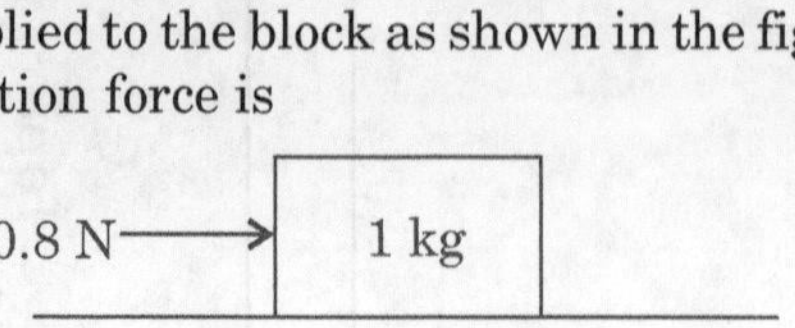

(a) 0 (b) 0.8 N
(c) 0.98 N (d) 1.2 N

[2011 : 2 Marks]

9. A block R of mass 100 kg is placed on a block S of mass 150 kg as shown in the figure. Block R is tied to the wall by a massless and inextensible string PQ. If the coefficient of static friction for all surfaces is 0.4, the minimum force F(in kN) needed to move the block S is

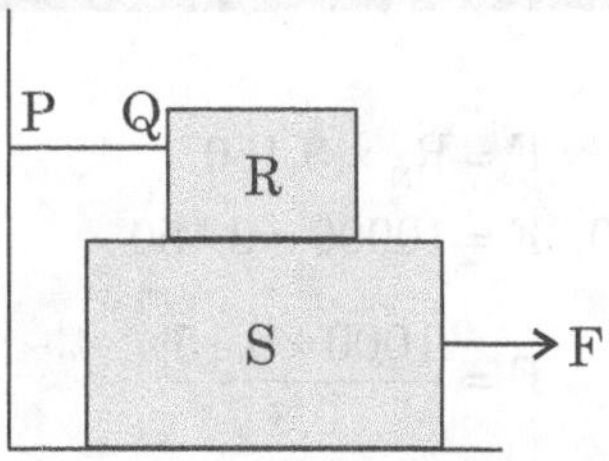

(a) 0.69 (b) 0.88
(c) 0.98 (d) 1.37

[2014 : 2 Marks, Set-1]

10. A block weighing 200 N is in contact with a level plane whose coefficients of static and kinetic friction are 0.4 and 0.2, respectively. The block is acted upon by a horizontal force (in newton) P = 10t, where t denotes the time in seconds. The velocity (in m/s) of the block attained after 10 seconds is ________.

[2014 : 2 Marks, Set-1]

11. A truck accelerates up a 10° incline with a crate of 100 kg. Value of static coefficient of friction between the crate and the truck surface is 0.3. The maximum value of acceleration (in m/s^2) of the truck such that the crate does not slide down is ______.

[2014 : 2 Marks, Set-2]

12. A body of mass (M) 10 kg is initially stationary on a 45° inclined plane as shown in figure. The coefficient of dynamic friction between the body and the plane is 0.5. The body slides down the plane and attains a velocity of 20 m/s. The distance travelled (in meter) by the body along the plane is ________.

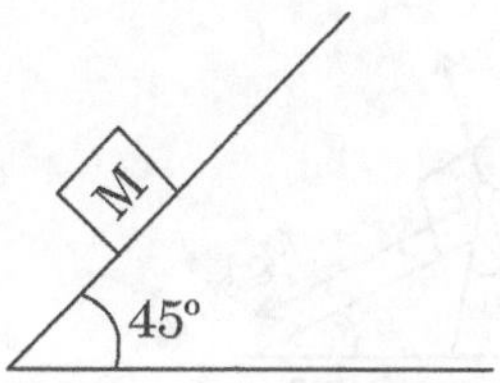

[2014 : 2 Marks, Set-3]

13. A wardrobe(mass 100 kg, height4m, width 2 m, depth 1 m), symmetric about the Y-Y axis, stands on a rough level floor as shown in the figure. A force P is applied at mid-height on the wardrobe so as to tip it about point Q without slipping. What are the minimum values of the force (in newton) and the static coefficient of friction μ between the floor and the wardrobe, respectively?

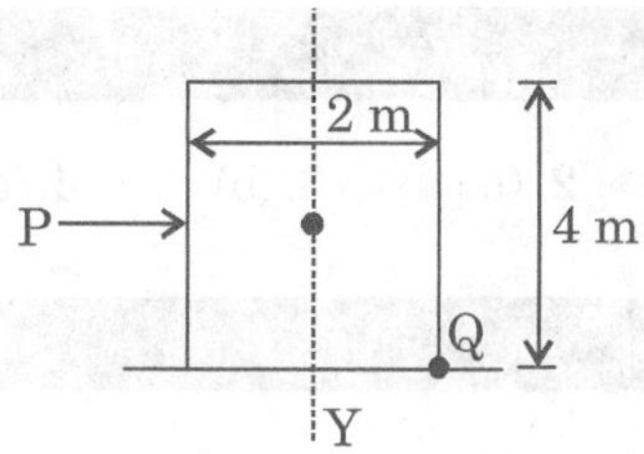

(a) 490.5 and 0.5 (b) 981 and 0.5
(c) 1000.5 and 0.15 (d) 1000.5 and 0.25

[2014 : 2 Marks, Set-4]

14. A block of mass m rests on an inclined plane and is attached by a string to the wall as shown in the figure. The coefficient of static friction between the plane and the block is 0.25. The string can with stand a maximum force of 20 N. The maximum value of the mass (m) for which the string will not break and the block will be in static equilibrium is ______ kg.

Take cos θ = 0.8 and sin θ = 0.6.

Acceleration due to gravity g = 10 m/s^2

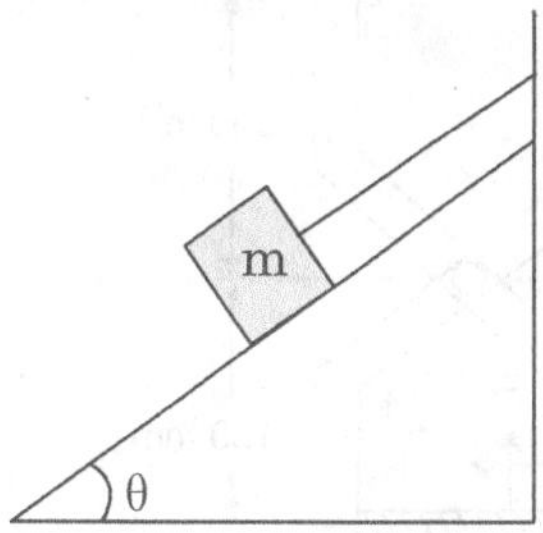

[2016 : 2 Marks, Set-1]

15. A force of 100 N is applied to the centre of a circular disc, of mass 10 kg and radius 1 m, resting on a floor as shown in the figure. If the disc rolls without slipping on the floor, the linear acceleration (in m/s^2) of the centre of the disc is ________ (correct to two decimal places).

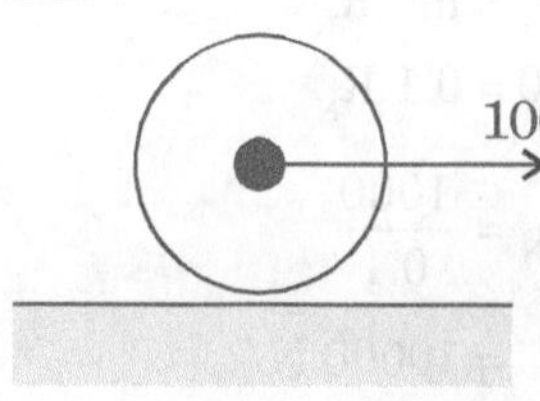

[2018 : 2 Marks, Set-2]

16. A block mass 10 kg rests on a horizontal floor. The acceleration due to gravity is 9.81 m/s^2. The coefficient of static friction between the floor and the block is 0.2. A horizontal force of 10 N is applied on the block as shown in the figure. The magnitude of force of friction (in N) on the block is ______.

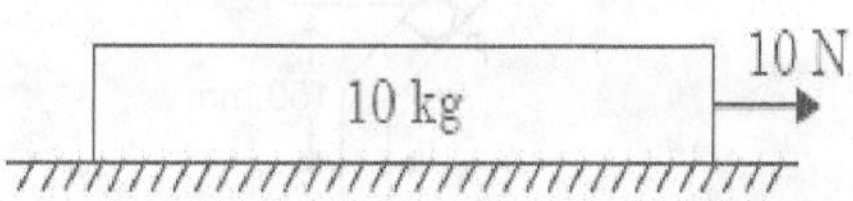

[2019 : 1 Mark, Set-1]

ANSWERS

1. (d) **2.** (d) **3.** (b) **4.** (d) **5.** (b) **6.** (a) **7.** (c) **8.** (b) **9.** (d) **13.** (b)

EXPLANATIONS

2. $r = \frac{600}{2} = 300$ mm.

W = 500 N

$\therefore$ F = μ × W

= 0.3 × 500

= 150 N

3. $\frac{T_1}{T_2} = e^{\mu\theta}$

4.

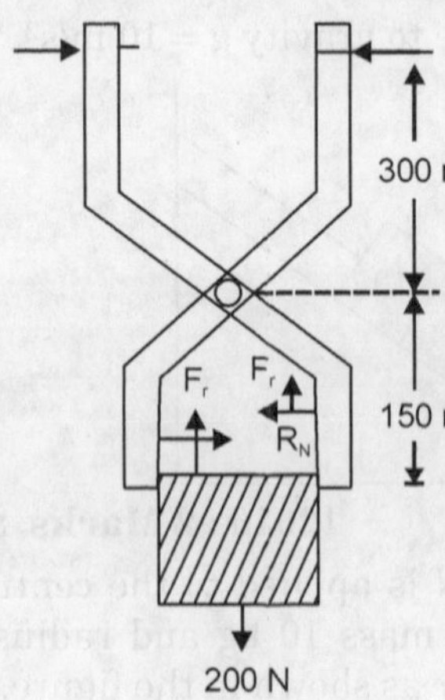

For vertical equilibrium $2F_r = 2000$ N

$F_r = 1000$ N

$F_r = m \cdot R_N$

$1000 = 0.1\ R_N$

$\therefore \quad R_N = \frac{1000}{0.1}$

= 10000 N.

Taking moment about P,

we have

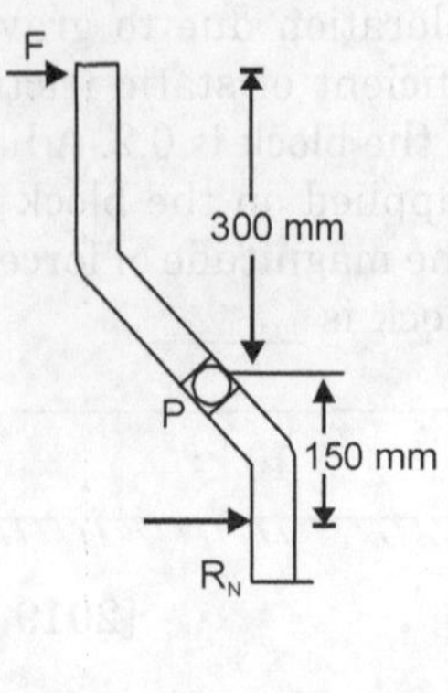

$0.3 \times F = R_N \times 0.150$

$0.3F = 10000 \times 0.150$

$F = \frac{1000 \times 0.15}{0.3}$

$= \frac{10,000 \times 15}{30}$

= 5000 N.

5. $[a_{max}]_A = ?$

$[a_{max}]_A \rightarrow [(f_s)_{max}]_{AB}$

[A] → 3N

$[(f_x)_{max}]_{AB} = \mu_s N_{AB}$

= 0.3 × 9.81 = 2.943 N

$(a_A)_{max} = \frac{[(f_s)_{max}]_{AB}}{m_A} = 2.943\ m/s^2$

At that condition,

2.943N ← [B] → a = 2.943

5.886N ← → F_{min} = ?

$F_{min} - 8.829 = m_B a_B$

$F_{min} = 8.829 + 1 \times 2.943$

$F_{min} = 11.772$ N

6.

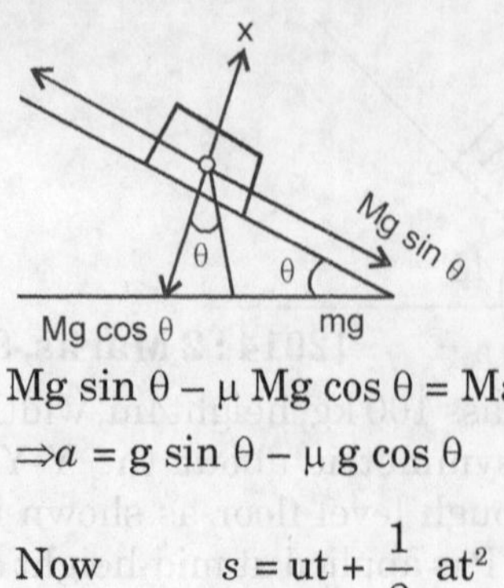

Mg sin θ – μ Mg cos θ = Ma

$\Rightarrow a = g \sin\theta - \mu g \cos\theta$

Now $s = ut + \frac{1}{2}at^2$

But $u = 0$

$\therefore \quad t = \sqrt{\frac{2s}{a}} = \sqrt{\frac{2s}{g\cos\theta(\tan\theta - \mu)}}$

7. Net normal force on block

N = W – T

= 981 – T

Frictional force, f = μN

Under equilibrium, i.e. when man is just able to move the block

$$\mu N = 100$$

$$\Rightarrow \mu(981 - T) = 100$$

$$\Rightarrow 0.2\,(981 - T) = 100$$

$$T = -\frac{100}{0.2} + 981$$

$$= 981 - 500$$

$$= 481\text{N}$$

8. Limiting friction force between block and the surface is 0.98 N. But applied force is 0.8 N which is less than the limiting friction force. Therefore friction force for the given case is 0.8 N.

9.

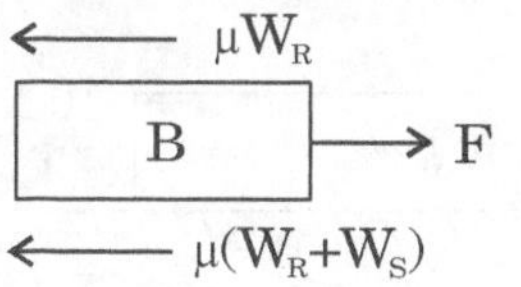

$F = \mu\,(W_s + W_R) + \mu\, W_R$

$F = \mu\,(W_s + 2W_R)$

$= 0.4\,(150 + 2 \times 1000)9.81 = 1.37$ kN

10. $\sum F_V = 0$

N = mg = 200 N

$(f_s)_{max} = \mu s\, N = 0.4 \times 200 = 8\;0$ N

As given in the question, upto 8 sec [P=10 t]→ no motion

⇒ From 8 sec to 10 sec

$f_k = \mu_K N = 0.2 \times 200 = 40$ N

Average applied force is given by,

$$P_{avg} = \frac{P_{max} + P_{min}}{2} = \frac{100 + 80}{2} = 90\text{N}$$

Average Resultant force = Avg. app. force – kinetic friction

= 90 – 40 = 50 N

$$\text{Average acceleration} = \frac{50 \times 9.8}{200}$$

⇒ V = u + at [∴ u = O]

$$a = \frac{50 \times 9.8}{200}$$

t = 2 second

$$\therefore V = \frac{50 \times 9.8}{200} \times 2 = 4.9\frac{m}{s}$$

11.

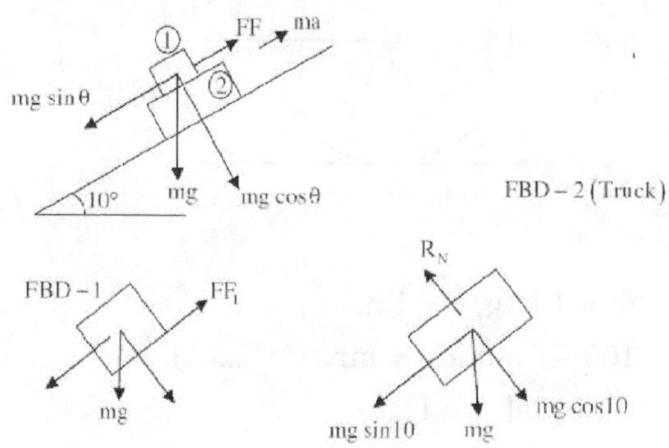

$$\Sigma F_{along} = ma$$

$$F_F - mg \sin\theta = ma$$

$$\Rightarrow \mu\, mg \cos 10 - mg \sin 10 = ma$$

$$\Rightarrow a = [0.3 \cos 10 - \sin 10]\, 9.81 = 1.19.$$

12. FBD of block

Net force = ma

mg sin 45 – μmg cos 45 = ma

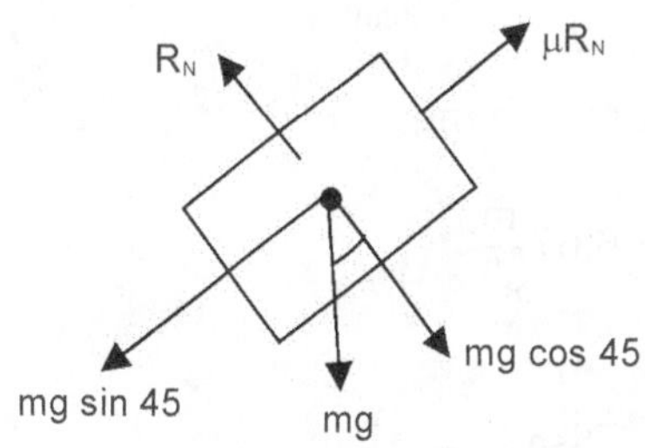

$$\Rightarrow a = 3.46 \text{ m/s}^2$$

$$\Rightarrow V^2 - u^2 = 2as$$

$$\Rightarrow S = \frac{V^2}{2a} = \frac{20^2}{2 \times 3.46} = 57.8 \text{ m.}$$

13. Taking moments about Q, $\Sigma M_Q = 0$

$$\Rightarrow W \times 1\text{m} = P \times 2\text{m}$$

$$\Rightarrow P = \frac{100 \times 9.81}{2} = 490.5 \text{ N}$$

$$\Sigma H = 0$$

$$\Rightarrow F_F = P = 490.5 \text{ N}$$

$$\Sigma V = 0$$

$$\Rightarrow R_N = mg = 981 \text{ N}$$

Friction Force = μR_N

$$\mu = \frac{490}{981} = 0.5$$

14.

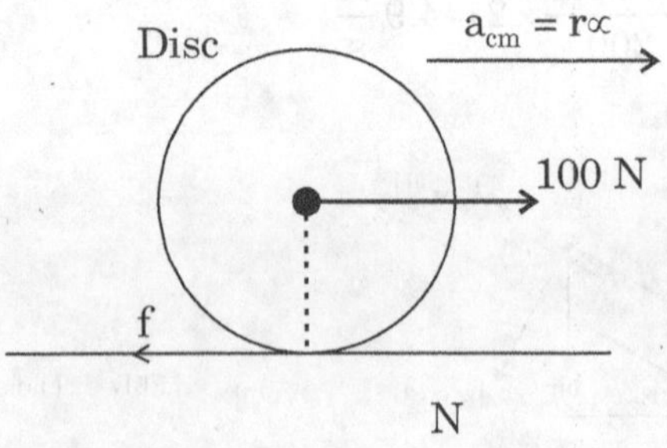

m = 10 kg, r = 1 m

$100 - f = ma_{CM} = mr\propto$ ...(A)

$[\text{Tarque}]_{CM} = I_{CM}' \propto$

$$f \times r = \frac{mr^2}{2} \times \propto$$

$2f = mr\propto$

$2f = ma_{CM}$

$f = \frac{m}{2} \times a_{cm}$ ...(B)

By (A) & (B) we get

$$100 - \frac{ma_{cm}}{2} = ma_{cm}$$

we get on solving

$$a_{CM} = 6.67 \frac{m}{s^2}$$

15.

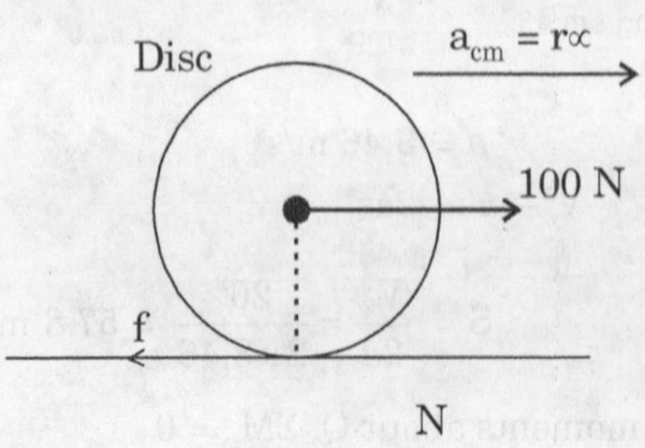

m = 10 kg, r = 1 m

$100 - f = ma_{CM} = mr\propto$ ...(A)

$[\text{Tarque}]_{CM} = I_{CM}' \alpha$

$$f \times r = \frac{mr^2}{2} \times \alpha$$

$2f = mr\,\alpha$

$2f = ma_{CM}$

$f = \frac{m}{2} \times a_{cm}$ ...(B)

By (A) & (B) we get

$$100 - \frac{ma_{cm}}{2} = ma_{cm}$$

we get on solving

$$a_{CM} = 6.67 \frac{m}{s^2}$$

16.

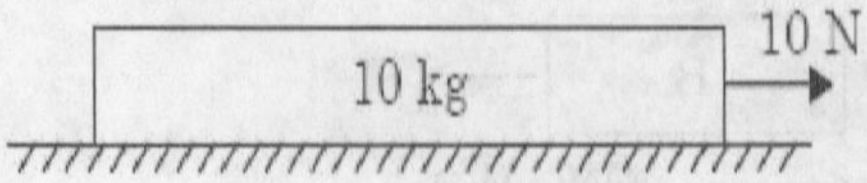

Here, $\mu = 0.2$

$g = 9.81\ m/s^2$

$M = 10\ kg$

$(f_s)_{max} = \mu Mg$

$= 0.2 \times 10 \times 9.81$

$= 19.62\ N$

As, $(F_s)_{max} >$ Force applied

So, magnitude of force of friction on block is,

$(F_s) = 10\ N$

4 CHAPTER

Impulse and Momentum, Work and Energy

Topic	Year/Set → Marks Distribution ↓		2019 Set 1	2019 Set 2	2018 Set 1	2018 Set 2	2017 Set 1	2017 Set 2	2016 Set 1	2016 Set 2	2016 Set 3	2015 Set 1	2015 Set 2	2015 Set 3
Impact and Collision	1 Mark	MCQ Type												
		Numerical Type											1	
	2 Marks	MCQ Type		1										
		Numerical Type										1		
	Total			2								2	1	
Impulse and Momentum	1 Mark	MCQ Type							1	1				
		Numerical Type												
	2 Marks	MCQ Type					1			1				
		Numerical Type			1						1			1
	Total				2		2		1	3	1			2
Work and Energy	1 Mark	MCQ Type					1							
		Numerical Type												
	2 Marks	MCQ Type												
		Numerical Type								1				
	Total						1			2				

Table: Analysis of Previous GATE Papers

Impact and Collision

1. A ball A of mass m falls under gravity from a height h and strikes another ball B of mass m which is supported at rest on a spring of stiffness k. Assume perfect elastic impact. Immediately after the impact

(a) the velocity of ball A is $\frac{1}{2}\sqrt{2gh}$

(b) the velocity of ball A is zero

(c) the velocity of both balls is $\frac{1}{2}\sqrt{2gh}$

(d) None of the above

[1996 : 1 Mark]

2. A bullet of mass m travels at a very high velocity V (as shown in the figure) and gets embedded inside the block of mass M initially at rest on a rough horizontal floor. The block with the bullet is seen to move a distance 's' along the floor. Assuming μ to be the coefficient of kinetic friction between the block and the floor and g the acceleration due to gravity. What is the velocity V of the bullet?

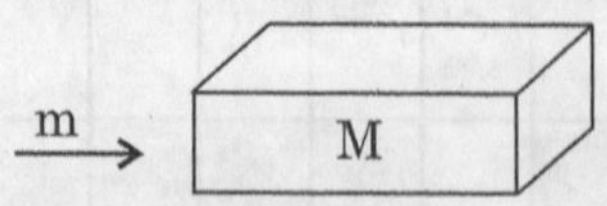

(a) $\frac{M+m}{m}\sqrt{2\mu gs}$ (b) $\frac{M-m}{m}\sqrt{2\mu gs}$

(c) $\frac{\mu(M-m)}{mm}\sqrt{2\mu gs}$ (d) $\frac{M}{m}\sqrt{2\mu gs}$

[2003 : 1 Mark]

3. A 1 kg mass of clay, moving with a velocity of 10 m/s, strikes a stationary wheel and sticks to it. The solid wheel has a mass of 20 kg and a radius of 1 m. Assuming that the wheel is set into pure rolling motion, the angular velocity of the wheel immediately after the impact is approximately

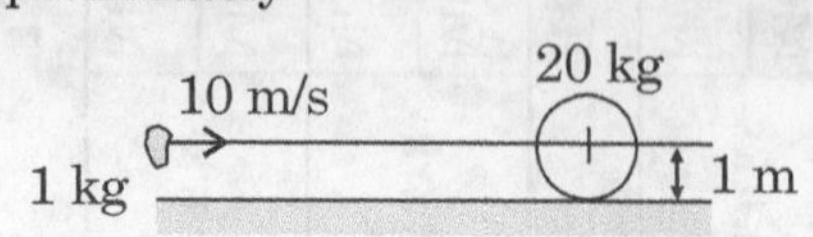

(a) zero (b) $\frac{1}{3}$ rad/s

(c) $\sqrt{\frac{10}{3}}$ rad/s (d) $\frac{10}{3}$ rad/s

[2005 : 2 Marks]

4. The coefficient of restitution of a perfectly plastic impact is

(a) 0 (b) 1

(c) 2 (d) ∞

[2011 : 1 Mark]

5. A ball of mass 0.1 kg, initially at rest, dropped from height of 1 m. Ball hits the ground and bounces off the ground. Upon impact with the ground, the velocity reduces by 20%. The height (in m) to which the ball will rise is ______.

[2015 : 2 Marks, Set-1]

6. A small ball of mass 1 kg moving with a velocity of 12 m/s undergoes a direct central impact with a stationary ball of mass 2 kg. The impact is perfectly elastic. The speed (in m/s) of 2 kg mass ball after the impact will be ______.

[2015 : 1 Mark, Set-2]

7. A ball of mass 3 kg moving with a velocity of 4 m/s undergoes a perfectly-elastic directcentral impact with a stationary ball of mass m. After the impact is over, the kinetic energy of the 3 kg ball is 6 J. The possible value(s) of m is/are

(a) 1 kg, 6 kg (b) 1 kg, 9 kg

(c) 6 kg only (d) 1 kg only

[2019 : 2 Marks, Set-2]

Impulse and Momentum

8. Match 4 correct pairs between **List-I** and **List-II** for questions.

List-I

A. Collision of particles
B. Stability
C. Satellite motion
D. Spinning top

List-II

1. Euler's equation of motion
2. Minimum kinetic energy
3. Minimum potential energy
4. Impulse momentum principle
5. Conservation of moment of momentum

Codes:

	A	B	C	D
(a)	1	3	5	1
(b)	2	4	3	5
(c)	4	3	2	5
(d)	4	3	1	5

[1996 : 2 Marks]

9. During inelastic collision of two particles, which one of the following is conserved?

(a) Total linear momentum only
(b) Total kinetic energy only
(c) Both linear momentum and kinetic energy
(d) Neither linear momentum nor kinetic energy

[2007 : 1 Mark]

10. A single degree of freedom system having mass 1 kg and stiffness 10 kN/m initially at rest is subjected to an impulse force of magnitude 5 kN for 10^{-4} seconds. The amplitude in mm of the resulting free vibration is

(a) 0.5 (b) 1.0
(c) 5.0 (d) 10.0

[2013 : 2 Marks]

11. A mass m_1 of 100 kg travelling with a uniform velocity of 5 m/s along a line collides with a stationary mass m_2 of 1000 kg. After the collision, both the masses travel together with the same velocity. The coefficient of restitution is

(a) 0.6 (b) 0.1
(c) 0.01 (d) 0

[2014 : 1 Mark, Set-3]

12. A bullet spins as the shot is fired from a gun. For this purpose, two helical slots as shown in the figure are cut in the barrel. Projections A and B on the bullet engage in each of the slots.

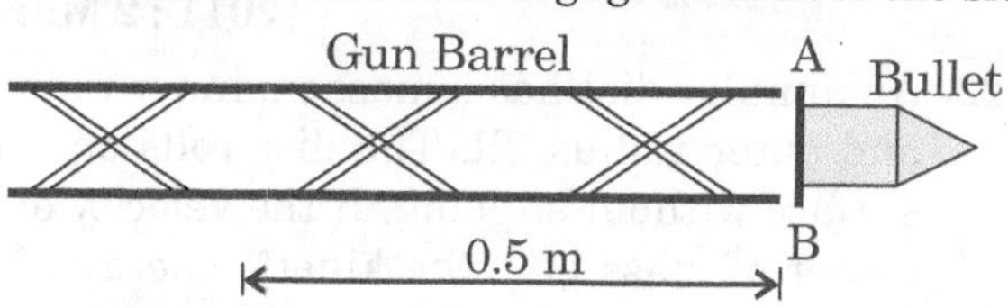

Helical slots are such that one turn of helix is completed over a distahce of 0.5 m. If velocity of bullet when it exits the barrel is 20 m/s, its spinning speed in rad/s is ______.

[2015 : 2 Marks, Set-3]

13. A point mass M is released from rest and slides down a spherical bowl of radius R from a height H as shown in the figure below. The surface of the bowl is smooth (no friction). The velocity of the mass at the bottom of the bowl is

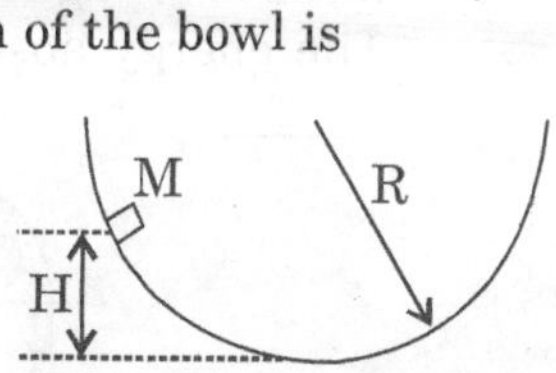

(a) $\sqrt{gH}$
(b) $\sqrt{2gR}$
(c) $\sqrt{2gH}$
(d) 0

[2016 : 1 Mark, Set-1]

14. A point mass having mass M is moving with a velocity V at an angle θ to the wall as shown in the figure. The mass undergoes a perfectly elastic collision with the smooth wall and rebounds. The total change (final minus initial) in the momentum of the mass is

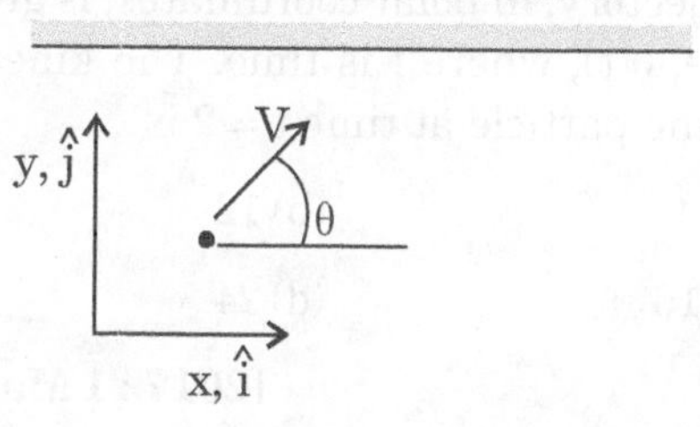

(a) $-2MV\cos\theta\,\hat{j}$ (b) $2MV\sin\theta\,\hat{j}$
(c) $2MV\cos\theta\,\hat{j}$ (d) $-2MV\sin\theta\,\hat{j}$

[2016 : 1 Mark, Set-2]

15. A system of particles in motion has mass center G as shown in the figure. The particle i has mass m_i and its position with respect to a fixed point O is given by the position vector r_i The position of the particle with respect to G is given by the vector p_i, The time rate of change of the angular momentum of the system of particles about G is (The quantity p_i indicates second derivative of p_i with respect to time and likewise for r_i).

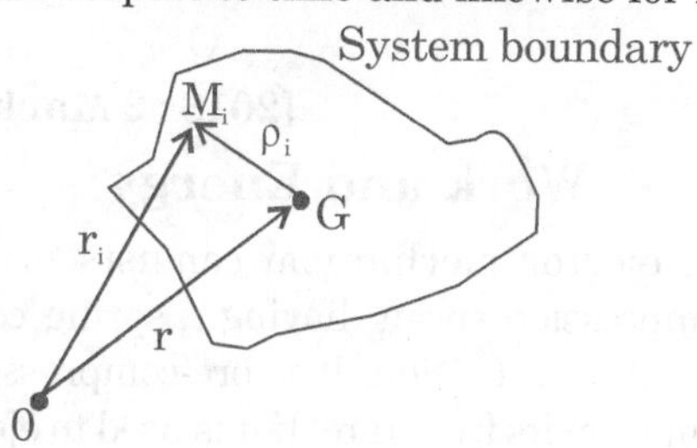

(a) $\sum_i r_i \times m_i p_i$ (b) $\sum p_i \times m_i r_i$
(c) $\sum_i r_i \times m_i r_i$ (d) $\sum_i p_i \times m_i p_i$

[2016 : 2 Marks, Set-2]

16. Acircular disc of radius 100 mm and mass 1 kg, initially at rest at position A, rolls without slipping down a curved path as shown in figure. The speed v of the disc when it reaches position B is ________ m/s.

Acceleration due to gravity, g = 10 m/s².

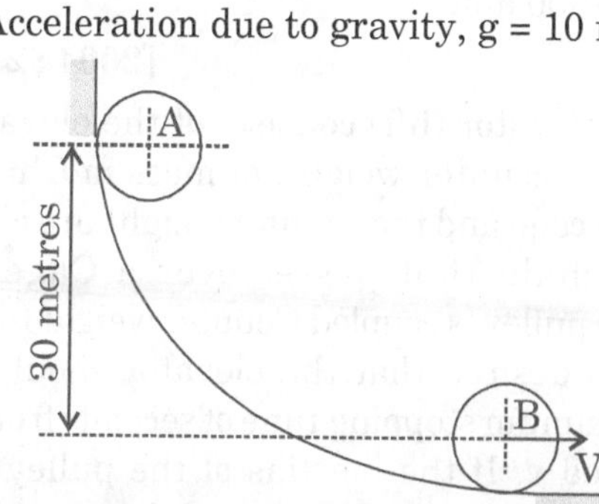

[2016 : 2 Marks, Set-3]

17. Two disc A and B with identical mass (m) and radius (R) are initially at rest. They roll down from the top of identical inclined planes without slipping. Disc A has all of its mass concentrated at the rim, while Disc B has its mass uniformly distributed. At the bottom of the plane, the ratio of velocity of the center of disc A to the velocity of the center of disc B is

(a) $\sqrt{\frac{3}{4}}$ (b) $\sqrt{\frac{3}{2}}$
(c) 1 (d) $\sqrt{2}$

[2017 : 2 Marks, Set-1]

18. Block P of mass 2 kg slides down the surface and has a speed 20 m/s at the lowest point, Q where the local radius of curvature is 2 m as shown in the figure. Assuming g = 10 m/s², the normal force (in N) at Q is ______ (correct to two decimal places).

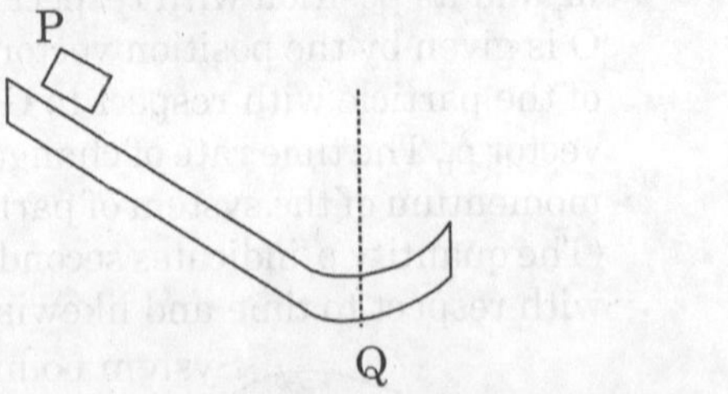

[2018 : 2 Marks, Set-1]

Work and Energy

19. An ejector mechanism consists of a helical compression spring having a spring constant of K = 981 × 10³ N/m. It is pre-compressed by 100 mm from its free state. If it is used to eject a mass of 100 kg held on it, the mass will move up through a distance of

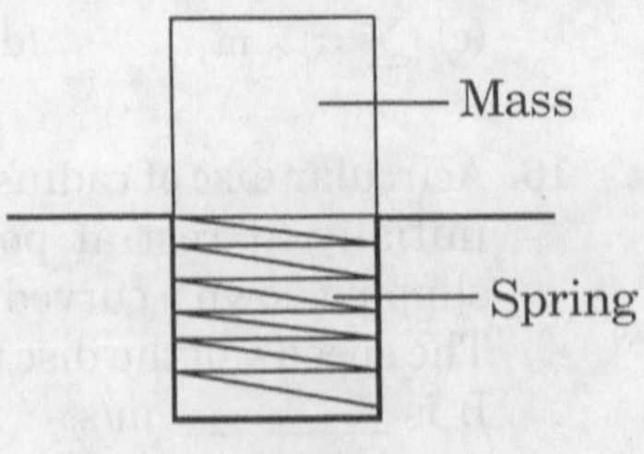

(a) 100 mm
(b) 5000 mm
(c) 581 mm
(d) 1000 mm

[2004 : 2 Marks]

20. An elevator (lift) consists of the elevator cage and a counter weight, of mass m Chain each. The cage and the counter weight are connected by chain that passes over a Cage pulley. The pulley is coupled Counter weight to a motor. It is desired that the elevator should have a maximum stopping time of seconds from a peak speed v. If the inertias of the pulley and the chain are neglected, the minimum power that the motor must have is

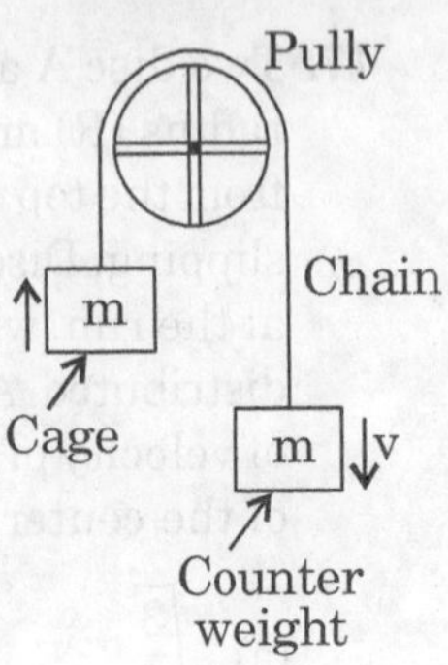

(a) $\frac{1}{2}mv^2$

(b) $\frac{mv^2}{2t}$

(c) $\frac{mv^2}{t}$

(d) $\frac{2mv^2}{t}$

[2005 : 2 Marks]

21. A stone with mass of 0.1 kg is catapulted as shown in the figure. The total force F_x (in N) exerted by the rubber band as a function of distance x (in m) is given by $F_x = 300x^2$. If the store is displaced by 0.1 m from the un-stretched position (x = 0) of the rubber band, the energy stored in the rubber band is

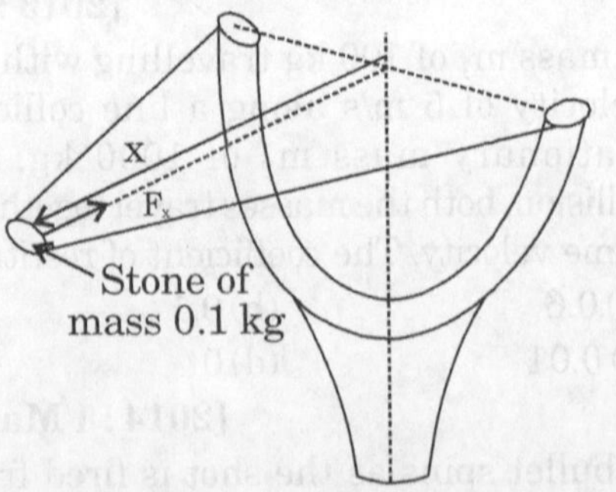

(a) 0.01 J (b) 0.1 J
(c) 1 J (d) 10 J

[2011 : 2 Marks]

22. An annular disc has a mass m, inner radius R and outer radius 2R. The disc rolls on a flat surface without slipping. If the velocity of the center of mass is v, the kinetic energy of the disc is

(a) 9/16 mv² (b) 11/16 mv²
(c) 13/16 mv² (d) 15/16 mv²

[2014 : 2 Marks, Set-3]

23. A mass of 2000 kg is currentiy being lowered at a velocity of 2 m/s from the drum as shown in the figure. The mass moment of inertia of the drum is 150 kg-m². On applying the brake, the mass is brought to rest in a distance of 0.5 m. The energy absorbed by the brake (in kJ) is ______.

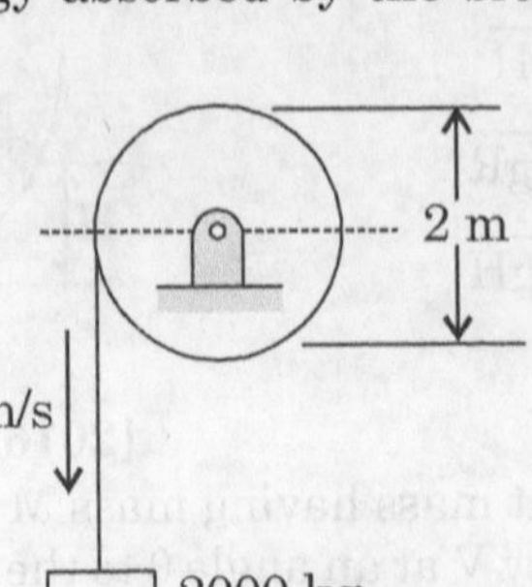

[2016 : 2 Marks, Set-2]

24. A particle of unit mass is moving on a plane. Its trajectory, in polar coordinates, is given by r(t) = t², θ(t), where f is time. The kinetic energy of the particle at time t = 2 is

(a) 4 (b) 12
(c) 16 (d) 24

[2017 : 1 Mark, Set-1]

ANSWERS

1. (b)	**2.** (a)	**3.** (b)	**4.** (a)	**7.** (b)	**8.** (d)	**9.** (a)	**10.** (c)	**11.** (d)	**13.** (c)
14. (d)	**15.** (b)	**17.** (a)	**19.** (a)	**20.** (c)	**21.** (b)	**22.** (c)	**24.** (c)		

EXPLANATIONS

1. In a perfectly elastic collision between equal masses of two bodies, velocities exchange on impact.

Velocity just before impect = Velocity immediately after impact.

Before impact	**After impact**
$V_A = \sqrt{2gh}$	$V_A = 0$
$V_B = 0$	$V_B = \sqrt{2gh}$

2. Let, m = mass of the bullet

v = velocity of bullet

M = mass of the block

u = final velocity of system (block + bullet)

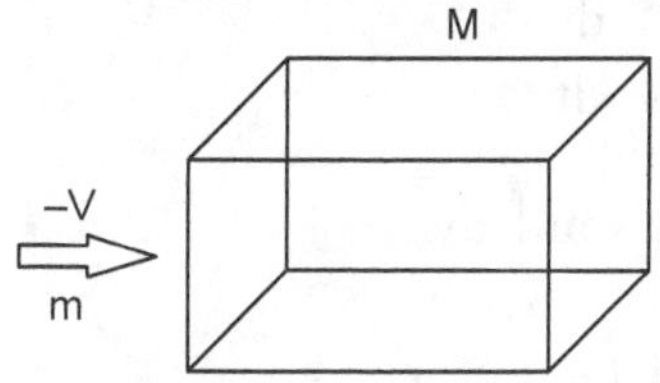

From conservation of linear momentum

$$mv = (M + m)u$$

$$\Rightarrow u = \frac{mv}{m + M}$$

For horizontal equilibrium of body

$$(M + m)g = N$$

F_r = Frictional force = $\mu N = u(M + m)g$

$$\text{Frictional retardation} = \frac{\mu(M+m)g}{(M+m)} = -\mu g$$

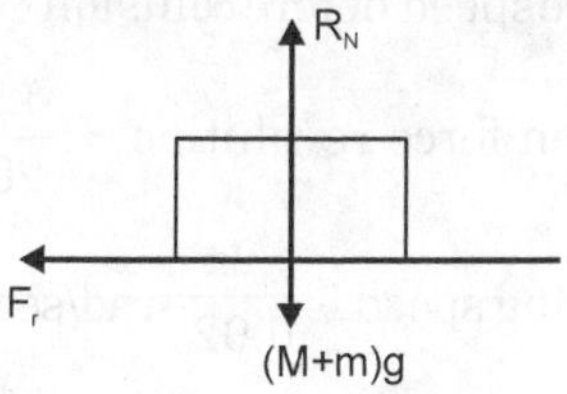

(–ve sign shows that acceleration is opposite to motion)

Let v_f = final velocity of system = 0

$$v_f^2 = u^2 + 2as$$

$$u^2 + 2as = 0$$

$$u^2 - 2\mu gs = 0$$

From equation (i)

$$\left(\frac{mv}{M+m}\right)^2 = 2\mu gs$$

$$\frac{mv}{M+m} = \sqrt{2\mu gs}$$

$$\Rightarrow v = \frac{M+m}{m}\sqrt{2\mu gs}$$

3. Taking both clay & wheel as system

$L_i = L_f$ (about A)

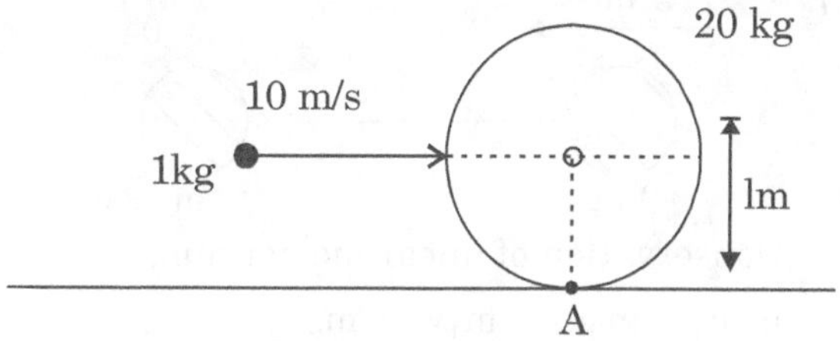

$$L_i = mvr = 1 \times 10 \times 1 = 10$$

$$L_f = mv_{cm}r + I\omega$$

$$= 20 \times r\omega \times r + \frac{mr^2\omega}{2} = 20\,\omega + 10\,\omega$$

$$L_f = 30\,\omega$$

$$L_i = Lf$$

$$10 = 30\,\omega$$

$$\omega = \frac{1}{3}\text{rad/sec}$$

4. For perfectly plastic impact,

Coefficient of restitution

$$= \frac{\text{Relative velocity of separation}}{\text{Relative velocity of approach}} = 0$$

5. $v = \sqrt{2gh} = \sqrt{2 \times 9.01 \times 1} \Rightarrow 4.4294$ m/s

$v' = 0 \times v = 3.5432$ m/s

$$h = \frac{v^2}{2g} = 0.64\text{m}$$

6. For elastic collision

$$m_1u_1 + m_1u_2 = m_1v_1 + m_2v_2 \quad ...(i)$$

By Moment conservation principle,

$m_1 = 1$ kg $u_1 = 12$ m/s

$m_2 = 2$ kg $u_2 = 0$ m/s

$$\frac{1}{2}m_1u_1^2 + \frac{1}{2}m_2u_2^2 = \frac{1}{2}m_1v_1^2 + \frac{1}{2}m_2v_2^2 \quad ...(ii)$$

By Energy conservation principle,

From equation (*i*)

$$12 = v_1 + 2v_2 \qquad ...(iii)$$

From equation (*ii*)

$$\frac{1}{2}\times 1\times 144+\frac{1}{2}\times 2\times 0 = \frac{1}{2}\times 1\times v_1^2+\frac{1}{2}\times 2\times v_2^2$$

$$\Rightarrow \quad 144 = v_1^2 + 2v_2^2 \qquad ...(iv)$$

From equations (*iii*) and (*iv*)

$$144 = 144 + 4v_2^2 - 48v_2 + 2v_2^2$$

$$\Rightarrow \quad 6v_2^2 - 48v_2 = 0$$

$$6v_2(v_2 - 8) = 0$$

$$\Rightarrow \quad v_2 = 8 \text{ m/s}$$

7. 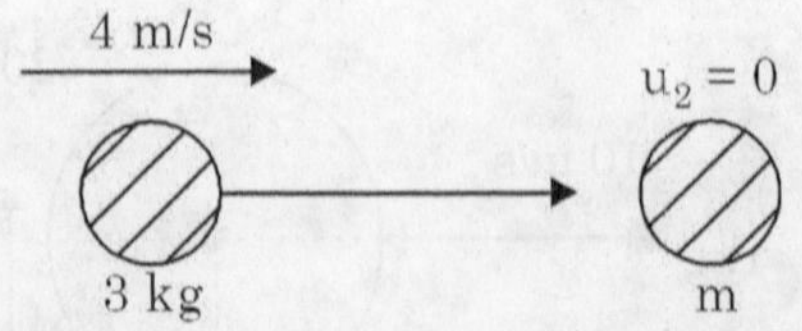

Conservation of linear momentum,

$m_1u_1 + m_2u_2 = m_1v_1 + m_2v_2$

$\Rightarrow u_2 = 0$

$3 \times 4 = 3 \times v_1 + mv_2 \qquad ...(i)$

$\because$ Coefficient of restitution, e = 1 for perfectly elastic collision,

$v_2 - v_1 = (u_1 - u_2)$

$\Rightarrow v_2 - v_1 = u_1 = 4$

$\Rightarrow v_2 - v_1 = 4 \qquad ...(ii)$

Conservation of energy,

$$\frac{1}{2}m_1u_1^2 + \frac{1}{2}m_2u_2^2 = \frac{1}{2}m_1v_1^2 + \frac{1}{2}m_2v_2^2$$

$$\Rightarrow \frac{1}{2}\times 3\times (4)^2 + 0 = 6 + \frac{1}{2}mv_2^2$$

$\Rightarrow (24 - 6)2 = mv_2^2$

$\Rightarrow mv_2^2 = 36 \qquad ...(iii)$

$\because v_1 = v_2 - 4$

Putting in equation (i)

$\Rightarrow 12 = 3v_2 - 12 + mv_2$

$\Rightarrow v_2(3 + m) = 24 \qquad ...(iv)$

Putting in equation (iii),

$$\Rightarrow m.\left(\frac{24}{3+m}\right)^2 = 36$$

$\Rightarrow (3 + m)^2 = 16m$

$\Rightarrow 9 + m^2 + 6m - 16m = 0$

$\Rightarrow m^2 - 10m + 9 = 0$

$\Rightarrow m^2 - 9m - m + 9 = 0$

$(m - 9)(m - 1) = 0$

$\boxed{m = 1,9 \text{ kg}}$

8. (*a*) –4, (*b*) –3, (*c*) –2, (*d*) –1

9. Linear momentum and Kinetic Energy is conserved in elastic collision and in inelastic collision Kinetic Energy (heat, sound etc.) is conserved.

10. 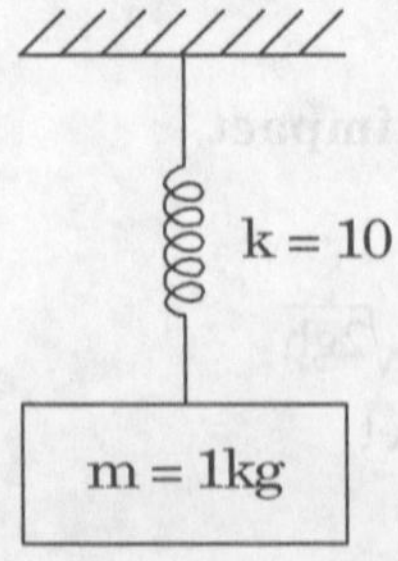

At t = 0, mass is at rest, F = 5 kN for 10^{-4} seconds & amplitude = x

$$F = m \times \frac{dv}{dt}$$

$$F.\int_{t_1}^{t_2} dt = m\int_{v_1}^{v_2} dv$$

$F(t_2 - t_1) = m \times (v_2 - v_1)$

$5 \times 10^{-3} \times 10^{-4} = 1 \times (v_2 - 0)$

$0.5 \text{ m/s} = v_2$

$$\frac{1}{2}mv_2^2 = \frac{1}{2}kx^2$$

$$\Rightarrow x = \sqrt{\frac{mv_2^2}{k}} = \sqrt{\frac{1\times 0.5\times 0.5}{10\times 1000}} = 5\text{mm}$$

11. Coefficient of restitution

$$= \frac{\text{Relative speed after collision}}{\text{Relative speed before collision}} = 0$$

12. Time taken force revolution = $\frac{0.5}{20}$ = 0.025 sec.

The spinning speed is $\frac{2\pi}{0.02s}$ rad/sec = 251.3 rad/sec

13. 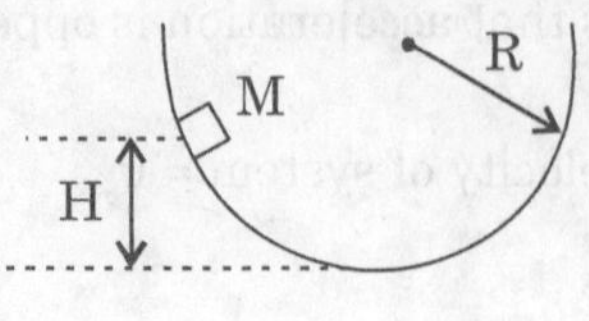

Loss in potential energy = Gain in kinetic energy

$$Mgh = \frac{1}{2}MV^2$$

$$V = \sqrt{2gH}$$

14. $P_i = MV\cos\theta\hat{i} + MV\sin\theta\hat{j}$

$P_f = MV\cos\theta\hat{i} - MV\sin\theta\hat{j}$

$P_f - P_i = -2MV\sin\theta\hat{j}$

16. Conservation of energy

Total (KE + PE) = const.

$$(KE + PE)_A = (KE + PE)_B$$

$$0 + 30 \times mg = \frac{1}{2}mv^2 + \frac{1}{2}I\omega^2 + 0$$

$$30 \times 10 = \frac{1}{2} \times 1 \times v^2 + \frac{1}{2} \times \frac{1}{2} \times 1 \times v^2$$

$$I = \frac{mr^2}{2},\ \omega = \frac{V}{I}$$

$$30 \times 10 = \frac{3}{4}V^2$$

$$V = 20 \text{ m/s}$$

17. Disc A $\rightarrow$ I= $mR^2 + mR^2 = 2mR^2$

Disc B $\rightarrow$ I= $\frac{mR^2}{2} + mR^2 = \frac{3}{2}mR^2$

For A $\rightarrow$ $mgx = \frac{1}{2}(2mR^2)\cdot\left(\frac{V^2}{R^2}\right)$

$$\sqrt{gx} = V_A$$

For B $\rightarrow$ $mgx = \frac{1}{2}\left(\frac{3}{2}mR^2\right)\cdot\left(\frac{V^2}{R^2}\right)$

$$\sqrt{\frac{4}{3}gx} = V_B$$

$$\frac{V_A}{V_B} = \sqrt{\frac{3}{4}}$$

18.

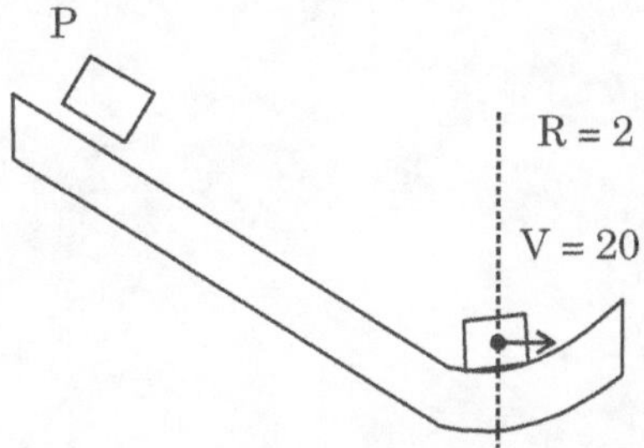

m = 2 kg, g = 10 m/s^2

FBD of mass P at point Q

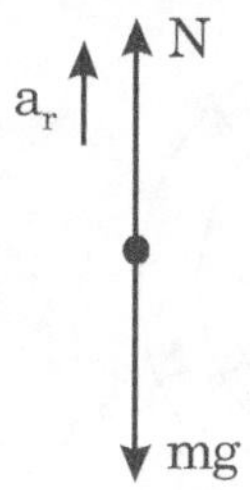

$N - mg = ma_r$ at no slip condition.

$$N = mg + \frac{mv^2}{R} = 2 \times 10 + \frac{2 \times 20 \times 20}{2} = 420 \text{ N}$$

19. From the conservation of energy after precompression,

Initial energy = Final energy

or $\frac{1}{2}k(0.1)^2 = \frac{1}{2}kx^2 + mg(x + 0.1)$

or $k(0.1)^2 = kx^2 + 2mg(x + 0.1)$

or $981 \times 10^3 \times 10^{-2} = 981 \times 10^3 \times x^2 + 2 \times 100 \times 9.81(x + 0.1)$

or $100x^2 + 2x - 0.8 = 0$

$$\therefore\ x = \frac{-2 \pm \sqrt{(2)^2 - 4 \times 100 \times 0.8}}{200}$$

Taking + ve sign, x = 100 mm

20. Power = $\frac{dW}{dt}$ (rate of doing work or rate of change of K.E. energy of the system)

Initial Kinetic energy of the system

$$= \frac{1}{2}mv^2 + \frac{1}{2}mv^2 = mv^2$$

Final Kinetic energy = 0

Time duration it occurs = t

$$\Rightarrow \quad \text{Power} = \frac{mv^2}{t}$$

21. Energy stored in the bar

= Work done by the stone

$$= \int_0^{0.1} F_x\, dx$$

$$= \int_0^{0.1} 300 \times x^2\, dx$$

$$= 300 \times \frac{x^3}{3}$$

$$= 100 \times 0.1^3$$

$$= 0.1 \text{J}$$

22.

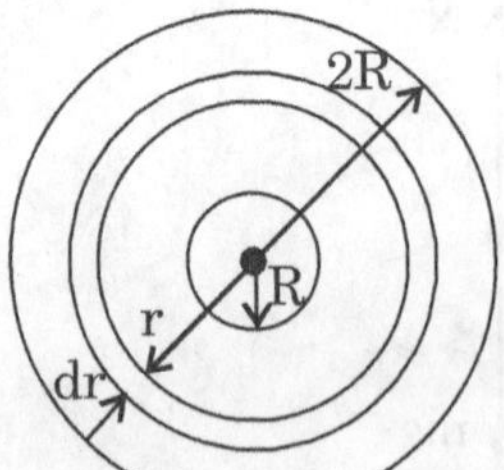

$$I = \frac{5mr^2}{2}$$

K.E. = (K.E.)$_{translation}$ + (K.E.)$_{Rotation}$

$$= \frac{1}{2}mv^2 + \frac{1}{2}I\omega^2$$

where $V = 2R\omega$

$$\Rightarrow \omega = \frac{V}{2R}$$

$$\therefore K.E = \frac{1}{2}mV^2 + \frac{5mV^2}{16}$$

$$K.F = \frac{13mV^2}{16}$$

23.

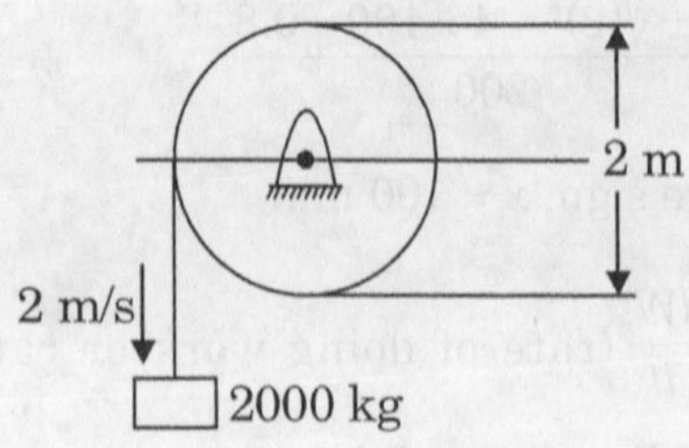

$$v = r\omega$$

$$\omega_i = \frac{V_u}{r} = \frac{2}{1} = 2 \text{ rad/sec}$$

$$\text{Loss in KE of drum} = \frac{1}{2}J(\omega_i^2 - \omega_f^2)$$

$$= \frac{1}{2} \times 150(2^2 - 0^2)$$

$$= 300 \text{ Joule}$$

$$\text{Loss in KE of block} = \frac{1}{2}m(V_i^2 - V_f^2)$$

$$= \frac{1}{2} \times 2000(2^2 - 0^2)$$

$$= 4000 \text{ Joule}$$

Loss in potential energy of block

$$= mgh$$

$$= 2000 \times 9.81 \times 0.5$$

$$= 9810 \text{ Joul}$$

Total loss of energy

$$= 300 + 4000 + 9810$$

$$= 14110 \text{ J}$$

$$= 14.11 \text{ kJ}$$

24. $r = t^2$

$\theta = t$

$x = r\cos\theta = t^2 \cos t$

$y = r\sin\theta = t^2 \sin t$

$$u = \frac{dx}{dt} = 2t\cos t - t^2 \sin t$$

at $t = 2$

$u = 4\cos 2 - 4\sin 2 = -5.301$

$$v = \frac{dy}{dt} = 2t\sin t + t^2\cos t$$

at $t = 2$

$v = 4\sin 2 + 4\cos 2 = 1.972$

$$V = \sqrt{u^2 + v^2} = 5.656$$

$$\text{K.E.} = \frac{1}{2}mv^2$$

$$= \frac{1}{2} \times 1 \times 5.656^2 = 15.99$$

5
CHAPTER

Plane Motion

Topic	Year/Set → Marks Distribution ↓		2019 Set 1	2019 Set 2	2018 Set 1	2018 Set 2	2017 Set 1	2017 Set 2	2016 Set 1	2016 Set 2	2016 Set 3	2015 Set 1	2015 Set 2	2015 Set 3
Plane Motion	1 Mark	MCQ Type									1			
		Numerical Type				1	1					1		
	2 Marks	MCQ Type			1									
		Numerical Type									1			
	Total				2	1	1				3	1		

Analysis of Previous GATE Papers

Plane Motion

1. Instantaneous centre of a body rolling without sliding on a stationary curved surface lies.

(a) at the point of contact

(b) on the common normal at the point of contact

(c) on the common tangent at the point of contact

(d) at the centre of curvature of the stationary surface

[1992 : 1 Mark]

Common Data Questions 2 and 3:

A reel of mass m and radius of gyration k is rolling down smoothly from rest with one end of the thread wound on it held in the ceiling as depicted in the figure. Consider the thickness of the thread and its mass negligible in comparison with the radius r of the hub and the reel mass m. Symbol g represents the acceleration due to gravity.

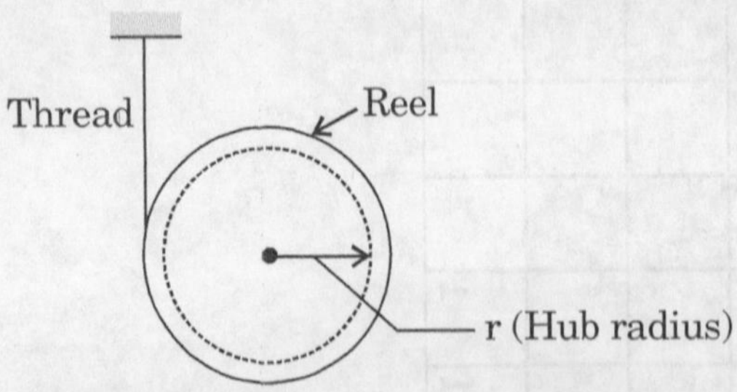

2. The linear acceleration of the reel is

(a) $\frac{gr^2}{(r^2+k^2)}$ (b) $\frac{gk^2}{(r^2+k^2)}$

(c) $\frac{grk}{(r^2+k^2)}$ (d) $\frac{mgr^2}{(r^2+k^2)}$

[2003 : 1 Mark]

3. The tension in the thread is

(a) $\frac{mgr^2}{(r^2+k^2)}$ (c) $\frac{mgrk}{(r^2+k^2)}$

(c) $\frac{mgk^2}{(r^2+k^2)}$ (d) $\frac{mg}{(r^2+k^2)}$

[2003 : 1 Mark]

4. A shell is fired from a cannon. At the instant the shell is just about to leave the barrel, its velocity relative to the barrel is 3 m/s, while the barrel is swinging upwards with a constant angular velocity of 2 rad/s. The magnitude of the absolute velocity of the shell is

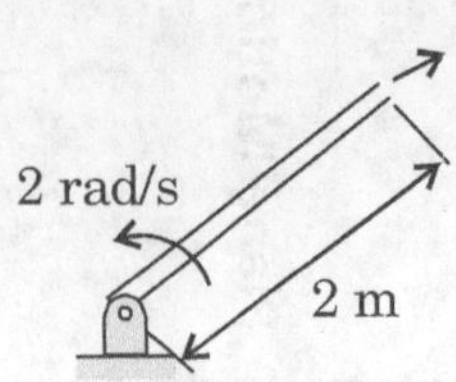

(a) 3 m/s

(b) 4 m/s

(c) 5 m/s

(d) 7 m/s

[2005 : 2 Marks]

5. A straight rod length L(t), hinged at one end freely extensible at the other end, rotates through an angle θ(t) about the hinge. At time t, L(t) = 1 m, L(t) = 1 m/s, θ(t) = π/4 rad and θ (t) = 1 rad/s. The magnitude of the velocity at the other end of the rod is

(a) 1 m/s (b) $\sqrt{2}$ m/s

(c) $\sqrt{3}$ m/s (d) 2 m/s

[2008 : 1 Mark]

6. A circular object of radius /rolls without slipping on a horizontal level floor with the center having velocity V. The velocity at the point of contact between the object and the floor is

(a) zero

(b) V in the direction of motion

(c) V opposite to the direction of motion

(d) V vertically upward from the floor

[2014 : 1 Mark, Set-1]

7. A wheel of radius r rolls without slipping on a horizontal surface shown below. If the velocity of point P is 10 m/s in the horizontal direction, the magnitude of velocity of point Q (in m/s) is _____.

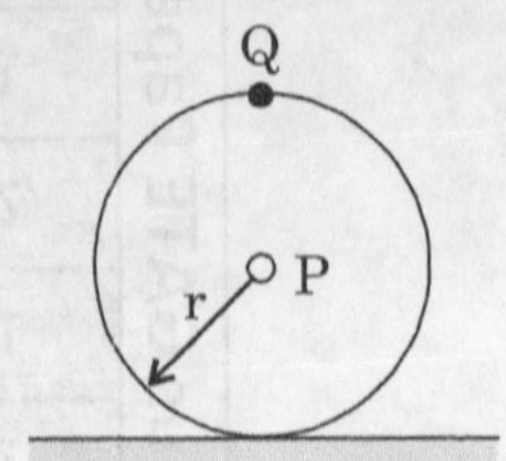

[2015 : 1 Mark, Set-1]

8. A rigid rod (AB) of length L = $\sqrt{2}$ m is undergoing translational as well as rotational motion in the x-y plane (see the figure). The point A has the velocity $V_1 = \hat{i} + 2\hat{j}$ m/s. The end B is constrained to move only along the x direction.

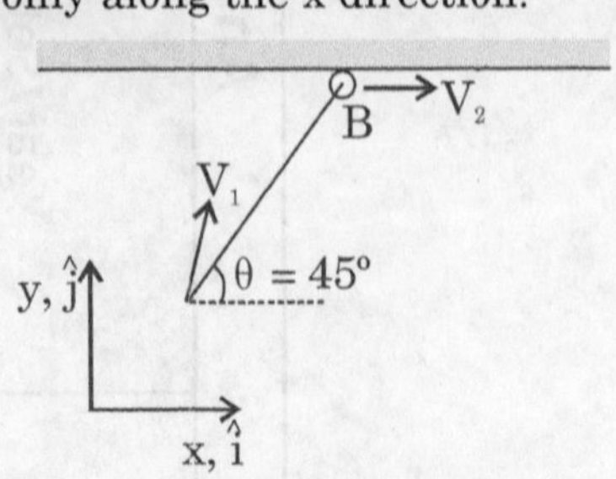

The magnitude of the velocity V_2 (in m/s) at the end B is __________.

[2016 : 2 Marks, Set-3]

9. A rigid link PQ is undergoing plane motion as shown in the figure (V_p and V_Q are non-zero). V_{QP} is the relative velocity of point Q with respect to point P.

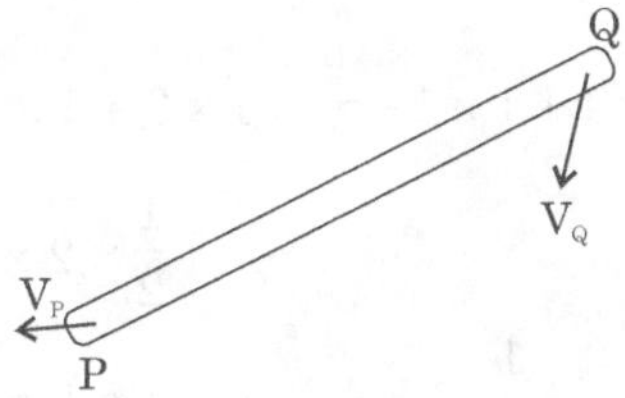

Which one of the following is TRUE?

(a) V_{QP} has components along and perpendicular to PQ

(b) V_{QP} has only one component directed from P to Q

(c) V_{QP} has only one component directed from Q toP

(d) V_{QP} has only one component perpendicular to PQ **[2016 : 1 Mark, Set-3]**

10. The following figure shows the velocity time plot for a particle traveling along a straight line. The distance covered by the particle from t = 0 to t = 5s is _______m.

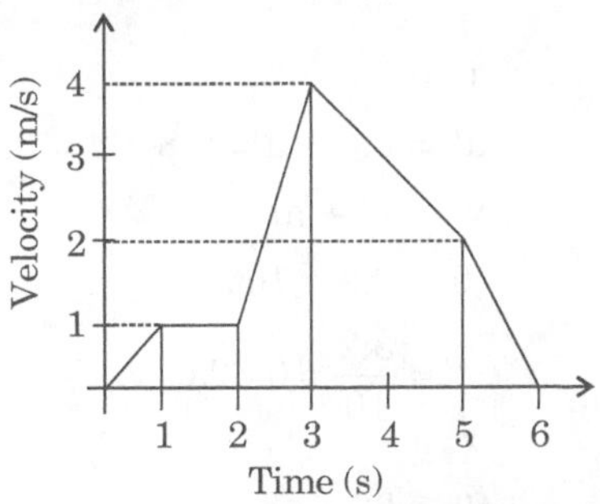

[2017 : 1 Mark, Set-1]

11. A point mass is shot vertically up from ground level with a velocity of 4 m/s at time, t = 0. It loses 20% of its impact velocity after each collision with the ground. Assuming that the acceleration due to gravity is 10 m/s² and that air resistance is negligible, the mass stops bouncing and comes to complete rest on the ground after a total time (in seconds) of

(a) 1

(b) 2

(c) 4

(d) ∞

[2018 : 2 Marks, Set-1]

12. A ball is dropped from rest from a height of 1 min a frictionless tube as shown in the figure. If the tube profile is approximated by two straight lines (ignoring the curved portion), the total distance travelled (in m) by the ball is (correct to two decimal places).

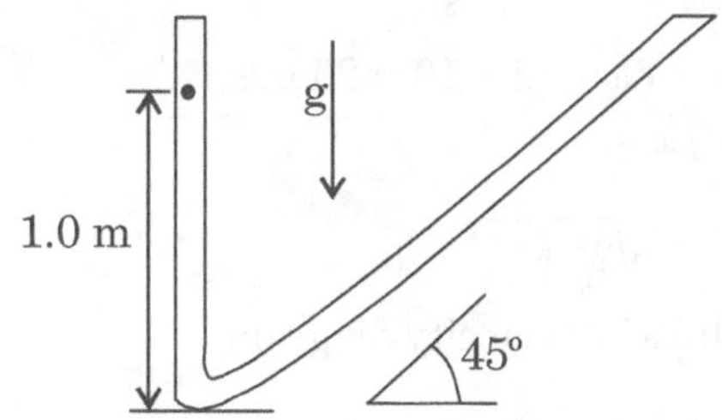

[2018 : 1 Mark, Set-2]

ANSWERS

1. (a) **2.** (a) **3.** (c) **4.** (c) **5.** (b) **6.** (a) **9.** (d) **11.** (c)

EXPLANATIONS

2. Given m = mass of the reel

k = radius of gyration

r = radius of the hub

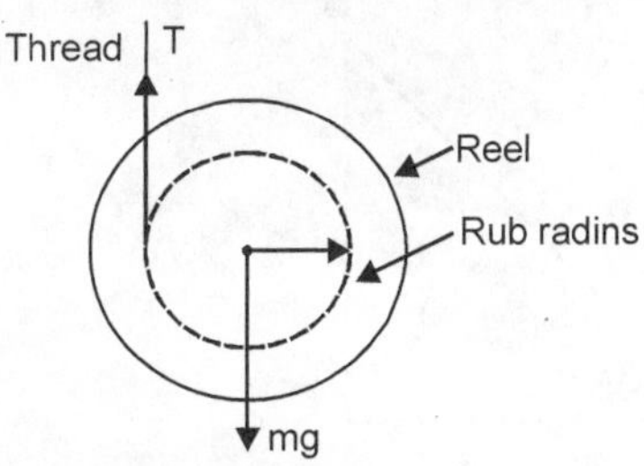

For vertical translation motion

$$mg - T = ma_t \quad ...(i)$$

where T = tension in thread

a_t = tangential acceleration of the point of contact of thread and hub.

For rotation motion

$$T \times r = I\alpha$$

$$T = \frac{I\alpha}{r}$$

$$T = I\frac{a_t}{r^2}$$

$$T = mk^2 \cdot \frac{a_t}{r^2} \quad ...(ii)$$

From equation (*ii*) and (*iii*), we get

$$mg - \frac{mk^2 a_t}{r^2} = ma_t$$

$$\Rightarrow mg = ma_t\left(1 + \frac{k^2}{r^2}\right) = ma_t\left(\frac{r^2 + k^2}{r^2}\right)$$

$$\Rightarrow a_f = \frac{gr^2}{(k^2 + r^2)}$$

3. Tension in thread,

$$T = \frac{mk^2}{r^2} \times \frac{gr^2}{(k^2+r^2)} = \frac{mgk^2}{(k^2+r^2)}$$

4.

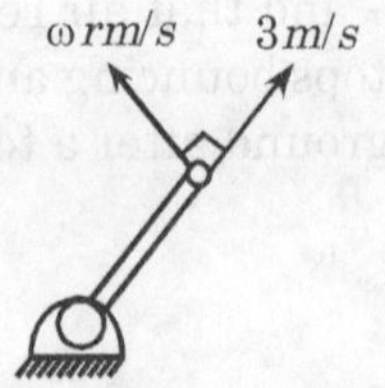

$$\omega r = 2 \times 2 = 4\text{m/sec}$$

$$v_{net} = \sqrt{3^2 + 4^2} = 5\text{ m/s}$$

5.

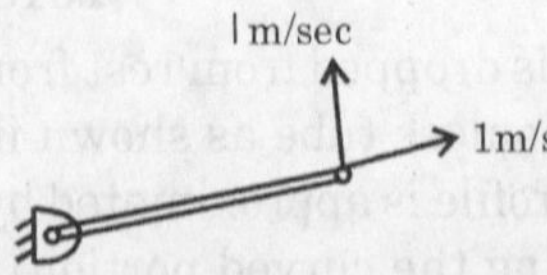

$$v = \sqrt{1^2 + 1^2} = \sqrt{2}\text{ m/s}$$

7. $V_p = r\omega = 10$ m/s

$V_q = 2r\omega = 2 \times 10 = 20$ m/s

8. $V_1 = i + 2j$

$$|V_1| = \sqrt{2^2 + 1^2} = \sqrt{5}$$

By drawing velocity diagram

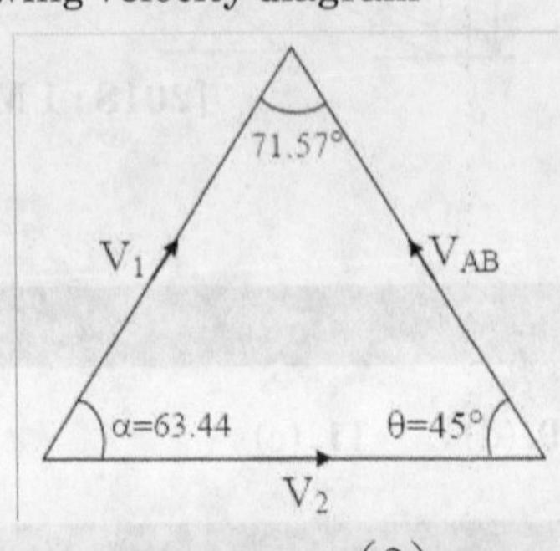

$$\alpha = \tan^{-1}\left(\frac{2}{1}\right) = 63.44$$

By applying sine rule

$$\frac{\sqrt{5}}{\sin 45} = \frac{V_2}{\sin 71.57}$$

$\therefore$ $V_2 = 3$ m/sec

9.

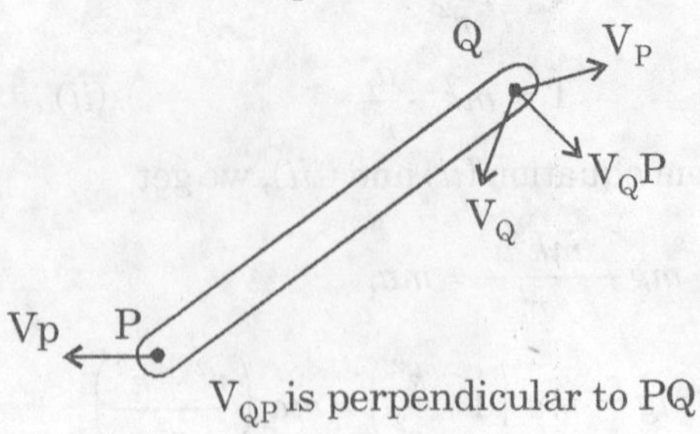

V_{QP} is perpendicular to PQ

10. Distance = Velocity × Time

Distance from $t = 0$ to $t = 5$s

= Area under the curve from $t = 0$ to $t = 5$

$$= \frac{1}{2} \times 1 \times 1 + 1 \times 1 + \frac{1}{2} \times 1 \times 3 + 1 \times 1 + \frac{1}{2} \times 2 \times 2 + 2 \times 2$$

$$= \frac{1}{2} + 1 + \frac{3}{2} + \frac{1}{1} + 2 + 4 = 4 + 2 + 1 + 1 + 2 = 10$$

11.

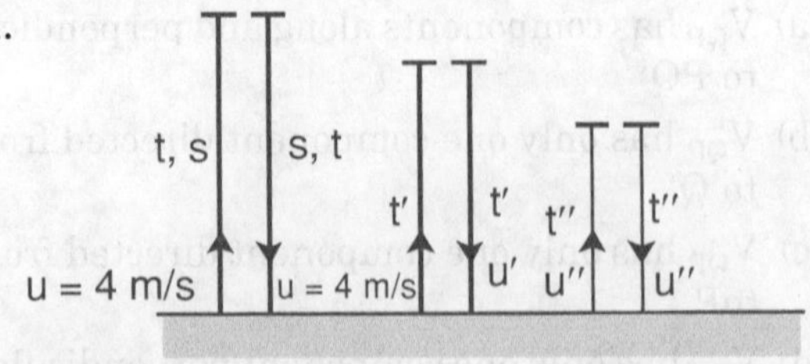

(1) →

$t = ?$

$v = u + at$

$0 = 4 - 10t$

$t = \frac{4}{10} = 0.4$s

(2) →

$t' = ?$

$u' = 0.8 \times u = 0.8 \times 4 = 3.2$ m/s

$v' = u' + at'$

$0 = 3.2 - 10t'$

$t' = \frac{3.2}{10} = 0.32$s

(3) →

$t'' = ?$

$u'' = 0.8\,u' = 0.8 \times 3.2 = 2.56$ m/s

$v'' = u'' + at''$

$0 = 2.56 - 10t''$

$t'' = 0.256$s

So, t, t', t'' are forming a GP series

So, total time = 2(t + t' + t'' + ... 0)

= 2[0.4 + 0.32 + 0.256 + ...0]

= 2 × 2 = 4 seconds

12.

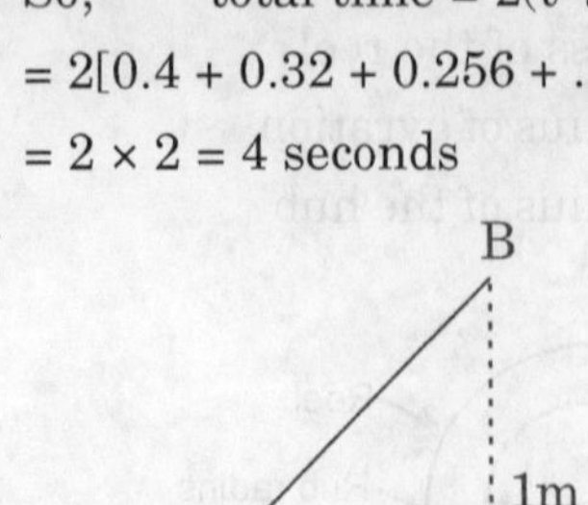

$$\frac{BC}{AB} = \sin 45°$$

$$AB = \frac{BC}{\sin 45°} = \frac{1}{\sin 45°} = 1.414\text{m}$$

Total travel, OA + AB = 1+1.414 = 2.414 m

Unit - III

Strength of Materials

Syllabus

Stress and strain, elastic constants, Poisson's ratio; Mohr's circle for plane stress and plane strain; thin cylinders; shear force and bending moment diagrams; bending and shear stresses; deflection of beams; torsion of circular shafts; Euler's theory of columns; energy methods; thermal stresses; strain gauges and rosettes; testing of materials with universal testing machine; testing of hardness and impact strength.

Contents

CHAPTER 1

Stress and Strain

Analysis of Previous GATE Papers

Topic	Year/Set →Marks Distribution ↓		2019 Set 1	2019 Set 2	2018 Set 1	2018 Set 2	2017 Set 1	2017 Set 2	2016 Set 1	2016 Set 2	2016 Set 3	2015 Set 1	2015 Set 2	2015 Set 3
Stress-Strain Relationship and Hooke's law	1 Mark	MCQ Type					1							1
		Numerical Type	1											
	2 Marks	MCQ Type						1						
		Numerical Type					2		1					
	Total		**1**				**5**	**2**	**2**					**1**
Poisson's Ratio	1 Mark	MCQ Type					1							
		Numerical Type												
	2 Marks	MCQ Type												
		Numerical Type									1			
	Total						**1**				**2**			
Elastic constants	1 Mark	MCQ Type												
		Numerical Type											1	
	2 Marks	MCQ Type												
		Numerical Type					1							
	Total						**2**						**1**	
Principle of Superposition	1 Mark	MCQ Type												
		Numerical Type												
	2 Marks	MCQ Type				1								
		Numerical Type							1					
	Total					**2**			**2**					
Thermal Stresses	1 Mark	MCQ Type												
		Numerical Type	1					1						
	2 Marks	MCQ Type												
		Numerical Type								1				
	Total		**1**					**1**		**2**				

Stress-strain Relationship, Hooke's law

1. Two identical circular rods of same diameter and same length are subjected to same magnitude of axial tensile force. One of the rods is made out of mild steel having the modulus of elasticity of 206 GPa. The other rod is made out of east iron having the modulus of elasticity of 100 GPa. Assume both the materials to be homogeneous and isotropic and the axial force causes the same amount of uniform stress in both the rods. The stresses developed are within the proportional limit of the respective materials. Which of the following observations is correct?

(a) Both rods elongate by the same amount
(b) Mild steel rod elongates more than the cast iron
(c) Cast iron rod elongates more than the mild steel rod
(d) As the stresses are equal strains are also equal in both the rods

[2003 : 1 Mark]

2. A rod of length L having uniform cross-section area A is subjected to a tensile force P as shown in the figure below. If the Young's modulus of the material varies linearly from E_1 to E_2 along the length of the rod, the normal stress developed at the section-SS at

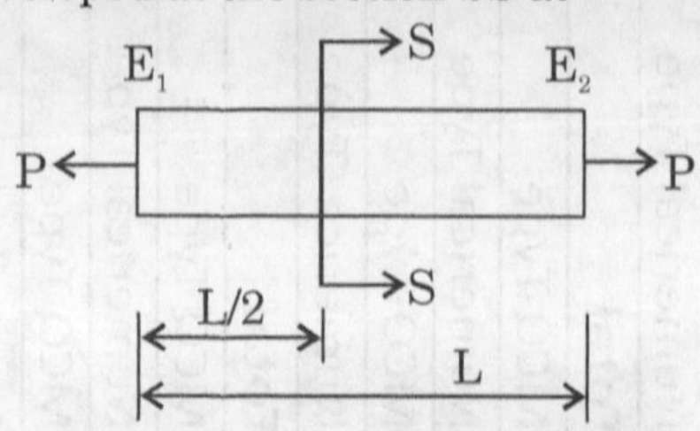

(a) $\frac{P}{A}$ (b) $\frac{P(E_1 - E_2)}{A(E_1 - E_2)}$

(c) $\frac{PE_2}{AE_1}$ (d) $\frac{PE_1}{AE_2}$

[2013 : 1 Mark]

3. A thin plate of uniform thickness is subject to pressure as shown in the figure below

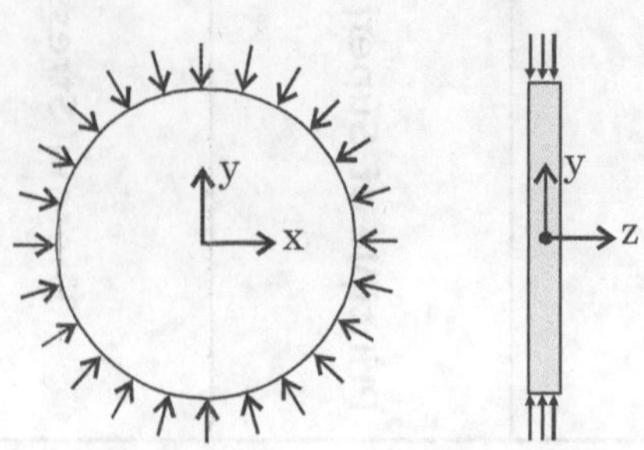

Under the assumption of plane stress, which one of the following is correct?

(a) Normal stress is zero in the z-direction
(b) Normal stress is tensile in the z-direction
(c) Normal stress is compressive in the z-direction
(d) Normal stress varies in the z-direction

[2014 : 1 Mark, Set-2]

4. The stress-strain curve for mild steel is shown in figure given below. Choose the correct option referring to both figure and table.

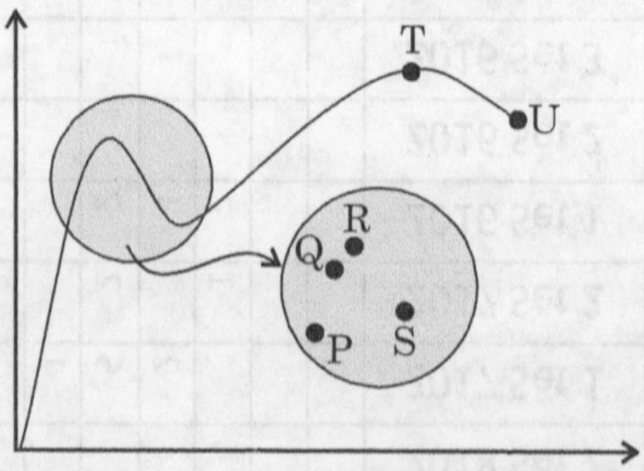

Pointonthegraph	Description of the point
P.	1. Upper yield point
Q.	2. Ultimate tensile strength
R.	3. Proportionality limit
S.	4. Elastic limit
T.	5. Lower yield point
U.	6. Failure

(a) P–1, Q–2, R–3, S–4, T–5, U–6
(b) P–3, Q–1, R–4, S–2, T–6, U–5
(c) P–3, Q–4, R–1, S–5, T–2, U–6
(d) P–4, Q–1, R–5, S–2, T–3, U–6

[2014 : 1 Mark, Set-3]

5. Which one of the following types of stress-strain relationship best describes the behavior of brittle materials, such as ceramics and thermosetting plastics, (σ = stress and ε - strain)?

(a) 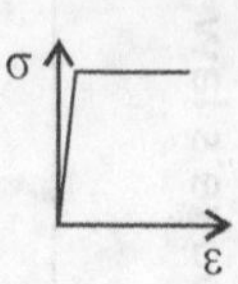

(b) 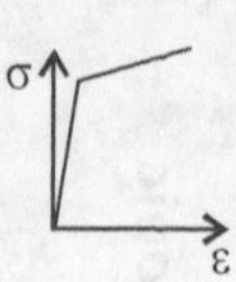

(c) 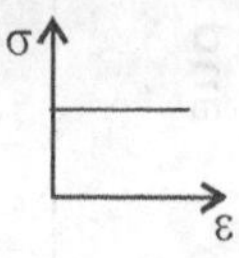

(d) 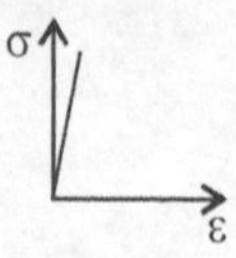

[2015 : 1 Mark, Set-3]

6. A hypothetical engineering stress-strain curve shown in the figure has three straight lines PQ, QR, RS with coordinates P(0, 0), Q(0.2, 100), R(0.6,140)andS(0.8,130). Q is the yield point, R is the UTS point and S the fracture point.

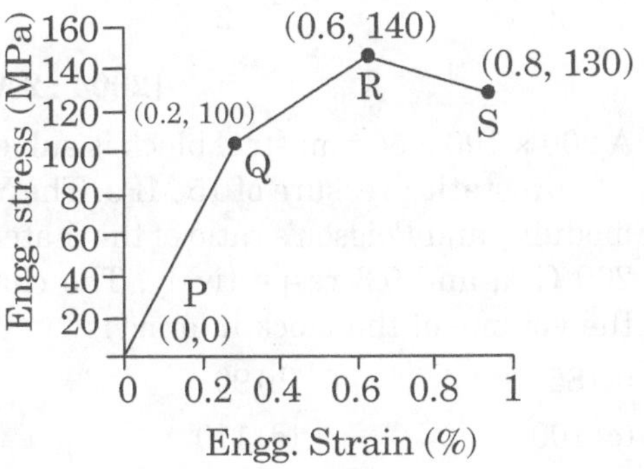

The toughness of the material (in MJ/m^3) is ______.

[2016 : 2 Marks, Set-1]

7. A point mass of 100 kg is dropped onto a massless elastic bar (cross-sectional area = 100 mm^2, length = 1 m, Young's modulus = 100 GPa) from a height H of 10 mm as shown (Figure is not to scale). If g = 10 m/s^2, the maximum compression of the elastic bar is______ mm.

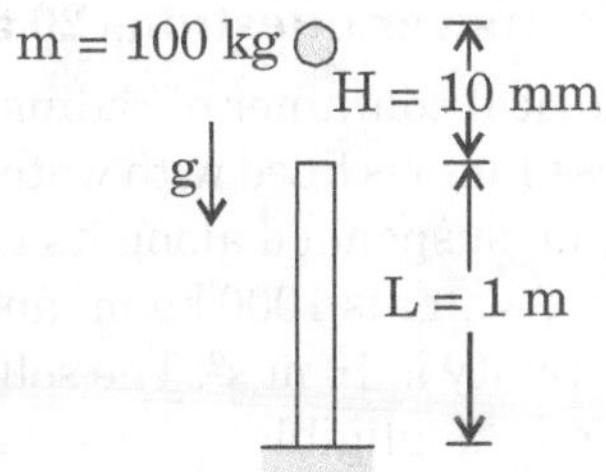

[2017 : 2 Marks, Set-1]

8. A rod of length 20 mm is stretched to make a rod of length 40 mm. Subsequently, it is compressed to make a rod of final length 10 mm. Consider the longitudinal tensile strain as positive and compressive strain as negative. The total true longitudinal strain in the rod is

(a) –0.5 (b) –0.69

(c) –0.75 (d) –1.0

[2017 : 2 Marks, Set-2]

9. In the engineering stress-strain curve for mild steel, the Ultimate Tensile Strength (UTS) refers to

(a) Yield stress

(b) Proportional limit

(c) Maximum stress

(d) Fracture stress

[2017 : 1 Mark, Set-1]

10. A rectangular region in a solid is in a state of plane strain. The (x, y) coordinates of the corners of the undeformed rectangle are given by P(0, 0), Q(4, 3), S(0, 3). The'rectangle is subjected to uniform strain ε_{xx} = 0.001, ε_{yy} = 0.002, γ_{xy} = 0.003. The deformed length of the elongated xy diagonal, upto three decimal places, is ______ units.

[2017 : 2 Marks, Set-1]

11. Consider the stress-strain curve for an ideal elastic-plastic strain hardening metal as shown in the figure. The metal was loaded in uniaxial tension starting from O. Upon loading, the stress-strain curve passes through initial yield point at P, and then strain hardens to point Q, where the loading was stopped. From point Q, the specimen was unloaded to point R, where the stress is zero. If the same specimen is reloaded in tension from point R, the value of stress τ which the material yields again is ______MPa

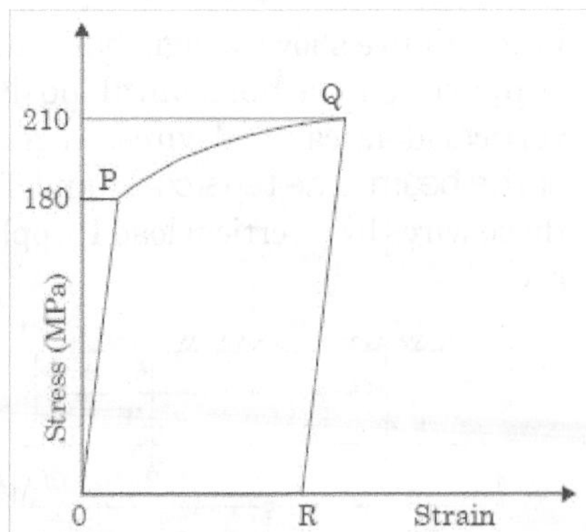

[2019 : 1 Mark, Set-1]

Poisson's Ratio

12. A metallic rod of 500 mm length and 50 mm diameter, when subjected to a tensile force of 100 kN at the ends, experiences an increase in its length by 0.5 mm and a reduction in its diameter by 0.015 mm. The Poisson's ratio of the rod material is ______

[2014: 1 Mark, Set-1]

13. If the Poisson's ratio of an elastic material is 0.4, the ratio of modulus of rigidity to Young's modulus is ______

[2014: 1 Mark, Set-4]

14. A square plate of dimension L × L is subjected to a uniform pressure load p = 250 MPa on its edges as shown in the figure. Assume plane stress conditions. The Young's modulus E= 200 GPa.

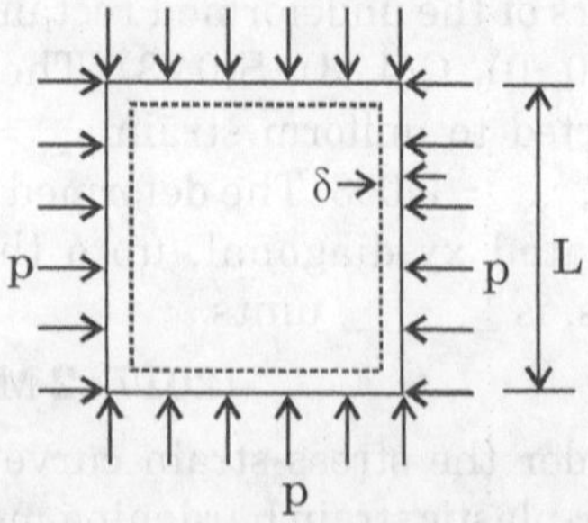

The deformed shape is a square of dimension L – 2δ. If L = 2 m and δ = 0.001 m, the Poisson's ratio of the piate material is ______

[2016 : 2 Marks, Set-3]

15. The Poisson's ratio for a perfectly incompressible linear elastic material is

(a) 1 (b) 0.5
(c) 0 (d) infinity

[2017 : 1 Mark, Set-1]

Elastic constants

16. Below figure shows a rigid bar hinged at A and supported in a horizontal position by two vertical identical steel wires. Neglect the weight of the beam. The tension T_1 and T_2 induced in these wires by a vertical load P applied as shown are

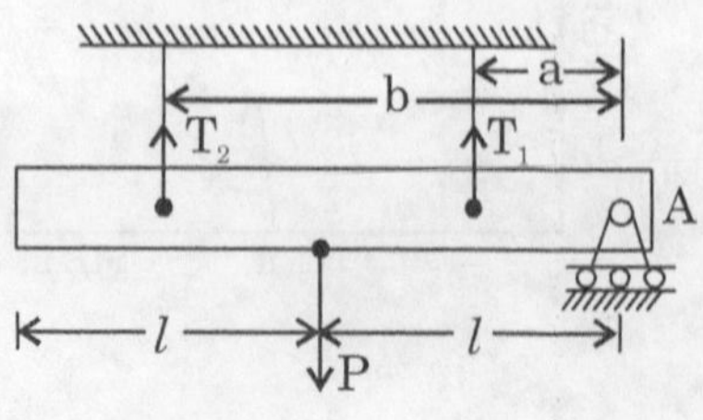

(a) $T_1 = T_2 = \frac{P}{2}$

(b) $T_1 = \frac{Pal}{(a^2+b^2)}, T_2 = \frac{Pbl}{(a^2+b^2)}$

(c) $T_1 = \frac{Pbl}{(a^2+b^2)}, T_2 = \frac{Pal}{(a^2+b^2)}$

(d) $T_1 = \frac{Pal}{2(a^2+b^2)}, T_2 = \frac{Pal}{2(a^2+b^2)}$

[1994 : 2 Marks]

17. In terms of Poisson's ratio (μ) the ratio of Young's modulus (E) to Shear modulus (G) of elastic materials is

(a) 2(1 + μ) (b) 2 (1– μ)

(c) $\frac{1}{2}(1+\mu)$ (d) $\frac{1}{2}(1-\mu)$

[2004 : 2 Marks]

18. A 200 × 100 × 50 mm steel block is subjected to a hydrostatic pressure of 15 MPa. The Young's modulus and Poisson's ratio of the material are 200 GPa and 0.3 respectively. The change in the volume of the block in mm³ is

(a) 85 (b) 90
(c) 100 (d) 110

[2007 : 2 Marks]

19. A rod of length L and diameter D is subjected to a tensile load P. Which of the following is sufficient to calculate the resulting change in diameter?

(a) Young's modulus
(b) Shear modulus
(c) Poisson's ratio
(d) Both Young's modulus and shear modulus

[2008 : 1 Mark]

Linked answer questions 20 and 21:

A cylindrical container of radius R = 1 m, wall thickness 1 mm is filled with water up to a depth of 2 m and suspended along its upper rim. The density of water is 1000 kg/m³ and acceleration due to gravity is 10 m/s². The self-weight of the cylinder is negligible. The formula for hoop stress in a thin-walled cylinder can be used at all points along the height of the cylindrical container.

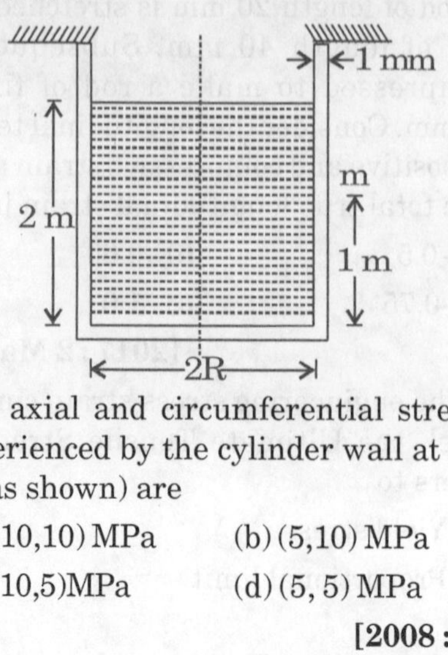

20. The axial and circumferential stress (σ_a, σ_c) experienced by the cylinder wall at mid-depth (m as shown) are

(a) (10,10) MPa (b) (5,10) MPa
(c) (10,5)MPa (d) (5, 5) MPa

[2008 : 2 Marks]

21. If the Young's modulus and Poisson's ratio of the container material' are 100 GPa and 0.3, respectively, the axial strain in the cylinder wall at mid-depth is

(a) 2×10^{-5} (b) 6×10^{-5}
(c) 7×10^{-5} (d) 1.2×10^{-4}

[2008 : 2 Marks]

22. The number of independent elastic constants required to define the stress-strain relationship for an isotropic elastic solid is _____.

[2014 : 1 Mark, Set-4]

23. A rod is subjected to a uni-axial load within linear elastic limit. When the change in the stress is 200 MPa, the change in the strain is 0.001. If the Poisson's ratio of the rod is 0.3, the modulus of rigidity (in GPa) is

[2015 : 1 Mark, Set-2]

24. A horizontal bar, fixed at one end (x = 0), has a length of 1 m, and cross-sectional area of 100 mm^2. Its elastic modulus varies along its length as given by $E(x) = 100e^{-x}$ GPa, where x is the length coordinate (in m) along the axis of the bar. An axial tensile load of 10 kN is applied at the free end (x = 1). The axial displacement of the free end is _____ mm.

[2017 : 2 Marks, Set-1]

Principle of superposition

25. The figure below shows a steel rod of 25 mm^2 cross sectional area. It is loaded at four points. K, L, M and N. Assume E_{steel} = 200 GPa. The total change in length of the rod due to loading is

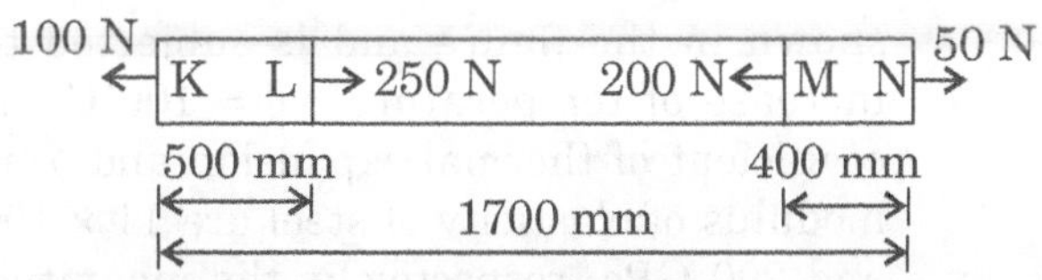

(a) 1 μm (b) −10 μm
(c) 16 μm (d) −20 μm

[2004 : 2 Marks]

26. A bar having a cross-sectional area of 700 mm^2 is subjected to axial loads at the positions indicated.

The value of stress in the segment QR is

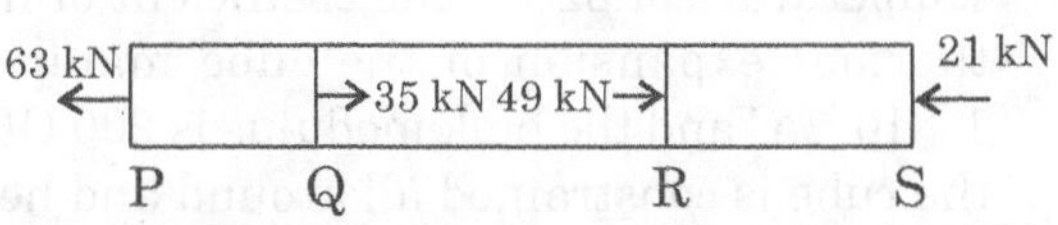

(a) 40MPa (b) 50 MPa
(c) 70 MPa (d) 120 MPa

[2006 : 2 Marks]

27. A horizontal bar with a constant cross-section is subjected to loading as shown in the figure. The Young's moduli for the sections AB and BC are 3E and E, respectively.

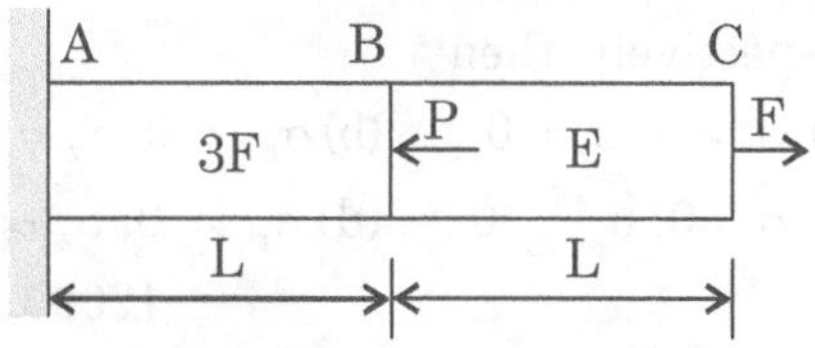

For the deflection at C to be Zero, the ratio P/F is _____

[2016 : 2 Marks, Set-1]

28. A bimetallic cylindrical bar of cross sectional area 1 m^2 is made by bonding Steel (Young's modulus = 210 GPa) and Aluminium (Young's modulus = 70 GPa) as shown in the figure. To maintain tensile axial strain of magnitude 10^{-6} in steel bar and compressive axial strain of magnitude 10^{-6} in Aluminum bar, the magnitude of the required force P(in kN) along the indicated direction is

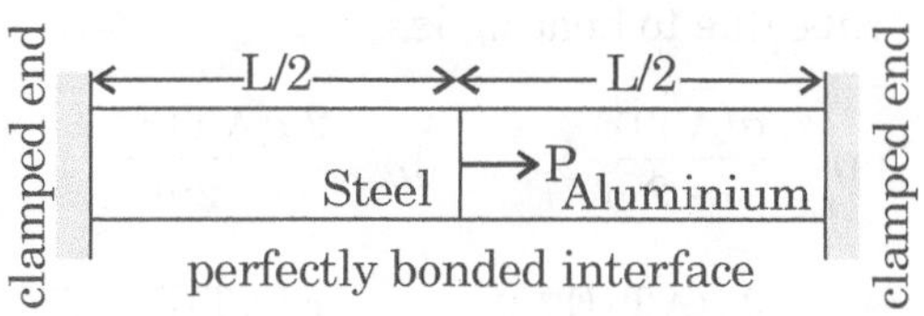

(a) 70 (b) 140
(c) 210 (d) 280

[2018 : 2 Marks, Set-2]

Thermal stresses

29. Determine the temperature rise necessary to induce buckling in a 1 m long circular rod of diameter 40 mm shown in the figure below. Assume the rod to be pinned at its ends and the coefficient of thermal expansion as $20 \times 10^{-6}/°C$. Assume uniform heating of the bar.

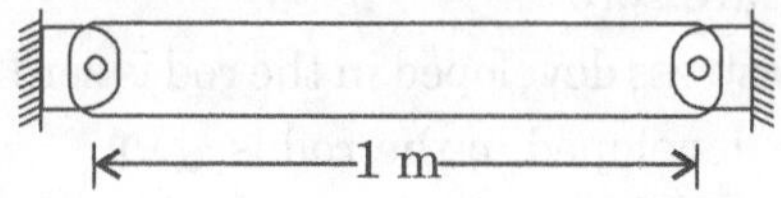

[1993 : 2 Marks]

30. A free bar of length l is uniformly heated from 0°C to a temperature t°C, α is the coefficient of linear expansion and E the modulus of elasticity. The stress in the bar is

(a) αtE (b) αtE/2
(c) Zero (d) None of these

[1995 : 1 Mark]

31. A uniform, slender cylindrical rod is made of a homogeneous and isotropic material. The rod rests on a frictionless surface. The rod is heated uniformly. If the radial and longitudinal thermal stresses are represented by σ_r and σ_z respectively, then

(a) $\sigma_r = 0, \sigma_z = 0$ (b) $\sigma_r \neq 0, \sigma_z = 0$

(c) $\sigma_r = 0, \sigma_z \neq 0$ (d) $\sigma_r \neq 0, \sigma_z \neq 0$

[2005 : 1 Mark]

32. A steel rod of length L and diameter D, fixed at both ends, is uniformly heated to a temperature rise of ΔT. The Young's modulus is E and the coefficient of linear expansion is α. The thermal stress in the rod is

(a) 0 (b) $\alpha\Delta T$

(c) $E\alpha\Delta T$ (d) $E\alpha\Delta TL$

[2007 : 1 Mark]

33. A solid steel cube constrained on all six faces is heated so that the temperature rises uniformly by ΔT. If the thermal coefficient of the material is α, Young's modulus is E and the Poisson's ratio is ν, the thermal stress developed in the cube due to heating is

(a) $-\frac{\alpha(\Delta T)E}{(1-2\nu)}$ (b) $-\frac{2\alpha(\Delta T)E}{(1-2\nu)}$

(c) $-\frac{3\alpha(\Delta T)E}{(1-2\nu)}$ (d) $-\frac{\alpha(\Delta T)E}{3(1-2\nu)}$

[2012 : 2 Marks]

34. A circular rod of length L and area of cross-section A has a modulus of elasticity E and coefficient of thermal expansion α. One end of the rod is fixed and other end is free. If the temperature of the rod is increased by ΔT, then

(a) stress developed in the rod is $E\alpha\Delta T$ and strain developed in the rod is a ΔT

(b) both stress and strain developed in the rod are zero

(c) stress developed in the rod is zero and strain developed in the rod is $\alpha\Delta T$

(d) stress developed in the rod is $E\alpha\Delta T$ and strain developed in the rod is zero

[2014 : 1 Mark, Set-1]

35. A steel cube, with all faces free to deform, has Young's modulus, E, Poisson's ratio, v, and coefficient of thermal expansion, a. The pressure (hydrostatic stress) developed within the cube, when it is subjected to a uniform increase in temperature, ΔT, is given by

(a) 0 (b) $\frac{\alpha(\Delta T)E}{1-2v}$

(c) $-\frac{\alpha(\Delta T)E}{1-2v}$ (d) $\frac{\alpha(\Delta T)E}{3(1-2v)}$

[2014 : 1 Mark, Set-2]

36. A 200 mm long, stress free rod at room temperature is held between two immovable rigid walls. The temperature of the rod is uniformly raised by 250°C. If the Young's modulus and coefficient of thermal expansion are 200 GPa and 1×10^{-5}/°C, respectively, the magnitude of the longitudinal stress (in MPa) developed in the rod is ______

[2014 : 2 Marks, Set-1]

37. A circular metallic rod of length 250 mm is placed between two rigid immovable walls as shown in the figure. The rod is perfect contact with the wall on the left side and there is a gap of 0.2 mm between the rod and the wall on the right side. If the temperature of the rod is increased by 200°C, the axial stress developed in the rod is ______ MPa.

Young's modulus of the material of the rod is 200 GPa and the coefficient of thermal expansion is10^{-5}per°C.

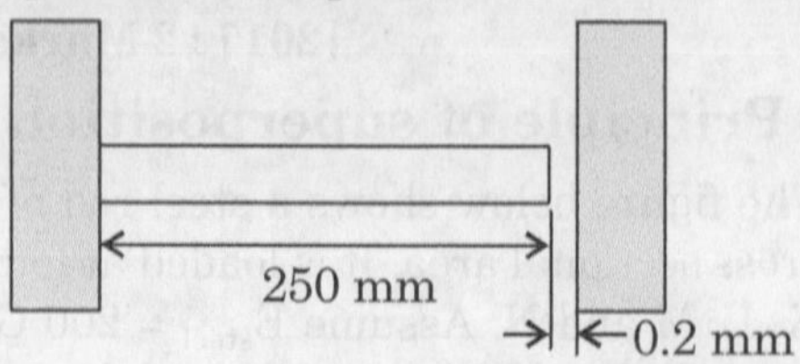

[2016 : 2 Marks, Set-2]

38. A steel bar is held by two fixed supports as shown in the figure and is subjected to an increase of temperature ΔT = 100°C. If the coefficient of thermal expansion and Young's modulus of elasticity of steel are 11×10^{-6}/°C and 200 GPa, respectively, the magnitude of thermal stress (in MPa) induced in the bar is ______.

[2017 : 1 Mark, Set-2]

39. A solid cube of side 1 m is kept at a room temperature of 32°C. The coefficient of linear thermal expansion of the cube material is 1×10^{-5}/°C and the bulk modulus is 200 GPa. If the cube is constrained all around and heated uniformly to 42°C, then the magnitude of volumetric (mean) stress (in MPa) induced due to heating is ______.

[2019 : 1 Mark, Set-1]

ANSWERS

1. (c)	2. (a)	3. (a)	4. (c)	5. (d)	8. (b)	9. (c)	15. (b)	16. (b)	17. (a)
18. (b)	19. (d)	20. (b)	21. (a)	22. (2)	25. (b)	26. (a)	28. (d)	30. (c)	31. (a)
32. (c)	33. (a)	34. (c)	35. (a)						

EXPLANATIONS

1. $E_{\text{mild steel}} = 206\text{ GPa}$

$E_{\text{cast iron}} = 100\text{ GPa}$

Now, Elongation in cast iron,

$$\Delta_{CI} = \left(\frac{Pl}{AE}\right)_{C.I.}$$

Elongation in mild steel,

$$\Delta_{m.s.} = \left(\frac{Pl}{AE}\right)_{M.S.}$$

$$\Rightarrow \frac{\Delta_{C.I.}}{\Delta_{M.S.}} = \frac{\dfrac{Pl}{AE_{C.I.}}}{\dfrac{Pl}{AE_{M.S.}}} = \frac{E_{M.S.}}{E_{C.I.}} = \frac{206}{100} = 2.06$$

$$\therefore \quad \Delta_{C.I.} > \Delta_{M.S.}$$

2. At section 55 : -

The left side of rod

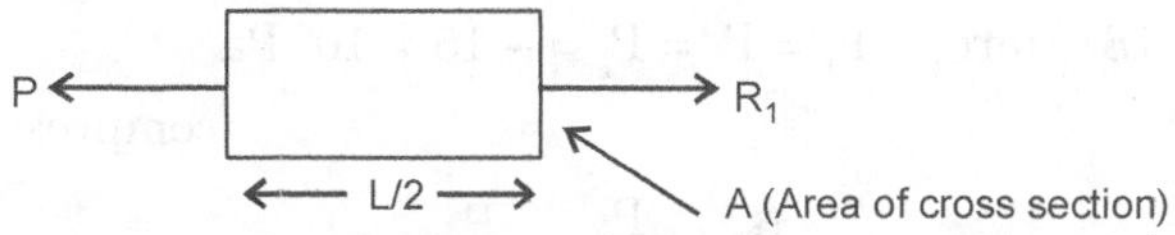

Normal stress $\sigma_n = \frac{R}{A} = \frac{P}{A}\left(\because \begin{matrix}\sum F_x = 0\\ R - P = 0\end{matrix}\right)$

The right side of rod

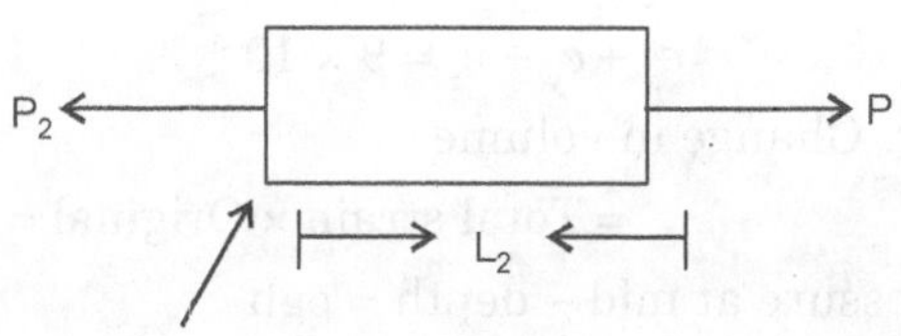

$$\Sigma fx = 0$$

$$\Rightarrow \quad R_2 - P = 0$$

$$\Rightarrow \quad R_2 = P$$

$$\sigma_n = \frac{P_2}{A} = \frac{P}{A}$$

$\therefore$ Hence normal $= \frac{P}{A}$

3. For a plane stress criteria.

Normal stress in Z direction = 0.

5. Brittle material breaks without plastic deformation.

6. Toughness = Area under diagram

$$= \frac{1}{2}\times\frac{0.2}{100}\times 100 + \frac{1}{2}\left(\frac{0.2}{100}\right)(100+140) + \frac{1}{2}\left(\frac{0.2}{100}\right)(140+130)$$

$$T = 0.1 + 0.48 + 0.27 = 0.85\text{ MJ/m}^3$$

7. m = 100 kg h = 10 mm, L = 1m

E = 100 GPa, g = 10m/s² A = 100 mm²

$$\delta = \frac{WL}{AE}\left[1 + \sqrt{\frac{2EAh}{wL}}\right]$$

$$\delta = \frac{100\times 10\times 1}{100\times 10^{-6}\times 100\times 10^{9}} \times \left[1 + \sqrt{1 + \frac{2\times 100\times 10^{9}\times 100\times 10^{-6}\times 0.01}{1000}}\right]$$

$$\delta = 1.517\text{ mm}$$

8. Volume remain same

$L_1 = 20$ mm, $L_2 = 40$ mm, $L_3 = 10$mm

$$A_1L_1 = A_2L_2 = A_3L_3$$

$$\frac{A_1}{A_3} = \frac{L_3}{L_1}$$

$$\text{True strain} = \ln\left(\frac{A_1}{A_3}\right) = \ln\left(\frac{L_3}{L_1}\right) = -0.693$$

10.

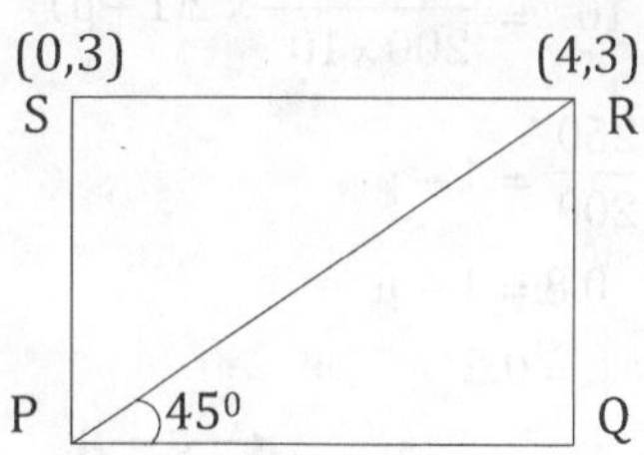

Length of diagonal PR $= \sqrt{4^2 + 3^2} = 5$

Strain at an angle 45°

$$\varepsilon = \varepsilon_x \cos^2\theta + \varepsilon_y \sin^2\theta + \gamma_{xy}\sin\theta\cos\theta$$

$$= 0.001\times\cos^2 45 + 0.002\sin^2 45 + 0.003\sin 45\cos 45$$

$$= 0.003$$

$$\varepsilon = \frac{\Delta l}{l} = 0.003$$

$\Delta l = 5 \times 0.003 = 0.015$ units

Deformed length = 5 + 0.015 = 5.015 units

11. As the point loaded beyond the yield point is unloaded completely and reloaded then because of strain hardening (residual strain remains in body) increase the yield strength.

In this case material with start yielding from point 'R' instead of 'P'.

12. l = 500 mm, d = 50 mm, p = 100 KN

δl = 0.5 mm, δd = 0.015

Poission's Ratio

$$\left(\frac{l}{m}\right) = \frac{\text{Lateral strain}}{\text{longitudinal strain}}$$

$$\frac{\delta d/d}{\delta l/l} = \frac{0.015/50}{0.5/500} = 0.3.$$

13. From E = 2G(1 + μ)

$$\frac{G}{E} = \frac{1}{2(1+\mu)}$$

$$= \frac{1}{2(1+0.4)} = 0.357$$

14. $$\frac{\Delta V}{V} = \frac{\sigma}{E}(1-\nu)$$

$$\varepsilon_v = \varepsilon_x + \varepsilon_y + \varepsilon_z$$

$$\varepsilon_v = \left(\frac{\sigma}{E} - \frac{\mu\sigma}{E}\right) + \left(\frac{\sigma}{E} - \frac{\mu\sigma}{E}\right) - \frac{\mu(\sigma+\sigma)}{E}$$

Due to plane stress condition $\varepsilon_z = 0$

$$\varepsilon_v = \frac{\sigma}{E} \times 2 \times (1-\mu)$$

$$\frac{(2-2\times0.001)^2 - 2^2}{2^2} = -\frac{P}{E} \times 2(1-\mu)$$

$$-2 \times 10^{-3} = \frac{250}{200\times10^3} \times 2(1-\mu)$$

$$\frac{250}{200} = 1-\mu$$

$$0.8 = 1 - \mu$$

$$\therefore \quad \mu = 0.2$$

16.

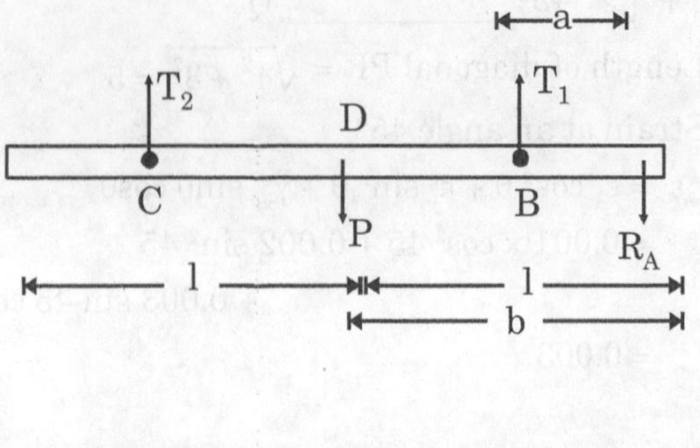

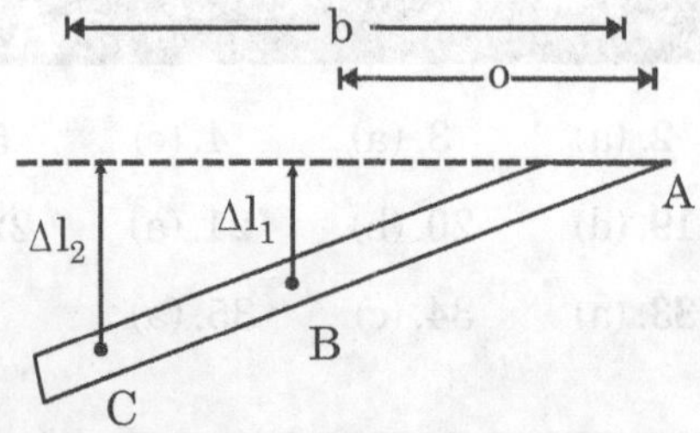

$\Sigma y = O$

$T_1 + T_2 - P - R_A = O$–(i)

$\Sigma M_A = O$

T_1 (a)+ T_2 (b) = P (l)–(ii)

Using similar Δ's in fig (2)

$$\frac{\Delta l_1}{a} = \frac{\Delta l_2}{b} \Rightarrow \left[\Delta l_1 = \frac{T_1 L}{AE} \Delta l_2 = \frac{T_2 L}{AE}\right]$$

$$\frac{T_1}{a} = \frac{T_2}{b}$$

Putting above eqⁿ in eqⁿ (ii)

$$T_1(a) + \frac{b^2}{a} T_1 = P(l)$$

$$T_1 = \frac{pla}{a^2+b^2} \quad T_2 = \frac{plb}{a^2+b^2}$$

17. We know E = 2G[1 + v]

$$\frac{E}{G} = 2[1+v].$$

18. Here, $P_x = P_y = P_z = -15 \times 10^6$ Pa (compression)

Now, $$e_x \frac{P_x}{E} - \frac{P_y}{E.m} - \frac{P_z}{Em} = 3\times10^{-5}$$

where, m = Poisson's ratio

$e_x = e_y = e_z = 3 \times 10^{-5}$

$e_x + e_y + e_z = 9 \times 10^{-5}$

∴ Change in volume

= Total strain × Original = 90 m³

20. Pressure at mid – depth = ρgh

= $10^3 \times 10 \times 1 = 10^4$ N/m²

$$\therefore \quad \sigma_a = \frac{pd}{4t} = \frac{10^4 \times 2}{4 \times 10^{-3}}$$

= 5×10^6 N/mm² = 5 MPa

$$\sigma_c = \frac{pd}{2t} = \frac{10^4 \quad 2}{2 \quad 10^{-3}} = 10\text{MPa}$$

21. Axial strain in the cylinder would be produced due to the axial (longitudinal) stress and due to the lateral stress

i.e.hoop stress, $e = \frac{1}{Y}\left(u - \frac{c}{m}\right)$

$$= \frac{1}{100 \quad 10^3}(5 - 10 \times 0.3)$$

$$= 2 \times 10^{-5}$$

22. Elastic constants are required to obtain relationship between stress & strain, There are 4 elastic constants.

E, μ, G, & E

But E & μ are independent.

23. $E\frac{\sigma}{t} = \frac{200 \times 10^6}{0.001} = 200\text{GPa}$

$$G = \frac{200}{2(1+0.3)} = 76.92\text{GPa}$$

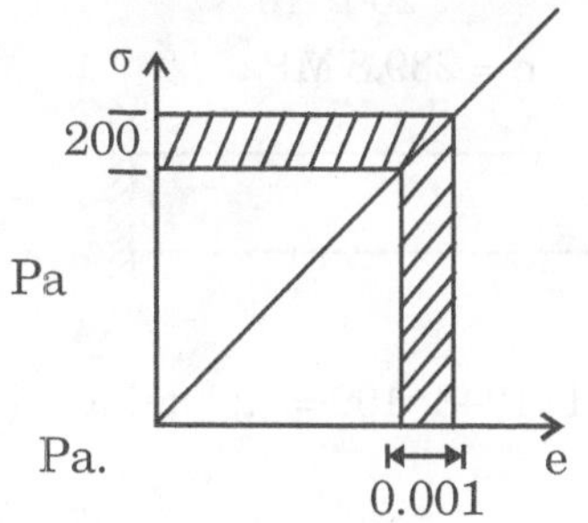

24.

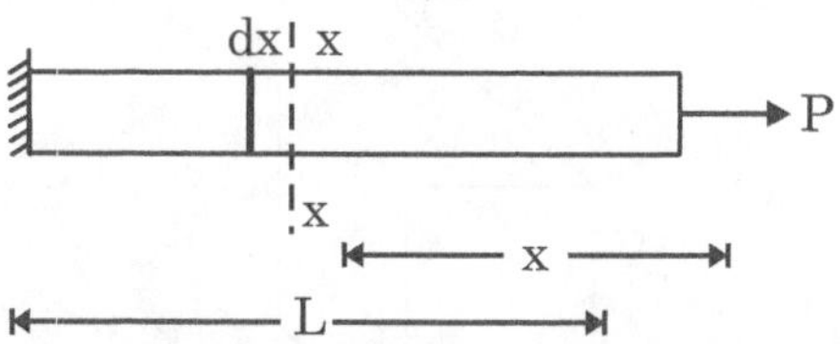

Change in length of small strip

$$\delta = \frac{Px - x(dx)}{Ax - x\,Ex - x}$$

Total change in length of bar

$$\delta_{total}\int_O^L \delta = \int_O^L \frac{Px - x\,dx}{Ax - x\,Ex - x}$$

$$= \frac{P}{A}\int_O^L \frac{dx}{100^{e-x}} = \frac{P}{A(100)}\int_O^L e^x d_x$$

$$= \frac{10 \times 10^3}{100 \times 10^3 \times 100}\left(e^x\right)_O^L = 1.7182 \times 10^{-3}\,\text{m}$$

$$\delta = 1.718\text{ mm}$$

25.

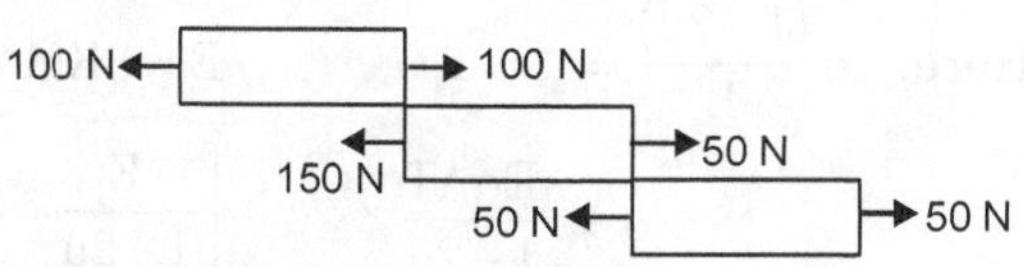

Total change in length

$$= \frac{\Sigma Pl}{AE} = \frac{P_1 l_1 + P_2 + P_3 l_3}{AE}$$

$$= \frac{100 \times 0.5}{AE} + \frac{(-150) \times (0.8)}{AE} + \frac{0.5 \times 0.4}{AE}$$

$$= \frac{1}{25 \times 10^{-6} \times 200 \times 10^9}[50 - 120 + 20]$$

$= -1 \times 10^{-5}$ m $= -10$μm

26.

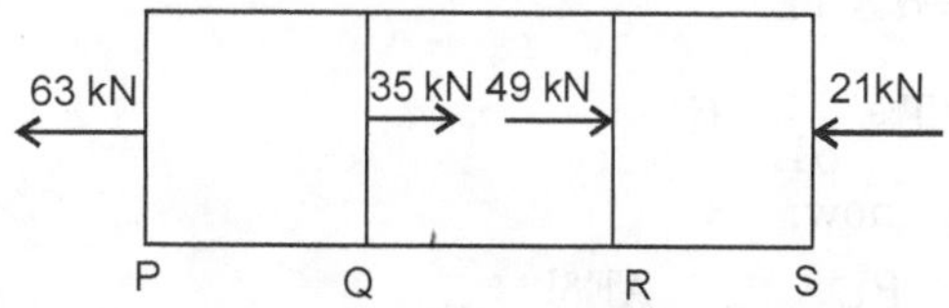

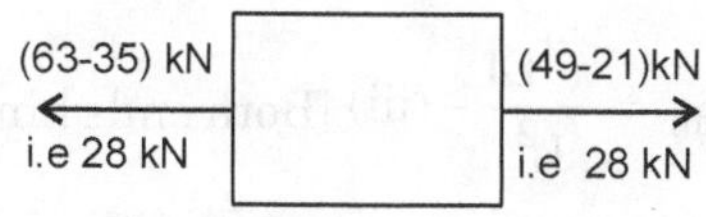

$$\text{Stress} = \frac{\text{Force}}{\text{Area}}$$

$$= \frac{28 \times 10^3}{700 \times 10^{-6}} N/m^2$$

$= 4 \times 10^{-5}$ N/m² = 40 MPa

27. $(\delta)_c = O$

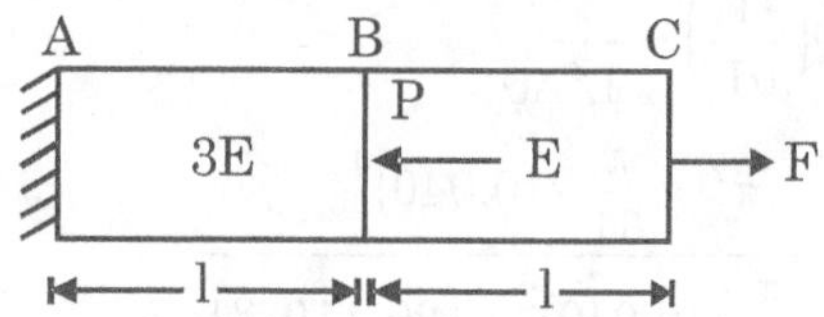

$$\delta l_{AB} + \delta l_{BC} = O$$

$$\frac{(F-P)}{A(3E)} + \frac{FL}{AE} = 0$$

$$\frac{(F-P)}{3} + \frac{F}{1} = 0$$

$$3F = P - F$$

$$\frac{P}{F} = 4$$

28.

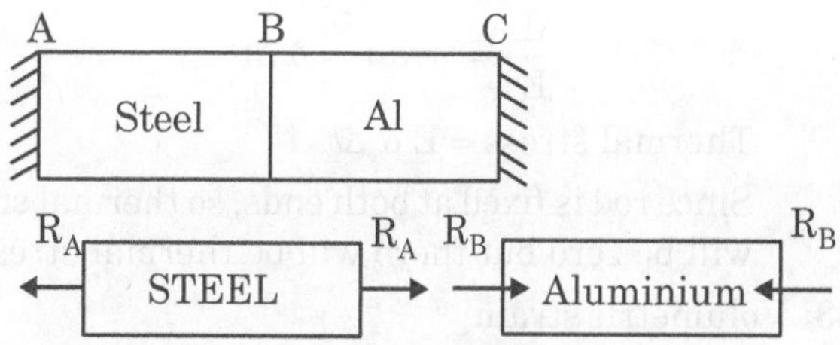

Let R_A & R_B be the reation at the supports A & B . For the equilibrium of the bar these reaction must act towards left. so that; $\boxed{R_A + R_B = P}$

$\sigma_A = \in E s \qquad \sigma B = \in E_{AL}$

$R_A = A \in E s \qquad R_B = A \in E_{AL}$

$R_A = 210\text{KN} \quad R_B = 70\text{ KN}$

$\therefore P = R_A + R_B = 70 + 210\text{uN}$

P = 280 KN.

29.

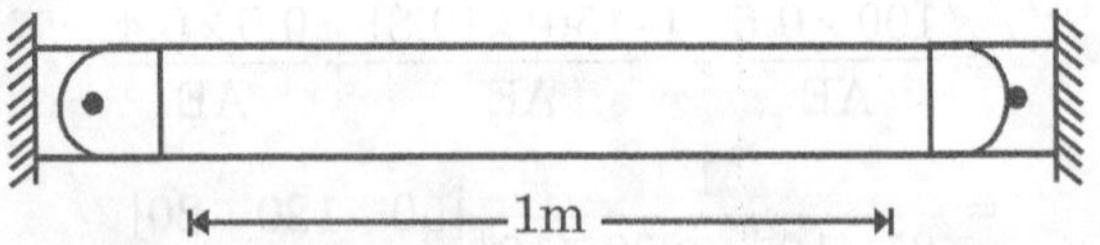

Let Bucking load be P

$\delta l = \alpha \Delta TL$

$\therefore \Delta T = \dfrac{\delta l}{\alpha L}$ – (i)

We know:

$\delta l = \dfrac{PL}{AE} \Rightarrow P = \dfrac{AE\delta l}{L}$ – (ii)

$P_{column} = \dfrac{\pi^2 EI}{L^2}$ – (iii) [Both ends hinged]

Compare equation (ii) + (iii)

$AE\dfrac{\delta l}{L} = \dfrac{\pi^2 EI}{L^2}$

$Sl = \dfrac{\pi^2 I}{LA}$ – (iv)

Putting the value of equation (iv) in equation (i)

$$\Delta T = \frac{\pi^2 I}{LA}\left(\frac{1}{\alpha L}\right) = \frac{\pi^2 I}{L^2 A\alpha}$$

$$\Delta T = \frac{\pi^2 \times \frac{\pi}{64} \times (0.040)^4}{1^2 \times \frac{\pi}{4} \times (0.040)^2 \times \left(20 \times 10^{-6}\right)}$$

$\Delta T = 49.35°C$

30. For thermal stress to be developed there must be constraint in the system to oppose. So strain develops but there is no thermal stress.

31. Rod is not restrained but completely free

32. Since rod is free to expand, therefore

ΔL = elongation = $L\alpha\ \Delta t$

$\therefore \quad \dfrac{\Delta L}{L}$ = strain = $\alpha\ \Delta t$

Thermal stress = $E\ \alpha\ \Delta t$

Since rod is fixed at both ends, so thermal strain will be zero but there will be thermal stresses.

33. Volumetric strain

$$\varepsilon_V = \frac{(1-2\mu)}{E}\left[\sigma_x + \sigma_y + \sigma_z\right]$$

$$\alpha\Delta T = \frac{(1-2\mu)}{E}(\sigma)$$

$$\sigma = \frac{\alpha(\Delta T)E}{(1-2\mu)}$$

34. Since one end of the rod is fixed and other is free to expand. Hence the temperature stresses is zero

$$\in = \frac{\delta L}{L} \frac{\alpha\Delta TL}{L} = \alpha\Delta T$$

35. Since all the faces are free to expand the stresses due to temperature rise is equal to 0.

36. $l = 200, \Delta T = 250°C, \alpha = 1 \times 10^{-5}\ /°C$

$E = 200\ GPa = 200 \times 10^3\ MPa$

$\sigma = \alpha\ \Delta TE = 1 \times 10^{-5} \times 250 \times 200 \times 10^3$

$= 500\ MPa.$

37. $\ell\ \alpha\ t = 250 \times (10^{-5})(200)$

$= \dfrac{1}{2} = 0.5\ mm > gap$

Deformation prevented $= 0.5 - 0.2 = 0.3 = \dfrac{P\ell}{AE}$

$\Rightarrow \quad 0.3 = \dfrac{\sigma(250 + 0.2)}{200 \times 10^3}$

$\sigma = 239.8\ MPa$

38.

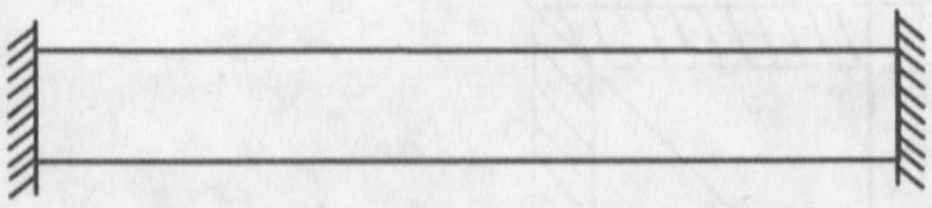

$\in = -\alpha\Delta T$

$\sigma = E\in$ = -200×11×10–6×100 = –220mPa

39.

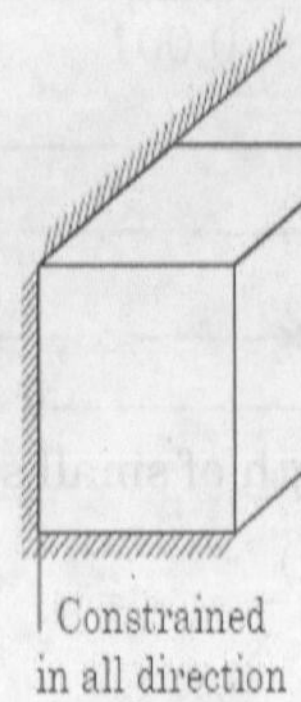

$a = 1\ m$

$T_a = 32°C$

$\alpha = 1 \times 10^{-5}/°C$

$K = 200\ GPa$

$\Delta T = 10°C$

$\in_v = \dfrac{(1-2\mu)}{E}(3\sigma) = 3\alpha T$

As $\quad \sigma_1 = \sigma_2 = \sigma_3 = \sigma$

(mean volumetric stress)

Here, $\sigma\dfrac{(1-2\mu)}{E} = \alpha\Delta\ T, \qquad E = 3K(1-2\mu)$

$\sigma = \dfrac{E\alpha\Delta T}{1-2\mu} \qquad \left[\dfrac{E}{1-2\mu} = 3K\right]$

$\sigma = 3K\alpha\Delta T$

$= 3 \times 200 \times 10^3 \times 1 \times 10^{-5} \times 10$

$\sigma = 60\ MPa$

2 CHAPTER

Principal Stress and Strain

Analysis of Previous GATE Papers

Topic	Year/Set → Marks Distribution ↓		2019 Set 1	2019 Set 2	2018 Set 1	2018 Set 2	2017 Set 1	2017 Set 2	2016 Set 1	2016 Set 2	2016 Set 3	2015 Set 1	2015 Set 2	2015 Set 3
Principal Stress	1 Mark	MCQ Type									1			
		Numerical Type						1						
	2 Marks	MCQ Type			1								1	
		Numerical Type												
	Total				2			1			1		2	
Principal Strain	1 Mark	MCQ Type												
		Numerical Type												
	2 Marks	MCQ Type												
		Numerical Type												
	Total													
Theories of Failure	1 Mark	MCQ Type					1							
		Numerical Type												
	2 Marks	MCQ Type						1	1					
		Numerical Type												
	Total						1	2	2					

Principal stress

1. An elastic body is subjected to a tensile stress X in a particular direction and a compressive stress Y in its perpendicular direction. X and Y are unequal in magnitude. On the plane of maximum shear stress in the body there will be

   (a) no normal stress

   (b) also the maximum normal stress

   (c) the minimum normal stress

   (d) both normal stress and shear stress

   [1989 : 1 Mark]

2. The three-dimensional state of stress at a point is given by

$$[\sigma] = \begin{bmatrix} 30 & 10 & -10 \\ 10 & 0 & 20 \\ -10 & 20 & 0 \end{bmatrix} \text{ MN/m}^2$$

   The shear stress on the x-face in y-direction at the same point is then equal to

   (a) zero MN/m^2 (b) -10 MN/m^2

   (c) 10 MN/m^2 (d) 20 MN/m^2

   [1990 : 2 Marks]

3. A large uniform plate containing a rivet hole is subjected to uniform uniaxial tension of 95 MPa. The maximum stress in the plate is

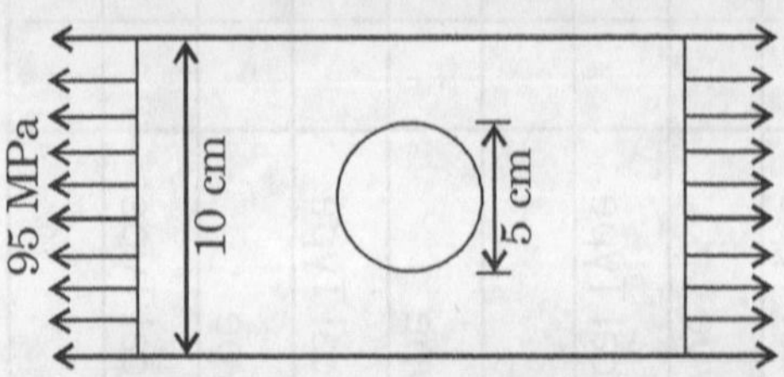

   (a) 100 MPa (b) 285 MPa

   (c) 190 MPa (d) Indetermine

   [1992 : 1 Mark]

4. At a point in a stressed body the state of stress on two planes 45° apart is as shown below. Determine the two principal stresses in MPa.

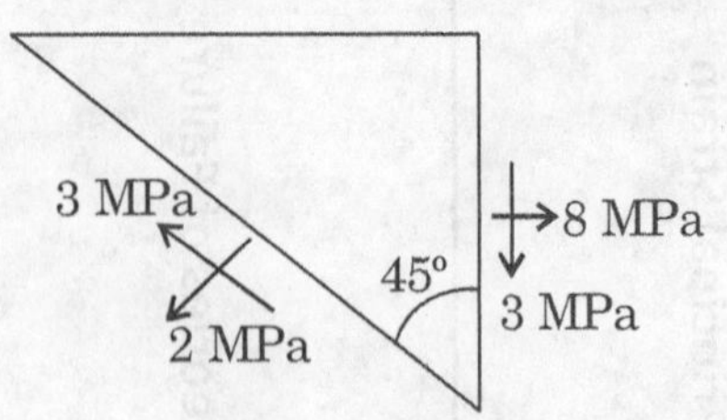

   (a) 8.242, 0.658 (b) 9.242, 0.758

   (c) 9.242, 0.758 (d) 8.242, 0.758

   [1993 : 2 Marks]

5. A shaft subjected to torsion experiences a pure shear stress x on the surface. The maximum principal stress on the surface which is at 45° to the axis will have a value

   (a) $\tau \cos 45°$ (b) $2\tau \cos 45°$

   (c) $\tau \cos^2 45°$ (d) $2\tau \sin 45° \cos 45°$

   [2003 : 2 Marks]

6. If the principal stresses in a plane stress problem are σ_1 = 100 MPa, σ_2 = 40 MPa, the magnitude of the maximum shear stress (in MPa) will be

   (a) 60 (b) 50

   (c) 30 (d) 20

   [2009 : 1 Mark]

7. The state of plane-stress at a point is given by σ_x =200 MPa, σ_y =100 MPa and τ_{xy} = 100 MPa. The maximum shear stress

   (a) 111.8 (b) 150.1

   (c) 180.3 (d) 223.6

   [2010 : 1 Mark]

8. The state of stress at a point is given by σ_x = –6 MPa, σ_y = 4 MPa, and τ_{xy} = – 8 MPa. The maximum tensile stress (in MPa) at the point is ______

   [2014 : 2 Marks, Set-1]

9. In a plane stress condition, the components of stress at a point are σ_x = 20 MPa, σ_y = 80 MPa and σ_{xy} = 40 MPa. The maximum shear stress (in MPa) at the point is

   (a) 20 (b) 25

   (c) 50 (d) 100

   [2015 : 2 Marks, Set-2]

10. The state of stress at a point on an element is shown in figure (a). The same state of stress is shown in another coordinate system in figure (b).

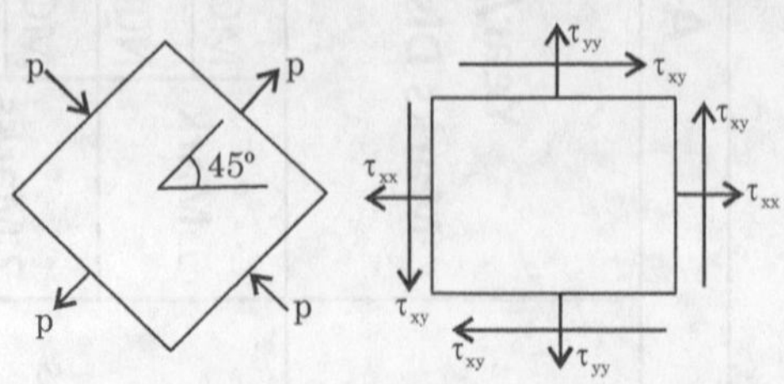

   The components (τ_{xx}, τ_{yy}, τ_{xy},) are given by

   (a) $\left(\frac{p}{\sqrt{2}}, -\frac{p}{\sqrt{2}}, 0\right)$ (b) (0,0, p)

   (c) $\left(p, -p, \frac{p}{\sqrt{2}}\right)$ (d) $\left(0, 0, \frac{p}{\sqrt{2}}\right)$

   [2016 : 1 Mark, Set-3]

11. The state of stress at a point is $\sigma_x = \sigma_y = \sigma_z = \tau_{xz} = \tau_{zx} = \tau_{yx} = \tau_{zy} = 0$ and $\tau_{xy} = \tau_{yx} = 50$ MPa. The maximum normal stress (in MPa) at that point is ______.

[2017 : 1 Mark, Set-2]

12. The State of stress at a point, for a body in place stress, is shown in the figure below. If the minimum principal stress is 10 kPa, then the normal stress σ_s (in kPa) is

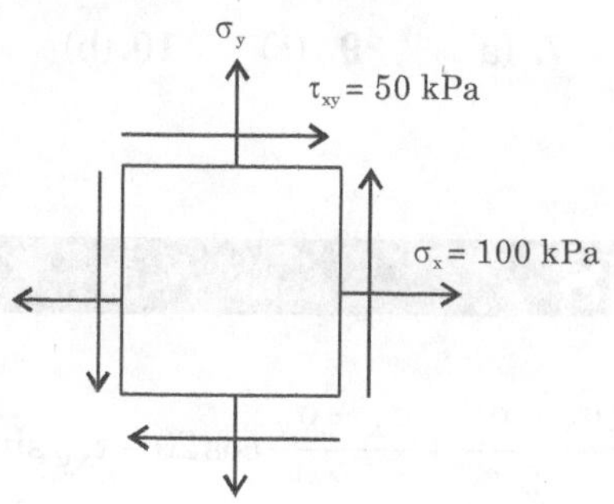

(a) 9.45 (b) 18.88

(c) 37.78 (d) 75.50

[2018 : 2 Marks, Set-1]

Principal strain

13. If the two principal strains at a point are 1000×10^{-6} and -600×10^{-6}, then the maximum shear strain is

(a) 800×10^{-6} (b) 500×10^{-6}

(c) 1600×10^{-6} (d) 200×10^{-6}

[1996 : 1 Mark]

Theories of failure

14. According to Von-Mises' distortion energy theory, the distortion energy under three dimensional stress state is represented by

(a) $\frac{1}{\varepsilon E}[\sigma_1^2 + \sigma_2^2 + \sigma_3^2 - 2m(\sigma_1\sigma_2 + \sigma_3\sigma_2 + \sigma_1\sigma_3)]$

(b) $\frac{1-2\mu}{6E}[\sigma_1^2 + \sigma_2^2 + \sigma_3^2 - 2m(\sigma_1\sigma_2 + \sigma_3\sigma_2 + \sigma_1\sigma_3)]$

(c) $\frac{1+\mu}{3E}[\sigma_1^2 + \sigma_2^2 + \sigma_3^2 - 2m(\sigma_1\sigma_2 + \sigma_3\sigma_2 + \sigma_1\sigma_3)]$

(d) $\frac{1}{3E}[\sigma_1^2 + \sigma_2^2 + \sigma_3^2 - 2m(\sigma_1\sigma_2 + \sigma_3\sigma_2 + \sigma_1\sigma_3)]$

[2006 : 2 Marks]

15. Match the following criteria of material failure, under biaxial stresses σ_1 and σ_2 and yield stress σ_y, with their corresponding graphic representations:

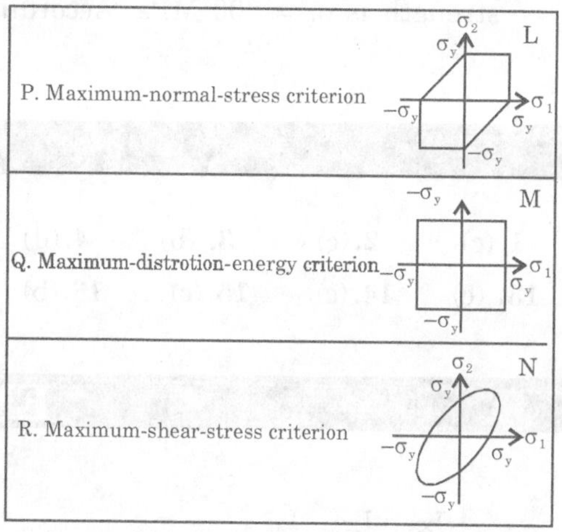

(a) P–M, Q–L, R–N (c) P–M, Q–N, R–L

(b) P–N, Q–M, R–L (d) Q–N, Q–L, R–M

[2011 : 1 Mark]

16. The homogenous state of stress for a metal part undergoing plastic deformation is

$$T = \begin{pmatrix} 10 & 5 & 0 \\ 5 & 20 & 0 \\ 0 & 0 & -10 \end{pmatrix}$$

Where the stress component values are in MPa. Using Von Mises yield criterion, the value of estimated shear yield stress, in MPa is

(a) 9.50 (b) 16.07

(c) 28.52 (d) 49.41

[2012 : 2 Marks]

17. The principal stresses at a point inside a solid object are $\sigma_1 = 100$ MPa, $\sigma_2 = 100$ MPa and $\sigma_3 = 0$ MPa. The yield strength of the material is 200 MPa. The factor of safety calculated using Tresca (maximum shear stress) theory is n_T and the factor of safety calculated using Von Mises(maximum distortional energy) theory is n_v. Which one of the following relations is TRUE?

(a) $n_T = \left(\frac{\sqrt{3}}{2}\right)n_v$ (b) $n_T = \left(\sqrt{3}\right)n_v$

(c) $n_T = n_v$ (d) $n_v = \left(\sqrt{3}\right)n_T$

[2016 : 2 Marks, Set-1]

18. In a metal forming operation when the material has just started yielding, the principal stresses $\sigma_1 = +180$ MPa, $\sigma_2 = -100$ MPa, $\sigma_3 = 0$. Following von Mises criterion, the yield stress is _____, MPa.

[2017 : 1 Mark, Set-1]

19. The principal stresses at a point in a critical - section of a machine component are $\sigma_1 = 60$ MPa, $\sigma_2 = 5$ MPa and $\sigma_3 = -40$ MPa. For the material of the component, the tensile yield strength is $\sigma_y = 200$ MPa. According to the maximum shear stress theory, the factor of safety is

(a) 1.67 (b) 2.00

(c) 3.60 (d) 4.00

[2017 : 2 Marks, Set-2]

ANSWERS

1. (c)	2. (c)	3. (b)	4. (d)	5. (d)	6. (c)	7. (a)	9. (c)	10. (b)	12. (c)
13. (c)	14. (c)	15. (c)	16. (b)	17. (c)	19. (b)				

EXPLANATIONS

2. $\sigma = \begin{bmatrix} \sigma_{xx} & \tau_{xy} & \tau_{xz} \\ \tau_{yx} & \sigma_{yy} & \tau_{yz} \\ \tau_{zx} & \tau_{zy} & \sigma_{zz} \end{bmatrix}$

3.

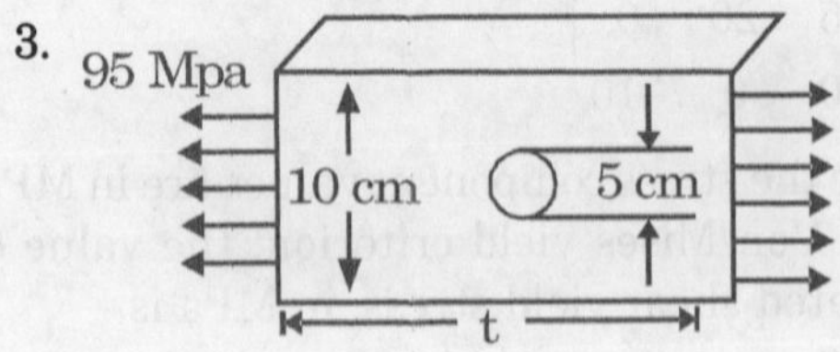

$\sigma = \frac{P}{A}$

$P = \sigma A = 95\times10^6\times10\times10^{-2}t$

$\sigma = \frac{P}{A} = \frac{95\times10^6\times10\times10^{-2}t}{(10\times10^{-2}-5\times10^{-2})t}$

$\sigma = 190$ MPa

4.

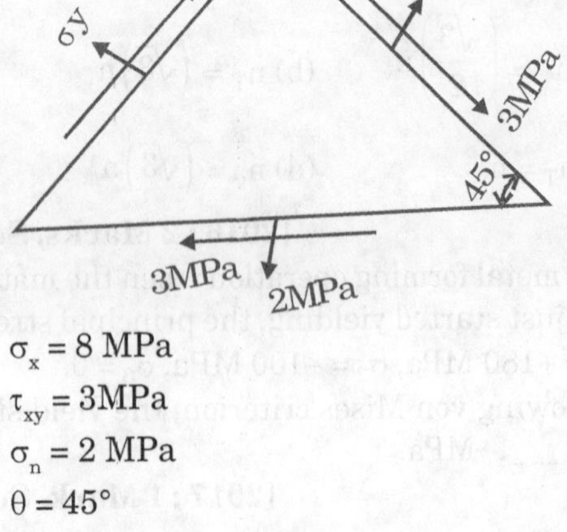

$\sigma_x = 8$ MPa

$\tau_{xy} = 3$MPa

$\sigma_n = 2$ MPa

$\theta = 45°$

$$\sigma_n = \frac{\sigma_x + \sigma_y}{2} + \frac{\sigma_x - \sigma_y}{2}\cos 2\theta - \tau_{xy}\sin 2\theta$$

$$2 = \frac{8+\sigma_y}{2} + \frac{8-\sigma_y}{2}(0) - 3\sin(90)$$

$(\sigma_y = 2\text{MPa})$

$$\sigma_{1,2} = \frac{\sigma_x + \sigma_y}{g} + \sqrt{\left(\frac{\sigma_x - \sigma_y}{2}\right)^2 + \tau^2 xy}$$

$$= \frac{8+2}{2} \pm \sqrt{\left(\frac{8-2}{2}\right)^2 + 3^2}$$

$= 5 \pm 4.242$

$\sigma_{1/2} = 9.242, 0.758$ MPa

5. $\sigma_1 = \frac{\sigma_x + \sigma_y}{2} + \sqrt{\left(\frac{\sigma_x - \sigma_y}{2}\right)^2 + \tau_{xy}^2}$

$\sigma_x = \sigma_y = 0, \tau_{zy} = \tau$

$\sigma_1 = \tau$

$\sigma_1 = 2\tau\sin(45)\cos(45) = \tau$

6. Maximum shear stress,

$$\tau_{max} = \frac{\sigma_1 - \sigma_2}{2}$$

$$= \frac{100-40}{2} = \frac{60}{2} = 30 \text{ M Pa}$$

7. $\sigma_x = 200\text{MPa}\ \sigma_y = 100\text{MPa}\ z_{xy} = 100\text{MPa}$

$$\tau_{max} = \sqrt{\left(\frac{\sigma_x - \sigma_y}{2}\right)^2 + \tau^2_{xy}}$$

$$= \sqrt{\left(\frac{200-100}{2}\right)^2 + 100^2} = \sqrt{50^2 + 100^2}$$

$= 111.80$ MPa.

8. $$\sigma_1 = \frac{\sigma_x + \sigma_y}{2} + \sqrt{\left(\frac{\sigma_x - \sigma_y}{2}\right)^2 + (\tau_{xy})^2}$$

$$= \frac{-6+4}{2} + \sqrt{\left(\frac{-6-4}{2}\right)^2 + (-8)^2} = 8.43.$$

9. $$\tau_{max} = \sqrt{\left(\frac{\sigma_x - \sigma_y}{2}\right)^2 + \tau^2_{xy}}$$

$$= \sqrt{\left(\frac{20-80}{2}\right)^2 + 40^2} = 50\text{MPa}$$

10. 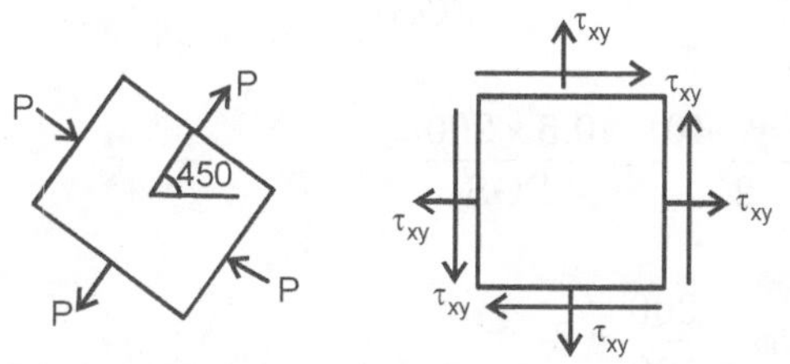

$$\sigma_\theta = \left(\frac{\sigma_x + \sigma_y}{2}\right) + \left(\frac{\sigma_x - \sigma_y}{2}\right)\cos 2\theta + \tau_{xy}\sin 2\theta$$

Here $\theta = -45$

$\sigma_\theta = Txy$

$\sigma_x = p$

$\sigma_y = -p$

$$\sigma_\theta = T_{xy} = \left(\frac{P-P}{2}\right) + \left(\frac{P+P}{2}\right)\cos 90°$$

$= 0$

When $\theta = +45$

$$\sigma_\theta = \left(\frac{P-P}{2}\right) + \left(\frac{P+P}{2}\right)\cos 90°$$

When $\theta = 45°$, $T_\theta = T_{xy}$

$$T_\theta = \left(\frac{\sigma_x - \sigma_y}{2}\right)\sin 2\theta - T_{xy}\cos 2\theta$$

$$T_\theta = T_{xy} = \left(\frac{P+P}{2}\right)\sin 90° = P$$

$\therefore\ T_{xx}, T_{yy}, T_{xy} = 0, 0, p$

11. The state of stress is pure shear

$\therefore\ \sigma_1 = -\sigma_2 = \tau = 50$ MPa.

$\sigma_{max} = 50$ MPa

12. $\sigma_x = 100$ kPa, $\tau_{xy} = 50$ kPa

Minimum principal stress

$$\sigma_{1,2} = \frac{\sigma_x + \sigma_y}{2} \pm \sqrt{\left(\frac{\sigma_x - \sigma_y}{2}\right)^2 + \tau^2_{xy}}$$

$$10 = \frac{100 + \sigma_y}{2} - \sqrt{\left(\frac{100 - \sigma_y}{2}\right)^2 + 50^2}$$

$$\therefore\ \sqrt{\left(50 - \frac{\sigma_y}{2}\right)^2 + 50^2} = 50 + \frac{\sigma_y}{2} - 10 = 40 + \frac{\sigma_y}{2}$$

By squaring

$$2500 + \frac{\sigma_y^2}{4} - 50\sigma_y + 2500 = 1600 + \frac{\sigma_y^2}{4} + 40\sigma_y$$

$\therefore\ 90\ \sigma_y = 3400$

$\sigma_y = 37.78$ MPa

13. 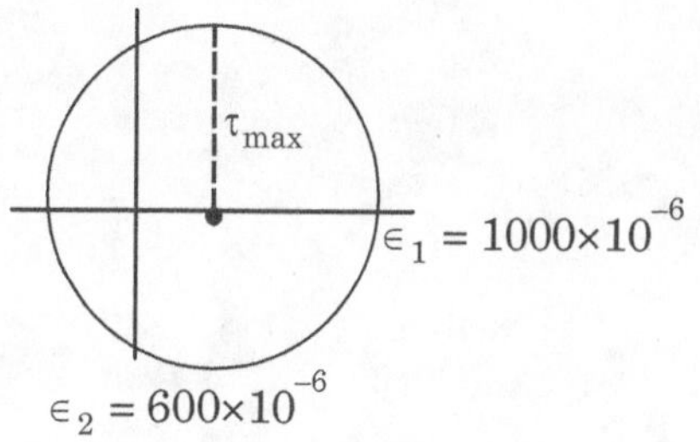

$$\tau_{max} = \frac{\phi_{max}}{2}, \text{radius} = \frac{\epsilon_1 - \epsilon_2}{2}$$

$\phi_{max} = \epsilon_1 = \epsilon_2$

$= 1000\times10^{-6} - (-600\times10^{-6})$

$\phi_{max} = 1600\times10^{-6}$

14. $\frac{1+\mu}{3E}\left(\sigma_1^2+\sigma_2^2+\sigma_3^2-(\sigma_1\sigma_2+\sigma_2\sigma_3+\sigma_3\sigma_1)\right)$

16. $\sigma_{eq} = \sqrt{\frac{1}{2}\left\{(\sigma_{11}-\sigma_{22})^2+(\sigma_{11}-\sigma_{22})^2+6\left[\sigma_{12}^2+\sigma_{23}^2+\sigma_{13}^2\right]\right\}}$

We know, $\sigma_{11} = 10, \sigma_{22} = 20,$

$\sigma_{33} = -10;$

$\sigma_{12} = 5;$

$\sigma_{23} = \sigma_{13} = 0$

$\therefore \quad \sigma_{eq} = 27.839$ MPa

Shear stress at yield, $\tau_y = \frac{\sigma_{eq}}{\sqrt{3}} = 16.07$ MPa

17. $\sigma_1 = 100$ MPa $\sigma_2 = 100$ MPa, $\sigma_3 = 0$ MPa

$S_{yt} = 200$ MPa

→ Tresca theory

$$\tau_{max} = \frac{0.5S_{ut}}{FOS}$$

$$\frac{\sigma_1-\sigma_3}{2} = \frac{0.5S_{yt}}{n_T}$$

$$n_T = \frac{0.5\times2\times200\times_2^6}{100\times10^6} = 2$$

→ Von mises theory

$$\sqrt{\frac{1}{2}(\sigma_1-\sigma_2)^2+(\sigma_2-\sigma_3)^2+(\sigma_3-\sigma_1)} = \frac{S_{yt}}{FOS}$$

$$\frac{1}{2}\left(0+100^2+100^2\right) = \frac{200\times10^6}{n_V}$$

$$n_V = \frac{200\times10^6}{100\times10^6} = 2$$

$\therefore n_T = n_V$

18. $\sigma_1 = 180$ MPa, $\sigma_2 = -100$MPa, $\sigma_3 = 0$

$$\sigma_1^2+\sigma_2^2-\sigma_1\sigma_2 \le \sigma_{yt}{}^2$$

$$\sigma_{yt} = \sqrt{180^2+100^2+180\times100}$$

$\sigma_{yt} = 245.76$ MPa.

19. $\sigma_1 = 60$ MPa, $\sigma_2 = 5$MPa $\sigma_3 = -40$ MPa

$$\tau_{max} = \frac{0.5S_{yt}}{FOS}$$

$$\tau_{max} = \frac{\sigma_1-\sigma_2}{2} = \frac{0.5S_{ut}}{FOS}$$

$$\frac{60-(-40)}{2} = \frac{0.5\times200}{FOS}$$

$$FOS = \frac{200}{100} = 2$$

3 CHAPTER

Mohr's Circle

Analysis of Previous GATE Papers

Topic	Year/Set → Marks Distribution ↓		2019 Set 1	2019 Set 2	2018 Set 1	2018 Set 2	2017 Set 1	2017 Set 2	2016 Set 1	2016 Set 2	2016 Set 3	2015 Set 1	2015 Set 2	2015 Set 3
Mohr's circle	1 Mark	MCQ Type												
		Numerical Type		1										
	2 Marks	MCQ Type												
		Numerical Type												
	Total			1										

Mohr's circle for plane stress

Common Data Questions Q.1 and Q.2

The state of stress at a point P in a two dimensional loading is such that the Mohr's circle is a point located at 175 MPa on the positive normal stress axis.

1. The maximum and minimum principal stresses respectively from the Mohr's circle are

(a) +175 MPa,-175 MPa

(b) +175 MPa,+175 MPa

(c) 0,-175 MPa

(d) 0, 0

[2003 : 2 Marks]

2. The directions of maximum and minimum principal stresses at the point P from the Mohr's circle are

(a) 0,90° (b) 90°, 0

(c) 45°, 135° (d) All directions

[2003 : 2 Marks]

3. The figure shows the state of stress at a certain point in a stressed body. The magnitudes of normal stresses in the x and y directions are 100 MPa and 20 MPa respectively. The radius of Mohr's stress circle representing this state of stress is

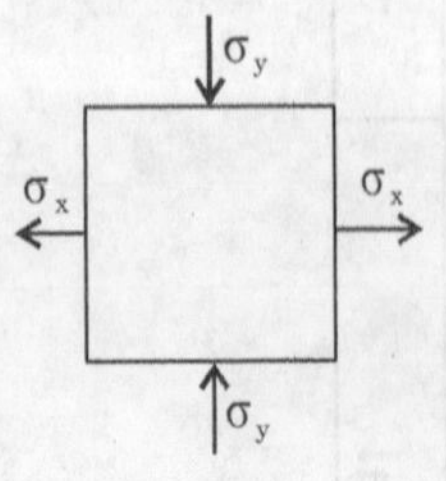

(a) 120 MPa (b) 80 MPa

(c) 60 MPa (d) 40 MPa

[2004 : 1 Mark]

4. The Mohr's circle of plane stress for a point in a body is shown. The design is to be done on the basis of the maximum shear stress theory for yielding. Then, yielding will just begin if the designer chooses a ductile material whose yield strength is

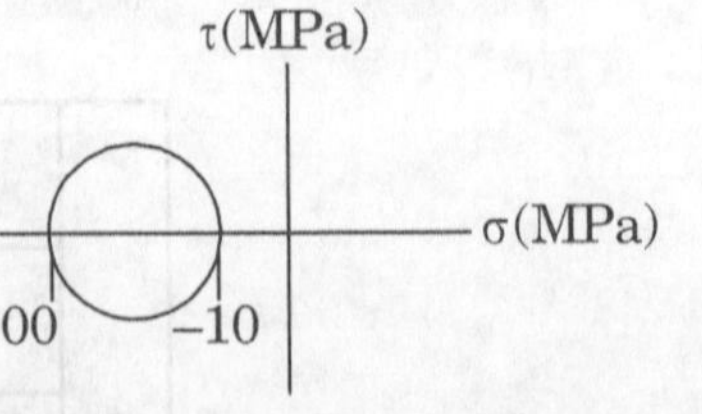

(a) 45 MPa (b) 50 MPa

(c) 90 MPa (d) 100 MPa

[2005 : 2 Marks]

5. A two dimensional fluid element rotates like a rigid body. At a point within the element, the pressure is 1 unit. Radius of the Mohr's circle, characterizing the state at that point, is

(a) 0.5 unit (b) 0 unit

(c) 1 unit (d) 2 units

[2008 : 2 Marks]

6. The state of stress at a point under plane stress condition is

σ_{xx} = 40 MPa, σ_{yy} = 100 MPa and τ_{xy} = 40 MPa

The radius of Mohr's circle representing the given state of stress in MPa is

(a) 40 (b) 50

(c) 60 (d) 100

[2012 : 2 Marks]

7. The state of stress at a point in a component is represented by a Mohr's circle of radius 100 MPa centered at 200 MPa on the normal stress axis. On a plane passing through the same point, the normal stress is 260 MPa. The magnitude of the shear stress on the same plane at the same point is ____ MPa.

[2019 : 1 Mark, Set-2]

ANSWERS

1. (b) 2. (d) 3. (c) 4. (c) 5. (b) 6. (b)

EXPLANATIONS

1. Mohr's Circle equation

$$\left(\sigma - \frac{\sigma_x + \sigma_y}{2}\right)^2 + (\tau)^2 = \left(\sqrt{\left(\frac{\sigma_x - \sigma_y}{2}\right)^2 + (\tau_{xy})^2}\right)^2$$

where σ = principle stress; τ = shear stress

According to question,

$$\frac{\sigma_x + \sigma_y}{2} = 175$$

$\Rightarrow \sigma_x = \sigma_y = 175$ MPa

2. Since Mohr's circle is a point, hence the direction of maximum and minimum principal stress of point P is in all direction.

3. $r = \dfrac{\sigma_x + \sigma_y}{2}$

$$= \frac{100 + 20}{2} = 60 \text{ MPa}$$

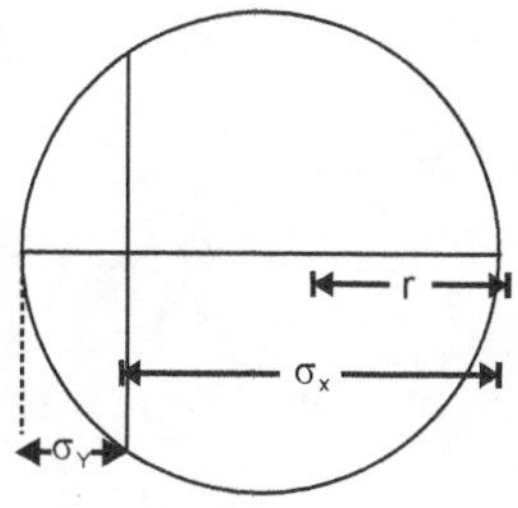

$\therefore$ Radius of Mohr's circle = 60 MPa

4. Applying maximum shear stress theory

$$\tau_{max} = \frac{\sigma_{Y.P}}{2} = \frac{\sigma_x - \sigma_y}{2}$$

$$\therefore \quad \sigma_{Y.P.} = \sigma_x - \sigma_y = -100 - (-10)$$

$$= -90 \text{ MPa}$$

5. Since the pressure in fluid is of hydrodynamic type

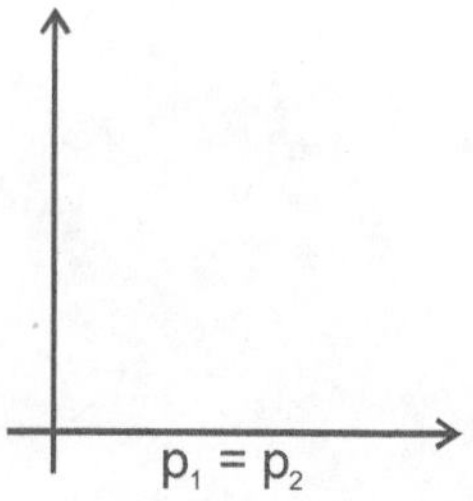

$\therefore p_1 = p_2 = p_3 = 1$

Hence radius of Mohr's circle is zero.

6. Here, $\sigma_{xx} = 40$ MPa; $\sigma_{yy} = 100$ MPa,

$\tau_{xy} = 40$ MPa

$$\text{Radius of Mohr's circle} = \sqrt{\left[\frac{\sigma_{xx} - \sigma_{yy}}{2}\right]^2 + (\tau_{xy})^2}$$

$$= \sqrt{30^2 + 40^2} = 50 \text{ MPa}$$

7.

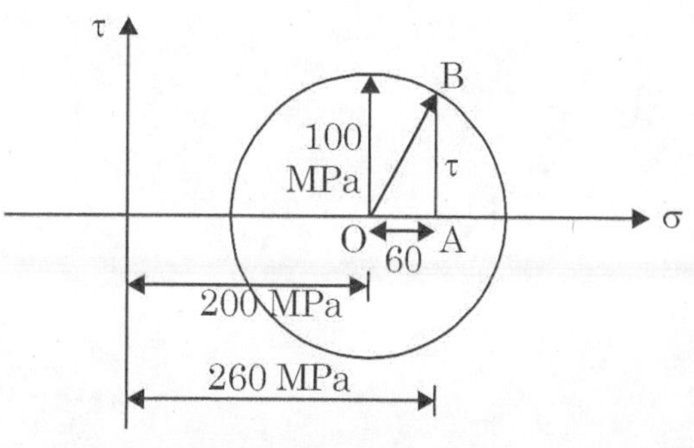

In triangle OAB,

$OB^2 = OA^2 + AB^2$

$\Rightarrow 100^2 = 60^2 + AB^2$

$\Rightarrow AB^2 = 6400$

AB = 80 MPa

$\because$ AB represents shear stress at the same point.

4

CHAPTER

Thin Cylinder

Analysis of Previous GATE Papers

Topic	Marks Distribution ↓	Year/Set →	2019 Set 1	2019 Set 2	2018 Set 1	2018 Set 2	2017 Set 1	2017 Set 2	2016 Set 1	2016 Set 2	2016 Set 3	2015 Set 1	2015 Set 2	2015 Set 3
Thin Cylindrical pressure vessel	1 Mark	MCQ Type											1	
		Numerical Type										1		
	2 Marks	MCQ Type		1										
		Numerical Type												
	Total			2								1	1	
Thin Spherical pressure vessel	1 Mark	MCQ Type												
		Numerical Type												
	2 Marks	MCQ Type												
		Numerical Type												
	Total													

Thin cylindrical pressure vessel

1. A thin cylinder of 100 mm internal diameter and 5 mm thickness is subjected to an internal pressure of 10 MPa and a torque of 2000 Nm. Calculate the magnitudes of the principal stresses.

[1996 : 2 Marks]

2. A thin cylinder of inner radius 500 mm and thickness 10 mm subjected to an internal pressure of 5 MPa. The average circumferential (hoop) stress in MPa is

(a) 100 (b) 250
(c) 500 (d) 1000

[2011 : 1 Mark]

3. A long thin walled cylindrical shell, closed at both the ends, is subjected to an internal pressure. The ratio of the hoop stress (circumferential stress) to longitudinal stress developed in the shell is

(a) 0.5 (b) 1.0
(c) 2.0 (d) 4.0

[2013 : 1 Mark]

4. A thin gas cylinder with an internal radius of 100 mm is subject to an internal pressure of 10 MPa. The maximum permissible working stress is restricted to 100 MPa. The minimum cylinder wall thickness (in mm) for safe design must be ______.

[2014 : 2 Marks, Set-4]

5. A gas is stored in a cylindrical tank of inner radius 7 m and wall thickness 50 mm. The gauge pressure of the gas is 2 MPa. The maximum shear stress (in MPa) in the wall is

(a) 35 (b) 70
(c) 140 (d) 280

[2015 : 1 Mark, Set-2]

6. A cylindrical tank with closed ends is filled with compressed air at a pressure of 500 kPa. The inner radius of the tank is 2 m, and it has wall thickness of 10 mm. The magnitude of maximum in-plane shear stress (in MPa) is ______.

[2015 : 1 Mark, Set-3]

7. Consider two concentric circular cylinders of different materials M and N in contact with each other at r = b, as shown below. The interface at r = b is frictionless. The composite cylinder system is subjected to internal pressure P. Let $\left(u_r^M, u_\theta^M\right)$ and $\left(\sigma_{rr}^M, \sigma_{\theta\theta}^M\right)$ denote the radial and tangential displacement and stress components, respectively, in material M. Similarly $\left(u_r^N, u_\theta^N\right)$ and $\left(\sigma_{rr}^N, \sigma_{\theta\theta}^N\right)$ denote the radial and tangential displacement and stress components, respectively, in material N. The boundary condition that need to be satisfied at the frictionless interface between the two cylinders are :

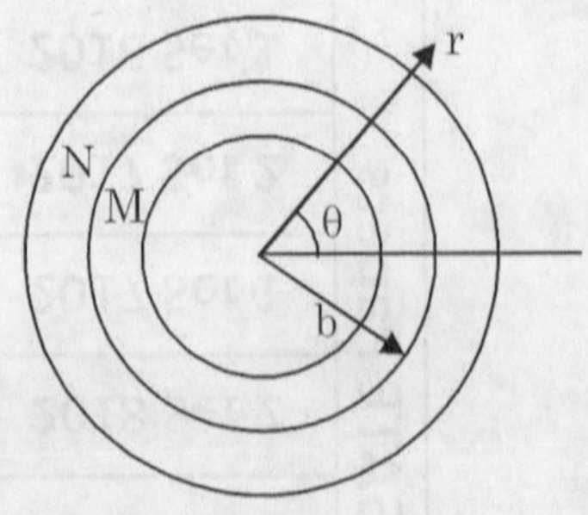

(a) $u_r^M = u_r^N$ and $\sigma_{rr}^M = \sigma_{rr}^N$ and $u_\theta^M = u_\theta^N$ and $\sigma_{\theta\theta}^M = \sigma_{\theta\theta}^N$

(b) $u_\theta^M = u_\theta^N$ and $\sigma_{\theta\theta}^M = \sigma_{\theta\theta}^N$ only

(c) $u_r^M = u_r^N$ and $\sigma_{rr}^M = \sigma_{rr}^N$ only

(d) $\sigma_{rr}^M = \sigma_{rr}^N$ and $\sigma_{\theta\theta}^M = \sigma_{\theta\theta}^N$ only

[2019 : 2 Marks, Set-2]

Thin spherical pressure vessel

8. A thin walled spherical shell is subjected to an internal pressure. If the radius of the shell is increased by 1% and the thickness is reduced by 1%, with the internal pressure remaining the same, the percentage change in the circumferential (hoop) stress is

(a) 0 (b) 1
(c) 1.08 (d) 2.02

[2012 : 1 Mark]

ANSWERS

2. (b) 3. (c) 5. (c) 7. (c) 8. (d)

EXPLANATIONS

1. $d_i = 0.1$ m, $d_o = di + 2t$

$t = 0.005$ m, $d_o = 0.11$ m

T = 2000 Nm

P = 10 MPa

$$\sigma_c = \frac{pd_i}{4t} = \frac{10 \times 10^6 \times 0.1}{4 \times 0.005} = 50\text{MPa}$$

$$\sigma_c = \frac{pd_i}{2t} = \frac{10 \times 10^6 \times 0.1}{2 \times 0.005} = 100\text{MPa}$$

$$\frac{T}{J} = \frac{\tau}{r}$$

$$\tau = \frac{Tr}{J}$$

$$= \frac{2000 \times \frac{0.11}{2}}{\frac{\pi}{32}\left(0.11^4 - 0.14\right)}$$

$\tau = 24.14$ MPa

$$\sigma_{1,2} = \frac{\sigma_x + \sigma_y}{2} \pm \sqrt{\left(\frac{\sigma x - \sigma y}{2}\right)^2 + \tau_{xy}^2}$$

$$= \frac{50+100}{2} \pm \sqrt{\left(\frac{50-100}{2}\right)^2 + 24.14^2}$$

$= 75 \pm 34.75$

= 109.75 & 40.25 MPa

2. Given : $p = 5$ MPa

$d = 1000$ mm

$t = 10$ mm

Hoop stress, $\sigma_{Hoop} = \dfrac{pd}{2t}$

= 250 MPa

3. $\sigma_h = \dfrac{PD}{2t}$ where P = internal pressure

$\sigma_l = \dfrac{PD}{4t}$, D = inner dia of cylinder.

t = thickness of cylinder.

$$\therefore \frac{\sigma_h}{\sigma_l} = \frac{\frac{PD}{2t}}{\frac{PD}{4t}} = 2$$

4. $\sigma_1 = \dfrac{pd}{2t}$

$$\Rightarrow 100 = \frac{100 \times 2 \times 100}{2 \times t}$$

$\Rightarrow t = 10$ mm

5. $\sigma_c = \dfrac{pd}{2t} = \dfrac{2 \times 14}{2 \times 0.05} = 280\text{MPa}$

$$\sigma_1 = \frac{pd}{4t} = \frac{2 \times 14}{4 \times 0.05} = 140\text{MPa}$$

Maximum shear stress,

$$\tau_{max} = \frac{\sigma_1 - \sigma_2}{2} = \frac{\sigma_c - \sigma_1}{2} = 70 \text{ MPa}$$

Note : This stress τ_{max} is the maximum in-plane shear stress.

Absolute τ_{max} will be $\dfrac{\sigma_c}{2} = 140$ MPa

6. Maximum in-plane shear stress

$$\tau_{max} = \frac{pd}{8t}$$

$$= \frac{500 \times 4}{8 \times 10}\text{MPa}$$

= 25 MPa

7. 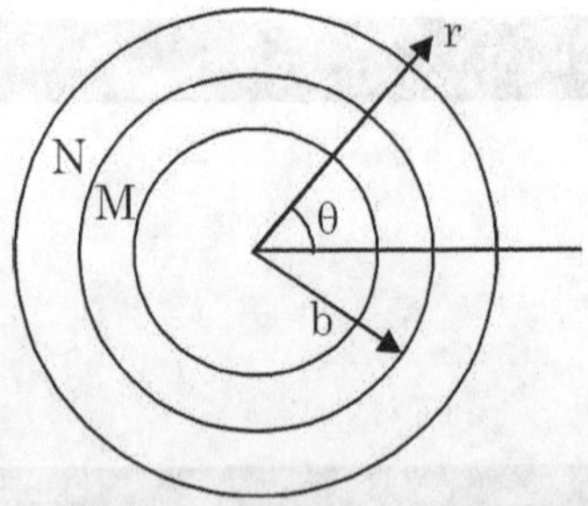

Due to internal pressure inside the cylinder, stresses are generated both in radial and circumferential direction. As interface of r = b is frictionless, so stress in radial direction must be equal otherwise relative motion will occur and parallely velocity in radial direction should be equal.

i.e. $u_r^M = u_r^N$ and $\sigma_{rr}^M = \sigma_{rr}^N$

In circumferential direction i.e. only 'θ' direction velocity cannot be equal because it depends upon radius i.e. V = rω and 'r' is varying.

So, $u_\theta^M \neq u_\theta^N$

Same is for stress in circumferential direction.

i.e. $\sigma_{\theta\theta}^M \neq \sigma_{\theta\theta}^N$

8. Hoop stress for a thin spherical shell

$$\sigma_h = \frac{Pr}{2t}$$

Applying logarithm on both sides, we get

$$\log(\sigma_h) = \log\left(\frac{P}{2}\right) + \log(r) - \log(t)$$

Differentiating above equation,

$$\frac{d\sigma_h}{\sigma_h} = 0 + \frac{dr}{r} - \frac{dt}{t}$$

Given : $\frac{dr}{r} = 0.01$ and $\frac{dt}{t} = -0.01$

Substituting, we get

$$\frac{d\sigma_h}{\sigma_h} = 0.02$$

Hence percentage increase will be 2%.

5 CHAPTER

Shear-force and Bending Moment Diagrams

Analysis of Previous GATE Papers

Topic	Year/Set → Marks Distribution ↓		2019 Set 1	2019 Set 2	2018 Set 1	2018 Set 2	2017 Set 1	2017 Set 2	2016 Set 1	2016 Set 2	2016 Set 3	2015 Set 1	2015 Set 2	2015 Set 3
Shear Force Diagram	1 Mark	MCQ Type												
		Numerical Type						1						
	2 Marks	MCQ Type												
		Numerical Type												
	Total							1						
Bending Moment Diagram	1 Mark	MCQ Type												
		Numerical Type												
	2 Marks	MCQ Type												
		Numerical Type											1	1
	Total												2	2

Shear force diagram

1. A block of steel is loaded by a tangential force on its top surface while the bottom surface is held rigidly. The deformation of the block is due to

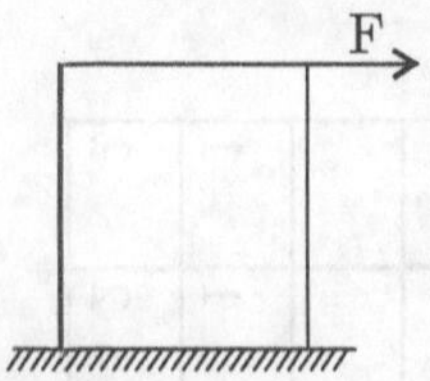

(a) shear only
(b) bending only
(c) shear and bending
(d) torsion

[1992 : 1 Mark]

2. A mass of 35 kg is suspended from a weightless bar AB which is supported by a cable CB and a pin at A as shown in figure. The pin reactions at A on the bar AB are

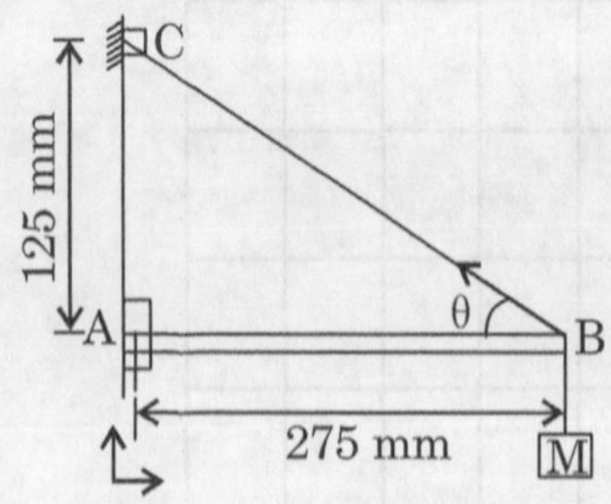

(a) R_x = 343.4 N, R_y =755.4 N
(b) R_x = 343.4 N, R_y = 0
(c) R_x = 755.4 N, R_y = 343.4 N
(d) R_x = 755.4 N, R_y = 0

[1997 : 2 Marks]

3. A uniformly loaded propped cantilever beam and its free body diagram are shown below. The reactions are

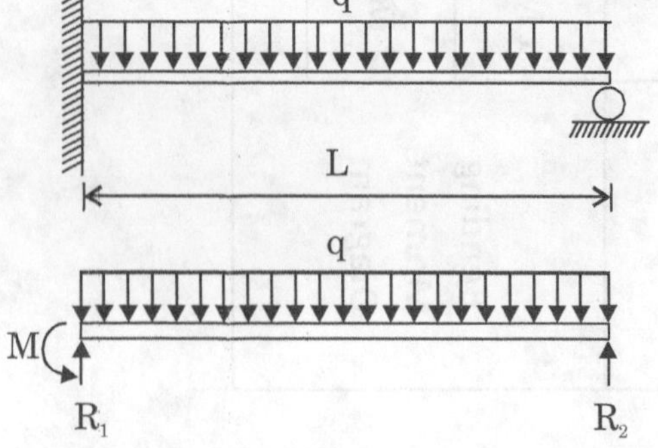

(a) $R_1 = \frac{5qL}{8}, R_2 = \frac{3qL}{8}, M = \frac{qL^2}{8}$

(b) $R_1 = \frac{3qL}{8}, R_2 = \frac{5qL}{8}, M = \frac{qL^2}{8}$

(c) $R_1 = \frac{5qL}{8}, R_2 = \frac{3qL}{8}, M = 0$

(d) $R_1 = \frac{3qL}{8}, R_2 = \frac{5qL}{8}, M = 0$

[2007 : 2 Marks]

4. A simply supported beam PQ is loaded by a moment of 1 kNm at the mid-span of the beam as shown in the figure. The reaction forces R_P and R_Q at supports P and Q respectively are

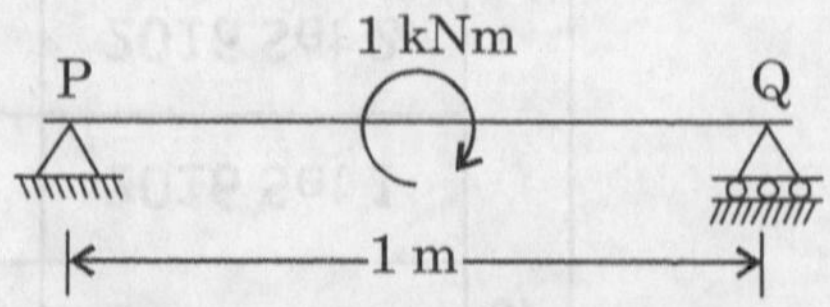

(a) 1 kN downward, 1 kN upward
(b) 0.5 kN upward, 0.5 kN downward
(c) 0.5 kN downward, 0.5 kN upward
(d) 1 kN upward, 1 kN upward

[2011 : 1 Mark]

5. A simply supported beam of length L is subjected to a varying distributed load sin (3πx/L) Nm^{-1}, where the distance x is measured from the left support. The magnitude of the vertical reaction force in N at the left support is

(a) zero
(b) L/3π
(c) L/π
(d) 2 L/π

[2013 : 2 Marks]

6. For a loaded cantilever beam of uniform cross-section, the bending moment (in Nmm) along the length is $M(x) = 5x^2 + 10x$, where x is the distance (in mm) measured from the free end of the beam. The magnitude of shear force (in N) in the cross- section at x = 10 mm is _____.

[2017 : 1 Mark, Set-2]

Bending moment diagram

7. For a simply supported beam on two end supports, the bending moment is maximum

(a) usually on the supports
(b) always at mid span
(c) where there is no shear force
(d) where the deflection is maximum

[1989 : 1 Mark]

8. A concentrated load P acts on a simply supported beam of span L at a distance L/3 from the left support. The bending moment at the point of application of the load is given by

(a) $\frac{PL}{3}$ (b) $\frac{2PL}{3}$

(c) $\frac{PL}{9}$ (d) $\frac{2PL}{9}$

[1993 : 1 Mark]

9. A concentrated load P acts on a simply supported beam of span L at a distance L/3 from the left support. The bending moment at the point of application of the load is given by

(a) $\frac{PL}{3}$ (b) $\frac{2PL}{3}$

(c) $\frac{PL}{9}$ (d) $\frac{2PL}{9}$

[2003 : 1 Mark]

Common Data for Q. 10 and 11

A steel beam of breadth 120 mm and height 750 mm is loaded as shown in the figure. Assume E_{steel} = 200 GPa.

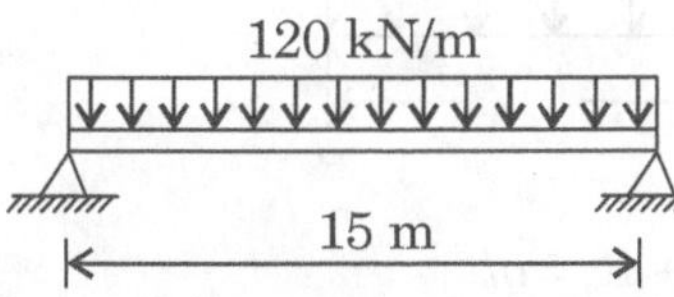

10. The beam is subjected to a maximum bending moment of

(a) 3375 kNm (b) 4750 kNm

(c) 6750 kNm (d) 8750 kNm

[2004 : 2 Marks]

11. The value of maximum deflection of the beam is

(a) 93.75 mm (b) 83.75 mm

(c) 73.75 mm (d) 63.75 mm

[2004 : 2 Marks]

12. A cantilever beam carries the antisymmetric load shown, where W_0 is the peak intensity of the distributed load. Qualitatively, the correct bending moment diagram for this beam is

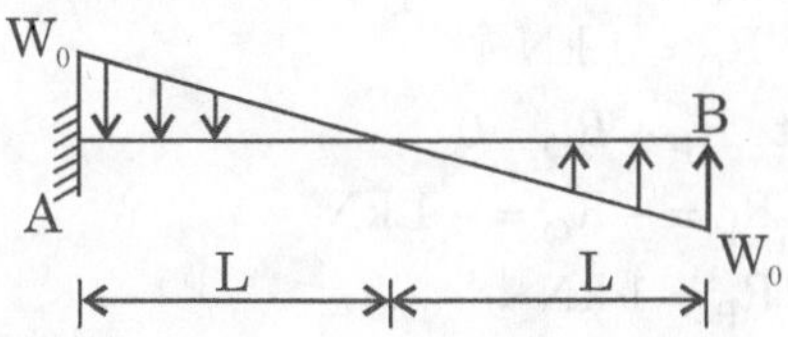

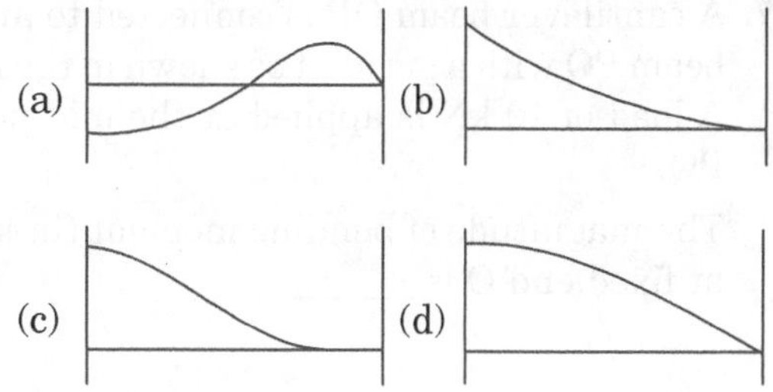

[2005 : 2 Marks]

13. A beam is made up of two identical bars AB and BC, by hinging them together at B. The end A is built-in (cantilevered) and the end C is simply-supported. With the load P acting as shown, the bending moment at A is

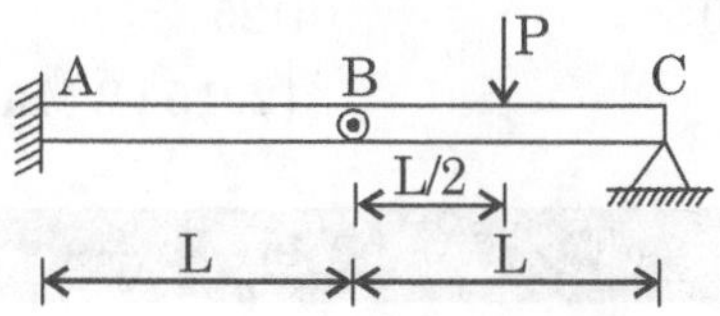

(a) Zero (b) PL/2

(c) 3PL/2 (d) Indeterminate

[2005 : 2 Marks]

Linked Data Questions 14 and 15:

A simply supported beam of span length 6 m and 75 mm diameter carries a uniformly distributed load of 1.5 kN/m.

14. What is the maximum value of bending moment?

(a) 9 kNm (b) 13.5 kNm

(c) 6.75 kNm (d) 125 kNm

[2006 : 2 Marks]

15. What is the maximum value of bending stress?

(a) 162.98 MPa (b) 325.95 MPa

(c) 625.95 MPa (d) 651.90 MPa

[2006 : 2 Marks]

16. In a simply-supported beam loaded as shown below, the maximum bending moment in Nm is

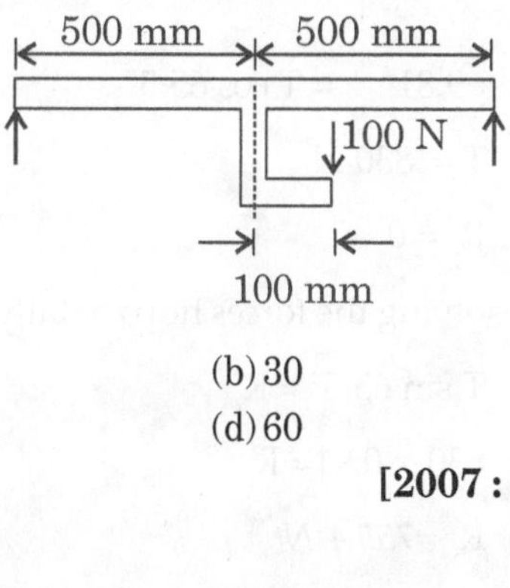

(a) 25 (b) 30

(c) 35 (d) 60

[2007 : 1 Mark]

17. A cantilever beam OP is connected to another beam PQ with a pin joint as shown in the figure. A load of 10 kN is applied at the mid-point of PQ.

The magnitude of bending moment (in kN–m) at fixed end O is ________

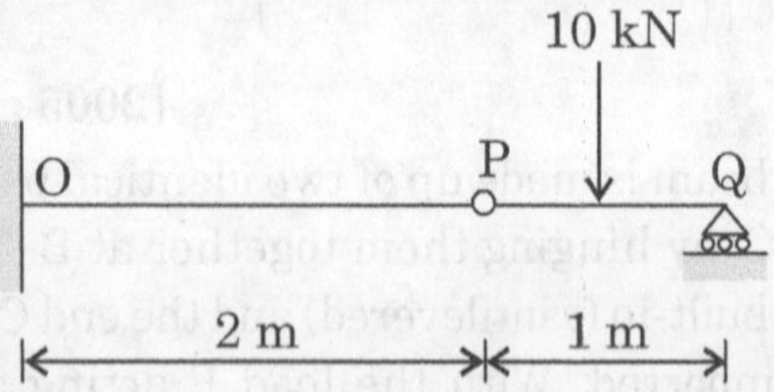

(a) 2.5 (b) 5
(c) 10 (d) 25

[2015 : 2 Marks, Set-2]

18. For the overhanging beam shown in figure, the magnitude of maximum bending moment (in kNm) is ________.

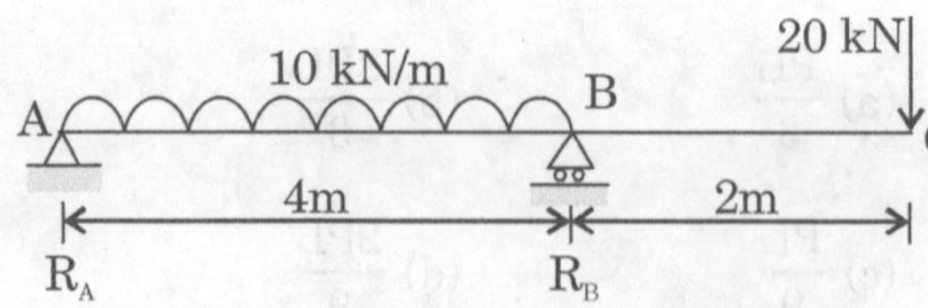

[2015 : 2 Marks, Set-3]

ANSWERS

1. (c) **2.** (d) **3.** (a) **4.** (a) **5** (b) **7.** (c) **8.** (d) **9.** (d) **10.** (a) **11.** (a)
12. (c) **13.** (b) **14.** (c) **15.** (a) **16.** (b) **17.** (c)

EXPLANATIONS

2. $\cos\theta = \dfrac{125}{302}$

$\Rightarrow \theta = 65.5°$

Resolving the forces vertically,

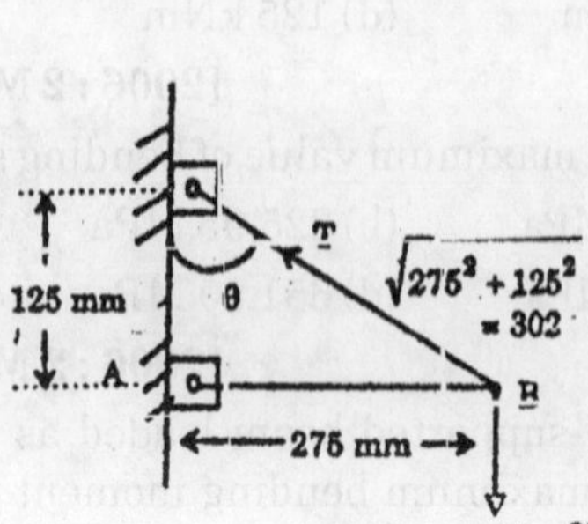

$35 \times 9.81 = T\cos 65.5°$

or $T = 830$ N

$R_y = 0$

Resolving the forces horizontally,

$T\sin 65.5° = R_x$

or $830 \times 0.91 = R_x$

or $R_x = 755.4\ N$

3.

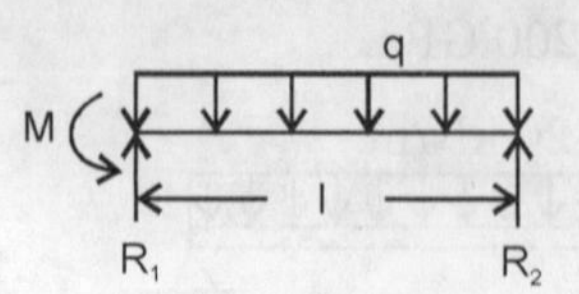

$R_1 + R_2 = ql$ ...(*i*)

Moment about (1),

$R_2 l + M - \dfrac{ql^2}{2} = 0$ ...(*ii*)

Moment about (2),

$R_1 l - \dfrac{ql^2}{2} - M = 0$ ...(*iii*)

From (*i*), (*ii*), (*iii*), we get $R_1 = \dfrac{5ql}{8}$,

$R_2 = \dfrac{3ql}{8}$ and $M = \dfrac{ql^2}{8}$

4. Take moments about 'Q'

$R_Q \times 1 - 1 = 0$

$\Rightarrow R_Q = 1\text{ kN}\uparrow$

But $R_P + R_Q = 0$

$\therefore R_P = -R_Q = -1$ kN

$\Rightarrow R_P = 1\text{ kN}\downarrow$

5.

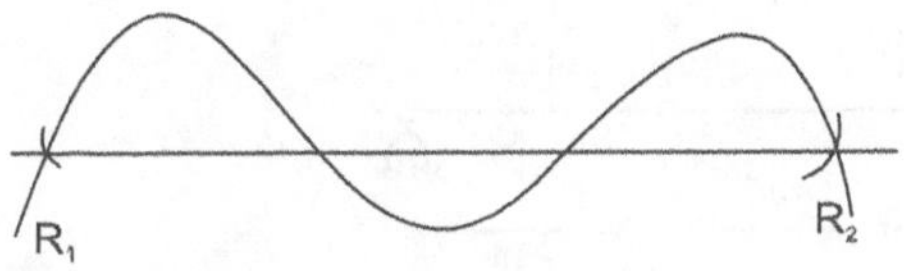

$$\text{Total load} = \int_0^L \sin\left(\frac{3\pi x}{L}\right) dx$$

$$= -\frac{L}{3\pi}\left[\cos\left(\frac{3\pi x}{L}\right)\right]_0^L = \frac{2L}{3\pi}$$

By symmetry, $R_1 = R_2 = P/2$

6. $M(x) = 5x^2+10x$

$$\text{Shear force}(F) = \frac{dM(x)}{dx}$$

$F_x = 10x+10$

$F_{(x=10)} = 10(10) + 10 = 110N$

7. Bending moment is maximum where shear force is zero.

8.

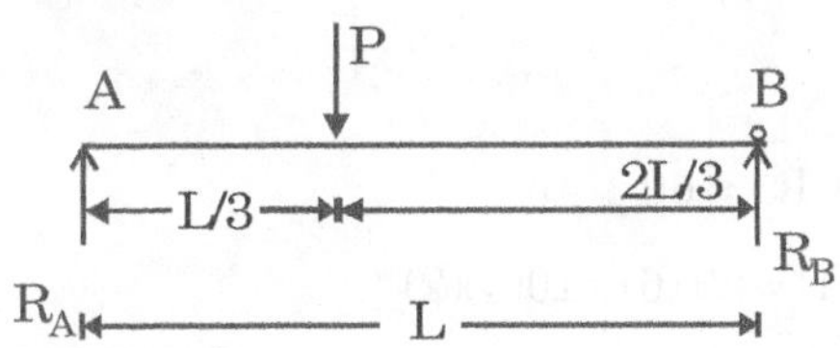

$R_A + R_B = P$

$$R_A = \frac{2P}{3}$$

$$R_B = \frac{P}{3}$$

$BM_x = R_A\, x - P\,(x-L/3)$

$$BM_{x=L/3} = R_A\frac{L}{3} - P\left(\frac{L}{3}-\frac{L}{3}\right) = R_A\frac{L}{3}$$

$$BM = \frac{2PL}{9}$$

9. For horizontal equilibrium

$R_1 + R_2 = P$

Taking moment about R_1, we have

$R_2 \times L = P \times L/3$

$$R_2 = \frac{P}{3}$$

$$\therefore \quad R_1 = \frac{2P}{3}$$

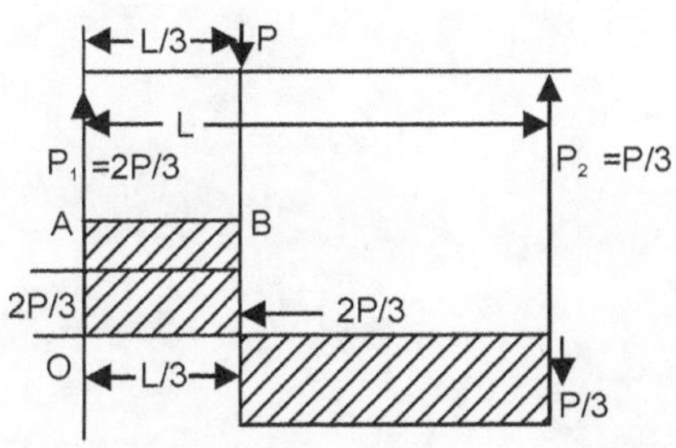

The bending moment at the point of application of load = Area under OABC

$$= \frac{2P}{3}\times\frac{L}{3} = \frac{2PL}{9}$$

10. Maximum bending moment occurs at the centre

$$= \frac{wL^2}{8} = \frac{120 \times (15)^2}{8} = 3375 \text{ kNm}$$

11. Moment of inertia

$$I = \frac{bh^3}{12} = \frac{120 \times (750)^3}{12} = 42.1875 \times 10^8 \text{ mm}^4$$

Maximum deflection in the beam is given by

$$\delta = \frac{5}{384}\frac{wL^4}{EI}$$

$$= \frac{5}{384} \times \frac{120 \times 10^3}{10^3} \times \frac{(15 \times 10^3)^4}{200 \times 10^3 \times 42.1875 \times 10^8}$$

$= 93.75$ mm

12.

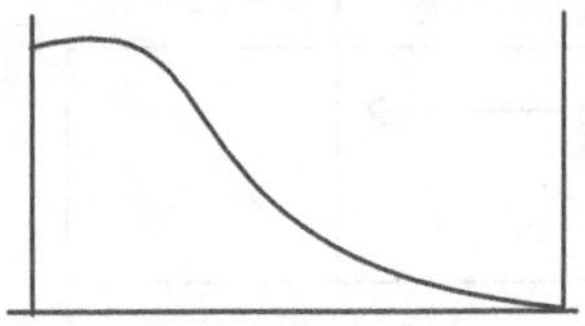

Here at both the ends, slope is zero and SF is zero. As moving form left to right, rate of shear force decreases due to varying uniform distributed load.

13.

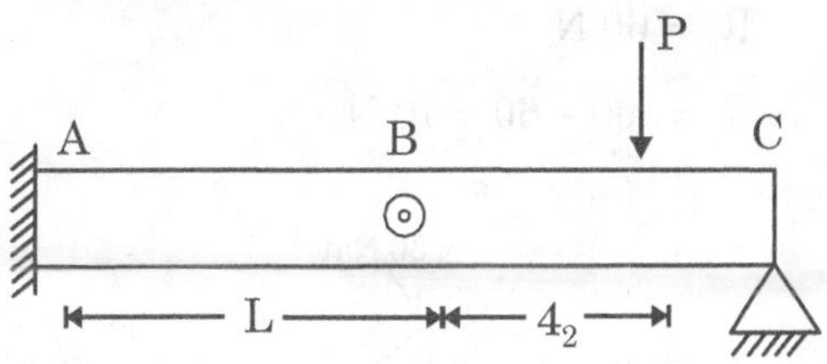

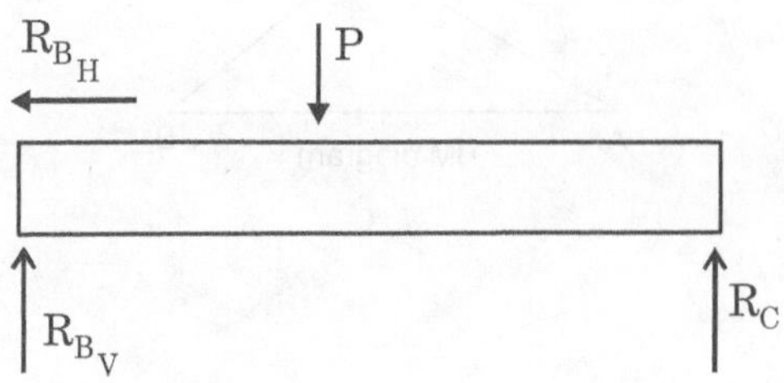

Consider Bar BC

$\Sigma Mc = 0$

$P\,(L/2) = R_{BV}(L)$

$R_{BV} = P/Z$

Now, moment at A

$$M_A = R_{B_V} XL = \frac{PL}{2}$$

14. $l = 6\text{m}, w = 1.5\text{KN/m}$

$$M = \frac{wl^2}{8} = \frac{1.5 \times 10^3 \times 6^2}{8} = 6.75\text{kNm}$$

15. Bending stress

$$\sigma_b = \frac{\text{(Maximum load} \times \text{distance) or Bending moment}}{\text{Section modulus}}$$

$$\sigma_b = \frac{9 \times 10^3 (\text{N}m)}{\frac{\pi}{32} \times (.075)^3} = 211.26 \text{ MPa}$$

Hence closest option is 162.98 MPa

16.

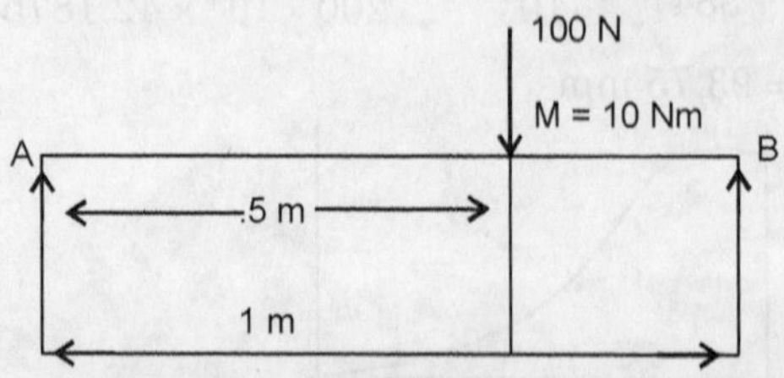

Taking moment about A,

$100 \times .5 + 10 = R_B$

$R_B = 60$ N

$\therefore$ $R_A = 100 - 60 = 40$ N

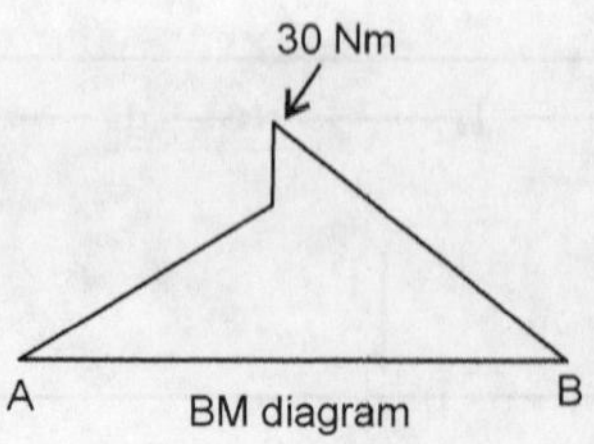

17.

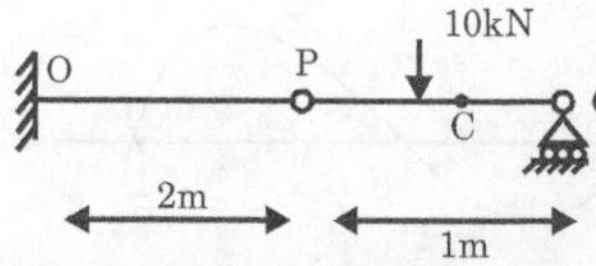

P-being internal hinge

$\Sigma M_P = 0$ ...(1)

Condition 'PQ', reaction at Q i.e. $R_Q = 10$ kN

Now from (1), $\Sigma M_P = 0$

$M_0 - R_Q \times 1 + R_C \times 0.5 = 0$

$\Rightarrow M_0 = 10\text{kN}$

18.

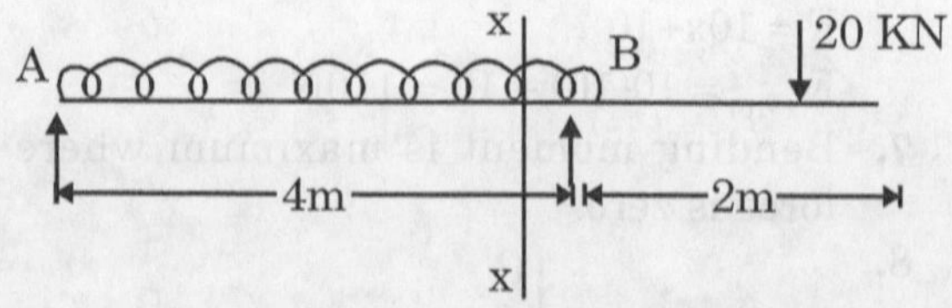

$R_A + R_B = 60$

$R_B (4) = 20 (6) + 10(4)(2)$

$R_B = 50\text{CN}$

$R_A = 10$ KN

$SF = R_A - 10x$

$SF_{x=0} = 10\text{KN}\ SF_{x=4} = -30\text{KN}$

Bending moment is maximum where shear force is zero

$\therefore$ $SF = R_A\ 10x = 0$

$R_A = 10x$

$x = 1$ m

$$BM_{x-x} = R_A x - \frac{10x^2}{2}$$

$$BM_{x=1} = 10(1) - \frac{10(1)^2}{2} = 5\text{kNm}$$

6 CHAPTER

Bending of Beams

Topic	Year/Set → Marks Distribution ↓		2019 Set 1	2019 Set 2	2018 Set 1	2018 Set 2	2017 Set 1	2017 Set 2	2016 Set 1	2016 Set 2	2016 Set 3	2015 Set 1	2015 Set 2	2015 Set 3
Bending stress in beam and MOI	1 Mark	MCQ Type					1							
		Numerical Type		1										
	2 Marks	MCQ Type							1					1
		Numerical Type	1		1				2					
	Total		2	1	2		1		5					2
Shear Stress in beam	1 Mark	MCQ Type									1			
		Numerical Type	1											
	2 Marks	MCQ Type												
		Numerical Type												
	Total		1								1			

Table: Analysis of Previous GATE Papers

Bending stress in beam and moment of Inertia

1. A 6 meter long supported wooden beam of rectangular section 10 cm × 20 cm deep is strengthened by mild steel plates 0.5 cm x 10 cm wide at the top and bottom fibre over the entire length. Find the minimum supportable uniformly distributed load considering failures in steel and wood due to flexure. Weakening of wood due to screws and weakening of the steel plates due to drilled holes may be ignored
Permissible tensile stress for steel
= 156.8 N/mm^2
Permissible tensile stress for wood
= 14.89 N/mm^2
Young's modulus of mild steel
= 1.96 × 10^5 N/mm^2
Young's modulus of wood
= 0.117 × 10^5 N/mm^2

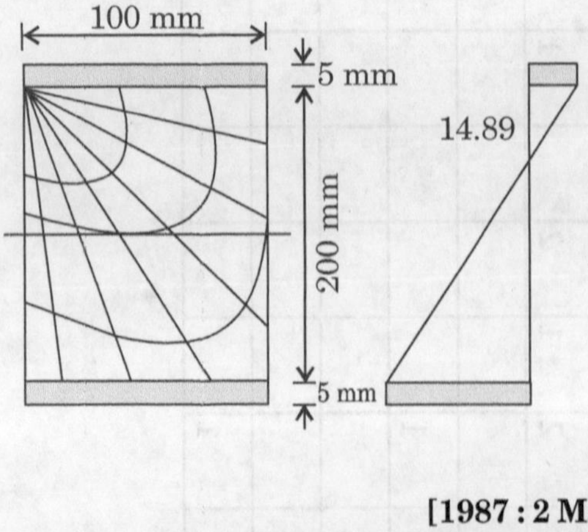

[1987 : 2 Marks]

2. A tapered cantilever beam of constant thickness is loaded as shown in the sketch below. The bending stress will be

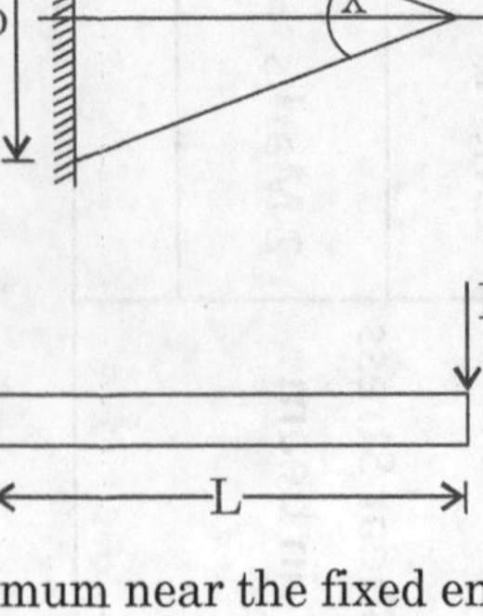

(a) maximum near the fixed end

(b) maximum at x = $\frac{1}{2}$L

(c) maximum at x = $\frac{2}{3}$L

(d) uniform throughout the length

[1988 : 2 Marks]

3. Which one of the following diagrams shows correctly the distribution of transverse shear stress across the depth h of a rectangular beam subjected to varying bending moment along its length?

(a) 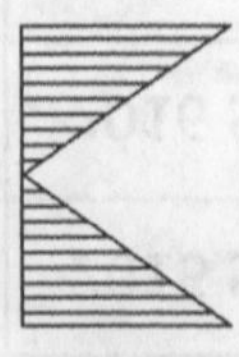 (b) 

(c)  (d) 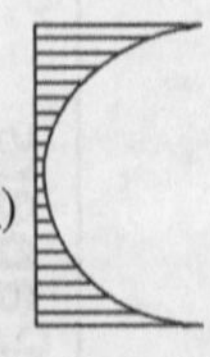

[1990 : 1 Mark]

4. The second moment of a circular area about the diameter is given by (D is the diameter)

(a) $\frac{\pi D^4}{4}$ (b) $\frac{\pi D^4}{16}$

(c) $\frac{\pi D^4}{32}$ (d) $\frac{\pi D^4}{64}$

[2003 : 1 Mark]

5. The beams, one having square cross-section and another circular cross-section, are subjected to the same amount of bending moment. If the cross sectional area as well as the material of both the beams are the same then

(a) maximum bending stress developed in both the beams is the same

(b) the circular beam experience more bending stress than the square one

(c) the square beam experiences more bending stress than the circular one

(d) as the material is same both beams will experience same deformation

[2003 : 1 Mark]

6. A cantilever beam has the square cross section of 10 mm × 10 mm. It carries a transverse load of 10 N. Considering only the bottom fibres of the beam, the correct representation of the longitudinal variation of the bending stress is

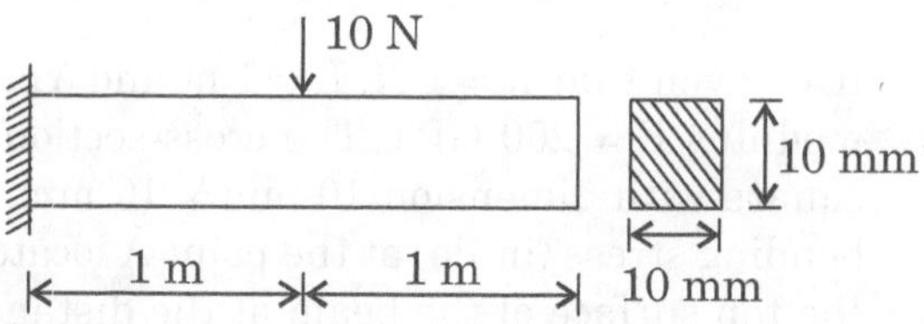

(a) 60 MPa 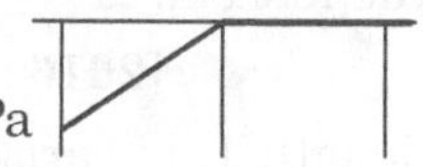

(b) 60 MPa 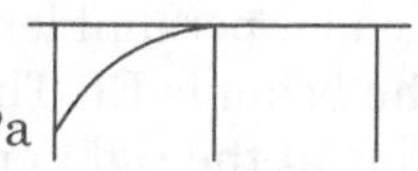

(c) 400 MPa 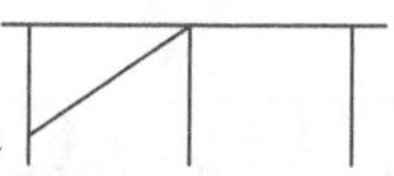

(d) 400 MPa 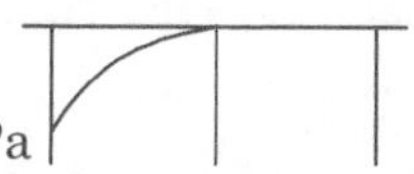

[2005 : 2 Marks]

Linked Data Questions 7 and 8

A machine frame shown in the figure below is subjected to a horizontal force of 600 N parallel to z-direction.

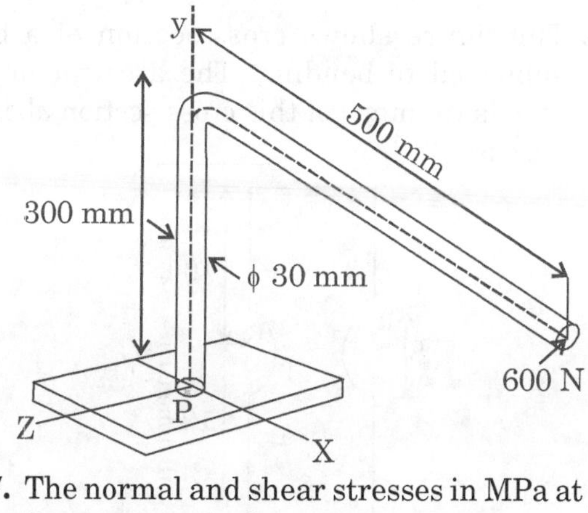

7. The normal and shear stresses in MPa at point P are respectively
   (a) 67.9 and 56.6 (b) 56.6 and 67.9
   (c) 67.9 and 0.0 (d) 0.0 and 56.6

[2007 : 2 Marks]

8. The maximum principal stress in MPa and the orientation of the corresponding principal plane in degrees are respectively
   (a) –32.0 and –29.52 (b) 100.0 and 60.48
   (c) –32.0 and 60.48 (d) 100.0 and –29.52

[2007 : 2 Marks]

Linked Data Question 9 and 10.

A massless beam has a loading pattern as shown in the figure. The beam is of rectangular cross-section with a width of 30 mm and height of 100 mm.

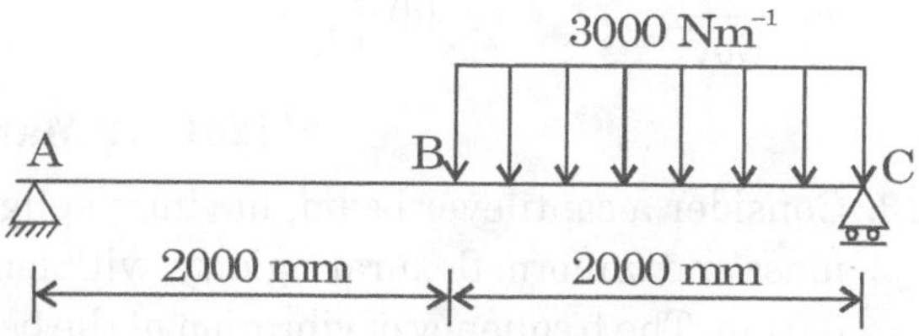

9. The maximum bending moment occurs at
   (a) location B
   (b) 2675 mm to the right of A
   (c) 2500 mm to the right of A
   (d) 3225 mm to the right of A

[2010 : 2 Marks]

10. The maximum magnitude of bending stress (in MPa) is given by
    (a) 60.0 (b) 67.5
    (c) 200.0 (d) 225.0

[2010 : 2 Marks]

Linked Data Question 11 and 12.

A triangular-shaped cantilever beam of uniform-thickness is shown in the figure. The Young's modulus of the material of the beam is E. A concentrated load P is applied at the free end of the beam.

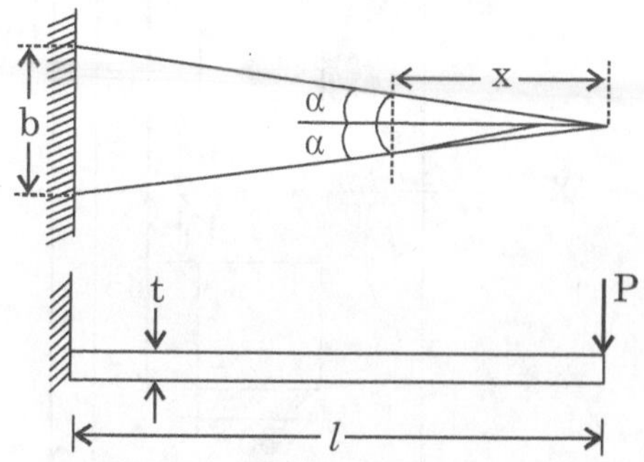

11. The area moment of inertia about the neutral axis of a cross-section at a distance x measured from the free end is

(a) $\frac{bxt^3}{6l}$ (b) $\frac{bxt^3}{12l}$

(c) $\frac{bxt^3}{24l}$ (d) $\frac{xt^3}{12l}$

[2011 : 2 Marks]

12. The maximum deflection of the beam is

(a) $\frac{24Pl^3}{Ebt^3}$ (b) $\frac{12Pl^3}{Ebt^3}$

(c) $\frac{8Pl^3}{Ebt^3}$ (d) $\frac{6Pl^3}{Ebt^3}$

[2011 : 2 Marks]

13. Consider a cantilever beam, having negligible mass and uniform flexural rigidity, with length 0.01 m. The frequency of vibration of the beam, with a 0.5 kg mass attached at the free tip, is 100 Hz. The flexural rigidity (in Nm^2) of the beam is ______.

[2014 : 2 Marks, Set-1]

14. Consider a simply supported beam of length, 50h, with a rectangular cross-section of depth, h, and width, 2h. The beam carries a vertical point load, P, at its mid-point. The ratio of the maximum shear stress to the maximum bending stress in the beam is

(a) 0.02 (b) 0.10
(c) 0.05 (d) 0.01

[2014 : 2 Marks, Set-3]

15. The value of moment of inertia of the section shown in the figure about the axis-XX is

All dimensions are in mm

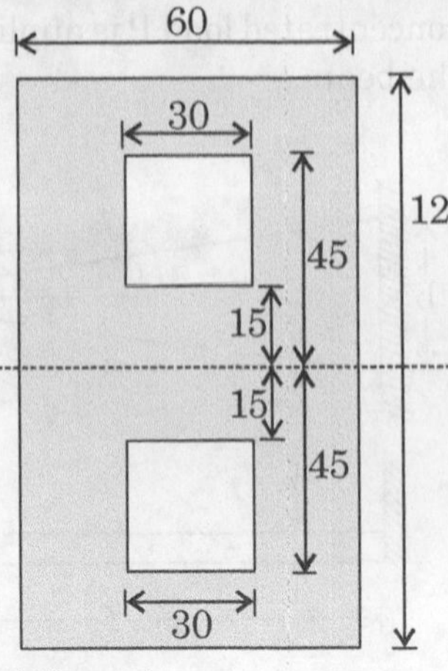

(a) 8.5050×10^6 mm^4
(b) 6.8850×10^6 mm^4
(c) 7.7625×10^6 mm^4
(d) 8.5725×10^6 mm^4

[2015 : 2 Marks, Set-3]

16. A simply supported beam of length 3L is subjected to the loading shown in the figure.

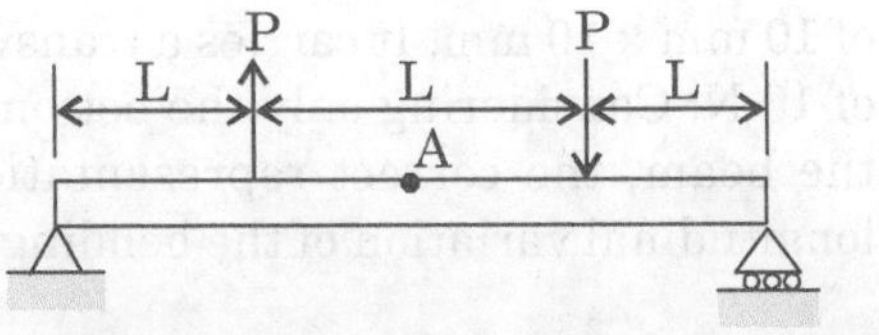

It is given that P = 1 N, L = 1 m and Young's modulus E = 200 GPa. The cross-section is a square with dimension 10 mm × 10 mm. The bending stress (in Pa) at the point A located at the top surface of the beam at the distance of 1.5 L from the left end is ______.

[2016 : 2 Marks, Set-1]

17. A beam of length L is carrying a uniformly distributed load w per unit length. The flexural rigidity of the beam is EI. The reaction at the simple support at the right end is

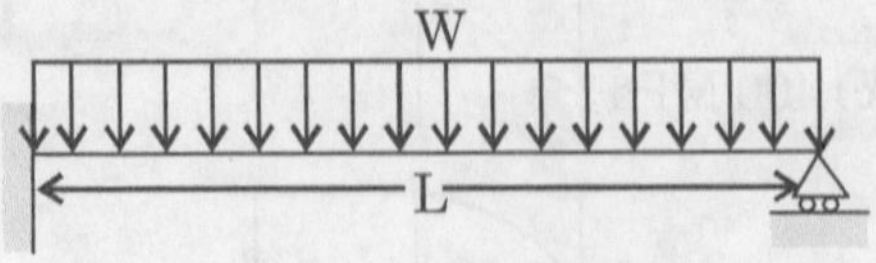

(a) $\frac{wL}{2}$ (b) $\frac{3wL}{8}$

(c) $\frac{wL}{4}$ (d) $\frac{wL}{8}$

[2016 : 2 Marks, Set-1]

18. The figure shows cross-section of a beam subjected to bending. The area moment of inertia (in mm^4) of this cross-section about its base is ______.

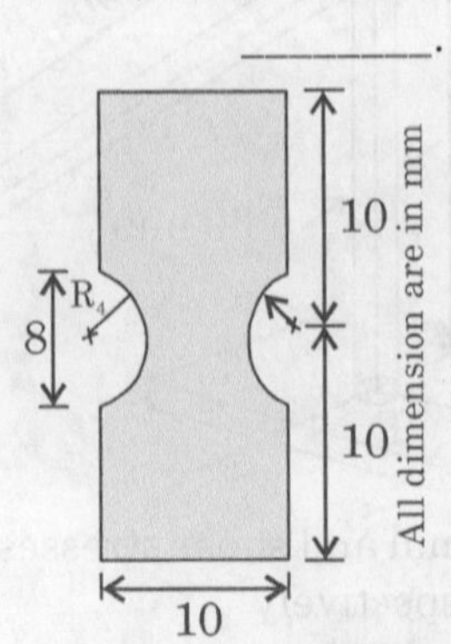

[2016 : 2 Marks, Set-1]

19. Consider a beam with circular cross-section of diameter d. The ratio of the second moment of area about the neutral axis to the section modulus of the area is

(a) $\frac{d}{2}$ (b) $\frac{\pi d}{2}$

(c) d (d) πd

[2017 : 1 Mark, Set-1]

20. A simply supported beam of width 100 mm, height 200 mm and length 4 m is carrying a uniformly distributed load of intensity 10 kNm. The maximum bending stress (in MPa) in the beam is ________ (correct to one decimal place)

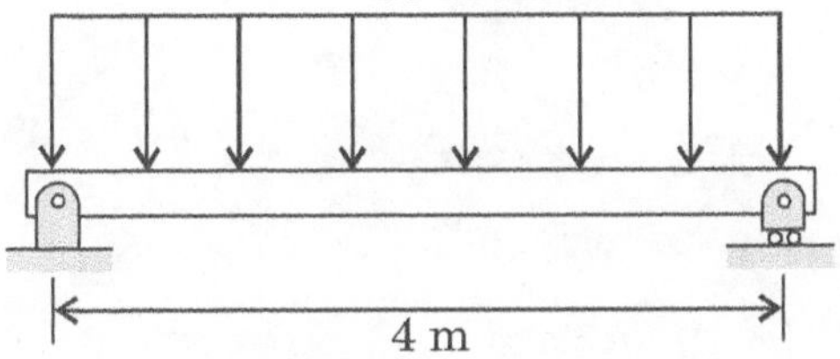

[2018 : 2 Marks, Set-1]

21. Consider an elastic straight beam of length L = 10 πm, with square cross-section of side a = 5 mm, and Young's modulus E = 200 GPa. This straight beam was bent in such a way that the two ends meet, to form a circle of mean radius R. Assuming that Euler-Bernoulli beam theory is applicable to this bending problem, the maximum tensile bending stress in the bent beam is ______ MPa.

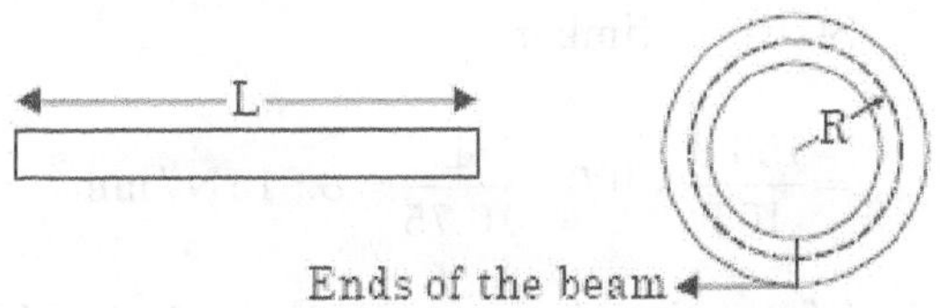

[2019 : 2 Marks, Set-1]

22. A wire of circular cross-section of diameter 1.0 mm is bent into a circular arc of radius 1.0 m by application of pure bending moments at its ends. The Young's modulus of the material of the wire is 100 GPa. The maximum tensile stress developed in the wire is ____ MPa.

[2019 : 1 Mark, Set-2]

Shear stress in Beam

23. The transverse shear stress acting in a beam of rectangular cross-section, subjected to a transverse shear load, is

(a) variable with maximum at the bottom of the beam

(b) variable with maximum at the top of the beam

(c) uniform

(d) variable with maximum on the neutral axis

[2008 : 1 Mark]

24. An axial residual compressive stress due to a manufacturing process is present on the outer surface of a rotating shaft subjected to bending. Under a given bending load, the fatigue life of the shaft in the presence of the residual compressive stress is

(a) decreased

(b) increased or decreased, depending on the external bending load

(c) neither decreased nor increased

(d) increased

[2008 : 1 Mark]

25. For the component loaded with a force F as shown in the figure, the axial stress at the corner point P is

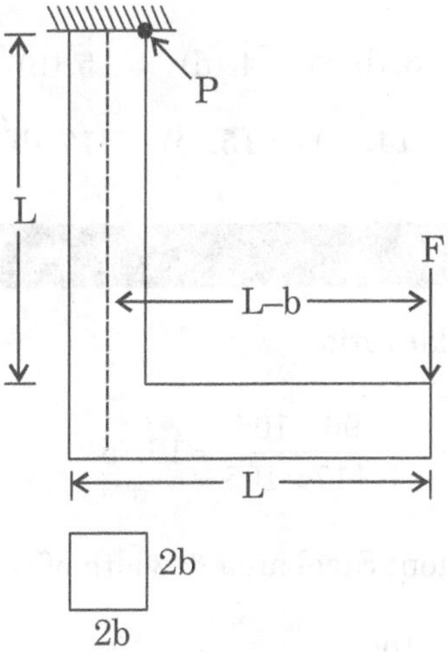

(a) $\dfrac{F(3L-b)}{4b^3}$ (b) $\dfrac{F(3L-4b)}{4b^3}$

(c) $\dfrac{F(3L-4b)}{4b^3}$ (d) $\dfrac{F(3L-2b)}{4b^3}$

[2008 : 2 Marks]

26. A solid circular shaft of diameter 100 mm is subjected to an axial stress of 50 MPa. It is further subjected to a torque of 10 kNm. The maximum principal stress experienced on the shaft is closest to

(a) 41 MPa

(b) 82 MPa

(c) 164 MPa

(d) 204 MPa

[2008 : 2 Marks]

27. The cross-sections of two solid bars made of the same material are shown in the figure. The square cross-section has flexural (bending) rigidity I_1 while the circular cross-section has flexural rigidity I_2. Both sections have the same cross-sectional area. The ratio I_1/I_2 is

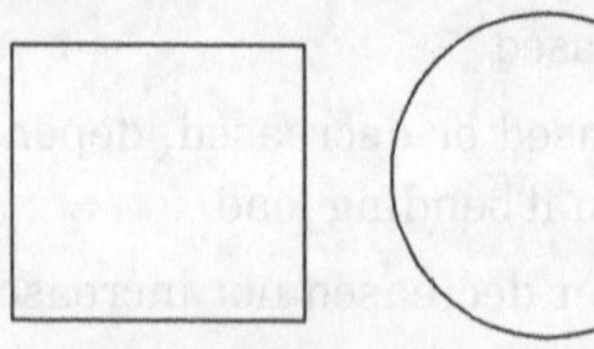

(a) $1/\pi$ (b) $2/\pi$

(c) $\pi/3$ (d) $\pi/6$

[2016 : 1 Mark, Set-3]

28. A cylindrical rod of diameter 10 mm and length 1.0 m fixed at one end. The other end is twisted by angle of 10° by applying a torque. If the maximum shear strain in the rod is $p \times 10^{-3}$, then p is equal to ______ (round off to two decimal places).

[2019 : 1 Mark, Set-1]

ANSWERS

2. (d)	**3.** (b)	**4.** (d)	**5.** (b)	**6.** (a)	**7.** (a)	**8.** (b)	**9.** (c)	**10.** (b)	**11.** (b)
12. (d)	**14.** (d)	**15.** (b)	**17.** (b)	**19.** (a)	**23.** (d)	**24.** (d)	**25.** (d)	**26.** (b)	**27.** (c)

EXPLANATIONS

1. Modular ratio

$$M = \frac{E_s}{E_t} = \frac{1.96 \times 105}{0.117 \times 105} = 16.75$$

Equivalent Steel area & width of timber

$$= \frac{200 \times 100}{16.75 \times 200} = 5.97\text{mm}$$

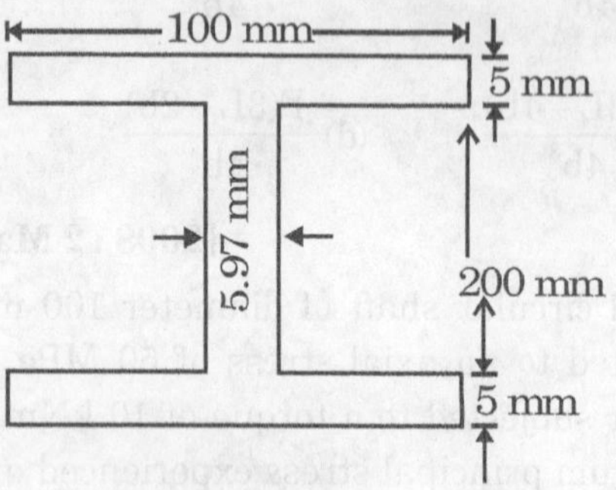

$$I = \frac{100 \times 210^3}{12}$$

$I = 14.5 \times 10^6 \text{mm}^4$

Stress in timber

$$= \frac{156.8}{105} \times 100 \times \frac{1}{16.75} = 8.915\text{N/mm}^2$$

(If permissible stress in steal = 156.8 MPa)

Stress in steel

$$= \frac{14.89}{100} \times 105 \times 16.75 = 261.88\text{N/mm}^2$$

(If permissible stress in wood = 14.8 MPa)

$I_{max} = 156.8$ MPa

$$\frac{M}{I} = \frac{\sigma}{y}$$

$$M = \frac{156.8 \times 14.5 \times 10^6}{105} = 21.65\text{Nm}$$

$$\frac{wl^2}{8} = M = 2165 \times 10^6$$

w = 4.81 kN/m.

2. 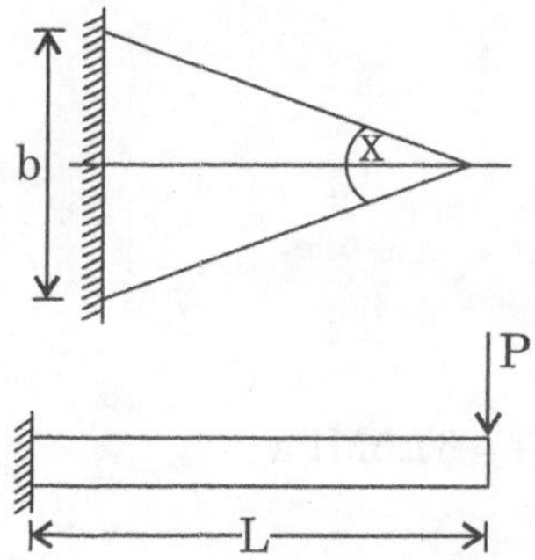

$$\frac{m}{I} = \frac{\sigma}{y}$$

$$\frac{\frac{PL}{\frac{bd^3}{12}}} = \frac{\sigma}{\frac{D}{2}}$$

$$\sigma = \frac{6PL}{bd^2}$$

3. Transverse shear stress distribution on a rectangular cross section due to varying bending moment is

$$q = \frac{F}{2l}\left[\frac{d^2}{4} - y^2\right]$$

4. Polar Moment of Inertia perpendicular to the plane of paper.

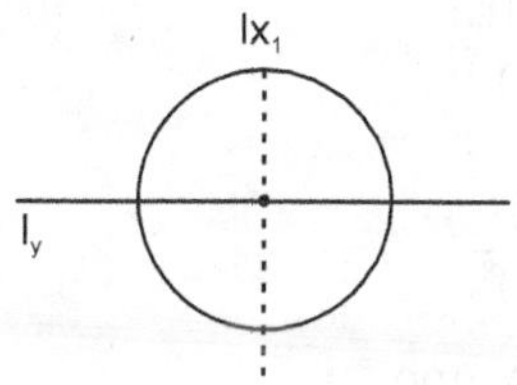

$$I_p = \frac{\pi D^4}{32}$$

Now, applying "Perpendicular Axes" Theorem.

$$I_x + I_y = I_p$$

$$\Rightarrow 2I_x = I_p$$

$$\Rightarrow I_x = \frac{I_p}{2} \; [\because I_x = I_y]$$

$$= \frac{\pi D^4}{64}$$

5. Since cross-sectional area of square and circular cross-sections are equal

Let d = diameter of circular cross-section

a_2 = side of square

$$\therefore \quad \frac{\pi}{4}d^2 = a^2$$

Now, σ for circular cross section

$$= \frac{M.C.}{I}$$

$$= \frac{M\cdot\frac{d}{2}}{M\cdot\frac{d^2}{6^4}} = \frac{32M}{\pi d^3}$$

σ for square cross section

$$= \frac{M.C.}{I} = \frac{M\frac{a}{2}}{\frac{a^4}{12}} = \frac{6M}{a^3}$$

$$\frac{\sigma \text{ for circular cross-section}}{\sigma \text{ for square cross-section}} = \frac{\frac{32M}{\pi d^3}}{\frac{6M}{d^3}}$$

$$= \frac{32M}{\pi d^3} \times \frac{a^3}{6M}$$

$$= \frac{32}{\pi}\cdot\left(\frac{a^3}{d^3}\right) = \frac{32}{\pi}\left(\sqrt{\frac{\pi}{4}}\right)^3 \cdot \frac{d^3}{d^3} \ldots$$

$$\text{Since, } a = \frac{\pi}{4}d^2 = 7$$

∴ σ for circular cross-section > σ for square cross section.

6. Here, $I = \frac{bh^3}{12} = \frac{0.01 \times 0.01^3}{12} m^4$

$y = 0.005$ m

10_x N-m

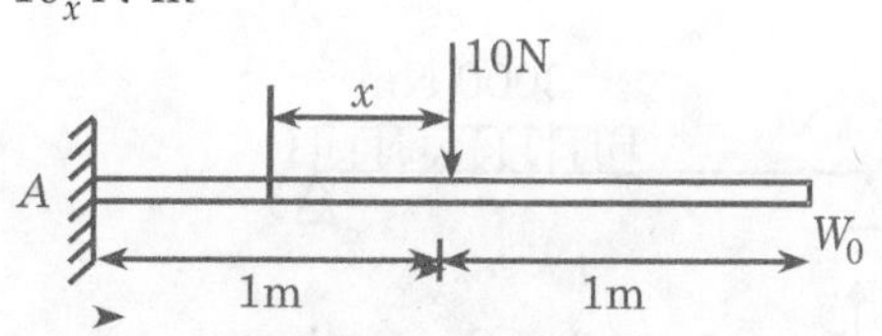

At left extreme of cantilever

$\sigma_{bending} = 60 \times 1$ MPa = 60 MPa after 1 m

$\sigma_{bending} = 60 \times 0$ MPa = 0 MPa

and it will continue over whole span.

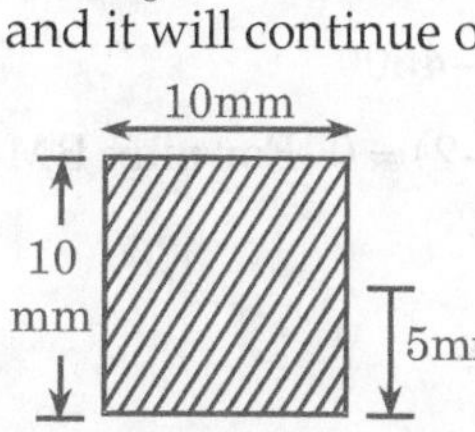

Bottom most fiber

$$\sigma_{(\text{bending stress})} = \frac{Mxy}{I} = \frac{(10x)(0.005)}{(0.01)^4/12} N/m^2$$

$\sigma_{bending}$ =60 x MPa (straight line equation)

7. Here, Twisting moment,

$T = 600 \times 500 \times 10^{-3}$

$= 300$ Nm

Bending moment,

$M = 600 \times 300 \times 10^{-3}$

$= 180$ Nm

τ_s = shear stress.

σ = Normal stress.

$$T = \frac{\pi}{16}\tau_s d^3$$

$\Rightarrow \tau_s = 56.58$ MPa

$$M = \frac{\pi}{16}\tau_s d^3 = \sigma$$

=67.9 MPa

8. Maximum principal stress

$$= \frac{\sigma - 0}{2} + \sqrt{\frac{(\sigma-0)^2 + \tau^2}{2}}$$

= 99.95 MPa.

9. R_A= 1500 N

R_B = 4500 N

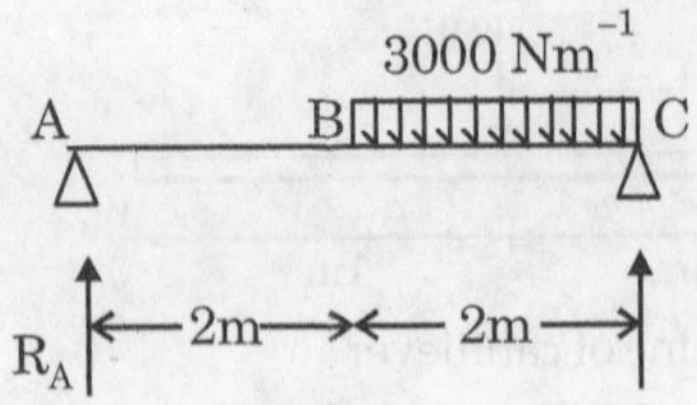

SF = R_A –3000 (x–2)

$SF_{x=2}$ = 1500 $SF_{x=4}$ = –4500

SF = 1500 – 3000 (x–2) = 0 [For max BM]

x = 2.5 m

x = 2500 mm from A.

10. $M_{X=2.5} = 1500\,(2.5) - \dfrac{3000(2.5-2)^2}{2}$

M = 3375 Nm.

$$\frac{M}{I} = \frac{\sigma}{y}$$

$$\sigma = \frac{33752 \times \frac{0.1}{2}}{\frac{0.03 \times 0.13}{12}} = 67.5\text{MPa}$$

11. At a distance of *x* from the free end width

$$b' = \frac{bx}{l}$$

$\therefore$ Moment of Inertia $I_x = \dfrac{bxt^3}{12l}$

12. Maximum deflection of the beam

$$y_{max} = \frac{Pl^3}{3EI} = \frac{6Pl^3}{Ebt^3}$$

where, $I = \dfrac{bt^3}{18l}$

13. $S = \dfrac{FL^3}{3EI}$

$$k = \frac{F}{S} = \frac{3EI}{L^3}$$

$$k = \frac{3EI}{(0.01)^3}$$

k = 3, 000, 000 EI

$$\omega_n = \sqrt{\frac{k}{m}} = \sqrt{\frac{3000000EI}{0.5}}$$

$= 2449.48\sqrt{EI}$

$$f_n = \frac{\omega_n}{2\pi} \Rightarrow 100 = \frac{2449.48\sqrt{EI}}{2\pi}$$

EI = 0.065 N.m^2.

14. $$\sigma_{max} = \frac{M_{max}}{z} = \frac{\frac{PL}{4}}{\frac{h^3}{3}} = \frac{\frac{Px50h}{4}}{\frac{h^3}{3}} = 37.5\frac{P}{h^2}$$

$$\tau_{max} = \frac{3}{2}\frac{P_{max}}{A} = \frac{3P_{max}}{4h^2} = 0.75\frac{P_{max}}{h^2}$$

$$\tau_{max} = \frac{0.75P}{h^2 2} = \frac{0.375P}{h^2}\left[P_{max} = \frac{P}{2}\right]$$

$$\frac{\tau_{max}}{\sigma_{max}} = \frac{0.375}{37.5} = 0.01$$

15. Moment of Inertia,

$$I_{xx} = \frac{1}{12}\left[(120)^3 \times 60\right] - 2\left[\frac{1}{12} \times (30)^4 + 30 \times 30 \times 30\right]$$

$= 6.885 \times 10^6$ mm^4

16.

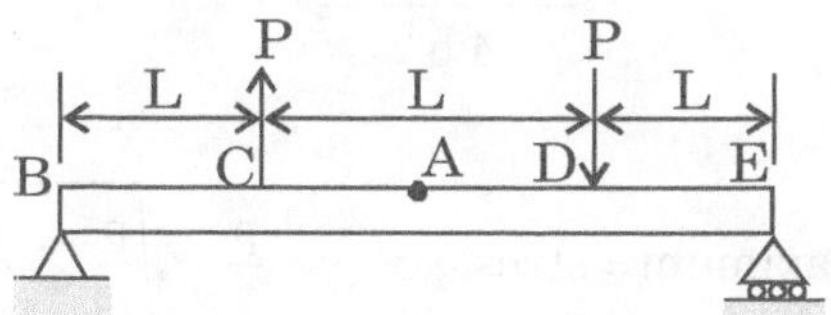

R_E = P/3N

R_B = – P/3 N

$$BM_{X=1.51} = \frac{P}{3}X - P(X - L)$$

$$= \frac{P}{3}\left(\frac{3L}{2}\right) - P\left(\frac{3L}{2} - L\right) = 0$$

BM = 0 0 = 0

17.

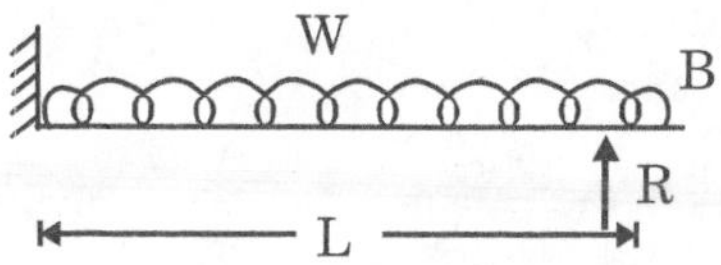

$\Delta_B = 0$

$$\frac{WL^4}{8EI} - \frac{RL^3}{3EI} = 0$$

$$\frac{WL^4}{8EI} - \frac{RL^3}{3EI} = 0$$

$$R = \frac{3WL}{8}$$

18. $I_{base} = I_{rect\ abt\ base} - 2I_{semicircle\ abt\ base}$

$I = ((I_{C.G} + AY^2) - 2(I_{C.G} + AY^2)$

$$= \left[\frac{bd_3}{12} + (bd)y^2\right] - 2\left[\frac{\pi}{64}\frac{d^2}{(2)} + \left(\frac{1}{2} \times \pi r^2 y^2\right)\right]$$

$$= \left[\frac{10 \times 20^3}{12} + 10 \times 20 \times 10^2\right] - 2$$

$$\left[\frac{\pi}{64} \times \frac{8^4}{4} + \frac{1}{2} \times \pi \times 4^2 \times 10^2\right]$$

$= 21439.06$ mm^4.

19. Second moment of area or MOI of circular beam = I

Section modulus $Z = \frac{I}{y}$

$$\frac{MOI}{Z} = \frac{I}{I_y} = y = \frac{d}{2}$$

20. For 0.01 maximum

Bending moment, M_{max}

$$M_{max} = \frac{Wl^2}{8}$$

and from bending equation:-

$$\sigma_{max} = \frac{6\,M_{max}}{bd^2} = \frac{6 \times 10 \times (4)^2 \times 1000}{8(0.1) \times (.2)^2}$$

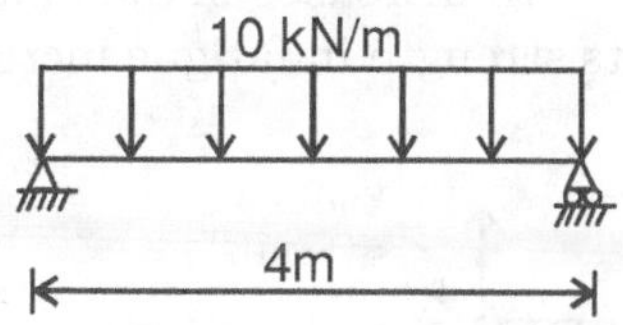

σ_{max} = 30MPa

21. By bending equation,

$$\frac{\sigma}{y} = \frac{M}{I} = \frac{E}{R}$$

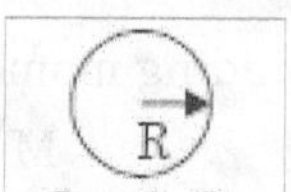

$$\sigma = \frac{Ey}{R}$$

$$= \frac{200 \times 10^3 \times 2.5 \times 10^{-3}}{5} = 100 \text{ Mpa}$$

22.

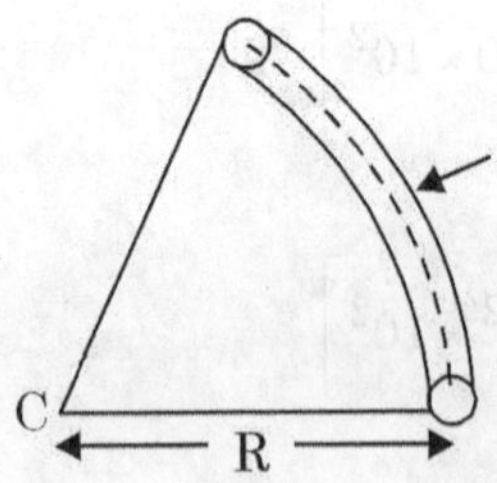

$$y = \frac{\text{Diameter of wire}}{2} = \frac{1}{2} = 0.5\text{mm}$$

$$\Rightarrow \frac{\sigma}{y} = \frac{E}{R}$$

$$\Rightarrow \sigma = \frac{E}{R}y = \frac{100 \times 10^3 \times 0.5}{1000}$$

Max tensile stress $\boxed{\sigma_{max} = 50 \text{ MPa}}$

23.

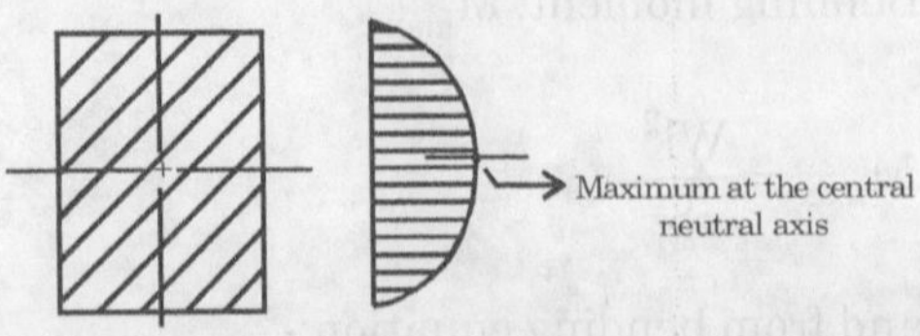

24. As we can see from the Gerber's parabola, the endurance strength remains constant at maximum value irrespective of the compressive load. Whereas for tensile load it decreases sharply with increases in σm and stress in the shaft, its strength in fatigue increases.

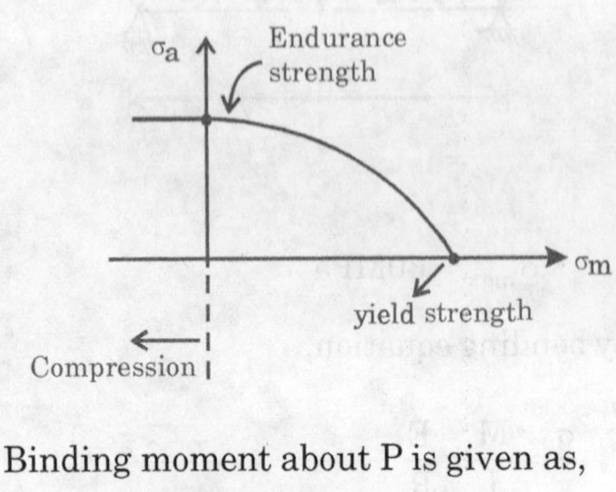

25. Binding moment about P is given as,

$$M = F(L - b)$$

$$\Rightarrow \frac{M}{z} = f$$

$$\Rightarrow f_P = \frac{F(L-b)}{\frac{1}{12}.2b.8b^2} = \frac{3F(L-b)}{4b^3}$$

Axial stress due to direct load F,

$$\sigma = \frac{F}{4b^2}$$

$$\Rightarrow \text{Axial stress} = f_p + \sigma$$

$$\therefore \quad f_p + \sigma = \frac{3F(L-b)}{4b^3} + \frac{F}{4b^2}$$

$$= \frac{1}{4b^3}[3FL - 2Fb]$$

$$= \frac{F(3L - 2b)}{4b^3}$$

26. Maximum= stress, $\sigma_{max} = \frac{p}{2} \sqrt{\frac{p^2}{4} q^2}$

$$q_{max} = \frac{T}{\frac{\pi}{16}D^3} = \frac{10 \times 10^3 \times 10^3}{\frac{\pi}{16} \times (100)^3}$$

$$= \frac{160}{3.14} \text{ N/mm}^2 = 51 \text{ MPa}$$

$$\therefore \quad \sigma_{max} = \frac{50}{2} + \sqrt{(25)^2 + (51)^2}$$

$$= 25 + 57 = 82 \text{ MPa}$$

28. θ = 10°

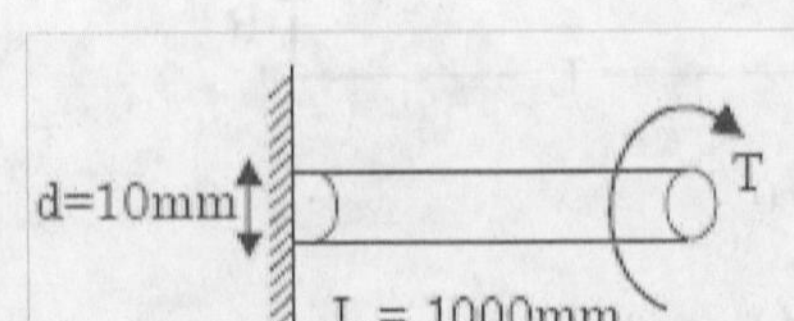

Torsion equation

$$\frac{T}{J} = \frac{\tau_{max}}{R} = \frac{G\theta}{L}$$

$$\frac{\tau_{max}}{G} = \gamma_{max} = \frac{R\theta}{L} = \frac{5 \times 10 \times \frac{\pi}{180}}{1000}$$

$$= 0.87 \times 10^{-3}$$

$$\Rightarrow \quad P \times 10^{-3} = 0.87 \times 10^{-3}$$

$$\therefore \quad P = 0.87$$

7 CHAPTER

Torsion of shafts

Analysis of Previous GATE Papers

Topic	Year/Set → Marks Distribution ↓		2019 Set 1	2019 Set 2	2018 Set 1	2018 Set 2	2017 Set 1	2017 Set 2	2016 Set 1	2016 Set 2	2016 Set 3	2015 Set 1	2015 Set 2	2015 Set 3
Shear stress in shaft	1 Mark	MCQ Type								2				
		Numerical Type					1							
	2 Marks	MCQ Type												
		Numerical Type												
	Total						1			2				
Composite Shaft	1 Mark	MCQ Type							1					
		Numerical Type										1		
	2 Marks	MCQ Type									1		1	
		Numerical Type											1	
	Total								1		2	1	4	
Torsional Strain Energy	1 Mark	MCQ Type												
		Numerical Type												
	2 Marks	MCQ Type												
		Numerical Type												
	Total													

Shear stress in shaft

1. A cyiindrical elastic body subjected to pure torsion about its axis develops
(a) tensile stress in a direction 45° to the axis
(b) no tensile or compressive stress
(c) maximum shear stress along the axis of the shaft
(d) maximum shear stress at 45° to the axis
[1989 : 1 Mark]

2. A circular rod of diameter d and length 3d is subjected to a compressive force F acting at the top point as shown below. Calculate the stress at the bottom most support point A

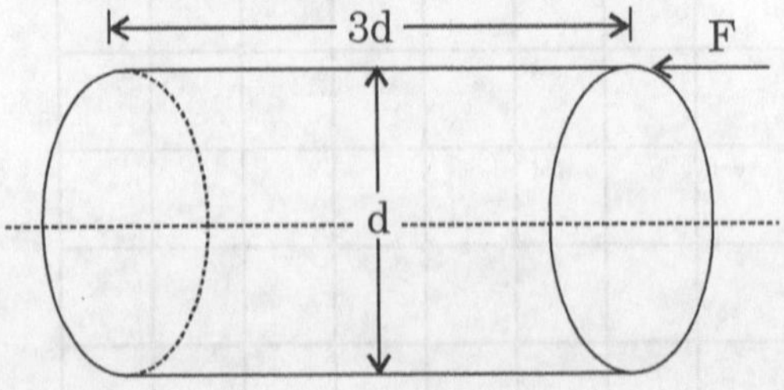

(a) $\frac{12F}{\pi d^2}$ (b) $\frac{16F}{\pi d^2}$
(c) $-\frac{4F}{\pi d^2}$ (d) $-\frac{12F}{\pi d^2}$
[1993 : 2 Marks]

3. Maximum shear stress developed on the surface of a solid circular shaft under pure torsion is 240 MPa. If the shaft diameter is doubled then the maximum shear stress developed corresponding to the same torque will be
(a) 120 MPa (b) 60 MPa
(c) 30 MPa (d) 15 MPa
[2003 : 1 Mark]

4. A solid circular shaft of 60 mm diameter transmits a torque of 1600 Nm. The value of maximum shear stress developed is
(a) 37.72 MPa (b) 47.72 MPa
(c) 57.72 MPa (d) 67.72 MPa
[2004 : 2 Marks]

5. For a circular shaft of diameter d subjected to torque T, the maximum value of the shear stress is
(a) $\frac{64T}{\pi d^3}$ (b) $\frac{32T}{\pi d^3}$
(c) $\frac{16T}{\pi d^3}$ (d) $\frac{8T}{\pi d^3}$
[2006 : 1 Mark]

6. A solid shaft of diameter d and length L is fixed at both the ends. A torque, T_0 is applied at a distance, L/4 from the left end as shown in the figure given below.

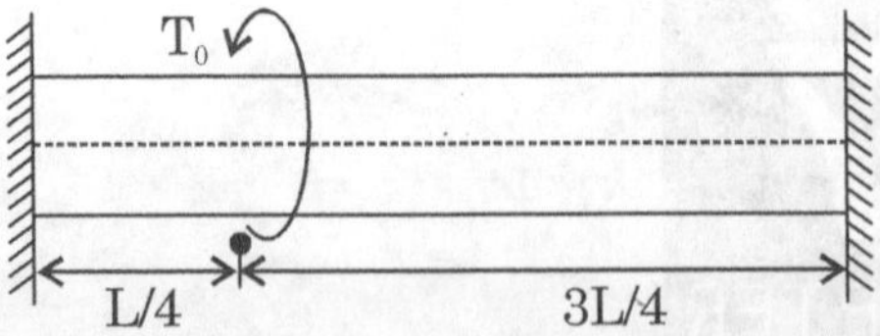

The maximum shear stress in the shaft is
(a) $\frac{16T_0}{\pi d^3}$ (b) $\frac{12T_0}{\pi d^3}$
(c) $\frac{8T_0}{\pi d^3}$ (d) $\frac{4T_0}{\pi d^3}$
[2009 : 2 Marks]

7. A solid circular shaft of diameter d is subjected to a combined bending moment M and torque, T. The material property to be used for designing the shaft using the relation $\frac{16}{\pi d^3}\sqrt{M^2+T^2}$ is
(a) ultimate tensile strength (S_u)
(b) tensile yield strength (S_y)
(c) torsional yield strength (S_{sy})
(d) endurance strength (S_e)
[2009 : 1 Mark]

8. A shaft with a circular cross-section is subjected to pure twisting moment. The ratio of the maximum shear stress to the largest principal stress is
(a) 2.0 (b) 1.0
(c) 0.5 (d) 0
[2016 : 1 Mark, Set-2]

9. A machine element XY, fixed at end X, is subjected to an axial load P, transverse load F, and a twisting moment T at its free end Y. The most critical point from the strength point of view is

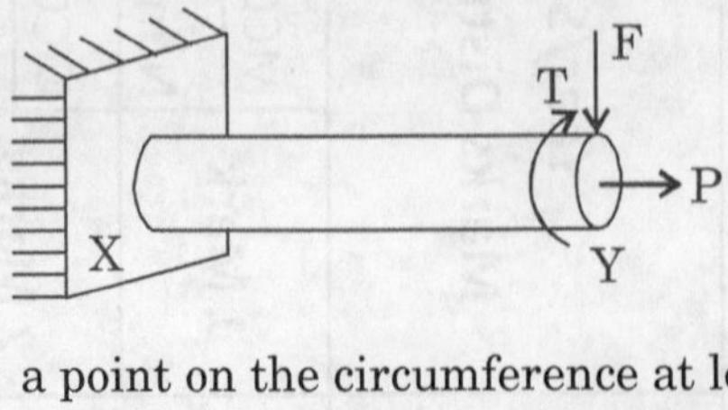

(a) a point on the circumference at location Y
(b) a point at the center at location Y
(c) a point on the circumference at location X
(d) a point at the center at location X
[2016 : 1 Mark, Set-2]

10. A motor driving a soiid circular steel shaft transmits 40 kW of power at 500 rpm. If the diameter of the shaft is 40 mm, the maximum shear stress in the shaft is ______ MPa.
[2017 : 1 Mark, Set-1]

11. A hollow circular shaft of inner radius 10 mm outer radius 20 mm and length 1 m is to be used as a torsional spring. If the shear modulus of the material of the shaft is 150 GPa, the torsional stiffness of the shaft (in kN-m/rad) is ______ (correct to two decimal places).

[2018 : 1 Mark, Set-2]

Composite shafts

12. The compound shaft shown is built-in at the two ends. It is subjected to a twisting moment T at the middle. What is the ratio of the reaction torques T_1, and T_2 at the ends?

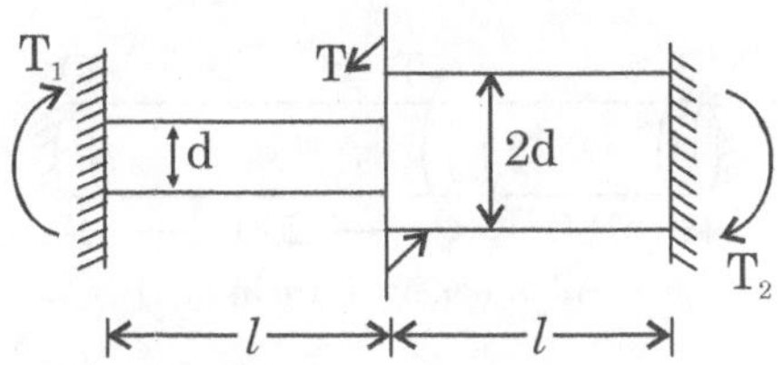

(a) $\frac{1}{16}$ (b) $\frac{1}{8}$

(c) $\frac{1}{4}$ (d) $\frac{1}{2}$

[1993 : 2 Marks]

13. Two shafts A and B are made of the same material. The diameter of shaft B is twice that of shaft A. The ratio of power which can be transmitted by shaft A to that of shaft B is (If maximum shear stress remains the same)

(a) $\frac{1}{2}$ (b) $\frac{1}{4}$

(c) $\frac{1}{8}$ (d) $\frac{1}{16}$

[1994 : 1 Mark]

14. A torque of 1.0 Nm is transmitted through a stepped shaft as shown in figure. The torsional stiffnesses of individual sections of lengths MN, NO and OP are 20 Nm/rad, 30 Nm/rad and 60 Nm/rad respectively. The angular deflection between the ends M and P of the shaft is

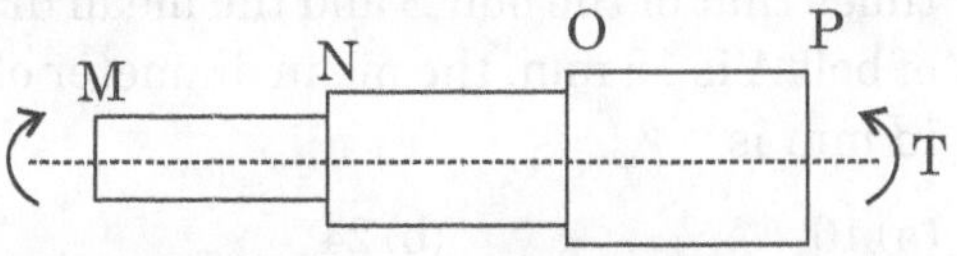

(a) 0.05 rad (b) 0.1 rad

(c) 0.5 rad (d) 1.0 rad

[2004 : 1 Mark]

15. Two shafts AB and BC, of equal length and diameters d and 2d, are made of the same material. They are joined at B through a shaft coupling, while the ends A and C are built-in (cantilevered). A twisting moment T is applied to the coupling. If T_A and T_c represent the twisting moments at the ends A and C, respectively, then

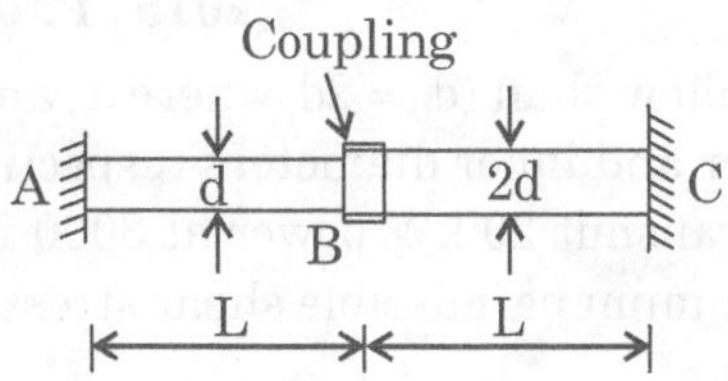

(a) $T_C = T_A$ (b) $T_C = 8T_A$

(c) $T_C = 16\ T_A$ (d) $T_A = 16T_C$

[2005 : 2 Marks]

16. A torque T is applied at the free end of a stepped rod of circular cross-sections as shown in the figure. The shear modulus of the material of the rod is G. The expression for d to produce an angular twist θ at the free end is

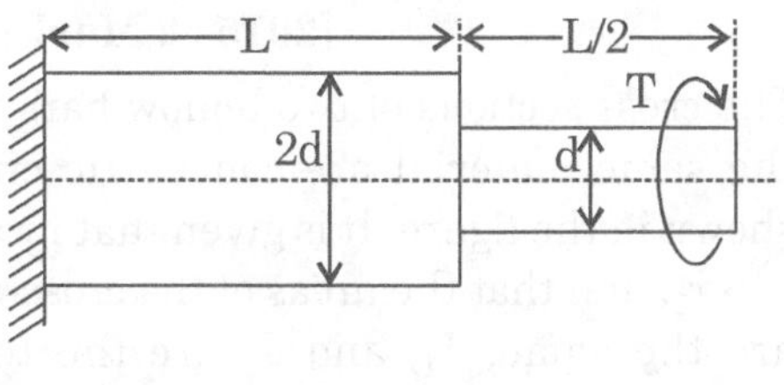

(a) $\left(\frac{32TL}{\pi\theta G}\right)^{\frac{1}{4}}$ (b) $\left(\frac{18TL}{\pi\theta G}\right)^{\frac{1}{4}}$

(c) $\left(\frac{16TL}{\pi\theta G}\right)^{\frac{1}{4}}$ (d) $\left(\frac{2TL}{\pi\theta G}\right)^{\frac{1}{4}}$

[2011 : 2 Marks]

17. Two solid circular shafts of radii R_1 and R_2 are subjected to same torque. The maximum shear stresses developed in the two shafts are τ_1 and τ_2. If $R_1/R_2 = 2$, then τ_2/τ_1 is ______.

[2014 : 1 Mark, Set-3]

18. Consider a stepped shaft subjected to a twisting moment applied at B as shown in the figure. Assume shear modulus, G = 77 GPa. The angle of twist at C (in degree) is ______.

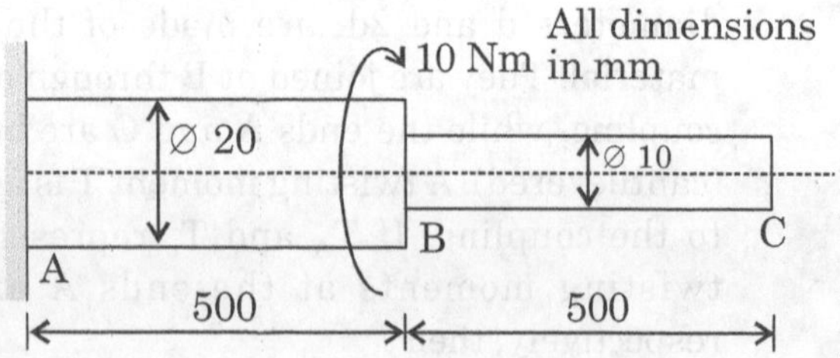

[2015 : 1 Mark, Set-1]

19. A hollow shaft ($d_0 = 2d_i$ where d_0 and d_i are the outer and inner diameters respectively) needs to transmit 20 kW power at 3000 RPM. If the maximum permissible shear stress is 30 MPa, d_0 is

(a) 11.29 mm (b) 22.58 mm
(c) 33.87 mm (d) 45.16 mm

[2015 : 2 Marks, Set-2]

20. A hollow shaft of 1 m length is designed to transmit a power of 30 kW at 700 rpm. The maximum permissible angle of twist in the shaft is 1°. The inner diameter of the shaft is 0.7 times the outer diameter. The modulus of rigidity is 80 GPa. The outside diameter (in mm) of the shaft is _____.

[2015 : 2 Marks, Set-2]

21. The cross sections of two hollow bars made of the same material are concentric circles as shown in the figure. It is given that $r_3 > r_1$, and $r_4 > r_2$, and that the areas of the cross-sections are the same, J_1, and J_2 are the torsional rigidities of the bars on the left and right, respectively. The ratio J_2/J_1 is

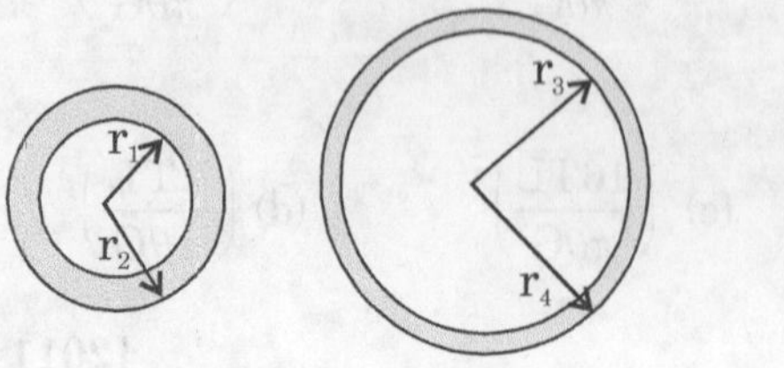

(a) > 1 (b) < 0.5
(c) = 1 (d) between 0.5 and 1

[2016 : 1 Mark, Set-1]

22. Two circular shafts made of same material, one solid (S) and one hollow (H), have the same length and polar moment of inertia. Both are subjected to same torque. Here, θ_S is the twist and τ_s is the maximum shear stress in the solid shaft, whereas θ_H is the twist and τ_H is the maximum shear stress in the hollow shaft. Which one of the following is TRUE?

(a) $\theta_S = \theta_H$ and $\tau_s = \tau_H$
(b) $\theta_S > \theta_H$ and $\tau_s > \tau_H$
(c) $\theta_S < \theta_H$ and $\tau_s < \tau_H$
(d) $\theta_S = \theta_H$ and $\tau_s < \tau_H$

[2016 : 2 Marks, Set-3]

23. A bar of circular cross section is clamped at ends P and Q as shown in the figure. A torsional moment T = 150 Nm is applied at a distance of 100 mm f rom end P. The torsional reactions (T_P T_Q) in Nm at the ends P and Q respectively are

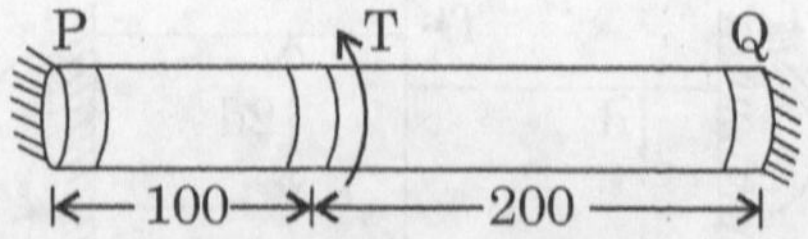

(All dimensions are in mm)

(a) (50,100) (b) (75,75)
(c) (100,50) (d) (120,30)

[2018 : 2 Marks, Set-2]

Torsional strain energy

24. A stepped steel shaft shown below is subjected to 10 N-m torque. If the modulus of rigidity is 80 GPa, the strain energy in the shaft in N-mm is

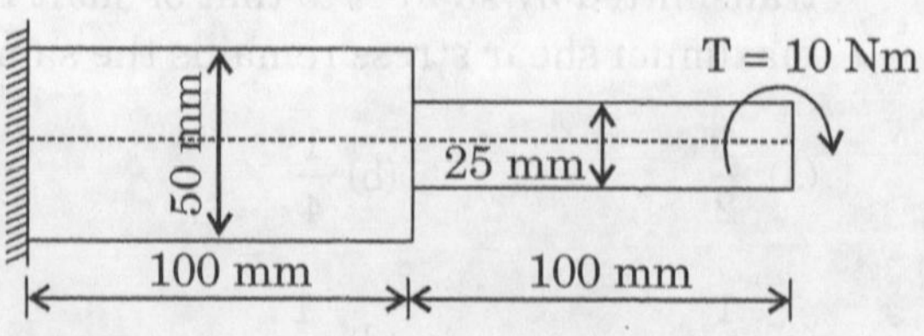

(a) 4.12 (b) 3.46
(c) 1.73 (d) 0.86

[2007 : 2 Marks]

25. Two threaded bolts A and B of same material and length are subjected to identical tensile load. If the elastic energy stored in bolt A is 4 times that of the bolt S and the mean diameter of bolt A is 12 mm, the mean diameter of bolt B in mm is

(a) 16 (b) 24
(c) 36 (d) 48

[2013 : 1 Mark]

ANSWERS

1. (a) **2.** (a) **3.** (c) **4.** (a) **5.** (c) **6.** (a) **7.** (c) **8.** (b) **9.** (c) **12.** (a) **13.** (c) **14.** (b) **15.** (c) **16.** (b) **19.** (b) **21.** (a) **22.** (b) **23.** (c) **24.** (c) **25.** (b)

EXPLANATIONS

1.

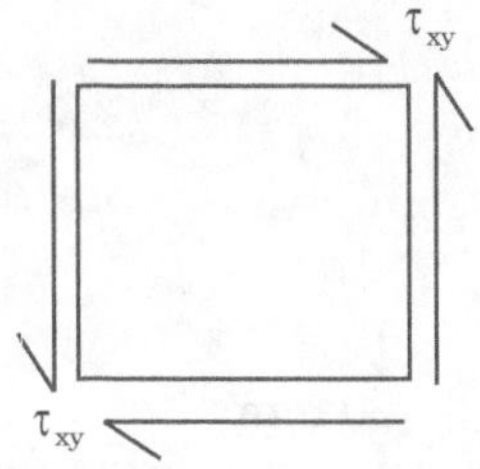

Pure torsion

$\sigma_x = \sigma_y = 0,\ \tau_{xy} = \tau$

principle stress

$$\sigma_{1,2} = \frac{\sigma_x + \sigma_y}{2} \pm \sqrt{\left(\frac{\sigma_x - \sigma_y}{2}\right)^2 + \tau^2_{xy}}$$

$$\sigma_{1,2} = \pm\tau_{xy}$$

So, angle of maximum principle stress

$$\tan(2\theta_p) = \frac{2\tau_{xy}}{\sigma_x - \sigma_y} = \infty$$

$$2\theta_P = \frac{\pi}{2}$$

$$\theta_P = \frac{\pi}{4} \text{ or } 45°$$

2.

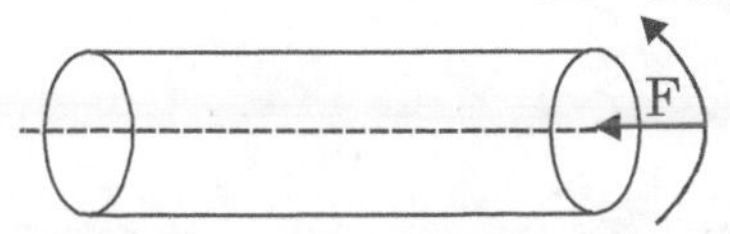

Stress due to Axial force

$$\sigma_a = \frac{F}{A} = \frac{F}{\frac{\pi}{4}d^2} = \frac{4F}{\pi d^2} \text{(compressive)}$$

Stress due to bending:

$$\sigma_b = \frac{My}{I} = \frac{\frac{Fd}{2} \times \frac{d}{2}}{\frac{\pi}{64}d^4} = \frac{16F}{\pi d^2} \text{(tensile)}$$

Combined stress:

$$\sigma_r = \sigma_a + \sigma_b$$

$$\sigma_r = \frac{-4F}{\pi d^2} + \frac{16F}{\pi d^2} = \frac{12F}{\pi d^2}$$

$$\sigma_r = \frac{12F}{\pi d^2}$$

3. $\tau_1 = 240$ MPa

When shaft diameter is doubled, maximum shear stress = τ_2

Let d be the original diameter

$$\frac{\tau_1}{\tau_2} = \frac{\left(\frac{T_{r_1}}{I_{P_1}}\right)}{\left(\frac{T_{r_2}}{I_{P_2}}\right)}$$

... (since torque in both the cases is constant)

$$\frac{\tau_1}{\tau_2} = \left(\frac{r_1}{I_{P_1}} \times \frac{I_{P_2}}{r_2}\right) = \frac{d}{2 \times \left(\frac{\pi d^2}{32}\right)} \times \frac{\pi(2d)^4}{32 \times \frac{2d}{2}}$$

$$= \frac{\pi d \times 16d^4}{\pi d^4 \cdot d \cdot 2} = 8$$

$$\Rightarrow \quad \tau_2 = \frac{\tau_1}{8} = \frac{240}{8} = 30 \text{ MPa}$$

4. $$f_s = \frac{16T}{\pi d^3} = \frac{16 \times 1600}{\pi \times (0.06)^3}$$

$$= 37.72 \text{ MPa}$$

5. Let T = torque; d = diameter of shaft and τ = maximum value of shear stress,

$$\therefore \quad T = \frac{\pi}{16}d^3 \times \tau \text{ or } \tau = \frac{16T}{\pi d^3}$$

6. $$\tau_{max} = \frac{16}{\pi d^3}\left[\sqrt{M^2 + T^2}\right]$$

But $M = 0$

$$\therefore \quad \tau_{max} = \frac{16T_0}{\pi d^3}$$

Alternately

We know, $\frac{T}{J} = \frac{\tau}{r}$

where, $J = \frac{\pi d^4}{32}$

$$\therefore \quad \tau = \frac{Tr}{J} = \frac{T_o \times R \times 32}{\pi d^4} = \frac{16T_o}{\pi d^3}$$

7. For a circular shaft of diameter d,

$$\frac{T'}{J} = \frac{\tau}{\left(\frac{d}{2}\right)}$$

where, T' = net torsional moment

τ = shear stress (torsional shear strength)

J = polar moment of inertia

$$\therefore \quad t = \frac{T'\left(\frac{d}{2}\right)}{J}$$

But for a solid circular shaft

$$J = \frac{\pi}{32} d^4$$

$$\therefore \quad \tau = \frac{16T'}{\pi d^3}$$

$$\text{But } T' = \sqrt{M^2 + T^2}$$

$$\therefore \quad \tau = \frac{16}{\pi d^3}\sqrt{M^2 + T^2}$$

8. 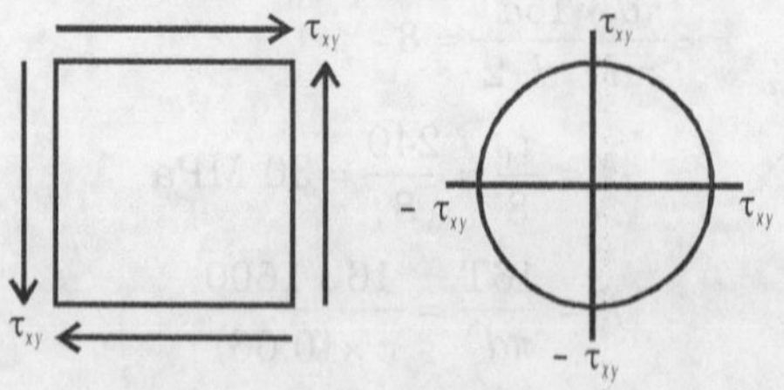

Mohr's circle

$$\tau_{max} = \tau_{xy}$$

$$\sigma_1 = \tau_{xy}$$

$$\therefore \quad \frac{\tau_{max}}{\sigma_1} = 1$$

9. 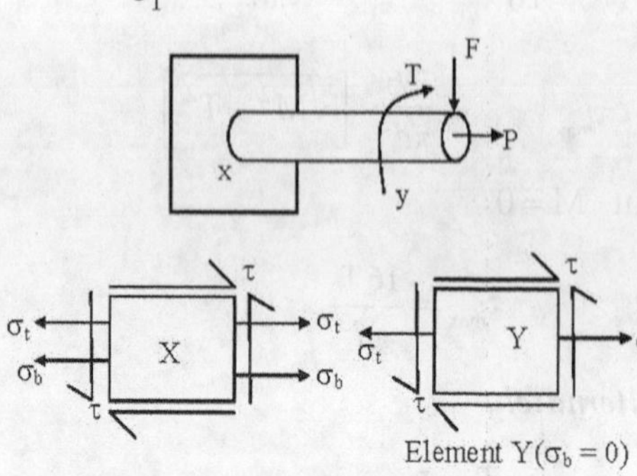

Element Y($\sigma_b = 0$)

At centre $\sigma_b = 0$ and torsional shear stress are zero.

10. We know that, Power is given by

$$P = \frac{2\pi NT}{60}$$

$$40 \times 10^3 = \frac{2\pi \times 500 \times T}{60}$$

$$T = 763.94 \text{ Nm}$$

$$T = \frac{\pi}{16} d^3 \tau$$

$$763.94 \times 10^3 = \frac{\pi}{16} \times 40^3 \times \tau$$

$$\tau = 60.79 \times 10^6 \text{ Pascal}$$

$$\tau = 60.79 \text{ MPa}$$

11. 35.34 KN- m/rad

G = 1504 Pa

$$\frac{T}{\theta} = \frac{GJ}{1}$$

$$\frac{I}{\theta} = \frac{150 \times 10^3}{1} \times \frac{\pi}{32} \times \left(0.04^4 - 0.02^4\right)$$

$$\frac{I}{\theta} = 35.34$$

12. 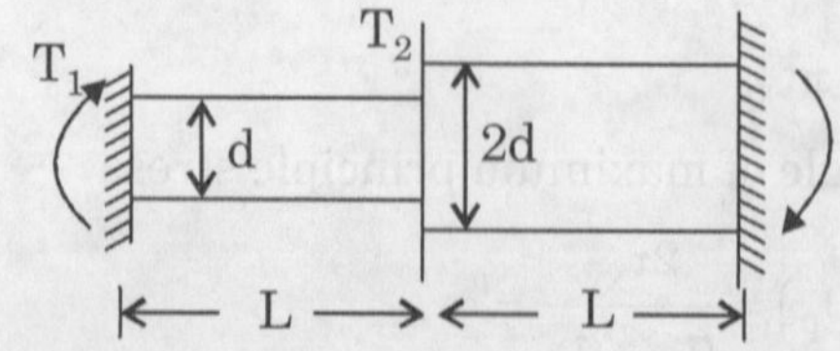

$$\theta_1 = \theta_2.$$

$$\frac{T_1 l}{J_1 G} = \frac{T_2 l}{J_2 G}$$

$$\frac{T_1}{T_2} = \frac{1}{16}$$

13. $d_B = 2d_A$.

$$\tau_A = \tau_B.$$

$$\left(\frac{T_r}{J}\right)_A = \left(\frac{T_r}{J}\right)_B$$

$$\frac{T_A}{T_B} = \frac{J_A}{J_B}\frac{r_B}{r_A}$$

$$\frac{2}{2^4} = \frac{1}{2^3} = \frac{1}{8}$$

14. 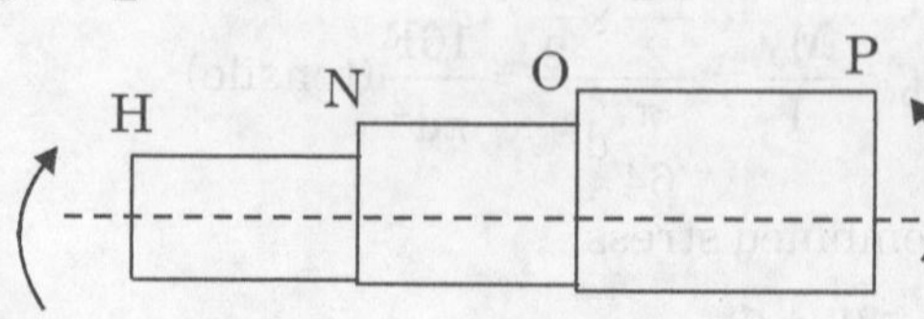

$$\frac{T}{\theta_{MN}} = 20 \text{ Nm/rad}$$

$$T = 1 \text{ N.m}$$

$$\theta_{MN} = \frac{1}{20}\text{rad}$$

$$\theta_{NO} = \frac{1}{30}\text{rad}$$

$$\theta_{OP} = \frac{1}{60}\text{rad}$$

The part MN, NO & OP are connected in series

$$\theta_{NP} = \theta_{MN} + \theta_{NO} + \theta_{OP} = \frac{1}{20} + \frac{1}{30} + \frac{1}{60} = 0.1 \text{ rad.}$$

$\theta = 0.1$ rad

15.

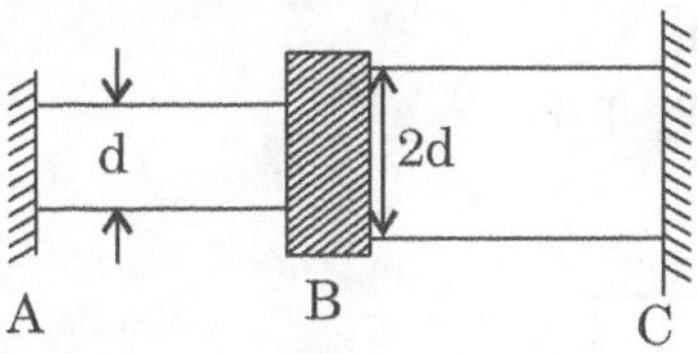

$\theta_A = \theta_C$.

$$\frac{T_A l}{J_A G} = \frac{T_C l}{T_C G}$$

$$\frac{T_A}{\frac{\pi}{32}d^4} = \frac{T_C}{\frac{\pi}{32}(2d)^4}$$

$T_C = 16\ T_A$.

16.

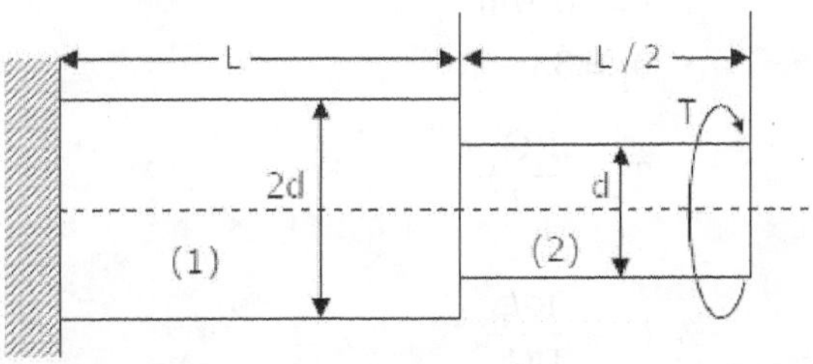

Angular twist at the free end

$$\theta = \Sigma\left(\frac{TL}{GJ}\right)$$

$$= \left(\frac{TL}{GJ}\right)_1 + \left(\frac{TL}{GJ}\right)_2$$

$$= \frac{T}{G}\left[\left(\frac{L}{J}\right)_1 + \left(\frac{L}{J}\right)_2\right]$$

$$= \frac{T}{G}\left[\frac{L}{\frac{\pi}{32}(2d)^4} + \frac{L/2}{\frac{\pi}{32}d^4}\right]$$

$$= \frac{TL}{\pi G d^4}\left[\frac{32}{16} + \frac{32}{2}\right]$$

$$= \frac{18TL}{\pi G d^4}$$

$$\Rightarrow d = \left[\frac{18TL}{\pi G\theta}\right]^{1/4}$$

17. $\frac{R_1}{R_2} = 2$

$T_1 = T_2$.

$$\left(\frac{\tau J}{r}\right)_1 = \left(\frac{\tau J}{r}\right)_2$$

$$\frac{\tau_1 {d_1}^4}{d_1} = \frac{\tau_2 {d_2}^4}{d_2}$$

$$\frac{\tau_1}{\tau_2} = \frac{{d_2}^3}{{d_1}^3} = \frac{1}{8}$$

$$\frac{\tau_2}{\tau_1} = 8$$

18. Angle of twist at (C) = Angle of twist at (B)

$$\Rightarrow \theta = \frac{TL}{GJ}$$

$$\Rightarrow \frac{10 \times 0.5 \times 32}{77 \times 10^9 \times \pi \times .024}$$

$\Rightarrow 0.236050$

19. $P = T\omega$

$$20 \times 10^3 = T \times \frac{2\pi \times 3000}{60}$$

$\therefore \quad T = 63.662$ N– m

Now $\frac{T}{I_p} = \frac{\tau}{r}$,

So, $\frac{63.662}{\frac{\pi}{32}(15d_1^4)} = \frac{30 \times 10^6}{d_1}$ $(r_0 = d)$,

$\therefore \quad d_1 = 11.295$ mm

$\therefore \quad d_0 = 2d_1 = 22.59$ mm

20. $P = T\omega 30 \times 1000 = T \times \frac{2\pi \times 700}{60}$

$\therefore$ T =409.256 N–m

$$\frac{T}{I_p} = \frac{G\theta}{l}$$

$$\frac{409.256}{\frac{\pi}{32}(1 - 0.7^4)d_0^4} = \frac{80 \times 10^9}{1} \times \frac{\pi}{180}\left(\begin{array}{c} l = 1\text{m} \\ \theta = \frac{\pi}{180}\text{radians}\end{array}\right)$$

Solving, we get , $d_0 = 44.5213$ mm

21. Cross-sectional area is same so

$$A_1 = A_2$$

$$\Rightarrow \pi\left[r_2^2 - r_1^2\right] = \pi\left[r_4^2 - r_3^2\right]$$

$$\therefore r_2^2 - r_1^2 = r_4^2 - r_3^2$$

$$\frac{J_2}{J_1} = \frac{\frac{\pi}{2}\left[r_4^4 - r_3^4\right]}{\frac{\pi}{2}\left[r_2^4 - r_1^4\right]} = \frac{\left(r_4^2 + r_3^2\right)\left(r_4^2 - r_3^2\right)}{\left(r_2^2 + r_1^2\right)\left(r_2^2 - r_1^2\right)}$$

$$\frac{J_2}{J_1} = \frac{\left(r_4^2 + r_3^2\right)}{\left(r_2^2 + r_1^2\right)}$$

$$\frac{J_2}{J_1} > 1 \qquad \left[\because r_4 > r_2 \;\&\; r_3 > r_1\right]$$

22. $\theta_S = \dfrac{TL}{CJ}$ (solid)

$\theta_H = \dfrac{TL}{CJ}$ (hollow)

$\theta_S = \theta_H = \theta$

Since material length , polar moment of inertia and applied torque all are same.

$$\tau_S = \frac{T}{J} \times r_S \text{ (or) } \tau_S \propto r_S$$

$$\tau_H = \frac{T}{J} \times r_H \text{ (or) } \tau_H \propto r_H$$

$$\therefore J_H = J_S$$

$$\frac{\pi}{32}(D_H^4 - D_H^4) = \frac{\pi}{32} D_S^4$$

$$D_H^4 - d_S^4 = D_S^4$$

$$1 - \left(\frac{d_S}{D_H}\right) = \left(\frac{D_S}{D_H}\right)^4$$

Since $\dfrac{d_S}{d_H} < 1$

$\therefore D_S < D_H$

$\therefore r_H > r_S$

$\therefore \tau_H > \tau_S$

23.

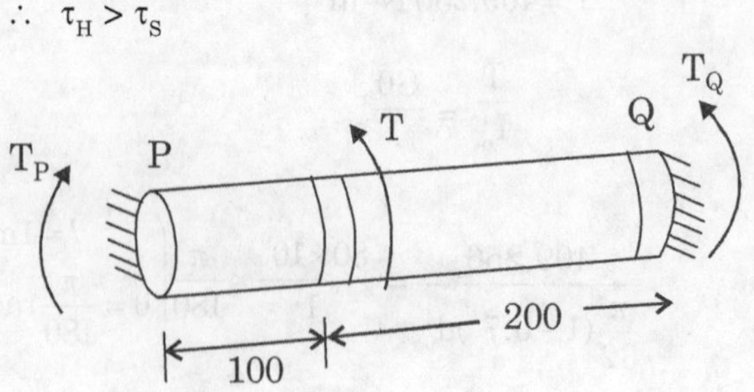

T = 150 Nm

T_P & T_Q = 150 Nm

$\theta_A = \theta_B$

$$\frac{T_P\,(0.1)}{\frac{\pi}{32} \times d^4 \times G} = \frac{T_Q\,(0.2)}{\frac{\pi}{32} \times d^4 \times G}$$

$T_P = 2T_Q$.

$3T_Q = 150$

$T_Q = 50$ Nm

$T_P = 100$ Nm

$(100, 50) = (T_P, T_Q)$

24.

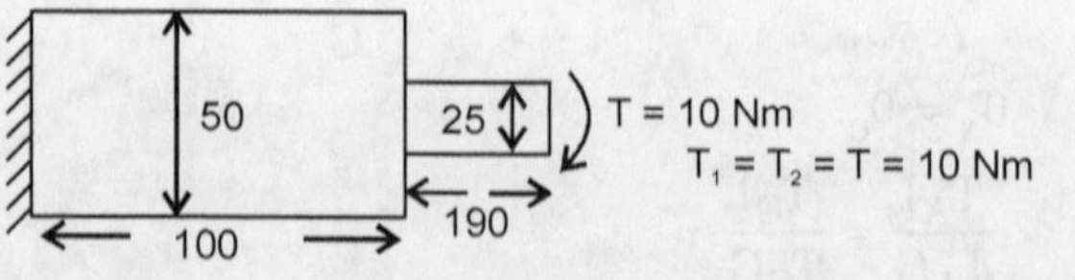

Here, $\quad J_1 = \dfrac{\pi}{32} d^4 = \left(\dfrac{\pi}{32}\right) \times (25)^4 \times 10^{-12}$

$G_1 = 80 \times 10^9$

$l = 0.1/\text{m}$

$\theta_1 = .33 \times 10^{-3}$

$$T_1 = \frac{J_1 G_1 \theta_1}{l_1}$$

$$\Rightarrow \theta_1 = \frac{T_1 l_1}{J_1 G_1}$$

$$\therefore \frac{\theta_1}{\theta_2} = \left(\frac{J_2}{J_1}\right)\left(\frac{1_1}{1_2}\right)\left(\frac{G_2}{G_1}\right)$$

Now $\quad \theta_1 = 4\theta_2$

$$\therefore \theta = \theta_1 + \theta_2 = \frac{5}{4} \times .33 \times 10^{-3}$$

Now, $\quad$ Strain energy $= \dfrac{1}{2} T\theta$

$$= \frac{1}{2} \times 10 \times \frac{5}{4} \times .33 \times 10^{-3} = 1.73$$

8 CHAPTER

Springs

Analysis of Previous GATE Papers

Topic	Year/Set → Marks Distribution ↓		2019 Set 1	2019 Set 2	2018 Set 1	2018 Set 2	2017 Set 1	2017 Set 2	2016 Set 1	2016 Set 2	2016 Set 3	2015 Set 1	2015 Set 2	2015 Set 3
Springs	1 Mark	MCQ Type							1					
		Numerical Type			1									
	2 Marks	MCQ Type												
		Numerical Type						1						
	Total				1			2	1					

Springs

1. The figure shows arrangements of springs. They have stiffness k_1, and k_2 as marked. Which of the following arrangements offers a stiffness $= \frac{2k_1k_2}{k_1 + 2k_2}$?

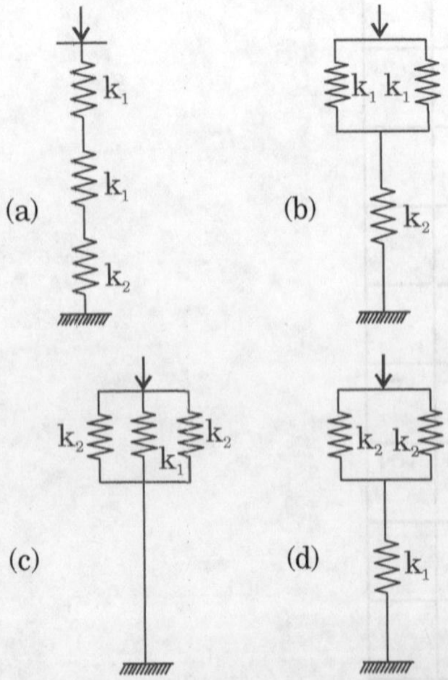

[1987 : 1 Mark]

2. The deflection of a spring with 20 active turns under a load of 1000 N is 10 mm. The spring is made into two pieces each of 10 active coils and placed in parallel under the same load. The deflection of this system is
   (a) 20 mm (b) 10 mm
   (c) 5 mm (d) 2.5 mm

[1995 : 2 Marks]

3. A weighing machine consists of a 2 kg pan resting on a spring. In this condition, with the pan resting on the spring, the length of the spring is 200 mm. When a mass of 20 kg is placed on the pan, the length of the spring becomes 100 mm. For the spring, the undeformed length L and the spring constant k (stiffness) are
   (a) L = 220 mm, k= 1862 N/m
   (b) L = 210 mm, k= 1960 N/m
   (c) L = 200 mm, k = 1960 N/m
   (d) L = 200 mm, k= 2156 N/m

[2005 : 2 Marks]

4. A compression spring is made of music wire of 2 mm diameter having a shear strength and shear modulus of 800 MPa and 80 GPa respectively. The mean coil diameter is 20 mm, free length is 40 mm and the number of active coils is 10. If the mean coil diameter is reduced to 10 mm, the stiffness of the spring is approximately
   (a) decreased by 8 times
   (b) decreases by 2 times
   (c) increased by 2 times
   (d) increased by 8 times

[2008 : 2 Marks]

5. The spring constant of a helical compression spring DOES NOT depend on
   (a) coil diameter
   (b) material strength
   (c) number of active turns
   (d) wire diameter

[2016 : 1 Mark, Set-1]

6. A helical compression spring made of a wire of circular cross-section is subjected to a compressive load. The maximum shear stress induced in the cross-section of the wire is 24 MPa. For the same compressive load, if both the wire diameter and the mean coil diameter are doubled, the maximum shear stress (in MPa) induced in the cross-section of the wire is ___.

[2017 : 2 Marks, Set-2]

7. If the wire diameter of a compressive helical spring is increased by 2%, the change in spring stiffness (in %) is _______ (correct to two decimal places.)

[2018 : 1 Mark, Set-1]

ANSWERS

1. (d) 2. (d) 3. (b) 4. (d) 5. (b)

EXPLANATIONS

1. $k = \dfrac{2k_1k_2}{k_1 + 2k_2}$

$$\frac{1}{k} = \frac{k_1 + 2k_2}{(2k_1)k_2} = \frac{1}{2k_2} + \frac{1}{k_1}$$

$$\frac{1}{k} = \frac{1}{k_1} + \frac{1}{2k_2}$$

$$\therefore \quad k = \frac{2k_1\,k_2}{k_1 + 2k_2}$$

2. Where;

$n = 20$

$P = 1000$ N

$\Delta x = 10$ mm

$F = K\Delta x$

$$K = \frac{P}{\Delta x} = 100 \text{N/mm}$$

When a spring is cut into two pieces, no of coils get halved

∴ Stiffness of each half coiled get doubled

The stiffness when connected in parallel

$= 2k + 2k = 4k$

$P = K\Delta x$

$$\Delta x = \frac{P}{k} = \frac{1000}{4 \times 100} = 2.5\text{mm}$$

3. Let initial length is δ without load, and stiffness is k

$\therefore \quad 2g = k(\delta - 0.2)$ ...(*i*)

$22g = k(\delta - 0.1)$ ...(*ii*)

Solving equations (i) and (ii), we get

$\delta = 210$ mm and $k = 1960$ N/m

4. $k_1 = \dfrac{G\,d_1^{\,4}}{8D_1^{\,3}\,n_1}$ and $k_2 = \dfrac{G\,d_2^{\,4}}{8D_2^{\,3}\,n_2}$

Here, $d_1 = d_2 = 2$ mm (diameter of spring wire)

$G = 80$ GPa

$D_1 = 20,\ D_2 = 10$

$n_1 = n_2 = 10$

$$\therefore \quad \frac{k_1}{k_2} = \left(\frac{D_2}{D_1}\right)^3 = \frac{1}{8}$$

$\Rightarrow k_2 = 8\,k_1$

5. Stiffness of helical spring

$$k = \frac{Gd^4}{64R^3n}$$

where

G is Shear modulus

d is Wire diameter

R is Wire radius

n is No. of active turns.

6. $\tau_{max} = \dfrac{8WD}{\pi d^2}k_w$

$$\tau_{max} \propto \frac{1}{d^2}$$

$$\left[\because W_1 = W_2 = W;\ C_1 = C_2;(K_w)_1 = (K_w)_2 = (K_w)\right]$$

$$\frac{\tau_2}{\tau_1} = \left(\frac{d_1}{d_2}\right)^2$$

$$\frac{\tau_2}{24} = \left(\frac{d}{2d}\right)^2$$

$\Rightarrow \tau_2 = 6$ MPa

7. Stiffness of helical spring, $k = \dfrac{Gd^4}{64R^3n}$

$\therefore k \propto d^4$

$$\frac{k'}{k} = \left(\frac{d'}{d}\right)^4$$

$$k' = \left(\frac{1.02d}{d}\right)^4 k$$

$k' = 1.08243\ k$

$$\% \text{ increase in } k = \frac{k' - k}{k} \times 100\%$$

$$= \left(\frac{1.08243K - k}{k}\right) \times 100\%$$

$= 8.243\%$

9 CHAPTER

Euler's Theory of column

Analysis of Previous GATE Papers

Topic	Year/Set → Marks Distribution ↓		2019 Set 1	2019 Set 2	2018 Set 1	2018 Set 2	2017 Set 1	2017 Set 2	2016 Set 1	2016 Set 2	2016 Set 3	2015 Set 1	2015 Set 2	2015 Set 3
Axial loading	1 Mark	MCQ Type				1								
		Numerical Type			1							1		
	2 Marks	MCQ Type					1							
		Numerical Type												
	Total				1	1	2					1		
Slenderness Ratio	1 Mark	MCQ Type												
		Numerical Type												
	2 Marks	MCQ Type												
		Numerical Type	1											
	Total		2											

Axial loading

1. If the length of a column is doubled, the critical load becomes

(a) 1/2 of the original value

(b) 1/4 of the original value

(c) 1/8 of the original value

(d) 1/16 of the original value

[1988 : 1 Mark]

2. For the case of a slender column of length l, and flexural rigidity El built in at its base and free at the top, the Euler's critical buckling load is

(a) $\frac{4\pi^2 El}{l^2}$ (b) $\frac{2\pi^2 El}{l^2}$

(c) $\frac{\pi^2 El}{l^2}$ (d) $\frac{\pi^2 El}{4l^2}$

[1994 : 1 Mark]

3. A pin-ended column of length L, modulus of elasticity E and second moment of the cross-sectional area I is loaded concentrically by a compressive load P. The critical buckling load (P_{cr}) is given by

(a) $P_{cr} = \frac{EI}{\pi L^2}$ (b) $P_{cr} = \frac{\pi^2 EI}{3L^2}$

(c) $P_{cr} = \frac{\pi EI}{L^2}$ (d) $P_{cr} = \frac{\pi^2 EI}{L^2}$

[2006 : 1 Mark]

4. The rod PQ of length L and with flexural rigidity EI is hinged at both ends. For what minimum force F is it expected to buckle?

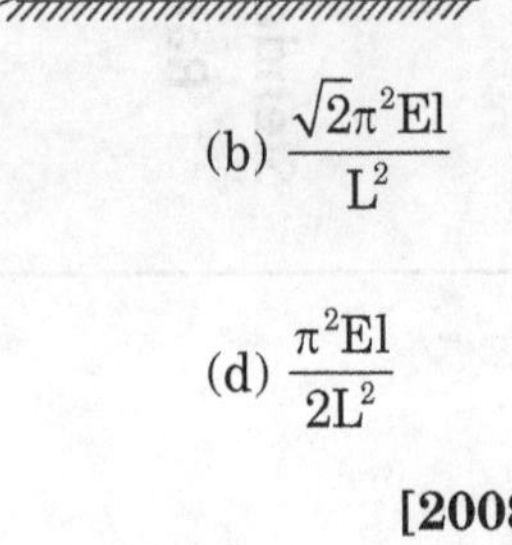

(a) $\frac{\pi^2 El}{L^2}$ (b) $\frac{\sqrt{2}\pi^2 El}{L^2}$

(c) $\frac{\pi^2 El}{\sqrt{2}L^2}$ (d) $\frac{\pi^2 El}{2L^2}$

[2008 : 2 Marks]

5. For a long slender column of uniform cross section, the ratio of critical buckling load for the case with both ends clamped to the case with both ends hinged is

(a) 1 (b) 2

(c) 4 (d) 8

[2012 : 1 Mark]

6. Consider a steel (Young's modulus E = 200 GPa) column hinged on both sides. Its heights is 1.0 m and cross-section is 10 mm x 20 mm. The lowest Euler critical buckling load (in N) is ___.

[2015 : 1 Mark, Set-1]

7. An initially stress-free massless elastic beam of length L and circular cross-section with diameter d(d<< L) is held fixed between two walls as shown. The beam material has Young's modulus E and coefficient of thermal expansion α.

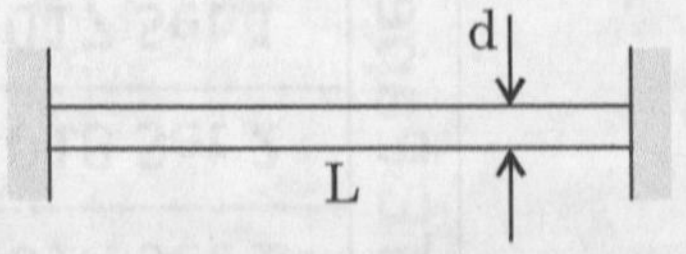

If the beam is slowly and uniformly heated, the temperature rise required to cause the beam to buckle is proportional to

(a) d (b) d^2

(c) d^3 (d) d^4

[2017 : 2 Marks, Set-1]

8. A steel column of rectangular section (15 mm × 10 mm) and length 1.5 m is simply supported at both ends. Assuming modulus of elasticity, E = 200 GPa for steel, the critical axial load (in kN) is _____ (correct to two decimal places)

[2018 : 1 Mark, Set-1]

9. The minimum axial compressive load, P required to initiate buckling for a pinned-pinned slender column with bending stiffness EI and length L is

(a) $P = \frac{\pi^2 EI}{4L^2}$ (b) $P = \frac{\pi^2 EI}{L^2}$

(c) $P = \frac{3\pi^2 EI}{4L^2}$ (d) $P = \frac{4\pi^2 EI}{L^2}$

[2018 : 1 Mark, Set-2]

Slenderness Ratio

10. A column has a rectangular cross-section of 10 mm × 20 mm and a length of 1 m. The slenderness ratio of the column is close to

(a) 200 (b) 346

(c) 477 (d) 1000

[2011 : 1 Mark]

11. Consider a prismatic straight beam of length L = π m, pinned at the two ends as shown in the figure. The beam has a square cross-section of side p = 6 mm. The Young's modulus E = 20 GPa, and the coefficient of thermal expansion $\alpha = 3 \times 10^{-6}\ K^{-1}$. The minimum temperature rise required to cause Euler buckling of the beam is ______ K.

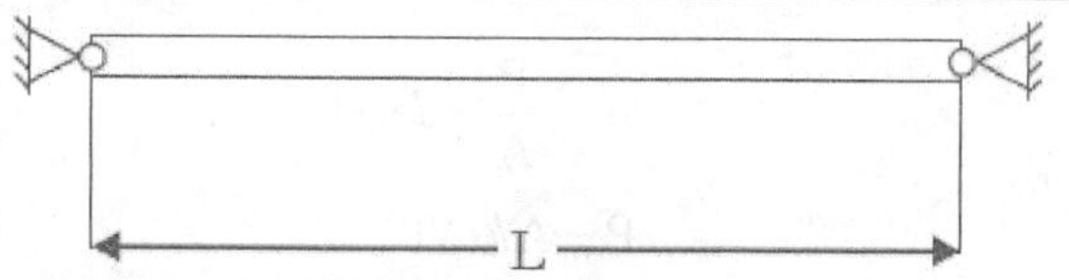

[2019 : 2 Marks, Set-1]

ANSWERS

1. (b) **2.** (d) **3.** (d) **4.** (c) **5.** (c) **7.** (b) **9.** (b) **10.** (b)

EXPLANATIONS

1. $Pe = \dfrac{\pi^2 EI}{l_e^2}$

$l_e = 2l$

$P_e' = \dfrac{\pi^2 EI}{4l_e^2}$

$P_e' = \dfrac{Pe}{4}$

2. $P = \dfrac{\pi^2 EI}{(2l)^2} = \dfrac{\pi^2 EI}{4l_e^2}$

3. The critical buckling load (P_{cr}) for a column having length L, modulus of elasticity E and second moment of cross sectional area I is loaded centrically.

Condition: Both ends are having pin joint.

i.e., hinged

so $n = 1$

$\therefore\ P_{cr} = n\dfrac{\pi^2 EI}{L^2} = \dfrac{\pi^2 EI}{L^2}$

4. Since both ends hinged, therefore, $L_e = L$

$\therefore$ Buckling load, $W = \dfrac{\pi^2 EI}{L^2}$,

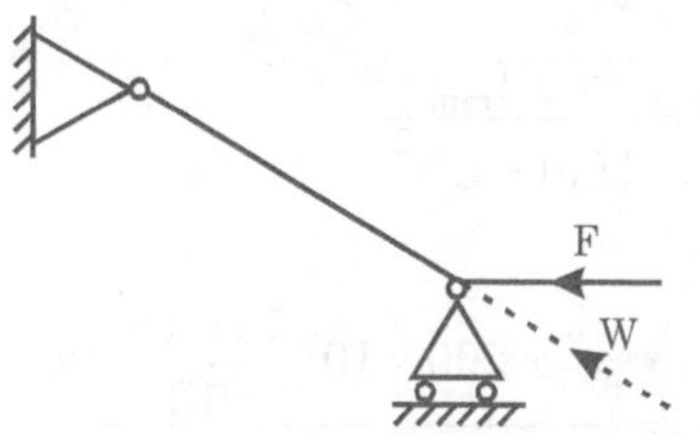

Also $W = F\cos 45°$

$\therefore\ F = \dfrac{\pi^2 EI}{L^2} / \cos 45° = \dfrac{\sqrt{2}\,\pi^2 EI}{L^2}$

5. Critical Buckling load for column fixed at both ends

$= \dfrac{4\pi^2 EI}{L^2}$

Critical Bucking load for a column hinged at both ends $= \dfrac{\pi^2 EI}{L^2}$

Hence, $\dfrac{(P_{cr})_1}{(P_{cr})_2} = 4$

6. Euler's critical load $= \dfrac{\pi^2 EI}{1^2}$

$\Rightarrow \dfrac{\pi^2 \times 200 \times 10^9 \times .02 \times .01^3}{12} = 3289.8681\ N$

7. By Buckling load,

$$P = \frac{4\pi^2 EI}{l^2}$$

Force due to change in temperature,

$$\sigma = \frac{P}{A} = l\alpha\Delta T$$

$$P = Al\alpha\Delta T$$

$$\frac{4\pi^2 EI}{l^2} = Al\alpha\Delta T$$

$$\frac{I}{A} \propto \Delta T$$

$$\Delta T \propto \frac{d^4}{d^2}$$

$$\Delta T \propto d^2$$

8. Steel column supported at both ends = hinged at both ends.

so for hinged condition:-

Effective length, $L_e = L$

so buckling load

$$(Pe) = \frac{\pi^2 EI_{min}}{\{Le(=L)\}^2}$$

$$= \frac{\pi^2 \times 200 \times 10^3 \frac{15 \times 10^3}{12}}{1500^2}$$

$$= 1.097 \text{ kN}$$

10. Slenderness ratio

$$= \frac{\text{Effective length of column}}{\text{least radius of gyration}}$$

$$= \frac{L_e}{K}$$

But $K = \sqrt{\frac{I_{min}}{A}}$

where I_{min} = minimum area moment of inertia i.e. I_{xx} or I_{yy}, whichever is less.

For given section

$$I_{min} = \frac{20 \times 10^3}{12} = 1667 \text{ mm}^3$$

$$\therefore \quad K = \sqrt{\frac{1667}{20 \times 10}} = 2.89$$

$$\text{and ratio} = \frac{1000}{2.89} = 346$$

11. Length of prismatic beam,

$$L = \pi m$$

$$P_{th} = P_{axial}$$

$$\text{or} \quad \alpha TEA = \frac{\pi^2 EI}{L^2}$$

$$\Rightarrow \; 3 \times 10^{-6} \times \Delta T \times 6 \times 6 \times 10^{-6}$$

$$= \frac{\pi^2 \times \frac{1}{12} \times \left(6 \times 10^{-3}\right)^4}{\pi^2}$$

$$\Delta T = 1 \text{ K}$$

10 CHAPTER

Strain Energy and thermal Stresses

Analysis of Previous GATE Papers														
Topic	**Year/Set →Marks Distribution ↓**		**2019 Set 1**	**2019 Set 2**	**2018 Set 1**	**2018 Set 2**	**2017 Set 1**	**2017 Set 2**	**2016 Set 1**	**2016 Set 2**	**2016 Set 3**	**2015 Set 1**	**2015 Set 2**	**2015 Set 3**
Deflection	**1 Mark**	MCQ Type							1					
		Numerical Type												
	2 Marks	MCQ Type								2				
		Numerical Type		1								1		
	Total			**2**					**1**	**4**		**2**		
Slope	**1 Mark**	MCQ Type												
		Numerical Type												
	2 Marks	MCQ Type												
		Numerical Type											1	
	Total												**2**	
Strain Energy	**1 Mark**	MCQ Type						1						
		Numerical Type												
	2 Marks	MCQ Type												
		Numerical Type												
	Total							**1**						

Deflection

1. A simply supported laterally loaded beam was found to deflect more than a specified value. Which of the following measures will reduce deflection?
   (a) Increase the area moment of inertia
   (b) Increase the span of the beam
   (c) Select a different material having lesser modulus of elasticity
   (d) Magnitude of the load to be increased

[2003 : 2 Marks]

2. Two identical cantilever beams are supported as shown, with their free ends in contact through a rigid roller. After the load P is applied, the free ends will have

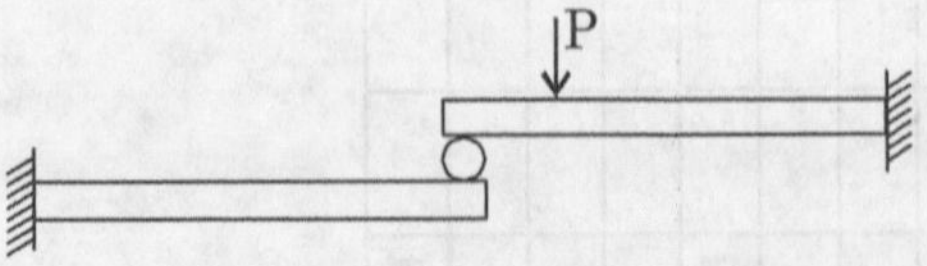

   (a) equal deflections but not equal slopes
   (b) equal slopes but not equal deflections
   (c) equal slopes as well as equal deflections
   (d) neither equal slopes nor equal deflections

[2005 : 1 Mark]

3. A frame of two arms of equal length L is shown in the adjacent figure. The flexural rigidity of each arm of the frame is EI. The vertical deflection at the point of application of load P is

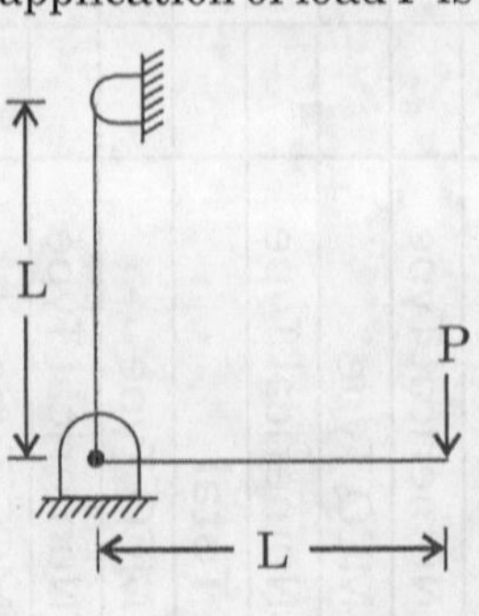

   (a) $\frac{PL^3}{3EI}$
   (b) $\frac{2PL^3}{3EI}$
   (c) $\frac{PL^3}{EI}$
   (d) $\frac{4PL^3}{3EI}$

[2009 : 2 Marks]

4. A cantilever beam of length L is subjected to a moment M at the free end. The moment of inertia of the beam cross section about the neutral axis is I and the Young's modulus is E The magnitude of the maximum deflection is

   (a) $\frac{ML^2}{2EI}$ (b) $\frac{ML^2}{EI}$
   (c) $\frac{2ML^2}{2EI}$ (d) $\frac{4ML^2}{EI}$

[2012 : 1 Mark]

5. A cantiever beam of length, L with uniform crossection and flexural rigidity, EI is loaded uniformly by a vertical load, w per unit length. The maximum vertical deflection of the beam is given by

   (a) $\frac{wL^4}{8EI}$ (b) $\frac{wL^4}{16EI}$
   (c) $\frac{wL^4}{4EI}$ (d) $\frac{wL^4}{24EI}$

[2014 : 2 Marks, Set-3]

6. A frame is subjected to a load P as shown in the figure. The frame has a constant flexural rigidity EI. The effect of axial load is neglected. The deflection at point A due to the applied load P is

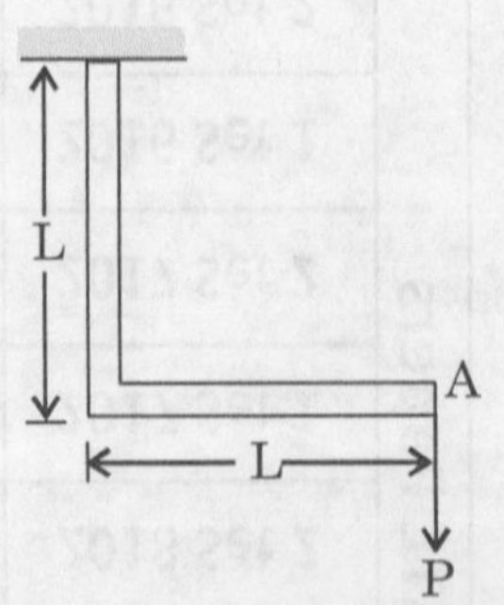

   (a) $\frac{1PL^3}{3EI}$ (b) $\frac{2PL^3}{3EI}$
   (c) $\frac{PL^4}{EI}$ (d) $\frac{4PL^3}{3EI}$

[2014 : 2 Marks, Set-2]

7. A cantilever beam with flexural rigidity of 200 Nm^2 is loaded as shown in the figure. The deflection (in mm) at the tip of the beam is ____.

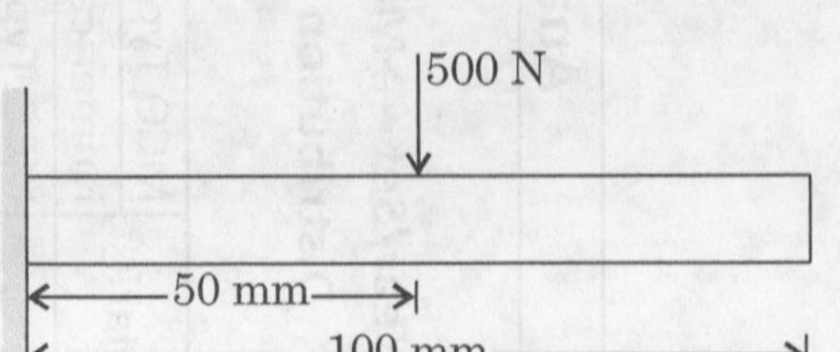

[2015 : 2 Marks, Set-1]

8. A cantilever beam having square cross-section of side a is subjected to an end load, If a is increased by 19%, the tip deflection decreases approximately by
   (a) 19% (b) 29%
   (c) 41% (d) 50%

[2016 : 1 Mark, Set-1]

9. A simply supported beam of length 2L is subjected to a moment M at the mid-point x = 0 as shown in the figure. The deflection in the domain $0 \le x \le L$ is given by

$$w = \frac{-Mx}{12EIL}(L-x)(x+c)$$

where E is the Young's modulus, I is the area moment of inertia and C is a constant (to be determined).

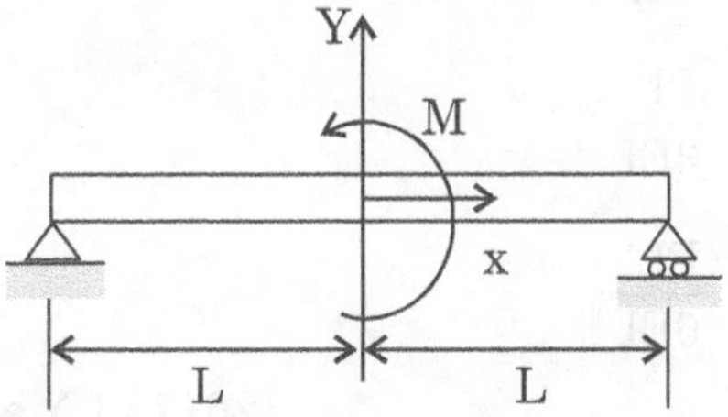

The slope at the center x = 0 is

(a) $\frac{ML}{(2EI)}$ (b) $\frac{ML}{(3EI)}$

(c) $\frac{ML}{(6EI)}$ (d) $\frac{ML}{(12EI)}$

[2016 : 2 Marks, Set-2]

10. A rigid horizontal rod of length 2L is fixed to a circular cylinder of radius R as shown in the figure. Vertical forces of magnitude P are applied at the two ends as shown in the figure. The shear modulus for the cylinder is G and the Young's modulus is E.

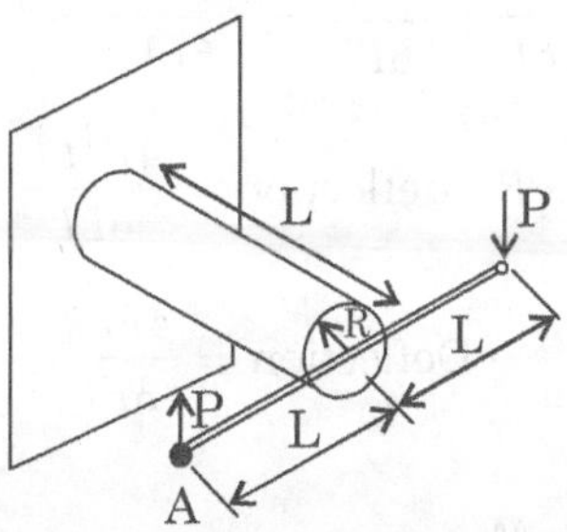

The vertical deflection at point A is

(a) $\frac{PL^3}{(\pi R^4 G)}$ (b) $\frac{PL^3}{(\pi R^4 E)}$

(c) $\frac{2PL^3}{(\pi R^4 E)}$ (d) $\frac{4PL^3}{(\pi R^4 G)}$

[2016 : 2 Marks, Set-2]

11. A horizontal cantilever beam of circular cross-section, length 1.0 m and flexural rigidity EI = 200 N-m^2 is subjected to an applied moment MA = 1.0 N-m at the free end as shown in the figure. The magnitude of the vertical deflection of the free end is ____ mm (round off to one decimal place)

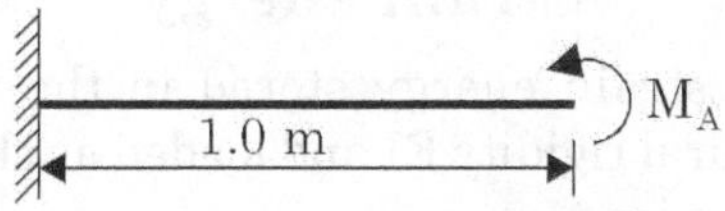

[2019 : 2 Marks, Set-2]

Slope

12. The flexural rigidity (EI) of a cantilever beam is assumed to be constant over the length of the beam shown in figure. If a load P and bending moment PL/2 are applied at the free end of the beam then the value of the slope at the free end is

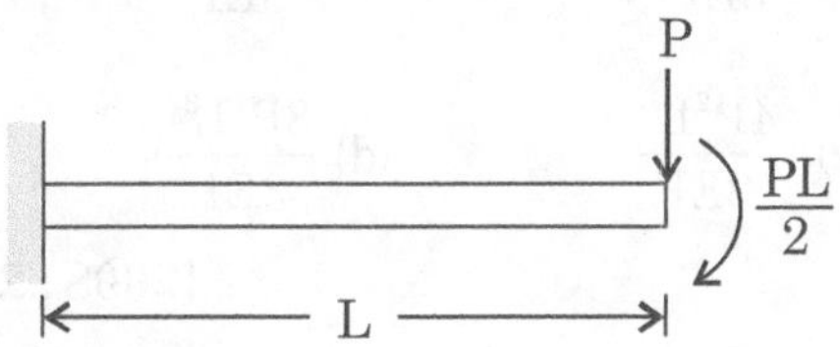

(a) $\frac{1PL^2}{2EI}$ (b) $\frac{PL^2}{EI}$

(c) $\frac{3PL^2}{2EI}$ (d) $\frac{5PL^2}{2EI}$

[2014 : 2 Marks, Set-3]

13. A force P is applied at a distance x from the end of the beam as shown in the figure. What would be the value of x so that the displacement at A is equal to zero?

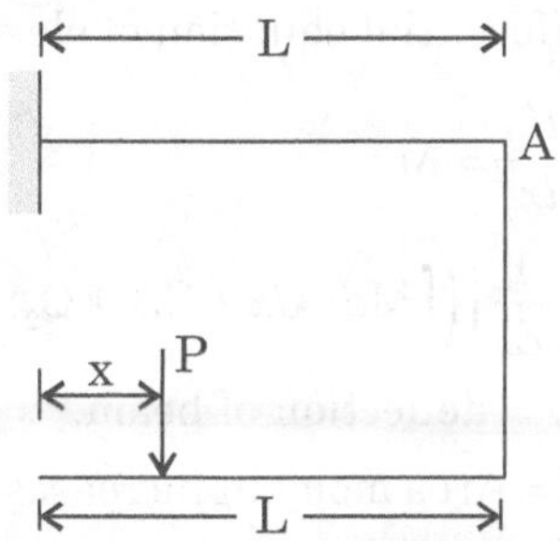

(a) 0.5 L (b) 0.25 L

(c) 0.33 L (d) 0.66 L

[2014 : 2 Marks, Set-3]

14. A cantilever beam with square cross-section of 6 mm side is subjected to a load of 2 kN normal to the top surface as shown in the figure. The Young's modulus of elasticity of the material of the beam is 210 GPa. The magnitude of slope (in radian) at Q (20 mm from the fixed end) is _____.

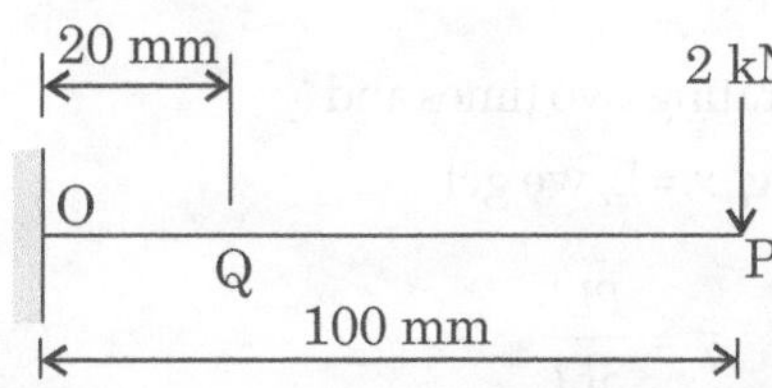

[2015 : 2 Marks, Set-2]

Strain energy

15. The strain energy stored in the beam with flexural rigidity EI and loaded as shown in the figure is

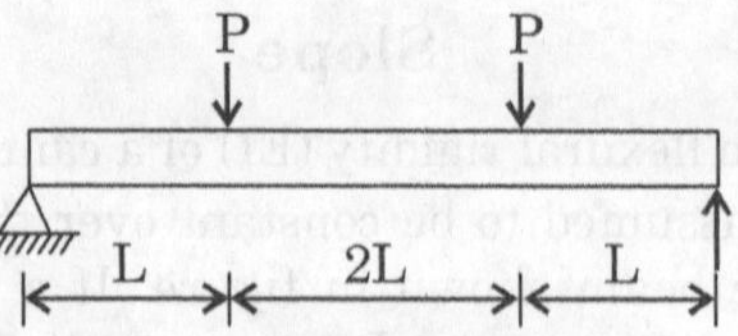

(a) $\frac{P^2L^3}{3EI}$ (b) $\frac{2P^2L^3}{3EI}$

(c) $\frac{4P^2L^3}{3EI}$ (d) $\frac{8P^2L^3}{3EI}$

[2008 : 2 Marks]

16. A cantilever beam of length L and flexural modulus EI is subjected to a point load Pat the free end. The elastic strain energy stored in the beam due to bending (neglecting transverse shear) is

(a) $\frac{P^2L^3}{6EI}$

(b) $\frac{P^2L^3}{3EI}$

(c) $\frac{PL^3}{3EI}$

(d) $\frac{PL^3}{6EI}$

[2017 : 1 Mark, Set-2]

ANSWERS

1. (a) **2.** (a) **3.** (d) **4.** (a) **5.** (a) **6.** (d) **8.** (d) **9.** (c) **10.** (d) **12.** (b) **13.** (c) **15.** (c) **16.** (a)

EXPLANATIONS

1. From differential equation of elastic curve

$$EI\frac{d^2y}{dx^2} = M$$

$$\Rightarrow y = \frac{1}{EI}\int\left(\int M dx\right)dx + C_1x + C_2$$

where y = deflection of beam.

As the I = Area moment increases, the deflection will reduce.

2. As it is rigid roller deflection must be same because after deflection they also will be in contact but slope unequal.

3. $EI\frac{d^2y}{dx^2} = -M_x$

or $M_x = -P(L-x)$

$$\therefore EI\frac{d^2y}{dx^2} = P(L-x)$$

Integrating two times and puting $x = L$, we get

$$y = \frac{PL^3}{3EI}$$

Total strain energy,

$$U = \frac{P^2L^3}{2EI} + \frac{P^2L^3}{6EI} = \frac{4P^2L^3}{6EI}$$

$$\therefore \quad \frac{1}{2}\times P\times \text{deflection} = \frac{4P^2L^3}{6EI}$$

or $\quad \text{Deflection} = \frac{4PL^3}{3EI}$

4. $E_I\frac{d^2y}{dx^2} = M$

On Integration, we get

$$EI\frac{dy}{dx} = Mx + C_1$$

Again integrating, we get

$$EI\,y = \frac{Mx^2}{2} + C_1x + C_2$$

For cantilever beam at $x = 0$,

$$\frac{dy}{dx} = 0$$

and $\quad x = 0;\ y = 0$

From this we get $C_1 = C_1 = 0$

$$\therefore \quad y = \frac{Mx^2}{2EI}$$

Maximum deflection, $y_{max} = \dfrac{ML^2}{2EI}$

5.

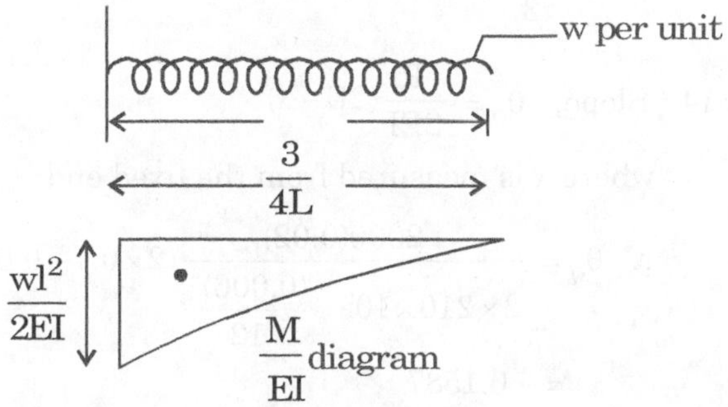

$$\delta_{max} = \frac{1}{3} \times L \times \frac{Wl^2}{2EI} \times \frac{3}{4} L$$

$$\delta_{max} = \frac{Wl^4}{8EI}$$

6. Strain energy (t) $= \displaystyle\int_0^L \frac{M_x^2 dx}{2EI} + \int_0^L \frac{M_y^2 dx}{2EI}$

$M_x = P_x \quad M_y = P_y$.

$$\frac{1}{2} P\delta = \int_0^L \frac{(Px)^2\, dx}{2EI} + \int_0^L \frac{(Px)^2\, dx}{2EI}$$

$$= \frac{2}{3} \frac{P^2L^2}{EI} = \frac{1}{2} P \times \frac{4}{3} \frac{PL^3}{EI}$$

$$\delta = \frac{4}{3} \frac{PL^3}{EI}$$

7. Deflection = Deflection at load + Slope × Distance

$$\Rightarrow \frac{Wl^3}{3EI} + \frac{Wl^2}{2EI} \times .05$$

W = 500

1 = .05

EI = 200

δ = 0.2604 mm

8. For cantilever beam with an end load.

$$\delta = \frac{Pl^3}{3EI}$$

$$\delta \propto \frac{1}{I}$$

$$\frac{\delta_1}{\delta_2} = \frac{I_2}{I_1}$$

$$\frac{\delta_1}{\delta_2} = \frac{I(1.19a)}{a^4} = 2.005$$

$$\delta_2 = \frac{\delta 1}{2.005} = 0.5\delta 1$$

$$\frac{\delta_1 - \delta_2}{\delta_1} \times 100 = 50\%$$

9. $\Delta_{BA} = \dfrac{\text{(first moment area APB)}}{EI}$

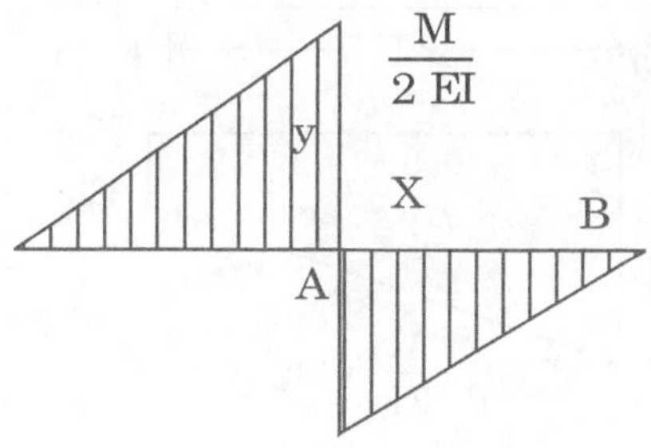

$$= \frac{-1}{EI}\left[\frac{M}{2} \times \frac{1}{2} \times L \times 2\frac{L}{3}\right] \frac{M}{2EI}$$

$$= \frac{ML^2}{6EI}$$

Slope at A $= \dfrac{\Delta BA}{L} = \dfrac{ML}{6EI}$

10.

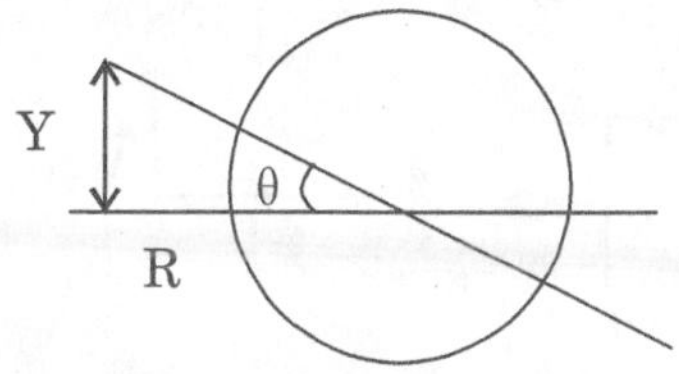

T = P (2L) = 2PL

$$\theta = \frac{TL}{GJ} = \frac{1PL(L)}{G\frac{\pi}{32}(2R)^4}$$

$$\theta = \frac{32 \times 2PL^2}{\pi GR^4 16} = \frac{4PL^2}{\pi GR^4}$$

$$y = R\theta = \frac{L\left(4PL^2\right)}{\pi GR^4}$$

$$= \frac{4PL^3}{\pi GR^4}$$

11.

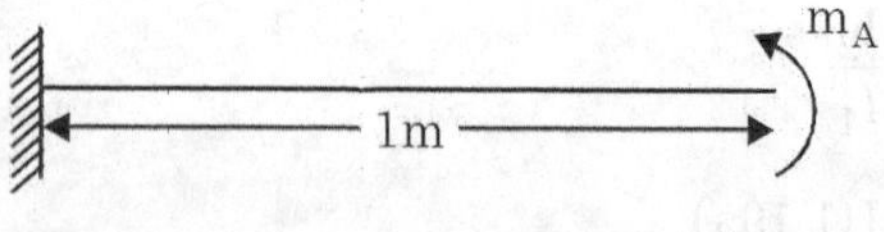

Deflection : $y_B = \dfrac{m_A \times L^2}{2EI}$ (standard result)

$= \dfrac{1 \times 1^2}{2 \times 200} = 2.5 \times 10^{-3} \times 10^3$

$y_{B=2.5\,mm}$

12.

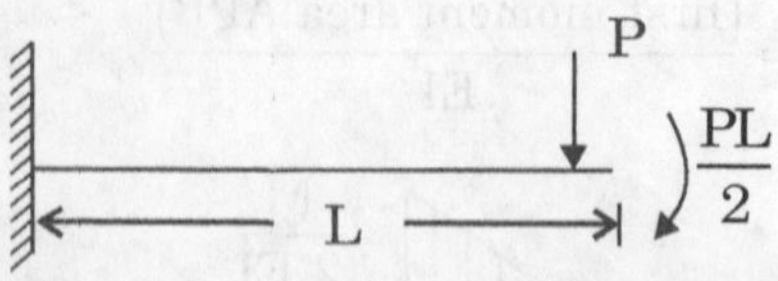

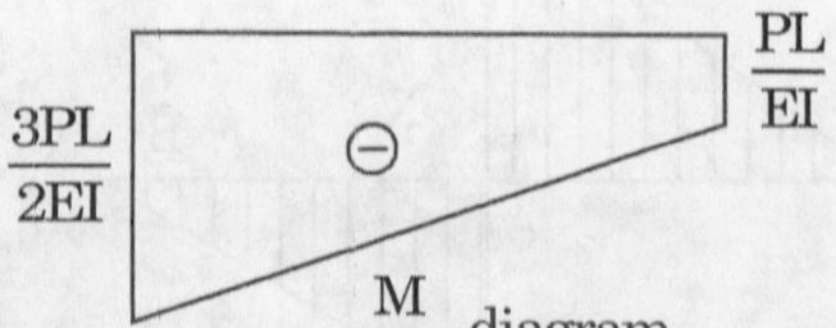

$\dfrac{M}{EI}$ diagram

Slope at free end

= Area under $\dfrac{M}{EI}$ diagram

$= \dfrac{1}{2} : \left[\dfrac{PL}{2EI} + \dfrac{3PL}{2EI}\right]$

$= \dfrac{PL^2}{EI}$

13.

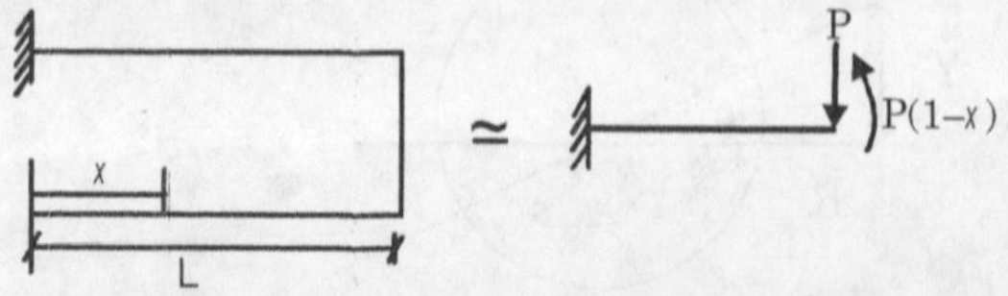

$\delta_1 = \dfrac{Pl^3}{3EI}$ due to point load (downward)

$\delta_2 = \dfrac{ML^2}{2EI}$ due to moment (upward)

$\Rightarrow \delta = \delta_1 - \delta_2 = \dfrac{Pl^3}{3EI} - \dfrac{P(l-x)l^2}{2EI} = 0$

$\Rightarrow \dfrac{Pl^3}{3EI} - \dfrac{Pl^3}{2EI} + \dfrac{P \times l^2 x}{2EI} = 0$

$\Rightarrow \left[\dfrac{-1}{3} l + x\right]\dfrac{Pl^2}{2EI} = 0$

$\Rightarrow x = \dfrac{1}{3} l = 0.33L.$

14. Slope, $\theta_Q = \dfrac{-Px}{2EI}(2L - x)$

where x is measured from the fixed end

$\therefore \theta_Q = \dfrac{-2000(0.02)}{2 \times 210 \times 10^9 \times \dfrac{(0.006)^4}{12}}(2 \times 0.1 - 0.02)$

$= -0.1587$

The magnitude of slope is 0.1587 radian

15. Total stain energy streed

$= 2\int_0^L \dfrac{(Px)^2\,dx}{2EI} + \dfrac{PL^2.2L}{2EI}$

$= \dfrac{2P^2}{2EI}\int_0^L x^2 dx + \dfrac{P^2L^3}{EI}$

$= \dfrac{P^2}{EI} \times \dfrac{L^3}{3} + \dfrac{P^2L^3}{EI} = \dfrac{4}{3}\dfrac{P^2L^3}{EI}$

16.

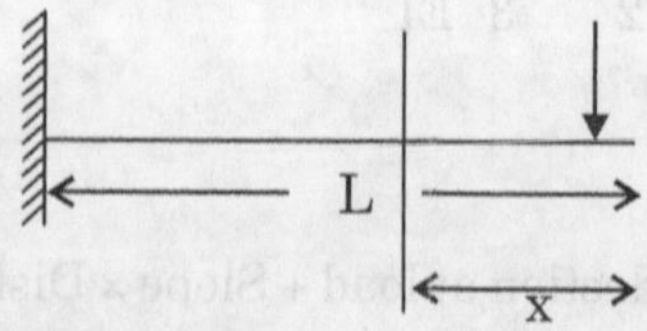

$U = \dfrac{M^2 dx}{2EI}$

$U = \int_0^L \dfrac{(Px)^2\,dx}{2EI}$

$U = P^2 L^3.$

Unit - IV

Theory of Machines

Syllabus

Displacement, velocity and acceleration analysis of plane mechanisms; dynamic analysis of linkages; cams; gears and gear trains; flywheels and governors; balancing of reciprocating and rotating masses; gyroscope. Vibrations: Free and forced vibration of single degree of freedom systems, effect of damping; vibration isolation; resonance; critical speeds of shafts.

Contents

1 CHAPTER

Planar Mechanisms

Analysis of Previous GATE Papers

Topic	Year/Set → Marks Distribution ↓		2019 Set 1	2019 Set 2	2018 Set 1	2018 Set 2	2017 Set 1	2017 Set 2	2016 Set 1	2016 Set 2	2016 Set 3	2015 Set 1	2015 Set 2	2015 Set 3
Kinematic pair, Kinematic Chain and Inversions	1 Mark	MCQ Type												
		Numerical Type												
	2 Marks	MCQ Type												
		Numerical Type												
	Total													
Mechanism(4Bar, Single Slider and Double Slider)	1 Mark	MCQ Type			1							1		
		Numerical Type			1									
	2 Marks	MCQ Type	1	1										
		Numerical Type					1							
	Total		2	2	2		2					1		
Degree of Freedom	1 Mark	MCQ Type									1			
		Numerical Type												
	2 Marks	MCQ Type												1
		Numerical Type												
	Total										1			2

Kinematic pair, Kinematic chain and Inversions

1. For the planar mechanism shown in figure select the most appropriate choice for the motion of link 2 when link 4 is moved upwards.

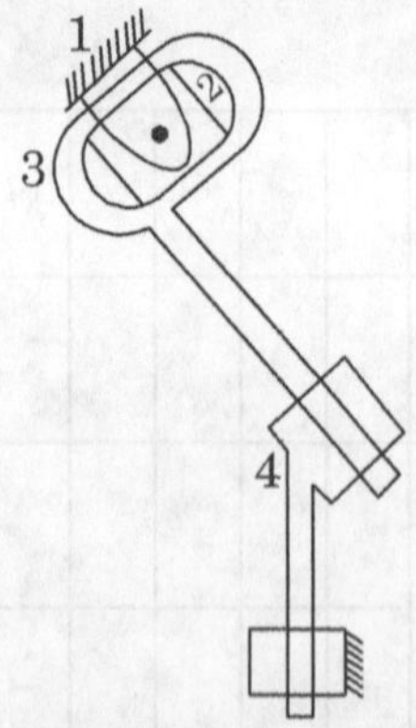

(a) Link 2 rotates clockwise
(b) Link 2 rotates counter clockwise
(c) Link 2 does not move
(d) Link 2 motion cannot be determined

[1999 : 1 Mark]

2. For the audio Cassette mechanism shown in figure given below where is the instantaneous centre of rotation (P) of the two spools?

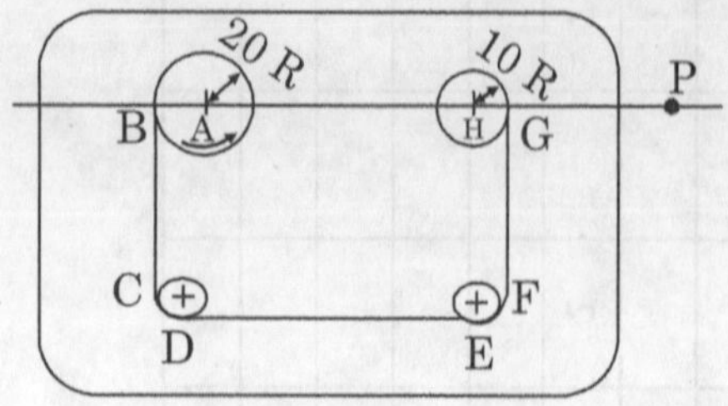

(a) Point P lies to the left of both the spools but at infinity along the line joining A and H
(b) Point P lies in between the two spools on the line joint A and H, such that PH = 2AP
(c) Point P lies to the right of both the spools on the line joining A and H, such that AH = HP
(d) Point P lies at the intersection of the line joining B and C and the line joining G and F

[1999 : 2 Marks]

3. The number of inversions for a slider crank mechanism is

(a) 6 (b) 5
(c) 4 (d) 3

[2006 : 1 Mark]

4. Match the items in columns I and II.

Column-I
P. Higher kinematic pair
Q. Lower kinematic pair
R. Quick return mechanism
S. Mobility of a linkage

Column-II
1. Grubler's equation
2. Line contact
3. Euler's equation
4. Planer
5. Shaper
6. Surface contact

(a) P–2, Q–6, R–4, S–3
(b) P–6, Q–2, R–4, S–1
(c) P–6, Q–2, R–5, S–3
(d) P–2, Q–6, R–5, S–1

[2006 : 2 Marks]

5. Match the approaches given below to perform stated kinematics/dynamics analysis of machine.

Analysis	Approach
P. Continuous relative rotation	1. D' Alembert's principle
Q. Velocity and acceleration	2. Grubler's criterion
R. Mobility	3. Grashof's law
S. Dynamic - static analysis	4. Kennedy's theorem

(a) P-1, Q-2, R-3, S-4
(b) P-3, Q-4, R-2, S-1
(c) P-2, Q-3, R-4, S-1
(d) P-4, Q-2, R-1, S-3

6. Which of the following statements is INCORRECT?

(a) Grashofs rule states that for a planar crank-rocker four bar mechanism, the sum of the shortest and longest link lengths cannot be less than the sum of the remaining two link lengths.
(b) Inversions of a mechanism are created by fixing different links one at a time.
(c) Geneva mechanism is an intermittent motion device.
(d) Grubler's criterion assumes mobility of a planar mechanism to be one.

[2010 : 1 Mark]

7. A planar closed kinematic chain is formed with rigid links PQ = 2.0 m, QR = 3.0 m, RS = 2.5 m and SP= 2.7 m with all revolute joints. The link to be fixed to obtain rocker (rocker-rocker) mechanism is

(a) PQ (b) QR

(c) RS (d) SP

[2013 : 1 Mark]

8. In a statically determinate plane truss, the number of joints (j) and the number of members (m) are related by

(a) j = 2m – 3 (b) m = 2j + 1

(c) m = 2j – 3 (d) m = 2j – 1

[2014 : 1 Mark, Set 4]

Mechanism (Four Bar, Single Slider and Double Slider)

9. Figure shows a quick return mechanism. The cranks OA rotates clockwise uniformly. OA = 2 cm, OO' = 4 cm. The ratio of time for forward motion to that for return motion is

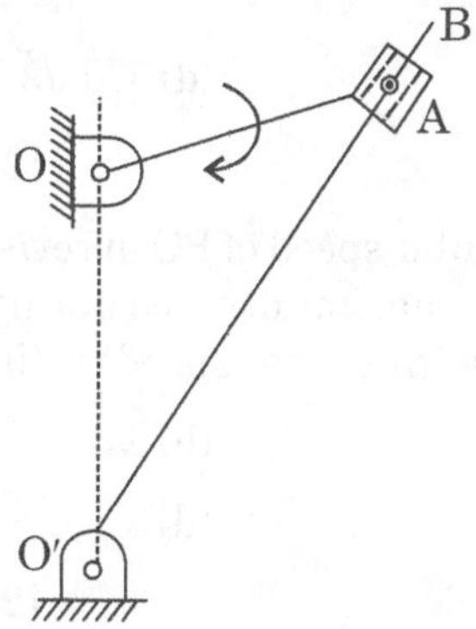

(a) 0.5 (b) 2.0

(c) $\sqrt{2}$ (d) 1

[1995 : 2 Marks]

10. A rod of length 1 m is sliding in a corner as shown in figure. At an instant when the rod makes an angle of 60 degrees with the horizontal plane., the velocity of point A on the rod is 1m/s. The angular velocity of the rod at this instant is

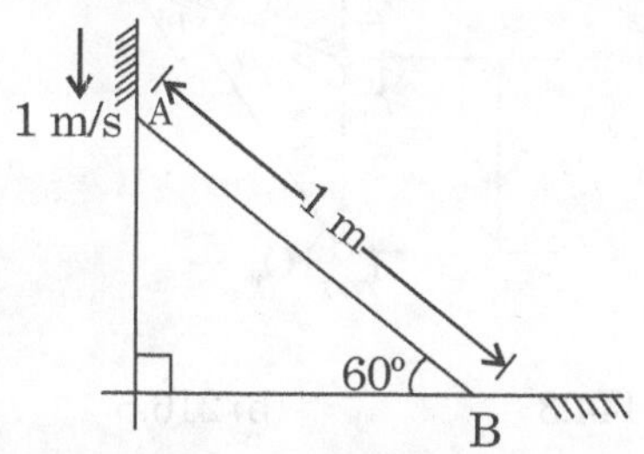

(a) 2 rad/s (b) 1.5 rad/s

(c) 0.5 rad/s (d) 0.75 rad/s

[1996 : 2 Marks]

11. The mechanism used in a shaping machine is

(a) a closed 4-bar chain having 4 revolute pairs

(b) a closed 6-bar chain having 6 revolute pairs

(c) a closed 4-bar chain having 2 revolute and 2 sliding pairs

(d) an inversion of the single slider-crank chain

[2003 : 1 Mark]

12. The lengths of the links of a 4-bar linkage with revolute pairs only are p, q, r and s units. Given that p < q< r< s. Which of these links should be the fixed one, for obtaining a "double crank" mechanism?

(a) links of length p (b) links of length q

(c) link of length r (d) link of length s

[2003 : 1 Mark]

13. Match the following:

Type of Mechanism

P. Scott - Russel mechanism

Q. Geneva mechanism

R. Off-set slider-crank mechanism

S. Scotch Yoke mechanism

Motion achieved

1. Intermittent motion
2. Quick return motion
3. Simple harmonic motion
4. Straight line motion

(a) P-2 Q-3 R-1 S-4

(b) P-3 Q-2 R-4 S-1

(c) P-4 Q-1 R-2 S-3

(d) P-4 Q-3 R-1 S-2

[2004 : 2 Marks]

Common Data for Q. 14 to 16:

An instantaneous configuration of a four-bar mechanism, whose plane is horizontal, is shown in the figure below. At this instant, the angular velocity and angular acceleration of link O_2A are ω = 8 rad/s and α = 0, respectively, and the driving torque (τ) is zero. The link O_2A is balanced so that its centre of mass falls at O_2.

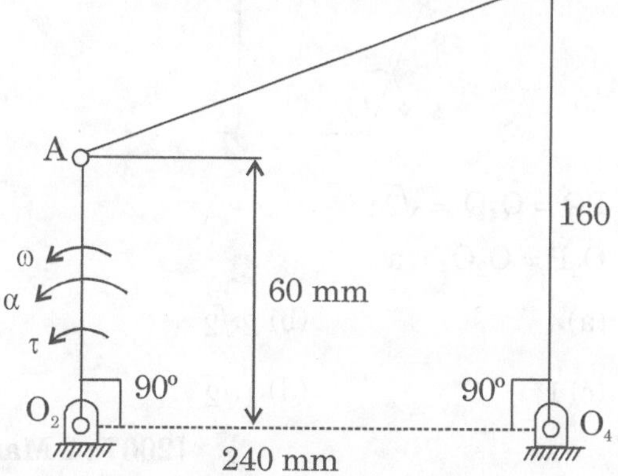

14. Which kind of 4-bar mechanism is O_2ABO_4?

(a) Double crank mechanism

(b) Crank-rocker mechanism

(c) Double rocker mechanism

(d) Parallelogram mechanism

[2005 : 2 Marks]

15. At the instant considered, what is the magnitude of the angular velocity of O_4B?

(a) 1 rad/s (b) 3 rad/s

(c) 8 rad/s (d) 64/3 rad/s

[2005 : 2 Marks]

16. At the same instant, if the component of the force at Joint A along AB is 30N, then the magnitude of the Joint reaction at O_2

(a) is zero

(b) is 30 N

(c) is 78 N

(d) cannot be determined from the given data

[2005 : 2 Marks]

17. In a four-bar linkage, S denotes the shortest link length, L is the longest link length, P and Q are the lengths of other two links. At least one of the three moving links will rotate by 360° if

(a) $S + L \leq P + Q$

(b) $S + L > P + Q$

(c) $S + P \leq L + Q$

(d) $S + P > L + Q$

[2006 : 2 Marks]

18. The input link O_2P of a four bar linkage is rotated at 2 rad/s in counter clockwise direction as shown below. The angular velocity of the coupler PQ in rad/s, at an instant when $\angle O_4O_2P = 180°$, is

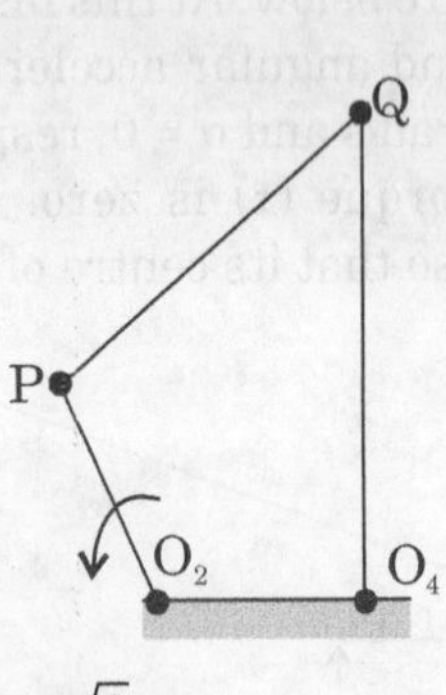

$PQ = Q_4Q = \sqrt{2}a$

$O_2P = O_2O_4 = a$

(a) 4 (b) $2\sqrt{2}$

(c) 1 (d) $1\sqrt{2}$

[2007 : 2 Marks]

Linked Answer Questions 19 and 20:

A quick return mechanism is shown below. The crank OS is driven at 2 rev/s in counter clockwise direction.

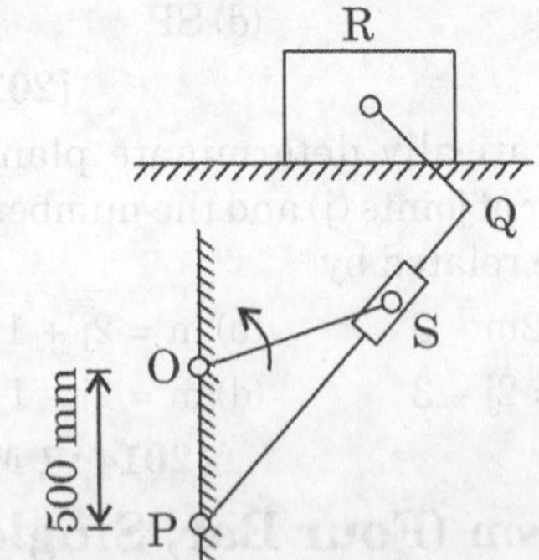

19. If the quick return ratio is 1 : 2, then the length of the crank in mm is

(a) 250 (b) $250\sqrt{3}$

(c) 500 (d) $500\sqrt{3}$

[2007 : 2 Marks]

20. The angular speed of PQ in rev/s when the block R attains maximum speed during forward stroke (stroke with slower speed) is (in rad/s)

(a) 1/3 (b) 2/3

(c) 2 (d) 3

[2007 : 2 Marks]

21. A simple quick return mechanism is shown in the figure. The forward to return ratio of the quick return mechanism is 2 :1. If the radius of the crank O, P is 125 mm, then the distance d(in mm) between the crank centre to lever pivot centre point should be

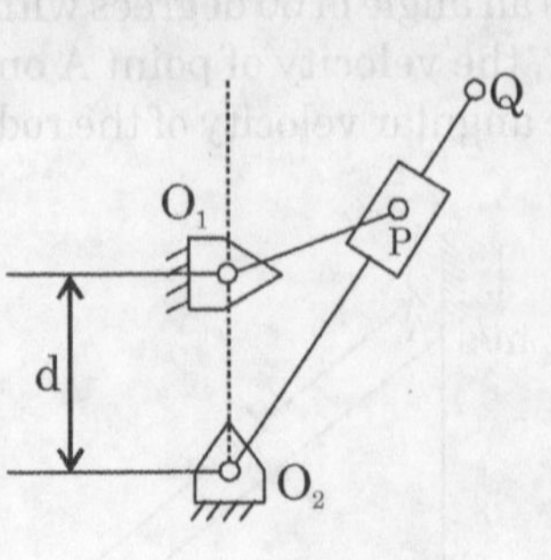

(a) 144.3 (b) 216.5

(c) 240.0 (d) 250.0

[2009 : 1 Mark]

22. For the four-bar linkage shown in the figure, the angular velocity of link AB is 1 rad/s. The length of link CD is 1.5 times the length of link AB. In the configuration shown, the angular velocity of link CD in rad/s is

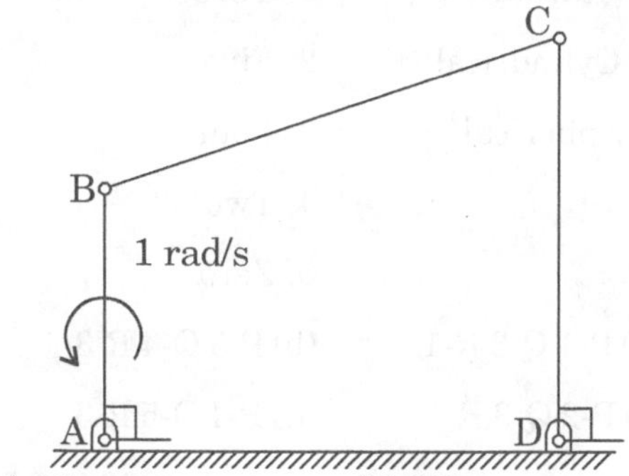

(a) 3 (b) 3/2
(c) 1 (d) 2/3

[2011 : 2 Marks]

23. A 4-bar mechanism with all revolute pairs has link lengths l_f = 20 mm, l_{in} = 40 mm, l_{co} = 50 mm and l_{out} = 60 mm. The suffixes 'f, in', 'co' and 'out' denote the fixed link, the input link, the coupler and output link respectively. Which one of the following statements is true about the input and output links?

(a) Both links can execute full circular motion
(b) Both links cannot execute full circular motion
(c) Only the output link cannot execute full circular motion
(d) Only the input link cannot execute full circular motion

[2014 : 1 Mark, Set-2]

24. Consider a slider crank mechanism with non zero masses and inertia. A constant torque τ is applied on the crank as shown in the figure. Which of the following plots best resembles variation of crank angle θ versus time

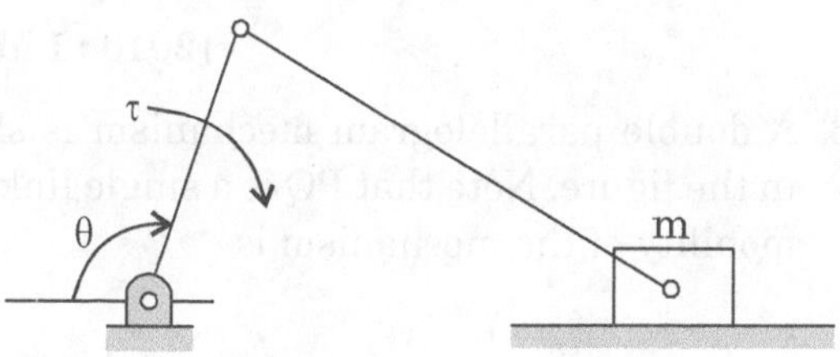

(a) 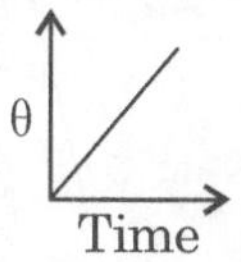

(b) 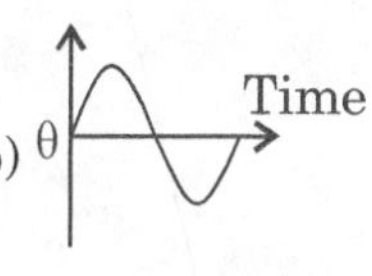

(c) 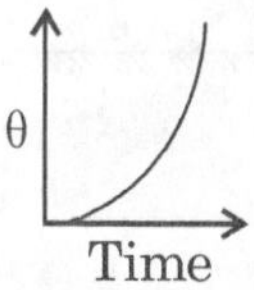

(d) 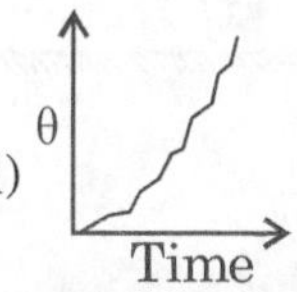

[2015 : 1 Mark, Set-1]

25. For an inline slider-crank mechanism, the length of the crank and connecting rod are 3 m and 4 m, respectively. At the instant when the connecting rod is perpendicular to the crank, if the velocity of the slider is 1 m/s, the magnitude of angular velocity (upto 3 decimal points accuracy) of the crank is ______ radian/s.

[2017 : 2 Marks, Set-1]

26. In a slider-crank mechanism, the lengths of the crank and the connecting rod are 100 mm and 160 mm, respectively. The crank is rotating with an angular velocity of 10 radian/s counter clockwise. The magnitude of linear velocity (in m/s) of the piston at the instant corresponding to the configuration shown in the figure is ___.

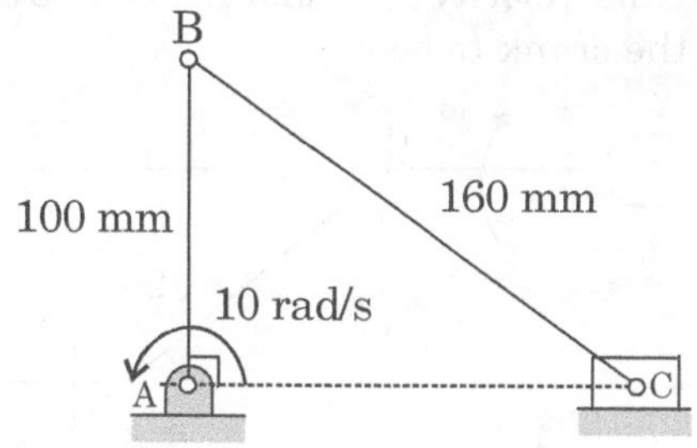

[2017 : 1 Mark, Set-2]

27. A four bar mechanism is made up of links of length 100, 200, 300 and 350 mm. If the 350 mm link is fixed, the number of links that can rotate fully is ________.

[2018 : 1 Marks, Set-1]

28. For an Oldham coupling used between two shafts, which among the following statements are correct?

I. Torsional load is transferred along shaft axis.
II. A velocity ratio of 1 : 2 between shaft is obtained without using gears.
III. Bending load is transferred transverse to shaft axis.
IV. Rotation is transferred along shaft axis.

(a) I and III (b) I and IV
(c) II and III (d) II and IV

[2018 : 1 Marks, Set-1]

29. In a four bar planar mechanism shown in the figure, AB = 5 cm, AD = 4 cm and DC = 2 cm. In the configuration shown, both AB and DC are perpendicular to AD. The bar AB rotates with an angular velocity of 10 rad/s. The magnitude of angular velocity (in rad/s) of bar DC at this instant is

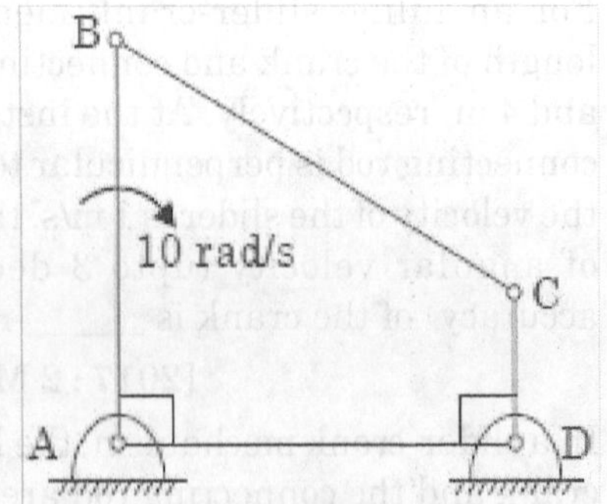

(a) 0 (b) 25
(c) 10 (d) 15

[2019 : 2 Marks, Set 1]

30. The crank of a slider-crank mechanism rotates counter-clockwise (CCW) with a constant angular velocity ω, as shown. Assume the length of the crank to be r.

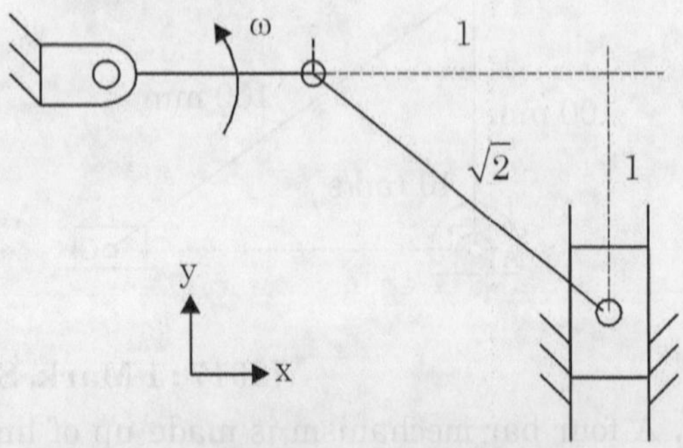

Using exact analysis, the acceleration of the slider in the y-direction, at the instant shown, where the crank is parallel to x-axis, is given by

(a) $-\omega^2 r$ (b) $2\omega^2 r$
(c) $\omega^2 r$ (d) $-2\omega^2 r$

[2019 : 2 Marks, Set-2]

Degrees of freedom

31. The number of degrees of freedom of a five link plane mechanism with five revolute pairs as shown in the figure is

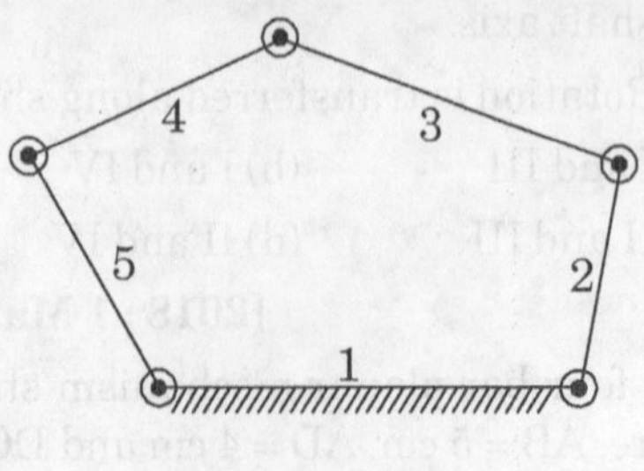

(a) 3 (b) 4
(c) 2 (d) 1

[1993 : 2 Marks]

32. Match the following with respect to spatial mechanisms.

Type of Joint	Degrees of constraint
P. Revolute	1. Three
Q. Cylindrical	2. Five
R. Spherical	3. Four
	4. Two
	5. Zero

(a) P-1 Q-3 R-1 (b) P-5 Q-4 R-3
(c) P-2 Q-3 R-1 (d) P-4 Q-5 R-3

[2004 : 2 Marks]

33. The number of degrees of freedom of a planar linkage with 8 links and 9 simple revolute joints is

(a) 1 (b) 2
(c) 3 (d) 4

[2005 : 1 Mark]

34. A planar mechanism has 8 links and 10 rotary joints. The number of degrees of freedom of the mechanism, using Grubler's criterion, is

(a) 0 (b) 1
(c) 2 (d) 3

[2008 : 1 Mark]

35. Mobility of a statically indeterminate structure is

(a) ≤ -1 (b) 0
(c) 1 (d) ≥ 2

[2010 : 1 Mark]

36. A double parallelogram mechanism is shown in the figure. Note that PQ is a single link. The mobility of the mechanism is

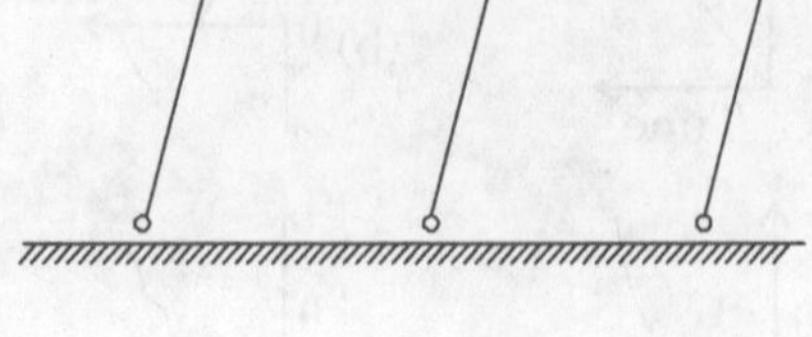

(a) -1 (b) 0
(c) 1 (d) 2

[2011 : 1 Mark]

37. The number of degrees of freedom of the linkage shown in the figure is

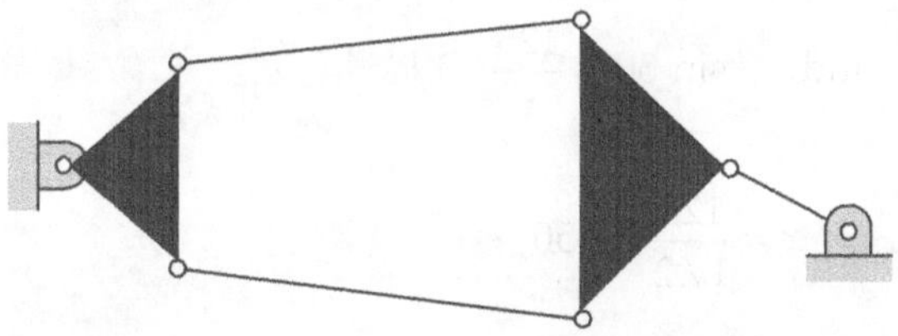

(a) –3 (b) 0
(c) 1 (d) 2

[2015 : 2 Marks, Set-3]

38. The number of degrees of freedom in a planar mechanism having n links and j simple hinge joints is

(a) 3(n – 3) – 2j (b) 3(n – 1) – 2j
(c) 3n – 2j (d) 2j – 3n + 4

[2016 : 1 Mark, Set-3]

ANSWERS

1. (b)	**2.** (c)	**3.** (c)	**4.** (d)	**5.** (b)	**6.** (a)	**7.** (c)	**8.** (c)	**9.** (b)	**10** (a)
11. (d)	**12.** (a)	**13.** (c)	**14.** (b)	**15.** (b)	**16.** (d)	**17.** (a)	**18.** (c)	**19.** (a)	**20.** (b)
21. (d)	**22.** (d)	**23.** (a)	**24.** (d)	**28.** (b)	**29.** (b)	**30.** (c)	**31.** (c)	**32.** (c)	**33.** (c)
34. (b)	**35.** (a)	**36.** (c)	**37.** (c)	**38.** (b)					

EXPLANATIONS

3. There are four number of inversions for a slider crank mechanism.

7. The link opposite to shortest link is fixed.

9. Ratio of time of forward motion to return motion

$$= \frac{180 + 2\alpha}{180 - 2\alpha} = \frac{180 + 60}{180 - 60} = \frac{240}{120} = 2$$

(Given OD = 2 cm, OO′ = 4 cm, $\sin \alpha = \frac{2}{4} = 0.5$ $\Rightarrow \alpha = 30°$)

10.

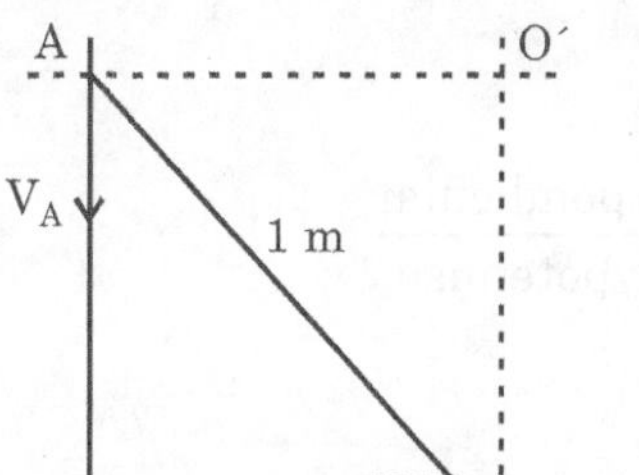

$$V_A = O'A \times \omega$$
$$1 = 1 \cos 60° \times \omega$$
$$\omega = 2 \text{ rad/s}$$

12. To obtain a "DOUBLE CRANK MECHANISM" shortest link is always fixed. While obtaining a "DOUBLE LEVER MECHANISM", the link opposite to the "SHORTEST LINK" is fixed.

14.
$$AB = \sqrt{(240)^2 + (160 - 60)^2}$$
$$= 260 \text{ mm}$$
$$= l_{AO_2} + l_{O_2O_4} < l_{AB} + l_{BO_4}$$

and O_2O_4 is fixed link. It will act as a crank-rocker mechanism

15.
$$\frac{\omega_4}{\omega_2} = \frac{I_{12}I_{24}}{I_{14}I_{24}} = \frac{144}{240 + 144}$$
$$\frac{\omega_4}{8} = \frac{144}{384}$$
$$\therefore \quad \omega_4 = \frac{144}{384} \times 8 = 3 \text{ rad/s}$$

17. According to Grashoff's Criteria.

$S + L \leq P + Q$

18.

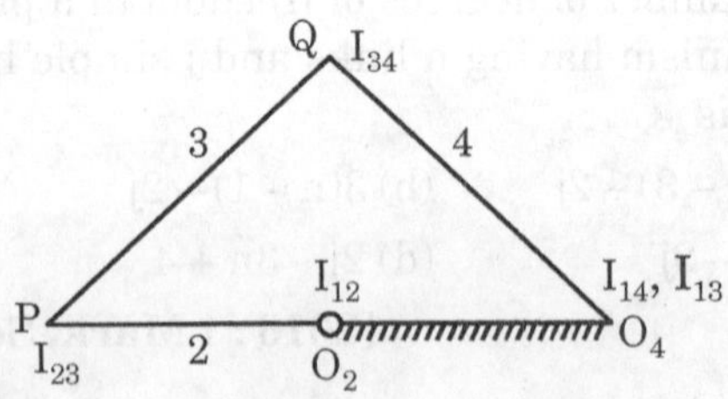

When $\angle O_4O_2P = 180°$

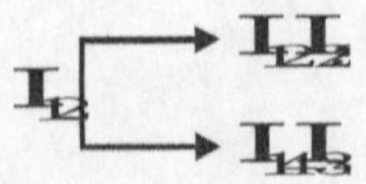

Now, $\frac{\omega_3}{\omega_2} = \frac{I_{12}I_{23}}{I_{13}I_{23}} = \frac{a}{2a}$

$\frac{\omega_3}{2} = \frac{1}{2}$

$\omega_3 = 1$ rad/s

20. Maximum speed during forward stroke occur when QR & QS are perpendicular.

$V = 250 \times 2 = 750 \times \omega_{PQ}$

$\omega_{PQ} = \frac{2}{3}$ rad/sec

21.

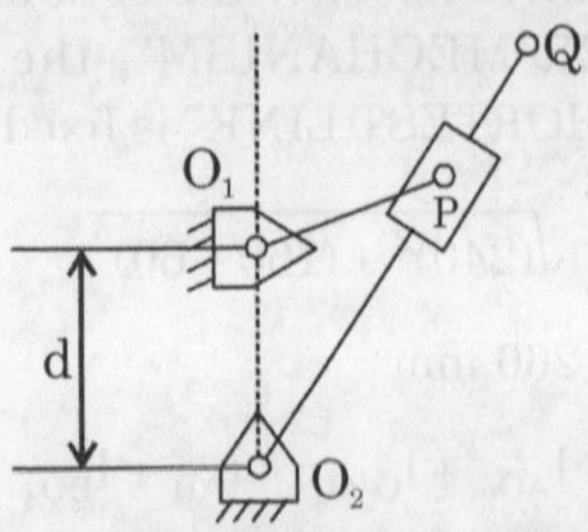

$\frac{2}{1} = \frac{360 - \alpha}{\alpha}$

or $2\alpha = 360 - \alpha$

or $\alpha = 120°$

$\therefore \frac{\alpha}{2} = 60°$

i.e. $\angle PO_2O_1 = 30°$

In ΔPO_2O_1

$\tan 30° = \frac{O_1P}{d}$

$\Rightarrow d = \frac{125}{\tan 30°} = 219.29$

and $\sin 30° = \frac{125}{x}$

$\Rightarrow x = \frac{125}{1/2} = 250$

22. For the given configuration

$V_{AB} = V_{CD}$

$\therefore \omega_{AB} AB = \omega_{CD} CD$

$\Rightarrow \omega_{CD} = \omega_{AB} \frac{AB}{CD}$

$= 1 \times \frac{1}{1.5}$

$= \frac{2}{3}$ rad/s

23. $S + L < P + Q$

$20 + 60 < 40 + 50$

$\Rightarrow 80 < 90$

If smaller link is fixed both input and Output link execute full circular motion.

25.

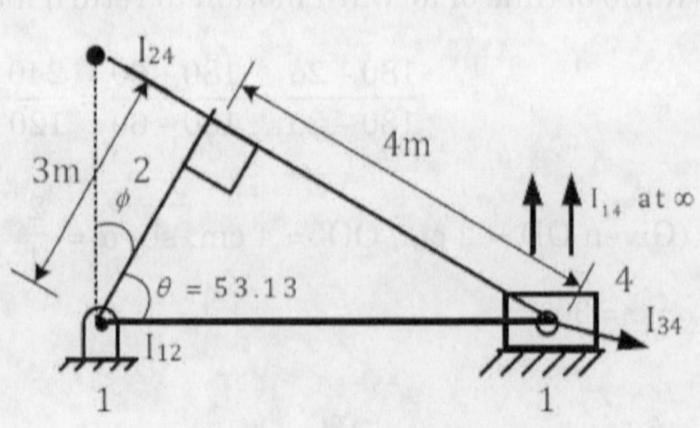

Here, $\tan\theta = \frac{\text{Perpendicular}}{\text{Hypotenuse}}$

$\tan \theta = \frac{4}{3}$

$\Rightarrow \theta = 53.13$

$\phi = 90 - \theta = 36.87$

$$\cos\phi = \frac{3}{I_{12} - I_{24}}$$

$$\cos 36.87 = \frac{3}{I_{12} - I_{24}}$$

$I_{12} - I_{24} \quad = 3.75$

$V_{I_{24}}$ = velocity of slider

$= \omega_2(I_{12} - I_{24})$

$I = \omega_2 \times 3.75$

$\omega_2 \; = 0.267$ radian/second

26.

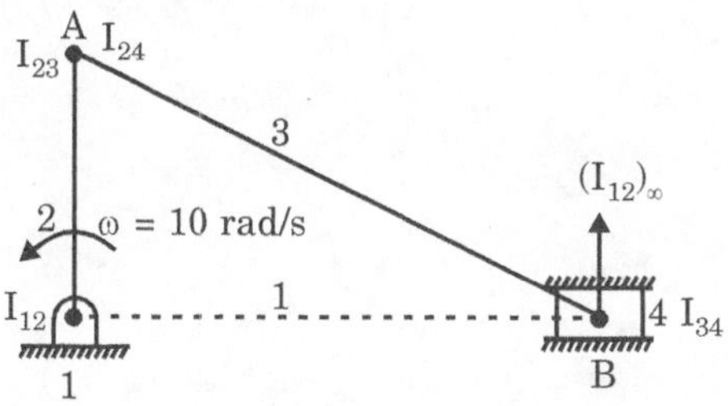

$$\omega_2(I_{24}I_{12}) = V_4 = V_B$$

$$V_B = \omega_2(I_{24}I_{12}) = 10 \times 0.1$$

$$V_B = 1 \text{ m/s}$$

27. Here the link, that is fixed is adjacent to the smaller one.

so in this condition we will get on fully rotation (crank)

⇒ if shortest link is fixed

⇒ Double crank (full rotation)

⇒ if opposite to shortest link is fixed

⇒ No Crank.

28. The oldham coupling connects parallel shafts having small offset and in this coupling the motion (rotation) and torsion load transfer is along axes of shafts.

29. $V_C = V_B$

$= AB \times \omega_{AB}$

$= (0.05 \times 10)$ m/s

Now, $\omega_{CD} = \dfrac{V_C}{CD} = \dfrac{0.05 \times 10}{0.02} = 25$ rad/s

30. Exactly analysis means exact dimensions have to be taken. No approximation is allowed. It means velocity diagram and acceleration diagram with exact geometry (dimensions) need to be drawn.

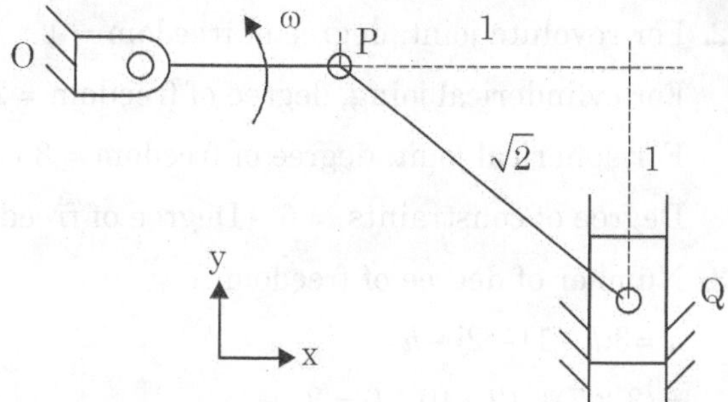

The velocity diagram of the above problem is as follows :

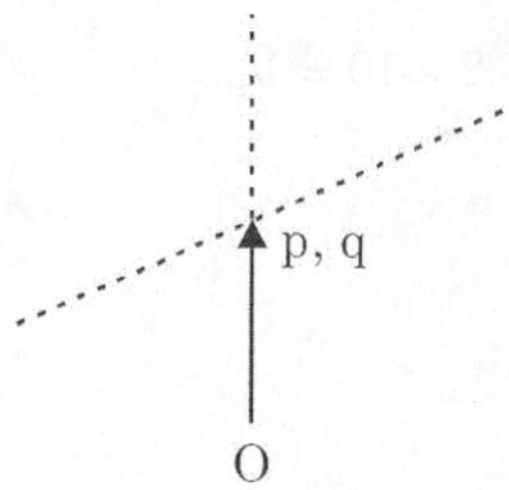

$\vec{op}$ perpendicular to OP

$\vec{oq}$ in the direction of movement of slider.

Since, p, q are coinciding i.e. no relative motion between p and q

$V_P = V_Q = r\omega$

and, $V_{qp} = 0$ and $\omega_{qp} = 0$

∴ There will be no radial acc. of link PQ.

Only tangential acc. $\left(a^t_{qp}\right)$ will be there.

Also for link OP, $\alpha = 0$ i.e. no tangential acceleration

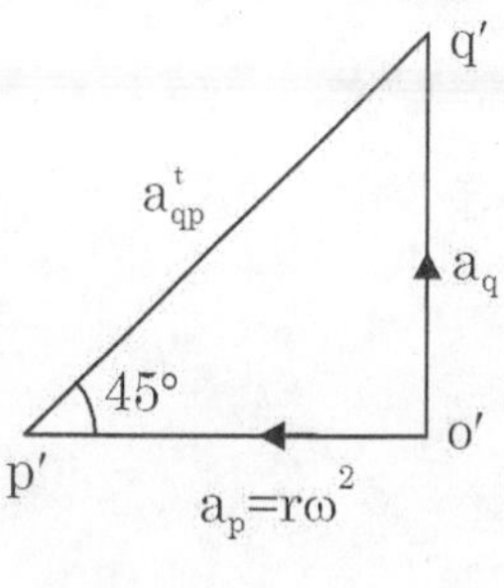

$$\tan 45° = \frac{a_q}{r\omega}$$

$$\Rightarrow a_q = r\omega^2$$

31. DOF $= 3(n-1) - 2j - h$

$n = 5, j = 5$

DOF $= 3(5-1) - 2 \times 5 = 2$

32. For revolute joint, degree of freedom = 1

For cylinderical joing, degree of freedom = 2

For spherical joint, degree of freedom = 3

Degree of constraints = 6 – Degree of freedom

33. Number of degree of freedom,

$n = 3(l - 1) - 2j - h$

$= (3 \times 7) - (2 \times 9) - 0 = 3$

34. By Gruebler's criterion,

$F = 3(l - 1) - 2j$

$= 3(8 - 1) - 2 \times 10 = 1$

35. Mobility (DOF) for a statically indeterminate structure is always less than zero.

36. There is one link which is redundant in nature. Therefore, DOF = 1.

37. Number of links, N = 6

Total number as binary joints, $j = 7$

$F = 3(N - 1) - 2j = 15 - 14 = 1.$

38. DOF= $3(n - 1) - 2j$

2 CHAPTER Velocity & Acceleration Analysis

Analysis of Previous GATE Papers														
Topic	Year/Set → Marks Distribution ↓		2019 Set 1	2019 Set 2	2018 Set 1	2018 Set 2	2017 Set 1	2017 Set 2	2016 Set 1	2016 Set 2	2016 Set 3	2015 Set 1	2015 Set 2	2015 Set 3
Velocity Analysis	1 Mark	MCQ Type		1							1			
		Numerical Type												
	2 Marks	MCQ Type				1								
		Numerical Type		1					1	1				
	Total			3		2			2	2	1			
Acceleration Analysis	1 Mark	MCQ Type												1
		Numerical Type												
	2 Marks	MCQ Type				1								
		Numerical Type						1						
	Total					2		2						1
Instantaneous Centre of Rotation(ICR)	1 Mark	MCQ Type												
		Numerical Type												
	2 Marks	MCQ Type												
		Numerical Type												
	Total													

Velocity analysis

Data for Question Q. 1

The circular disc shown in its pian view in the figure rotates in a plane parallel to the horizontal plane about the point O at a uniform angular velocity ω. Two other points A and B are located on the line OZ at distances r_A and r_B from O respectively.

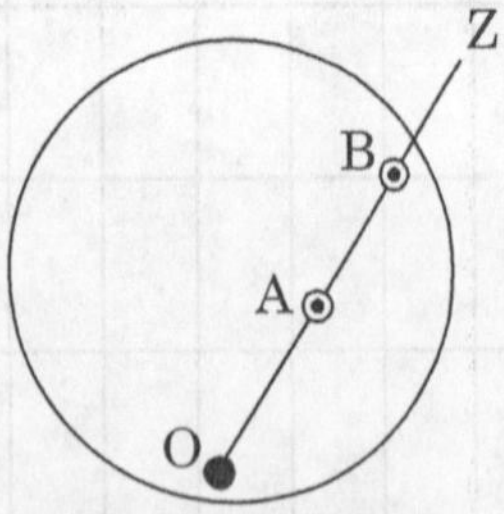

1. The velocity of point B with respect to point A is a vector of magnitude

(a) 0

(b) $\omega (r_B - r_A)$ and direction opposite to the direction of motion of point B

(c) $\omega (r_B - r_A)$ and direction same as the direction of motion of point B.

(d) $\omega^2 (r_B - r_A)$ and direction being from O to Z

[2003 : 2 Marks]

2. There are two points Pand O on a planar rigid body. The relative velocity between the two points

(a) should always be along PQ

(b) can be oriented along any direction

(c) should always be perpendicular to PQ

(d) should be along QP when the body undergoes pure translation

[2010 : 1 Mark]

3. For the configuration shown, the angular velocity of link AB is 10 rad/s counter clockwise. The magnitude of the relative sliding velocity (in ms^{-1}) of slider B with respect to rigid link CD is

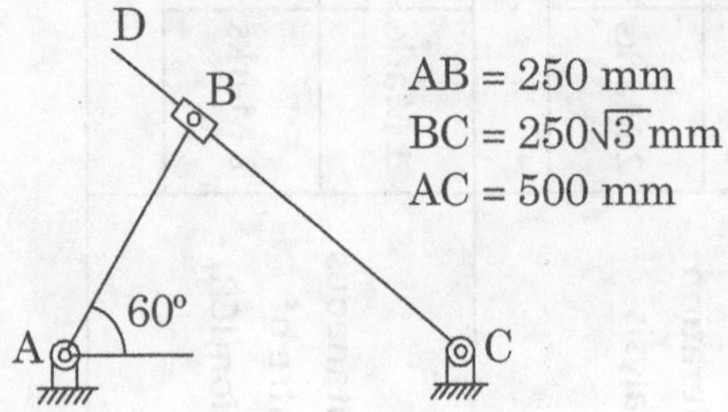

(a) 0 (b) 0.86

(c) 1.25 (d) 2.50

[2010 : 2 Marks]

4. A rigid link PQ is 2 m long and oriented at 20° to the horizontal as shown in the figure. The magnitude and direction of velocity V_Q, and the direction of velocity V_p are given. The magnitude of V_p (in m/s) at this instant is

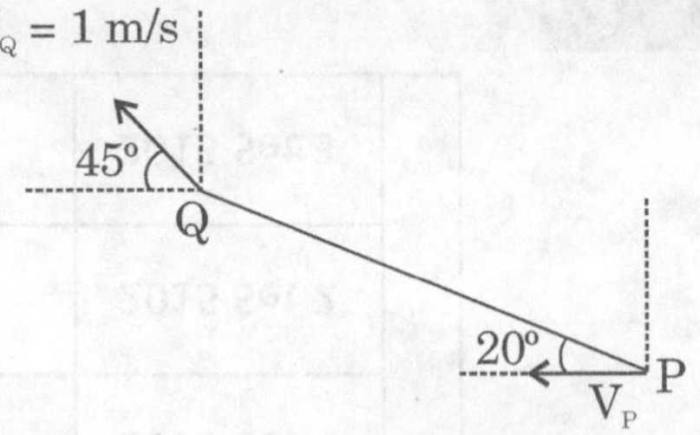

(a) 2.14 (b) 1.89

(c) 1.21 (d) 0.96

[2014 : 1 Mark, Set-1]

5. The rod AB, of length 1 m, shown in the figure is connected to two sliders at each end through pins. The sliders can slide along QP and QR. If the velocity V_A of the slider at A is 2 m/s, the velocity of the midpoint of the rod at this instant is _____ m/s.

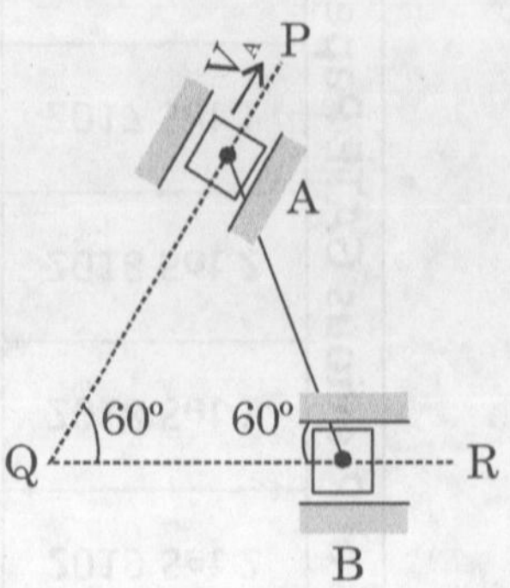

[2016 : 2 Marks, Set-2]

6. A slider crank mechanism with crank radius 200 mm and connecting rod length 800 mm is shown. The crank is rotating at 600 rpm in the counterclockwise direction. In the configuration shown, the crank makes an angle of 90° with the sliding direction of the slider, and a force of 5 kN is acting on the slider. Neglecting the inertia forces, the turning moment on the crank (in kNm) is _______.

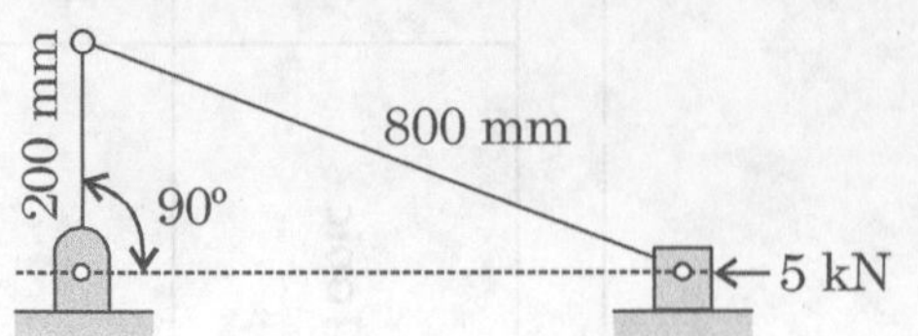

[2016 : 2 Marks, Set-1]

7. A rigid link PQ is undergoing plane motion as shown in the figure (V_P and V_Q are non-zero). V_{QP} is the relative velocity of point Q with respect to point P

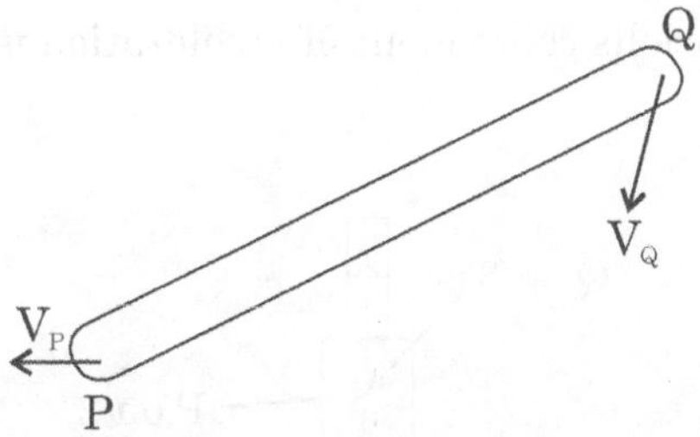

Which one of the following is TRUE?

(a) V_{QP} has components along and perpendicular to PQ

(b) V_{QP} has only one component directed from P to Q

(c) V_{QP} has only one component directed from Q to P

(d) V_{QP} has only one component perpendicular to PQ

[2016 : 1 Mark, Set-3]

8. A rigid rod of length 1 m is resting at an angle $\theta = 45°$ as shown in the figure. The end P is dragged with a velocity of U = 5 m/s to the right. At the instant shown, the magnitude of the velocity V(in m/s) of point Q as it moves along the wall without losing contact is

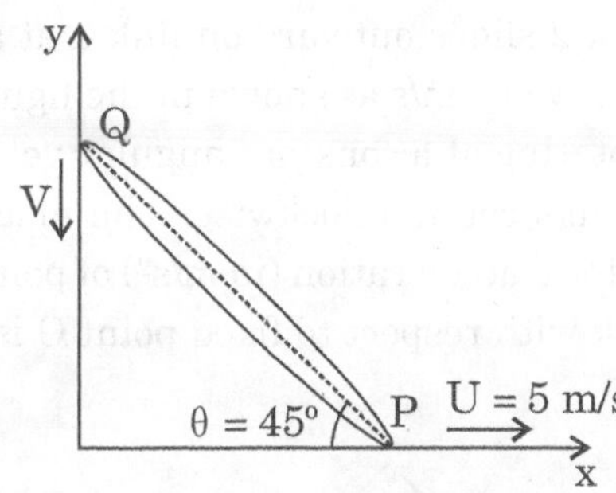

(a) 5 (b) 6

(c) 8 (d) 10

[2018 : 2 Marks, Set-2]

9. A rigid triangular body, PQR, with sides of equal length of 1 unit moves on a flat plane.

At the instant shown, edge QR is parallel to the x-axis, and the body moves such that velocities of points P and R are VP and VR, in the x and y directions, respectively. The magnitude of the angular velocity of the body is

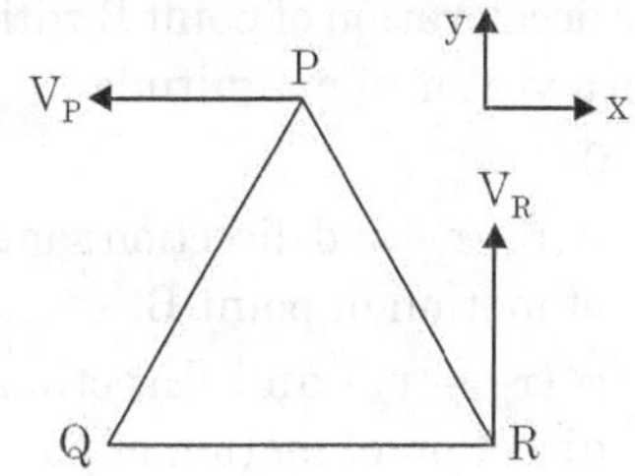

(a) $V_P/\sqrt{3}$ (b) $2V_R$

(c) $2V_P$ (d) $V_R/\sqrt{3}$

[2019 : 1 Mark, Set 2]

10. A four bar mechanism is shown in the figure. The link numbers are mentioned near the links. Input link 2 is rotating anticlockwise with a constant angular speed ω_2. Length of different links are

$O_2O_4 = O_2A = L,$

$AB = O_4B = \sqrt{2}L$

The magnitude of the angular speed of the output link 4 is ω_4 at the instant when link 2 makes an angle of 90° with O_2O_4 as shown. The ratio $\frac{\omega_4}{\omega_2}$ is _____ (round off to two decimal places).

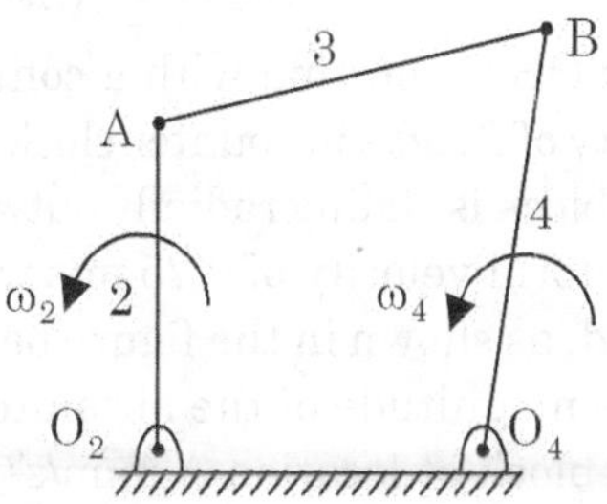

[2019 : 2 Marks, Set 2]

Acceleration analysis

Data for Question Q. 11

The circular disc shown in its plan view in the figure rotates in a plane parallel to the horizontal plane about the point O at a uniform angular velocity ω. Two other points A and B are located on the line OZ at distances r_A and r_B from O respectively.

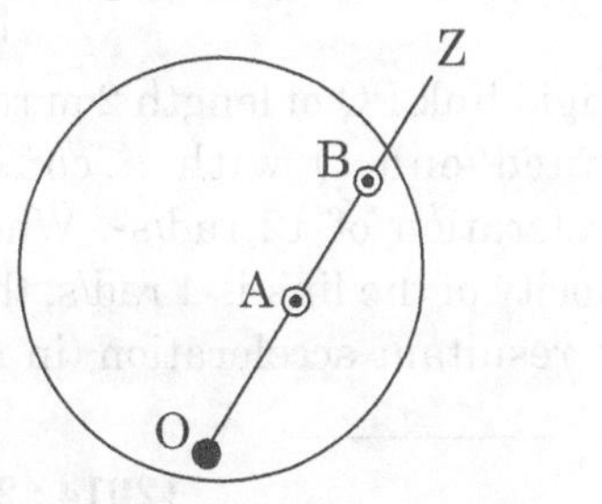

11. The acceleration of point B with respect to point A is a vector of magnitude

(a) 0

(b) $\omega\,(r_B - r_A)$ and direction same as the direction of motion of point B.

(c) $\omega\,(r_B - r_A)$ and direction opposite to be direction of motion of point B.

(d) $\omega^2\,(r_B - r_A)$ and direction being from Z to O.

[2003 : 2 Marks]

12. In the figure shown, the relative velocity of link 1 with respect of link 2 is 12 m/s. Link 2 rotates at a constant speed of 120 rpm. The magnitude of Coriolis component of acceleration of link 1 is

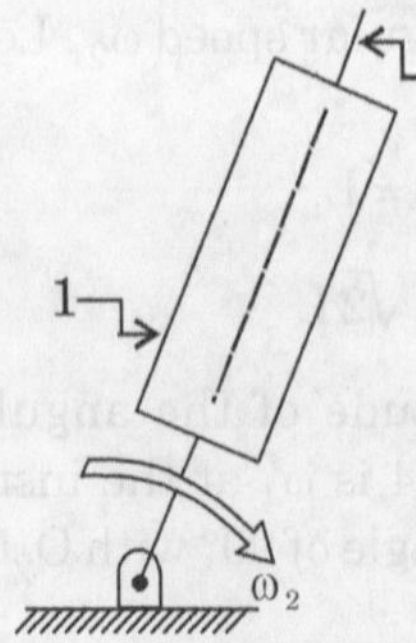

(a) 302 m/s² (b) 906 m/s²
(c) 604 m/s² (d) 1208 m/s²

[2004 : 2 Marks]

13. A link OB is rotating with a constant angular velocity of 2 rad/s in counter clockwise direction and a block is sliding radially outward on it with an uniform velocity of 0.75 m/s with respect to the rod, as shown in the figure below. If OA = 1 m, the magnitude of the absolute acceleration of the block at location A in m/s² is

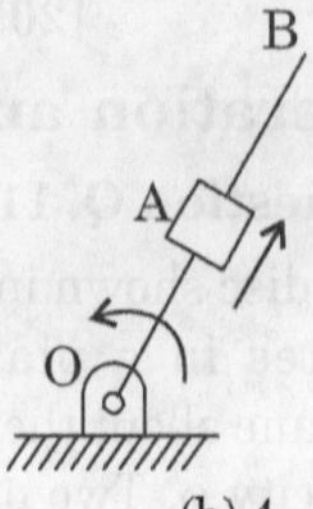

(a) 3 (b) 4
(c) 5 (d) 6

[2013 : 1 Mark]

14. A rigid link PQ of length 2 m rotates about the pinned end Q with a constant angular acceleration of 12 rad/s². When the angular velocity of the link is 4 rad/s, the magnitude of the resultant acceleration (in m/s²) of the end P is ________.

[2014 : 2 Marks, Set-2]

15. In the figure, link 2 rotates with constant angular velocity ω_2. A slider link 3 moves outwards with a constant relative velocity $V_{Q/P}$, where Q is a point on slider 3 and P is a point on link 2. The magnitude and direction of Coriolis component of acceleration is given by

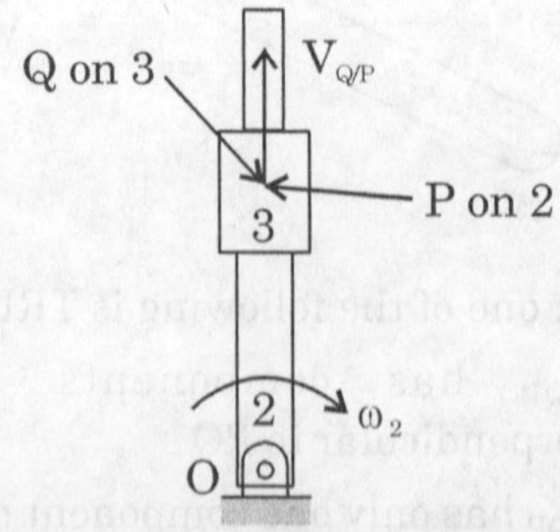

(a) $2\omega_2 V_{Q/P}$, direction of $V_{Q/P}$ rotated by 90° in the direction ω_2

(b) $\omega_2 V_{Q/P}$; direction of V_{QP} rotated by 90° in the direction ω_2

(c) $2\omega_2 V_{Q/P}$, direction of $V_{Q/P}$ rotated by 90° opposite to the direction of ω_2

(d) $\omega_2 V_{Q/P}$; direction of $V_{Q/P}$ rotated by 90° opposite to the direction ω_2

[2015 : 1 Mark, Set-3]

16. Block 2 slides outward on link 1 at a uniform velocity of 6 m/s as shown in the figure. Link 1 is rotating at a constant angular velocity of 20 radian/s counterclockwise. The magnitude of the total acceleration (in m/s²) of point P of the block with respect to fixed point O is ______.

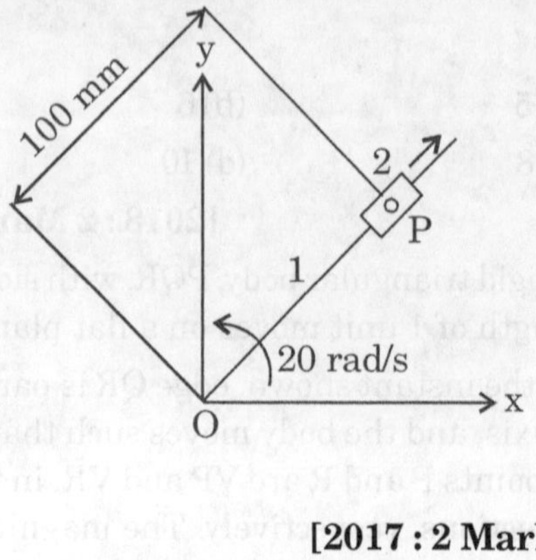

[2017 : 2 Marks, Set-2]

17. In a rigid body in plane motion, the point R is accelerating with respect to point P at $10 \angle 180^\circ$ m/s^2. If the instantaneous acceleration of point O is zero, the acceleration (in m/s^2) of point R is

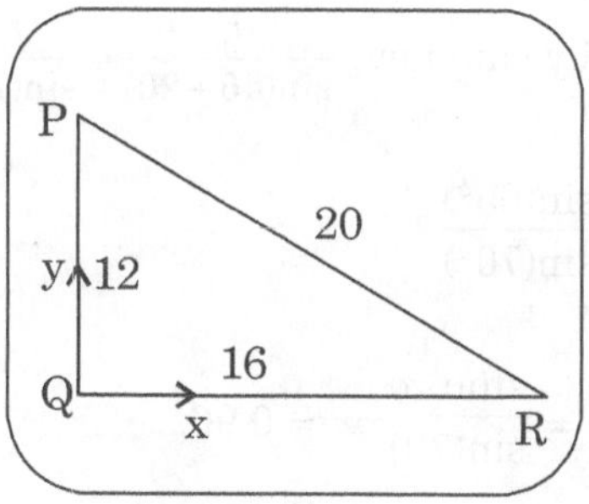

(a) $8 \angle 233^\circ$ (b) $10 \angle 225^\circ$

(c) $10 \angle 217^\circ$ (d) $8 \angle 217^\circ$

[2018 : 2 Marks, Set-2]

Instantaneous centre

18. Instantaneous centre of a body rolling with sliding on a stationary curved surface lies

(a) at the point of contact

(b) on the common normal at the point of contact

(c) on the common tangent at the point of contact

(d) at the center of curvature of the stationary surface

[1992 : 2 Marks]

19. The figure below shows a planar mechanism with single degree of freedom. The instant center 24 for the given configuration is located at a position

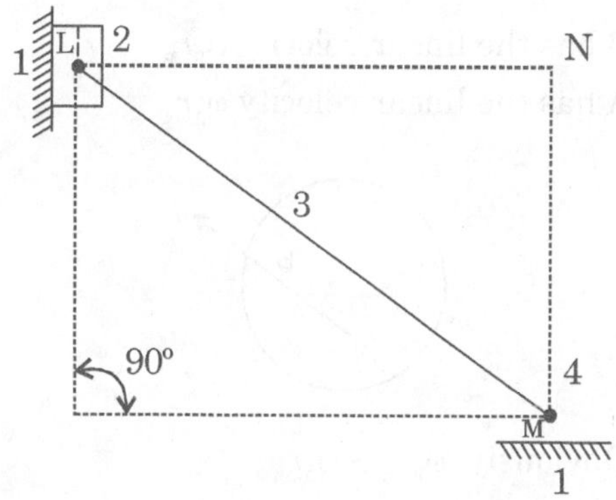

(a) L (b) M

(c) N (d) ∞

[2004 : 2 Marks]

20. Match the items in columns I and II.

Column I

P. Addendum

Q. Instantaneous center of velocity

R. Section modulus

S. Prime circle

Column II

1. Cam
2. Beam
3. Linkage
4. Gear

(a) P–4, Q–2, R–3, S–1

(b) P–4, Q–3, R–2, S–1

(c) P–3, Q–2, R–1, S–4

(d) P–3, Q–4, R–1, S–2

[2006 : 2 Marks]

ANSWERS

1. (c)	**2.** (b)	**3.** (d)	**4.** (d)	**7.** (d)	**8.** (a)	**9.** (b)	**11.** (d)	**12.** (a)	**13.** (c)
15. (a)	**17.** (d)	**18.** (a)	**19.** (d)	**20.** (b)					

EXPLANATIONS

1. B has the linear velocity $\omega_B r_B$

A has the linear velocity $\omega_A r_A$

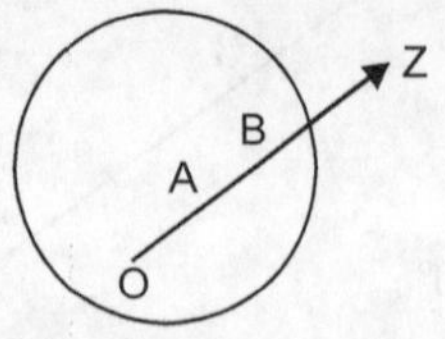

Obviously, $\omega_B r_B > \omega_A r_A$

because $\omega_B = \omega_B = \omega$ and $r_B > r_A$

$\therefore \omega r_B > \omega r_A$

So relative velocity = $\omega(r_B - r_A)$ in the direction of motion of point B.

and, it will experience a centripetal acceleration of $\omega^2(r_B - r_A)$ towards "O"

3. Given : ω_{AB} = 10 rad/sec (CCW)

AB = 250 mm, BC = $250\sqrt{3}$, AC = 500

Direction of velocity of AB is perpendicular to link AB and direction of velocity of AB is parallel to line CD.

Also, direction of relative velocity of slides B with respect to C is in line with link BC.

$\therefore \quad V_C = 0$

But $V_B = r_{AB} \times W_{AB} = 250 \times 10 \times 10^{-3} = 2.5$ m/sec

$\therefore \quad V_{BC} = V_B - V_C = 2.5$ m/sec

4. By Instantaneous center method; say ω = angular velocity of rod

$v_q = r_q.\omega \quad ...(i)$

$v_p = r_p.\omega ...(ii)$

Dividing equation (*i*) with equation (*ii*) we get

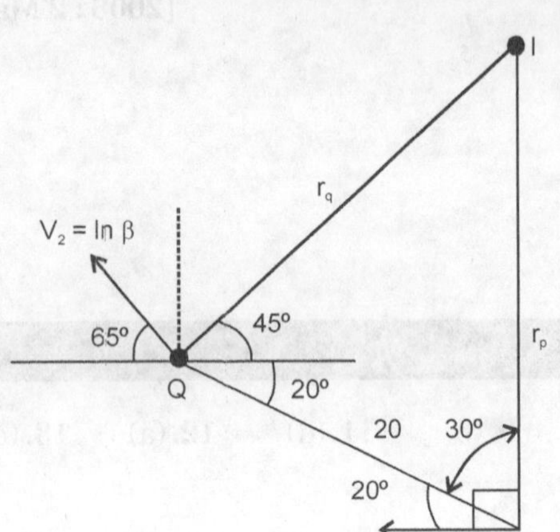

$$\frac{v_q}{v_p} = \frac{r_q}{r_p}; v_p = \frac{r_p}{r_q}.v_q$$

Now by sine law, $\dfrac{r_p}{\sin(45+20)} = \dfrac{r_q}{\sin(70°)}$

$$\frac{r_p}{r_q} = \frac{\sin(65°)}{\sin(70°)}$$

$$\text{So, } v_p = \frac{\sin(65)}{\sin(70)} v_q = 0.96.$$

5. Locate the I-centre for the link AB as shown in figure M is the mid point of AB

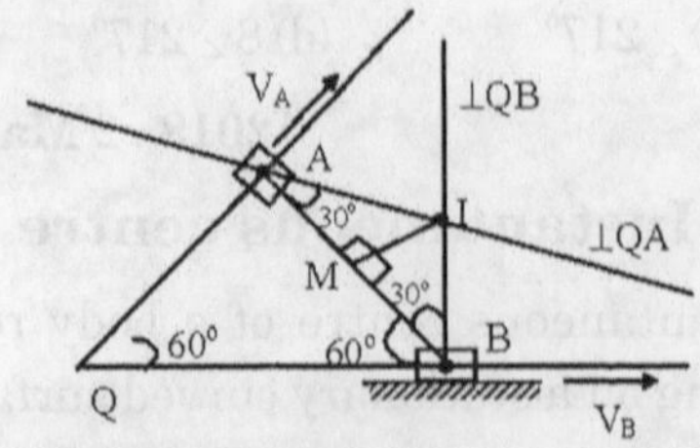

Given V_A = 2 m/sec

$$V_A = IA.\omega \Rightarrow \omega = \frac{V_A}{IA}$$

$$V_M = IM.\omega = IM\frac{V_A}{IA} = \frac{IM}{IA}.V_A$$

$$= \sin 30°.\ V_A = \frac{1}{2}.2 = 1 \text{ m/sec}$$

6. Turning moment on crankshaft,

$$T = F_T \times r$$

$$= \frac{F}{\cos\beta} \times \cos\beta \times r$$

$$= F \times r = 5 \times 0.2 = 1 \text{ kN-m}$$

7.

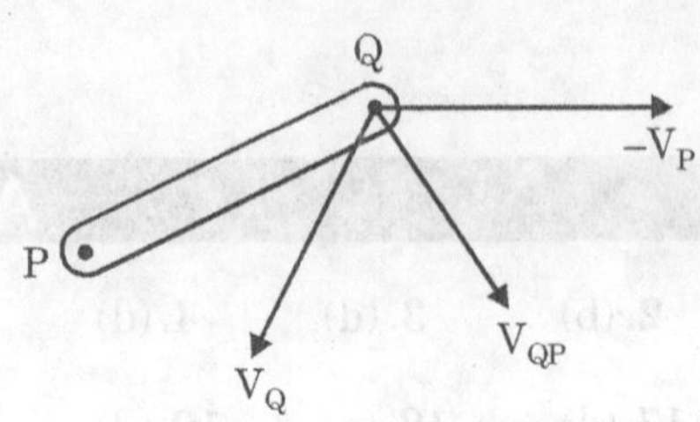

Resultant V_{QP} is $\perp_r$ to link PQ.

8.

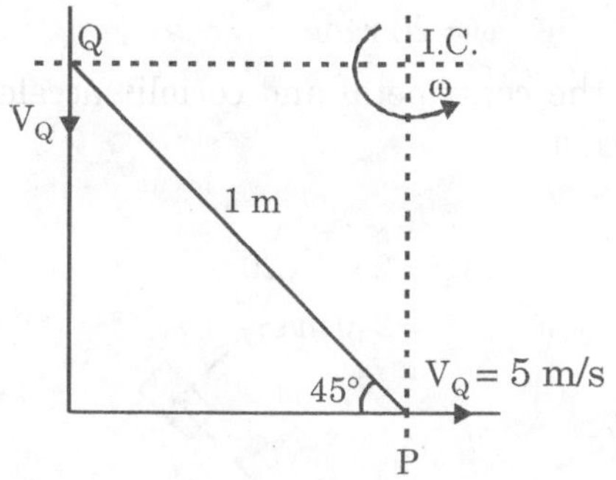

$$V_P = r_P\omega$$

$$\Rightarrow \quad 5 = 1 \times \sin 45° \times \omega$$

$$\Rightarrow \quad \omega = \frac{5}{1 \times \sin 45°}$$

$$V_Q = r_Q\omega$$

$$\Rightarrow \quad V_Q = 1 \times \cos 45° \times \frac{5}{1 \times \sin 45°}$$

$$= 5 \text{ m/s}$$

9.

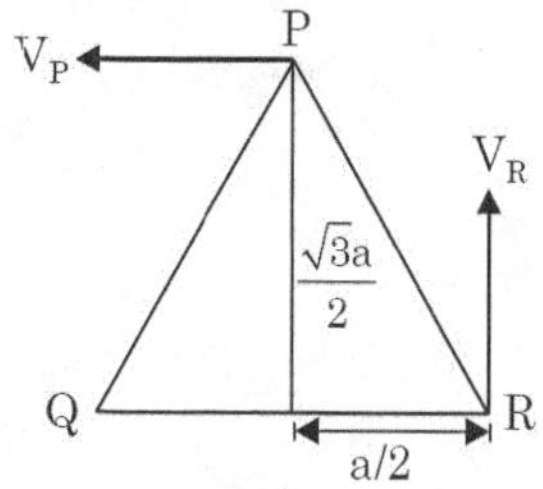

$$\frac{2V_P}{\sqrt{3}a} = \frac{2V_R}{a} = \omega_{body}$$

$$V_P = \sqrt{3}V_R \text{ (a = 1 unit)}$$

$$\omega_{body} = 2V_R$$

10.

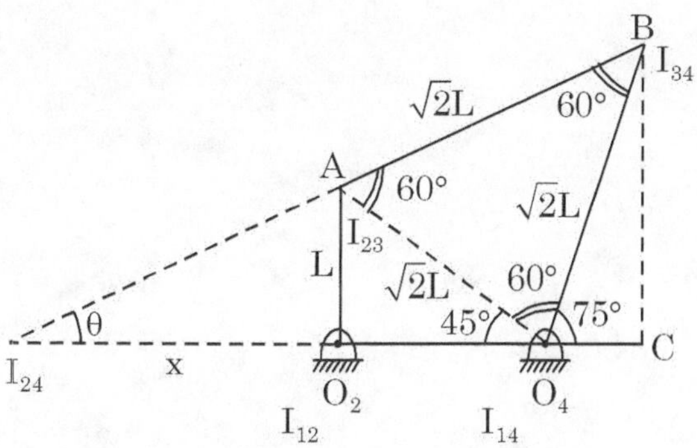

$$\frac{\theta_4}{\omega_2} = \frac{I_{24}I_{12}}{I_{24}I_{14}} = \frac{x}{x+L} = \frac{1}{1+\frac{L}{x}}$$

Now, $O_4C = \cos 75°$

$O_4C = \sin 75°$

$$\Delta AO_2I_{24} \approx \Delta BI_{24}C$$

$$\frac{L}{x} = \frac{\sqrt{2}L\sin 75°}{x + L + \sqrt{2}\cos 75°}$$

$$\frac{L}{x} = \frac{\sqrt{2}\frac{L}{x}\sin 75°}{1 + \frac{L}{x} + \sqrt{2}\frac{L}{x}\cos 75°}$$

$$\Rightarrow \quad \frac{L}{x} = 0.2679$$

$$\frac{\omega_4}{\omega_2} = \frac{1}{1+0.267} = 0.7887 \approx 0.79$$

11. Acceleration of point B with respect to point A

$$a_{B/A} = \bar{\omega} \times \bar{v}_{B/A}$$

$$= \omega \cdot \omega\,(r_B - r_A)$$

$$= \omega^2(r_B - r_A)$$

and direction from Z to O.

12. Velocity of link 1 with respect to link 2,

$V_{1/2}$ = 12 m/s

$$\omega = \frac{2\pi N}{60} = \frac{2\pi \times 120}{60} = 4\pi$$

∴ Corioli's component of acceleration

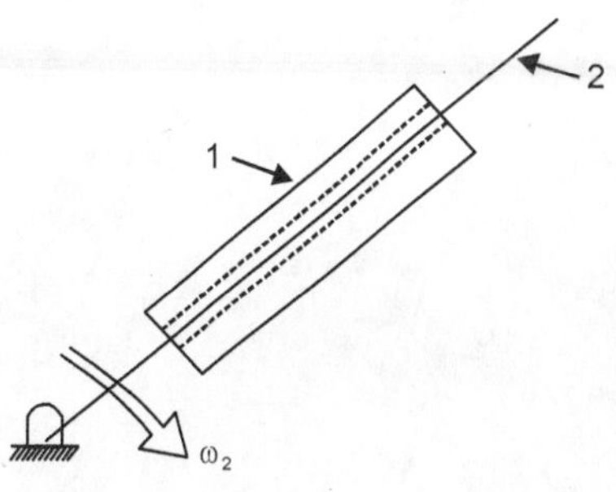

$$= 2V_{1/2}\omega = 2 \times 12 \times 4\pi$$

$$= 301.59 \approx 302 \text{ m/s}^2$$

13. $\alpha_t = 2v\omega = 2 \times 0.75 \times 2 = 3$

$$\alpha_r = \frac{v^2}{r} = 4;$$

$$a = \sqrt{a_t^2 + a_r^2} = \sqrt{3^2 + 4^2}$$

$$a = 5 \text{ m/s}^2$$

14.

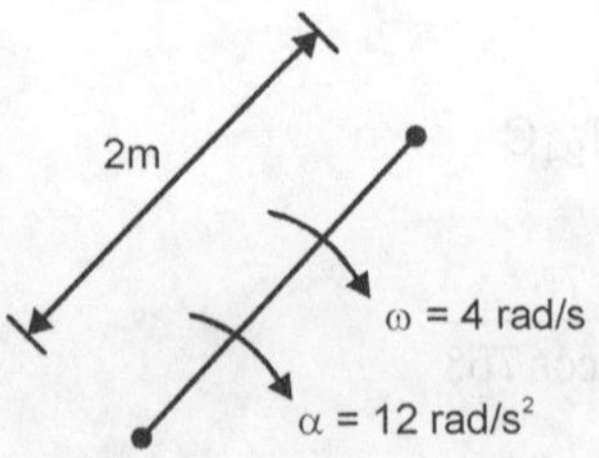

$a_t = R\alpha = 2 \times 12 = 24$

$a_R = \omega^2 R = 16 \times 2 = 32$

$a = \sqrt{a_t^2 + a_R^2}$

$= \sqrt{24^2 + 32^2} = 40 \text{ m/s}^2.$

15. Direction is obtained by rotating velocity vector through 90° in the direction of rotation of the link.

16.

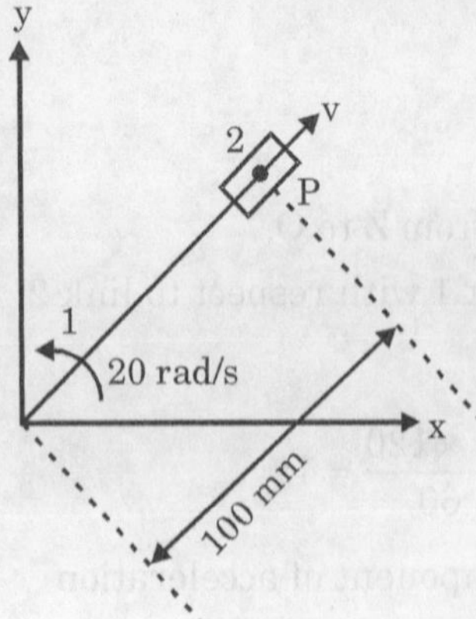

v = 6 m/s

ω = 20 rad/s

Here the centripetal and coriolis acceleration will exist.

$a_p^c = 2\,v\omega = 2 \times 6 \times 20$

$= 240 \text{ m/s}^2$

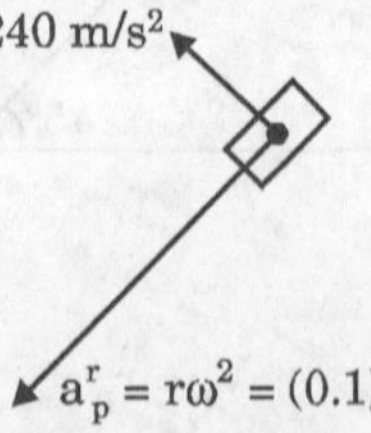

$a_p^r = r\omega^2 = (0.1)\,(20)^2$

$= 40 \text{ m/s}^2$

Total acceleration $\Rightarrow Q_P = \sqrt{40^2 + 240^2}$

$= 243.310 \text{ m/s}^2$

3 CHAPTER

Cams and follower

Analysis of Previous GATE Papers

Topic	Year/Set → Marks Distribution ↓		2019 Set 1	2019 Set 2	2018 Set 1	2018 Set 2	2017 Set 1	2017 Set 2	2016 Set 1	2016 Set 2	2016 Set 3	2015 Set 1	2015 Set 2	2015 Set 3
Motion of follower	1 Mark	MCQ Type	1											
		Numerical Type												
	2 Marks	MCQ Type				1								
		Numerical Type												
	Total		1			2								
Terminology	1 Mark	MCQ Type												
		Numerical Type												
	2 Marks	MCQ Type												
		Numerical Type												
	Total													

Motion of follower

1. In a cam-follower mechanism, the follower needs to rise through 20 mm during 60° of cam rotation, the first 30° with a constant acceleration and then with a deceleration of the same magnitude. The initial and final speeds of the follower are zero. The cam rotates at uniform speed of 300 rpm. The maximum speed of the follower is

(a) 0.60 m/s (b) 1.20 m/s
(c) 1.68 m/s (d) 2.40 m/s

[2005 : 2 Marks]

2. In the mechanism given below, if the angular velocity of the eccentric circular disc is 1 rad/s, the angular velocity (rad/s) of the follower link for the instant shown in the figure is
[Note : All dimensions are in mm.]

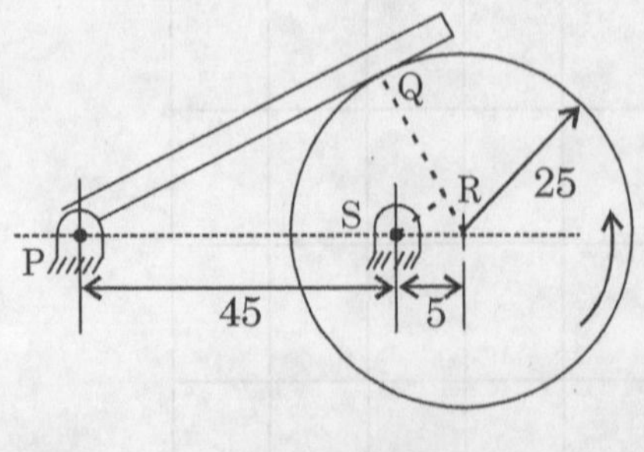

(a) 0.05 (b) 0.1
(c) 5.0 (d) 10.0

[2012 : 1 Mark]

3. Consider a rotating disk cam and a translating roller follower with zero offset. Which one of the following pitch curves, parameterized by t, lying in the interval 0 to 2π, is associated with the maximum translation of the follower during one full rotation of the cam rotating about the center at (x, y) = (0, 0)?

(a) x(t) = cost, y(t) = sint
(b) x(t) = cost, y(t) = 2 sint
(c) $x(t) = \frac{1}{2} + \cos t,\ y(t) = 2\sin t$
(d) $x(t) = \frac{1}{2} + \cos t,\ y(t) = \sin t$

[2014 : 2 Marks, Set-3]

4. In a cam-follower, the follower rises by h as the cam rotates by δ (radians) at constant angular velocity ω (radians/s). The follower is uniformly accelerating during the first half of the rise period and it is uniformly decelerating in the later half of the rise period. Assuming that the magnitudes of the acceleration and deceleration are same, the maximum velocity of the follower is

(a) $\frac{4h\omega}{\delta}$ (b) hω
(c) $\frac{2h\omega}{\delta}$ (d) 2hω

[2018 : 2 Marks, Set-2]

5. A flat-faced follower is driven using a circular eccentric can rotating at a constant angular velocity ω. At time t = 0, the vertical position of the follower is y(0) = 0, and the system is in the configuration shown below.

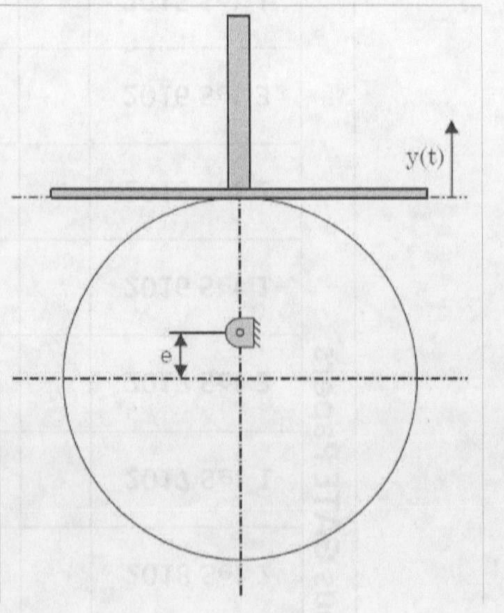

The vertical position of the follower face, y(t) is given by

(a) e sin ωt (b) e(1 + cos 2ωt)
(c) e sin 2 ωt (d) e(1 – cos ωt)

[2019 : 1 Mark, Set-1]

Terminology

6. In a cam design, the rise motion is given by a Simple Harmonic Motion (SHM) $s = \frac{h}{2}\left(1 - \cos\frac{\pi\theta}{\beta}\right)$ where h is total rise, θ is camshaft angle, β is the total angle of the rise interval. The jerk is given by

(a) $\frac{h}{2}\left(1 - \cos\frac{\pi\theta}{\beta}\right)$ (b) $\frac{\pi}{\beta}\frac{h}{2}\sin\left(\frac{\pi\theta}{\beta}\right)$

(c) $\frac{\pi^2}{\beta^2}\frac{h}{2}\cos\left(\frac{\pi\theta}{\beta}\right)$ (d) $-\frac{\pi^3}{\beta^3}\frac{h}{2}\sin\left(\frac{\pi\theta}{\beta}\right)$

[2008 : 2 Marks]

ANSWERS

1. (*d*) **2.** (*b*) **3.** (*c*) **5.** (d) **6.** (*d*)

EXPLANATIONS

1. $\omega = \dfrac{2\pi N}{60} = 10\,\pi$

Time taken to move 30°

$$\frac{\frac{\pi}{180}\times 30}{\omega} = \frac{\frac{\pi}{6}}{10\,\pi} = \frac{1}{60}\text{s}$$

During this period follower move 20 mm = .02 m and u = 0

From $s = ut + \dfrac{1}{2}at^2$

$$0.02 = \frac{1}{2}\times a \times \left(\frac{1}{60}\right)^2$$

$a = 144$ m/sec²

$$v = 0 + \left(144 \times \frac{5}{300}\right) = 2.4 \text{ m/sec}$$

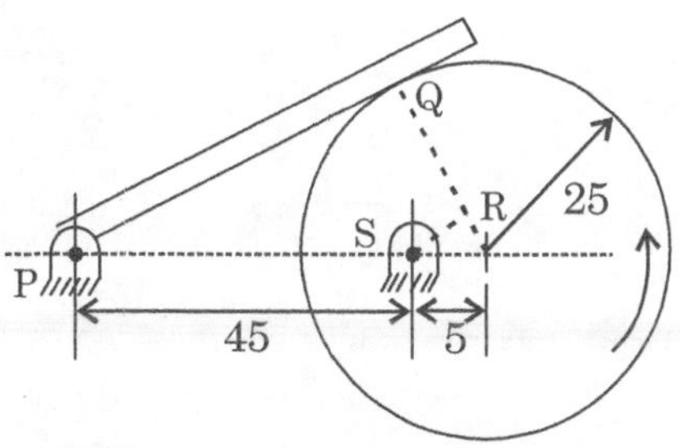

2. $\Delta PQR \sim \Delta SRO$

$$\therefore \quad \frac{PQ}{SR} = \frac{PO}{SO} \quad ...(1)$$

$$(PO)^2 = (PQ)^2 + (OQ)^2$$

$$\therefore \quad PQ = \sqrt{(PO)^2 - (OQ)^2}$$

$$= \sqrt{50^2 - 25^2} = 43.3 \text{ mm}$$

From (1),

$$\frac{43.3}{SR} = \frac{45+5}{5}$$

$$SR = 4.33 \text{ mm}$$

$$V_Q = V_R$$

$$= SR \times \omega$$

$$= 4.33 \times 1 = 4.33 \text{ m/s}$$

$$\omega_{PQ} = \frac{V_{PQ}}{PQ} = \frac{4.33}{43.3} = 0.1 \text{ rad/sec}$$

3. From all the four options the maximum amplitudes is in point 'C' as t = 0.

$(x)_{t=0}\,(y)_{t=0}$

(a) $x = 1 \quad y = 0$

(b) $x = 1 \quad y = 0$

(c) $x = \dfrac{3}{2} \quad y = 0$

(d) $x = \dfrac{3}{2} \quad y = 0$

Option 'c' has maximum net amplitude.

4. Here, out stroke angle

$$\theta_o = \delta$$

and stroke length = h

Angular velocity = ω

$$V = u + at$$

$$(V_O)_{max} = O + a \times \frac{t_o}{2} = \frac{a\times\delta}{2\omega} \quad ...(1)$$

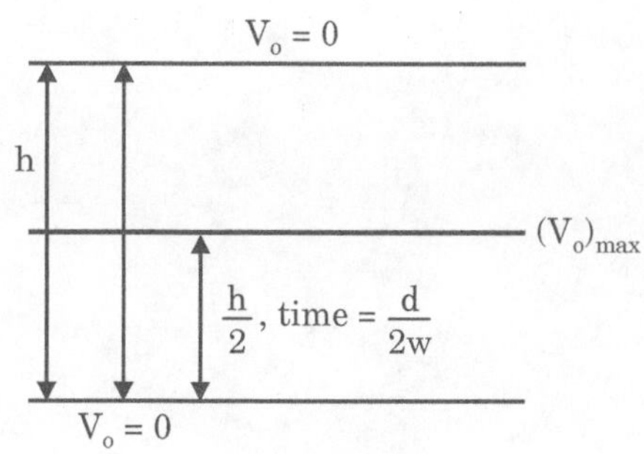

$$\frac{h}{2} = O + \frac{1}{2}a\left(\frac{t_o}{2}\right)^2$$

$$h = \frac{a\delta^2}{4\,\omega^2}$$

$$a = \frac{4\,\omega^2 h}{\delta^2} \qquad ...(2)$$

From (1) & (2), we get

$$(V_o)_{max} = \frac{a\delta}{2\omega} = \frac{2\,\omega h}{\delta}$$

5.

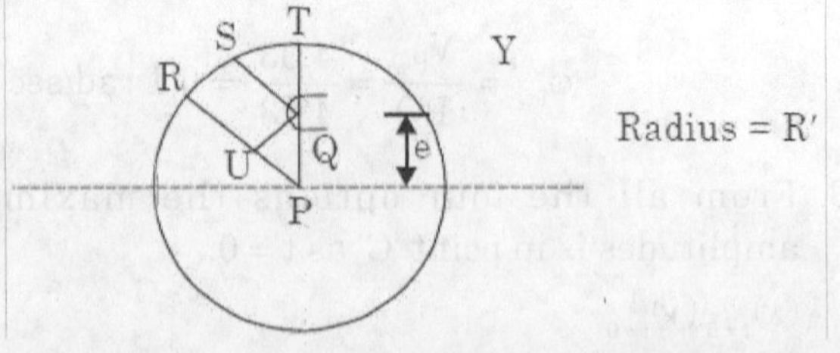

As the follower starts moving, the displacement Y can be shown as marked in the diagram

$$Y = QS - QT$$
$$= RU - QT \qquad (\because QS \| RU)$$
$$= (PR - PU) - QT \qquad ...(i)$$
$$PR = R'$$
$$\Rightarrow \quad PU = e\cos\theta$$
$$QT = (R' - e) \quad \text{from geometry}$$

Now, putting the above values in equation (i)

$$Y = (R' - e\cos\theta) - (R' - e)$$
$$= e - e\cos\theta$$
$$= e(1 - \cos\theta)$$
$$\Rightarrow \quad Y = e(1 - \cos\omega t) \text{ (In time 't', } \theta = \omega t)$$

6. The jerk is given by $\frac{d^3s}{dt^3}$

if $S = \frac{h}{2}\left(1 - \cos\frac{\pi\theta}{\beta}\right)$

$$\therefore \quad \frac{d^3s}{dt^3} = \frac{-\pi^3}{\beta^3}\frac{h}{2}\sin\left(\frac{\pi\theta}{\beta}\right)$$

4 CHAPTER

Dynamic Analysis of Slider-crank Mechanism

Analysis of Previous GATE Papers

Topic	Year/Set → Marks Distribution ↓		2019 Set 1	2019 Set 2	2018 Set 1	2018 Set 2	2017 Set 1	2017 Set 2	2016 Set 1	2016 Set 2	2016 Set 3	2015 Set 1	2015 Set 2	2015 Set 3
Motion of Links	1 Mark	MCQ Type												
		Numerical Type												
	2 Marks	MCQ Type												
		Numerical Type											1	
	Total												**2**	
Dynamic force analysis	1 Mark	MCQ Type												
		Numerical Type												
	2 Marks	MCQ Type												
		Numerical Type			1									
	Total				**2**									

Motion of links

1. The cross head velocity in the slider crank mechanism, for the position shown in figure

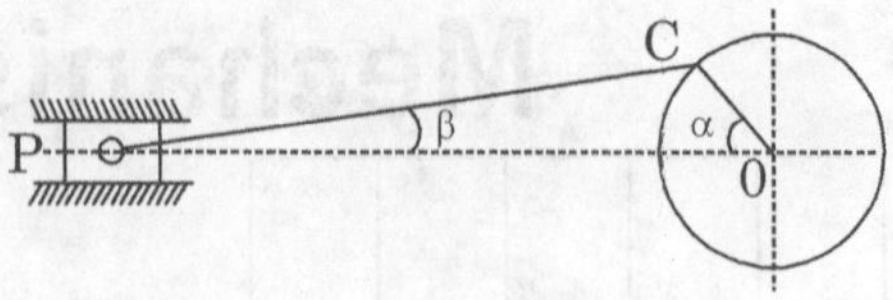

(a) $V_c \cos\left(90-\overline{\alpha+\beta}\right)\cos\beta$

(b) $V_c \cos\left(90-\overline{\alpha+\beta}\right)\sec\beta$

(c) $V_c \cos\left(90-\overline{\alpha-\beta}\right)\cos\beta\beta$

(d) $V_c \cos\left(90-\overline{\alpha-\beta}\right)\sec\beta$

[1997 : 1 Mark]

2. Consider the triangular formed by the connecting rod and the crank of an IC engine as the two sides of the triangle. If the maximum area of this triangle occurs when the crank is 75°, the ratio of connecting rod length to crank radius to

(a) 5 (b) 4
(c) 3.73 (d) 3

[1998 : 1 Mark]

3. An offset slider-crank mechanism is shown in the figure at an instant. Conventionally, the Quick Return Ratio (QRR) is considered to be greater than one. The value of QRR is ______.

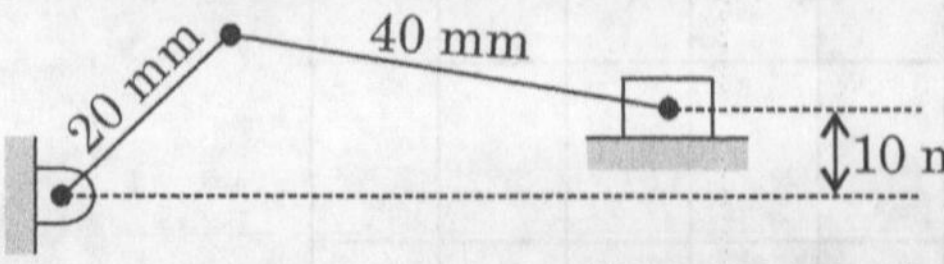

[2014: 2 Marks, Set-1]

4. A slider-crank mechanism with crank radius 60 mm and connecting rod length 240 mm is shown in figure. The crank is rotating with a uniform angular speed of 10 rad/s, counter clockwise. For the given configuration, the speed (in m/s) of the slider is _______.

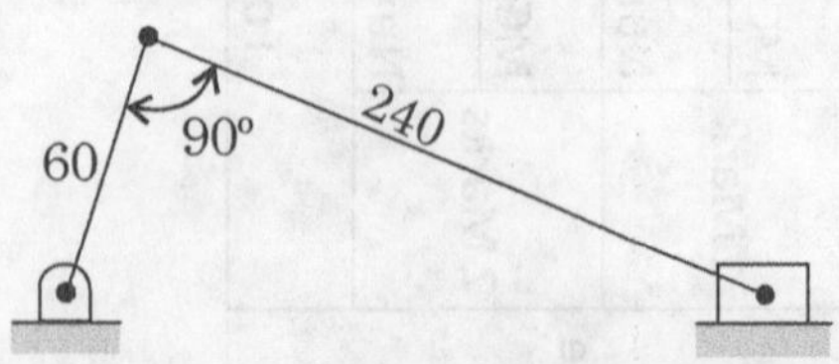

[2014 : 2 Marks, Set-3]

5. In a certain slider-crank mechanism, lengths of crank and connecting rod are equal. If the crank rotates with a uniform angular speed of 14 rad/s and the crank length is 300 mm, the maximum acceleration of the slider (in m/s^2) is _________.

[2015 : 2 Marks, Set-2]

Dynamic force Analysis

6. For a mechanism shown below, the mechanical advantage for the given configuration is

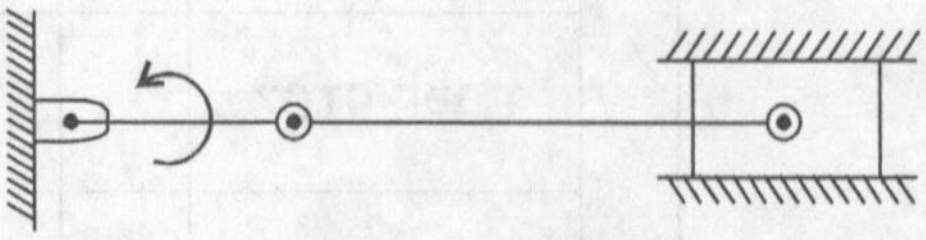

(a) 0.0 (b) 0.5
(c) 1.0 (d) ∞

[2004 : 1 Mark]

7. For a four bar linkage in toggle position, the value of mechanical advantage is

(a) 0.0 (b) 0.5
(c) 1.0 (d) ∞

[2006 : 1 Mark]

8. A slider crank mechanism has slider mass of 10 kg, stroke of 0.2 m and rotates with a uniform angular velocity of 10 rad/s. The primary inertia forces of the slider are partially balanced by a revolving mass of 6 kg at the crank, placed at a distance equal to crank radius. Neglect the mass of connecting rod and crank. When the crank angle (with respect to slider axis) is 30°, the unbalanced force (in newton) normal to the slider axis is _________.

[2014 : 2 Marks, Set-1]

9. A slider crank mechanism is shown in the figure. At some instant, the crank angle is 45° and a force of 40 N is acting towards the left on the slider. The length of the crank is 30 mm and the connecting rod is 70 mm. Ignoring the effect of gravity, friction and inertial forces, the magnitude of the crankshaft torque (in Nm) needed to keep the mechanism in equilibrium is ________ (correct to two decimal places).

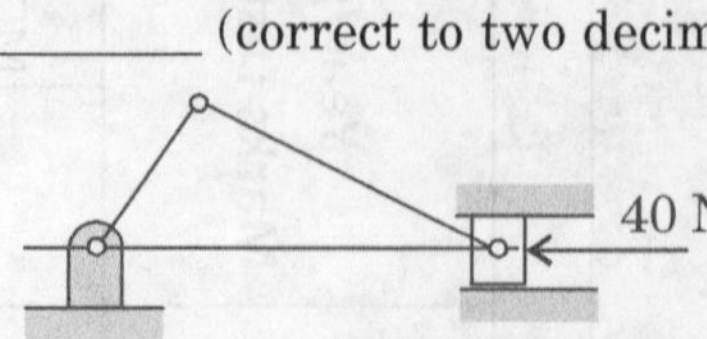

[2018 : 2 Marks, Set-1]

ANSWERS

1. (b) 2. (c) 6. (a) 7. (d)

EXPLANATIONS

2. $\Delta = \frac{1}{2}(PQ)(PR)\sin A$

Area will be maximum when A = 90°

i.e. PQR is a right angled triangle.

∴ Ratio of connecting rod length to crank radius,

$\frac{\ell}{r}$ =tan 75° = 3.732;

ℓ = 3.732 r

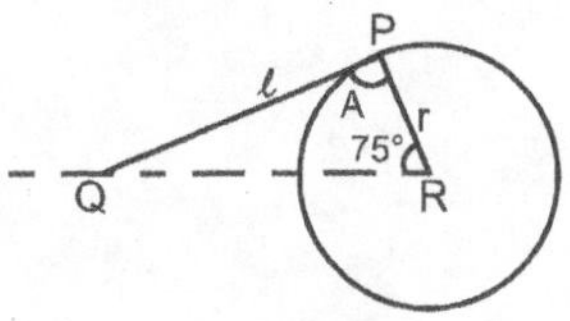

3. AB = stroke length

AO = 40 – 20 = 20

BO = 40 + 20 = 60

∠AOB = ∠BOC – ∠AOC

= 80.41 – 60 = 20.41

$$QRR = \frac{180+\phi}{180-\phi} = \frac{180+20.41}{180-20.41} = 1.255.$$

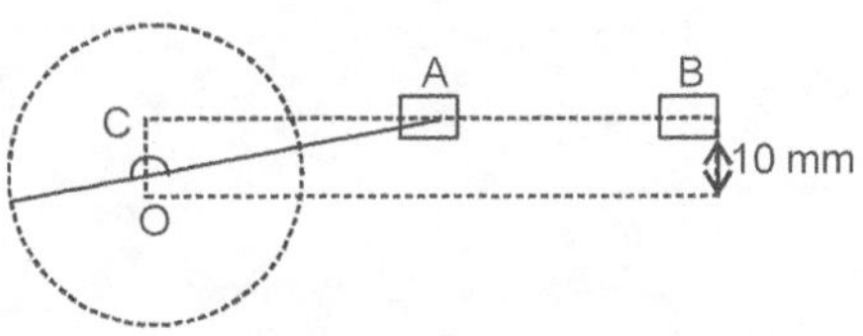

4.

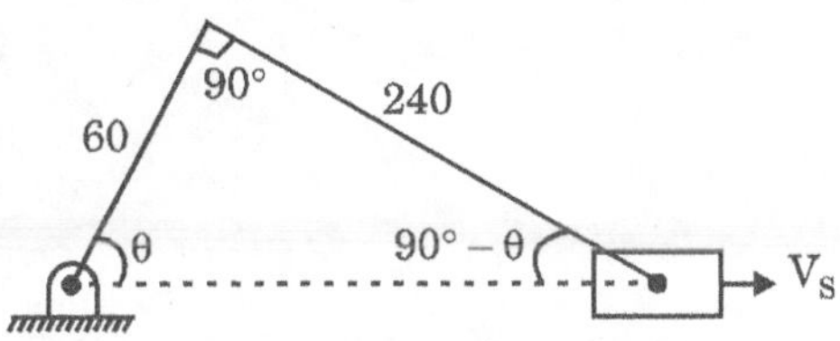

L = 240 mm

r = 60 mm

$n = \frac{L}{r} = 4$

ω = 10 rad/s

$\cos\theta = \frac{\sin\theta}{n}$

$\theta = \tan^{-1}(4) = 75.96°$

$$\therefore \quad V_S = \omega r\left[\sin\theta + \frac{\sin 2\theta}{2n}\right]$$

$$V_S = 10 \times 0.06\left[\sin 75.96° + \frac{\sin 151.92°}{2\times 4}\right]$$

= 0.6 × 1.0288

= 0.6 m/s

5. $a_{max} = 2r\omega^2$

(where θ = 0) i.e at Inner dead centre

$= 2 \times 0.3 \times 14^2$

$a_{max} = 117.6\ m/s^2$

6. Mechanical advantage

$$= \frac{\text{Load (output force)}}{\text{Effort (input force)}}$$

For a four bar linkage in toggle position, effort = 0

∴ Mechanical Advantage = ∞

7. Mechanical advantages

$$= \frac{\text{Load (output force)}}{\text{Effort (input force)}}$$

For a four bar linkage in toggle position, effort = 0

∴ Mechanical Advantage = ∞

8. $r = \frac{0.2}{2} = 0.1$

Now m= 6kg

$F = mr\,(10^2)\sin\theta$

= 6 × 0.1 × 100 × sin 30° = 30 N

9.

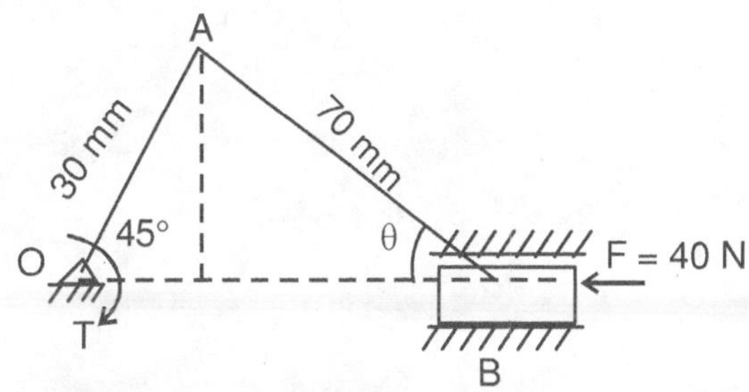

The Torque required to hold the mechanism in equilibrium is equal & opposite of turning moment.

$$T = r\times F\left\{\sin\theta + \frac{\sin 2\theta}{2\sqrt{x^2 - \sin^2\theta}}\right\}$$

$$= 1\times 2\left\{\frac{1}{\sqrt{2}} + \frac{1}{2\sqrt{\frac{49}{2} - \frac{1}{2}}}\right\}$$

$\boxed{T = 1.12\ Nm}$

5 CHAPTER

Gear and Gear Train

Analysis of Previous GATE Papers

Topic	Year/Set → Marks Distribution ↓		2019 Set 1	2019 Set 2	2018 Set 1	2018 Set 2	2017 Set 1	2017 Set 2	2016 Set 1	2016 Set 2	2016 Set 3	2015 Set 1	2015 Set 2	2015 Set 3
Gear Terminology, profile and Classification	1 Mark	MCQ Type	1	1										
		Numerical Type												
	2 Marks	MCQ Type												
		Numerical Type												
	Total		1	1										
Gear Train	1 Mark	MCQ Type				1							1	1
		Numerical Type												
	2 Marks	MCQ Type			1				1					
		Numerical Type					1	1						
	Total				2	1	2	2	2				1	1

Gear Terminology and profile, classification

1. In spur gears having involute teeth, the product of circular pitch and diametral pitch is ______.

[1994 : 1 Mark]

2. **List-I (Gear types)**

A. Worm gears
B. Cross helical gears
C. Bevel gears
D. Spur gears

List-II (Application)

1. Parallel shafts
2. Nonparallel, intersecting shafts
3. Nonparallel, non intersecting shafts
4. Large speed ratios

Codes:

	A	B	C	D
(a)	2	3	1	4
(b)	4	1	3	2
(c)	4	3	2	1
(d)	3	1	2	4

[1995 : 2 Marks]

3. A 1.5 kW motor is running at 1440 rev/min. It is to be connected to a stirrer running at 36 rev/min. The gearing arrangement suitable for this application is

(a) differential gear
(b) helical gear
(c) spur gear
(d) worm gear

[2000 : 1 Mark]

Common Data Questions Q. 4 – 5 : A compacting machine shown in the figure below is used to create a desired thrust force by using a rack and pinion arrangement. The input gear is mounted on the motor shaft. The gears have involute teeth of 2 mm module.

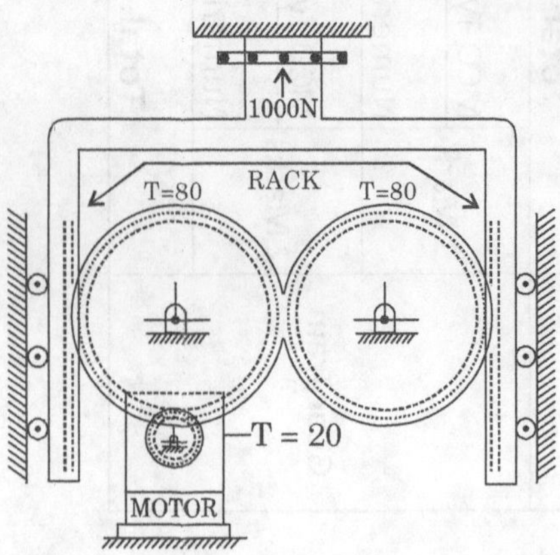

4. If the drive efficiency is 80%, the torque required on the input shaft to create 1000 N output thrust is

(a) 20 Nm (b) 25 Nm
(c) 32 Nm (d) 50 Nm

[2004 : 2 Marks]

5. If the pressure angle of the rack is 20°, the force acting along the line of action between the rack and the gear teeth is

(a) 250 N (b) 342 N
(c) 532 N (d) 600 N

[2004 : 2 Marks]

6. Tooth interference in an external involute spur gear pair can be reduced by

(a) decreasing center distance between gear pair
(b) decreasing module
(c) decreasing pressure angle
(d) increasing number of gear teeth

[2010 : 1 Mark]

7. The following are the data for two crossed helical gears used for speed reduction:

Gear I : Pitch circle diameter in the plane of rotation 80 mm and helix angle 30°

Gear II : Pitch circle diameter in the plane of rotation 120 mm and helix angle 22.5°

If the input speed is 1440 rpm, the output speed in rpm is

(a) 1200 (b) 900
(c) 875 (d) 720

[2012 : 1 Mark]

8. It is desired to avoid interference in a pair of spur gears having a 20° pressure angle. With increase in pinion to gear speed ratio, the minimum number of teeth on the pinion

(a) increases
(b) decreases
(c) first increases and then decreases
(d) remains unchanged

[2014 : 2 Marks, Set-4]

9. A spur gear with 20° full depth teeth is transmitting 20 kW at 200 rad/s. The pitch circle diameter of the gear is 100 mm. The magnitude of the force applied on the gear in the radial direction is

(a) 0.36 kN (b) 2.78 kN
(c) 0.73 kN (d) 1.39 kN

[2019 : 1 Mark, Set 1]

10. A spur gear has pitch circle diameter D and number of teeth T. The circular pitch of the gear is

(a) $\frac{D}{T}$ (b) $\frac{\pi D}{T}$

(c) $\frac{2\pi D}{T}$ (d) $\frac{T}{D}$

[2019 : 1 Mark, Set 2]

Gear Train

11. The sun gear in the figure is driven clockwise at 100 rpm. The ring gear is held stationary. For the number of teeth shown on the gears, the arm rotates at

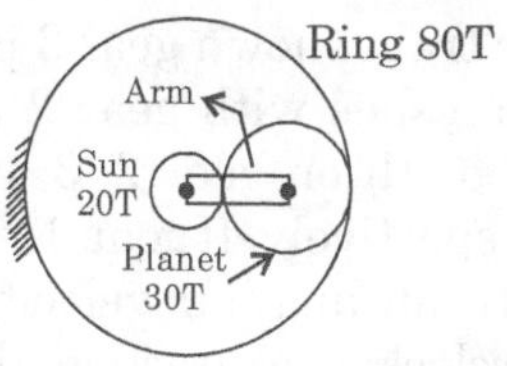

(a) 0 rpm (b) 20 rpm

(c) 33.33 rpm (d) 666.67 rpm

[1993 : 2 Marks]

12. The arm OA of a epicyclic gear train shown in figure revolves counter clockwise about Owith an angular velocity of 4 rad/s. Both gear are of same size. The angular velocity of gears C, if the sun gear B is fixed is

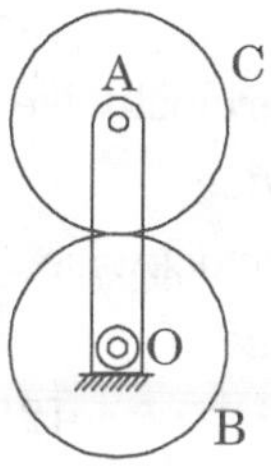

(a) 4 rad/s (b) 8 rad/s

(c) 10 rad/s (d) 12 rad/s

[1995 : 2 Marks]

Common Data for Questions 13 and 14:

A planetary gear train has four gears and one carrier. Angular velocities of the gears are ω_1, ω_2, ω_3, and ω_4, respectively. The carrier rotates with angular velocity ω_5.

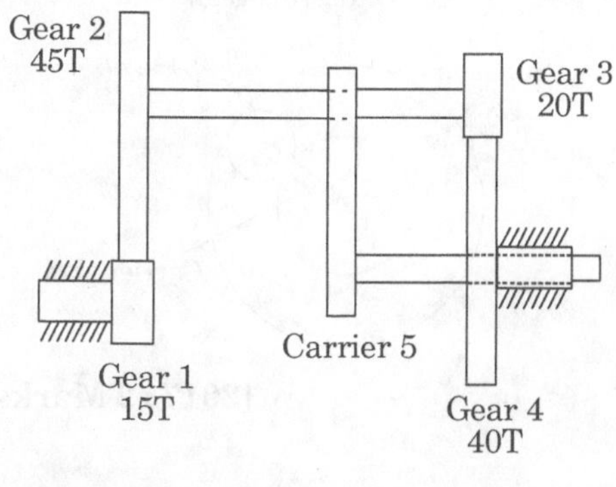

13. What is the relation between the angular velocities of Gear 1 and Gear 4?

(a) $\frac{\omega_1 - \omega_5}{\omega_4 - \omega_5} = 6$ (b) $\frac{\omega_4 - \omega_5}{\omega_1 - \omega_5} = 6$

(c) $\frac{\omega_1 - \omega_2}{\omega_4 - \omega_5} = -\left(\frac{2}{3}\right)$ (d) $\frac{\omega_2 - \omega_5}{\omega_4 - \omega_5} = \frac{8}{9}$

14. For ω_1 = 60 rpm clockwise (cw) when viewed from the left, what is the angular velocity of the carrier and its direction. so that Gear 4 rotates in counter clockwise (ccw) direction at twice the angular velocity of Gear 1 when viewed from the left?

(a) 130 rpm, cw (b) 223 rpm, ccw

(c) 256 rpm, cw (d) 156 rpm, ccw

[2006 : 2 Marks]

15. An epicyclic gear train is shown schematically in the adjacent figure. The sun gear 2 on the input shaft is a 20 teeth external gear. The planet gear 3 is a 40 teeth external gear. The ring gear 5 is a 100 teeth internal gear. The ring gear 5 is fixed and the gear 2 is rotating at 60 rpm ccw(ccw-counterclockwise and cw- clockwise)

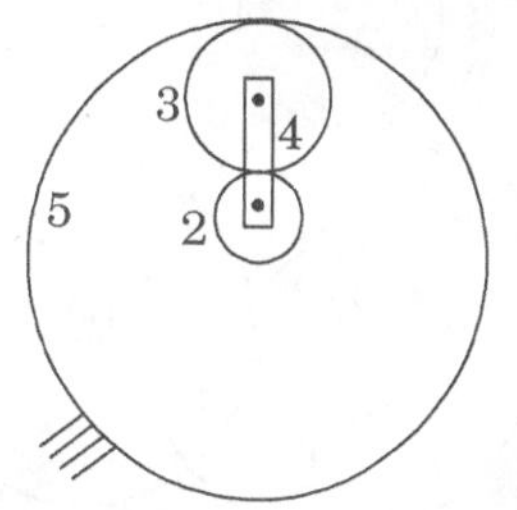

The arm 4 attached to the output shaft will rotate at

(a) 10 rpm ccw (b) 10 rpm cw

(c) 12 rpm cw (d) 12 rpm ccw

[2009 : 2 Marks]

16. For the epicyclic gear arrangement shown in the figure, ω_2 = 100 rad/s clockwise (cw) and ω_{arm} = 80 rad/s counter clockwise (ccw). The angular velocity ω_5 (in rad/s) is

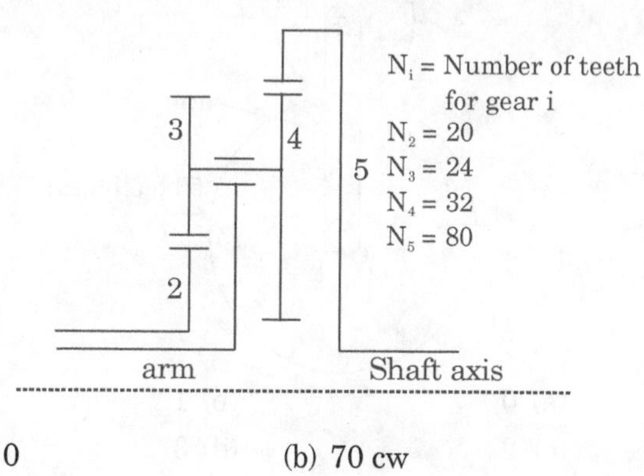

(a) 0 (b) 70 cw

(b) 140 ccw (d) 140 cw

[2010 : 2 Marks]

17. A compound gear train with gears P, Q, R and S has number of teeth 20,40,15 and 20 respectively. Gears Q and P are mounted on the same shaft as shown in the figure below. The diameter of the gear Q is twice that of the gear R. If the module of the gear R is 2 mm, the centre distance in mm between gears Pand S is

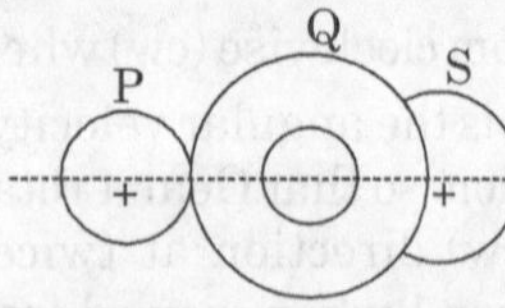

(a) 40 (b) 80
(c) 120 (d) 160

[2013 : 2 Marks]

18. Gear 2 rotates at 1200 rpm is counter clockwise direction and engages with Gear 3 and Gear 4 mounted on the same shaft. Gear 5 engages with Gear 4. The numbers of teeth on Gears 2, 3, 4 and 5 are 20, 40, 15 and 30, respectively. The angular speed of Gear 5 is

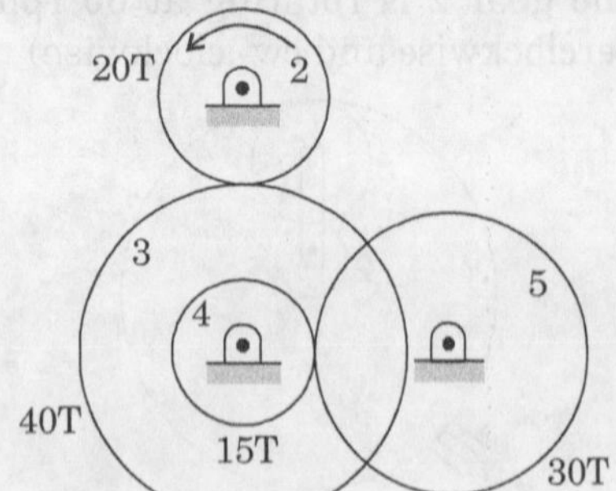

(a) 300 rpm counter clockwise
(b) 300 rpm clockwise
(c) 4800 rpm counter clockwise
(d) 4800 rpm clockwise

[2014 : 1 Mark, Set-3]

19. The number of degree of freedom of the planetary gear train shown in the figure is

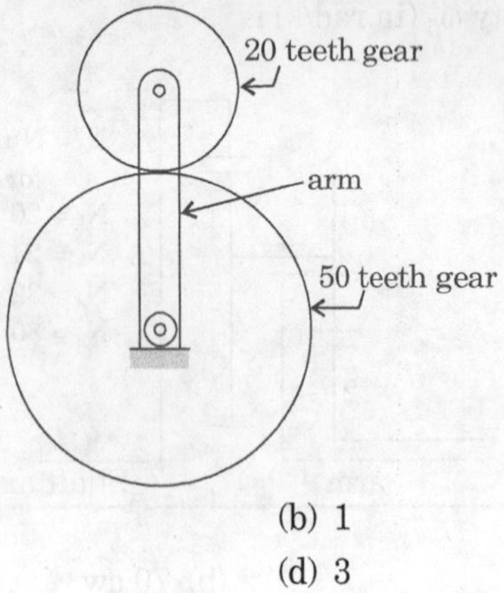

(a) 0 (b) 1
(c) 2 (d) 3

[2015 : 1 Mark, Set-2]

20. A gear train is made up of five spur gears as shown in the figure. Gear 2 is driver and gear 6 is driven member. N_2, N_3, N_4, N_5 and N_6 represent number of teeth on gears 2,3,4,5 and 6 respectively. The gear(s) which act(s) as idler(s) is/are

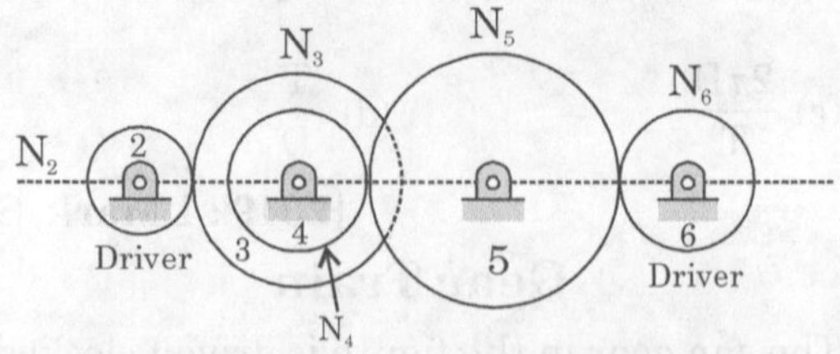

(a) only 3 (b) only 4
(c) only 5 (d) Both 3 and 5

[2015 : 1 Mark, Set-3]

21. In the gear train shown gear 3 is carried on arm 5, Gear 3 meshes with gear 2 and gear 4. The number of teeth on gear 2, 3 and 4 are 60, 20, and 100, respectively. If gear 2 is fixed and gear 4 rotates with an angular velocity of 100 rpm in the counterclockwise direction, the angular speed of arm 5 (in rpm) is

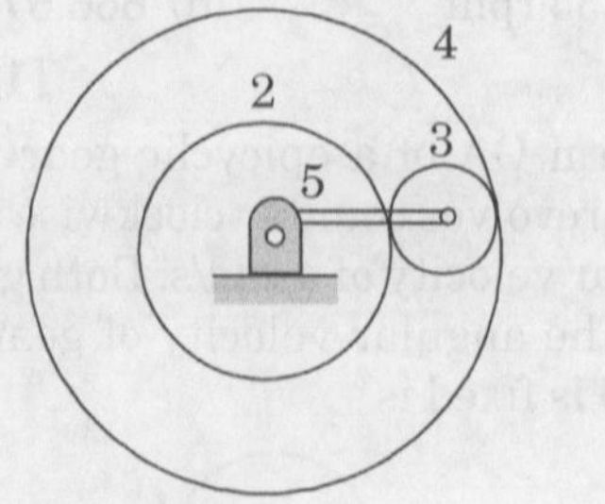

(a) 166.7 counterclockwise
(b) 166.7 clockwise
(c) 62.5 counterclockwise
(d) 62.5 clockwise

[2016 : 2 Marks, Set-1]

22. In an epicyclic gear train, shown in the figure, the outer ring gear is fixed, while the sun gear rotates counterclockwise at 100 rpm. Let the number of teeth on the sun planet and outer gears to be 50, 25 and 100, respectively. The ratio of magnitude of angular velocity of the planet gear to the angular velocity of the carrier arm is __________.

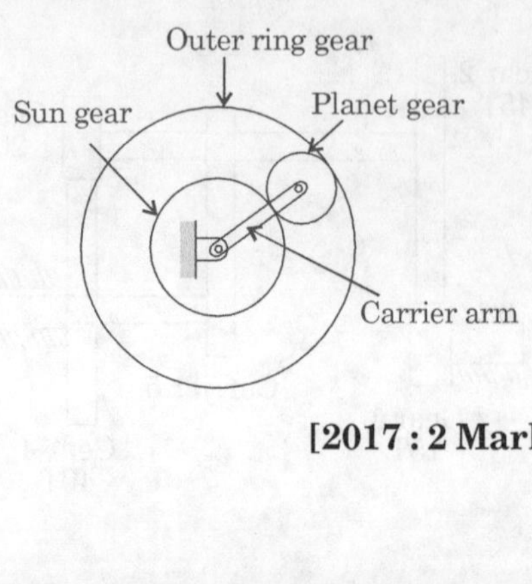

[2017 : 2 Marks, Set-1]

23. A gear train shown in the figure consists of gears P, Q, R and S. Gear Q and gear R are mounted on the same shaft. All the gears are mounted on parallel shafts and the number of teeth of P, Q, R and S are 24,45, 30 and 80, respectively. Gear P is rotating at 400 rpm. The speed (in rpm) of the gear S is _______.

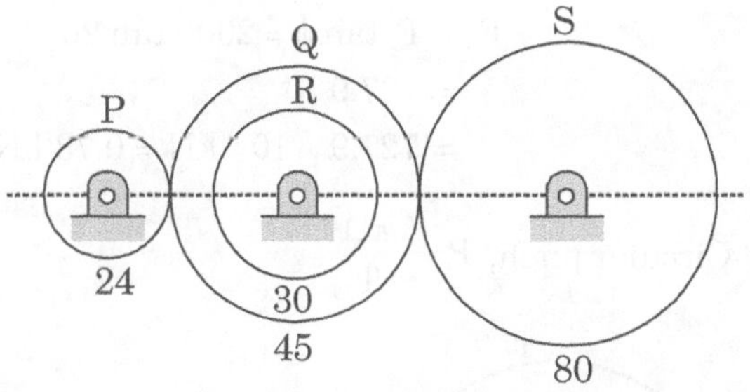

[2017 : 2 Marks, Set-2]

24. An epicyclic gear train is shown in the figure below. The number of teeth on the gears A, B and D are 20, 30 and 20 respectively. Gear C has 80 teeth on the inner surface and 100 teeth on the outer surface. If the carrier arm AB is fixed and the sun gear A rotates at 300 rpm in the clockwise direction, then the rpm of D in the clockwise direction is

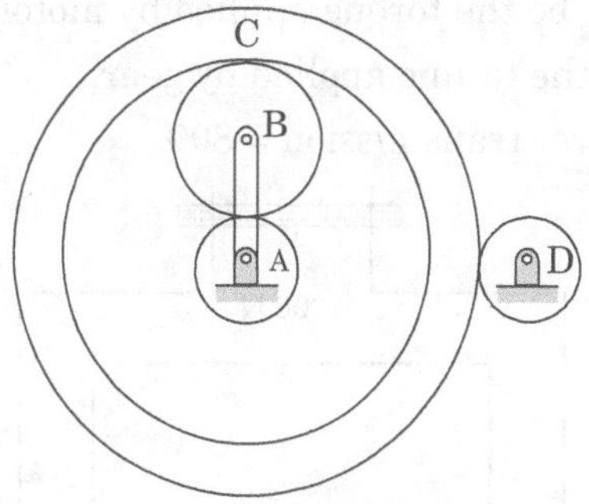

(a) 240 (b) –240

(c) 375 (d) –375

[2018 : 2 Marks, Set-1]

25. A frictionless gear train is shown in the figure. The left most 12-teeth gear is given a torque of 100 Nm. The output torque from the 60-teeth gear on the right in Nm is

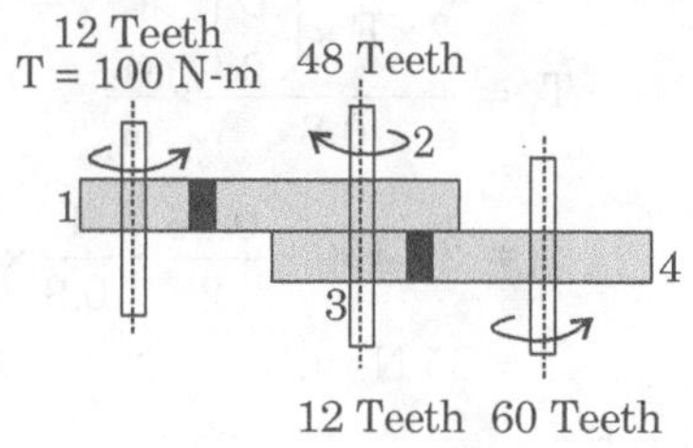

(a) 5 (b) 20

(c) 500 (d) 2000

[2018 : 1 Mark, Set-2]

ANSWERS

2. (c)	**3.** (d)	**4.** (b)	**5.** (c)	**6.** (d)	**7.** (b)	**8.** (b)	**9.** (c)	**10.** (b)	**11.** (b)
12. (b)	**13.** (a)	**14.** (d)	**15.** (b)	**16.** (b)	**17.** (b)	**18.** (a)	**19.** (c)	**20.** (c)	**21.** (c)
24. (c)	**25.** (d)								

EXPLANATIONS

1. $P_c \times P_d = \Pi$

2. $A \rightarrow 4, B \rightarrow 3, C \rightarrow 2, D \rightarrow 1$

3. Power of motor = 1.5 kW

N = 1440 rev/min

Speed of stirrer = 36 rev/min

$$\text{S.R.D} = \frac{1440}{36} = 40$$

$\therefore$ Worm gear is used.

4. Given : Module m = 2,

$$\frac{D}{T} = 2$$

$\therefore$ D = 80 × 2 = 160 mm

2F = 1000,

or F = 500 N

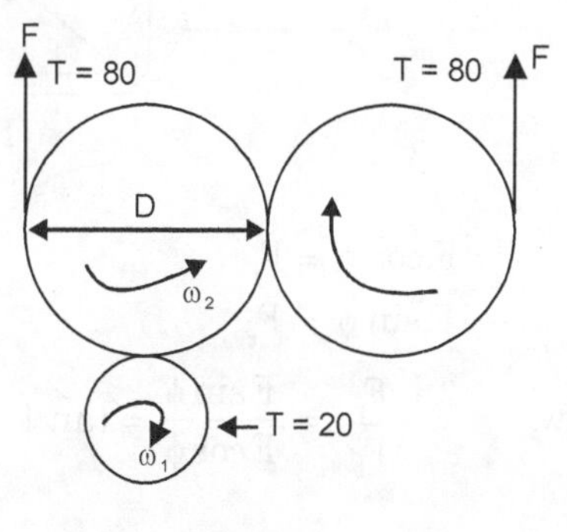

Let T_1 be the torque applied by motor.

T_2 be the torque applied by gear.

$\therefore$ Power transmission = 80%

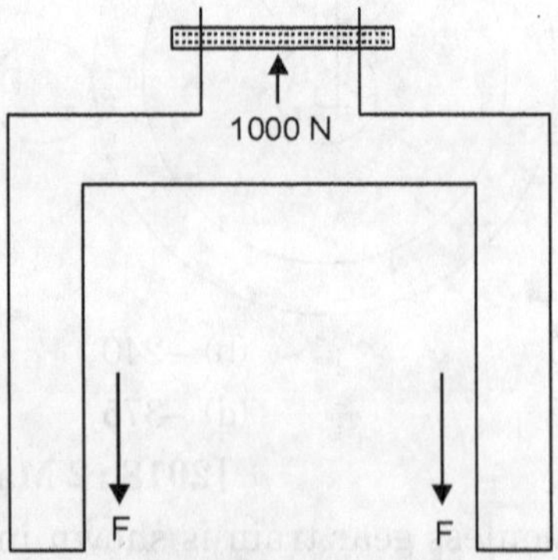

Now, $T_1\omega_1 = \dfrac{2T_2 \times \omega_2}{0.8}$

or $T_1 = \dfrac{2 \times F \times \left(\dfrac{D}{2}\right)}{0.8} \times \dfrac{\omega_2}{\omega_1}$

$= 2 \times 500 \times \dfrac{0.16}{2} \times \dfrac{1}{0.8} \times \dfrac{1}{4}$

$= 25$ N-m.

5. $P \cos\phi = F$

$\therefore$ Force acting along the line of action,

$P = \dfrac{F}{\cos\phi} = \dfrac{500}{\cos 20^\circ} = 532\text{N}$

7. For helical gears

velocity ratio$= \dfrac{d_1 \cos\phi}{d_2 \cos\phi}$

$= \dfrac{80 \cos 30^\circ}{120 \cos 22.5} = \dfrac{N_2}{N_1}$

$\therefore$ $N_2 = 1440 \times 0.625 = 900$ rpm

9. Given : $\phi = 20^\circ$

$P = 20$ kW, $\omega = 200$ rad/s

Pitch Circle Diameter = 100 mm

Now, force on radial direction (F_r),

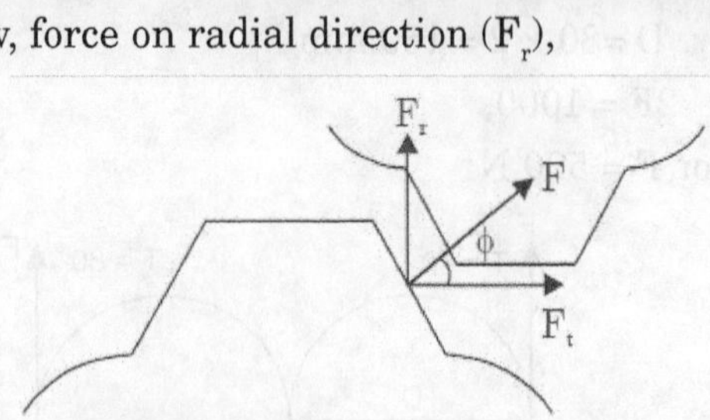

$F \cos\phi = F_t$

$F \sin\phi = F_r$

Now, $\dfrac{F_r}{F_t} = \dfrac{F\sin\phi}{F\cos\phi} = \tan\phi$

$\Rightarrow$ $F_r = F_t \tan\phi$

Also, $T = \dfrac{P}{\omega} = \dfrac{20 \times 1000}{200} = 100\text{N.m.}$

$T = F_t \times r$

$\Rightarrow$ $F_t = \dfrac{T}{r} = \dfrac{100}{\dfrac{100}{2} \times 10^{-3}} = 2000\text{N}$

$F_r = F_t \tan\phi = 2000 \tan 20$

$= 727.9$ N

$= 727.9 \times 10^{-3}$ kN $= 0.73$ kN

10. Circular pitch, $P_c = \dfrac{\pi D}{T}$

11.

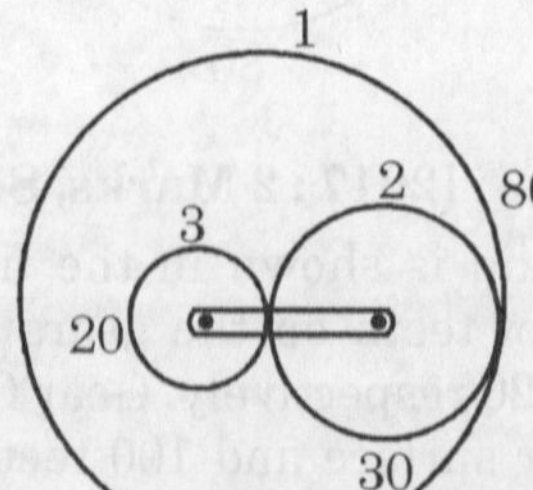

Gear	1	2	3
No. of teeth	80	30	20(sun)
N(rpm)	+N	$+\dfrac{80N}{30}$	$-\dfrac{80N \times 30}{30 \times 20}$
Adding 'y' rpm	N + y	$\dfrac{80N}{30} + y$	$-4N + y$

Given :

$N + y = 0$

$y = -N$

Sun gear means "3"

$\dfrac{-N \times 80 \times 30}{30 \times 20} + y = 0$

$-4N - N = 100$

$-5N = 100$

$N = -20$

$\therefore$ $y = -N = +20$ rpm

12.

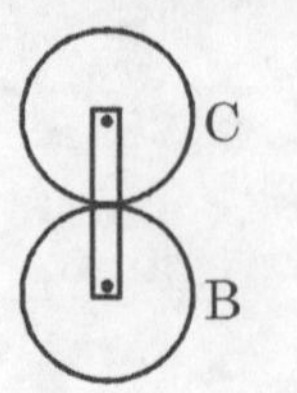

Given : y = – 4 [CCW]

N + y = 0 [Sun gear B is fixed]

$\therefore$ N = – y

= 4 rad/sec

Speed of gear C = – 4 – 4 = – 8 rad/sec

13. $\dfrac{\omega_1-\omega_5}{\omega_2-\omega_5}=3$ (with respect to arm 5 or carrier 5)

$\dfrac{\omega_3-\omega_5}{\omega_4-\omega_5}=2$ (with respect to carrier 5)

As, $\omega_3=\omega_2$

$\therefore\ \dfrac{\omega_1-\omega_5}{\omega_4-\omega_5}=6$

14. $\omega_1=60$ rpm (Clockwise)

$\omega_4=120$ rpm (Counter clockwise)

$\dfrac{60-\omega_5}{-120-\omega_5}=6$

$\therefore\ \omega_5=-156$ i.e. counter clockwise

15. $T_2=20$; $T_3=40$; $T_5=100$

Condition	Arm	Gear 2	Gear 3	Gear 5
1. Arm fixed– gear 2 rotates with + 1 rotation	0	+1	$-\dfrac{T_2}{T_3}$	$-\dfrac{T_2}{T_3}\times\dfrac{T_3}{T_5}$
2. Arm fixed –gear 2 rotates with + x rotations	0	$+x$	$-x\dfrac{T_2}{T_3}$	$-x\dfrac{T_2}{T_5}$
3. Adding + y rotations	y	$y+x$	$y-x\dfrac{T_2}{T_3}$	$y-x\dfrac{T_2}{T_5}$

Given $y-x\dfrac{T_2}{T_5}=0$

$\Rightarrow\ x=5y$ ....(1)

and $y+x=60$ ....(2)

Solving (1) and (2), we get

$y+5y=60$

$\Rightarrow$ y = 10 rpm ccw

16.

Condition	Arm/a	2	3/4	5
Arm fixed 2-rotates +1 revolution	0	1	$-\dfrac{20}{26}=\dfrac{5}{6}$	$-\dfrac{20}{26}\times\dfrac{32}{80}=-\dfrac{1}{3}$
Arm fixed 2-rotates +X revolution	0	X	$-\dfrac{5}{6}X$	$-\dfrac{X}{3}$
Add Y (for arm rotation)	Y	X+Y	$-\dfrac{5}{6}X+Y$	$-\dfrac{X}{3}+Y$

Given : $\omega_{arm}=-80$ rad/sec = Y (CCW)

$\omega_2=100$ rad/sec = X + Y (CW)

$\therefore$ X = 100 – Y

= 100 + 8

= 108 rad/sec (CW)

Also, $\omega_s=-\dfrac{1}{3}X+Y=-\dfrac{1}{3}(180)+(-80)$

= – 140 rad/sec (CCW)

$\therefore\ \omega_s=140$ rad/sec (CW)

Note : CCW→ Counter Clock Wise

CW → Clock Wise

17. $m=\dfrac{D}{T}$ and $m_p=m_Q$

$m_R=m_s$ for gear meshing condition

18. $\dfrac{N_5}{N_2}=\dfrac{T_2\times T_4}{T_3\times T_5}=\dfrac{20\times15}{40\times30}$

$\Rightarrow N_5=1200\times0.25=300$ rpm ccw.

19. A planetary gear train has 2 DOF and hence requires two input to get desired output.

20. $\dfrac{\omega_2}{\omega_6}=\dfrac{\omega_2}{\omega_3}\dfrac{\omega_3}{\omega_5}\dfrac{\omega_5}{\omega_6}=\dfrac{N_3}{N_2}\dfrac{N_5}{N_4}\dfrac{N_6}{N_5}=\dfrac{N_3N_6}{N_2N_4}$

$\therefore$ only Gear(5) is Idle

21. Given $T_2=60$

$N_2=0$

$T_3=20$

$T_4 = 100$
$N_2 = 100$ rpm (ccw)

Relative velocity equation

$$\frac{N_4 - N_a}{N_2 - N_a} = \frac{T_2}{T_4} = \frac{100 - N_a}{0 - N_a} = \frac{-60}{100}$$

$1.6N_a = 100$

$N_a = 62.5$ rpm (ccw)

22.

	Arm	S	P	R
Arm → Fix S → +1	0	1	−2	−0.5
$\times x$	0	x	$-2x$	$0.5x$
Add y	Y	$x+y$	$-2x+y$	$0.5x+y$

$N_S = 100 \Rightarrow x + y = 100$ ...(*i*)

$N_R = 0 \Rightarrow 0.5x + y = 0$ ...(*ii*)

By solving eqn (*i*) and (*ii*), we get

$$x = \frac{200}{3},$$

$$y = \frac{100}{3}$$

$$\frac{N_P}{N_{arm}} = \frac{-2x + y}{y}$$

$$= \frac{-2\left(\frac{200}{3}\right) + \frac{100}{3}}{\frac{100}{3}} = -3$$

23. $T_P = 24$
$T_Q = 45$
$T_R = 30$
$T_S = 80$
$N_P = 400$ rpm
$N_S = ?$

Here gear G is not meshing,

$$\frac{N_P}{N_Q} = \frac{T_Q}{T_P} = \frac{45}{24} \quad ...(i)$$

$$\frac{N_Q}{N_S} = \frac{T_S}{T_Q} = \frac{80}{45} \quad ...(ii)$$

By (i) and (ii)

$$\frac{N_P}{N_S} = \frac{45}{24} \times \frac{80}{45}$$

$$N_S = N_P \times \frac{24}{80} = \frac{400 \times 24}{80}$$

$N_S = 120$ r.p.m.

24. $T_A = 20$, $T_B = 30$, $T_D = 20$, $T_C = 80$ (Inner)
$T_D = 100$ (outer)

Arm is fixed, no epicyclic nature. Taking clockwise direction as positive.

$N_A = +300$

$$(A, B)_{N_B} = \frac{-300 \times 20}{30} = -200$$

$$(C, C)_{N_C} = -200 \times \frac{30}{80} = -75$$

$$(C, D)_{N_D} = +75 \times \frac{100}{20} = +375$$

25. $(1, 2) \Rightarrow N_2 = N_1 \times \frac{Z_1}{Z_2} = N_1 \times \frac{12}{48} = \frac{N_1}{4}$

$$N_3 = N_2 = \frac{N_1}{4}$$

$$(3, 4) \Rightarrow N_4 = N_3 \times \frac{Z_3}{Z_4} = \frac{N_1}{4} \times \frac{12}{60}$$

$$N_4 = \frac{N_1}{20}$$

By power conservation,

$$T_1 \times N_1 = T_4 \times N_4$$

$$100 \times N_1 = T_4 \times \frac{N_1}{20}$$

$$\boxed{T_4 = 2000 \text{N-m}}$$

6 CHAPTER

Fly wheel

Analysis of Previous GATE Papers

Topic	Year/Set → Marks Distribution ↓		2019 Set 1	2019 Set 2	2018 Set 1	2018 Set 2	2017 Set 1	2017 Set 2	2016 Set 1	2016 Set 2	2016 Set 3	2015 Set 1	2015 Set 2	2015 Set 3
Mean Speed of rotation	1 Mark	MCQ Type												
		Numerical Type												
	2 Marks	MCQ Type												
		Numerical Type												
	Total													
Energy Fluctuation	1 Mark	MCQ Type												
		Numerical Type												
	2 Marks	MCQ Type												
		Numerical Type												
	Total													
Turning moment diagram	1 Mark	MCQ Type												
		Numerical Type												
	2 Marks	MCQ Type												
		Numerical Type												1
	Total													2

Mean speed of rotation

1. A flywheel of moment of inertia 9.8 kg m^2 fluctuates by 30 rpm for a fluctuation in energy of 1936 Joules. The mean speed of the flywheel is (in rpm)

(a) 600 (b) 900
(c) 968 (d) 29470

[1998 : 2 Marks]

2. Which of the following statements is correct?

(a) Flywheel reduces speed fluctuations during a cycle for a constant load, but flywheel does not control the mean speed of the engine if the load changes.
(b) Flywheel does not reduce speed fluctuations during a cycle for a constant load, but flywheel does control the mean speed of the engine if the load changes
(c) Governor control a speed fluctuations during a cycle for a constant load, but governor does not control the mean speed of the engine if the load changes.
(d) Governor controls speed fluctuations during a cycle for a constant load, and governor also controls the mean speed of the engine if the load changes.

[2001 : 1 Mark]

3. If C_f is the coefficient of speed fluctuation of a flywheel then the ratio of $\omega_{max}/\omega_{min}$ will be

(a) $\frac{1-2C_f}{1+2C_f}$ (b) $\frac{2-C_f}{2+C_f}$

(c) $\frac{1+2C_f}{1-2C_f}$ (d) $\frac{2+C_f}{2-C_f}$

[2006 : 2 Marks]

Energy fluctuation

4. For a certain engine having an average speed of 1200 rpm, a flywheel approximated as a solid disc, is required for keeping the fluctuation of speed within 2% about the average speed. The fluctuation of kinetic energy per cycle is found to be 2 kJ. What is the least possible mass of the flywheel if its diameter is not to exceed 1 m?

(a) 40 kg (b) 51kg
(c) 62 kg (d) 73 kg

[2003 : 2 Marks]

5. The speed of an engine varies from 210 rad/s to 190 rad/s. During cycle the change in kinetic energy is found to be 400 Nm. The inertia of the flywheel in kg-m^2 is

(a) 0.10 (b) 0.20
(c) 0.30 (d) 0.40

[2007 : 2 Marks]

6. A flywheel connected to a punching machine has to supply energy of 400 Nm while running at a mean angular speed of 20 rad/s. If the total fluctuation of speed is not to exceed ± 2% the mass moment of inertia of the flywheel in kg-m^2 is

(a) 25 (b) 50
(c) 100 (d) 125

[2013 : 2 Marks]

7. Maximum fluctuation of kinetic energy in an engine has been calculated to be 2600 J. Assuming that the engine runs at an average speed of 200 rpm, the polar mass moment of inertia (in kg.m^2) of a flywheel to keep the speed fluctuation within ±0.5% of the average speed is ________.

[2014 : 2 Marks, Set-2]

Turning moment diagram

8. Torque and angular speed data over one cycle for a shaft carrying a flywheel are shown in the figures. The moment of inertia (in kgm^2) of the flywheel is ______ is.

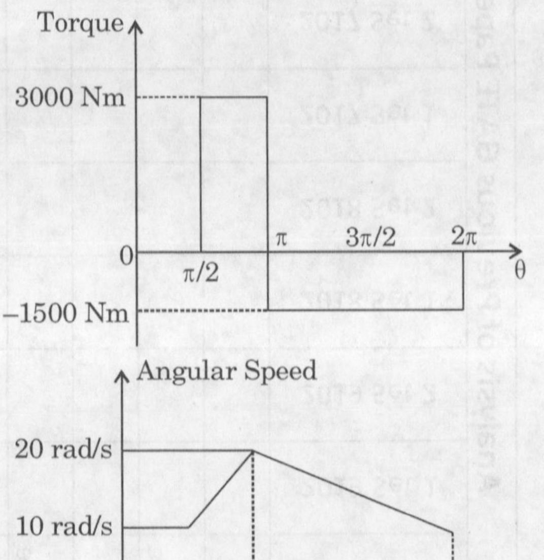

[2014 : 2 Marks, Set-4]

9. The torque (in Nm) exerted on the crank shaft of a two stroke engine can be described as T = 10000 + 1000 sin 2θ - 1200 cos 2θ, where θ is the crank angle as measured from inner dead center position. Assuming the resisting torque to be constant, the power (in kW) developed by the engine at 100 rpm is ________.

[2015 : 2 Marks, Set-3]

ANSWERS

1. (a) **2.** (a) **3.** (d) **4.** (b) **5.** (a) **6.** (a)

EXPLANATIONS

1 . $\Delta E = mR^2\omega^2\Delta S$

$= mR^2\omega(\omega_1 - \omega_2)$

$\therefore\ 1936 = 9.8 \times \omega\left(\frac{2\pi}{60} \times 30\right)$

or Mean speed of flywheel, ω = ***600 rpm***

3. $C_f = \frac{\omega_{max} - \omega_{min}}{\frac{\omega_{max} + \omega_{min}}{2}}$

$\Rightarrow \frac{\omega_{max}}{\omega_{min}} = \frac{2 + C_f}{2 - C_f}$

4. Average speed, N = 1200 rpm

Co-efficient of fluctuation of speed

$= c_s = \frac{\omega_1 - \omega_2}{\omega} = 2\% = 0.02$

Fluctuation of kinetic energy

$= \Delta E = 2 \times 10^3 J$

Now $\Delta E = \frac{1}{2}I\omega_1^2 - \frac{1}{2}\omega_2^2$

$= \frac{1}{2}I(\omega_1^2 - \omega_2^2)$

Since $\frac{\omega_1 + \omega_2}{2} = \omega$

$= I\left(\frac{\omega_1 + \omega_2}{\omega}\right)(\omega_1 - \omega_2)$

$= I\omega \frac{(\omega_1 + \omega_2)}{\omega} \cdot \omega$

$= I\omega^2 c_s$.

$\Rightarrow 2 \times 10^2 = \frac{1}{2}MR^2 \cdot \omega^2 \cdot c_s^2,$

where R = Radius of disc

$= \frac{1}{2}M \times \left(\frac{1}{2}\right)^2 \times \left(\frac{2176200}{60}\right) \times 0.02$

$\therefore\ M = \frac{2 \times 10^3 \times 60 \times 60 \times 8}{0.02 \times (2 \times \pi \times 1200)^2}$

$= 50.65 \approx 51$ kg.

5. We know, $\Delta E = I\omega^2 C_S$

Here, $\omega = 200$; $\Delta E = 400$

$C_S = \frac{\omega_{max} - \omega_{min}}{\omega_{max} + \omega_{min}}$

$= \frac{20}{200} = 0.1$

$\Delta E = 400$ N – m

$\therefore\ I = \frac{\Delta E}{\omega^2 C_S} = \frac{400}{(200)^2 \times .1}$

$= .10$

6. $\Delta E = 400$ Nm

$\omega_{mean} = 20$ rad/s

$C_S = \pm 2\%$

$= 2 \times \frac{2}{100} = 0.04$

$\Delta E = I\omega_{mean}^2 C_S$

$400 = I(20)^2 \times 0.04$

$I = 25$ kg-m²

7. $\Delta E = 2600$ J

$N_{average} = 200$ rpm

I = ?

$C_S = \pm 0.5\%$

$= 2 \times \frac{0.5}{100} = 0.1$

$$\Delta E = I\omega^2 C_S$$

$$2600 = I \times \left(\frac{2\pi}{60} \times 200\right)^2 \times 0.01$$

$$I = 592.73 \text{ kg-m}^2$$

8. I = ?

$$E = 3000 \times \left[\pi - \frac{\pi}{2}\right] - 1500[2\pi - \pi]$$

$$= 0$$

$$T_{mean} \times 2\pi = E$$

$$T_{mean} = 0$$

$$\omega_{min} = 10 \text{ rad/s at } \theta = \frac{\pi}{2}$$

$$\omega_{max} = 20 \text{ rad/s at } \theta = \pi$$

$$C_S = \frac{\omega_{max} - \omega_{min}}{\frac{\omega_{max} + \omega_{min}}{2}} = \frac{20 - 10}{\frac{20 + 10}{2}}$$

$$= \frac{10 \times 2}{30}$$

$$= 0.67$$

$$\Delta E = 3000 \times \frac{\pi}{2} = 1500\pi = I\omega^2 C_S$$

$$1500\,\pi = I \times \omega_{mean}^2 \times 0.67$$

$$1500\,\pi = I \times 15^2 \times 0.67$$

$$I = 31.26 \text{ kg-m}^2$$

9. T = 10000 + 1000 sin 2θ – 1200 cos 2θ

N = 100 rpm

$$\omega = \frac{2\pi \times 100}{60} = 10.472 \text{ rad/s}$$

$$T_{mean} = \frac{1}{2\pi}\int_0^{2\pi}[10000 + 1000\sin 2\theta - 1200\cos 2\theta]\,d\theta$$

By solving, we get

$$T_{mean} = 10000 \text{ N-m}$$

$$\text{Power} = 10000 \times \frac{2\pi \times 100}{60}$$

$$= 104719.75 \text{ W}$$

$$= 104.719 \text{ kW}$$

7 CHAPTER

Vibration

Analysis of Previous GATE Papers														
Topic	**Year/Set →** **Marks Distribution ↓**		**2019 Set 1**	**2019 Set 2**	**2018 Set 1**	**2018 Set 2**	**2017 Set 1**	**2017 Set 2**	**2016 Set 1**	**2016 Set 2**	**2016 Set 3**	**2015 Set 1**	**2015 Set 2**	**2015 Set 3**
Damped and Undamped Vibration	**1 Mark**	MCQ Type												
		Numerical Type												
	2 Marks	MCQ Type												
		Numerical Type												
	Total													
Forced Vibration	**1 Mark**	MCQ Type			1		1			1				1
		Numerical Type							1				1	
	2 Marks	MCQ Type							1				1	1
		Numerical Type									1	1		
	Total				**1**		**1**		**3**	**1**	**2**	**2**	**3**	**3**
Transmissibility	**1 Mark**	MCQ Type				1								
		Numerical Type												
	2 Marks	MCQ Type												
		Numerical Type			1									
	Total				**2**	**1**								
Torsional Vibration and Critical Speed of Shaft	**1 Mark**	MCQ Type												
		Numerical Type												
	2 Marks	MCQ Type												
		Numerical Type												
	Total													

Damped & undamped vibration

1. A mass of 1 kg is suspended by means of 3 springs as shown in figure. The spring constant K_1, K_2 and K_3 are respectively 1 kN/m, 3 kN/m and 2 kN/m. The natural frequency of the system is approximately

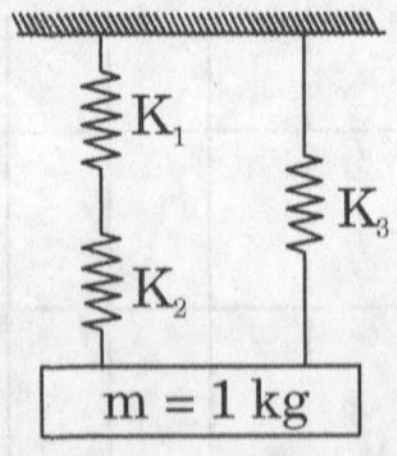

(a) 46.90 rad/s (c) 60.55 rad/s

(b) 52.44 rad/s (d) 77.46 rad/s

[1996 : 2 Marks]

2. Consider the system of two wagons shown in figure. The natural frequencies of this system are

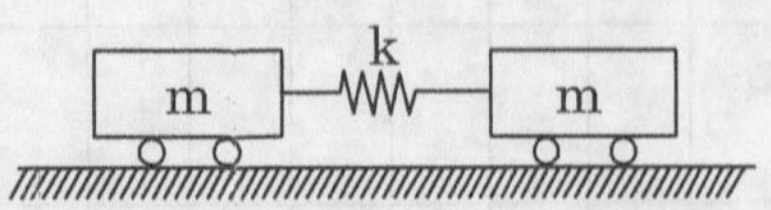

(a) $0, \sqrt{\frac{2k}{m}}$

(b) $\sqrt{\frac{k}{m}}, \sqrt{\frac{2k}{m}}$

(c) $\sqrt{\frac{k}{m}}, \sqrt{\frac{k}{2m}}$

(d) $0, \sqrt{\frac{k}{2m}}$

[1999 : 2 Marks]

3. As shown in figure, a mass of 100 kg is held between two springs. The natural frequency of vibration of the system in cycle/s, is

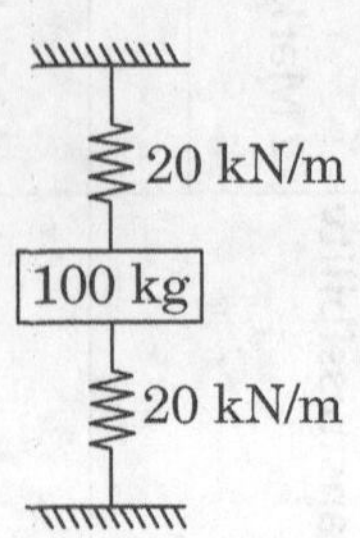

(a) $\frac{1}{2\pi}$ (b) $\frac{5}{\pi}$

(c) $\frac{10}{\pi}$ (d) $\frac{20}{\pi}$

[2000 : 2 Marks]

4. In the figure shown, the spring deflects by δ to position A (the equilibrium position) when a mass m is kept on it. During free vibration, the mass is at position B at some instant. The change in potential energy of the spring mass system from position A to position B is

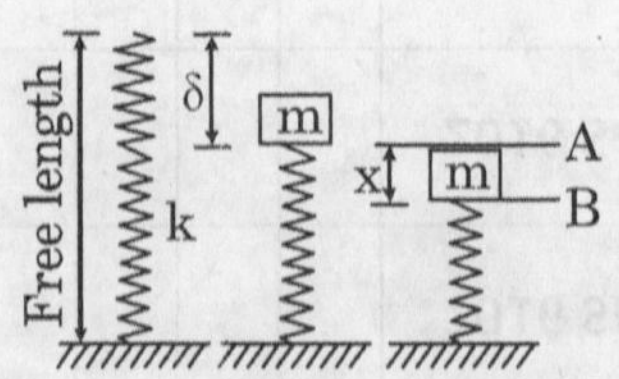

(a) $\frac{1}{2}kx^2$ (b) $\frac{1}{2}kx^2 - mgx$

(c) $\frac{1}{2}k(x+\delta)$ (d) $\frac{1}{2}kx^2 + mgx$

[2001 : 1 Mark]

5. The assembly shown in the figure is composed of two massless rods of length l with two particles, each of mass m. The natural frequency of this assembly for small oscillations is

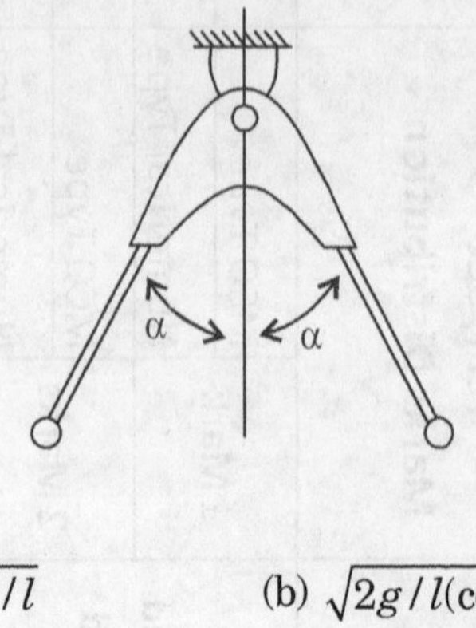

(a) $\sqrt{g/l}$ (b) $\sqrt{2g/l(\cos\alpha)}$

(c) $\sqrt{g/l(\cos\alpha)}$ (d) $\sqrt{(g\cos\alpha)/l}$

[2001 : 2 Marks]

6. A uniform stiff rod of length 300 mm and having a weight of 300 N is pivoted at one end and connected to a spring at the other end. For keeping the rod vertical in a stable position the minimum value of spring constant K needed is

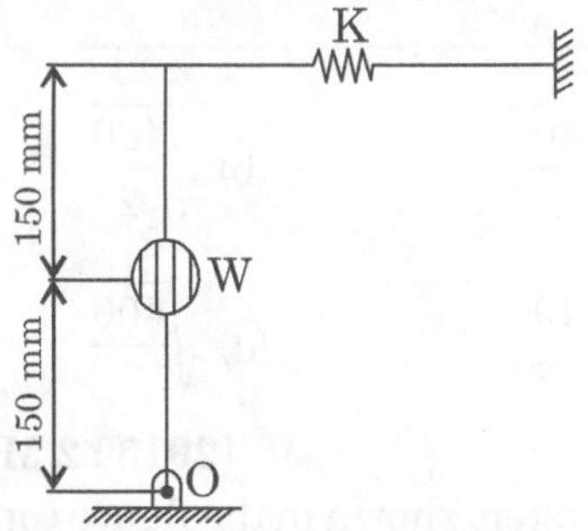

(a) 300 N/m (c) 500 N/m
(b) 400 N/m (d) 1000 N/m

[2004 : 2 Marks]

7. For an under damped harmonic oscillator, resonance

(a) occurs when excitation frequency is greater than undamped natural frequency
(b) occurs when excitation frequency is less than undamped natural frequency
(c) occurs when excitation frequency is equal to undamped natural frequency
(d) never occurs

[2007 : 1 Mark]

8. The natural frequency of the system shown below

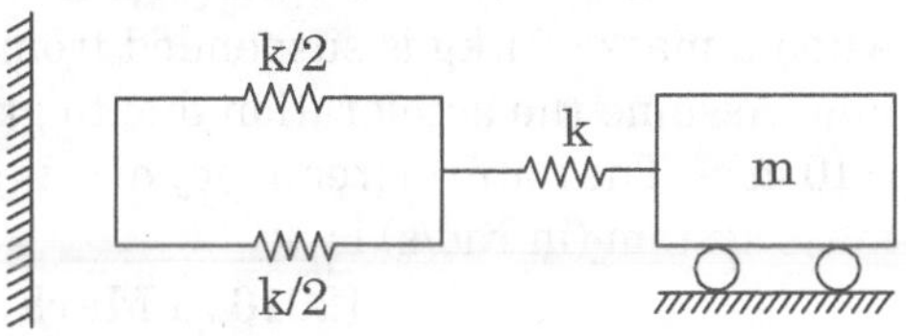

(a) $\sqrt{\frac{k}{2m}}$ (b) $\sqrt{\frac{k}{m}}$

(c) $\sqrt{\frac{2k}{m}}$ (d) $\sqrt{\frac{3k}{m}}$

[2007 : 2 Marks]

9. The natural frequency of the spring mass system shown in the figure is closest to

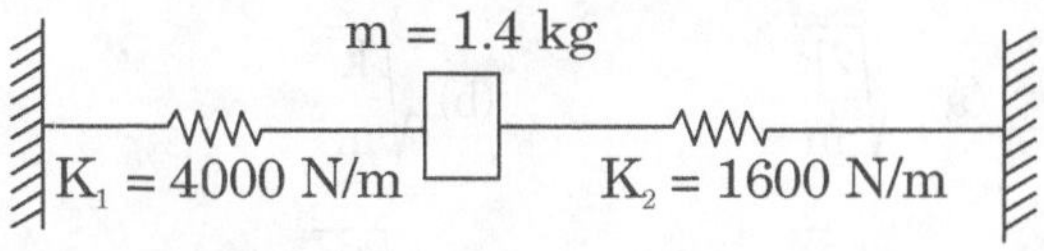

(a) 8Hz (b) 10Hz
(c) 12 Hz (d) 14 Hz

[2008 : 2 Marks]

10. A uniform rigid rod of mass m = 1 kg and length L = 1 m is hinged at its centre & laterally supported at one end by a spring of spring constant k = 300 N/m. The natural frequency ω_n in rad/s is

(a) 10 (b) 20
(c) 30 (d) 40

[2008 : 2 Marks]

11. The natural frequency of a spring-mass system on earth is ω_n. The natural frequency of this system on the moon ($g_{moon} = g_{earth}/6$) is

(a) ω_n (b) 0.408 ω_n
(c) 0.204 ω_n (d) 0.167ω_n

[2010 : 1 Mark]

12. A disc of mass m is attached to a spring of stiffness k as shown in the fig. The disc rolls without slipping on a horizontal surface. The natural frequency of vibration of the system is

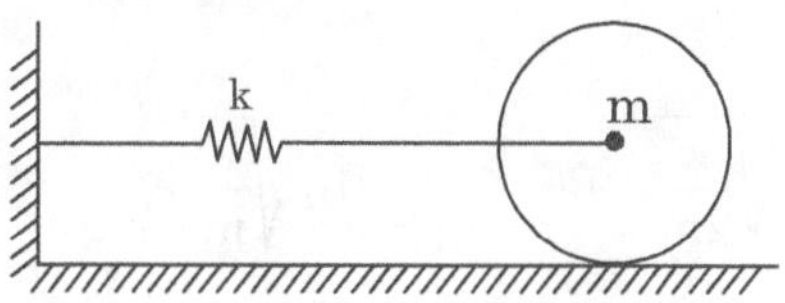

(a) $\frac{1}{2\pi}\sqrt{\frac{k}{m}}$ (b) $\frac{1}{2\pi}\sqrt{\frac{2k}{m}}$

(c) $\frac{1}{2\pi}\sqrt{\frac{2k}{3m}}$ (d) $\frac{1}{2\pi}\sqrt{\frac{3k}{2m}}$

[2011 : 2 Marks]

13. A concentrated mass m is attached at the centre of a rod of length 2L as shown in the figure. The rod is kept in a horizontal equilibrium position by a spring of stiffness k. For very small amplitude of vibration, neglecting the weights of the rod and spring, the undamped natural frequency of the system is

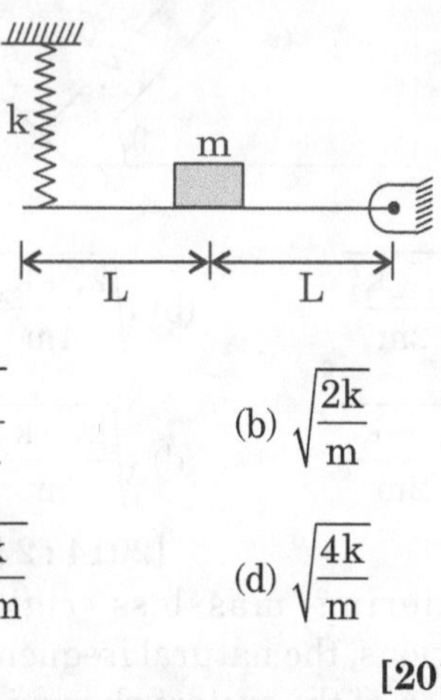

(a) $\sqrt{\frac{k}{m}}$ (b) $\sqrt{\frac{2k}{m}}$

(c) $\sqrt{\frac{k}{2m}}$ (d) $\sqrt{\frac{4k}{m}}$

[2012 : 2 Marks]

14. A point mass is executing simple harmonic motion with an amplitude of 10 mm and frequency of 4 Hz. The maximum acceleration (m/s^2) of the mass is ______

[2014 : 1 Mark, Set-4]

15. A rigid uniform rod AB of length L and mass m is hinged at C such that AC = L/3, CB = 2L/3. Ends A and B are supported by springs of spring constant k. The natural frequency of the system is given by

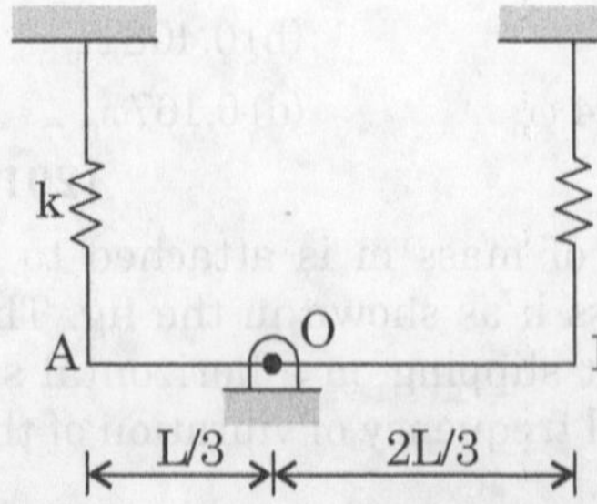

(a) $\sqrt{\frac{k}{2m}}$ (b) $\sqrt{\frac{k}{m}}$

(c) $\sqrt{\frac{2k}{m}}$ (d) $\sqrt{\frac{5k}{m}}$

[2014 : 2 Marks, Set-1]

16. What is the natural frequency of the spring mass system shown below? The contact between the block and the inclined plane is frictionless. The mass of the block is denoted by m and the spring constants are denoted by k_1 and k_2 as shown below

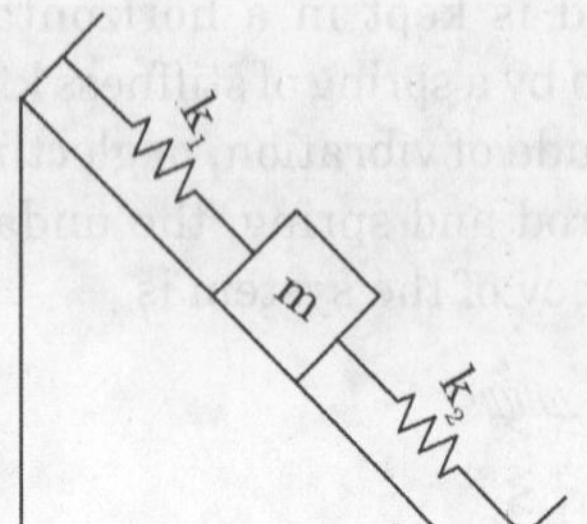

(a) $\sqrt{\frac{k_1 + k_2}{2m}}$ (b) $\sqrt{\frac{k_1 + k_2}{4m}}$

(c) $\sqrt{\frac{k_1 - k_2}{2m}}$ (d) $\sqrt{\frac{k_1 + k_2}{m}}$

[2014 : 2 Marks, Set-2]

17. Considering massless rigid rod small oscillations, the natural frequency (in rad/s) of vibration of the system shown in the figure is

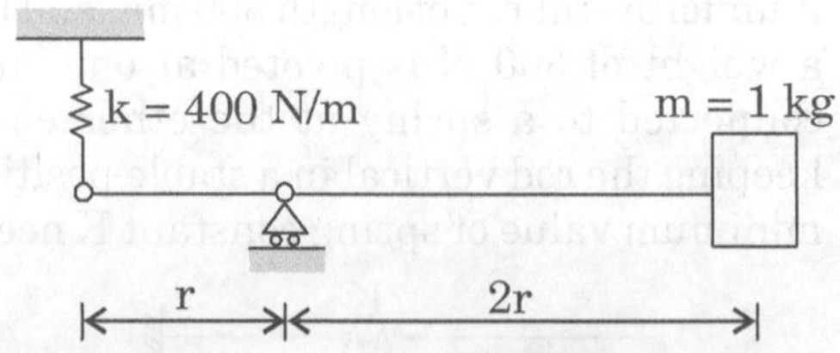

(a) $\sqrt{\frac{400}{1}}$ (b) $\sqrt{\frac{400}{2}}$

(c) $\sqrt{\frac{400}{3}}$ (d) $\sqrt{\frac{400}{4}}$

[2015 : 2 Marks, Set-1]

18. The system shown in the figure consists of block A of mass 5 kg connected to a spring through a massless rope passing over pulley B of radius r and mass 20 kg. The spring constant k is 1500 N/m. If there is no slipping of the rope over the pulley, the natural frequency of the system is ______ rad/s.

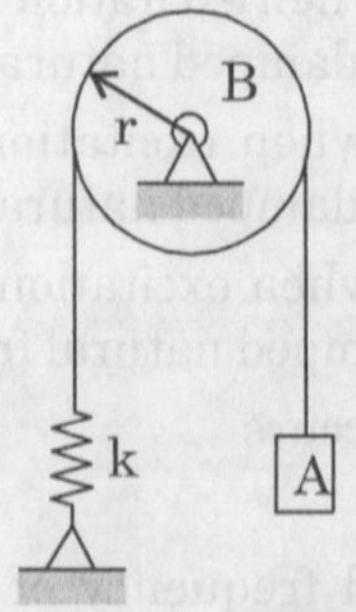

[2016 : 2 Marks, Set-2]

19. The static deflection of a spring under gravity, when a mass of 1 kg is suspended from it, is 1 mm. Assume the acceleration due to gravity g = 10 m/s^2. The natural frequency of this spring-mass system (in rad/s) is ______

[2016 : 1 Mark, Set-3]

20. A mass m is attached to two identical springs having constant k as shown in the figure. The natural frequency ω of this single degree of freedom system is

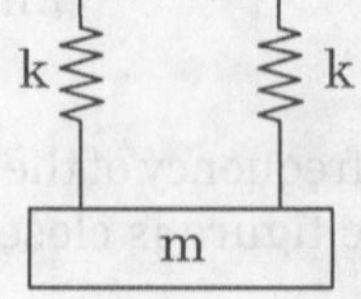

(a) $\sqrt{\frac{2k}{m}}$ (b) $\sqrt{\frac{k}{m}}$

(c) $\sqrt{\frac{k}{2m}}$ (d) $\sqrt{\frac{4k}{m}}$

[2017 : 1 Mark, Set-2]

21. A thin uniform rigid bar of length L and M is hinged at point O, located at a distance of L/3 from one of its ends. The bar is further supported using springs, each of stiffness k located at the two ends. A particle of mass m = M/4 is fixed at one end of the bar, as shown in the figure. For small rotations of the bar about O, the natural frequency of the system is

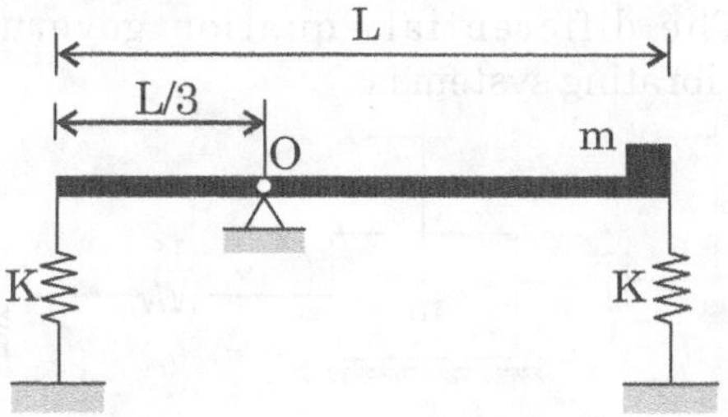

(a) $\sqrt{\frac{5k}{M}}$ (b) $\sqrt{\frac{5k}{2M}}$

(c) $\sqrt{\frac{3k}{2M}}$ (d) $\sqrt{\frac{3k}{M}}$

[2017 : 2 Marks, Set-1]

22. The radius of gyration of a compound pendulum about the point of suspension is 100 mm. The distance between the point of suspension and the centre of mass is 250 mm. Considering the acceleration due to gravity is 9.81 m/s^2, the natural frequency (in radian/s) of the compound pendulum is ________.

[2017 : 2 Marks, Set-2]

23. The natural frequencies corresponding to the spring-mass system I and II are ω_I and ω_{II}, respectively. The ratio $\frac{\omega_I}{\omega_{II}}$ is

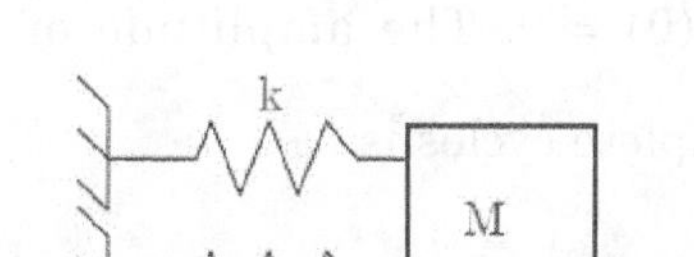

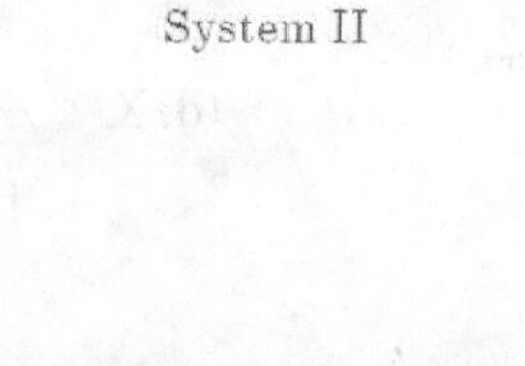

(a) 4 (b) 2

(c) $\frac{1}{4}$ (d) $\frac{1}{2}$

[2019 : 1 Mark, Set 1]

24. A uniform thin disk of mass 1 kg and radius 0.1 m is kept on a surface as shown in the figure. The spring of stiffness k_1 = 400 N/m is connected to the disk center A and another spring of stiffness k_2 = 100 N/m is connected at point B just above point A on the circumference of the disk. Initially, both the springs are unstretched. A assume pure rolling of the disk. For small disturbance from the equilibrium, the natural frequency of vibration of the system is _____ rad/s (round off to one decimal place).

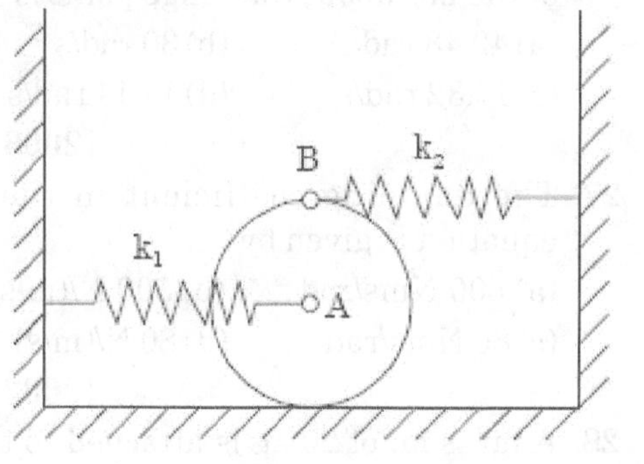

[2019 : 2 Marks, Set 1]

25. A slender uniform rigid bar of mass m is hinged at O and supported by two springs, with stiffnesses 3k and k, and a damper with damping coefficient c, as shown in the figure. For the system to be critically damped, the ratio $c/\sqrt{km}$ should be

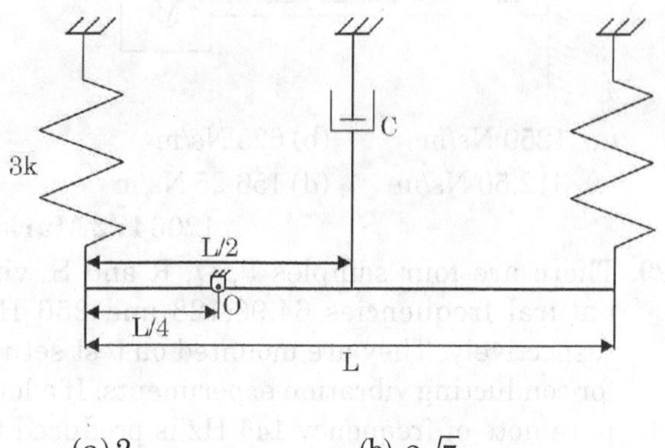

(a) 2 (b) $2\sqrt{7}$

(c) 4 (d) $4\sqrt{7}$

[2019 : 2 Marks, Set 2]

Forced vibration

Common Data Questions 26 and 27

A uniform rigid slender bar of mass 10 kg, hinged at the left end is suspended with the help of spring and damper arrangement as shown in the figure where K= 2 kN/m, C = 500 Ns/m and the stiffness of the torsional spring k_θ is 1 kN /m/rad. Ignore the hinge dimensions.

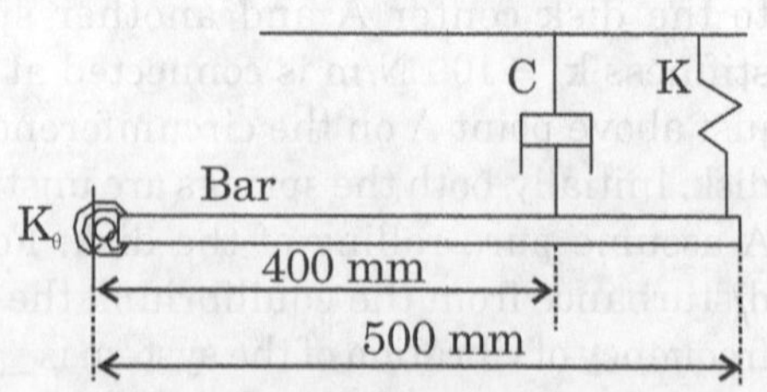

26. The undamped natural frequency of oscillations of the bar about the hinge point is

(a) 42.43 rad/s (b) 30 rad/s
(c) 17.32 rad/s (d) 14.14 rad/s

[2003 : 2 Marks]

27. The damping coefficient in the vibration equation is given by

(a) 500 Nms/rad (b) 500 N/(m/s)
(c) 80 Nms/rad (d) 80 N/(m/s)

[2003 : 2 Marks]

28. A mass m, of 20 kg is attached to the free end of a steel cantilever beam of length 1000 mm having a cross-section of 25 × 25 mm. Assume the mass of the cantilever to be negligible and E_{steel} = 200 GPa. If the lateral vibration of this system is critically damped using a viscous damper, the damping constant of the damper is

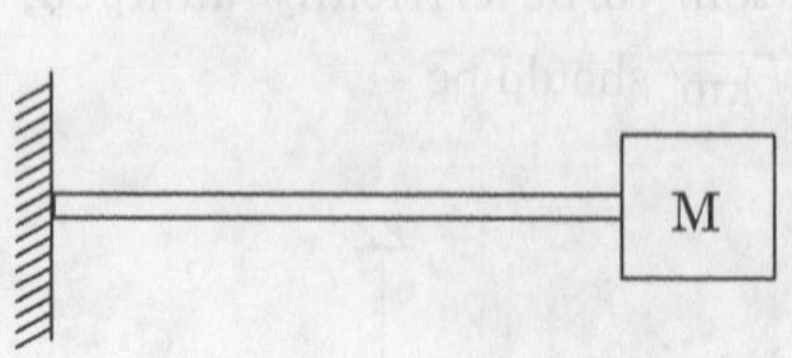

(a) 1250 Ns/m (b) 625 Ns/m
(c) 312.50 Ns/m (d) 156.25 Ns/m

[2004 : 2 Marks]

29. There are four samples P, Q, R and S with natural frequencies 64,96,128 and 256 Hz, respectively. They are mounted on test setups for conducting vibration experiments. If a loud pure note of frequency 144 Hz is produced by some instrument, which of the samples will show the most perceptible induced vibration?

(a) P (b) Q
(c) R (d) S

[2005 : 1 Mark]

30. In a spring-mass system, the mass is 0.1 kg and the stiffness of the spring is 1 kN/m. By introducing a damper, the frequency of oscillation is found to be 90% of the original value. What is the damping coefficient of the damper?

(a) 1.2 Ns/m (b) 3.4 Ns/m
(c) 8.7 Ns/m (d) 12.0 Ns/m

[2005 : 2 Marks]

31. The differential equation governing the vibrating system is

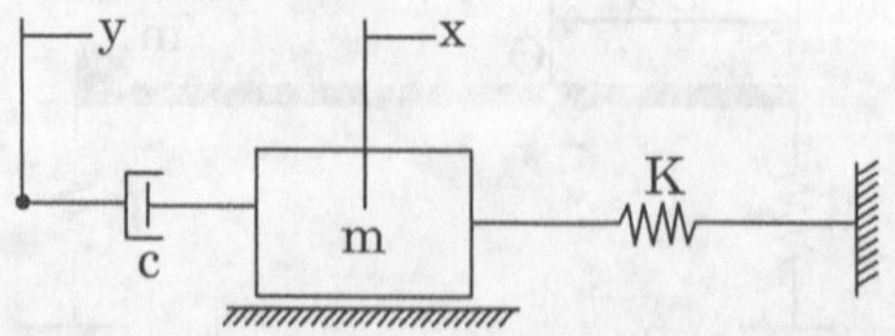

(a) $m\ddot{x} + c\dot{x} + k(x - y) = 0$

(b) $m(\ddot{x} - \ddot{y}) + c(\dot{x} - \dot{y}) + kx = 0$

(c) $m\ddot{x} + c(\dot{x} - \dot{y}) + kx = 0$

(d) $m(\ddot{x} - \ddot{y}) + c(\dot{x} - \dot{y}) + k(x - y) = 0$

[2006 : 1 Mark]

Linked Answer Questions 32 and 33

A vibratory system consists of a mass 12.5 kg, a spring of stiffness 1000 N/m, and a dashpot with damping coefficient of 15 Ns/m.

32. The value of critical damping of the system is

(a) 0.223 Ns/m (b) 17.88 Ns/m
(c) 71.4 Ns/m (d) 223.6 Ns/m

[2006 : 2 Marks]

33. The value of logarithmic decrement is

(a) 1.35 (b) 1.32
(c) 0.68 (d) 0.66

[2006 : 2 Marks]

34. The equation of motion of a harmonic oscillator is given by

$$\frac{d^2x}{dt^2} + 2\xi\omega_n \frac{dx}{dt} + \omega_n^2 x = 0$$

and the initial conditions at t = 0 are x(0) = X, $\frac{dx}{dt}(0) = 0$. The amplitude of x(t) after n complete cycles is

(a) $Xe^{-2n\pi\left(\frac{\xi}{\sqrt{1-\xi^2}}\right)}$ (b) $Xe^{2n\pi\left(\frac{\xi}{\sqrt{1-\xi^2}}\right)}$

(c) $Xe^{-2n\pi\left(\frac{\sqrt{1-\xi^2}}{\xi}\right)}$ (d) X

[2007 : 2 Marks]

35. An automotive engine weighing 240 kg is supported on four springs with linear characteristics. Each of the front two springs have a stiffness of 16 MN/m while the stiffness of each rear spring is 32 MN/m. The engine speed (in rpm), at which resonance is likely to occur, is

(a) 6040 (b) 3020

(c) 1424 (d) 955

[2009 : 2 Marks]

36. A vehicle suspension system consists of a spring and a damper. The stiffness of the spring is 3.6 kN/m and the damping constant of the damper is 400 Ns/m. If the mass is 50 kg, then the damping factor (d) and damped natural frequency (f_n), respectively, are

(a) 0.471 and 1.19 Hz

(b) 0.471 and 7.48 Hz

(c) 0.666 and 1.35 Hz

(d) 0.666 and 8.50 Hz

[2009 : 2 Marks]

37. A mass m attached to a spring is subjected to a harmonic force as shown in figure. The amplitude of the forced motion is observed to be 50 mm. The value of m (in kg) is

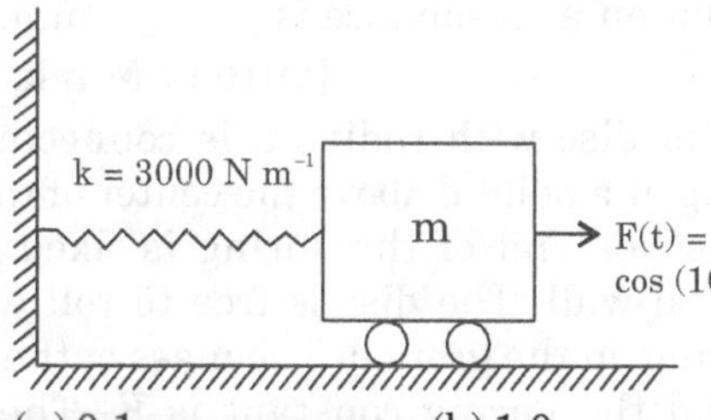

(a) 0.1 (b) 1.0

(c) 0.3 (d) 0.5

[2010 : 2 Marks]

38. A mass of 1 kg is attached to two identical springs each with stiffness k = 20 kN/m as shown in the figure. Under frictionless condition, the natural frequency of the system in Hz is close to

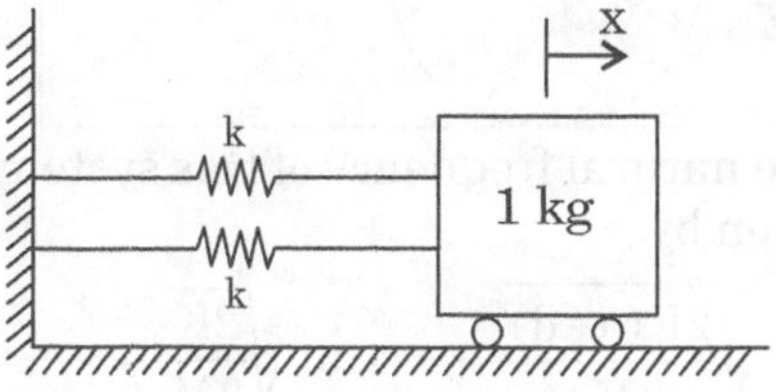

(a) 32 (b) 23

(c) 16 (d) 11

[2011 : 2 Marks]

39. Critical damping is the

(a) largest amount of damping for which no oscillation occurs in free vibration

(b) smallest amount of damping for which no oscillation occurs in free vibration

(c) largest amount of damping for which the motion is simple harmonic in free vibration

(d) smallest amount of damping for which the motion is simple harmonic in free vibration

[2014 : 1 Mark, Set-1]

40. Consider a single degree of freedom system with viscous damping excited by a harmonic force. At resonance, the phase angle (in degree) of the displacement with respect to the exciting force is

(a) 0 (b) 45

(c) 90 (d) 135

[2014 : 1 Mark, Set-3]

41. The damping ratio of a single degree of freedom spring mass damper system with mass of 1 kg, stiffness 100 N/m and viscous damping coefficient of 5 Ns/m is______.

[2014 : 2 Marks, Set-3]

42. A mass-spring-dashpot system with mass m = 10 kg, spring constant k = 6250 N/m is excited by a harmonic excitation of 10 cos(25t) N. At the steady state, the vibration amplitude of the mass is 40 mm. The damping coefficient (c, in Ns/m) of the dashpot is __________.

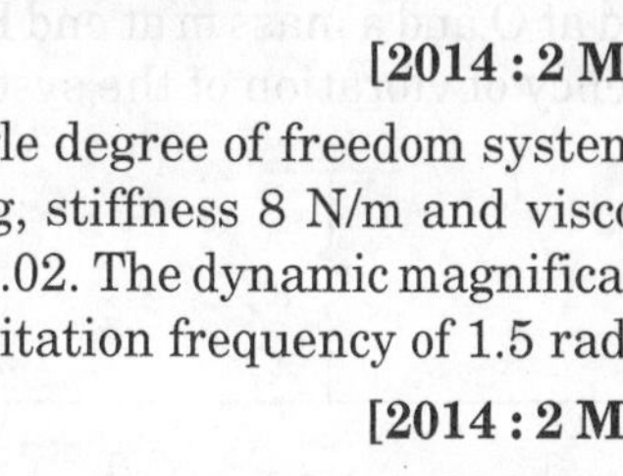

[2014 : 2 Marks, Set-3]

43. A single degree of freedom system has a mass of 2 kg, stiffness 8 N/m and viscous damping ratio 0.02. The dynamic magnification factor at an excitation frequency of 1.5 rad/s is ______

[2014 : 2 Marks, Set-4]

44. In a spring-mass system, the mass is m and the spring constant is k. The critical damping coefficient of the system is 0.1 kg/s. In another spring-mass system, the mass is 2m and the spring constant is 8k. The critical damping coefficient (in kg/s) of this system is ______.

[2015 : 1 Mark, Set-2]

45. Which of the following statements are TRUE for damped vibrations?

P. For a system having critical damping, the value of damping ratio is unity and system does not undergo a vibratory motion.

Q. Logarithmic decrement method is used to determine the amount of damping in a physical system.

R. In case of damping due to dry friction between moving surfaces resisting force of constant magnitude acts opposite to the relative motion.

S. For the case of viscous damping, drag force is directly proportional to the square of relative velocity.

(a) P and Q only (b) P and S only
(c) P, Q and R only (d) Q and S only

[2015 : 1 Mark, Set-3]

46. A precision instrument package (m = 1 kg) needs to be mounted on a surface vibrating at 60 Hz. It is desired that only 5% of the base surface vibration amplitude be transmitted to the instrument. Assuming that the isolation is designed with its natural frequency significantly lesser than 60 Hz, so that the effect of damping may be ignored. The stiffness (in N/m) of the required mounting pad is ________.

[2015 : 2 Marks, Set-1]

47. A single-degree-freedom spring-mass system is subjected to a sinusoidal force of 10 N amplitude and frequency ω along the axis of the spring. The stiffness of the spring is 150 N/m, damping factor is 0.2 and the undamped natural frequency is 10 ω. At steady state, the amplitude of vibration (in m) is approximately

(a) 0.05 (b) 0.07
(c) 0.70 (d) 0.90

[2015 : 2 Marks, Set-2]

48. Figure shows a single degree of freedom system. The system consists of a massless rigid bar OP hinged at O and a mass m at end P. The natural frequency of vibration of the system is

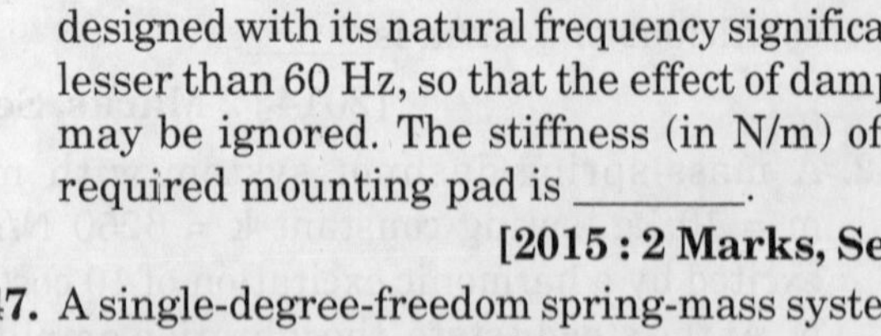

(a) $f_n = \frac{1}{2\pi}\sqrt{\frac{k}{4m}}$ (b) $f_n = \frac{1}{2\pi}\sqrt{\frac{k}{2m}}$

(c) $f_n = \frac{1}{2\pi}\sqrt{\frac{k}{m}}$ (d) $f_n = \frac{1}{2\pi}\sqrt{\frac{2k}{m}}$

[2015 : 2 Marks, Set-3]

49. A single degree of freedom spring-mass system is subjected to a harmonic force of constant amplitude. For an excitation frequency of $\sqrt{\frac{3k}{m}}$, the ratio of the amplitude of steady state response to the static deflection of the spring is______ .

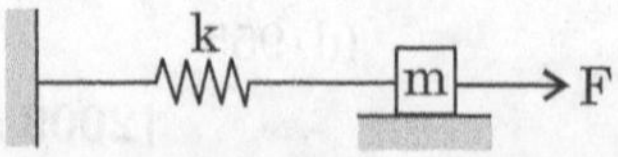

[2016 : 2 Marks, Set-3]

50. A single degree of freedom mass-spring-viscous damper system with mass m, spring constant k and viscous damping coefficient q is critically damped. The correct relation among m, k, and q is

(a) $q = \sqrt{2km}$ (b) $q = 2\sqrt{km}$

(c) $q = \sqrt{\frac{2k}{m}}$ (d) $q = 2\sqrt{\frac{k}{m}}$

[2016 : 1 Mark, Set-2]

51. A single degree of freedom spring mass system with viscous damping has a spring constant of 10 kN/m. The system is excited by a sinusoidal force of amplitude 100 N. If the damping factor (ratio) is 0.25, the amplitude of steady state oscillation at resonance is ______ mm.

[2016 : 1 Mark, Set-1]

52. A solid disc with radius a is connected to a spring at a point d above the center of the disc. The other end of the spring is fixed to the vertical wall. The disc is free to roll without slipping on the ground. The mass of the disc is M and the spring constant is K. The polar moment of inertia for the disc about its centre is $J = \frac{Ma^2}{2}$

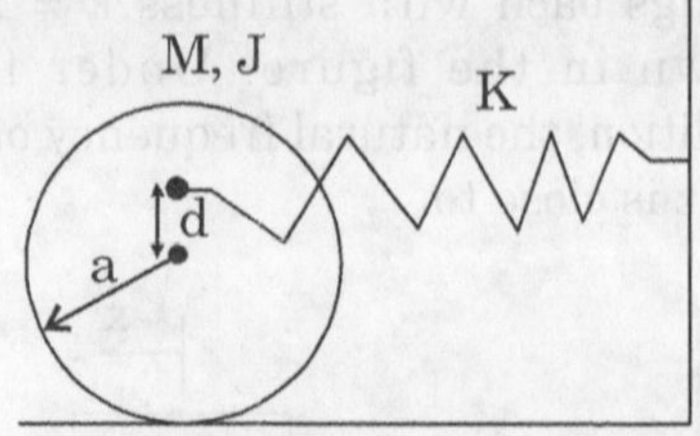

The natural frequency of this system in rad/s is given by

(a) $\sqrt{\frac{2K(a+d)^2}{3Ma^2}}$ (b) $\sqrt{\frac{2K}{3M}}$

(c) $\sqrt{\frac{2K(a+d)^2}{Ma^2}}$ (d) $\sqrt{\frac{K(a+d)^2}{Ma^2}}$

[2016 : 2 Marks, Set-1]

53. The damping ratio for a viscously damped spring mass system, governed by the relationship

$$m\frac{d^2x}{dt^2}+c\frac{dx}{dt}+kx = F(t), \text{ is given by}$$

(a) $\sqrt{\frac{c}{mk}}$ (b) $\frac{c}{2\sqrt{km}}$

(c) $\frac{c}{\sqrt{km}}$ (d) $\sqrt{\frac{c}{\sqrt{2km}}}$

[2017 : 1 Mark, Set-1]

54. The equation of motion for a spring-mass system excited by a harmonic force is

$M\ddot{x} + Kx = F\cos(\omega t)$

where M is the mass, K is the spring stiffness, F is the force amplitude and ω is the angular frequency of excitation. Resonance occurs when ω is equal to

(a) $\sqrt{\frac{M}{K}}$ (b) $\frac{1}{2\pi}\sqrt{\frac{K}{M}}$

(c) $2\pi\sqrt{\frac{K}{M}}$ (d) $\sqrt{\frac{K}{M}}$

[2018 : 1 Mark, Set-1]

Transmissibility

55. A vibrating machine is isolated from the floor using springs. If the ratio of excitation frequency of vibration of machine to the natural frequency of the isolation system is equal to 0.5, the transmissibility ratio of isolation is

(a) $\frac{1}{2}$ (b) $\frac{3}{4}$

(c) $\frac{4}{3}$ (d) 2

[2004 : 1 Mark]

56. A machine of 250 kg mass is supported on springs of total stiffness 100 kN/m. Machine has an unbalanced rotating force of 350 N at speed of 3600 rpm. Assuming a damping factor of 0.15, the value of transmissibility ratio is

(a) 0.0531 (b) 0.9922

(c) 0.0162 (d) 0.0028

[2006 : 2 Marks]

57. In vibration isolation, which one of the following statements is **NOT** correct regarding Transmissibility (T)?

(a) T is nearly unity at small excitation frequencies

(b) T can be always reduced by using higher damping at any excitation frequency

(c) T is unity at the frequency ratio of $\sqrt{2}$

(d) T is infinity at resonance for undamped systems

[2014 : 1 Mark, Set-2]

58. A machine of mass m = 200 kg is supported on two mounts, each of stiffness k = 10 kN/m. The machine is subjected to an external force (in N) F(t) = 50 cos 5t. Assuming only vertical translatory motion, the magnitude of the dynamic force (in N) transmitted from each mount to the ground is ______ (correct to two decimal places).

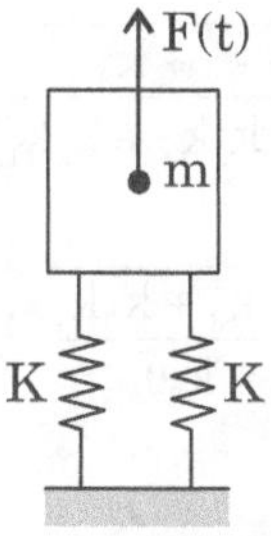

[2018 : 2 Marks, Set-1]

59. In a single degree of freedom underdamped spring-mass-damper system as shown in the figure, an additional damper, is added in parallel such that the system still remains underdamped. Which one of the following statements is ALWAYS true?

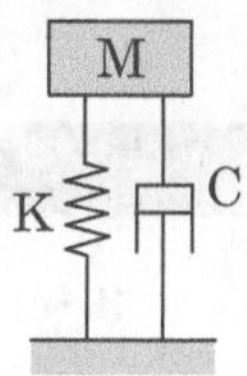

(a) Transmissibility will increase

(b) Transmissibility will decrease

(c) Time period of free oscillations will increase

(d) Time period of free oscillations will decrease

[2018 : 1 Mark, Set-2]

Torsional vibration & Critical Speed of Shaft

60. Consider the arrangement shown in the figure below where J is the combined polar mass moment of inertia of the disc and the shafts. k_1 k_2, k_3 are the torsional stiffness of the respective shafts. The natural frequency of torsional oscillation of the disc is given by

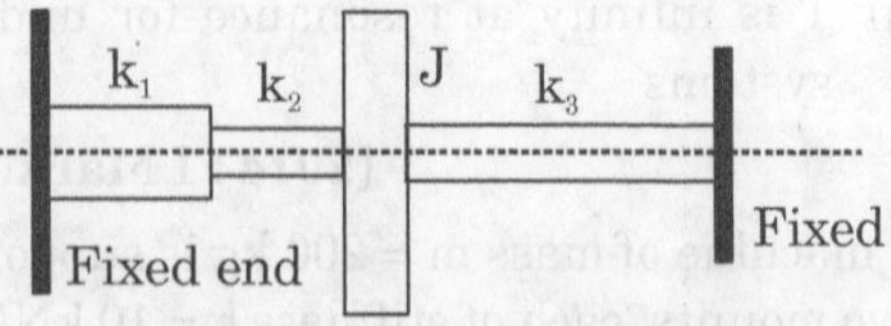

(a) $\sqrt{\dfrac{k_1 + k_2 + k_3}{J}}$

(b) $\sqrt{\dfrac{k_1k_2 + k_2k_3 + k_3k_1}{J(k_1 + k_2)}}$

(c) $\sqrt{\dfrac{k_1 + k_2 + k_3}{J(k_1k_2 + k_2k_3 + k_3k_1)}}$

(d) $\sqrt{\dfrac{k_1k_2 + k_2k_3 + k_3k_1}{J(k_2 + k_3)}}$

[2003 : 1 Mark]

61. A flexible rotor-shaft system comprises of a 10 kg rotor disc placed in the middle of a massless shaft of diameter 30 mm and length 500 mm between bearings (shaft is being taken massiess as the equivalent mass of the shaft is included in the rotor mass) mounted at the ends. The bearings are assumed to simulate simply supported boundary conditions. The shaft is made of steel for which the value of E is 2.1 × 10^{11} Pa. What is the critical speed of rotation of the shaft?

(a) 60 Hz (b) 90 Hz

(c) 135 Hz (d) 180 Hz

[2003 : 2 Marks]

62. A rotating disc of 1 m diameter has two eccentric masses of 0.5 kg each at radii of 50 mm and 60 mm at angular positions of 0° and 150°, respectively. A balancing mass of 0.1 kg is to be used to balance the rotor. What is the radial position of the balancing mass?

(a) 50 mm (b) 120 mm

(c) 150 mm (d) 280 mm

[2005 : 2 Marks]

63. The rotor shaft of a large electric motor supported between short bearings at both the ends shows a deflection of 1.8 mm in the middle of the rotor. Assuming the rotor to be perfectly balanced and supported at knife edges at both the ends, the likely critical speed (in rpm) of the shaft is

(a) 350 (b) 705

(c) 2810 (d) 4430

[2009 : 1 Mark]

64. If two nodes are observed at a frequency of 1800 rpm during whirling of simply supported long slender rotating shaft, the first critical speed of the shaft in rpm is

(a) 200 (b) 450

(c) 600 (d) 900

[2013 : 1 Mark]

ANSWERS

1. (b)	**2.** (a)	**3.** (c)	**4.** (a)	**5.** (d)	**6.** (c)	**7.** (c)	**8.** (a)	**9.** (b)	**10.** (c)
11. (a)	**12.** (c)	**13.** (d)	**15.** (d)	**16.** (d)	**17.** (d)	**20.** (a)	**21.** (b)	**23.** (d)	**25.** (d)
26. (a)	**27.** (c)	**28.** (a)	**29.** (c)	**30.** (c)	**31.** (c)	**32.** (d)	**34.** (a)	**35.** (a)	**36.** (a)
37. (a)	**38.** (a)	**39.** (b)	**40.** (c)	**45.** (c)	**47.** (b)	**48.** (a)	**50.** (b)	**52.** (a)	**53.** (b)
54. (d)	**55.** (c)	**56.** (c)	**57.** (b)	**59.** (c)	**60.** (b)	**61.** (b)	**62.** (c)	**63.** (b)	**64.** (a)

EXPLANATIONS

1. K_1 and K_2 are in series,

$$\therefore \quad \frac{1}{K_{eq1}} = \frac{1}{K_1} + \frac{1}{K_2} = \frac{1}{1} + \frac{1}{3} = \frac{4}{3}$$

K_{eq1} and K_3 are in parallel arrangement,

$$K_{eq} = K_{eq1} + K_3 = \frac{3}{4} + 2 = \frac{11}{4} \text{ kN/m}$$

$$\text{Natural frequency, } \omega = \sqrt{\frac{K_{eq}}{m}} = \sqrt{\frac{11 \times 10^3}{4 \times 1}}$$

$$= 52.44 \text{ rad/s}$$

2. Natural frequencies will be $\left(0, \sqrt{\frac{2K}{m}}\right)$

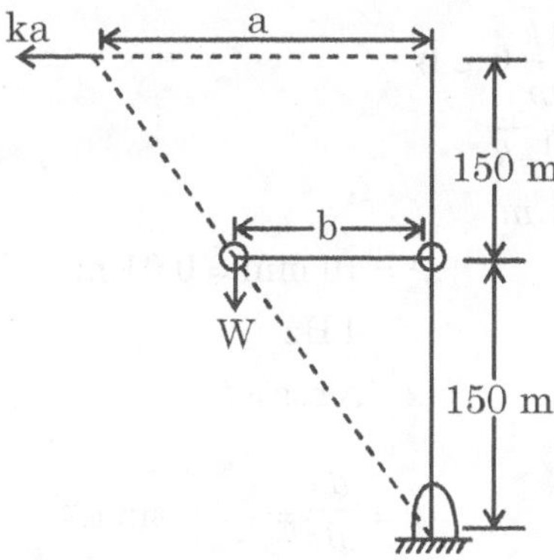

For small displacement

$$\theta = \frac{a}{300} = \frac{b}{150}$$

Taking moment about hinged point 'O'

k.a (0.3) = Wb

$\Rightarrow$ K$(0.3)^2$ = 300 × .150

$$K = \frac{300 \times 0.150}{0.3^2} = 500 \text{ N/m}$$

3. $S = S_1 + S_2 = 20 + 20$

= 40 kN/m = 40,000 N/m

$\therefore$ Natural frequency of vibration of the system,

$$f_n = \frac{1}{2\pi}\sqrt{\frac{S}{m}} = \frac{1}{2\pi}\sqrt{\frac{40 \times 1000}{100}} = \frac{20}{2\pi} = \frac{10}{\pi}$$

4. Potential energy at A = mg $(l - \delta)$

Total energy at B = mg $[l - (\delta + x)] + \frac{1}{2}kx^2$

$\therefore$ Change in energy

$$= mgl - mg(\delta + x) + \frac{1}{2}kx^2 - mgl + mg\delta.$$

$$= \frac{1}{2}kx^2 - mgx \cdot \delta$$

5. Net restoring torque when displaced by a small angle θ

$\tau = mg \cos(\alpha - \theta)\, l - mg\,(\alpha + \theta) l$

$= 2$ mgl $\cos\alpha \sin\theta$

For very small θ, $\sin\theta \approx \theta$

$\therefore \tau = 2mgl \cos\alpha\, \theta$ (restorative)

Now, $$I\frac{d^2\theta}{dt^2} + 2\, mgl \cos\alpha\, \theta = 0$$

But $I = 2ml^2$

$$\therefore \quad 2ml^2 \frac{d^2\theta}{dt^2} + 2mgl\cos\alpha\,\theta = 0$$

$$\text{or} \quad \frac{d^2\theta}{dt^2} + \frac{g\cos\alpha}{l}\theta = 0$$

$$\therefore \quad \omega_n = \sqrt{\frac{g\cos\alpha}{l}}$$

7. For underdamped harmonic oscillator,

$\omega_d = \quad \omega_n$

8. Equivalent k,

$$k_{equ} = \frac{1}{k} + \frac{1}{k/2 + k/2} = \frac{k}{2}$$

$$\therefore \quad \omega_n = \sqrt{\frac{k}{2m}}$$

9. $k_{eq} = k_1 + k_2 = 5600$ N/m

$$\therefore \quad f = \frac{1}{2}\sqrt{\frac{k}{m}} = 10 \text{ Hz}$$

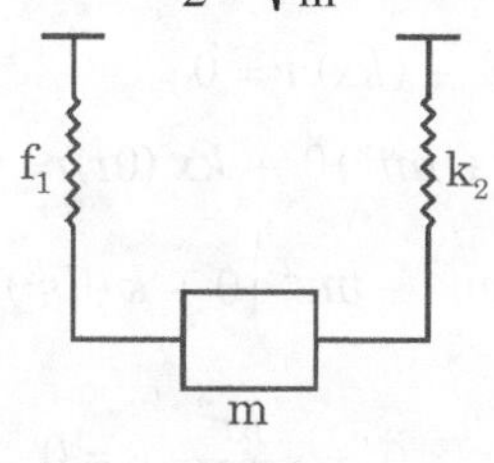

10. 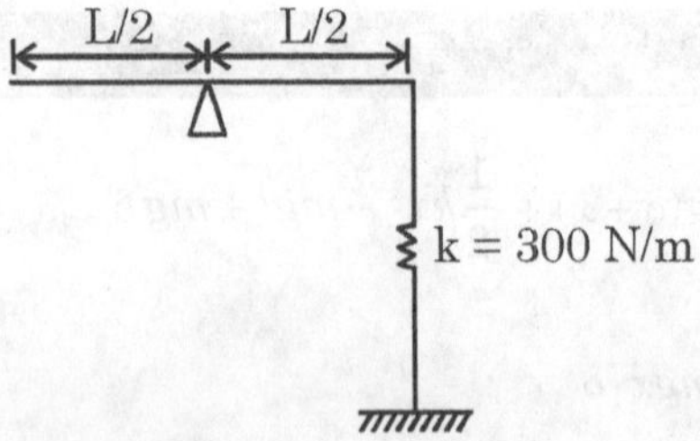

$$\frac{1}{2}I\dot{\theta}^2 + \frac{1}{2}kx^2 = \text{Constant}$$

$$\frac{1}{2}I\dot{\theta}^2 + \frac{1}{2}k\left(\frac{L}{2}\theta\right)^2 = \text{Constant}$$

Differentiating above equation w.r.t. t, we get

$$\frac{1}{2}I\ddot{\theta} + \frac{kL^2\theta}{4} = 0$$

$$I = \frac{mL^2}{12}$$

$$\Rightarrow \frac{mL^2}{12}\ddot{\theta} + \frac{kL^2}{4}\theta = 0$$

$$\ddot{\theta} + \frac{3k}{m}\theta = 0$$

$$\omega_n = \sqrt{\frac{3k}{m}} = \sqrt{\frac{3\times 300}{1}} = 30 \text{ rad/s}$$

11. Natural frequency, $\omega_n = \sqrt{\dfrac{K}{m}}$

It depends only on mass but not on weight. Therefore it is independent of gravity of a system, so it will remain ω_n.

12. 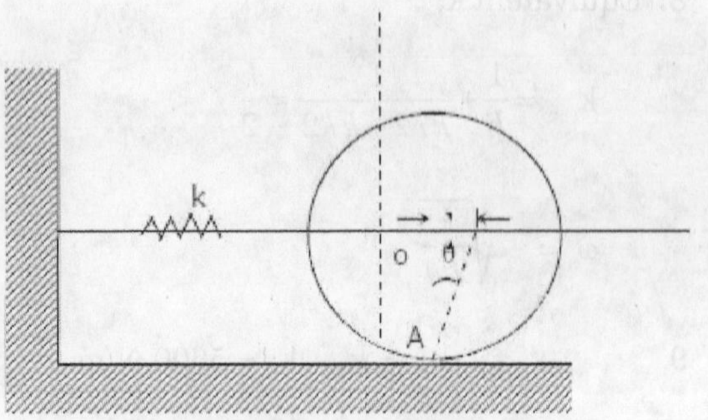

Taking moments about instantaneous centre 'A',

$$I_a\ddot{\theta} + (kx)\,r = 0$$

$$\Rightarrow (I_O + mr^2)\ddot{\theta} + kx\,(\theta r)r = 0$$

$$\Rightarrow \left(\frac{1}{2}mr^2 + mr^2\right)\theta + k\,(\theta r^2) = 0$$

$$\Rightarrow \ddot{\theta} + \frac{kr^2}{\frac{3}{2}mr^2}\theta = 0$$

$$\Rightarrow \ddot{\theta} + \frac{2k}{3m}\theta = 0$$

$$\therefore \omega_n = \frac{1}{2\pi}\sqrt{\frac{2k}{3m}}$$

13. 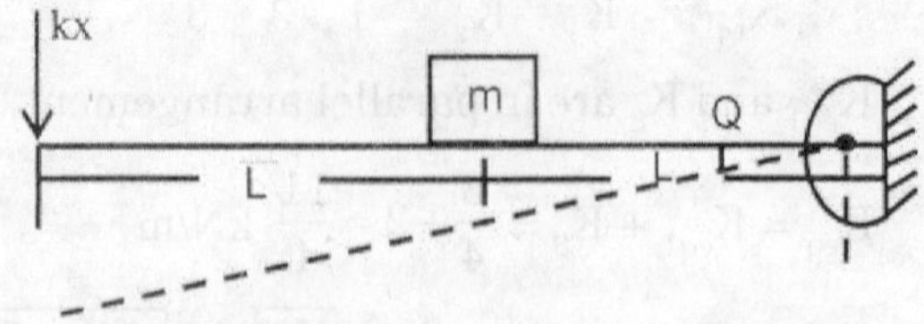

Equation of motion, is

$$m\left(L\ddot{\theta}\right).L\bar{\theta} + kx\times 2\text{L} = 0$$

$$\Rightarrow m\left(L\ddot{\theta}\right).L\bar{\theta} + \text{k}(2\text{L}.\theta)\times 2\text{L} = 0$$

$$\Rightarrow mL^2\ddot{\theta} + 4\text{kL}^2\theta = 0$$

$$\Rightarrow \ddot{\theta} + \sqrt{\frac{4k}{m}}\theta = 0$$

$$\therefore \omega_n = \sqrt{\frac{4k}{m}}$$

14. Given

$$A = 10 \text{ mm} = 0.01 \text{ m}$$

$$f = 4 \text{ Hz}$$

$$x = A\cos\omega t$$

$$v = \frac{dx}{dt} = -A\omega\sin\omega t$$

$$\frac{d^2x}{dt^2} = a = -A\omega^2\cos\omega t$$

For maximum, $t = 0$

$$\therefore a = A\omega^2 = 0.01(2\pi f)^2$$

$$\because \omega = 2\pi f = 0.01(2\times\pi\times 4)^2 = 6.32 \text{ m/s}^2$$

15. $$\text{K}\theta\left(\frac{\text{L}}{3}\right)^2 + \text{K}\theta\left(\frac{2\text{L}}{3}\right)^2 + \frac{m\text{L}^2}{9}\ddot{\theta} = 0$$

$$\Rightarrow \left[\frac{m\text{L}^2}{9}\right]\ddot{\theta} + \left[\frac{5\text{kL}^2}{9}\right]\theta = 0$$

$$\ddot{\theta} + \left(\frac{5\text{k}}{\text{m}}\right)\theta = 0$$

$$\omega_n = \sqrt{\frac{5\text{k}}{\text{m}}}$$

16. It is parallel

$$k_{eq} = k_1 + k_2$$

$$\omega_n = \sqrt{\frac{k_{eq}}{m}}$$

$$\omega_n = \sqrt{\frac{k_1 + k_2}{m}}$$

17. Form $\Rightarrow dx = 2r\theta$

$$\frac{d^2x}{dt^2} = 2r.\ddot{\theta}$$

Taking moments

$$2r\,\theta.m.\,2r.\ddot{\theta} + 400 \times (r\theta) \times r\theta = 0$$

$$\Rightarrow 4m.\ddot{\theta} + 400 \times \theta = 0$$

$$\Rightarrow \omega = \sqrt{\frac{400}{4}}$$

18. $KE = \frac{1}{2}m\dot{x}^2 + \frac{1}{2}I\dot{\theta}^2$

$m = 5$ kg, $\quad \theta = \frac{x}{r}$

$$I = \frac{20 \times r^2}{2} = 10r^2$$

$$KE = \frac{1}{2}5\dot{x}^2 + \frac{1}{2}10r^2.\frac{\dot{x}^2}{r^2} = \frac{1}{2}(15)\dot{x}^2$$

$$\therefore \quad m_{eq} = 15$$

$$PE = \frac{1}{2}kx^2$$

$$\therefore \quad k_{eq} = k = 1500 \text{ N/m}$$

Natural frequency

$$\omega_n = \sqrt{\frac{k_{eq}}{m_{eq}}} = \sqrt{\frac{1500}{15}} = 10 \text{ rad/sec}$$

19. $g = 10$ m/s², $m = 1$ kg

Deflection : $\Delta = 1$ mm

$$\omega_n = \sqrt{\frac{g}{\Delta}} = \sqrt{\frac{10}{1 \times 10^{-3}}} = 100 \text{ rad/s}$$

20. $K_{eq} = k + k = 2k$

$$\omega_n = \sqrt{\frac{Keq}{m}} = \sqrt{\frac{2k}{m}}$$

21.

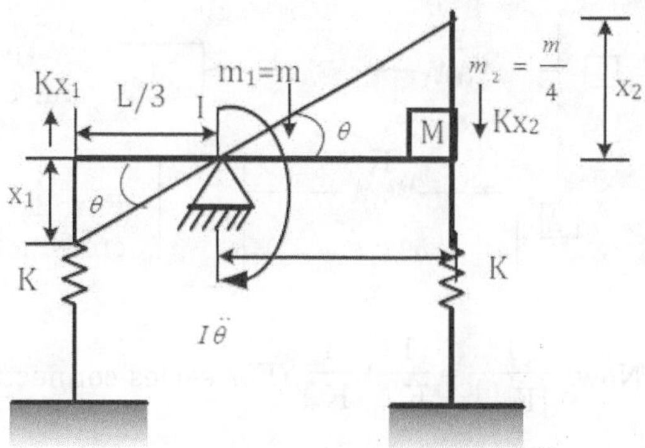

$$I = (I_{\text{about I}})_{rod} + (I_{\text{about I}})_{\text{point mass}}$$

$$I = \left[\frac{ml^2}{12} + m\left(\frac{2l}{3} - \frac{l}{2}\right)^2\right]\left[\frac{m}{4}\left(\frac{2l}{3}\right)^2\right]$$

$$I = \frac{ml^2}{9} + \frac{ml^2}{9} = \frac{2ml^2}{9}$$

$$\Sigma M_I = 0$$

$$I\ddot{\theta} + Kx_2 + \frac{2l}{3} + Kx_1 \times \frac{l}{3} = 0$$

$$\frac{2ml^2}{9} + k\left(\frac{5}{9}l^2\right)\theta = 0$$

$$\omega_n = \sqrt{\frac{k \times \frac{5}{9}}{\frac{2m}{9}}} = \sqrt{\frac{5k}{2m}}$$

22. For compound pendulum,

$$\ddot{\theta} + \left(\frac{mg}{I}a\right)\theta = 0$$

$$(\because I = mk^2 = m(0.1)^2)$$

$$I = m(0.01) \text{ kg m}^2$$

$$a = 0.250 \text{ m}$$

$$g = 9.81 \text{ m/s}^2$$

$$\ddot{\theta} + \left(\frac{m \times 9.81 \times 0.250}{(0.01)m}\right) = 0$$

$$\omega_n = \sqrt{245.25}$$

$$= 15.660 \text{ rad/sec}$$

23. 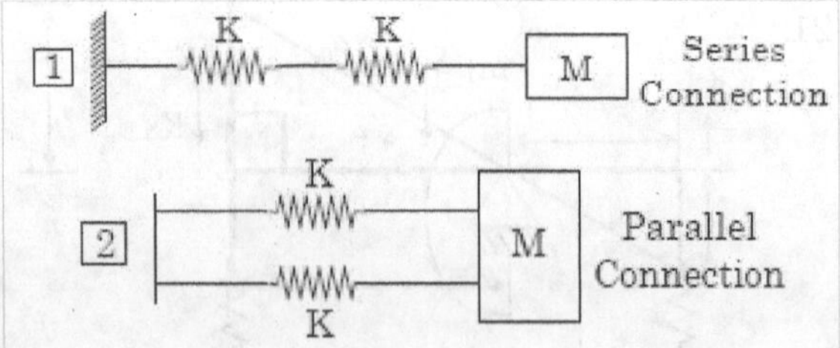

Now, $\frac{1}{\{K_{eq}\}} = \frac{1}{K_1} + \frac{1}{K_2}$ (For series connection)

$$\Rightarrow \left(K_{eq1} = \frac{K_1K_2}{K_1+K_2}\right) = \frac{K}{2}$$

$\left(K_{eq}\right)_2 = K_1 + K_2 = 2K$ (For parallel connection)

Natural frequency $\left(\omega = \sqrt{\frac{K}{m}}\right)$

$$\frac{\omega_1}{\omega_2} = \frac{\sqrt{\frac{K}{2m}}}{\sqrt{\frac{2K}{m}}} = \frac{1}{2}$$

24. Given, Disc mass m = 1 kg, R = 0.1 m
$K_1 = 400$ N/m, $K_2 = 100$ N/m

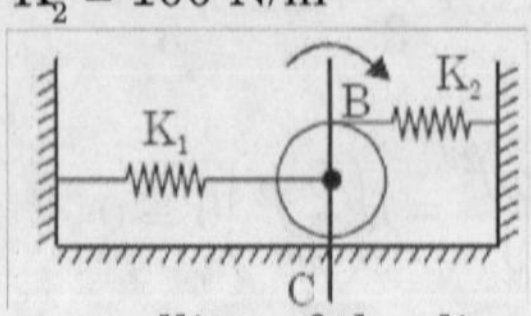

By assuming pure rolling of the disc,
$M_c = 0$ at I.C. at point (C)

$$I_c\ddot{\theta} + K_2(2R)\theta \times 2R + K_1R\theta \times R = 0$$

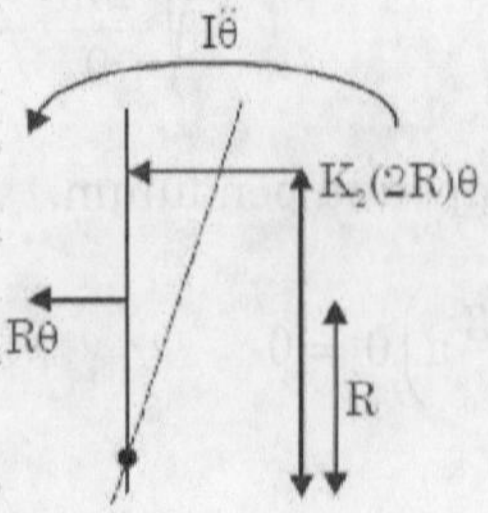

$$I_c = I_{disc} + mR^2 = \frac{mR^2}{2} + mR^2$$

$$= \frac{3}{2}mR^2 = \frac{3}{2} \times 0.1^2 = 1.5 \times 10^{-2}$$

$$1.5 \times 10^{-2}\ddot{\theta} + 4 \times 100 \times 0.1^2\theta + 400 \times 0.1^2\theta = 0$$

$$1.5 \times 10^{-2}\ddot{\theta} + 800 \times 0.1^2\theta = 0$$

$$\omega_n = \sqrt{\frac{800 \times 0.1^2}{1.5 \times 0.1^2}} = \sqrt{533.3}$$

$$\Rightarrow \omega_n = \sqrt{533.3} \text{ rad/second}$$

$$\omega_n = 23.094 \text{ rad/s}$$

$$\omega_n = 23.1$$

25. 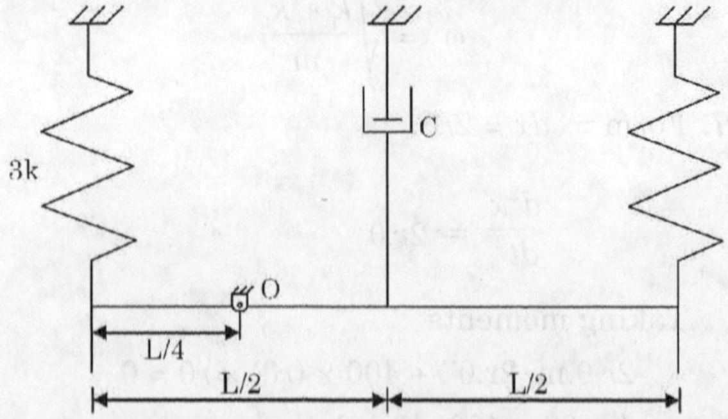

Displacing rod by angle θ about hinge,

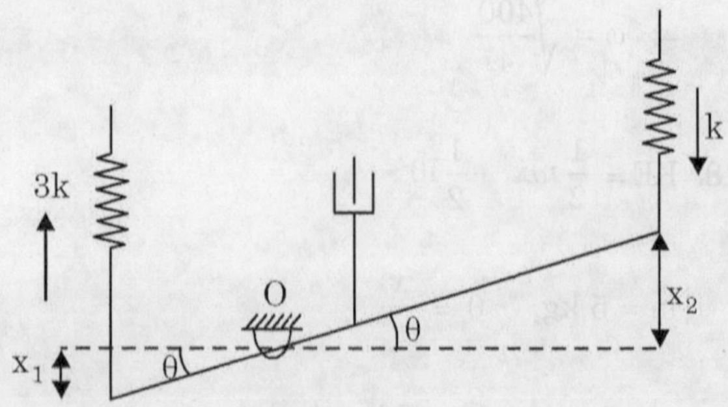

$$I_o = \frac{mL^2}{12} + m\left(\frac{L}{4}\right)^2$$

$$= \frac{mL^2}{12} + \frac{mL^2}{16} = \frac{7}{48}mL^2$$

Torque about hinge 'O'

$$= \frac{7mL^2}{48}\ddot{\theta} + 3k.\left(\frac{L^2}{16}\right)\theta + k.\frac{9L^2}{16}.\theta + C.\frac{L^2}{16}.\dot{\theta} = 0$$

$$\Rightarrow \frac{7mL^2}{48}\ddot{\theta} + \frac{12kL^2}{16}\theta + \frac{CL^2}{16}\dot{\theta} = 0$$

$\Rightarrow$ Natural frequency, $\omega_n = \sqrt{\frac{k}{m}}$

$$\omega_n = \sqrt{\frac{12kL^2.48}{16.7.mL^2}} = \sqrt{\frac{36}{7}\frac{k}{m}}$$

$$\omega_n = 6\sqrt{\frac{k}{7m}}$$

$$2\xi\omega_n = \frac{C \times 48}{16 \times 7m} = \frac{3C}{7m}$$

$$\Rightarrow 2\sqrt{km} = \frac{3}{7} \times \frac{\sqrt{7} \times C}{6}$$

$$\boxed{\frac{C}{\sqrt{km}} = 4\sqrt{7}}$$

26. $x_1 = l_1 Q$

$x_2 = Q \cdot l_2$

For small deflection, after equilibrium

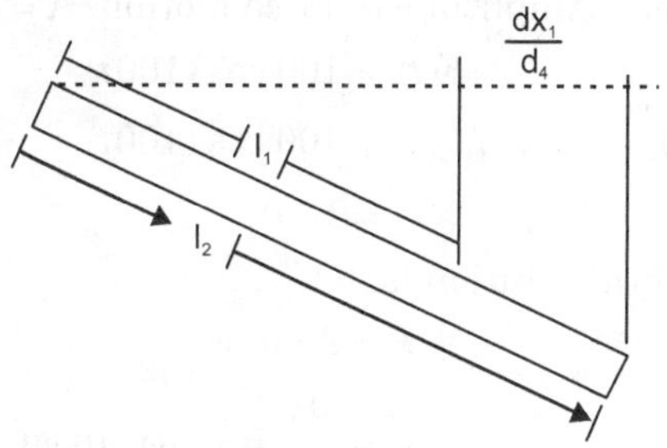

Taking momentum about O

$$C\frac{dx_1}{dt} \cdot l_1 + kx_2 l_2 + I\alpha + k_\theta \cdot Q = 0 \quad ...(i)$$

$$w_n = \sqrt{\frac{kl_2^2 + kw}{\frac{ml^2}{3}}} = \sqrt{\frac{2 \times 10^3 \times (0.5)^2 \times 1 \times 10^3}{\frac{10 \times (0.05)^2}{3}}}$$

$$= \sqrt{\frac{1.5 \times 10^3 \times 3}{10 \times (.5)^2}} = 42.43 \text{ rad/sec.}$$

27. For static equilibrium

$k_t \cdot \theta_s + k_t \cdot \theta_s L^2 = mgl^2$

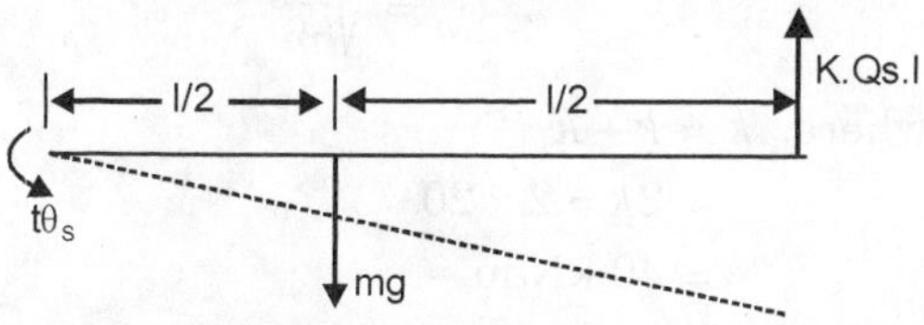

For additional deflection θ, equation will be.

$$I \cdot \ddot{\theta} = \frac{mgl}{2} - k_t(\theta_s + \theta) - C.l_1^2\theta - kl^2(\theta_s + \theta)$$

$$\Rightarrow I \cdot \ddot{\theta} + C.l_1^2\theta + (k_t + kl^2)\theta = 0$$

where, $I = ml^2/3$

∴ Un-damped frequency,

$$\omega = \sqrt{\frac{(k_t + kl^2) \times 3}{ml^2}}$$

$$= \sqrt{\frac{(1000 + 2000 \times 0.5)^2 \times 3}{10 \times 0.5^2}}$$

$= 42.426$

$= 42.43$ rad/sec.

and, Equivalent damping coefficient

$= Cl_1^2$ Nm/rad

$= 500 \times (0.04)^2$

$= 80$ Nms/rad

28. Given length of cantilever beam,

$l = 1000$ mm $= 1$ m

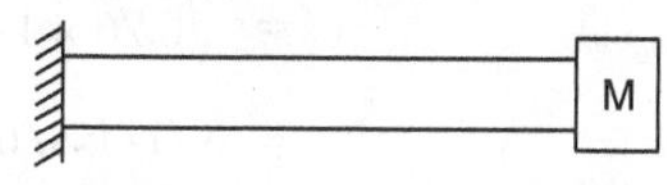

Moment of inertia of the shaft,

$$I = \frac{1}{12}bd^3 = \frac{25 \times (25)^3}{12}$$

$= 3.25 \times 10^{-8}$ m^4

$E_{steel} = 200 \times 10^9$ Pa.

Mass, $M = 20$ kg

∴ Weight W = 20 × 9.81 = 196.2

Now, deflection of free end,

$$\delta = \frac{Wl^3}{3EI} = \frac{196.2 \times (1)^3}{3 \times 200 \times 10^9 \times 3.25 \times 10^{-8}}$$

$= 0.01$ m

$$\text{Now, } \omega_n = \sqrt{\frac{g}{\delta}} = \sqrt{\frac{9.81}{0.01}} = 31.22 \text{ rad/s}$$

As the system is critically damped,

$$\left(\frac{C_c}{2m}\right)^2 = \omega_n^2$$

Critical damping constant,

$C_c = 2m\omega_n = 2 \times 20 \times 31.22$

$= 1248.99$ N-s/m^2.

29. R produces vibration close to natural frequency 144 Hz.

$$-2n\pi\left(\frac{\xi}{\sqrt{1-\xi^2}}\right)$$

30. $d = 0.9 f_n$

$\Rightarrow \quad \omega_d = 0.9\,\omega_n$

$\Rightarrow \quad \sqrt{\omega_n^2 - a^2} = 0.9\,\omega_n$

Squaring $\quad \omega_n^2 - a^2 = 0.81\,\omega_n^2$

$\Rightarrow \quad 0.19\omega_n^2 = a^2$

$\Rightarrow \quad 0.19 \times \frac{s}{m} = \frac{c^2}{(2m)^2}$

$\Rightarrow \quad c^2 = 0.76\ sm$

$\Rightarrow \quad c = \sqrt{0.76 \times 1 \times 10^3 \times 0.1}$

$= 8.717$ Ns/m

31.

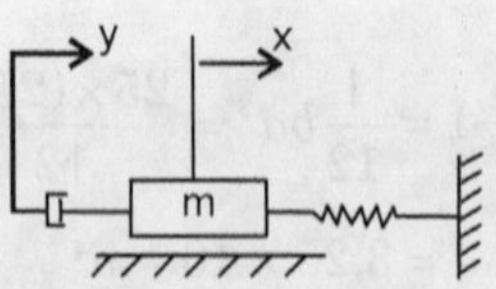

$$\therefore\ m\frac{d^2x}{dt^2} + c\left(\frac{dx}{dt} - \frac{dy}{dt}\right) + kx = 0$$

This is the differential equation governing the above vibrating system.

32. For critical damping,

$$C_c = 2m \times \sqrt{\frac{S}{M}} = 223.6\,\text{Ns}/m$$

33. Logarithmic decrement,

$$\delta = \frac{2\pi \times C}{\sqrt{C_c^2\ C^2}} = 0.42$$

35. At resonance, $\omega_n = \sqrt{\frac{k_{eq}}{m}}$

$k_{eq} = 2k_1 + 2k_2$

$= 2(k_1 + k_2)$

$= 2(32 + 16)$

$= 96 \times 10^6$ N/m

$w = \sqrt{\frac{k}{m}}$ rad/s

$= \frac{60}{2\pi}\sqrt{\frac{k}{m}}\ rpm$

$= \frac{30}{\pi}\sqrt{\frac{96 \times 10^6}{240}} = 6019.1\ rpm$

36. Damping factor $= \frac{C}{2\sqrt{sm}}$

$= \frac{400}{2\sqrt{50 \times 3.6 \times 10^3}} = 0.741$

$\omega_d = \left(\sqrt{1-\xi^2}\right)\omega_n$

$\omega_d = \sqrt{1-(471)^2} \times 8.481$

$\omega_d = 7.48$

$\therefore \quad N_d = \frac{\omega_d}{2\pi} = \frac{7.48}{2 \times 3.14} = 1.19$ Hz

37. Given : Amplitude of force motion, A = 50 mm

$\therefore \quad F(t) = 100\cos(100t) \quad \ldots(i)$

Also, $\quad m\ddot{x} + kx = 100\cos(100t)$

$\omega = 100$

General solution is,

$x = A\cos\omega t$

$\Rightarrow \quad \dot{x} = \frac{\partial x}{\partial t} = -A\omega\sin\omega t$

and $\quad \ddot{x} = \frac{\partial^2 x}{\partial t^2} = -A\omega^2\cos\omega t$

Putting in equation (*i*), we have

$m(-A\omega^2\cos\omega t) + k(A\cos\omega t) = 100\cos(\omega t)$

$\Rightarrow \quad -m\omega^2 A + kA = 100$

Here $k = 3000$ N/m

$\therefore \quad -m(0.05)(100)^2 + 3000(0.05) = 100$

$\Rightarrow \quad m = \frac{50}{100} = 0.1$

38. Natural frequency of the system

$\omega_n = \sqrt{\frac{k_e}{m}}$

where, $k_e = k + k$

$= 2k = 2 \times 20$

$= 40$ kN/m

$\therefore \quad \omega_n = \sqrt{\frac{40 \times 1000}{1}}$

$= 200$ rad/s $= 32$ Hz

41. Damping ratio.

$$(\xi) = \frac{C}{C_c} = \frac{C}{2\sqrt{km}} = \frac{C}{2\sqrt{1 \times 100}} = \frac{5}{20} = 0.25$$

42. $$X = \frac{F}{\sqrt{(k - m\omega^2)^2 + (C\omega)^2}} = 40 \text{ mm} = 0.04$$

$$F = 10 \text{ N}$$
$$\omega = 25$$

$$\Rightarrow \quad 0.04 = \frac{10}{\sqrt{(6250 - 10 \times 25^2)^2 + (C \times 25)^2}}$$

$$C = 10 \text{Ns/m}$$

43. $$\text{M.F.} = \frac{1}{\sqrt{\left[1 - \left(\frac{\omega}{\omega_n}\right)^2\right]^2 + \left[2\xi \frac{\omega}{\omega_n}\right]^2}}$$

$\omega = 1.5$ rad/s

$$\omega_n = \sqrt{\frac{k}{m}} = \sqrt{\frac{8}{2}} = 2 \text{ rad/s}$$

$\xi = 0.02$

$$\text{M.F.} = \frac{1}{\sqrt{\left[1 - (0.75)^2\right]^2 + (2 \times 0.02 \times 0.75)^2}}$$

$\therefore$ M.F. = 2.28

44. $$x = \frac{C}{2\sqrt{mk}}$$

$$\Rightarrow \quad C = 2\xi\sqrt{mk}$$

$$\frac{C_2}{C_1} = \frac{2\xi_2\sqrt{m_2k_2}}{2\xi_1\sqrt{m_1k_1}}$$

$$\frac{C_2}{0.1} = \sqrt{16}$$

$C_2 = 0.4$ kg/s

46. $\omega = 2\pi N = 2 \times \pi \times 60 = 376.99$ rad/second

$$.05 = \frac{1}{1 - \left(\frac{\omega}{\omega_n}\right)^2}$$

$$\Rightarrow \quad -20 = 1 - \left(\frac{\omega}{\omega_n}\right)^2 \Rightarrow \left(\frac{\omega}{\omega_n}\right)^2 = 21$$

$$\omega_2 = 21\omega_n^2$$
$$\omega_n = 82.266 \text{ rad/s}$$

$$\omega_n = \sqrt{\frac{k}{m}} = 82.266$$

1 = 6767.6005 N/m

47. Amplitude of vibration,

$$A = \frac{f_0 / s}{\sqrt{1 - \left(\frac{\omega}{\omega_n}\right)^2 + \left(\frac{2\xi\omega}{\omega_n}\right)^2}}$$

$$= \frac{10/150}{\sqrt{\left[1 - \left(\frac{1}{10}\right)\right]^2 + \left(2 \times 0.2 \times \frac{1}{10}\right)^2}} = 0.06605 \simeq 0.07$$

48. Force in the spring F = 2mg [from equilibrium]
Deflection as mass at P,

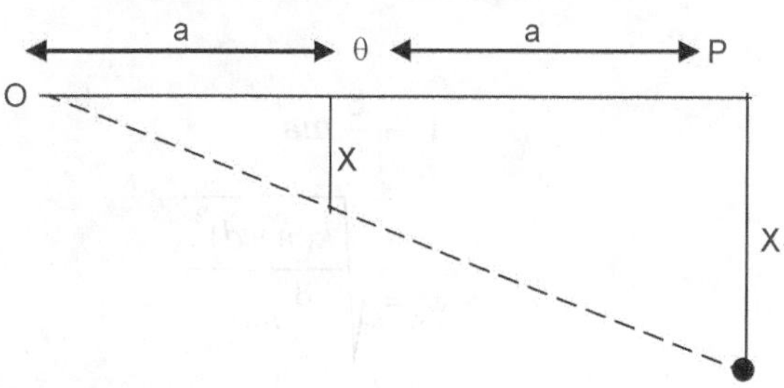

$$x_1 = \frac{x}{a} \times 2a = 2x = 2 \times \frac{2mg}{k} = \frac{4mg}{k}$$

$$w_n = \sqrt{\frac{g}{s}} = \sqrt{\frac{g}{x_1}} = \sqrt{\frac{k}{4m}};$$

$$f_n = \frac{1}{2\pi}\omega_n = \frac{1}{2\pi}\sqrt{\frac{k}{4m}}$$

49. $$\text{MF} = \pm\frac{1}{1 - \left(\frac{\omega}{\omega_n}\right)^2} = \pm\frac{1}{1-3} = 0.5$$

50. $$w_n = \frac{q}{2m}$$

$$q = 2mw_n = 2m\sqrt{\frac{k}{m}}$$

$$q = 2\sqrt{km}$$

51. $k = 10$ kN/m
$F_0 = 100$ N
$\xi = 0.25$

$$X = \frac{(F_0/k)}{\sqrt{\left(1-\left(\frac{\omega}{\omega_n}\right)^2\right)^2+\left(2\zeta\frac{\omega}{\omega_n}\right)^2}}$$

$\frac{\omega}{\omega_n} = 1$ at resonance

$$X = \frac{F_0}{2k\zeta} = \frac{100}{2\times100\times.25\times10^3} = 20\text{ mm}$$

52.

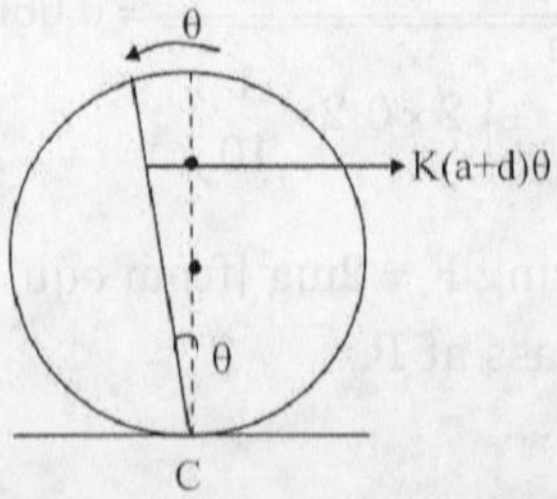

$$I_c = \frac{3}{2}ma^2$$

$$\omega_c = \sqrt{\frac{k(a+d)^2}{\frac{3}{2}ma^2}}$$

$$\omega_n = \sqrt{\frac{2k(a+d)^2}{3ma^2}}$$

53. $\xi = \frac{C}{C_c}$

$$\xi = \frac{C}{2\sqrt{km}}$$

55. Transmissibility ratio,

$$\xi = \frac{\sqrt{1+\left(\frac{2c\omega}{\omega_n c_c}\right)^2}}{\sqrt{\left(\frac{2c\omega}{\omega_n c_c}\right)^2+\left(1-\frac{\omega^2}{w_n^2}\right)}}$$

If $c = 0$, then

$$\xi = \frac{1}{\left(1-\frac{\omega^2}{w_n^2}\right)} = \frac{1}{1-(0.5)^2} = \frac{4}{3}$$

56. $TR = \frac{\sqrt{1+(2\xi r)^2}}{\sqrt{(1-r^2)^2+(2\xi r)^2}}$

$$\left[\omega_n = \sqrt{\frac{k}{m}} = \sqrt{\frac{100\times1000}{250}}\right]$$

$$r = \frac{\omega}{\omega_n} = \frac{377}{20} = 18.85$$

$$TR = \frac{\sqrt{1+(2\times0.15\times18.85)^2}}{\sqrt{(1-18.85^2)^2+(2\times0.15\times18.85)^2}}$$

TR = .0162

58. m = 200 kg, k = 10 kN/m = 10000 N/m

$K_{eq} = K + K = 10000 + 10000 = 20000$ N/m

$F_t = 50\cos 5t$

$F_0 = 50$ N

$\omega = 5$ rad/s

$$\omega_n = \sqrt{\frac{K_{eq}}{m}} = \sqrt{\frac{20000}{200}} = \sqrt{100} = 10\text{ rad/s}$$

$$\frac{\omega}{\omega_n} = \frac{5}{10} = \frac{1}{2}$$

No damping $\Rightarrow c = 0 \Rightarrow \xi = 0$

Transmissibility

$$\in = \frac{\sqrt{1+\left(\frac{2\xi\omega}{\omega_n}\right)^2}}{\sqrt{\left\{1-\left(\frac{\omega}{\omega_n}\right)^2\right\}^2+\left\{\frac{2\xi\omega}{\omega_n}\right\}^2}} \qquad (\xi = 0)$$

$$= \frac{1}{\sqrt{\left\{1-\left(\frac{\omega}{\omega_n}\right)^2\right\}^2}} = \frac{1}{\sqrt{\left\{1-\left(\frac{1}{2}\right)^2\right\}^2}}$$

$$= \frac{1}{1-\frac{1}{4}} = \frac{1}{\frac{3}{4}} = \left(\frac{4}{3}\right)$$

$$\in = \frac{F_t}{F_0}$$

$$\frac{4}{3} = \frac{F_t}{50}$$

$$F_t = \frac{200}{3} = 66.666\text{ N}$$

Force from each mount

$$= \frac{F_t}{2} = \frac{66.666}{2} = 33.333\text{ N}$$

60. $K_{eg} = \dfrac{1}{\dfrac{1}{K_1}+\dfrac{1}{K_2}} + \dfrac{1}{K_3}$

$$= \frac{K_1K_2}{K_1K_2} + K_3$$

$$= \frac{K_1K_2 + K_1K_3 + K_2K_3}{K_1 + K_2}$$

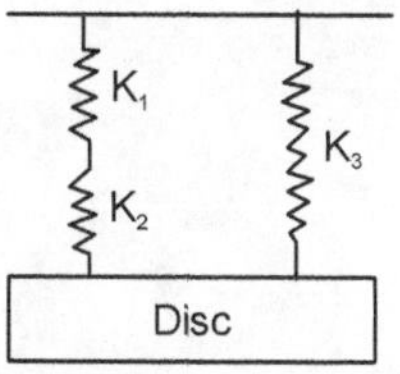

∴ Natural frequency

$$= \sqrt{\frac{K_1K_2 + K_1K_3 + K_2K_3}{J(K_1 + K_2)}}$$

61. Here, m = 10 kg = mass of rotor

d = diameter of shaft = 30 × 10^5 m

l = length of shaft = 500 × 10^{-3} m

E for steel = 2.1 × 10^{11} N/m²

$$\Delta = \text{deflection of shaft} = \frac{mgl^3}{4gEI}$$

$$I = \frac{\pi}{64}d^4 = \frac{\pi}{64} \times (30 \times 10^{-3})^4$$

$$= 3.976 \times 10^{-8}\ m^4$$

$$\Delta = \frac{mgl^3}{48EI}$$

$$= \frac{10 \times 9.81 \times (500 \times 10^{-3})^3}{48 \times 2.1 \times 10^{11} \times 3.976 \times 10^{-8}}$$

$$= 3.06 \times 10^{-5}\ m$$

$$\omega_n = \sqrt{\frac{g}{\Delta}} = \sqrt{\frac{9.81}{3.06 \times 10^{-5}}} = 566.24\ \text{rad/s}$$

$$f_n = \frac{\omega_n}{2\pi}$$

$$= \frac{566.24}{2 \times 3.142} = 90\ \text{Hz}$$

62. Along x-axis ,

$0.5\,(-60 \times 10^{-3} \cos 30° + 50 \times 10^{-3})\,\omega^2 = 0.1\omega^2 \times x \times 10^{-3}$

∴ $x = -9.8076$ mm

Along y-axis, $0.5\,(60 \times 10^{-3} \sin 30°)\,\omega^2 = 0.1\omega^2 y$

∴ $y = 150$ mm

∴ $r = \sqrt{x^2 + y^2} = 150.32$ mm

63.
$$N_c = \frac{0.4985}{\sqrt{\delta}}\ \text{rps}$$

$$= \frac{0.4985 \times 60}{\sqrt{1.8 \times 10^{-3}}}\ \text{rpm}$$

$$= 705\ \text{rpm}$$

Alternately

When speed of the shaft (ω) is equal to natural speed of vibration, it is called critical speed (ω_n)

i.e. $\omega = \omega_n = \sqrt{\dfrac{g}{A}}$

$$= \sqrt{\frac{9.81}{1.8 \times 10^{-3}}} = 73.82$$

$$= \frac{60 \times 73.82}{2\pi}$$

$$= 705\ \text{rpm}$$

64.

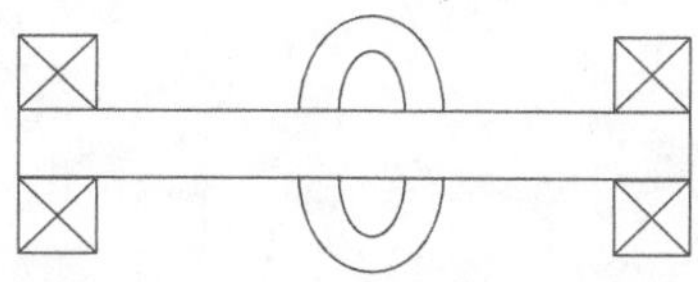

$n = 1$

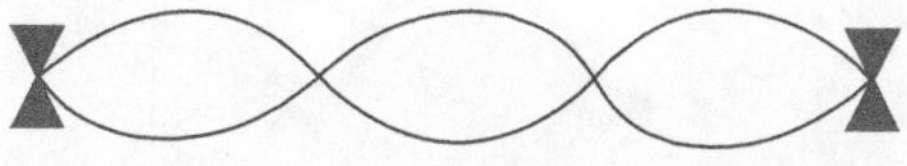

$n = 3$

The whiring frequency of shaft is given by

$$f = \frac{\pi}{2}n^2\sqrt{\frac{gEI}{wL^4}}$$

$$f_1 = \frac{\pi}{2}\sqrt{\frac{gEI}{wL^4}}$$

$n \Rightarrow$ node number

frequency of 1st mode

2 nodes are present in 3rd mode.

$$f_3 = 3^2 f_1 = 1800\ (\text{rpm})$$

$$\Rightarrow \quad f_1 = \frac{1800}{9} = 200(rpm)$$

8 CHAPTER

Balancing and Gyroscope

Analysis of Previous GATE Papers

Topic	Marks	Year/Set → Marks Distribution ↓	2019 Set 1	2019 Set 2	2018 Set 1	2018 Set 2	2017 Set 1	2017 Set 2	2016 Set 1	2016 Set 2	2016 Set 3	2015 Set 1	2015 Set 2	2015 Set 3
Balancing	1 Mark	MCQ Type												
		Numerical Type												
	2 Marks	MCQ Type												
		Numerical Type		1				1			1			
		Total		2				2			2			
Gyroscope	1 Mark	MCQ Type												
		Numerical Type							1					
	2 Marks	MCQ Type	1											
		Numerical Type		1										
		Total	2	2					1					

Balancing

1. A cantilever type gate hinged at Q is shown in the figure. P and R are the centers of gravity of the cantilever part and the counterweight respectively. The mass of the cantilever part is 75 kg. The mass of the counterweight, for static balance, is

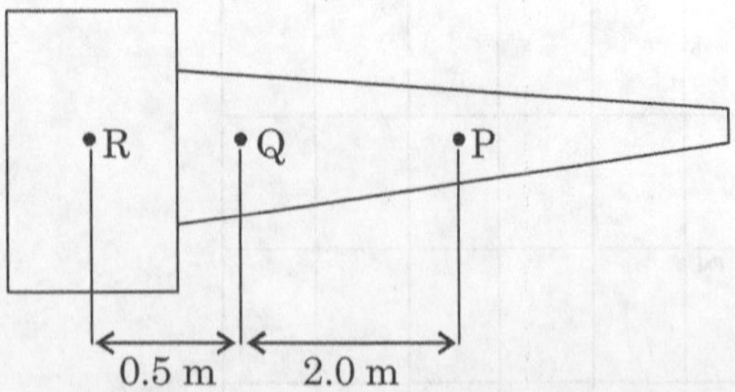

(a) 75 kg (c) 225 kg

(b) 150 kg (d) 300 kg

[2008 : 1 Mark]

2. Two masses m are attached to opposite sides of a rigid rotating shaft in the vertical plane. Another pair of equal masses m_1 is attached to the opposite sides of the shaft in the vertical plane as shown in figure. Consider m = 1 kg, e = 50 mm, e_1 = 20 mm, b = 0.3 m, a = 2 m and a_1 = 2.5 m. For the system to be dynamically balanced, m_1 should be ______ kg.

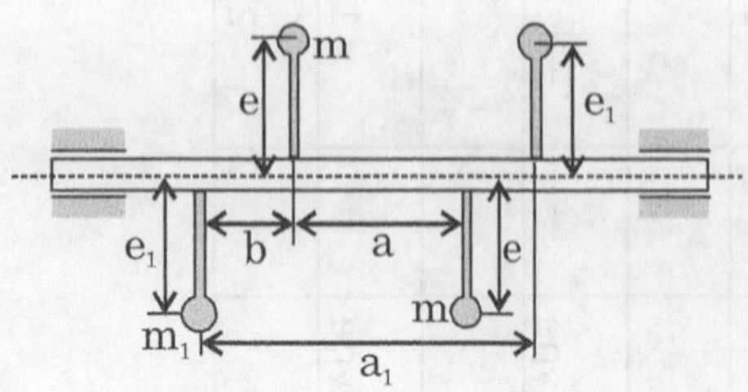

[2016 : 2 Marks, Set-3]

3. Three masses are connected to a rotating shaft supported on bearing A and B as shown in the figure. The system is in a space where the gravitational effect is absent. Neglect the mass of shaft and rods connecting the masses. For m_1 = 10 kg, m_2 = 5 kg and m_3 = 2.5 kg and for a shaft angular speed of 1000 radian/s, the magnitude of the bearing reaction (in N) at location be is ________.

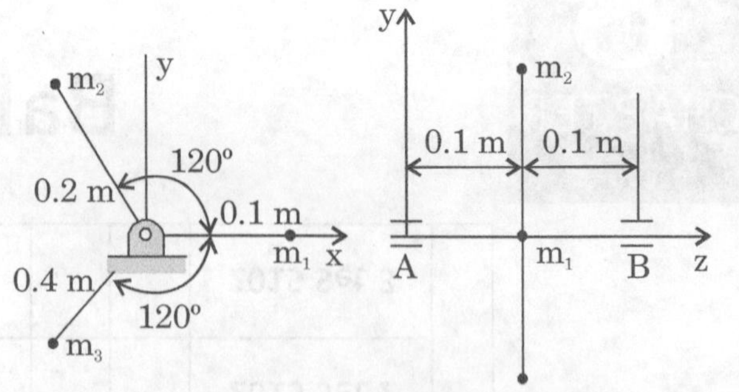

[2017 : 2 Marks, Set-2]

4. Two masses A and B having mass m_a and m_b, respectively, lying in the plane of the figure shown, are rigidly attached to a shaft which revolves about an axis through O perpendicular to the plane of the figure. The radii of rotation of the masses m_a and m_b are r_a and r_b, respectively. The angle between lines OA and OB is 90°. If m_a = 10 kg, m_b = 20 kg r_a = 200 mm and r_b = 400 mm, then the balance mass to be placed at a radius of 200 mm is ____ kg (round off to two decimal places)

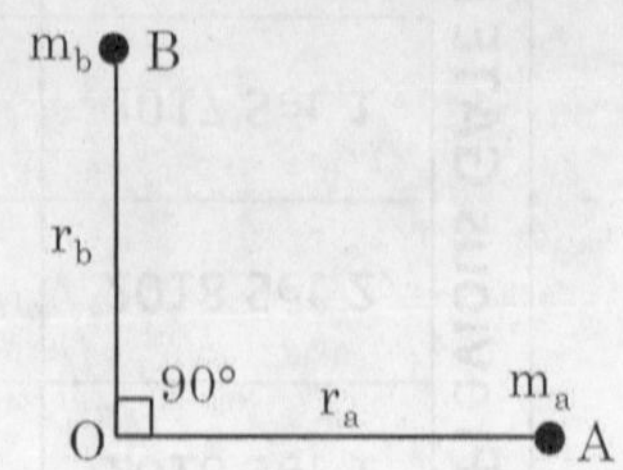

[2019 : 2 Marks, Set 2]

Gyroscope

5. A car is moving on a curved horizontal road of radius 100 m with a speed of 20 m/s, The rotating masses of the engine have an angular speed of 100 rad/s in clockwise direction when viewed from the front of the car. The combined moment of inertia of the rotating masses is 10 kg-m^2. The magnitude of the gyroscopic moment in (N-m) is __________.

[2016 : 1 Mark, Set-1]

6. The rotor of turbojet engine of an aircraft has a mass 180 kg and polar moment of inertia 10 kg.m^2 about the rotor axis. The rotor rotates at a constant speed of 1100 rad/s in the clockwise direction when viewed from the front of the aircraft. The aircraft while flying at a speed of 800 km per hour takes a turn with a radius of

1.5 km to the left. The gyroscopic moment exerted by the rotor on the aircraft structure and the direction of motion of the nose when the aircraft turns, are

(a) 1629.6 N.m and the nose goes up
(b) 162.9 N.m and the nose goes up
(c) 162.9 N.m and the nose goes down
(d) 1629.6 N.m and the nose goes down

[2019 : 2 Marks, Set 1]

7. A uniform disc with radius r and a mass of m kg is mounted centrally on a horizontal axle of negligible mass and length of 1.5r. The disc spins counter-clockwise about the axle with angular speed ω, when viewed from the right-hand side bearing, Q. The axle precesses about a vertical axis at $\omega_p = \omega/10$ in the clockwise direction when viewed from above. Let R_P and R_Q (positive upwards) be the resultant reaction forces due to the mass and the gyroscopic effect, at bearing P and Q, respectively. Assuming $\omega^2 r = 300\ m/s^2$ and $g = 10\ m/s^2$, the ratio of the larger to the smaller bearing reaction force (considering appropriate signs) is ____.

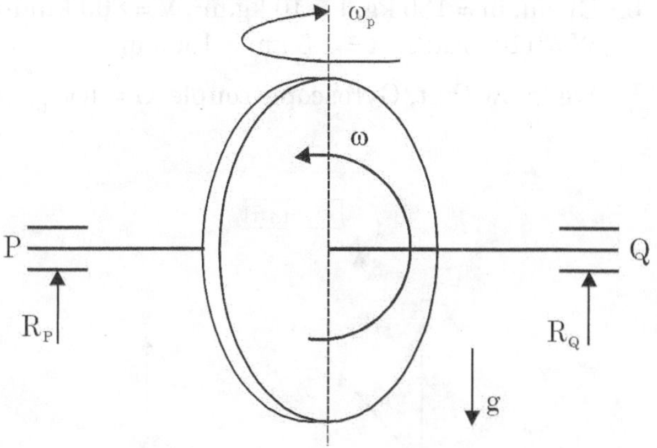

[2019 : 2 Marks, Set 2]

ANSWERS

1. (d) **2.** (2) **6.** (d)

EXPLANATIONS

1. For balancing the moments about the hinge Q,

$$75 \times 2 = m_R \times 0.5$$

$$\Rightarrow \quad m_R = 300\ kg$$

2.

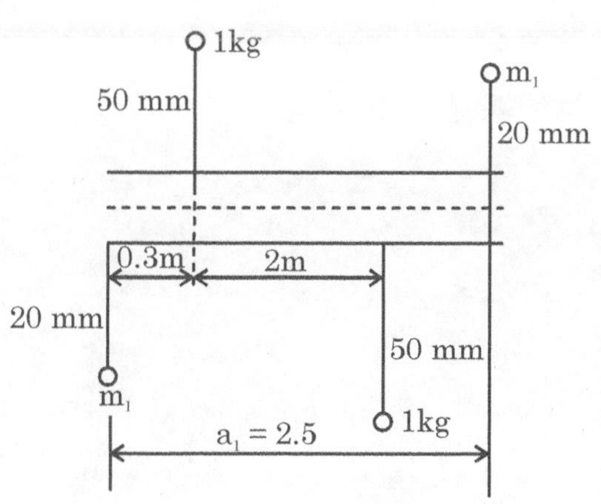

Balance moment of all forces

$$\omega^2(m_1 \times 0.02 \times 2.5) + (1 \times .05 \times .3)\omega^2$$

$$= (m_1 \times 0.02 \times 0)\ \omega^2 + (1 \times .05 \times 2.3)\omega^2$$

$$m_1 = 2\ kg$$

3. Reaction at B is zero.

4. Let balancing mass = m_c

$$m_c r_c = \sqrt{\Sigma(mr\cos\theta)^2 + \Sigma(mr\sin\theta)^2}$$

$$= \sqrt{\begin{matrix}(10\times200\times\cos0° + 20\times400\times\cos90°)^2 \\ +(10\times200\times\sin0° + 20\times400\times\sin90°)^2\end{matrix}}$$

$$= \sqrt{(2000)^2 + (20\times400)^2}$$

$$= \sqrt{(2000)^2 + (8000)^2}$$

$$m_c r_c = 8246.211$$

$$m_c = \frac{8246.211}{200} = 41.23$$

$$\boxed{m_c = 41.23\,kg}$$

5. R = 100m, V = 200 m/sec,

$$\omega_p = \frac{V}{R} = 0.2\ rad/sec$$

$\omega_s = 100$ rad/sec

$I = 10$ kg-m²

Gyroscopic moment $= I\omega_s\omega_p$

$= 10 \times 0.2 \times 100$

$= 200$ N-m

6. Given, m = 180 kg, I = 10 kg.m², V = 800 km/hr., W = 1100 rad/s, R = 1.5 km = 1500 m

We know that, Gyroscopic couple $G = I\omega\omega_p$

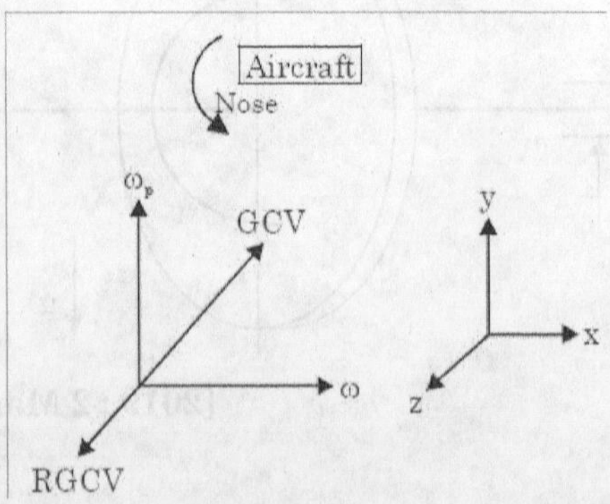

Effect ⇒ Nose goes down

$$G = 10 \times 1100 \times \frac{800 \times 5}{1800 \times 18}$$

$= 1629.63$ Nm

$G = 1629.63$ N.m

7. Gyroscopic couple, $J = I.\,\omega\,.\omega_p$

$$= \frac{mr^2}{2}.\omega.\frac{\omega}{10} = \frac{mr^2.\omega^2}{20}$$

$$= \frac{m.r.300}{20} = 15\,mr \qquad (\because r\omega^2 = 300\,m/s^2)$$

Gyroscopic force on each bearing

$$= \frac{15mr}{1.5r} = 10m$$

Force on each bearing due to weight of disc

$$= \frac{mg}{2} = \frac{m \times 10}{2} = 5m$$

Now,

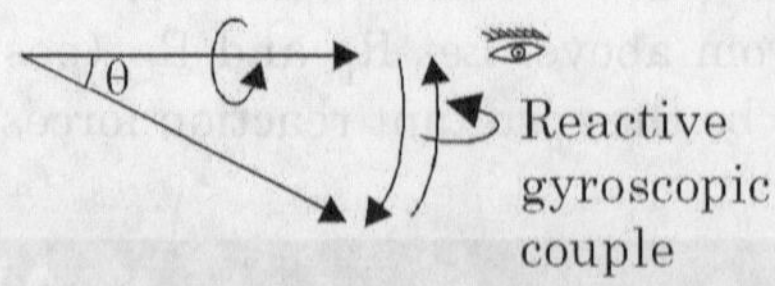

Total force on bearing,

$Q = 10m + 5m = 15m$

Total force on bearing,

$P = -10m + 5m = -5m$

R_Q (upward) = 15 m

R_P (downward) = –5 m

$$\frac{R_Q}{R_P} = \frac{15m}{-5m} = -3$$

Unit - V

Machine Design

Syllabus

Design for static and dynamic loading; failure theories; fatigue strength and the S-N diagram; principles of the design of machine elements such as bolted, riveted and welded joints; shafts, gears, rolling and sliding contact bearings, brakes and clutches, springs.

Contents

1 CHAPTER

Static, Dynamic Loading and Failure Theories

Analysis of Previous GATE Papers

Topic	Year/Set → Marks Distribution ↓		2019 Set 1	2019 Set 2	2018 Set 1	2018 Set 2	2017 Set 1	2017 Set 2	2016 Set 1	2016 Set 2	2016 Set 3	2015 Set 1	2015 Set 2	2015 Set 3
Static loading	1 Mark	MCQ Type			1									
		Numerical Type												
	2 Marks	MCQ Type			1									
		Numerical Type												
	Total				3									
Dynamic Loading	1 Mark	MCQ Type					1							1
		Numerical Type	1					1						
	2 Marks	MCQ Type												
		Numerical Type												
	Total		1				1	1						1
Failure Theories	1 Mark	MCQ Type												
		Numerical Type											1	
	2 Marks	MCQ Type										1		
		Numerical Type	2											
	Total		4									2	1	

Static loading

1. Torque to weight ratio for a circular shaft transmitting power is directly proportional to the
(a) square root of the diameter
(b) diameter
(c) square of the diameter
(d) cube of the diameter

[1991 : 1 Mark]

2. A large uniform plate containing a rivet hole subjected to uniform uniaxial tension of 95 MPa.
The maximum stress in the plate is

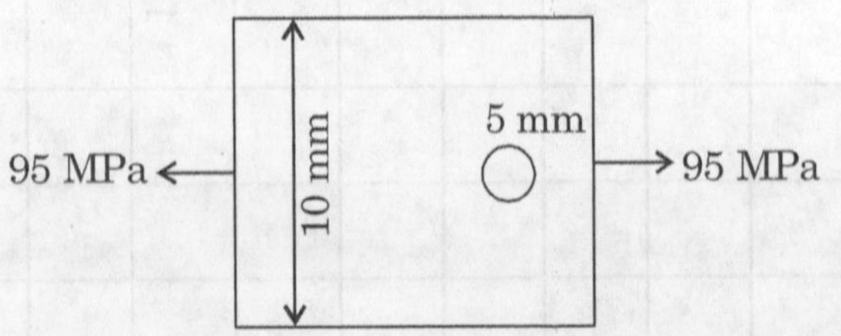

(a) 100 MPa (b) 285 MPa
(c) 190 MPa (d) Indeterminate

[1992 : 1 Mark]

3. The outside diameter of a hollow shaft that is twice its inside diameter the ratio of its torque carrying capacity to that of a solid shaft of the same material and the same outside diameter is
(a) 15/16 (b) 3/4
(c) 1/2 (d) 1/16

[1993 : 1 Mark]

4. Two shafts A and B are made of the same material. The diameter of shaft B is twice that of shaft A The ratio of power which can be transmitted by shaft A to that of shaft B is
(a) 1/2 (b) 1/4
(c) 1/8 (d) 1/16

[1994 : 1 Mark]

5. A solid shaft can resist a bending moment of 3.0 kNm and a twisting moment of 4.0 kNm together, then the maximum torque that can be applied is
(a) 7.0 kNm (b) 3.5 kNm
(c) 4.5 kNm (d) 5.0 kNm

[1996 : 1 Mark]

6. A pin jointed uniform rigid rod of weight W and length L is supported horizontally by an external force F as shown in the figure below. The force F is suddenly removed. At the instant of force removal, the magnitude of vertical reaction developed at the support is

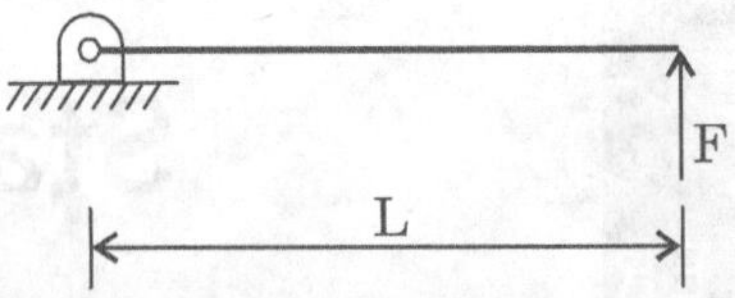

(a) zero (b) W/4
(c) W/2 (d) W

[2013 : 2 Marks]

7. If σ_1 and σ_3 are the algebraically largest and smallest principal stresses respectively, the value of the maximum shear stress is

(a) $\frac{\sigma_1 + \sigma_3}{2}$ (b) $\frac{\sigma_1 - \sigma_3}{2}$

(c) $\sqrt{\frac{\sigma_1 + \sigma_3}{2}}$ (d) $\sqrt{\frac{\sigma_1 - \sigma_3}{2}}$

[2018 : 1 Marks, Set-1]

8. A carpenter glues a pair of cylindrical wooden logs by bonding their end faces at an angle of $\theta = 30°$ as shown in the figure.

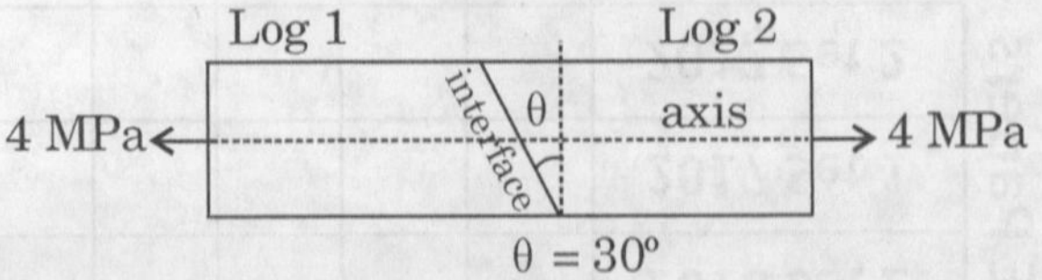

The glue used at the interface fails if
Criterion 1 : the maximum normal stress exceeds 2.5 MPa
Criterion 2 : the maximum shear stress exceeds 1.5 MPa
Assume that the interface fails before the logs fail. When a uniform tensile stress of 4 MPa is applied, the interface
(a) fails only because of criterion 1
(b) fails only because of criterion 2
(c) fails because of both criteria 1 and 2
(d) does not fail.

[2018 : 2 Marks, Set-1]

Dynamic loading

9. A thin spherical pressure vessel of 200 mm diameter and 1 mm thickness is subjected to an internal pressure varying from 4 to 8 MPa. Assume that the yield, ultimate, and endurance strength of material are 600,800 and 400 MPa respectively. The factor of safety as per Goodman's relation is
(a) 2.0 (b) 1.6
(c) 1.4 (d) 1.2

[2007 : 2 Marks]

10. The piston rod of diameter 20 mm and length 700 mm in a hydraulic cylinder is subjected to a compressive force of 10 kN due to the internal pressure. The end conditions for the rod may be assumed as guided at the piston end and hinged at the other end. The Young's modulus is 200 GPa. The factor of safety for the piston rod is

(a) 0.68 (b) 2.75
(c) 6.32 (d) 11.0

[2007 : 2 Marks]

11. A forged steel link with uniform diameter of 30 mm at the centre is subjected to an axial force that varies from 40 kN in compression to 160 kN in tension. The tensile (S_u), yield (S_y) and corrected endurance (S_e) strengths of the steel material are 600 MPa, 420 MPa and 240 MPa respectively. The factor of safety against fatigue endurance as per Soderberg's criterion is

(a) 1.26 (b) 1.37
(c) 1.45 (d) 2.00

[2009 : 2 Marks]

12. A bar is subjected to fluctuating tensile load from 20 kN to 100 kN. The material has yield strength of 240 MPa and endurance limit in reversed bending is 160 MPa. According to the Soderberg principle, the area of cross-section in mm² of the bar for a factor of safety of 2 is

(a) 400 (b) 600
(c) 750 (d) 1000

[2013 : 2 Marks]

13. A rotating steel shaft is supported at the ends. It is subjected to a point load at the center. The maximum bending stress developed is 100 MPa. If the yield, ultimate and corrected endurance strength of the shaft material are 300 MPa, 500 MPa and 200 MPa, respectively, then the factor of safety for the shaft is ______.

[2014 : 1 Mark, Set-3]

14. In a structure subjected to fatigue loading, the minimum and maximum stresses developed in a cycle are 200 MPa and 400 MPa respectively. The value of stress amplitude (in MPa) is ______.

[2014 : 1 Mark, Set-2]

15. For the given fluctuating fatigue load, the values of stress amplitude and stress ratio are respectively

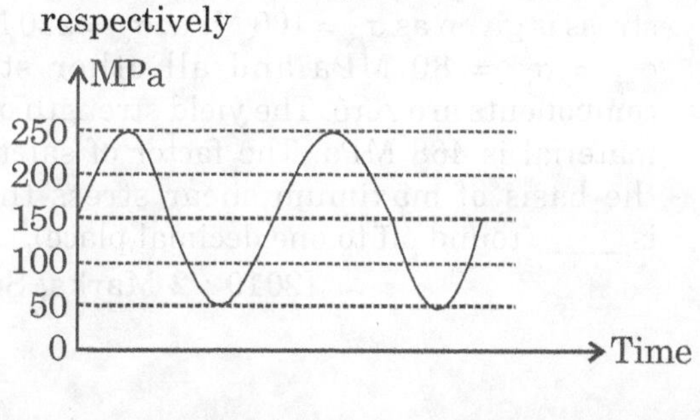

(a) 100MPa and 5
(b) 250 MPa and 5
(c) 100 MPa and 0.20
(d) 250 MPa and 0.20

[2015 : 1 Mark, Set-3]

16. Consider the schematic of a riveted lap Joint subjected to tensile load F, as shown below. Let d be the diameter of the rivets, and S_f be the maximum permissible tensile stress in the plates. What should be the minimum value for the thickness of the plates to guard against tensile failure of the plates? Assume the plates to be identical.

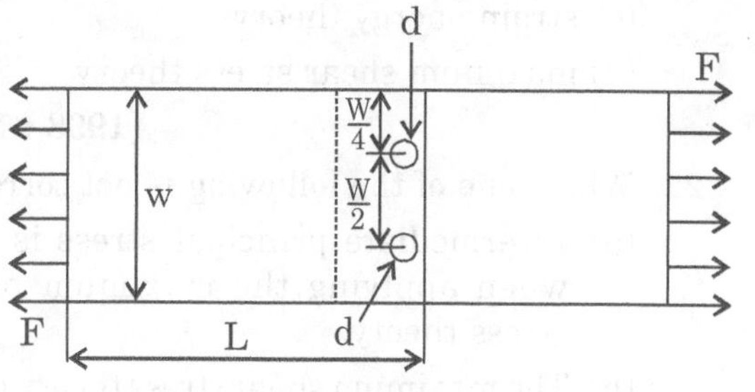

(a) $\frac{F}{S_f(w-2d)}$ (b) $\frac{F}{S_f w}$

(c) $\frac{F}{S_f(w-d)}$ (d) $\frac{2F}{S_f w}$

[2017 : 1 Mark, Set-1]

17. A machine component made of a ductile material is subjected to a variable loading with $\sigma_{min} = -50$ MPa and $\sigma_{max} = 50$ MPa. If the corrected endurance limit and the yield strength for the material are $\sigma_e = 100$ MPa and for the material are $\sigma_y = 300$ MPa, respectively, the factor of safety is ________.

[2017 : 1 Mark, Set-2]

18. During a high cycle fatigue test, a metallic specimen is subjected to cyclic loading with a mean stress of + 140 MPa, and a minimum stress of –70 MPa. The R-ratio (minimum stress to maximum stress) for this cycle loading is _______ (round off to one decimal place).

[2019 : 1 Mark, Set-1]

Failure Theory

19. In the design of shafts made of ductile materials subjected to twisting moment and bending moment, the recommended theory of failure is

(a) maximum principal stress theory
(b) maximum principal strain theory
(c) maximum shear stress theory
(d) maximum strain-energy theory

[1988 : 1 Mark]

20. A small element at the critical section of a component in biaxial state of stress with the two principal stresses being 360 MPa and 140 MPa. The maximum working stress according to distortion energy theory is

(a) 220 MPa (b) 110 MPa
(c) 314 MPa (d) 330 MPa

[1997 : 2 Marks]

21. Which theory of failure will you use for aluminium components under steady loading is

(a) principal stress theory
(b) principal strain theory
(c) strain energy theory
(d) maximum shear stress theory

[1998 : 1 Mark]

22. Which one of the following is not correct?

(a) Intermediate principal stress is ignored when applying the maximum principal stress theory.
(b) The maximum shear stress theory gives the most accurate results amongst all the failure theories.
(c) As per the maximum strain energy theory, failure occurs when the strain energy per unit volume exceeds a critical value.
(d) As per the maximum distortion energy theory, failure occurs when the distortion energy per unit volume exceeds a critical value.

[2014 : 1 Mark, Set-3]

23. Consider the two states of stress as shown in configurations I and II in the figure below. From the standpoint of distortion energy (von-Mises) criterion, which one of the following statements is true?

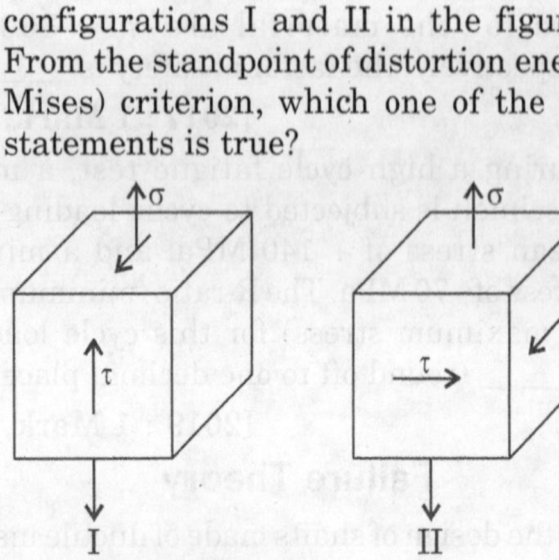

(a) I yields after II
(b) II yields after I
(c) Both yield simultaneously
(d) Nothing can be said about their relative yielding

[2014 : 2 Marks, Set-2]

24. A shaft is subjected to pure torsional moment. The maximum shear stress developed in the shaft is 100 MPa. The yield and ultimate strengths of the shaft material in tension are 300 MPa and 450 MPa, respectively. The factor of safety using maximum distortion energy (Von-Mises) theory is

[2014 : 2 Marks, Set-4]

25. The uniaxial yield stress of a material is 300 MPa. According to Von Mises criterion, the shear yield stress (in MPa) of the material is ______

[2015 : 1 Mark, Set-2]

26. A machine element is subjected to the following bi-axial state of stress; σ_x = 80 MPa; σ_y = 20 MPa; τ_{xy} = 40 MPa. If the shear strength of the material is 100 MPa, the factor of safety as per Tresca's maximum shear stress theory is

(a) 1.0 (b) 2.0
(c) 2.5 (d) 3.3

[2015 : 2 Marks, Set-1]

27. A plane-strain compression (forging) of a block is shown in the figure. The strain in the z-direction is zero. The yield strength (S_y) in uniaxial tension/compression of the material of the block is 300 MPa and its follows the Tresca (maximum shear stress) criterion. Assume that the entire block has started yielding. At a point where σ_x = 40 MPa (compressive) and τ_{xy} = 0, the stress component σ_y is

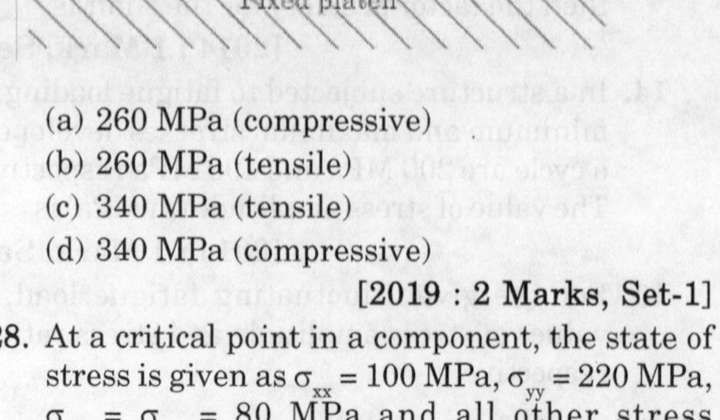

(a) 260 MPa (compressive)
(b) 260 MPa (tensile)
(c) 340 MPa (tensile)
(d) 340 MPa (compressive)

[2019 : 2 Marks, Set-1]

28. At a critical point in a component, the state of stress is given as σ_{xx} = 100 MPa, σ_{yy} = 220 MPa, σ_{xy} = σ_{yx} = 80 MPa and all other stress components are zero. The yield strength of the material is 468 MPa. The factor of safety on the basis of maximum shear stress theory is_____ (round off to one decimal place).

[2019 : 2 Marks, Set-1]

ANSWERS

1. (b) **2.** (a) **3.** (a) **4.** (c) **5.** (d) **6.** (b) **7.** (b) **8.** (c) **9.** (b) **10.** (c)
11. (a) **12.** (d) **15.** (c) **16.** (a) **19.** (c) **20.** (c) **21.** (d) **22.** (b) **23.** (c) **26.** (b)
27. (b)

EXPLANATIONS

1. $T = \frac{\pi}{16} d^3 \tau$

$w = \pi \frac{d^2}{4} \rho l$

$\frac{T}{w} \propto d$

2. $\text{Max. Stress} = \frac{\text{Avg. Force}}{\text{Min. Area}}$

$= \frac{95 \times 10 \times t}{(10 - 0.5)t} = 100\text{MPa}$

3. $T_{hollow} = \frac{\pi}{32}\left(d^4 - \frac{d^4}{16}\right)\frac{2\tau}{d}$ ...(1)

$T_{solid} = \frac{\pi}{16} d^3 \tau$ ...(2)

By dividing (1) with (2)

$\frac{T_{hollow}}{T_{solid}} = \frac{15}{16}$

4. $T_A = \frac{\pi}{16} \times d^3 \tau$

$T_B = \frac{\pi}{16} \times (2d)^3 \tau$

$\frac{P_A}{P_B} = \frac{T_A}{T_B} = \frac{1}{8}$

5. Maximum tensile stress due to bending,

$f_t = \frac{M}{Z}$

$= \frac{3 \times 10^3}{\left(\frac{\frac{\pi}{64} \times d^4}{\frac{d}{2}}\right)} = \frac{30.558}{d^3}\text{kPa.}$

Maximum shear stress due to torsion, $f_s = \frac{TR}{J}$

$= \frac{4 \times 10^3 \times d/2}{\frac{\pi}{32} \times d^4} = \frac{20.7}{d^3}\text{kPa.}$

When both the stress acts simultaneously, maximum induced shear stress,

$$f_{s\,max.} = \frac{1}{2}\left[\sqrt{(f_t)^2 + 4(f_s)^2}\right]$$

Maximum torque that can be applied,

$$T = \frac{\pi d^3}{16} f_{s\,max.}$$

$$= \frac{\pi d^3}{16} \times \frac{1}{2}\sqrt{\left(\frac{30.558}{d^3}\right)^2 + 4\left(\frac{20.37}{d^3}\right)^2}$$

= 5 kNm

6.

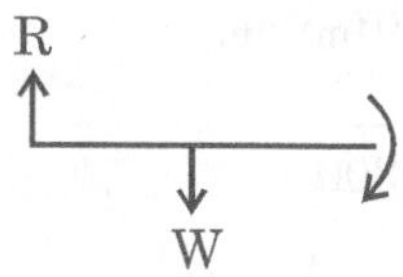

$I = \frac{1}{3}\frac{W}{g}L^2$; $W \times \frac{L}{2} = I\alpha \Rightarrow \alpha = \frac{3g}{2L}$

$\therefore$ Linear acceleration at centre $= \alpha \times \frac{L}{2} = \frac{3g}{4}$

$\therefore$ Inertial force at centre $= \frac{3}{4}W$

$\therefore$ Reaction at support $= W - \frac{3W}{4} = \frac{W}{4}$

7. Maximum shear stress = Radius of Mohr's Circle (for plane stress condition)

$$\boxed{\tau_{max} = \frac{\sigma_1 - \sigma_2}{2}}$$

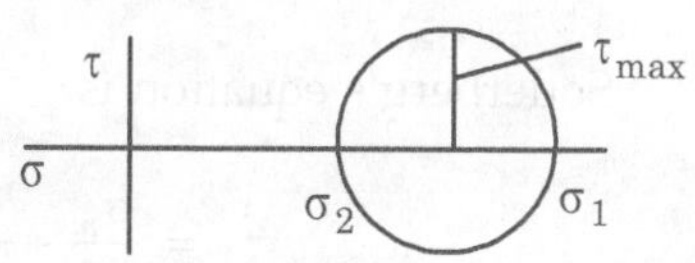

8. Normal stress on inclined plane

$\sigma = \sigma_x \cos^2\theta$

$= 4 \times \cos^2 30 = 3$ MPa

Shear stress on inclined plane $\tau = \frac{\sigma_x}{2} \sin 2\theta$

$= 2 \times \sin 60° = 1.73$ MPa

Since both the stress exceeds the given limits, answer is option (*c*).

9. Stress induced,

$$\sigma_1 = \sigma_2 = \frac{pr}{2t}$$

$$\sigma_{1max} = \frac{8 \times 100}{2 \times 1} = 400\text{MPa}$$

$$\sigma_{1min} = \frac{4 \times 100}{2 \times 1} = 200\text{MPa}$$

$\sigma_{2max} = 400$ MPa

$\sigma_{2min} = 200$ MPa

$\sigma_{1m} = 300$ MPa

$\sigma_{1a} = 100$ MPa

$\sigma_{2m} = 300$ MPa

$\sigma_{2a} = 100$ MPa

Equivalent stress are as follows:

$$\sigma_{me} = \sqrt{\sigma_{1m}{}^2 + \sigma_{2m}{}^2 - \sigma_{1m}\sigma_{2m}}$$

$$= \sqrt{300^2 + 300^2 - 300 \times 300}$$

$= 300$ MPa

Similarly, $\sigma_{ae} = 100$ MPa

From Goodman equation,

$$\frac{\sigma_{ae}}{S_e} + \frac{\sigma_{me}}{S_{ut}} = \frac{1}{n}$$

$$\Rightarrow \frac{100}{400} + \frac{300}{800} = \frac{1}{n}$$

$= n = 1.6$

11. Fatigue endurance, $\sigma_{en} = 240$ MPa

$$A = \frac{\pi}{4} \times (30)^2 = 706.5 \times 10^{-6}\, m^2$$

Soderberg's equation is

$$\frac{1}{h} = \frac{\sigma_m}{\sigma_y} + \frac{\sigma_a}{\sigma_{en.f}}$$

Now $\quad F_m = \frac{160 + 40}{2} = 100$ kN

$$\therefore \quad \sigma_m = \frac{F_m}{A} = \frac{100kN}{706.5 \times 10^{-6}} = 141.54 \text{ MPa}$$

and $\quad F_a = \frac{160 - 40}{2} = 60$ kN

$$\therefore \quad \sigma_a = \frac{60 \times 10^3}{706.5} = 84.93 \text{ MPa}$$

$$\therefore \quad \frac{1}{h} = \frac{141.54}{420} + \frac{84.93}{240}$$

or $\quad h = 1.45$

12. Soderberg criteria is:

$$\frac{\sigma_a}{\sigma_e} + \frac{\sigma_m}{s_{y+}} = \frac{1}{n} \text{ and } \sigma_a = \frac{100 - 20}{2A}$$

$$\Rightarrow \frac{40}{160} + \frac{60}{240} = \frac{1}{2} \times A \times 10^3 \quad \sigma_m = \frac{100 + 20}{2A}$$

$A = 1000$ mm^2

13. $\sigma = \min$ of $\frac{(S_{yt}, S_{ut}, S_e)}{FOS}$

$$FOS = \frac{200}{100} = 2.$$

14. $\sigma_{min} = 200$ MPa

$\sigma_{max} = 400$ MPa

$$\sigma_a = \frac{\sigma_{max} - \sigma_{min}}{2}$$

$$= \frac{400 - 200}{2}$$

$= 100$ MPa

15. Stress amplitude $= \frac{\sigma_{max} - \sigma_{min}}{2}$

$$= \frac{250 - 50}{2} = 100 \text{ MPa}$$

Stress ratio $= \frac{\sigma_{min}}{\sigma_{max}} = \frac{50}{250} = 0.2$

16. Tensile stress

$$S_f = \frac{F}{(w - 2d)t}$$

$$t = \frac{F}{S_f(w - 2d)}$$

17. $N = \dfrac{\text{Endurance limit}}{\text{Max. stress}}$

$= \dfrac{100}{50} = 2.$

18. Here, $\sigma_{mean} = +140$ MPa

$\sigma_{min} = -70$ MPa

Now, we know

$$\sigma_{mean} = \frac{\sigma_{max} + \sigma_{min}}{2} = 140$$

$$\frac{\sigma_{max} - 70}{2} = 140$$

$\Rightarrow \quad \sigma_{max} = 350$ MPa

Now, minimum stress to maximum stress ratio is,

$$\frac{\sigma_{min}}{\sigma_{max}} = \frac{-70}{350} = -0.2$$

20. According to distorsion energy theorem

$$f_1^2 + f_2^2 - f_1 f_2 = (\text{Working stress})^2$$

$$\text{Working stress} = \sqrt{(360)^2 + (140)^2 - (360 \times 140)}$$

$$= 314.32 \text{ MPa}$$

23. Both yields simultaneously.

24. $\sigma_1 = 100$ MPa

$\sigma_2 = 100$ MPa

As per Von-mises yield theory.

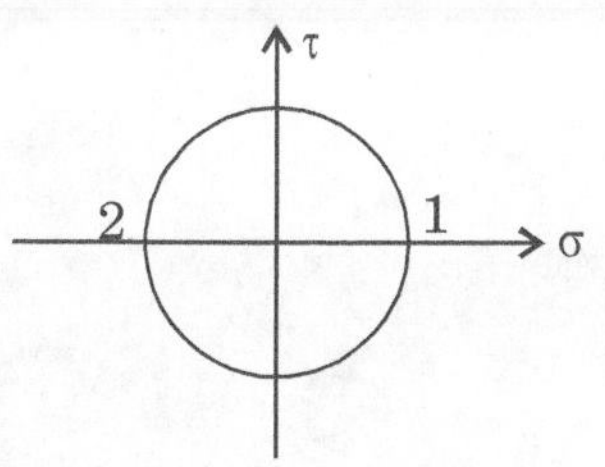

$$n^2\left[(\sigma_1 - \sigma_2)^2 + (\sigma_2 - \sigma_3)^2 + (\sigma_1 - \sigma_3)^2\right] \le 2\sigma_y^{\,2}$$

In this case, $\sigma_3 = 0$

$$2n^2\left[\sigma_1^{\,2} + \sigma_2^{\,2} - \sigma_1\sigma_2\right] \le 2\sigma_y^{\,2}$$

$$\Rightarrow n^2\left[(100)^2 + (100)^2 + (100)^2\right] \le (300)^2$$

$\therefore \quad n = 1.732$

25. If there is uniaxial loading yield stress is σ_y

As per Von Mises failure theory

$$\sigma_y^2 = \sigma_1^2 + \sigma_1\sigma_2 + \sigma_2^2$$

Under pure shear stress loading

$$\sigma_1 = -\sigma_2 = \tau$$

then $\quad \sigma_y^2 = \tau^2 + \tau^2 + \tau^2 = 3\tau^2$

$\Rightarrow \quad \tau = \dfrac{\sigma_y}{\sqrt{3}}$

Hence $\quad \tau = \dfrac{\sigma_y}{\sqrt{3}} = 0.5577\sigma_y = 173.28$

where τ is shear yield stress

26. $\sigma_1 = \dfrac{80+20}{2} + \sqrt{\left(\dfrac{80-20}{2}\right)^2 + 40^2}$

$= 50 + \sqrt{50^2} = 100$

$\sigma_2 = 0$

$\tau = \dfrac{\sigma_1 - \sigma_2}{7} = 50$

$\text{FOS} = \dfrac{100}{50} = 2$

27. σ_x, σ_y & σ_z are principal stresses.

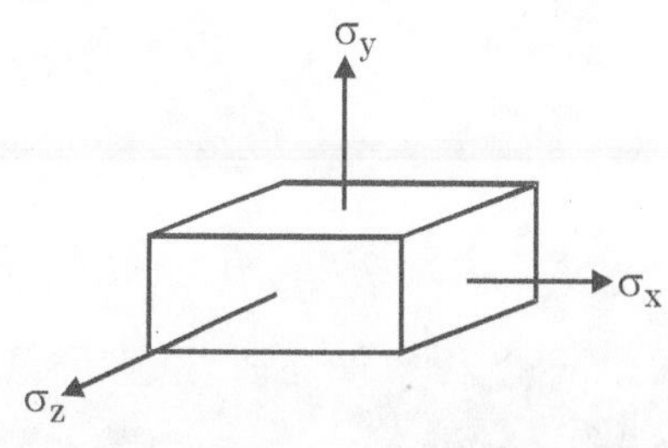

Under plane strain condition, $\epsilon_z = 0$

$$\Rightarrow \frac{\sigma_z}{E} - \frac{\mu\sigma_y}{E} - \frac{\mu\sigma_x}{E} = 0$$

$$\Rightarrow \sigma_z = (\sigma_x + \sigma_y) = \mu(\sigma_x + \sigma_y)$$

but $\sigma_x = -40$

$$\Rightarrow \sigma_z = \mu(-40 + \sigma_y)$$

From Tresca theory,

$$\tau_{max} = \max\left[\frac{|\sigma_x - \sigma_y|}{2}, \frac{|\sigma_y - \sigma_z|}{2}, \frac{|\sigma_z - \sigma_x|}{2}\right]$$

$$= \frac{S_{yt}}{2N}$$

$$\Rightarrow \max\left[\frac{|-40 - \sigma_y|}{2}, \frac{(\sigma_y - \mu(\sigma_y - 40))}{2}, \frac{|\mu(\sigma_y - 40) - (-40)|}{2}\right]$$

$$\Rightarrow \max\left[|40 + \sigma_y|, |(1-\mu)\sigma_y + 40\mu|, |\mu\sigma_y + 40(1-\mu)|\right]$$

$$= 300$$

But μ is not given

Hence if we neglect σ_z then

$\max [|40 + \sigma_y|, |\sigma_y|, |+40|] = 300$

$\Rightarrow |40 + \sigma_y| = 300$

$40 + \sigma_y = 300$

$\Rightarrow \sigma_y = 260$ MPa

28. Given, $\sigma_x = 100$ MPa

$\sigma_y = 220$ MPa

$\sigma_{xy} = 80$ MPa

$$\sigma_{1,2} = \frac{\sigma_x + \sigma_y}{2} \pm \sqrt{\left(\frac{\sigma_x - \sigma_y}{2}\right)^2 + \sigma_{xy}^2}$$

$$= \frac{100 + 200}{2} \pm \sqrt{\left(\frac{100 - 220}{2}\right)^2 + 80^2}$$

$$= 160 \pm \sqrt{55^2 + 80^2}$$

$$\sigma_{1,2} = 160 \pm 97.0824$$

$\Rightarrow$ $\sigma_1 = 257.08$ Mpa, $\sigma_2 = 62.9$ MPa

As per maximum shear stress theory

$$\frac{S_{yt}}{2N} = \text{Max}\left[\left|\frac{\sigma_1 - \sigma_2}{2}\right|, \frac{\sigma_1}{2}, \frac{\sigma_2}{2}\right]$$

$$\frac{S_{yt}}{2N} = \frac{\sigma_1}{2} = \frac{257.08}{2}$$

$$N = \frac{468}{257.08} = 1.8$$

2
CHAPTER

Fatigue Strength and S-N Diagram

Analysis of Previous GATE Papers														
Topic	**Year/Set →** **Marks Distribution ↓**		**2019 Set 1**	**2019 Set 2**	**2018 Set 1**	**2018 Set 2**	**2017 Set 1**	**2017 Set 2**	**2016 Set 1**	**2016 Set 2**	**2016 Set 3**	**2015 Set 1**	**2015 Set 2**	**2015 Set 3**
Fatigue Strength	**1 Mark**	MCQ Type										1		
		Numerical Type												
	2 Marks	MCQ Type												
		Numerical Type				1								
	Total					2						1		
S-N Diagram	**1 Mark**	MCQ Type		1										
		Numerical Type				1								
	2 Marks	MCQ Type												
		Numerical Type					1							
	Total			1		1	2							

Fatigue strength

1. A ductile material having an endurance limit of 196 N/mm^2 and the yield point at 294 N/mm^2 is stressed under variable load. The maximum and minimum stresses are 147 N/mm^2 and 49 N/mm^2. The fatigue stress concentration factor is 1.32. The available factor of safety for this loading is

(a) 3.0 (b) 1.5
(c) 1.33 (d) 4.0

[1987 : 2 Marks]

2. Stress concentration in a machine component of a ductile material is not so harmful as it is in a brittle material because

(a) in ductile material local yielding may distribute stress concentration
(b) ductile material has larger young's material
(c) Poisson's ratio is larger in ductile materials
(d) Modulus of rigidity is larger in ductile materials

[1989 : 1 Mark]

3. The process of shot peening increases the fatigue life of steel springs mainly because it results in

(a) surface hardening
(b) increased stiffness of the material
(c) structural changes in the material
(d) residual compression at the surface

[1990 : 1 Mark]

4. In a shaft with a transverse hole, as the hole to the shaft diameter ratio_____(increase/ decrease), the torsional stress concentration factor ______ (increases/ decreases)

[1991 : 1 Mark]

5. Fatigue strength of rod subjected to cyclic axial force is less than that of a rotating beam of the same dimensions subjected to steady lateral force.

[True / False]

[1992 : 1 Mark]

6. The yield strength of a steel shaft is twice its endurance limit. Which of the following torque fluctuations represent the most critical situation according to Soderberg criterion?

(a) – T to + T (b) –T/2 to + T
(c) 0 to + T (d) + T/2 to + T

[1993 : 1 Mark]

7. A static load is mounted at the centre of a shaft rotating at uniform angular velocity. This shaft will be designed for

(a) the maximum compressive stress (static)
(b) the maximum tensile stress (static)
(c) the maximum bending moment (static)
(d) fatigue loading

[2002 : 1 Mark]

8. Which one of the following is the most conservation fatigue failure criterion?

(a) Soderberg (b) Modified Goodman
(c) ASME Elliptic (d) Gerber

[2015 : 1 Mark, Set-1]

9. A bar is subjected to a combination of a steady load of 60 kN and a load fluctuating between -10 kN and 90 kN. The corrected endurance limit of the bar is 150 MPa. the yield strength of the material is 480 MPa and the ultimate strength of the material is 600 MPa. The bar cross-section is square with side a. If the factor of safety is 2, the value of a (in mm), according to the modified Goodman's criterion, is ________ (correct to two decimal places).

[2018 : 2 Marks, Set-2]

S-N Diagram

10. In terms of theoretical stress concentration factor k_t and fatigue stress concentration factor k_f, the notch sensitivity q is expressed as

(a) $\frac{(k_f - 1)}{(k_t - 1)}$ (b) $\frac{(k_f - 1)}{(k_t + 1)}$

(c) $\frac{(k_t - 1)}{(k_f - 1)}$ (d) $\frac{(k_f + 1)}{(k_t + 1)}$

[2004 : 1 Mark]

11. The S-N curve for steel becomes asymptotic nearly at

(a) 10^3 cycles (b) 10^4 cycles
(c) 10^6 cycles (d) 10^9 cycles

[2004 : 1 Mark]

12. A cylindrical shaft is subjected to an alternating stress of 100 MPa. Fatigue strength to sustain 1000 cycles is 490 MPa. If the corrected endurance strength is 70 MPa, estimated shaft life will be

(a) 1071 cycles (b) 15000 cycles
(c) 281914 cycles (d) 928643 cycles

[2006 : 2 Marks]

13. A machine element has an ultimate strength (σ_u) of 600 N/mm², and endurance limit (σ_{en}) of 250 N/mm². The fatigue curve for the element on a log-log plot is shown below. If the element is to be designed for a finite life of 10000 cycles, the maximum amplitude of a completely reversed operating stress is ______N/mm².

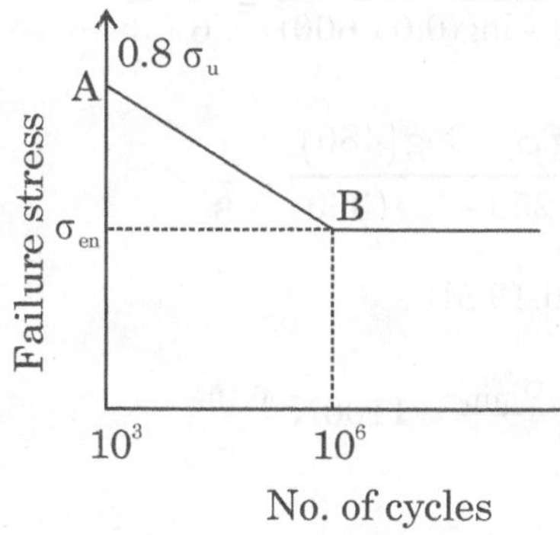

[2017 : 2 Marks, Set-1]

14. Fatigue life of a material for a fully reversed loading condition is estimated from $\sigma_a = 1100\ N^{-0.15}$ where σ_a is the stress amplitude in MPa and N is the failure life in cycles. The maximum allowable stress amplitude (in MPa) for a life of 1×10^5 cycles under the same loading condition is______(correction to two decimal places).

[2018 : 1 Marks, Set-2]

15. Endurance limit of a beam subjected to pure bending decreases with
(a) increase in the surface roughness and increase in the size of the beam
(b) decrease in the surface roughness and increase in the size of the beam
(c) decrease in the surface roughness and decrease in the size of the beam
(d) increase in the surface roughness and decrease in the size of the beam

[2019 : 1 Mark, Set 2]

ANSWERS

1. (b) **2.** (a) **3.** (d) **6.** (a) **7.** (d) **8.** (a) **10.** (a) **11.** (c) **12.** (c) **15.** (a)

EXPLANATIONS

1. Consider Soderberg equation

$$\frac{\tau_a}{S_e} + \frac{\tau_m}{S_y} = \frac{1}{FOS}$$

$$\tau_a = \frac{147-49}{2} = 49 \text{ MPa}$$

$$\tau_m = \frac{147+49}{2} = 98 \text{ MPa}$$

$$S_e = \frac{196}{1.32} = 147.48 \text{ MPa}$$

$$\frac{49}{147.48} + \frac{98}{294} = \frac{1}{FOS}$$

FOS = 1.5

2. Local yielding reduces stress concentration effect.
3. Shot preening increases fatigue life of springs
4. Increase, decreases
5. True
6. $$\frac{\tau_a}{S_e} + \frac{\tau_m}{S_y} = \frac{1}{Fos}$$

$$\frac{\tau_a}{1} + \frac{\tau_m}{2} = \frac{S_e}{Fos}$$

τa is highest in (a)

7. Shaft will be designed for fatigue loading.

9. $$P_m = \frac{P_{max} + P_{min}}{2}$$

$$P_a = \frac{P_{max} - P_{min}}{2}$$

$P_m = 100$ kN; $P_a = 50$ kN

$$\sigma_m = \frac{100\times10^3}{a^2}\text{MPa};$$

$$\sigma_a = \frac{50\times10^3}{a^2}\text{MPa}$$

By Goodman's equation,

$$\frac{\sigma_m}{Sut}+\frac{\sigma_a}{\sigma_e}=\frac{1}{N};1000\left[\frac{1000}{a^2\times600}+\frac{50}{150a^2}\right]=\frac{1}{2}$$

a = 31.62 mm

By yield line equation,

$$\frac{\sigma_m}{Syt}+\frac{\sigma_a}{Syt}=\frac{1}{N};1000\left[\frac{100}{480a^2}+\frac{50}{480a^2}\right]=\frac{1}{2}$$

a = 25 mm

Final answer is 31.62 mm

12.

$\log_{10} 490 = 2.69$

$\log_{10} 70 = 1.84$

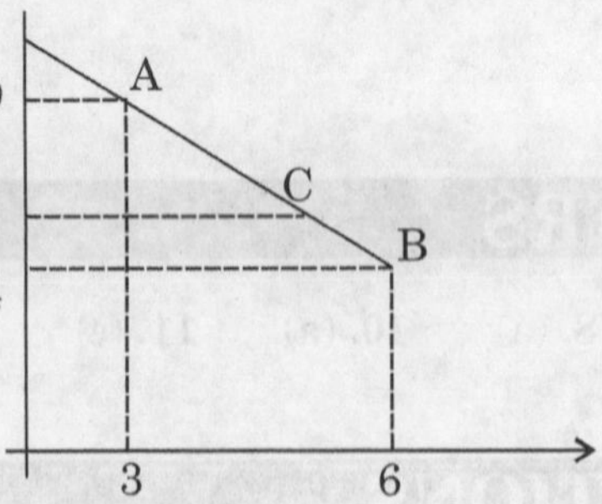

By concept of straight line,

$$\frac{y-1.8451}{x-6}=\frac{1.84-2.6902}{6-3}$$

At y = 2

$\Rightarrow$ x = 5.4501

$\log 10_N = 5.4501$, N = 281914

13. $$\frac{\log S_f-\log(0.8\times600)}{\log250-\log(0.6\times600)}=\frac{4-3}{6-3}$$

$$\Rightarrow \frac{\log S_f-\log(480)}{\log250-\log(360)}=\frac{1}{3}$$

S_f = 386.19 MPa

14. $$\frac{\sigma_{max}-\sigma_{min}}{2}=1100N^{-0.15}$$

$$\frac{\sigma_{max}-(-\sigma_{max})}{2}=1100N^{-0.15}$$

$$\frac{2\sigma_{max}}{2}=1100N^{-0.15}(N=10^5)$$

$\therefore$ we get, σ_{max} = 195.61N

3 CHAPTER Bolted, Riveted and Welded Joint

Analysis of Previous GATE Papers														
Topic	Year/Set → Marks Distribution ↓		2019 Set 1	2019 Set 2	2018 Set 1	2018 Set 2	2017 Set 1	2017 Set 2	2016 Set 1	2016 Set 2	2016 Set 3	2015 Set 1	2015 Set 2	2015 Set 3
Bolted Joint	1 Mark	MCQ Type				1								
		Numerical Type												
	2 Marks	MCQ Type						1						
		Numerical Type									1			1
	Total					1		2			2			2
Riveted Joint	1 Mark	MCQ Type												
		Numerical Type												
	2 Marks	MCQ Type												
		Numerical Type										1		
	Total											2		
Welded Joint	1 Mark	MCQ Type												
		Numerical Type												
	2 Marks	MCQ Type												
		Numerical Type												
	Total													

Bolted joint

1. The bolts in a rigid flanged coupling connecting two shafts transmitting power are subjected to
   (a) shear force and bending moment
   (b) axial force
   (c) torsion
   (d) torsion and bending moment

   [1996 : 1 Mark]

2. Bolts in the flanged end of pressure vessel are usually pre-tensioned. Indicate which of the following statements is true
   (a) Pre-tensioning helps to seal the pressure vessel
   (b) Pre-tensioning increase the fatigue life of the bolts
   (c) Pre-tensioning reduces the maximum tensile stress in the bolts
   (d) Pre-tensioning helps to reduce the effect of pressure pulsations in the pressure vessels

   [1998 : 1 Mark]

3. In a bolted Joint two smooth members are connected with an axial tightening force of 2200 N. If the bolt used has metric threads of 4 mm pitch, the torque required for achieving the tightening force is

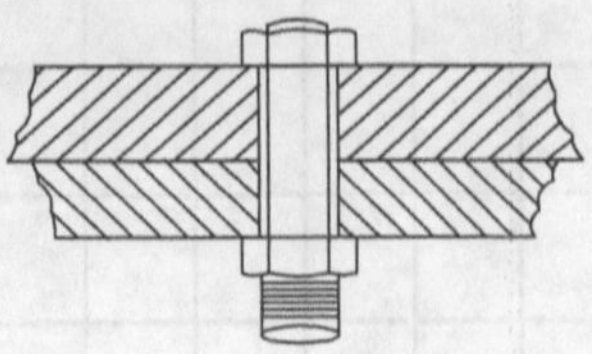

   (a) 0.7 Nm (b) 1.0 Nm
   (c) 1.4 Nm (d) 2.8 Nm

   [2004 : 2 Marks]

4. A bolted Joint is shown below. The maximum shear stress, in MPa, in the bolts at A and B, respectively are

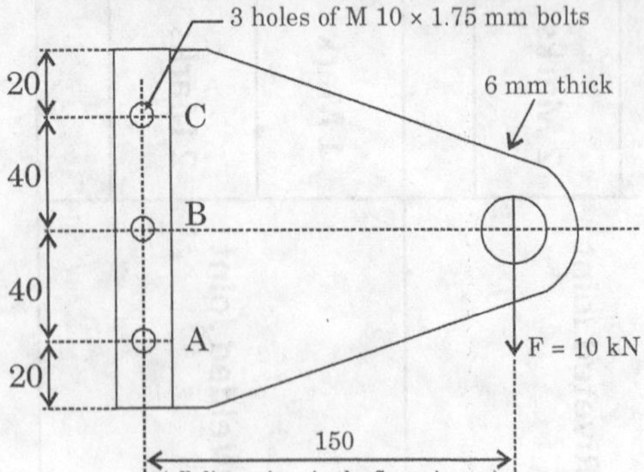

   (a) 242.6, 42.5 (b) 42.5, 242.6
   (c) 42.5,42.5 (d) 60.15,343.64

   [2007 : 2 Marks]

Linked Answer Questions 5 and 6

A steel bar of 10 × 50 mm is cantilevered with two M12 bolts (P and Q) to support a static load of 4 kN as shown in the figure.

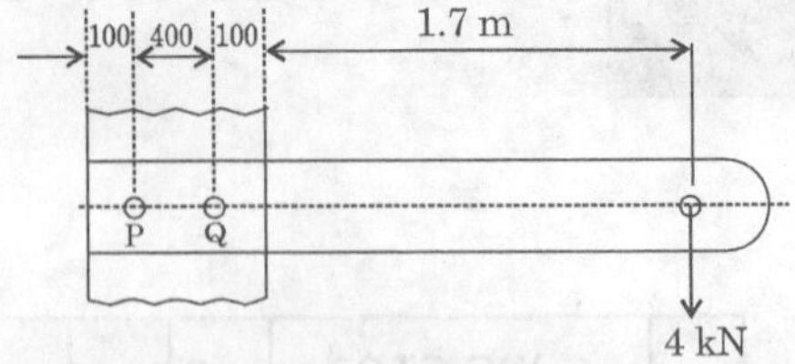

5. The primary and secondary shear loads on bolt P, respectively, are
   (a) 2 kN, 20 kN (b) 20 kN, 2 kN
   (c) 20 kN, 0 kN (d) 0 kN, 20 kN

   [2008 : 2 Marks]

6. The resultant shear stress on bolt P is closest to:
   (a) 132 MPa (b) 159 MPa
   (c) 178 MPa (d) 195 MPa

   [2008 : 2 Marks]

7. For the three bolt system shown in the figure, the bolt material has shear yield strength of 200 MPa. For a factor of safety of 2, the minimum metric specification required for the bolt is
   (a) M8
   (b) M10
   (c) M12
   (d) M16

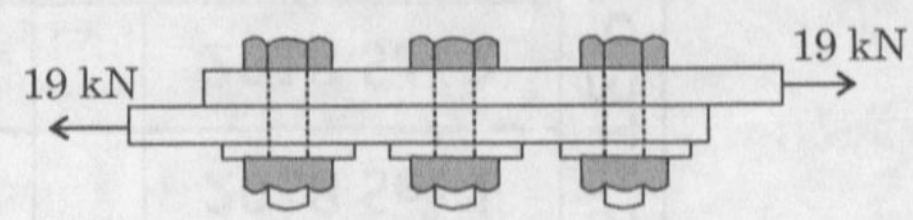

   [2014 : 2 Marks, Set-2]

8. A bolt of major diameter 12 mm is required to clamp two steel plates. Cross sectional area of the threaded portion of the bolt is 84.3 mm^2. The length of the threaded portion in grip is 30 mm, while the length of the unthreaded portion in grip is 8 mm. Young's modulus of material is 200 GPa. The effective stiffness (in MN/m) of the bolt in the clamped zone is _____. **[2014 : 2 Marks, Set-4]**

9. A cantilever bracket is bolted to a column using three M12 × 1.75 bolts, P, Q and R. The value of maximum shear stress developed in the bolt P (in MPa) is ______.

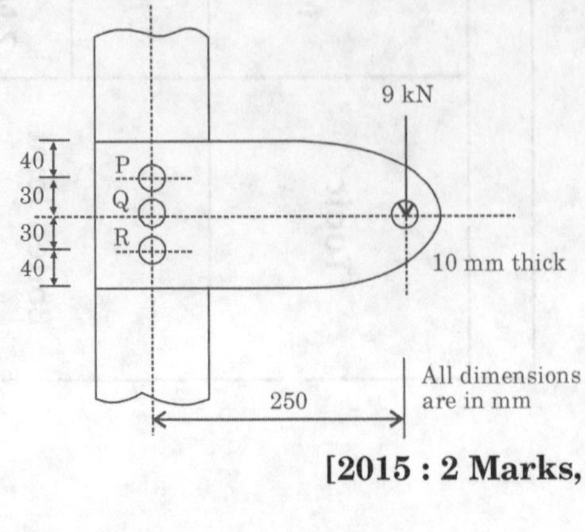

   [2015 : 2 Marks, Set-3]

10. A bolted Joint has four bolts arranged as shown in figure. The cross sectional area of each bolt is 25 mm². A torque T= 200 Nm is acting on the Joint. Neglecting friction due to clamping force, maximum shear stress in a bolt is ________ MPa.

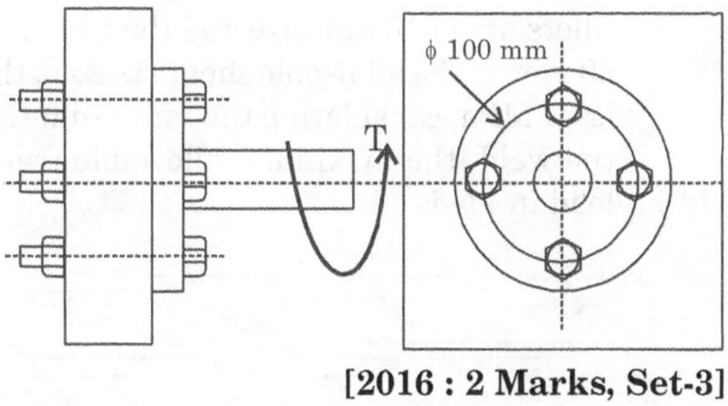

[2016 : 2 Marks, Set-3]

11. A steel plate, connected to a fixed Channel using three identical bolts A, B and C, carries a load of 6 kN as shown in the figure. Considering the effect of direct load and moment, the magnitude of resultant shear force (in kN) on bolt C is

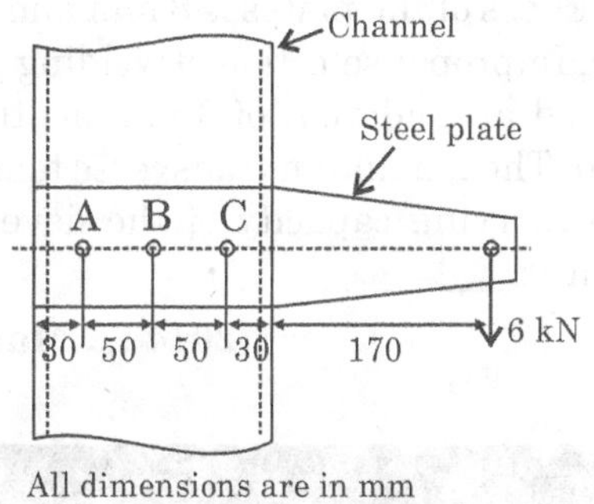

All dimensions are in mm

(a) 13 (b) 15
(c) 17 (d) 30

[2017 : 2 Marks, Set-2]

12. Pre-tensioning of a bolted Joint is used to

(a) strain harden the bolt head
(b) decrease stiffness of the bolted Joint
(c) increase stiffness of the bolted Joint
(d) prevent yielding of the thread root

[2018 : 1 Mark, Set-2]

Riveted joint

13. If the ratio of the diameter of rivet hole to the pitch of rivets is 0.25, then the tearing efficiency of the Joint is

(a) 0.50 (b) 0.75
(c) 0.25 (d) 0.87

[1996 : 1 Mark]

14. A bracket (shown in figure) is rigidly mounted on wall using four rivets. Each rivet is 6 mm in diameter and has an effective length of 12 mm.

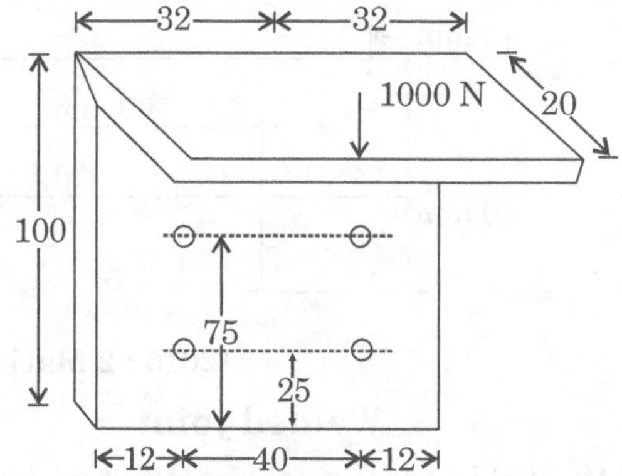

Direct shear stress (in MPa) in the most heavily loaded rivet is

(a) 4.4 (b) 8.8
(c) 17.6 (d) 35.2

[2010 : 2 Marks]

Common Data Questions 15 and 16

A single riveted lap Joint of two similar plates as shown in the figure below has the following geometrical and material details.

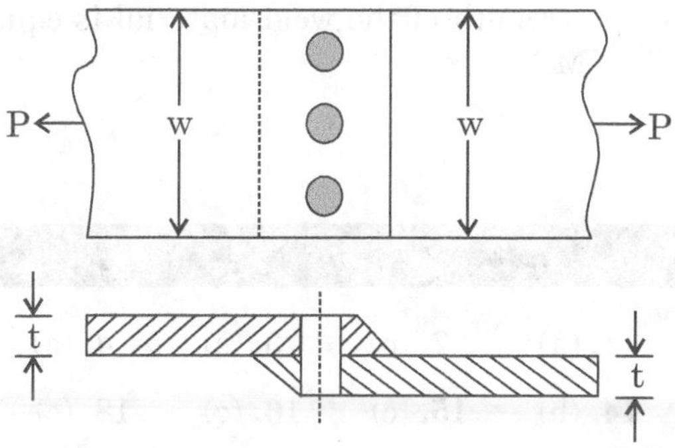

width of the plate ω = 200 mm, thickness of the plate t = 5 mm, number of rivets n = 3, diameter of the rivet d_r = 10 mm, diameter of the rivet hole dr = 11 M mm, allowable tensile stress of the plate σ_p = 200 MPa, allowable shear stress of the rivet σ_s = 100 MPa and allowable bearing stress of the rivet σ_c = 150 MPa.

15. If the rivets are to be designed to avoid crushing failure, the maximum permissible load P in kN is

(a) 7.50 (b) 15.00
(c) 22.50 (d) 30.00

[2013 : 2 Marks]

16. If the plates are to be designed to avoid tearing failure, the maximum permissible load P in kN is

(a) 83 (b) 125
(c) 167 (d) 501

[2013 : 2 Marks]

17. A horizontal plate has been joined to a vertical to a vertical post using four rivets arranged as shown in the figure. The magnitude of the load on the worst loaded rivet (in N) is ______.

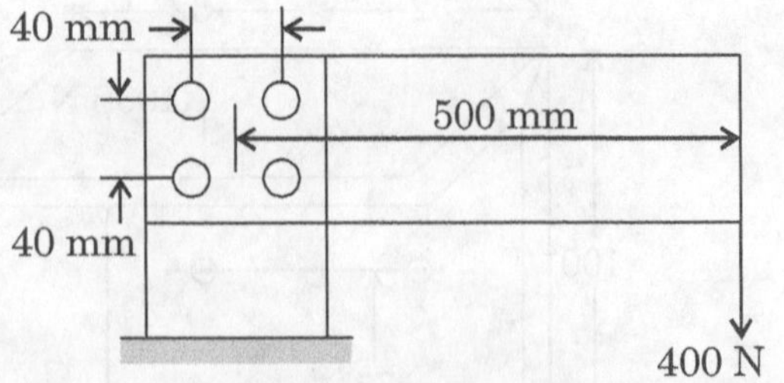

[2015 : 2 Marks, Set-1]

Welded joint

18. Weldments in fabricated steel beams are designed for

(a) bending stresses at the flange

(b) shear stresses in transverse plane

(c) combination of bending and shear

(d) none of these because in fabricated beams welds not to get stressed

[1987 : 1 Mark]

19. A 60 mm long and 6 mm thick fillet weld carries a steady load of 15 kN along the weld. The shear strength of the weld material is equal to 200 MPa.

The factor of safety is

(a) 2.4 (b) 3.4

(c) 4.8 (d) 6.8

[2006 : 2 Marks]

20. A fillet welded Joint is subjected to transverse loading F as shown in figure. Both legs of the fillets are of 10 mm size and the weld length is 30 mm. If the allowable shear stress of the weld is 94 MPa, considering the minimum throat of the weld, the maximum allowable transverse load in kN is

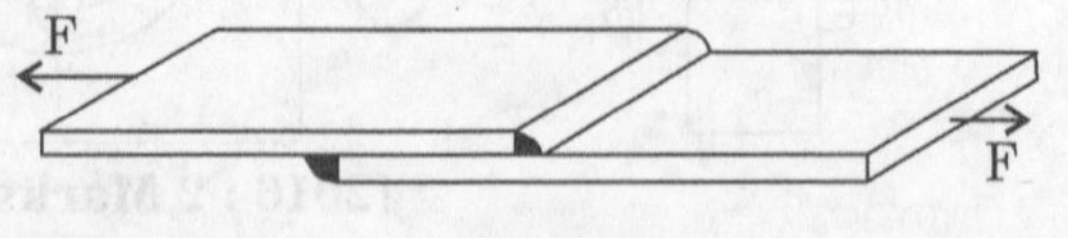

(a) 14.44 (b) 17.92

(c) 19.93 (d) 22.16

[2012 : 2 Marks]

21. A butt weld Joint is developed on steel plates having yield and ultimate tensile strength of 500 MPa and 700 MPa, respectively. The thickness of the plates is 8 mm and width is 20 mm. Improper selection of welding parameters caused an undercut of 3 mm depth along the weld. The maximum transverse tensile load (in kN) carrying capacity of the developed weld Joint is ______.

[2014 : 2 Marks, Set-4]

ANSWERS

1. (a) **2.** (b) **3.** (c) **4.** (a) **5.** (a) **6.** (b) **7.** (b) **11.** (c) **12.** (c) **13.** (b)

14. (b) **15.** (c) **16.** (c) **18.** (c) **19.** (b) **20.** (c)

EXPLANATIONS

3. Tightening force = 2200 N

Pitch = 4 mm = 0.004 m

$$\text{Torque} = \text{Tightening force} \times \frac{0.004}{2\pi}$$

$$= 1.4 \text{ N-m}$$

4. $F_A = F_B = F_C = \frac{10}{3}$ kN

$$F_A = \frac{P.e.r}{r_A^2 + r_B^2 + r_C^2} = \frac{10 \times 10^3 \times 150 \times 40}{40^2 + 0^2 + 40^2}$$

$$F_A' = 18.75\text{kN} = F_C'$$

Shear load at A and C

$$F_R = \sqrt{(F_A)^2 + (F_A')^2} = \sqrt{3.33^2 + 18.75^2}$$

$$F_R = 19.04 \text{ KN}$$

$$\tau_{max} = \frac{19.04 \times 10^3}{\frac{\pi}{4} \times 10^2}$$

$$(\tau_{max})_{A,C} = 242.5\text{MPa}$$

$$(\tau_{max})_B = 42.44\text{MPa}$$

5. Let F_1 and F_2 resist the 4 kN eccentric load. The primary shear load is due to a direct shear stress of 4kN, which is equally divided amongst the two bolts, i.e. 2 k N,

$\therefore$ $F_{primary}$ on P = 2 kN

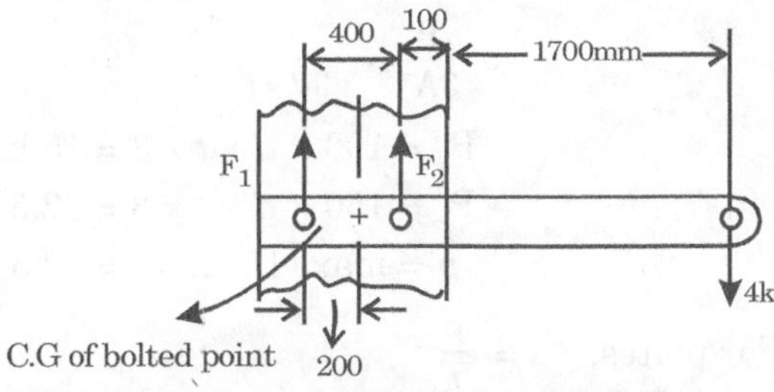

The secondary load is to resist the moment produced by the eccentric load 4 kN about the C. G. of the joint

Moment $M = 4 \text{ kN} \times 2000 \text{ mm}$

$= 8000 \text{ Nm}$

Now this moment must be balanced by a moment produced by equal and opposite forces i.e. secondary loads on the bolts

i.e. $F \times 400 = 800 \text{ Nm}$

$\Rightarrow$ $F = 20 \text{ kN}$

6. Resultant load on $P = F_{sec} - F_{prim}$

$= 20 - 2 = 18 \text{ kN}$

$$\therefore \quad \text{Stress} = \frac{18 \times 10^3}{\frac{\pi}{4}(12)^2} = 159 \text{ MPa}$$

7. $\sigma_y = 200$ MPa

$P = 19 \times 10^3$ N

$$\sigma = \frac{P}{\frac{\pi}{4}d^2} \Rightarrow d = 10.9 \text{ mm}$$

M10

8. $d_1 = 12$ mm;

$l_1 = 8$ mm

$A_2 = 84.33 \text{ mm}^2$

$$A_1 = \frac{\pi}{4}(d_1)^2$$

$l_2 = 30$ mm

$$K_1 = \frac{A_1E_1}{l_1};$$

$$K_2 = \frac{A_2E_2}{l_2}$$

$$= \frac{84.33 \times 200}{30} = 562.2$$

$$= \frac{\frac{\pi}{4}(12)^2 \times 200}{8} = 2827.4$$

$$\frac{1}{K} = \frac{1}{K_1} + \frac{1}{K_2}$$

$$= \frac{\pi}{2} \times 3000 + 1500 \times \pi$$

$$= \frac{1}{2827.4} + \frac{1}{562.2}$$

$\Rightarrow$ $K = 468.9$ MN/m

9. $P = 9$ kN, $e = 250$ mm

Primary shear force

$$P'_p = P'_q = P'_r = \frac{p}{3} = \frac{q}{3} = 3 \text{ kN}$$

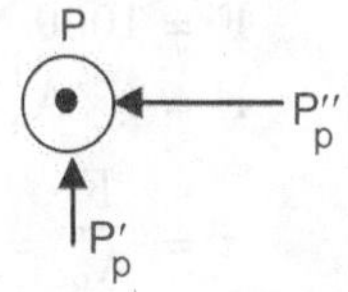

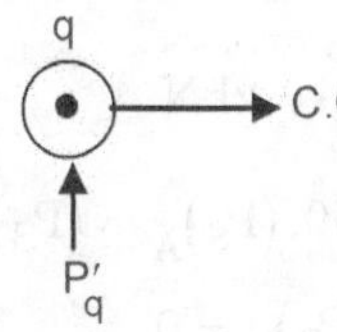

Secondary shear force:

By symmetry C.G lies at the centre of bolt Q.

$\therefore$ $r_p = 30$ mm

$r_r = 30$ mm

$r_q = 0$

$$C = \frac{Pe}{r_q^2 + r_q^2 + r_r^2}$$

$$= \frac{9000 \times 250}{30^2 + 0^2 + 30^2} = 1250$$

$P''_p = P''_r = C$

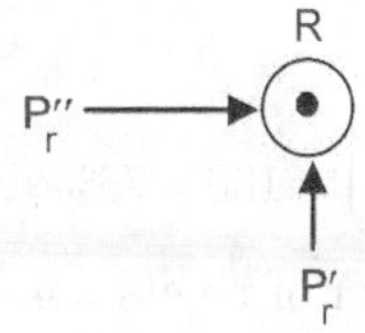

$P''_p = Cr_p = 37.5$ kN

$P'_r = Cr_r = 37.5$ kN

$P''_q = 0$

Resultant shear force

Due to symmetry stress in P Q R will be equal in magnitude

$$P_p = P_r = \sqrt{(3)^2 + (37.5)^2}$$

$= 38.5648$ kN

$$\tau = \frac{P}{A}$$

$$= \frac{38.5648}{\frac{\pi}{4}(0.012)^2}$$

$= 332$ MPa $\simeq$

10. Let Resisting force in each bolt = FN

Net Resisting torque = TR = $4F \times 50 \times 10^{-3}$

Applied torque (T) = Resisting torque

$$T = 4F \times 50 \times 10^{-3}$$

$$F = \frac{200}{4 \times 50} \times 1000$$

$$F = 1000 \text{ N}$$

$$F = \tau \times A_R$$

$$\tau = \frac{F}{A_R} = \frac{1000}{25}$$

$$= 40 \text{ MPa}$$

11. $P_P = \frac{P}{n} = \frac{P}{3} = 2\text{kN}.$

$r_A = r_C = 50, (P_S)_A = (P_S)_C$

$r_B = 0 \Rightarrow (P_S)_B = 0$

$$\frac{(P_S)_C}{r_C}\left[r_C^2 + r_B^2 + r_A^2\right] = P \times e$$

$(P_S)_C \times 2 \times r_C = P \times e$

$$(P_S)_C = \frac{6 \times 250}{2 \times 50} = 15\text{KN}$$

$Q_A = 180°$, $Q_C = 0°$

Resultant shear force $(R_C) = P_P + (P_S)_C$.

$R_C = 2 + 15 = 17$ KN

12. Pre-tensioning increases the stiffness of the bolt.

13. $\frac{d}{p} = 0.25$

$$\eta_{tearing} = \left(\frac{P - d}{P}\right) \times 100 = 75\%$$

14. Given : Number of rivets = 4,

Diameter of each rivet, $d = 6 \times 10^{-3}$ m

Direct load on the bracket = 1000 N

since there is no eccentricity

$\therefore$ Load on each rivet $= \frac{1000}{4} = 250$ N

$\therefore$ Shear stress $= \frac{P}{A} = \frac{250}{\frac{\pi}{4}(6)^2} = 8.8$ MPa

15. For rivet,

$$\sigma \frac{P}{3A_r} = \frac{P}{3d \times t}$$

$\therefore$ $P_s = 100 \times d \times t \times 3 = 15$ kN

and $P_c = 150 \times d \times t \times 3 = 22.5$ kN

$\therefore$ $p = \max(15, 22.5) = 22.5$ kN

16. For plates, $\sigma = \frac{P}{A}$

$$\Rightarrow \quad 200 = \frac{P}{(W - 3d_h)t}$$

$$P = 167\text{kN}$$

17. Primary force $(P_1) = 100$ N

$$P_1'' = P_2'' = P_3'' = P_4'' = \frac{P_e \times r}{4r^2}$$

$$= \frac{400 \times 500 \times \left(\frac{40\sqrt{2}}{2}\right)}{4 \times \left(\frac{40\sqrt{2}}{2}\right)^2}$$

$P_1'' = 1767.766$N

1 & 4 are worst loaded

$$\therefore (P_{net})^2 = (P_1)^2 + (P_1'')^2 + 2(P_1)(P_1'')\cos 45°$$

$P_{net} = 1839.83$ N

19. $P = .707\, t \times L_e \times \tau_S$.

$$\frac{15 \times 10^3}{.707 \times 6 \times 60} = \tau_S$$

$\therefore$ $\tau_S = 58.94$ MPa

Given shear strength of material = 200 MPa

$$\text{FOS} = \frac{200}{58.94} = 3.4$$

20. $P = 0.707$ sl. $\tau_{allowable}$

$= 0.707 \times 10 \times 30 \times 94$

$= 19937.4$N

$= 19.934$ kN

21. $P_t = (s_t)(t_{under\ cut})\, L$

$= 700 \times 20 \times (8 - 3)$

$= 700 \times 100$

$= 70$ kN

4
CHAPTER

Bearings, Shafts and Keys

Analysis of Previous GATE Papers

Topic	Marks Distribution		2019 Set 1	2019 Set 2	2018 Set 1	2018 Set 2	2017 Set 1	2017 Set 2	2016 Set 1	2016 Set 2	2016 Set 3	2015 Set 1	2015 Set 2	2015 Set 3
Bearing	1 Mark	MCQ Type									1			
		Numerical Type												
	2 Marks	MCQ Type			1									
		Numerical Type												1
	Total				2						1			2
Shaft Design	1 Mark	MCQ Type												
		Numerical Type												
	2 Marks	MCQ Type												
		Numerical Type												
	Total													
Keys	1 Mark	MCQ Type												
		Numerical Type												
	2 Marks	MCQ Type												
		Numerical Type												
	Total													

Bearing

1. The expected life of a ball bearing subjected to a load of 9800 N and working at 1000 rpm is 3000 hours. What is the expected life of the same bearing for a similar load of 4900 N and speed of 2000 rpm?

(a) Unchanged (b) 12000 hours

(c) 1500 hours (d) 6000 hours

[1987 : 1 Mark]

2. If the load on a ball bearing is reduced to half, the life of the ball bearing will

(a) increase 8 times (b) increase 4 times

(c) increase 2 times (d) not change

[1988 : 1 Mark]

3. Match the rolling element bearings with the most appropriate loading condition

Bearing-type

(a) Ball bearing

(b) Roller bearing

(c) Needle bearing

(d) Taper roller bearing

Loading condition

(p) Tangential load

(q) Radial load

(r) Heavy radial load with impact

(s) Light radial load with space limitation

(t) Heavy radial and axial load

(u) Fatigue load

[1992 : 1 Mark]

4. Starting friction is low in

(a) hydrostatic lubrication

(b) hydrodynamic lubrication

(c) mixed (or semi fluid) lubrication

(d) boundary lubrication

[1992 : 1 Mark]

5. Spherical roller bearings are normally used

(a) for increased radial load

(b) for increased thrust load

(c) when there is less radial load

(d) to compensate for angular misalignment

[1992 : 1 Mark]

6. In thick film hydrodynamic Journal bearings, the coefficient of friction

(a) increase with increase in load

(b) is independent on load

(c) decreases with increase in load

(d) may increase or decrease with increase in load

[1996 : 1 Mark]

7. To restore stable operating condition in a hydrodynamic Journal bearing, when it encounters higher magnitude of loads

(a) oil viscosity is to be decreased

(b) oil viscosity is to be increased

(c) oil viscosity index is to be increased

(d) oil viscosity index is to be decreased

[1997 : 1 Mark]

8. The dynamic load capacity of 6306 bearing is 22 kN. The maximum radial load it can sustain to operate at 600 rev/min, for 2000 hours is

(a) 4.16 kN (b) 3.60 kN

(c) 6.25 kN (d)5.29 kN

[1997 : 2 Marks]

9. The basic load rating of a ball bearing is

(a) the maximum static radial load that can be applied without causing any plastic deformation of bearing components

(b) the radial load at which 90% of the group of apparently identical bearings run for on million revolutions before the first evidence of failure

(c) the maximum radial load that can be applied during operation without any plastic deformation of bearing components

(d) a combination of radial and axial loads that can be applied without any plastic deformation

[2005 : 1 Mark]

10. Which one of the following is criterion in the design of hydrodynamic Journal bearings?

(a) Sommerfield number

(b) Rating life

(c) Specific dynamic capacity

(d) Rotation factor

[2005 : 1 Mark]

11. A ball bearing operating at a load F has 8000 hours of life. The life of the bearing, in hours, when the load is doubled to 2F is

(a) 8000 (b) 6000

(c) 4000 (d) 1000

[2007 : 1 Mark]

12. A natural feed Journal bearing of diameter 50 mm and length 50 mm operating at 20 revolution/sec. carries a load of 2.0 kN. The lubricant used has a viscosity of 20 mPa.s. The radial clearance is 50 μm. The Sommerfield number for the bearings is

(a) 0.062 (b) 0.125

(c) 0.250 (d) 0.785

[2007 : 2 Marks]

13. A Journal bearing has a shaft diameter of 40 mm and a length of 40 mm. The shaft is rotating at 20 rad/s and the viscosity of the lubricant is 20 mPa.s. The clearance is 0.020 mm. The loss of torque due to the viscosity of the lubricant is approximately

(a) 0.040 Nm (b) 0.252 Nm

(c) 0.400 Nm (d) 0.652 Nm

[2008 : 2 Marks]

14. A lightly loaded full Journal bearing has Journal diameter of 50 mm, bush bore of 50.05 mm and bush length of 20 mm. If rotational speed of Journal is 1200 rpm and average viscosity of liquid lubricant is 0.03 Pa s, the power loss (in W) will be

(a) 37 (b) 74

(c) 118 (d) 237

[2010 : 2 Marks]

15. Two identical ball bearings Pand Q are operating at loads 30 kN and 45 kN respectively. The ratio of the life of bearing P to the life of bearing Q is

(a) 81/16 (b) 27/8

(c) 9/4 (d) 3/2

[2011 : 2 Marks]

16. Ball bearings are rated by a manufacturer for a life of 10^6 revolutions. The catalogue rating of a particular bearing is 16 kN. If the design load is 2 kN, the life of the bearing will be P x 10^6 revolutions, where P is equal to ________

[2014 : 1 Mark, Set-4]

17. A hydrodynamic Journal bearing is subjected to 2000 N load at a rotational speed of 2000 rpm. Both bearing bore diameter and length are 40 mm. If radial clearance is 20 mm and bearing is lubricated with an oil having viscosity 0.03 Pa.s, the Sommerfeld number of the bearing is ________.

[2014 ; 2 Marks, Set-1]

18. For ball bearings, the fatigue life L measured in number of revolutions and the radial load P are related by $FL^{1/3} = K$, where K is a constant. It withstands a radial load of 2 kN for a life of 540 million revolutions. The load (in kN) for a life of one million revolutions is ______.

[2015 : 2 Marks, Set-3]

19. Which of the bearings given below SHOULD NOT be subjected to a thrust load?

(a) Deep groove ball bearing

(b) Angular contact ball bearing

(c) Cylindrical (straight) roller bearing

(d) Single row tapered roller bearing

[2016 : 1 Mark, Set-3]

20. A self-aligning ball bearing has a basic dynamic load rating (C_{10}, for 10^6 revolutions) of 35 kN. If the equivalent radial load on the bearing is 45 kN, the expected life (in 10^6 revolutions) is

(a) below 0.5 (b) 0.5 to 0.8

(c) 0.8 to 1.0 (d) above 1.0

[2018 : 2 Marks, Set-1]

Shaft Design

21. A solid circular shaft needs to be designed to transmit a torque of 50 N.m. If the allowable shear stress of the material is 140 MPa, assuming a factor of safety of 2, minimum allowable design diameter in mm is

(a) 8 (b) 16

(c) 24 (d) 34

[2012 : 2 Marks]

Keys

22. Square key of side d/4 each and length l is used to transmit torque T from the shaft of diameter d to the hub of a pulley. Assuming the length of the key to be equal to the thickness of the pulley, the average shear stress developed in the key is given by

(a) $\frac{4T}{ld}$ (b) $\frac{16T}{ld^2}$

(c) $\frac{8T}{ld^2}$ (d) $\frac{16T}{\pi d^3}$

[2003 : 2 Marks]

ANSWERS

1. (b) **2.** (a) **4.** (a) **5.** (d) **6.** (c) **7.** (b) **8.** (d) **9.** (b) **10.** (a) **11.** (d)

12. (b) **13.** (a) **14.** (a) **15.** (b) **19.** (c) **20.** (a) **21.** (b) **22.** (c)

EXPLANATIONS

1. $L(P)^3$ = constant

$$\frac{L_1}{L_2} = \left(\frac{P_2}{P_1}\right)^3$$

$$\Rightarrow \frac{1000 \times 60 \times 3000}{2000 \times 60 \times t_2} = \left(\frac{4900}{9800}\right)^3$$

t_2 = 12000 hours

2. $LP^3 = C$

$$\Rightarrow L_1 P_1^3 = L_2\left(\frac{P_1}{2}\right)^3$$

$\Rightarrow L_2 = 8L_1$.

3. a – q, b – r, c – s, d – t

4. It is due to thick fluid friction from the start

6. $f \propto \frac{\mu N}{P}$

7.

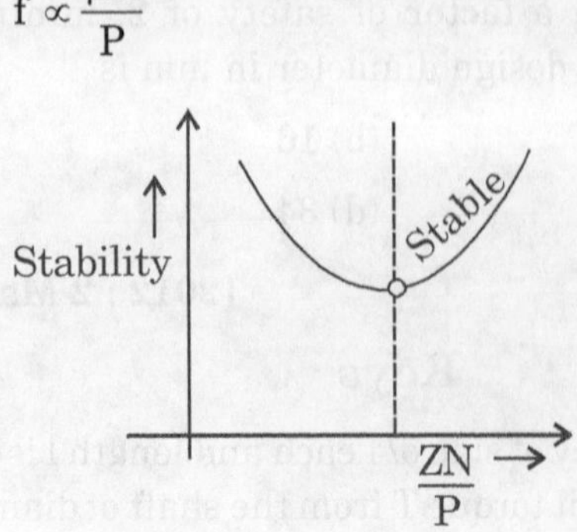

For stability $\left(\frac{ZN}{P}\right) \uparrow$ & $Z \uparrow$ & $P \downarrow$

8. Number of revolutions in life = 2000 × 60 × 600

= 72×10^5 revolutions

∴ L = 72

$$\text{Maximum radial load} = \frac{22}{\sqrt[3]{L}} = \frac{22}{\sqrt[3]{72}}$$

= 5.29 kN

10. Sommerfeld Number, also known as Bearing Characteristic Number,

$$S = \frac{Zn}{P} \cdot \left(\frac{D}{C_d}\right)^2$$

11. Life $\propto \frac{1}{(\text{Load})^3}$

$$\therefore \quad \text{New life} = \frac{8000}{(2)^3} = 1000 \text{ hours}$$

12. $S = \left(\frac{r}{c}\right)^2 \times \frac{\mu N}{P}$

$$P = \frac{P}{d \times l} = \frac{2000}{50 \times 50} = 0.8 \text{N/mm}^2$$

$$S = \left(\frac{25}{50 \times 10^{-3}}\right)^2 \times \frac{20 \times 20 \times 10^{-3}}{0.8 \times 10^6} = .125$$

13. $\tau = \frac{\mu v}{c} = \frac{20 \times 10^{-3} \times 20 \times 20 \times 10^{-3}}{.02 \times 10^{-3}}$

= 400 N/m².

Force = dA = 400 × 3.14 × .04 × .04

Torque = Force × .02

= .040 N-m

14. Given : d_j = 50 mm = 0.05 m

d_B = 50.05 = 0.05005 m

l_B = 20 mm = 0.02 m

N = 1200 rpm

μ = 0.03 Pa–sec

Frictional factor, f = Shear stress × Area

$$= \frac{\mu U}{h} \times \pi d_j l_B$$

Here, $h = r_B - r_j$

$= 25.025 - 25 = 0.025 = 0.000025\text{m}$

$$U = \frac{\pi d_j N}{60} = \frac{\pi \times 0.05 \times 1200}{60} = 3.141$$

$$\therefore \quad f = \frac{0.03 \times 3.141}{0.000025} \times \pi(0.05)(0.02)$$

$= 11.841$

Torque, $T = f \times r_j$

$= 11.841 \times 0.025$

$= 0.2960$

Power loss $= T \times \omega$

$$= \frac{2\pi NT}{60}$$

$$= \frac{2\pi \times 1200 \times 0.2960}{60}$$

$= 37.196 \text{ W} \approx 37.2 \text{ W}$

15. For ball bearing

$$P.(L)^{1/3} = C$$

Given: $P_P = 30$ kN

$P_2 = 45$ kN

$$\frac{L_P}{L_2} = \left(\frac{P_2}{P_P}\right)^3$$

$$= \left(\frac{45}{30}\right)^3 = \left(\frac{3}{2}\right)^3 = \frac{27}{8}$$

16. $L_{10} = \left(\frac{C}{W}\right)^3$

where

L_{10} = Basic life rating in millions of revolutions

C = Basic dynamic load rating

P = Equivalent dynamic radial load

$$\therefore L_{10} = \left(\frac{16}{2}\right)^3 = \frac{5}{2}$$

17. $S = \left(\frac{r}{c}\right)^2 \frac{\mu D_s}{P}$

$d = 40$ mm $\Rightarrow r = 20$ mm

$c = 20\ \mu\text{m}$ $\Rightarrow c = 20 \times 10^{-3}$ mm

$\mu = 0.03$ PaS $\Rightarrow \mu = 0.03 \times 10^{-6}$ MPa.S

$D_s = 2000$ rpm $\Rightarrow D_s = \frac{2000}{60}$ rps

$$P = \frac{p}{1 \times d} = \frac{2000}{40 \times 40} = 1.25 \text{ MPa}$$

$$S = \left(\frac{20}{20 \times 10^{-3}}\right)^2 \times \frac{0.03 \times 10^{-6} \times \frac{2000}{60}}{1.25} = 0.8.$$

18. $FL^{\frac{1}{3}} = k$

$$2 \times (450)^{\frac{1}{3}} = F \times (1)^{\frac{1}{3}}$$

$F = 16.2865$ kN

20. Life (in Million Revolution) $= \left(\frac{C}{P_e}\right)^q$

$q \rightarrow 3$ = ball bearing

$q \rightarrow \frac{10}{3} \Rightarrow$ Roller bearing

$$\boxed{\text{Life} = \left(\frac{35}{45}\right)^3 = 0.4705 \textbf{ Million Revolution}}$$

C → Dynamic load capacity = 35 KN

P_e equivalent load = 45 KN

21. Here, $T = 50\text{N} - m$

$\tau_{allowable} = 140$ MPa

FOS = 2;

$$\tau_{safe} = \frac{\tau_{allowable}}{FOS} = 70 \text{ MPa}$$

We know, $\frac{T}{Z_p} = \tau_{safe}$

$$Z_P = \frac{\pi d^3}{16}$$

$$\Rightarrow \quad d^3 = \frac{16\,T}{\pi\,\tau_{safe}}$$

$$\Rightarrow \quad d = \sqrt[3]{\frac{16 \times 50 \times 10^3}{\pi \times 70}}$$

$$= 15.38 \text{ mm} \approx 16 \text{ mm}$$

22. Force of the shaft circumference is given by

$$P = \frac{T}{d/2}$$

$$\Rightarrow \quad P = \frac{2T}{d}$$

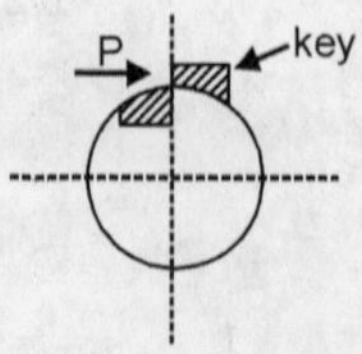

Shearing area = width × length

$$= \frac{d}{4} \times l = \frac{ld}{4}$$

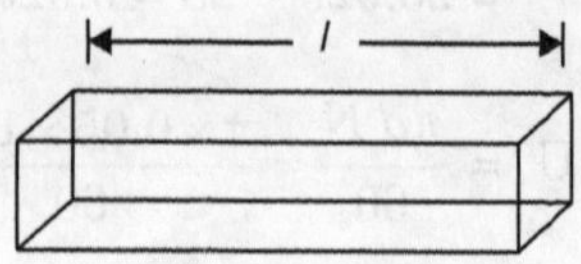

$$\text{Average shear stress} = \frac{P}{\text{Shearing area}}$$

$$= \frac{\frac{2T}{d}}{\frac{ld}{4}} = \frac{8T}{ld^2}$$

5 CHAPTER Gears

Analysis of Previous GATE Papers

Topic	Year/Set → Marks Distribution ↓		2019 Set 1	2019 Set 2	2018 Set 1	2018 Set 2	2017 Set 1	2017 Set 2	2016 Set 1	2016 Set 2	2016 Set 3	2015 Set 1	2015 Set 2	2015 Set 3
Classification	1 Mark	MCQ Type												
		Numerical Type												
	2 Marks	MCQ Type												
		Numerical Type												
	Total		0						0				0	
Standard systems of Gear teeth	1 Mark	MCQ Type												
		Numerical Type												
	2 Marks	MCQ Type												
		Numerical Type												
	Total		0						0				0	
Dynamic Loading in Gears	1 Mark	MCQ Type												
		Numerical Type												
	2 Marks	MCQ Type												
		Numerical Type												
	Total		0						0				0	

Classification

1. Match the Following

List-I (Gear Types)	**List-II (Applications)**
A. Worm gears	1. Parallel shafts
B. Cross helical gears	2. Nonparallel, intersecting shafts
C. Bevel gears	3. Nonparallel, non-intersecting shafts
D. Spur gears	4. Large speed ratios

Codes:

	A	B	C	D
(a)	2	3	4	1
(b)	4	3	2	1
(c)	3	4	2	1
(d)	1	3	4	2

[1995 : 1 Mark]

2. Match the following

Type of gears	**Arrangement of shafts**
P. Bevel gears	1. Nonparallel offset shafts
Q. Worm gears	2. Nonparallel intersecting shafts
R. Herringbone gears	3. Nonparallel, non-intersecting shafts
S. Hypoid gears	4. Parallel shafts

(a) P – 4, Q – 2, R –1, S – 3

(b) P – 2, Q – 3, R – 4, S –1

(c) P – 3, Q – 2, R – 1, S – 4

(d) P– 1, Q – 3, R – 4, S – 2

[2004 : 2 Marks]

3. Match the type of gears with their most appropriate description.

Type of gear	**Description**
P. Helical	1. Axes non parallel and non intersecting
Q. Spiral Bevel	2. Axes parallel and teeth are inclined to the axis
R. Hypoid	3. Axes parallel and teeth are parallel to the axis
S. Rack and pinion	4. Axes are perpendicular and intersecting, and teeth are inclined to the axis
	5. Axes are perpendicular and used for large speed reduction
	6. Axes parallel and one of the gears has infinite radius

(a) P–2, Q–4, R–1, S–6

(b) P–1, Q–4, R–5, S–6

(c) P–2, Q–6, R–4, S–2

(d) P–6, Q–3, R–1, S–5

[2008 : 2 Marks]

4. For the given statements:

I. Mating spur gear teeth is an example of higher pair

II. A revolute Joint is an example of lower pair

Indicate the correct answer.

(a) Both I and II are false

(b) I is true and II is false

(c) I is false and II is true

(d) Both I and II are true

[2014 : 1 Mark, Set-1]

5. Which one of the following is used to convert a rotational motion into a translational motion?

(a) Bevel gears (b) Double helical gears

(c) Worm gears (d) Rack and pinion gears

[2014 : 1 Mark, Set-4]

Sandard systems of gear teeth

6. For full depth of involute spur gears, minimum number of teeth of pinion to avoid interference depends upon

(a) pressure angle (b) speed ratio

(c) circular pitch (d) pitch diameter

[1988 : 1 Mark]

7. Modern gear tooth profile is given involute shape because

(a) this is a very easy curve for manufacturing

(b) sliding does not takes place anywhere on meshing teeth

(c) involute is the only profile that gives conjugate action

(d) change in the centre distance does not change gear ratio

[1989 : 1 Mark]

8. For a pinion of 15 teeth, under cutting ______ (increases/decreases) with _______(increase/ decrease) of pressure angle.

[1991 : 1 Mark]

9. Interference in a pair of gears is avoided, if the addendum circles of both the gears intersect common tangent to the base circles with in the points of tangency.

(a) True

(b) False

[1995 : 1 Mark]

10. In spur gears, the circle on which the involute is generated is called the

(a) pitch circle (b) clearance angle

(c) base circle (d) addendum circle

[1996 : 1 Mark]

11. To make a worm drive reversible, it is necessary to increase

(a) centre distance (b) worm diameter factor

(c) number of starts (d) reduction ratio

[1997 : 1 Mark]

12. A 1.5 kW motor is running at 1440 rev/min. It is to be connected to a stirrer running at 36 rev/min. The gearing arrangement suitable for this application is

(a) differential gear (b) helical gear

(c) spur gear (d) worm gear

[2000 : 1 Mark]

13. Large speed reductions (greater than 20) in one stage of a gear train are possible through

(a) Spur gearing (b) Worm gearing

(c) Bevel gearing (d) Helical gearing

[2002 : 1 Mark]

14. The minimum number of teeth on the pinion to operate without interference in standard full height involute teeth gear mechanism with 20° pressure angle is

(a) 14 (b) 12

(c) 18 (d) 32

[2002 : 1 Mark]

Common Data Questions 15 and 16

The overall gear ratio in a 2 stage speed reduction gear box (with all spur gears) is 12. The input and output shafts of the gear box are collinear. The countershaft which is parallel to the input and output shafts has a gear (Z_2 teeth) and pinion (Z_3 = 15 teeth) to mesh with pinion (Z_1 = 16 teeth) on the input shaft and gear (Z_4 teeth) on the output shaft respectively. It was decided to use a gear ratio of 4 with 3 module in the first stage and 4 module in the second stage.

15. Z_2 and Z_4 are

(a) 64 and 45 (b) 45 and 64

(o) 48 and 60 (d) 60 and 48

[2003 : 2 Marks]

16. The centre distance in the second stage is

(a) 90 mm (b) 120 mm

(c) 160 mm (d) 240 mm

[2003 : 2 Marks]

17. Twenty degree full depth involute profiled 19-teeth pinion and 37-teeth gear are in mesh. If the module is 5 mm, the center distance between the gear pair will be

(a) 140 mm (b) 150 mm

(c) 280 mm (d) 300 mm

[2006 : 2 Marks]

18. One tooth of a gear having 4 module and 32 teeth is shown in the figure. Assume that the gear tooth and the corresponding tooth space make equal intercepts on the pitch circumference. The dimensions a and b, respectively, are closest to.

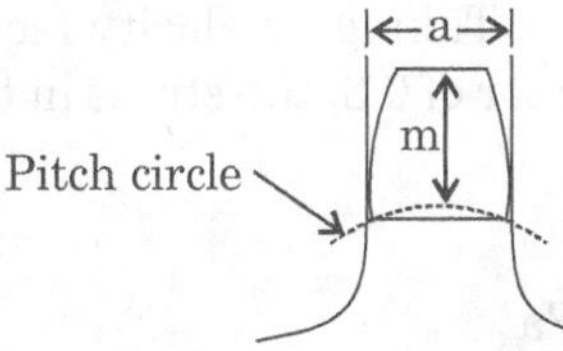

(a) 6.08 mm, 4 mm (b) 6.48 mm, 4.2 mm

(c) 6.28 mm, 4.3 mm (d) 6.28 mm, 4.1 mm

[2008 : 2 Marks]

19. A pair of spur gears with module 5 mm and a center distance of 450 mm is used for a speed reduction of 5 :1. The number of teeth on pinion is ______ .

[2014 : 2 Marks, Set-1]

Dynamic loading in Gears

20. In order to test the efficiency of reducer gear train 1 kW input was given at the input end at a speed of 1440 rpm and at the output end the measured torque was 56.36 Nm. If the ratio of speed reduction in this unit is 10 : 1, the efficiency is about

(a) 78% (b) 85%

(c) 63% (d) 96% **[1989 : 2 Mark]**

21. Two mating spur gears have 40 and 120 teeth respectively. The pinion rotates at 1200 rpm and transmits a torque of 20 N.m. The torque transmitted by gear is

(a) 6.6 Nm (b) 20 Nm

(c) 40 Nm (d) 60 Nm

[2004 : 1 Mark]

Linked Answer Questions 22 and 23:

A gear set has a pinion with 20 teeth and a gear with 40 teeth. The pinion runs at 30 rev/s and transmits a power of 20 kW. The teeth are on the 20° full-depth system and have a module of 5 mm. The length of the line of action is 19 mm.

22. The center distance for the above gear set in mm is

(a) 140 (b) 150

(c) 160 (d) 170 **[2007 : 2 Marks]**

23. The contact ratio of the contacting tooth is

(a) 1.21 (b) 1.25

(c) 1.29 (d) 1.33

[2007 : 2 Marks]

24. The resultant force on the contacting gear tooth is N is

(a) 77.23 (b) 212.20

(c) 2258.1 (d) 289.43

[2007 : 2 Marks]

25. A spur gear has a module of 3 mm, number of teeth 16, a face width of 36 mm and a pressure angle of 20°. It is transmitting a power of 3 kW at 20 rev/s. Taking a velocity factor of 1.5 and a form factor of 0.3, the stress in the gear tooth is about

(a) 32 MPa

(b) 46 MPa

(c) 58 MPa

(d) 70 MPa

[2008 : 2 Marks]

Linked Answer Questions 26 and 27

A 20° full depth involute spur pinion of 4 mm module and 21 teeth is to transmit 15 kW at 960 rpm. Its face width is 25 mm.

26. The tangential force transmitted (in N) is

(a) 3552 (b) 2611

(c) 1776 (d) 1305

[2009 : 2 Marks]

27. Given that the tooth geometry factor is 0.32 and the combined effect of dynamic load and allied factors intensifying the stress is 1.5; the minimum allowable stress (in MPa) for the gear material is

(a) 242.0 (b) 166.5

(c) 121.0 (d) 74.0

[2009 : 2 Marks]

28. A spur pinion of pitch diameter 50 mm rotates at 200 rad/s and transmits 3 kW power. The pressure angle of the tooth of the pinion is 20°. Assuming that only one pair of the teeth is in contact, the total force (in Newton) exerted by a tooth of the pinion on the tooth on a mating gear is ______.

[2014 : 2 Marks, Set-2]

ANSWERS

1. (b)	**2.** (b)	**3.** (a)	**4.** (d)	**5.** (d)	**6.** (a)	**7.** (d)	**9.** (a)	**10.** (c)	**11.** (c)
12. (d)	**13.** (b)	**14.** (c)	**15.** (a)	**16.** (b)	**17.** (a)	**18.** (d)	**20.** (b)	**21.** (d)	**22.** (b)
23. (c)	**24.** (c)	**25.** (b)	**26.** (a)	**27.** (b)					

EXPLANATIONS

3.

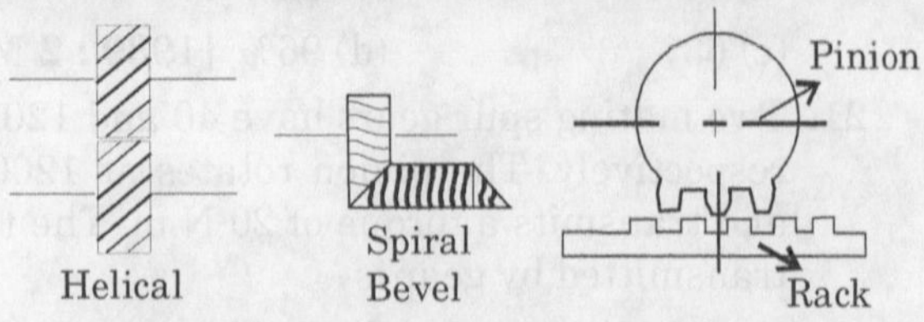

4. Since higher pair has a line or point contact and lower pair has a surface of Area contact. Hence both are true.

i.e., spur gear has line contact (Higher pair) and Revolute joint has surface contact (lower pair).

8. Interference decreases with increase in pressure

12. Velocity reduction $= \frac{36}{1440} = \frac{1}{40}$

For this kind of reduction, worm gear arrangement is most suitable.

13. Worm gearing is used for reduction ratios greater than 20.

14. Z_{min} = Minimum number of teeth on pinion to avoid interference.

$$Z_{min} = \frac{2 \times A_W}{\sin^2 \phi} = \frac{2 \times 1}{\sin^2 20°} = 17.09 \simeq 18 \text{ teeth}$$

15. Given, $\frac{N_1}{N_4} = 12$, $\frac{N_1}{N_2} = 4 = \frac{D_2}{D_1}$

$m_1 = 3$, $m_2 = 4$

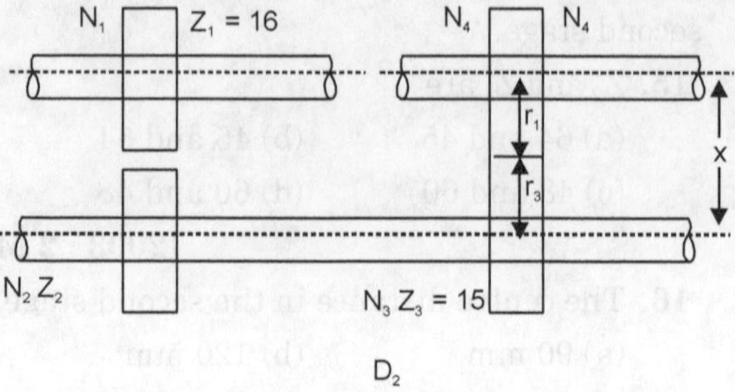

Now, $\dfrac{D_1}{Z_1} = \dfrac{D_2}{Z_2}$

$\Rightarrow \dfrac{Z_1}{Z_2} = \dfrac{D_1}{D_2} = \dfrac{N_2}{N_1} = \dfrac{1}{4}$

$\Rightarrow Z_2 = Z_1 \times 4 = 64$

and $\dfrac{N_1}{N_4} = \dfrac{D_2 \times D_4}{D_1 \times D_3}$

$\Rightarrow 12 = \dfrac{D_4}{D_3}$

$\Rightarrow \dfrac{D_4}{D_3} = 3$

Also, $\dfrac{Z_3}{Z_4} = \dfrac{D_3}{D_4}$

$\Rightarrow Z_4 = Z_3 \cdot \dfrac{D_4}{D_3} = Z_3 \times 3 = 15 \times 3 = 45$

16. Now, $x = r_4 + r_3 = \dfrac{D_4 + D_3}{2}$

But $\dfrac{D_4}{Z_4} = \dfrac{D_3}{Z_3} = 4$

$\Rightarrow D_4 = 180,\ D_3 = 60$

$\therefore\ x = \dfrac{180 + 60}{2} = 120 \text{ mm}$

17. Centre distance $= \dfrac{D_1 + D_2}{2} = \dfrac{mT_1 + mT_2}{2}$

$= \dfrac{5(19+37)}{2} = 140 \text{ mm}$

18. $a \simeq \dfrac{\pi D}{2Z} \ldots (D = mZ)$

$a \simeq \dfrac{3.14 \times m}{2} = 6.28$

$\Rightarrow$ b = addendum + PR

The addendum = module for standard addendum

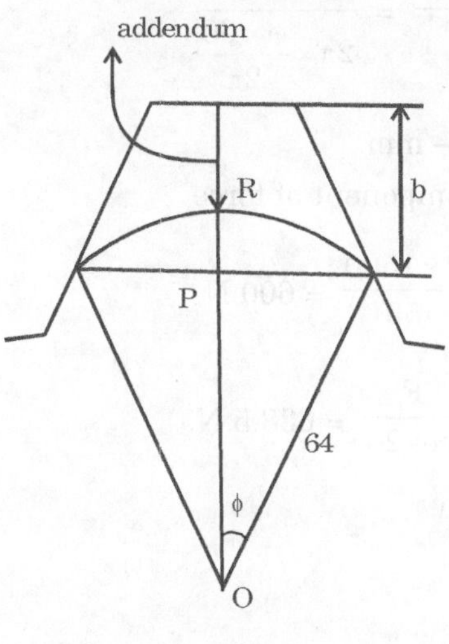

Now $\phi = \dfrac{360°}{32 \times 4} = 2.8°$

$\therefore$ OP = 64 cos ϕ = 63.9

$\Rightarrow$ b = m + PR

= 4 + (64 – 63.9) = 4.1 mm

19. Given speed Ratio = 5 : 1

$\dfrac{5}{1} = \dfrac{T_2}{T_1} = \dfrac{d_2}{d_1}$

$\Rightarrow d_2 = 5d_1$

centre distance $= \dfrac{d_1 + d_2}{2} = 450$

$\Rightarrow d_1 + d_2 = 900$

$\Rightarrow 5d_1 + d_1 = 900$

$\Rightarrow d_1 = 150$

$m = \dfrac{d_1}{T_1} \Rightarrow T_1 = \dfrac{150}{5} = 30$.

20. Speed reduction = 10 : 1

Output speed $= \dfrac{1440}{10} = 144$ rpm

Output Torque $= T.\omega = 56.36 \times \dfrac{2\pi \times 144}{60}$

= 849.88 N – m

% efficiency $= \dfrac{\text{Output}}{\text{Input}} \times 100$

$= \dfrac{849.88}{1000} \times 100 = 85\%$ (Approximately)

21. We know

$\dfrac{N_p}{N_G} = \dfrac{T_G}{T_p}$

where, N_p = speed of pinion,

N_G = speed of gear wheel

T_G = number of teeth of gear,

T_p = number of teeth of pinion

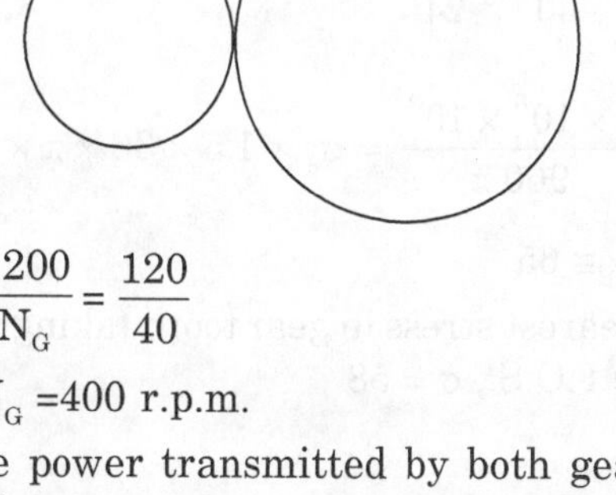

$\therefore\ \dfrac{1200}{N_G} = \dfrac{120}{40}$

or N_G = 400 r.p.m.

Since power transmitted by both gear will be equal

i.e. $T_p\omega_p = T_G\omega_G$

where,

T_p = torque transmited by pinion

T_G = torque transmitted by gear wheel

$$\therefore \frac{20 \times 2\pi \times 1200}{60} = \frac{T_G \times 2\pi \times 400}{60}$$

$\therefore$ Torque transmitted by gear, T_G = 60 N-m.

22. Centre distance

$$= \frac{m}{2}(T_p + T_g) = \frac{5(20+40)}{2} = 150 \text{ mm}$$

23. Line of action (LOA) = 19 mm

$$\text{Arc of contact (AOC)} = \frac{19}{\cos 20°} = 20.21 \text{ mm}$$

$$\text{Circular pitch} = \frac{\pi d}{z} = \pi m = \pi \times 5 = 5\pi \text{mm}$$

$$\text{Contact ratio} = \frac{\text{LOA}}{\text{Circcular pitch}} = \frac{20.21}{5.\pi} = 1.29$$

24. P = Tω

$\therefore$ 20×10^3 =T.2π × 30

⇒ T =106. 1 Nm

Now, T = F_t. r

$$\text{Here, r} = \frac{mTp}{2} = \frac{5 \times 20}{2} = 50 \text{ nm} = .05 \text{ m}$$

$$F_t = \frac{106.1}{r} = \frac{106.1}{.05} = 2122\text{N}$$

$\therefore$ Resultant force,

$$F = \frac{F_t}{\cos\theta} = \frac{2122}{\cos 20°} = 2258 \text{ N.}$$

25. Lewis equation is given as

$F = \sigma_d C_v b\pi ym$

F can be calculated as $\frac{P}{\omega r}$ ...[ω = 20 × 2π rad /sec]

D = mZ = 3 × 16 = 48mm

C_v = 1.5, b = 36, y = 0.3, m = 3

$$\therefore F = \frac{3 \text{ kW}}{40 \quad 24}$$

$$\Rightarrow \frac{3 \times 10^3 \times 10^3}{960\,\pi} = \sigma_d \times 1.5 \times 36 \times \pi \times 0.3 \times 3$$

⇒ $\sigma_d \cong 65$

⇒ nearest stress in gear tooth taking in account some F.O.S., σ = 58

26. P = 15 kW

m = 4 mm

Z_p = 21

N = 960 rpm

b = 25mm

$D_p = mZ_p = 4 \times 21 = 84$ mm

$$\text{Speed of pinion, } v = \frac{\pi D_P N}{60000} \text{ m/s}$$

$$= \pi \times 84 \times \frac{960}{60000}$$

= 4.22 m/s

Also, P = $F_t v$

$$\text{or } F_t = \frac{P}{v} = \frac{15 \times 1000}{4.22}$$

≈ 3552 N

27. Toot hgeomet r yf act or Y=0.32

f_s =1.5

$$f_t = \frac{\sigma_b.b.Y_m}{f_s}$$

$$\Rightarrow 3552 = \frac{\sigma_b(25\text{mm})(0.32)(4\text{mm})}{1.5}$$

⇒ σ_b = 166.5MPa

28. Diameter= 50 mm = 0.05 m

ω = 200 rad/s

$$\Rightarrow N = \frac{\omega \times 60}{2\pi}$$

α = 20°

P = 3 kW

$$T = \frac{p \times 10^6 \times 60}{2\pi N} = \frac{3 \times 10^6 \times 60}{2\pi \times \frac{\omega \times 60}{2\pi}}$$

T = 15000 N – mm.

Tangential component of force

$$F_t = \frac{2T}{d} = \frac{2 \times 1500}{50} = 600 \text{ N}$$

$$\text{Total force} = \frac{F_t}{\cos 20} = 638.5 \text{ N.}$$

6 CHAPTER Brakes, Clutches and Ropes

Analysis of Previous GATE Papers

Topic	Year/Set → Marks Distribution ↓		2019 Set 1	2019 Set 2	2018 Set 1	2018 Set 2	2017 Set 1	2017 Set 2	2016 Set 1	2016 Set 2	2016 Set 3	2015 Set 1	2015 Set 2	2015 Set 3
Brakes	1 Mark	MCQ Type								1				
		Numerical Type								1				
	2 Marks	MCQ Type												
		Numerical Type	1	1	1									
	Total		2	2	2					2				
Clutches	1 Mark	MCQ Type												
		Numerical Type												
	2 Marks	MCQ Type												
		Numerical Type						1						
	Total							2						
Ropes	1 Mark	MCQ Type												
		Numerical Type												
	2 Marks	MCQ Type												
		Numerical Type												
	Total													

Brakes

1. In a band brake the ratio of tight side band tension to the tension on the slack side is 3. If the angle of overlap of band on the drum is 180°, the coefficient of friction required between drum and the band is

(a) 0.20 (b) 0.25
(c) 0.30 (d) 0.35

[2003 : 2 Marks]

Linked Answer Questions 2 and 3

A band brake consists of a lever attached to one end of the band. The other end of the band is fixed to the ground. The wheel has a radius of 200 mm and the wrap angle of the band is 270°. The braking force applied to the lever is limited to 100 N, and the coefficient of friction between the band and the wheel is 0.5. No other information is given.

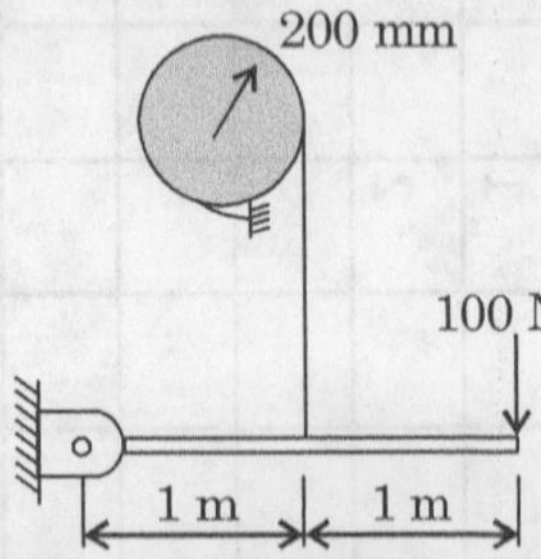

2. The maximum tension that can be generated in the band during braking is

(a) 1200 N (b) 2110 N
(c) 3224 N (d) 4420 N

[2005 : 2 Marks]

3. The maximum wheel torque that can be completely braked is

(a) 200 N.m (b) 382 N.m
(c) 604 N.m (d) 844 N.m

[2005 : 2 Marks]

4. A block-brake shown below has a face width of 300 mm and a mean coefficient of friction of 0.25. For an activating force of 400 N, the braking torque in Nm is

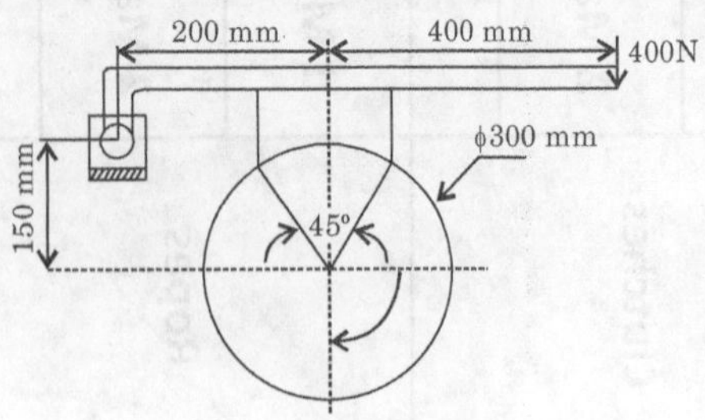

(a) 30 (b) 40
(c) 45 (d) 60

[2007 : 2 Marks]

5. A band brake having bandwidth of 80 mm, drum diameter of 250 mm, coefficient of friction of 0.25 and angle of wrap of 270 degrees is required to exert a friction torque of 1000 N m. The maximum tension (in kN) developed in the band is

(a) 1.88 (b) 3.56
(c) 6.12 (d) 11.56

[2010 : 2 Marks]

6. A force of 400 N is applied to the brake drum of 0.5 m diameter in a band brake system as shown in figure, where the wrapping angle is 180°. If the coefficient of friction between the drum and the band is 0.25, the braking torque applied, in Nm is

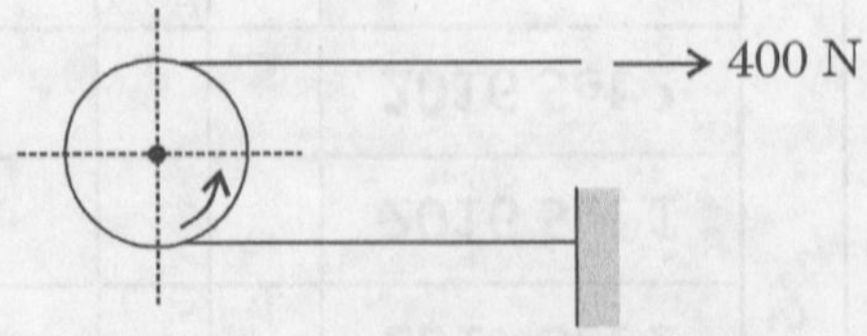

(a) 100.6 (b) 54.4
(c) 22.1 (d) 15.7

[2012 : 2 Marks]

7. A drum brake is shown in the figure. The drum is rotating in anticlockwise direction. The coefficient of friction between drum and shoe is 0.2. The dimensions shown in the figure are in mm. The braking torque (in Nm) for the brake shoe is______.

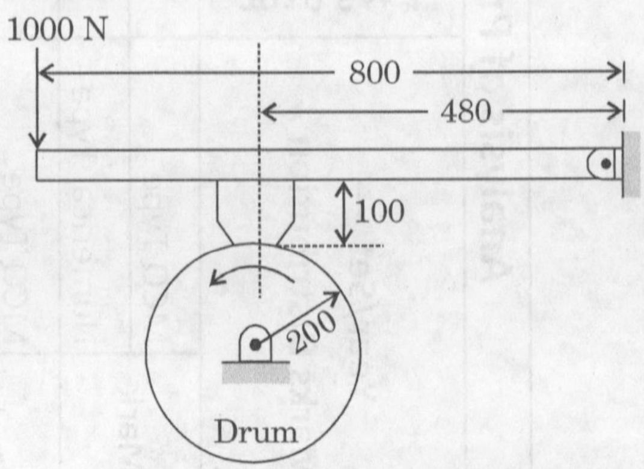

[2014 : 2 Marks, Set-3]

8. A four-wheel vehicle of mass 1000 kg moves uniformly in a straight line with the wheels revolving at 10 rad/s. The wheels are identical, each with a radius of 0.2 m. Then a constant braking torque is applied to all the wheels and the vehicle experience a uniform deceleration. For the vehicle to stop in 10s, the braking torque (in Nm) on each wheel is ______.

[2014 : 2 Marks, Set-3]

9. The forces F_1 and F_2 in a brake band and the direction of rotation of the drum are as shown in the figure. The coefficient of friction is 0.25. The angle of wrap is $3\pi/2$ radians. It is given that $R = 1$ m and $F_2 = 1$ N. The torque (in Nm) exerted on the drum is _______ .

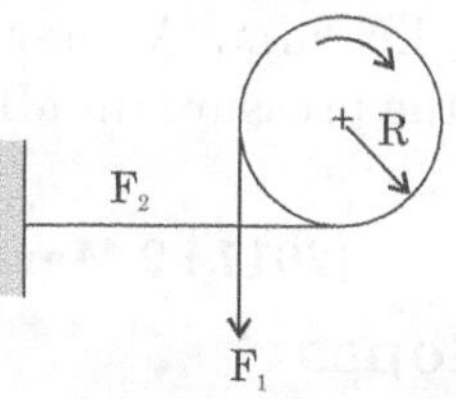

[2016 : 1 Mark, Set-2]

10. For the brake shown in the figure, which one of the following is TRUE?

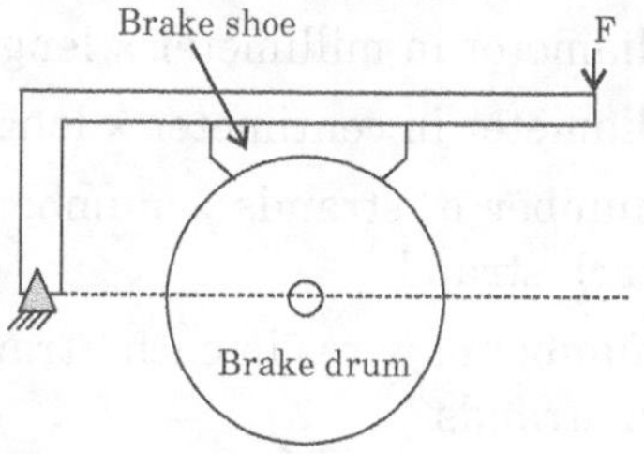

(a) Self energizing for clockwise rotation of the drum
(b) Self energizing for anti-clockwise rotation of the drum
(c) Self energizing for rotation in either direction of the drum
(d) Not of the self energizing type

[2016 : 1 Mark, Set-2]

11. The schematic of an external drum rotating clockwise engaging with a short shoe is shown in the figure. The shoe is mounted at point Y on a rigid lever XYZ hinged at point X. A force F = 100 N is applied at the free end of the lever as shown. Given that the coefficient of friction between the shoe and the drum is 0.3, the braking torque (in Nm) applied on the drum is ____ (correct to two decimal places).

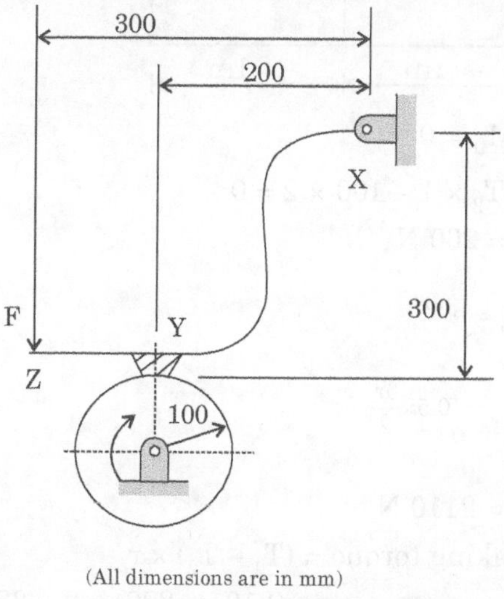

(All dimensions are in mm)

[2018 : 2 Marks, Set-1]

12. A single block brake with a short shoe and torque capacity of 250 N-m is shown. The cylindrical brake drum rotates anticlockwise at 100 rpm and the coefficient of friction is 0.25. The value of a, in mm (round off to one decimal place), such that the maximum actutating force P is 2000 N, is_____.

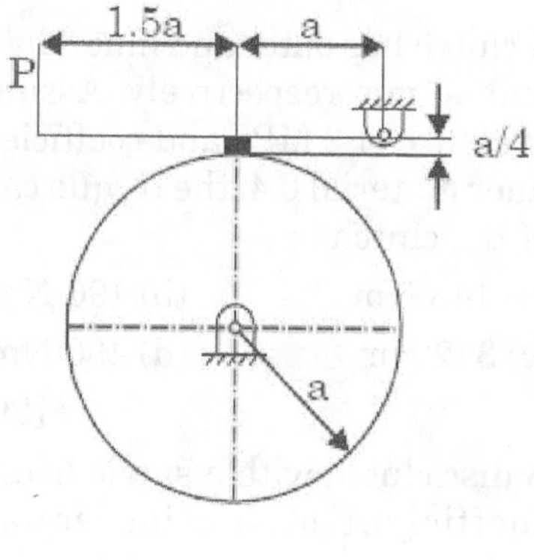

[2019 : 2 Marks, Set-1]

13. A short shoe external drum brake is shown in the figure. The diameter of the brake drum is 500 mm. The dimensions a = 1000 mm, b = 500 mm and c = 200 mm. The coefficient of friction between the drum and the shoe is 0.35. The force applied on the lever F = 100 N as shown in the figure. The drum is rotating anti-clockwise. The braking torque on the drum is ____ N-m (roundoff to two decimal places)

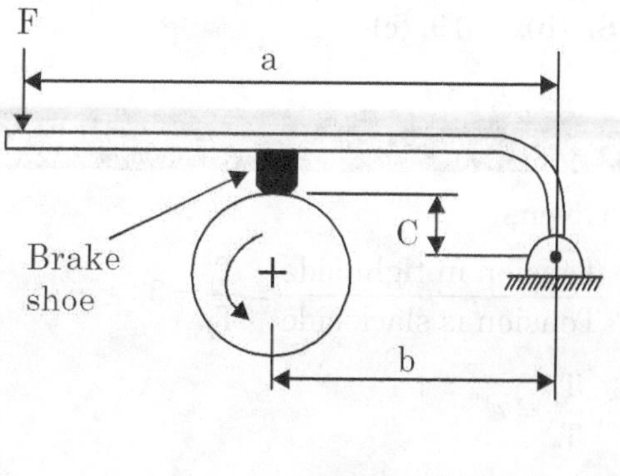

[2019 : 2 Marks, Set-2]

Clutches

14. Axial operation claw clutches having self-locking tooth profile

(a) can be disengaged at any speed
(b) can be disengaged onlyunloaded
(c) can be engaged only when unloaded
(d) can work only with load

[1987 : 1 Mark]

15. A disk clutch is required to transmit 5 kW at 2000 rpm. The disk has a friction lining with coefficient of friction equal to 0.25. Bore radius of friction lining is equal to 25 mm. Assume uniform contact pressure of 1 MPa. The value of outside radius of the friction lining is

(a) 39.4 mm (b) 49.5 mm

(c) 97.9 mm (d) 142.9 mm

[2006 : 2 Marks]

16. A clutch has outer and inner diameters 100 mm and 40 mm respectively. Assuming a uniform pressure of 2 MPa and coefficient of friction of liner material 0.4, the torque carrying capacity of the clutch is

(a) 148 Nm (b) 196 Nm

(c) 372 Nm (d) 490 Nm

[2008 : 2 Marks]

17. A disc clutch with a single friction surface has coefficient of friction equal to 0.3. The maximum pressure which can be imposed on the friction material is 1.5 MPa. The outer diameter of the clutch plate is 200 mm and its internal diameter is 100 mm. Assuming uniform wear theory for the clutch plate, the maximum torque (in Nm) that can be transmitted is ________.

[2014 : 2 Marks, Set-2]

18. A single-plate clutch has a friction disc with inner and outer radii of 20 mm and 40 mm, respectively. The friction lining in the disc is made in such a way that the coefficient of friction μ varies radially as μ = 0.01 r, where r is in mm. The clutch needs to transmit a friction torque of 18.85 kN mm. As per uniform pressure theory, the pressure (in MPa) on the disc is ________.

[2017 : 2 Marks, Set-2]

Ropes

19. A wire rope is designated as 6 x 19 standard hoisting. The numbers 6 × 19 represent

(a) diameter in millimeter × length in meter

(b) diameter in centimeter × length in meter

(c) number of strands × number of wires in each strand

(d) number of wires in each Strand × number of strands

[2003 : 1 Mark]

ANSWERS

1. (d) **2.** (b) **3.** (b) **4.** (c) **5.** (d) **6.** (b) **8.** (10) **10.** (a) **14.** (c) **15.** (a)
16. (b) **19.** (c)

EXPLANATIONS

1. Given,

$$\frac{\text{Tension in tight side}}{\text{Tension is slack side}} = \frac{T_1}{T_2} = 3$$

$$\frac{T_1}{T_2} = e^{\mu\theta}$$

where, θ = angle of overlap = 180 = π radians

$\therefore\ 3 = e^{\mu\theta}$

$\ln(3) = \mu\pi.$

$\therefore\ \mu = \frac{1}{\pi}\ln(3) = 0.35$

2. θ = 270°

μ = 0.5

Taking moment about 0,

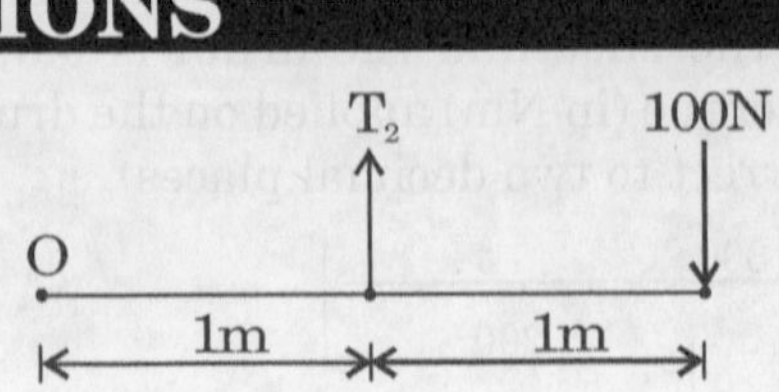

$\Sigma M_0 = 0$

$\Rightarrow T_2 \times 1 - 100 \times 2 = 0$

$T_2 = 200$ N

$$\frac{T_1}{T_2} = e^{\mu\theta}$$

$$\frac{T_1}{T_2} = e^{0.5\times\frac{3\pi}{2}}$$

$T_1 = 2110$ N

3. Braking torque = $(T_1 - T_2) \times r$

= (2101 – 200) × 2 = 382 N–m

4. Taking moment about A,

$R_N \times 200 = 400 \times 600$

$\Rightarrow R_N = 1200$ N

$\therefore$ Braking torque

$T_B = \mu \cdot R_N \times D/2 = 45\,\text{Nm}$

5. Given : $b = 80$ mm $= 0.08$ m

D = 250 mm = 0.25 m

$R = \frac{D}{2}$

$\mu = 0.25$

$\theta = 270° = 270 \times \frac{\pi}{180} = 4.712$ radians

$F_t = 1000$ N-m.

As $\frac{p_1}{p_2} = e^{\mu\theta}$

$\Rightarrow p_1 = e^{\mu\theta} \cdot p_2 = e^{0.25 \times 4.712} \cdot p_2 = 3.25 p_2$

Now $F_t = (p_1 - p_2)$ R

$\therefore 1000 = \left(p_1 - \frac{p_1}{3.25}\right) 0.125$

Solving, we get maximum tension in the belt,

$p_1 = 11.56$ kN

6.

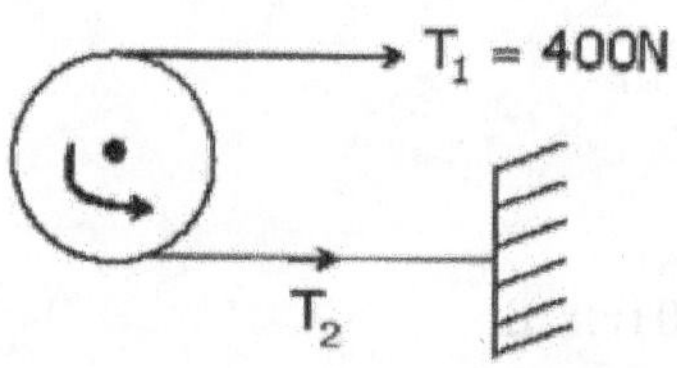

As the drum is rotating in anti clock wise direction, T_1 will be right side & T_2 will be clock side.

$\frac{T_1}{T_2} = e^{\mu\theta} = e^{0.25 \times \pi} = 2.19$

$\Rightarrow T_2 = 182.375$ N

Braking torque $= (T_1 - T_2)r$

$= 54.4$ N – m

7. Taking moments about pin joint (SM)

$\Rightarrow 1000 \times 0.8 = R_N \times 0.48 + \mu R_N \times 0.1$

$\Rightarrow R_N = \frac{800}{0.5} = 1600$

$T_B = \mu\, R_N \times r$

$= 0.2 \times 1600 \times \frac{200}{1000} = 64$ N – *m*.

8. m = 1000 kg, $\omega = 10$ Rad/sec, R = 0.2 m, $t = 10$ sec when $\omega = 0$

$\therefore \omega_F = \omega_i - \alpha t$

$\therefore 0 = 10 - \alpha \cdot 10$

$\therefore \alpha = 1\ \text{rad}/\text{sec}^2$

Now $T = I \cdot \alpha$

$\therefore (T)_{\text{each wheel}} = \left(\frac{1000}{4}\right) \times 0.2^2 \times 1$

$\therefore (T)_{\text{each wheel}} = 10$ N – m

9. $\frac{F_1}{F_2} = e^{\mu\theta} = e^{0.25 \times \frac{3\pi}{2}} = 3.248$

$\therefore \quad F_1 = 3.248$N

Torque, $\quad T = (F_1 - F_2) \times r$

$= (3.248 - 1) \times 1 = 2.248$ N-m

10.

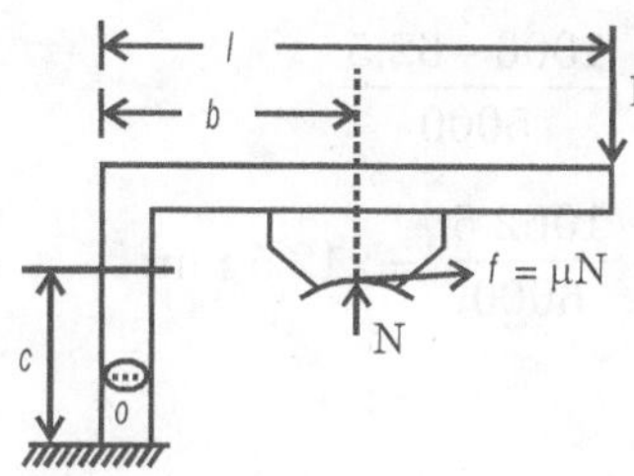

Taking moment about hinge in clockwise

$N_b - Fl - \mu Nc = 0$

$F = \frac{N(b - \mu c)}{l}$

Self energizing if $b > \mu c$ so for clockwise rotation of drum the brake is self energizing.

11.

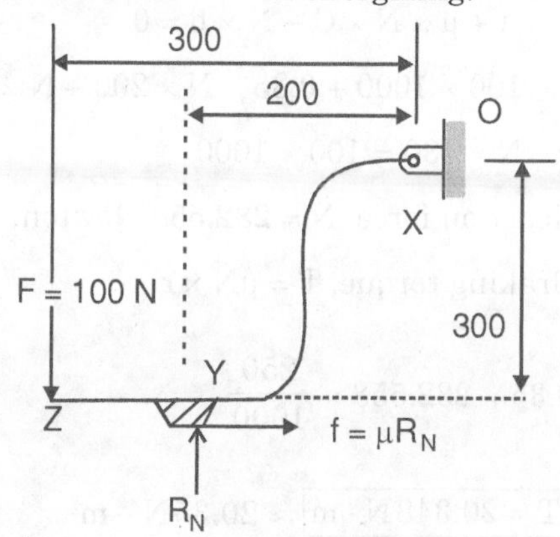

$\Sigma M_0 = 0$

$F \times 300 \curvearrowright + f \times 300 \curvearrowright = RN \times 200 \curvearrowleft$

$100 \times 300 + \mu R_N \times 300 = R_N \times 200$

$100 \times 300 + 0.3 \times R_N \times 300 = R_N \times 200$

$300 = 1.1\, R_N$

$R_N = 272.72$ N

Braking Torque $= \mu R_N \times R$

$= 0.3 \times 272.72 \times 0.100 = 8.18$ Nm

12. $T = 250$ Nm

$F \times r = 250$

$\Rightarrow \quad F = \dfrac{250}{a}$

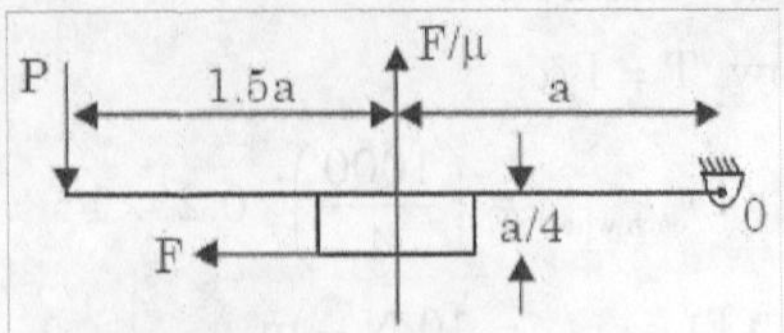

$\Sigma M_0 = 0$

$P \times 2.5a = F \times \dfrac{a}{4} + \dfrac{F}{\mu} \times a$

$2000 \times 2.5a = \dfrac{250 \times a}{a \times 0.25} + \dfrac{250}{a} \times \dfrac{a}{4}$

$a = \dfrac{1000 + 62.5}{5000}$

$= \dfrac{1062.5}{5000} = 212.5$ mm

13. 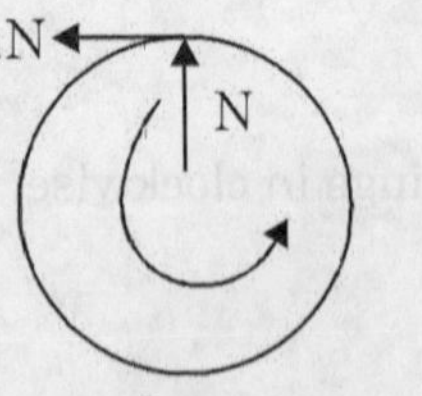

drum

Taking moment about hinge point,

$F \times a + \mu \times N \times C - N \times b = 0$

$\Rightarrow 100 \times 1000 + 0.35 \times N \times 200 - N \times 500 = 0$

$\Rightarrow N \times 430 = 100 \times 1000$

Reaction force, N = 232.558 Newton.

Braking torque, $T = \mu N \times r$

$0.35 \times 232.558 \times \dfrac{250}{1000}$

$\boxed{T = 20.348\,\text{N - m}} \approx 20.35\,\text{N} - \text{m}$

15. Torque, $T = \dfrac{P \times 60}{2\pi \times N} = 23.87$ Nm

= Axial thrust,

$W = P \times \pi \left(r_1^2 - r_2^2\right)$

But $T = \dfrac{2}{3}\mu \times P \times \pi (r_1^2 - r_2^2)$

$\dfrac{(r_1^3 - r_2^3)}{(r_1^2 - r_2^2)} = \mu w r$

$\therefore r_2 = 39.4$ mm

16. $T = \int r.\ \mu.\ p.\ 2\pi r.\ dr$

$= \int r.\ 0.4.\ 2\ .\ 2\pi r.\ dr$

$= 1.6\,\pi \displaystyle\int_{20}^{50} r^2\ .\ dr = \dfrac{1.6\,\pi}{3}\dfrac{\left[50^3 - 20^3\right]}{10^3} = 196$ Nm

17. As per uniform wear theory

$\text{Torque} = \dfrac{\pi \mu P_a d}{8}\left[D^2 - d^2\right]$

$= \dfrac{\pi \times 0.3 \times 1.5 \times 10^6}{8}\left[0.2^2 - 0.1^2\right] = 530.14$ N – m

18. $T_f = 2\pi P \displaystyle\int_{R_i}^{R_0} \mu r^2 dr$

$T_f = 2\pi P \displaystyle\int_{R_i}^{R_0} (0.01r) r^2 dr$

$T_f = (2\pi p)(0.01)\left[\dfrac{R_0^4 - R_i^4}{4}\right]$

$18.85 \times 10^3 = 2\pi P(0.01)\left[\dfrac{40^4 - 20^4}{4}\right]$

$P = 0.5$ MPa

Unit - VI

Fluid Mechanics and Hydraulic Machinery

Syllabus

Fluid properties; fluid statics, manometry, buoyancy, forces on submerged bodies, stability of floating bodies; control-volume analysis of mass, momentum and energy; fluid acceleration; differential equations of continuity and momentum; Bernoulli's equation; dimensional analysis; viscous flow of incompressible fluids, boundary layer, elementary turbulent flow, flow through pipes, head losses in pipes, bends and fittings.

Turbomachinery: Impulse and reaction principles, velocity diagrams, Pelton-wheel, Francis and Kaplan turbines.

Contents

1 CHAPTER Fluid Properties

Analysis of Previous GATE Papers

Topic	Year/Set → Marks Distribution ↓		2019 Set 1	2019 Set 2	2018 Set 1	2018 Set 2	2017 Set 1	2017 Set 2	2016 Set 1	2016 Set 2	2016 Set 3	2015 Set 1	2015 Set 2	2015 Set 3
Basic Properties	1 Mark	MCQ Type								1				
		Numerical Type												
	2 Marks	MCQ Type												
		Numerical Type												
	Total									1				
Viscosity	1 Mark	MCQ Type												
		Numerical Type												
	2 Marks	MCQ Type												
		Numerical Type	1											
	Total		2											

Basics Properties

1. A fluid is said to be Newtonian fluid when the shear stress is

(a) directly proportional to the velocity gradient
(b) inversely proportional to the velocity gradient
(c) independent of the velocity gradient
(d) none of these

[1995 : 1 Mark]

2. A fluid is one which can be defined as a substance that

(a) has that same shear stress at all points
(b) can deform indefinitely under the action of the smallest shear force
(c) has the small shear stress in all directions
(d) is practically incompressible

[1996 : 1 Mark]

3. The dimension of surface tension is

(a) ML^{-1} (b) L^2T^{-1}
(c) $ML^{-1}T^{-1}$ (d) MT^{-2}

[1996 : 1 Mark]

4. If P is the gauge pressure within a spherical droplet, then gauge pressure within a bubble of the same fluid and of same size will be

(a) P/4 (b) P/2
(c) P (d) 2P

[1999 : 1 Mark]

5. For a Newtonian fluid

(a) shear stress is proportional to shear strain
(b) rate of shear stress is proportional to shear strain
(c) shear stress is proportional to rate of shear strain
(d) rate of shear stress is proportional to rate of shear strain

[2006 : 1 Mark]

6. Assuming constant temperature condition and air to be an ideal gas, the variation in atmospheric pressure with height calculated from fluid statics is

(a) linear (b) exponential
(c) quadratic (d) cubic

[2016 : 1 Mark, Set-2]

Viscosity

7. Kinematic viscosity of air at 20°C is given to be 1.6×10^{-5} m²/s. Its kinematic viscosity at 70°C will be varying approximately

(a) 2.2×10^{-5} m²/s (b) 1.6×10^{-5} m²/s
(c) 1.2×10^{-5} m²/s (d) 3.2×10^{-5} m²/s

[1999 : 1 Mark]

8. The SI unit of kinematic viscosity ν is

(a) m²/s (b) kg/ms
(c) m/s² (d) m³/s²

[2001 : 1 Mark]

9. Two immiscible, incompressible, viscous fluids having same densities but different viscosities are contained between two infinite horizontal parallel plates. 2 m apart as shown below. The bottom plate is fixed and the upper plate move to the right with a constant velocity of 3 m/s. With the assumptions of Newtonian fluid steady, and fully developed laminar flow with zero pressure gradient in all directions, the momentum equations simplify to $\frac{d^2u}{dy^2} = 0$

If the dynamic viscosity of the lower fluid, μ_2 is twice that of the input fluid, μ_1, then the velocity at the interface (round off to two decimal places) is_____m/s.

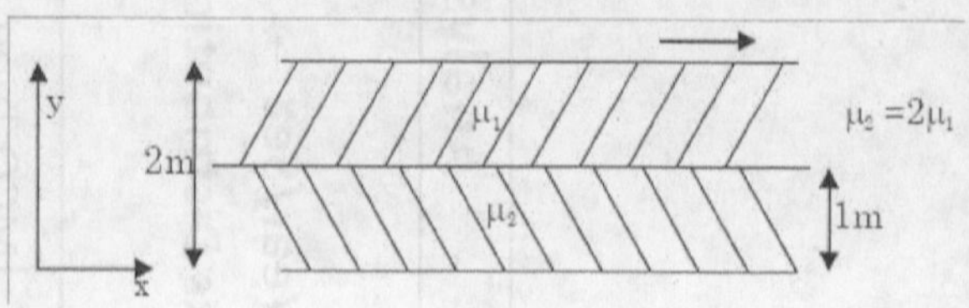

[2019 : 2 Marks, Set 1]

ANSWERS

1. (a) **2.** (b) **3.** (d) **4.** (d) **5.** (c) **6** (b) **7** (a) **8.** (a)

EXPLANATIONS

1. Newtonian fluid obeys Newton's law of viscosity.

3. Surface tension (σ)

$$\sigma = \frac{F}{L} = \frac{N}{m} = \frac{kg - m}{s^2 \times m} = MT^{-2}$$

4. P = Gauge pressure within a spherical droplet

$$P_{droplet} = \frac{4\sigma}{d} \text{ for droplet}$$

$$P_{bubble} = \frac{8\sigma}{d} \text{ for bubble}$$

$$P_{bubble} = 2\ P_{droplet}$$

5. For a Newtonian fluid, shear stress is proportional to rate of shear strain.

7. Viscosity of gas increases with increasing temperature.

8. Kinematic Viscosity:

$$\upsilon = \frac{\text{Dynamic viscosity}}{\text{Density}} = \frac{\mu}{\rho}$$

$$\upsilon = \frac{\mu}{\rho} = \frac{Ns/m^2}{kg/m^3}$$

$$= \frac{kg.m}{s^2} \times \frac{s}{m^2} \times \frac{1}{kg/m^3}$$

$$= m^2/s$$

9.

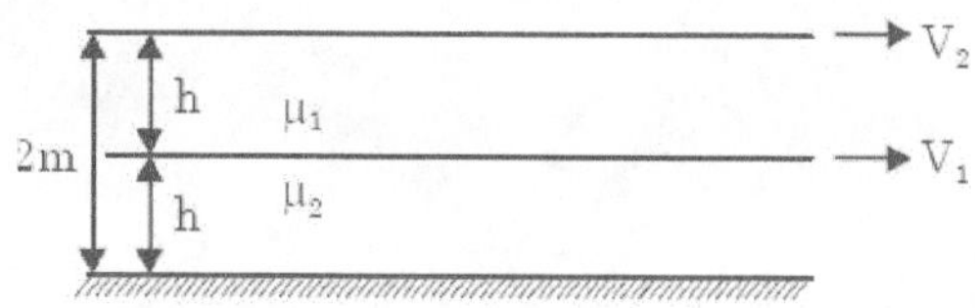

$$\mu_2 = 2\mu_1$$

$$V_2 = 3 \text{ m/s}$$

Shear stress value at intermediate plate will be

$$\tau = \frac{\mu_2 V_1}{h} = \frac{\mu_1 (V_2 - V_1)}{h}$$

Since $\mu_2 = 2\mu_1$

$\therefore$ $2V_1 = V_2 - V_1$

$\Rightarrow$ $V_2 = V_1 + 2V_1$

$V_2 = 3V_1$ $[V_2 = 3 \text{ m/s}]$

$\Rightarrow$ $3 = 3V_1$

$\Rightarrow$ $V_1 = \frac{3}{3} = 1 \text{ m/s}$

2 CHAPTER Fluid Statics

Analysis of Previous GATE Papers

Topic	Year/Set → Marks Distribution ↓		2019 Set 1	2019 Set 2	2018 Set 1	2018 Set 2	2017 Set 1	2017 Set 2	2016 Set 1	2016 Set 2	2016 Set 3	2015 Set 1	2015 Set 2	2015 Set 3
Fluid Pressure and its measurement	1 Mark	MCQ Type												
		Numerical Type												
	2 Marks	MCQ Type												
		Numerical Type							1					
	Total								2					
Hydrostatic forces	1 Mark	MCQ Type												
		Numerical Type												
	2 Marks	MCQ Type							1					
		Numerical Type												
	Total								2					
Buoyancy and Floatation	1 Mark	MCQ Type						1	1					
		Numerical Type												
	2 Marks	MCQ Type												
		Numerical Type	1							1				
	Total		2					1	1	2				

Fluid pressure and Its measurement

1. The cross-sectional area of one limb of a U-tube manometer (figure shown below) is made 500 times larger than the other, so that the pressure difference between the two limbs can be determined by measuring h on one limb of the manometer. The percentage error involved is

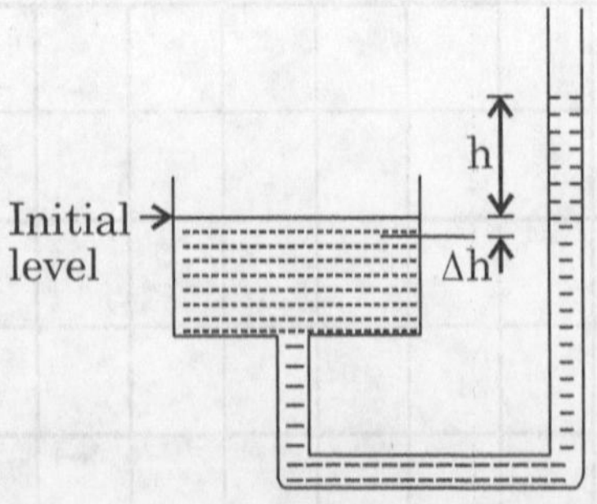

(a) 1.0
(b) 0.5
(c) 0.2
(d) 0.05

[1990 : 2 Marks]

2. The force F needed to support the liquid of density ρ and the vessel on top (fig.) is

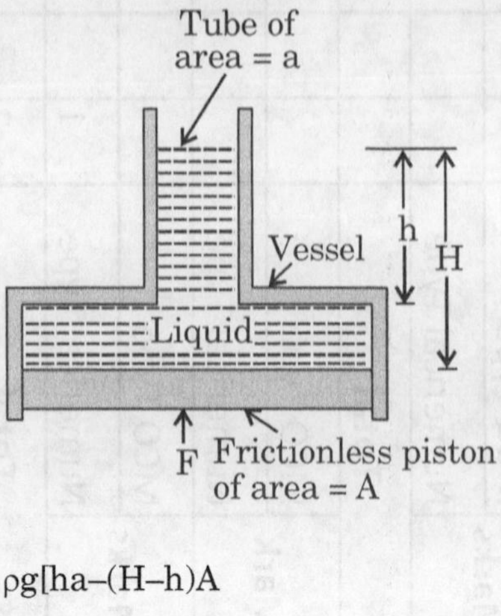

(a) ρg[ha–(H–h)A
(b) ρgHa
(c) ρgHA
(d) ρg(H–h)A

[1995 : 2 Marks]

3. A mercury manometer is used to measure the static pressure at a point in a water pipe as shown in figure. The level difference of mercury in the two limbs is 10 mm. The gauge pressure at that point is

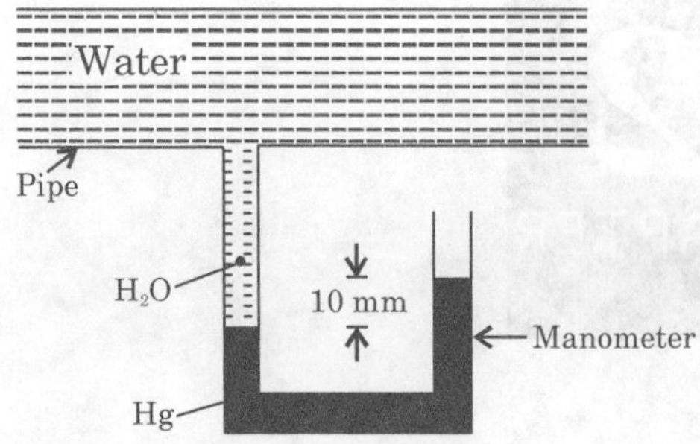

(a) 1236 Pa
(b) 1333Pa
(c) Zero
(d) 98 Pa

[1996 : 1 Mark]

4. Refer to figure, the absolute pressure of gas A in the bulb is

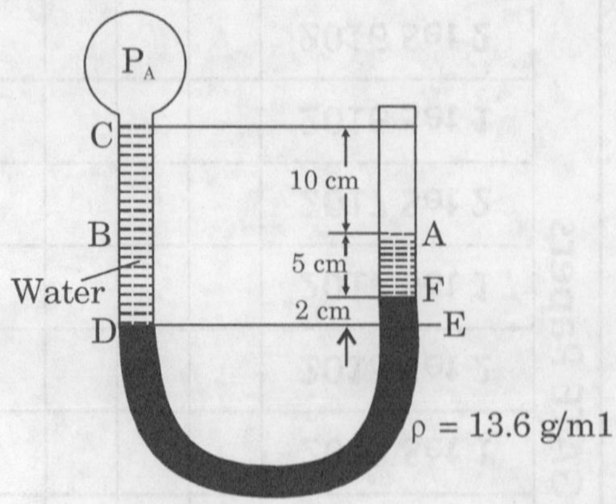

(a) 771.2 mm Hg
(c) 767.35 mm Hg
(b) 752.65 mm Hg
(d) 748.8 mm Hg

[1997 : 2 Marks]

5. In given figure, if the pressure of gas in bulb A is 50 cm Hg vacuum and p_{atm} = 76 cm Hg, then height of column H is equal to

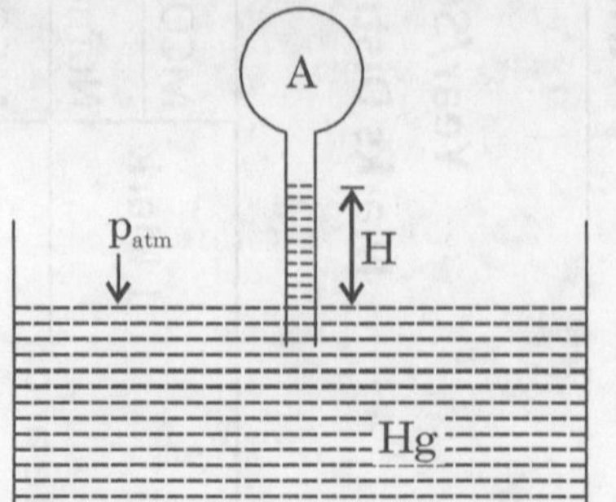

(a) 26 cm
(c) 76 cm
(b) 50 cm
(d) 126 cm

[2000 : 2 Marks]

6. The pressure gauges G_1 and G_2 installed on the system show pressure of $p_{G_1} = 5.00$ bar and $p_{G_2} = 1.00$ bar. The value of unknown pressure P is [Atmospheric pressure = 1.01 bar]

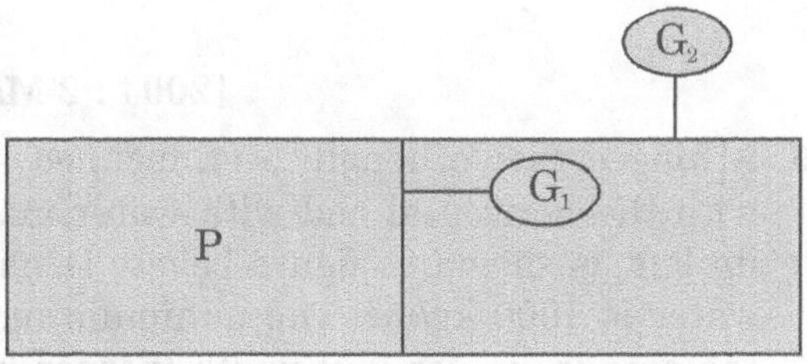

(a) 1.01 bar (b) 2.01 bar
(c) 5.00 bar (d) 7.01 bar

[2004 : 2 Marks]

7. Oil in a hydraulic cylinder is compressed from an initial volume 2 m³ to 1.96 m³. If the pressure of oil in the cylinder changes from 40 MPa to 80 MPa during compression, the bulk modulus of elasticity of oil is

(a) 1000 MPa (b) 2000 MPa
(c) 4000 MPa (d) 8000 MPa

[2007 : 2 Marks]

8. A large tank with a nozzle attached contains three immiscible, inviscid fluids as shown. Assuming that the changes in h_1, h_2 and h_3 are negligible, the instantaneous discharge velocity is

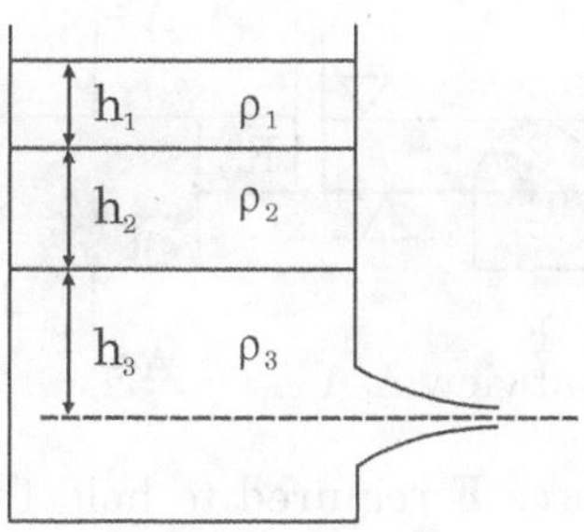

(a) $\sqrt{2gh_3\left(1+\frac{\rho_1}{\rho_3}\frac{h_1}{h_3}+\frac{\rho_2}{\rho_3}\frac{h_2}{h_3}\right)}$

(b) $\sqrt{2g(h_1+h_2+h_3)}$

(c) $\sqrt{2g\left(1+\frac{\rho_1h_1+\rho_2h_2+\rho_3h_3}{\rho_1+\rho_2+\rho_3}\right)}$

(d) $\sqrt{2g\left(\frac{\rho_2h_2h_2+\rho_2h_3h_1+\rho_3h_1h_2}{\rho_1h_1+\rho_2h_2+\rho_3h_3}\right)}$

[2012 : 2 Marks]

9. The difference in pressure (in N/m²) across an air bubble of diameter 0.001 m immersed in water (surface tension = 0.072 N/m) is_____.

[2014 : 1 Mark, Set-2]

10. An inverted U-tube manometer is used to measure the pressure difference between two pipes A and B, as shown in the figure. Pipe A is carrying oil (specific gravity = 0.8) and pipe B is carrying water. The densities of air and water are 1.16 kg/ m³ and 1000 kg/m³, respectively. The pressure difference between pipes A and B is________kPa.

Acceleration due to gravity: g = 10 m/s²

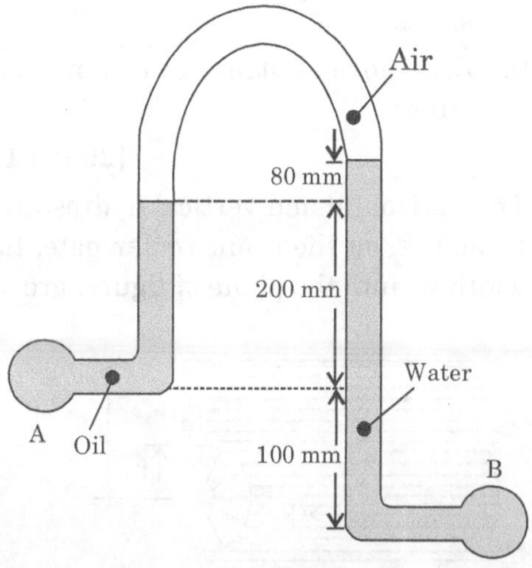

[2016 : 2 Marks, Set-1]

Hydrostatic forces

11. A circular plate 1 m in diameter is submerged vertically in water such its upper edge is 8 m below the free surface of water. The total hydrostatic pressure force on one side of plate is

(a) 6.7 kN (b) 65.4 kN
(c) 45.0 kN (d) 77.0 kN

[1988 : 2 Marks]

12. Shown below are three cylindrical gates which restrain water in a 2–D Channel. Which gate experiences the maximum vertical component, the minimum vertical component and the maximum horizontal component of the hydrostatic force?

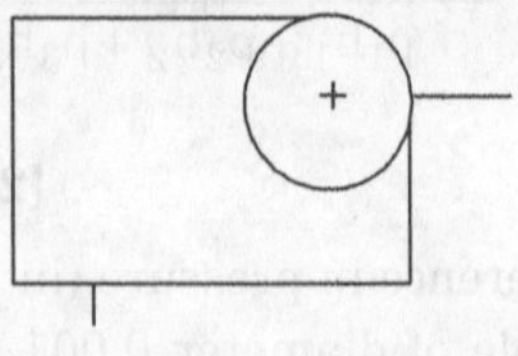

(a)

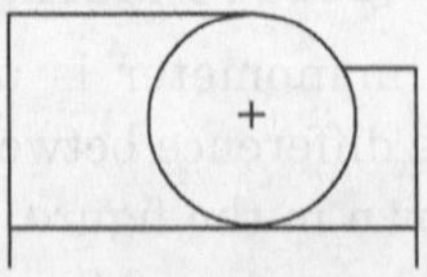

(b)

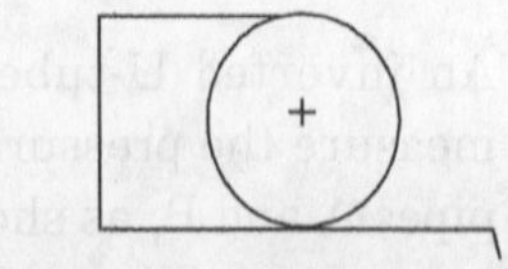

(c)

[1993 : 1 Mark]

13. A static fluid can have

(a) Non-zero normal and shear stress

(b) Negative normal stress and zero shear stress

(c) Positive normal stress and zero shear stress

(d) Zero normal stress and non-zero shear stress

[2001 : 1 Mark]

14. The horizontal and vertical hydrostatic forces F_x and F_y on the semicircular gate, having a width w into the plane of figure, are

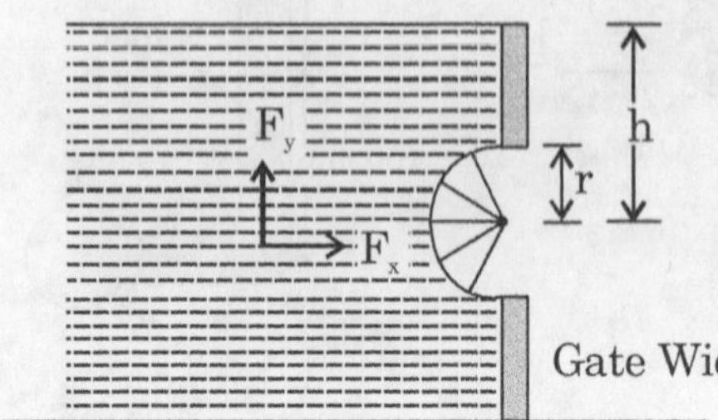

(a) $F_x = \rho ghrw$ and $F_y = 0$

(b) $F_x = 2\rho ghrw$ and $F_y = 0$

(c) $F_x = \rho ghrw$ and $F_y = \rho gwr^2$

(d) $F_x = 2\rho ghrw$ and $F_y = \pi\rho gwr^2/2$

[2001 : 2 Marks]

15. A closed cylinder having a radius R and height H is filled with oil of density ρ. If the cylinder is rotated about its axis at an angular velocity of ω, the thrust at the bottom of the cylinder is

(a) $\pi R^2\rho gH$

(b) $\pi R^2 \dfrac{\rho\omega^2R^2}{4}$

(c) $\pi R^2(\rho\omega^2R^2 + \rho gH)$

(d) $\pi R^2\left(\dfrac{\rho\omega^2R^2}{4} + \rho gh\right)$

[2004 : 2 Marks]

16. A hinged gate of length 5 m, inclined at 30° with the horizontal and with water mass on its left, is shown in figure below. Density of water is 1000 kg/m³. The minimum mass of the gate in kg per unit width (perpendicular to the plane of paper), required to keep it closed

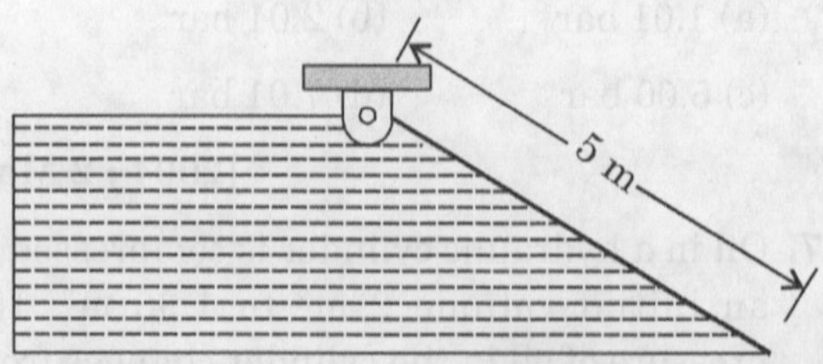

(a) 5000 (b) 6600

(c) 7546 (d) 9623

[2013 : 2 Marks]

17. Consider a frictionless, massless and leak-proof plug blocking a rectangular hole of dimensions 2R × L at the bottom of an open tank as shown in the figure. The head of the plug has the shape of a semi-cylinder of radius R. The tank is filled with a liquid of density ρ up to the tip of the plug.

The gravitational acceleration is g. Neglect the effect of the atmospheric pressure.

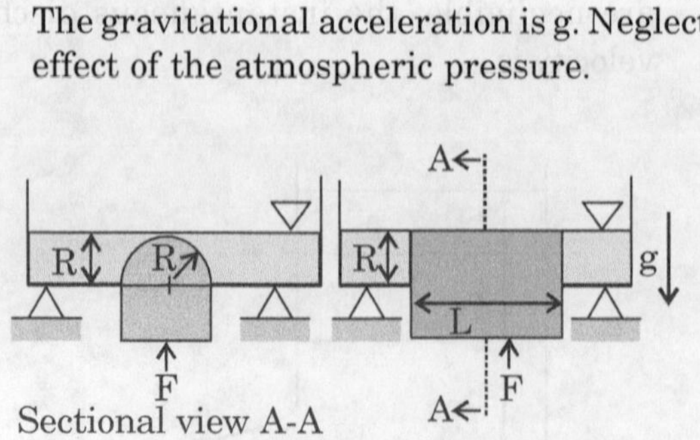

Sectional view A-A

The force F required to hold the plug in its position is

(a) $2\rho R^2gL\left(1-\dfrac{\pi}{4}\right)$ (b) $2oR^2gL\left(1+\dfrac{\pi}{4}\right)$

(c) $\rho R^2\delta gL$ (d) $\dfrac{\pi}{2}\rho R^2gL$

[2016 : 2 Marks, Set-1]

Buoyancy and floatation

18. Bodies in flotation to be in stable equilibrium the necessary and sufficient condition is that the centre of gravity is located below the ____.

[1994 : 1 Mark]

19. A cylindrical body of cross-sectional area A, height H and density ρ_s, is immersed to a depth h in a liquid of density ρ, and tied to the bottom with a string. The tension in the string is ____.

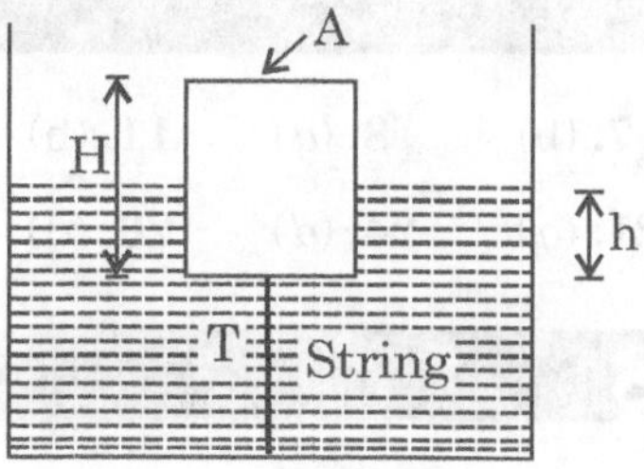

(a) ρghA (b) $(\rho_s - \rho)ghA$

(c) $(\rho - \rho_s)ghA$ (d) $(\rho h - \rho_s H)gA$

[2003 : 1 Mark]

20. For the stability of a floating body, under the influence of gravity alone, which of the following is TRUE?

(a) Metacenter should be below centre of gravity.

(b) Metacenter should be above centre of gravity.

(c) Metacenter and centre of gravity must lie on the same horizontal line.

(d) Metacenter and centre of gravity must lie on the same vertical line.

[2010 : 1 Mark]

21. For a completely submerged body with centre of gravity G and centre of buoyancy B, the condition of stability will be

(a) G is located below B

(b) G is located above B

(c) G and B are coincident

(d) Independent of the locations of G and B

[2014 : 1 Mark, Set-1]

22. An aluminium alloy (density 2600 kg/m^3) casting is to be produced. A cylindrical hole of 100 mm diameter and 100 mm length is made in the casting using sand core (density 1600 kg/m^3). The net buoyancy force (in Newton) acting on the core is ______.

[2014 : 2 Marks, Set-1]

23. A spherical balloon with a diameter of 10 m, shown in the figure below is used for advertisements. The balloon is filled with helium (R_{He} = 2.08 kJ/kgK) at ambient conditions of 15°C and 100 kPa. Assuming no disturbances due to wind, the maximum allowable weight (in Newton) of balloon material and rope required to avoid the fall of the balloon (R_{air} = 0.289 kJ/kgK) is______.

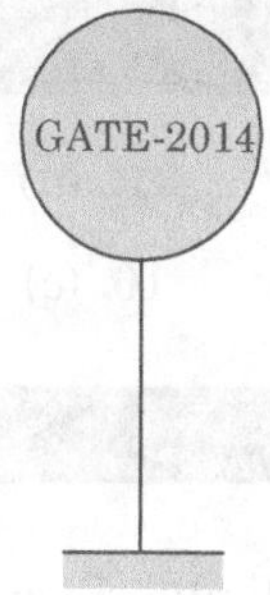

[2014 : 2 Marks, Set-2]

24. For a floating body, buoyant force acts at the

(a) centroid of the floating body

(b) center of gravity of the body

(c) centroid of the fluid vertically below the body

(d) centroid of the displaced fluid

[2016 : 1 Mark, Set-1]

25. The large vessel shown in the figure contains oil and water. A body is submerged at the interface of oil and water such that 45 percent of its volume is in oil while the rest is in water. The density of the body is ______ kg/m^3.

The specific gravity of oil is 0.7 and density of water is 1000 kg/m^3.

Acceleration due to gravity: g = 10 m/s^2.

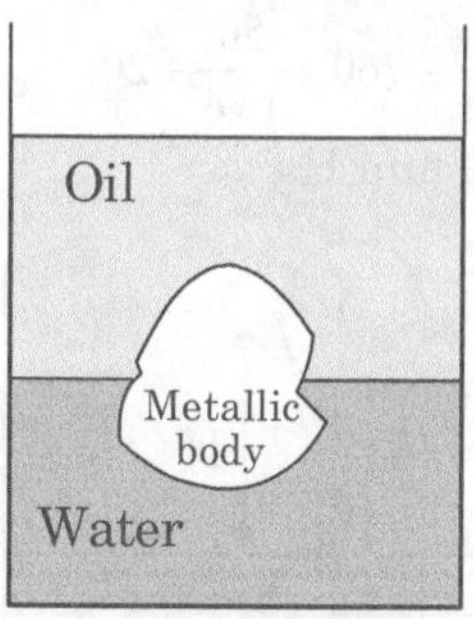

[2016 : 2 Marks, Set-2]

26. For the stability of a floating body the
(a) centre of buoyancy must coincide with the centre of gravity
(b) centre of buoyancy must be above the centre of gravity
(c) centre of gravity must be above the centre of buoyancy
(d) metacentre must be above the centre of gravity

[2017 : 1 Mark, Set-2]

27. A cube of side 100 mm is placed at the bottom of an empty container on one of its faces. The density of the material of the cube is 800 kg/m³. Liquid of density 1000 kg/m³ is now poured into the container. The minimum height to which the liquid needs to be poured into the container for the cube to just lift up is _____ mm.

[2019 : 2 Marks, Set-1]

ANSWERS

1. (c)	**2.** (c)	**3.** (*a*)	**4.** (*a*)	**5.** (*b*)	**6.** (*d*)	**7.** (b)	**8.** (*a*)	**11.** (b)	**13.** (c)
14. (d)	**15.** (*d*)	**16.** (*c*)	**17.** (a)	**19.** (*d*)	**20.** (*b*)	**21.** (*a*)	**24.** (*d*)	**26.** (d)	

EXPLANATIONS

1. Fall in larger limb = Rise in smaller limb

$A_1 \, \Delta h = A_2 h$

or $\dfrac{\Delta h}{h} = \dfrac{A_2}{A_1} = \dfrac{1}{500}$

$\%\text{ error} = \dfrac{\Delta h}{h} \times 100 = \dfrac{1}{500} \times 100 = 0.2\%$

2. Gauge pressure = $(\delta m - \delta w)gh$
$= (13600 - 1000) \times 9.81 \times 10 \times 10^{-3} = 1236$ Pa

3. Height of water column over plate is H.
$F = P \times A = \rho gHA$

4. $P_{atm} = 760$ mm Hg

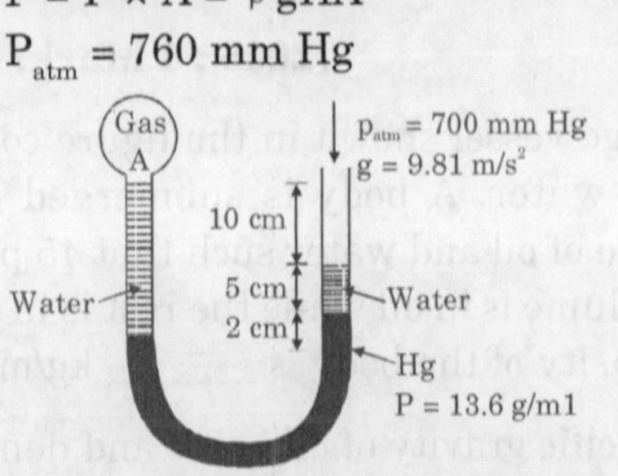

Let P_A be the absolute pressure at 'A'
Pressure at D = Pressure at E

or $P_A + P_{CD} = P_{atm} + P_{AF} + P_{FE}$

or $P_A + \dfrac{170}{13.6} = 760 + \dfrac{50}{13.6} + 20$

$P_A = 771.2$ mm Hg

5.

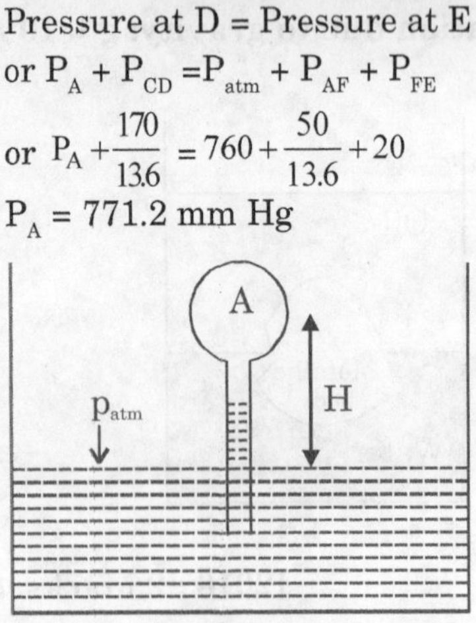

$p_A = 50$ cm Hg vacuum $= p_H$

6.

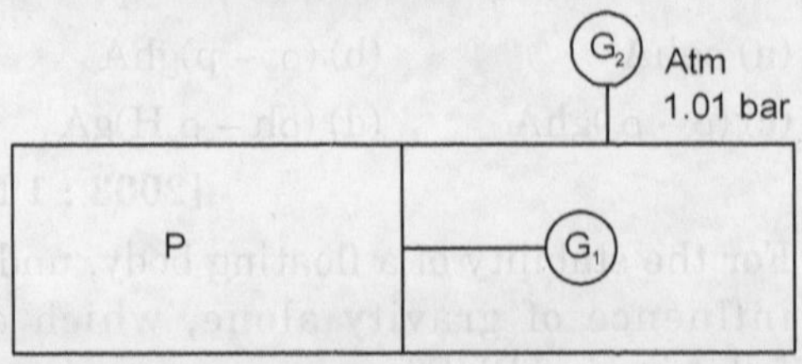

$P = P_{G1} + P_{G2} + P_{atm}$
$P = 5 + 1 + 1.01 = 7.01$ bar

7. Bulk modules of elasticity, $K = \dfrac{-dh}{\dfrac{dv}{V_1}}$

$K = -V_1 \dfrac{-dp}{dv}$

$K = \dfrac{-2 \times 40}{-0.04}$

$K = 2000$ MPa

8. Applying Bernoulli's equation at exit, we get

$\dfrac{P_1}{\rho_3} + gz_1 + \dfrac{V_1^2}{2} = \dfrac{P_2}{\rho_3} + gz_2 + \dfrac{V_2^2}{2}$

We know $Z_1 = Z_2$, $V_1 = 0$ & $P_2 = P_{atm}$
Hence it reduce to

$\dfrac{P_1}{\rho_3} = \dfrac{V_2^2}{2}$

$V_2 = \sqrt{\dfrac{2P_1}{\rho_3}}$

But $P_1 = \rho_1 gh_1 + \rho_2 gh_2 + \rho_3 gh_3$
Upon substituting, we get

$V_2 = \sqrt{2gh_3\left[1 + \dfrac{\rho_1 h_1}{\rho_3 h_3} + \dfrac{\rho_2 h_2}{\rho_3 h_3}\right]}$

9.

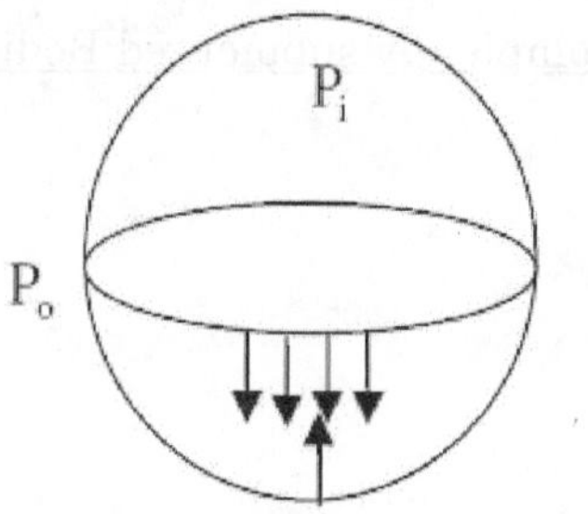

Surface tension in a bubble = $\frac{\Delta p.r}{4}$

we know, $P_i - P_o = \frac{4T}{r}$

$\Delta p = \frac{4 \times 0.072}{0.001} = 288 \text{ N/m}^2$.

10. $P_A - (\rho_{oil} \times g \times 0.2) - (\rho_{air} \times g \times 0.08) + (\rho_w \times g \times 0.38) - P_B = 0$

$P_A - P_B = -2.2$ kPa

11.

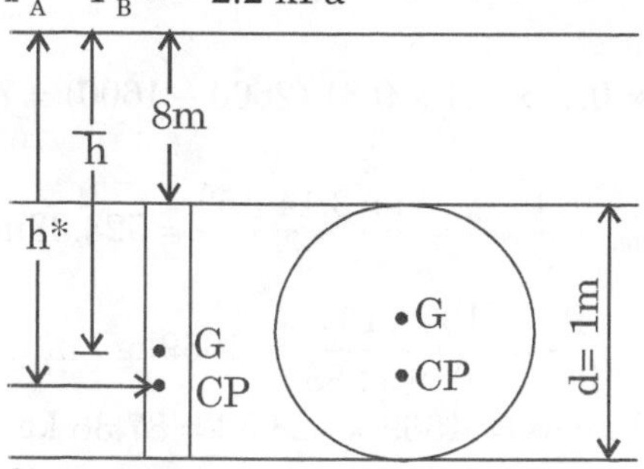

diameter, d = 1 m

$\bar{h} = 8 + \frac{d}{2}$

$= 8 + \frac{1}{2} = 8.5$ m

Surface area, $A = \frac{\pi}{4} d^2 = \frac{3.14}{4} \times 1^2 = 0.785 \text{ m}^2$

Total hydrostatic pressure force, $F = \rho g \bar{h} A$

$F = 1000 \times 9.81 \times 8.5 \times 0.785 = 65458$ N

or

F = 65.45 KN

12. A → Experiences Maximum vertical hydrostatic force component

B → Experiences Maximum horizontal hydrostatic force component

C → Experiences minimum vertical component of force

14. $F_x = 2\rho ghrw$

$F_y = w_2 - w_1 = \frac{\pi r^2}{2} \cdot pg \times w = \frac{\pi r^2 pgw}{2}$

15. We know that

$$\frac{\partial p}{\partial r} = \frac{\rho v^2}{r} = \frac{\rho \cdot \omega^2 r^2}{r} = \rho \omega^2 r$$

$[\because v = \omega \times r]$

$\therefore \int_0^p \partial p = \int_0^r \rho \omega^2 r \, dr$

$p = \frac{\rho}{2} \omega^2 r^2$

Area of circular ring = $2\pi r dr$

Force on elementry ring

= Intensity of pressure × Area of ring

$= \frac{\rho}{2} \omega^2 r^2 2\pi r dr$

∴ Total force on the top of the cylinder

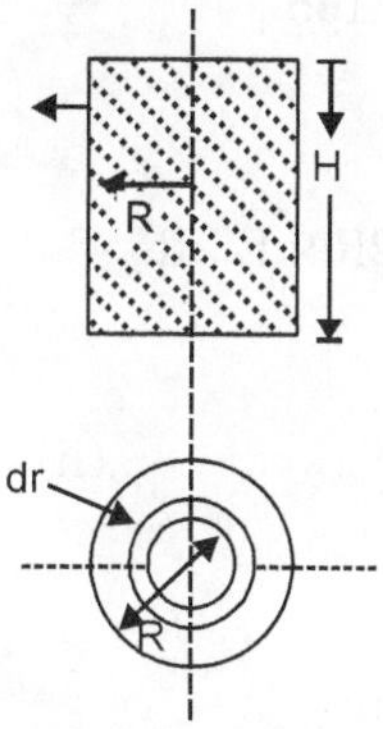

$$= \int_0^R \frac{\rho}{2} \omega^2 r^2 2\pi r \, dr = \frac{\rho}{2} \omega^2 r^2 2\pi \int_0^R r^3 \, dr$$

$$= \frac{\rho}{2} \cdot \omega^2 2\pi \frac{R^4}{4} = \frac{\rho}{4} \omega^2 \times \pi R^4$$

Thrust at the bottom of the cylinder

= Weight of water in cylinder + Total force on the top of cylinder

$$= \rho g \times \pi R^2 \times H + \frac{\rho}{4} \omega^2 \times \pi R^4$$

$$= \pi R^2 \left[\frac{\rho \omega^2 R^2}{4} + \rho g H \right]$$

16. Given data

l = 5 m

θ = 30°

ρ = 1000 kg/m³

$\bar{h} = 2.5 \sin 30° = 1.25$m

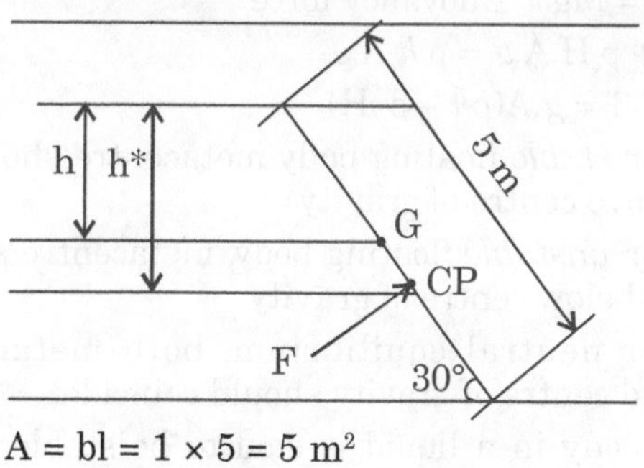

$A = bl = 1 \times 5 = 5 \text{ m}^2$

$I_G = \frac{bl^3}{12} = \frac{1 \times (5)^3}{12} = 10.41 \text{m}^4$

$F = \rho g A \bar{h} = 1000 \times 9.81 \times 5 \times 1.25 = 61312.5$N

$$h^* = \bar{h} + \frac{I_G}{A\bar{h}}.\sin^2\theta$$

$$= 1.25 + \frac{10.41}{5 \times 1.25}\sin^2 30° = 1.67m$$

Taking moment about hinge

$$F \times \frac{h^*}{\sin 30°} = W \times 2.5 \times \cos 30°$$

$$61312.5 \times \frac{1.66}{0.5} = Mg \times 2.165$$

$$\boxed{M = 9623 \text{ kg}}$$

17. Total volume $\Rightarrow V_{total} = 2R \times L \times R$
$= 2R^2L$ ...(i)

volume : $v_1 = \frac{\pi R_2}{2} \times L$ ...(ii)

F = Weight of fluit
$= [V_{total} - V_1]\rho g$

$$= \left[2R^2L - \frac{\pi R^2 L}{2}\right]\rho g$$

$$F = 2G\rho R^2 gL\left(1\frac{-\pi}{4}\right)$$

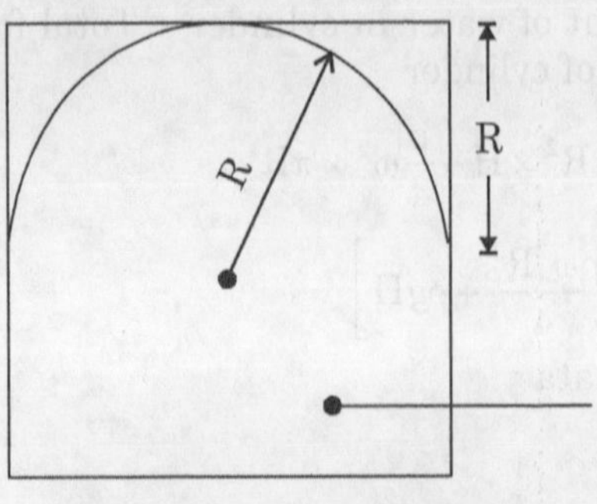

18. Located below the metacentre.
19. Let T = Tension in string
$T_1 = Mg$ = Buoyancy force
$T + \rho_s H.A.g = \rho.h\, A.g$
$\therefore\; T = g.A(\rho h - \rho_s.H)$
20. For *stable* floating body metacentre should be above centre of gravity.
For *unstable* floating body, metacentre should be below centre of gravity.
For neutral equilibrium, both metacentre and centre of gravity should coincide.
21. A body in a liquid is said to be stable, when given small displacement, it returns to its original position.

Stability of completely submerged Bodies

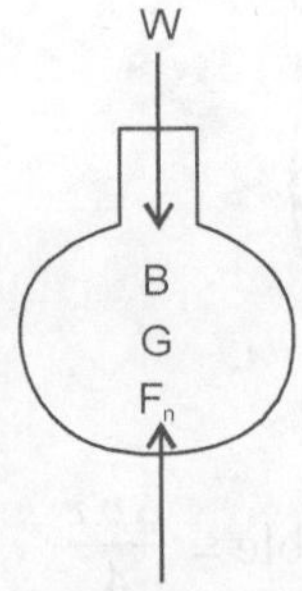

The center of gravity 'G' is below the center of Buoyancy 'B'.

22. $d = 1600$ kg/m³
$\rho = 2600$ kg/m³
Net buouancy force = weight of liquid displaced − weight of solid body
$= v \times \rho \times g - v \times d \times g = vg(\rho - d)$
$= \frac{\pi}{4} d^2 L \times g(\rho - d)$
$= \frac{\pi}{4} \times 0.1^2 \times 0.1 \times 9.81\,(2600 - 1600) = 7.7$ N.

23. $V_{ballon} = \frac{4}{3}\pi r^3 = \frac{4 \times 3.14 \times 5^3}{3} = 523.33m^3$

$$\rho_{He} = \frac{P}{RT} = \frac{100 \times 10_3}{2080 \times 288} = .1669 kg/m^3$$

Total mass = .1669 × 523.33 = 87.36 kg
mass of air :

$$\rho = \frac{P}{RT} = \frac{100 \times 10^3}{289 \times 288} = 1.2 \text{ kg/m}^3$$

Mass of displaced volume of air
= 1.2 × 523.33 = 628 kg
Net Buoyant mass = 628 – 87. 36 = 540.64 kgl
Allowable Mass, $M_{allowable}$ = 540.64 × 9.81
= 5303.67 N

24. For floating body Buoyancy force acts through the center of buoyancy which is C.G for displaced volume.

27. Given, side of cube (a) = 100 m
Density of material of cube = 800 kg/m³
Poured liquid density = 1000 kg/m³
Now, Minimum height to which the liquid needs to be poured into the container for the cube to just lift up, should satisfy the condition

$$F_B \geq (mg)_{block}$$

$$\rho_f g V_{fd} \geq (\rho V)_{block} \times g$$

$$1000 \times 0.1 \times 0.1 \times h \geq 800 \times 0.1 \times 0.1 \times 0.1$$

$$h \geq 0.08 \text{ m}$$

$$h \geq 80 \text{ mm}$$

3 CHAPTER

Fluid Kinematics

Analysis of Previous GATE Papers

Topic	Year/Set → Marks Distribution ↓		2019 Set 1	2019 Set 2	2018 Set 1	2018 Set 2	2017 Set 1	2017 Set 2	2016 Set 1	2016 Set 2	2016 Set 3	2015 Set 1	2015 Set 2	2015 Set 3
Fluid Velocity	**1 Mark**	MCQ Type												
		Numerical Type			1									
	2 Marks	MCQ Type												
		Numerical Type												
	Total				1									
Fluid Acceleration	**1 Mark**	MCQ Type					1							
		Numerical Type	1											
	2 Marks	MCQ Type												
		Numerical Type												
	Total		1				1							
Types of Fluid flow	**1 Mark**	MCQ Type			1									
		Numerical Type												
	2 Marks	MCQ Type			1							1		
		Numerical Type												
	Total				3							2		
Fluid flow lines and application	**1 Mark**	MCQ Type		1						1	1		1	
		Numerical Type												
	2 Marks	MCQ Type					1				1			
		Numerical Type										1		
	Total			1			2			1	3	2		1

Fluid velocity

1. The stream function in a two dimensional flow field is given by $\psi = x^2 - y^2$ The magnitude of the velocity at point (1, 1) is

(a) 2 (b) $2\sqrt{2}$

(c) 4 (d) 8

[1989 : 2 Marks]

2. Shown below are three pipe sections through which water flows as shown. Option a to d below pertain to the direction of the net force on the pipe section due to the flow of water.

(a) 45° to both +ve x and +ve y axes

(b) 45° to both -ve x and +ve y axes

(c) 45° to both +ve x and -ve y axes

(d) 45° to both -ve x and -ve y axes

State which of the options indicate the correct direction of force for pipe 1, pipe 2 and pipe 3

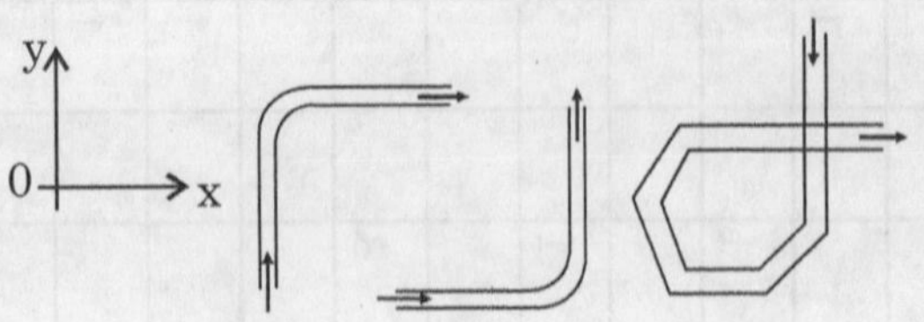

Pipe 1 Pipe 2 Pipe 3

[1993 : 1 Mark]

3. Circulation is defined as line integral of tangential component of velocity about a ________ (fill in the blanks)

[1994 : 1 Mark]

4. The velocity components in the x and y directions are given by $u = \lambda xy^3 - x^2y$, $v = xy^2 - \frac{3}{4}y^4$. The value of λ for a possible flow field involving an incompressible fluid is

(a) $-\frac{3}{4}$ (b) $-\frac{4}{3}$

(c) $\frac{4}{3}$ (d) 3

[1995 : 2 Marks]

5. The velocity components in the x and y directions of a two dimensional potential flow are u and v, respectively. Then $\frac{\partial u}{\partial x}$, is equal to

(a) $\frac{\partial v}{\partial x}$ (b) $-\frac{\partial v}{\partial x}$

(c) $\frac{\partial v}{\partial y}$ (d) $-\frac{\partial v}{\partial y}$

[2005 : 1 Mark]

6. A leaf is caught in a whirlpool. At a given instant, the leaf is at a distance of 120 m from the centre of the whirlpool. The whirlpool can be described by the following velocity distribution: $V_r = -\left(\frac{60 \times 10^3}{2\pi r}\right)$ m/s and $V_\theta = \frac{300 \times 10^3}{2\pi r}$ m/s, where r (in meters) is the distance from the centre of the whirlpool. What will be the distance of the leaf from the centre when it has moved through half a revolution?

(a) 48 m (b) 64 m

(c) 120 m (d) 142 m

[2005 : 2 Marks]

7. In a steady flow through a nozzle, the flow velocity on the nozzle axis is given by $v = u_0(1 + 3x/L)i$, where x is the distance along the axis of the nozzle from its inlet plane and L is the length of the nozzle. The time required for a fluid particle on the axis to travel from the inlet to the exit plane of the nozzle is

(a) $\frac{L}{u_0}$ (b) $\frac{L}{3u_0}\ln 4$

(c) $\frac{L}{4u_0}$ (d) $\frac{L}{2.5u_0}$

[2007 : 1 Mark]

Linked Answer Questions 8 and 9:

The gap between a moving circular plate and a stationary surface is being continuously reduced, as the circular plate comes down at a uniform speed V towards the stationary bottom surface, as shown in the figure. In the process, the fluid contained between the two plates flows out radially. The fluid is assumed to be incompressible and inviscid.

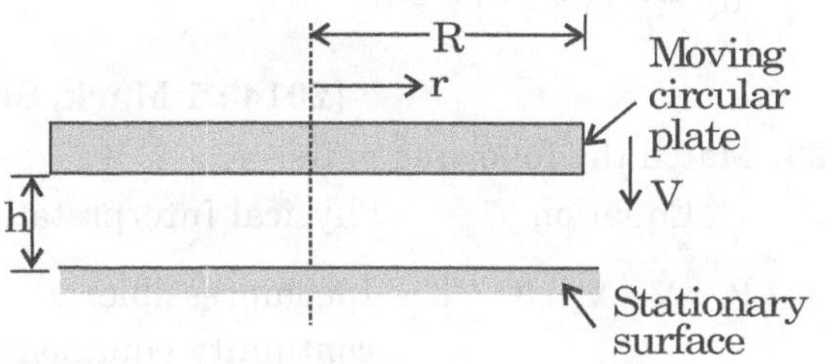

8. The radial velocity V_r at any radius r, when the gap width is h, is

(a) $V_r = \frac{Vr}{2h}$ (b) $V_r = \frac{2Vr}{h}$

(c) $V_r = \frac{2Vh}{r}$ (d) $V_r = \frac{Vh}{r}$

[2008 : 2 Marks]

9. The radial component of the fluid acceleration at r = R is

(a) $\frac{3V^2R}{4h^2}$ (b) $\frac{V^2R}{4h^2}$

(c) $\frac{V^2R}{2h^2}$ (d) $\frac{V^2h}{2R^2}$

[2008 : 2 Marks]

10. A flat plate of width L = 1 m is pushed down with a velocity U = 0.01 m/s towards a wall resulting in the drainage of the fluid between the plate and the wall as shown in the figure. Assume two- dimensional incompressible flow and that the plate remains parallel to the wall. The average velocity, u_{avg} of the fluid (in m/s) draining out at the instant shown in the figure is _____ (correct to three decimal places).

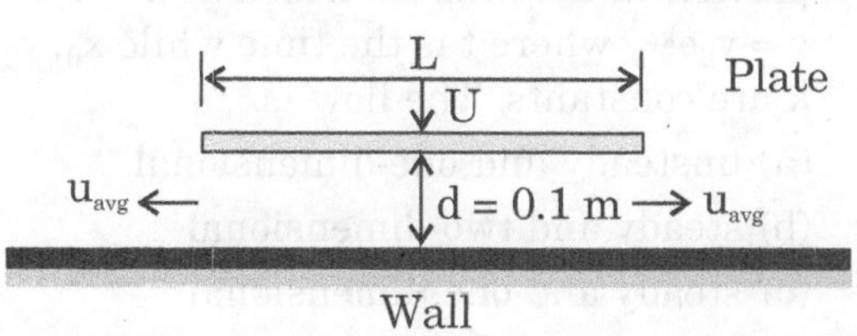

[2018 : 1 Mark, Set-1]

Fluid acceleration

11. In a two-dimensional velocity field with velocities u and v along the x and y directions respectively, the convective acceleration along the x-direction is given by

(a) $u\frac{\partial V}{\partial x} + v\frac{\partial u}{\partial y}$ (b) $u\frac{\partial u}{\partial x} + v\frac{\partial v}{\partial y}$

(c) $u\frac{\partial u}{\partial x} + v\frac{\partial u}{\partial y}$ (d) $u\frac{\partial u}{\partial x} + u\frac{\partial u}{\partial y}$

[2006 : 1 Mark]

12. A flow field which has only convective acceleration is

(a) a steady uniform flow

(b) an unsteady uniform flow

(c) a steady non-uniform flow

(d) an unsteady non-uniform flow

[2014 : 1 Mark, Set-4]

13. Consider the two-dimensional velocity field given by $\vec{V} = (5 + a_1x + +b_1y)\hat{i} + (4 + a_2x + b_2y)\hat{j}$, where a_1, b_1, a_2 and b_2 are constants. Which one of the following conditions needs to be satisfied for the flow to be incompressible?

(a) $a_1 + b_1 = 0$ (b) $a_1 + b_2 = 0$

(c) $a_2 + b_2 = 0$ (d) $a_2 + b_1 = 0$

[2017 : 1 Mark, Set-1]

14. Water flows though a pipe with a velocity given by $\vec{V} = \left(\frac{4}{t} + x + y\right)\hat{j}$ m/s, where $\hat{j}$ is the unit vector in the y direction, t(>0) is in seconds, and x and y are in meters. The magnitude of total acceleration at the point (x, y) = (1, 1) at t = 2 s is _____ m/s².

[2019 : 1 Mark, Set-1]

Types of fluid flow

15. A velocity field is given as

$$\vec{V} = 3x^2y\hat{i} - 6xyz\hat{k}$$

where x, y, z are in m and V in m/s. Determine if

(i) It represents an incompressible flow

(ii) The flow is irrotational

(iii) The flow is steady

[1993 : 2 Marks]

16. Streamlines, path lines and streak lines are virtually identical for

(a) Uniform flow (b) Flow of ideal fluids

(c) Steady flow (d) Non uniform flow

[1994 : 1 Mark]

17. For a fluid element in a two dimensional flow field (x-y plane), if it will undergo

(a) Translation only

(b) Translation and rotation

(c) Translation and deformation

(d) Deformation only

[1994 : 1 Mark]

18. Existence of velocity potential implies that

(a) Fluid is in continuum

(b) Fluid is irrotational

(c) Fluid is ideal

(d) Fluid is compressible

[1994 : 1 Mark]

19. The 2-D flow with, velocity $\vec{V} = (x + 2y + 2)\hat{i} + (4 - y)\hat{j}$ is

(a) Compressible and irrotational

(b) Compressible and not irrotational

(c) Incompressible and irrotational

(d) Incompressible and not irrotational

[2001 : 2 Marks]

20. Which combination of the following statements about steady incompressible forced vortex flow is correct?

P : Shear stress is zero at all points in the flow.

Q : Vorticity is zero at all points in the flow.

R : Velocity is directly proportional to the radius from the centre of the vortex.

S : Total mechanical energy per unit mass is constant in the entire flow field. Select the correct answer using the codes given below:

(a) P and Q (b) R and S

(c) P and R (d) P and S

[2007 : 2 Marks]

21. For the continuity equation given by $\vec{\Delta} \cdot \vec{V} = 0$ to be valid, where $\vec{V}$ is the velocity vector, which one of the following is a necessary condition?

(a) Steady flow

(b) Irrotational flow

(c) Inviscid flow

(d) Incompressible flow

[2008 : 1 Mark]

22. For an incompressible flow field, V, which one of the following conditions must be satisfied?

(a) $\nabla \cdot \vec{V} = 0$

(b) $\nabla \times \vec{V} = 0$

(c) $(\vec{V} \cdot \nabla)\vec{V} = 0$

(d) $\dfrac{\partial \vec{V}}{\partial t} + (\vec{V} \cdot \nabla)\vec{V} = 0$

[2014 : 1 Mark, Set-2]

23. Match the following pairs:

	Equation		Physical Interpretation
P.	$\nabla \times \vec{V} = 0$	I.	Incompressible continuity equation
Q.	$\nabla \cdot \vec{V} = 0$	II.	Steady flow
R.	$\dfrac{\Delta \vec{V}}{\Delta t} = 0$	III.	Irrotational flow
S.	$\dfrac{\partial \vec{V}}{\partial t} = 0$	IV.	Zero acceleration of fluid particle

(a) P-IV, Q-I, R-II, S-III

(b) P-IV, Q-III, FRS-II

(c) P-III, Q-I, R-IV, S-II

(d) P-III, Q-I, R-II, S-IV

[2015 : 2 Marks, Set-1]

24. For a two-dimensional incompressible flow field given by $\vec{u} = A(x\hat{i} - y\hat{j})$, where $A > 0$, which one of the following statements is FALSE?

A. It satisfies continuity equation

B. It is unidirectional when $x \to 0$ and $y \to \infty$.

C. Its streamlines are given by x = y.

D. It is irrotational

(a) A (b) B

(c) C (d) D

[2018 : 1 Mark, Set-1]

25. In a Lagrangian system, the position of a fluid particle in a flow is described as $x = x_0 e^{-kt}$ and $y = y_0 e^{kt}$ where t is the time while x_0, y_0, and k are constants. The flow is

(a) unsteady and one-dimensional

(b) steady and two-dimensional

(c) steady and one-dimensional

(d) unsteady and two-dimensional

[2018 : 2 Marks, Set-1]

Fluid flow lines and application

26. The velocity potential function for a source varies with the distance r as

(a) 1/r (b) $1/r^2$
(c) e^r (d) In r

[1987 : 1 Mark]

27. A streamlined body is defined as a body about which

(a) The flow is laminar
(b) The flow is along the streamlines
(c) The flow separation is suppressed
(d) The drag is zero

[1987 : 1 Mark]

28. The Newtonian fluid has the following velocity field:

$\vec{V} = x^2y\hat{i} + 2xy^2z\hat{j} - yz^3\hat{k}$

The rate shear deformation ε_{yz} at the point x = –2, y = –1 and z = 2 for the given flow is

(a) –6 (b) –2
(c) –12 (d) 4

[1988 : 2 Marks]

29. In a flow field the stream lines and equipotential lines

(a) are parallel
(b) Cut at any angle
(c) Are orthogonal every where in the field
(d) Cut orthogonal except at the stagnation points

[1994 : 1 Mark]

30. A fluid flow is represented by the velocity field $\vec{V} = ax\hat{i} + ay\hat{j}$, where a is a constant. The equation of stream line passing through a point (1, 2) is

(a) x – 2y = 0 (b) 2x + y = 0
(c) 2x – y = 0 (d) x + 2y = 0

[2004 : 1 Mark]

31. A two-dimensional flow field has velocities along the x and y directions given by $u = x^2t$ and $v = -2xyt$ respectively, where t is time. The equation of streamline is

(a) x^2y = constant
(b) xy^2 = constant
(c) xy = constant
(d) not possible to determine

[2006 : 2 Marks]

32. Velocity vector of a flow field is given as $\vec{V} = 2xy\hat{i} - x^2z\hat{j}$. The vorticity vector at (1, 1, 1) is

(a) $4\hat{i} - \hat{j}$ (b) $4\hat{i} - \hat{k}$
(c) $\hat{i} - 4\hat{j}$ (d) $\hat{i} - 4\hat{k}$

[2010 : 2 Marks]

33. A streamline and an equipotential line in a flow field

(a) are parallel to each other
(b) are perpendicular to each other
(c) intersect at an acute angle
(d) are identical

[2011 : 1 Mark]

34. Consider the following statements regarding streamline(s):

(i) It is a continuous line such that the tangent at any point on it shows the velocity vector at that point
(ii) There is no flow across streamlines
(iii) $\frac{dx}{u} = \frac{dy}{v} = \frac{dz}{w}$ is the differential equation of a streamline, where u, v and w are velocities in directions x, y and z, respectively
(iv) In an unsteady flow, the path of a particle is a streamline

Which one of the following combinations of the statements is true?

(a) (i), (ii), (iv) (b) (ii), (iii), (iv)
(c) (i), (iii), (iv) (d) (i), (ii), (iii)

[2014 : 2 Marks, Set-4]

35. Consider a velocity field $\vec{V} = K(y\hat{i} + x\hat{k})$, where K is a constant. The vorticity, Ω_z, is

(a) –K (b) K
(c) –K/2 (d) K/2

[2014 : 2 Marks, Set-4]

36. If the fluid velocity for a potential flow is given by V(x, y) = u(x, y)i + v(x, y)j with usual notations, then the slope of the potential line at (x, y) is

(a) $\frac{v}{u}$ (b) $-\frac{u^2}{v}$

(c) $-\frac{v}{u^2}$ (d) $\frac{u}{v}$

[2015 : 1 Mark, Set-2]

37. The velocity field of an incompressible flow is given by

$V = (a_1x + a_2y + a_3z)i + (b_1x + b_2y + b_3z)j + (c_1x + c_2y + c_3z)k$, where $a_1 = 2$ and $c_3 = -4$. The value of b_2 is ______.

[2015 : 2 Marks, Set-1]

38. The volumetric flow rate (per unit depth) between two streamlines having stream function ψ_1 and ψ_2 is

(a) $|\psi_1 + \psi_2|$ (b) $\psi_1\psi_1$

(c) ψ_1/ψ_2 (d) $|\psi_1 - \psi_2|$

[2016 : 1 Mark, Set-2]

39. For a certain two-dimensional incompressible flow, velocity field is given by $2xy\,\hat{i} - y^2\,\hat{j}$. The streamlines for this flow are given by the family of curves

(a) x^2y^2 = constant

(b) xy^2 = constant

(c) $2xy - y^2$ = constant

(d) xy – constant

[2016 : 1 Mark, Set-3]

40. For a two-dimensional flow, the velocity field is $\vec{u} = \frac{x}{x^2+y^2}\hat{i} + \frac{y}{x^2+y^2}\hat{j}$ where $\hat{i}$ and $\hat{j}$ are the basis vectors in the x – y Cartesian coordinate system. Identify the CORRECT statements from below.

1. The flow is incompressible
2. The flow is unsteady
3. y-component of acceleration,
$$a_y = \frac{-y}{(x^2+y^2)^2}$$
4. x-component of acceleration,
$$a_x = \frac{-(x+y)}{(x^2+y^2)^2}$$

(a) 2 and 3 (b) 1 and 3

(c) 1 and 2 (d) 3 and 4

[2016 : 2 Marks, Set-3]

41. For a steady flow, the velocity field is $\vec{V} = (-x^2 + 3y)\,\hat{i} + (2xy)\,\hat{j}$. The magnitude of the acceleration of a particle at (1, - 1) is

(a) 2 (b) 1

(c) $2\sqrt{5}$ (d) 0

[2017: 2 Marks, Set-1]

42. A two-dimensional incompressible frictionless flow field is given by $\vec{u} = x\hat{i} - y\hat{j}$. If ρ is the density of the fluid, the expression for pressure gradient vector at any point in the flow field is given as

(a) $\rho(x\hat{i} - y\hat{j})$ (b) $\rho(x\hat{i} + y\hat{j})$

(c) $-\rho(x^2\hat{i} + y^2\hat{j})$ (d) $-\rho(x\hat{i} + y\hat{j})$

[2019 : 1 Mark, Set-2]

ANSWERS

1. (b)	**4.** (d)	**5.** (d)	**6.** (b)	**7.** (b)	**8.** (a)	**9.** (a)	**11.** (c)	**12.** (c)	**13.** (b)
16. (c)	**17.** (c)	**18.** (b)	**19.** (d)	**20.** (b)	**21.** (d)	**22.** (a)	**23.** (c)	**24.** (c)	**25.** (b)
26. (d)	**27.** (c)	**28.** (a)	**29.** (d)	**30.** (c)	**31.** (a)	**32.** (d)	**33.** (b)	**34.** (d)	**35.** (a)
36. (b)	**38.** (d)	**39.** (b)	**40.** (b)	**41.** (c)	**42.** (d)				

EXPLANATIONS

1. Stream function,

$\psi = x^2 - y^2$

$$u = \frac{-\partial \psi}{\partial y} = \frac{-\partial}{\partial y}\left(x^2 - y^2\right) = 2y$$

$$\text{or } V = \frac{-\partial \psi}{\partial x} = \frac{\partial}{\partial x}\left(x^2 - y^2\right) = 2x$$

$$\vec{V} = u\hat{i} + v\hat{j} \text{ at } (1,1) = 2\hat{i} + 2\hat{j}$$

$$V = \sqrt{(2)^2 + (2)^2} = 2\sqrt{2}$$

2. 1 – d, 2 – d, 3 – b

3. In a flow field, closed contour exists.

4. $u = \lambda xy^3 - x^{2y}; \frac{\partial u}{\partial x} = \lambda y^3 - 2xy$

$$v = xy^2 \frac{-3}{4} y^4; \frac{\partial v}{\partial y} = 2xy - 3y^3$$

For 2D incompressible flow,

$$\frac{\partial u}{\partial x} + \frac{\partial v}{\partial y} = 0$$

$$\Rightarrow xy^3 - 2xy + 2xy - 3y^3 = 0$$

$$\boxed{\lambda = 3}$$

5. $\frac{\partial^2 \phi}{\partial x^2} + \frac{\partial^2 \phi}{\partial y^2} = 0 \ldots (i)$ (Laplace Equation should be satisfied)

$$\therefore u = -\frac{\partial \phi}{\partial x} \quad \text{and} \quad v = -\frac{\partial \phi}{\partial y}$$

$$\Rightarrow \frac{\partial u}{\partial x} = -\frac{\partial^2 \phi}{\partial x^2}, \text{ and } \quad \frac{\partial v}{\partial y} = -\frac{\partial^2 \phi}{\partial y^2}$$

Putting values in equation (i), we get

$$\left(-\frac{\partial u}{\partial x}\right) + \left(-\frac{\partial v}{\partial y}\right) = 0$$

$$\Rightarrow \frac{\partial u}{\partial x} = -\frac{\partial v}{\partial y}$$

6. Radial distance = 120 m

$$V_r = \frac{-60 \times 10^3}{2\pi r} \text{m/s}$$

$$\& \ V_\theta = \frac{300 \times 10^3}{2\pi r} \text{m/s}$$

$$\& \ \theta = \pi$$

$$\frac{V_R}{V_\theta} = \frac{-1}{5}$$

we know that

$$V_r = \frac{dr}{dt} \ldots \text{(i)}$$

$$\& \ V_\theta = rw = r\frac{d\theta}{dt} \quad -5V_r = r\frac{d\theta}{dt}$$

$$V_r = \frac{-r}{5}\frac{d\theta}{dt} \ldots \text{(ii)}$$

By equating (i) & (ii), we get

$$\frac{dr}{dt} = \frac{-r}{5}\frac{d\theta}{dt}$$

Integrating both sides, we get

$$\int_{120}^{r} \frac{dr}{r} = \frac{-1}{5}[\theta]_0^\pi$$

$$\ln r]_{120}^{r} = \frac{-1}{5} \times \pi$$

$$\ln\left(\frac{r}{120}\right) = \frac{-\pi}{5}$$

By solving above, we get

r = 64 m

7.

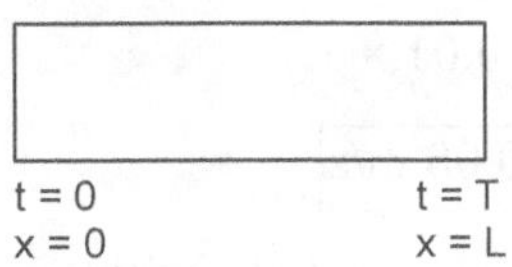

$$v = u_0\left(1 + \frac{3x}{L}\right)$$

$$\therefore \frac{dx}{dt} = u_0\left(1 + \frac{3x}{L}\right)$$

$$\Rightarrow \frac{dx}{u_0\left(1 + \frac{3x}{L}\right)} = dt$$

$$\text{Integrating } \int_0^L \frac{dx}{u_0\left(1 + \frac{3x}{L}\right)} = \int_0^T dt$$

$$\Rightarrow \frac{L}{3u_0}\left[\ln\left(1 + \frac{3x}{L}\right)\right]_0^L = T$$

$$\Rightarrow T = \frac{L}{3u_0}\ln(4) = \frac{L}{3u_0}\ln(4)$$

8. $V \times \pi r^2 = V_r \times 2\pi rh$

$$V_r = \frac{v.r}{2h}$$

9. Radial acceleration

$$a_r = V_r \times \frac{\partial Vr}{\partial r} + \frac{\partial V_r}{\partial t}$$

$$a_r = \frac{V.r}{2h} \times \frac{\partial}{\partial r}\left(\frac{Vr}{2h}\right) + \frac{\partial}{\partial t}\left(\frac{Vr}{\partial t}\right)$$

$$\frac{-\partial h}{\partial t} = V$$

$$\therefore a_r = \frac{Vr}{2h} \times \frac{Vr}{2h} + \frac{Vr}{2} \times \left(\frac{-1}{h^2}\frac{\partial h}{\partial t}\right)$$

$$\therefore a_r = \frac{V^2 r}{4h^2} + \frac{2V^2 r}{4h^2}$$

$$a_r = \frac{3V^2 r}{4h^2}$$

10.

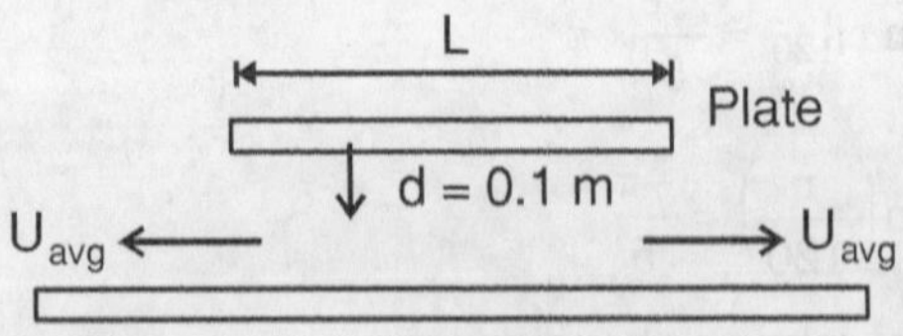

By mass conservation:-

$2(s \times u_{avg} \times db) = (s.\ u \times Lb)$

$\Rightarrow 2 \times v_{avg} \times 0.1 = 0.01 \times 1$

$$\boxed{u_{avg} = 0.05 \text{ m/s}}$$

11. Two dimensional velocity field with velocities u, v and along x and y direction.

$\therefore$ Acceleration along x direction, $a_x = a_{convective} + a_{temporal\ or\ local}$

$$= u\frac{\partial u}{\partial x} + v\frac{\partial u}{\partial y} + w\frac{\partial u}{\partial z} + \frac{\partial u}{\partial t}$$

Convective acceleration   temporal acceleration

Sicne, $\frac{\partial u}{\partial z} = 0$ for 2-dimensional field, therefore

Convective acceleration

$$= u\frac{\partial u}{\partial x} + v\frac{\partial u}{\partial y}$$

12. Convective acceleration is the effect of time independent acceleration of fluid with respect to space that means flow is steady non-uniform flow.

13. For continuous and incompressible flow

$u_x + u_y = 0$

$a_1 + b_2 = 0$

14. Given: $\bar{V} = \left(\frac{4}{t} + x + y\right)\hat{j}\,\frac{m}{s}$

$$\bar{V} = u\hat{i} + v\hat{j} + w\hat{k}$$

$$u = 0,\ w = 0$$

$$V = \left(\frac{4}{t} + x + y\right)\hat{j}$$

$$\vec{a} = \vec{a}_x\hat{i} + \vec{a}_y\hat{j} + \vec{a}_z\hat{k}$$

$$\vec{a}_x = u\frac{\partial u}{\partial x} + v\frac{\partial u}{\partial y} + w\frac{\partial u}{\partial z} + \frac{\partial u}{\partial t} = 0$$

$$\vec{a}_z = u\frac{\partial w}{\partial x} + v\frac{\partial w}{\partial y} + w\frac{\partial w}{\partial z} + \frac{\partial w}{\partial t} = 0$$

$$\vec{a}_y = u\frac{\partial v}{\partial x} + v\frac{\partial v}{\partial y} + w\frac{\partial v}{\partial z} + \frac{\partial v}{\partial t}$$

$$= v\frac{\partial v}{\partial y} + \frac{\partial v}{\partial t}$$

$$= \left(\frac{4}{t} + x + y\right) \times 1 + \left(-\frac{4}{t^2}\right)$$

$$\Rightarrow \quad \vec{a}_y = \left(\frac{4}{t} + x + y - \frac{4}{t^2}\right)\hat{j}$$

Now, At (x,y) = 1,1 and t = 2 sec.

Total acceleration is given by,

$$\vec{a} = \vec{a}_y = \left(\frac{4}{2} + 1 + 1 - \frac{4}{4}\right) = 3\frac{m}{s^2}$$

15. $\bar{V} = 3x^{2y}\hat{i} - 6xyz\hat{k}$

$$\frac{\partial u}{\partial x} + \frac{\partial v}{\partial y} + \frac{\partial w}{\partial z} = 6xy + 0 - 6xy = 0$$

Flow is incompressible

$$w_y = \frac{1}{2}\left[\frac{\partial u}{\partial z} \frac{-\partial w}{\partial x}\right]$$

$$= \frac{1}{2}[0 - (-6yz)]$$

$w_y = 3yz$

Flow is rotational.

Flow field is independent of time since terms does not include time 't'.

19. $\bar{v} = (x+2y+2)\hat{i} + (4-y)\hat{j}$

$u = x + 2y + 2,\ v = 4 - y$

$$\therefore \frac{\partial v}{\partial x} = 1, \frac{\partial v}{\partial y} = -1$$

$$\therefore \frac{\partial v}{\partial x} + 1, \frac{\partial v}{\partial y} = 0,$$

hence incompressible.

$$\text{Again, } \omega = \frac{1}{2}\left(\frac{\partial v}{\partial x} - \frac{\partial v}{\partial y}\right)$$

$$= \frac{1}{2}(0-2) = -1.$$

hence not irrotational.

20. Clearly zero shear stress and vortex.

21. The basic equation of continuity for fluid flow is given by

$$\frac{\rho u}{u} \quad \frac{\rho v}{y} \quad \frac{\rho w}{z} \quad \frac{\rho}{t}$$

Now if ρ remains constant, then only we can write

$$\vec{\nabla} \bullet \vec{V} = 0$$

$$\text{i.e. } \frac{\partial u}{\partial x} + \frac{\partial v}{\partial y} + \frac{\partial w}{\partial z} = 0$$

hence incompressible flow

22. Incompressible flow condition

$$\nabla \cdot \vec{V} = 0$$

24. C is the false statement

2D incompressible flow continuity equation.

$$\frac{\partial u}{\partial x} + \frac{\partial v}{\partial y} = 0$$

$$\frac{\partial (Ax)}{\partial x} + \frac{\partial (-Ay)}{\partial y} = 0$$

A – A = 0 it satisfies continuity equation.

⇒ As $\vec{V} = Ax\hat{i} - Ay\hat{j}$

As $y \to \infty$ velocity vector field will not be defined along y axis.

So flow will be along x-axis i.e. 1-D flow.

⇒ Stream line equation for 2D

$$\frac{dx}{u} = \frac{dy}{v}$$

$$\frac{dx}{Ax} = \frac{dy}{-Ay}$$

$\ln x = -\ln y + \ln c$

$\ln xy = \ln c$

$xy = c \to$ streamline equation

25. x direction scalar of velocity field,

$$u = \frac{dx}{dt}$$

$$u = -kx.e^{-kt}$$

y direction scalar of velocity field

$$v = \frac{dy}{dt}$$

$$v = ky_0e^{kt}$$

$$\vec{V} = u\hat{i} + v\hat{j}$$

$$\vec{V} = -kx_0e^{-kt}\hat{i} + ky_0e^{kt}\hat{j}$$

u & v are non zero scalar $t \geq 0$ so it is 2D flow.

2D possible flow field

$$\frac{\partial u}{\partial x} + \frac{\partial v}{\partial y} = 0$$

$$\frac{\partial}{\partial x}\left(-kx_0e^{-kt}\right) + \frac{\partial}{\partial y}\left(ky_0e^{kt}\right) = 0$$

0 + 0 = 0 continuity satisfied.

$$\frac{\partial u}{\partial t} = +k^2x_0e^{-kt}$$

$$\frac{\partial v}{\partial t} = k^2y_0e^{kt}$$

$$\frac{\partial u}{\partial t} \neq 0$$

$$\frac{\partial v}{\partial t} \neq 0$$

So, flow is unsteady.

26. $U_r = \dfrac{\partial \phi}{\partial r} = \dfrac{-q}{2\pi r}$

$$\phi = \frac{-q}{2\pi}\ln r + c$$

(2) 1 –d, 2 – d, 3 - c

(3) Closed contour in a flow field.

28. $\vec{V} = x^2y\hat{i} + 2xy^2z\hat{j} - yz^2\hat{k}$

Shear strain

$$\sum_{yz} = \frac{1}{2}\left(\frac{\partial v}{\partial z} + \frac{\partial w}{\partial y}\right) = \frac{1}{2}\left(2xy^2 - z^3\right)$$

at $(-2\pi, 2)$, we get

$$\sum_{yz} = \frac{1}{2}(-4 - 8) = -6$$

30. Given: $u_x = ax$ and $u_y = ay$

Equation of steam line is,

$$\frac{du}{u_x} = \frac{dy}{u_y}$$

$$\Rightarrow \frac{du}{ax} = \frac{dy}{ay}$$

Integrating both sides, we have

$\log(ax) = \log(ay) + \log c$

or $ax = c \cdot ay$

or $x = cy$

Since the steam line is passed through point (1, 2), therefore

$1 = 2c$

$$\Rightarrow c = \frac{1}{2}$$

$$\therefore x = \frac{y}{2}$$

Hence equation of steam line is

$2x - y = 0.$

31. Given, $u = x^2t$ and $v = -2xyt$

We know, $\frac{\partial\psi}{\partial y} = v = -2xyt$ ...(*i*)

$\frac{\partial\psi}{\partial y} = -u = -x^2t$ ...(*ii*)

Integrating equation (*i*), we get

$\psi = -x^2yt + f(y)$ ...(*iii*)

Differentiating equation (*iii*) with respect to *y*, we get

$\frac{\partial\psi}{\partial y} = -x^2t + f(y)$ ...(*v*)

Equating the value of $\frac{\partial\psi}{\partial y}$ from equations (*ii*) and (*iv*), we get

$-x^2t = -x^2t + f'(y)$

Since, $f'(y) = 0$, thus $f(y) = C$

(where 'C' is constant of integration)

$\psi = -x^2yt + C$

C is a numerical constant so it can be taken as zero

$\psi = -x^2yt$

For equation of stream lines,

ψ = constant

$-x^2yt$ = constant

For a particular instance,

x^2y = constant

32. $\bar{v} = 2xyi - x^2zj$

Velocity of a vector = $\bar{\nabla} \times \bar{v}$

$$= \begin{vmatrix} i & j & k \\ \frac{\partial}{\partial x} & \frac{\partial}{\partial y} & \frac{\partial}{\partial z} \\ 2xy & -x^2z & 0 \end{vmatrix}$$

$$= \hat{i}\left(0 - \frac{\partial}{\partial z}(-x^2z)\right) - \hat{j}\left(0 - \frac{\partial}{\partial z}(2xy)\right) + \hat{k}\left(\frac{\partial}{\partial x}(-x^2z) - \frac{\partial}{\partial y}(2xy)\right)$$

$$= \hat{i}(-x^2) + \hat{k}(-2\times z - 2x)$$

At (1, 1, 1),

$\bar{\nabla} \times \bar{v} = \hat{i} + \hat{k}(-2-2) = i - 4k$

33. $\left(\frac{dy}{dx}\right)_\phi \times \left(\frac{dy}{dx}\right)_\psi = -1$

Slope of equipotential Line

× slope of stream function = – 1

They are orthogonal to each other.

36. Here $V(x, y) = u(x, y)\hat{i} + v(x, y)\hat{j}$

As we know that $u = \frac{-\partial\phi}{\partial x}$

$\therefore \quad \frac{-\partial\phi}{\partial x} = -u(x, y) \qquad ...(i)$

Similarly $v = \frac{-\partial\phi}{\partial y}$

$\therefore \quad \frac{\partial\phi}{\partial y} = -v(x, y) \qquad ...(ii)$

From equations (*i*) and (*ii*) we get,

$$\frac{\partial y}{\partial x} = \frac{\partial\phi}{\partial x} \times \frac{\partial y}{\partial\phi} = \frac{+u(x, y)}{-v(x, y)} = \frac{-u}{v}$$

37. $\frac{\partial u}{\partial x} + \frac{\partial v}{\partial y} + \frac{\partial w}{\partial z} = 0$

$a_1 + b_2 + c_3 = 0$

$2 - 4 + b_2 = 0$

$b_2 = 2$

39. $v = 2xy\, i - y^2 j$

$u = \frac{\partial x}{\partial y},\ v = -\frac{\partial x}{\partial x}$

$2xy = \frac{\partial x}{\partial y} = 2$

$2xy\partial y = \partial\psi$

on integrating

$\psi = xy^2 + f(x)$

Now $\frac{\partial x}{\partial x} = y^2 + f'(x) - v = y^2 + f'(x) - (-y^2)$

$= y^1 + f'(x)$

$f'(x) = 0$

$\Rightarrow f(x) = \text{constant}$

so $\psi = xy^2 + \text{constant}$

40. $a_x = u\frac{\partial u}{\partial x} + v\frac{\partial u}{\partial y}$

$$= \frac{x}{x^2 + y^2}\left(\frac{x^2 + y^2 - x \times 2x}{(x^2 + y^2)^2}\right) - \frac{y}{(x^2 + y^2)} \times x \times \frac{1}{(x^2 + y^2)} \times 2y$$

$$= \frac{x(x^2 + y^2 - 2x^2) - 2xy^2}{(x^2 + y^2)(x^2 + y^2)} = \frac{-x^3 - xy^2}{(x^2 + y^2)}$$

$$\therefore \quad a_x = \frac{x}{(x^2 + y^2)^2}$$

$$a_y = u\frac{\partial v}{\partial x} + v\frac{\partial v}{\partial y}$$

$$= \frac{x}{(x^2 + y^2)} \times \frac{-y}{(x^2 + y^2)^2} \times 2x \times \frac{y}{(x^2 + y^2)} \times \left(\frac{(x^2 + y^2) - y \times 2y}{(x^2 + y^2)^2}\right)$$

$$= \frac{-2x^2 y + yx^2 - y^3}{(x^2 + y^2)^3} = \frac{-y}{(x^2 + y^2)^2}$$

The velocity components are not functions of time, so flow is steady according to continuity equation,

$$\frac{\partial u}{\partial x} + \frac{\partial v}{\partial y} = \frac{-(x^2 - y^2)}{(x^2 + y^2)^2} + \frac{(x^2 - y^2)}{(x^2 + y^2)^2} = 0$$

Since it satisfies the above continuity equation for 2D and incompressible flow.

∴ The flow is incompressible.

41. $\vec{V} = (-x^2 + 3y)i + (2xy)j$

$u = -x^2 + 3y,\ v = 2xy$

$$\frac{\partial u}{\partial x} = -2x, \quad \frac{\partial v}{\partial x} = 2y$$

$$\frac{\partial u}{\partial y} = 3, \quad \frac{\partial v}{\partial y} = 2x$$

Now magnitude of particle at (1, –1)

$$u = -4,\ v = -2,\ \frac{\partial u}{\partial x} = -2,\ \frac{\partial u}{\partial y} = 3,\ \frac{\partial v}{\partial x}$$

$$= -2,\ \frac{\partial v}{\partial y} = 2$$

$$a_x = u\frac{\partial u}{\partial x} + v\frac{\partial u}{\partial y} = (-4)\times(-2) + (-2)\times 3$$

$$a_x = 8 - 6 = 2$$

$$a_y = u\frac{\partial v}{\partial x} + v\frac{\partial v}{\partial y} = [(-4)\times(-2) + (-2)\times 2]$$

$$a_y = 8 - 4 = 4$$

$$a = \sqrt{a_x^2 + a_y^2} = \sqrt{2^2 + 4^2} = 2\sqrt{5}$$

42. From Navier stokes equation,

$$\vec{V} = x\hat{i} - y\hat{j}$$

$u = x,\ v = -y$

x-momentum

$$\cancelto{0}{\frac{\partial(\rho u)}{\partial t}} + \frac{\partial(\rho u^2)}{\partial x} + \frac{\partial(\rho uv)}{\partial y} + \cancelto{0}{\frac{\partial(\rho uw)}{\partial z}}$$

$$= \frac{-\partial p}{\partial x} + \frac{1}{R_{cr}}\left[\cancelto{0}{\frac{\partial \tau_{xx}}{\partial x}} + \cancelto{0}{\frac{\partial \tau_{yy}}{\partial y}} + \cancelto{0}{\frac{\partial \tau_{zz}}{\partial t}}\right]$$

$$\frac{-\partial p}{\partial x} = \frac{\partial(\rho.x^2)}{\partial x} + \frac{\partial(\rho.x.-y)}{\partial y}$$

$$= 2\rho x - \rho x$$

$$\frac{\partial p}{\partial x} = -\rho x$$

Similarly, $\frac{\partial p}{\partial y} = -\rho y$

So, pressure gradient vector

$$= \frac{\partial p}{\partial y}\hat{i} + \frac{\partial p}{\partial y}\hat{j}$$

$$= -\rho x\hat{i} - \rho y\hat{j}$$

$$\boxed{\Delta p = -\rho(x\hat{i} + y\hat{j})}$$

4
CHAPTER

Fluid Dynamics

Analysis of Previous GATE Papers

Topic	Year/Set → Marks Distribution ↓		2019 Set 1	2019 Set 2	2018 Set 1	2018 Set 2	2017 Set 1	2017 Set 2	2016 Set 1	2016 Set 2	2016 Set 3	2015 Set 1	2015 Set 2	2015 Set 3
Bernoulli's equation, conservation of energy and mass	1 Mark	MCQ Type											1	
		Numerical Type												
	2 Marks	MCQ Type				1								
		Numerical Type												
	Total					2							1	
Flow measuring devices	1 Mark	MCQ Type												
		Numerical Type												
	2 Marks	MCQ Type												
		Numerical Type						1				1		1
	Total							2				2		2
Flow through orifice and mouthpiece	1 Mark	MCQ Type												
		Numerical Type												
	2 Marks	MCQ Type			1									
		Numerical Type				1					1			
	Total				2	2					2			

Bernoulli's equation, conservation of energy and mass

1. Bernoulli's equation can be applied between any two points on a stream line for a rotational flow field. State : (T/F)

[1994 : 1 Mark]

2. Navier Stoke's equation represents the conservation of

(a) Energy

(b) Mass

(c) Pressure

(d) Momentum

[2000 : 1 Mark]

3. The following data about the flow of liquid was observed in a continuous Chemical process plant:

Flow rate (litres/sec.)	7.5 to 7.7	7.7 to 7.9	7.9 to 8.1	8.1 to 8.3	8.3 to 8.5	8.5 to 8.7
Frequency	1	5	35	17	12	10

Mean flow rate of the liquid is

(a) 8.00 liters/s

(b) 8.06 liters/s

(c) 8.16 liters/s

(d) 8.26 liters/s

[2004 : 2 Marks]

4. Consider steady, incompressible and irrotational flow through a reducer in a horizontal pipe where the diameter is reduced from 20 cm to 10 cm. The pressure in the 20 cm pipe just upstream of the reducer is 150 kPa. The fluid has a vapour pressure of 50 kPa and a specific weight of 5 kN/m^3. Neglecting frictional effects, the maximum discharge (in m^3/s) that can pass through the reducer without causing cavitation is

(a) 0.05

(b) 0.16

(c) 0.27

(d) 0.38

[2009 : 2 Marks]

5. Consider steady flow of water in a situation where two pipe lines (pipe 1 and pipe 2) combine into a single pipe line (pipe 3) as shown in figure. The cross-sectional'area of all three pipelines are constant. The following data is given

Pipe number	Area (m^2)	Velocity (m/s)
1	1	1
2	2	2
3	2.5	?

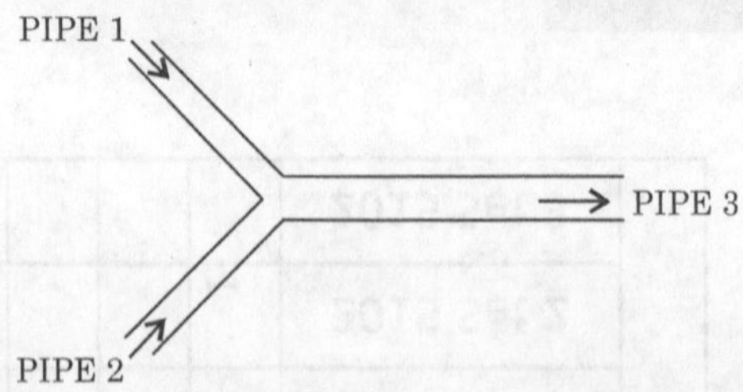

Assuming the water properties and the velocities to be uniform across the cross-section of the inlets and the outlet, the exit velocity (in m/s) in pipe 3 is

(a) 1

(b) 1.5

(c) 2

(d) 2.5

[2009 : 2 Marks]

6. Within a boundary layer for a steady incompressible flow, the Bernoulli equation

(a) holds because the flow is steady

(b) holds because the flow is incompressible

(c) holds because the flow is transitional

(d) does not hold because the flow is frictional

[2015 : 1 Mark, Set-2]

7. Air flows at the rate of 1.5 m^3/s through a horizontal pipe with a gradually reducing cross-section as shown in the figure. The two cross-sections of the pipe have diameters of 400 mm and 200 mm. Take the air density as 1.2 kg/m^3 and assume inviscid incompressible flow. The change in pressure ($p_2 - p_1$) (in kPa) between sections 1 and 2 is

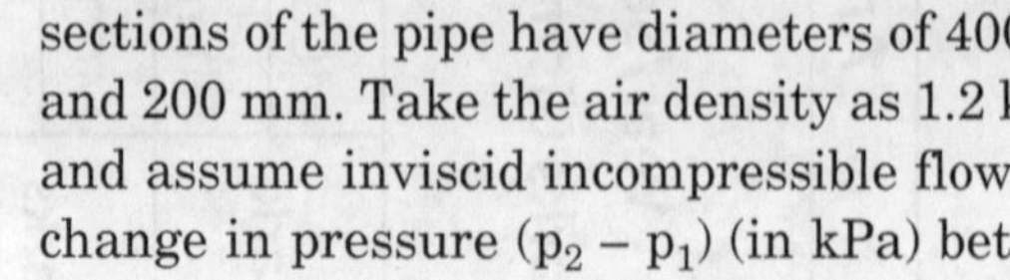

(a) –1.28

(b) 2.56

(c) –2.13

(d) 1.28

[2018 : 2 Marks, Set-2]

Flow measuring devices

8. A mercury manometer is attached to a section of the pipe shown in the figure. Mercury levels are indicated when there is no water flowing through the pipe. When water starts flowing through the pipe continuously at constant rate in the direction of the arrow, the level of mercury at B.

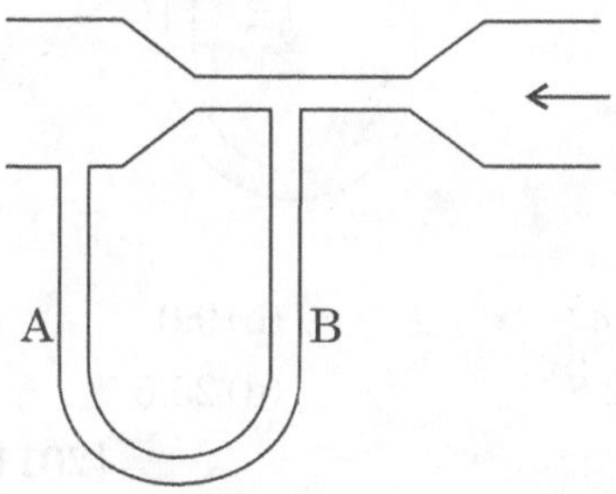

[1987 : 2 Marks]

9. Water flows through a pipe of diameter 0.30 m. What would be the velocity V for the conditions shown in the figure below?

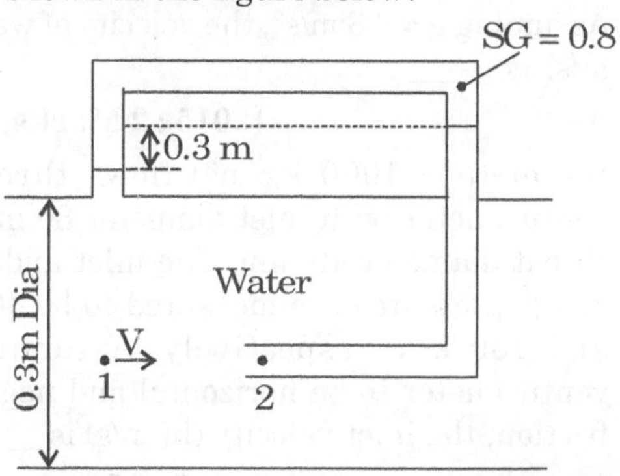

[1988 : 2 Marks]

10. In a hand operated liquid sprayer (figure shown below) the liquid from the container rises to the top of the tube because of

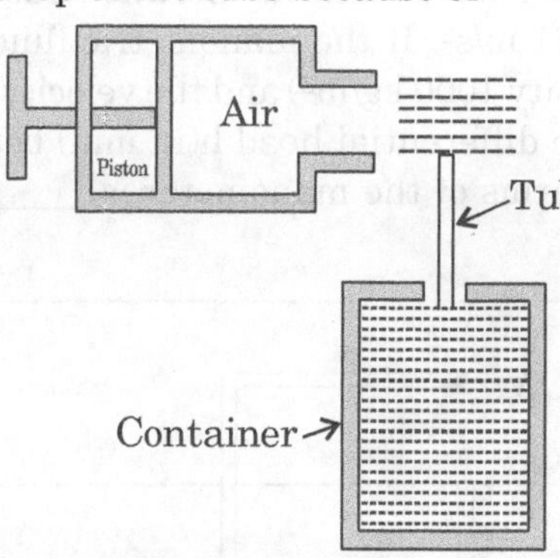

(a) Capillary effect

(b) Suction produced by the air jet at the top end of tube

(c) Suction produced by the piston during the backward stroke

(d) Pumping of the air into the container

[1990 : 1 Mark]

11. In a venturimeter, the angle of the diverging section is more than that of converging section. State: (T/F)

[1994 : 1 Mark]

12. Water flows though a vertical contraction from a pipe of diameter d to another of diameter d/2 (see fig.) The flow velocity at the inlet to the contraction is 2 m/s and pressure 200 kN/m². If the height of the contraction measures 2m, then pressure at the exit of the contraction will be very nearly

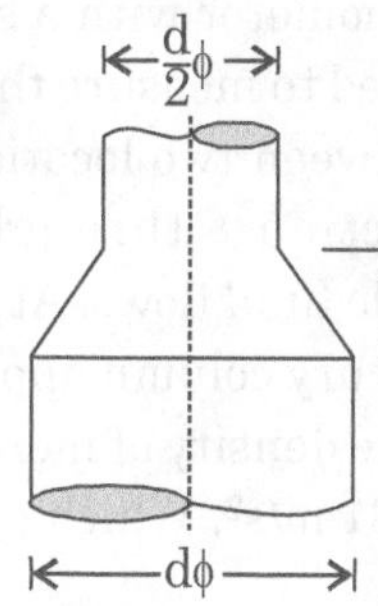

(a) 168 kN/m² (c) 150 kN/m²

(b) 192 kN/m² (d) 174 kN/m²

[1999 : 2 Marks]

13. Air flows through a venturi and into atmosphere. Air density is ρ_a; atmospheric pressure is ρ_a; throat diameter is D_t; exit diameter is D and exit velocity is U. The throat is connected to a cylinder containing a frictionless piston attached to a spring. The spring constant is k. The bottom surface of the piston is exposed to atmosphere. Due to the flow, the piston moves by distance x. Assuming incompressible frictionless flow, x is

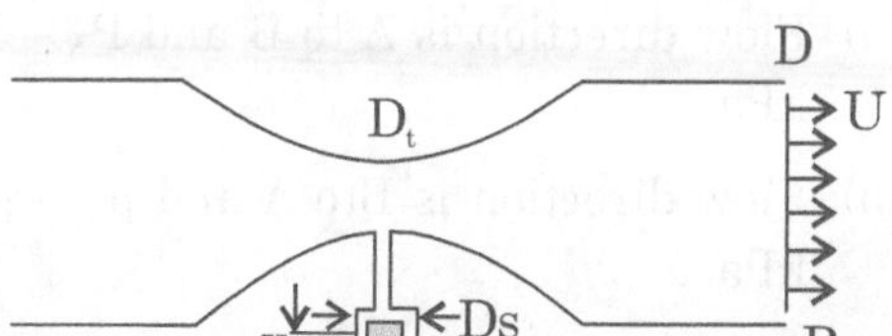

(a) $(\rho U^2/2k)\pi D_s^2$

(b) $(\rho U^2/8k)\left(\frac{D^2}{D_t^2}-1\right)\pi D_s^2$

(c) $(\rho U^2/2k)\left(\frac{D^2}{D_t^2}-1\right)\pi D_s^2$

(d) $(\rho U^2/8k)\left(\frac{D^4}{D_t^4}-1\right)\pi D_s^2$

[2003 : 2 Marks]

14. A venturimeter of 20 mm throat diameter is used to measure the velocity of water in a horizontal pipe of 40 mm diameter. If the pressure difference between the pipe and throat sections is found to be 30 kPa then, neglecting frictional losses, the flow velocity is

(a) 0.2 m/s (b) 1.0 m/s
(c) 1.4 m/s (d) 2.0 m/s

[2005 : 2 Marks]

15. A U-tube manometer with a small quantity of mercury is used to measure the static pressure difference between two locations A and B in a conical section through which an incompressible fluid flows. At a particular flow rate, the mercury column appears as shown in the figure. The density of mercury is 13600 kg/m^3 and g - 9.81 m/s^2. Which of the following is correct?

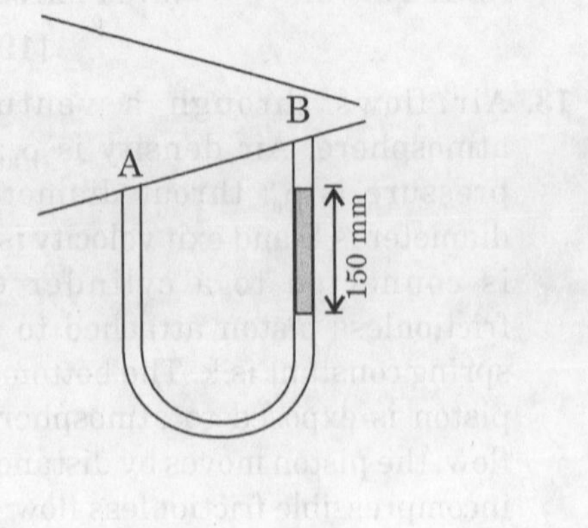

(a) Flow direction is A to B and $P_A - P_B = 20$ kPa

(b) Flow direction is Bto A and $p_A - p_B = 1.4$ kPa

(c) Flow direction is A to B and $p_B - p_A = 20$ kPa

(d) Flow direction is B to A and $p_B - p_A = 1.4$ kPa

[2005 : 2 Marks]

16. Figure shows the schematic for the measurement of velocity of air (density = 1.2 kg/m^3) through a constant-area duct using a pitot tube and a water tube manometer. The differential head of water (density = 1000 kg/m^3) in the two columns of the manometer is 10 mm. Take acceleration due to gravity as 9.8 m/s^2. The velocity of air in m/s is

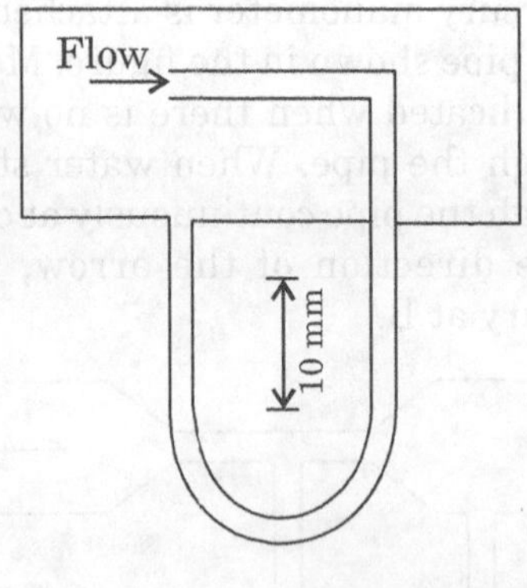

(a) 6.4 (b) 9.0
(c) 12.8 (d) 25.6

[2011 : 2 Marks]

17. A Prandtl tube (Pitot-static tube with C = 1) is used to measure the velocity of water. The differential manometer reading is 10 mm of liquid coiumn with a relative density of 10. Assuming g = 9.8 m/s^2, the velocity of water (in m/s) is _____.

[2015 : 2 Marks, Set-3]

18. Water (ρ = 1000 kg/m^3) flows through a venturimeter with inlet diameter 80 mm and throat diameter 40 mm. The inlet and throat gauge pressures are measured to be 400 kPa and 130 kPa respectively. Assuming the venturimeter to be horizontal and neglecting friction, the inlet velocity (in m/s) is _____.

[2015 : 2 Marks, Set-1]

19. The arrangement shown in the figure measures the velocity V of a gas of density 1 kg/m^3 flowing through a pipe. The acceleration due to gravity is 9.81 m/s^2. If the manometric fluid is water (density 1000 kg/m^3) and the velocity V is 20 m/s, the differential head h(in mm) between the two arms of the manometer is _____.

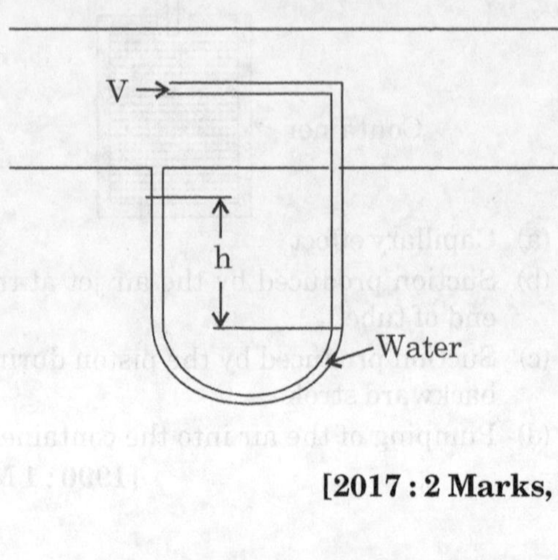

[2017 : 2 Marks, Set-2]

Flow through orifice and mouthpiece

20. A jet of water issues from a nozzle with a velocity 20 m/s and it impinges normally on a flat plate moving away from it at 10 m/s. The cross-sectional area of the jet is 0.01 m^2, and the density of water = 1000 kg/m^3. The force developed on the plate is

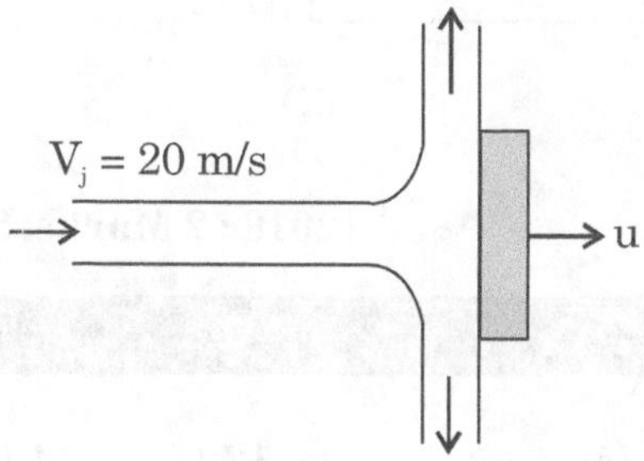

(a) 1000 N (b) 100 N
(c) 10 N (d) 2000 N

[1990 : 2 Marks]

21. A fan in the duct shown below sucks air from the ambient and expels it as a jet at 1 m/s to the ambient. Determine the gauge pressure at the point marked as A. Take the density of air as 1 kg/m^3.

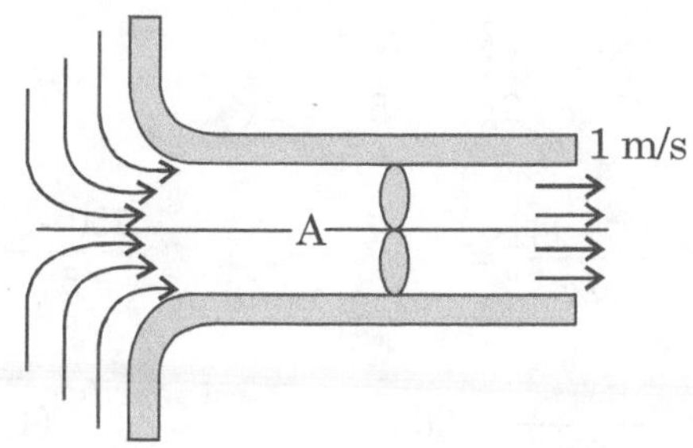

[1993 : 2 Marks]

22. A water container is kept on a weighing balance. Water from a tap is falling vertically into the container with a volume flow rate of Q; the velocity of the water when it hits the water surface is U. At a particular instant of time the total mass of the container and water is m. The force registered by the weighing balance at this instant of time is

(a) $mg + \rho QU$ (b) $mg + 2\rho QU$
(c) $mg + \rho QU^2/2$ (d) $\rho QU^2/2$

[2003 : 2 Marks]

23. Water is coming out from a tap and a falls vertically downwards. At thc tap opening, the stream diameter is 20 mm with uniform velocity of 2 m/s. Acceleration due to gravity is 9.81 m/s^2. Assuming steady, inviscid flow, constant atmospheric pressure everywhere and neglecting curvature and surface tension effects, the diameter in mm of the stream 0.5 m below the tap is approximately

(a) 10 (b) 15
(c) 20 (d) 25

[2013 : 2 Marks]

24. The water jet exiting from a stationary tank through a circular opening of diameter 300 mm impinges on a rigid wall as shown in the figure. Neglect all minor losses and assume the water level in the tank to remain constant. The net horizontal force experienced by the wall is ______ kN. Density of water is 1000 kg/m^3.

Acceleration due to gravity g = 10 m/s^2.

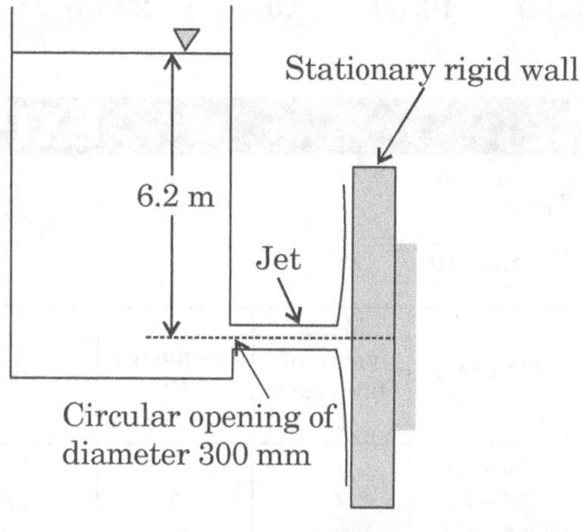

[2016 : 2 Marks, Set-3]

25. A tank open at the top with a water level of 1 m, as shown in the figure, has a hole at a height of 0.5 m. A free jet leaves horizontally from the smooth hole. The distance X(in m) where the jet strikes the floor is

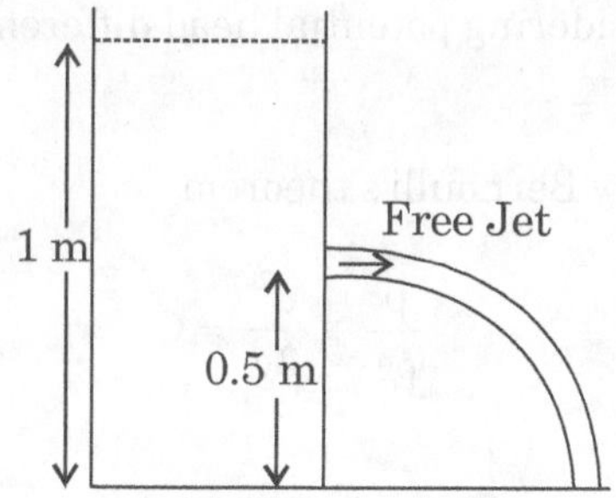

(a) 0.5 (b) 1.0
(c) 2.0 (d) 4.0

[2018 : 2 Marks, Set-1]

26. A frictionless circular piston of area $10^{-2}m^2$ and mass 100 kg sinks into a cylindrical container of the same area filled with water of density 1000 kg/m³ as shown in the figure.

The container has a hole of area $10^{-3}m^2$ at the bottom that is open to the atmosphere. Assuming there is no leakage from the edges of the piston and considering water to be incompressible, the magnitude of the piston velocity (in m/s) at the instant shown is ______ (correct to three decimal places).

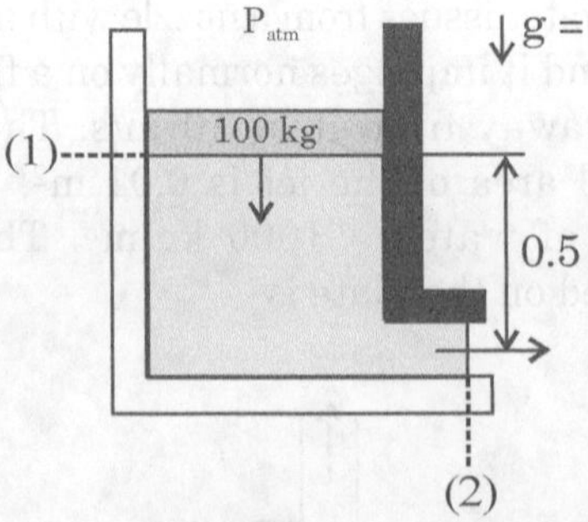

[2018 : 2 Marks, Set-2]

ANSWERS

2. (d)	**3.** (c)	**4.** (b)	**5.** (c)	**6.** (d)	**7.** (a)	**10.** (c)	**12.** (c)	**13.** (d)	**14.** (d)
15. (a)	**16.** (c)	**20.** (a)	**22.** (a)	**23.** (b)	**25.** (b)				

EXPLANATIONS

1. True
2. Momentum
3.

Flow rate (litre/sec)	Mean value of flow rate (x)	Frequency (f)	fx
7.5 – 7.7	7.6	1	7.6
7.7 – 7.9	7.8	5	39
7.9 – 8.1	8.0	35	280
8.1 – 8.3	8.2	17	139.4
8.3 – 8.5	8.4	12	100.8
8.5 – 8.7	8.6	10	86
		$\Sigma f = 80$	$\Sigma fx = 652.8$

$\therefore$ Mean flow rate $= \dfrac{\Sigma fx}{\Sigma f} = \dfrac{652.8}{80} = 8.16.$

4. Considering potential head difference = 0,

i.e $z_1 = z_2$

Apply Bernoulli's theorem

$$\frac{p}{\rho g} + \frac{v^2}{2g} = C$$

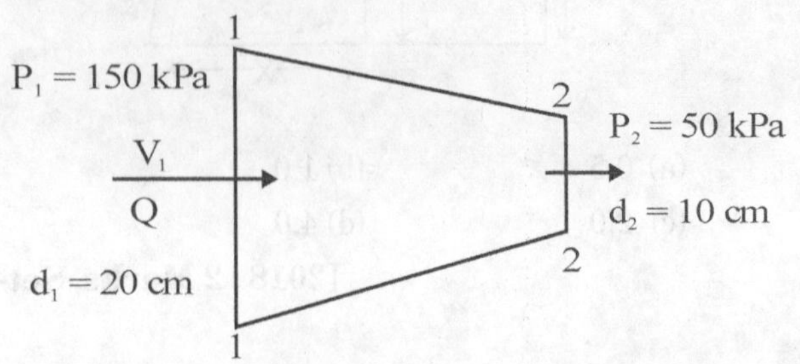

$$\frac{p_1}{w_1} + \frac{v_1^2}{2g} = \frac{p_2}{w_2} + \frac{v_2^2}{2g}$$

But $w_1 = w_2 = w = 5$ (incompressible flow)

$$\therefore \quad \frac{150}{5} + \frac{v_1^2}{2g} = \frac{50}{5} + \frac{v_2^2}{2g}$$

$$\text{or} \quad \frac{v_2^2 - v_1^2}{2g} = \frac{150-5}{5}$$

$$\text{or} \quad \frac{v_2^2 - v_1^2}{2g} = 20\ m \qquad ....(1)$$

Also, discharge, $Q = \left(\dfrac{\pi}{4}d_1^2\right)v_1 = \left(\dfrac{\pi}{4}d_2^2\right)v_2$

$$\text{or} \quad \frac{v_1}{v_2} = \left(\frac{d_2}{d_1}\right)^2 = \left(\frac{10}{20}\right)^2$$

$$\text{or} \quad v_1 = \frac{v_2}{4} \qquad ...(2)$$

From equations (1) and (2)

$$\frac{v_2^2 - \frac{v_2^2}{16}}{2g} = 20$$

$$\text{or} \quad \frac{15v_2^2}{32g} = 20$$

or $v_2 = \sqrt{\frac{20 \times 32g}{15}} = 20.45$ m/s

$\therefore$ Discharge, $Q = \left(\frac{\pi}{4} d_2^2\right) v_2$

$= \frac{\pi}{4} (0.1)^2 \times 20.45 = 0.16$ m³/sec

5. Q1 + Q2 = Q3

$A_1V_1 + A_2V_3 = A_3V_3$

$1 + 4 = 2.5\, V_3$

or $V_3 = 2$ m/s

6. Bernoulli equation does not hold because the flow is frictional.

7. $\frac{p_1}{\delta g} + \frac{V_1{}^2}{2g} + Z_1 = \frac{P_2}{\delta g} + \frac{V_2{}^2}{2g} + Z_2\,[Z = \text{cons tan t}]$

$$A_1 = \frac{\pi}{4} \times 0.4^2 = 0.1256 \text{ m}^2$$

$$A_2 = \frac{\pi}{4} \times 0.2^2 = 0.0314 \text{ m}^2$$

$$\frac{P_2 - P_1}{\delta g} = \frac{V_1^2 - V_2^2}{2g}$$

$$\frac{P_2 - P_1}{1.2} = \frac{1}{2}\left[\frac{Q^2}{A_1^2} - \frac{Q^2}{A_2^2}\right]$$

$$P_2 - P_1 = 0.6 \times (1.5)^2 \left[\frac{1}{A_1^2} \frac{-1}{A_2^2}\right]$$

$$P_2 - P_1 = 1.35\left[\frac{1}{0.01577} - \frac{1}{0.00985}\right]$$

$= 1.35\,[63.41 - 1015.22]$

$= -1.28 \times 10^3$ Pascal

$= -1.28$ kPa

8. When flow starts, pressure at B will be lower than pressure at A, therefore level of mercury at B rises.

9. $h = x\left[1 - \frac{\rho_{mano}}{\rho_{pipe}}\right]$

$$= 0.3\left[1 - \frac{800}{1000}\right]$$

= 0.06 m of water

$V_1 = \sqrt{2gh} = \sqrt{2 \times 9.8 \times 0..6}$

= 1.084 m/s

11. False

12. From continuity equation,

$A_1 v_1 = A_2 v_2$

or $\frac{\pi}{4} \times d^2 \times 2 = \frac{\pi}{4}\left(\frac{d}{2}\right)^2 \times v_2$

or $v_2 = 8$ m/s.

Applying Bernoulli's theorem,

$$\frac{p_1}{w} + \frac{v_1^2}{2g} + z_1 = \frac{p_2}{w} + \frac{v_2^2}{2g} + z_2$$

or $\frac{200 \times 1000}{9810} + \frac{(2)^2}{2 \times 9.81} + 0$

$= \frac{p_2}{w} + \frac{(8)^2}{2 \times 9.81} + 2$

or $p_2 = 150.38$ kN/m²

13.

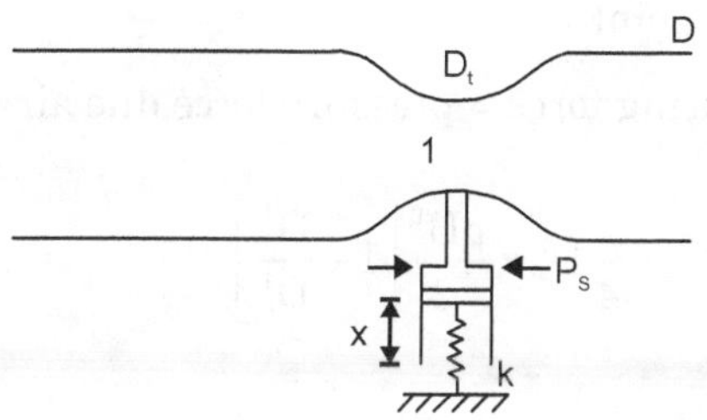

Applying Bernoulli's equation at points (1) and (2), we have

$$\frac{P_1}{\rho g} + \frac{v_1^2}{2g} + z_1 = \frac{P_2}{\rho g} + \frac{v_2^2}{2g} + z_2$$

Since venturi is horizontal

$z_1 = z_2$

Now $\left(\frac{P_1}{\rho g} - \frac{P_2}{\rho g}\right) = \frac{v_2^2}{2g} - \frac{v_1^2}{2g}$

$\Rightarrow (P_1 - P_2) = \frac{\rho g}{2g}(v_2^2 - v_1^2) = \frac{\rho}{2}(v_2^2 - v_1^2)$

Since $P_2 = P_a$ = atmospheric pressure

$\therefore (P_1 - P_a) = \frac{\rho}{2}\left(v_2^2 - v_1^2\right)$ ...(*i*)

Applying continuity equation at points (*i*) and (*ii*), we have

$A_1 v_1 = A_2 v_2$

$\Rightarrow v_1 = \left(\frac{A_2}{A_1}\right) v_2$ since $V_2 = U$

$$v_1 = \left(\frac{\frac{\pi}{4}D^2}{\frac{\pi}{4}D_t^2}\right) U$$

$$\Rightarrow v_1 = \left(\frac{D}{D_t}\right)^2 U$$

From equation (*i*),

$$P_1 - P_a = \frac{\rho}{2}\left[v^2 - \left(\frac{D}{D_t}\right)^2 U^2\right]$$

$$= \frac{\rho}{2} U^2 \left[1 - \frac{D^4}{D_t^4}\right]$$

At point P

Spring force = pressure force due air

$$-kx = \frac{\pi}{4} D_s^2 \times \frac{\rho U^2}{2}\left[1 - \frac{D^4}{D_t^4}\right]$$

$$\Rightarrow x = \frac{\pi}{8}\frac{D_s^2 \rho U^2}{k}\left[1 - \frac{D^4}{D_t^4}\right]$$

14. We know, $A_1 V_1 = A_2 V_2$

Þ $V_2 = \frac{D_1^2}{D_2^2} V_1 = \frac{16}{4} V_1$

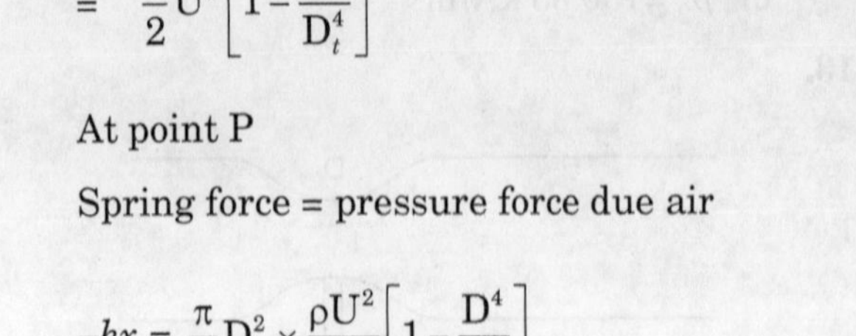

$\therefore$ $\boxed{V_2 = 4V_1}$

Applying Bernoulli's Equation

$$\frac{P_1}{\rho g} + \frac{V_1^2}{2g} + z_1 = \frac{P_2}{\rho g} + \frac{V_2^2}{2g} + z_2$$

$$\frac{P_1 - P_2}{eg} = \frac{V_2^2 - V_1^2}{2g}$$

Þ $\frac{15V_1^2}{2} = \frac{30 \times 10^3}{1000}$

Þ $V_1^2 = 4$

Þ $V_1 = 2.0$ m/s

So velocity of flow is 2.0 m/sec

15.

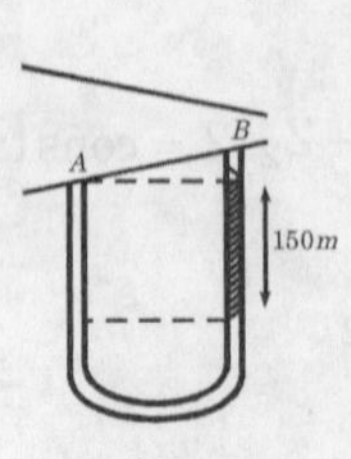

$P_A - P_B = rg\ .\ Dh$

$= 13600 \times 9.81 \times .15$

$= 20$ kPa

As pressure is decreasing from A to B, so flow direction is A to B.

16. From Bernoulli's equation

$$\frac{V_1^2 - V_2^2}{2g} = \frac{p_2 - p_1}{\rho_a g}$$

$$\Rightarrow V_1 = \sqrt{\frac{2(p_2 - p_1)}{\rho_a}}$$

But $p_2 - p_1 = (\rho g h)_{water}$

$= 9810 \times 10 \times 10^{-3}$

$= 98.1\ N/m^2$

$$\therefore V_1 = \sqrt{\frac{2 \times 98.1}{1.2}} = 12.8 \text{ m/s}$$

17. Velocity as water = $C_v\sqrt{2gh}$

$C_v = 1$ (Given)

$$h = x\left[\frac{s_g}{s_0} - 1\right]$$

$= 0.01\,(10 - 1) = 0.09$ m

$\therefore$ velocity of flow =

$\sqrt{2 \times 9.8 \times 0.09} = 1.328\,\text{m/s}$

18. $\dfrac{p_1}{\rho g} + \dfrac{v_1^2}{2g} = \dfrac{p_2}{\rho g} + \dfrac{v_2^2}{2g}$ $\begin{bmatrix} p_1 = 400000 \\ p_2 = 130000 \end{bmatrix}$

$v_1 \times 80^2 = v_2 \times 40^2$

$v_2 = 4v_1$

Substituting v_2 and solving for v_1 we get

$v_1 = 6$m/s

19. Dynamic pressure of gas = $(\delta gh)_{water}$

$$\frac{1}{2}\delta_{gas} V_2 = \delta_m \times 9.81 \times h$$

$$\frac{1}{2} \times 1 \times (20)2 = 1000 \times 9.81 \times h$$

h = 0.02038 m of water

h = 20.38 mm of water

20. The force developed on the plate,

$$F_x = \rho a (v_j - u)^2 = 1000 \times 0.01(20 - 10)^2$$

$F_x = 1000$ N

21. Applying Bernoulli's equation between sections (i) & A

$$P_1 = \frac{P_A}{\rho g} + \frac{V_A^2}{2g}$$

$$\frac{P_A - P_1}{\rho g} = \frac{-V_A^2}{2g}$$

P_1 = Pambient

$\therefore P_A - P_1 = (P_A)$gauge

$$\therefore \frac{(P_A)\text{gauge}}{\rho g} == \frac{-V_A^2}{2g}$$

$$(P_A)_{gauge} = \frac{-1 \times 1^2}{2} = 0.5\,\text{N/m}^2$$

22. Volume flow rate = Q

Mass of water strike = ρQ

Velocity of the water when it hit the water surface = U

Force on weighing balance due to water strike

= Initial momentum – final momentum

= ρQU – 0= ρQU

(since final velocity is perpendicular to initial velocity)

Now total force on weighing balance

= mg + ρQU

23. $Q = A_1 V_1 = A_2 V_2$

$$\Rightarrow \frac{\pi}{4}(20)^2 \times 2 = \frac{\pi}{4} \times d^2 \times \sqrt{2 \times 9.8 \times 0.5}$$

$d \approx 15$mm

24. Force exerted by a jet of water striking fixed wall = $\rho a v^2$

$$= 1000 \times \frac{\pi}{4} \times 0.3^2 \times V^2$$

$$= 1000 \times \frac{\pi}{4} \times 0.3^2 \times (2 \times 10 \times 6^2) \left[\because V = \sqrt{2gh}\right]$$

= 8.76 N

25. $u_x = \sqrt{2gh}$

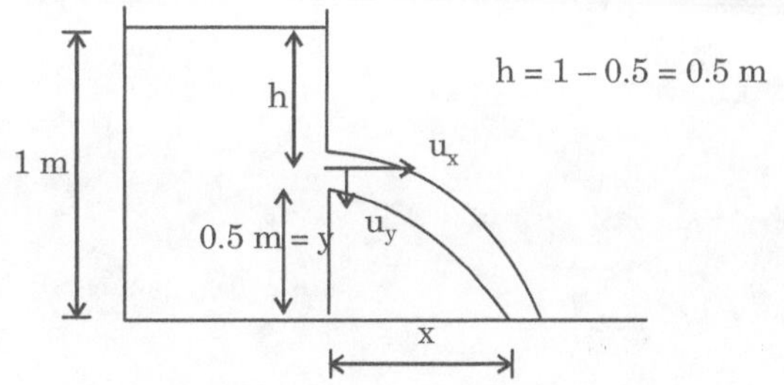

$$= \sqrt{2 \times 9.8 \times 0.5}$$

$u_x = 3.13$ m/s

Now for distance x

we require u_x so–

$$x = u_x \cdot t + \frac{1}{2} \times a \times b$$

$x = 3.13\ t$

($\because$ acceleration (a) = 0 in horizontal direction)

Now for $t \Rightarrow y = u_y t + \frac{1}{2} g t^2$

$\Rightarrow 0.5 = 0 + \frac{1}{2} \times 9.81 \times t^2 \qquad [\because u_y = 0]$

$t = 0.319$ sec.

so distance, $x = 3.13 \times 319 = 1$m

26. $A_1 V_1 = A_2 V_2$

$$V_2 = \left(\frac{A_2}{V_2}\right) V_1$$

By applying Bernoulli's equation between points (1) & (2), we get

$$\frac{P_1}{\delta g} + \frac{V_1^2}{2g} + Z_1 = \frac{P_2}{\delta g} + \frac{V_2^2}{2g} + Z_2$$

$$\left[P_{atm} + \frac{100 \times 10}{10^{-2}}\right] + \frac{V_1^2}{2g} + 0.5 = \frac{p_{atm}}{\delta g} + \frac{A_1^2}{A_2^2} \times \frac{V_1^2}{2y}$$

$$\frac{100 \times 10}{10^{-2} \times 1000 \times 10} + 0.5 = \left(\frac{A_1^2}{A_2^2} - 1\right)\frac{V_1^2}{2g}$$

$$10 + 0.5 = (10^2 - 1)\frac{V_1^2}{2g} \Rightarrow V_1 = 1.456\text{m/s}$$

5 CHAPTER

Viscous, Turbulent Flow and Boundary Layer Theory

Analysis of Previous GATE Papers

Topic	Year/Set → Marks Distribution ↓		2019 Set 1	2019 Set 2	2018 Set 1	2018 Set 2	2017 Set 1	2017 Set 2	2016 Set 1	2016 Set 2	2016 Set 3	2015 Set 1	2015 Set 2	2015 Set 3
Viscous flow	1 Mark	MCQ Type					1	1						1
		Numerical Type		1		1								
	2 Marks	MCQ Type												
		Numerical Type			1			1	1	1	1	1	1	
	Total			1	2	1	1	3	2	2	2	2	2	1
Turbulent flow	1 Mark	MCQ Type							1					
		Numerical Type												
	2 Marks	MCQ Type												
		Numerical Type												
	Total								1					
Boundary layer theory	1 Mark	MCQ Type					1							
		Numerical Type												
	2 Marks	MCQ Type												
		Numerical Type		1					1					
	Total			2			1		2					

Viscous flow

1. The discharge in m^3/s for laminar flow through a pipe of diameter 0.04 m having a centre line velocity of 1.5 m/s is

(a) $\frac{3\pi}{50}$

(b) $\frac{3\pi}{2500}$

(c) $\frac{3\pi}{5000}$

(d) $\frac{3\pi}{10000}$

[1988 : 2 Marks]

2. Shown below are three tanks, tank 1 without an orifice tube and tanks 2 and 3 with orifice tubes as shown. Neglecting losses and assuming the diameter of orifice to be much less than that of the tank, write expressions for the exit velocity in each of the three tanks.

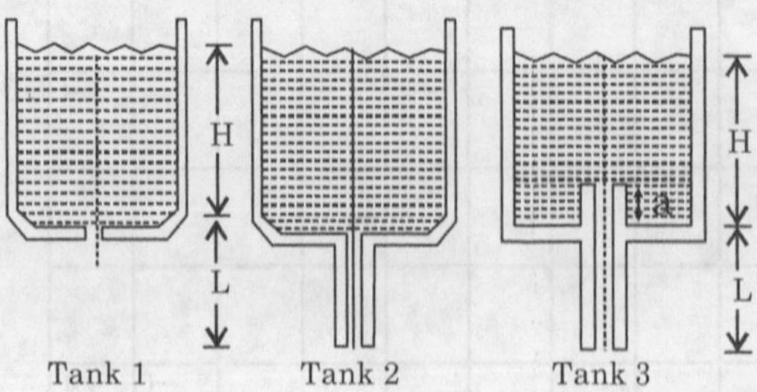

[1993 : 2 Marks]

3. For a fully developed flow through a pipe, the ratio of the maximum velocity to the average velocity is ________(fill in the blanks)

[1994 : 1 Mark]

4. As the transition from laminar to turbulent flow is induced in a cross flow past a circular cylinder the value of the drag coefficient drops. (True/False)

[1994 : 1 Mark]

5. Fluid is flowing with an average velocity of V through a pipe of diameter d. Over a length of L, the head loss is given by $\frac{fLV^2}{2gD}$. The friction factor, f for laminar flow in terms of Reynolds number (Re) is _______ (fill in the blanks)

[1994 : 2 Marks]

6. In fully developed laminar flow in the circular pipe, the head loss due to friction is directly proportional to _____ (mean velocity/square of the mean velocity)

[1995 : 1 Mark]

7. For laminar flow through along pipe, the pressure drop per unit length increases

(a) in linear proportion to the cross-sectional area

(b) in proportion to the diameter of the pipe

(c) in inverse proportion to the cross-sectional area

(d) in inverse proportional to the square of cross-sectional area

[1996 : 2 Marks]

8. The discharge velocity at the pipe exit in figure is

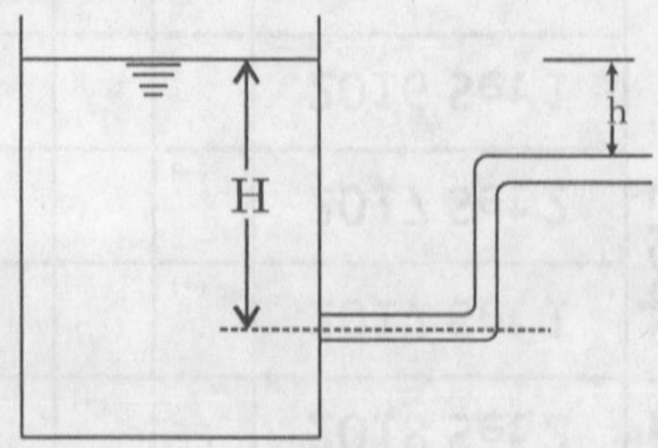

(a) $\sqrt{2gH}$ (b) $\sqrt{2gh}$

(c) $\sqrt{g(H+h)}$ (d) 0

[1998 : 2 Marks]

Common Data Questions 9 and 10:

A syringe with a frictionless plunger contains water and has at its end a 100 mm long needle of 1 mm diameter. The internal diameter of the syringe is 10 mm. Water density is 1000 kg/m^3. The plunger is pushed in at 10 mm/s and the water comes out as a jet.

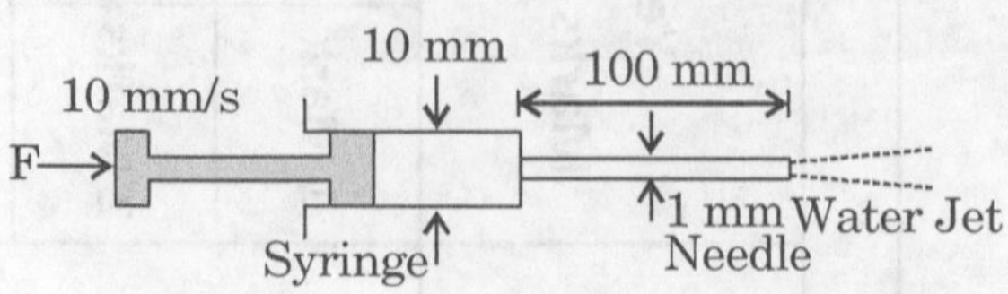

9. Assuming ideal flow, the force F in newtons required on the plunger to push out the water is

(a) 0 (b) 0.04

(c) 0.13 (d) 1.15

[2003 : 2 Marks]

10. Neglect losses in the cylinder and assume fully developed laminar viscous flow throughout the needle: the Darcy friction factor is 64/Re. where Re is the Reynolds number. Given that the viscosity of water is 1.0×10^{-3} kg/ms, the force F in newtons required on the plunger is

(a) 0.13 (b) 0.16
(c) 0.3 (d) 4.4

[2003 : 2 Marks]

11. A cubic block of side L and mass M is dragged over an oil film across table by a string connects to a hanging block of mass m as shown in figure. The Newtonian oil film of thickness h has dynamic viscosity μ and the flow condition is laminar. The acceleration due to gravity is g. The steady state velocity V of block is

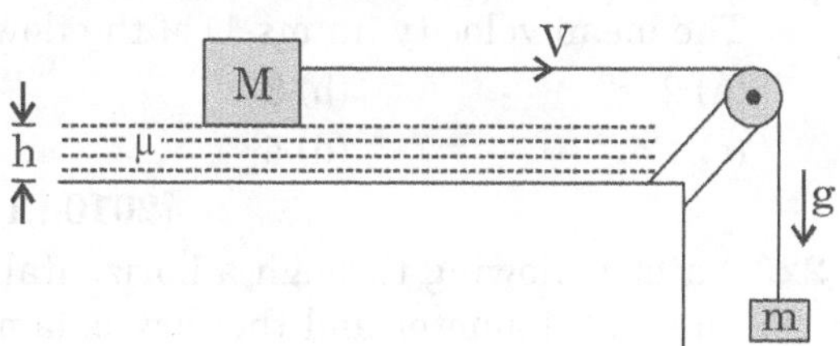

(a) $\frac{Mgh}{\mu L^2}$ (b) $\frac{Mgh}{\mu}$

(c) $\frac{mgh}{\mu L^2}$ (d) $\frac{mgh}{\mu}$

[2004 : 2 Marks]

12. An incompressible fluid (kinematic viscosity, 7.4×10^{-7} m²/s, specific gravity, 0.88) is held between two parallel plates. If the top plate is moved with a velocity of 0.5 m/s while the bottom one is held stationary, the fluid attains a linear velocity profile in the gap of 0.5 mm between these plates: the shear stress in Pascals on the surface of top plate is

(a) 0.651×10^{-3} (b) 0.651
(c) 6.51 (d) 0.651×10^3

[2004 : 1 Mark]

13. The velocity profile in fully developed laminar flow in a pipe of diameter D is given by $u = u_0 (1 - 4r^2/D^2)$, where r is the radial distance from the center. If the viscosity of the fluid is ρ, the pressure drop across a length L of the pipe is

(a) $\frac{\mu u_0 L}{D^2}$ (b) $\frac{4\mu u_0 L}{D^2}$

(c) $\frac{8\mu u_0 L}{D^2}$ (d) $\frac{16\mu u_0 L}{D^2}$

[2006 : 2 Marks]

14. Consider steady laminar incompressible axi-symmetric fully developed viscous flow through a straight circular pipe of constant cross-sectional area at a Reynolds number of 5. The ratio of inertia force to viscous force on a fluid particle is

(a) 5 (b) 1/5
(c) 0 (d) ∞

[2007 : 1 Mark]

Linked Answer Questions 15 and 16:

Consider a steady incompressible flow through a channel as shown below:

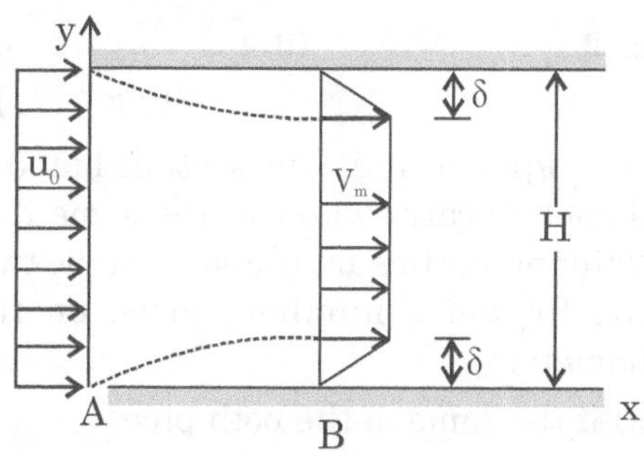

The velocity profile is uniform with a value of u_0 at the inlet section A. The velocity profile at section B downstream is

$$u = \begin{cases} V_m \frac{y}{\delta}, & 0 \le y \le \delta \\ V_m, & \delta \le y \le H - \delta \\ V_m \frac{H-y}{\delta}, & H - \delta \le y \le H \end{cases}$$

15. The ratio V_m/u_0 is

(a) $\frac{1}{1-2(\delta/H)}$ (b) 1

(c) $\frac{1}{1-(\delta/H)}$ (d) $\frac{1}{1+(\delta/H)}$

[2007 : 2 Marks]

16. The ratio $\frac{p_A - p_B}{\frac{1}{2}\rho u_0^2}$. (where p_A and p_B are the pressures at section A and B, respectively, and ρ is the density of the fluid), is

(a) $\frac{1}{[1+(\delta/H)]^2}$ (b) $\frac{1}{[1-(\delta/H)]^2}$

(c) $\frac{1}{[1-(2\delta/H)]^2}$ (d) $\frac{1}{1+(\delta/H)}$

[2007 : 2 Marks]

17. Oil is being pumped through a straight pipe, the pipe length, diameter and volumetric flow rate are all doubled in a new arrangement. The pipe friction factor, however, remains constant. The ratio of pipe frictional losses in the new arrangement to that in the original configuration would be

(a) $\frac{1}{4}$ (b) $\frac{1}{2}$

(c) 2 (d) 4

[2008 : 2 Marks]

18. Two pipes of uniform section but different diameters carry water at the same flow rate. Water properties are the same in the two pipes. The Reynolds number, based on the pipe diameter

(a) if the same in the both pipes

(b) is large in then arrow pipe

(c) is smaller in the narrower pipe

(d) depends on the pipe material

[2008 : 2 Marks]

19. The velocity profile of a fully developed laminar flow in a straight circular pipe, as shown in the figure, is given by the expression

$$u(r) = -\frac{R^2}{4\mu}\left(\frac{\partial p}{\partial x}\right)\left(1-\frac{r^2}{R^2}\right)$$

where $\frac{\partial p}{\partial x}$ is a constant

The average velocity of fluid in the pipe is

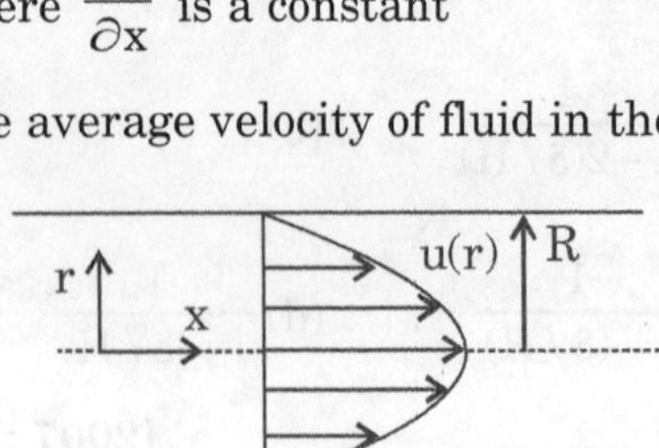

(a) $-\frac{R^2}{8\mu}\left(\frac{\partial p}{\partial x}\right)$ (b) $-\frac{R^2}{4\mu}\left(\frac{\partial p}{\partial x}\right)$

(c) $-\frac{R^2}{2\mu}\left(\frac{\partial p}{\partial x}\right)$ (d) $-\frac{R^2}{\mu}\left(\frac{\partial p}{\partial x}\right)$

[2009 : 2 Marks]

20. The pressure drop for laminar flow of a liquid in a smooth pipe at normal temperature and pressure is

(a) directly proportional to density

(b) inversely proportional to density

(c) Independent of density

(d) proportional to (density)$^{0.75}$

[2009 : 2 Marks]

21. The maximum velocity of a one-dimensional incompressible fully developed viscous flow, between two fixed parallel plates, is 6 ms-1. The mean velocity (in ms-1) of the flow is

(a) 2 (b) 3

(c) 4 (d) 5

[2010 : 1 Mark]

22. Water is flowing through a horizontal pipe of constant diameter and the flow is laminar. If the diameter of the pipe is increased by 50% keeping the volume flow rate constant, then the pressure drop in the pipe due to friction will decrease by

(a) 33% (b)50%

(c) 70% (d) 80%

[2011 : 2 Marks]

23. Water flows through a pipe having an inner radius of 10 mm at the rate of 36 kg/hr at 25°C. The viscosity of water at 25°C is 0.001 kg/ms. The Reynolds number of the flow is ______.

[2014 : 2 Marks, Set-1]

24. For a fully developed flow of water in a pipe having diameter 10 cm, velocity 0.1 m/s and kinematic viscosity 10^{-5} m²/s, the value of Darcy friction factor is ______.

[2014 : 2 Marks, Set-1]

25. In a simple concentric shaft-bearing arrangement, the lubricant flows in the 2 mm gap between the shaft and the bearing. The flow may be assumed to be a plane Couette flow with zero pressure gradient. The diameter of the shaft is 100 mm and its tangential speed is 10 m/s. The dynamic viscosity of the lubricant is 0.1 kg/ms. The frictional resisting force (in Newton) per 100 mm length of the bearing is ______.

[2014 : 2 Marks, Set-1]

26. Consider laminar flow of water over a flat plate of length 1 m. If the boundary layer thickness at a distance of 0.25 m from the leading edge of the plate is 8 mm, the boundary layer thickness (in mm), at a distance of 0.75 m, is ______

[2014 : 2 Marks, Set-2]

27. Couette flow is characterized by

(a) steady, incompressible, laminar flow through a straight circular pipe

(b) fully developed turbulent flow through a straight circular pipe

(c) steady, incompressible, laminar flow between two fixed parallel plates

(d) steady, incompressible, laminar flow between one fixed plate and the other moving with a constant velocity

[2015 : 1 Mark, Set-3]

28. Air (ρ = 1.2 kg/m³ and kinematic viscosity, $\upsilon = 2 \times 10^{-5}$ m²/s) with a velocity of 2 m/s flows over the top surface of a flat plate of length 2.5 m. If the average value of friction coefficient is $C_f = \frac{1.328}{\sqrt{Re_x}}$. the total drag force (in N) per unit width of the plate is ________.

[2015 : 2 Marks, Set-1]

29. For a fully developed laminar flow of water (dynamic viscosity 0.001 Pa-s) through a pipe radius 5 cm, the axial pressure gradient is –10 Pa/m. The magnitude of axial velocity (in m/s) at a radial location of 0.2 cm is _____.

[2015 : 2 Marks, Set-2]

30. Oil (kinematic viscosity, $\upsilon_{oil} = 1.0 \times 10^{-6}$ m²/s) flows through a pipe of 0.5 m diameter with a velocity of 10 m/s. Water (kinematic viscosity, $v_w = 0.89 \times 10^{-6}$ m²/s) is flowing through a model pipe of diameter 20 mm. For satisfying the dynamic similarity, the velocity of water (in m/s) is ______.

[2016 : 2 Marks, Set-1]

31. Consider fluid flow between two infinite horizontal plates when are parallel (the gap between them being 50 mm), The top plate is sliding parallel to the stationary bottom plate at a speed of 3 m/s.

The flow between the plates is solely due to the motion of thc top plate. The force per unit area (magnitude) required to maintain the bottom plate stationary is _______ N/m².

Viscosity of the fluid μ = 0.44 kg/ms and density p = 888 kg/m³.

[2016 : 2 Marks, Set-2]

32. Consider a fully developed steady laminar flow of an incompressible fluid with viscosity μ. through a circular pipe of radius R, Given that the velocity at a radial location of R/2 from the centerline of the pipe is U_1, the shear stress at the wall is $K\mu U_1/R$, where K is ______.

[2016 : 2 Marks, Set-3]

33. For steady flow of a vlscous Incompressible fluid through pipe of constant of diameter, the average veloolty In the fully developed region is constant, Whieh one of the following statements about the average velocity in the developing region is TRUE?

(a) It increases until the flow is fully developed,

(b) It is constant and is equal to the average velocity in the fully developed region,

(c) It decreases until the flow is fully developed.

(d) It is constant but is always lower than the average velocity in the fully developed region,

[2017 : 1 Mark, Set-1]

34. Consider a laminar flow at zero incidence over a flat plate, The shear stress at the wall is denoted by τ_W, The axial position x_1 and x_2 on the plate are measured from the leading edge in the direction of flow. If $x_2 > x_1$, then

(a) $\tau_W|_{x1} = \tau_W|_{x2} = 0$

(b) $\tau_W|_{x1} = \tau_W|_{x2} \neq 0$

(c) $\tau_W|_{x1} > \tau_W|_{x2}$

(d) $\tau_W|_{x1} < \tau_W|_{x2}$

[2017 : 1 Mark, Set-2]

35. For the laminar flow of water oyer a sphere, the drag coefficient C_F is defined as $C_F = F/(\rho U^2D^2)$, where F is the drag force, ρ is the fluid density, U is the fluid velocity and D is the diameter of the sphere. The density of water is 1000 kg/m³. When the diameter of the sphere is 100 mm and the fluid velocity is 2 m/s, the drag coefficient is 0.5. If water now flows over another sphere of diameter 200 mm under dynamically similar conditions, the drag force (in N) on this sphere is ______.

[2017: 2 Marks, Set-2]

36. A solid block of 2.0 kg mass slides steadily at a velocity V along a vertical wall as shown in the figure below. A thin oil film of thickness h = 0.15 mm provides lubrication between the block and the wall. The surface area of the face of the block in contact with the oil film is 0.04 m^2. The velocity distribution within the oil film gap is linear as shown in the figure. Take dynamic viscosity of oil as 7×10^{-3} Pa-s and acceleration due to gravity as 10 m/s^2. Neglect weight of the oil. The terminal velocity V (in m/s) of the block is ______(correct to one decimal place).

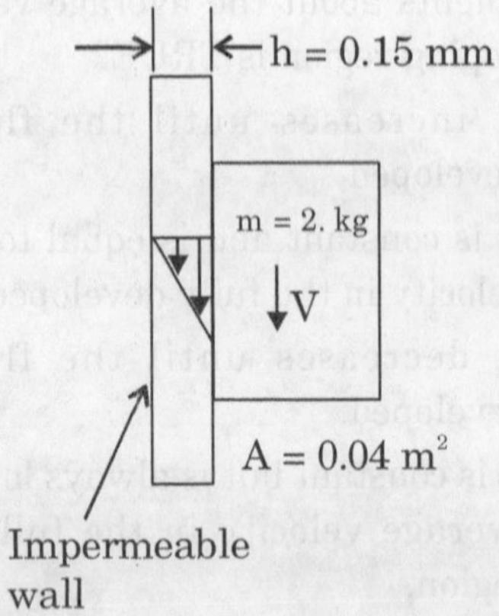

[2018 : 2 Marks, Set-1]

37. The viscous laminar flow of air over a flat plate results in the formation of a boundary layer. The boundary layer thickness at the end of the plate of length L is δ_L. When the plate length is increased to twice of its original length, the percentage change in laminar boundary layer thickness at the end of the plate (with respect to δ_L) is ________(correct to two decimal places).

[2018 : 1 Mark, Set-2]

38. Water enters a circular pipe of length L = 5.0 m and diameter D = 0.20 m with Reynolds number Re_D = 500. The velocity profile at the inlet of the pipe is uniform while it is parabolic at the exit. The Reynolds number at the exit of the pipe is_____

[2019 : 1 Mark, Set-2]

Turbulent flow

39. In the case of turbulent flow of a fluid though a circular tube (as compared to the case of laminar flow at the same flow rate) the maximum velocity is ________shear stress at the wall is ________, and the pressure drop across a given length is __________

The correct words for the blanks are, respectively

(a) higher, higher, higher
(b) higher, lower, lower
(c) lower, higher, higher
(d) lower, higher, lower

[1987 : 2 Marks]

40. The parameters which determines the friction factor for turbulent flow in a rough pipe are

(a) Froude number and relative roughness
(b) Froude number and Mach number
(c) Reynolds number and relative roughness
(d) Mach number and relative roughness

[1988 : 1 Mark]

41. Prandtl's mixing length in turbulent flow signifies

(a) the average distance perpendicular to the mean flow covered by the mixing particles
(b) the ratio of mean free path to characteristic length of the flow field
(c) the wavelength corresponding to the lowest frequency present in the flow field
(d) the magnitude of turbulent kinetic energy

[1994 : 1 Mark]

42. A phenomenon is modeled using-n dimensional variables with k primary dimensions. The number of non-dimensional variables is

(a) k (b) n
(c) n - k (d) n + k

[2010 : 1 Mark]

43. Match the following

P: Compressible flow	U: Reynolds number
Q: Free surface flow	V: Nusselt number
R: Boundary layer flow	W: Weber number
S: Pipe flow	X: Froude number
T: Heat convection	Y: Mach number
Z: Skin friction coefficient	

(a) P-U; Q-X; R-V; S-Z; T-W
(b) P-W; Q-X; R-Z; S-U; T-V
(c) P-Y; Q-W; R-Z; S-U; T-X
(d) P-Y; Q-X; R-Z; S-U; T-V

[2010 : 2 Marks]

44. Consider the turbulent flow of a fluid through a circular pipe of diameter D. Identify the correct pair of statements.

I. The fluid is well-mixed
II. The fluid is unmixed
III. Re_D < 2300
IV. Re_D > 2300

(a) I and III (b) II and IV
(c) II and III (d) I and IV

[2014 : 1 Mark, Set-3]

45. The instantaneous stream wise velocity of a turbulent flow is given as follows:

$u(x, y, z, t) = \bar{u}(x, y, z) + u'(x, y, z, t)$

The time average of the fluctuating velocity u (x, y, z, t) is

(a) $\frac{u'}{2}$ (b) $-\frac{\bar{u}}{2}$

(c) zero (d) $\frac{\bar{u}}{2}$

[2016 : 1 Mark, Set-1]

Boundary layer theory

46. The predominant forces acting on an element of fluid in the boundary layer over a flat plate in a uniform parallel stream are

(a) Viscous and pressure forces

(b) Viscous and inertia forces

(c) Viscous and body forces

(d) Inertial and pressure forces

[1990 : 1 Mark]

47. The necessary and sufficient condition which brings about separation of boundary layer is

$\frac{dp}{dx} > 0$ (True/False)

[1994 : 1 Mark]

48. Flow separations flow past a solid object is caused by

(a) A reduction of pressure to vapour pressure

(b) A negative pressure gradient

(c) A positive pressure gradient

(d) The boundary layer thickness reducing to zero

[2002 : 1 Mark]

49. If x is the distance measured from the leading edge of a flat plate, then laminar boundary layer thickness varies as

(a) 1/x (b) $x^{4/5}$

(c) x^2 (d) $x^{1/2}$

[2002 : 1 Mark]

50. For air flow over a flat plate, velocity (U) and boundary layer thickness (δ) can be expressed respectively, as

$$\frac{u}{u_\infty} = \frac{3}{2}\frac{y}{\delta} - \frac{1}{2}\left(\frac{y}{\delta}\right)^3 ; \delta = \frac{4.64x}{\sqrt{Re_x}}$$

If the free stream velocity is 2 m/s, and air has kinematic viscosity of 1.5×10^{-5} m²/s and density of 1.23 kg/m³, the wall shear stress at x = 1 m, is

(a) 2.36×10^2 N/m² (b) 43.6×10^{-3} N/m²

(c) 4.36×10^{-3} N/m² (d) 2.18×10^{-3} N/m²

[2004 : 2 Marks]

Linked Answer Questions 51 and 52:

A smooth flat plate with a sharp leading edge is placed along a gas stream flowing at U = 10 m/s. The thickness of the boundary layer at section r – s is 10 mm, the breadth of the plate is 1 m (into the paper) and the density of the gas, ρ = 1.0 kg/m³. Assume that the boundary layer is thin, two-dimensional, and follows a linear velocity distribution, u = U (y/δ), at the section r–s, where y is the height from plate.

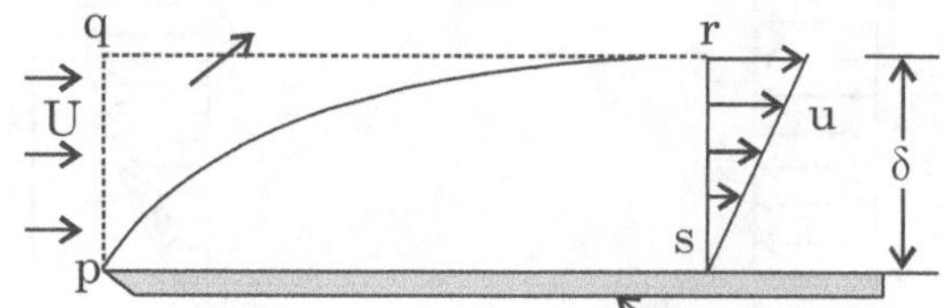

51. The mass flow rate (in kg/s) across the section q – r is

(a) zero (b) 0.05

(c) 0.10 (d) 0.15

[2006 : 2 Marks]

52. The integrated drag force (in N) on the plate, between p – s, is

(a) 0.67 (b) 0.33

(c) 0.17 (d) zero

[2006 : 2 Marks]

53. Consider an incompressible laminar boundary layer flow over a flat plate of length L, aligned with the direction of an incoming uniform free stream. If F is the ratio of the drag force on the front half of the plate to the drag force on the rear half, then

(a) F < 1/2 (b) F = 1/2

(c) F = 1 (d) F > 1

[2007 : 1 Mark]

54. An incompressible fluid flows over a flat plate with zero pressure gradient. The boundary layer thickness is 1 mm at a location where the Reynolds number is 1000. If the velocity of the fluid alone is increased by a factor of 4, then the boundary layer thickness at the same location, in mm will be

(a) 4 (b) 2

(c) 0.5 (d) 0.25

[2012 : 2 Marks]

55. A steady laminar boundary layer Is formed over a flate plate as shown in the figure, The free stream velocity of the fluid is U_0. The velocity profile at the inlet a-b is unlform, while that at a downstream location c–d is given by

$$u = \left[2\left(\frac{y}{\delta}\right) - \left(\frac{y}{\delta}\right)^2\right].$$

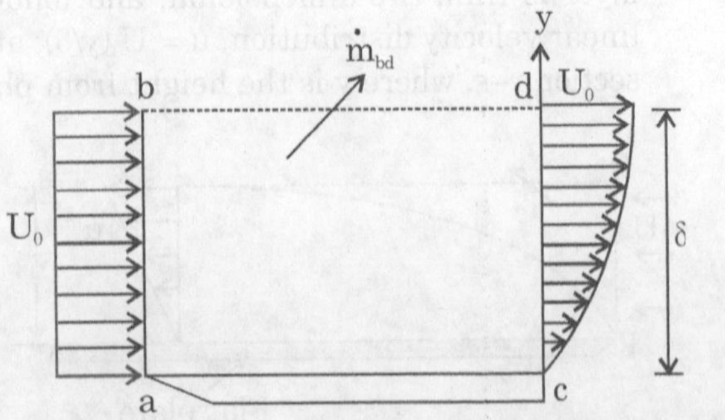

The ratio of the mass flow rate, $\dot{m}_{bd}$, leaving through the horizontal section b-d to that entering through the vertical section a-b is ____

[2016 : 2 Marks, Set-1]

56. The velocity profile inside the boundary layer for flow over a plate is given as $\frac{u}{U_\infty} = \sin\left(\frac{\pi}{2}\frac{y}{\delta}\right)$, where U_∞ is the free stream velocity and δ is the local boundary layer thickness. If δ* is the local displacement thickness, the value of $\frac{\delta^*}{\delta}$ is

(a) $\frac{2}{\pi}$ (b) $1 - \frac{2}{\pi}$

(c) $1 + \frac{2}{\pi}$ (d) 0

[2017 : 2 Marks, Set-1]

57. The aerodynamic drag on a sports car depends on its shape. The car has a drag coefficient of 0.1 with the windows and the roof closed. With the windows and the roof open, the drag coefficient becomes 0.8. The car travels at 44 km/h with the windows and roof closed. For the same amount of power needed to overcome the aerodynamic drag, the speed of the car with the windows and roof open (round off to two decimal places), is ____ km/h. (The density of air and the frontal area may be assumed to be constant.

[2019 : 2 Marks, Set-2]

ANSWERS

1. (d)	**7.** (c)	**8.** (b)	**9.** (b)	**10.** (c)	**11.** (c)	**12.** (b)	**13.** (d)	**14.** (a)	**15.** (c)
16. (b)	**17.** (a)	**18.** (b)	**19.** (a)	**20.** (c)	**21.** (c)	**22.** (d)	**27.** (d)	**33.** (b)	**34.** (c)
39. (c)	**40.** (c)	**41.** (a)	**42.** (c)	**43.** (d)	**44.** (d)	**45.** (c)	**46.** (b)	**48** (c)	**49.** (d)
50. (c)	**51.** (b)	**52.** (c)	**53.** (d)	**54.** (c)	**56.** (b)				

EXPLANATIONS

1. $u_{max} = 2\,u_{avg}$

$u_{max} = 2\,u_{avg}$

$1.5 = 2\,u_{avg}$

$u_{avg} = \frac{1.5}{2}\,m/s$

$Q = A \times \overline{V} = A \times u_{avg}$

$$= \frac{\pi}{4}d^2 \times u_{avg} = \frac{\pi}{4} \times (0.04)^2 \times \frac{1.5}{2} = \frac{3\pi}{10000}\,m^3/s$$

2. $V_1 = \sqrt{2gH}$

$V_2 = \sqrt{2g(H+L)}$

$V_3 = \sqrt{2g(H+L)}$

3. Two

4. True

5. For Laminar flow,

$$f = \frac{64}{Re}$$

6. $f \propto$ Mean velocity

7. $P_1 - P_2 = \dfrac{32\mu \bar{U} L}{D^2}$

8.

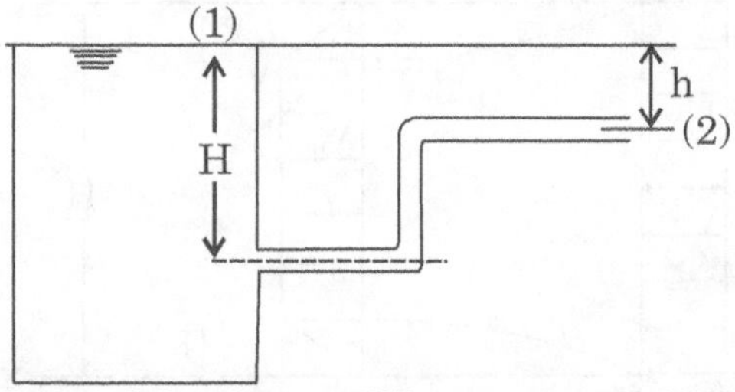

By applying Bernoulli's equation at (1) and (2), we get

$$\frac{P_1}{\rho g} + \frac{V_1^2}{2g} + Z_1 = \frac{P_2}{\rho g} + \frac{V_2^2}{2g} + Z_2$$

$P_1 = P_2 = P_{atm}$

$V_1 = 0$

$V_2 = ?$

$Z_2 = H - h$

$$\therefore H = \frac{V_2^2}{2g} + H - h$$

$$V_2 = \sqrt{2gh}$$

9. $\rho_{water} = 1000$ kg/m³

Velocity at points 1 = velocity of plunger = 10 mm/ s = 0.014 m/s

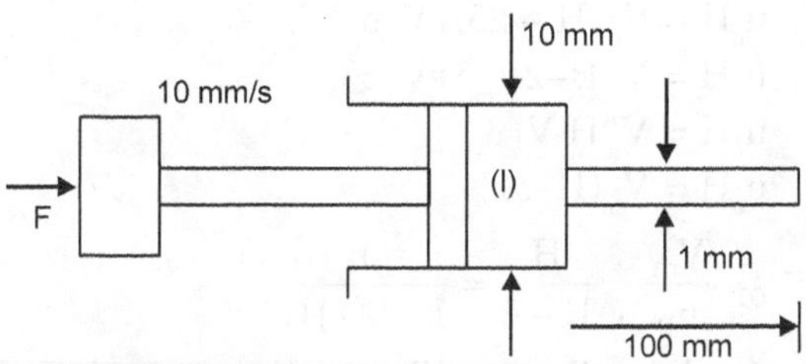

Applying Bernoulli's equation at points 1 and 2, we have

$$\frac{P_1}{\rho g} + \frac{v_1^2}{2g} + z_1 = \frac{P_2}{\rho g} + \frac{v_2^2}{2g} + z_2$$

Since $z_1 = z_2$ and $P_2 = 0$

$$\frac{P_1}{\rho q} = \frac{v_2^2}{2g} - \frac{v_1^2}{2g}$$

$$P_1 = \frac{\rho}{2}(v_2^2 - v_1^2) \quad ...(i)$$

Applying continuity equation at points (i) and (ii), we have

$$A_1 v_1 = A_2 v_2$$

$$\Rightarrow v_2 = \left(\frac{A_1}{A_1}\right) v_1$$

$$= \frac{\frac{\pi}{4} \times (0.01)^2}{\frac{\pi}{4} \times (0.001)^2} v_1$$

$= 100\, v_1$

$= 100 \times 0.01 = 1$ m/s

Now from equation (i),

$$P_1 = \frac{\rho}{2}\left[v_2^2 - v_1^2\right]$$

$$= \frac{1000}{2}\left[(1)^2 - (0.01)^2\right]$$

$= 499.95$ N/m²

Force required on plunger

$= P_1 \times v_1$

$$= 499.95 \times \frac{11}{4} \times (0.01)^2 = 0.04\text{N}$$

10. Given, v = velocity of water = 10 × 10⁻³ kg/sm

$$\text{Now Re} = \frac{\rho v_2 d}{v}$$

$$= \frac{1000 \times 1 \times 0.001}{1 \times 10^{-3}} \quad \text{... since } = v_2 = 1$$

Re = 1000

Now, Darcy's friction factor,

$$4f = \frac{64}{\text{Re}} = \frac{64}{1000} = 0.064$$

$$\text{Head loss in needle} = h_l = \frac{flv^2}{2gD}$$

$$= \frac{0.064 \times 0.1 \times (1)^2}{2 \times 9.8 \times 0.001} = 0.3265$$

Applying Bernoulli's equation at points 1 and 2, we have

$$\frac{P_1}{\rho g} + \frac{v_1^2}{2g} + z_1 = \frac{P_2}{\rho g} + \frac{v_2^2}{2g} + z_2 + h_l$$

Since $z_1 = z_2$ and $P_2 = 0$

$$\therefore \frac{P_1}{\rho g} = \frac{v_2^2 - v_1^2}{2g} + h_l$$

$$P_1 = \frac{\rho}{2}(v_2^2 - v_1^2) + \rho g h_l$$

$$= \frac{1000}{2}\left[(1)^2 - (0.01)^2\right] + 1000 \times 9.8 \times 0.3265$$

$= 499.95 + 3199.7 = 3699.65$

Now force required on plunger

$= P_1 \times A_1$

$$= 3699.65 \times \frac{\pi}{4} \times (0.01)^2$$

$= 0.3$ N

11. $\tau = \mu\left(\frac{du}{dy}\right)$

$du = V - O - V$

$dy = h$

$\tau = u\frac{V}{h}$ ...(1)

Also $\tau = \frac{F}{A}$

where $F = mg$

$A = L^2$

$\therefore \tau = \frac{mg}{L^2}$ ...(2)

By (1) (2), we get

$$V = \frac{mgh}{uL^2}$$

12. Given: $\nu = 7.4 \times 10^{-7}\ m^2/s$

$\rho = 0.88 \times 1000 = 880$

$v = 0.5\ m/s$

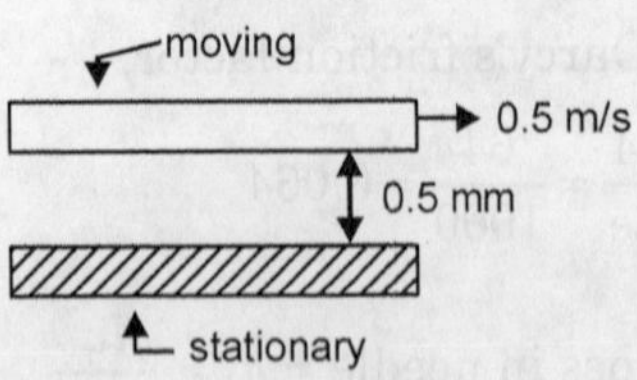

Shear stress, $\tau = \mu \cdot \frac{du}{dy} = (v \times \rho) \times \frac{du}{dy}$

$= 7.4 \times 10^{-7} \times 880 \times \frac{0.5}{(0.0005)} = 0.6512.$

13. By Hagen – Poiseuille law, for steady laminar flow in circular pipes

$$\tau = -\mu\frac{\partial u}{\partial r}$$

$$\tau = \frac{-\partial P}{\partial x}\cdot\frac{r}{2}$$

$$\mu\frac{\partial u}{\partial r} = \frac{\partial P}{\partial x}\cdot\frac{r}{2}$$

$$\mu u_0\left(\frac{-8r}{D^2}\right) = \frac{P}{L}\cdot\frac{r}{2}$$

$$\cdots\left[\because u = u_0\left(1 - \frac{4r^2}{D^2}\right)\right]$$

$$P = \frac{-16\mu L u_0}{D^2}$$

[(–) sign is due to drop]

14. Reynold's number $= \frac{\text{Inertia force}}{\text{Viscous force}} = 5$

15. Applying continuity equation at sections A and B, mass flow rate at inlet section A = mass rate at section B.

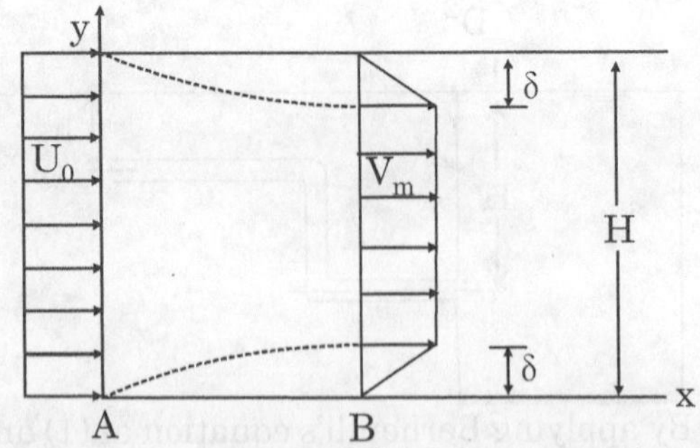

Hence, per unit width of the channel, we get

$$\rho u_o H \times 1 = \rho Vm(H - 2\delta) \times 1 + \int_o^\delta u dy \times 1$$

$$u_o H = V_m(H - 2\delta) + 2\int_o^\delta \frac{V_m Y}{\delta} dy$$

$$\left(\therefore u = \frac{V_m Y}{\delta}\right)$$

$$u_o H = V_m(H - 2\delta) + 2\frac{V_m}{\delta}\left[\frac{Y^2}{2}\right]_o^\delta$$

$$u_o H = V_m(H - 2\delta) + 2\frac{V_m}{\delta} \times \frac{\delta^2}{2}$$

$$u_o H = V_m(H - 2\delta) + V_m\delta$$

$$u_o H = V_m H - 2V_m\delta + V_m\delta$$

$$u_o H = V_m H - V_m\delta$$

$$u_o H = V_m(H - \delta)$$

or $\frac{V_m}{u_0} = \frac{H}{H-\delta} = \frac{1}{1-(\delta/H)}$

16. Applying Bernoulli's equation at sections A and B, we get

$$\frac{P_A}{\rho g} + \frac{u_0^2}{2g} + Z_A = \frac{P_A}{\rho g} + \frac{V_m^2}{2g} + Z_B$$

$$\frac{P_A}{\rho g} + \frac{u_0^2}{2g} = \frac{P_B}{\rho g} + \frac{V_m^2}{2g}$$

or $\frac{P_A}{Pg} + \frac{P_B}{\rho g} = \frac{V_m^2}{2g} + \frac{u_o^2}{2g}$

Now, the ratio

$$\frac{P_A - P_B}{\frac{1}{2}\rho_0^2} = \frac{\frac{\rho V_m^2}{2} - \frac{\rho u_0^2}{2}}{\frac{1}{2}\rho u_0^2}$$

$$= \frac{V_m^2 - u_0^2}{u_0^2} = \left(\frac{V_m}{u_0}\right)^2 - 1$$

Substituting the Value of $\frac{V_m}{u_o} = \frac{1}{1-(\delta/H)}$

we get

$$\frac{P_A - P_B}{\frac{1}{2}\rho u_0^2} = \frac{1}{[1-(\delta/H)]^2} - 1$$

17. For original configuration

$l_1 = l$

$d_1 = d$

$Q_1 = Q$

For new configuration

$l_2 = 2l$

$d_2 = 2d$

$Q_2 = 2Q$

Head loss due to friction,

$$h_f = \frac{flV^2}{2gb}$$

where f = friction factor

l = Length of pipe

v = Average velocity

d = Diameter pipe

Discharge: $Q = AV = \frac{\pi}{4}a^2V$

or $V = \frac{4Q}{\pi a^2}$

$$h_f = \frac{fl}{2gd} \times \left(\frac{4Q}{\pi a^2}\right)^2 = \frac{8flQ^2}{\pi g a^5}$$

For new Pipe,

$$h_{f2} = \frac{8fl_1Q_2^2}{\pi g d_2^5} = \frac{8f \times 2f \times 4Q^2}{\pi g \times 2^5 \times a^5} = \frac{2fQ^2}{\pi g a^5}$$

$$\frac{h_1}{h_2^1} = \frac{2flQ^2}{\pi g a^5} \times \frac{\pi g a^5}{2flQ^2} = \frac{1}{4}$$

18. Reynolds number, $R_e = \frac{Vd}{\upsilon}$

Discharge, $Q = AV = \frac{\pi}{4}d^2V$

From above, $V = \frac{4Q}{\pi d^2}$

$\therefore$ Reynolds No. $= \frac{4Q}{\pi d^2} \times \frac{d}{2}$

Reynolds No. $= \frac{4Q}{\pi d}$ ...(1)

From (1), $R_e \propto \frac{1}{2}$

Reynolds number is large in the Narrow pipe.

19. 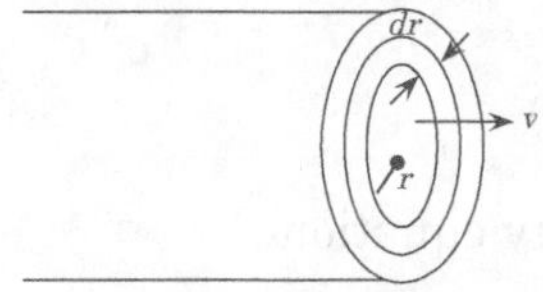

$$u = -\frac{R^2}{4\mu}\left(\frac{dp}{dx}\right)\left(1 - \frac{r^2}{R}\right)$$

$dQ = u\,(2\pi r dr)$

$$\therefore Q = 2\pi \int_o^R 4rdr$$

$$= 2\pi \int_o^R \frac{1}{4\mu} R^2 \left(-\frac{dp}{dx}\right)\left(1 - \frac{r^2}{R^2}\right) rdr$$

Solving, we get

$$Q = \frac{\pi R^4}{8\mu}\left(-\frac{dp}{dx}\right)$$

Now Q = Area ´ average velocity

$$\therefore u_{av} = \frac{Q}{Area} = \frac{Q}{\pi R^2}$$

$$\text{or } u_{av} = \frac{R^2}{8\mu}\left(\frac{dp}{dx}\right)$$

20. $\Delta P = \frac{32\mu\bar{U}L}{D^2}$

Above equation shows that ΔP is independent of velocity.

21. For fully developed incompressible viscous flow,

$$V_{mean} = \frac{2}{3} V_{max}$$

$$= \frac{2}{3} \times 6 = 4 \text{ m/s}$$

22. $\Delta P = \frac{32\mu\bar{U}L}{\Delta^2}$

$$Q = AV = \frac{\pi}{4}d^2V$$

$$V = \frac{4Q}{\pi d^2}$$

$$\Delta P = \frac{128\mu QL}{\pi D^4}$$

Above equation shows that

$$\Delta P \propto \frac{1}{D^4}$$

$$\Delta P_1 = \frac{C}{D_1^4} = \frac{C}{D^4} \text{ where} D_1 = D$$

$$\& \ \Delta P_2 = \frac{C}{D_2^4} = \frac{C}{(d_2^4)} \text{ where} D_2 = 1.5D$$

% change in pressure drop

$$\% = \frac{\Delta P_1 - \Delta P_2}{\Delta P_1} \times 100$$

$= 80.24\% \simeq 80\%$

23. $R_e = \frac{\rho VD}{\mu}$ ...(1)

From continuity equation :-

$\dot{m} = \rho AV$

$V = \frac{\dot{m}}{\rho A} = \frac{\dot{m}}{\rho \times \frac{\pi d^2}{4}} = \frac{4\dot{m}}{\rho \times \frac{\pi d^2}{4}} = \frac{4\dot{m}}{\rho \pi d^2}$ ...(2)

put (2) in (1), we get

$R_e = \frac{\rho \times 4\dot{m} \times d}{\rho \pi d_2 \times \mu}$

$R_e = \frac{4\dot{m}}{\pi \mu d} = \frac{4 \times .01}{3.14 \times .001 \times .02} = 636.94$

$\left[\dot{m} = 36 kg/hr = \frac{36}{3600} kg/s\right] = .01$ kg/s

24. Given, D = 10 cm = 0.1 m

V = 0.1 m/s

$\upsilon = 10^{-5}$ m²/s

$R_e = \frac{VD}{\upsilon} = \frac{0.1 \times 0.1}{10^{-5}}$

$R_e = 1000$

∴ flow is laminar

Darcy friction factor = $\frac{64}{R_e}$ (for laminar flow)

$= \frac{64}{1000} = 0.064$

25.

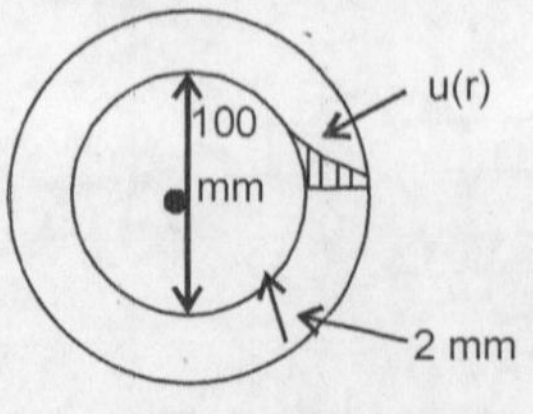

$\tau_w = \tau_{cylinder} = \mu \left.\frac{du}{dr}\right|$

$F = A \times \mu \frac{du}{dr} = \frac{\pi D\mu[u_t - u_w]}{(\Delta t)}$

u_t = tangential velocity

u_w = velocity at bearing

$F = \pi \times 0.1 \times 0.1 \times \frac{0.1[10-0]}{2 \times 10^{-3}}$

F = 15.707 N.

26. ∴ $\frac{\delta}{x} = \frac{k}{\sqrt{x}}$, here '$k$' is constant

∴ $\frac{\delta_1}{\sqrt{x_1}} = \frac{\delta_2}{\sqrt{x_2}}$

∴ $\frac{8}{\sqrt{0.25}} = \frac{\delta_2}{\sqrt{0.75}}$ ∴ $\delta_2 = 13.86$ mm.

∵ $\frac{\delta}{x} = \frac{0.60}{\sqrt{Re_x}}$

or $\frac{\delta}{x} = \frac{0.60}{\sqrt{xu_\infty/v}} = \frac{0.60\sqrt{v}}{\sqrt{x}\cdot\sqrt{u_\infty}}$

∵ $\sqrt{v}$ and $\sqrt{u_\infty}$ is constant in both the cases

27. Couette flow is steady incompressible, laminar flow between one fixed plate and other moving with constant velocity.

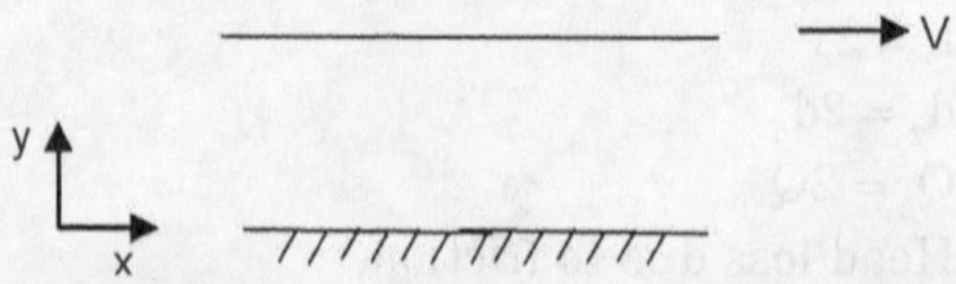

28. $C_f = \frac{1.320}{\sqrt{Re_x}}$

$R_{ex} = \frac{\rho v d}{\mu} = \frac{vd.}{\upsilon}$

v = 2m/s

1 = 2.5m

$\upsilon = 2 \times 10^{-5}$ m²/s

$F = \frac{1}{2} C_f \rho A v^2$

A = 2.5 × 1

On substituting we get

F = 0.0159N

29. $u = \frac{-1}{4\mu}\frac{\partial P}{\partial x}(R^2 - r^2)$

$= \frac{-1}{4 \times 0.001} \times (-10)((0.05)^2 - (0.002)^2)$

u = 6.24 m/s

30. oil water

$v = 1.0 \times 10^{-5}$ m²/s $\quad v = 0.89 \times 10^{-6}$ m²/s

d = 0.5 m $\quad d$ = 0.02 m

v = 10 m/sec $\quad v$ = ?

$[R_e]_{(oil)} = [R_e]_{(w)}$

$\frac{10 \times 0.5}{1.0 \times 10^{-5}} = \frac{v \times 0.02}{0.89 \times 10^{-6}}$

⇒ v = 22.25 m/s

31.

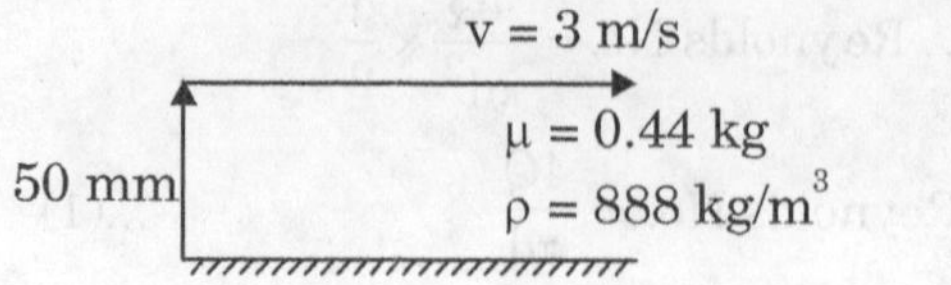

$\frac{du}{dy} = 7\frac{8.0}{0.05} = 60$

Shear strees= $\tau = \mu \dfrac{du}{dy}$

$= 0.44 \times 60 = 26.4 \text{ N/m}^2$

34. From Blausius equation,

$$C_{fx} = \frac{0.664}{\sqrt{Re_x}}$$

$$\& \; C_{fx} = \frac{\tau_w}{\frac{1}{2}\delta U_\infty^2}$$

$$\frac{.664}{\sqrt{\frac{U_{\infty x}}{\upsilon}}} = \frac{\tau_w}{\frac{1}{2}\delta U_\infty^2}$$

$$\tau_w \propto \frac{1}{\sqrt{x}}$$

& we know $x_2 > x_1$

$\therefore \tau_{w/x_1} = \tau_{w/x2}$

35. Drag coefficient,

$$C_f = \frac{F}{\rho U^2 D^2}$$

$F = C_F \, \rho \, U^2 D^2$

For sphere –1

$\rho_1 = 1000 \text{ kg/m}^3$

$D_1 = 100 \text{ mm} = 0.1\text{m}$

$U_1 = 2 \text{ m/s}$

$C_F = 0.5$

For sphere –2

$\rho_2 = 1000 \text{ kg/m}^3$

But $D_2 = 0.2\text{m}$

Due to dynamically similar conditions:

$(Re)_1 = (Re)_2$

$$\frac{U_1 D_1}{\upsilon_1} = \frac{U_2 D_2}{\upsilon_2}$$

$2 \times 0.1 = U_2 \times 0.2$

$U_2 = 1 \text{ m/s}$

$$F_2 = C_F \rho U_2^2 D_2^2$$

$$F_2 = 0.5 \times 1000 \times 1^2 \times .2^2$$

$= 20\text{N}$

36.

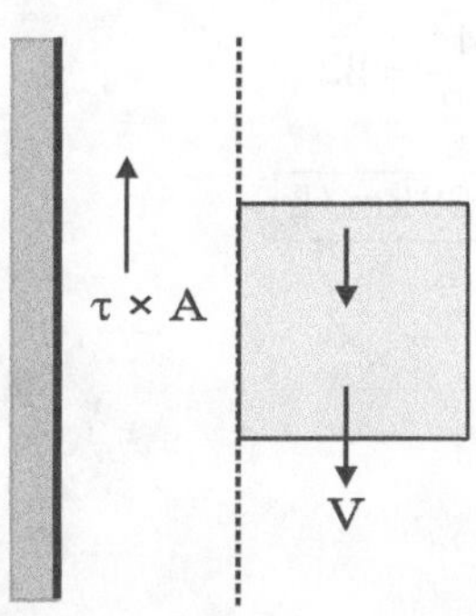

Terminal velocity is a constant velocity i.e. the net acceleration is zero.

So, $\Sigma F_{net} = ma$

$mg - \tau A = 0$

$\tau A = mg$

$$\mu \frac{V}{h} A = mg$$

$$7 \times 10^{-3} \times \frac{V}{0.15 \times 10^{-3}} \times 0.04 = 2 \times 10$$

$V = 10.714 \text{ m/s}$

37. As per Blausius equation,

$$\rho = \frac{5x}{\sqrt{Re_x}} = \frac{5x}{\sqrt{\frac{\rho V_\infty . x}{\mu}}}$$

$$\rho \propto \sqrt{x}$$

$$\rho_L \propto \sqrt{L}$$

$$\rho_{2L} \propto \sqrt{2L}$$

$$\frac{\rho_{2L} - \rho_L}{\rho_L} 100 = \left(\frac{\sqrt{2L} - \sqrt{L}}{\sqrt{L}}\right) \times 100$$

$$= \left(\sqrt{2} - 1\right) \times 100 = 41.42\%$$

38. Since discharge is not changing average velocity in constant dia pipe will not change

$$Re = \frac{\rho VD}{\mu}$$

= Reynold's no. will not change

Re at exit = 500

44. $Re_D > 2300$ means it is a turbulent flow. In turbulent flow, the fluid is well mixed. The fluid is unmixed, for a very-low Reynolds number laminar flow.

45. Time average of fluctuating velocity is zero.

49. $\rho = \dfrac{5x}{\sqrt{Re}}$ for laminar flow

$$Re_x = \frac{Vx}{\upsilon}$$

$$\therefore \rho = \frac{5\sqrt{x}}{\sqrt{V/\upsilon}}$$

$$\therefore \rho \propto \sqrt{x}$$

50. Given: $\rho = 1.23\ kg/m^3$,

$$\nu = \frac{\mu}{\rho} = 1.5 \times 10^{-5}\ m^2/s.$$

$$u = 2\ m/s,\ x = L = 1\ m$$

$$R_{ex} = \frac{\rho u L}{\mu} = \frac{2 \times 1}{\left(\frac{\mu}{\rho}\right)} = \frac{2 \times 1}{1.5 \times 10.5}$$

$$= 1.34 \times 10^5$$

Now, shear stress,

$$\tau_0 = \mu\left(\frac{du}{dy}\right)_{y=0}$$

where, $\frac{u}{U} = \frac{3}{2}\frac{y}{\delta} - \frac{y^3}{2\delta^3}$

$$\frac{du}{dy} = U\left[\frac{3}{2\delta} - \frac{3}{2}\frac{y^2}{\delta^3}\right]$$

Hence $\left(\frac{du}{dy}\right)_{y=0} = \frac{3U}{2\delta}$

Given: $\delta = \frac{4.64x}{\sqrt{Re_x}} = \frac{4.64 \times x}{\sqrt{\frac{\rho U x}{\mu}}}$

Putting $x = 1$,

$$(\delta)_{x=1} = \frac{4.64}{\sqrt{\frac{2 \times 1}{1.5 \times 10^{-5}}}} = 0.0127$$

$$\therefore \quad \tau_0 = \mu \cdot \frac{du}{dy} = \frac{3}{2}\frac{\mu U}{\delta}$$

$$= \frac{3}{2} \times \frac{(1.5 \times 10^{-5} \times 1.23) \times 2}{0.0127}$$

$$= 4.355 \times 10^{-3}.\ N/m^2$$

51. Mass entering from side *q-p* = Mass leaving from side *q-r* + Mass leaving the side *r-s*

52. By momentum equation, we can find drag force.

53. A, B, C can be used out easily.

54. Boundary layer thickness, $\delta \propto \frac{1}{\sqrt{Re}}$

$Re \propto V \Rightarrow \delta \propto \frac{1}{\sqrt{V}}$

If velocity of fluid is increased by 4 times then boundary layer thickness reduces by half

55. mass entering = mass leaving

$$\dot{m}_{ab} = \dot{m}_{bd} + \int_0^\delta \rho dy$$

$$\int_0^\delta u dy = u_0 \int_0^\delta 2\left(\frac{y}{\delta}\right) - \left(\frac{y}{\delta}\right)^2$$

$$= u_0\left[\frac{y^2}{\delta} - \frac{y^3}{3\delta^2}\right]_0^\delta$$

$$= u_0\left[\delta - \frac{\delta}{3}\right]$$

$$\int_0^\delta u dy = \frac{2}{3} u_0 \rho\delta$$

$$\dot{m}_{ab} = \dot{m}_{bd} + \frac{2}{3} u_0 \rho\delta$$

$$\frac{\rho u_0 \delta}{\rho u_0 \delta} = \frac{\dot{m}_{bd}}{\dot{m}_{ab}} + \frac{2 u_0 \rho\delta}{3 u_0 \rho\delta}$$

$$\frac{\dot{m}_{bd}}{\dot{m}_{ab}} = 1 - \frac{2}{3} = \frac{1}{3}$$

56. $$\frac{\delta^*}{\delta} = \frac{\int_0^\delta \left(1 - \frac{u}{u_\infty}\right) dy}{\delta} = \frac{\int_0^\delta \left(1 - \sin\frac{\pi y}{2\delta}\right) dy}{\delta}$$

By integrating and re-arranging we get,

$$\frac{\delta^*}{\delta} = 1 - \frac{2}{\pi}$$

57. Drag force, $F_D = C_d \times \frac{1}{2}\rho V^2 \times A$

Power, $P = F_D \times V$

$$= C_d \times \frac{1}{2}\rho V^2 \cdot A \cdot V$$

$$= C_d \times \frac{1}{2}\rho V^3 \cdot A$$

Power, $P = KC_d \times V_3$

where K = constant

$\Rightarrow$ Given, $(C_d)_1 = 0.1$; $V_1 = 44$ km/h

$(C_d)_2 = 0.8$, $V_2 = ?$

$\because\ P_1 = P_2$

$$\Rightarrow (C_d)_1 \times V_1^3 = (C_d)_2 \cdot V_2^3$$

$$\Rightarrow V_2^3 = \frac{(C_d)_1 \times V_1^3}{(C_d)_2} = \frac{0.1 \times (44)^3}{0.8}$$

$$V_2 = \frac{44}{2} = 22$$

$$\boxed{V_2 = 22\ km/h}$$

6 CHAPTER

Flow Through Pipes

Analysis of Previous GATE Papers

Topic	Year/Set → Marks Distribution ↓		2019 Set 1	2019 Set 2	2018 Set 1	2018 Set 2	2017 Set 1	2017 Set 2	2016 Set 1	2016 Set 2	2016 Set 3	2015 Set 1	2015 Set 2	2015 Set 3
Losses in pipes	1 Mark	MCQ Type										1		1
		Numerical Type									1			
	2 Marks	MCQ Type											1	
		Numerical Type					1							
	Total						2				1	1	2	1
Flow through Siphon	1 Mark	MCQ Type												
		Numerical Type												
	2 Marks	MCQ Type												
		Numerical Type												
	Total													
Power Transmission	1 Mark	MCQ Type												
		Numerical Type					1							
	2 Marks	MCQ Type												
		Numerical Type		1										
	Total			2			1							

Losses in pipes

1. A smooth pipe of diameter 200 mm carries water. The pressure in the pipe at section S1 (elevation: 10 m) ie 50 kPa, At section S2 (elevation : 12 m) the pressure is 20 kPa and velocity is 2 m/s. Density of water is 1000 kg/m^3 and acceleration due to gravity is 9:8 ms^{-2}, Which of the following is TRUE

(a) flow is from S1 to S2 and head loss is 0.53 m
(b) flow is from S2 to S1 and head loss is 0.53 m
(c) flow is from S1 to S2 and head loss is 1.06 m
(d) flow is from S2 to S1 and head loss is 1.06 m

[2010 : 2 Marks]

2. Oil flows through a 200 mm diameter horizontal cast iron pipe (friction factor, f = 0.0225) of length 500 m, The volumetric flow rate is 0.2 m^3/s, The head loss (in m) due to friction is (assume g = 9.81 m/s^2)

(a) 116.18 (b) 0.116
(c) 18.22 (d) 232.36

[2012 : 1 Mark]

3. Water flows through a 10 mm diameter and 250 m long smooth pipe at an average velocity of 0.1 m/s. The density and the viscosity of water are 997 kg/m^3 and 855 × 10^{-6} Ns/m^2, respectively. Assuming fully- developed flow, the pressure drop (in Pa) in the pipe is ____?

[2014 : 2 Marks, Set-2]

4. A fluid of dynamic viscosity 2 × 10^{-5} kg/ms and density 1 kg/m^3 flows with an average velocity of 1 m/s through a long duct of rectangular (25 mm × 15 mm) cross-section, Assuming laminar flow, the pressure drop (in Pa) in the fully developed region per meter length of the duct is _______.

[2014 : 2 Marks, Set-3]

5. Consider fully developed flow in a circular pipe with negligible entrance length effects. Assuming the mass flow rate, density and friction factor to be constant, if the length of the pipe is doubled and the diameter is halved, the head loss due to friction will increase by a factor of

(a) 4 (b) 16
(c) 32 (d) 64

[2015 : 1 Mark, Set-1]

6. Three parallel pipes connected at the two ends have flow-rates Q_1, Q_2 and Q_3 respectively, and the corresponding frictional head losses are h_{L1}, h_{L2} and h_{L3} respectively. The correct expressions for total flow rate (Q) and frictional head loss across the two ends (h_L) are

(a) $Q = Q_1 + Q_2 + Q_3$; $h_L = h_{L1} + h_{L2} + h_{L3}$
(b) $Q = Q_1 + Q_2 + Q_3$; $h_L = h_{L1} = h_{L2} = h_{L3}$
(c) $Q = Q_1 = Q_2 = Q_3$; $h_L = h_{L1} + h_{L2} + h_{L3}$
(d) $Q = Q_1 = Q_2 = Q_3$; $h_L = h_{L1} = h_{L2} = h_{L3}$

[2015 : 1 Mark, Set-3]

7. The head loss for a laminar incompressible flow through a horizontal circular pipe is h_1. Pipe length and fluid remaining the same, if the average flow velocity doubles and the pipe diameter reduces to half its previous value, the head loss is h_2. The ratio h_2/h_1. is

(a) 1 (b) 4
(c) 8 (d) 16

[2015 : 2 Marks, Set-2]

8. A channel of width 450 mm branches into two sub-channels having width 300 mm and 200 mm as shown in figure. If the volumetric flow rate (taking unit depth) of an incompressible flow through the main channel is 0.9 m^3/s and the velocity in the sub-channel of width 200 mm is 3 m/s, the velocity in the sub-channel of width 300 mm is ________m/s.

Assume both inlet and outlet to be at the same elevation.

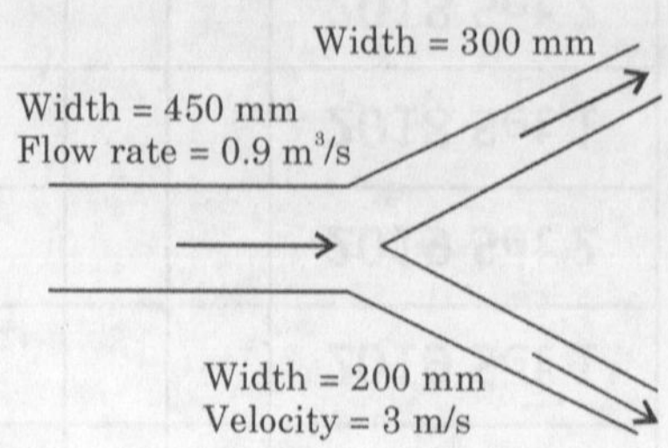

[2016 : 1 Marks, Set-3]

9. Consider steady flow of an incompressible fluid through two long and straight pipes of diameters d_1, and d_2 arranged in series. Both pipes are of equal length and the flow is turbulent in both pipes. The friction factor for turbulent flow though pipes is of the for, f = K(Re)$^{-n}$, Where K and n are known positive constants and Re is the Reynolds number. Neglecting minor losses, the ratio of the frictional pressure drop in pipe 1 to that in pipe 2. $\left(\frac{\Delta P_1}{\Delta P_2}\right)$, is given by

(a) $\left(\frac{d_2}{d_1}\right)^{(5-n)}$ (b) $\left(\frac{d_2}{d_1}\right)^{5}$

(c) $\left(\frac{d_2}{d_1}\right)^{(3-n)}$ (d) $\left(\frac{d_2}{d_1}\right)^{(5+n)}$

[2017 : 2 Marks, Set-1]

Flow through siphon

10. A siphon draws water from a reservoir and discharge it out at atmospheric pressure. Assuming ideal fluid and the reservoir is large, the velocity at point Pin the siphon tube is

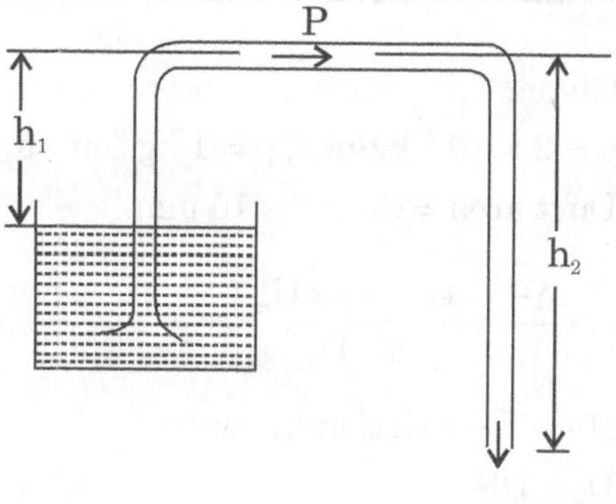

(a) $\sqrt{2gh_1}$ (b) $\sqrt{2g\,h_2}$

(c) $\sqrt{2g(h_2-h_1)}$ (d) $\sqrt{2g(h_2+h_1)}$

[2006 : 2 Marks]

11. A siphon is used to drain water from a large tank as shown in figure below. Assume that the level of water is maintained constant. Ignore frictional effect due to viscosity and losses at entry and exit. At the exit of the siphon, the velocity of water is

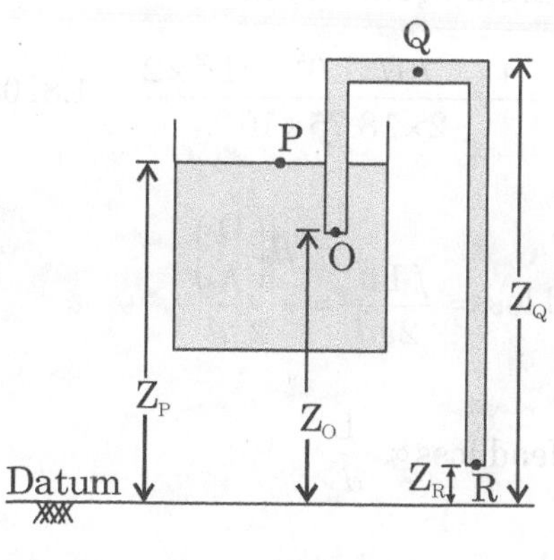

(a) $\sqrt{2g(Z_Q-Z_R)}$ (b) $\sqrt{2g(Z_P-Z_R)}$

(c) $\sqrt{2g(Z_O-Z_R)}$ (d) $\sqrt{2g\,Z_Q}$

[2014 : 2 Marks, Set-3]

Power transmission

12. For a fluid flow through a divergent pipe of length L having inlet and outlet radii of R_1, and R_2 respectively and a constant flow rate of Q, assuming the velocity to be axial and uniform at any cross-section, the acceleration at the exit is

(a) $\dfrac{2Q(R_1-R_2)}{\pi LR_2^3}$ (b) $\dfrac{2Q^2(R_1-R_2)}{\pi LR_2^3}$

(c) $\dfrac{2Q^2(R_1-R_2)}{\pi^2 LR_2^5}$ (d) $\dfrac{2Q^2(R_2-R_1)}{\pi^2 LR_2^5}$

[2004 : 2 Marks]

13. Water at 25°C is flowing throug a 1.0 km long G.I. pipe of 200 mm diameter at the rate of 0.07 m^3/s. If value of Darcy friction factor for this pipe is 0.02 and density of water is 1000 kg/m^3, the pumping power (in kW) required to maintain the flow is

(a) 1.8 (b) 17.4

(c) 20.5 (d) 41.0

[2009 : 2 Marks]

14. For steady, fully developed flow Inside straight pipe of diameter D, neglecting gravity effects, the pressure drop Δp over a length L and the wall shear stress τ_W are related by

(a) $\tau_W = \dfrac{\Delta pD}{4L}$ (b) $\tau_W = \dfrac{\Delta pD^2}{4L^2}$

(c) $\tau_W = \dfrac{\Delta pD}{2L}$ (d) $\tau_W = \dfrac{4\Delta pL}{D}$

[2013 : 1 Mark]

15. Water (density = 1000 kg/m^3) at ambient temperature flows through a horizontal pipe of uniform cross-section at the rate of 1 kg/s. If the pressure drop across the pipe is 100 kPa, the minimum power required to pump the water across the pipe, in Watt, is ______.

[2017 : 1 Mark, Set-1]

16. Water flows through two different pipes A and B of the same circular cross-section but at different flow rates. The length of pipe A is 1.0 m and that of pipe B is 2.0 m. The flow in both the pipes is laminar and fully developed. If the frictional head loss across the length of the pipes is same, the ratio of volume flow rates Q_B/Q_A is ____ (round off to two decimal places)

[2019 : 2 Marks, Set-2]

ANSWERS

1. (c) 2. (a) 5. (d) 6. (b) 7. (c) 9. (a) 10. (b) 11. (b) 12. (c)
13. (b) 14. (a)

EXPLANATIONS

1. At section S_1 — At section S_2

$P_1 = 50$ kPa — $P_2 = 20$ kPa

$z_1 = 10$ m — $z_2 = 12$ m

D = 200 mm,

$l = 1000$ kg/m^3

V = 2 m/s

$g = 9.8$ m/s^2

Applying Bernoulli's equation at two sections, we have

$$\frac{P_1}{\rho g} + z_1 = \frac{P_2}{\rho g} + z_2 + h_2$$

where h_2 = Loss of head from s_1 to s_2

$$\therefore \frac{50\times10^3}{10^3\times9.81} + 10 = \frac{20\times10^3}{10^3\times9.8} + 12 + h_2$$

$$\Rightarrow \quad h_2 = 1.06$$

2. $$H = \frac{fLV^2}{2gD}$$

$$= \frac{0.0225\times500\times\left(\frac{0.2}{\Pi\times0.2^2/4}\right)^2}{2\times9.81\times0.4}$$

$= 116.18$ m

3. $$Re = \frac{\rho VD}{\mu}$$

$$= \frac{997\times0.1\times10\times10^{-3}}{855\times10^{-6}} = 1166.081 \text{ (laminar flow)}$$

$$f = \frac{64}{Re} = \frac{64}{Re} = \frac{64}{1166.081} = 0.054$$

$$\Delta p = \frac{fl\rho V^2}{2D}$$

$$= \frac{0.054\times250\times997\times(0.1)^2}{2\times10\times10^{-3}} = 6840 \text{ pascal.}$$

4. Given,

$\mu = 2\times10^{-5}$ kg/m.s, $\rho = 1$ kg / m^3, $u_{av} = 1$m/sec

Duct area =25 mm × 15 mm

$$\therefore \frac{\Delta P}{L} = \frac{4\times\lambda\times\rho\times U_{av}^2}{2\times D_n} \quad ....(1)$$

Here λ = Friction factor

D_n = Dia

$$\therefore D_n = \frac{4\times A}{P} \text{ (A = Area, P = Parameter)}$$

$$\lambda = \frac{4\times25\times15}{2(25+15)} = 18.75 \text{ mm} \quad ...(2)$$

$$\lambda = \frac{16}{Re} \quad \therefore Re = \frac{\rho\cdot u_{av}\cdot D_n}{\pi}$$

$$\therefore Re = \frac{1\times1\times18.75\times10^{-3}}{2\times10^{-5}} = 937.5$$

$$\therefore \quad \lambda = \frac{16}{937.5} = 1.707\times10^{-2} \quad ...(3)$$

Here from equation (3),

$$\frac{\Delta P}{L} = \frac{4\times1.707\times10^{-2}\times(1)^2\times2}{2\times18.75\times10^{-3}} = 1.8208 \text{ Pa/m.}$$

5. Head loss= $$\frac{fLv^2}{2gd} = \frac{fl\left(\frac{Q}{A}\right)^2}{2gd}$$

$$\Rightarrow \text{Head loss} \propto \frac{L}{d^5}$$

$$h_1^* \alpha \frac{L}{d^5}$$

$$h_2^* \alpha \frac{2.L}{(d/2)^5} = \frac{L}{d^5}\times64$$

$$\frac{h_1}{h_2} = 1/64 \Rightarrow \frac{h_2}{h_1} = 64$$

6. Total flow rate $Q = Q_1 + Q_2 + Q_3$

Head loss $h = h_{L_1} = h_{L_2} = h_{L_3}$

7. Head loss, $h \propto \dfrac{u_{avg}}{D^2}$

$$\therefore \quad \frac{h_2}{h_1} = \left(\frac{D_1}{D_2}\right)^2 \frac{u_{avg,2}}{u_{avg,1}} = 2^2 \times 2$$

$$\frac{h_2}{h_1} = 8$$

8.

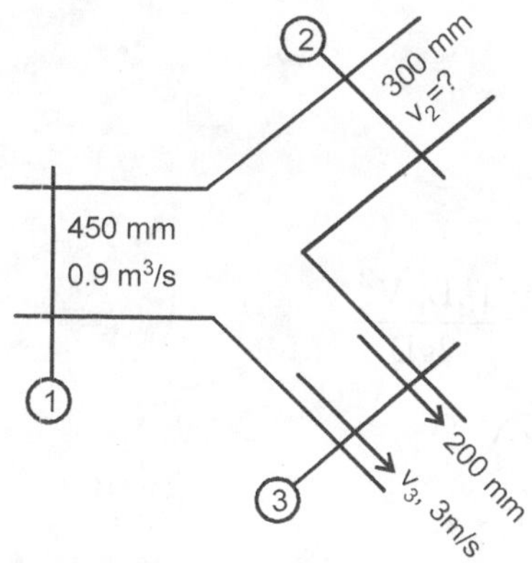

Given data:

$b_1 = 450 \text{ mm} = 0.45 \text{ m}$

$b_2 = 0.3 \text{ m}$

$b_3 = 0.2 \text{ m}$

Depth: $h = 1\text{m}$

$Q_1 = 0.9 \text{ m}^3/\text{s}$

$V_3 = 3\text{m/s}$

$V_2 = ?$

$Q_1 = Q_2 + Q_3$

$Q_1 = A_2V_2 + A_3V_3$

$Q_1 = b_2hV_2 + b_3hV_3$

$0.9 = 0.3 \times 1 \times V_2 + 0.2 \times 1 \times 3$

$0.9 = 0.3\, V_2 + 0.6$

$V_2 = 1 \text{ m/s}$

9. $\dfrac{h_{f1}}{h_{f2}} = \dfrac{f_1 lQ^2}{12d_1^5} \times \dfrac{12d_2^5}{f_2 lQ^2}$

$$\frac{h_{f1}}{h_{f2}} = \frac{f_1}{f_2} \times \left(\frac{d_2}{d_1}\right)^5$$

$$f \propto R_e^{-n}$$

$$f \propto \left(\frac{\rho V d}{\mu}\right)^{-n}$$

$$f \propto \left(\frac{\rho Q d}{\frac{\pi}{4} d^2 \mu}\right)^{-n}$$

$$f \propto d^n$$

$$\frac{h_{f1}}{h_{f2}} = \left(\frac{d_1}{d_2}\right)^n \times \left(\frac{d_2}{d_1}\right)^5$$

$$\frac{h_{f1}}{h_{f2}} = \left(\frac{d_2}{d_1}\right)^{5-n}$$

10. By energy conservation, velocity at point

$Q = \sqrt{2g(h_2 - h_1)}$

As there is a continuous and uniform flow, so velocity of liquid at point Q and P is same.

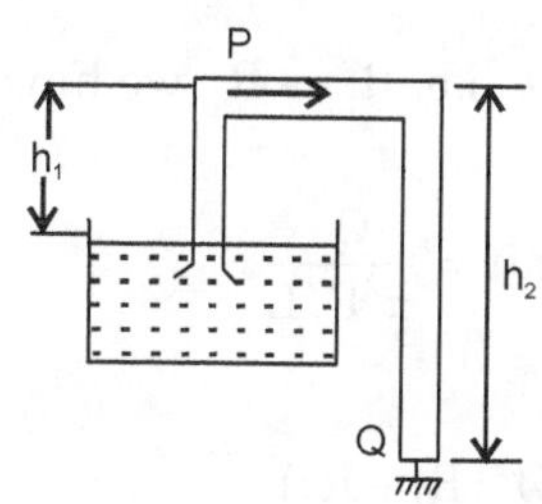

$V_P = \sqrt{2g(h_2 - h_1)}$

11. Applying Bernoulli's equation between the points 'P' and 'R'.

$$\frac{P_{atm}}{\rho g} + \frac{V_P^2}{2g} + Z_p = \frac{P_{atm}}{\rho g} + \frac{V_R^2}{2g} + Z_R$$

Since the level of liquid in tank remains the same, $V_P = 0$

$$\therefore\ V_R = \sqrt{2g(Z_P - Z_R)}.$$

12. Velocity at inlet,

$$u_1 = \frac{Q}{\pi R_1^2}$$

Velocity at outlet,

$$u_2 = \frac{Q}{\pi R_2^2}$$

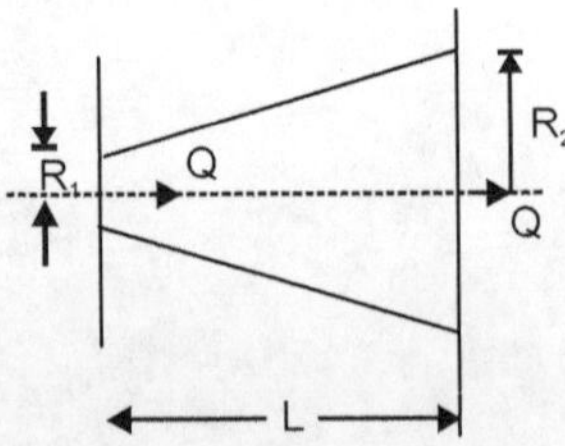

$$\text{Acceleration} = \frac{v \cdot dv}{dx}$$

$$\therefore \frac{dv}{dx} = \frac{u_2 - u_1}{L} = \frac{Q}{\pi L}\left(\frac{1}{R_2^2} - \frac{1}{R_1^2}\right)$$

$$\text{Acceleration at the exit} = u_2 \cdot \frac{dv}{dx}$$

$$= \frac{Q}{\pi R_2^2}\left[\frac{Q}{\pi L}\left(\frac{R_1^2 - R_2^2}{R_1^2 R_2^2}\right)\right]$$

$$= \frac{Q^2}{\pi^2 R_2^2 L}\left[\frac{(R_1 - R_2)(R_1 + R_2)}{R_1^2 R_2^2}\right]$$

Consider limiting case, i.e. $R_1 \to R_2$, we have

$$\text{Acceleration at the exit} = \frac{Q^2}{\pi^2 R_L^2 L}$$

$$\left[\frac{(R_1 - R_2)(2R_2)}{R_2^4}\right] = \frac{2Q^2(R_1 - R_2)}{\pi^2 R_2^5 L}$$

13. $v = \dfrac{0.07}{\dfrac{\pi}{4} \times (0.2)^2} = 2.23\text{m/s}$

$$\text{Head loss, } h_f = \frac{4fLv^2}{d \times 2g}$$

$$= \frac{4 \times 0.02 \times 1000 \times (2.23)^2}{0.2 \times 2 \times 9.81} = 25.35$$

$\therefore$ Power $= \rho g h_f$

$= 9.81 \times 0.07 \times 25.35$

$= 17.4$ kW

14. Shear stress on wall,

$$\tau_w = \frac{-\partial P}{\partial x} \cdot \frac{R}{2}$$

$$\text{where } \frac{-\partial P}{\partial x} \cdot \frac{\Delta P}{L}$$

$$\& \ R = \frac{D}{2}$$

$$\therefore \tau_w = \frac{\Delta P}{L} \times \frac{D}{2 \times 2} = \frac{\Delta PD}{4L}$$

15. $W = \dot{m} v (p_1 - p_2)$

$$W = 1 \times \frac{1}{1000} \times 100 \times 10^3 = 100 \text{ Watt}$$

16. $L_A = 1.0$ m

$L_B = 2.0$ m

$h_{L_A} = h_{L_B}$

$$\Rightarrow \frac{f_A L_A V^2}{2gD} = \frac{f_B L_B V^2}{2gD}$$

Now, $Q = A \times V$

$$\Rightarrow V = \frac{Q}{A} = \frac{Q \times 4}{\pi D^4}$$

$$f_A \times 1 \times \frac{Q_A^2 \times 16}{\pi^2 D^8} = f_B \times \frac{Q_B^2 \times 16}{\pi^2 D^8} \times 2$$

$$\frac{64}{Re_A} \times Q_A^2 = \frac{64}{Re_B} \times Q_B^2 \times 2$$

$$\frac{Q_A^2}{Q_B^2} = 2 \times \frac{Re_A}{Re_B}$$

$$\Rightarrow \frac{Q_B^2}{Q_A^2} = \frac{1}{2} \times \frac{Re_B}{Re_A} = \frac{\frac{1}{2} \times \rho \times \frac{Q_B}{D^2} \times \frac{D}{\mu}}{\rho \times \frac{Q_A}{D^2} \times \frac{D}{\mu}}$$

$$\frac{Q_B^2}{Q_A^2} = \frac{1}{2} \times \frac{Q_B}{Q_A}$$

$$\Rightarrow \frac{Q_B}{Q_A} = 0.5$$

7 CHAPTER Turbines and Pumps

Analysis of Previous GATE Papers

Topic	Year/Set → Marks Distribution ↓		2019 Set 1	2019 Set 2	2018 Set 1	2018 Set 2	2017 Set 1	2017 Set 2	2016 Set 1	2016 Set 2	2016 Set 3	2015 Set 1	2015 Set 2	2015 Set 3
Turbines	1 Mark	MCQ Type	1					1						
		Numerical Type			1				1		1			
	2 Marks	MCQ Type												1
		Numerical Type				1								
	Total		1		1	2		1	1		1			2
Pumps	1 Mark	MCQ Type					1							
		Numerical Type												
	2 Marks	MCQ Type												
		Numerical Type		1										
	Total			2			1							
Impact of Jet	1 Mark	MCQ Type												
		Numerical Type												
	2 Marks	MCQ Type												
		Numerical Type			1			1						
	Total				2			2						

Turbines

1. Cavitation in a hydraulic reaction turbine is most likely to occur at the turbine

(a) Entry (b) Exit
(c) Stator exit (d) Rotor exit

[1993 : 2 Marks]

2. Specific speed of a Kaplan turbine ranges between

(a) 30 to 60 (b) 60 to 300
(c) 300 to 600 (d) 600 to 1000

[1993 : 2 Marks]

3. Kaplan turbine is

(a) A high head mixed flow turbine
(b) A low head axial flow turbine
(c) An outward flow reaction turbine
(d) An impulse inward flow turbine

[1997 : 1 Mark]

4. If there are m physical quantities and n fundamental dimensions in a particular process, the number of non-dimensional parameters is

(a) m + n (b) m × n
(c) m – n (d) m/n

[2002 : 1 Mark]

5. Match List-I with List-II and select the correct answer using the codes given below the lists:

List-I	List-II
P. Curtis	1. Reaction steam turbine
Q. Rateau	2. Gas turbine
R. Kaplan	3. Velocity compounding
S. Francis	4. Pressure compounding
	5. Impulse water turbine
	6. Axial turbine
	7. Mixed flow turbine
	8. Centrifugal pump

Codes:

	P	Q	R	S
(a)	2	1	1	6
(b)	3	4	6	7
(c)	1	3	1	5
(d)	3	4	7	6

[2003 : 2 Marks]

6. At a hydro electric power plant site, available head and flow rate are 24.5 m and 10.1 m^3/s respectively. If the turbine to be installed is required to run at 4.0 revolution per second (rps) with an overall efficiency of 90%, the suitable type of turbine for this site is

(a) Francis (b) Kaplan
(c) Pelton (d) Propeller

[2004 : 2 Marks]

7. In a Pelton wheel, the bucket peripheral speed is 10 m/s, the water jet velocity is 25 m/s and volumetric flow rate of the jet is 0.1 m^3/s. If the jet deflection angle is 120° and the flow is ideal, the power developed is

(a) 7.5 kW (b) 15.0 kW
(c) 22.5 kW (d) 37.5 kW

[2006 : 1 Mark]

8. A large hydraulic turbine is to generate 300 kW at 1000 rpm under a head of 40 m. For initial testing, a 1 : 4 scale model of the turbine operates under a head of 10 m. The power generated by the model (in kW)will be

(a) 2.34 (b) 4.68
(c) 9.38 (d) 18.75

[2006 : 2 Marks]

9. The inlet angle of runner blades of a Francis turbine is 90°. The blades are so shaped that the tangential component of velocity at blade outlet is zero. The flow velocity remains constant through out the blade passage and is equal to half of the blade velocity at runner inlet. The blade efficiency of the runner is

(a) 25% (b) 50%
(c) 80% (d) 89%

[2007 : 2 Marks]

10. A model of a hydraulic turbine is tested at a head of 1/4th of that under which the full scale turbine works. The diameter of the model is half of that of the full scale turbine. If N is the RPM of the full scale turbine, then the RPM of the model will be

(a) N/4 (b) N/2
(c) N (d) 2N

[2007 : 2 Marks]

11. A hydraulic turbine develops 1000 kW power for a head of 40 m. If the head is reduced to 20 m, the power developed (in kW) is

(a) 177 (b) 354
(c) 500 (d) 707

[2010 : 1 Mark]

12. The velocity triangles at the inlet and exit of the rotor of a turbo machine are shown. V denotes the absolute velocity of the fluid, W denotes the relative velocity of the fluid and U denotes the blade velocity. Subscripts 1 and 2 refer to inlet and outlet respectively If $V_2 = W_1$ and $V_1 = W_2$, then the degree of reaction is

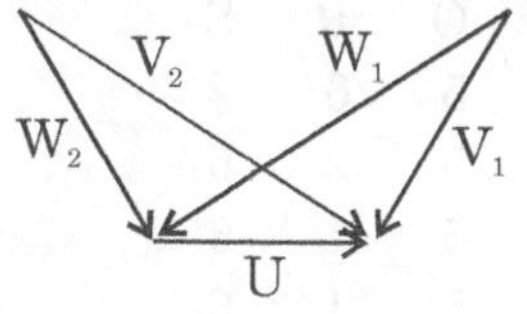

(a) 0 (b) 1

(c) 0.5 (d) 0.25

[2012 : 1 Mark]

13. In order to have maximum power from a Pelton turbine, the bucket speed must be

(a) equal to the jet speed

(b) equal to half of the jet speed

(c) equal to twice the jet speed

(d) independent of the jet speed

[2013 : 1 Mark]

14. Kaplan water turbine is commonly used when the flow through its runner is

(a) axial and the head available is more than 100 m

(b) axial and the head available is less than 10 m

(c) radial and the head available is more than 100 m

(d) mixed and the head available is about 50 m

[2014 : 1 Mark, Set-4]

15. At the inlet of an axial impulse turbine rotor, the blade linear speed is 25 m/s, the magnitude of absolute velocity is 100 m/s and the angle between them is 25°. The relative velocity and the axial component of velocity remain the same between the inlet and outlet of the blades. The blade inlet and outlet velocity triangles are shown in figure. Assuming no losses, the specific work (in J/kg) is ______.

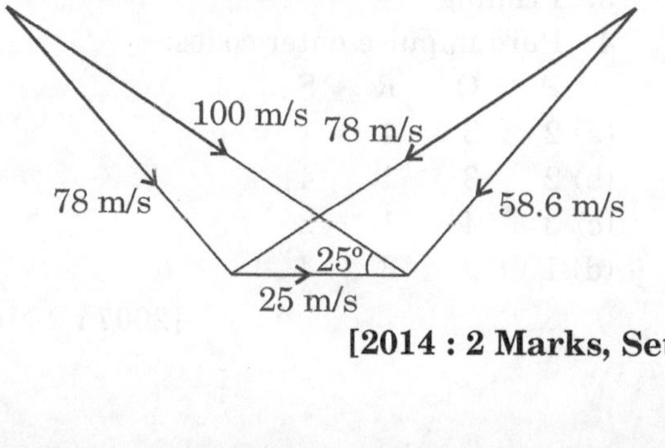

[2014 : 2 Marks, Set-3]

16. Which of the following statements are TRUE, when the cavitation parameter $\sigma = 0$?

i. the local pressure is reduced to vapor pressure

ii. cavitation starts

iii. boiling of liquid starts

iv. cavitation stops

(a) i, ii and iv (b) only ii and iii

(c) only i and iii (d) i, ii and iii

[2015 : 2 Marks, Set-3]

17. Consider two hydraulic turbines having identical specific speed and effective head at the inlet. If the speed ratio (N_1/N_2) of the two turbines is 2, then the respective power ratio (P_1/P_2) is ____.

[2016 : 1 Mark, Set-1]

18. The blade and fluid velocities for an axial turbine are as shown in the figure.

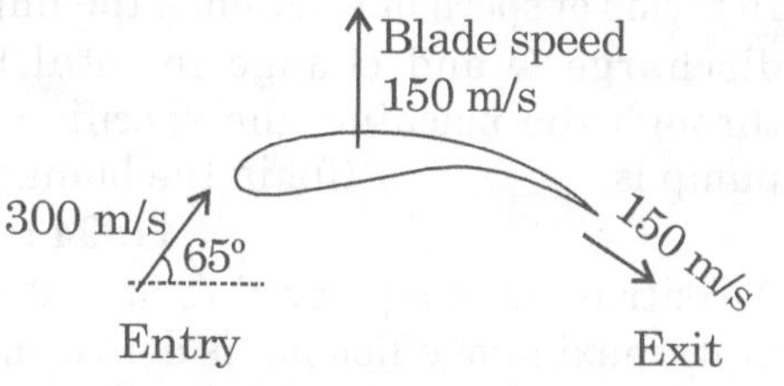

The magnitude of absolute velocity at entry is 300 m/s at an angle of 65° to the axial direction, while the magnitude of the absolute velocity at exit is 150 m/s. The exit velocity vector has a component in the downward direction. Given that the axial (horizontal) velocity is the same at entry and exit, the specific work (in kJ/kg) is ______.

[2016 : 1 Mark, Set-3]

19. Which one of the following statements is TRUE?

(a) Both Pelton and Francis turbines are impulse turbines.

(b) Francis turbine is a reaction turbine but Kaplan turbine is an impulse turbine.

(c) Francis turbine is an axial-flow reaction turbine.

(d) Kaplan turbine is an axial-flow reaction turbine.

[2017 : 1 Mark, Set-2]

20. For a Pelton wheel with a given water jet velocity, the maximum output power from the Pelton wheel is obtained when the ratio of the bucket speed to the water jet speed is ______ (correct to two decimal places).

[2018 : 1 Mark, Set-1]

21. A test is conducted on a one-fifth scale model of a Francis turbine under a head of 2 m and volumetric flow rate of 1 m^3/s at 450 rpm. Take the water density and the acceleration due to gravity as 10^3 kg/m^3 and 10 m/s^2, respectively. Assume no losses both in model and prototype turbines. The power (in MW) of a full sized turbine while working under a head of 30 m is ______ (correct to two decimal places).

[2018 : 2 Marks, Set-2]

22. As per common design practice, the three types of hydraulic turbines, in descending order of flow rate, are

(a) Pelton, Francis, Kaplan
(b) Francis, Kaplan, Pelton
(c) Pelton, Kaplan, Francis
(d) Kaplan, Francis, Pelton

[2019 : 1 Mark, Set-1]

Pumps

23. In terms of speed of rotation of the impeller N, discharge Q and change in total head H, through the machine, the specific speed for pump is ____ (fill in the blanks)

[1994 : 1 Mark]

24. A centrifugal pump running at 500 rpm and at its maximum efficiency is delivering a head of 30 m at a flow rate of 60 liters per minute. If the rpm is changed to 1000, then the head H in meters and flow rate Q in liters per minute at maximum efficiency are estimated to be

(a) H = 60, Q = 120 (b) H = 120, Q= 120
(c) H = 60, Q = 480 (d) H = 120, Q = 30

[2003 : 2 Marks]

25. A centrifugal pump is required to pump water to an open water tank situated 4 km away from the location of the pump through a pipe of diameter 0.2 m having Darcy's friction factor of 0.01. The average speed of water in the pipe is 2 m/s. If it is to maintain a constant head of 5 m in the tank, neglecting other minor losses, the absolute discharge pressure at the pump exit is

(a) 0.449 bar (b) 5.503 bar
(c) 44.911 bar (d) 55.203 bar

[2004 : 2 Marks]

26. Match List-I with List-II and select the correct answer using the codes given below the lists:

List-I
R. Reciprocating pump
Q. Axial flow pump
R. Microhydel plant
S. Backward curved vanes

List-II
1. Plant with power output below 100 kW
2. Plant with power output between 100 kW to 1 MW
3. Positive displacement
4. Draft tube
5. High flow rate, low pressure ratio
6. Centrifugal pump impeller

Codes:

	P	Q	R	S
(a)	3	5	6	2
(b)	3	5	2	6
(c)	3	5	1	6
(d)	4	5	1	6

[2004 : 2 Marks]

27. A horizontal - shaft centrifugal pump lifts water at 65°C. The suction nozzle is one meter below pump centerline. The pressure at this point equals 200 kPa gauge and velocity is 3 m/s. Steam tables show saturation pressure at 65°C is 25 kPa, and specific volume of the saturated liquid is 0.001020 m^3/kg. The pump Net Positive Suction Head (NPSH) in meters is

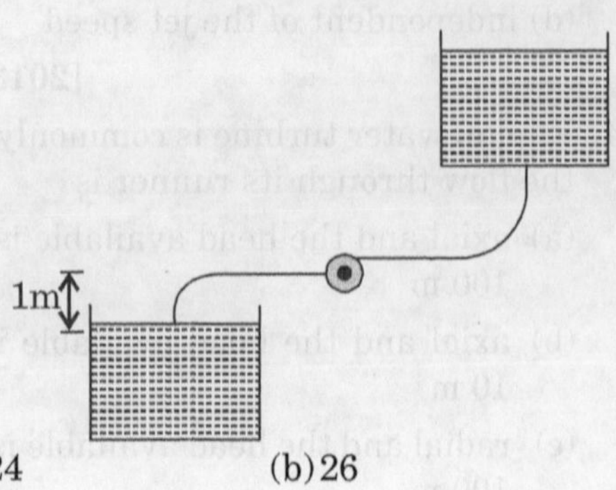

(a) 24 (b) 26
(c) 28 (d) 30

[2006 : 2 Marks]

28. Match List-I with List-II and select the correct answer using the codes given below the lists:

List-I
P. Centrifugal compressor
Q. Centrifugal pump
R. Pelton wheel
S. Kaplan turbine

List-II
1. Axial flow
2. Surging
3. Priming
4. Pure impulse enter codes:

	P	Q	R	S
(a)	2	3	4	1
(b)	2	3	1	4
(c)	3	4	1	2
(d)	1	2	3	4

[2007 : 2 Marks]

29. A pump handling a liquid raises its pressure from 1 bar to 30 bar. Take the density of the liquid as 990 kg/m^3. The isentropic specific work done by the pump in kJ/kg is

(a) 0.10 (b) 0.30
(c) 2.50 (d) 2.93

[2011 : 2 Marks]

30. Which one of the following is NOT a rotating machine?

(a) Centrifugal pump (b) Gear pump
(c) Jet pump (d) Vane pump

[2017 : 1 Mark, Set-1]

31. An idealized centrifugal pump (blade outer radius of 50 mm) consumes 2 kW power while running at 3000 rpm. The entry of the liquid into the pump is axial and exit from the pump is radial with respect to impeller. If the losses are neglected, then the mass flow rate of the liquid through the pump is ____ kg/s (round off to two decimal places).

[2019 : 2 Marks, Set-2]

Impact of jet

32. Water, having a density of 1000 kg/m^3, issues from a nozzle with a velocity of 10 m/s and the jet strikes a bucket mounted on a Pelton wheel. The wheel rotates at 10 rad/s. The mean diameter of the wheel is 1 m. The jet is split into two equal streams by the bucket, such that each stream is deflected by 120° as shown in the figure. Friction in the bucket may be neglected. Magnitude of the torque exerted by the water on the wheel, per unit mass flow rate of the incoming jet, is

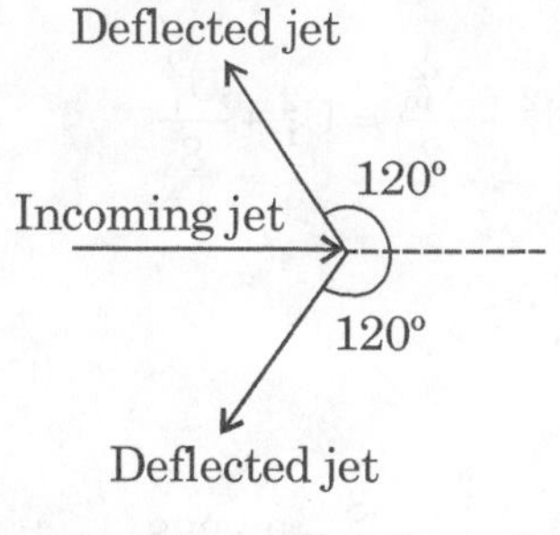

(a) 0 (Nm)/(kg/s) (b) 1.25 (Nm)/(kg/s)
(c) 2.5 (Nm)/(kg/s) (d) 3.75 (Nm)/(kg/s)

[2008 : 2 Marks]

33. An ideal water jet with volume flow rate of 0.05 m^3/s strikes a flat plate placed normal to its path and exerts a force of 1000 N. Considering the density of water as 1000 kg/m^3, the diameter (in mm) of the water jet is ______.

[2014 : 2 Marks, Set-1]

34. A 60 mm diameter water jet strikes a plate containing a hole of 40 mm diameter as shown in the figure. Part of the jet passes through the hole horizontally, and the remaining is deflected vertically. The density of water is 1000 kg/m^3. If velocities are as indicated in the figure, the magnitude of horizontal force (in N) required to hold the plate is ________.

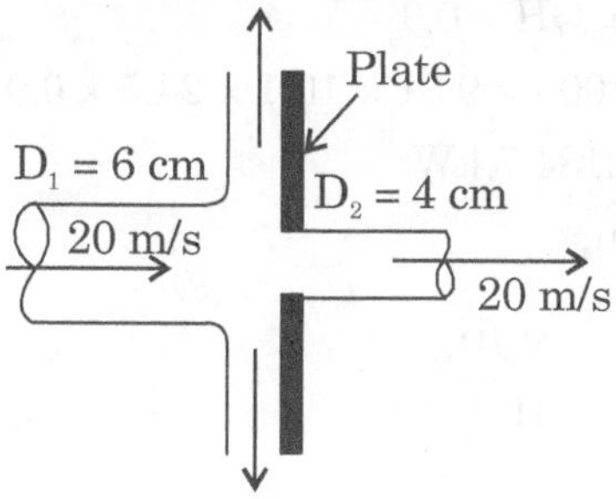

[2017 : 2 Marks, Set-2]

35. A sprinkler shown in the figure rotates about its hinge point in a horizontal plane due to water flow discharged through its two exit nozzles.

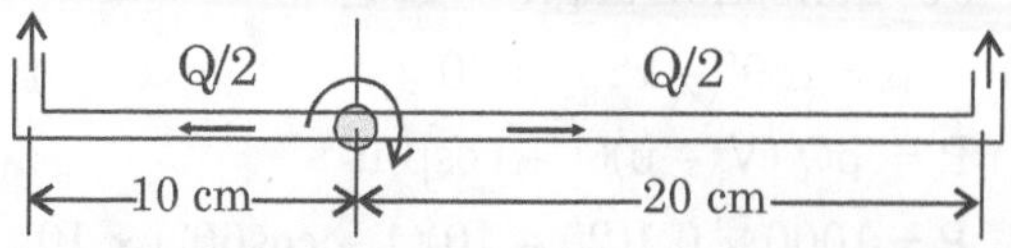

The total flow rate Q through the sprinkler is 1 litre/sec and the cross-sectional area of each exit nozzle is 1 cm^2. Assuming equal flow rate through both arms and a frictionless hinge, the steady state angular speed of rotation (rad/s) of the sprinkler is _______ (correct to two decimal places).

[2018 : 2 Marks, Set-1]

ANSWERS

1. (d) **2.** (d) **3.** (b) **4.** (c) **5.** (b) **6.** (a) **7.** (c) **8.** (a) **9.** (d) **10.** (c)
11. (b) **12.** (c) **13.** (b) **14.** (b) **16.** (d) **19.** (d) **22.** (d) **24.** (b) **25.** (b) **26.** (b)
27. (c) **28.** (a) **29.** (d) **30.** (c) **32.** (d)

EXPLANATIONS

6. Given:

$H = 24.5$ m, $Q = 10.1$ m³/s

$N = 4$rev/s $= 4 \times 60 = 240$ r.p.m.

$\eta_0 = 0.90$

∴ Power generated

$P = \rho g Q H \times 0.9$

$P = 1000 \times 9.81 \times 10.1 \times 24.5 \times 0.9$

$P = 2184.7$ kW

Again,

$$N_s = \frac{N\sqrt{P}}{H^{5/4}}$$

$$= \frac{240\sqrt{2184.7}}{(24.5)^{5/4}} = 205.80$$

$51 < N_s < 255$, hence turbine in Francis.

7. $u = 10$ m/s

$V_i = 25$ m/s

$Q = 0.1$ m³/s

Jet deflection angle = 120°

$\therefore \beta = 180° - 120° = 60°$

$P = \rho Q (V_i - u)(1 + \cos\beta)u$

$P = 1000 \times 0.1(25 - 10)(1 + \cos 60°) \times 10$

$P = 22500$ W $= 22.5$ kW

8. $$P_M = P_P\left(\frac{N_m}{Np}\right)^3\left(\frac{D_m}{D_p}\right)^5$$

$$= 300\left(\frac{2000}{1000}\right)^3\left(\frac{1}{5}\right)^5 = 2.34\text{ kW}$$

9. Given data:

$B_i = 90°$

$V_{wo} = 0$

$$V_{fi} = V_{fo} = \frac{U_i}{2}$$

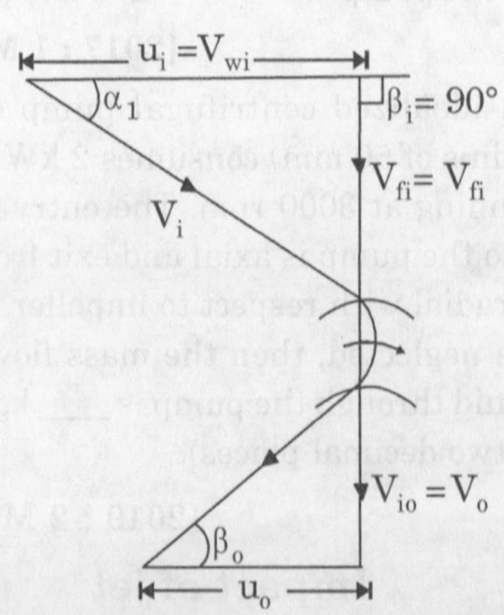

Blade Efficiency of the Runner,

Blade efficiency of the runner,

$$\eta_H = \frac{V_{wi}u_i}{gH} = \frac{U_i^2}{gH} \therefore V_{wi} = u_i$$

By energy balance equation,

$$\rho QgH = \rho QV_{wi}U_i + \frac{\rho QV_0^2}{2}$$

$$gH = U_i^2 + \frac{V_{fo}^2}{2} = U_i^2 + \frac{U_i^2}{8}$$

$$gH = \frac{9}{8}U_i^2$$

$$\therefore \eta_H = \frac{U_i^2}{\frac{9}{8}U_i^2} = \frac{8}{9} = 0.8888$$

$= 88.88\% \approx 98\%$

10. Given data :

Model : $H_m = \frac{H_p}{4}; D_m = \frac{D_p}{2}$

$N_m = ?$

Prototype (Full scale turbine)

$N_P = N$

$$C_H = \frac{H}{(ND)^2}$$

$$(C_H)_{model} = (C_H)_{prototype}$$

$$\frac{H_m}{N_m^2 D_m^2} = \frac{H_P}{N_p^2 D_P^2}$$

$$\frac{H_{P/4}}{N_m^2 \times \frac{D_p^2}{4}} = \frac{H_p}{N_2 D_p^2}$$

$$\boxed{N_m = N}$$

11. $P_u = \frac{P}{H^{3/2}}$

$$\therefore \quad \frac{P_1}{H_1^{3/2}} = \frac{P_2}{H_2^{3/2}}$$

$$\Rightarrow \quad \frac{1000}{40^{3/2}} = \frac{P_2}{20^{3/2}}$$

$\Rightarrow P_2 = 353.55$ kW.

12. $DOR = \frac{\frac{W_2^2 - W_1^2}{2}}{W_2^2 - W_1^2}$

$$= \frac{1}{2} = 0.5$$

13. For a pelton wheel, maximum power is obtained at,

$$u = \frac{v}{2}$$

14. Kaplan turbine is an axial flow turbine and works at low heads (<50).

15.

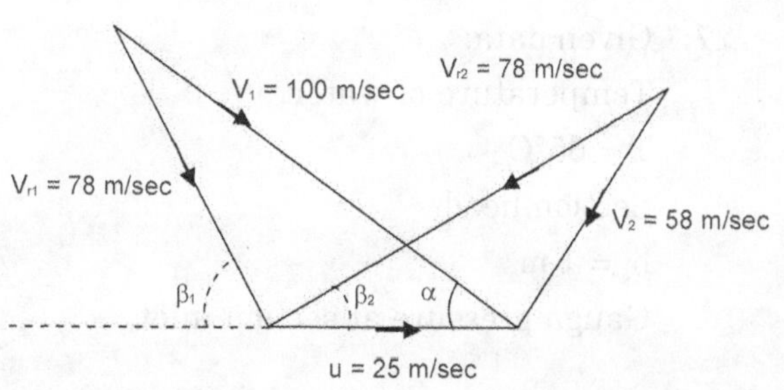

Given $\alpha = 25^\circ$

$V_{r_1} \sin\beta_1 = V_1 \sin\alpha$

$\therefore$ $78 \sin\beta_1 = 100 \sin 25^\circ$

$\therefore$ $\beta_1 = 32.81^\circ$

$\because$ $V_{r_1} = V_{r_2}$

$\therefore$ $\beta_1 = \beta_2 = 32.81^\circ$

Now $\therefore \Delta V_w = V_{r_1}\cos\beta_1 + V_{r_2}\cos\beta_2$

$= 2V_{r_1}\cos\beta_1$

$= 2 \times 78 \times \cos 32.81^\circ$

$\therefore \Delta V_w = 131.1136$ m/sec

$\therefore$ Sp. power $= \Delta V_w \cdot u = 131.1136 \times 25$

$= 3277.84$ J/kg.

16. $\sigma = 0$ implies (*i*), (*ii*) and (*iii*)

17. $N_s = \frac{N\sqrt{P}}{H^{\frac{5}{4}}}$

Given $N_{s1} = N_{s2}$, $H_1 = H_2$

$$\frac{N_1}{N_2} = 2, \frac{P_1}{P_2} = ?$$

$$N_1\sqrt{P_1} = N_2\sqrt{P_2}$$

$$\Rightarrow \quad \frac{P_1}{P_2} = \left[\frac{N_1}{N_2}\right]^2 = \left(\frac{1}{2}\right)^2 = 0.25$$

18. Given: $V_1 = 300$ m/sec

$u = 150$ m/sec

$V_{f1} = V_{f2}$

$(\alpha = 25^\circ)$

$V_2 = 150$ m/sec

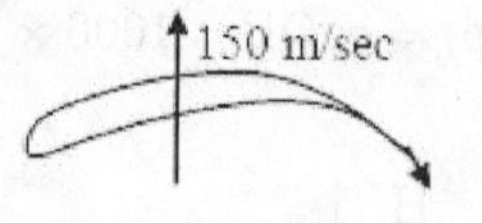

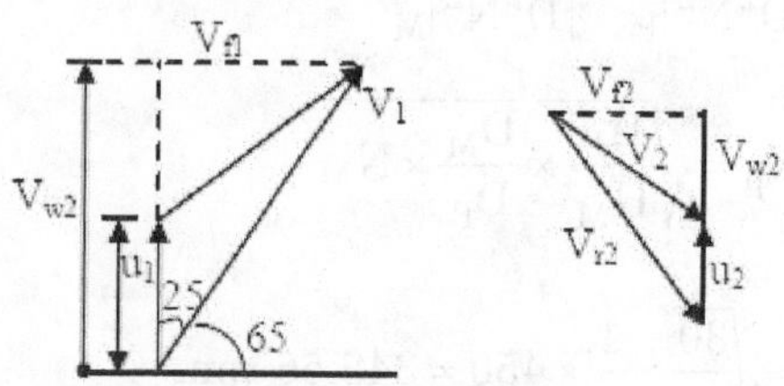

Specific work $= [V_{w1} + V_{w2}] \cdot u$

$V_{w1} = V_1 \cos 25^\circ$

$V_{w1} = 300 \cos 25^\circ$

$= 271.89$ m/s

$V_{f1} = V_{f2} = V_1 \sin 25$

$V_{f2} = 300 \sin 25$

$= 126.78$ m/s

$V_{w2} = \sqrt{V_2^2 - V_{f2}^2}$

$= \sqrt{150^2 - 126.78^2} = \sqrt{6426.83}$

$V_{w2} = 80.16$ m/s

Specific work $= [V_{w1} + V_{w2}] \cdot u$

$= [271.89 + 80.76] \times 150$

$= 52807.5$ J/kg

$= 52.81$ kJ/kg

19. Kaplan turbine is an axial-flow reaction turbine

20. In pelton wheel:-

$$\eta = \frac{2(vu - u^2)}{v^2} \text{ where,}$$

u = bucket speed

v = water jet speed

So, For maximum efficiency:-

$$\frac{dx}{du} = 0$$

$2(v - 2u) = 0$

$v = 2u$

$$\boxed{u = \frac{v}{2}}$$

21. $\frac{D_m}{D_p} = \frac{1}{5}$

Model : H = 2 m, Q = 1 m³/s

Prototype : H = 30 m, P = ?

N = 450 rpm, ρ = 1000 kg/m³ /g = 10 m/s²

For model, $p_m = \rho gQH = 1000 \times 10 \times 1 \times 2$

= 20 kW

$$\left|\frac{H}{D^2N^2}\right|_P = \left|\frac{H}{D^2N^2}\right|_M$$

$$N_P = \sqrt{\frac{H_P}{H_M} \times \frac{D_M}{D_P}} \times N_M$$

$$= \sqrt{\frac{30}{2}} \times \frac{1}{5} \times 450 = 348.56 \text{ rpm}$$

$$\left.\frac{P}{D^5N^3}\right|_P = \left.\frac{P}{D^5N^3}\right|_M$$

$$P_P = P_M \times \frac{D_P^5}{D_M^5} \times \frac{N_P^3}{N_M^3} = 20 \times 5^5 \times \frac{348.56^3}{450^3}$$

= 29045.24 kW

= 29.05 mW

22. In descending order of flow rate.

Kaplan > Francis > Pelton

23. $N_s = \frac{N\sqrt{Q}}{H^{3/4}}$

24. $N_1 = 500$ rpm, $H_1 = 30$ m $Q_1 = 60l$/minute

$N_2 = 1000$ rpm, $H_2 = ?$ $Q_2 = ?$

Since $\frac{\sqrt{H_1}}{DN_1} = \frac{\sqrt{H_2}}{DN_2}$

$$\therefore H_2 = \left(\frac{N_2}{N_1}\right)^2$$

$$H_1 = \left(\frac{1000}{500}\right)^2 \times 30 = 120 \text{ m}$$

$$\frac{Q_1}{D^3N_1} = \frac{Q_2}{D^3N_2}$$

$$\Rightarrow Q_2 = \left(\frac{N_2}{N_1}\right)$$

$$Q_1 = \left(\frac{1000}{500}\right) \times 60 = 120l/\text{minute}$$

25. Given: $d = 0.2$ m, L = 4000 m

$f = 0.01$, $v = 2$ m/s

Head loss due to friction,

$$h_f = \frac{fLv^2}{2gd} = \frac{0.01 \times 4000 \times (2)^2}{2 \times 9.81 \times 0.2}$$

= 40.77 m

Pressure corresponding to this head

$= \rho g(h_f + h + h_{atm})$

$= 1000 \times 9.81(40.77 + 5 + 10.3)$

$= 5.50 \times 10^5$ N/m² = 5.50 bar

27. Given data:

Temperature of water,

T = 65°C

Section head,

$h_s = 1$ m

Gauga pressure at section inlet,

$p_{1.g}$ = 200 kPa

∴ Absolute pressure,

$P_{1.abs} = P_{1.g} + P_{atm}$

= 200 + 101.325

= 301.325 kPa

= 301.325 × 10^3 pa

Velocity at section nozzle inlet,

V_1 = 3 m/s

Saturated pressure at 65°C,

$P_{v.abs}$ = 25 kPa, vapour pressure

= 25 × 10^3 Pa

Specific volume,

V = 0.001020 m³/kg

Specific volume,

V= 0.001020 m³/kg

$$\therefore \text{Density} : \rho = \frac{1}{0.001020} = 980.39\text{kg}/\text{m}^3$$

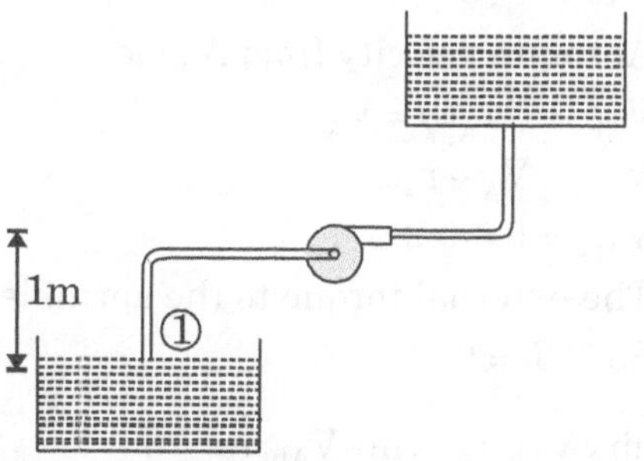

Net Positive Section Head

Net positive section head,

$$\text{NPSH} = \frac{P_{S,abs}}{\rho g} + \frac{V_s^2}{2g} - \frac{Pv.abs}{\rho g}$$

Applying Bernoulli's equation at section nozzle inlet and section of the pump

$$\frac{P_{1.abs}}{\rho g} + \frac{V_1^2}{2g} + \frac{P_{s.abs}}{\rho g} + \frac{V_s^2}{2g} + h_s$$

$$\frac{P_{1.abs}}{\rho g} + \frac{V^2}{2g} + 0 = \frac{P_{s.abs}}{\rho g} + \frac{V_s^2}{2g} + h_s$$

$$\text{or} \quad \frac{P_{s.abs}}{\rho g} + \frac{V_s^2}{2g} + \frac{P_{t.abs}}{\rho g} + \frac{V_1^2}{2g} - h_s$$

$$\therefore \text{NPSH} = \frac{P_{s.abs}}{\rho g} + \frac{V_s^2}{2g} - h_s \frac{P_{v.abs}}{\rho g}$$

$$= \frac{301.325 \times 10^3}{980.39 \times 9.81} + \frac{(3)^2}{2 \times 9.81} - 1 - \frac{25 \times 10^3}{980.39 \times 9.81}$$

= 31.33 + 0.45 – 0.45–1 – 2.56

= 28.22m = 28 m

28. Kaplan – Axial flow

Pelton – Impulse type

29. Work done by pump = $\upsilon(p_2 - p_1)$

$$= \frac{(30-1) \times 100}{990}$$

= 2.93 kJ/kg

31. D = 2 × 50 = 100 mm

N = 3000 rpm.

Power, P = m.u²

$$u = \frac{\pi DN}{60} = \frac{\pi \times 100 \times 3000}{60}$$

= 15707.963 mm/sec.

= 15.7079 m/sec.

$\Rightarrow$ 2000 = $\dot{m}(15.7079)^2$

$\Rightarrow$ $\dot{m}$ 8.105 kg/s ≈ 8.11 kg/s

32. ϕ = 60°, k = 1 (No friction)

v = 10 m/sec

Wheel speed, u = 10 × $\frac{1}{2}$ = 5 m/sec

Work done per sec by wheel

$$= \frac{W}{g} (v - u)(1 + k\cos\phi).\, u$$

$$\text{Power} = \frac{m.g}{g} (v - u)(1 + K\cos\phi).\, u$$

Now P = T. ω

$$\frac{T}{m} = \frac{(v-u)(1+k\cos\phi)u}{w}$$

$$= \frac{(10-5)(1+\cos 60°).5}{10}$$

= 3.75 N.m/(kg/sec)

33.

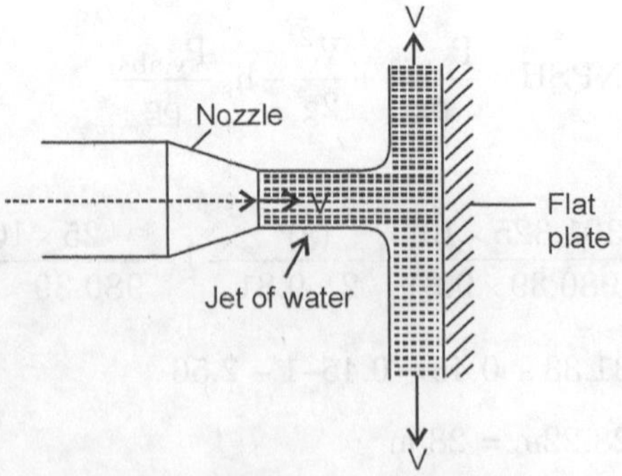

Given : $Q = 0.05\ m^3/s$

$F_x = 1000$ N

$\rho = 1000\ kg/m^3$

D = ?

F_x = Rate of change of momentum in the direction of force

$$= \frac{Mass}{Time} \times [\text{Initial velocity} - \text{Final velocity}]$$

$\frac{Mass}{Sec} = \rho aV$, velocity of jet after striking is equal to zero

$\therefore \quad \rho aV (V - 0) = F_x$

$$\left[\because Q = aV \Rightarrow V = \frac{Q}{a}\right]$$

$$1000 = 1000 \times a \times \frac{Q^2}{a^2}$$

$Q^2 = a;\ a = (0.05)^2$

$$\frac{\pi}{4}d^2 = 2.5 \times 10^{-3}$$

$d = 0.05641$ m = 56.41 mm.

34. $\rho = 1000\frac{kg}{m^3}$

$F = (Momentum)_i - (Momentum)_o$

$$F = \left(\dot{m}\,V\right)_i - \left(\dot{m}\,V\right)_{out}$$

$$F = \rho\left[\frac{\pi}{4}\left(D_1^2 - D_2^2\right)\right] \times V^2$$

$$= 1000 \times \frac{\pi}{4} \times \left[0.6^2 - 0.4^2\right] \times 20^2$$

= 628.32 N

35.

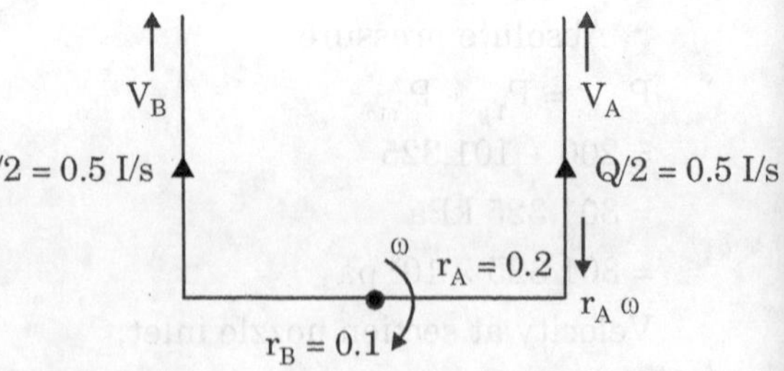

Relative velocities of water with sprinkler

$$V_A = \frac{Q/2}{A}$$

$$= \frac{1 \times 10^{-3}}{2 \times 10^{-4}} = 5 m/s$$

$V_B = 5$ m/s

Absolute velocity from B side

$$V'_{Abs} - (+r_B\omega) = V_B$$

$$V'_{Abs} = V_B + r_B\omega = 5 + 0.1\ \omega$$

Absolute velocity from A side

$V_{Abs} - (-r_A\omega) = V_A$

$V_{Abs} = V_A - r_A\omega$

$V_{Abs} = 5 - 0.2\ \omega$

The external torque to the sprinkler is zero.

So, $\Sigma T = 0$

$$\dot{m}_A V_{Abs} r_A - \dot{m}_B V'_{Abs} r_B = 0$$

$$\rho\left(\frac{Q}{2}\right)\{5 - 0.2\ \omega\}0.2 - \rho\frac{Q}{2}\{5 + 0.1\ \omega\}0.1 = 0$$

$1 - 0.04\ \omega - 0.5 - 0.01\ \omega = 0$

$0.05\ \omega = 0.5$

$\omega = 10$ rad/s

Unit - VII

Heat-Transfer

Syllabus

Modes of heat transfer; one dimensional heat conduction, resistance concept and electrical analogy, heat transfer through fins; unsteady heat conduction, lumped parameter system, Heisler's charts; thermal boundary layer, dimensionless parameters in free and forced convective heat transfer, heat transfer correlations for flow over flat plates and through pipes, effect of turbulence; heat exchanger performance, LMTD and NTU methods; radiative heat transfer, Stefan- Boltzmann law, Wien's displacement law, black and grey surfaces, view factors, radiation network analysis.

Contents

CHAPTER 1

Conduction

Analysis of Previous GATE Papers

Topic	Year/Set → Marks Distribution ↓		2019 Set 1	2019 Set 2	2018 Set 1	2018 Set 2	2017 Set 1	2017 Set 2	2016 Set 1	2016 Set 2	2016 Set 3	2015 Set 1	2015 Set 2	2015 Set 3
Basic Introduction	1 Mark	MCQ Type												
		Numerical Type												
	2 Marks	MCQ Type												
		Numerical Type												
	Total													
Fourier law of conduction	1 Mark	MCQ Type							1					
		Numerical Type									1			
	2 Marks	MCQ Type		1										1
		Numerical Type			1									
	Total			**2**	**2**				**1**		**1**			**2**
Thermal resistance	1 Mark	MCQ Type	1								1			
		Numerical Type												
	2 Marks	MCQ Type												
		Numerical Type												
	Total		**1**								**1**			
Critical radius of insulation	1 Mark	MCQ Type												
		Numerical Type											1	
	2 Marks	MCQ Type										1		
		Numerical Type												
	Total											**2**	**1**	
Heat Generation	1 Mark	MCQ Type												
		Numerical Type												
	2 Marks	MCQ Type											1	
		Numerical Type	1				1							
	Total		**2**				**2**						**2**	

Basic introduction

1. Thermal conductivity is lower for

(a) wood (b) air

(c) water at 100°C (d) steam at 1 bar

[1990 : 1 Mark]

2. Match the property with their units

Property	Units
A. Bulk modulus	1. W/s
B. Thermal conductivity	2. N/m^2
C. Heat transfer coefficient	3. N/m^3
D. Heat flow rate	4. W
	5. W/mK
	6. W/m^2K

Codes:

	A	B	C	D
(a)	1	2	6	5
(b)	2	5	6	1
(c)	2	6	4	1
(d)	1	5	3	2

[1991 : 2 Marks]

3. For a given heat flow and for the same thickness, the temperature drop across the material will be maximum for

(a) Copper (b) Steel

(c) Glass wool (d) Refractory brick

[1996 : 1 Mark]

4. In descending order of magnitude, the thermal conductivity (a) Pure iron (b) Liquid water (c) Saturated water vapour (d) Pure aluminum can be arranged as

(a) abcd (b) bcad

(c) dabc (d) dcba

[2001 : 1 Mark]

5. As the temperature increases, the thermal conductivity of a gas

(a) increases

(b) decreases

(c) remains constant

(d) increases up to a certain temperature and then decreases

[2014 : 1 Mark, Set-4]

6. Match Group with Group B:

Group A

A : Biot number

B : Grashoff number

C : Prandtl number

D : Reynolds number

Group B

1: Ratio of buoyancy to viscous force

2: Ratio of inertia force to viscous force

3: Ratio of momentum to thermal diffusivities

4: Ratio of internal thermal resistance to boundary layer thermal resistance

Codes:

	A	B	C	D
(a)	4	1	3	2
(b)	4	3	1	2
(c)	3	2	1	4
(d)	2	1	3	4

[2014: 1 Mark, Set-4]

Fourier's law

7. The temperature variation under steady heat conduction across a composite slab of two materials with thermal conductivities k_1 and k_2 is shown in figure then, which one of the following statements holds?

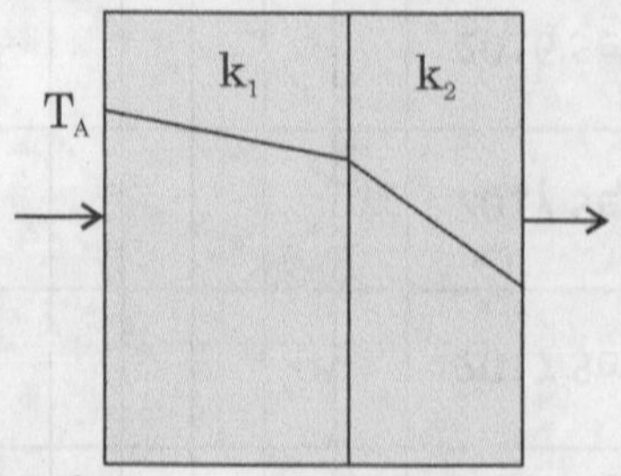

(a) $k_1 > k_2$ (b) $k_1 = k_2$

(c) $k_1 = 0$ (d) $k_1 < k_2$

[1998 : 2 Marks]

Common Data Questions Q.8 & Q.9:

Heat is being transferred by convection from water at 48°C to a glass plate whose surface temp. is exposed to the water is at 40°C. The thermal conductivity of water is 0.6 W/mK and the thermal conductivity of glass is 1.2 W/mK. The spatial gradient of temperature in the water at the water-glass interface is dT/dy = 1×10^4 K/m.

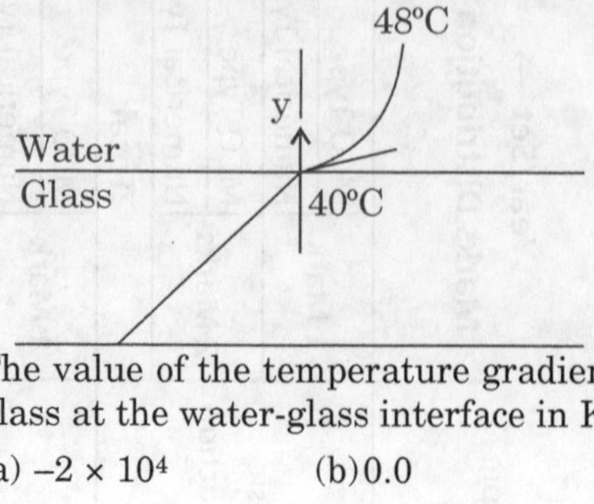

8. The value of the temperature gradient in the glass at the water-glass interface in K/m is

(a) -2×10^4 (b) 0.0

(c) 0.5×10^4 (d) 2×10^4

[2003 : 2 Marks]

9. The heat transfer coefficient h in w/m^2K is

(a) 0.0 (b) 4.8

(c) 6 (d) 750

[2003 : 2 Marks]

10. One dimensional unsteady state heat transfer equation for a sphere with heat generation at the rate of q can be written

(a) $\frac{1}{r}\frac{\partial}{\partial r}\left(r\frac{\partial T}{\partial r}\right)+\frac{q}{k}=\frac{1}{\alpha}\frac{\partial T}{\partial t}$

(b) $\frac{1}{r^2}\frac{\partial}{\partial r}\left(r^2\frac{\partial T}{\partial r}\right)+\frac{q}{k}=\frac{1}{\alpha}\frac{\partial T}{\partial t}$

(c) $\frac{\partial^2 T}{\partial r^2}+\frac{q}{k}=\frac{1}{\alpha}\frac{\partial T}{\partial t}$

(d) $\frac{\partial^2}{\partial r^2}(rT)+\frac{q}{k}=\frac{1}{\alpha}\frac{\partial T}{\partial t}$

[2004 : 1 Mark]

11. A stainless steel tube (k_s = 19 W/mK) of 2 cm ID and 5 cm OD is insulated with 3 cm thick asbestos (k_a = 0.2 W/mK). If the temperature difference between the inner most and outermost surfaces is 600°C, the heat transfer rate per unit length is

(a) 0.94 W/m (b) 9.44 W/m

(c) 944.72 W/m (d) 9447.21 W/m

[2004 : 2 Marks]

12. A well machined steel plate of thickness L is kept such that the wall temperature are T_h and T_c as seen in the figure below. A smooth copper plate of the same thickness L is now attached to the steel plate without any gap as indicated in the figure below. The temperature at the interface is T_t. The temperatures of the outer walls are still the same at T_h and T_c. The heat transfer rates are q_1 and q_2 per unit area in the two cases respectively in the direction shown. Which of the following statements is correct?

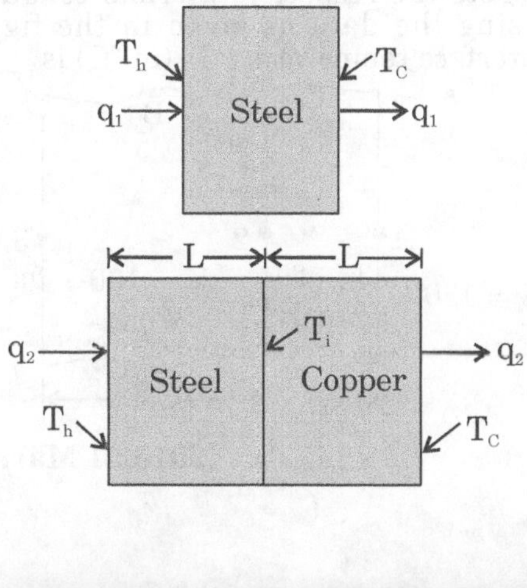

(a) $T_h > T_i > T_c$ and $q_1 < q_2$

(b) $T_h < T_i > T_c$ and $q_1 = q_2$

(c) $T_h = (T_i + T_c)/2$ and $q_1 > q_2$

(d) $T_i < (T_h + T_c)/2$ and $q_1 > q_2$

[2005 : 1 Mark]

13. In a case of one dimensional heat conduction in a medium with constant properties, T is the temperature at position x, at time t. Then $\frac{\partial T}{\partial t}$ is proportional to

(a) $\frac{T}{x}$ (b) $\frac{\partial T}{\partial x}$

(c) $\frac{\partial^2 T}{\partial x \partial t}$ (d) $\frac{\partial^2 T}{\partial x^2}$

[2005 : 1 Mark]

14. Heat flows through a composite slab, as shown below. The depth of the slab is 1 m. The k values are in W/mK. The overall thermal resistance in K/W is

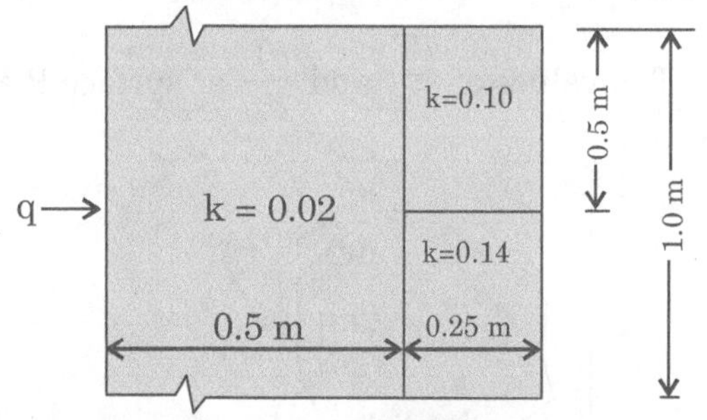

(a) 17.2 (b) 21.9

(c) 28.6 (d) 39.2

[2005 : 2 Marks]

15. In a composite slab, the temperature at the interface (T_{inter}) between two materials is equal to the average of the temperature at the two ends. Assuming steady one dimensional heat conduction, which of the following statements is true about the respective thermal conductivities?

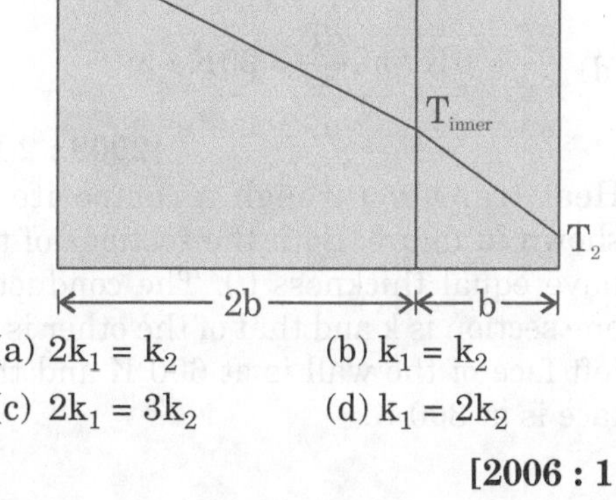

(a) $2k_1 = k_2$ (b) $k_1 = k_2$

(c) $2k_1 = 3k_2$ (d) $k_1 = 2k_2$

[2006 : 1 Mark]

16. Heat is being transferred conductively from a cylindrical nuclear reactor fuel rod of 50 mm diameter to water at 75°C, under steady state condition, the rate of heat generation within the fuel element is 106 W/m^3 and the convective heat transfer coefficient is 1 kW/m^{2}K, the outer surface temperature of the fuel element would be

(a) 700 K (b) 625 K
(c) 360 K (d) 400 K

[2007 : 2 Marks]

17. Steady two-dimensional heat conduction takes place in the body shown in the figure below. The normal temperature gradients over surface P and Q can be considered to be uniform. The temperature gradient $\frac{\partial T}{\partial x}$ at surface Q is equal to 10 K/m. Surfaces P and Q are maintained at constant temperatures as shown in the figure, while the remaining part of the boundary is insulated. The body has a constant thermal conductivity of 0.1 W/mK.

The values of $\frac{\partial T}{\partial y}$ and $\frac{\partial T}{\partial x}$ at surface P are

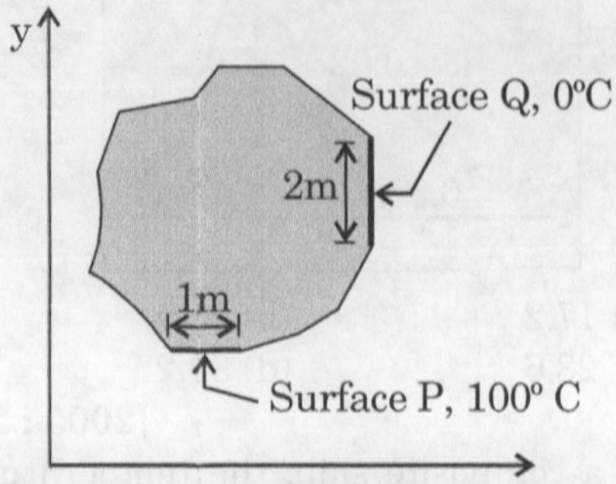

(a) $\frac{\partial T}{\partial x} = 20\,K/m, \frac{\partial T}{\partial y} = 0\,K/m$

(b) $\frac{\partial T}{\partial x} = 0\,K/m, \frac{\partial T}{\partial y} = 10\,K/m$

(c) $\frac{\partial T}{\partial x} = 10\,K/m, \frac{\partial T}{\partial y} = 10\,K/m$

(d) $\frac{\partial T}{\partial x} = 0\,K/m, \frac{\partial T}{\partial y} = 20\,K/m$

[2008 : 2 Marks]

18. Heat transfer through a composite wall is shown in figure. Both the sections of the wall have equal thickness (l). The conductivity of one section is k and that of the other is 2k. The left face of the wall is at 600 K and the right face is at 300 K.

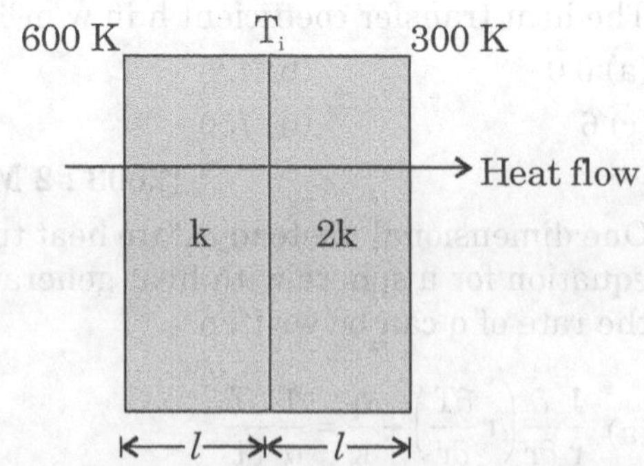

The interface temperature T_i (in K) of the composite wall is ________

[2014 : 2 Marks, Set-3]

19. A brick wall (k = 0.9 W/mK) of thickness 0.18 m separates the warm air in a room from the cold ambient air. On a particular Winter day, the outside air temperature is - 5°C and the room needs to be maintained at 27°C. The heat transfer coefficient associated with outside air is 20 W/m^{2}K. Neglecting the convective resistance of the air inside the room, the heat loss, in (W/m^2), is

(a) 88 (b) 110
(c) 128 (d) 160

[2015 : 2 Marks, Set-3]

20. A plastic sleeve of outer radius r_0 = 1 mm covers a wire (radius r = 0.5 mm) carrying electric current, Thermal conductivity of the plastic is 0.12 W/mK. The heat transfer coefficient on the outer surface of the sleeve exposed to air is 25 W/m^{2}K. Due to the addition of the plastic cover, the heat transfer from the wire to the ambient will

(a) increase (b) remain the same
(c) decrease (d) be zero

[2016 : 1 Mark, Set-1]

21. Steady one-dimensional heat conduction takes place across the faces 1 and 3 of a composite slab consisting of slabs A and B in perfect contact as shown in the figure, where k_A, k_B denote the respective thermal conductivities. Using the data as given in the figure, the interface temperature T_2 (in °C) is ________

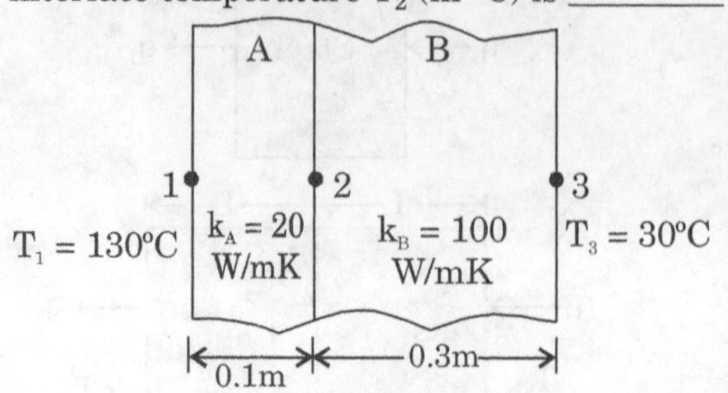

[2016 : 1 Mark, Set-3]

22. A plane slab of thickness L and thermal conductivity k is heated with a fluid on one side (P), and the other side (Q) is maintained at a constant temperature, T_Q of 25°C, as shown in the figure. The fluid is at 45°C and the surface heat transfer coefficient h, is 10 W/m²K. The steady state temperature. T_p (in °C) of the side which is exposed to the fluid is (correct to two decimal places).

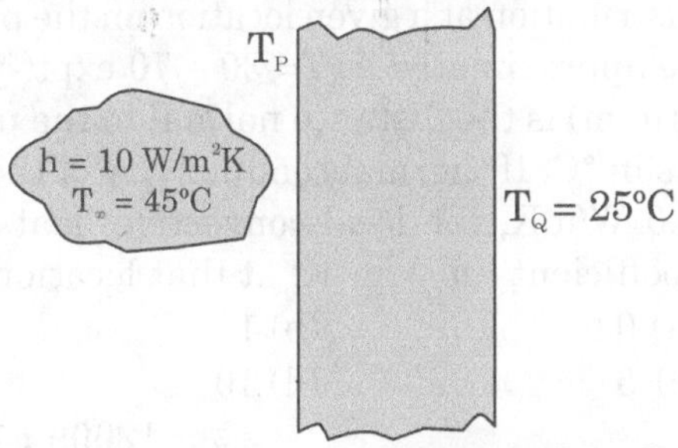

[2018 : 2 Marks, Set-1]

23. One-dimensional steady state heat conduction takes place through a solid whose cross-sectional area varies linearly in the direction of heat transfer. Assume there is no heat generation in the solid and the thermal conductivity of the material is constant and independent of temperature. The temperature distribution in the solid is

(a) Quadratic (b) Linear
(c) Logarithmic (d) Exponential

[2019 : 1 Mark, Set-2]

Thermal Resistance

24. A long glass cylinder of inner diameter = 0.03 m and outer diameter = 0.05 m carries hot fluid inside. If the thermal conductivity of glass = 1.05 W/mK, the thermal resistance (K/W) per unit length of the cylinder is

(a) 0.031 (b) 0.077
(c) 0.17 (d) 0.34

[2007 : 2 Marks]

25. Consider a long cylindrical tube of inner and outer radii, r_i and r_0, respectively, length L and thermal conductivity, k. Its inner and outer surfaces are maintained at T_i, and T_0, respectively ($T_i > T_0$). Assuming one-dimensional steady state heat conduction in the radial direction, the thermal resistance in the wall of the tube is

(a) $\frac{1}{2\pi kL}\ln\left(\frac{r_i}{r_o}\right)$ (b) $\frac{L}{2\pi r_i k}$

(c) $\frac{1}{2\pi kL}\ln\left(\frac{r_0}{r_i}\right)$ (d) $\frac{1}{4\pi kL}\ln\left(\frac{r_0}{r_i}\right)$

[2014 : 1 Mark, Set-3]

26. A hollow cylinder has length L, inner radius r_1 outer radius r_2, and thermal conductivity k. The thermal resistance of the cylinder for radial conduction is

(a) $\frac{\ln(r_2/r_1)}{2\pi k_L}$ (b) $\frac{\ln(r_2/r_1)}{2\pi kL}$

(a) $\frac{2\pi kL}{\ln(r_2/r_1)}$ (d) $\frac{2\pi kL}{\ln(r_1/r_2)}$

[2016 : 1 Mark, Set-3]

27. A slender rod of length L, diameter d (L >> d) and thermal conductivity k_1 is joined with another rod of identical dimensions, but of thermal conductivity k_2, to form a composite cylindrical rod of length 2L. The heat transfer in radial direction and contact resistance are negligible. The effective thermal conductivity of the composite rod is

(a) $\frac{2k_1k_2}{k_1+k_2}$ (b) $\sqrt{k_1k_2}$

(c) $\frac{k_1k_2}{k_1+k_2}$ (d) $k_1 + k_2$

[2019 : 1 Mark, Set-1]

Critical radius of insulation

28. For a current carrying wire of 20 mm diameter exposed to air (h = 20 W/m²K), maximum heat dissipation occurs when thickness of insulation (0.5 W/mK) is

(a) 30 mm (b) 25 mm
(c) 20 mm (d) 15 mm

[1993 : 2 Marks]

29. Two insulating material of thermal conductivity k and 2k are available for lagging a pipe carrying a hot fluid. If the radial thickness of each material is same

(a) Material with higher thermal conductivity should be used for inner layer and one with lower thermal conductivity for the outer
(b) Material with lower thermal conductivity should be used for inner layer and one with higher thermal conductivity for the outer
(c) It is immaterial in which sequence the insulating material are used
(d) It is not possible to judge unless numerical values of dimension are given

[1994 : 1 Mark]

30. It is proposed to coat a 1 mm diameter wire with enamel paint (k = 0.1 W/mK) to increase heat transfer with air. If the air side heat transfer coefficient is 100 W/m²K, the optimum thickness of enamel paint should be

(a) 0.25 mm (b) 0.5 mm
(c) 1 mm (d) 2 mm

[1999 : 2 Marks]

31. With an increase in the thickness of insulation around a circular pipe, heat loss to surroundings due to
(a) convection increases, while that due to conduction decreases
(b) convection decreases, while that due to conduction increases
(c) convection and conduction decreases
(d) convection and conduction increases

[2006 : 2 Marks]

32. A 10 mm diameter electrical conductor is covered by an insulation of 2 mm thickness. The conductivity of the insulation is 0.08 W/mK and the convection coefficient at the insulation surface is 10 W/m^{2}K. Addition of further insulation of the same material will
(a) increase heat loss continuously
(b) decrease heat loss continuously
(c) increase heat loss to a maximum and then decrease heat loss
(d) decrease heat loss to a minimum and then increase heat loss

[2015 : 2 Marks, Set-1]

33. If a foam insulation is added to a 4 cm outer diameter pipe as shown in the figure, the critical radius of insulation (in cm) is ________.

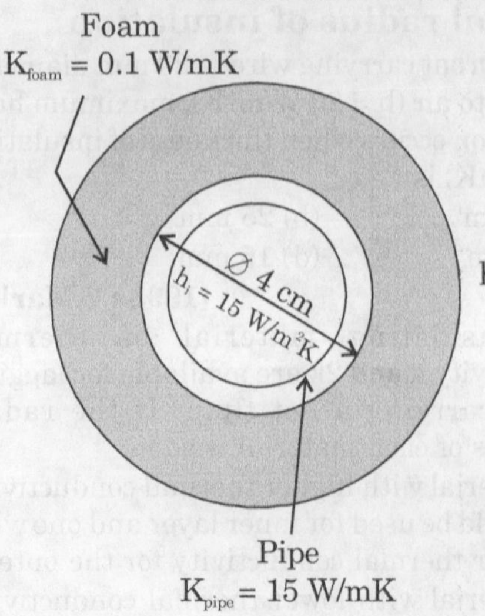

[2015 : 1 Mark, Set-2]

Heat Generation

Linked Answer Questions Q.34 and Q.35:
Consider steady one-dimensional heat flow in a plate of 20 mm thickness with a uniform heat generation of 80 MW/m^3. The left and right faces are kept at constant temperatures of 160°C and 120°C respectively. The plate has a constant thermal conductivity of 200 W/mK.

34. The location of maximum temperature within the plate from its left face is
(a) 15 mm (b) 10 mm
(c) 5 mm (d) 0 mm

[2007 : 2 Marks]

35. The maximum temperature within the plate in °C is
(a) 160 (b) 165
(c) 200 (d) 250

[2007 : 2 Marks]

36. A coolant fluid at 30°C flows over a heated flat plate maintained at a constant temperature of 100°C. The boundary layer temperature distribution at a given location on the plate may be approximated as T= 30 + 70 expt(-y) where y(in m) is the distance normal to the plate and fis in °C. If thermal conductivity of the fluid is 1.0 W/mK, the local convective heat transfer coefficient (in W/m^{2}K) at that location will be
(a) 0.2 (b) 1
(c) 5 (d) 10

[2009 : 1 Mark]

37. Consider one-dimensional steady state heat conduction, without heat generation, in a plane wall; with boundary conditions as shown in the figure below. The conductivity of the wall is given by $k = k_0 + bT$; where k_0 and b are positive constant, and T is temperature.

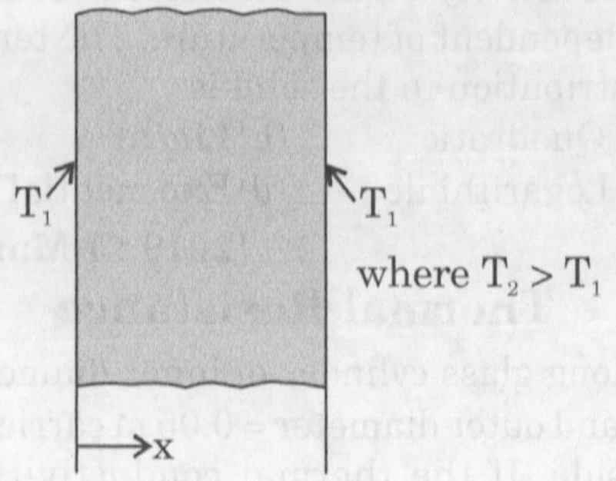

As x increases, the temperature gradient (dT/dx) will
(a) remains constant (b) be zero
(c) increase (d) decrease

[2013 : 1 Mark]

38. Consider one-dimensional steady state heat conducting along x-axis ($0 \le x \le L$), through a plane wall with the boundary surfaces (x = 0 and x = L) maintained at temperature of 0°C and 100°C. Heat is generated uniformly through out the wall.
Choose the CORRECT statement
(a) The direction of heat transfer will be from the surface at 100°C to the surface of 0°C.
(b) The maximum temperature inside the wall must be greater than 100°C
(c) The temperature distribution is linear within the wall
(d) The temperature distribution is symmetric about the mid-plane of the wall

[2013 : 1 Mark]

39. Consider one dimensional steady state heat conduction across a wall (as shown in figure below) of thickness 30 mm and thermal conductivity 15 W/mK. At x = 0, a constant heat flux, $q'' = 1 \times 10^5$ W/m^2 is applied. On the other side of the wall, heat is removed from the wall by convection with a fluid at 25°C and heat transfer coefficient of 250 W/m^{2}K. The temperature (in °C), at x = 0 is ______.

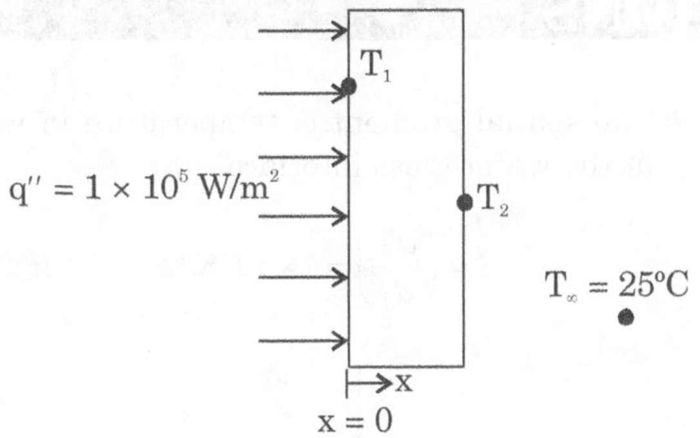

[2014 : 2 Marks, Set-1]

40. A material P of thickness 1 mm is sandwiched between two steel slabs, as shown in the figure below. A heat flux 10 kW/m^2 is supplied to one of the steel slabs as shown. The boundary temperatures of the slabs are indicated in the figure. Assume thermal conductivity of this steel is 10 W/mK. Considering one-dimensional steady state heat conduction for the configuration, the thermal conductivity (k, inW/mK) of material P is ______.

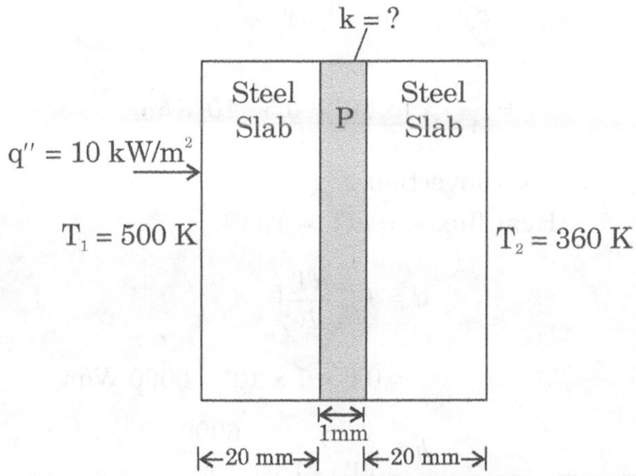

[2014 : 2 Marks, Set-2]

41. A plane wall has a thermal conductivity of 1.15 W/mK. If the inner surface is at 1100°C and the outer surface is at 350°C, then the design thickness (in meter) of the wall to maintain a steady heat flux of 2500 W/m^2 should be

[2014 : 2 Marks, Set-4]

42. A cylindrical uranium fuel rod of radius 5 mm in a nuclear is generating heat at the rate of 4×10^7 W/m^3. The rod is cooled by a liquid (convective heat transfer coefficient 1000 W/m^{2}K) at 25°C. At steady state, the surface temperature (in K) of the rod is

(a) 308 (b) 398
(c) 418 (d) 448

[2015 : 2 Marks, Set-2]

43. Heat is generated uniformly in a long solid cylindrical rod (diameter =10 mm) at the rate of 4×10^7 W/m^3. The thermal conductivity of the rod material is 25 W/mK. Under steady state conditions, the temperature ditference between the centre and the surface of the rod is 1°C.

[2017 : 2 Marks, Set-1]

44. Three slabs are joined together as shown in the figure. There is no thermal contact resistance at the interfaces. The center slab experience a nonuniform internal heat generation with an average value equal to 10000 Wm^{-3}, while the left and right slabs have no internal heat generation. All slabs have thickness equal to 1 m and thermal conductivity of each slab is equal to 5 Wm^{-1} K^{-1}. The two extreme faces are exposed to fluid with heat transfer coefficient 100 Wm^{-2}K^{-1} and bulk temperature 30°C as shown. The heat transfer in the slabs is assumed to be one dimensional and steady, and all properties are constant. If the left extreme face temperature T_1 is measured to be 100°C, the right extreme faced temperature T_2 is__________°C.

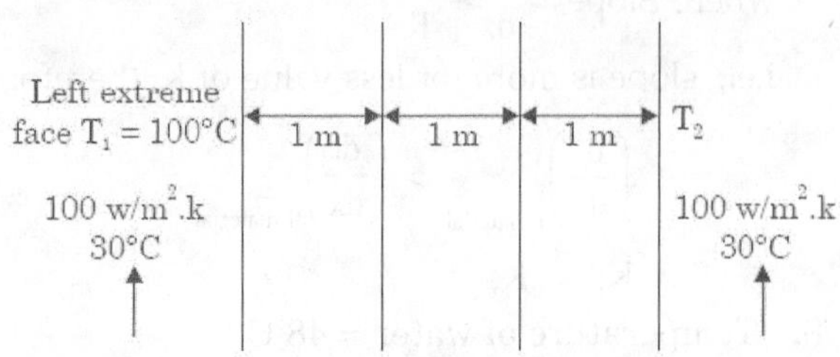

[2019 : 2 Mark, Set-1]

ANSWERS

1. (b) **2.** (b) **3.** (c) **4.** (c) **5.** (a) **6.** (a) **7.** (a) **8.** (c) **9.** (d) **10.** (b)
11. (c) **12.** (d) **13.** (d) **14.** (c) **15.** (d) **16.** (c) **17.** (d) **19.** (c) **20.** (a) **23.** (c)
24. (b) **25.** (c) **26.** (a) **27.** (a) **28.** (d) **29.** (b) **30.** (b) **31.** (a) **32.** (c) **34.** (c)
35. (b) **36.** (b) **37.** (c) **38.** (b) **42.** (b)

EXPLANATIONS

3. $dT \propto \frac{1}{K}$

So, glass wool has minimum thermal conductivity and therefore maximum temperature difference.

5. $K = \frac{nV\lambda C_V}{3N_A}$

Where K is thermal conductivity
V is mean particle speed
λ is mean free path
C_V is molar head capacity
N_A is Avogadro's number
n is particles per unit volume

Gases transfer heat by direct collisions between molecules. As the temperature increases, the thermal conductivity increases due to increase in speed, movement and collisions in the molecules. From the above expression, by increasing mean particle speed, the thermal conductivity increases.

7. $Q = KA\frac{dt}{dx}$

where Slope $= \frac{dt}{dx} = \frac{Q}{KA}$

i.e., slope is more for less value of K, therefore

$$\left(\frac{dt}{dx}\right)_{\text{I material}} > \left(\frac{dt}{dx}\right)_{\text{II material}}$$

$\therefore \quad K_1 < K_2$

8. Temperature of water = 48°C

Temperature of glass plate = 40°C

Thermal conductivity of water

$= k_w = 0.6$ W/mK

Thermal conductivity of glass

$= k_G = 1.2$ W/mK

The spatial gradient of temperature in water at the water-glass interface

$$= \left(\frac{dT}{dy}\right) = 1 \times 10^5 \text{ K/m}$$

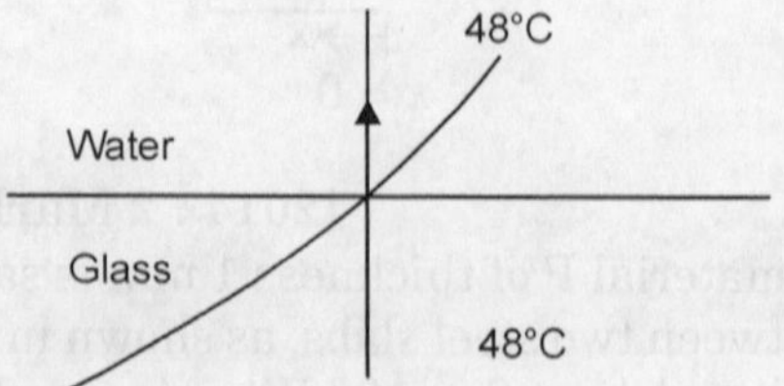

At water glass interface

$$q = k_w\left(\frac{dT}{dy}\right)_w = k_G\left(\frac{dT}{dy}\right)_G$$

$$\Rightarrow \left(\frac{dT}{dy}\right)_G = \frac{k_w}{k_G} \times \left(\frac{dT}{dy}\right)_w$$

$$= \frac{0.6}{1.1} \times 1 \times 10^4 = 0.5 \times 10^4 \text{ K/m}$$

9. For convection

Heat flux, $q = h(T_f - T_i)$

$$q = k_w\left(\frac{dT}{dy}\right)_w$$

$$= 0.6 \times 1 \times 10^4 = 6000 \text{ W/m}$$

$$\therefore \quad h = \frac{q}{(T_f - T_i)} = \frac{6000}{8}$$

$$= 750 \text{ W/m}^2\text{K}$$

11. Given: $d_1 = 2$ cm

$r_1 = 1$ cm

$d_2 = 5$ cm

$r_2 = 2.5$ cm

$d_3 = 5 + 2 \times 3 = 11$

$r_3 = 5.5$ cm

Equivalent thermal resistance

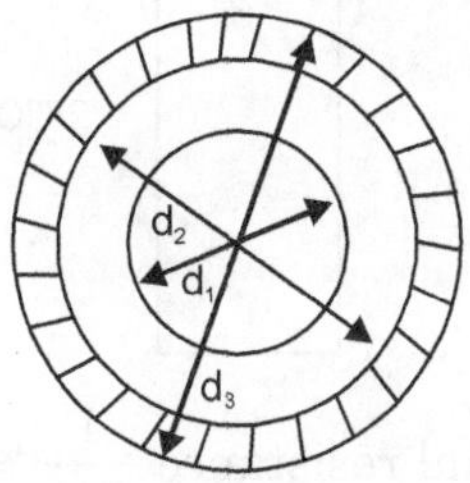

$$R_t = \frac{1}{2\pi k_{st}} \ln\frac{r_2}{r_1} + \frac{1}{2\pi k_a} \ln\frac{r_3}{r_2}$$

$$= \frac{1}{2\pi \times 19} \ln\frac{2.5}{1} + \frac{1}{2\pi \times 0.2} \times \ln\frac{5.5}{2.5}$$

$$= 7.67 \times 10^{-3} + 0.627 = 0.635$$

∴ Heat transfer rate per unit length

$$= \frac{t_1 - t_2}{R_t} = \frac{600}{0.635} = 944.88 \text{ W/m.}$$

12. Case -I:

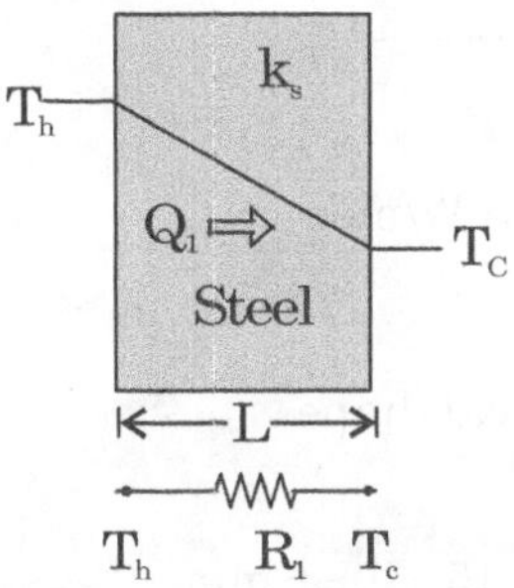

Case-II:

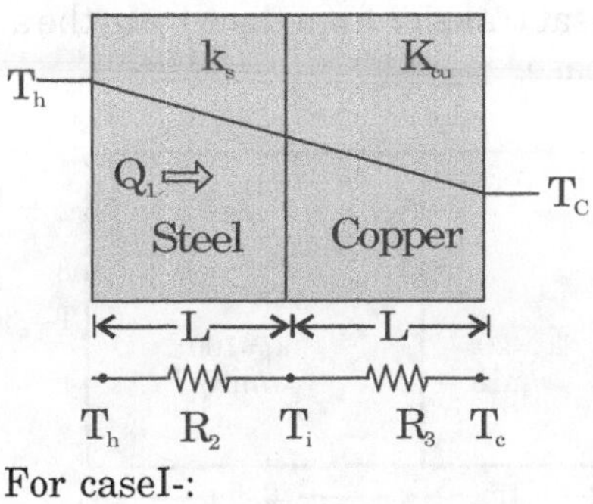

For caseI-:

$$Q_1 = \frac{T_h - T_c}{R_1}$$

where $R_1 = \frac{L}{k_S A}$

For Case-II:

$$Q_2 = \frac{T_h - T_c}{R_2 + R_3}$$

where $R_2 = \frac{L}{k_s A}$ and $R_3 = \frac{L}{k_{cu} A}$

As $R_2 + R_3 > R_1$ at constant temperature difference in both cases.

Then $q_1 \left(\text{or} \frac{Q_1}{A}\right) > q_2 \left(\text{or} \frac{Q_2}{A}\right)$

For Case-II:

$R_2 > R_3$ $\quad \because k_{cu} > k_s$

$$Q_2 = \frac{T_h - T_1}{Q_2}$$

$$\text{or} \quad R_2 = \frac{T_h - T_1}{Q_2}$$

$$\text{also } Q_2 = \frac{T_i - T_c}{R_3}$$

$$\text{or} \quad R_3 = \frac{T_i - T_c}{Q_2}$$

$$\therefore \frac{T_h - T_i}{Q_2} > \frac{T_i - T_c}{Q_2}$$

or $T_h - T_i > T_i - T_c$

or $T_h + T_c > 2T_i$

$$\text{or} \quad T_i < \frac{T_h + T_c}{2}$$

13. One dimensional heat conduction equation is,

$$\frac{\partial^2 T}{\partial x^2} = \frac{1}{\alpha}\frac{\partial T}{\partial t}$$

$$\therefore \quad \frac{\partial^2 T}{\partial x^2} \propto \frac{\partial T}{\partial t}$$

14. $\frac{L_1}{k_1} = \frac{0.5}{0.02} = 25$

$$\frac{2L_2}{k_2} = 5$$

$$\frac{2L_3}{k_3} = 12.5$$

$$R_{total} = 25 + 3.6 = 28.6 \text{ K/W}$$

15. $Q = \frac{-k_1 A}{2b}(T_{inter} - T_1) = \frac{-k_2 A(T_2 - T_{inter})}{b}$

$$T_{inter} = \frac{T_1 + T_2}{2}$$

$$\therefore \quad k_1 = 2k_2$$

16. By applying energy balance, we get

Rate of heat generation = Rate of convective heat transfer

$q_G \times$ Volume of fuel rod $= h_0 A (T_s - T_0)$

$$q_G \times \frac{\pi}{4} d^2 \times l = h_0 \times \pi d l (T_s - T_0)$$

$$\frac{d.q_G}{4} = h_0 (T_s - T_0)$$

$$\frac{0.05 \times 1 \times 10^3}{4} = (T_s - 75)$$

$T_s - 75 = 12.5$

$T_s = 87.5°C$ or $360.5K$

17. Direction of heat flow is always normal to surface of constant temperature

So for surface P_1, $\frac{dT}{dx} = 0$

From energy conservation,

Heat rate at P = Heat rate at Q

$$0.1 \times 1 \left(\frac{dT}{dy}\right)_P = 0.1 \times 2 \times \left(\frac{dT}{dx}\right)_Q$$

$$\Rightarrow \quad \frac{dT}{dy} = 20 \text{ K/m}$$

18. $R = R_1 + R_2$

$$= \frac{L}{KA} + \frac{L}{KA} = \frac{L}{KA} + \frac{L}{2KA} = \frac{3L}{2KA}$$

$$Q = \frac{T_L - T_R}{T_R}$$

$$= \frac{300}{\frac{3L}{2KA}} = 200 \times \frac{KA}{L} \quad ...(1)$$

At interface,

$$Q = \frac{T_L - T_i}{R_i}$$

$$= \frac{600 - T_i}{\frac{L}{KA}} = (600 - T_i)\frac{KA}{L} \quad ...(2)$$

Equating 1 and 2,

$$200\frac{KA}{L} = (600 - T_i)\frac{KA}{L}$$

$200 = 600 - T_i$

$T_i = 400$ K

19.

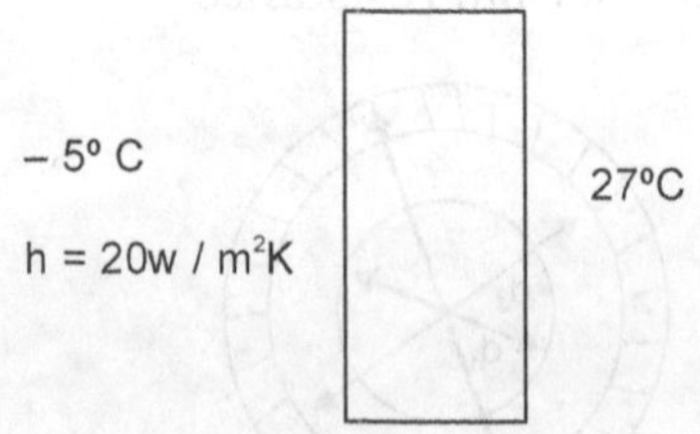

$$\text{Total thermal resistance } \frac{1}{R_{th}} = \frac{1}{\frac{1}{h} + \frac{1}{k}}$$

$$\frac{1}{R_{th}} = \frac{1}{\frac{1}{20} + \frac{0.18}{0.9}} = 4 \text{ W/m}^2\text{K}$$

$$Q = \frac{\Delta T}{R_{th}} = [27 - (-5)] \times 4 = 128 \text{ W/m}^2$$

20.

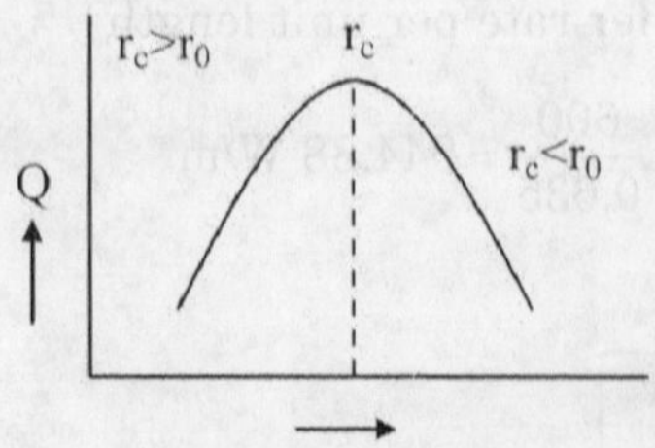

$r_0 = 1$ mm, $k = 0.15$ W/m-k

$h = 25$ W/m²-K

$r_c = \frac{k}{h_0}$ for cylindrical shape

$$= \frac{0.15}{25} \times 1000 = 0.15 \times 40 = 6 \text{ mm}$$

$\therefore r_c > r_0$

$\Rightarrow$ The heat transfer from the wire to the ambient will increase

21.

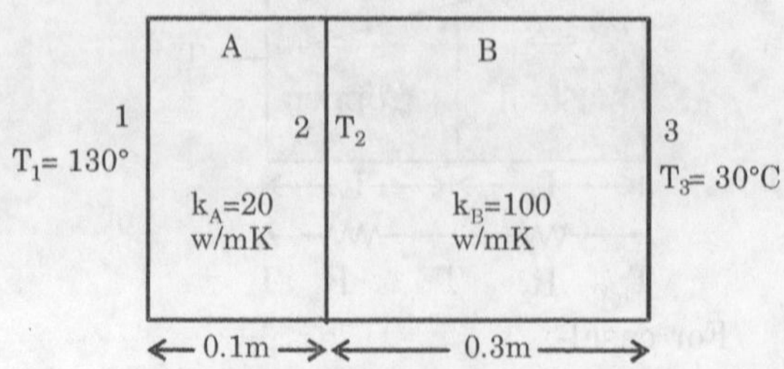

$$\frac{Q}{A} = \frac{kA.(T_1 - T_2)}{LA}$$

$$= \frac{kB(T_2 - T_3)}{LB}$$

$$\frac{20(130 - T_2)}{0.1} = \frac{100(T_2 - 30)}{0.3}$$

$\therefore \quad T_2 = 67.5°C$

22. For steady state condition

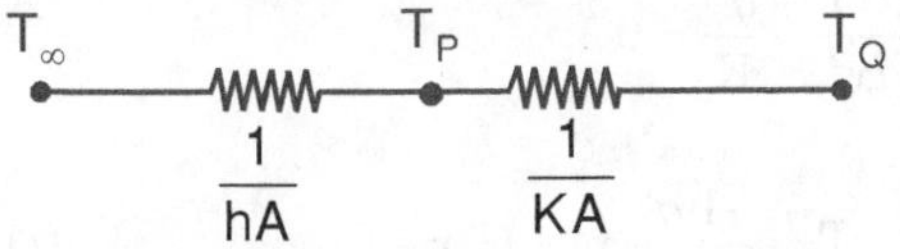

$$\frac{T_\infty - T_Q}{\frac{1}{hA}+\frac{L}{KA}} = \frac{T_P - T_Q}{\frac{L}{KA}} = \frac{T_\infty - T_P}{\frac{1}{h}A}$$

so from above equation:

$$\frac{45-25}{\frac{1}{10}+\frac{0.2}{2.5}} = \frac{T_P - 25}{\frac{0.2}{2.5}}$$

$$\boxed{T_P = 33.889°C}$$

23. We know that one-dimensional steady state heat conduction equation with no heat generation is given by,

$$\frac{d}{dx}\left(-kA\frac{dT}{dx}\right) = 0$$

$$\Rightarrow -kA\frac{dT}{dx} = C_1$$

$\because$ A = Cx + B (linear variation)

$$\Rightarrow \int_1^2 dT = \int_1^2 \frac{-C_1}{k(Cx+B)}dx$$

$$\Rightarrow T_2 - T_1 = \frac{-C_1}{k}\ln\left(\frac{Cx_2+B}{Cx_1+B}\right)$$

Logarithmic temperature distribution

24. Given data:

$d_1 = 0.03$ m

$$\therefore r_1 = \frac{d_1}{2} = \frac{0.3}{2} = 0.015\text{m}$$

$d_2 = .05$ m

$\therefore r_2 = .025$ m

$$K = 1.05\frac{W}{m\,K}$$

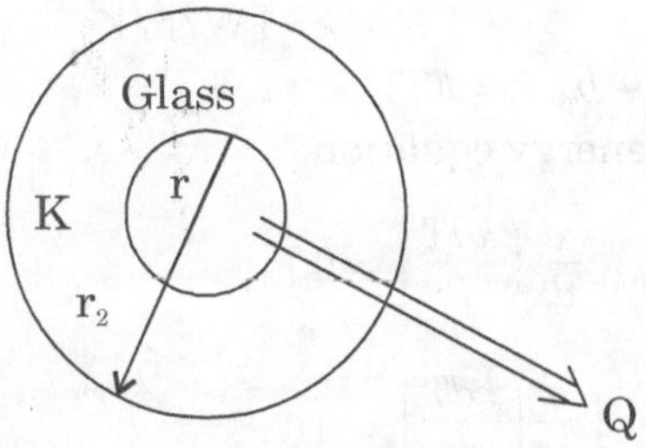

Thermal resistance

$$R_t = \frac{1}{2\pi Kl}\log_e\left(\frac{r_2}{r_1}\right)$$

$$= \frac{1}{2\times 3.14\times 1.05\times 1}\log_e\left(\frac{.025}{.015}\right)$$

$$= .077\frac{W}{m}$$

25. $A_r = 2\pi rL$

From Fourier 's Law

$$q_r = kA_r\frac{dT}{dr}$$

$$q_r = 2\pi kr L\frac{dT}{dr}$$

Boundary conditions:

$T = T_i$ at $r = r_i$

$T = T_0$ at $r = r_0$

$$q = \frac{2\pi k L(T_i - T_0)}{\ln(r_0/r_i)}$$

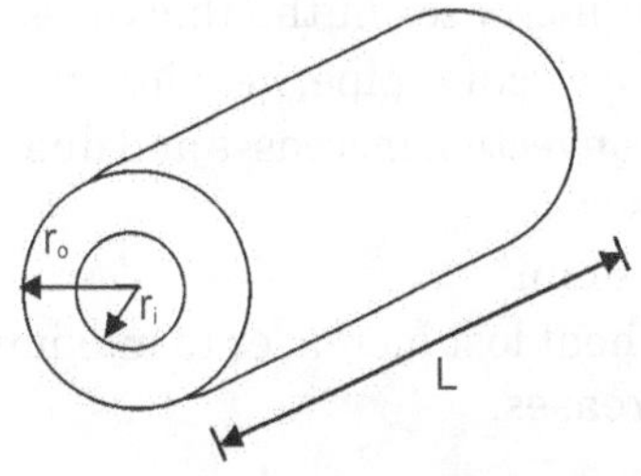

$$= \frac{T_i - T_0}{\frac{\ln(r_0/r_i)}{2\pi k L}} = \frac{T_i - T_0}{R_{th}}$$

$$R_{th} = \frac{\ln(r_0/r_i)}{2\pi k L}$$

27.

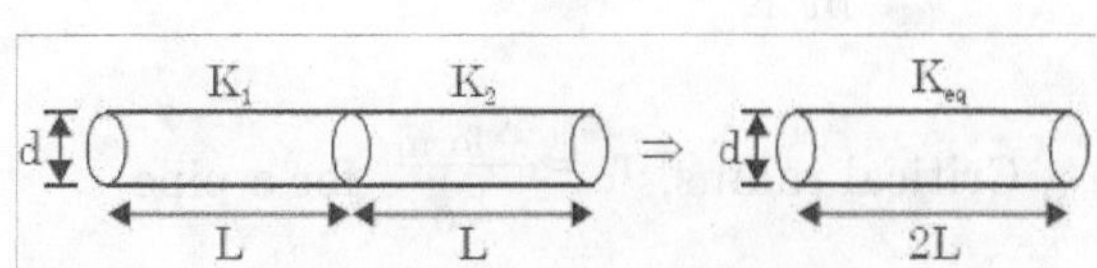

As per question, heat transfer in radial direction and contact resistance are negligible.

In series connection,

$$(R_{th})_{eq} = R_1 + R_2$$

$$\frac{2L}{Ak_{eq}} = \frac{L}{A\,K_1} + \frac{L}{A\,K_2}$$

$$\frac{2}{K_{eq}} = \frac{1}{K_1} + \frac{1}{K_2}$$

$$\Rightarrow \quad K_{eq} = \frac{2K_1K_2}{K_1+K_2}$$

28. For maximum heat dissipation,
Critical radius,

$$r_c = \frac{k}{h}$$

where $k = 0.5\frac{W}{m\,K}$

$$h = 20\frac{W}{m^2K},\ r = 10\text{ mm}$$

$$r_c = \frac{0.5}{20}$$

= 25 mm

Insulation thickness = $r_c - r$
= 25 – 10
= 15 mm

30. Critical radius of insulation,

$$r_c = \frac{k}{h} = \frac{0.1}{100} = 0.001 = 1\text{ mm.}$$

Radius of min = 0.5 mm.

∴ Thickness of coating = 1 – 0.5 = 0.5 mm.

31. With an increases in the thickness of insulation around a circular pipe, heat loss to surroundings due to convection increases and due to conduction decreases.

32. r_c = 8mm

∴ the heat lost increases to maximum and then decreases.

33. $K_{foam} = 0.1\frac{W}{mK}$

$$h_0 = 2\frac{W}{m^2K}$$

Critical radius, $r_c = \frac{K_{form}}{h_0}$ for a pipe

$$= \frac{0.1}{2} = 0.05\text{ m} = 5\text{cm}$$

34.

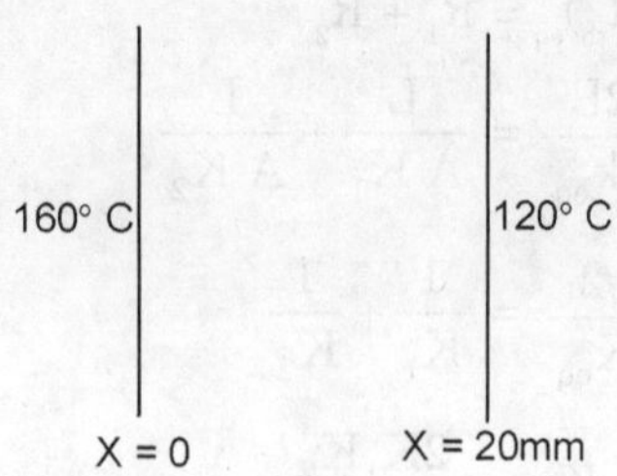

$$\frac{d^2T}{dx^2} + \frac{q^0}{K} = 0$$

$$\therefore\ T = \frac{-1}{2}\frac{q^0}{K}x^2 + C_1x + C_2 \qquad ...(i)$$

At $x = 0$, T = 433 K

∴ $C_2 = 433$

At $x = 20 \times 10^{-3}$ m, T = 393 K.

∴ C1 = 200

For maximum temperature,

$$\frac{dT}{dx} = 0$$

$$\therefore\ \frac{-q^0x}{K} + C_1 = 0$$

$$x = 5 \times 10^{-3} = 5\text{ mm}$$

35. T_{max} = 165° [putting x = 5 × 10^{-3} is equation (1)]

36. Under steady state conditions
Heat transfer by conduction
= Heat transfer by convection

$$\therefore\ -kA\frac{dT}{dy} = hA\Delta T$$

or $$-k\frac{dT}{dy} = h\Delta T$$

or $$-k\frac{dT}{dy}(30 + 70e^{-y}) = h\ (100 - 30°)$$

or $$-k\frac{\partial}{\partial y}(-70e^{-y}) = h \times 70$$

But $y = 0$

∴ $$-k(-70\overset{-o}{e}) = 70h$$

or $70k = 70h$

But $k = 1$

∴ $h = 1\text{W/m}^2\text{k}$

Alternately

$$hA\Delta t = kA\frac{dT}{dy}$$

or $h(100-30) = k \times (100 - 30)$

or $h = k = 1\text{W/m}^2\text{K}$

37. $k = k_0 + b_T$, $k = f(T)$

From energy equation:

$$\frac{d}{dx}\left(\frac{kdT}{dx}\right) = 0$$

$$\int\frac{d}{an}\left(\frac{kdT}{dx}\right) = \int 0$$

$$\Rightarrow \quad \frac{kdT}{dx}.C \qquad ...(1)$$

$$\int K(T)dT = \int C.dx$$

$$\int (k_0 + b_T)dT = Cx + A$$

A is integration constant

$$\left(K_o T + \frac{bT^2}{2}\right) = Cx + A$$

Using B.C's

$x = 0, T = 0$

$x = 1, T = 100$

$A = 0$

$$k_0 \times (100) + \frac{b(10)^4}{2} = C$$

$(100\ k_0 + 5000b) = C$

from equation (1)

$$\boxed{\frac{dT}{dx} = \left[\frac{100k_0 + 5000b}{k_0 + b_T}\right]} \qquad ...(2)$$

38. $\frac{d^2t}{dx^2} + \frac{q}{x} = 0$

$$a \quad \frac{dT}{dx} = \frac{-q}{k}x + c_1$$

$$1 = \frac{-q}{k}\frac{x^2}{2} + c_1 x + c_2 \text{ (parabolic)}$$

at $x = 0, T_1 = 0$

$x = l, T_2 = 100$

$0 = 0 + c_2 \Rightarrow \boxed{c_2 = 0}$

$$100 = \frac{-qt^2}{2k} + c_1 l$$

$$\Rightarrow \quad \boxed{\frac{100}{L} - \frac{qL}{2K} = c_1}$$

$$\therefore \quad T = \frac{-q}{k}\frac{x^2}{2} + \left(\frac{100}{l} - \frac{ql}{2k}\right)x$$

For maximum temp:

$$\frac{dT}{dx} - 0 \Rightarrow \frac{-q \times Lx}{2x} + \frac{100}{L}\frac{qL}{2x} = 0$$

39. $Q = \dfrac{T - T_\infty}{\dfrac{L}{kA} + \dfrac{1}{hA}}$

$$q = \frac{Q}{A} = \frac{T - T_\infty}{\dfrac{L}{k} + \dfrac{1}{h}}$$

$$1 \times 10^5 = \frac{T - 25}{\dfrac{30 \times 10^{-3}}{15} + \dfrac{1}{250}}$$

$600 = T - 25°C$

$T = 625°C.$

40. $t_1 = 20$ mm; $k_1 = 10$ W/mk

$t_2 = 1$ mm; $k_2 = ?$

$t_3 = 20$ mm; $k_3 = 10$ W/mk

$T_1 = 500\ k$; $T_2 = 360\ k$; $q_{in} = 10$ kW/m^2

Diagram

$$q_{in} = \frac{T_1 - T_2}{\dfrac{t_1}{k_1 A_1} + \dfrac{t_2}{k_2 A_2} + \dfrac{t_3}{k_3 A_3}}$$

Taking unit Area

$$10 \times 10^3 \text{ W/m}^2 = \frac{(500 - 360)}{\left(\dfrac{20 \times 10^{-3}}{10} + \dfrac{1 \times 10^{-3}}{k_2} + \dfrac{20 \times 10^{-3}}{10}\right)}$$

$$10\left(4 + \frac{1}{k_2}\right) = 140$$

$$\frac{1}{k_2} = 10$$

$$\Rightarrow \quad k_2 = 0.10 \text{ W/m.k.}$$

41. $Q = KA\dfrac{dT}{dx}$

$$q = \frac{Q}{A} = 2500 \text{ W/m}^2$$

$$2500 = K\frac{dT}{dx}$$

$$2500 = 1.15 \times \frac{(1100 - 350)}{x}$$

$x = 0.345$ m.

42. Given data: r = 0.005 m

d = 2r = 2 × 0.005 = 0.010 m

$$q_6 = 4 \times 10^7 \frac{W}{m^3}$$

$$h_0 = 1000\frac{W}{m^2K}$$

$T_0 = 25°c$

For steady state,

Rate of heat generation in the rod} = Rate of convection H.T. from rod surface to fluid

q_G × volume of rod = $h_0 A(T_s - T_0)$

$$q_G \times \frac{\pi}{4} d^2 l = h_0 \times \pi d l (T_s - T_0)$$

$$\frac{d.q_G}{4} = h_0 (T_s - T_0)$$

$$\frac{.010 \times 4 \times 10^7}{4} = 1000(T_s - 25)$$

$T_s - 25 = 100$

$T_s = 125°$ c or 398 K

43. $$\frac{d}{dr}\left(r\frac{dT}{dr}\right) + \frac{q_g.r}{k} = 0$$

$$\frac{d}{dr}\left(r\frac{dT}{dr}\right) + \frac{4 \times 10^7}{25} r = 0$$

$$\frac{d}{dr}\left(r\frac{dT}{dr}\right) + 16 \times 10^5 r = 0$$

$$r\frac{dT}{dr} + 8 \times 10^5 r^2 = C_1$$

at $r = 0, \frac{dT}{dr} = 0$

$C_1 = 0$

$$\therefore \quad r\frac{dT}{dr} + 8 \times 10^5 r^2 == 0$$

$$\frac{dT}{dr} + 8 \times 10^5 r = 0$$

$T + 4 \times 10^5 r^2 = C_2$

$T = C_2 - 4 \times 10^5 r^2$

$T_{r=R} - (T)_{r=0} = (C_2 - 4 \times 10^5 r^2) - C_2$

$(T)_{r=R} - (T)_{r=0} = -4 \times 10^5 r^2 = -4 \times 10^5 (0.005)^2$

$(T)_{r=R} - (T)_{r=0} = -10°C$

44.

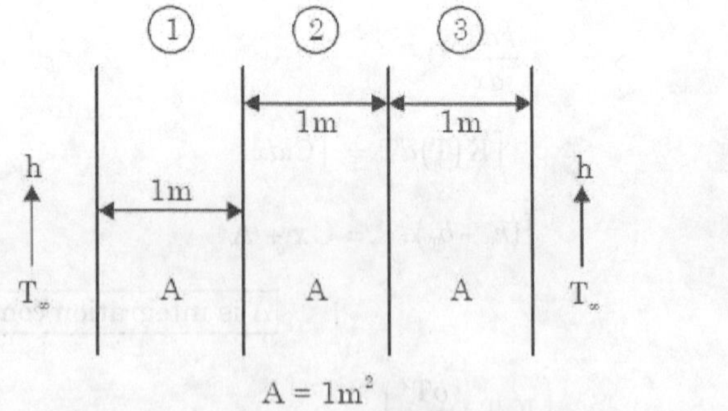

Heat generated in the slab (2) will be convected away from the slabs 1 and 3 i.e.

$$\dot{E}_{gen} = \dot{Q}_{1, conv} + \dot{Q}_{3, conv}$$

$$\dot{Q}_{1, conv} = h \times A \times (T_1 - T_\infty)$$

$$= 100 \times 1 \times (100 - 30)$$

$$= 100 \times 70 = 7000 \text{ W}$$

$$\dot{Q}_{3, conv} = h \times A \times (T_3 - T_\infty)$$

$$\dot{Q}_{3, conv} = 100 \times 1 \times (T_3 - 30)$$

Now $\dot{E}_{gen} = 10,000 \times A \times \text{width}$

$$= 10,000 \times 1 \times 1$$

$$\dot{E}_{gen} = 10,000\text{W}$$

$\Rightarrow \quad 10,000 = 7000 + 100 \times (T_3 - 30)$

$\Rightarrow \quad 3000 = 100 T_3 - 3000$

$\Rightarrow \quad 3000 + 3000 = 100 T_3$

$\Rightarrow \quad T_3 = \frac{6000}{100} = 60°C$

CHAPTER 2

Free and Forced Convection

Analysis of Previous GATE Papers

Topic	Year/Set →Marks Distribution ↓		2019 Set 1	2019 Set 2	2018 Set 1	2018 Set 2	2017 Set 1	2017 Set 2	2016 Set 1	2016 Set 2	2016 Set 3	2015 Set 1	2015 Set 2	2015 Set 3
Free convection	1 Mark	MCQ Type									1			1
		Numerical Type												
	2 Marks	MCQ Type												
		Numerical Type												
	Total										1			1
Forced convection	1 Mark	MCQ Type												
		Numerical Type												
	2 Marks	MCQ Type												
		Numerical Type												
	Total													
Boundary layer analogies	1 Mark	MCQ Type	1										1	
		Numerical Type												
	2 Marks	MCQ Type		1										
		Numerical Type							1					
	Total		1	2					2				1	

Free convection

1. Match List-I with List-ll and select the correct answer using the code given below the lists:

List-I	**List-II**
A. Grashof number	1. Mass diffusion
B. Schmidt number	2. Transient heat conduction
C. Weber number	3. Free convection
D. Fourier number	4. Forced convection
	5. Surface tension
	6. Radiation

Codes:

	A	B	C	D
(a)	4	2	3	5
(b)	3	1	5	2
(c)	5	1	2	4
(d)	1	2	5	4

[1996 : 2 Marks]

2. The ratio of momentum diffusivity (υ) to thermal diffusivity (α), is called

(a) Prandt number (b) Nusselt number
(c) Biot number (d) Lewis number

[2015 : 1 Mark, Set-3]

3. Grashof number signifies the ratio of

(a) inertia force to viscous force
(b) buoyancy force to viscous force
(c) buoyancy force to inertia force
(d) inertia force to surface tension force

[2016 : 1 Mark, Set-3]

Forced convection

4. In pool boiling the highest HTC occurs in

(a) Subcooled boiling zone
(b) Nucleate boiling zone
(c) Partial film boiling zone
(d) Film boiling zone

[1990 : 1 Mark]

5. Heat transfer coefficients for free convection in gases, forced convection in gases and vapours, and for boiling water lie, respectively, in the range of

(a) 5-15; 20-200 and 3000-50000 W/m^{2}K
(b) 20-50,200-500 and 50000-100000 W/m^{2}K
(c) 50-100, 500-1000 and 100000-100000 W/m^{2}K
(d) 20-100, 200-1000 and a constant 1000000 Wm^{2}K

[1998 : 1 Mark]

6. For the three-dimensional object shown in the figure below, five faces are insulated. The sixth face (PQRS), which is not insulated, interacts thermally with the ambient, with a convective heat transfer coefficient of 10W/m^{2}K. The ambient temperature is 30°C. Heat is uniformly generated inside the object at the rate of 100 W/m^3. Assuming the face PQRS to be at uniform temperature, its steady state temperature is

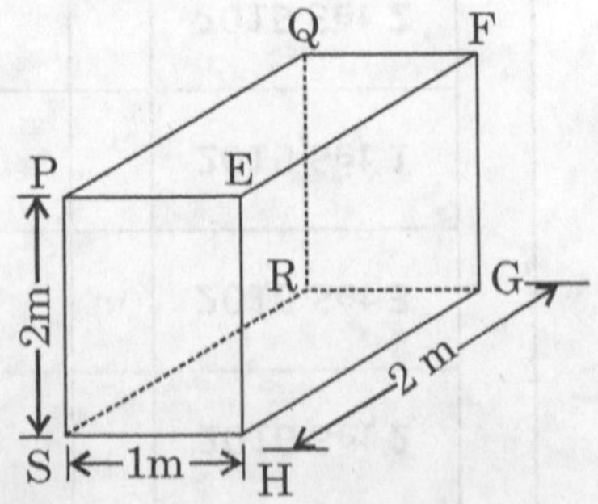

(a) 10°C (b) 20°C
(c) 30°C (d) 40°C

[2008 : 2 Marks]

Common Data Questions Q.7 and Q.8:

Water (specifie heat, c = 4.18 kJ/kgK) enters a pipe at a rate 0.01 kg/s and a temperature of 20°C. The pipe, of diameter 50 mm and length 3 m, is subjected to a wall heat flux q"w in W/m^2.

7. If q''_w = 2500x, where x is in m and in the direction of flow (x = 0 at the inlet), the bulk means temperature of the water leaving the pipe in °C is

(a) 42 (b) 62
(c) 74 (d) 104

[2013 : 2 Marks]

8. If q''_w = 5000 and the convection heat transfer coefficient at the pipe outlet is 1000 W/m^{2}K, the temperature in °C at the inner surface of the pipe at the outlet is

(a) 71 (b) 76
(c) 79 (d) 81

[2013 : 2 Marks]

9. For laminar forced convection over a flat plate, if the free stream velocity increases by a factor of 2, the average heat transfer coefficient

(a) remains same
(b) decreases by a factor of $\sqrt{2}$
(c) rises by a factor of $\sqrt{2}$.
(d) rises by a factor of 4

[2014 : 1 Mark, Set-2]

Boundary layer analogies

10. The properties of mercury at 300 K are: Density = 13529 kg/m^3, c_p = 0.1393 kJ/kgK, dynamic viscosity = 0.1523 × 10^{-2} Ns/m^2 and thermal conductivity = 8.540 W/mK. The Prandtl number of the mercury at 300 K is

(a) 0.0248 (b) 2.48

(c) 24.8 (d) 248 **[2002 : 2 Marks]**

11. In the laminar flow of air (Pr = 0.7) over a heated plate, if δ and δ_T denote, respectively, the hydrodynamic and thermal boundary layer thicknesses, then

(a) $\delta = \delta_T$ (b) $\delta > \delta_T$

(c) $\delta < \delta_T$ (d) $\delta = 0$ but $\delta_T = 0$

[2015 : 1 Mark, Set-2]

12. A fluid (Prandt/number, Pr = 1) at 500 K flows over a flat plate of 1.5 m length, maintained at 300 K. The velocity of the fluid is 10 m/s. Assuming kinematic viscosity. $\upsilon = 30 \times 10^{-6}$ m^2/s, the thermal boundary layer thickness (in mm) at 0.5 m from the leading edge is ______.

[2016 : 2 Marks, Set-1]

13. For a hydrodynamically and thermally fully developed laminar flow through a circular pipe of constant cross-section. The Nusselt number at constant wall heat flux (Nu_q) and that at constant wall temperature (Nu_T) are related as

(*a*) $Nu_q = Nu_T$ (*b*) $(Nu_q = Nu_T)^2$

(*c*) $Nu_q < Nu_T$ (*d*) $Nu_q > Nu_T$

[2019 : 1 Mark, Set-1]

14. The wall of a constant diameter pipe of length 1 m is heated uniformly with flux q" by wrapping a heater coil around it. The flow at the inlet to the pipe is hydrodynamically fully developed. The fluid is incompressible and the flow is assumed to be laminar and steady all through the pipe. The bulk temperature of the fluid is equal to 0°C at the inlet and 50°C at the exit. The wall temperatures are measured at three locations, P, Q and R as shown in the figure. The flow thermally develops after some distance from the inlet. The following measurements are made :

Point	P	Q	R
Wall Temp(°C)	50	80	90

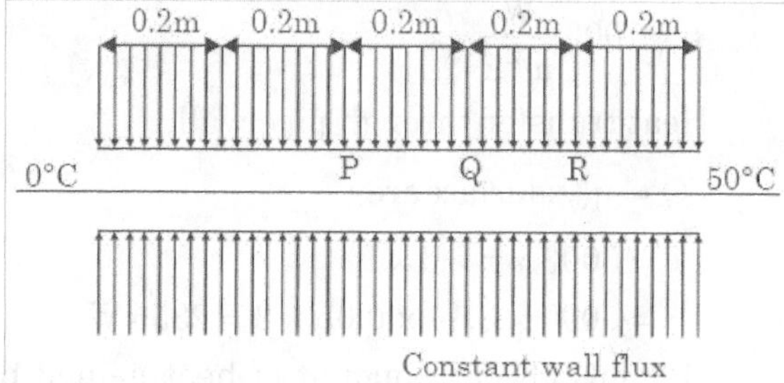

Among the locations P, Q and R, the flow is thermally at :

(a) R only

(b) P and Q only

(c) P, Q and R

(d) Q and R only

[2019 : 2 Mark, Set-2]

ANSWERS

1. (b) **2.** (a) **3.** (b) **4.** (b) **5.** (a) **6.** (d) **7.** (b) **8.** (d) **9.** (c) **10.** (a)
11. (c) **13.** (d) **14.** (d)

EXPLANATIONS

2. $$Pr = \frac{\mu C_p}{k} = \frac{\rho V C_p}{k} = \frac{V}{\frac{k}{\rho c_p}} = \frac{V}{\alpha}$$

3. Grashof number = inertia force × $\frac{\text{Buoyancy force}}{(\text{viscous force})^2}$

6. Given data;

$h_o = \frac{W}{m^2K}$

$T_O = 30°C$

$q_G = 100\frac{W}{m^3}$

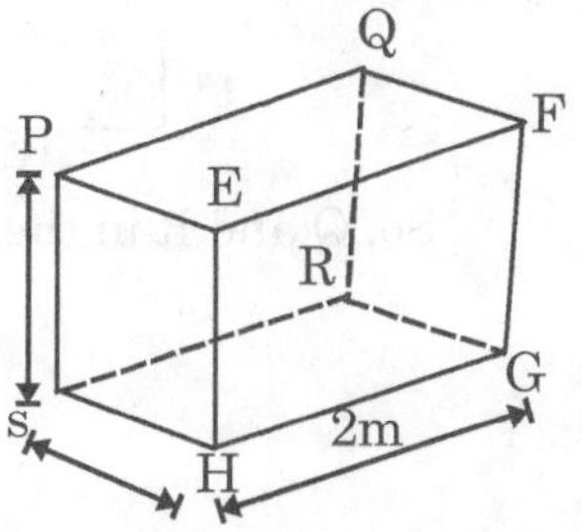

Volume,

V = 2×1×2=4 m^3

Heat generated,

$Q = q_G \times V = 100 \times 4 = 400$ W

Convection heat transfer from face PQRS,

$Q = h_O A (T_S - T_O)$

$400 = 10 \times 2 \times 2 (T_S - 30)$

$T_S = 40°C$

7. At the inlet of pipe

x = O

$q''_{wi} = 2500 \times 0 = 0$

At the exit of pipe,

x = l

$q''_{we} = 2500 \times 1 \times 3 = 7500\frac{W}{m^2}$

Average heat transfer,

$$q''_{avg} = \frac{q''_{wi} + q''_{we}}{2} = \frac{o + 7500}{2} = 3750 \frac{w}{m^2}$$

Net heat transfer

$Q = q''_{avg} \times$ surface area $= 3750 \times \pi$ dl

$= 3750 \times 3.14 \times 0.05 \times 3 = 1766.25$ W

Also $Q = mC_P (T_o - T_i)$

$1766.25 = 0.01 \times 4180 (T_o - 20)$

By solving above equation, we get

$T_o = 62°$ C

8. $q''_w = 5000 \frac{W}{m^2}$

$h = 1000 \frac{W}{m^2 k}$

Heat transfer through pipe wall,

$Q = q''_w$ x surface area

$= 5000 \times \pi \times d \times l$

$= 5000 \times 3.14 \times 0.05 \times 3 = 2355$ W

$Q = mC_P(T_o - T_i)$, amount of heat gained by the water

$2355 = 0.01 \times 4180 (T_o - 20)$

$T_o = 76.33°C$

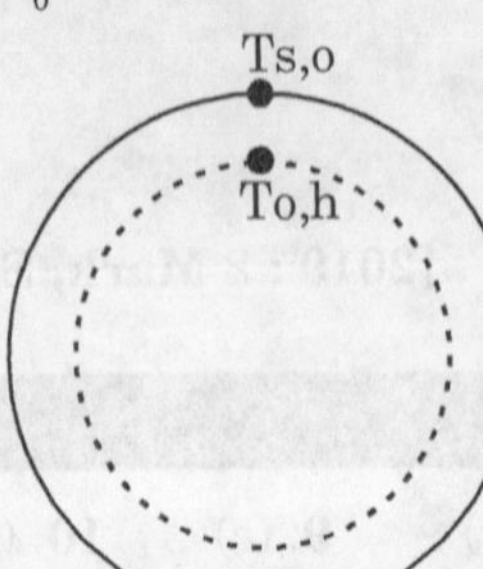

$q''_w = h(Tso - To)$

$5000 = 1000 (Tso - 76.33)$

$Tso = 81°C$

9. For laminar flow, $Nu = 0.664 (Re)^{0.5} (Pr)^{0.33}$

$$\frac{hL}{k} = 0.664 \left(\frac{\rho VD}{\mu}\right)^{0.5} (Pr)^{0.33}$$

$$h \propto V^{0.5};\ h \propto \sqrt{V}$$

So when free stream velocity increases by a factor of 2, then the average heat transfer coefficient rises by a factor of $\sqrt{2}$.

10. Prandtl number $= \frac{\mu C_p}{K}$

$$= 0.1523 \times 10^{-2} \frac{Ns}{m^2} \times \frac{0.1393 \times 1000}{8.54 / mK} \frac{kJ}{kgK} = 0.0248$$

11. When $P_r < 1 \Rightarrow \delta_T > \delta$

$P_r > 1 \Rightarrow \delta_T < \delta$

$P_r = 1 \Rightarrow \delta_T = \delta$

12. $Re_{x=0.5} = \frac{Ux}{\upsilon} = \frac{10 \times 0.5}{30 \times 10^{-6}} = 166666.67$

Since $Re_{x=0.5} \le 5 \times 10^5$

Then flow will be laminar.

$$\therefore \delta = \frac{5x}{\sqrt{Re_x}} = \frac{5 \times 0.5}{\sqrt{166666.67}} = 6.1237 \times 10^{-3} \text{ m}$$

$$\frac{\delta_t}{\delta} = (Pr)^{\frac{1}{3}} \Rightarrow \delta t = (1)^{\frac{1}{3}} \times \delta =$$

$1 \times 6.1237 \times 10^{-3}$ m $= 6.1237$mm

13. $(Nu)_q$ for constant wall heat flux and $(Nu)_T$ at constant wall temperature for a hydrodynamically and thermally fully developed laminar flow through a circular pipe of constant cross-section is 4.36 and 3.66 respectively.

Hence, $(Nu)_q > (Nu)_T$

14. In case of uniform heat flux, bulk mean temperature varies linearly. The difference between bulk mean temperature and wall temperature is constant in thermally developed region so bulk tempeature,

$T(x) = A + Bx$

At $x = 0$, $T = 0°C$

$\Rightarrow$ $A = 0°C$

At $x = 1$, $T = 50°C$

$\Rightarrow$ $B = 50°C$

$\therefore$ $T(x) = 50x$

Here,

Location	Wall temp. (T_w)	bulk temp T(x)	$\Delta T = T_w - T(x)$
P, x = 0.4	50°C	20°C	30°C
Q, x = 0.6	80°C	30°C	50°C
R, x = 0.8	90°C	40°C	50°C

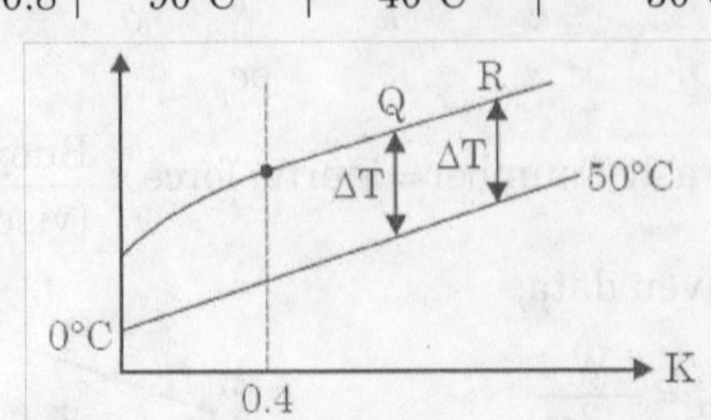

So, Q and R in thermally developed region.

3
CHAPTER

Heat Transfer in Flow Over Plates and Pipes

Analysis of Previous GATE Papers

Topic	Year/Set → Marks Distribution ↓		2019 Set 1	2019 Set 2	2018 Set 1	2018 Set 2	2017 Set 1	2017 Set 2	2016 Set 1	2016 Set 2	2016 Set 3	2015 Set 1	2015 Set 2	2015 Set 3
Flow over plates and pipes	1 Mark	MCQ Type										2		
		Numerical Type		1										
	2 Marks	MCQ Type										1		
		Numerical Type												
	Total			1								4		

Flow over plates and pipes with Prandtl numbers

1. For the fluid flowing over a flat plate with Prandtl number greater than unity, the thermal boundary layer for laminar forced convection.

(a) is thinner than the hydrodynamic boundary layer

(b) has thickness equal to zero

(c) is of same thickness as hydrodynamic boundary layer

(d) is thicker than the hydrodynamic boundary layer

[1988 : 1 Mark]

2. A fluid flowing over a flat plate has the following properties: dynamic viscosity = 25×10^{-6} kg/ms, specific heat = 2.0 kJ/kgK, thermal conductivity 0.05 W/mK. The hydrodynamic boundary layer thickness is measured to be 0.5 mm. The thickness of the thermal boundary layer would be

(a) 0.1 mm (b) 0.5 mm

(c) 1.0 mm (d) None of these

[1992 : 2 Marks]

3. For air near atmospheric condition flowing over a flat plate the laminar thermal boundary layer is thicker than hydrodynamic boundary layer (T/F).

[1994 : 1 Mark]

4. Water (Prandtl number = 6) flows over a flat plate which is heated over the entire length. Which one of the following relationship between the hydrodynamic boundary layer thickness (δ) and thermal boundary layer thickness (δ_t) is true

(a) $\delta_t = \delta$ (b) $\delta_t < \delta$

(c) $\delta_t > \delta$ (d) Can not be predicted

[2001 : 2 Marks]

5. Consider a laminar boundary layer over a heated flat plate. The free stream velocity is U_∞. At some distance x from the leading edge the velocity boundary layer thickness is δ_v and the thermal boundary layer thickness is δ_T. If the Prandtl number is greater than 1, then

(a) $\delta_v > \delta_T$ (b) $\delta_T > \delta_v$

(c) $\delta v \sim \delta_T \sim (U_\infty x)^{-1/2}$ (d) $\delta_v \sim \delta_T \sim x^{-1/2}$

[2003 : 2 Marks]

Linked Answer Questions Q.6 and Q.7: An un-insulated air conditioning duct of rectangular cross section 1 m × 0.5 m, carrying air at 20°C with a velocity of 10 m/s, is exposed to an ambient of 30°C. Neglect the effect of duct construction material. For air in the range of 20 – 30°C, data is as follows: thermal conductivity = 0.025 W/mK; viscosity = 18 μPa.s; Prandtl number = 0.73; density =1.2 kg/m³. For laminar flow Nusselt number is 3.4 for constant wall temperature conditions and , for turbulent flow, $Nu = 0.023\ Re^{0.8}Pr^{0.33}$.

6. The Reynolds number for the flow is

(a) 444 (b) 890

(c) 4.44×10^5 (d) 5.33×10^5

[2005 : 2 Marks]

7. The heat transfer per meter length of the duct, in watts, is

(a) 3.8 (b) 5.3

(c) 89 (d) 769

[2005 : 2 Marks]

8. The temperature distribution within the thermal boundary layer over a heated isothermal flat plate is given by

$$\frac{T - T_W}{T_\infty - T_W} = \frac{3}{2}\left(\frac{y}{\delta_t}\right) - \frac{1}{2}\left(\frac{y}{\delta_t}\right)^3$$ where T_W and T_∞

are the temperatures of plate and free stream respectively, and y is the normal distance measured from the plate. The local Nusselt number based on the thermal boundary layer thickness δ_t is given by

(a) 1.33 (b) 1.50

(c) 2.0 (d) 4.64

[2007 : 2 Marks]

9. For flow of fluid over a heated plate, the following fluid properties are known: viscosity = 0.001 Pa.s; specific heat at constant pressure = 1 kJ/kgK; thermal conductivity = 1 W/mK.

The hydrodynamic boundary layer thickness at a specified location on the plate is 1 mm. The thermal boundary layer thickness at the same location is

(a) 0.001 mm (b) 0.01 mm

(c) 1 mm (d) 1000 mm

[2008 : 1 Marks]

10. A pipe of 25 mm outer diameter carries steam. The heat transfer coefficient between the cylinder and surroundings is 5 W/m²K. It is proposed to reduce the heat loss from the pipe by adding insulation having a thermal conductivity of 0.05 W/mK. Which one of the following statements is TRUE?

(a) The outer radius of the pipe is equal to the critical radius.

(b) The outer radius of the pipe is less than the critical radius.

(c) Adding the insulation will reduce the heat loss.

(d) Adding the insulation will increase the heat loss.

[2011 : 1 Mark]

11. The ratios of the laminar hydrodynamic boundary layer thickness to thermal boundary layer thickness of flows of two fluids P and Q on a flat plate are 1/2 and 2 respectively. The Reynolds number based on the plate length for both the flows is 10^4. The Prandtl and Nusselt numbers for P are 1/8 and 35 respectively. The Prandlt and Nusselt numbers for Q are respectively

(a) 8 and 140 (b) 8 and 70 .

(c) 4 and 70 (d) 4 and 35

[2011 : 2 Marks]

12. Consider a two-dimensional laminar flow over a long cylinder as shown in the figure below.

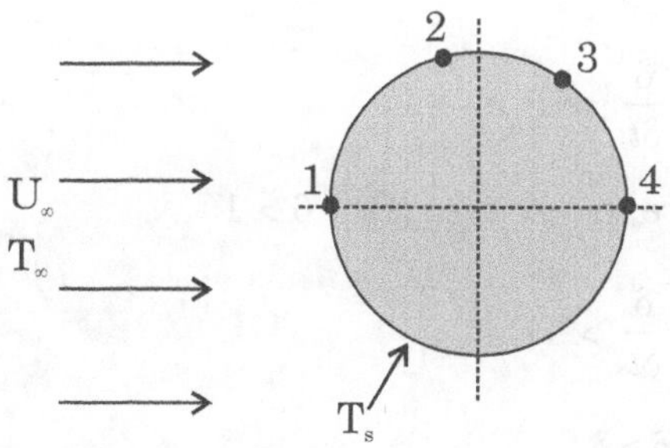

The free stream velocity is U_∞ and the freestream temperature T_∞ is lower than the cylinder surface temperature T_∞. The local heat transfer coefficient is minimum at point

(a) 1 (b) 2

(c) 3 (d) 4

[2014 : 1 Marks, Set-1]

13. The non-dimensional fluid temperature profile near the surface of a convectively cooled flat plate is given by $\frac{T_W - T}{T_W - T_\infty} = a + b\frac{y}{L} + c\left(\frac{y}{L}\right)^2$, where y is measured perpendicular to the plate, L is the plate length, and a, b and c are arbitrary constants. T_W and T_∞ are wall and ambient temperatures, respectively. If the thermal conductivity of the fluid is k and the wall heat flux is q', the Nusselt number $Nu = \frac{q''}{T_W - T_\infty}\frac{L}{k}$ is equal to

(a) a (b) b

(c) 2c (d) (b + 2c)

[2014 : 2 Marks, Set-1]

14. Water flows through a tube of diameter 25 mm at an average velocity of 1.0 m/s. The properties of water are ρ = 1000 kg/m³, μ = 7.25 × 104 Ns/m², k = 0.625 W/mK, Pr = 4.85. Using $Nu = 0.023\,Re^{0.8}\,Pr^{0.4}$, the convective heat transfer coefficient (inW/m²K) is ______

[2014 : 2 Marks, Set-4]

15. For flow of viscous fluid over a flat plate, if the fluid temperature is the same as the plate temperature, the thermal boundary layer is

(a) thinner than the velocity boundary layer

(b) thicker than the velocity boundary layer

(c) of the same thickness as the velocity boundary layer

(d) not formed at all

[2015 : 1 Mark, Set-1]

16. The Blausius equation related to boundary layer theory is a

(a) third-order linear partial differential equation

(b) third-order nonlinear partial differential equation

(c) second-order nonlinear ordinary differential equation

(d) third-order nonlinear ordinary differential equation

[2015 : 1 Mark, Set-1]

17. For flow through a pipe of radius R, the velocity and temperature distribution are as follows:

$u(r, x) = C_1$ and $T(r, x) = C_2\left[1-\left(\frac{r}{R}\right)^3\right]$, where

C_1 and C_2 are constants.

The bulk temperature is given by $T_m = \frac{2}{U_m R^2}\int_0^R u(r,x)T(r,x)rdr$ with U_m being the mean velocity of flow. The value of T_m is

(a) $\frac{0.5C_2}{U_m}$ (b) $0.5C_2$

(c) $0.6C_2$ (d) $\frac{0.6C_2}{U_m}$

[2015 : 2 Marks, Set-1]

18. A thin vertical flat plate of height L, and infinite width perpendicular to the plane of the figure, is losing heat to the surroundings by natural convection. The temperature of the plate and the surroundings, and the properties of the surrounding fluid, are constant. The relationship between the average Nusselt and Rayleigh numbers is given as $Nu = K\,Ra^{1/4}$, where K is a constant.

The length scales for Nusselt and Rayleigh numbers are the height of the plate. The height of the plate is increased to 16L keeping all other factors constant

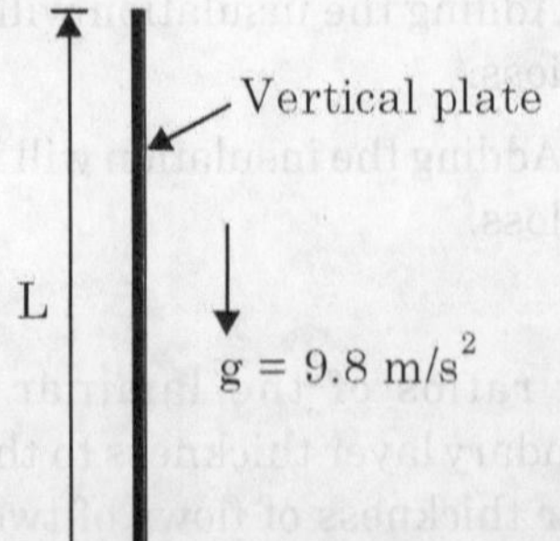

If the average heat transfer coefficient for the first plate is h_1 and that for the second plate is h_2, the value of the ratio h_1/h_2 is

[2019 : 1 Mark, Set-2]

ANSWERS

1. (a)	**2.** (b)	**4.** (b)	**5.** (a)	**6.** (c)	**7.** (d)	**8.** (b)	**9.** (c)	**10.** (c)	**11.** (a)
12. (b)	**13.** (b)	**15.** (d)	**16.** (d)	**17.** (c)					

EXPLANATIONS

1. $P_r^{1/3} = \frac{\delta}{\delta_t}$, if Pr > 1, it means $\delta < \delta_t$

2. $Pr = \frac{\mu C_P}{K}$

$$= \frac{25\times10^{-6}\times2\times10^3}{0.05} = 1$$

$$\Rightarrow \frac{\delta}{\delta t} = 1$$

Now, $\delta = 0.5$ mm

$\therefore\ \delta = \delta_t = 0.5$mm

3. The given statement is False.

At near atmospheric or ambient condition, $Pr = 1$, i.e., $\delta = \delta_t$

4. $\frac{\delta}{\delta t} = (P_r)^n$

Here, $P_r = 6 > 1$

or $\frac{\delta}{\delta t} > 1$

$\therefore\ \delta > \delta_t$

5. Prandtl number

$$= \frac{\text{Molecular diffusivity of momentum}}{\text{Molecular diffusivity of heat}}$$

From question, since prandtl number > 1

$\therefore$ Velocity boundry thickness (δ_υ) > thermal boundary thickness (δ_t)

6. $L_C = \frac{4A}{D} = 4 \times \frac{1 \times 0.5}{2 \times 1.5} = 0.666$

Reynolds number,

$$Re = \frac{\rho L_C V}{\mu} = \frac{1.2 \times 0.66 \times 10}{18 \times 10^{-6}} = 4.4 \times 10^5.$$

7. As Re > 4000, the flow is turbulent

$$Nu = .023\ R_e^{0.8}\ Pr^{0.33}$$

$$\frac{hL}{k} = 0.23 \times \left(4.44 \times 10^5\right) \times (0.73)^{0.33}$$

$$\frac{h \times .0667}{.025} = 683.173$$

$$h = 25.60 \frac{W}{m^2 K}$$

Surface area of duct

$A = 2 \times al + 2 \times bl$

where, l = length of duct

$\therefore\ A = 3l$

Heat transfer rate, $Q = h\,A(T_0 - T)$

$Q = 25.60 \times 3l \times (30 - 20)$

$$\frac{Q}{l} = 769W$$

8. Only (b) satisfies the conditions

9. Given:

$\mu = 0.001$ Pa – s

$$C_P = 1 \frac{kJ}{kg\ K}$$

$K_f = \frac{1W}{mK}$ (Fluid thermal conductivity)

Hydrodynamic boundary layer thickness, $\delta = 1$ mm

$$Pr = \frac{\mu C_P}{K_f} = \frac{.001 \times 1000}{1} = 1$$

$$\delta_t = \delta (Pr)^{\frac{-1}{3}}$$

$$= 1 \times (1)^{\frac{-1}{3}} = 1mm$$

10. Critical radius of insulation

$$= \left(\frac{k}{h}\right)$$

$$= \left(\frac{0.05}{5}\right) m$$

$$= 10\ mm$$

$(r_{outer}) > r_{critical}$

Thus, adding insulation shall decrease H.T. Rate.

11. $\frac{\delta_t}{\delta} = \frac{1}{1.026} \times P_r^{-1/3}$

For fluid Q: $\frac{1}{2} = \frac{\delta_t}{\delta} = \frac{1}{1.026} \times P_r^{-1/3}$

$\Rightarrow\quad P_r = 8$

For fluid P: Laminar flow over flat plate

$$N_u = 0.664\ Re_L^{1/2}\ P_r^{1/3}$$

$$= 35$$

Similarly for fluid Q:

$$Nu = 0.664\ Re_L^{1/2}\ P_r^{1/3}$$

$$= 0.664\ \left(10^4\right)^{1/2}\ 8^{1/3}$$

$$\simeq 140$$

12. For laminar flow, the heat transfer coefficient is minimum where the boundary layer thickness is maximum and vice versa. For turbulent-region boundary layer thickness is maximum at 3 but for laminar boundary layer thickness is maximum at 2 so minimum heat transfer coefficient.

13. $\frac{T_w - T}{T_w - T_\infty} = a + \frac{by}{L} + c\left(\frac{y}{L}\right)^2$

$$T = T_w + (T_\infty - T_w)\left[a + \frac{by}{L} + c\left(\frac{y}{L}\right)^2\right]$$

$$q'' = -k\frac{dT}{dy} = -k(T_\infty - T_w)\left[\frac{b}{L} + \frac{2Cy}{L^2}\right]$$

at y = 0

$$\Rightarrow \frac{q''L}{(T_w - T_\infty)k} = b = N_u$$

$\therefore\ N_u = b.$

14. $Re = \frac{\rho v d}{\mu}$

$= \frac{1000 \times 1 \times 25 \times 10^{-3}}{7.25 \times 10^{-4}}$

$= 34482.75$

$Pr = 4.85$

$Nu = 0.023\, Re^{0.8}\, Pr^{0.4}.$

$Nu = .0.23 \times (34482.758)^{0.8}\,(4.85)^{0.4}.$

$= 184.546$

$\&Nu = \frac{hd}{k}$

$h = \frac{Nu.k}{d}$

$= \frac{184.546 \times 0.625}{25 \times 10^{-3}}$

$= 4613.66 \frac{W}{m^2K}$

16. Blausius equation:

$$\frac{2d^3f}{d\eta^3} + \frac{fd^2f}{d\eta^2} = 0$$

This order non-linear D.E.

17. $T_m = \frac{2}{u_m R^2}\int_0^R C_1C_2\left(1-\left(\frac{r}{R}\right)^3\right) r\,dr$

$= \frac{2C_1C_2}{u_m R^2}\int_0^R \left[r - \frac{r^4}{R^3}\right] dr$

$= \frac{2C_1C_2}{u_m R^2}\left(\frac{3}{10}R^2\right)$

$= \frac{0.6C_1C_2}{u_m}$

Since $u(r, x)$ is constant, $u_m = C_1$

$\therefore \quad t_m = 0.6C_2$

18. $Nu = k.(Ra)^{1/4}$

$Ra = f(Gr, Pr)$

$= f\left(\frac{g\beta.\Delta T.L^3}{v^2}, \frac{\mu C_p}{k}\right)$

$\frac{h.L}{k} \propto (L)^{3/4}$

$h \propto (L)^{\frac{3}{4}-1}$

$h \propto \frac{1}{(L)^{1/4}}$

$\therefore \frac{h_1}{h_2} = \left(\frac{L_2}{L_1}\right)^{1/4} = \left(\frac{16L}{L}\right)^{1/4}$

$\Rightarrow \boxed{\frac{h_1}{h_2} = 2}$

4 CHAPTER

Radiation

Analysis of Previous GATE Papers

Topic	Year/Set → Marks Distribution ↓		2019 Set 1	2019 Set 2	2018 Set 1	2018 Set 2	2017 Set 1	2017 Set 2	2016 Set 1	2016 Set 2	2016 Set 3	2015 Set 1	2015 Set 2	2015 Set 3
Basic definitions and associated laws	1 Mark	MCQ Type												
		Numerical Type												
	2 Marks	MCQ Type												
		Numerical Type												
	Total													
Net heat radiation and emissivity	1 Mark	MCQ Type				1		1						
		Numerical Type												
	2 Marks	MCQ Type											1	
		Numerical Type				1	1		1		1			
	Total					**3**	**2**	**1**	**2**		**2**		**2**	
Shape Factor	1 Mark	MCQ Type		1						1				
		Numerical Type												
	2 Marks	MCQ Type												
		Numerical Type		1										1
	Total			**3**						**1**				**2**
Radiation Shields	1 Mark	MCQ Type												
		Numerical Type												
	2 Marks	MCQ Type												
		Numerical Type												
	Total													

Basic Definitions and law associated with radiation

1. For a glass plate transmissivity and reflectivity are specified as 0.86 and 0.08 respectively, the absorptivity of the glass plate is

(a) 0.86 (b) 0.08
(c) 1.00 (d) 0.06

[1988 : 1 Mark]

2. A diffuse radiation surface has

(a) Radiation intensity Independent of angle
(b) Emissive power Independent of angle
(c) Emissive power independent of wavelength
(d) Radiation intensity Independent of both angle and wavelength

[1991 : 1 Mark]

3. What is the value of the view factor for two inclined flat plates having common edge of equal length and wlth an angle of 20 degrees?

(a) 0.83 (b) 1.17
(c) 0.66 (d) 1.34

[2002 : 2 Marks]

4. The following figure was generated from experimental data relating spectral black body emissive power to wave length at the three temperatures T_1, T_2 and T_3($T_1 > T_2 > T_3$).

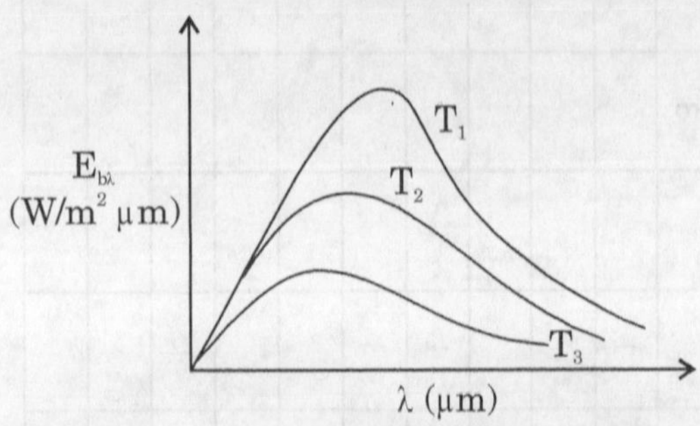

The conclusion in that the measurements are

(a) correct because the maxima in $E_{b\lambda}$ show the correct trend
(b) correct because planck's law is satisfied
(c) wrong because the Stefan Boltzmann law in not satisfied
(d) wrong because Wien's displacement law is not satisfied

[2005 : 1 Mark]

5. For an opaque surface, the absorptivity (α), transmissivity (τ) and reflectivity (ρ) are related by the equation

(a) $\alpha + \rho = \tau$ (b) $\rho + \alpha + \tau = 0$
(c) $\alpha + \rho = 1$ (d) $\alpha + \rho = 0$

[2012 : 1 Mark]

Net heat radiation and emissivity

6. The radiative heat transfer rate per unit area (W/m^2) between two plane parallel gray surfaces (emissivity = 0.9) maintained at 400 K and 300 K is (σ = Stefan Boltzmann constant = 5.67×10^{-8} W/m^2K^4)

(a) 992 (b) 812
(c) 464 (d) 567

[1993 : 2 Marks]

7. A plate having 10 cm^2 area each side is hanging in the middle of a room of 100 m^2 total surface area. The plate temperature and emissivity are respectively 800 K and 0.6. The temperature and emissivity valute for the surfaces of the room are 300 K and 0.3 respectively. Boltzmann's constant σ = 5.67×10^{-8} W/m^2K^4. The total heat loss from the two surfaces of the plate is

(a) 13.66 W (b) 27.32 W
(c) 27.87 W (d) 13.66 MW

[2003 : 1 Mark]

8. A100 W electric bulb was switched on in a 2.5 m × 3 m × 3 m size thermally insulated room having a temperature of 20°C. The room temperature at the end of 24 hours will be

(a) 321 °C (b) 341 °C
(c) 450°C (d) 470°C

[2006 : 2 Marks]

Common Data for Q.9 and Q.10: Radiative heat transfer is intercepted between the inner surfaces of two very large isothermal parallel metal plates. While the upper plate (designated as plate 1) is a black surface and is the warmer one being maintained at 727°C, the lower plate (plate 2) is a diffuse and gray surface with an emissivity of 0.7 and is kept at 227°C. Assume that the surfaces are sufficiently large to form a two-surface enclosure and steady-state conditions to exist. Stefan-Boltzmann constant is given as 5.67×10^{-8} W/m^2K^4.

9. The irradiation (in kW/m^2) for the plate (plate 1) is

(a) 2.5 (b) 3.6
(c) 17.0 (d) 19.5

[2009 : 2 Marks]

10. If plate 1 is also a diffuse and gray surface with an emissivity value of 0.8, the net radiatibn heat exchange (in kW/m^2) between plate 1 and plate 2 is

(a) 17.0 (b) 19.5
(c) 23.0 (d) 31.7

[2009 : 2 Marks]

11. Two large diffuse gray parallel plates, separated by a small distance, have surface temperature of 400 K and 300 K. If the emissivities of the surfaces are 0.8 and the Stefan-Boltzmann constant is 5.67×10^{-8} W/m^{2}K^4, the net radiation heat exchange rate in kW/m^2 between the two plates is

(a) 0.66 (b) 0.79
(c) 0.99 (d) 3.96

[2013 : 2 Marks]

12. A hemispherical furnace of 1 m radius has the inner surface (emissivity, $\varepsilon = 1$) of its roof maintained at 800 K, while its floor ($\varepsilon = 0.5$) is kept at 600 K. Stefan-Boltzmann constant is 5.668×10^{-8} W/m^{2}K^4. The net radiative heat transfer (in kW) from the roof to the floor is ______.

[2014 : 2 Marks, Set-2]

13. The total emissive power of a surface is 500 W/m^2 at temperature T_1 and 1200 W/m^2 at emperature T_2, where the temperature are in Kelvin. Assuming the emissivity of the surface to be constant, the ratio of the temperature T_1/T_2 is

(a) 0.0308 (b) 0.416
(c) 0.803 (d) 0.874

[2015 : 2 Marks, Set-2]

14. An infinitely long function 0.5 m × 0.4 m cross-section is shown in the figure below. Consider all surfaces of the furnace to be black. The top and bottom walls are maintained at temperature $T_1 = T_3 = 927°C$ while the side walls are attemperature $T_2 = T_4 = 527°C$. The view factor, F_{1-2} is 0.26.The net radiation heat loss or gain on side 1 is ______ W/m.
Stefan-Boltzmann constant = 5.67×10^{-8} W/m^{2}K^4.

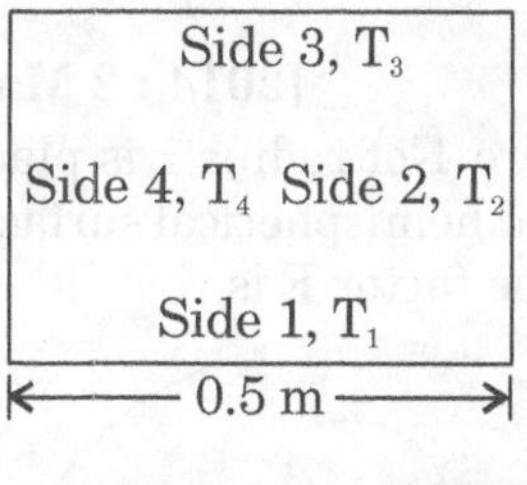

[2016 : 2 Marks, Set-1]

15. Two large parallel plates having a gap of 10 mm in between them are maintained at temperatures T_1, = 1000 K and T_2 = 400 K. Given emissivity values, $\varepsilon_1 = 0.5$, $\varepsilon_2 = 0.25$ and Stefan-Boltzmann constant $\sigma = 5.67 \times 10^{-8}$ W/m^{2}K^4, the heat transfer between the plates (in kW/m^2) is ______

[2016 : 2 Marks, Set-3]

16. Two black surfaces, AB and BC, of length 5m and 6 m, respectively, are oriented as shown. Both surfaces extend infinitely into the third dimension. Given that view factor $F_{12} = 0.5$. T_1 = 800 K, T_2 = 600 K, $T_{surrounding}$ = 300 K and Stefan Boltzmann constant, a = 5.67×10^{-8} W/(m^{2}K^4), the heat transfer rate from Surface 2 to the surrounding environment is _____kW.

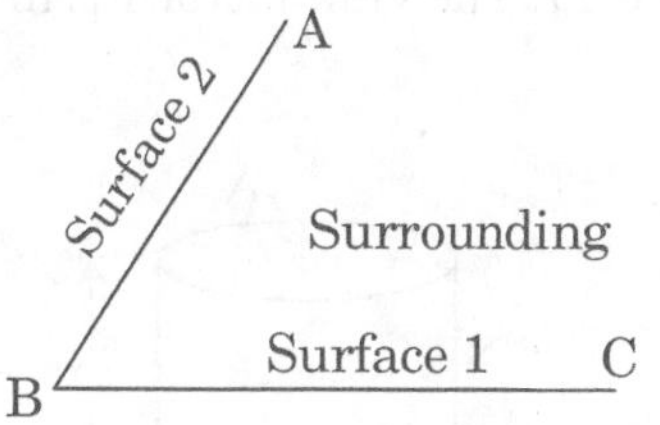

[2017 : 2 Marks, Set-1]

17. The emissive power of a blackbody is P. If its absolute temperature is doubled, the emissive power becomes

(a) 2P (b) 4P
(c) 8P (d) 16P

[2017 : 1 Mark, Set-2]

18. The peak wavelength of radiation emitted by a black body at a temperature of 2000 K is 1.45 μm. If the peak wavelength of emitted radiation changes to 2.90 μm , then the temperature (in K) of the black body is

(a) 500 (b) 1000
(c) 4000 (d) 8000

[2018 : 1 Mark, Set-2]

19. A 0.2 m thick infinite black plate having a thermal conductivity of 3.96 W/m-K is exposed to two infinite black surfaces at 300 K and 400 K as shown in the figure. At steady state the surface temperature of the plate facing the cold side is 350 K. The value of Stefan-Boltzmann constant a is 5.67×10^{-8} W/m^2 K^4. Assuming 1-D heat conduction, the magnitude of heat flux through the plate (in W/m^2) is ______(correct to two decimal places).

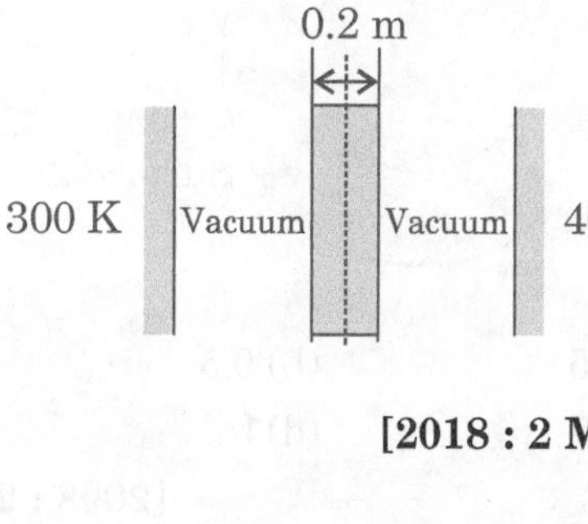

[2018 : 2 Marks, Set-2]

Shape factor

20. The shape factors with themselves of two infinitely long black body concentric cylinders with a dia ratio of 3 are ____ for the Inner and ____ for the outer

[1994 : 1 Mark]

21. For the circular tube of equal length and diameter shown in figure below, the view factor F_{13} is 0.17. The view factor F_{12} in this case will be

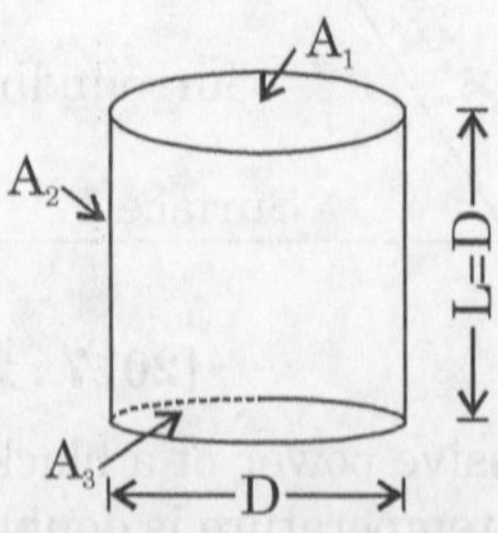

(a) 0.17 (b) 0.21
(c) 0.79 (d) 0.83

[2001 : 2 Marks]

22. A solid cylinder (surface 2) is located at the centre of a hollow sphere (surface 1). The diameter of the sphere is 1 m, while the cylinder has a diameter and length of 0.5 m each. The radiation configuration factor F_{11} is

(a) 0.375 (b) 0.625
(c) 0.75 (d) 1

[2005 : 2 Mark]

23. A hollow enclosure is formed between two inf initely long concentric cylinders of radii 1 m and 2 m, respectively. Radiative heat exchange takes place between the inner surface of the larger cylinder (surface-2) and the outer surface of the smaller cylinder (surface-1). The radiating surface are diffuse and the medium in the enclosure is non- participating. The fraction of the thermal radiation leaving the larger surface and striking itself is

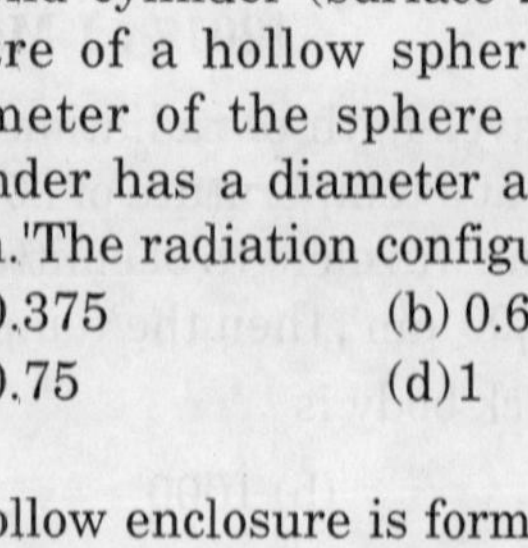

(a) 0.25 (b) 0.5
(c) 0.75 (d) 1

[2008 : 2 Marks]

24. Consider two infinitely long thin concentric tubes of circular cross section as shown in figure. If D_1 and D_2 are the diameters of the inner and outer tubes respectively, then the view factor F_{22} is given by

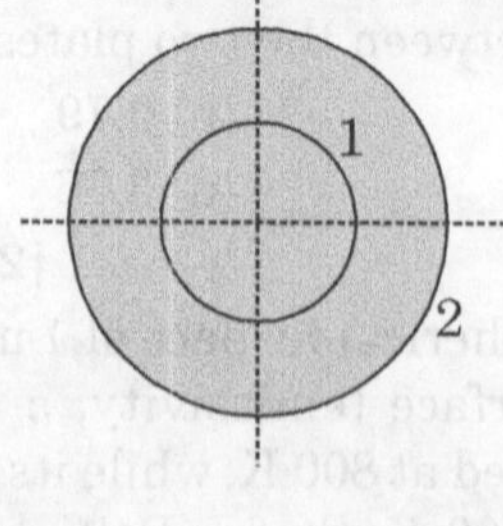

(a) $\left(\frac{D_2}{D_1}\right)-1$ (b) Zero

(c) $\left(\frac{D_1}{D_2}\right)$ (d) $1-\left(\frac{D_1}{D_2}\right)$

[2012 : 2 Marks]

25. A solid sphere of radius r_1 = 20 mm is placed concentrically inside a hollow sphere of radius r_2 = 30 mm as shown in the figure.

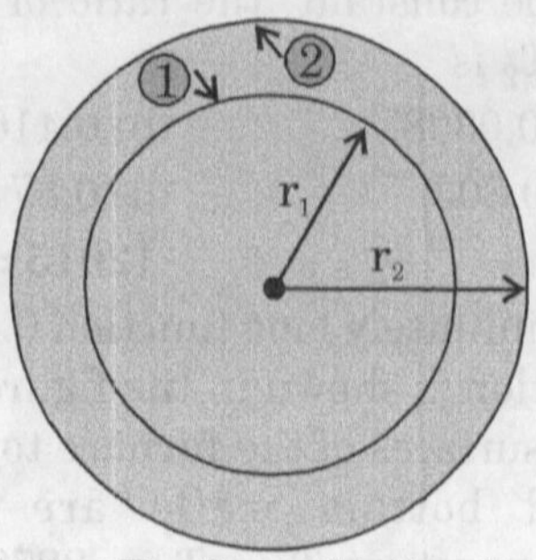

The view factor F_{21} for radiation heat transfer is

(a) $\frac{2}{3}$ (b) $\frac{4}{9}$

(c) $\frac{8}{27}$ (d) $\frac{9}{4}$

[2014 : 2 Marks, Set-3]

26. A solid sphere 1 of radius r is placed inside a hollow, closed hemispherical surface 2 of radius 4r. The shape factor F is,

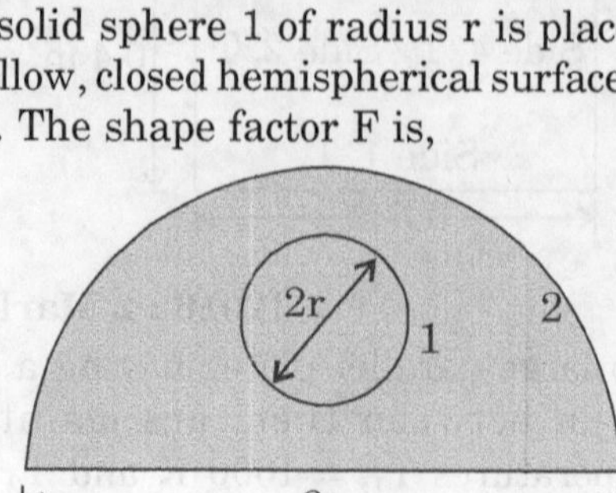

(a) $\frac{1}{12}$ (b) $\frac{1}{2}$

(c) 2 (d) 12

[2015 : 2 Marks, Set-3]

27. Consider the radiation heat exchange inside an annulus between two very long concentric cylinders. The radius of the outer cylinder is R_0 and that of the inner cylinder is R_i. The radiation view factor of the outer cylinder onto itself is

(a) $1-\sqrt{\frac{R_i}{R_0}}$ (b) $\sqrt{1-\frac{R_i}{R_0}}$

(c) $1-\left(\frac{R_i}{R_0}\right)^{1/3}$ (d) $1-\frac{R_i}{R_0}$

[2016 : 1 Mark, Set-2]

28. Sphere 1 with a diameter of 0.1 m is completely enclosed by another sphere 2 of diameter 0.4 m. The view factor F_{12} is

(a) 0.0625 (b) 0.25
(c) 0.5 (d) 1.0

[2019 : 1 Mark, Set-2]

29. Three sets of parallel plates LM, NR and PQ are given in Figures 1, 2 and 3. The view factor Fij is defined as the fraction of radiation leaving plate I that is intercepted by plate J. Assume that the values of FLM and FNR are 0.8 and 0.4, respectively. The value of FPQ (round off to one decimal place) is _____.

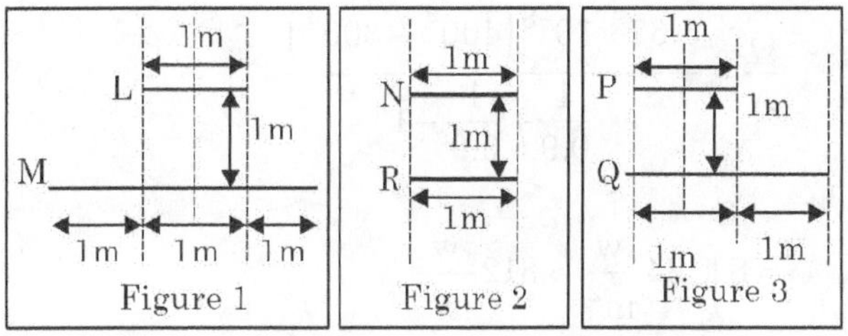

[2019 : 2 Marks, Set-2]

Radiation shields

30. Two infinite parallel plates are placed at a certain distance apart. An infinite radiation shield is inserted between the plates without touching any of them to reduce heat exchange between the plates. Assume that the emissivities of plates and radiation shield are equal. The ratio of the net heat exchange between the plates with and without the shield is

(a) 1/2 (b) 1/3
(c) 1/4 (d) 1/8

[2014 : 2 Marks, Set-4]

ANSWERS

1. (d)	**2.** (a)	**3.** (a)	**4.** (d)	**5.** (c)	**6.** (b)	**7.** (b)	**8.** (d)	**9.** (d)	**10.** (d)
11. (a)	**13.** (c)	**17.** (d)	**18.** (b)	**21.** (d)	**22.** (b)	**23.** (b)	**24.** (d)	**25.** (b)	**26.** (a)
27. (d)	**28.** (d)	**30.** (a)							

EXPLANATIONS

1. $\tau = 0.86$
$\rho = 0.08$
$\alpha = ?$
$\alpha + \tau + \rho = 1$
$\alpha + 0.86 + 0.08 = 1$
$\alpha = 0.06$

2. Radiation intensity is independent of angle in diffuse radiation.

3. $F_{12} = F_{21} = 1 - \sin\left(\frac{\theta}{2}\right)$

$= 1 - \sin\left(\frac{20°}{2}\right)$

$= 0.826$

4. Wien's displacement law is not satisfied, *i.e.*, $\lambda_{max} T = C$

which tells lower λ_{max} for higher temperature.

5. $\alpha + \tau + \rho = 1$

for opaque surface, $\tau = 0$

$\therefore \alpha + \rho = 1$

6. For plane surfaces

$$\frac{Q}{A} = \frac{\sigma\left(T_1^4 - T_2^4\right)}{\frac{1}{\epsilon_1} + \frac{1}{\epsilon_2} - 1}$$

Here $\epsilon_1 = \epsilon_2 = 0.9$

$$\frac{Q}{A} = \frac{5.67 \times 10^{-8}\left(400^4 - 300^4\right)}{\frac{1}{0.9} + \frac{1}{0.9} - 1}$$

$$= 8\,||\,.84\frac{w}{m^2} \approx 812\frac{w}{m^2}$$

7. Heal loss from one surface of the plate,

$$Q_{net} = \frac{\sigma\left(T_1^4 - T_2^4\right)}{\frac{\rho_1}{A_1 \in_1} + \frac{1}{A_1} + \frac{\rho_2}{A_2 \in_2}}$$

$$= \frac{\sigma\left(T_1^4 - T_2^4\right)}{\frac{1}{A_1}\left(\frac{1-\in_1}{\in_1} + 1\right) + \frac{1}{A_2}\left(\frac{1-\in_2}{\in_2}\right)}$$

Putting

$A_1 = 10 \times 10^{-4}\ m^2$, $T_1 = 800$

$\in_1 = 0.6$ $T_2 = 300$

$A_2 = 100\ m^2$

$\in_2 = 0.3$

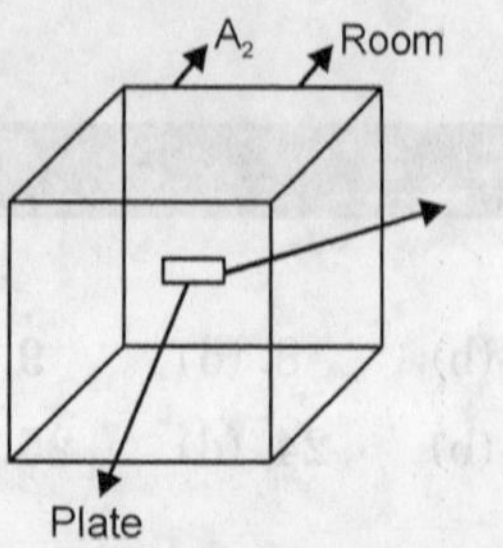

$$\therefore Q = \frac{5.67 \times 10^{-8}\left(800^4 - 300^4\right)}{\frac{10^4}{10}\left(\frac{1}{0.6}\right) + \frac{1}{100}\left(\frac{1}{0.3} - 1\right)}$$

$$= 13.66\ W$$

∴ Heat loss from both the surfaces

$= 2\ Q_{net}$

$= 2 \times 13.66 = 27.32\ W$

8. Heat generated by bulb = 100 × 24 × 60 × 60 Joule.

∴ Heat dissipated = $\rho \times V \times C_v \times (T - 20)$

⇒ $100 \times 24 \times 60 \times 60 = 1.24 \times 9 \times 2.5 \times 3 \times C_v\ (T - 20)$

⇒ $T = 450^\circ$

9. $T_1 = 1000\ K$

$T_2 = 500\ K$

$\in = 0.7$

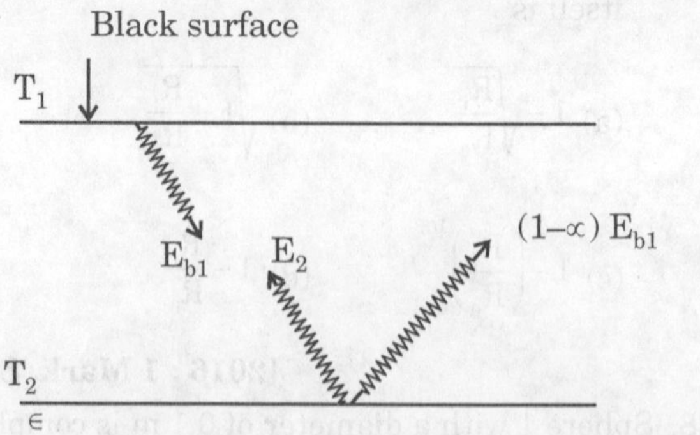

Energy indicent upon plate 1 due to irradiation = Radiosity of plate 2

Energy leaving surface 2 = $E_2 + (1-\alpha 2)\ E_{b1}$

$E_2 = \in\sigma T_2^{\ 4} = 0.7 \times 5.67 \times 10^{-8} \times 500^4$

$= 2480.62\frac{w}{m^2}$

$\sigma\,(1-\alpha)\ E_{b1}$ = Fraction of energy reflected from surface 2

$= (1-0.7) \times 5.67 \times 10^{-8} \times 1000^4$

$= 17010\frac{W}{m^2}$

∴ Total energy incident upon plate 1 = $19.49\frac{kW}{m^2}$

10. $f_{12} = \frac{1}{\frac{1}{\in_1} + \frac{1}{\in_2} - 1} = \frac{1}{\frac{1}{0.8} + \frac{1}{0.7} + 1} = \frac{1}{1.25 + 1.42 - 1}$

$= 0.59$

∴ $Q = f_{12}\ \sigma_b\left(T_1^4 - T_2^4\right)$

$= 0.59 \times 5.67 \times (10000 - 625)$

$= 31362.18\ W/m^2$

$= 31.36\ kW/m^2$

Alternately

$1000\ K = T_A$ A (Black), $\in_A = 0.8$

$500\ K = T_B$ B(Grey), $\in_B = 0.7$

Net heat exchange between A and B,

$$Q = \sigma \frac{\left(T_A^4 - T_B^4\right)}{\frac{1}{\epsilon_A} + \frac{1}{\epsilon_B} - 1}$$

$$= \frac{(1000)^4 - (500)^4}{\frac{1}{0.8} + \frac{1}{0.7} - 1} \times 5.67 \times 10^{-8}$$

$$= \frac{(1000)^4 - (500)^4}{1.25 + 1.4286 - 1}$$

$$= \frac{(1000)^4 - (500)^4}{1.6786} \times 5.67 \times 10^{-8}$$

$$= \frac{53156.25}{1.6786} \text{ W/m}^2$$

$$\approx 31.7 \text{ kW/m}^2$$

11. $$q_{T\text{-}2} = \frac{0\left(T_1^4 - T_2^4\right)}{1/\varepsilon_1 + 1/\varepsilon_2 - 1}$$

$$= \frac{5.67 \times 10^{-8}(400^4 - 300^4)}{\left(2/.8 - 1\right)}$$

$$= 0.66 \frac{\text{kW}}{\text{m}^2}$$

12. Let the base be (1) and hemispherical furnace be (2)

$\therefore \quad F_{11} + F_{12} = 1 \quad ...(1)$

$F_{21} + F_{22} = 1 \quad ...(2)$

$\because \quad F_{11} = 0$

$\therefore \quad F_{12} = 2$

$A_1F_{12} = A_2F_{21}$

$$\therefore \quad F_{21} = \frac{A_1}{A_2}F_{12} = \frac{\pi R^2}{2\pi R^2}F_{12} = 0.5F_{12}$$

$\therefore \quad F_{21} = 0.5 \times 2.0.5$

$\therefore \quad F_{22} = 0.5$

So, $F_{11} = 0$, $F_{12} = 1$, $F_{21} = 0.5 < F_{22} = 0.5$

Now Radiative heat transfer,

$\Rightarrow Q = A_1F_{12}\varepsilon_2 \times 6(800^4 - 600^4)$ watt

$\therefore \; Q = \pi \times 1^2 \times 1 \times 0.5 \times 5.668 \times 10^{-8}(800^4 - 600^4)$ watt

or $Q = 24.9$ kW.

13. Emissive power $\propto T^4$

$$\Rightarrow \quad \frac{500}{1200} = \left(\frac{T_1}{T_2}\right)^4$$

$$\therefore \quad \frac{T_1}{T_2} = 0.8034$$

14. $T_1 = 927°C = 1200$ K,

$T_2 = 527°C = 800$ K,

$F_{12} = F_{14} = 0.26$

$F_{11} + F_{12} + F_{13} + F_{14} = 1$

$F_{13} = 0.48$

$Q = Q_{12} + Q_{13} + Q_{14}$

$Q_{13} = 0$ since the temperatures are same

$$Q = Q_{12} + Q_{14} = 2 \times \sigma_b \times A \times F_{12} \times \left(T_1^4 - T_2^4\right)$$

$$Q = 2 \times 5.67 \times 10^{-8} \times (0.5 \times 1) \times 0.26 \times (1200^4 - 800^4)$$

$= 24530.688$

15. 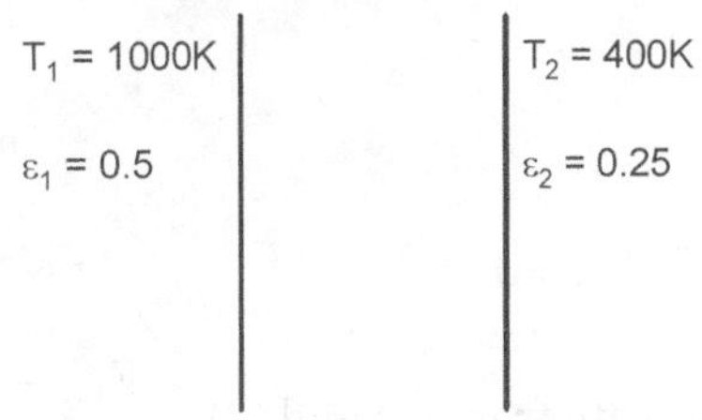

$$Q + 2 = \frac{\sigma\left(T_1^4 - T_2^4\right)}{\frac{1}{\varepsilon_1} + \frac{1}{\varepsilon_2} - 1}$$

$$= \frac{5.76 \times 10^{-8}\left(1000^4 - 400^4\right)}{\frac{1}{0.5} + \frac{1}{0.25} - 1}$$

$= 11049.696$ W/m^2

$= 11.049$ kW/m^2

16. $F_{12} = 0.5$

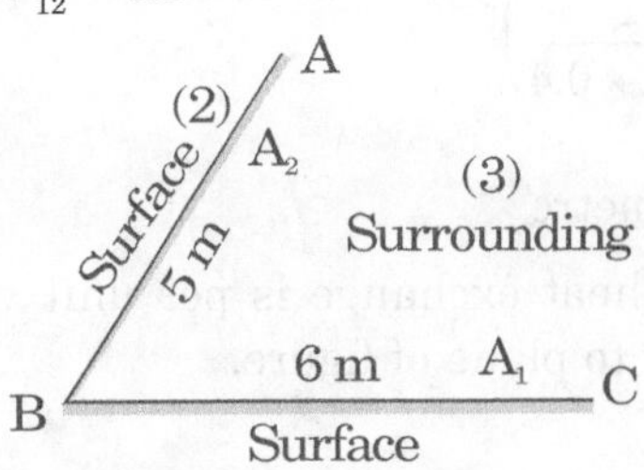

$T_1 = 800$ K, $T_2 = 600$ K

$T_{surrounding} = 300$ K

Let (3) denote the surrounding environment (Assume unit width perpendicular to plane of Fig.)

From summation rule:

$F_{21} + F_{22} + F_{23} = 1$

$F_{21} + F_{23} = 1$

$\therefore F_{22} = 0$

or $F_{23} = 1 - F_{21}$

$A_1F_{12} = A_2F_{21} = 6 \times 1 \times 0.5$

$= (5 \times 1) \times F_{21}$

$$F_{21} = \frac{6 \times 0.5}{5} = 0.6$$

$\therefore F_{23} = 1 - F_{21} = 0.4$

Since surrounding environment being large. No surface has a surface resistance

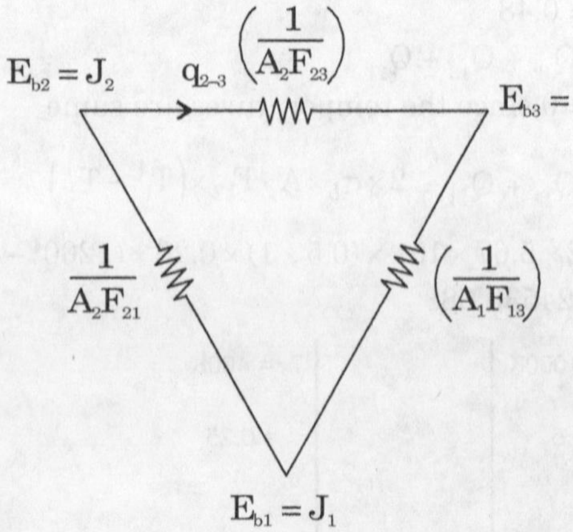

Heat transfer rate from (2) and (3) $= \dfrac{E_{b_2} - E_{b_3}}{\dfrac{1}{A_2F_{23}}}$

$$= \frac{\sigma\left(T_2^4 - T_3^4\right)}{\dfrac{1}{A_2F_{23}}}$$

$$= \frac{5.67 \times 10^{-8}\left(600^4 - 300^4\right)}{\left(\dfrac{1}{5 \times 1 \times 0.4}\right)} \text{ watt/metre}$$

= 13.778 kW/metre

Assume that heat exchange is per unit width perpendicular to plane of figure.

17. $E_b = \sigma T^4$

$$\frac{Eb_2}{Eb_1} = \frac{T_2^4}{T_1^4} = \left(\frac{2T}{T}\right)^4 = 16$$

$E_{b2} = 16\ E_{b1}$

$E_{b2} = 16\ P\ (\because E_{b1} = P)$

18. From wein's displacement law

$\lambda_{max}\ T$ = constant

$(\lambda_{max})_1\ T_1 = (\lambda_{max})_2\ T_2$

$1.45 \times 2000 = 2.90 \times T_2$

$$\therefore T_2 = \left(\frac{1.45}{2 \circ 90} \times 2000\right)$$

$T_2 = 1000$ K

19. Heat flux through the wall} = heat flux from wall to surface at 300 k

$$= \frac{\sigma(350^4 - 300^4)}{\dfrac{1}{1} + \dfrac{1}{1} - 1}$$

$$= 391.612 \frac{w}{m^2}\left(\because 6 = 5.67 \times 10^{-8} \frac{W}{m^2k^4}\right)$$

20. $\dfrac{d_2}{d_1} = 3$

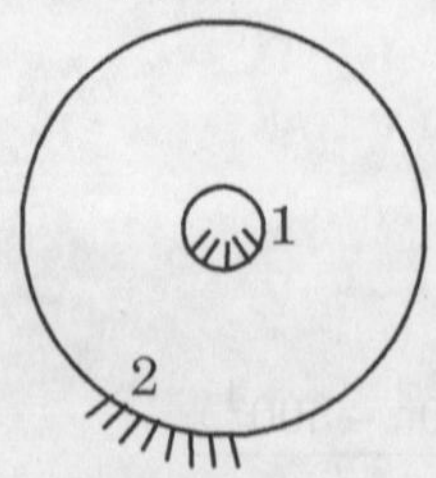

For Body – 1

$F_{11} + F_{12} = 1$

$F_{12} = 1\ (\because F_{11} = 0)$

For Body – 2

$F_{21} + F_{22} = 1 - (A)$

By applying reciprocity theorem,

$A_1F_{12} = A_2\ F_{21}$

$A_1 = A_2\ F_{21}$

$$\text{or } F_{21} = \frac{A_1}{A_2} = \frac{\pi d_1 l}{\pi d_2 l} = \frac{d_1}{d_2} = \frac{1}{3}$$

$$F_{21} = \frac{1}{3}$$

Substituting, $F_{21} = \dfrac{1}{3}$ in (A), we get

$$\frac{1}{3} + F_{22} = 1$$

$$F_{22} = \frac{2}{3}$$

21. $\frac{1}{r_1} = 2$

and $\frac{r_3}{l} = 0.5$

$F_{13} = 0.17,$

$F_{11} = 0,$

$F_{11} + F_{12} + F_{13} = 1$

$\therefore\ F_{12} = 1 - F_{13}$

$= 1 - 0 - 17 = 0.83$

22.

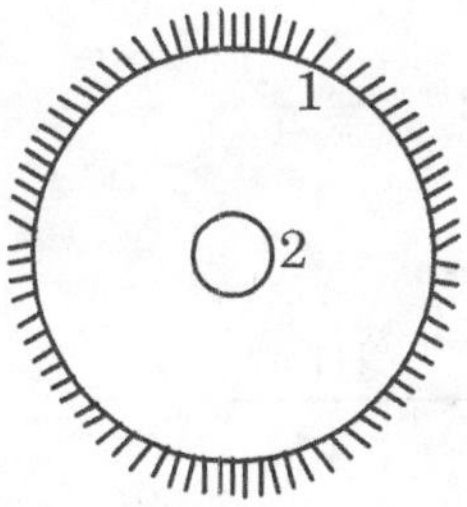

Diameter of sphere = 1 m

Diameter of aylinder = 0.5 m

$F_{21} + F_{22} = 1\ (\because\ F_{22} = 0)$

$F_{21} = 1$

we know that

$A_1 F_{12} = A_2 F_{21}$

$F_{12} = \frac{A_2}{A_1}$

Now, $F_{11} + F_{12} = 1$

$\therefore F_{11} = 1 - F_{12} = 1 \frac{-A_2}{A_1}$

$\therefore F_{11} = 1 - \frac{A_2}{A_1}$

$$F_{11} = 1 - \frac{\left[(\pi \times 0.5 \times 0.5) + \left(2 \times \frac{\lambda}{4} \times 0.5 \times 0.5\right)\right]}{4\pi \times 0.5 \times 0.5}$$

$= 0.625$

23. $F_{11} + F_{12} = 1$

$F_{21} + F_{22} = 1$

$F_{11} = 0$ due to concave surface

$F_{12} = 1$

Now $A_1 F_{12} = A_2 F_{21}$

$\Rightarrow\ F_{21} = F_{12} \frac{A_1}{A_2}$

$\Rightarrow\ F_{12} = 1$

$\therefore\ F_{21} = \frac{A_1}{A_2} \quad \frac{2\ r_1 l}{2\ r_2 l} \quad \frac{r_1}{r_2} \quad \frac{1}{2}$

$F_{21} = 0.5$

$F_{22} = 1 - F_{21} = 0.5$

24. $F_{22} = 1 - F_{21}$

$= 1 - \frac{A_1}{A_2} = 1 - \frac{D_1}{D_2}$

25. $F_{11} + F_{12} = 1$, here $F_{11} = 0$

$\therefore\ F_{12} = 1$

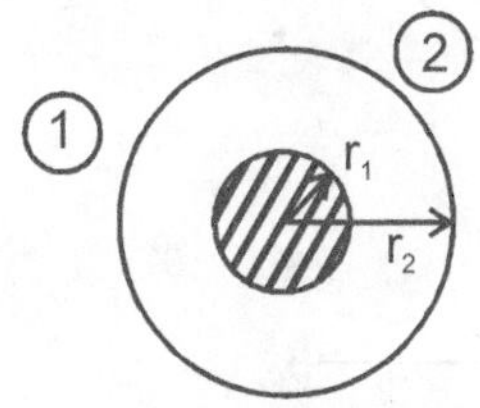

Now from reciprocating law

$A_1 F_{12} = A_2 F_{21}$

$\therefore\ 2\pi(20)^2 \times 1 = 2\pi(30)^2 \times F_{21}$

$\Rightarrow \therefore F_{21} = \left(\frac{10}{30}\right)^2 = \frac{4}{9}.$

26. $f_{11} + f_{12} = 1$

$\therefore\ f_{12} = 1$

$f_{21} A_2 = f_{12} A_1$

$\therefore\ f_{21} = \frac{f_{12} A_1}{A_2} = \frac{1 \times 4\pi r^2}{\frac{1}{2} 4\pi(4r)^2 + \pi(4r)^2} = \frac{1}{12}$

27.

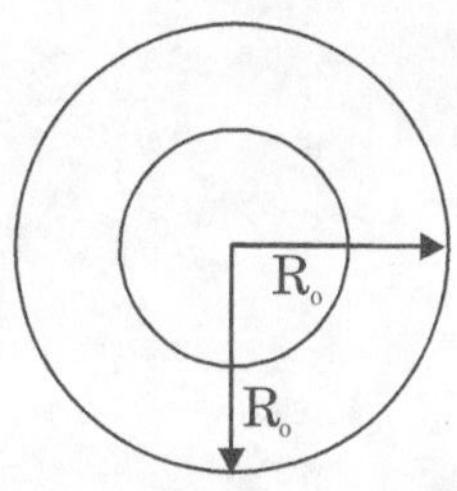

View factor

$F_{1-1} = 0$

$F_{1-2} = 1$

Reciprocacy theorem

$A_2 F_{2-1} = A_1 F_{1-2}$

$$\Rightarrow \qquad F_{2-1} = \frac{A_1}{A_2} F_{1-2}$$

$$= \frac{2\pi R_i L}{2\pi R_o L} = \frac{R_i}{R_o}$$

$$F_{2-1} + F_{2-2} = 1$$

$$\therefore \qquad F_{2-2} = 1 - F_{2-1} = 1 - \frac{R_i}{R_o}$$

28. 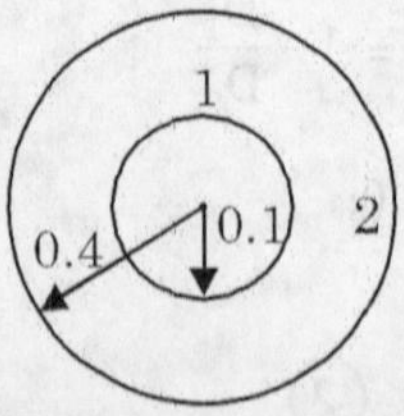

$F_{11} = 0$

$F_{12} = 1$

29. 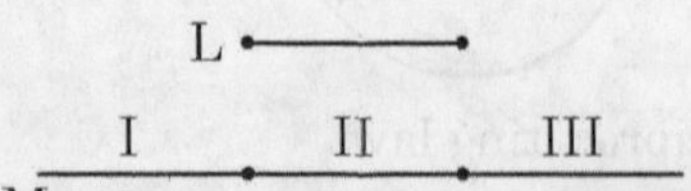

Fig. (1)

$F_{LM} = F(I) + F(II) + F(III)$

Where F is shape factor of surface area I, II and III of M with surface area of 'L'

$0.8 = F(I) + F(II) + F(III)$

$\because$ (I) and (III) are symmetrical in shape so, shape factor of these surface will be same

i.e. F(I) = F(III)

So, 0.8 = 2F(I) + F(II) ... (1)

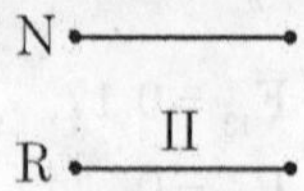

Fig. (2)

In this area (II) is same as fig. (I) area II.

i.e. F_{NR} = F(II) = 0.4 ... (2)

From equation (1),

$$F(III) = \frac{0.8 - 0.4}{2} = 0.2$$

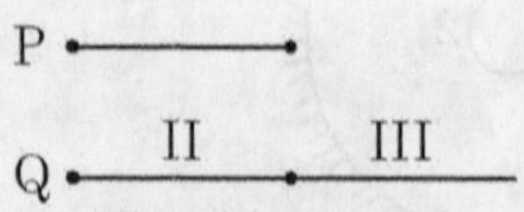

Fig. (3)

Here, F_{PQ} = F(II) + F(III)

= 0.4 + 0.2 = 0.6

30. $$\frac{q_{\text{with sheilds}}}{q_{\text{without sheilds}}} = \frac{1}{n+1}$$

here, $n = 1$

$$\therefore \qquad \frac{q}{q_{\text{without}}} = \frac{1}{2}.$$

5
CHAPTER

Heat Exchanger

Analysis of Previous GATE Papers

Topic	Year/Set → Marks Distribution ↓		2019 Set 1	2019 Set 2	2018 Set 1	2018 Set 2	2017 Set 1	2017 Set 2	2016 Set 1	2016 Set 2	2016 Set 3	2015 Set 1	2015 Set 2	2015 Set 3
Types of Heat Exchanger	1 Mark	MCQ Type												
		Numerical Type												
	2 Marks	MCQ Type												
		Numerical Type												
	Total													
Heat Transfer Coefficient	1 Mark	MCQ Type												
		Numerical Type												
	2 Marks	MCQ Type						1					1	
		Numerical Type				1				1				
	Total					2		2		2			2	
LMTD Method	1 Mark	MCQ Type									1			
		Numerical Type					1							
	2 Marks	MCQ Type												
		Numerical Type		1										
	Total			2			1				1			
NTU Method	1 Mark	MCQ Type												1
		Numerical Type												
	2 Marks	MCQ Type												
		Numerical Type												
	Total													1

Types of heat exchanger

1. In shell and tube heat exchanger, baffles are mainly used to

(a) increase the mixing of fluid

(b) increase the heat transfer area

(c) deflect the flow in desired direction

(d) reduce fouling of the tube surface

[1991 : 1 Mark]

2. The practice to use steam on the shell side and water on the tube side in condensers of steam power plant is because

(a) to increase overall HT coefficient, water side velocity can be increased if water is on the tube side

(b) condenser can act as a Storage unit for condensed steam

(c) rate of condensation of steam is invariably smaller than the mass flow rate of cooling water

(d) it is easier to maintain vacuum on the shell side than on the tube side

[1994 : 1 Mark]

3. Hot oil is cooled from 80 to 50°C in an oil cooler which uses air as the coolant. The air temperature rises from 30 to 40°C. The designer uses a LMTD value of 26°C. The type of heat exchanger is

(a) parallel flow (b) double pipe

(c) counter flow (d) cross flow

[2005 : 2 Marks]

4. In a heat exchanger, it is observed that $\Delta T_1 = \Delta T_2$ where ΔT_1 is the temperature difference between the two single phase fluid streams at one end and ΔT_2 is the temperature difference at the other end. This heat exchanger is

(a) a condenser

(b) an evaporator

(c) a counter flow heat exchanger

(d) a parallel flow heat exchanger

[2014 : 1 Mark, Set-2]

Heat transfer coefficient

5. In certain heat exchanger, both the fluids have identical mass flow rate specific heat product. The hot fluid enters at 76°C and leaves at 47°C, and the cold fluid entering at 28°C leave at 55°C, the effectiveness of the HE is

(a) 0.16 (b) 0.60

(c) 0.72 (d) 1.0

[1997 : 2 Marks]

6. In a counter flow heat exchanger, for the hot fluid the heat capacity = 2 kJ/kgK, mass flow rate = 5 kg/s, inlet temperature = 150°C, outlet temperature = 100°C. For the cold fluid, heat capacity = 4 kJ/kg K, mass flow rate = 10 kg/s, inlet temperature = 20°C. Neglecting heat transfer to the surroundings, the outlet temperature of the cold fluid in °C is

(a) 7.5 (b) 32.5

(c) 45.5 (d) 70.0

[2003 : 2 Marks]

7. In a condenser, water enters at 30°C and flows at the rate 1500 kg/hr. The condensing steam is at a temperature of 120°C and cooling water leaves the condenser at 80°C. Specific heat of water is 4.187 kJ/kgK. If the overall heat transfer coefficient is 2000 W/m^{2}K, the heat transfer area is

(a) 0.707 m^2 (b) 7.07 m^2

(c) 70.7 m^2 (d) 141.4 m^2

[2004 : 2 Marks]

8. An industrial gas (c = 1 kJ/kgK) enters a parallel flow heat exchanger at 250°C with a flow rate of 2 kg/s to heat a water stream. The water stream (c = 4 kJ/kgK) enters the heat exchanger at 50°C with a flow rate of 1 kg/s. The heat exchanger has an effectiveness of 0.75. The gas stream exit temperature will be

(a) 75°C (b) 100°C

(c) 125°C (d) 150°C

[2010 : 2 Marks]

9. Cold water flowing at 0.1 kg/s is heated from 20°C to 70°C in a counter flow type heat exchanger by a hot water stream flowing at 0.1 kg/s and entering at 90°C. The specific heat of water is 4200 J/(kgK) and density is 1000 kg/m^3. If the overall heat transfer coefficient U for the heat exchanger is 2000 W/(m^{2}K), the required heat exchange area (in m^2) is

(a) 0.052 (b) 0.525

(c) 0.151 (d) 0.202

[2011 : 2 Marks]

10. In a concentric counter flow heat exchanger, water flows through the inner tube at 25°C and leaves at 42°C. The engine oil enters at 100°C and flows in the annular flow passage. The exit temperature of the engine oil is 50°C. Mass flow rate of water and the engine oil are 1.5 kg/s and 1 kg/s, respectively. The specific heat of water and oil are 4178 J/kgK and 2130 J/kgK, respectively. The effectiveness of this heat exchanger is

[2014 : 2 Marks, Set-2]

11. A balanced counter flow heat exchanger has a surface area of 20 m^2 and overall heat transfer coefficient of 20 W/m^2K. Air (c_p = 1000 J/kgK) entering at 0.4 kg/s and 280 K is to be preheated by the air leaving the system at 0.4 kg/s and 300 K. The temperature (in K) of the preheated air is

(a) 290 (b) 300
(c) 320 (d) 350

[2015 : 2 Marks, Set-2]

12. Consider a parallel-flow heat exchanger with area A_p and a counter-flow heat exchanger with area A_c. In both the heat exchangers, the hot stream flowing at 1 kg/s cools from 80°C to 50°C. For the cold stream in both the heat exchangers, the flow rate and the inlet temperature are 2 kg/s and 10°C, respectively. The hot and cold streams in both the heat exchangers are of the same fluid. Also, both the heat exchangers have the same overall heat transfer coefficient. The ratio A_c/A_p is _____.

[2016 : 2 Marks, Set-2]

13. In a counter-flow heat exchanger, water is heated at the rate of 1.5 kg/s from 40°C to 80°C by an oil entering at 120°C and leaving at 60°C. The specific heats of water and oil are 4.2 kJ/kgK and 2 kJ/kgK, respectively. The overall heat transfer coefficient is 400 W/m^2K. The required heat transfer surface area (in m^2) is

(a) 0.104 (b) 0.022
(c) 10.4 (d) 21.84

[2017 : 2 Marks, Set-2]

14. Steam in the condenser of a thermal power plant is to be condensed at a temperature of 30°C with cooling water which enters the tubes of the condenser at 14°C and exits at 22°C. The total surface area of the tubes is 50 m^2, and the overall heat transfer coefficient is 2000 W/m^2. The heat transfer (in MW) to the condenser is _____(correct to two decimal places).

[2018 : 2 Marks, Set-2]

LMTD method

15. Air enters a counter flow HE at 70°C and leaves at 40°C. Water enters at 30°C and leaves at 50°C, the LMTD in degree C is

(a) 5.65 (b) 14.43
(c) 19.52 (d) 20.17

[2000 : 2 Marks]

16. For the same inlet and exit temperatures of the hot and cold fluids, the log mean temperature difference (LMTD) is

(a) Greater for parallel flow heat exchanger than the counter flow heat exchanger
(b) Greater for counter flow heat exchanger than the parallel and cross flow heat exchanger
(c) Same for both parallel and counter flow heat exchangers
(d) Depending on the properties of fluid

[2002 : 1 Mark]

17. In a counterflow heat exchanger, hot fluid enters at 60°C and cold fluid leaves at 30°C. Mass flow rate of the hot fluid is 1 kg/s and that the cold fluid is 2 kg/s. Specific heat of the hot fluid is 10 kJ/kgK and that of the cold fluid is 5 kJ/kgK. The Log Mean Temperature Difference (LMTD) for the heat exchanger in °C is

(a) 15 (b) 30
(c) 35 (d) 45

[2007 : 2 Marks]

18. The logarithmic mean temperature difference (LMTD) of a counter flow heat exchanger is 20°C. The cold fluid enters at 20°C and the hot fluid enters at 100°C. Mass flow rate of the cold fluid is twice that of the hot fluid. Specific heat at constant pressure of the hot fluid is twice that of the cold fluid. The exit temperature of the cold fluid

(a) is 40°C (b) is 60°C
(c) is 80°C (d) cannot be determined

[2008 : 2 Marks]

19. In a condenser of a power plant, the steam condenses at a temperature of 60°C. The cooling water enters at 30°C and leaves at 45°C. The Logarithmic Mean Temperature Difference (LMTD) of the condenser is

(a) 16.2°C (b) 21.6°C
(c) 30°C (d) 37.5*C

[2011 : 1 Mark]

20. Water (c = 4.18 kJ/kgK) at 80°C enters a counter flow heat exchanger with a mass flow rate of 0.5 kg/s. Air (c = 1 kJ/kgK) enter at 30°C with a mass flow rate 2.09 kg/s. If the effectiveness of the heat exchanger is 0.8, the LMTD (in °C) is

(a) 40 (b) 20

(c) 10 (d) 5

[2012 : 2 Marks]

21. A double pipe counter flow heat exchanger transfers heat between two water streams. Tube side water at 19 litre/s is heated frim 10°C to 38°C. Shell side water at 25 litre/s is entering at 46°C Assume constant properties of water, density is 1000 kg/m³ and specific heat is 4186 J/kgK. The LMTD (in °C) is ____.

[2014 : 2 Marks, Set-3]

22. For a heat exchanger, ΔT_{max} is the maximum temperature difference and ΔT_{min} is the minimum temperature difference between the two fluids. LMTD is the log mean temperature difference. C_{min} and C_{max} are the minimum and the maximum heat capacity rates. The maximum possible heat transfer (Q_{max}) between the two fluids is

(a) C_{min}LMTD (b) $C_{min}\Delta T_{max}$

(c) $C_{max}\Delta T_{max}$ (d) $C_{max}\Delta T_{min}$

[2016 : 1 Mark, Set-3]

23. Saturated steam at 100°C condenses on the outside of a tube. Cold enters the tube at 20°C and exits at 50°C. The value of the Log Mean Temperature Difference (LMTD) is ____°C.

[2017 : 1 Mark, Set-1]

24. Hot and cold fluids enter a parallel flow double tube heat exchanger at 100°C and 15°C, respectively. The heat capacity rates of hot and cold fluids are C_h = 2000 W/K and C_c = 1200 W/K, respectively. If the outlet temperature of the cold fluid is 45°C, the log mean temperature difference (LMTD) of the heat exchanger is ____ K (round off to two decimal places).

[2019 : 2 Marks, Set-2]

NTU method

25. In a parallel flow heat exchanger operating under steady state, the heat capacity rates (product of specific heat at constant pressure and mass flow rate) of the hot and cold fluid are equal. The hot fluid, flowing at 1 kg/s with cp = 4 kJ/kgK, enters the heat exchanger at 102°C while the cold fluid has an inlet temperature of 15°C. The overall heat transfer coefficient for the heat exchanger is estimated to be 1 kW/m²K and the corresponding heat transfer surface area is 5 m². Neglect heat transfer between the heat exchanger and the ambient. The heat exchanger is characterized by the following relation:

2ε = 1 – expt(–2NTU). The exit temperature (in °C) for the cold fluid is

(a) 45 (b) 55

(c) 65 (d) 75

[2009 : 2 Marks]

26. Saturated vapor is condensed to saturated liquid in a condenser. The heat capacity ratio is $C_r = \frac{C_{min}}{C_{max}}$ The effectiveness (ε) of the condenser is

(a) $\frac{1-\exp[-NTU(1+C_r)]}{1+C_r}$

(b) $\frac{1-\exp[-NTU(1+C_r)]}{1+C_r[-NTU(1+C_r)]}$

(c) $\frac{NTU}{1+NTU}$

(d) 1-exp(–NTU)

[2015 : 1 Mark, Set-3]

ANSWERS

1. (c) **2.** (d) **3.** (*d*) **4.** (c) **5.** (*b*) **6.** (*b*) **7.** (*a*) **8.** (b) **9.** (b) **11.** (*a*)

13. (d) **15.** (*b*) **16.** (*b*) **17.** (*b*) **18.** (c) **19.** (*b*) **20.** (*c*) **22.** (*b*) **25.** (*b*) **26.** (*d*)

EXPLANATIONS

1. Baffles are used to deflect the flow.
2. Air leaked into the condensor can be easily extracted.

3. $\theta = \dfrac{\theta_1 - \theta_2}{\ln\dfrac{\theta_1}{\theta_2}}$

$\theta_1 = 50°,\ \ \theta_2 = 10°$
$\theta_1 = 40°,\ \ \theta_2 = 20°$

As given value is between these two, so cross flow is the best option.

5. $\in = \dfrac{C_h(T_{h1} - T_{h2})}{C_{min}(T_{h1} - T_{c1})}$

Here, $C_h = C_c$.

$$\in = \frac{T_{h1} - T_{h2}}{T_{h1} - T_{c1}} = \frac{76 - 47}{76 - 28} = 0.604$$

6. Heat lost by fluid = Heat gained by cold fluid

$\dot{m}_h\, C_{ph}\,(T_{hi} - T_{ho}) = \dot{m}_c\, C_{PC}\,(T_{co} - T_{ci})$

$5 \times 2000 \times 150 - 100) = 10 \times 4000\,(T_{co} - T_{ci})$

$\therefore$ Tco = 32.5°c

7. Here, $\theta_1 = 120 - 30 = 90°C$
$\theta_2 = 120 - 80 = 40°C$

Log mean temperature difference,

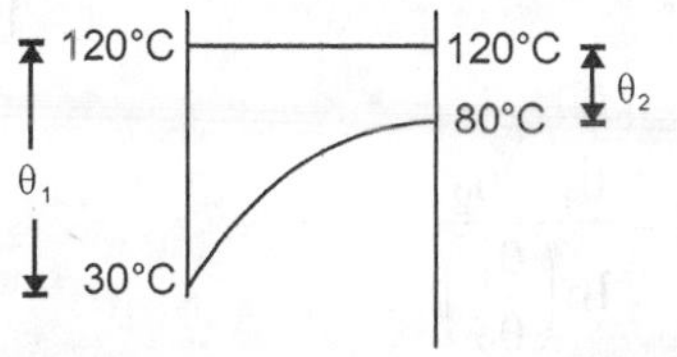

$$\theta_m = \frac{\theta_1 - \theta_2}{\ln\dfrac{\theta_1}{\theta_2}} = \frac{(90 - 40)}{\ln\dfrac{90}{40}} = 61.65$$

$Q = mc_{pw}\Delta T$

$$= 1500 \times \frac{1}{3600} \times 4.187 \times (50) \times 10^3$$

$= 87.22 \times 10^3$ Joule/s

Again

$Q = UA\theta_m$

or $87.22 \times 10^3 = 2000 \times A \times 61.65$

$\therefore$ $A = 0.707\ m^2$

8. $C_h = \dot{m}_c\, C_{ph} = 2 \times 1 \times 10^3 = 2$ kJ/Ks

$C_c = \dot{m}_c\, C_{pc} = 1 \times 4 \times 10^3 = 4$ kJ/Ks

$$\in = \frac{C_h(T_{h1} - T_{h2})}{C_{min}(T_{hi} - T_{ci})}$$

$$0.75 = \frac{2 \times (250 - T_{h2})}{2 \times (250 - 50)}$$

$Th_2 = 100°C$

9.

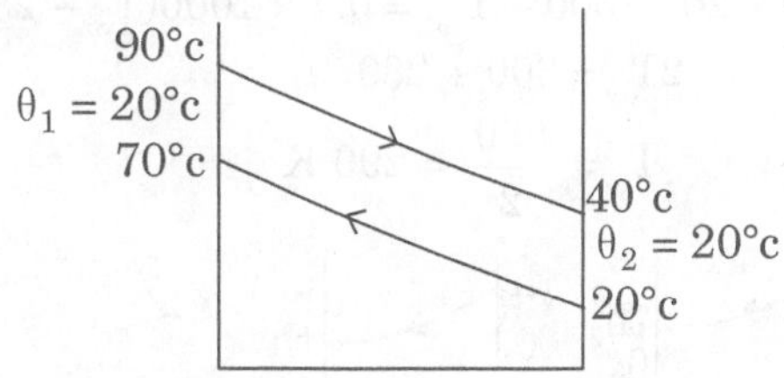

$C_h = C_c = 0.1 \times 4200 = 420$ J/s
$C_c\,(T_{c1} - T_{c2}) = C_h\,(Th_1 - Th_2)$
$(70 - 20) = 90°\ T_{h2}$
LMTD $= \theta_1 = \theta_2\ (\because \theta_1 = \theta_2)$
$C_c\,(T_{c2} - T_{c1}) = UA(LMTD)$
or $A = 0.525\ m^2$.

10. $T_{c,i} = 25°C$;
$T_{c,o} = 42°C$;
$m_c = 1.5$ kg/s;
$C_c = 4.178$ kJ/kg.k
$T_{h,i} = 100°C$;
$T_{h,o} = 50°C$;
$m_h = 1$ kg/s;
$C_h = 2.130$ kJ/kg.k
$m_c C_c = 1.5 \times 4.178$
$= 6.267\ kW/°C = C_{max}$
$m_h C_h = 1 \times 2.130$
$= 2.130\ kW/°C = C_{min}$

effectiveness $= \varepsilon = \dfrac{C_h\left(T_{h,i} - T_{h,o}\right)}{C_{min}\left(T_{h,i} - T_{c,i}\right)}$

$$= \frac{100 - 50}{100 - 25}$$

$= 0.666$

11. Counter flow heat exchanged

Surface Area $A = 20\ m^2$, mass flow rate = 0.4 kg/s

$u = \dfrac{20W}{m^2K}$; Temperature $T_{ci} = 280K$

C_p of air = $1000 \frac{1}{kgK}$

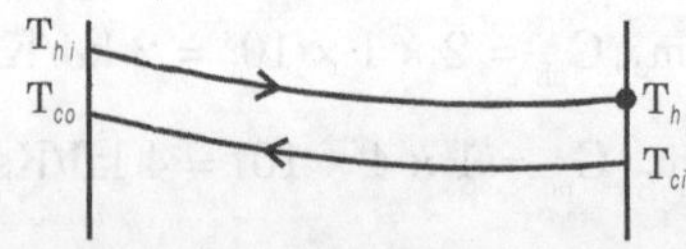

Since m is same for both flow = 0.4 kg/s

Assume C_p is same = 100 J/kg.K

Hence $\Delta T_1 = T_{ci} - T_{co} = \Delta T_2 = T_{h0} - T_{ci}$

$\Delta T_1 = 300 - T_{co} = T_{ho} - 280$

$\Delta T_m = \Delta T_1 = \Delta T_2$

$uA\Delta T_m = mC_p(T_{co} - T_{ci})$

$20 \times 20 \times (300 - T_{co}) = 0.4 \times 1000(T_{co} - 280)$

$2T_{co} = 300 + 280$

$T_{co} = \frac{580}{2} = 290 \text{ K}$

12.

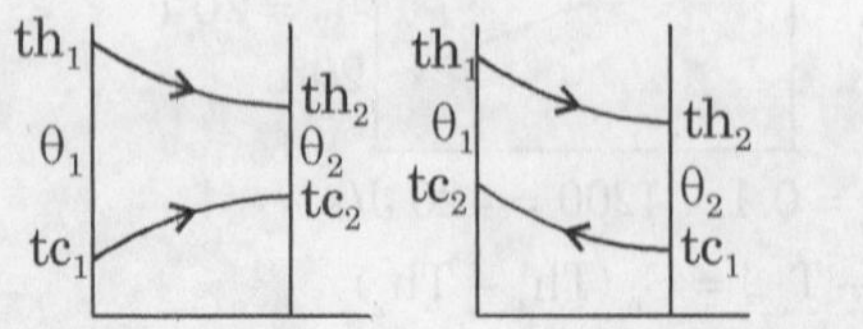

Parallel flow

$\theta_1 = th_1 - tc_1$

$= 80 - 10$

$= 70°C$

$\theta_2 = th_2 - tc_2$

$= 50° - 25°$

$= 25°$

$\theta_m = \frac{\theta_1 - \theta_2}{\ln\left(\frac{\theta_1}{\theta_2}\right)}$

$\theta_m = \frac{70 - 25}{\ln\left(\frac{70}{23}\right)}$

$\theta_m = 43.705°C$

Counter flow

$\theta_1 = th_1 - tc_2$

$= 80 - 25$

$= 55°C$

$\theta_2 = th_2 - tc_1$

$= 50 - 10$

$= 40°$

$\theta_m = \frac{\theta_1 - \theta_2}{\ln\left(\frac{\theta_1}{\theta_2}\right)}$

$\theta_m = \frac{50 - 40}{\ln\left(\frac{55}{40}\right)}$

$\theta_m = 47.1°C$

$\theta = UA_P\theta m_P = UA_c\,\theta m_c$

$\frac{A_C}{A_P} = \frac{\theta m_P}{\theta m_c} = \frac{43.705}{47.1} = 0.927$

13. Heat exchanger is a counterflow type.

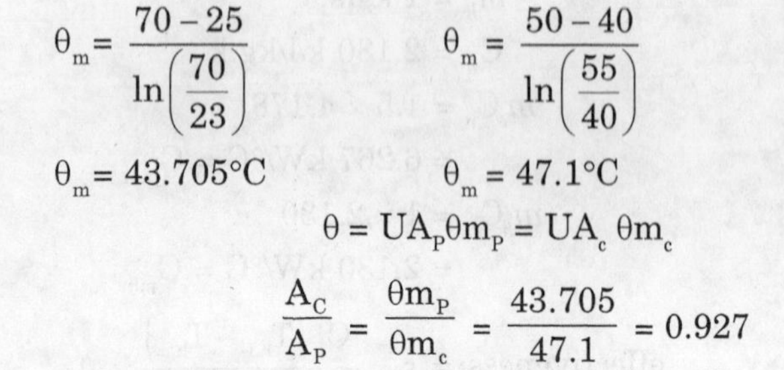

$\Delta T_i = 120 - 80 = 40°C\ (\theta_1)$

$\Delta T_e = 60 - 40 = 20°C\ (\theta_2)$

$\text{LMTD} = \frac{\theta_1 - \theta_2}{\ln\left(\frac{\theta_1}{\theta_2}\right)}$

$= \frac{40 - 20}{\ln\left(\frac{40}{20}\right)} = 28.86°C$

$Q = \dot{m}_c\, C_{pc}\left(T_{ce} - T_{ci}\right)$

$= 1.5 \times 4.2 \times 10^3\ (80 - 40)$ watt

$Q = 252000$ W

$Q = UA\Delta T_m$.

$A = \frac{Q}{U\Delta T_m} = \frac{1.5 \times 4.2 \times 10^3 \times 40}{400 \times 28.86}$

$= 21.83\ m^2$.

14. Condenser is given.

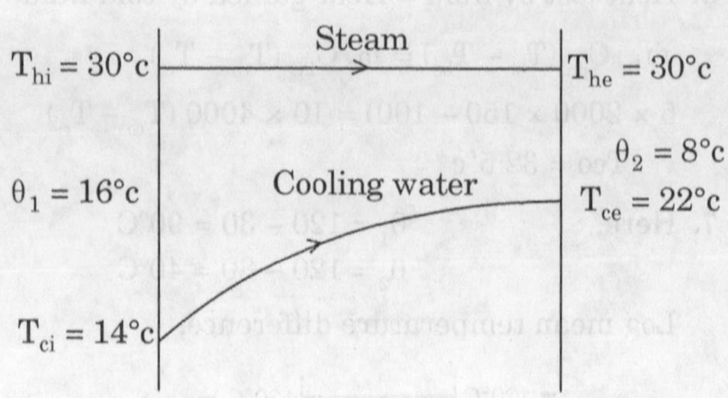

$\text{LMTD} = \frac{\theta_1 - \theta_2}{\ln\left(\frac{\theta_1}{\theta_2}\right)}$

$= \frac{16 - 8}{\ln\left(\frac{16}{8}\right)} = 11.54°c$

$Q = UA\ (\text{LMTD})$

$= 2000 \times 50 \times 11.54$

$= 1.154$ MW

15. $\therefore \Delta T_1 = 70 - 50 = 20°C, \quad \Delta T_2 = 40 - 30 = 10°C$

$\text{LMTD} = \frac{\Delta T_1\ \Delta T_2}{\log\left(\frac{\Delta T_1}{\Delta T_2}\right)}$

$= \frac{20 - 10}{\log\left(\frac{20}{10}\right)} = \frac{10}{\log 2} = \mathbf{14.43}$

17. Heat capacity of hot fluid

$= 1 \times 10 = 10$ kJ / k – s

Heat capacity of cold fluid

$= 2 \times 5 = 10$ kJ / k – s

Since heat capacity is same, so LMTD is difference of temperature at either end

i.e. LMTD = 60° – 30° = 30° C

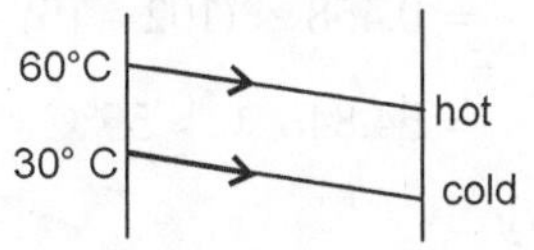

18. $\theta_1 = \theta_2$

$\therefore \theta_m = 20°C$

$T_{h1} - T_{c2} = 20°C$

$100 - T_{c2} = 20°C$

$T_{c2} = 80°C$

19. Flow configuration in condenser as shown below.

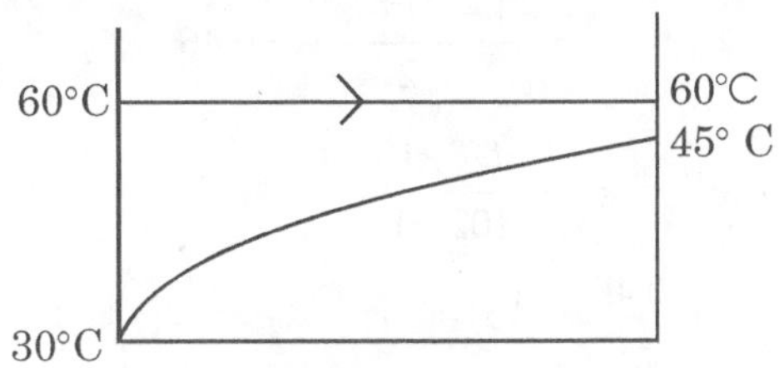

$\Delta T_1 = 30°C\ (Q_1)$

$\Delta T_2 = 15°C\ (Q_2)$

$$\text{LMTD} = \frac{\Delta T_1 - \Delta T_2}{\ln\left(\frac{\Delta T_1}{\Delta T_2}\right)} = \frac{30-15}{\ln\left(\frac{30}{15}\right)}$$

$= 21.6°$ C

20. $$\varepsilon = \frac{C_h}{C_{min}} \times \frac{(t_{h_1} - t_{h_2})}{(t_{h_1} - t_{c_1})}$$

$$\therefore \quad 0.8 = \frac{4.18 \times 0.5}{2.09 \times 1} \times \frac{(80 - t_{h_2})}{(80-30)}$$

$\Rightarrow \quad t_{h_2} = 40°C$

$m_h c_{p_h}(t_{h_1} - t_{h_2}) = m_c c_{p_c}(t_{c_2} - t_{c_1})$

$\therefore\ 0.5 \times 4.18 \times 40 = 2.09 \times 1 \times (t_{c_2} - 30)$

$\Rightarrow \quad t_{c_2} = 70°C$

$\theta_1 = \theta_2 = 10°C$

$\therefore \quad$ LMTD = 10°C

21. Given: $\dot{m}_h = 25$ L/S; $T_{h,i} = 46°C$, $T_{h,o} = ?$

$\dot{m}_c = 19$ L/S; $T_{c,i} = 10°C$, $T_{c,o} = 38°C$

$\rho = 1000$ kg/m³, C = 4186 J/kg.k

Energy balance,

$\dot{m}_c = C(T_{c,o} - T_{c,i}) = m_n C(T_{h,i} - T_{h,o})$

$19\,(38 - 10) = 25\,(46 - T_{h,o})$

$T_{h,o} = 24.72$ °C

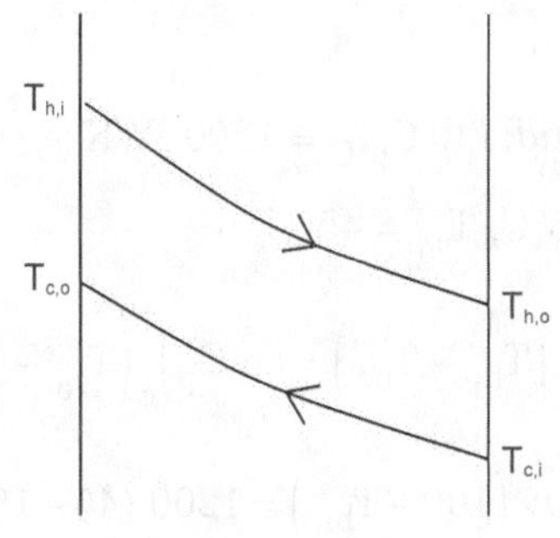

$$\text{LMTD} = \frac{\theta_1 - \theta_2}{\ln \theta_1 / \theta_2}$$

$\theta_1 = T_{h,i} - T_{c,o} = 46 - 38 = 8°C$

$\theta_2 = T_{h,o} - T_{c,i} = 24.72 - 10 = 14.72$

$$\text{LMTD} = \frac{8 - 14.72}{\ln(8/14.72)} = 11.0206°C.$$

22. The temperature difference is not for a given fluid but across the fluids and maximum heat transfer occurs for C_{min} and the temperature difference is equal to $(T_{h_i} - T_{c_i})$.

23.

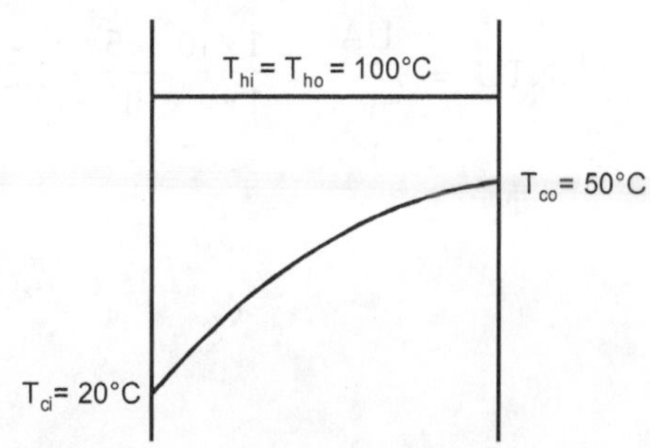

Here

$T_{ci} \rightarrow$ cold fluid which enters the tube (temperature)

$T_{co} \rightarrow$ cold fluid which exits at tube (temperature)

$\theta_1 = T_{hi} - T_{ci} = 100 - 20 = 80°C$

$\theta_2 = T_{ho} - T_{co} = 100 - 50 = 50°C$

$$\theta_{im} = \frac{\theta_1 - \theta_2}{ln\frac{\theta_1}{\theta_2}} = \frac{80° - 50°}{ln\left(\frac{80}{50}\right)} = 63.83°C$$

24. Hot fluid: $(\dot{m}C_p)_h = 2000$ W/K

$T_{hi} = 100°C$

Cold fluid: $(\dot{m}C_p)_C = 1200$ W/K

$T_{c_i} = 15°C, T_{c_o} = 45°C$

$$(\dot{m}C_p)_h (T_{h_i} - T_{h_o}) = (\dot{m}C_p)_c (T_{c_o} - T_{c_i})$$

$$\Rightarrow 2000 \times (100 - T_{h_o}) = 1200\ (45 - 15)$$

$$\Rightarrow T_{h_o} = 82°C$$

$$(\Delta T_i) = T_{h_i} - T_{c_i} = 100 - 15 = 85°C$$

$$(\Delta T_o) = T_{h_o} - T_{c_o} = 82 - 45 = 37°C$$

$$(LMTD)_{Parallel} = \frac{85 - 37}{\ell n\left(\frac{85}{37}\right)} = 57.71°C \text{ or } K$$

25. $t_{h_1} = 102°\ C$

$t_{c_2} = 15°C$

$$NTU = \frac{UA}{C_{min}} = \frac{1 \times 10^3 \times 5}{1 \times 4 \times 10^3} = 1.25$$

$$\in = \frac{1 - \exp(-2NTU)}{2}$$

$$= \frac{1 - \exp(-2.5)}{2} = 0.458$$

$$\therefore \quad 0.458 = \frac{t_{c_1} - t_{c_2}}{t_{h1} - t_{c_2}} = \frac{t_{c_1} - 15}{102 - 15}$$

or $t_{c_1} = 0.458 \times (102 - 15) + 15$

$= 54.846°\ C \approx 55°C$

Alternately

$$C = M_c C_c = M_h C_h = 1 \times 4\ kJ/kg\ K$$

$$\therefore \quad C = \frac{M_c C_c}{M_h C_h} = 1$$

$$NTU = \frac{U.A}{C_{min}} = \frac{1 \times 5}{4} = \frac{5}{4}$$

From the relation

$$\in = \frac{1 - e^{-(2 \times 5/4)}}{2} = 0.46$$

Now, $\in = \dfrac{t_{c_2} - 15}{102 - 15}$

$\therefore \quad 0.46 = t_{c_2}$

or $t_{c_2} = 54.9$

26. ε of condenser is given by 1-exp(-NTU)

Because $C_r = \dfrac{C_{min}}{C_{max}} = 0.$ (as $C_{max} \to \infty$)

6 CHAPTER Fins and Unsteady Heat Transfer (Conduction and Convection)

Analysis of Previous GATE Papers

Year/Set →	Fins: 1 Mark MCQ Type	Fins: 1 Mark Numerical Type	Fins: 2 Marks MCQ Type	Fins: 2 Marks Numerical Type	Fins: Total	Unsteady heat transfer: 1 Mark MCQ Type	Unsteady heat transfer: 1 Mark Numerical Type	Unsteady heat transfer: 2 Marks MCQ Type	Unsteady heat transfer: 2 Marks Numerical Type	Unsteady heat transfer: Total
2019 Set 1										
2019 Set 2										
2018 Set 1										
2018 Set 2										
2017 Set 1										
2017 Set 2		1			1				1	2
2016 Set 1									1	2
2016 Set 2									1	2
2016 Set 3									1	2
2015 Set 1										
2015 Set 2										
2015 Set 3										

Fins

1. Two rods, one of length L and the other of length 2L are made of the same material and have the same diameter. The two ends of the longer rod are maintained at 100°C. One end of the shorter rod is maintained at 100°C while the other end is insulated. Both the rods are exposed to the same environment at 40°C. The temp at the insulated end of the shorter rod is measured to be 55°C. The temperature at the mid point of the longer rod would be.

(a) 40°C (b) 50°C
(c) 55°C (d) 100°C

[1992 : 2 Marks]

2. A fin has 5 mm diameter and 100 mm length. The thermal conductivity of fin material is 400 $Wm^{-1}K^{-1}$. One end of the fin is maintained at 130°C and its remaining surface is exposed to ambient air at 30°C. If the convective heat transfer coefficient is 40 $Wm^{-2}K^{-1}$, the heat loss (in W) from the fin is

(a) 0.08 (b) 5.0
(c) 7.0 (d) 7.8 **[2010 : 2 Marks]**

3. Which one of the following configurations has the highest fin effectiveness?

(a) Thin, closely spaced fins
(b) Thin, widely spaced fins
(c) Thick, widely spaced fins
(d) Thick, closely spaced fins **[2012 : 1 Mark]**

4. The heat loss from a fin is 6 W. The effectiveness and efficiency of the fin are 3 and 0.75 respectively. The heat loss (in W) from the fin, keeping the entire fin surface at base temperature is _____. **[2017 : 1 Mark, Set-2]**

Unsteady heat transfer

5. The heat transfer process between body and its ambient is governed by an Intemal Conductive Resistance (ICR) and an External Convective Resistance (ECR). The body can be considered to be a lumped heat capacity system is

(a) ICR > ECR
(b) ICR is marginally smaller than ECR
(c) ICR = ECR
(d) ICR is negligible

[1989 : 1 Mark]

6. Biot number signifies

(a) The ratio of heat conducted to heat convected
(b) The ratio of heat convected to heat conducted
(c) The ratio of external convective resistance to internal conductive resistance
(d) The ratio of internal conductive resistance to external convective resistance

[1991 : 1 Mark]

7. When the fluid velocity is doubled the thermal time constant of a thermometer used of measuring the fluid temperature reduces by a factor of 2 (T/F)

[1994 : 1 Mark]

8. Lumped heat transfer analysis of a solid object suddenly exposed to a fluid medium at a different temp is valid when

(a) Biot number < 0.1
(b) Biot number > 0.1
(c) Fourier number < 0.1
(d) Fourier number > 0.1 **[2001 : 1 Mark]**

9. The value of Biot number is very small (less than 0.01),when

(a) The convective resistance of fluid is negligible
(b) The conductive resistance of fluid is negligible
(c) The conductive resistance of solid is negligible
(d) None of the above

[2002 : 1 Mark]

10. A spherical thermocouple junction of diameter 0.706 mm is to be used for the measurement of temperature of a gas stream. The convective heat transfer coefficient on bead surface is 400 W/m^2K, Thermophysical properties of thermocouple material are k = 20 W/mK, c = 400 J/kgK and ρ = 8500 kg/m^3. If the thermocouple initially at 30°C is placed in a hot stream of 300°C, the time taken by the bead to reach 298°C, is

(a) 2.35 s (b) 4.9 s
(c) 14.7 s (d) 29.4 s

[2004 : 2 Marks]

11. A small copper ball of 5 mm diameter at 500 K is dropped into an oil bath whose temperature is 300 K. The thermal conductivity of copper is 400 W/mK, its density 9000 kg/m^3 and its specific heat 385 J/kgK. If the heat transfer coefficient is 250 W/m^2K and lumped analysis is assumed to be valid, the rate of fall of the temperature of the ball at the beginning of cooling will be, in K/s,

(a) 8.7 (b) 13.9
(c) 17.3 (d) 27.7

[2005 : 2 Marks]

12. The average heat transfer coefficient on a thin hot vertical -plate suspended in still air can be determined from observations of the change in plate temperature with time as it cools. Assume the plate temperature to be uniform at any instant of time and radiation heat exchange with the surroundings negligible. The ambient temperature is 25°C, the plate has a total surface area of 0.1 m^2 and a mass of 4 kg. The specific heat of the plate material is 2.5 kJ/kgK. The convective heat transfer coefficient in W/m^2K, at the instant when the plate temperature is 225°C and the change in plate temperature with time dT/dt = – 0.02 K/s, is

(a) 200 (b) 20
(c) 15 (d) 10

[2007 : 2 Marks]

13. Consider steady-state heat conduction across the thickness in a plane composite wall (as shown in the figure) exposed to convection conditions on both sides.

Given: h_i = 20 W/m^2K; h_0 = 50 W/m^2K; $T_{\infty i}$ = 20°C; $T_{\infty,0}$ = –2°C; k_1 = 20 W/mK; k_2 = 50 W/mK; L_1 = 0.30 m and L_2 = 0.15 m. Assuming negligible contact resistance between the wall surfaces, the interface termperature, T(in °C), of the two walls will be

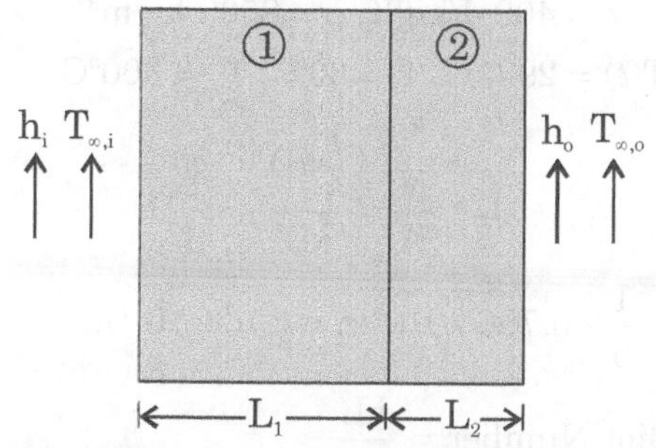

(a) – 0.50 (b) 2.75
(c) 3.75 (d) 4.50

[2009 : 2 Marks]

14. A spherical steel ball of 12 mm diameter is initially at 1000 K. It is slowly cooled in a surrounding of 300 K. The heat transfer coefficient between the steel ball and the surroundings is 5 W/m^2K. The thermal conductivity of steel is 20 W/mK. The ternperature difference between the centre and the surface of the steel ball is

(a) large because conduction resistance is far higher than the convective resistance.
(b) large because conduction resistance is far less than the convective resistance.
(c) small because conduction resistance is far higher than the convective resistance.
(d) small because conduction resistance is far less than the convective resistance.

[2011 : 2 Marks]

15. A steel ball of diameter 60 mm is initially in thermal equilibrium at 1030°C in a furnace. It is suddenly removed from the furnace and cooled in ambient air at 30°C, with convective heat transfer coefficient h = 20 W/m^2K. The thermo-physical properties of steel are : density ρ = 7800 kg/m^2, conductivity k =40 W/mK and specific heat c - 600 J/kgK. The time required in seconds to cool the steel ball in air from 1030°C to 430°C is

(a) 519 (b) 931
(c) 1195 (d) 2144 **[2013 : 2 Marks]**

16. Biot number signifies the ratio of

(a) convective resistance in the fluid to conductive resistance in the solid
(b) conductive resistance in the solid to convective resistance in the fluid
(c) inertia force to viscous force in the fluid
(d) buoyancy force to viscous force in the fluid

[2014 : 1 Mark, Set-1]

17. A steel ball of 10 mm diameter at 1000 K is required to be cooled to 350 K by immersing it in a water environment at 300 K. The convective heat transfer coefficient is 1000 W/m^2K. Thermal conductivity of steel is 40 W/m K. The time constant for the cooling process is 16 s. The time required (in s) to reach the final temperature is

[2016 : 2 Marks, Set-1]

18. Two cylindrical shafts A and B at the same initial temperature are simultaneously placed in a furnace. The surfaces of the shafts remain at the furnace gas temperature at all times after they are introduced into the furnace. The termperature variation in the axial direction of the shafts can be assumed to be negligible. The data related to shafts A and B is given in the following Table.

Quantity	Shaft A	Shaft B
Diameter (m)	0.4	0.1
Thermal Conductivity (W/mK)	40	20
Volumetric heat capacity (J/m^3K)	2×10^6	2×10^7

The temperature at the centerline of the shaft A reaches 400°C after two hours. The time required (in hours) for the centerline of the shaft B to attain the temperature of 400°C is ____.

[2016 : 2 Marks, Set-2]

19. A cylindrical steel rod, 0.01 m in diameter and 0.2 m in length is first heated to 750°C and then immersed in a water bath at 100°C. The heat transfer coefficient is 250 W/m²K. The density, specific heat and thermal conductivity of steel are ρ = 7801 kg/m³, c = 473 J/kgK, and k= 43 W/mK, respectively. The time required for the rod to reach 300°C is ____ seconds.

[2016 : 2 Marks, Set-3]

20. A metal ball of diameter 60 mm is initially at 220°C. The ball is suddenly cooled by an air jet of 20°C. The heat transfer coefficient is 200 W/m²K. The specific heat, thermal conductivity and density of the metal ball are 400 J/kgK, 400 W/mK and 9000 kg/m³, respectively. The ball temperature (in °C) after 90 seconds will be approximately

(a) 141 (b) 163
(c) 189 (d) 210

[2017 : 2-Marks, Set-2]

ANSWERS

1. (c) **2.** (*b*) **3.** (*a*) **5.** (d) 6. (d) **8.** (a) **9.** (*c*) **10.** (*b*) **11.** (*c*) **12.** (*a*)
13. (*c*) **14.** (*d*) **15.** (*d*) **16.** (*b*) **20.** (a)

EXPLANATIONS

2. Fin diameter, d = 5mm

Fin length, L = 100 mm

$$k = 400 \text{ W/mK}$$
$$T_o = 130°C$$
$$T_\infty = 30°C$$
$$h = 40 \text{ W/m}^2\text{K}$$

$\therefore$ Heat loss from fin

$$Q = \sqrt{kAhP}\,(T_o - T_\infty)\ \tanh ml$$

where $P = \pi d = 3.14 \times 5 \times 10^{-3} = 0.015$

$$A = \frac{\pi}{4}d^2 = \frac{\pi}{4}\times(5\times10^{-3})^2$$
$$= 1.96\times10^{-5}\text{ m}^2$$

$$\therefore \quad m = \sqrt{\frac{hp}{kA}} = \sqrt{\frac{40\times0.0157}{400\times1.96\times10^{-5}}} = 8.949$$

$$\therefore \quad ml = 8.949 \times 0.100 = 0.895$$

$$\therefore \quad Q = \sqrt{40\times0.0157\times400\times1.96\times10^{-5}}\ \tanh(0.895)$$
$$= 5.0 \text{ W}$$

4. $$\eta_{fin} = \frac{q_{actual}}{q_{maximum\ possible}}$$

$$q_{max} = \frac{6}{0.75} = 8\text{W}$$

5. Lumped heat capacity can be applied when Bi < 0.1

$\text{Bi} = \dfrac{\text{hl}}{\text{k}}$, for desired condition i.e. Bi < 0

k > h, ICR is small

6. Biot number $= \dfrac{R_{conduction}}{R_{convective}}$

7. It is not necessary that h would become a half if velocity becomes double.

8. For lumed system analysis,

Bi < 0.1

9. When conductive resistance of solid is negligible, then biot number is very small.

10. Given

$$h = 400 \text{ W/m}^2\text{K},\ k = 20 \text{ W/mK}$$
$$c = 400 \text{ J/kg.K},\ \rho = 8500 \text{ kg/m}^3$$
$$T(t) = 298°\text{ C},\ T_i = 30°C,\ T_\infty = 300°C$$

$$\text{Now,} \quad L = \frac{V}{A} = \frac{\frac{1}{6}\pi D^3}{\pi D^2} = \frac{1}{6}D$$

$$= \frac{1}{6}\times0.706\times10^{-3}\text{ m} = 1.76\times10^{-4}\text{ m}$$

$$\text{Biot Number} = \frac{hL}{k}$$

$$= \frac{400\times1.176\times10^{-4}}{20} = 0.0023$$

Since B_i < 0.1, lumped system analysis can be used

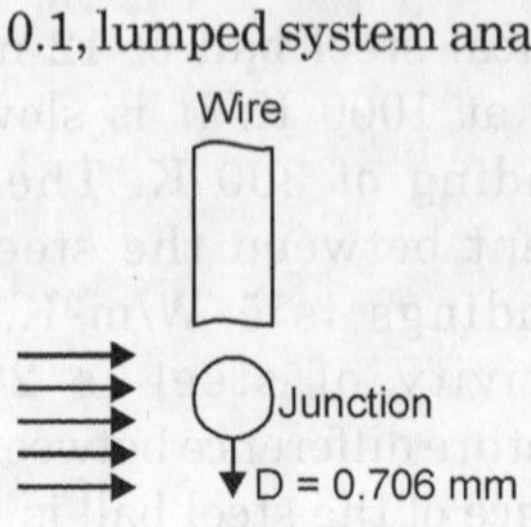

$$\frac{T(t)}{T_i - T_\infty} = e^{\frac{hAT}{\rho cv}}$$

$$\frac{298-300}{30-300} = e^{\frac{hAT}{\rho cv}}$$

$$\frac{-2}{270} = e^{-\frac{hAT}{\rho cv}}$$

$$\ln\left(\frac{270}{2}\right) = \frac{hAT}{\rho cV}$$

$$\Rightarrow T = \frac{\rho cV}{cA}\ln\left(\frac{270}{2}\right) = \frac{\rho c}{h}\left(\frac{V}{A}\right)\ln 135$$

$$= \frac{8500 \times 400 \times 1.176 \times 10^{-4} \times 4.90}{400}$$

$\approx$ 4.9 sec.

11. Given data:

$K = 400$ W/mK; $T_i = 500$ K; $\rho = 9000$ Kg/m³.

$T_\infty = 300$ K; $C_p = 385$ J/KG K; $h = 250\frac{W}{m^2K}$

$$L_c = \frac{V}{A} = \frac{D}{6} = \frac{(.005)}{6} = 8.33 \times 10^{-4}$$

$$T = T_\infty + (T_i - T_\infty)e^{\frac{-hAt}{\rho cV}}$$

$$\frac{dT}{dt} = \frac{-hA}{\rho CV}(T_i - T_\infty)e^{\frac{-hAt}{\rho VC}}$$

$$\left.\frac{dT}{dt}\right|_{t=0} = \frac{hA}{\rho VC}(T_i - T_\infty)$$

$$= \frac{250 \times 200}{9000 \times 385 \times 8.33 \times 10^{-4}} = 17.31\,K/S$$

12. From heat balance equation,

$$-hA(T - T_0) = mc\frac{dT}{dt}$$

where, T_0 = Ambient temperature

$$\therefore \quad h = \frac{4 \times 2.5 \times 10^3 \times .2}{.1 \times 200} = 10\ W/m^2K$$

13. $$R_{eq} = \frac{1}{h_2A} + \frac{l_1}{k_1A} + \frac{l_2}{k_2A} + \frac{1}{h_2A}$$

$$= \frac{1}{A}\left(\frac{1}{20} + \frac{0.30}{20} + \frac{0.15}{50} + \frac{1}{50}\right) = \frac{0.088}{A}$$

Now $$Q_2 = \frac{(t_1 + t_2)}{R_{eq}} = \frac{t_1 - t}{\frac{1}{h_iA} + \frac{L_1}{k_1A}}$$

$$\therefore \quad \frac{22}{\frac{0.88}{A}} = \frac{t_1 - t}{\frac{0.065}{A}}$$

$$\Rightarrow \quad \frac{2.95 \times 0.065}{0.088} = 293 - t$$

$$\Rightarrow \quad t = 3.75\ °C$$

Alternately

Under steady state conditions

$$\frac{(20 - t_i)}{\frac{1}{h_iA} + \frac{l_1}{k_1A}} = \frac{t_1 - (-2)}{\frac{l_2}{k_2A} + \frac{1}{h_0A}}$$

$$\therefore \quad \frac{20 - t_i}{\frac{1}{20} + \frac{0.3}{20}} = \frac{t_i + 2}{\frac{0.15}{50} + \frac{1}{50}}$$

or $$\frac{20 - t_i}{\frac{1.3}{20}} = \frac{T_i + 2}{\frac{1.15}{50}}$$

or $$20 - t_i = \frac{1.3}{1.15} \times \frac{50}{20}\{t_i + 2\}$$

or $$20 - t_i = 2.826\,(t_i + 2)$$

or $$3.826\,t_i = 20 - 2 \times 2.826$$

or $$t_i = 3.75°C$$

14. $$Bi = \frac{hL}{K} = \frac{5 \times 0.002}{20} = 0.0005$$

For the given condition, Biot number tends to zero, i.e. conduction resistance is far less than convection resistance. Therefore temperature between centre and surface is very small.

15. Diameter of steel ball,

$d = 0.06$ m

Initial temperature,

$T_i = 1030°C$

Ambient temperature

$T_0 = 30°C$

Convective heat transfer coefficient

$$h = 20\frac{W}{m^2K}$$

$$\rho = 7800\frac{kg}{m^3}$$

$$k = 40\frac{W}{mK}$$

$$C = 600\frac{J}{kgK}$$

Final temperature of steel ball, T = 430°c

Time : t = ?

$$\frac{T_i - T_0}{T_i - T_0} = e^{\frac{-ht}{\rho Cl}} \qquad \text{...(i)}$$

$$l = \frac{V}{A} = \frac{\pi d^3}{6 \times \pi d^2} = \frac{d}{6} = \frac{0.6}{6} = .01m$$

From (i)

$$\frac{430 - 30}{1030 - 30} = e^{\frac{-20 \times t}{7800 \times 600 \times 0.1}}$$

t = 2144.16 seconds

16. Biot - number :

Biot number provides a way to compare the conduction resistance within a solid body to the convection resistance external to the body (offered by the surrounding fluid) for heat transfer:

$$Bi = \frac{hs}{k}; \quad s = \frac{\text{Volume of the body}}{\text{Surface area}}$$

Where 's' is a characteristic dimension of the solid

'h' is convective heat transfer coefficient

'k' is thermal conductivity of the body.

17. $\text{Biot Number} = \frac{hL_C}{K}$

For sphere $L_C = \frac{\text{Volume}}{\text{surface area}} = \frac{d}{6}$

$$\therefore \quad Bi = \frac{hd}{6K} = \frac{1000 \times 0.01}{6 \times 40}$$

$$= 0.0416 < 0.1$$

Hence lumped heat analysis is used

$$\frac{T - T_\infty}{T_i - T_\infty} = e^{\frac{-hA_S t}{\rho VC_p}} = e^{\frac{-t}{t^*}}$$

Thermal time constant,

$$t^* = \frac{\rho VC_p}{hA_S} = 16\,\text{sec}$$

$$\therefore \quad \frac{350 - 300}{1000 - 300} = e^{\frac{-t}{16}}$$

$$\Rightarrow \quad t = 42.2249\ \text{sec}$$

18. $\frac{T - T_\infty}{T_i - T_\infty} = e^{\frac{-hA_s t}{\rho VC_p}}$

$(\rho C_P)_A = 2 \times 10^6$ J/m³K,

$(\rho C_P)_B = 2 \times 10^7$ J/m³K,

$d_A = 0.4$ m, $d_B = 0.1$ m

$t_A = 2\text{hrs} = 2 \times 3600 = 7200$ sec

$h_A = h_B$

$$\left(\frac{hA_s}{\rho VC_p} \times t\right)_A = \left(\frac{hA_s}{\rho VC_p} \times t\right)_B$$

$$\Rightarrow \left(\frac{\pi \times 0.4 \times L \times 7200}{2 \times 10^6 \times \frac{\pi}{4} \times (0.4)^2 \times L}\right)_A$$

$$= \left(\frac{\pi \times 0.1 \times L \times t}{2 \times 10^7 \times \frac{\pi}{4} \times (0.1)^2 \times L}\right)_B$$

$$\Rightarrow t_B = 2.5\ \text{hrs}$$

19. d = 0.01 m

L = 0.2 m, R = 7801 kg/m³

$t_i = 750°C$, $t_a = 100°C$

C = 473 J/Kg K, K = 43 W/mK

h = 250 W/m²K

$$L_c = \frac{V}{A} = \frac{\frac{\pi}{4}d^2 l}{\pi dl} = \frac{d}{4}$$

$$\left(\frac{t - t_a}{t_i - t_a}\right) = \exp\left[\frac{-hA\tau}{\rho VC}\right] = \exp\left[\frac{-h\tau}{\rho CL_c}\right]$$

$$\frac{300 - 100}{750 - 100} = \exp\left[\frac{-250 \times 4 \times \tau}{7801 \times 473 \times 0.01}\right]$$

$$\Rightarrow \tau = 43.49\ \text{s}$$

20. $C_P = 400 \frac{J}{KgK}$

$K = 400 \frac{W}{mK}$

$\rho = 9000 \frac{kg}{m^3}$

Time (τ) = 90 sec

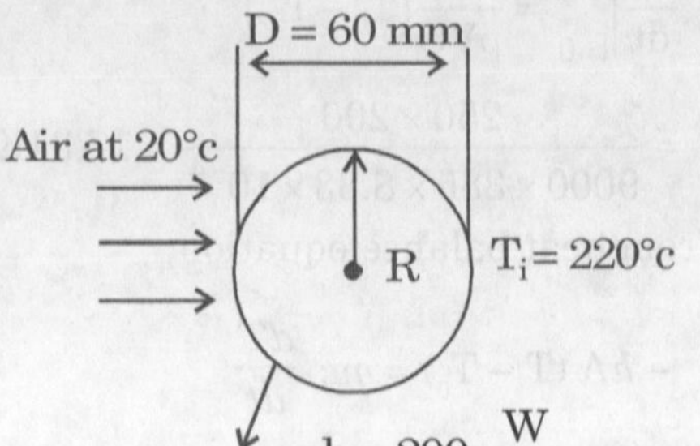

$h = 200 \frac{W}{m^2K}$

$$\left(\frac{hA}{\rho VC_P}\right)\tau$$

$$\frac{T_i - T_\infty}{T - T_\infty} = e$$

Put $\frac{A}{v} = \frac{3}{R} = 100$

$$\left(\frac{hA}{\rho VC_P}\right)\tau = \left(200 \times \frac{3}{0.03} \times \frac{1}{9000} \times \frac{1}{400}\right) \times 90$$

$= 0.5$

$$\frac{220 - 20}{T - 20} = e^{0.5}$$

By solving, we get T = 141.3°C

Unit - VIII

Thermodynamics, Power Plant Engineering and Internal Combustion Engines

Syllabus

Thermodynamic systems and processes; properties of pure substances, behaviour of ideal and real gases; zeroth and first laws of thermodynamics, calculation of work and heat in various processes; second law of thermodynamics; thermodynamic property charts and tables, availability and irreversibility; thermodynamic relations.

Applications: Power Engineering: Air and gas compressors; vapour and gas power cycles, concepts of regeneration and reheat. I.C. Engines: Air-standard Otto, Diesel and dual cycles.

Contents

1 CHAPTER

Thermodynamic System and Processes

Analysis of Previous GATE Papers														
Topic	Year/Set → Marks Distribution ↓		2019 Set 1	2019 Set 2	2018 Set 1	2018 Set 2	2017 Set 1	2017 Set 2	2016 Set 1	2016 Set 2	2016 Set 3	2015 Set 1	2015 Set 2	2015 Set 3
Thermodynamic System	1 Mark	MCQ Type												
		Numerical Type												
	2 Marks	MCQ Type												
		Numerical Type												
	Total													
Processes and Cycle	1 Mark	MCQ Type	1											
		Numerical Type												
	2 Marks	MCQ Type						1						
		Numerical Type												
	Total		**1**					**2**						
State and Properties	1 Mark	MCQ Type							1					
		Numerical Type												
	2 Marks	MCQ Type												
		Numerical Type												
	Total								**1**					

Thermodynamic system

1. An isolated thermodynamic system executes a process. Choose the correct statement (s) from the following
   (a) No heat is transferred
   (b) No work is done
   (c) No mass flows across the boundary of the system
   (d) No Chemical reaction takes place within the system

[1999 : 2 Marks]

Processes and cycle

2. The following four figures have been drawn to represent a fictitious thermodynamic cycle, on the p-v and T-s planes.

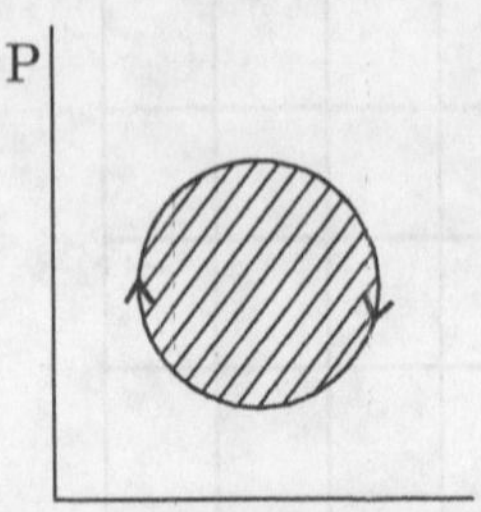

Fig. 1

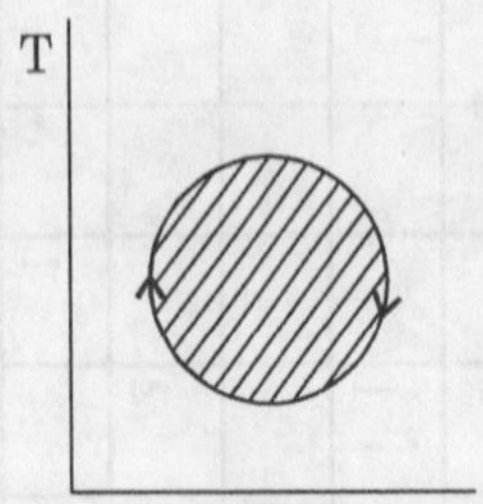

Fig. 2

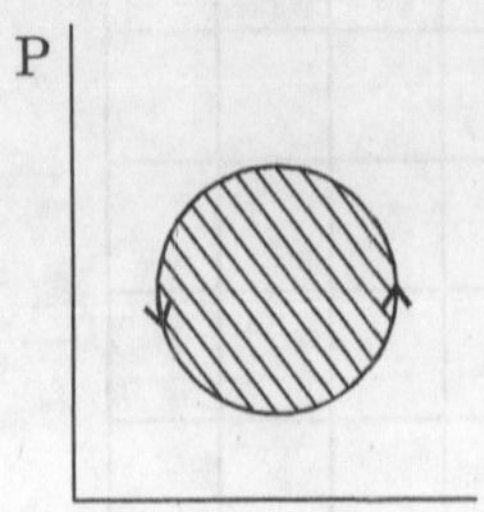

Fig. 3

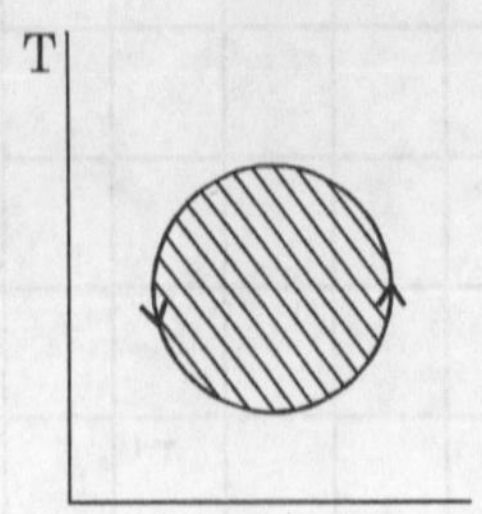

Fig. 4

According to the first law of thermodynamics, equal areas are enclosed by
(a) Figures 1 and 2 (b) Figures 1 and 3
(c) Figures 1 and 4 (d) Figures 2 and 3

[2005 : 1 Mark]

3. A reversible thermodynamic cycle containing only three processes and producing work is to be constructed. The constraints are :
   (i) there must be one isothermal process,
   (ii) there must be one isentropic process,
   (iii) the maximum and minimum cycle pressures and the clearance volume are fixed, and
   (iv) polytropic processes are not allowed. Then the number of possible cycles are
   (a) 1 (b) 2
   (c) 3 (d) 4

[2005 : 2 Marks]

Common Data for Questions 4 and 5

A thermodynamic cycle with an ideal gas as working fluid is shown below.

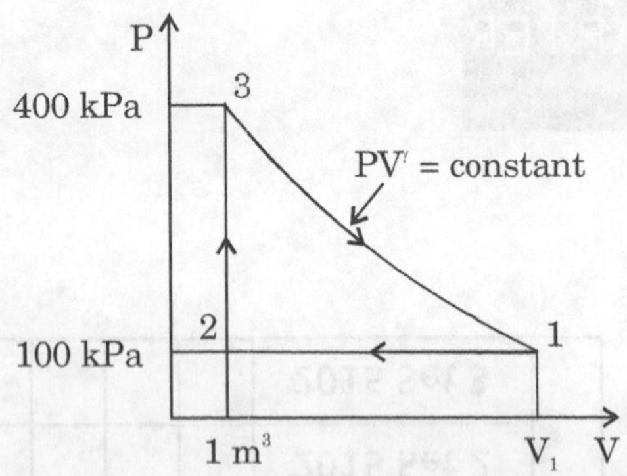

4. The above cycle is represented on T-s plane by

(a)

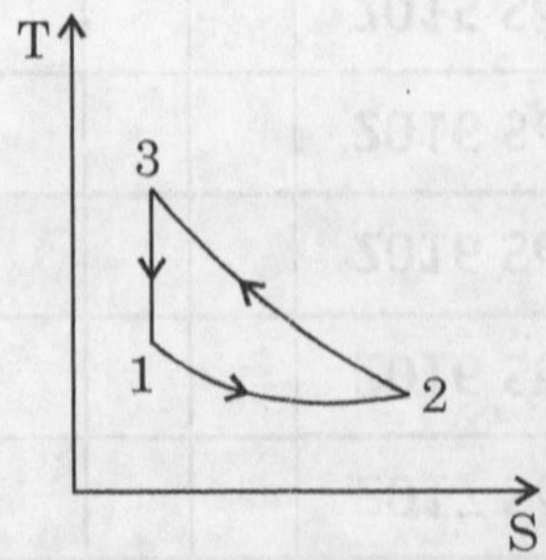

(b)

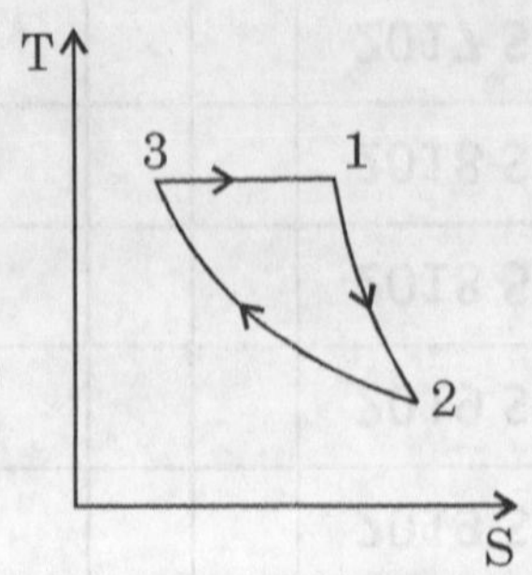

(c)

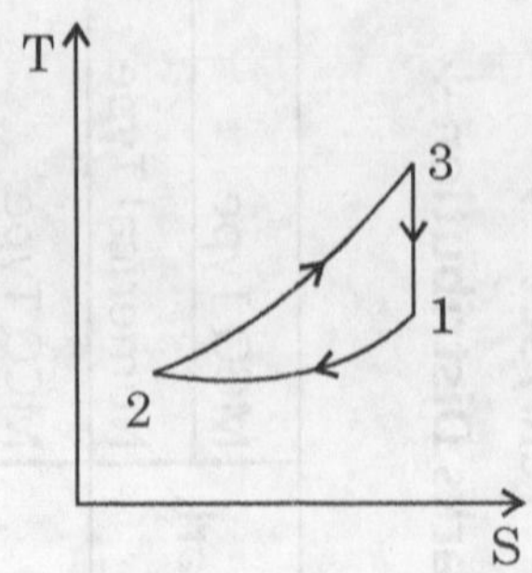

(d)

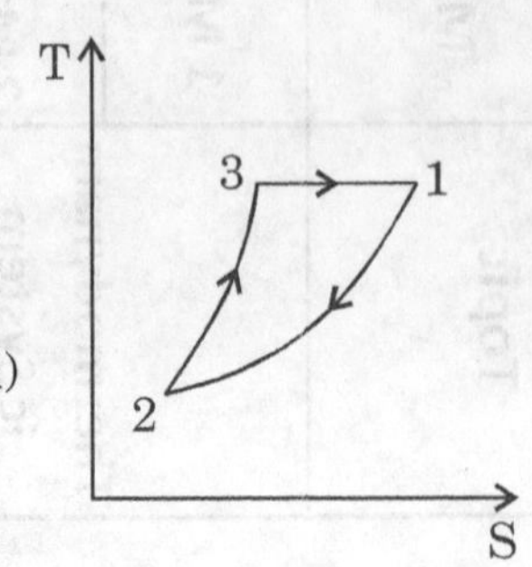

[2007 : 2 Marks]

5. If the specific heats of the working fluid are constant and the value of specific heat ratio γ is 1.4, the thermal efficiency (%) of the cycle is

(a) 21 (b) 40.9
(c) 42.6 (d) 59.7

[2007 : 2 Marks]

6. A certain amount of an ideal gas is initially at a pressure p_1 and temperature T_1 First, it undergoes a constant pressure process 1-2 such that $T_2 = 3T_1/4$. Then, it undergoes a constant volume process 2-3 such that $T_3 = T_1/2$. The ratio of the final volume to the initial volume of the ideal gas is

(a) 0.25 (b) 0.75
(c) 1.0 (d) 1.5

[2014 : 2 Marks, Set-3]

7. Two identical metal blocks L and M(specific heat = 0.4 kJ/kg K), each having a mass of 5 kg, are initially at 313 K. A reversible refrigerator extracts heat from block L and rejects heat to block M until the temperature of block L reaches 293 K. The final temperature (in K) of block M is _____ .

[2014 : 2 Marks, Set-4]

8. The volume and temperature of air (assumed to be an ideal gas) in a closed vessel is 2.87 m^3 and 300 K, respectively. The gauge pressure indicated by a manometer fitted to the wall of the vessel is 0.5 bar. If the gas constant of air is R = 287 J/kgK and the atmospheric pressure is 1 bar, the mass of air (in kg) in the vessel is

(a) 1.67 (b) 3.33
(c) 5.00 (d) 6.66

[2017 : 2 Marks, Set-2]

9. During a non-flow thermodynamic process (1–2) executed by a perfect gas, the heat interaction is equal to the work interaction ($Q_{1\text{-}2} = W_{12}$) when the process is

(a) Adiabatic (b) Polytropic
(c) Isothermal (d) Isentropic

[2019 : 1 Mark, Set-1]

State and properties

10. An insulated rigid vessel contains a mixture of fuel and air. The mixture is ignited by a minute spark. The contents of the vessel experience

(a) Increase in temperature, pressure and energy
(b) Decrease in temperature, pressure and energy
(c) Increase in temperature and pressure but no change in energy
(d) Increase in temperature and pressure but decrease in energy

[1993 : 1 Mark]

11. The definition of 1 K as per the internationally accepted temperature scale is

(a) 1/100th the difference between normal boiling point and normal freezing polnt of water
(b) 1/273.15th the normal freezing point of water
(c) 100 times the difference between the triple point of water and the normal freezing point of water
(d) 1 /273.16th of the triple point of water

[1994 : 1 Mark]

12. The specific heats of an ideal gas depend on its

(a) Temperature
(b) Pressure
(c) Volume
(d) Molecular weight and structure

[1996 : 1 Mark]

13. Match items from groups I, II, III, IV and V.

Group I	Group II When added to the system, is	Group III Differential	Group IV Function	Group V Phenomenon
E Heat	G Positive	I Exact	K Path	M Transient
F Work	H Negative	J Inexact	L Point	N Boundary

(a) F–G–J–K–M (b) F–G–I–K–M
E–G–I–K–N F–H–I–K–N
(c) F–H–J–L–N (d) E–G–J–K–N
E–H–I–L–M F–H–J–K–M

[2006 : 2 Marks]

14. Heat and work are

(a) intensive properties
(b) extensive properties
(c) point functions
(d) path functions

[2011 : 1 Mark]

15. Which of the following statements are TRUE with respect to heat and work?

(i) They are boundary phenomena
(ii) They are exact differentials
(iii) They are path functions

(a) both (i) and (ii) (b) both (i) and (iii)
(c) both (ii) and (iii) (d) only(iii)

[2016 : 1 Mark, Set-1]

ANSWERS

1. (a, b, c) **2.** (a) **3.** (a) **4.** (c) **5.** (a) **6.** (b) **8.** (c) **9.** (c) **10.** (c)
11. (d) **12.** (a) **13.** (d) **14.** (d) **15.** (b)

EXPLANATIONS

1. For an isolated system, Q = 0, W = 0 and mass does not flow through the system.

4. Since 3 – 1 is adiabatic process, it will be represented by a straight line on T – s plane. 1 – 2 is isobaric process, so with decreases of volume, temperature will also decrease.

6. For (1 – 2) process : Constant pressure process $\{P_2 = P_1\}$

$$\frac{T_1}{V_1} = \frac{T_2}{V_2} \Rightarrow V_2 = \frac{T_2}{T_1} \times V_1$$

For (2 – 3) process : Constant volume process $\{V_3 = V_2\}$

Given $T_2 = \frac{3T_1}{4}$

$$V_2 = \frac{3T_1}{4T_1} \times V_1 = \frac{3}{4} V_1$$

$$\Rightarrow \frac{V_3}{V_1} = \frac{V_2}{V_1} = \frac{3}{4} = 0.75$$

7. Given $m_1 = m_2 = 5\text{kg}$

$$c_1 = c_2 = 0.4\text{kJ / kg.K}$$
$$T_{L,i} = T_{M,i} = 313\text{K}$$
$$T_{M.f} = 293\text{K},\ T_{L.f} = ?$$
$$m_1 c_1 \Delta T_1 = m_2 c_2 \Delta T_2$$
$$\Delta T_1 = \Delta T_2$$
$$T_{L,i} - T_{L.f} = T_{M,i} - T_{M,f}$$
$$313 - T_{L.f} = 313 - 293$$
$$T_{L.f} = 333\text{k}$$

8. P = 1 + 0.5 = 1.5 bar

$$Pv = MRT;\quad M = \frac{PV}{RT}$$

$$M = \frac{1.5 \times 10^5 \times 2.87}{287 \times 300} = 5 \text{ kg}$$

9. For isothermal process

$$\delta U = 0$$

As per 1st law of thermodynamics

$$\delta Q = \delta U^{\,0} + \delta W$$

$$\delta Q = \delta W$$

10. Increase in temperature and pressure but no change in energy

Temperature increases

Volume = constant; p – decreases

$\Delta Q = 0$ (insulated)

12. Temperature C is the amount of heat required to increase the temperature of a substance by °C or 1K per kg.

13.

When added to the system	Differential	Function	Phenomenon
Heat Positive	Inexact	Path	Boundary
Work Negative	Inexact	Path	Transient

14. Heat and work are path functions.

Since Q and W are dependent on path followed between two given end states of a thermodynamic process undergone by system.

15. Since heat and work are energy in transit i.e. heat and work has meaning only when there is process. So both are boundary phenomena and are inexact i.e. path function.

2 CHAPTER

First Law of Thermodynamics, Heat, Work and Energy Interactions

Analysis of Previous GATE Papers

Topic	Year/Set → Marks Distribution ↓		2019 Set 1	2019 Set 2	2018 Set 1	2018 Set 2	2017 Set 1	2017 Set 2	2016 Set 1	2016 Set 2	2016 Set 3	2015 Set 1	2015 Set 2	2015 Set 3
Heat, Work and Energy interactions	1 Mark	MCQ Type								1		1		
		Numerical Type												
	2 Marks	MCQ Type												
		Numerical Type								1				1
	Total									3		1		2
First law of Thermodynamics	1 Mark	MCQ Type						1						
		Numerical Type					1							
	2 Marks	MCQ Type												
		Numerical Type				1			1			2		
	Total					2	1	1	2			4		
Steady flow system	1 Mark	MCQ Type												
		Numerical Type				1								
	2 Marks	MCQ Type	1					1						
		Numerical Type			1						1		1	1
	Total		2		2	1		2			2		2	2

Heat, Work and Energy interactions

1. A balloon containing an ideal gas is initially kept in an evacuated and insulated room. The balloon ruptures and the gas fills up the entire room. Which one of the following statements is TRUE at the end of above process?

(a) The internal energy of the gas decreases from its initial value, but the enthalpy remains constant.

(b) The internal energy of the gas increases from its initial value, but the enthalpy remains constant.

(c) Both internal energy and enthalpy of the gas remain constant

(d) Both internal energy and enthalpy of the gas increase

[2008 : 2 Marks]

2. A frictionless piston-cylinder device contains a gas initially at 0.8 MPa and 0.015 m^3. It expands quasistatically at constant temperature to a final volume of 0.030 m^3. The work output (in kJ) during this process will be

(a) 8.32 (b) 12.00

(c) 554.67 (d) 8320.00

[2009 : 1 Mark]

3. For an ideal gas with constant values of specific heats, for calculation of the specific enthalpy,

(a) it is sufficient to know only the temperature

(b) both temperature and pressure are required to be known

(c) both temperature and volume are required to be known

(d) both temperature and mass are required to be known

[2015 : 1 Mark, Set-1]

4. A mixture of ideal gases has the following composition by mass:

N_2	O_2	CO_2
60%	30%	10%

If the universal gas constant is 8314 J/kmol-K, the characteristic gas constant of the mixture (in J/kgK) is ______.

[2015 : 2 Marks, Set-3]

5. The internal energy of an ideal gas is a function of

(a) temperature and pressure

(b) volume and pressure

(c) entropy and pressure

(d) temperature only

[2016 : 1 Mark, Set-2]

6. A piston-cylinder device initially contains 0.4 m^3 of air (to be treated as an ideal gas) at 100 kPa and 80°C. The air is now isothermally compressed to 0.1 m^3. The work done during this process is ______kJ.

(Take the sign convention such that work done on the system is negative)

[2016 : 2 Marks, Set-2]

First law of Thermodynamics

7. The first law of thermodynamics takes the form $W = -\Delta H$ when applied to

(a) A closed system undergoing a reversible adiabatic process

(b) An open system undergoing an adiabatic process with negiigible changes in kinetic and potential energies

(c) A closed system undergoing a reversible constant volume process

(d) A closed system undergoing a reversible constant pressure process.

[1993 : 1 Mark]

8. A steel ball of mass 1 kg of specific heat 0.4 kJ/kgK is at a temperature of 60°C. It is dropped into 1 kg water at 20°C. The final steady state temperature of water is

(a) 23.5°C (b) 30°C

(c) 35°C (d) 40°C

[1993 : 1 Mark]

9. A vertical cylinder with a freely floating piston contains 0.1 kg air at 1.2 bar and a small electrical resistor. The resistor of is wired to an external 12 volt battery. When a current of 1.5 amps is passed through the resistor for 90 secs, the piston sweeps a volume of 0.01 m^3. Assume (i) piston and the cylinder are insulated and (ii) air behaves as an ideal gas with c_v = 700 J/kgK, Find the rise in temperature of air.

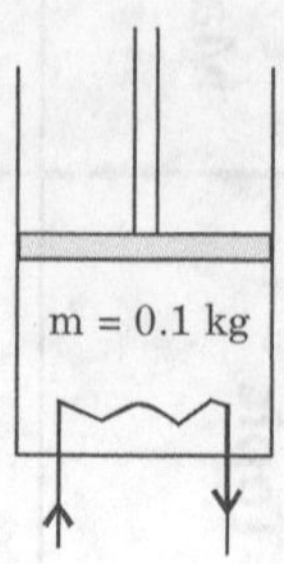

[1993 : 2 Marks]

10. A 2 kW, 40 litres water heater is switched on for 20 minutes. The heat capacity c_p for water is 4.2 kJ/kgK. Assuming all the electrical energy has gone into heating the water, increase of the water temperature in degree centigrade is

(a) 2.7 (b) 4.0

(c) 14.3 (d) 25.25 **[2003 : 1 Mark]**

Common Data Questions Q.11 and Q.12

Nitrogen gas (molecular weight 28) is enclosed in a cylinder by a piston, at the initial condition of 2 bar, 298 K and 1 m³. In a particular process, the gas slowly expands under isothermal condition, until the volume becomes 2 m³. Heat exchange occurs with the atmosphere at 298 K during this process.

11. The work interaction for the nitrogen gas is

(a) 200kJ (b) 138.6 kJ

(c) 2 kJ (d) –200 kJ

[2003 : 2 Marks]

Linked Answer Questions 12 and 13

A football was inflated to a gauge pressure of 1 bar when the ambient temperature was 15°C. When the game started next day, the air temperature at the stadium was 5°C. Assume that the volume of the football remains constant at 2500 cm³.

12. The entropy changes for the universe during the process in kJ/K is

(a) 0.4652 (b) 0.0067

(c) 0 (d) –0.6711

[2003 : 2 Marks]

13. A gas contained in a cylinder is compressed, the work required for compression being 5000 kJ. During the process, heat interaction of 2000 kJ causes the surroundings to be heated. The changes in internal energy of the gas during the process is

(a) –7000 kJ (b) –3000 kJ

(c) +3000 kJ (d) +7000 kJ

[2004 : 1 Mark]

14. The amount of heat lost by the air in the football and the gauge pressure of air in the football at the stadium respectively equal

(a) 30.6 J, 1.94 bar (b) 21.8 J, 0.93 bar

(c) 61.1 J, 1.94 bar (d) 43.7 J, 0.93 bar

[2006 : 2 Marks]

15. Gauge pressure of air to which the ball must have been originally inflated so that it would be equal 1 bar gauge at the stadium is

(a) 2.23 bar (b) 1.94 bar

(c) 1.07 bar (d) 1.00 bar

[2006 : 2 Marks]

16. Which of the following relationships is valid only for reversible processes undergone by a closed system of simple compressible substance (neglect changes in kinetic and potential energy)?

(a) $\delta Q = dU + \delta W$ (b) $TdS = dU + pdV$

(c) $TdS = dU + \delta W$ (d) $\delta Q = dU + pdV$

[2007 : 1 Mark]

17. A gas expands in a frictionless piston-cylinder arrangement. The expansion process is very slow, and is resisted by an ambient pressure of 100 kPa. During the expansion process, the pressure of the system (gas) remains constant at 300 kPa. The change in volume of the gas is 0.01 m³. The maximum amount of work that could be utilized from the above process is.

(a) 0 kJ (b) 1 kJ

(b) 2 kJ (d) 3 kJ

[2008 : 2 Marks]

18. The contents of a well-insulated tank are heated by a resistor of 23 Ω in which 10 A current is flowing. Consider the tank along with its contents as a thermodynamic system. The work done by the system and the heat transfer to the system are positive. The rates of heat (Q), work (W) and change in internal energy (ΔU) during the process (kW) are

(a) $Q = 0, W = -2.3, \Delta U = +2.3$

(b) $Q = +2.3, W = 0, \Delta U = +2.3$

(c) $Q = -2.3, W = 0, \Delta U = -2.3$

(d) $Q = 0, W = +2.3, \Delta U = -2.3$

[2011 : 1 Mark]

19. Temperature of nitrogen in a vessel of volume 2 m³ is 288 K. A U-tube manometer connected by the vessel shows a reading of 70 cm of mercury (level higher in the end open to atmosphere). The universal gas constant is 8314 J/kmol-K, atmospheric pressure is 1.01325 bar, acceleration due to gravity is 9.81 m/s² and density of mercury is 13600 kg/m³. The mass of nitrogen (in kg) in the vessel is _______.

[2015 : 2 Marks, Set-1]

20. A well insulated rigid container of volume 1 m³ contains 1.0 kg of an ideal gas [c_p = 1000 J/(kgK) and c_v = 800 J/(kgK)] at a pressure of 10^5 Pa. A stirrer is rotated at constant rpm in the container for 1000 rotations and the applied torque is 100 Nm. The final temperature of the gas (in K) is______.

[2015 : 2 Marks, Set-1]

21. An ideal gas undergoes a reversible process in which the pressure varies linearly with volume. The conditions at the start (subscript 1) and at the end (subscript 2) of the process with usual notation are : p_1 = 100 kPa, V_1 = 0.2 m^3 and p_2 = 200 kPa, V_2 = 0.1 m^3 and the gas constant, R = 0.275 kJ/kgK. The magnitude of the work required for the process (in kJ) is______.

[2016 : 2 Marks, Set-1]

22. The molar specific heat at constant volume of an ideal gas is equal to 2.5 times the universal gas constant (8.314 J/mol-K). When the temperature increases by 100 K, the change in molar specific enthalpy is ____ J/mol.

[2017 : 1 Mark, Set-1]

23. A mass m of a perfect gas at pressure p_1 and volume V_1 undergoes an isothermal process. The final pressure is p_2 and volume is V_2. The work done on the system is considered positive. If R is the gas constant and T is the temperature, then the work done in the process is

(a) $p_1V_1 \ln \frac{V_2}{V_1}$ (b) $-p_1V_1 \ln \frac{p_1}{p_2}$

(c) $RT \ln \frac{V_2}{V_1}$ (d) $-mRT \ln \frac{p_1}{p_2}$

[2017 : 1 Mark, Set-2]

24. Air is held inside a non-insulated cylinder using a piston (mass M = 25 kg and area A = 100 cm^2) and stoppers (of negligible area), as shown in the figure. The initial pressure P_1 and temperature T_1 of air inside the cylinder are 200 kPa and 400° C_1 respectively, The ambient pressure P_∞ and temperature T_∞ are 100 kPa and 27° C respectively. The temperature of the air inside the cylinder (°C) at which the piston will begin to move is ______(correct to two deeimal places).

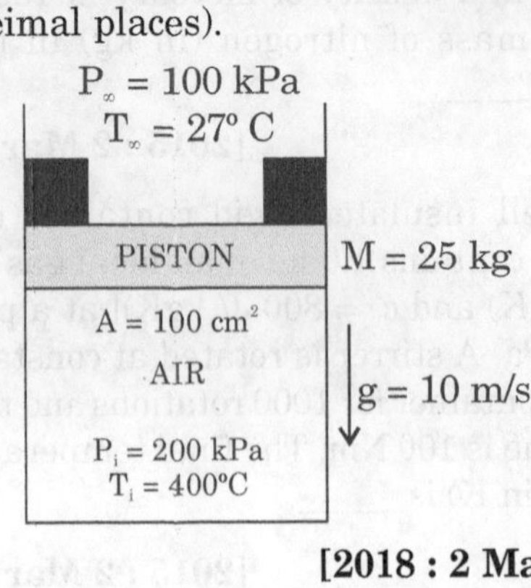

[2018 : 2 Marks, Set-2]

Steady flow system

25. For reversible adiabatic compression in a steady flow process, the work transfer per unit mass is

(a) $\int pdv$ (b) $\int vdp$

(c) $\int Tds$ (d) $\int sdT$

[1996 : 1 Mark]

26. A steam turbine receives steam steadily at 10 bar with a enthalpy of 3000 kJ/kg and discharges at 1 bar with an enthalpy of 2700 kJ/kg. The work output is 250 kJ/kg. The changes in kinetic and potential energies are negligible. The heat transfer from the turbine casing to the surroundings is equal to

(a) 0 kJ (b) 50 kJ

(c) 150 kJ (d) 250 kJ

[2000 : 1 Mark]

27. When an ideal gas with constant specific heats is throttled adiabatically, with negligible changes in kinetic and potential energies.

(a) $\Delta h = 0, \Delta T = 0$ (b) $\Delta h > 0, \Delta T = 0$

(c) $\Delta h > 0, \Delta S > 0$ (d) $\Delta h = 0, \Delta S > 0$

[2000 : 2 Marks]

28. A small steam whistle (perfectly insulated and doing no shaft work) causes a drop of 0.8 kJ/kg in enthalpy of steam from entry to exit. If the kinetic energy of the steam at entry is negligible, the velocity of the steam at exit is

(a) 4 m/s (b) 40 m/s

(c) 80 m/s (d) 120 m/s

[2001 : 2 Marks]

29. A rigid, insulated tank is initially evacuated. The tank is connected with a supply line through which air (assumed to be ideal gas with constant specific heats) passes at 1 MPa, 350°C.

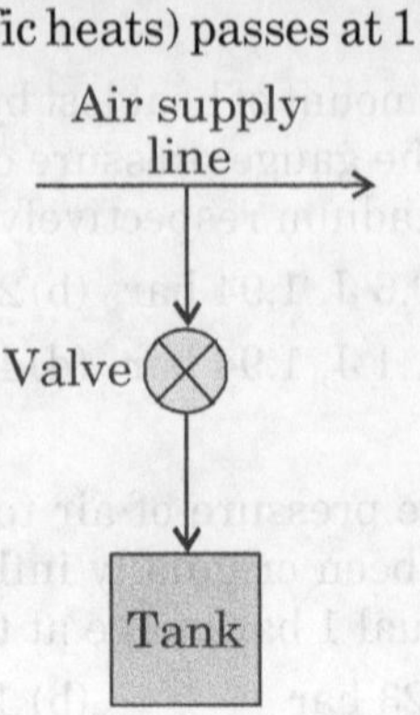

A valve connected with the supply line is opened and the tank is charged with air until the final pressure inside the tank reaches 1 MPa. The final temperature inside the tank

(a) is greater than 350°C

(b) is less than 350°C

(c) is equal to 350°C

(d) may be greater than, less than, or equal to 350°C, depending on the volume of the tank.

[2008 : 2 Marks]

30. In a steady state steady flow process taking place in a device with a single inlet and a single outlet, the work done per unit mass flow rate is given by $w = -\int_{inlet}^{outlet} v\,dp$, where v is the specific volume and p is the pressure, The expression for w given above

(a) is valid only if the process is both reversible and adiabatic

(b) is valid only if the process is both reversible and isothermal

(c) is valid for any reversible process.

(d) is incorrect; it must be $-\int_{inlet}^{outlet} v\,dp$

[2008 : 2 Marks]

31. A compressor undergoes a reversible, steady flow process. The gas at inlet and outlet of the compressor is designated as state 1 and state 2 respectively. Potential and kinetic energy changes are to be ignored. The following notations are used: v = specific volume and P = pressure of the gas. The specific work required to be supplied to the compressor for this gas compression process is

(a) $\int_1^2 Pdv$ (b) $\int_1^2 vdP$

(c) $v_1(P_2-P_1)$ (d) $-P_2(v_1-v_2)$

[2009 : 1 Mark]

Common Data for Question 32 and 33.

The inlet and the outlet conditions of steam for an adiabatic steam turbine are as indicated in the notations are as usually followed.

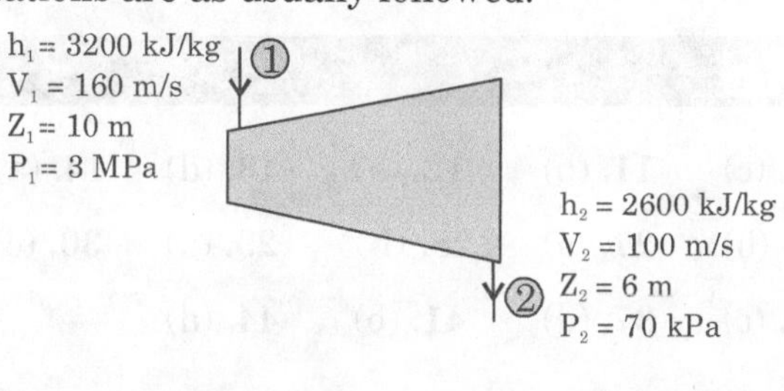

32. If mass flow rate of steam through the turbine is 20 kg/s, the power output of the turbine (in MW) is

(a) 12.157 (b) 12.941

(c) 168.001 (d) 168.785

[2009 : 2 Marks]

33. Assume the above turbine to be part of a simple Rankine cycle. The density of water at the inlet to the pump is 1000 kg/m³. Ignoring kinetic and potential energy effects, the specific work (in kJ/kg) supplied to the pump is

(a) 0.293 (b) 0.351

(c) 2.930 (d) 3.510

[2009 : 2 Marks]

Linked Data Questions 34 and 35

The temperature and pressure of air in a large reservoir are 400 K and 3 bar respectively. A converging diverging nozzle of exit are 0.005 m² is fitted to the wall of the reservoir as shown in the figure. The static pressure of air at the exit section for isentropic flow through the nozzle is 50 kPa. The characteristic gas constant and the ratio of specific heats of air are 0.287 kJ/kgK and 1.4 respectively.

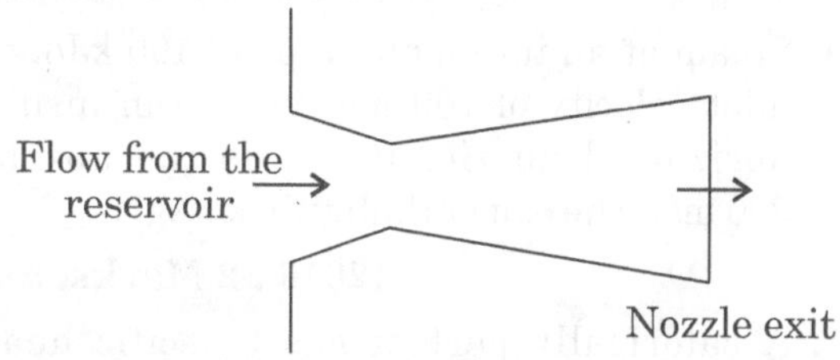

34. The density of air in kg/m³ at the nozzle exit is

(a) 0.560 (b) 0.600

(c) 0.727 (d) 0.800

[2011 : 2 Marks]

35. The mass flow rate of air through the nozzle in kg/s is

(a) 1.30 (b) 1.77

(c) 1.85 (d) 2.06

[2011 : 2 Marks]

Linked Data Questions 36 and 37

Air enters an adiabatic nozzle at 300 kPa, 500 K with a velocity of 10 m/s. It leaves the nozzle at 100 kPa with a velocity of 180 m/s. The inlet area is 80 cm². The specific heat of air c is 1008 J/kgK

36. The exit temperature of the air is

(a) 516 K (b) 532K

(c) 484 K (d) 468 K

[2012 : 2 Marks]

37. The exit area of the nozzle in cm^2 is

(a) 90.1 (b) 56.3

(c) 4.4 (d) 12.9

[2012 : 2 Marks]

38. Work is done on an adiabatic system due to which its velocity changes from 10 m/s to 20 m/s, elevation increases by 20 m and temperature increases by 1 K. The mass of the system is 10 kg. c_v = 100 J/(kgK) and gravitational acceleration is 10 m/s^2. If there is no change in any other component of the energy of the system, the magnitude of total work done (in kJ) on the system is ______.

[2015 : 2 Marks, Set-2]

39. Steam enters a turbine at 30 bar, 300°C (u = 2750 kJ/kg, h = 2993 kJ/kg) and exits the turbine as saturated liquid at 15 kPa (u = 225 kJ/kg, h = 226 kJ/kg). Heat loss to the surrounding is 50 kJ/kg of steam flowing through the turbine. Neglecting changes in kinetic energy and potential energy, the work output of the turbine (in kJ/kg of steam) is______.

[2015 : 2 Marks, Set-3]

40. Steam at an initial enthalpy of 100 kJ/kg and inlet velocity of 100 m/s, enters an insulated horizontal nozzle. It leaves the nozzle at 200 m/s. The exit enthalpy (in kJ/kg) is ______

[2016 : 2 Marks, Set-3]

41. A calorically perfect gas (specific heat at constant pressure 1000 J/kgK) enters and leaves a gas turbine with the same velocity. The emperature of the gas at turbine entry and exit are 1100 K and 400 K, respectively. The power produced is 4.6 MW and heat escapes at the rate of 300 kJ/s through the turbine casing. The mass flow rate of the gas (in kg/s) through the turbine is

(a) 6.14 (b) 7.00

(c) 7.50 (d) 8.00

[2017 : 2 Marks, Set-2]

42. Steam flows through a nozzle at mass flow rate of m = 0.1 kg/s with a heat loss of 5 kW. The enthalpies at inlet and exit are 2500 kJ/kg and 2350 kJ/kg, respectively. Assuming negliglble velocty at inlet (C_1~ 0), the velocity (C_2) of steam (in m/s) at the nozzle exit is ____(correct to two deeimal places)

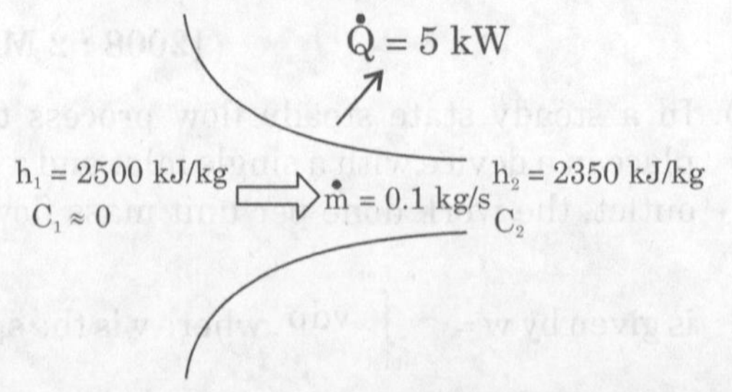

[2018 : 2 Marks, Set-1]

43. An engine operates on the reversible cycles as shown in the figure. The work output from the engine (in kJ/cycle) is ______(correct to two decimals)

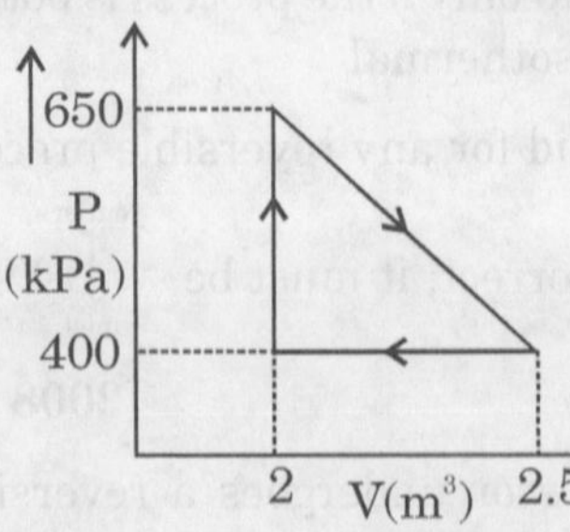

[2018 : 1 Marks, Set-2]

44. A gas is heated in a duct as it flows over a resistance heater. Consider a 101 kW electric heating system. The gas enters the heating section of the duct at 100 kPa and 27°C with a volume flow rate of 15 m^3/s. If heat is lost from the gas in the duct to the surroundings at a rate of 51 kW, the exit temperature of the gas is (Assume constant pressure, ideal gas, negligible change in kinetic and potential energies and constant specific heat : C_p = 1 kJ/kg. K; R = 0.5 kJ/kg K).

(a) 37°C (b) 76°C

(c) 53°C (d) 32°C

[2019 : 2 Marks, Set 1]

ANSWERS

1. (c)	**2.** (a)	**3.** (a)	**5.** (d)	**7.** (b)	**10.** (c)	**11.** (b)	**12.** (a)	**13.** (d)	**14.** (a)
15. (c)	**16.** (b)	**17.** (c)	**18.** (a)	**25** (b)	**26** (b)	**27.** (a)	**28.** (b)	**29.** (a)	**30.** (c)
31. (b)	**32.** (a)	**33.** (c)	**34.** (c)	**35.** (d)	**36.** (c)	**37.** (d)	**41.** (b)	**44.** (d)	

EXPLANATIONS

1. Since the kinetic energy of the molecules won't change in vacuum ⇒ temperature remains constant ⇒ internal energy remains constant.

Now $\frac{pV}{T}$ = constant

⇒ pV = constant

⇒ H = U + pV = constant

2. $W = P_1V_1 \, ln \frac{V_2}{V_1} = 8.317 kJ$

4. Gas constant of mixture,

$$R_m = \frac{\text{Universal gas constant}}{\text{Average molar mass}}$$

Average molar mass

$$= \frac{100}{\frac{60}{28}+\frac{30}{32}+\frac{10}{44}} = 30.233 \text{ kg/ kmol}$$

$$R_m = \frac{8314}{30.233} = 274.996 \text{ J / kg – K}$$

6. For isothermal process

$$W = p_1v_1 \ln\left(\frac{v_2}{v_1}\right)$$

$$= 100 \times 0.4 \ln\left(\frac{0.1}{0.4}\right) = -55.45 \text{ kJ}$$

7. Steady flow equation is given be,

$H_1 + q = H_2 + W$

Adiabatic q = 0

$H_1 - H_2 = W$

$W = -\Delta H$

9. m = 0.1 kg

P = 1.2 bar

Voltage

E = 12 volt

I = 1.5 A

t = 90 sec

$\Delta v = 0.01/m^3$

$C_v = 700 J/kg - K$

Insulated condition $\delta q = 0$

$\delta q = \delta w + \Delta U$

$0 = mCv\Delta T + \delta w_{air} - \delta w_e$

$$\Delta T = \frac{\delta w_{air} - \delta w_e}{mCv}$$

$$\Delta T = \frac{1.2\times10^5 \times 0.01 - 12\times1.5\times90}{0.1\times700}$$

10. Energy given by water = Energy taken by heater

$P.t = mc_p\Delta T = 2 \times 20 \times 60 = 40 \times 4.2 \times \Delta t$

$\Delta t = 14.3$ °C

11. Work interaction,

$$W = \int_{V_1}^{V_2} PdV = \int_{V_1}^{V_2} \frac{c}{V} dV = c\ln\left(\frac{V_2}{V_1}\right) = P_1V_1 \ln\left(\frac{V_2}{V_1}\right)$$

$= 2 \times 10^5 \times \ln(2)$

$= 2 \times 10^5 \times 0.693$

$= 1.386 \times 10^5 J = 138.6$ kJ

12. $dQ = dU + dW$

For isothermal process,

$dU = 0$

$\therefore dQ = dW$

$$ds = \frac{dQ}{T} - \frac{dW}{T} = \frac{138.6\times10^3}{298} = 0.4652 \text{ kJ/K}$$

13. q = u + w

2000 = u – 5000

u = 7000

14. $P_1V_1 = mRT_1$

$m = 6.049 \times 10^{-3}$kg

$Q = mC_vdT$

$Q = -43.43J$

$P_2V_1 = mRT_2$

$P_2 = 1.93$ bar (absolute)

$P_2 = 1 - 1.93$

$P_2 = 0.93$ bar

15. $P_2V_1 = mRT_2$

m = 0.6266 kg

$P_1V_1 = mRT_1$

$$P_1 = \frac{(0.6266)(281)(288)}{0.75}$$

P_1, abs = 2.07 bar

P_1, gauge = 2.07 – 1 = 1.07 bar

16. Q = dU + PdV

Reversible closed system

$\frac{\delta Q}{T} = ds \leftarrow$ Reversibe

TdS = du + Pdv

17. +2kJ

P = 300 kPa

P = 100 kPa

dv = 0.01

w → (300 - 100)dv = (200) (0.01)

[W = 2kJ]

18. $W_{electric} = i^2R$

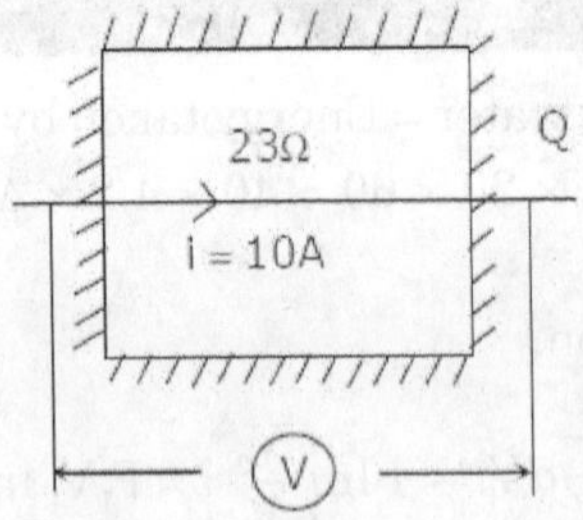

$W_{electric} = i^2R$

$= (10^2 \times 20)$ watts $= -2.3$ kW (on system)

First law : $Q - w = \Delta U$

$\therefore 0 - (-W_{electric}) = \Delta u$

19. $p = \rho gh = 0.7 \times 1360 \times 9.81 = 9339.12$Pa

Actual pressure = atmospheric pressure

= 110664.12 Pa

$pv = mRT \Rightarrow m = pv/RT$

$$m = \frac{110664.12 \times 2}{288 \times 8314} \Rightarrow m = 4.55\text{kg}$$

20.

T = 100 N– m

$$T_1 = \frac{p_1v_1}{mR} = \frac{p_1v_1}{m(c_p - c_v)}$$

$T_1 \Rightarrow 500$k

$w = 2\pi NT = -2\pi \times 10^5$ J

Q = U + W (insulated)

W = –U

$U = 2\pi \times 10^5$

$Cv\,(T - 500) = 2\pi \times 10^5$

T = 1295.4K

21.

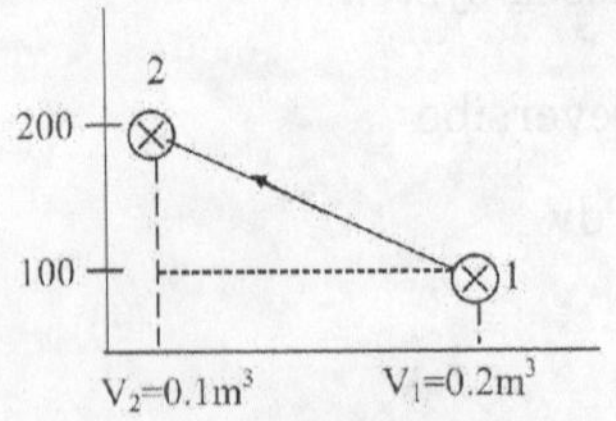

$$_1W_2 = \frac{1}{2}(P_1 + P_2)(V_1 - V_2)$$

$$= \frac{1}{2}(100 + 200)(0.2 - 0.1)$$

15 kJ

22. $C_V = 2.5$ R (R = 8.314)

$C_V = 2.5 \times 8.314 = 20.785$ J/molK

$C_P = R + C_V$, n = no. of moles

$= 8.314 + 20.785 = 29.099$

$\Delta H = nC_P\Delta T = 1 \times 29.099 \times 100 = 2909.9$ J/mol

23. $W_{iso} = p_1v_1 \ln \frac{V_2}{V_1} = mRT \ln \frac{V_2}{V_1}$

$$= mRT \ln \frac{p_1}{p_2} = p_1v_1 \ln \frac{p_1}{p_2}$$

24.

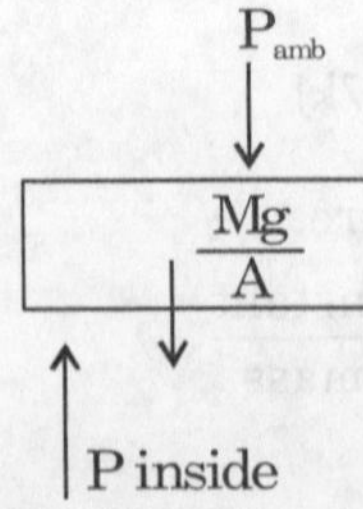

$$P_{in} = P_{amb} + \frac{mg}{A} = 100 + \frac{25 \times 4.81}{100 \times 10^{-4}} \times 10^{-3}$$

$P_2 = 124.525$

$$\frac{P_1}{T1} = \frac{P_2}{T_2}$$

$T_2 = 146.03$ °C

25. $\left[\frac{w}{m} = \int vdp\right]$ for flow process

26.

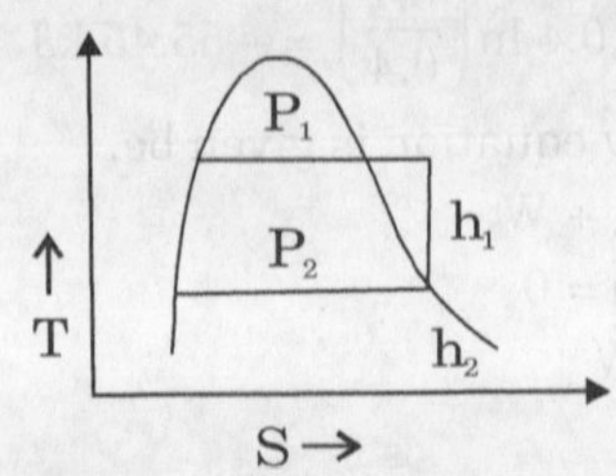

$\Delta h = \Delta Q - \Delta W$

$\therefore \Delta Q = (-300 + 250)$ kJ/kg $= -50$ kJ/kg

27. $\Delta h = 0$, $\Delta T = 0$

Throttling is known as is isenthalpic process and as for ideal gas

H = f(T)

$\boxed{\Delta T = 0}$ also.

28 $\frac{v_1^2}{2} + h_1 = \frac{v_2^2}{2} + h_2$

Heres $v_1 = 0$, therefore

$$h_1 = \frac{v_2^2}{2} + h_2$$

or $v_2 = \sqrt{2(h_1 - h_2)} = \sqrt{2 \times 0.8 \times 1000} = 40$ m/sec.

29. Greater than 350°c $(\because T_f = \gamma T_i)$

30. For steady flow process,

$\Delta W = -\Delta H + \Delta Q$

$= -\Delta H + T.\,\Delta S$

But for reversible process,

$T.\,\Delta S = \Delta H - V.\,\Delta p$

$\Rightarrow \Delta W - \Delta H + \Delta H = -V.\,\Delta p$

$\Rightarrow \Delta W = -V.\,\Delta p$

For steady flow reversible process,

$$W = -\int_{\text{inlet}}^{\text{outlet}} v\,dp$$

31. $h = u + Pv$

$\Rightarrow dh = du + Pdv + vdP$

But $dQ = du + Pdv$

$\therefore du = dQ - Pdv$

$\therefore dh = dQ - Pdv + Pdv + vdP$

$\Rightarrow dh = dQ + vdP$

$\Rightarrow -vdP = d(P.E) + d(K.E) + dW$

Since K.E and P.E are constant

$$\therefore dW = \int_1^2 vdP$$

32. SFEE $h_1 + z_1 g + \dfrac{c_1^2}{2} + Q = h_2 + z_2 g + \dfrac{c_2^2}{2} + W$

For adiabatic process, $Q = 0$

$$\therefore h_1 + z_1 g + \frac{c_1^2}{2} = h_2 + z_2 g + \frac{c_2^2}{2} + W$$

$$\text{or } 3200 \times 10^3 + 10 \times 9.81 + \frac{160 \times 160}{2}$$

$$= 2600 \times 10^3 + 6 \times 9.81 + \frac{100 \times 100}{2} + W$$

$$\Rightarrow W = 600 \times 10^3 + 39.24 + \frac{25600 - 10000}{2}$$

$$= 607839.24 \times 20 \text{ J/s} = 12.156 \text{ MW}$$

Alternately

$h_1 = 3200$ kJ/kg

$V_1 = 160$ m/s

$Z_1 = 10$m

$P_1 = 3$mPa

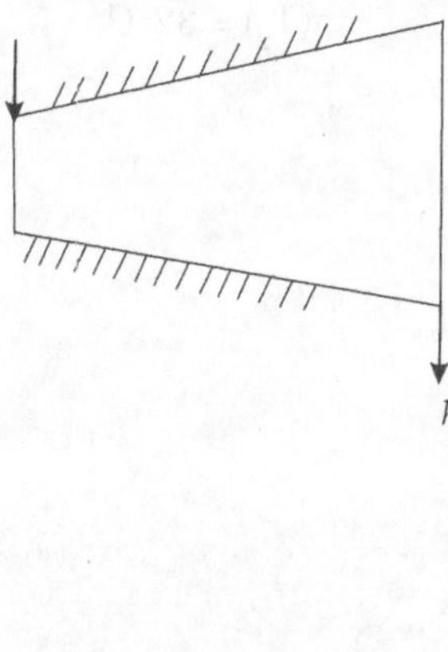

$h_2 = 2600$ kJ/kg

$V_2 = 100$ m/s

$Z_2 = 6$m

$P_2 = 70$kPa

Applying S.F.E.E

$$w_1\left(h_1 + \frac{V_1^2}{2} + gz_1\right) + \frac{dQ}{d\tau} \quad w_2\left(h_2 + \frac{V_2^2}{2} + gz_2\right) + \frac{\partial w}{\partial \tau}$$

Here, w_1 = mass flow rate

$\dfrac{dQ}{d\tau} = 0$ for insulated turbine

$$\therefore \; 20\left(3200 + \frac{160^2}{2 \times 10^3} \times 9.81 \times 10\right)$$

$$= 20\left(2600 + \frac{160^2}{2 \times 10^3} + \frac{9.81 \times 6}{10^3}\right) + \frac{dw}{d\tau}$$

$$\text{or } \frac{dw}{d\tau} = 20\left\{(3200 - 2600) + \frac{1}{10^3}\left(\frac{160^2}{2} - \frac{100^2}{2}\right) + \frac{9.81(10-6)}{10^3}\right\}$$

$= 20 \times \{600 + 7.8 + 0.03924\}$

$= 20 \times 607.83924$ kW

$= 12156.7848$ kW $\simeq 12.157$ MW

33. Pump work

$h_{\text{inlet}} + q = h_{\text{outlet}} + W_{\text{pump}}$

here, $q = 0$

$W_{\text{pump}} = h_{\text{inlet}} - h_{\text{outlet}}$

$\therefore W = \Delta h_{\text{pump}}$

Reversible adiabatic

$dq = dh - Vdp$

$\therefore dh = Vdp$

$\therefore dh_{\text{pump}} = V(p_2 - p_1) = \dfrac{1}{\rho}(3000\text{–}70)$ kPa $= 2.930$kPa

34. Given : $T_1 = 400$ K, $P_1 = 300$ kPa,

$P_2 = 50$ kPa, $R = 0.289$ kJ/kgK

$v = 1.4$, $A_2 = 0.005$ m^2

Process from entrance to exit is isentropic process, therefore

$$\frac{T_2}{T_1} = \left(\frac{P_2}{P_1}\right)^{\frac{\gamma-1}{\gamma}}$$

$$\Rightarrow T_2 = 400\left(\frac{50}{300}\right)^{\frac{0.4}{1.4}} = 239.5 \text{ K}$$

From the perfect gas equation,

$$\rho = \frac{P}{RT}$$

$$\Rightarrow \rho_2 = \frac{P_2}{RT_2} = \frac{50}{0.287 \times 239.5} = 0.727 \text{ kg/m}^3$$

35. Mass flow rate, $m = \rho Q$

where, $Q = A_2 V_2$

But $V_2 = \sqrt{2C_P(T_1 - T_2)}$

$= \sqrt{2 \times 1.005 \times (400 - 239.5)} = 568$ m/s

$\therefore m = 0.727 \times 0.005 \times 568 = 2.06$ kg/s

36. $h_1 + \frac{V_1^2}{2} = h_2 + \frac{V_2^2}{2}$

$$\Rightarrow \frac{V_2^2 - V_1^2}{2} = h_1 - h_2 = C_p(T_1 - T_2)$$

$$\therefore \frac{180^2 - 10^2}{2} = 1008 \times (500 - T_2)$$

$$\Rightarrow T_2 = 484 \text{ K}$$

37. From continuity equation, we get

$$\rho_1 A_1 V_1 = \rho_2 A_2 V_2$$

$$\therefore \frac{P_1}{RT_1} \times A_1 V_1 = \frac{P_2}{RT_2} \times A_2 V_2$$

$$\Rightarrow A_2 = \frac{P_1 T_2}{P_2 T_1} \times \frac{V_1}{V_2} \times A_1 = \frac{300 \times 484}{500 \times 100} \times \frac{10}{180} \times 80$$

$$= 12.9 \text{ cm}^2$$

38. Using SFEE

$$m\left[\left(\frac{v_1^2}{2} - \frac{v_2^2}{2}\right) + (z_1 - z_2)g + (h_1 - h_2)\right] = W$$

$$10\left[\frac{10^2}{2} - \frac{20^2}{2} + (-20) \times 10 + (100 \times -1)\right] = W$$

$$W = -4.5 \text{ kJ}$$

$\therefore$ Work done on the system is 4.5 kJ.

39. Work output = (2993 – 226.50) kJ/kg = 2717 kJ/kg

40. h_1 = 100 KJ/Kg

V_1 = 100 m/s

V_2 = 200 m/s

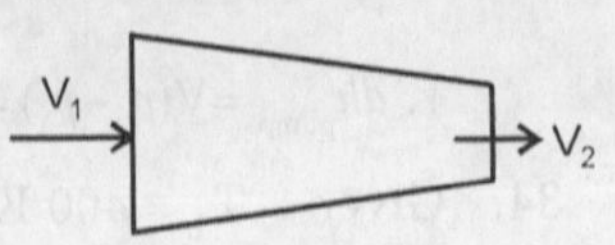

$$h_1 + \frac{V_1^2}{2} = h_2 + \frac{V_2^2}{2}$$

$$100 \times 10^3 + \frac{(100)^2}{2} = h_2 + \frac{(200)^2}{2}$$

$$h_2 = 85 \times 10^3 \text{ J/Kg}$$

41. $m\left(h_1 + \frac{C_1^2}{2} + gz_1\right) + Q = m\left(h_2 + \frac{C_2^2}{2} + gz_2\right) + w$

$m(h_1 - h_2) + Q = w$

$m(700)(1000) = (4600 + 300) \times 10^3$ J/s = 7kg/s

42. $\dot{m} = 0.1$ kg/s, $\dot{Q} = 5$ kW(heat loss)

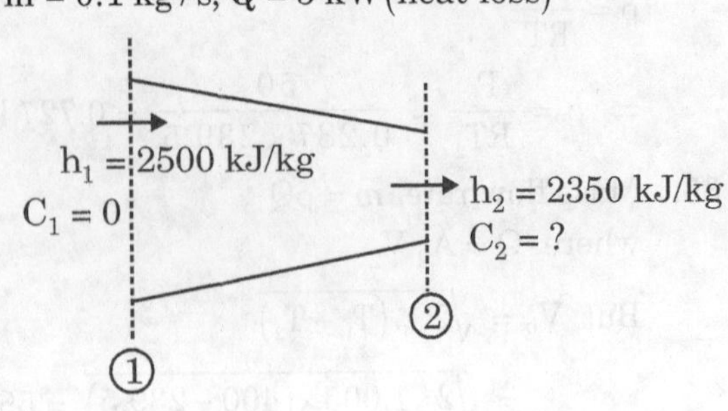

Applying SFEE

$$\dot{m}\left(h_1 + \frac{1}{2}C_1^2 + gz_1\right) + \dot{Q} = \dot{m}\left(h_2 + \frac{1}{2}C_2^2 + gz_2\right) + \dot{W}_{CV}$$

$C_1 = 0$ and $\dot{W}_{CV} = 0$

$z_1 = z_2$ (assume)

$$\dot{m}h_1 + \dot{Q} = \dot{m}h_2 + \dot{m}\frac{1}{2}C_2^2$$

$$\Rightarrow \dot{m} \times \frac{1}{2}C_2^2 = \dot{m}(h_1 - h_2) + \dot{Q}$$

$$\Rightarrow 0.1 \times \frac{1}{2}C_2^2 \times 10^{-3} = 0.1(2500 - 2350) - 5$$

$$C_2 = 447.213 \text{ m/s}$$

43. Work output = $\int pdv$

$$W = \frac{1}{2} \times (2 - 5 - 2)(650 - 900)$$

$$W = \frac{1}{2}(0.5)(250)$$

$$W = 62.5 \text{ kJ/cycle}$$

44.

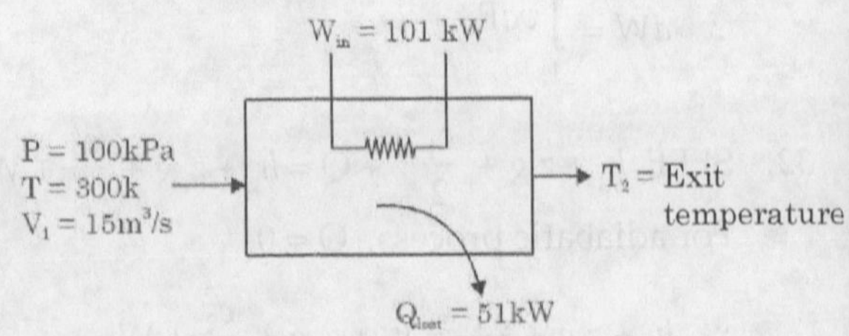

$$C_p = 1 \frac{\text{kJ}}{\text{kgK}}$$

$$R = 0.5 \frac{\text{kJ}}{\text{kgK}}$$

$$\dot{m} = \frac{P_1 \dot{V}_1}{RT_1} = \frac{100 \times 15}{0.5 \times 300} = 10 \frac{\text{kg}}{\text{s}}$$

By applying steady flow energy equation, (SFEE)

$$\dot{m}h_1 + \dot{Q} = \dot{m}h_2 + \dot{W}$$

$$10 \times 1 \times 300 + (-51) = 10 \times 1 \times T_2 + (-101)$$

$$T_2 = 305 \text{ K} = 32°\text{C}$$

Hence, Exit temperature

$(T_2) = 32°$C.

3 CHAPTER

Second Law of Thermodynamics, Carnot Cycle and Entropy

Analysis of Previous GATE Papers

Topic	Year/Set → Marks Distribution ↓		2019 Set 1	2019 Set 2	2018 Set 1	2018 Set 2	2017 Set 1	2017 Set 2	2016 Set 1	2016 Set 2	2016 Set 3	2015 Set 1	2015 Set 2	2015 Set 3
Second law of Thermodynaics	1 Mark	MCQ Type										1		
		Numerical Type					1			1				
	2 Marks	MCQ Type		1										
		Numerical Type								1				
	Total			2			1			3		1		
Carnot Cycle	1 Mark	MCQ Type												
		Numerical Type												
	2 Marks	MCQ Type												
		Numerical Type												
	Total													
Entropy	1 Mark	MCQ Type												
		Numerical Type	1		1									
	2 Marks	MCQ Type												
		Numerical Type					1	1					1	
	Total		1		1		2	2					2	

Second law of thermodynamics

1. A condenser of a refrigeration system rejects heat at a rate of 120 kW, while is compressor consumes a power of 30 kW. The coefficient of performance of the system would be

(a) 1/4 (b) 4

(c) 1/3 (d) 3

[1992 : 1 Mark]

2. A reversible heat transfer demands:

(a) The temperature difference causing heat transfer tends to zero

(b) The system receiving heat must be at a constant temperature

(c) The system transferring out heat must be at a constant temperature

(d) Both interacting systems must be at constant temperatures

[1993 : 1 Mark]

3. Figure below shows a reversible heat engine E_R having heat interactions with three constant temperature systems. Calculate the thermal efficiency of the heat engine

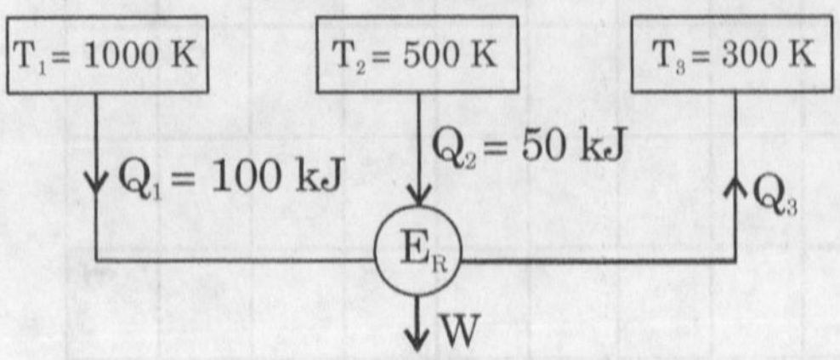

[1993 : 2 Marks]

4. When a system executes an irreversible cycle

(a) $\oint \frac{\delta Q}{T} < 0$ (b) $\oint dS > 0$

(c) $\oint \frac{dQ}{T} = 0$ (d) $\oint \frac{\delta Q}{T} > 0$

[1994 : 1 Mark]

5. Any thermodynamic cycle operating between two temperature limits is reversible if the product of the efficiency when operating as a heat engine and the COP when operating as a refrigerator is equal to l. [TRUE/FALSE]

[1994 : 1 Mark]

6. In the case of a refrigeration system undergoing an irreversible cycle, $\oint \frac{\delta Q}{T}$ is.... (< 0 / $= 0$ / > 0)

[1995 : 1 Mark]

7. A solar energy based heat engine which receives 80 kJ of heat at 100°C and rejects 70 kJ of heat to the ambient at 30°C is to be designed. The thermal efficiency of the heat engine is

(a) 70% (b) 18.8%

(c) 12.5% (d) Indetermine

[1996 : 2 Marks]

8. For two cycles coupled in series, the topping cycle has an efficiency of 30% and the bottoming cycle has an efficiency of 20%. The overall combined cycle efficiency is

(a) 50% (b) 44%

(c) 38% (d) 55%

[1996 : 2 Marks]

9. A cycle heat engine does 50 kJ of work per cycle. If the efficiency of the heat engine is 75%. The heat rejected per cycle is

(a) $16\frac{2}{3}$ kJ (b) $33\frac{1}{3}$ kJ

(d) $37\frac{1}{2}$ kJ (c) $66\frac{1}{2}$ kJ

[2001 : 2 Marks]

10. A solar collector receiving solar radiation at the rate of 0.6 kW/m^2 transforms it to the internal energy of a fluid at an overall efficiency of 50%. The fluid heated to 350 K is used to run a heat engine which rejects heat at 313 K. If the heat engine is to deliver 2.5 kW power, the minimum area of the solar collector required would be

(a) 83.33 m^2 (b) 16.66 m^2

(c) 39.68 m^2 (d) 79.36 m^2

[2004 : 2 Marks]

11. A heat transformer is device that transfers a part of the heat, supplied to it at an intermediate temperature, to a high temperature reservoir while rejecting the remaining part to a low temperature heat sink. In such a heat transformer, 100 kJ of heat is supplied at 350 K. The maximum amount of heat in kJ that can be transferred to 400 K, when the rest is rejected to a heat sink at 300 K is

(a) 12.50 (b) 14.29

(c) 33.33 (d) 57.14

[2007 : 2 Marks]

12. A cyclic device operates between three thermal reservoirs, as shown in the figure. Heat is transferred to/from the cycle device. It is assumed that heat transfer between each thermal reservoir and the cyclic device takes place across negligible temperature difference. Interactions between the cyclic device and the respective thermal reservoirs that are shown in the figure are all in the form of heat transfer.

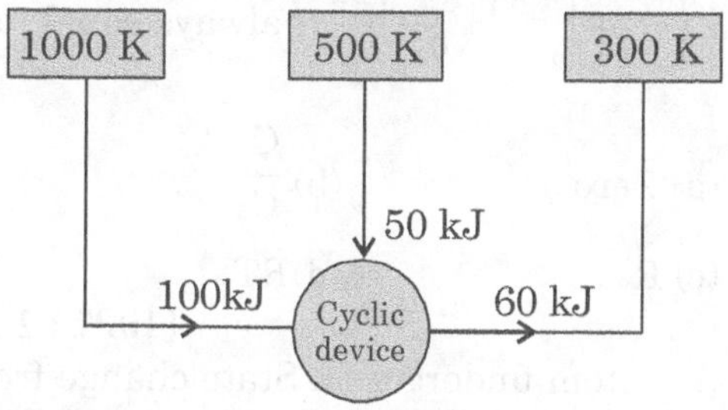

The cyclic device can be
(a) a reversible heat engine
(b) a reversible heat pump or a reversible refrigerator
(c) an irreversible heat engine
(d) an irreversible heat pump or an irreversible refrigerator.

[2008 : 2 Marks]

13. An irreversible heat engine extracts heat from a high temperature source at a rate of 100 kW and rejects heat to a sink at a rate of 50 kW. The entire work output of the heat engine is used to drive a reversible heat pump operating between a set of independent isothermal heat reservoirs at 17°C and 75°C. The rate (in kW) at which the heat pump delivers heat to its high temperature sink is
(a) 50 (b) 250
(c) 300 (d) 360

[2009 : 2 Marks]

14. Consider the following two processes;
I. A heat source at 1200 K loses 2500 kJ of heat to a sink at 800 K
II. A heat source at 800 K loses 2000 kJ of heat to a sink at 500 K
Which of the following statements is true?
(a) Process I is more irreversible than Process II
(b) Process II is more irreversible than Process I
(c) Irreversibility associated in both the processes are equal
(d) Both the processes are reversible

[2010 : 2 Marks]

15. A reversible heat engine receives 2 kJ of heat from a reservoir at 1000 K and a certain amount of heat from a reservoir at 800 K. It rejects 1 kJ of heat to a reservoir at 400 K. The net work output (in kJ) of the cycle is
(a) 0.8 (b) 1.0
(c) 1.4 (c) 2.0

[2014 : 2 Marks, Set-1]

16. A source at a temperature of 500 K provides 1000 kJ of heat. The temperature of environment is 27°C. The maximum useful work (in kJ) that can be obtained from the heat source is_______.

[2014 : 1 Mark, Set-3]

17. A Carnot engine (CE-1) works between two temperature reservoirs A and B, where T_A = 900 K and T_B = 500 K. A second Carnot engine (CE-2) works between temperature reservoirs B and C, where T_c = 300 K. In each cycle of CE-1 and CE-2, all the heat rejected by CE-1 to reservoir B is used by CE-2. For one cycle of operation, if the net Q absorbed by CE-1 from reservoir A is 150 MJ, the net heat rejected to reservoir C by CE-2 (in MJ) is ______.

[2015 : 1 Mark, Set-1]

18. The heat removal rate from a refrigerated space and the power input to the compressor are 7.2 kW and 1.8 kW, respectively. The coefficient of performance (COP) of the refrigerator is____.

[2016 : 1 Mark, Set-2]

19. A reversible cycle receives 40 kJ of heat from one heat source at a temperature of 127°C and 37 kJ from another heat source at 97°C. The heat rejected (in kJ) to the heat sink at 47°C is_______

[2016 : 2 Marks, Set-2]

20. A heat pump absorbs 10 kW of heat from outside environment at 250K while absorbing 15 kW of work. It delivers the heat to a room that must be kept warm at 300 K. The Coefficient of Performance (COP) of the heat pump is____.

[2017 : 1 Mark, Set-1]

21. The figure shows a heat engine (HE) working between two reservoirs. The amount of heat (Q_2) rejected by the heat engine is drawn by a heat pump (HP). The heat pump receives the entire work output (W) of the heat engine. If temperatures, $T_1 > T_3 > T_2$, then the relation between the efficiency (η) of the heat engine and the coefficient and the coefficient of performance (COP) of the heat pump is

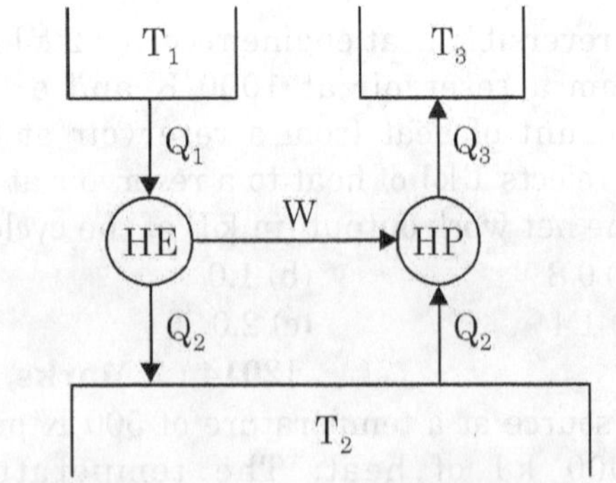

(a) COP = 1 + η (b) COP = $\eta^{-1} - 1$

(c) COP = η^{-1} (d) COP = η

[2019 : 2 Marks, Set-2]

Carnot cycle

22. Consider a refrigerator and a heat pump working on the reversed Carnot cycle between the same temperature limits. Which of the following is correct?

(a) COP of refrigerator = COP of heat pump

(b) COP of refrigerator = COP of heat pump + 1

(c) COP of refrigerator = COP of heat pump –1

(d) COP of refrigerator = Inverse of the COP of heat pump

[1995 : 1 Mark]

23. A Carnot cycle is having an efficiency of 0.75. If the temperature of the high temperature reservoir is 727°C. What is the temperature of low temperature reservoir?

(a) 23°C (b) –23°C

(c) 0°C (d) 250°C

[2001 : 2 Marks]

Entropy

24. Which among the following relations is/are valid only for reversible process undergone by a pure substance?

(a) δQ = dU + δW (b) TdS = du + δW

(c) TdS = dU + PdV (d) δQ = PdV + dU

[1993 : 1 Mark]

25. The slopes of constant volume and constant pressure lines in the T-s diagram are ________ and ________ respectively.

[1994 : 1 Mark]

26. A 1500 W electrical heater is used to heat 20 kg of water (c_p = 4186 J/kgK) in an insulated bucket, from a temperature of 30°C to 80°C. If the heater temperature is only infinitesimally larger than the water temperature during the process, the change in entropy for the heater is ________ J/K and for water J/K.

[1994 : 1 Mark]

27. One kilomole of an ideal gas is throttled from an initial pressure of 0.5 MPa to 0.1 MPa. The initial temperature is 300 K. The entropy change of the universe is

(a) 13.38 kJ/K (b) 4014.3 kJ/K

(c) 0.0446 kJ/K (d) 0.0445 kJ/K

[1995 : 2 Marks]

28. For an ideal gas the expression

$$\left[T\left(\frac{\delta s}{\delta T}\right)_p - T\left(\frac{\delta s}{\delta T}\right)_v\right] \text{ is always equal to}$$

(a) Zero (b) $\frac{C_p}{C_v}$

(c) R (d) RT

[1997 : 1 Mark]

29. A system undergoes a State change from 1 to 2. According to the second law of thermodynamics, for the process to be feasible, the entropy change, $S_2 - S_1$ of the system

(a) is positive or zero

(b) is negative or zero

(c) is zero

(d) can be positive, negative or zero

[1997 : 1 Mark]

30. If a closed system is undergoing an irreversible process, the entropy of the system

(a) must increase

(b) always remains constant

(c) must decrease

(d) can increase, decrease or remain constant

[2009 : 1 Mark]

31. One kilogram of water at room temperature is brought into contact with a high temperature thermal reservoir. The entropy change of the universe is

(a) equal to entropy change of the reservoir

(b) equal to entropy change of water

(c) equal to zero

(d) always positive

[2010 : 1 Mark]

Common Data Question 32 and 33

In an experimental set-up, air flows between two stations P and Q adiabatically. The direction of flow depends on the pressure and temperature conditions maintained at P and Q. The conditions at station P are 150 kPa and 350 K. The temperature at station Q is 300 K. The following are the properties and relations pertaining to air:

Specific heat at constant pressure,

c_p = 1.005 kJ/kgK;

Specific heat at constant volume,

c_v = 0.718 kJ/kgK;

Characteristic gas constant,
R = 0.287 kJ/kgK.
Enthalpy, $h = c_pT$.
Internal energy, $u = c_vT$.

32. If the air has to flow from station P to station Q, the maximum possible value of pressure in kPa at station Q is close to
(a) 50 (b) 87
(c) 128 (d) 150

[2011 : 2 Marks]

33. If the pressure at station Q is 50 kPa, the change in entropy $(s_Q - s_p)$ in kJ/kgK is
(a) –0.155 (b) 0
(c) 0.160 (d) 0.355

[2011 : 2 Marks]

34. An ideal gas of mass m and temperature T_1 undergoes a reversible isothermal process from an initial pressure P_1 to final pressure P_2. The heat loss during the process is Q. The entropy change ΔS of the gas is

(a) $mR\ln\left(\frac{P_2}{P_1}\right)$ (b) $mR\ln\left(\frac{P_1}{P_2}\right)$

(c) $mR\ln\left(\frac{P_2}{P_1}\right) - \frac{Q}{T_1}$ (d) zero

[2012 : 1 Mark]

35. Which one of the following pairs of equations describes an irreversible heat engine?

(a) $\oint \delta Q > 0$ and $\oint \frac{\delta Q}{T} < 0$

(b) $\oint \delta Q < 0$ and $\oint \frac{\delta Q}{T} < 0$

(c) $\oint \delta Q > 0$ and $\oint \frac{\delta Q}{T} > 0$

(d) $\oint \delta Q < 0$ and $\oint \frac{\delta Q}{T} > 0$

[2014 : 1 Mark, Set-3]

36. An amount of 100 kW of heat is transferred through a wall in steady State. One side of the wall is maintained at 127°C and the other side at 27°C. The entropy generated (in W/K) due to the heat transfer through the wall is ______.

[2014 : 2 Marks, Set-3]

37. A closed system contains 10kg of saturated liquid ammonia at 10°C. Heat addition required to convert the entire liquid into saturated vapour at a constant pressure is 16.2 MJ. If the entropy of the saturated liquid is 0.88 kJ/kgK, the entropy (in kJ/kgK) of saturated vapour is____

[2014 : 2 Marks, Set-4]

38. One kg of air (R = 287 J/kgK) undergoes an irreversible process between equilibrium state 1 (20°C, 0.9 m³) and equilibrium state 2 (20°C, 0.6 m³). The change in entropy $s_2 - s_1$ (in J/kgK) is ______.

[2015 : 2 Marks, Set-2]

39. One kg of an ideal gas (gas constant, R= 400 J/kgK; specific heat at constant volume, c_p = 1000 J/kgK) at 1 bar, and 300 K is contained in a sealed rigid cylinder. During an adiabatic process, 100 kJ of work is done on the system by a stirrer. The increase in entropy of the system is ______ J/K.

[2017 : 2 Marks, Set-1]

40. One kg of an ideal gas (gas constant R = 287 J/kgK) undergoes an irreversible process from state-1 (1 bar, 300 K) to state-2 (2 bar, 300 K). The change in specific entropy $(s_2 - s_1)$ of the gas (in J/kgK) in the process is ______.

[2017 : 2 Marks, Set-2]

41. An ideal gas undergoes a process from state 1 (T_1 = 300 K, p_1 = 100 kPa) to state 2 (T_2 = 600 K, p_2 = 500 kPa). The specific heats of the ideal gas are: c_p = 1 kJ/kg-K and c_v = 0.7 kJ/kg-K. The change in specific entropy of the ideal gas from state 1 to state 2 (in kJ/kg-K) is ______ (correct to two decimal places)

[2018 : 1 Mark, Set-1]

42. For an ideal gas with constant properties undergoing a quasi-static process, which one of the following represents the change of entropy (Δs) from state 1 to 2?

(a) $\Delta s = c_p\ln\left(\frac{T_2}{T_1}\right) - R\ln\left(\frac{P_2}{R_1}\right)$

(b) $\Delta s = c_v\ln\left(\frac{T_2}{T_1}\right) - c_p\ln\left(\frac{V_2}{V_1}\right)$

(c) $\Delta s = c_p\ln\left(\frac{T_2}{T_1}\right) - c_v\ln\left(\frac{P_2}{P_1}\right)$

(d) $\Delta s = c_v\ln\left(\frac{T_2}{T_1}\right) + R\ln\left(\frac{V_1}{V_2}\right)$

[2018 : 1 Mark, Set-2]

43. Air of mass 1 kg. initially at 300 K and 10 bar, is allowed to expand isothermally till it reaches a pressure of 1 bar. Assuming air as an ideal gas with gas constant of 0.287 kJ/kg K, the change in entropy of air (in kJ/kg K, round off to two decimal places) is ______.

[2019 : 1 Mark, Set-1]

ANSWERS

1. (d)	**2.** (a)	**4.** (a)	**7.** (c)	**8.** (b)	**9.** (a)	**10.** (a)	**11.** (d)	**12.** (a)	**13.** (c)
14. (b)	**15.** (c)	**21.** (c)	**22.** (c)	**23.** (b)	**24.** (d)	**27.** (a)	**28.** (c)	**29.** (d)	**30.** (a)
31. (d)	**32.** (b)	**33.** (c)	**34.** (b)	**35.** (a)	**42.** (a)				

EXPLANATIONS

1. $Q_1 = 120$ kW

$W = 30 \times$ kW

$Q_2 = 120 - 30 = 90$kW

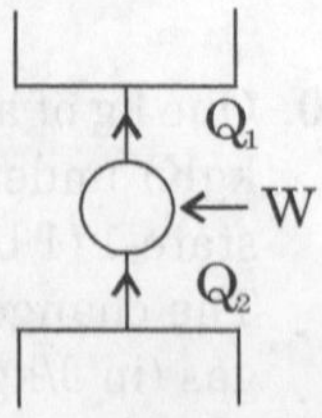

$$(COP)_{ref} = \frac{Q_2}{W} = \frac{90}{30} = 3$$

2. $\Delta S = 0$

$$\Rightarrow \frac{\delta Q}{T_1} - \frac{\delta Q}{T_2} = 0$$

$$\Rightarrow \boxed{T_1 = T_2}$$

3. $Q_1 + Q_2 - Q_3 = W$ ... (i)

$$\frac{Q_1}{1000} + \frac{Q_2}{500} - \frac{Q_3}{300} = 0 \quad \text{... (ii)}$$

From eq^n (ii)

$$\frac{Q_1}{1000} + \frac{Q_2}{500} - \frac{Q_3}{300} = 0$$

$Q_3 = 60$ kJ

$W \Rightarrow 100 + 50 - 60$

$W = 90$ kJ

$$\eta = \frac{W}{Q_1 + Q_2} = \frac{90}{150} = 60\%$$

4. $\oint \frac{\delta Q}{T} < 0$ clausius inequality

5. False

$\eta_{HE} \times COP_{HP} = 1$

Heat pump

6. $\oint \frac{\delta Q}{T} < 0$

Clausius inequality

7. $\eta_{th} = \frac{(80 - 70)}{80} = \frac{1}{8}$

$= 12.5\ \%$

8. $\eta_0 = \eta_1 + \eta_2 - \eta_1 \eta_2$

$= 0.3 + 0.2 - (0.3)(0.2) = 0.44 = 44\%$

9. $\eta = 0.75$

$$\therefore \eta_{HE} = \frac{W}{Q_1} = 0.75$$

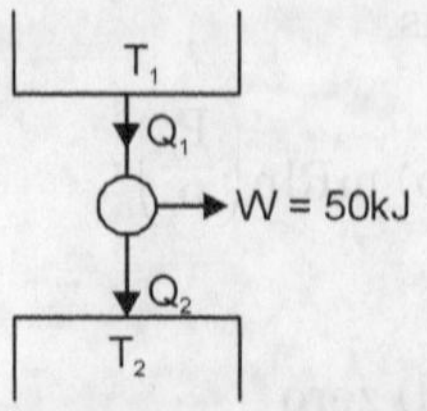

or, Net heat input,

$$Q_1 = \frac{50}{0.75} = \frac{200}{3}$$

Also, $W = Q_1 - Q_2$

or $50 = \frac{200}{3} - Q_2$

$$\therefore \text{Net heat required, } Q_2 = \frac{50}{3} = 16.66 \text{ kJ}$$

10. Given; Receiving solar radiation at the rate of 0.6 kW/m²

Internal energy of fluid after absorbing solar radiation

$$= 0.6 \times \frac{1}{2} \text{kW/m}^2 = 0.3\text{kW/m}^2$$

$$\eta_{engine} = 1 - \frac{315}{350} = 0.1$$

$$0.1 = \frac{W}{Q_1}$$

$$\therefore Q_1 = \frac{2.5}{0.1} = 25 \text{ kW}$$

Let A be minimum area of collector

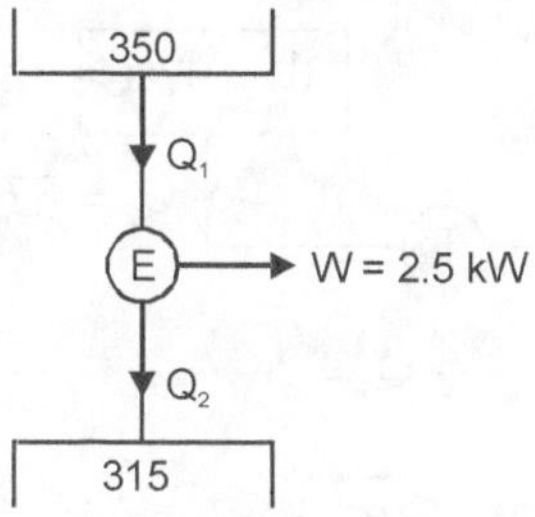

$\therefore\ Q_1 = 0.3 \times A$

or 25 kW =0.3 kW/m^2

or $A = \frac{25}{0.3} = 83.33\ m^2$

11. $\frac{Q_H}{Q_L} = \frac{400}{300} - \frac{4}{3}$

$Q_H + Q_L = Q = 100$

$\therefore\ \frac{4}{3}Q_L + Q_L = 100$

$\Rightarrow Q_L = 42.85$ kJ

$\therefore\ Q_H = 100 - 42.85$

= 57.14 kJ

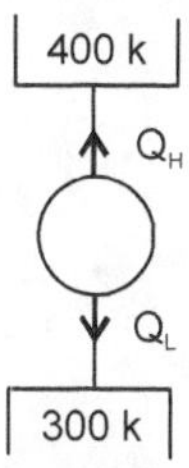

12. Since heat is taken from the high temperature sources and rejected to low temperature sink, hence the device is a heat engine not heat pump. Since, the temperature differences are negligible, the engine is reversible.

13.

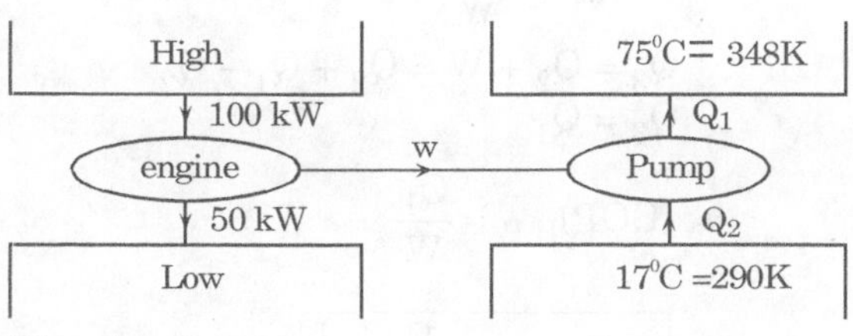

$W = Q_1 - Q_2$

= 100 – 50

= 50 kW

$C.O.P = \frac{348}{348 - 290} = \frac{Q_1}{W}$

Q_1 = 300 kW

Alternately

Work output from irreversible heat engine,

$W = Q_1 - Q_2$

= 100 –50 = 50 kW

Reversible heat pump

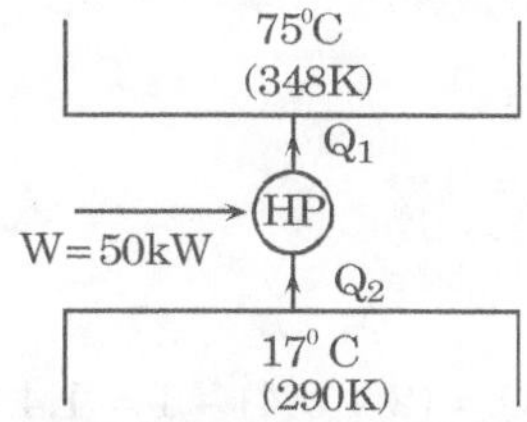

Also for irreversible heat engine

$Q_1 + W = Q_2$

$\therefore T_1\ \Delta S + 50 = T_2\ \Delta S$

or $(T_2 - T_1)\ \Delta S = 50$

or $(348 - 290)\ \Delta S = 50$

or $\Delta S = \frac{50}{58}$

$\therefore$ Rate of heat rejection,

$Q_2 = T_2\ \Delta S$

$= 348 \times \frac{50}{58} = 300$ kW

14. From Clausius inequality,

$\oint \frac{\partial Q}{T} < 0$ for irreversible process

i.e., $\frac{Q_1}{T_1} + \frac{Q_2}{T_2} + \frac{Q_3}{T_3} + \ldots < 0$

For process I

$\oint \frac{\delta Q}{T} = \frac{2500}{1200} - \frac{2500}{800} = -1.042 < 0$

For process II

$\oint \frac{\delta Q}{T} = \frac{2000}{800} - \frac{2000}{500} = -1.5 < 0$

Hence process II is more irreversible than process I.

15.

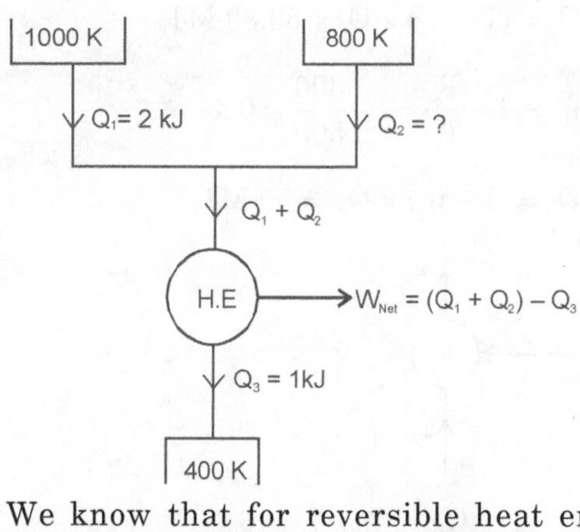

We know that for reversible heat engine, change in entropy is always zero

That is $\Delta S = 0$

$$\frac{Q_3}{T_3} - \left(\frac{Q_1}{T_1} + \frac{Q_2}{T_2}\right) = 0$$

$$\frac{1}{400} - \frac{2}{1000} - \frac{Q_2}{800} = 0$$

$Q_2 = 0.4$ kJ

$W_{Net} = (Q_1 + Q_2) - Q_3 = (2 + 0.4) - 1 = 1.4$ kJ.

16. $\eta = \dfrac{W_{net}}{Q_{in}} = 1 - \dfrac{T_{sink}}{T_{source}}$

$$1 - \frac{300}{500} = \frac{W_{net}}{1000}$$

$W_{net} = 1000 \times 0.4 = 400$ kJ.

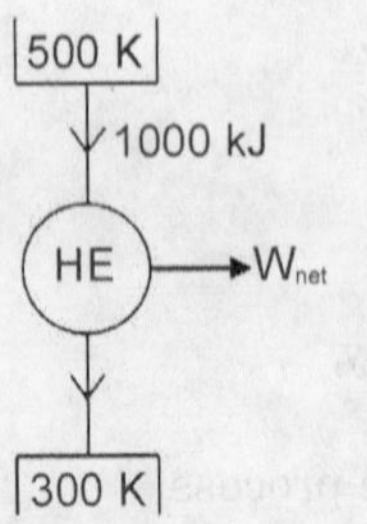

17.

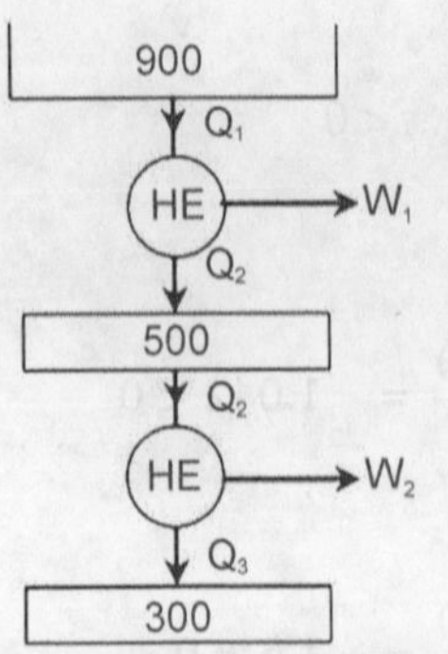

$$\eta_1 = \frac{1 - T_1}{T_1} = \frac{1 - 500}{900} = 4.44$$

$Q_2 = (1 - \eta_1) \times Q_1 = 53.33$ MJ

$$\eta_2 = 1 - \frac{T_3}{T_2} = 1 - \frac{300}{500} = 0.4$$

$Q_3 = (1 - \eta_2) \times Q_2 = 50$ MJ

18.

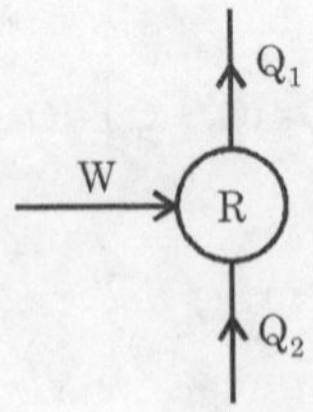

$$COP = \frac{Q_2}{W} = \frac{7.2}{1.8}$$

COP = 4

19.

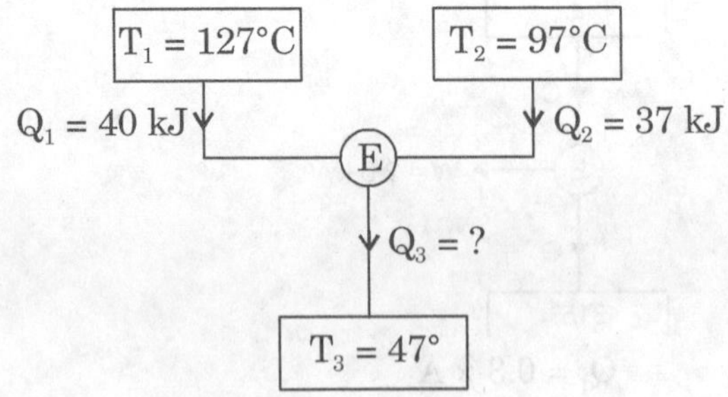

$$\frac{Q_1}{T_1} + \frac{Q_2}{T_2} - \frac{Q_3}{T_3} > 0$$

$$\frac{40}{127 + 273} + \frac{37}{97 + 273} - \frac{Q_3}{47 + 273} = 0$$

$Q_3 = 64$ kJ

20.

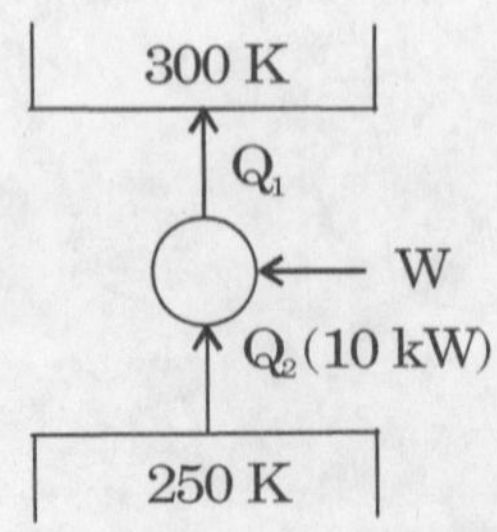

By question, given:

$Q_2 = 10$kW ; W = 15kW

$Q_1 = Q_2 + W = 25$kW

$$COP = \frac{Q_1}{W} = 1.67$$

21. Efficiency of heat engine,

$$\eta = \frac{W}{Q_1} = \frac{Q_1 - Q_2}{Q_1}$$

$$(COP)_{H.P.} = \frac{Q_3}{W}$$

$\because Q_3 = Q_2 + W = Q_2 + Q_1 - Q_2$

$Q_3 = Q_1$

so, $(COP)_{H.P} = \dfrac{Q_1}{W}$

$$\boxed{(COP)_{H.P} = \frac{1}{\eta} = \eta^{-1}}$$

23.

$$1 - \frac{T_2}{T_1} = 0.75$$

$$\frac{T_2}{273 + 727} = 0.25$$

$T_2 = 250$K

$T_2 = 273 - 250$

$[T_2 - 23°C]$

25. Tds = du + pdv

Constant volume

$Tds = dv \qquad (dv = 0)$

$Tds = C_v dt$

$$\left[\frac{dt}{ds} = \frac{C_V}{T}\right] \qquad \ldots \text{(i)}$$

$Tds = du - vdp$

Constant pressure

$$\left(\frac{dt}{ds} = \frac{C_P}{T}\right) \qquad \ldots \text{(ii)}$$

26. $ds = mC\ln\frac{T_h}{T_i}$

$= (20)\,(4186)\ln\frac{353}{303}$

$ds_{water} = 12786.99$ J/kg

$ds_{rod} = -12786.99$ J/kg

27. Throttling, $T_i = T_f$

Hence

$ds = R\ln\frac{P_i}{P_f}$

$= 8.314\ln\frac{0.5}{0.1}$

8.314 ln 5

(ds) = 13.388 kJ/K

29. Entropy of irreversible process always increases.

30. $(ds)_{uni} \geq 0$

31.

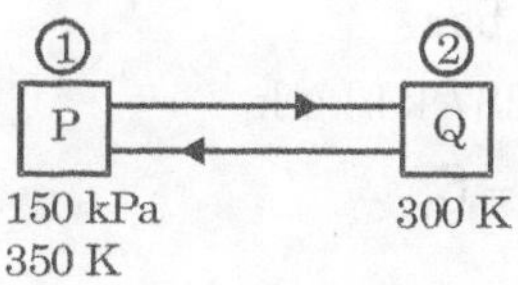

$$\frac{P_2}{P_1} = \left(\frac{T_2}{T_1}\right)^{\frac{\gamma}{\gamma-1}}$$

$$\Rightarrow P_2 = \left(\frac{300}{350}\right)^{\frac{0.4}{1.4}} \times 150$$

$P_2 = 87.4$ kPa

$P_2 = 87$kPa

32. $S_Q - S_P = C_v\ln\left(\frac{T_2}{T_1}\right) + R\ln\left(\frac{V_2}{V_1}\right)$

From perfect gas law,

$$\frac{V_2}{V_1} = \frac{P_1}{P_2}\frac{T_2}{T_1}$$

$$= \frac{150\times300}{300\times350} = 2.57$$

$\therefore S_Q - S_p = -0.1107 + 0.287\ln 2.57$

= 0.16 kJ/kgK

33. $\Delta s = \frac{\delta q}{T}$ for reversible process

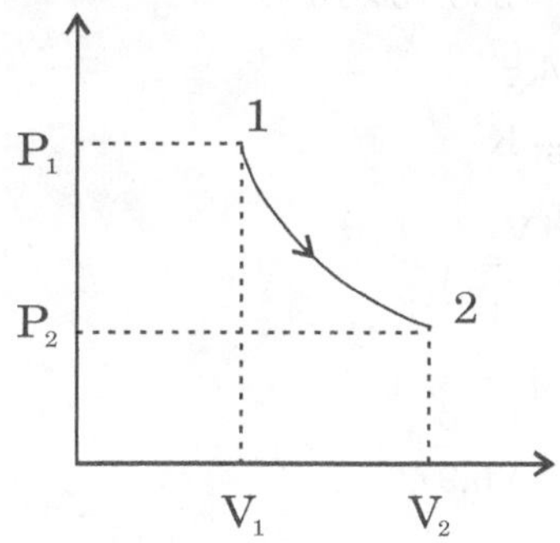

$$\delta Q = \delta W = \mathbf{P_1V_1\ ln}\frac{V_2}{V_1}$$

$$\Delta s = \frac{\delta q}{T} = \mathbf{mRln}\frac{p_1}{p_2}$$

$$\left[\Delta s = mR\ln\frac{p_1}{p_2}\right].$$

34. For clausius theorem, $\oint\frac{\delta Q}{T} < 0$ (for irreversible Heat engine).

36. $\Delta S_1 = \frac{Q}{T_1}$

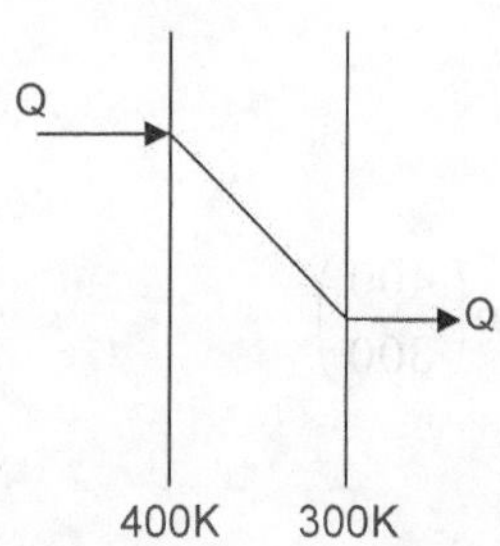

$$\Delta S_2 = \frac{Q}{T_2}$$

$$(\Delta S)_{generated} = \Delta S_1 + \Delta S_2$$

$$= \frac{Q}{400} - \frac{Q}{300}$$

$$= \frac{100\times 10^3}{100}\left(\frac{1}{4}-\frac{1}{3}\right) = 10^3 \times -0.0833$$

$$= -83.33 \text{ W/K}$$

Entropy generated = 83.33 W/K.

37. Tds = du + pdv (closed system)

$$283.15 \times 10(S_2 - S_1) = 16.2 \times 10^3$$

$$\therefore\ S_2 = 0.88 + \frac{16.2\times 10^3}{283.15\times 10}$$

$$= 6.6013 \text{ kJ/kg}$$

38. –116.36 J/kg–K

$$\Delta S = mR\ln\frac{V_2}{V_1}$$

$$= (287)(1)\ln\left(\frac{0.6}{0.9}\right)$$

$$\Rightarrow -116.36 \text{ J/kgK}$$

39. Here,

$m = 1$ kg

R = 400 J/kgK

$C_V = 1000$ J/kgK

$P_1 = 1$ bar

$T_1 = 300$ K

W = – 100 kJ = – 100 × 10^3 J

Q = 0

$$0 = -100\times 10^3 + mC_V(T_2 - T_1)$$

$$100\times 10^3 = 1\times 1000\,(T_2 - 300)$$

$$T_2 = 400 \text{ K}$$

$$V_1 = V_2$$

$$\Rightarrow \ln\frac{V_2}{V_1} = 0$$

$$S_2 - S_1 = mC_V\ln\frac{T_2}{T_1}$$

$$S_2 - S_1 = 1\times 1000\times\ln\left(\frac{400}{300}\right)$$

$$S_2 - S_1 = 287.68 \text{ J/K}$$

40. $\Delta S = mC_p\ln\frac{T_2}{T_1} - R\ln\frac{P_2}{P_1}$

$$\Delta S = -287\ln 2$$

$$\Delta S = -198.93 \text{ J/kg K}$$

41. Ideal gas

State - 1 : $T_1 = 300$ K, $P_1 = 100$ kPa

State - 2 : $T_2 = 600$ K, $P_2 = 500$ kPa,

$c_p = 1$ kJ/kg-K, $c_p - c_v = R$, $c_v = 0.7$ kJ/kg-K

$\Rightarrow c_p - c_v = 1 - 0.7 = R$

R = 0.3 kJ/kg–K

Change in specific entropy

$$s_2 - s_1 = c_p\ln\frac{T_2}{T_1} - R\ln\frac{P_2}{P_1}$$

$$= 1\times\ln\frac{600}{300} - 0.3\ln\frac{500}{100} = 0.21 \text{ kJ/kg-K}$$

42. $\Delta S = C_p\ln\frac{T_2}{T_1} - R\ln\frac{P_2}{P_1}$

Tds = dh – vdp

$= C_p dt$ – vdp

$$ds = \frac{C_p dt}{T} - \frac{R}{p}dp$$

$$\left[ds = C_P\ln\frac{T_2}{T_1} - R\ln\frac{P_2}{P_1}\right]$$

43. Air, m = 1 kg

$$T = 300K \xrightarrow{\text{Isothermally}} T = 300K,\ P_2 = 1 \text{ bar}$$

$P_1 = 10$ bar

R = 0.287 kJ/kg k

Now, $\Delta s = S_2 - S_1$

$$= mc_p\ell n\left(\frac{T_2}{T_1}\right)^{0} - mR\ell n\left(\frac{P_2}{P_1}\right)$$

$$= -1\times 0.287\ln\left(\frac{1}{10}\right)$$

$$= 0.66 \text{ kJ/kg k}$$

$$\therefore\ \Delta s = 0.66 \text{ kJ/kg k}$$

4 CHAPTER Availability and Irreversibility

Analysis of Previous GATE Papers

Topic	Year/Set → Marks Distribution ↓		2019 Set 1	2019 Set 2	2018 Set 1	2018 Set 2	2017 Set 1	2017 Set 2	2016 Set 1	2016 Set 2	2016 Set 3	2015 Set 1	2015 Set 2	2015 Set 3
Availability	1 Mark	MCQ Type												
		Numerical Type												
	2 Marks	MCQ Type												
		Numerical Type												
	Total													
Irreversibility	1 Mark	MCQ Type												
		Numerical Type												
	2 Marks	MCQ Type												
		Numerical Type		1										1
	Total			2										2

Availability

1. A heat reservoir at 900 K is brought into contact with the ambient at 300 K for a short time. During this period 9000 kJ of heat is lost by the heat reservoir. The total loss in availability due to this process is

(a) 18000 kJ (b) 9000 kJ
(c) 6000 kJ (d) None of these

[1995 : 2 Marks]

2. Availability of a system at any given state is

(a) a property of the system
(b) the maximum work obtainable as the system goes to dead state
(c) the total energy of the system
(d) the maximum useful work obtainable as the system goes to dead state

[2000 : 1 Mark]

3. Considering the relationship TdS = dU + pdV between the entropy (S), internal energy (U), pressure (p), temperature (T) and volume (V), which of the following statements is correct?

(a) It is applicable only for a reversible process
(b) For an irreversible process, TdS > dU+ pdV
(c) It is valid only for an ideal gas
(d) It is equivalent to 1st law, for a reversible process

[2003 : 2 Marks]

4. A steel billet of 2000 kg mass is to be cooled from 1250 K to 450 K. The heat released during this process is to be used as a source of energy. The ambient temperature is 303 K and specific heat of steel is 0.5 kJ/kgK. The available energy of this billet is

(a) 490.44 MJ (b) 30.95 MJ
(c) 10.35 MJ (d) 0.10 MJ

[2004 : 2 Marks]

5. The pressure, temperature and velocity of air flowing in pipe are 5 bar, 500 K and 50 m/s, respectively. The specific heats of air at a constant pressure and at constant volume are 1.005 kJ/kgK and 0.718 kJ/kgK, respectively. Neglect potential energy. If the pressure and temperature of the surroundings are 1 bar and 300 K, respectively, the available energy in kJ/kg of the air stream is

(a) 170 (b) 187
(c) 191 (d) 213

[2013 : 2 Marks]

6. The maximum theoretical work obtainable, when a system interacts to equilibrium with a reference environment, is called

(a) Entropy (b) Enthalpy
(c) Exergy (d) Rothalpy

[2014 : 1 Mark, Set-1]

Irreversibility

7. One side of a wall is maintained at 400 K and the other at 300 K. The rate of heat transfer through the wall is 1000 W and the surrounding temperature is 25°C. Assuming no generation of heat within the wall, the irreversibility (in W) due to heat transfer through the wall is.

[2015 : 2 Marks, Set-3]

8. Water flowing at the rate of 1 kg/s through a system is heated using an electric heater such that the specific enthalpy of the water increases by 2.50 kJ/kg and the specific entropy increases by 0.007 kJ/kg.K. The power input to the electric heater is 2.50 kW. There is no other work or heat interaction between the system and the surroundings. Assuming an ambient temperature of 300 K, the irreversibility rate of the system is ____ kW (round off to two decimal places)

[2019 : 2 Marks, Set-2]

ANSWERS

1. (c) **2.** (*d*) **3.** (*d*) **4.** (*a*) **5.** (*b*) **6.** (*c*)

EXPLANATIONS

1. Entropy change for hot reservoir

$$\Delta S_h = \frac{Q}{T} = \frac{-9000}{900} = -10 kJ/k$$

Energy gain in cold reservoir

$$\Delta S_C = \frac{Q}{T} = \frac{-9000}{300} = +30 kJ/k$$

Loss in availability = $T_0[\Delta S_c + \Delta S_h]$

$\Rightarrow 300(30 - 10)$

$= 300(20)$

$= 6000$ kJ

2. Availability of system of any given state is when no maximum useful work obtainable as the system goes to dead state.

3. The equation is valid for between reversible or inversible process of a closed system. Because it is the relation between properties which are independent to path.

4. Heat lost by steel = (0.5)(2000)(450 – 1250)

Gained = 800MJ

$$\Delta S_{Lost} \text{ by steel} = (2000)(0.5))\ln\left(\frac{450}{1250}\right)$$

$\Rightarrow -1.021$ MJ

(ΔS) gained = 1.021MJ

AE (W) = $a - T_0 d_s$

$= 800 - (303)(1.021)$

$= 490.7$ MJ

5. For flow stream,

A.E. $= (h_2 - h_1) + K.E. - T_0(S_2 - S_1)$

$$= c_p(T_2 - T_1) + \frac{1}{2}v^2 - T_0\left(C_p \ln\frac{T_2}{T_1} - R/n\frac{p_2}{p_1}\right)$$

$= 187$ kJ/kg

6. Exergy (or) Available Energy :

The maximum portion of energy which could be converted into useful work by ideal processes which reduce the system to dead state (a state in equilibrium with the earth and its atmosphere).

7. $T_0 = 273 + 25 = 298$ K

$AE_1 = a_1 - T_0 S_1$

$$AE_1 = 1000 - 298 \times \frac{1000}{400} = 255 \text{ watts}$$

$AE_2 = a_2 - T_0 S_2$

$$AE_2 = 1000 - 298 \times \frac{1000}{300} = 6.67 \text{ watts}$$

$\Delta AE = AE_2 - AE_1$

$= 248.33$ watt.

8. Irreversibility of system

$= T_0 .(\Delta S)_{syst}$

$= 300[0.007]$

$I = 2.1$ kJ/kg

$\dot{I} = \dot{m} \times 2.1 = 1 \times 2.1$

$\boxed{\dot{I} = 2.1 \text{ kW}}$

5

CHAPTER

Properties of Pure Substances

Analysis of Previous GATE Papers

Topic	Year/Set → Marks Distribution ↓		2019 Set 1	2019 Set 2	2018 Set 1	2018 Set 2	2017 Set 1	2017 Set 2	2016 Set 1	2016 Set 2	2016 Set 3	2015 Set 1	2015 Set 2	2015 Set 3
Pure Substances	1 Mark	MCQ Type			1						1			1
		Numerical Type												
	2 Marks	MCQ Type												
		Numerical Type			1				1					
	Total				3				2		1			1
Properties of gas mixture and Ideal gas equation of state	1 Mark	MCQ Type		1									1	
		Numerical Type												
	2 Marks	MCQ Type												
		Numerical Type	1											
	Total		2	1									1	

Pure substances

1. During the phase change of a pure substance
(a) dG = 0 (b) dP = 0
(c) dH = 0 (d) dU= 0
[1993 : 1 Mark]

2. At the triple point of a pure substance, the number of degrees of freedom is
(a) 0 (b) 1
(c) 2 (d) 3
[1993 : 1 Mark]

3. A vessel of volume 1.0 m^3 contains a mixture of liquid water and steam in equilibrium at 1.0 bar. Given that 90% of the volume is occupied by the steam, find the fraction of the mixture. Assume at 1.0 bar, v_f = 0.001 m^3/kg and v_g = 1.7 m^3/kg.
[1993 : 2 Marks]

4. In the vicinity of the triple point, the vapour pressures of liquid and solid ammonia are respectively given by
In P= 15.16 – 3063/T and
In P= 18.70 – 3754/T
where P is in atmospheric and T is in kelvin.
What is the temperature at the triple point?
[1993 : 2 Marks]

Common Data for Q. 5, 6 and 7

In the figure shown, the system is a pure substance kept in a piston-cyllnder arrangement. The system is initially a two-phase mixture containing 1 kg of liquid and 0.03 kg of vapour at a pressure of 100 kPa. Initially, the piston rests on a set of stops, a shown in the figure. A pressure of 200 kPa is required to exactly balance the weight of the piston and the outside atmospheric pressure. Heat transfer takes place into the system until its volume increases by 50%. Heat transfer to the system occurs in such a manner that the piston, where it is allowed to move, does so in a very slow (quasi-Static/quasi-equilibrium) process. The thermal reservoir from which heat is transferred to the system has a temperature of 400°C. Average temperature of the system boundary can be taken as 175°C. The heat transfer to the system is 1 kJ, during which its entropy increases by 10 J/K. Specific volumes of liquid (v_f) and vapour (v_g) phases, as well as values of saturation temperatures, are given in the table below.

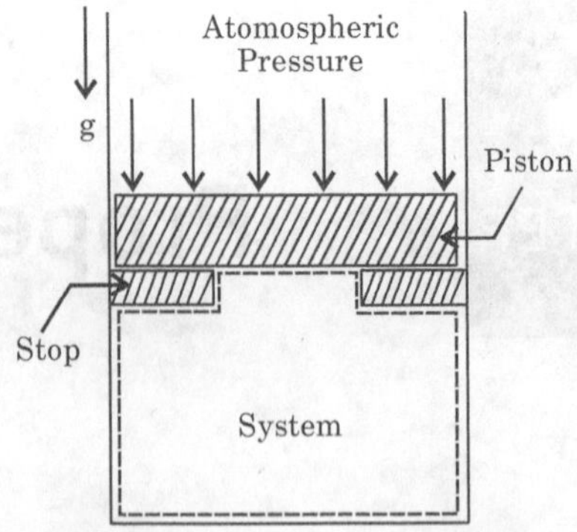

Pressure (kPa)	Saturation temp. T_{sat}(°C)	v_f(m^3/kg)	v_g(m^3/kg)
100	100	0.001	0.1
200	200	0.0015	0.002

5. At the end of the process, which one of the following situations will be true?
(a) superheated vapour will be left in the system
(b) no vapour will be left in the system
(c) a liquid + vapour mixture will be left in the system
(d) the mixture will exist at a dry saturated vapour state
[2008 : 2 Marks]

6. The work done by the system during the process is
(a) 0.1 kJ (b) 0.2 kJ
(c) 0.3 kJ (d) 0.4 kJ
[2008 : 2 Marks]

7. The net entropy generation (considering the system and the thermal reservoir together) during the process is closest to
(a) 7.5 J/K (b) 7.7 J/K
(c) 8.5 J/K (d) 10 J/K
[2008 : 2 Marks]

8. 1.5 kg of water is in saturated liquid state at 2 bar (v_f = 0.001061 m^3/kg, u_f= 504.0 kJ/kg, h_f = 505 kJ/kg). Heat is added in a constant pressure process till the temperature of water reaches 400°C (v = 1.5493 m^3/kg, u = 2967.0 k J/kg, h = 3277.0 kJ/kg). The heat added (in kJ) in the process is______.
[2014 : 2 Marks, Set-1]

9. A pure substance at 8 MPa and 400°C is having a specific internal energy of 2864 kJ/kg and a specific volume of 0.03432 m^3/kg. Its specific enthalpy (in kJ/kg) is, ________.
[2014 : 1 Mark, Set-2]

10. A rigid container of volume 0.5 m³ contains 1.0 kg of water at 120°C (v_f = 0.00106 m³/kg, v_g = 0.8908 m³/kg). The state of water is

(a) compressed liquid
(b) saturated liquid
(c) a mixture of saturated liquid and saturated vapor
(d) superheated vapor

[2015 : 1 Mark, Set-3]

11. For water at 25°C, dp_s/dT_s = 0.189 kPa/K (p_s is the saturation pressure in kPa and T_s is the saturation temperature in K) and the specific volume of dry saturated vapour is 43.38 m³/kg.
Assume that the specific volume of liquid is negligible in comparison with that of vapour. Using the Clausius-Clapeyron equation, an estimate of the enthalpy of evaporation of water at 25°C (in kJ/kg) is

[2016 : 2 Marks, Set-1]

12. The INCORRECT statement about the characteristics of critical point of a pure substance is that

(a) there is no constant temperature vaporization process
(b) it has point of inflection with zero slope
(c) the ice directly converts from solid phase to vapor phase
(d) saturated liquid and saturated vapor states are identical

[2016 : 1 Mark, Set-3]

13. Which one of the following statements is correct for a superheated vapour?

(a) Its pressure is less than the saturation pressure at a given temperature.
(b) Its temperature is less than the saturation temperature at a given pressure.
(c) Its volume is less than the volume of the saturated vapour at a given temperature.
(d) Its enthalpy is less than enthalpy of the saturated vapour at a given pressure.

[2018 : 1 Mark, Set-1]

14. A tank of volume 0.05 m³ contains a mixture of saturated water and saturated steam at 200°C. The mass of the liquid present is 8 kg. The entropy (in kJ/kgK) of the mixture is _______ (correct of two decimal places)
Property data for saturated steam and water are:
At 200°C [P_{sat} = 1.5538 MPa]
v_f = 0.001157 m³/kg, v_g = 0.12736 m³/kg
s_{fg} = 4.1014 kJ/kgK, s_f = 2.3309 kJ/kgK

[2018 : 2 Marks, Set-1]

Properties of gas mixture and ideal gas equation of state

15. The relationship $(\partial T/\partial P)_h = 0$ holds good for

(a) An ideal gas at any state
(b) A real gas at any state
(c) Any gas at its critical state
(d) Any gas at its inversion point

[1993 : 1 Mark]

16. Constant pressure lines in the superheated region of the Mollier diagram will have

(a) A positive slope
(b) A negative slope
(c) Zero slope
(d) Both positive and negative slope

[1995 : 1 Mark]

17. When wet steam flows through a throttle valve and remains wet at exit

(a) Its temperature and quality increase
(b) Its temperature decreases but quality increases
(c) Its temperature increases but quality decreases
(d) Its temperature and quality decrease

[1996 : 1 Mark]

18. Which combination of the following statements is correct?

P : A gas cools upon expansion only when its Joule-Thomson coefficient is positive in the temperature range of expansion.

Q : For a system undergoing a process, its entropy remains constant only when the process is reversible.

R : The work done by a closed system in an adiabatic is a point function.

S : A liquid expands upon freezing when the slope of its fusion curve on pressure-Temperature diagram is negative.

(a) R and S (b) P and Q
(c) Q, R and S (d) P, Q and R

[2007 : 2 Marks]

19. Water has a critical specific volume of 0.003155 m^3/kg. A closed and rigid steel tank of volume 0.025 m^3 contains a mixture of water and steam at 0.1 MPa. The mass of the mixture is 10 kg. The tank is now slowly heated. The liquid level inside the tank

(a) will rise

(b) will fall

(c) will remain constant

(d) may rise or fall depending on the amount of heat transferred

[2007 : 1 Mark]

20. 2 moles of oxygen are mixed adiabatically with another 2 moles of oxygen in a mixing chamber, so that the final total pressure and temperature of the mixtures become same as those of the individual constituents at their initial states. The universal gas constant is given as R. The change in entropy due to mixing, per mole of oxygen, is given by

(a) - Rln2 (b) 0

(c) Rln2 (d) Pln4

[2008 : 1 Mark]

21. The Vander Waals equation of state is $\left(p+\frac{a}{v^2}\right)(v-b)=RT$, where p is pressure, v is specific volume, T is temperature and R is characteristic gas constant. The SI unit of a is

(a) J/kgK (b) m^3/kg

(c) $m^5/kg\text{-}s^2$ (d) Pa/kg

[2015 : 1 Mark, Set-2]

22. If one mole of H_2 gas occupies a rigid container with a capacity of 1000 liters and the temperature is raised from 27°C to 37°C, the change in pressure of the contained gas (round off to two decimal places), assuming ideal gas behavior, is _____ Pa. (R = 8.314 J/mol.K).

[2019 : 2 Marks, Set-1]

23. For a simple compressible system, v, s, p and T are specific volume, specific entropy, pressure and temperature, respectively. As per Maxwell's relation, $\left(\frac{\partial v}{\partial s}\right)_p$ is equal to

(a) $\left(\frac{\partial p}{\partial v}\right)_T$ (b) $\left(\frac{\partial T}{\partial p}\right)_s$

(c) $\left(\frac{\partial s}{\partial T}\right)_p$ (d) $-\left(\frac{\partial T}{\partial v}\right)_p$

[2019 : 1 Mark, Set-2]

ANSWERS

1. (b)	2. (a)	5. (a)	6. (d)	7. (c)	10. (c)	12. (c)	13. (a)	15. (a & b)	16. (a)
17. (b)	18. (a)	19. (a)	20. (b)	21. (c)	23. (b)				

EXPLANATIONS

1. dP = 0

During phase change

$\Delta P = 0, \Delta T = 0$

2. According to Gibb's phase rule

$P + F = C + 2$

at triple point C = 1, P = 3

$3 + F = 1 + 2$

$F = 0$

3. $v_f = 0.001\ m^3/kg$

$v_g = 1.7\ m^3/kg$

$$\text{mass of mixture} = \frac{0.9}{v_g}+\frac{0.1}{v_f}$$

$$=\frac{0.9}{1.7}+\frac{0.1}{0.001} = 100.53\ kg$$

Specific volume $v = \frac{1}{m} = 9.947 \times 10^{-3}\ m^3/kg$

$v = v_f + x(v_f - v_g)$

$9.947 \times 10^{-3} = 0.001 + x(1.7 - 0.001)$

$x = 5.266 \times 10^{-3}$

4. At triple point the vapour pressure of solid ammonia is equal to vapor pressure of liquid ammonia

$$-\frac{3063}{T} = 18.7 - \frac{3754}{T}$$

$$T = \frac{3754-3063}{18.7-15.16}$$

$T = 195.2K$

5. Given at 100 kPa

$m_l = 1$ kg, $m_v = 0.03$ kg

$V_1 = m_l v_1 + m_v v_g$

$= 1 \times .001 + 0.03 \times 0.1$

$V_1 = 0.004$ m^3

$V_2 = 1.5V_1$ (Heat addition increase volume by 50%)

$= 0.006$ m^3

Now at 2kPa

$v_g = 0.002$ m^3/kg

Total volume

$= v_g \times$ m(at 200kPa) ($m = m_1 + m_v$)

$= (0.002) \times (1.03)$

$= 0.00206$ m^3

Since final volume V_2 is more than that of saturated total volume. So, the system will be in superheated state.

6. w = pdv

$= 200\ (v_2 - v_1)$

$= 200\ (0.5 \times v_1)$

$= 200 \times 0.5 \times 0.004$

[w = 0.4 kJ]

7. $S_{gen} = \Delta S - \frac{\delta q}{T}$

$$= 10 - \frac{1000}{673} = 8.51$$

$S_{gen} = 8.51$ J/k

8. Given, m = 1.5 kg

$h_1 = h_f = 505$ kJ/kg $\qquad h_2 = 3277.0$

From Ist Law,

$dQ = du + pdv = dh - vdp$

$dQ = dh$ (as $vdp = 0$)

$Q_{add} = dQ = m\ (h_2 - h_1) = (3277.0 - 505) \times 1.5$

$Q_{added} = 4158$ kJ.

9. $h = u + pv$

$h = 2864 + (8000\text{ kPa}) \times (0.03432)$

$= 2864 + 274.56$

$= 3138.56$

$= 3139$

10. $V = \frac{0.5}{1}$ m^3 / kg = 0.5 m^3 / kg

Since $V_f < V < V_g$ the state of water is mixture of saturated water and saturated vapour.

11. $\frac{dP_s}{dT_s} = 0.189$ kPa/K

$T_{sat} = 273 + 25 = 298$ K

$v_g = 43.38$ m^3/kg

$v_f = 0$

$v_{fg} = v_g - v_f$

$= 43.38 - 0 = 43.38$ m^3/kg

$$\frac{dP_s}{dT_s} = \frac{h_{fg}}{T_{sat} \times v_{fg}}$$

$$= \frac{h_{fg}}{298 \times 43.38}$$

$h_{fg} = 2443.25$ kJ/kg

12. Because it is sublimation.

13.

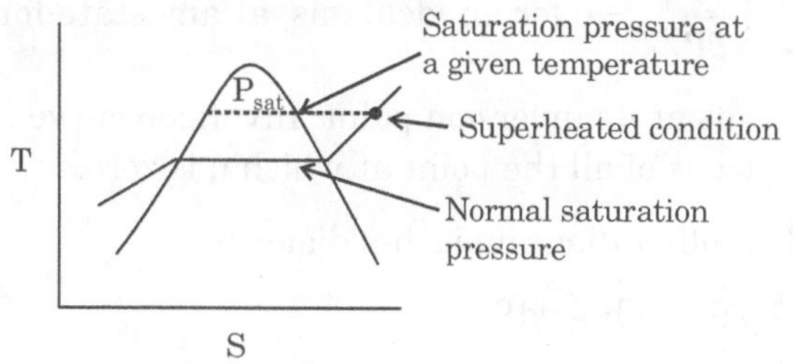

14. Total volume of tank (V) = 0.05 m^3

Means of liquid (m_2) = 8 kg

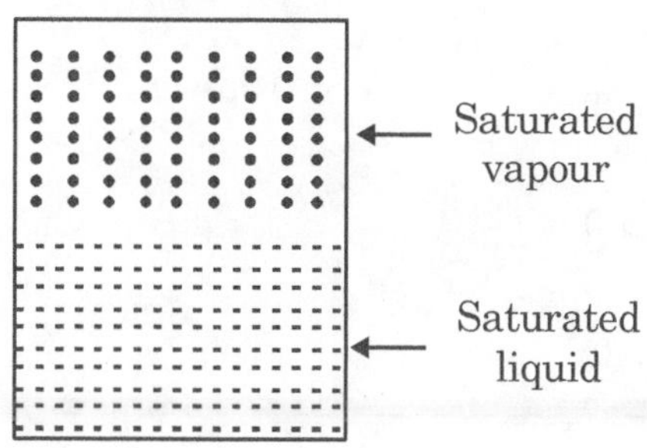

$$x = \frac{m_V}{m_V + m_L} \text{(dryness fraction)}$$

$$V = V_L + V_V \quad \begin{bmatrix} V_L - v \text{ of vapour in tank} \\ V_V - v \text{ of vapour in tank} \end{bmatrix}$$

$V_V = m_V V_V$

$V_V = V_g$ @ 200°C

$= 0.12736$ m^3/kg

$V_V = m_L V_L$

$V_L = V_f$ @ 200°C

$= 0.001157$ m^3/kg

$V_L = 8 \times 0.001157$

$= 9.256 \times 10^{-3}$ m^3

$0.05 = 9.256 \times 10^{-3} + V_V$

$V_V = 0.0474\ m^3$

$0.04074 = m_V \times 0.12736$

$m_V = 0.3198\ kg$

$$x = \frac{0.3198}{0.3198 + 8}$$

$\Rightarrow x = 0.0384$

Specific entropy of mixture(s)

$s = s_f + xs_{fg}$

$s = 2.3309 + 0.0384 \times 4.1014$

$s = 2.4884$ kJ/kg-K

15. $\left(\frac{\partial T}{\partial P}\right)_h$ is Joule-Thompson coefficient (μ)

$\left(\frac{\partial T}{\partial P}\right)_h = 0$ for an ideal gas, at any state for any gas at its inversion point. Inversion curve is the locus of all the point at which μ is zero.

16. Mollier diagram is h-s diagram

$Tds = dh - vdP$

$dP = 0$

$Tds = dh$

$$\frac{dh}{ds} = T$$

$\because T > 0$

$$\frac{dh}{ds} > 0$$

slope > 0

17. Throttling always takes place in direction of decreasing pressure

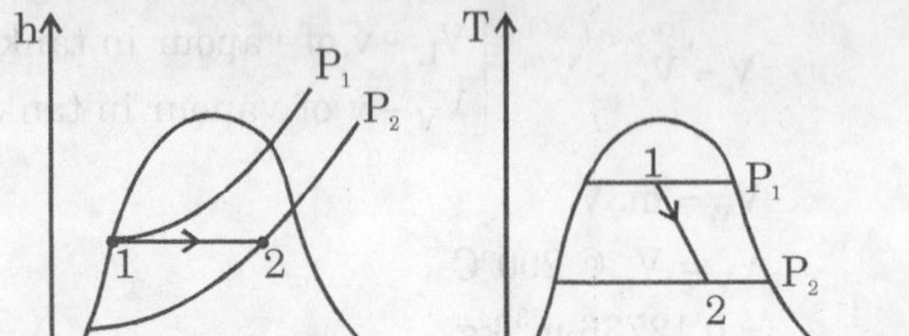

Hence we can say during throtting temperature also decreases quality increases.

18. First law of thermodymnamics,

1. $\delta Q = dU + \delta W$

Adiabatic, $\delta Q = 0$

$dU = -\delta W$

Hence work acts as the point function.

2. Liquid expands upon freezing when the liquid is water. In case of water the slope of fusion curve is negative.

19. $v_c = 0.0031\ m^3/kg$

$$v_2 = \frac{0.025}{10} = 0.0025\ m^3\ kg$$

as ($v_c > v$)

As heat is supplied the amount of steam in mixture will decrease and pressure will increase, because volume of mixture is less than critical volume. The final state will move towards saturated liquid line. Hence level of liquid in tank will rise.

20. Since, $\Delta Q = 0$, therefore $S = \int \frac{dQ}{T} = 0$

21. $p + \frac{a}{v^2}$ both term should gave same unit since they are getting added

$$\Rightarrow \frac{N}{m^2} = a\left(\frac{kg}{m^3}\right)^2$$

$$\Rightarrow a(\text{unit}) = \frac{m^6}{kg^2}.kg.\frac{m}{s^2 m^2}$$

$$= \frac{m^5\ kg}{kg^2 s^2} = \frac{m^5}{kgs^2}$$

22. $P_1 V = n\bar{R}T_1$

$P_2 V = n\bar{R}T_2$

Now, Changs in pressure,

$$P_2 - P_1 = n\bar{R}(T_2 - T_1)$$

$$\Rightarrow \Delta P \times 1 = 1 \times 8.314(37 - 27)$$

$$\Rightarrow \Delta P = 83.14 \text{ Pascal}$$

23. $dH = TdS + vdP$

$$\Rightarrow \left(\frac{\partial T}{\partial P}\right)_S = \left(\frac{\partial V}{\partial S}\right)_P$$

So, $\Rightarrow \left(\frac{\partial V}{\partial S}\right)_P = \left(\frac{\partial T}{\partial P}\right)_S$

6 CHAPTER

Power Plant Engineering

Analysis of Previous GATE Papers

Topic	Year/Set → Marks Distribution ↓		2019 Set 1	2019 Set 2	2018 Set 1	2018 Set 2	2017 Set 1	2017 Set 2	2016 Set 1	2016 Set 2	2016 Set 3	2015 Set 1	2015 Set 2	2015 Set 3
Boiler	1 Mark	MCQ Type												
		Numerical Type												
	2 Marks	MCQ Type												
		Numerical Type												
	Total													
Steady flow through nozzle	1 Mark	MCQ Type												
		Numerical Type												
	2 Marks	MCQ Type												
		Numerical Type												
	Total													
Rankine and Brayton cycles	1 Mark	MCQ Type		1									1	
		Numerical Type								1				
	2 Marks	MCQ Type												
		Numerical Type	1									1	1	
	Total		2	1						1		2	3	
Steam Turbines	1 Mark	MCQ Type				1								
		Numerical Type												
	2 Marks	MCQ Type												
		Numerical Type	1				1	1	1					
	Total		1			1	2	2	2					
Air compressors	1 Mark	MCQ Type												
		Numerical Type												
	2 Marks	MCQ Type												
		Numerical Type									1			
	Total										2			

Boiler

1. Boiler rating is usually defined in terms of

(a) Maximum temperature of steam in kelvin

(b) Heat transfer rate in kJ/hr

(c) Heat transfer area in meter^2

(d) Steam output in kg/hr

[1992 : 2 Marks]

2. The equivalent evaporation (kg/hr) of a boiler producing 2000 kg/hr of steam with enthalpy content of 2426 kJ/kg from feed water at temperature 40°C (liquid enthalpy =168 kJ/kg) is (enthalpy of vaporization of water at 100°C = 2258 kJ/kg)

(a) 2000 (b) 2149

(c) 186 (d) 1649

[1993 : 2 Marks]

3. Which among the following is the boiler mounting?

(a) Blow off cock (b) Feed pump

(c) Economizer (d) Super heater

[1997 : 1 Mark]

Steam flow through nozzle

4. When initially dry and saturated steam flows through a nozzle, the ratio of actual discharge to calculated discharge is

(a) Equal to 1.0

(b) Greater than 1.0

(c) Less than 1.0

(d) Independent of inlet conditions

[1990 : 2 Marks]

5. Subsonic and supersonic diffusers have the following geometry.

(a) Divergent and convergent respectively

(b) Both divergent

(c) Both convergent

(d) Convergent and divergent respectively

[1992 : 2 Marks]

6. In adiabatic flow with friction, the stagnation temperature along a streamline .... (increase/ remain constant)

[1995 : 1 Mark]

7. **List-I**

A. Steam nozzle

B. Compressible

C. Surface tension

D. Heat conduction

List-II

1. Mach number
2. Reaction turbine
3. Biot number
4. Nusselt number
5. Super saturation
6. Weber number

[1997 : 2 Marks]

8. The isentropic heat drop in the nozzle of an impulse steam turbine with a nozzle efficiency 0.9, blade velocity ratio 0.5 and mean blade velocity 150 m/s in kJ/kg is

(a) 50 (b) 40

(c) 60 (d) 75

[1998 : 2 Marks]

9. For a compressible fluid, sonic velocity is

(a) a property of the fluid

(b) always given by $\sqrt{\gamma RT}$ where γ, R and T are respectively the ratio of specific heats, gas constant and temperature in K

(c) always given by $\left(\frac{\partial p}{\partial \rho}\right)_s^{1/2}$, where p, ρ and s are respectively pressure, density and entropy

(d) always greater then the velocity of fluid at any location

[2000 : 2 Marks]

Rankine and Brayton cycles

10. The fundamental objective of employing the condenser in a steam power plant is to _____ (reduce/increase) the dryness fraction of steam.

[1989 : 2 Marks]

11. In a gas turbine power plant intercooler are used to cool the ____ (hot gases/compressed air) in order to decrease the ______ (expansion work/ compression work)

[1989 : 2 Marks]

12. In the Rankine cycle when superheated steam is used

(a) Thermal efficiency increases

(b) Steam consumption decreases

(c) Steam dryness after expansion increases

(d) All of the above

[1990 : 2 Marks]

13. In a Rankine cycle heat is added
(a) Reversibly at constant volume
(b) Reversibly at constant temperature
(c) Reversibly at constant pressure and temperature
(d) Reversibly at constant pressure

[1991 : 2 Marks]

14. In steam and other vapour cycle, the process of removing non-condensable is called
(a) Scavenging process
(b) Deaeration process
(c) Exhaust process
(d) Condensation process

[1992 : 2 Marks]

15. Air expands steadily through a turbine from 6 bar, 800 K to 1 bar, 520 K. During the expansion, heat transfer from air to the surroundings at 300 K is 10 kJ/kg air. Neglect the changes in kinetic and potential energies and evaluate the irreversibility per kg air. Assume air to behave as an ideal gas with c_p = 1.0 kJ/kgK and R = 0.3 kJ/kgK.

[1993 : 2 Marks]

16. In problem above, find the actual work and maximum work per kg air.

[1993 : 2 Marks]

17. A gas turbine cycle with heat exchange and reheating improves
(a) Only the thermal efficiency
(b) Only the specific power output
(c) Both thermal efficiency and specific power output
(d) Neither thermal efficiency nor specific power output

[1993 : 2 Marks]

18. For a given set of operating pressure limits of a Rankine cycle the highest efficiency occurs for
(a) Saturated cycle
(b) Superheated cycle
(c) Reheat cycle
(d) Regenerative cycle

[1994 : 1 Mark]

19. A gas turbine cycle with infinitely large number of stages during compression and expansion leads to
(a) Stirling cycle (b) Atkinson cycle
(c) Ericson cycle (d) Brayton cycle

[1994 : 1 Mark]

20. Consider a Rankine cycle with superheat. If the maximum pressure in the cycle is increased without changing the maximum temperature and the minimum pressure, the dryness fraction of steam after the isentropic expansion will increase [TRUE/FALSE]

[1995 : 1 Mark]

21. Consider an actual regenerative Rankine cycle with one open feed water heater. For each kg steam entering the turbine, if m kg steam with a specific enthalpy of h_1 is blade from the turbine, and the specific enthalpy of liquid water entering the heater is h_2, then h_3 the specific enthalpy of saturated liquid leaving the heater is equal to
(a) $mh_1 - (h_2 - h_1)$ (b) $h_1 - m(h_2 - h_1)$
(c) $h_2 - m(h_2 - h_1)$ (d) $mh_2 - (h_2 - h_1)$

[1997 : 2 Marks]

22. Select statement from List-II matching the processes in List-I. Enter your answer as A, B is the correct choice for (1) is (A) and (2) and (B).

List-I
1. Intercooling
2. Isothermal compression

List-Il
A. No heat transfer during compression
B. Reduce low pressure compressor work
C. Heat rejection during compression
D. Reduce high pressure compressor work

[1999 : 2 Marks]

23. The efficiency of superheat Rankine cycle is higher than that of simple Rankine cycle because
(a) the enthalpy of main steam is higher for superheat cycle
(b) the mean temperature of heat addition is higher for superheat cycle
(c) the temperature of steam in the condenser is high
(d) the quality of steam in the condenser is low

[2002 : 2 Marks]

24. In Rankine cycle, regeneration results in higher efficiency because
(a) pressure inside the boiler increases
(b) heat is added before steam enters the low pressure turbine
(c) average temperature of heat addition in the boiler increases
(d) total work delivered by the turbine increases

[2003 : 1 Mark]

25. In a gas turbine, hot combustion products with the specific heats c_p = 0.98 kJ/kgK, and c_v = 0.7538 kJ/kgK enters the turbine at 20 bar, 1500 K and exits at 1 bar. The isentropic efficiency of the turbine is 0.94. The work developed by the turbine per kg of gas flow is

(a) 689.64 kJ/kg (b) 794.66 kJ/kg
(c) 1009.72 kJ/kg (d) 1312.00 kJ/kg

[2003 : 2 Marks]

26. The compression ratio of a gas power plant cycle corresponding to maximum work output for the given temperature limits of T_{min} and T_{max} will be

(a) $\left(\frac{T_{max}}{T_{min}}\right)^{\frac{\gamma}{2(\gamma-1)}}$ (b) $\left(\frac{T_{min}}{T_{max}}\right)^{\frac{\gamma}{2(\gamma-1)}}$

(c) $\left(\frac{T_{max}}{T_{min}}\right)^{\frac{\gamma-1}{\gamma}}$ (d) $\left(\frac{T_{min}}{T_{max}}\right)^{\frac{\gamma-1}{\gamma}}$

[2004 : 1 Mark]

Common Data Questions Q. 27 and Q. 28

Consider a steam power plant using a reheat cycle as shown. Steam leaves the boiler and enters the turbine at 4 MPa, 350°C (h_3 = 3095 kJ/kg). After expansion in the turbine to 400 kPa (h_4 = 2609 kJ/kg), the steam is reheated to 350°C (h_5 = 3170 kJ/kg), and then expanded in a low pressure turbine to 10 kPa (h_6 = 2165 kJ/kg) the specific volume of liquid handled by the pump can be assumed to be

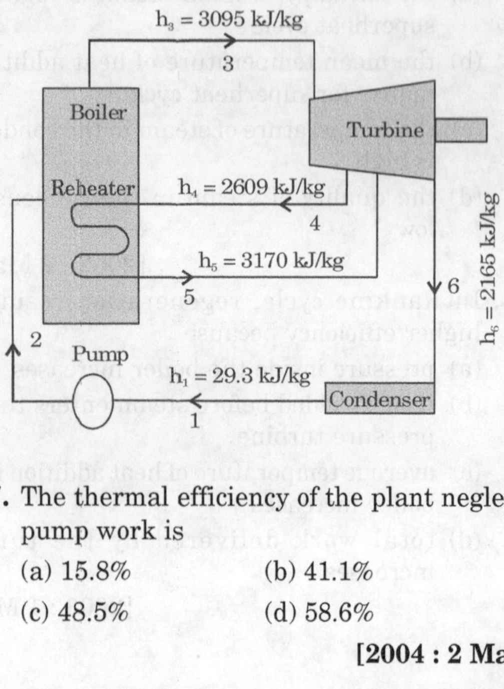

27. The thermal efficiency of the plant neglecting pump work is

(a) 15.8% (b) 41.1%
(c) 48.5% (d) 58.6%

[2004 : 2 Marks]

28. The enthalpy at the pump discharge (h_2) is

(a) 0.33 kJ/kg (b) 3.33 kJ/kg
(c) 4.0 kJ/kg (d) 33.3 kJ/kg

[2004 : 2 Marks]

Common Data for Questions 29 and 30

In two air standard cycles - one operating in the Otto and the other on the Brayton cycle - air is isentropically compressed from 300 to 450 K. Heat is added to raise the temperature to 600 K in the Otto cycle and to 550 K in the Brayton cycle.

29. In η_0 and η_B are the efficiencies of the Otto and Brayton cycles, then

(a) $\eta_0 = 0.25, \eta_B = 0.18$
(b) $\eta_0 = \eta_B = 0.33$
(c) $\eta_0 = 0.5, \eta_B = 0.45$
(d) it is not possible to calculate the efficiencies unless the temperature after the expansion is given

[2005 : 2 Marks]

30. If W_0 and W_B are work outputs per unit mass, then

(a) $W_0 > W_B$
(b) $W_0 < W_B$
(c) $W_0 = W_B$
(d) it is not possible to calculate the work outputs unless the temperature after the expansion is given

[2005 : 2 Marks]

Linked Answer Questions 31 and 32

The following table of properties was printed out for saturated liquid and saturated vapour of ammonia. The title for only the first two columns are available. All that we know that the other columns (columns 3 to 8) contain data on specific properties, namely, internal energy (kJ/ kg), enthalpy (kJ/kg) and entropy (kJ/kgK).

t (°C)	p (kPa)						
-20	190.2	88.76	0.3657	89.05	5.6155	1299.5	1418.0
0	429.6	179.69	0.7114	180.36	5.3309	1318.0	1442.2
20	857.5	272.89	1.0408	274.30	5.0860	1332.2	1460.2
40	1554.9	368.74	1.3574	371.43	4.8662	1341.0	1470.2

31. The specific enthalpy data are in columns

(a) 3 and 7 (b) 3 and 8
(c) 5 and 7 (d) 5 and.8

[2005 : 2 Marks]

32. when saturated liquid at 40°C is throttled to – 20°C, the quality at exit will be

(a) 0.189 (b) 0.212

(c) 0.231 (d) 0.788

[2005 : 2 Marks]

33. Given below is an extract from steam tables.

Temperature (°C)	P_{sat} (bar)	Specific volume (m^3/kg)		Enthalpy (KJ/kg)	
		Saturated liquid	Saturated vapour	Saturated liquid	Saturated vapour
45	0.09593	0.001010	15.26	188.45	2394.8
342.24	150	0.001658	0.010337	1610.5	2610.5

Specific enthalpy of water in kJ/kg at 150 bar and 45°C is

(a) 203.60 (b) 200.53

(c) 196.38 (d) 188.45

[2006 : 2 Marks]

34. Determine the correctness or otherwise Assertion (A) and the Reason (R).

Assertion (A): In a power plant working on a Rankine cycle, the regenerative feed water heating improves the efficiency of the steam turbine.

Reason (R): The regenerative feed water heating raises the average temperature of heat addition in the Rankine cycle.

(a) Both (A) and (R) are true and (R) is the correct reason for (A)

(b) Both (A) and (R) are true but (R) is NOT the correct reason for (A)

(c) Both (A) and (R) are false

(d) (A) is false but (R) is true

[2006 : 2 Marks]

35. Determine the correctness or otherwise of the following Assertion (A) and the Reason (R).

Assertion (A) : Condenser is an essential equipment in a steam power plant.

Reason (R): For the same mass flow rate and the same pressure rise, a water pump requires substantially less power than a steam compressor.

(a) Both (A) and (R) are true and (R) is the correct reason for (A)

(b) Both (A) and (R) are true but (R) is NOT the correct reason for (A)

(c) Both (A) and (R) are false

(d) (A) is false but (R) is true

[2006 : 2 Marks]

36. Group-I shows different heat addition processes in power cycles. Likewise, Group-II shows different heat removal processes. Group-III lists power cycles. Match items from Groups I, II and III.

Group I	Group II	Group III
P. Pressure constant	S. Pressure constant	1. Rankine cycle
Q. Volume constant	T. Volume constant	2. Otto cycle
R. Temperature	U. Temperature	3. Carnot cycle
		4. Diesel cycle
		5. Brayton cycle

(a) P–S–5, R–U–3, P–S–1, Q–T–2

(b) P–S–1, R–U–3, P–S–4, P–T–2

(c) R–T–3, P–S–1, P–T–4, Q–S–5

(d) P–T–4, R–S–3, P–S–1, P–S–5

[2006 : 2 Marks]

37. Which combination of the following statements is correct?

The incorporation of reheater in a steam power plant:

P : Always increases the thermal efficiency of the plant.

Q : Always increases the dryness fraction of steam at condenser inlet.

R : Always increases the mean temperature of heat addition.

S : Always increases the specific work output.

(a) P and S (b) Q and S

(c) P, R and S (d) P, Q, R and S

[2007 : 2 Marks]

38. A thermal power plant operates on a regenerative cycle with a single open feedwater heater, as shown in the figure. For the state points shown, the specific enthalpies are: h_1 = 2800 kJ/kg and h_2 = 200 kJ/kg. The bleed to the feedwater heater is 20% of the boiler steam generation rate. The specific enthalpy at state 3 is

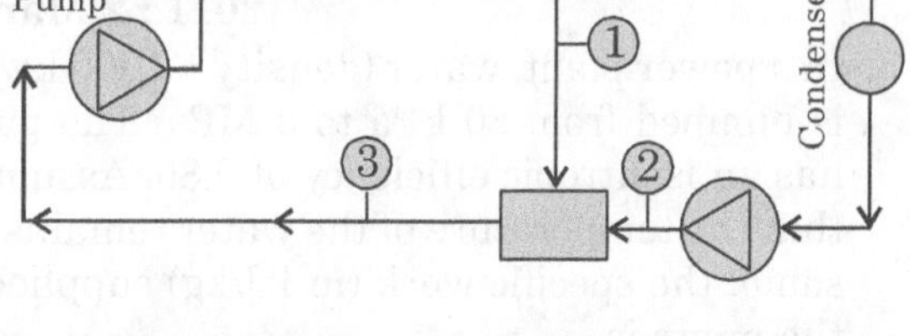

(a) 720 kJ/kg (b) 2280 kJ/kg

(c) 1500 kJ/kg (d) 3000 kJ/kg

[2008 :2 Marks]

Common Data for Questions 39 and 40.

In a steam power plant operating on the Rankine cycle, steam enters the turbine at 4 MPa, 350°C and exists at a pressure of 15 kPa. Then it enters the condenser and exists as saturated water. Next, a pump feeds back the water to the boiler. The adiabatic efficiency of the turbine is 90%. The thermodynamic states of water and steam are given in the table.

State	$h(kJkg^{-1})$		$s(kJkg^{-1}K^{-1})$		$v(m^3kg^{-1})$	
Steam : 4 MPa, 350°C	3092.5		6.5821		0.06645	
Water: 15kPa	h_f	h_g	s_f	s_g	v_f	v_g
	225.94	2599.1	0.7549	8.0085	0.001014	10.02

h is specific enthalpy, s is specific entropy and v the specific volume; subscript f and g denote saturated liquid state and saturated vapour state.

39. The net work output ($kJ\ kg^{-1}$) of the cycle is

(a) 498 (b) 775
(c) 860 (d) 957

[2010 : 2 Marks]

40. Heat supplied ($kJ\ kg^{-1}$) to the cycle is

(a) 2372 (b) 2576
(c) 2863 (d) 3092

[2010 : 2 Marks]

41. The values of enthalpy of steam at the inlet and outlet of a steam turbine in a Rankine cycle are 2800 kJ/kg and 1800 kJ/kg respectively. Neglecting pump work, the specific steam consumption in kg/kW hour is

(a) 3.60 (b) 0.36
(c) 0.06 (d) 0.01

[2011 : 2 Marks]

42. An ideal Brayton cycle, operating between the pressure limits of 1 bar and 6 bar, has minimum and maximum temperatures of 300 K and 1500 K. The ratio of specific heats of the working fluid is 1.4. The approximate final temperature in Kelvin at the end of the compression and expansion processes are respectively

(a) 500 and 900 (b) 900 and 500
(c) 500 and 5400 (d) 900 and 900

[2011 : 2 Marks]

43. In a power plant, water (density = 1000 kg/m^3) is pumped from 80 kPa to 3 MPa. The pump has an isentropic efficiency of 0.85. Assuming that the temperature of the water remains the same, the specific work (in kJ/kg) supplied to the pump is

(a) 0.34 (b) 2.48
(c) 2.92 (d) 3.43

[2014 : 1 Mark, Set-1]

44. An ideal reheat Rankine cycle operates between the pressure limits of 10 kPa and 8 MPa, with reheat being done at 4 MPa. The temperature of steam at the inlets of both turbines is 500°C and the enthalpy of steam is 3185 kJ/kg at the exit of the high pressure turbine and 2247 kJ/kg at the exit of low pressure turbine. The enthalpy of water at the exit from the pump is 191 kJ/kg. Use the following table for relevant data.

Superheated steam temperature (°C)	Pressure (MPa)	V (m^3/kg)	h (kJ/kg)	s (kJ/kg.K)
500	4	0.08644	3446	7.0922
500	8	0.04177	3399	6.7266

Disregarding the pump work, the cycle efficiency (in percentage) is ______.

[2014 : 2 Marks, Set-1]

45. The thermal efficiency of an air-standard Brayton cycle in terms of pressure ratio r_p and $\gamma(= c_p/c_v)$ is given by

(a) $1-\frac{1}{r_p^{\gamma-1}}$ (b) $1-\frac{1}{r_p^{\gamma}}$

(c) $1-\frac{1}{r_p^{1/\gamma}}$ (d) $1-\frac{1}{r_p^{(\gamma-1)/\gamma}}$

[2014 : 1 Mark, Set-2]

46. In an ideal Brayton cycle, atmospheric air (ratio of specific heats, c_p/c_v = 1.4, specific heat at constant pressure = 1.005 kJ/kgK) at 1 bar and 300 K is compressed to 8 bar. The maximum temperature in the cycle is limited to 1280 K. If the heat is supplied at the rate of 80 MW, the mass flow rate (in kg/s) of air required in the cycle is ______.

[2014 : 2 Marks, Set-2]

47. For a gas turbine power plant, identify the correct pair of statements.

P. Smaller in size compared to steam power plant for same power output

Q. Starts quickly compared to steam power plant

R. Works on the principle of Rankine cycle

S. Good compatibility with solid fuel

(a) P and Q (b) R and S
(c) Q and R (d) P and S

[2014 : 1 Mark, Set-3]

48. Steam with specific enthalpy (h) 3214 kJ/kg enters an adiabatic turbine operating at steady state with a flow rate 10 kg/s. As it expands, at a point where his 2920 kJ/kg, 1.5 kg/s is extracted for heating purposes. The remaining 8.5 kg/s further expands to the turbine exit, where h = 2374 kJ/kg.

Neglecting changes in kinetic and potential energies, the net power output (in kW) of the turbine is_____.

[2014 : 2 Marks, Set-4]

49. Steam enters a well insulated turbine and expands isentropically throughout. At an intermediate pressure, 20 percent of the mass is extracted for process heating and the remaining steam expands isentropically to 9 kPa.

Inlet to turbine: P = 14 MPa, T = 560°C,

h = 3486 kJ/kg, s = 6.6 kJ/(kgK)

Intermediate stage: h = 276 kJ/kg

Exit of turbine: P = 9 kPa, h_f = 174 kJ/kg,

h_g = 2574 kJ/kg, s_f = 0.6 kJ/(kgK), s_g = 8.1 kJ/(kgK)

If the flow rate of steam entering the turbine is 100 kg/s, then the work output (in MW) is_____.

[2015 : 2 Marks, Set-1]

50. Which of the following statements regarding a Rankine cycle with reheating are TRUE?

(i) increase in a average temperature of heat addition

(ii) reduction in thermal efficiency

(iii) drier steam at the turbine exit

(a) only (i) and (ii) are correct

(b) only (ii) and (iii) are correct

(c) only (i) and (iii) are correct

(d) (i), (ii) and (iii) are correct

[2015 : 1 Mark, Set-2]

51. In a Rankine cycle, the enthalpies at turbine entry and outlet are 3159 kJ/kg and 2187 kJ/kg, respectively. If the specific pump work is 2 kJ/kg, the specific steam consumption (in kg/kWh) of the cycle based on net output is.

[2015 : 2 Marks, Set-2]

52. The INCORRECT statement about regeneration in vapor power cycle is that

(a) it increases the irreversibility by adding the liquid with higher energy content to the steam generator

(b) heat is exchanged between the expanding fluid in the turbine and the compressed fluid before heat addition

(c) the principle is similar to the principle of Stirling gas cycle

(d) it is practically implemented by providing feed water heaters

[2016 : 1 Mark, Set-1]

53. Which one of the following modifications of the simple ideal Rankine cycle increases the thermal efficiency and reduces the moisture content of the steam at the turbine outlet?

(a) Increasing the turbine inlet temperature

(b) Decreasing the condenser pressure

(c) Increasing the boiler pressure

(d) Decreasing the boiler pressure

[2019 : 1 Mark, Set-2]

54. Consider a simple gas turbine (Brayton) cycle and a gas turbine cycle with perfect regeneration. In both the cycles, the pressure ratio is 6 and the ratio of the specific heats of the working medium is 1.4. The ratio of minimum to maximum temperatures is 0.3 (with temperatures expressed in K) in the regenerative cycle. The ratio of the thermal efficiency of the simple cycle to that of the regenerative cycle is._____.

[2016 : 1 Mark, Set-2]

55. A gas turbine with air as the working fluid has an isentropic efficiency of 0.70 when operating at a pressure ratio of 3. Now, the pressure ratio of the turbine is increased to 5, while maintaining the same inlet conditions. Assume air as a perfect gas with specific heat ratio γ = 1.4. If the specific work output remains the same for both the cases, the isentropic efficiency of the turbine at the pressure ratio of 5 is___(round off to two decimal places).

[2019 : 2 Marks, Set-1]

Steam turbine

56. An economizer in a steam generator performs the function of

(a) Preheating the combustion air

(b) Preheating the feed water

(c) Preheating the input fuel

(d) Raising the temperature of steam

[1989 : 2 Marks]

57. The current level of the maximum temperature of steam turbine inlet.is much lower than that at gas turbine inlet because

(a) The fuel combustion temperature in a steam generator is lower than that in a gas turbine engine

(b) Of the corrosive nature of high temperature steam on super heater tubes

(c) The materials used for the gas turbine blades are not suitabie for the steam turbine blades

(d) Unlike the gas turbine blades the steam turbine blades cannot be cooled

[1989 : 2 Marks]

58. Velocity compounded steam turbine known as _____(curtis/rateau) turbine uses the principle of converting entire _____(pressure/velocity) energy before entering the first stage runner itself.

[1991 : 2 Marks]

59. A steam turbine operating with less moisture is _____(more/less) efficiency and _____ (less/more) prone to blade damage.

[1992 : 2 Marks]

60. For a single stage impulse turbine with rotor diameter of 2 m and a speed of 3000 rpm when the nozzle angle is 20°, the optimum velocity of steam in m/s is

(a) 334 (b) 356

(c) 668 (d) 711

[1994 : 1 Mark]

61. A steam power plant has the boiler efficiency of 92%, turbine efficiency (mechanical) of 94%, generator efficiency of 95% and cycle efficiency of 44%. If 6% of the generated power is used to run the auxiliaries, the overall plant efficiency is

(a) 34% (b) 39%

(c) 45% (d) 30%

[1996 : 2 Marks]

62. The following data pertain to a single stage impulse steam turbine:

Nozzle angle = 20°

Blade velocity = 200 m/s

Relative steam velocity at entry = 350 m/s

Blade inlet angle = 30°

Blade exit angle = 25°

If blade friction is neglected the work done per kg steam is

(a) 124 kJ (b) 164 kJ

(c) 169 kJ (d) 174 kJ

[1997 : 2 Marks]

63. If V_N and α are the nozzle exit velocity and nozzle angle in an impulse turbine, the optimum blade velocity is given by

(a) $V_N \cos 2\alpha$ (b) $V_N \sin 2\alpha$

(c) $V_N \cos\left(\frac{\alpha}{2}\right)$ (d) $\frac{V_N \cos\alpha}{2}$

[1998 : 1 Mark]

64. A Curtis stage, Rateau stage and a 50% reaction stage in a steam turbine are examples of

(a) different types of impulse stages

(b) different types of reaction stages

(c) a simple impulse stage, a velocity compounded impulse stage and reaction stage

(d) a.velocity compounded impulse stage, a pressure compounded impulse stage and a reaction stage

[1998 : 1 Mark]

65. Which of the following is a pressure compounded turbine?

(a) Parsons (b) Curtis

(c) Rateau (d) All of three

[2000 : 1 Mark]

66. The Rateau turbine belongs to the category of

(a) pressure compounded turbine

(b) reaction turbine

(c) velocity compounded turbine

(d) radial flow turbine

[2001 : 1 Mark]

67. Considering the variation of static pressure and absolute velocity in an impulse steam turbine, across one row of moving blades

(a) both pressure and velocity decrease

(b) pressure decreases but velocity increases

(c) pressure remains constant, while velocity increases

(d) pressure remains constant, while velocity decreases

[2003 : 1 Mark]

68. In the velocity diagram shown below, u = blade velocity, C = absolute fluid velocity and W = relative velocity of fluid and the subscripts 1 and 2 refer to inlet and outlet. This diagram is for

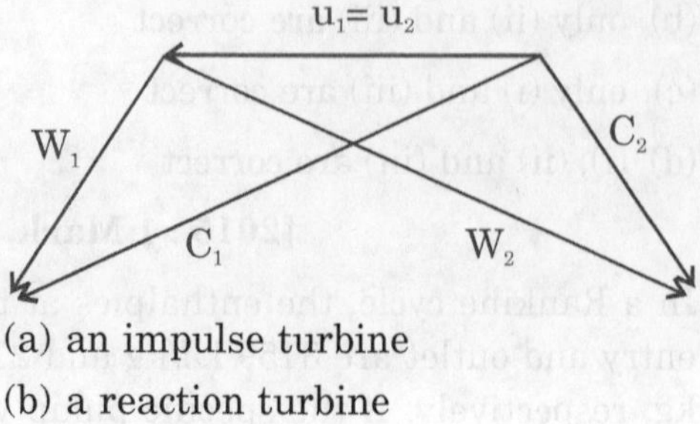

(a) an impulse turbine

(b) a reaction turbine

(c) a centrifugal compressor

(d) an axial flow compressor

[2005 : 2 Marks]

69. Specific enthalpy and velocity of steam at inlet and exit of a steam turbine, running under steady state, are as given below:

	Specific enthalpy (kJ/kg)	Velocity (m/s)
Inlet steam condition	3250	180
Exit steam condition	2360	5

The rate of heat loss from the turbine per kg of steam flow rate is 5 kW. Neglecting changes in potential energy of steam, the power developed in kW by the steam turbine per kg of steam flow rate, is

(a) 901.2 (b) 911.2
(c) 17072.5 (d) 17082.5

[2014 : 2 Marks]

70. Steam at a velocity of 10 m/s enters the impulse turbine stage with symmetrical blading having blade angle 30°. The enthalpy drop in the stage is 100 kJ. The nozzle angle is 20°. The maximum blade efficiency (in percent) is ____.

[2014: 2 Marks, Set-2]

71. In a steam power plant operating on a ideal Rankine cycle, superheated steam enters the turbine at 3 MPa and 350°C. The condenser pressure is 75 kPa. The thermal efficiency of the cycle is ________%.

Given data:

For saturated liquid, at p = 75 kPa, h_f = 384.39 kJ/kg, v_f = 0.001037 m^3/kg, s_f = 1.213 kJ/kgK.

At 75 kPa, h_{fg} = 2278.6 kJ/kg,

s_{fg} = 6.2434 kJ/kgK

At P = 3 MPa and T = 350°C (superheated steam), h = 3115.3 kJ/kg, s = 6.7428 kJ/kgK.

[2016 : 2 Marks, Set-1]

72. The pressure ratio across a gas turbine (for air, specific heat at constant pressure, c_p = 1040 J/kg K and ratio of specific heats γ = 1.4) is 10. If the inlet temperature to the turbine is 1200 K and the isentropic efficiency is 0.9, the gas temperature at turbine exit is ________K.

[2017 : 2 Marks, Set-1]

73. In the Rankine cycle for a steam power plant the turbine entry and exit enthalpies are 2803 kJ/kg and 1800 kJ/kg, respectively. The enthalpies of water at pump entry and exit are 121 kJ/kg and 124 kJ/kg respectively. The specific steam consumption (in kg/kWh) of the cycle is_____.

[2017 : 2 Marks, Set-2]

74. Select the correct statement for 50% reaction stage in a steam turbine.

(a) The rotor blade is symmetric

(b) The stator blade is symmetric

(c) The absolute inlet flow angle is equal to absolute exit flow angle

(d) The absolute exit flow angle is equal to inlet angle of rotor blade.

[2018 : 1 Marks, Set-2]

75. A steam power cycle with regeneration as shown below on the T-s diagram employs a single open feedwater heater for efficiency improvement. The fluids mix with each other in an open feedwater heater. The turbine is isentropic and the input (bleed) to the feedwater heater from the turbine is at state 2 as shown in the figure. Process 3-4 occurs in the condenser. The pump work is negligible. The input to the boilder is at state 5. The following information is available from the steam tables:

State	1	2	3	4	5	6
Enthalpy(kJ/kg)	3350	2800	2300	175	700	1000

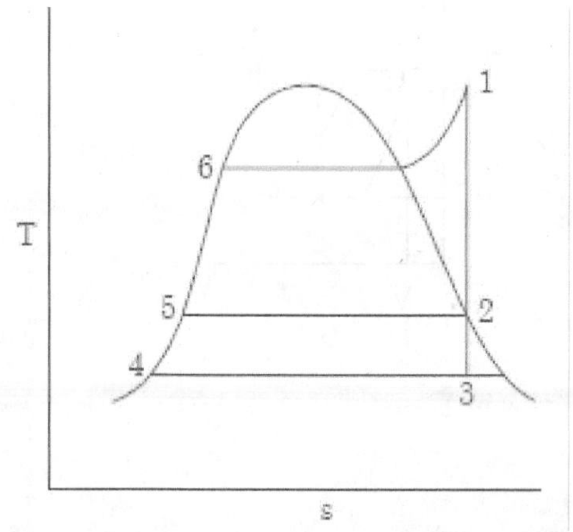

The mass flow rate of steam bled from the turbine as a percentage of the total mass flow rate at the inlet to the turbine at state 1 is _____.

[2019 : 2 Marks, Set-1]

Air Compressors

76. A single acting two stage compressor with complete inter cooling delivers air at 16 bar. Assuming an intake state of 1 bar at 15°C, the pressure ratio per stage is

(a) 16 (b) 8
(c) 4 (d) 2

[2001 : 2 Marks]

77. A p-V diagram has been obtained from a test on a reciprocating compressor. Which of the following represents that diagram?

(a)

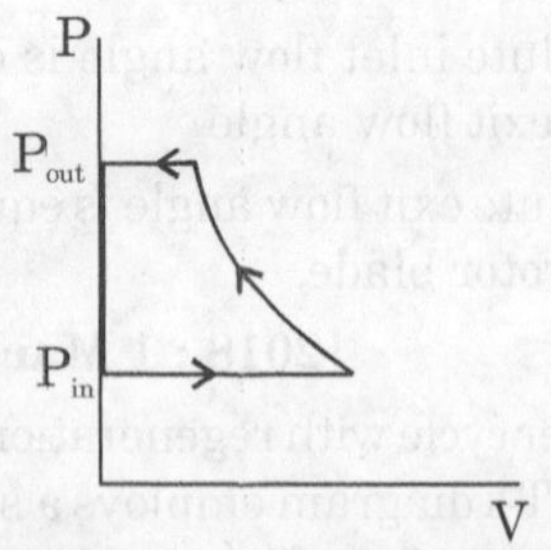

(b)

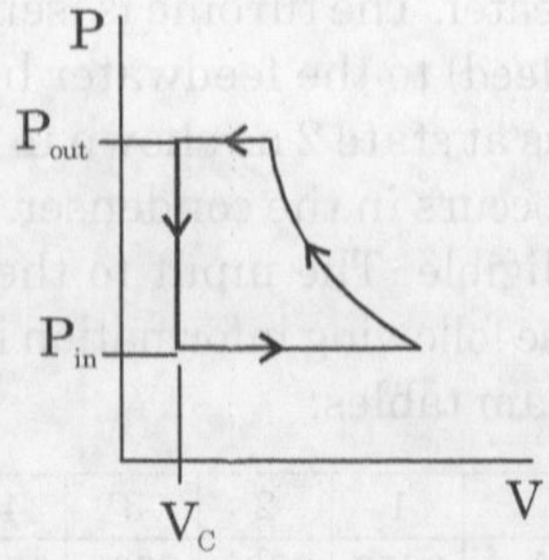

(c)

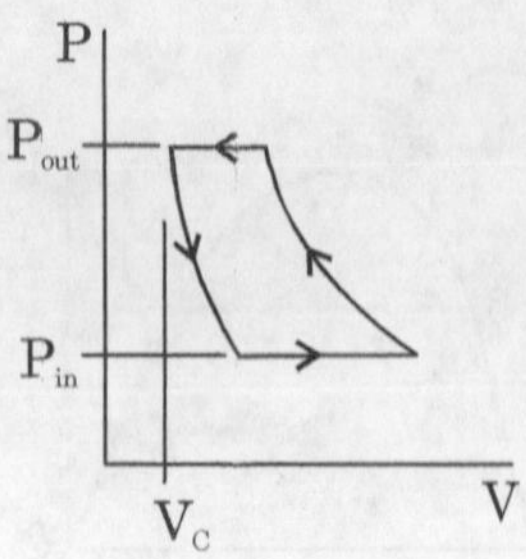

(d)

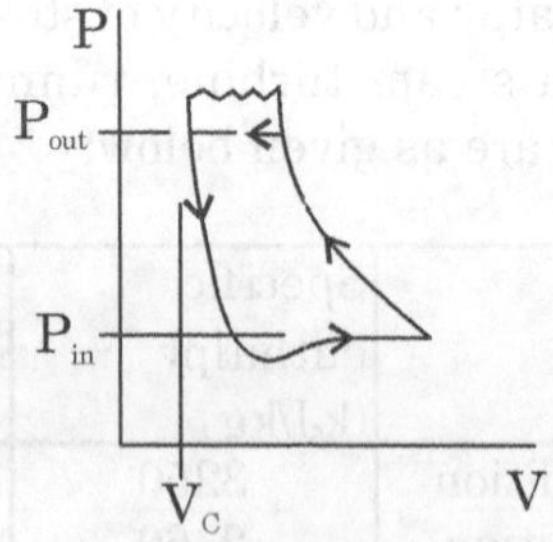

[2005 : 1 Mark]

Common Data Questions 78 and 79

In a simple Brayton cycle, the pressure ratio is 8 and temperature at the entrance of compressor and turbine are 300 K and 1400 K, respectively. Both compressor and gas turbine have isentropic efficiencies equal to 80%. For the gas, assume a constant value of c_p (specific heat at constant pressure) equal to 1 kJ/kgK and ratio of specific heats as 1.4. Neglect changes in kinetic and potential energies.

78. The power required by the compressor in kW/kg of gas flow are is

(a) 194.7 (b) 243.4
(c) 304.3 (d) 378.5

[2013 : 2 Marks]

79. The thermal efficiency of the cycle in percentage (%) is

(a) 24.8 (b) 38.6
(c) 44.8 (d) 53.1

[2013 : 2 Marks]

80. In a 3-stage air compressor, the inlet pressure is p1, discharge pressure is p_4 and the intermediate pressures are p_2 and p_3 ($p_2 < p_3$). The total pressure ratio of the compressor is 10 and the pressure ratios of the stages are equal. If p_1 = 100 kPa, the value of the pressure p_3 (in kPa) is ______.

[2016 : 2 Marks, Set-3]

ANSWERS

1. (d)	**2.** (a)	**3.** (a)	**4.** (b)	**5.** (d)	**8.** (a)	**9.** (a, b)	**12.** (d)	**13.** (d)	**14.** (b)
17. (c)	**18.** (d)	**19.** (c)	**21.** (c)	**22.** (c, b)	**23.** (b)	**24.** (c)	**25.** (a)	**26.** (a)	**27.** (b)
28. (d)	**29.** (b)	**30.** (a)	**31.** (d)	**32.** (b)	**33.** (d)	**34.** (a)	**35.** (b)	**36.** (a)	**37.** (b)
38. (a)	**39.** (a)	**40.** (d)	**41.** (a)	**42.** (a)	**43.** (d)	**45.** (d)	**47.** (a)	**50.** (c)	**52.** (c)
53. (a)	**56.** (b)	**57.** (d)	**60.** (c)	**61.** (a)	**62.** (a)	**63.** (d)	**64.** (d)	**65.** (c)	**66.** (a)
67. (d)	**68.** (b)	**69.** (a)	**74.** (d)	**76.** (c)	**77.** (d)	**78.** (c)	**79.** (a)		

EXPLANATIONS

1. Boiler rating is usually defined in term of steam output in kg/hr

2. Equivalent evaporation $= \left(\frac{h_1 - h_{feed}}{h_{fgam}}\right) \times \dot{m}$

$$= \frac{(2000)(2426 - 168)}{2258}$$

$= 2000$ kg/hr

3. Blow-off cock

6. Remain constant

7. A-5, B-1, C-6, D-3

8. Blade velocity ratio $= \frac{u}{v_1} = 0.5$

or $u = 0.5\ v_1$

where u = mean blade velocity, and

v_1 = Absolute velocity of steam at inlet

$$= \frac{150}{0.5} = 300 \text{ m/s}$$

K.E. at inlet $= \frac{1}{2} \times m \times v_1^2 = \frac{m}{2}(300)^2$ J

Nozzle efficiency = 0.9

∴ Isentropic heat drop in nozzle

$$= \frac{(300)^2}{2 \times 0.9} = \textbf{50kJ / kg.}$$

10. Reduce

11. Compressed air; compression work

Intercoolers are used to cool compressed air so that the compression work can be reduced and more air means can be supplied to turbine.

15. $I = T_0(\Delta S)_{un}$

$$= 70\ (\Delta S)_{system}\ (\Delta S)_{surr}$$

$$= \left[S_2 - S_1 + \frac{q}{T_0}\right)$$

$$= 300\left[C_p \ln\frac{T_2}{T_1} - R\ln\frac{P_2}{T_1}\right) + \frac{10}{300}$$

16. 312.04KJ/kg-maximum work

270KJ/Kg-Actual work

19. Infinitally large number of stages during compression and expansion in the gas turbine cycle i.e Brayton cycle leads to constant temperature and constant pressure cycle i.e. Ericsson cycle.

20. *False*

21. From energy balance

$(1 - m)h_2 + mh_1 = h_3$

$h_3 = h_2 - m(h_2 - h_1)$

22. Intercooling provides significant benefits to the brayton cycle gas turbine power plant through dereating the work of compression for the high pressure compressor.

Heat rejection taken place during isothermal compression.

25.

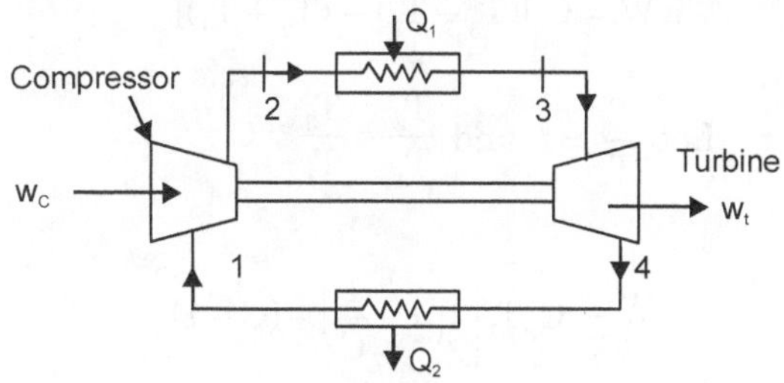

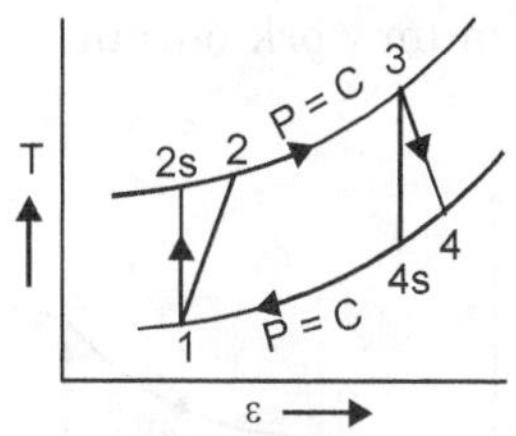

Given,, $c_p = 0.98$ kJ/kg k.

$c_p = 0.7538$ kJ/kg k

$P_3 = 20$ bar $= 20 \times 10^5$ N/m²

$T_3 = 1500$ K

$P_4 = 1$ bar $= 1 \times 10^5$ N/m

$h_{turbine} = 0.94$

For process, 3 – 4s

$$\left(\frac{T_3}{T_{4s}}\right) = \left(\frac{P_3}{P_4}\right)^{\frac{q-1}{q}}$$

where, $q = \frac{c_P}{c_V} = \frac{0.98}{0.7538} = 1.3$

$$\frac{1500}{T_{4s}} = \left(\frac{20 \times 10^5}{1 \times 10^5}\right)^{\frac{1.3-1}{1.3}}$$

$$\Rightarrow \quad T_{4s} = \frac{1500}{(20)^{\frac{0.3}{1.3}}}$$

$$\Rightarrow \quad T_{4s} = 751.37 \text{ K}$$

Now turbine efficiency, $\eta_t = \frac{T_3 - T_4}{T_3 - T_{4s}}$

$$\Rightarrow \quad 0.94 = \frac{1500 - T_4}{1500 - 751.37}$$

$$\therefore \quad T_4 = 796.3 \text{ K}$$

Now turbine work,

$$W_t = h_3 - h_4$$
$$= c_p(T_3 - T_4)$$
$$= 0.08(1500 - 796.3)$$
$$= 698.64 \text{ kJ/kg}$$

26. W = Turbine work – compressor work

$$W = C_p[(T_3 - T_4) - (T_2 - T_1)]$$

Let $\frac{T_3}{T_1} = t$ and $\frac{T_2}{T_4} = \frac{T_3}{T_4} = C$

$$W = C_p T_1 \left[t\left(1 - \frac{1}{C}\right) - (C - 1) \right]$$

For maximum work output

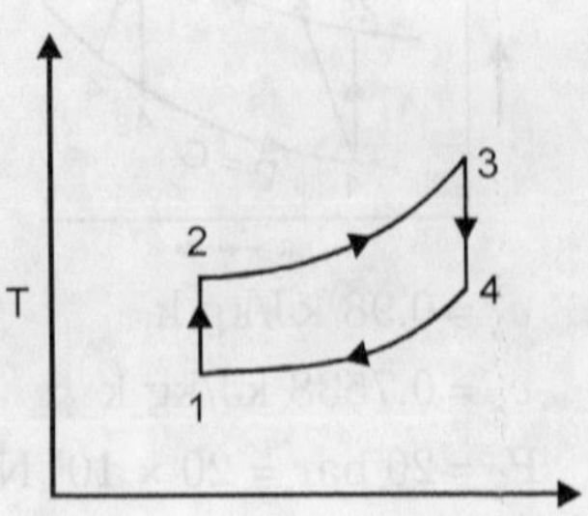

$$\frac{dW}{dc} = C_p T_1 \left[T\left(\frac{1}{C^2}\right) - 1 \right] = 0$$

gives $\quad C^2 = \sqrt{t}$

$$\therefore \quad \frac{T_2}{T_1} = \sqrt{\frac{T_3}{T_1}}, \left(\frac{p_2}{p_1}\right)^{\frac{r-1}{r}} = \left(\frac{T_{max}}{T_{min}}\right)^{½}$$

$$\frac{p_2}{p_1} = \left(\frac{T_{max}}{T_{min}}\right)^{\frac{\gamma}{2(\gamma-1)}}$$

27. Given: $h_1 = 29.3$ kJ/kg

$h_2 = ?$

$h_3 = 3095$ kJ/kg

$h_4 = 2609$ kJ/kg

$h_5 = 3170$ kJ/kg

$h_6 = 2165$ kJ/kg

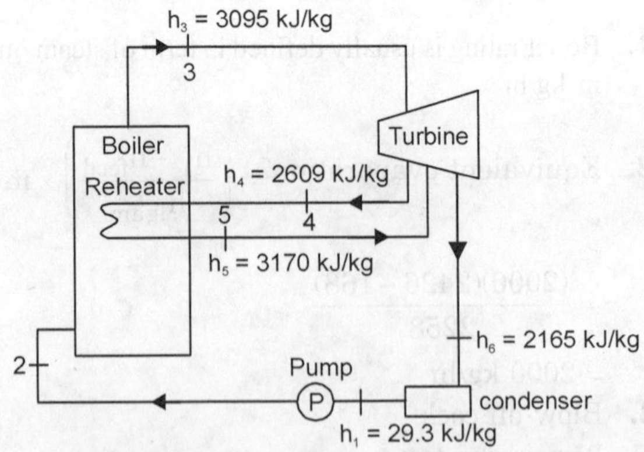

Turbine work

$$W_T = (h_3 - h_4) + (h_5 - h_6)$$
$$= (3095 - 2609) + (3170 - 2165)$$
$$= 1491 \text{ kJ/kg}$$
$$Q = \text{Heat input} = (h_3 - h_1) + (h_5 - h_4)$$
$$= (3095 - 29.3) + 3(3170 - 2609)$$
$$= 3626.7 \text{ kJ/kg}$$

∴ Thermal efficiency of the plant

$$= \frac{1491}{3226.7} = 41.11\%$$

28. Enthalpy at the pump discharge will be greater than 29.3 kJ/kg

Hence from given choice clearly we can say

$$h_2 = 33.3 \text{ kJ/kg.}$$

29. $\eta_0 = \eta_0 = 1 - \frac{T_1}{T_2} = 1 - \frac{300}{450}$

$$= \frac{150}{450} = 0.333$$

$$[\eta_B = \eta_0 = 33.3\%]$$

30. $W_0 = \eta_0 C_V \times 150 = 35.54 \text{ kJ/kg}$

$W_B = \eta_0 C_P \times 100 = 33.165 \text{ kJ/kg}$

34. In a regenerative the heat added in low temperature range is reduced which increase heat addition temperature.

37. Reheater always increases the dryness fractions of steam at condensor inlet. Always increases the specific work output.

38. $h_3 = 0.2h_1 + h_2 \times 0.8$
$$= 560 + 160$$
$$= 720 \text{ kJ/kg}$$

39 & 40.

Given : Rankine cycle on T-S diagram

$P_3 = 4$ MPa

$T_3 = 350°C$

$\eta_t = 0.9$

$h_3 = 3092.5$ kJ/kg

$s_3 = s_4 = 6.5821$ kJ/kgk

$h_1 = 225.94$ kJ/kg $= h_f$

$h_g = 2599.1$ kJ/kg

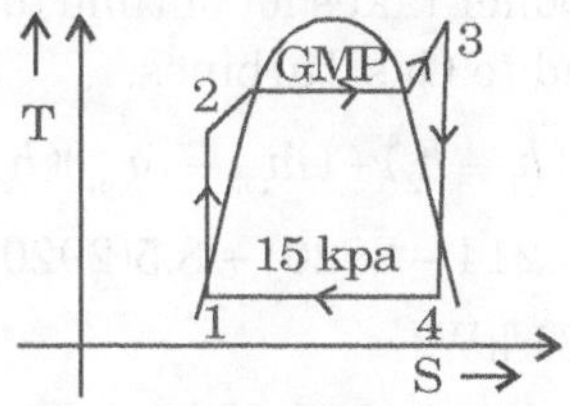

$h_{fg} = h_g - h_f = 2373.16$ kJ/kg

$s_1 = s_f = 0.7549$ kJ/kgK

$s_g = 8.0085$ kJ/kgK

$s_{fg} = 8.0085 - 0.7949$

$= 7.2536$ kJ/kg K

Also, $s_3 = s_4 = 6.5821$

$\therefore$ $s_4 = s_1 + x\, s_{fg}$

$= 0.7549 + x\,(7.2536)$

$= 6.5821$

$\Rightarrow$ $x = 0.8$

$h_4 = h_1 + x h_{fg}$

$= 225.94 + (0.8 \times 2375.16)$

$= 2132.4$ kJ/kg

Net work output of the cycle

$= \eta_t (h_3 - h_4)$

$= 0.9\,(3092.5 - 2132.4)$

$= 864$ kJ/kg

≈ 860 kJ/kg

Heat supplied to the cycle $= h_3 - h_2$

Since we do not have information regarding h_2, therefore neglect compressor work and take

$h_2 = h_1$

$\therefore$ $Q_s = 3092.5 - 2866$ kJ/kg $= 2863$ kJ/kg

41. Work done by the turbine

W = 2800 – 1800

= 1000 kJ/kg

= 1000 kW-s/kg

$$\text{Specific fuel consumption} = \frac{1}{1000} \times 3600$$

= 3.6 kg/kW – hr

42. Ideal Brayton cycle,

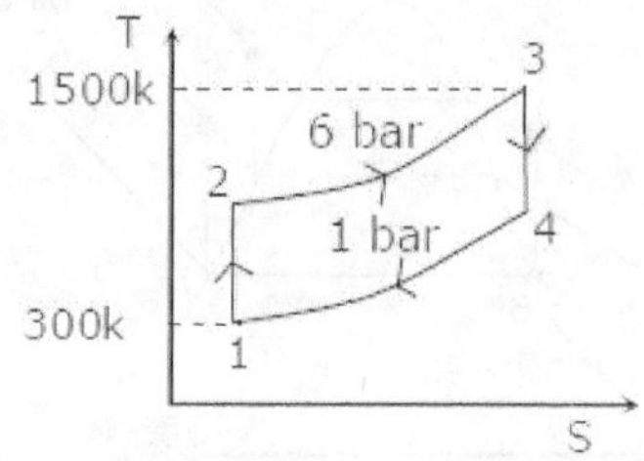

At the end of compression,

$$\text{Temperature, } T_2 = T_1\left(\frac{P_3}{P_4}\right)^{\frac{\gamma-1}{\gamma}}$$

$$= 300 \times 6^{\frac{0.4}{1.4}} = 500 \text{ K}$$

At the end of expansion;

$$\text{Temperature, } T_4 = \frac{T_3}{\left(\frac{P_3}{P_4}\right)^{\frac{\gamma-1}{\gamma}}}$$

$$= \frac{1500}{6^{\frac{0.4}{1.4}}} = 900 \text{ K}$$

43. $$\text{Specific volume} = \frac{\text{volume of fluid}}{\text{Mass of fluid}}$$

$$= \frac{1}{\rho} = \frac{1}{1000} = 10^{-3} \text{ m}^3/\text{kg}$$

$$\eta = \frac{\text{Isentropic compressor work}}{\text{Actual compressor work}}$$

$$\text{Actual compressor work} = \frac{V(\Delta P)}{\eta}$$

$$= \frac{10^{-3} \times (3000 - 80)\text{KPa}}{0.85}$$

$$= \frac{2.92}{0.85} = 3.43 \text{ kJ/kg.}$$

44. $W_{HP} = h_2 - h_3$

$W_{LP} = h_4 - h_5$

Net work

$W_T = W_{HP} + W_{LP} = (h_2 - h_3) + (h_4 - h_5)$

Given :

$h_1 = 191$ kJ/kg

$h_2 = 3399$ kJ/kg

$h_3 = 3185$ kJ/kg

$h_4 = 3446$ kJ/kg

$h_5 = 2247$ kJ/kg

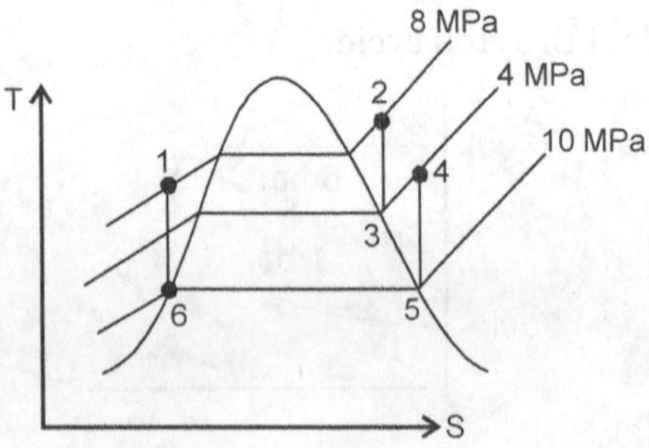

$$\eta_{\text{cycle}} = \frac{w_{\text{net}}}{Q_{\text{added}}}$$

$$\eta_{\text{cycle}} = \frac{w_1 - w_{\text{pump}}}{Q_{\text{added}}}$$

$$\eta_{\text{cycle}} = \frac{(h_2 - h_3) + (h_4 - h_5)}{(h_2 - h_1) + (h_4 - h_3)}$$

$$\eta_{\text{cycle}} = 0.407 = 40.7\%.$$

45. Thermal efficiency of air standard efficiency

$$= 1 - \frac{1}{(r_p)^{\frac{\gamma-1}{\gamma}}}$$

46.

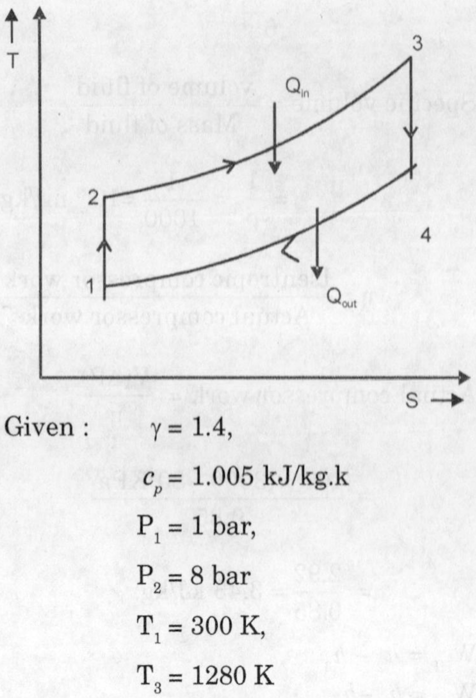

Given : $\gamma = 1.4,$

$c_p = 1.005$ kJ/kg.k

$P_1 = 1$ bar,

$P_2 = 8$ bar

$T_1 = 300$ K,

$T_3 = 1280$ K

$$r = \frac{P_2}{P_1} = 8$$

$Q_{in} = 80000$ kW.

(1-2) isentropic process

$$\frac{T_2}{T_1} = (r_p)^{\gamma - \frac{1}{\gamma}}$$

$$T_2 = 300 \times (8)^{\frac{1.4-1}{1.4}}$$

$$= 543.43 \text{ k}$$

$$Q_{in} = mc_p (T_3 - T_2)$$

$$80{,}000 = m \times 1.005 \,(1280 - 300)$$

$$m = 108.07 \text{ kg/s.}$$

47. Steam power plants are bulky due to presence of boiler and condenser. Gas turbines are compact, as compressors and turbines are coupled on a common shaft. In steam power plants, boiler takes lot of time to get started, as compared to Gas Turbines.

48. $\dot{w} = \dot{m}_{\text{tot}}(h_1 - h_2) + (\dot{m}_{\text{tot}} - \dot{m}_{\text{out}})(h_2 - h_3)$

$\dot{w} = 10(3214 - 2920) + 8.5(2920 - 2374)$

$\dot{w} = 7581$ kW.

49. $h_1 = 3486$ kJ/kg

$h_2 = 2776$ kJ/kg

$s_1 = s_3 = 6.6$

$6.6 = .6 + x(0.1 - .6)$

$x = 0.8$

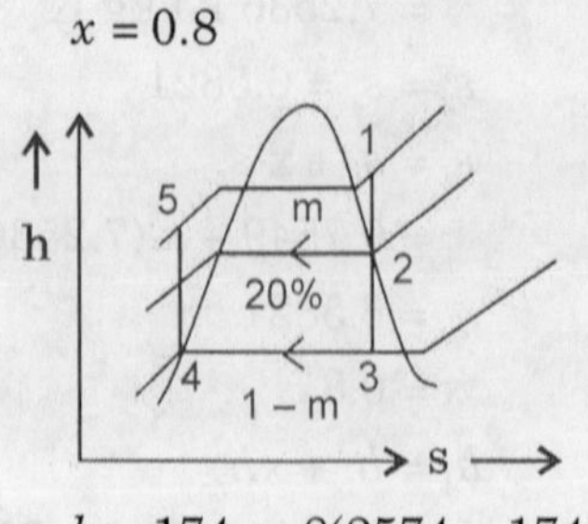

$h = 174 + .8(2574 - 174)$

$= 2094$ kJ/kg

$w = (3486 - 2776) + .8\,(2776 - 2094)$

$= 1255.6$ kJ/kg $= 125.56$ MW

50. With reheat average temperature of heat addition increases. Hence, efficiency of cycle also increases. Further, the quality of steam is higher at turbine exit.

51. Specific steam consumption $= \dfrac{3600}{W_T - W_P}$

Now, $W_T = h_2 - h_1$

$= 3159 - 2187$ kJ/kg

$\therefore$ $W_T = 972$ kJ/kg

and $W_P = 20$ kJ/kg

Thus specific steam consumption

$$= \frac{3600}{972 - 20} \text{ kg/kW-h} = 3.7815 \text{ kg/kW.h}$$

53. $\eta = f(T_m)$

$T_m \uparrow \eta \uparrow$

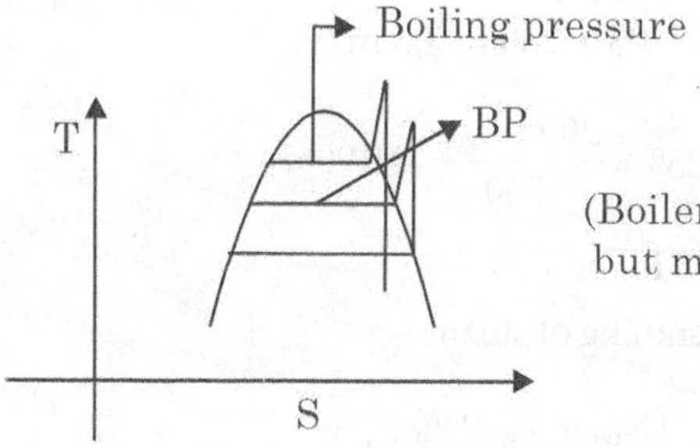

(Boiler pressure ↑ but moisture % ↑)

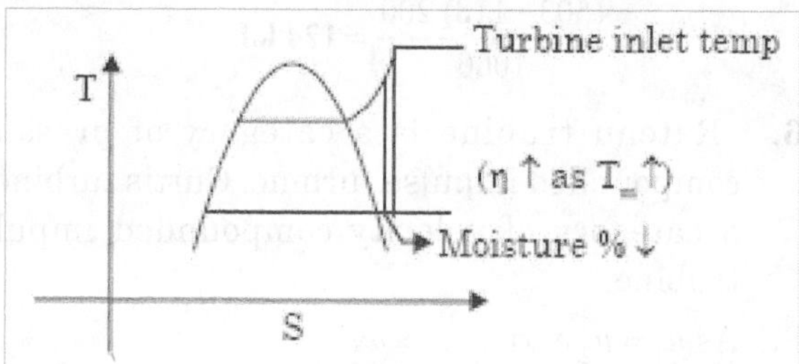

54. Brayton cycle

$$\eta_{Brayton} = 1 - \frac{1}{(r_p)^{\frac{\gamma-1}{\gamma}}}$$

given $r_p = 6, \gamma = 1.4$

$$\therefore \quad \eta_{Brayton} = 1 - \frac{1}{(6)^{\frac{1.4-1}{1.4}}} = 0.4006$$

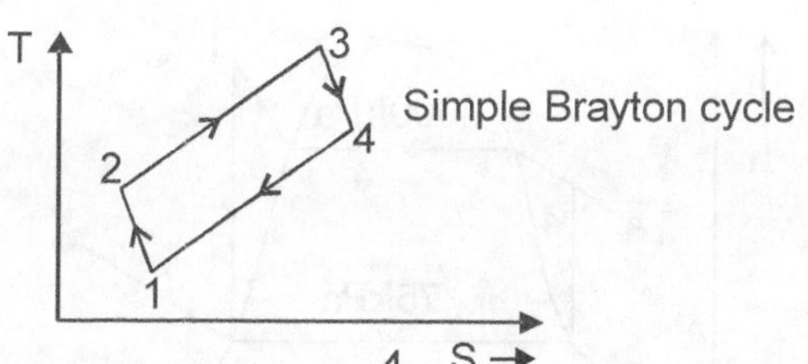

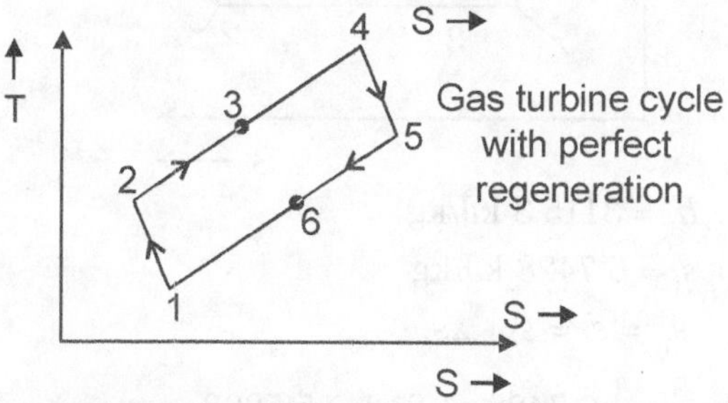

$r_p = 6, \gamma = 1.4$

$$\text{effectiveness } \varepsilon = \frac{C_p(T_5 - T_2)}{C_p(T_3 - T_2)} = 1$$

$$T_5 = T_3$$

$$\text{Given } \frac{T_{min}}{T_{max}} = 0.3 = \frac{T_1}{T_4}$$

$$\begin{aligned} W_{net} &= W_T - W_C \\ &= C_p(T_4 - T_5) - C_p(T_2 - T_1) \\ &= C_p(T_4 - T_3) - C_p(T_2 - T_1)[\because T_4 = T_3] \end{aligned}$$

$$\eta_{Regenerative} = \frac{W_{net}}{\text{Heat-supplied}}$$

$$= \frac{C_p[(T_4 - T_3) - (T_2 - T_1)]}{C_p(T_4 - T_3)}$$

$$= 1 - \left(\frac{T_2 - T_1}{T_4 - T_3}\right)$$

$$= \frac{1 - T_1\left(\frac{T_2}{T_1} - 1\right)}{1 - T_4\left(1 - \frac{T_3}{T_4}\right)}$$

$$\eta_{Regenerative} = 1 - \frac{T_1}{T_4}\left(\frac{\frac{T_2}{T_1} - 1}{1 - \frac{T_5}{T_4}}\right)$$

$$= 1 - \frac{T_1}{T_4}\left(\frac{(r_p)^{\frac{\gamma-1}{\gamma}} - 1}{1 - \frac{1}{(r_p)^{\frac{\gamma-1}{\gamma}}}}\right)$$

$$= 1 - \frac{T_1}{T_4}(r_p)^{\frac{\gamma-1}{\gamma}}$$

$$= 1 - 0.3(6)^{\frac{1.4-1}{1.4}} = 0.4994$$

$$\frac{\eta_{brayton}}{\eta_{Regenerative}} = \frac{0.4006}{0.4994} = 0.8021$$

55. $\eta = 0.7, (r_p)_1 = 3$

$\eta_2 = ?\ (r_p)_2 = 5$

$r = 1.4$

$(W_T)_1 = (W_T)_2$

As Inlet condition same,

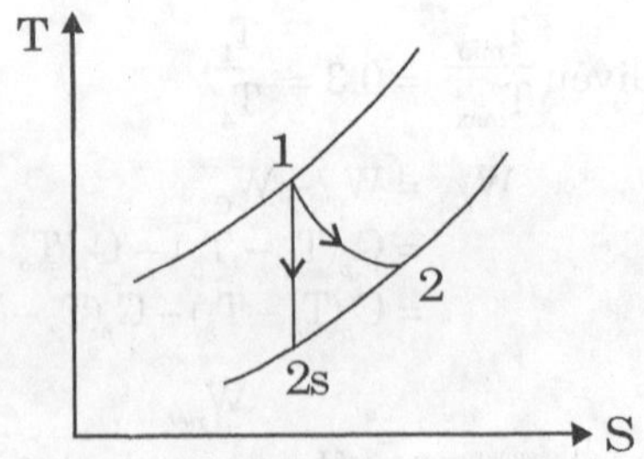

Here, $\eta_1 = \frac{h_1 - h_2}{h_1 - h_{2s}}$

$$= \frac{T_1 - T_2}{T_1 - T_{2s}} = \frac{T_1 - T_2}{T_1 - \frac{T_1}{x}}$$

$$\eta_1\left(T_1 - \frac{T_1}{x}\right) = \eta_2\left(T_1 - \frac{T_2}{y}\right)$$

$$0.7\left(1 - \frac{1}{(3)^{2/7}}\right) = \eta_2\left(1 - \frac{1}{(5)^{2/4}}\right)$$

$\Rightarrow$ $\eta_2 = 0.51$

Hence, Isentropic efficiency of turbine is 0.51.

56. Economiser is used for preheating the feed water heater using exhaust fuel gasses.

57. In gas turbine cooling is provided to blades which keeps the blade material at much lower temperature.

58. Curtis Pressure

59. Less, more more, less

60. $\frac{u}{v_1} = \frac{\text{Cos}\alpha}{2}$

$$v_1 = \frac{2v}{\cos\alpha}$$

$$= \frac{2 \times \left(\frac{\pi DN}{60}\right)}{\cos\alpha} = \frac{\pi \times 3000}{30 \times \cos 20}$$

$[V_1 = 668.6 \text{ m/s}]$

61. $\eta_0 = \eta_b \times \eta_m \times \eta_s \times \eta_c \,|1 - \eta a)$

$= 0.92 \times 0.94 \times 0.95 \times 0.44r \times (1 - 0.06)$

$\eta_0 = 34\%$

62.

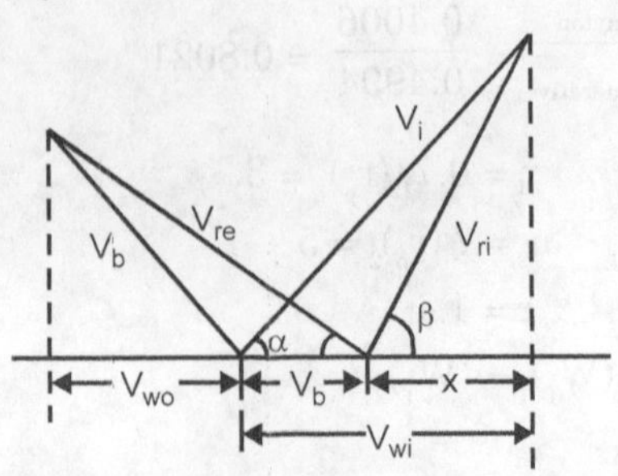

Given :$\alpha = 20°$

$\beta = 30°$

$x = 350 \cos 30° = 303 \text{ m/s}$

$\therefore \quad V_{wi} = 303 + V_b$

$= 303 + 200 = 503 \text{ m/s}$

$$\cos 25° = \frac{200 + V_{wo}}{350} = 0.9063$$

$V_{wo} = 118 \text{ m/s}$

Work done/kg of steam

$$= \frac{(V_{wi} + V_{wo})V_b}{1000} \text{ kJ}$$

$$= \frac{(503 + 118)\,200}{1000} = \mathbf{124\ kJ}$$

66. Rateau trubine is a category of pressure compounded impulse turbine. Curtis turbine is a category of velocity compounded impulse trubine.

68. As $u_1 = u_2$ and $w_2 >> w_1$

69. $\dot{Q} - \dot{W} = \dot{m}\left[(h_2 - h_1) + \left(\frac{V_2^2 - V_1^2}{2}\right)\right]$

and $\dot{Q} = -5$ kW, $\dot{m} = 1 kg/s$

70. 85.1 to 89.9

Maximum blade efficiency

$\eta_{max} = \cos^2 \alpha$

$= .883$

$= 88.3\%$

71.

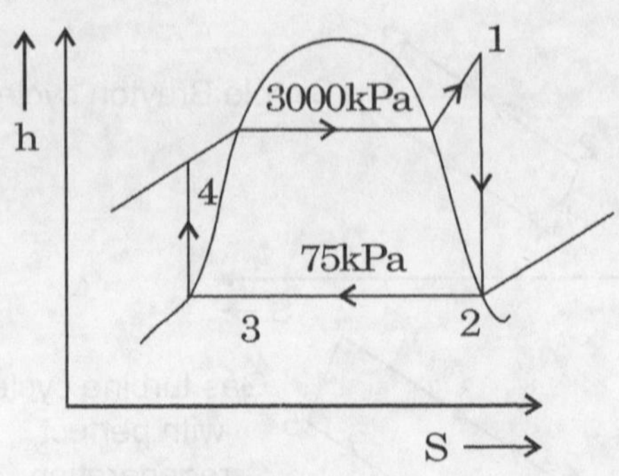

$h_1 = 3115.3$ kJ/kg

$s_1 = 6.7428$ kJ/kg

$s_1 = s_2 = s_f + xs_{fg}$

$$x = \frac{6.7428 - 1.213}{6.2434} = \frac{5.5298}{6.2434} = 0.8857$$

$h_2 = h_f + xh_{fg}$

$= 348.39 + 0.8857 \times 2278.6$

$= 348.39 + 2018.16$

$= 2402.55$ kJ/kg

Pump work $= W_p = v_f(p_4 - p_3) = 3.033$ kJ/kg

$h_4 = h_3 + v_f(p_4 - p_3)$

$= 384.34 + 0.001037(3000 - 75)$

$= 384.34 + 3.033$

$= 387.37$ kJ/kg

$W_{net} = W_T - W_P$

$= (h_1 - h_2) - W_P$

$= (3115.3 - 2402.55) - 3.033$

$= 709.72$ kJ/kg

Q_s = heat supplied $= h_1 - h_4$

$= 3115.3 - 387.37 = 2727.93$ kJ/kg

$$\eta_{th} = \frac{W_{net}}{Q_s}$$

$$= \frac{709.72}{2727.93} \times 100 = 0.26 \text{ or } 26\%$$

72.

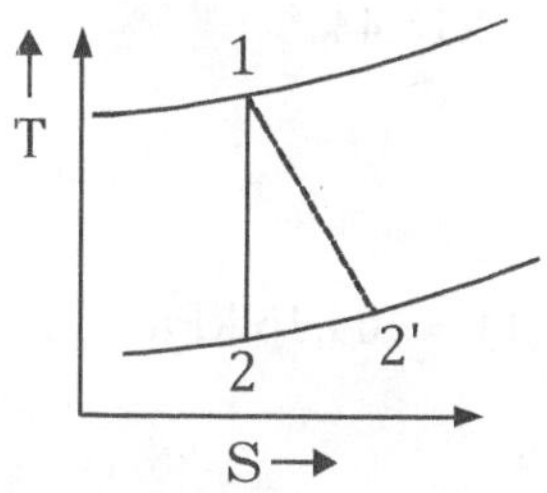

By question

$T_1 = 1200$ K

$$r_p = 10 = \frac{P_1}{P_2}$$

$\eta_T = 0.9$

$T_2 = ?$

$$\frac{T_2}{T_1} = \left(\frac{P_2}{P_1}\right)^{\frac{\gamma-1}{\gamma}}$$

$$\frac{T_2}{1200} = \left(\frac{1}{10}\right)^{\frac{1.4-1}{1.4}}$$

$T_2 = 621.53$ K

$$\eta_T = \frac{T_1 - T_{2'}}{T_1 - T_2}$$

$$0.9 = \frac{1200 - T_{2'}}{1200 - 621.53}$$

$T_2 = 679.4$ K

73. Pump work

$W_p = h_2 - h_1 = 124 - 121 = 3$kJ/kg

Power output

$w_t = h_u - h_s = 2803 - 1800 = 1003$ kJ/kg

Net output

$W_{net} = w_t - w_p = 1003 - 3 = 1000$ k

Specific steam consumption

$$= \frac{3600}{W_{net}} (\text{kg/kW-h}) = 3.6 \text{ kg/kW-h}$$

75. From energy balance equation,

$$\dot{m}h_2 + (1 - \dot{m})h_4 = h_5$$

$$\dot{m}(h_2 + h_4) = h_2 - h_4$$

$$\Rightarrow \dot{m} = \frac{h_5 - h_4}{h_2 - h_4} = \left(\frac{700 - 175}{2800 - 175}\right) \times 100$$

$= 0.2$ or 20%

∴ 20% of mass flow rate of steam bled from the turbine of total mass flow rate.

76. For perfect inter cooling, pressure ratio are same in each stage

$$r^2 = \frac{16}{1}$$

$r = 4$

77. At the time of intake, intake valve operates only when the pressure inside the cylinder is lower than the atmospheric pressure. And appreciable large pressure difference at the time of suction starting, causing fluttering at initial.

At the time of compression, process followed is Polytropic and continues till pressure exceeds the delivery line pressure for opening the delivery valve.

78.

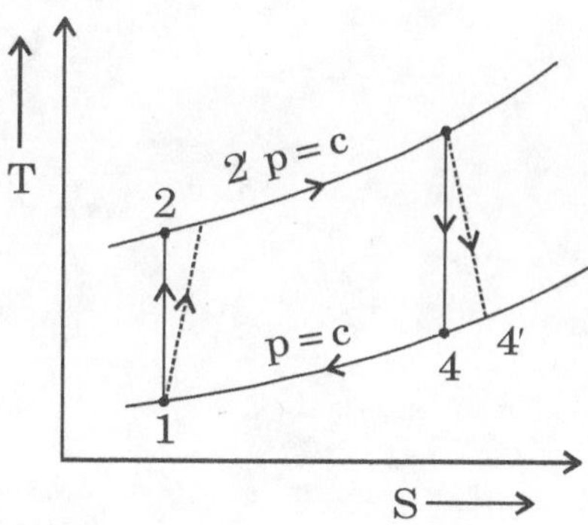

$$T_2 = T_1 (r_p)^{\frac{r-1}{r}}$$

$$T_2 = 300\ (8)^{\frac{0.4}{1.4}}$$

T2 = 543.43 K

$$\eta_c = \frac{T_2 - T_1}{T_2' - T_1}$$

$$\eta_c = \eta_T = 0.8$$

$r_p = 8$

$T_1 = 300$ K

$T_3 = 1400$ K

$$\eta_c = \frac{543.43 - 300}{T_2' - 300}$$

$$0.8 = \frac{543.43 - 300}{T_2' - 300}$$

$$\boxed{T_2' = 604.3\text{K}}$$

$$\rho_c = mC_p\left(T_2' - T_1\right) = C_p\left(T_2' - T_1\right)\frac{\text{KW}}{\text{kg}}$$

$= 1\ (604.3 - 300)$

$$= 304.3\ \frac{\text{kW}}{\text{kg}}$$

79. $T_4 = \dfrac{T_3}{8^{.2857}} = 772.9\text{K}$

$$\eta_T = \frac{T_3 - T_4'}{T_3 - T_4} \Rightarrow 0.8 = \frac{1400 - T_4'}{1400 - 772.9}$$

$\Rightarrow T_4' = 898.32\text{K}$

$$\eta_{cycle} = \frac{(W_T)_{actual} - (W_C)_{actual}}{Q_s}$$

$$= \frac{\left(T_3 - T_4'\right) - \left(T_2' - T_1\right)}{T_3 - T_2'}$$

$$\eta_{cycle} = \frac{(1400 - 898.32) - (604.3 - 300)}{1400 - 604.3}$$

$= .248$ or 24.8%

80. $(r_p)_{OPT} = \left(\dfrac{P_4}{P_1}\right)^{\frac{1}{3}} = (10)^{\frac{1}{3}} = 2.1544$

$$\frac{P_2}{P_1} = \frac{P_3}{P_2} = \frac{P_4}{P_3} = 2.1544$$

$P_2 = 2.1544 \times 100 = 215.44$

$$\frac{P_3}{P_2} = 2.1544$$

$P_3 = 2.1544 \times 215.44 = 464.15$ kPa

7 CHAPTER Internal Combustion Engine

Analysis of Previous GATE Papers														
Topic	Year/Set →Marks Distribution ↓		2019 Set 1	2019 Set 2	2018 Set 1	2018 Set 2	2017 Set 1	2017 Set 2	2016 Set 1	2016 Set 2	2016 Set 3	2015 Set 1	2015 Set 2	2015 Set 3
Air Standard Cycles	1 Mark	MCQ Type												1
		Numerical Type										1		
	2 Marks	MCQ Type											1	
		Numerical Type		1	1	1								
	Total			2	2	2						1	2	1
Combustion in SI and CI	1 Mark	MCQ Type												
		Numerical Type							1					
	2 Marks	MCQ Type												
		Numerical Type					1							
	Total						2		1					
Performance Parameter Analysis	1 Mark	MCQ Type												
		Numerical Type											1	
	2 Marks	MCQ Type												
		Numerical Type												
	Total												1	

Air standard cycles

1. In air standard Otto cycle the terminal pressures at the end of compression, heat release and expansion are respectively P_2, P_3 and P_4. If the corresponding values are P_2, P_3, and P_4, taking into account the effect of variable specific heat and dissociation of the working fluid, then

(a) $P_2 < P_2'$ and $P_3 > P_3'$

(b) $P_3 < P_3'$ and $P_4 > P_4'$

(c) $P_2 > P_2'$, $P_3 > P_3'$ and $P_4 < P_4'$

(d) $P_2 > P_2'$, $P_3 > P_3'$

[1989 : 2 Marks]

2. An air standard diesel cycle consists of

(a) Two adiabatic and two constant volume processes

(b) Two constant volume and two isothermal processes

(c) One constant pressure, one constant volume and two adiabatic processes

(d) One constant pressure, one constant volume and two isothermal processes

[1990 : 2 Marks]

3. The figure below shows a thermodynamic cycle undergone by a certain system. Find the mean effective pressure in N/m^2.

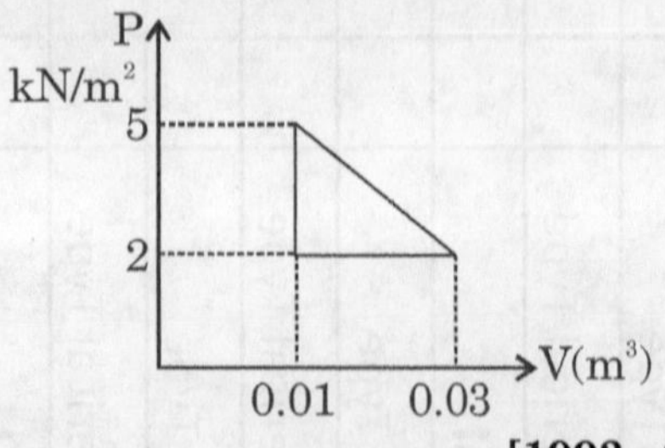

[1993 : 2 Marks]

4. With increasing temperature of intake air, CI engine efficiency

(a) decreases

(b) increases

(c) remains same

(d) depends on other factors

[1998 : 1 Mark]

5. An IC engine has a bore and stroke of 2 units each. The area to calculate heat loss can be taken as

(a) 4π (b) 5π

(c) 6π (d) 4

[1998 : 2 Marks]

6. An air breathing air craft is flying at an altitude where the air density is half the value at ground level. With reference to the ground level, the air- fuel ratio at this altitude will be

(a) $3\sqrt{2}$ (b) $\sqrt{\frac{1}{2}}$

(c) 2 (d) 4

[1998 : 2 Marks]

7. In a spark ignition engine working on the ideal Otto cycle, the compression ratio is 5.5. The work output per cycle (i.e., area of the P – V diagram) is equal to $23.625 \times 10^5 \times V_c J$, where V_C is the clearance volume in m^3. The indicated mean effective pressure is

(a) 4.295 bar (b) 5.250 bar

(c) 86.870 bar (d) 106.300 bar

[2001 : 2 Marks]

8. An ideal air standard Otto cycle has a compression ratio of 8.5. If the ratio of the specific heats of air (γ) is 1.4, what is the thermal efficiency (in percentage) of the Otto cycle?

(a) 575 (b) 45.7

(c) 52.5 (d) 95

[2002 : 2 Marks]

9. For an engine operating on air standard Otto cycle, the clearance volume is 10% of the swept volume. The specific heat ratio of air is 1.4. The air standard cycle efficiency is-

(a) 38.3% (b) 39.8%

(c) 60.2% (d) 61.7%

[2003 : 2 Marks]

10. An engine working on air standard Otto cycle has a cylinder diameter of 10 cm and stroke length of 15 cm. The ratio of specific heats for air is 1.4. If the clearance volume is 196.3 cc and the heat supplied per kg of air per cycle is 1800 kJ/kg, the work output per cycle per kg of air is

(a) 879.1 kJ (b) 890.2 kJ

(c) 895.3 kJ (d) 973.5 kJ

[2004 : 2 Marks]

11. The stroke and bore of a four stroke spark ignition engine are 250 mm and 200 mm respectively. The clearance volume is 0.001 m^3. If the specific heat ratio $\gamma = 1.4$, the air-standard cycle efficiency of the engine is

(a) 46.40% (b) 56.10%

(c) 58.20% (d) 62.80%

[2007 : 2 Marks]

12. Which one of the following is NOT a necessary assumption for the air-standard Otto cycle?

(a) All processes are internally reversible.

(b) Intake and exhaust processes are constant volume heat rejection processes.

(c) The combustion process is a constant volume heat addition process.

(d) The working fluid is an ideal gas with constant specific heats.

[2008 : 1 Mark]

13. In an air-standard Otto cycle, the compression ratio is 10. The condition at the beginning of the compression process is 100 kPa and 27°C. Heat added at constant volume is 1500 kJ/kg, while 700 kJ/kg of heat is rejected during the other constant volume process in the cycle. Specific gas constant for air = 0.287 kJ/kgK. The mean effective pressure (in kPa) of the cycle is

(a) 103 (b) 310

(c) 515 (d) 1032

[2009 : 2 Marks]

14. The crank radius of a single-cylinder IC engine is 60 mm and the diameter of the cylinder is 80 mm. The swept volume of the cylinder in cm^3 is

(a) 48 (b) 96

(c) 302 (d) 603

[2011 : 2 Marks]

15. In an air-standard Otto cycle, air is supplied at 0.1 MPa and 308 K. The ratio of the specific heats (γ) and the specific gas constant (R) of air are 1.4 and 288.8 J/kgK, respectively. If the compression ratio is 8 and the maximum temperature in the cycle is 2660 K, the heat (in kJ/kg) supplied to the engine is ______.

[2014 : 2 Marks, Set-1]

16. A diesel engine has a compression ratio of 17 and cut-off takes place at 10% of the stroke. Assuming ratio of specific heats (γ) and the air standard efficiency (in %) is –.

[2014 : 2 Marks, Set-3]

17. In a compression ignition engine, the inlet air pressure is 1 bar and the pressure at the end of isentropic compression is 32.42 bar. The expansion ratio is 8. Assuming ratio of specific heats (γ) as 1.4, the air standard efficiency (in percent) is

[2014 : 2 Marks, Set-4]

18. Air enters a diesel engine with a density of 1.0 kg/m^3. The compression ratio is 21. At steady state, the air intake is 30×10^{-3} kg/s and the net work output is 15 kW. The mean effective pressure (kPa) is_____.

[2015 : 1 Mark, Set-1]

19. For the same values of peak pressure, peak temperature and heat rejection, the correct order of efficiencies for Otto, Duel and Diesel cycles is

(a) $\eta_{otto} > \eta_{Dual} > \eta_{Diesel}$

(b) $\eta_{Diesel} > \eta_{Dual} > \eta_{Otto}$

(c) $\eta_{Dual} > \eta_{disel} > \eta_{Otto}$

(d) $\eta_{Diesel} > \eta_{Otto} > \eta_{Dual}$

[2015 : 2 Marks, Set-2]

20. An air-standard Diesel cycle consists of the following processes:

1-2 : Air is compressed isentropically.

2-3 : Heat is added at constant pressure.

3-4 : Air expands isentropically to the original volume.

4-1 : Heat is rejected at constant volume. If γ and T denote the specific heat ratio and temperature, respectively, the efficiency of the cycle is

(a) $1-\frac{T_4-T_1}{T_3-T_2}$ (b) $1-\frac{T_4-T_1}{\gamma(T_3-T_2)}$

(c) $1-\frac{\gamma(T_4-T_1)}{T_3-T_2}$ (d) $1-\frac{T_4-T_1}{(\gamma-1)(T_3-T_2)}$

[2015 : 1 Mark, Set-3]

21. An engine working on air standard Otto cycle is supplied with air at 0.1 MPa and 35°C. The compression ratio is 8. The heat supplied is 500 kJ/kg. Property data for air: c_p = 1.005 kJ/kgK, c_v = 0.718 kJ/kgK, R = 0.287 kJ/kgK. The maximum temperature (in K) of the cycle is ______(correct to one decimal place).

[2018: 2 Marks, Set-1]

22. A vehicle powered by a spark ignition engine follows air standard Otto cycle (γ = 1.4). The engine generates 70 kW while consuming 10.3 kg/hr of fuel. The calorific value of fuel is 44000 kJ/kg. The compression ratio is (correct to two decimal places).

[2018 : 2 Marks, Set-2]

23. An air standard Otto cycle has thermal efficiency of 0.5 and the mean effective pressure of the cycle is 1000 kPa. For air, assume specific heat ratio $\gamma = 1.4$ and specific gas constant R = 0.287 kJ/kg.K. If the pressure and temperature at the beginning of the compression stroke are 100 kPa and 300 K, respectively, then the specific net work output of the cycle is ___ kJ/kg (round off to two decimal places).

[2019 : 2 Marks, Set-2]

Combustion in SI and CI

24. The power output from a spark ignition engine is varied by

(a) Changing the ignition timing

(b) Regulating the amount of air-fuel inducted

(c) Regulating the amount of air fuel mixture

(d) Regulating the amount of fuel

[1990 : 2 Marks]

25. For determining the ignition quality of compression ignition engine fuels, the reference fuels used are

(a) Isooctane and n-heptane

(b) Cetane and α-methylnapthalene

(c) Hexadecane and n-heptane

(d) Cetane and isooctane

[1991 : 2 Marks]

26. If air fuel ratio of the mixture in petrol engine is more than 15:1

(a) NO_x is reduced

(b) CO_2 is reduced

(c) HC is reduced

(d) CO reduced

[1991 : 2 Marks]

27. Alcohols are unsuitable at diesel engine fuels because

(a) The cetane number of alcohol fuels is very low which prevents their ignition by compression

(b) The cetane number of alcohol fuels is very high which prevents their ignition by compression

(c) The cetane number of alcohol fuels is very low which prevents good combustion

(d) None of these

[1992 : 2 Marks]

28. Knocking tendency in a Sl engine reduces with increasing

(a) Compression ratio

(b) Wall temperature

(c) Supercharging

(d) Engine speed

[1993 : 2 Marks]

29. In order to burn 1 kilogram of CH_4 completely, the minimum number of kilograms of oxygen needed is (take atomic weights of H, C and O as 1, 12 and 16 respectively)

(a) 3 (b) 4

(c) 5 (d) 6

[1995 : 2 Marks]

30. The silencer of an internal combustion engine

(a) reduces noise

(b) decrease break specific fuel consumption (BSFC)

(c) increase BSFC

(d) has no effect on its efficiency

[1999: 2 Marks]

31. For a spark ignition engine, the equivalence ratio (ϕ) of mixture entering the combustion chamber has values

(a) $\phi < 1$ for idling and $\phi > 1$ for peak power conditions

(b) $\phi > 1$ for both idling and peak power conditions

(c) $\phi > 1$ for idling and $\phi < 1$ for peak power conditions

(d) $\phi < 1$ for both idling and peak power conditions

[2003 : 1 Mark]

32. A diesel engine is usually more efficient than a spark ignition engine because

(a) diesel being a heavier hydrocarbon, releases more heat per kg than gasoline

(b) the air standard efficiency of diesel cycle is higher than the Otto cycle, at a fixed compression ratio

(c) the compression ratio of a diesel engine is higher than that of an SI engine

(d) self ignition temperature of diesel is higher than that of gasoline

[2003 : 1 Mark]

33. At the time of starting, idling and low speed operation, the carburettor supplies a mixture which can be termed as

(a) Lean

(b) slightly leaner than stoichiometric

(c) stoichiometric

(d) rich

[2004 : 1 Mark]

34. Propane (C_3H_8) is burned in an oxygen atmosphere with 10% deficit oxygen with respect to the stoichiometric requirement. Assuming no hydrocarbons in the products, the volume percentage of CO in the products is___.

[2016 : 1 Mark, Set-1]

35. Air contains 79% N_2 and 21% O_2 on a molar basis. Methane (CH_4) is burned with 50% excess air than required stoichiometrically. Assuming complete combustion of methane, the molar percentage of N_2 in the products is______.

[2017 : 2 Marks, Set-1]

Performance parameter analysis

36. BHP of a diesel engine can be increased by

(a) increasing the pressure of intake air

(b) increasing the temperature of intake air

(c) increasing the density of intake air

(d) decreasing the density of intake air

[1991 : 2 Marks]

37. Brake thermal efficiency of the three basic types of reciprocating engines commonly used in road vehicles are given in the increasing order as

(a) 2 Stroke SI engine, 4 Stroke SI engine, 4 Stroke CI engine

(b) 2 Stroke SI engine, 4 Stroke CI engine, 4 Stroke SI engine

(c) 4 Stroke SI engine, 2 Stroke SI engine, 4 Stroke CI engine

(d) 4 Stroke CI engine, 4 Stroke SI engine, 2 Stroke SI engine

[1992: 2 Marks]

38. An automobile engine operates at a fuel air ratio of 0.05, volumetric efficiency of 90% and indicated thermal efficiency of 30%. Given that the calorific value of the fuel is 45 MJ/kg and the density of air at intake in 1 kg/m^3, the indicated mean effective pressure for the engine is

(a) 6.075 bar (b) 6.75 bar

(c) 67.5 bar (d) 243 bar

[2003 : 2 Marks]

39. During a Morse test on a 4 cylinder engine, the following measurements of brake power were taken at constant speed.

All cylinders firing	3037 kW
Number 1 cylinder not firing	2102 kW
Number 2 cylinder not firing	2102 kW
Number 3 cylinder not firing	2100 kW
Number 4 cylinder not firing	2098 kW

The mechanical efficiency of the engine is

(a) 91.53% (b) 85.07%

(c) 81.07% (d) 61.22%

[2004 : 2 Marks]

40. A turbocharged four-stroke direct injection diesel engine has a displacement volume of 0.0259 m^3 (25.9 liters). The engine has an output of 950 kW at 2200 rpm. The mean effective pressure (in MPa) is closest to

(a) 2 (b) 1

(c) 0.2 (d) 0.1

[2010 : 1 Mark]

41. A rope-brake dynamometer attached to the crank shaft of an I.C. engine measures a brake power of 10 kW when the speed of rotation of the shaft is 400 rad/s. The shaft torque (in Nm) sensed by the dynamometer is _____.

[2015 : 1 Mark, Set-2]

ANSWERS

1. (d)	**2.** (c)	**4.** (a)	**5.** (a)	**6.** (c)	**7.** (c)	**8.** (a)	**9.** (d)	**10.** (d)	**11.** (c)
12. (b)	**13.** (d)	**14.** (d)	**19.** (b)	**20.** (b)	**24.** (c)	**25.** (b)	**26.** (a)	**27.** (c)	**28.** (d)
29. (b)	**30.** (a)	**31.** (b)	**32.** (c)	**33.** (d)	**36.** (a & c)	**37.** (a)	**38.** (a)	**39.** (c)	**40.** (a)

EXPLANATIONS

3. Work done = Area under the cycle

$$= \frac{1}{2} \times 3 \times 0.02 = 0.03\text{kNm}$$

$$\text{mep} = \frac{\text{Work done}}{\text{Volume}} = \frac{0.03}{0.02}$$

mep = 1.5 kPa or 1500 N/m^2

5. Total area for heat loss = surface area of cylinder

$$= \pi dl = \pi \times 2 \times 2 = 4\pi$$

6. $\text{A/F ratio} = \dfrac{m_a}{m_f}$

where m_a – mass of air,

m_f – mass of fuel

$$\frac{(\text{A / F ratio})_{\text{ground}}}{(\text{A / F ratio})_{\text{altitude}}} = \frac{(m_a)_{\text{ground}}}{(m_a)_{\text{altitude}}}$$

as m_f is same at both places.

$$\frac{(m_a)_{\text{ground}}}{(m_a)_{\text{altitude}}} = \frac{V \times \rho_{\text{ground}}}{V \times \rho_{\text{altitude}}}$$

where 'V' is the volume of air taken in.

$$= \frac{1}{(1\backslash 2)} = 2$$

7. $\dfrac{V_1}{V_C} = 5.5$

$V_1 = 5.5\, v_c$

$V_s = V_1 - V_c$

$V_s = 4.5\, V_c$

work output = $23.625 \times 10^5 \times V_C$ Joules

$$\text{mep} = \frac{23.625 \times 10^5 V_C}{4.5 V_C}$$

$= 5.25 \times 10^5$ Pa

$= 5.250$ bar

8. $\eta = 1 - \dfrac{1}{r^{\gamma-1}} = 1 - \dfrac{1}{(8.5)^{1.4-1}}$

$$= 1 - \frac{1}{(8.5)^{0.4}} = 57.5$$

9. V_C = clearance volume = 10% of swept volume

$\therefore \quad V_c = 10\%\, V_s$

$= 0.1\, V_s$

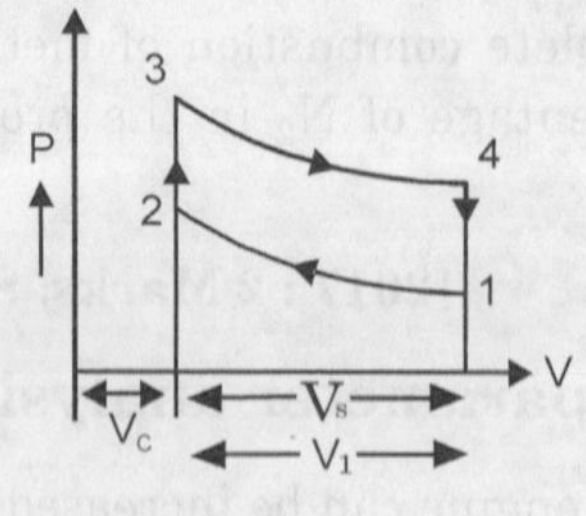

Now r = compression ratio = $\dfrac{V_c + V_s}{V_c} \approx \dfrac{V_1}{V_c}$

$$= \frac{0.1V_s + V_s}{0.1V_s}$$

$$= \left(\frac{1.1}{.1}\right)\frac{V_s}{V_s} = 11$$

Now standard cycle efficiency,

$$\eta = 1 - \frac{1}{r^{\gamma-1}}$$

$$= 1 - \frac{1}{(11)^{1.4-1}} = 61.7\%$$

10. Given: $V_c = V_2 = 196.3$ cm^3 = 0.0001963 m^3 stroke volume,

$$V_s = \frac{\pi}{4} D^2 \times L = \frac{\pi}{4}(10)^2 \times 15$$

$= 1178.09$ cc

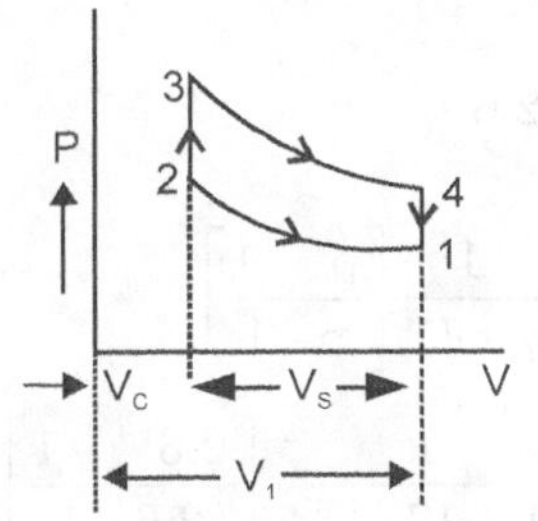

Compression ratio,

$$(r) = \frac{V_s + V_c}{V_c} = \frac{1178.09 + 196.3}{196.3} = 7$$

$$\eta_{otto} = 1 - \frac{1}{r^{\gamma-1}} = 1 - \frac{1}{7^{1.4-1}} = 1 - \frac{1}{7^{0.4}}$$

$= 0.5408$

Now,

$$\eta_{otto} = \frac{\text{work output}}{\text{Heat supplied}} = \frac{W}{Q_s}$$

or $0.5408 = \dfrac{\text{work output}}{1800}$

or Work output = 973.5 kJ

11. Here, $V_C = .001\ m^3$

$$V_S \frac{\pi}{4} d^2 l = \frac{\pi}{4} \times (.2)^2 \times (.25)^2$$

$\therefore$ Compression ratio, $r = 1 + \dfrac{V_S}{V_C} = 0.85$

$\therefore n = 1 - \dfrac{1}{r^{\gamma-1}} = 58.20\%$

12. Intake process is the constant volume heat addition process

Hence (b) is not the right assumption

13.

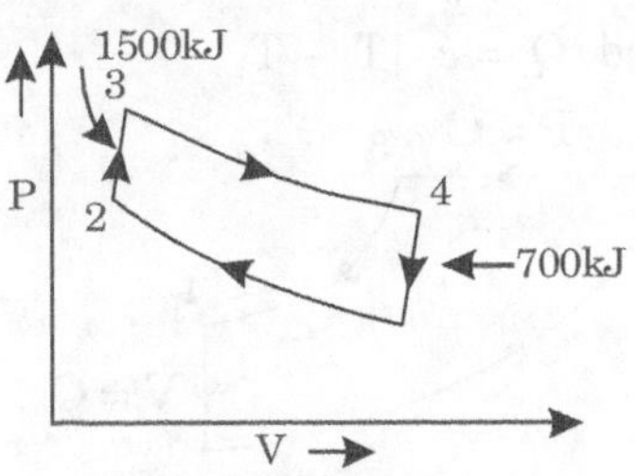

Here, $P_1 = 100$ kPa

$T_1 = 27°C$

$$\frac{V_4}{V_3} = \frac{V_1}{V_2} = 10$$

$C_v = 0.287$ kJ/kgK

$P_m \times (V_1 - V_2) =$ work done

$= h_1 - h_2$

$= 1500 - 700$

$= 800$ kJ

Now $P_1V_1 = mRT_1$

or $100 \times 10^3 \times V_1 = 1 \times 0.287 \times 10^3 \times 300$

or $V_1 = 0.861$

$\therefore$ $V_2 = 0.0861$

Now $P_m\ (0.861 - 0.0861) = 800$

$\therefore$ $P_m = 1032$ kPa

14. Stroke of the cylinder, $l = 2r = 2 \times 60 = 120$ mm

$$\text{Swept volume} = \frac{\pi}{4} d^2 \times l$$

$$= \frac{\pi}{4} \times 80^2 \times 120 = 603\ cm^3$$

15. Otto cycle

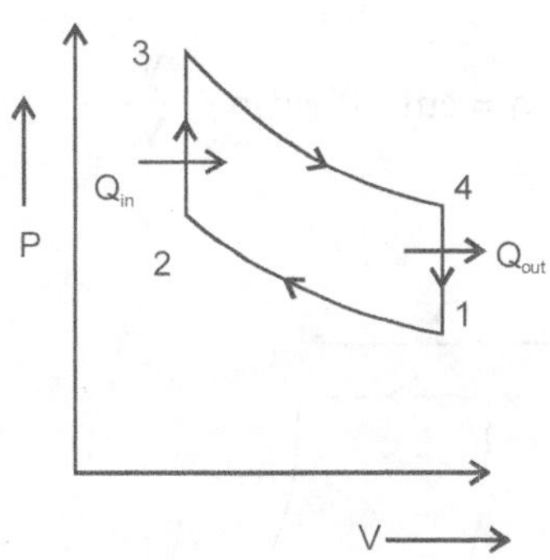

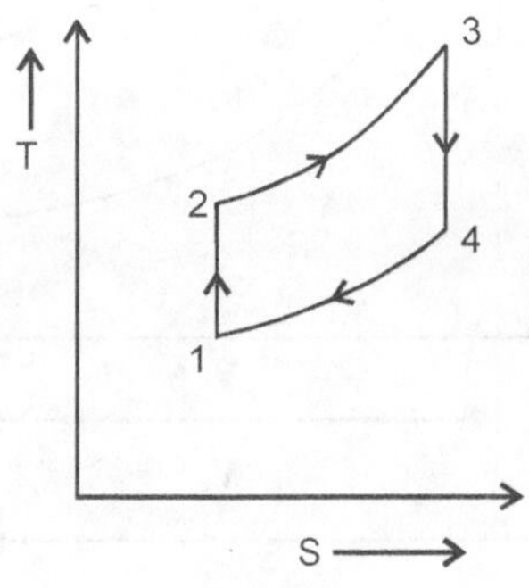

Given : $P_1 = 0.1$ MPa

$T_1 = 308$ K

$\gamma = 1.4$

$R = 0.2888$ kJ/kg.k

$T_3 = 2660$ K

(1-2) Isentropic compression

PV^r = constant

$$\frac{T_2}{T_1} = (r)^{\gamma-1}$$

$T_2 = 308 \times 8^{0.4} = 698.40$ k

(2-3) Isochoric Heat addition process

$$\frac{C_p}{C_v} = 1.4$$

R = 0.2888 kJ/kg

$C_p - C_v = 0.2888$

$$C_v\left(\frac{C_p}{C_v} - 1\right) = 0.2888$$

$$C_v = \frac{0.2888}{1.4-1} = 0.722 \text{ kJ/kg}$$

$Q_{in} = C_v (T_3 - T_2) = 0.722 (2660 - 698.40)$
$= 1416.27$ kJ/kg

16. Compression ratio $= 17 = \frac{V_1}{V_2} = V_1 = 17V_2$

where ρ = cut-off ratio $= \frac{V_3}{V_2}$

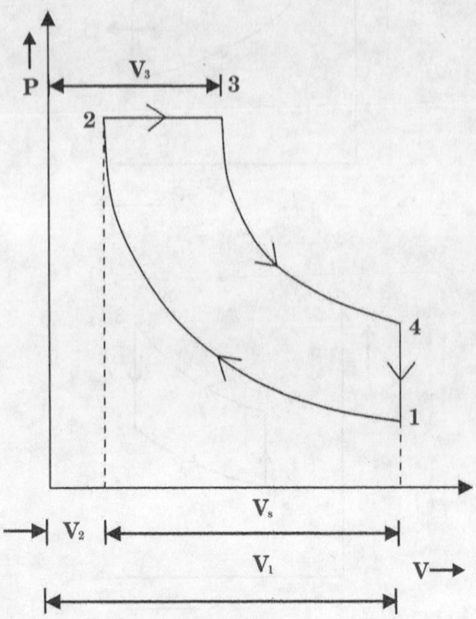

$$V_3 - V_2 = \frac{10}{100} V_s$$

$V_3 - V_2 = 0.1(V_1 - V_2)$
$= 0.1(17V_2 - V_2)$
$= 0.1(16V_2)$
$V_3 = 1.6V_2 + V_2$

$V_3 = 2.6V_2$

$$\frac{V_3}{V_2} = \rho = 2.6$$

$$\eta = 1 - \frac{1}{\gamma(r)^{\gamma-1}}\left[\frac{\rho^{\gamma}-1}{\rho-1}\right]$$

$$= 1 - \frac{1}{1.4(17)^{1.4-1}} \times \left[\frac{2.6^{1.4}-1}{2.6-1}\right]$$

$= 0.5960 = 59.60\%$.

17. Compression ratio,

$$r_k = \frac{V_1}{V_2} = \left(\frac{p_2}{p_1}\right)^{\frac{1}{\gamma}} = (32.42)^{\frac{1}{1.4}} = 11.9999 \simeq 12$$

$$\text{cutoff ratio}, r_k = \frac{\text{compression ratio}}{\text{expansion ratio}} = \frac{12}{8} = 1.5$$

$$\eta = 1 - \frac{1}{r_k^{\gamma-1}}\left(\frac{r_e^{\gamma}-1}{\gamma(r_c-1)}\right)$$

$$= 1 - \frac{1}{12^{0.4}}\left(\frac{(15)^{1.4}-1}{1.4(1.5-1)}\right)$$

$\eta = 59.599\%$.

18.
$$\text{mep} = \frac{\text{work output}}{\text{swept volume}}$$

$$= \frac{12}{30 \times 10^{-3}\left(1 - \frac{1}{21}\right)}$$

$= 525$ kPa

19. For same values of peak pressure and temperature. Diesel cycle is most efficient and Otto cycle is least. Efficiency of dual cycle lies in between. $\eta_{Diesel} > \eta_{Dual} > \eta_{Otto}$

20. Heat applied, $Q_s = c_p (T_3 - T_2)$

Heat rejected, $Q_r = c_V (T_4 - T_1)$

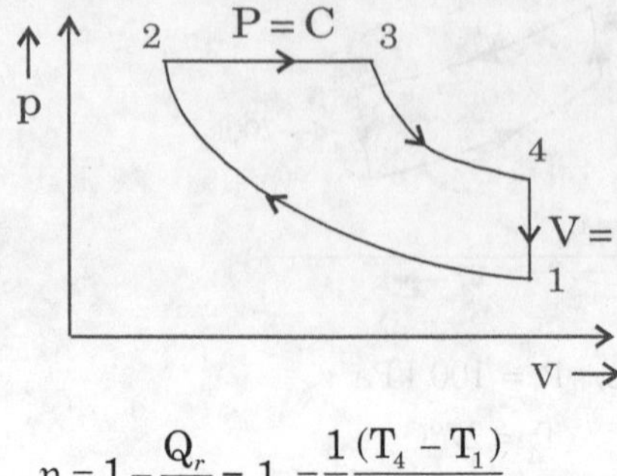

$$\eta = 1 - \frac{Q_r}{Q_s} = 1 - \frac{1}{\gamma}\frac{(T_4 - T_1)}{(T_3 - T_2)}$$

21. $P_1 = 0.1$ MPa

$T_1 = 35°C = 35 + 273 = 308$ K

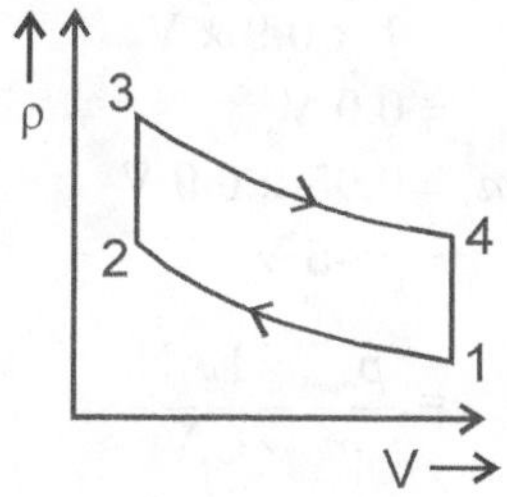

$$V_2 = \frac{V_1}{8}$$

Heat Supplied $Q_s = C_v(T_3 - T_2)$

$C_v(T_3 - T_2) = 500$ kJ/kg ...(*i*)

Given, $\left(\frac{V_2}{V_1}\right)^{\gamma-1} = \frac{T_1}{T_2}$

$$\Rightarrow \quad \left(\frac{1}{8}\right)^{1.4-1} = \frac{308}{T_2} = 707.60\text{K}$$

$$\Rightarrow \quad T_2 = 707.60 \text{ K}$$

After putting the value of T_2 is equation(i):-

$0.718\,(T_3 - 707.6) = 500$

$$\Rightarrow \quad \boxed{T_3 = 1403.98 \text{ K}}$$

22. $r = 14$

BP = 70 kW

$\dot{M}_f = 10.3$ kg/hr

C.V. = 44000 kJ/kg

$$\eta = \frac{BP}{\dot{M}_f \times C.V.} = \frac{70}{\frac{10.3}{3600} \times 44000}$$

$\eta = 0.556$

$$\eta_{otto} = 1 - \frac{1}{(r)^{\gamma-1}} = 0.556$$

$\Rightarrow \gamma = 7.61$

23.

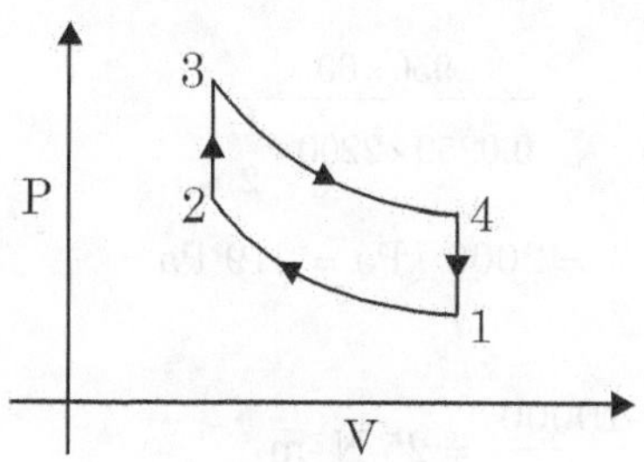

Otto cycle

$$\eta = 0.5 = 1 - \frac{1}{(r)^{\gamma-1}}$$

$$\because r = \text{compression ratio} = \frac{V_T}{V_C} = 1 + \frac{V_S}{V_C}$$

$\because$ Before compression,

Total volume, $V_T = \frac{m.RT}{P}$ [let m = 1 kg]

$$V_T = \frac{RT}{P} = \frac{0.287 \times 300}{100}$$

$V_T = 0.861$ m³/kg

From equation (1)

$\Rightarrow r = 5.656$

$$\Rightarrow \frac{V_T}{V_C} = 5.656$$

$$\Rightarrow V_C = \frac{0.861}{5.656} = 0.152 \text{m}^3/\text{kg}$$

Swept volume, $V_S = (r - 1).V_C$

$= 4.656 \times 0.152$

$= 0.70877$ m³/kg

Specific net work output of cycle,

$W = P_{mep} \times V_S$

$= 1000 \times 0.70877$

$\boxed{W_{net} = 708.77 \text{ kJ/kg}}$

29. $CH_4 + 2C_2 \rightarrow CO_2 + 2 4_2O$

16 kg 24 kg 44 kg 36 kg

1kg $\frac{64}{16}$kg

$CH_4 = \frac{64}{16}$kg [from above balance reaction mass of O_2 for complete combustion of 1 kg $\rightarrow$ mass of oxygen = 4 kg]

33. Very rich mixture is provided during starting, idling and peak power to ensure burning due to low supply of air, low temperature etc.

34. $C_3H_8 + xO_2 \rightarrow aCO_2 + bH_2O$

Carbon balance:

$a = 3$

Hydrogen balance:

$2b = 8 \rightarrow b = 4$

Oxygen balance:

$$2x = 2a + b$$

$$\rightarrow \quad x = a + \frac{b}{2} = 3 + \frac{4}{2} = 5$$

For chemical correct or stoichiometric burning, no of moles of O_2 required are = 5.

As it is burnt with 10% deficient oxygen, it will generate CO.

The new equation is

$$C_3H_8 + 0.9 \times 5O_2 \rightarrow aCO_2 + bCO + cH_2O$$

Carbon balance:

$$a + b = 3$$

Hydrogen balance:

$$2c = 8 \rightarrow c = 4$$

Oxygen balance:

$$2a + b + c = 0.9 \times 5 \times 2 = 9$$

$$2a + b + c = 9$$

$$\Rightarrow 2a + b + c \Rightarrow 2a + b = 5$$

$$a + b = 3$$

By solving these two equations

$a = 2$ and $b = 1$

In the exhaust products the no, of moles of CO are 1. % by volume of CO in exhaust,

$$= \frac{b}{a+b+c} \times 100 = \frac{1}{2+1+4} \times 100$$

$$= \frac{1}{7} \times 100 = 14.29\%$$

36. The BHP of diesel engine is increasing burning more fuel. And burning more fuel can be achieved by increasing density of intake air and increasing the pressure of intake air.

38. Given, $\frac{\text{Fuel}}{\text{Air}} = 0.05$

$$\eta_{\text{volumetric}} = 90\%$$

$$\eta_{\text{thermal}} = 30\%$$

Calorific value = 45 MJ/kg

$$\frac{m_f}{m_a} = 0.05$$

$$\rho_{\text{air}} = 1 \text{ kg/m}^3$$

Volumetric efficiency,

$$\eta_V = \frac{\text{actual volume}}{\text{Swept volume}} = \frac{V_{ac}}{V_s}$$

$$\therefore \quad V_{ac} = 0.9\, V_s$$

$$m_a = \rho_{\text{air}} V_{ac} = 1 \times 0.9 \times V_s = 0.9\, V_s$$

$$m_f = 0.05 \times 0.9\, V_s = .045\, V_s$$

$$\eta_{\text{thermal}} = \frac{p_{mep} \times \text{LAN}}{m_f \times \text{C.V}}$$

$$0.3 = \frac{p_{mep} \times V_s}{.045 V_s \times 45 \times 10^6}$$

... since LAN = V_s

$p_{mep} = 0.3 \times 0.045 \times 45 \times 10^6 = 6.075 \times 10^5 \text{N/m}^2$ = 6.075 bar

39. Given: B.P. = 3037 kW

$(\text{I.P.})_1 = 3037 - 2102 = 935$ kW

$(\text{I.P.})_2 = 3037 - 2102 = 935$ kW

$(\text{I.P.})_3 = 3037 - 2100 = 937$ kW

$(\text{I.P.})_4 = 3037 - 2098 = 937$ kW

Total I.P. of the cylinder

$= (\text{I.P})_1 + (\text{I.P})_2 + (\text{I.P})_3 + (\text{I.P})_4$

$= 935 + 935 + 937 + 937 = 3746$ kW

∴ Mechnical efficiency of the engine

$$= \frac{\text{B.P.}}{\text{I.P}} = \frac{3037}{3746} = 81.07\%$$

40. $\text{B.P.} = \frac{n \cdot p_m \text{ VN}.k}{60}$

where p_m = mean effective pressure

n = number of cylinders = 1.

V = volume of displacement = 0.025^{10} cm^3

N = r.p.m

$k = \frac{1}{2}$ for 4-stroke

= 1 for 2-stroke

$$\therefore \quad p_m = \frac{\text{B.P} \times 60}{n.\text{V.N}.k}$$

$$= \frac{950 \times 60}{0.0259 \times 2200 \times \frac{1}{2}}$$

= 2000 kPa = 219 Pa

41. P = Tω

$$\therefore \quad T = \frac{P}{\omega} = \frac{10000}{400} = 25 \text{ N–m}$$

42. (0.51)

Unit - IX

Refrigeration and Air-conditioning

Syllabus

Vapour and gas refrigeration and heat pump cycles; properties of moist air, psychrometric chart, basic psychrometric processes.

Contents

CHAPTER 1

Vapour Refrigeration

Analysis of Previous GATE Papers

Topic	Year/Set → Marks Distribution ↓		2019 Set 1	2019 Set 2	2018 Set 1	2018 Set 2	2017 Set 1	2017 Set 2	2016 Set 1	2016 Set 2	2016 Set 3	2015 Set 1	2015 Set 2	2015 Set 3
Vapour Compression Cycles	1 Mark	MCQ Type	1											
		Numerical Type												
	2 Marks	MCQ Type												
		Numerical Type				1				1	1			1
	Total		1			2				2	2			2
Refrigerants	1 Mark	MCQ Type												
		Numerical Type												
	2 Marks	MCQ Type												1
		Numerical Type												
	Total													2

Vapour Compression cycles

1. Round the clock cooling of an apartment having a load of 300 MJ/day requires and air conditioning plant of capacity about

(a) 1 ton (b) 5 tons

(c) 10 tons (d) 100 tons

[1993 : 2 Marks]

2. In a vapour compression refrigeration system, liquids to suction heat exchanger is used to

(a) Keep the COP constant

(b) Prevent the liquid refrigerant from entering the compressor

(c) Sub-cool the liquid refrigerant leaving the condenser

(d) Sub-cool the vapour refrigerant from the evaporator

[2000 : 1 Mark]

Common Data for Q. 3 & 4

A refrigerator based on ideal vapour compression cycle operates between the temperature limits of –20°C and 40°C. The refrigerant enters the condenser as saturated vapour and leaves as saturated liquid. The enthalpy and entropy values for saturated liquid and vapour at these temperatures are given in the table below.

T (°C)	h_f (kJ/kg)	h_g (kJ/kg)	s_f (kJ/kgK)	s_g (kJ/kgK)
–20	20	180	0.07	0.7366
40	80	200	0.3	0.67

3. If refrigerant circulation rate is 0.025 kg/s, the refrigeration effect is equal to

(a) 2.1 kW (b) 2.5 kW

(c) 3.0 kW (d) 4.0 kW

[2003 : 2 Marks]

4. The COP of the refrigerator is

(a) 2.0 (b) 2.33

(c) 5.0 (d) 6.0

[2003 : 2 Marks]

5. In the window air conditioner, the expansion device used is

(a) capillary tube

(b) thermostatic expansion valve

(c) automatic expansion valve

(d) float valve

[2004 : 1 Mark]

6. A R-12 refrigerant reciprocating compressor operates between the condensing temperature of 30° and evaporator temperature of –20°C. The clearance volume ratio of the compressor is 0.03. Specific heat ratio of the vapour is 1.15 and the specific volume at the suction is 0.1089 m³/kg. Other properties at various states are given in the figure. To realize 2 tons of refrigeration, the actual volume displacement rate considering the effect of clearance is

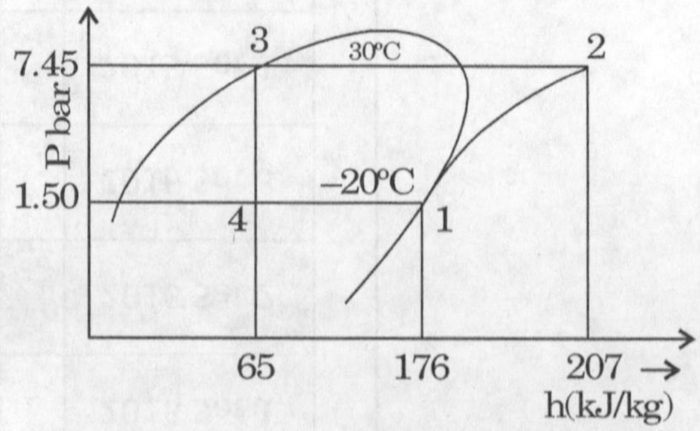

(a) 6.35×10^{-3} m³/s (b) 63.5×10^{-3} m³/s

(c) 635×10^{-3} m³/s (d) 4.88×10^{-3} m³/s

[2004 : 2 Marks]

7. The vapour compression refrigeration cycle is represented as shown in the figure below, with state 1 being the exit of the evaporator. The coordinate system used in this figure is

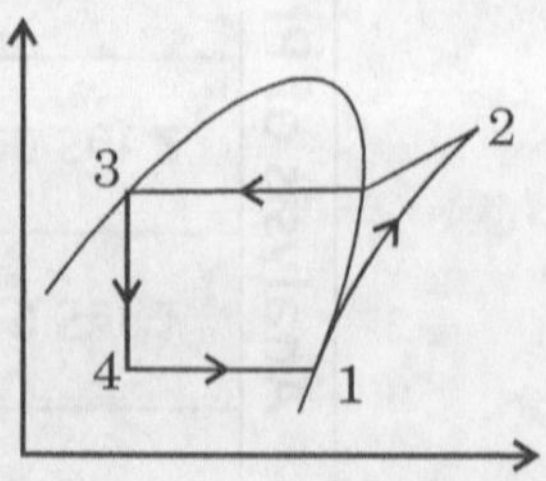

(a) p – h (b) T – s

(c) p – s (d) T – h

[2005 : 2 Marks]

8. In an ideal vapour compression refrigeration cycle, the specific enthalpy of refrigerant (in kJ/kg) at the following states is given as:

Inlet of condenser : 283

Exit of condenser : 116

Exit of evaporator : 232

The COP of this cycle is

(a) 2.27 (b) 2.75

(c) 3.27 (d) 3.75

[2009 : 1 Mark]

Common Data Questions 9 and 10:

A refrigerator operates between 120 kPa and 800 kPa in an ideal vapor compression cycle with R-134a as the refrigerant. The refrigerant enters the compressor as saturated vapor and leaves the condenser as saturated liquid. The mass flow rate of the refrigerant is 0.2 kg/s. Properties for R -134a are as follows:

Saturated R-134a

Saturated R-134a

P(kPa)	T(°C)	h_f(kJ/kg)	h_s(kJ/kg)	s_f(kJ/kgK)	s_g(kJ/kgK)
120	–22.32	22.5	237	0.093	0.95
800	31.31	95.6	267.3	0.354	0.918

Superheated R-134a

P(kPa)	T(°C)	h(kJ/kg)	s(kJ/kgK)
800	40	276.45	0.95

9. The rate at which heat is extracted, in kJ/s from the refrigerated space is

(a) 28.3 (b) 42.9

(c) 34.4 (d) 14.6

[2012 : 2 Marks]

10. The power required for compressor in kW is

(a) 5.94 (b) 1.83

(c) 7.9 (d) 39.5

[2012 : 2 Marks]

11. Refrigerant vapor enters into the compressor of a standard vapor compression cycle at –10°C (h = 402 J/kg) and leaves the compression at 50°C (h = 432 kJ/kg). It leaves the condenser at 30°C (h = 237 kJ/kg). The COP of the cycle is ________.

[2015 : 2 Marks, Set-3]

12. A refrigerator uses R134a as its refrigerant and operates on a ideal vapour-compression refrigeration cycle between 0.14 MPa and 0.8 MPa. If the mass flow rate of the refrigerant is 0.05 kg/s, the rate of heat rejection to the environment is ______ kW.

Given data:

At P = 0.14 MPa, h = 236.04 kJ/kg, s = 0.9322 kJ/kgK

At P= 0.8 MPa, h = 272.05 kJ/kg (superheated vapour)

At P= 0.8 MPa, h = 93.42 kJ/kg (saturated liquid)

[2016 : 2 Marks, Set-2]

13. In the vapour compression cycle shown in the figure, the evaporating and condensing temperatures are 260 K and 310 K, respectively. The compressor takes in liquid-vapour mixture (state 1) and isentropically compresses it to a dry saturated vapour condition (state 2). The specific heat of the liquid refrigerant is 4.8 kJ/kgK and may be treated as constant. The enthalpy of evaporation for the refrigerant at 310K is 1054 kJ/kg.

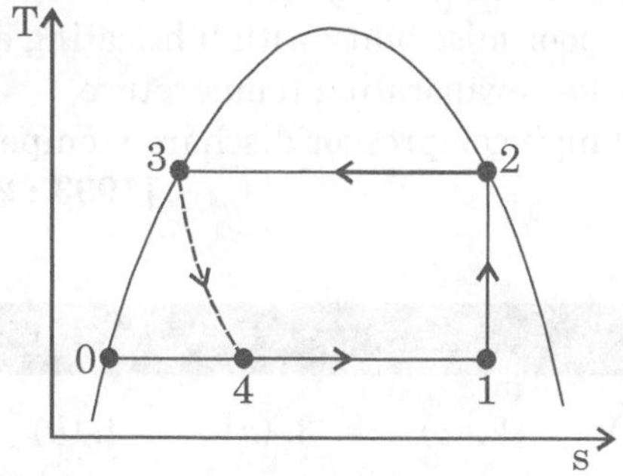

The difference between the enthalpies at state points 1 and 0 (in kJ/kg) is ________

[2016 : 2 Marks, Set-3]

14. A standard vapour compression refrigeration cycle operating with a condensing temperature of 35° C and an evaporating temperature of -10° C develops 15 kW of cooling. The P-h diagram shows the enthalpies at various states. If the isentropic efficiency of the compressor is 0.75, the magnitude of compressor power (in kW) is ________(correct to two decimal places).

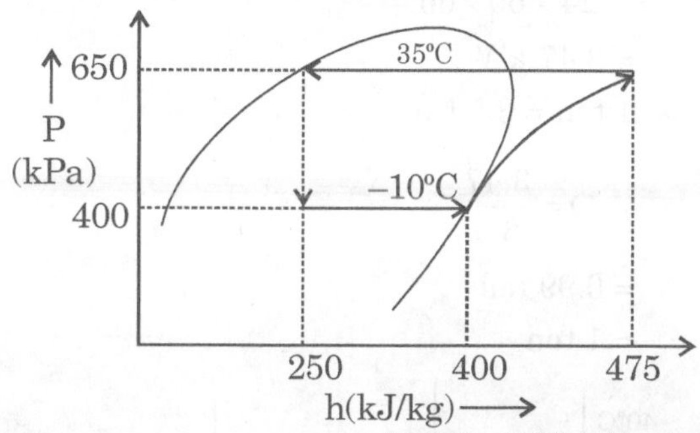

[2018 : 2 Marks, Set-2]

15. Consider an ideal vapor compression refrigeration cycle. If the throttling process is replaced by an isentropic expansion process, keeping all the other processes unchanged, which one of the following statements is true for the modified cycle?

(*a*) Coefficient of performance is higher than that of the original cycle.

(*b*) Coefficient of performance is the same as that of the original cycle.

(*c*) Coefficient of performance is lower than that of the original cycle.

(*d*) Refrigerating effect is lower than that of the original cycle.

[2019 : 1 Mark, Set-1]

Refrigerants

16. A refrigeration compressor, designed to operate with R 22 ______(can/cannot) be operated with R 12 because the condensing pressure of R 22 at any given temperature is ____(higher/ lower) than that of R 12.

[1992 : 2 Marks]

17. The use of Refrigerant R–22 for temperature below -30°C is not recommended due to its
(a) good miscibility with lubricating oil
(b) poor miscibility with lubricating oil
(c) low evaporating temperature
(d) high compressor discharge temperature

[1993 : 2 Marks]

18. Global warming is caused by
(a) Ozone (b) Carbon dioxide
(c) Nitrogen (d) Carbon monoxide

[2000 : 1 Mark]

19. Environment friendly refrigerant R134a is used in the new generation domestic refrigerators. Its Chemical formula is
(a) $CH\,Cl\,F_2$ (b) $C_2\,Cl_3\,F_3$
(c) $C_2\,Cl_2\,F_4$ (d) $C_2\,H_2\,F_4$

[2004 : 1 Mark]

20. Which one of the following is a CFC refrigerant?
(a) R744 (b) R290
(c) R502 (d) R718

[2015 : 2 Marks, Set-3]

ANSWERS

1. (a)	**2.** (c)	**3.** (a)	**4.** (b)	**5.** (a)	**6.** (a)	**7.** (d)	**8.** (a)	**9.** (a)	**10.** (c)
15. (a)	**16.** (b)	**17.** (b)	**18.** (b)	**19.** (d)	**20.** (c)				

EXPLANATIONS

1. Capacity= $\dot{Q}$= 300 MJ/day

$$= \frac{300 \times 10^3}{24 \times 60 \times 60} \text{ kJ/S}$$

= 3.47 kW

1 ton = 3.5 kW

$$\therefore \dot{Q} = \frac{3.47}{3.5}$$

= 0.99 ton

= 1 ton

3.

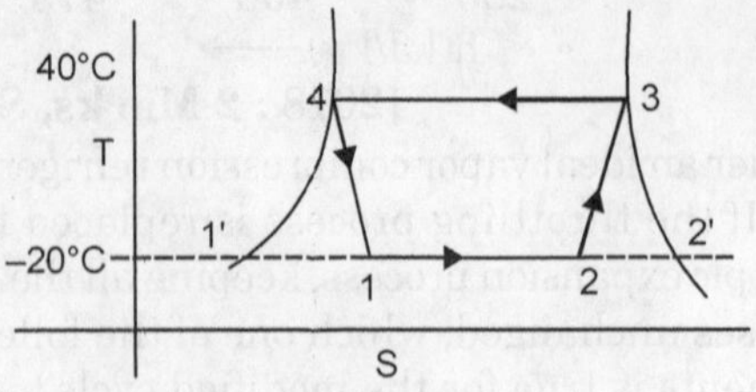

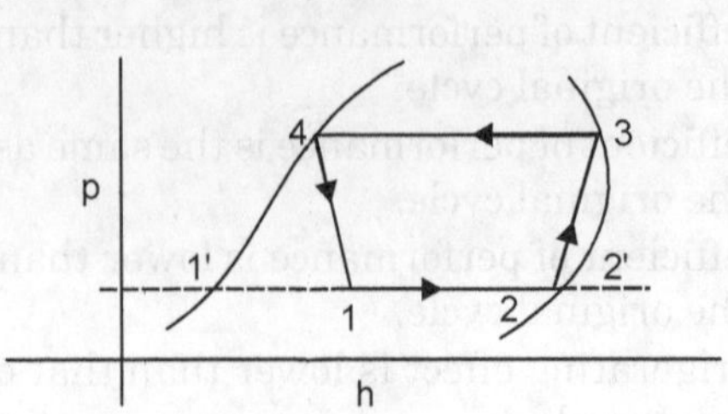

$h_1 = h_4 = 80$ kJ/kg $s_3 = 0.67$

$h_3 = 200$ kJ/kg $s_2 = 0.67$

$h_1' = 20$ kJ/kg $s_2' = 0.7366$

$h_2' = 20$ kJ/kg $s_1' = 0.07$

$$s_2 = s_1' + x(s_2' - s_1')$$

$$= 0.07 + x + (0.7366 - 0.07)$$

$$\Rightarrow \quad 0.67 = 0.07 + x + 0.6666$$

$$\Rightarrow \quad x = \frac{0.067 - 0.07}{0.6666} = 0.90$$

$$h_2 = h_1' + x \times (h_2' - h_1')$$

$$= 20 + 0.9(180 - 20)$$

$$= 20 + 0.9 \times 160 = 165 \text{ kJ/kg}$$

Refrigerating effect per kg = $h_2 - h_1$

= 164 – 80 = 84 kJ/kg

Hence refrigerating effect for 0.025 kg/sec circulation

= 84 × 0.025 = 2.1 kW

4. $$\text{C.O.P.} = \frac{h_2 - h_1}{h_3 - h_2}$$

$$= \frac{164 - 80}{200 - 164} = \frac{84}{36} = 2.33$$

5. The capillary tube is used as expansion device

6. Given: Clearance volume ratio,

$C = 0.03$

Specific volume at suction,

$v_1 = 0.1089\ m^3/kg$

Net refrigerating effect

$= 2$ ton $= 2 \times 3.516$ kJ

$= 7.032$ kJ/s

Specific heat ratio,

$c = 1.15$

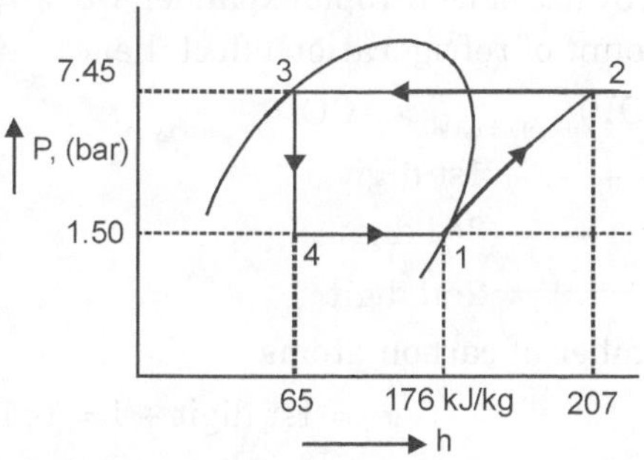

$\therefore$ Net refrigerating effect

$= \dot{m}(h_1 - h_4)$

or $7.032 = \dot{m}(176 - 65)$

or $\dot{m} = 0.063$ kg/s

$\therefore$ Volume $= 0.063 \times 0.1089$

$= 6.89 \times 10^{-3}\ m^3/s$

Volumetric efficiency

$$= 1 + C - C\left(\frac{p_2}{p_1}\right)^{1/n}$$

$$= 1 + 0.03 - 0.03\left(\frac{7.45}{1.50}\right)^{\frac{1}{1.15}}$$

$= 0.909$

$\therefore$ Volume displacement rate considering clearance effect

$= 6.89 \times 10^{-3} \times 0.909$

$= 6.26 \times 10^{-3}\ m^3/s$

7. T-h curve

8.

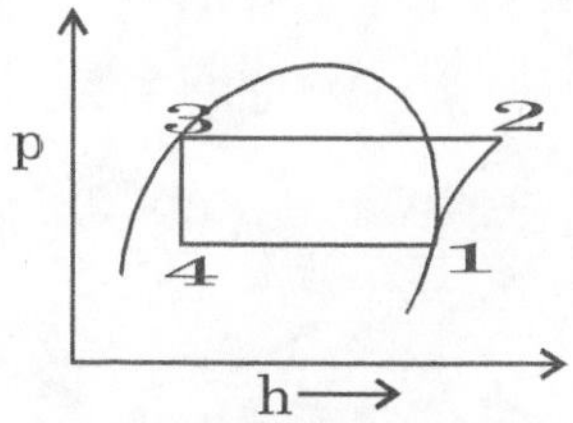

1–2 → work done by the compressor

2 – 3 → condenser heat rejected at constant pressure

3 – 4 → throttling

4 – 1 → heat addition in evaporator

Given : $h_2 = 283$ kJ/ kg

$h_3 = 116$ kJ/kg $= h_4$ (from ph curve)

$h_1 = 232$ kJ/kg

We know

$$COP = \frac{\text{desired effect}}{\text{work input}}$$

$$= \frac{\text{cooling effect}}{\text{work done by compressor}}$$

$$= \frac{h_1 - h_4}{h_2 - h_1}$$

$$= \frac{232 - 116}{283 - 232} = 2.27$$

9. Rate at which heat is extracted

$= \dot{m}(h_1 - h_4)$

$= 0.2\ (237 - 95.5)$

$= 28.3$ kW

10. Power required by compressor

$p = \dot{m}(h_2 - h_1) = 0.2\ (276.45 - 237) = 7.9$ kW

11. Work done $= 432 - 402 = 30$ kJ/kg

Refrigerating effect $= 402 - 237 = 165$ kJ/kg.

$$COP = \frac{165}{30} = 5.5$$

12.

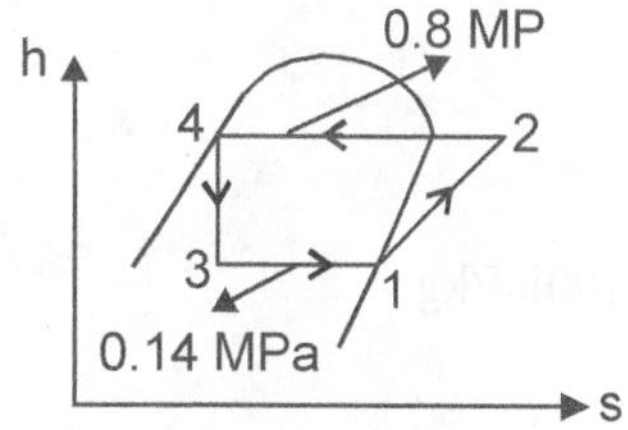

$h_1 = 236.04$ kJ/kg

$h_2 = 272.05$ kJ/kg

$h_3 = 93.42$ kJ/kg

$s_1 = 0.9322$ kJ/kg

$\dot{m} = 0.05$ kg/s

Heat Rejection $= m(h_2 - h_3)$

$= 0.05(272.05 - 93.42)$

$= 8.931$ kW

13.

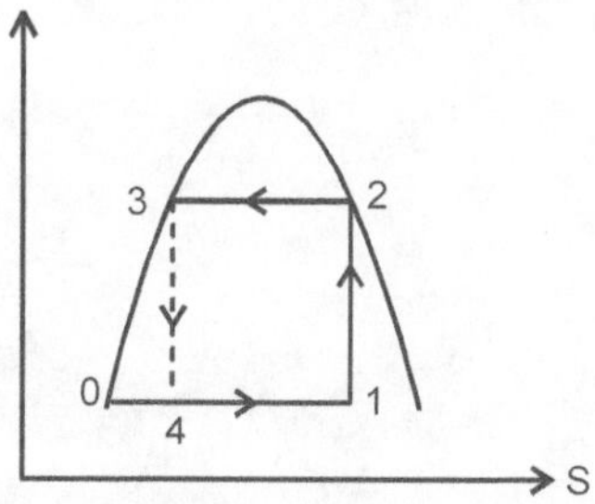

$$dQ = C_p dT$$

$$S_2 - S_0 = \int_0^3 \frac{C_p dT}{T} + \frac{h_{fg}}{310}$$

$$S_1 - S_0 = 4.8 \ln\left(\frac{310}{260}\right) + \frac{1054}{310} \qquad [\because S_1 = S_2]$$

$$\frac{h_1 - h_0}{260} = 4.8 \ln\left(\frac{310}{260}\right) + \frac{1054}{310}$$

$$h_1 - h_0 = 1103.51 \text{ kJ/kg}$$

14.

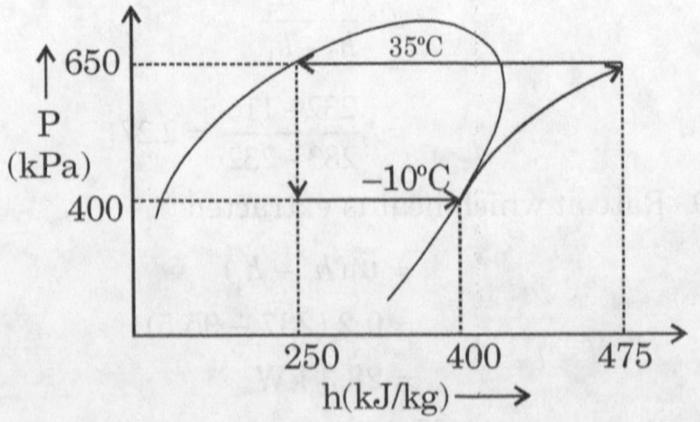

RC = 15 kW

$RC = \dot{m}(h_1 - h_4)$

$15 = \dot{m} \times (400 - 250)$

Mass flow rate, $\dot{m} = 0.1$ kg/s

$W_{isentropic} = (h_2 - h_1) = (475 - 400)$

$= 75$ kJ/kg

$$\eta_c = \frac{W_{isentropic}}{W_{actual}}$$

$$W_{actual} = \frac{75}{0.75} = 100 \text{kJ/kg}$$

$P = \dot{m} \times W_{actual}$

$= 0.1 \times 100 = 10$ kW

15. Vapor compression refrigeration cycle :

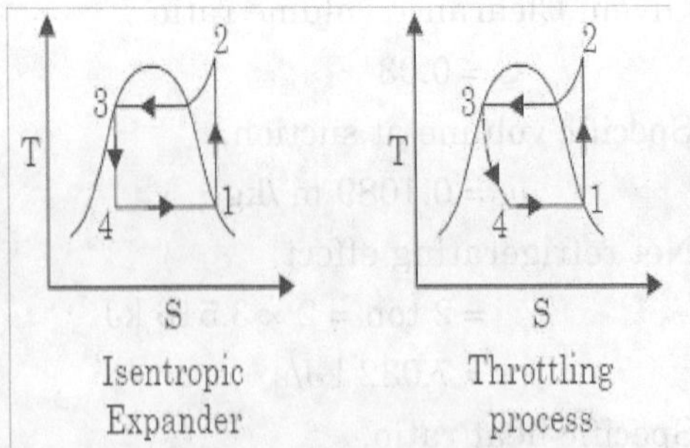

Both cycle required same amount of work input but by use of Isentropic expander, we can get more amount of refrigeration effect, hence

$((COP)_{\text{modified cycle}} > ((COP)_{\text{original cycle}})$

19. R134 — 1st digit, 3rd digit, 2nd digit

Number of carbon atoms

= 1st digit + 1 = 1+1 = 2

Number of hydrogen atoms

= 2nd digit – 1 = 3 – 1 = 2

Number of fluorine atoms = 3rd digit = 4

So chemical formula is $C_2H_2F_4$

20. R744–CO_2

R502 – $CHClF_3$ + $CCl\,F_2CF_3$

2 CHAPTER

Psychrometric Process

Analysis of Previous GATE Papers

Topic	Year/Set → Marks Distribution ↓		2019 Set 1	2019 Set 2	2018 Set 1	2018 Set 2	2017 Set 1	2017 Set 2	2016 Set 1	2016 Set 2	2016 Set 3	2015 Set 1	2015 Set 2	2015 Set 3
Psychrometric Processes and Applications	1 Mark	MCQ Type						1						
		Numerical Type												
	2 Marks	MCQ Type												
		Numerical Type				1				1	1			1
	Total					2		1		2	2			2

Psychrometric processes and application

1. Wet bulb depression, under saturated ambient air conditions:
   (a) is always positive
   (b) is always negative
   (c) is always zero
   (d) May have a value depending upon the dew point temperature **[1989 : 2 Marks]**

2. Select statement from List-II matching the processes in List-I. Enter you answer as D, C if the correct choice for (1) is (D) and that for (2) is (C)

   List-I
   1. Cooling and dehumidification
   2. Chemical dehumidification

   List-II
   A. Dry-bulb temperature increase, but dew-point temperature decreases.
   B. Dew-point temperature increase and dry bulb temperature remains unchanged.
   C. Dry-bulb and wet-bulb temperature decrease
   D. Dry-bulb temperature decreases, but dew-point temperature increases

   [1999 : 2 Marks]

3. During the Chemical dehumidification process of air
   (a) dry bulb temperature and specific humidity decrease
   (b) dry bulb temperature increases and specific humidity decreases
   (c) dry bulb temperature decreases and specific humidity increases
   (d) dry bulb temperature and specific humidity increases **[2004 : 1 Mark]**

4. Water at 42°C is sprayed into a stream of air at atmospheric pressure, dry bulb temperature of 40°C and a wet bulb temperature of 20°C. The air leaving the spray humidifier is not saturated. Which of the following statements is true?
   (a) Air gets cooled and humidified
   (b) Air gets heated and humidified
   (c) Air gets heated and dehumidified
   (d) Air gets cooled and dehumidified

   [2005: 1 Mark]

5. Various psychrometric processes are shown in the figure below.

Process in Figure	Name of the process
P. 0-1	(i) Chemical dehumidification
Q. 0-2	(ii) Sensible heating
R. 0-3	(iii) Cooling and dehumidification
S. 0-4	(iv) Humidification with steam injection
T. 0-5	(v) Humidification with water injection

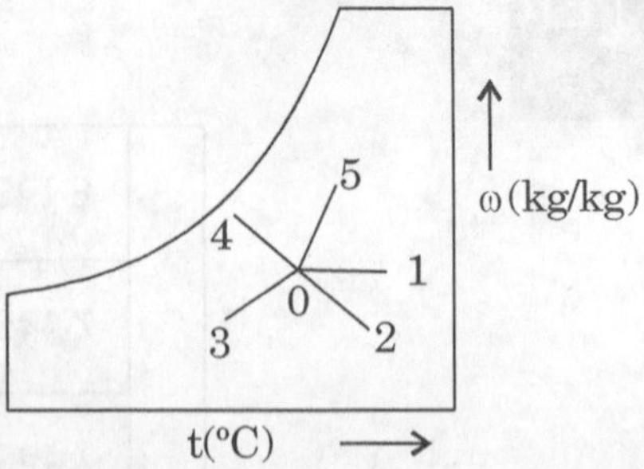

The matching pair are
   (a) P-i, Q-ii, R-iii, S-iv, T-v
   (b) P-ii, Q-i, R-iii, S-v, T-iv
   (c) P-ii, Q-i, R-iii, S-iv, T-v
   (d) P-iii, Q-iv, R-v, S-i, T-ii **[2005 : 2 Marks]**

6. Dew point temperature is the temperature at which condensation begins when the air is cooled at constant
   (a) volume (b) entropy
   (c) pressure (d) enthalpy

   [2006 : 1 Mark]

7. The statements concern psychrometric chart.
   1. Constant relative humidity lines are uphill straight lines to the right.
   2. Constant wet bulb temperature lines are downhill straight lines to the right.
   3. Constant specific volume lines are downhill straight lines to the right.
   4. Constant enthalpy lines are coincident with constant wet bulb temperature lines.

   Which of the statements are correct?
   (a) 2 and 3 (b) 1 and 2
   (c) 1 and 3 (d) 2 and 4

   [2006 : 2 Marks]

8. Atmospheric air at a flow rate of 3 kg/s (on dry basis) enters a cooling and dehumidifying coil with an enthalpy of 85 kJ/kg of dry air and a humidity ratio of 19 grams/kg of dry air. The air leaves the coil with an enthalpy of 43 kJ/kg of dry air and a humidity ratio of 8 grams/kg of dry air. If the condensate water leaves the coil with an enthalpy of 67 kJ/kg, the required cooling capacity of the coil in kW is
   (a) 75.0 (b) 123.8
   (c) 128.2 (d) 159.0

   [2007 : 2 Marks]

9. Air (at atmospheric pressure) at a dry bulb temperature of 40°C and wet bulb temperature of 20°C is humidified in an air washer operating with continuous water recirculation. The wet bulb depression, the difference between the dry and wet bulb temperatures) at the exit is 25% of that at the inlet. The dry bulb temperature at the exit of the air washer is closest to

(a) 10°C (b) 20°C

(c) 25°C (d) 30°C

[2008 : 2 Marks]

10. A sample of moist air at a total pressure of 85 kPa has a dry bulb temperature of 30°C (saturation vapour pressure of water = 4.24 kPa). If the air sample has a relative humidity of 65%, the absolute humidity (in gram) to water vapour per kg of dry air is _____.

[2014 : 1 Mark, Set-3]

11. Moist air at 35°C and 100% relative humidity is entering a psychrometric device and leaving at 25°C and 100% relative humidity. The name of the device is

(a) Humidifier (b) Dehumidifier

(c) Sensible heater (d) Sensible cooler

[2014 : 1 Mark, Set-4]

12. Air in a room is at 35°C and 60% relative humidity (RH). The pressure in the room is 0.1 MPa. The saturation pressure of water at 35°C is 5.63 kPa. The humidity ratio of the air (in gram/kg of dry air) is_________.

[2015 : 2 Marks, Set-3]

13. The partial pressure of water vapour in a moist air sample of relative humidity 70% is 1.6 kPa, the total pressure being 101.325 kPa. Moist air may be treated as an ideal gas mixture of water vapour and dry air. The relation between saturation temperature (T_s in K) and saturation pressure (P_s in kPa) for water is given by

$$\text{in}\left(\frac{p_s}{p_0}\right) = 14.317 - \frac{5304}{T_s}$$

where $p_0 = 101.325$ kPa. The dry bulb temperature of the moist air sample (in °C) is_______.

[2016 : 2 Marks, Set-2]

14. In a mixture of dry air and water vapor at a total pressure of 750 mm of Hg, the partial pressure of water vapor is 20 mm of Hg. The humidity ratio of the air in grams of water vapor per kg of dry air (g_w/kg_{da}) is _______

[2016 : 2 Marks, Set-3]

15. If a mass of moist air contained in a closed metallic vessel is heated, then its

(a) relative humidity decreases

(b) relative humidity increases

(c) specific humidity increases

(d) specific humidity decreases

[2017 : 1 Mark, Set-2]

16. Ambient air is at a pressure of 100 kPa, dry bulb temperature of 30° C and 60% relative humidity, The saturation pressure of water at 30° C is 4.24 kPa. The specific humidity of air (in g/kg of dry air) is (correct to two decimal places).

[2018 : 2 Marks, Set-2]

ANSWERS

1. (c) **2.** (*c, d*) **3.** (*b*) **4.** (*b*) **5.** (*b*) **6.** (*c*) **7.** (*a*) **8.** (*b*) **9.** (c) **11.** (*b*)
15. (a)

EXPLANATIONS

1. WBD = 0 for saturated air

∴ DBT = WBT

2. 1 – C, 2 – A

3.

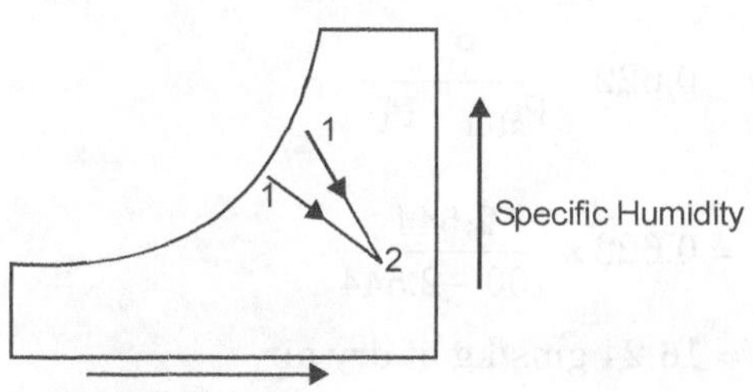

4. Here, $t_{DBT} = 40°$, $t_{WBT} = 20°$

Water sprayed at temperature = 42°

Since, $t_{water\ spray} > t_{DBT}$ so heating and humidification.

5.

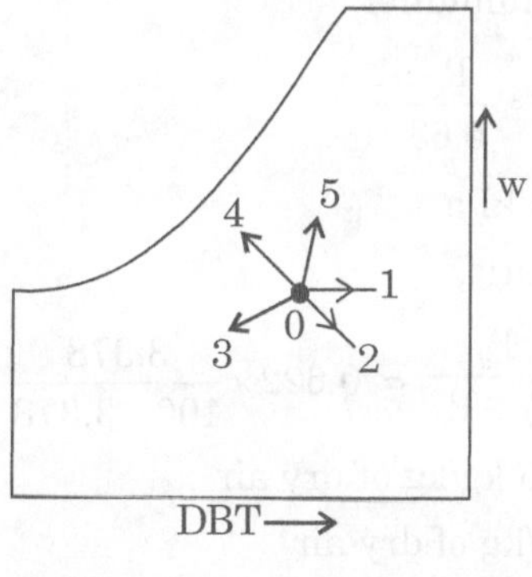

Process 0 – 1 : DBT↑, w = C; sensible heating

Process 0 – 2 : DBT↑, w↓; heating & dehumidification

Process 0 – 3 ; Cooling and dehumidification

Process 0 – 4 ; Humidification with water injection (DBT↓, w↑)

Process 0 – 5; DBT↑, w↑: Humidification with steam injection.

6.

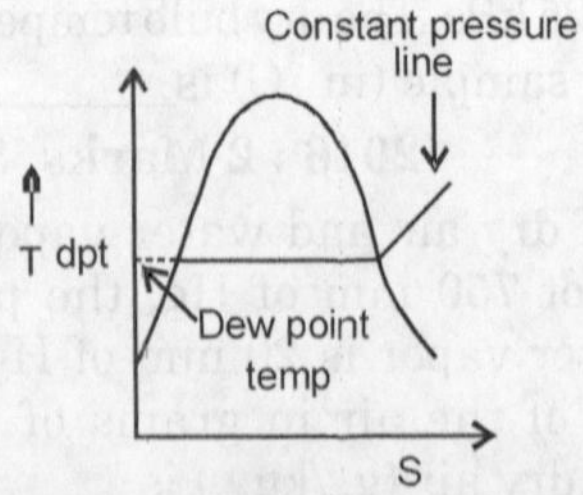

Air is cooled at constant pressure to make unsaturated air to saturated one.

8. Mass flow rate of condensate,

$\dot{m}_w = \dot{m}_a(w_1 - w_2) = 3 \times (.019 - .008) = .033\,kg/s$

By applying energy balance equation we get

$\dot{m}_a h_1 = \dot{m}_a h_2 + \dot{m}_w h_{fg} + Q$

$3 \times 85 = 3 \times 43 + 0.033 \times 67 + Q$

$Q = 123.8$ kW

10. $P_T = 85$ KPa, DBT = 30ºC

$P_{w.s} = 4.24$ KPa, RH = 65%

$\therefore\; P_w = P_{w.s} \times \dfrac{RH}{100} = 4.24 \times 0.65 = 2.756$ kPa

now $\omega = 622\,\dfrac{P_w}{P_T}$ gm (or water) / mg.d.a

$\therefore\; \omega = \dfrac{622 \times 2.756}{85} = 20.17$ gm (of water)/mg.d.a

11.

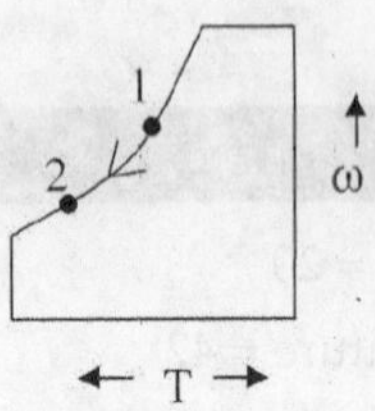

Since water content in moist air is reducing. The device is dehumidifer.

12. $\phi = \dfrac{P_w}{P_s} = 0.6 = \dfrac{P_w}{5.63}$

$\therefore P_w = 3.378$ kPa

Humidity Ratio,

$w = 0.622\,\dfrac{P_w}{P_a - P_w} = 0.622 \times \dfrac{3.378}{100 - 3.378}$

$= 0.021745$ kg/kg of dry air

or 21.745 g/kg of dry air

13. $\phi = 0.7,\ P_0 = 101.325$

$\phi = \dfrac{P_v}{P_{sat}} \Rightarrow P_{sat} = \dfrac{P_v}{\phi}$

$P_{sat} = \dfrac{1.6}{0.7} = 2.28571$ kPa

Corresponding to saturation pressure whatever is the temperature is the DBT

$$\ln\left(\frac{P_S}{P_0}\right) = 14.317 - \frac{5304}{T_{sat}}$$

$$\ell n\left(\frac{2.28571}{101.325}\right) = 14.317 - \frac{5304}{T_{sat}} - 3.7916$$

$$= 14.317 - \frac{5304}{T_{sat}}$$

$$T_{sat} = \frac{5304}{18.1086} = 292.898\text{ K} = 19.89°\text{C}$$

14. $w = 0.622\,\dfrac{P_v}{P_t - P_v}$

$= 0.622 \times \left(\dfrac{20}{750 - 20}\right)$

$= 0.017$ Kgw/Kgda $= 17$ gw/Kgda

15.

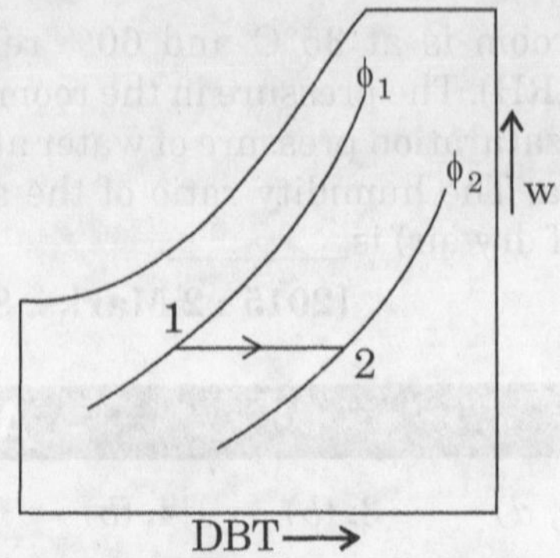

16. $P_{atm} = 100$ kPa

DBT = t = 30°C → $P_{VS} = 4.24$ k Pa

$\phi = 60° = 0.6$

$\phi = \dfrac{P_V}{P_{VS}} \Rightarrow 0.6 = \dfrac{P_V}{4.24}$

$P_V = 2.544$ kPa

$w = 0.622 \times \dfrac{P_V}{P_{atm} - P_V}$

$w = 0.622 \times \dfrac{2.544}{100 - 2.544}$

w = 16.24 gms/kg of dry air

CHAPTER 3 Heat Pumps and Cycles

Analysis of Previous GATE Papers

Topic	Year/Set → Marks Distribution ↓		2019 Set 1	2019 Set 2	2018 Set 1	2018 Set 2	2017 Set 1	2017 Set 2	2016 Set 1	2016 Set 2	2016 Set 3	2015 Set 1	2015 Set 2	2015 Set 3
Heat Pump	1 Mark	MCQ Type												1
		Numerical Type											1	
	2 Marks	MCQ Type												
		Numerical Type												
	Total												1	1
Refrigeration Cycles	1 Mark	MCQ Type												
		Numerical Type												
	2 Marks	MCQ Type												
		Numerical Type												
	Total													

Heat pump

1. Clearance volume of a reciprocating compressor is 100 ml, and the volume of the cylinder at bottom dead centre is 1.0 litre. The clearance ratio of the compressor is

(a) 1/11 (b) 1/10

(c) 1/9 (d) 1/1.1

[1997 : 2 Marks]

2. **List-I**

A. Liquid to suction heat exchanger
B. Constant volume heat addition
C. Normal shock
D. Ammonia-water

List-II

1. Vapour absorption refrigeration
2. Vapour compression refrigeration
3. Diesel cycle
4. Otto cycle
5. Converging nozzle
6. Converging diverging nozzle

[1997 : 2 Marks]

3. An industrial heat pump operates between the temperatures of 27°C and -13°C. The rates of heat addition and heat rejection are 750 W and 1000 W, respectively. The COP for the heat pump is

(a) 7.5 (b) 6.5

(c) 4.0 (d) 3.0

[2003 : 1 Mark]

4. A vapour absorption refrigeration system is a heat pump with three thermal reservoirs as shown in the figure. A refrigeration effect of 100 W is required at 250 K when the heat source available is at 400 K. Heat rejection occurs at 300 K. The minimum value of heat required (in W) is

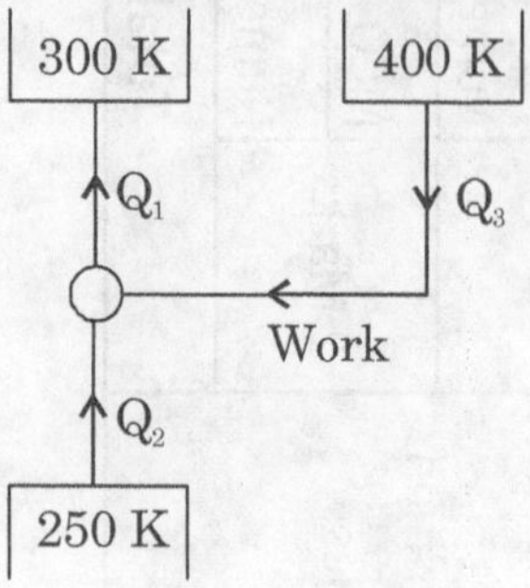

(a) 167 (b) 100

(c) 80 (d) 20

[2005 : 2 Marks]

5. A heat pump with refrigerant R22 is used for space heating between temperature limits of –20°C and 25°C. The heat required is 200 MJ/h. Assume specific heat of vapour at the time of discharge as 0.98 kJ/kgK. Other relevant properties are given below. The enthalpy (in kJ/ kg) of the refrigerant at isentropic compressor discharge is________.

Saturation temperature	Pressure	Specific enthalpy		Specific entropy	
T_{sat}(°C)	P(MN/m²)	h_f(kJ/kg)	h_g(kJ/kg)	s_f(kJ/kgK)	s_g(kJ/kgK)
–20	0.2448	177.21	397.53	0.9139	1.7841
25	1.048	230.27	413.02	1.1047	1.7183

[2014 : 2 Mark, Set-2]

6. The COP of a Carnot heat pump operating between 6°C and 37°C is

[2015 : 1 Mark, Set-2]

7. The thermodynamic cycle shown in figure (T-s diagram) indicates

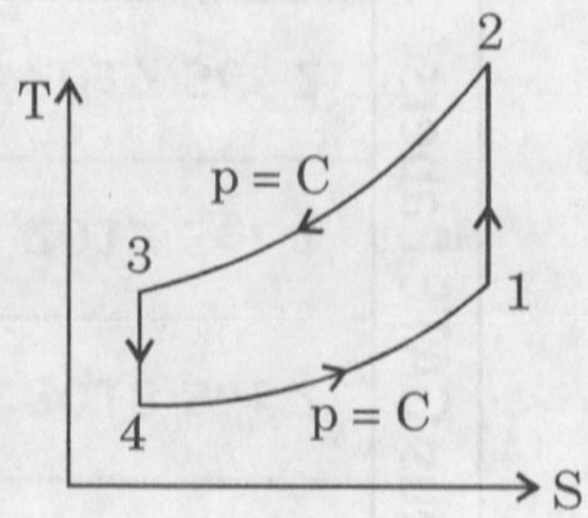

(a) reversed Carnot cycle
(b) reversed Brayton cycle
(c) vapor compression cycle
(d) vapor absorption cycle

[2015 : 1 Mark, Set-3]

Refrigeration cycles

8. A heat engine having an efficiency of 70% is used to drive a refrigerator having a coefficient of performance of 5. The energy absorbed from low temperature reservoir by the refrigerator for each kJ of energy absorbed from high temperature source by the engine is

(a) 0.14kJ (b)0.71kJ

(c) 3.5 kJ (d) 7.1 kJ

[2004 : 2 Marks]

9. A thin layer of water in a field is formed after a farmer has watered it. The ambient air conditions are: temp. 20°C and relative humidity 5%. An extract of steam tables is given below.

Temp.(°C)	-15	-10	-5	0.01	5	10	15	20
Saturation Pressure(kPa)	0.10	0.26	0.40	0.61	0.87	1.23	1.71	2.34

Neglecting the heat transfer between the water and the ground, the water temperature in the field after phase equilibrium is reached equals

(a) 10.3°C (b) –10.3°C

(c) –14.5°C (d) 14.5°C

[2006 : 2 Marks]

10. A reversed Carnot cycle refrigerator maintains a temperature of –5°C. The ambient air temperature is 35°C. The heat gained by the refrigerator at a continuous rate is 2.5 kJ/s. The power (in watt) required to pump this heat out continuously is _______

[2014 : 1 Mark, Set-4]

ANSWERS

1. (c) 3. (c) 4. (c) 7. (b) 8. (c) 9. (c)

EXPLANATIONS

1. Clearance ratio $= \dfrac{\text{Clearance volume}}{\text{Swept volume}}$

$$= \frac{100}{(1000-100)} = \frac{1}{9}$$

2. A – 2, B – 4, C – 6, D – 1

3. Heat addition, $Q_2 = 750$ W

Heat rejected, $Q_1 = 1000$ W

$$[\text{C.O.P}]_{\text{H.P.}} = \frac{Q_1}{W} = \frac{Q_1}{Q_1 - Q_2}$$

$$= \frac{1000}{1000-750} = \frac{1000}{250} = 4$$

$T_1 = 273 + 27° = 300°K$

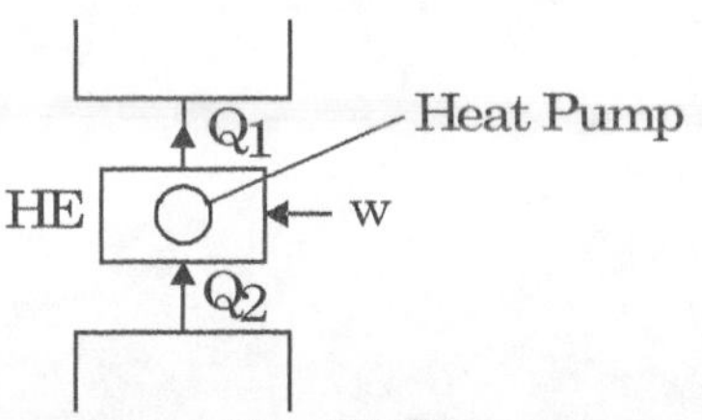

$T_2 = 273 - 13 = 260K$

$$[\text{C.O.P}]_{\text{H.P.}} = \frac{Q_1}{Q_1 - Q_2} \text{ for actual heat pump}$$

$$= \frac{1000}{1000-750} = \frac{1000}{250} = 4$$

4. From question, since refrigeration effect of 100 W is required

$$\text{So, } \frac{100}{\text{Work obtained}} = \frac{250}{300-250}$$

$$\Rightarrow \text{Work obtained} = \frac{100 \times 50}{250} = 20\text{W}$$

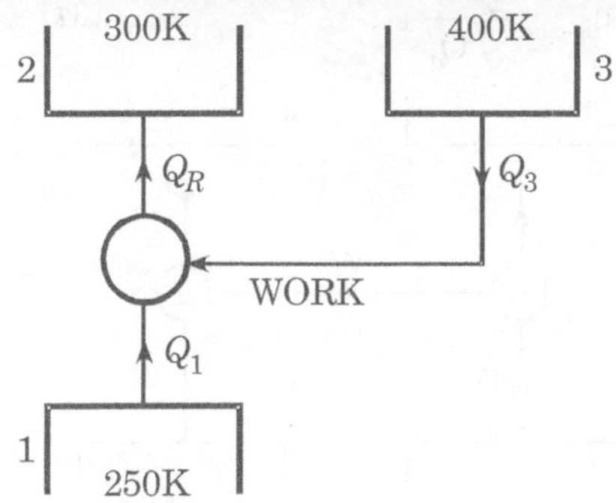

Now for this amount of work, heat is absorbed fromreservoir 3 and rejected to sink 2.

$$\therefore \eta = \frac{W}{Q_3} = 1 - \frac{300}{400} = \frac{1}{4}$$

$\Rightarrow Q_3 = 4W = 4 \times 20\ W = 80\ W$

5.

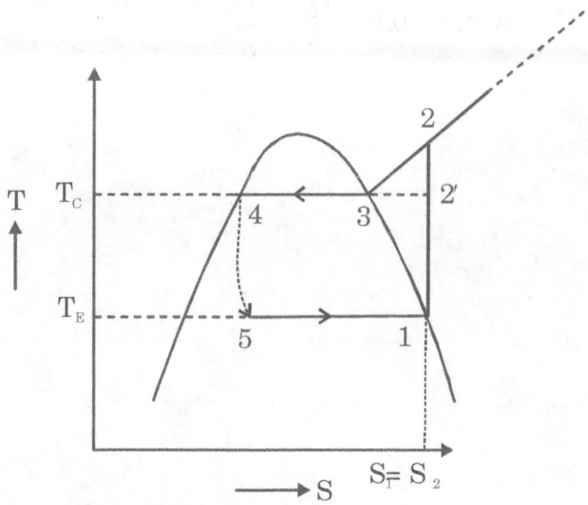

Given : $T_1 = 253$ K

$T_2' = 298$ K

$S_1 = S_2 = S_2'$

$$\therefore\ S_1 = S_2 = S_2' + C_P \ln\frac{T_2}{T_2'}$$

$$1.7841 = 1.7183 + 0.98 \ln \frac{T_2}{298}$$

$\therefore T_2 = 318.69$ K

$\therefore$ Enthalpy of discharge of compressor

$\Rightarrow h_2 = h_2' + C_P(T_2' - T_1')$

$\therefore h_2 = 413.02 + 0.98\,(318.69 - 298)$

$\therefore h_2 = 433.3 \dfrac{kJ}{kg}$.

6. $(COP)_{c.p.} = \dfrac{T_1}{T_1 - T_2} = \dfrac{310}{31} = 10$

7. T-S diagram represent a reversed brayton cycle used in air conditioning of aeroplanes where air is used as a refrigerant.

8. Given: $\eta_{engine} = 0.7$, $(C.O.P)_R = 5$

$$\eta_{engine} = \frac{W}{Q_1} \quad ...(i)$$

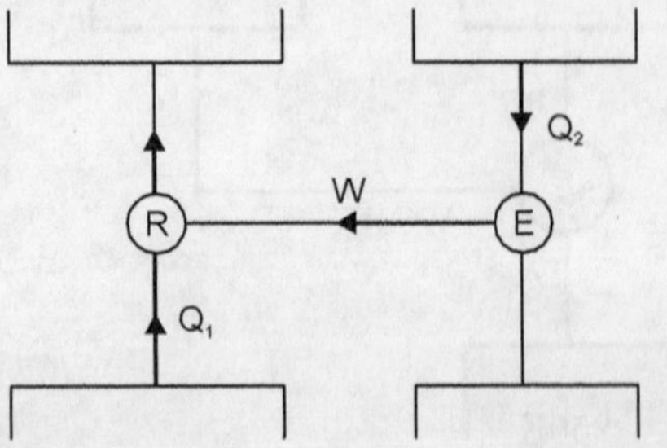

Now,

$$(C.O.P)_R = \frac{Q_2}{W}$$

$$\therefore W = \frac{Q_2}{5} \quad ...(ii)$$

Again, $0.7 = \dfrac{Q_2}{5} \times \dfrac{1}{Q_1}$

or $\dfrac{Q_2}{Q_1} = 3.5$

Energy absorbed from low temperature reservoir by the refrigerator for each kJ of energy absorbed from high temperature source by the engine = 3.5 kJ.

9. Given data:

$\phi = 5\% = 0.05$

$T_{db} = 20°C$

From table

At 20°C, $P_s = 2.34$ kPa

$$\phi = \frac{P_V}{P_s}$$

$$0.05 = \frac{P_V}{2.34}$$

$P_V = .117$ k Pa

By Interpolation method,

$$T_{dp} = -15 + \frac{(-10+15)}{0.26 - 0.10} \times (.117 - .1)$$

$$= -15 + 0.531 = -14.5° \text{ C}$$

10. $\dfrac{Q_1}{268} = \dfrac{Q_2}{308}$

$Q_2 = \dfrac{2.5 \times 308}{268}$ $[Q_1 = 2.5]$

$Q_2 = 2.873$ kW

$w = Q_2 - Q_1 = 0.373$ kw = 373.13 Watt.

CHAPTER 4 Properties of Moist Air

Analysis of Previous GATE Papers

Topic	Year/Set → Marks Distribution ↓		2019 Set 1	2019 Set 2	2018 Set 1	2018 Set 2	2017 Set 1	2017 Set 2	2016 Set 1	2016 Set 2	2016 Set 3	2015 Set 1	2015 Set 2	2015 Set 3
Properties of moist air and air-conditioning	1 Mark	MCQ Type												
		Numerical Type										1		
	2 Marks	MCQ Type												
		Numerical Type					1							
	Total						2					1		

Properties of moist air and airconditioning

1. Atmospheric air from 40°C and 60 percent relative humidity can be brought to 20°C and 60 percent relative humidity by

(a) Cooling and dehumidification process
(b) Cooling and humidification process
(c) Adiabatic saturation process
(d) Sensible cooling process

[1990 : 2 Marks]

2. If moist air is cooled by sensible heat removal, which of the following is true?

(a) Neither relative humidity nor specific humidity changes
(b) Specific humidity changes but not relative humidity
(c) Both relative humidity and specific humidity change
(d) None of these

[1991 : 2 Marks]

3. For air at a given temperature, as the relative humidity is increased isothermally,

(a) the wet bulb temperature and specific enthalpy increase
(b) the wet bulb temperature and specific enthalpy decrease
(c) the wet bulb temperature increases and specific enthalpy decreases
(d) the wet bulb temperature decreases and specific enthalpy increases

[2001 : 2 Marks]

4. For air with a relative humidity of 80%

(a) the dry bulb temperature is less than the wet bulb temperature
(b) the dew point temperature is less than wet bulb temperature
(c) the dew point and web bulb temperature are equal
(d) the dry bulb and dew point temperatures are equal

[2003 : 1 Mark]

5. Dew point temperature of air at one atmospheric pressure (1.013 bar) is 18°C. The air dry bulb temperature is 30°C. The saturation pressure of water at 18°C and 30°C are 0.02062 bar and 0.04241 bar respectively. The specific heat of air and water vapour respeetively are 1.005 and 1.88 kJ/kgK and the latent heat of vaporization water of 0°C is 2500 kJ/kg. The specific humidity (kg/kg of dry air) and enthalpy (kJ/kg of dry air) of this moist air respectively, are

(a) 0.01051, 52.64 (b) 0.01291, 63.15
(c) 0.01481, 78.60 (d) 0.01532, 81.40

[2004 : 2 Marks]

6. For a typical sample of ambient air (at 30°C, 75% relative humidity and standard atmospheric pressure), the amount of moisture in kg per kg of dry air will be approximately.

(a) 0.002 (b) 0.027
(c) 0.25 (d) 0.75

[2005 : 1 Mark]

7. A building has to be maintained at 21 °C (dry bulb) and 14.5°C(wet bulb). The dew point temperature under these conditions is 10.17°C. The outside temperature is –23°C (dry bulb) and the internal and external surface heat transfer coefficients are 8 W/m^{2}K and 23 W/m^{2}K respectively. If the building wall has a thermal conductivity of 1.2 W/mK, the minimum thickness (in m) of the wall required to prevent condensation is

(a) 0.471 (b) 0.407
(c) 0.321 (d) 0.125

[2007 : 2 Marks]

8. Moist air at a pressure of 100 kPa is compressed to 500 kPa and then cooled to 35°C in an aftercooler. The air at the entry to the aftercooler is unsaturated and becomes just saturated at the exit of the aftercooler. The saturation pressure of water at 35°C is 5.628 kPa. The partial pressure of water vapour (in kPa) in the moist air entering the compressor is closest to

(a) 0.57 (b) 1.13
(c) 2.26 (d) 4.52

[2008 : 2 Marks]

9. A moist air sample has dry bulb temperature of 30°C and specific humidity of 11.5 g water vapour per kg dry air. Assume molecular weight of air as 28.93. If the saturation vapour pressure of water at 30°C is 4.24 kPa and the total pressure is 90 kPa, then the relative humidity (in %) of air sample is

(a) 50.5 (b) 38.5
(c) 56.5 (d) 68.5

[2010 : 2 Marks]

10. If a mass of moist air in an airtight vessel is heated to a higher temperature, then

(a) specific humidity of the air increases
(b) specific humidity of the air decreases
(c) relative humidity of the air increases
(d) relative humidity of the air decreases

[2011 : 1 Mark]

11. A room contains 35 kg of dry air and 0.5 kg of water vapor. The total pressure and temperature of air in the room are 100 kPa and 25°C respectively. Given that the saturation pressure for water at 25°C, is 3.17 kPa, the relative humidity of the air in the room is

(a) 67% (b) 55%

(c) 83% (d) 71% **[2012 : 2 Marks]**

12. The pressure, dry bulb temperature and relative humidity of air in a room are 1 bar, 30°C and 70%, respectively. If the saturated steam pressure at 30°C is 4.25 kPa, the specific humidity of the room air in kg of water vapour/kg dry air is

(a) 0.0083 (b) 0.0101

(c) 0.0191 (d) 0.0232

[2013 : 1 Mark]

13. A stream of moist air (mass flow rate = 10.1 kg/s) with humidity ratio of 0.01 kg/kg dry air mixes with a second stream of superheated water vapour flowing at 0.1 kg/s. Assuming proper and uniform mixing with no condensation, the humidity ratio of the final stream in (kg/kg dry air) is _____.

[2015 : 1 Mark, Set-1]

14. Moist air is treated as an ideal gas mixture of water vapor and dry air (molecular weight of air = 28.84 and molecular weight of water = 18). At a location, the total pressure is 100 kPa, the temperature is 30°C and the relative humidity is 55%. Given that the saturation pressure of water at 30°C is 4246 Pa, the mass of water vapor per kg of dry air is _____ grams.

[2017 : 2 Marks, Set-1]

ANSWERS

1. (a) 2. (d) 3. (a) 4. (*b*) 5. (*b*) 6. (*b*) 7. (*b*) 8. (*b*) 9. (*b*) 10. (*d*) 11. (*d*) 12. (*c*)

EXPLANATIONS

1. Cooling and dehumidification process,

DBT↓, w↓

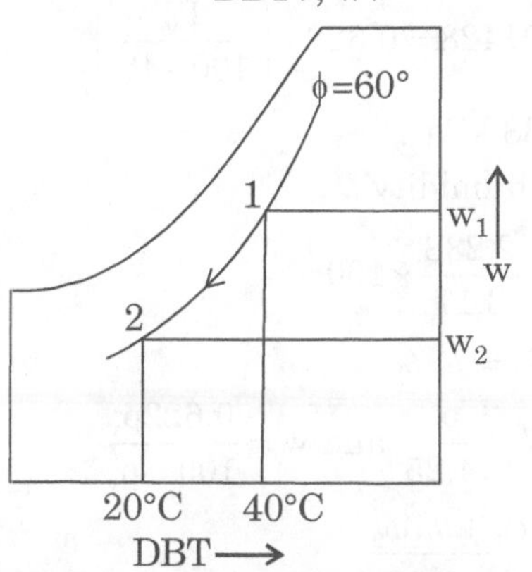

2. Sensible cooling process

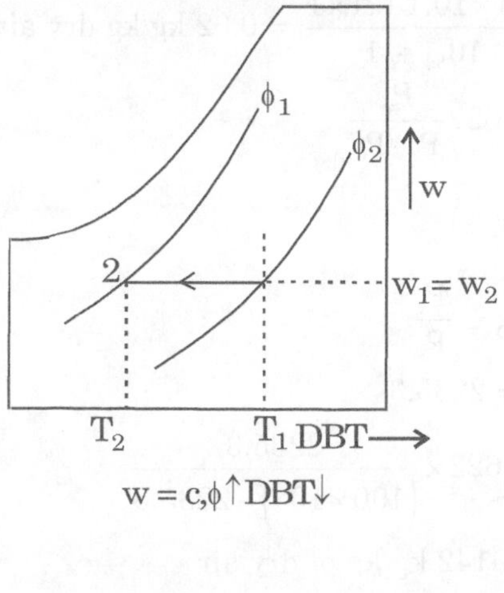

w = c, ϕ↑ DBT↓

3.

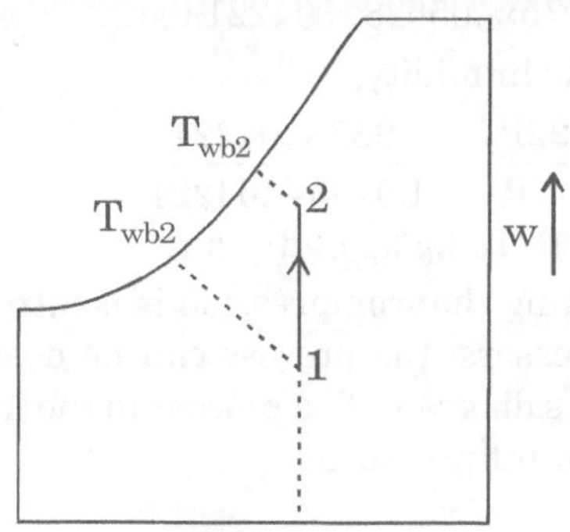

$DBT = DBT_2$

DBT⟶

ϕ↑, Twb↓

Relative humidity is increased isothermally.

4.

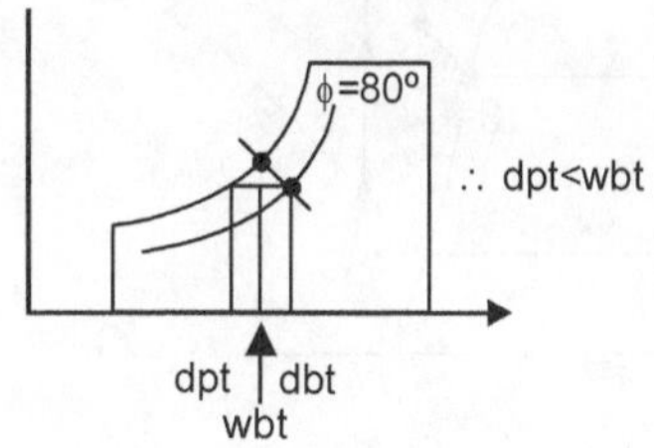

5. Given: $p = 1.013$ bar, $(p_v)_{tdp} = 0.02062$

$$\text{Specific humidity} = 0.622\frac{p_v}{p - p_v}$$

$$= \frac{0.622 \times 0.02062}{1.013 - 0.02062} = 0.01291 \text{ kg/kg of dry air}$$

$$h = 1.022\, t_d + w(h_{fgdp} + 2.3 t_{dp})$$

$$= 63.15 \text{ kJ/kg of dry air.}$$

Alternative method

p_v = Saturation pressure = 0.02062 bar

p_a = Partial pressure of air = 0.04241 bar

t_d = Dry bulb temperature = 30°

t_w = Wet bulb temperature = 18°

Specific humidity,

$$w = \frac{p_v}{p - p_v} \times 0.622 = 0.01292 \text{ kg/kg of dry air.}$$

Total enthalpy,

$$h = {}^{c}p_a t_d + wh_s + w_g p_s (t_d - t_\omega)$$

$$= 1.005 \times 30 + 0.0129 \times 2500 + 0.01292 \times 1.88(30 - 18)$$

$$= 63.15 \text{ kJ/kg of dry air.}$$

6. Given data:

Dry bulb temperature:

T_{db} = 35°C

Relative humidity

ϕ = 75° = 0.75

At 35°C, P_s = .05628 bar

$$\phi = \frac{P_V}{P_S} \Rightarrow 0.75 = \frac{P_V}{.05628}$$

P_v = 0.75× .05628 =.04221 bar

Specific humidity,

$$\omega = \frac{.622 P_V}{P - P_V} = \frac{.622 \times .04221}{1.0135 - .04221}$$

= 0.0270 kg/kg of dry air.

8. Assuming that compression is isentropic in air compressors, the process can be described on the T - s diagram. The process in the intercooler is constant pressure

$p_2 = p_3$

But $p_2 = 5p_1$

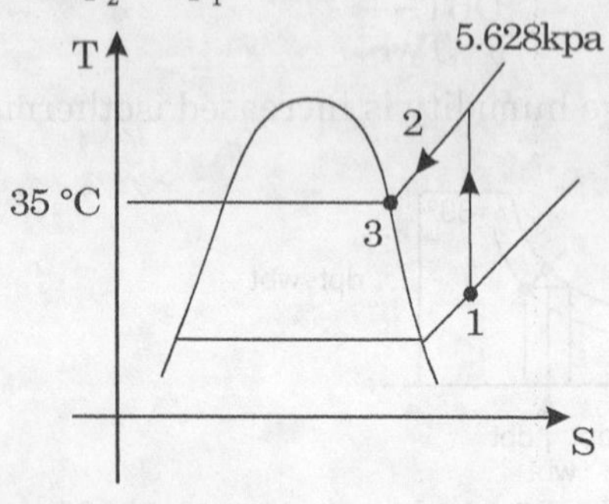

$$p_1 = \frac{p_2}{5} = \frac{p_3}{5}$$

$$p_1 = \frac{5.628}{5} = 1.13 \text{kPa}$$

9. Given : DBT = 30°C, w = 11.5 g/kg

∴ Relative humidity

$$= \frac{\text{Specific humidity}}{0.622} \times \frac{\text{Total pressure}}{\text{Saturation vapour pressure}}$$

$$\phi = \frac{w}{0.622} \times \frac{P_a}{P_{vs}}$$

$$\phi = \frac{11.5 \times 90}{0.622 \times 1000 \times 4.24} = 0.385$$

$\phi \approx 38.5\%$

10. R.H. decreases

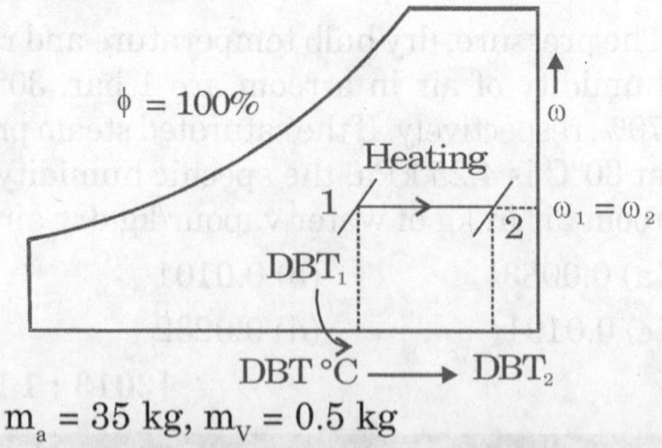

11. m_a = 35 kg, m_v = 0.5 kg

$$w = \frac{m_V}{m_a} = \frac{0.5}{35} = 0.01428$$

$$\text{also } w = 0.622 \frac{P_V}{P_a}$$

P_t = 100 kPa

$$\therefore w = 0.622 \times \frac{P_V}{P_t - P_V}$$

$$\text{Now, } 0.01428 = 0.622 \times \left(\frac{P_V}{100 - P_V}\right)$$

P_V = 2.238 kPa

Relative humidity

$$\phi = \frac{P_V}{P_S} = \frac{2.238}{3.17} \times 100$$

ϕ = 70.6% = 71 %

12. $$\text{RH} = 0.7 = \frac{P_V}{4.25} \text{ and } w = \frac{0.622 p_V}{100 - p_V}$$

13. $$\omega_{new} = \frac{m_1\omega_1 + m_2\omega_2}{m_1 + m_2}$$

$$\Rightarrow \frac{0.1 \times 10.1 + .1 \times 1}{10.1 + .1} = 0.02 \text{ kg/kg dry air}$$

14. $$w = 0.622 \frac{P_V}{P - P_V}$$

$$\phi = \frac{P_V}{P_s}$$

$$\Rightarrow 0.55 = \frac{P_V}{P_S}$$

$\Rightarrow P_v$ = 2235.3

$$w = 0.622 \times \frac{2235.3}{(100 \times 10^3) - 2235.3}$$

w = 0.0142 kg/kg of dry air

w = 14.2 gm/kg of dry air

Unit - X

Materials Science and Manufacturing Engineering

Syllabus

Engineering Materials: Structure and properties of engineering materials, phase diagrams, heat treatment, stress-strain diagrams for engineering materials.

Casting, Forming and Joining Processes: Different types of castings, design of patterns, moulds and cores; solidification and cooling; riser and gating design.

Plastic deformation and yield criteria; fundamentals of hot and cold working processes; load estimation for bulk (forging, rolling, extrusion, drawing) and sheet (shearing, deep drawing, bending) metal forming processes; principles of powder metallurgy. Principles of welding, brazing, soldering and adhesive bonding.

Machining and Machine Tool Operations: Mechanics of machining; basic machine tools; single and multi-point cutting tools, tool geometry and materials, tool life and wear; economics of machining; principles of non-traditional machining processes; principles of work holding, design of jigs and fixtures.

Metrology and Inspection: Limits, fits and tolerances; linear and angular measurements; comparators; gauge design; interferometry; form and finish measurement; alignment and testing methods; tolerance analysis in manufacturing and assembly.

Computer Integrated Manufacturing: Basic concepts of CAD/CAM and their integration tools.

Contents

CHAPTER 1 Engineering Materials

Analysis of Previous GATE Papers														
Topic	**Year/Set → Marks Distribution ↓**		**2019 Set 1**	**2019 Set 2**	**2018 Set 1**	**2018 Set 2**	**2017 Set 1**	**2017 Set 2**	**2016 Set 1**	**2016 Set 2**	**2016 Set 3**	**2015 Set 1**	**2015 Set 2**	**2015 Set 3**
Heat Treatment	**1 Mark**	MCQ Type				1				1				
		Numerical Type						1						
	2 Marks	MCQ Type		1						1				
		Numerical Type												
		Total		**2**		**1**		**1**		**3**				
Iron Carbon Diagram	**1 Mark**	MCQ Type				1				1				
		Numerical Type						1						
	2 Marks	MCQ Type		1						1				
		Numerical Type												
		Total		**2**		**1**		**1**		**3**				
Mechanical Properties	**1 Mark**	MCQ Type		1					1					
		Numerical Type												
	2 Marks	MCQ Type												
		Numerical Type												
	Total			1					1					
Crystal Structure	**1 Mark**	MCQ Type			1			1						
		Numerical Type											1	
	2 Marks	MCQ Type												
		Numerical Type												
	Total				1			1					1	

Heat Treatment

1. When 1.0% plane carbon steel is slowly cooled from the molten state to 740°C, the resulting structure will contain

(a) Pearlite and cementite

(b) Ferrite and cementite

(c) Austenite and ferrite

(d) Austenite and cementite **[1990 : 1 Mark]**

2. During heat treatment of steel, the hardness of various structures in increasing order is

(a) martensite, fine pearlite, coarse pearlite, spherodite

(b) fine pearlite, martensite, spherodite, coarse pearlite

(c) martensite, coarse pearlite, fine pearlite, spherodite

(d) spherodite, coarse pearlite, fine pearlite, martensite **[ME 2003 : 1 Mark]**

3. Cold working of steel is defined as working

(a) at its recrystallisation temperature

(b) above its recrystallisation temperature

(c) below its recrystallisation temperature

(d) at two thirds of the melting temperature of the metal **[ME 2003 : 1 Mark]**

4. Hardness of steel greatly improves with

(a) annealing (b) cyaniding

(c) normalising (d) tempering

[ME 2003 : 2 Marks]

5. From the lists given below, choose the most appropriate set of heat treatment process and the corresponding process characteristies

Process

P. Tempering

Q. Austempering

R. Martempering

Characteristics

1. Austenite is converted into bainite
2. Austenite is converted into martensite
3. Cementite is converted into globular structure
4. Both hardness and brittleness are reduced
5. Carbon is absorbed into the metal

(a) P-3, Q-1, R-5 (b) P-4, Q-3, R-2

(c) P-4, Q-1, R-2 (d) P-1, Q-5, R-4

[ME 2004 : 2 Marks]

6. When the temperature of a solid metal increases,

(a) strength of the metal decreases but ductility increases

(b) both strength and ductility of the metal decreases

(c) both strength and ductility of the metal increases

(d) strength of the metal increases but ductility decreases

[ME 2005 : 1 Mark]

7. The main purpose of spheroidising treatment is to improve

(a) hardenability of low carbon steels

(b) machinability of low carbon steels

(c) hardenability of high carbon steels

(d) machinability of high carbon steels

[ME 2006 : 1 Mark]

8. Match the items in columns I and II.

Column-I	Column-II
P. Charpy test	1. Fluidity
Q. Knoop test	2. Microhardness
R. Spiral test	3. Formability
S. Cupping test	4. Toughness
	5. Permeability

(a) P-4, Q-5, R-3, S-2 (b) P-3, Q-5, R-1, S-4

(c) P-2, Q-4, R-3, S-5 (d) P-4, Q-2, R-1, S-3

[ME 2006 : 2 Marks]

9. During normalizing process of steel, the specimen is heated

(a) between the upper and lower critical temperature and cooled in still air

(b) above the upper critical temperature and cooled in furnace

(c) above the upper critical temperature and cooled in still air

(d) between the upper and lower critical temperature and cooled in furnace

[ME 2012 : 1 Mark]

10. The process of reheating the martensitic steel to reduce its brittleness without any significant loss in its hardness is

(a) normalising (b) annealing

(c) quenching (d) tempering

[ME 2014 : 1 Mark, Set-1]

11. Match the heat treatment processes (Group A) and their associated effects on properties (Group B) of medium carbon steel

Group A	Group B
P. Tempering	1. Strengthening and grain refinement
Q. Quenching	2. Inducing toughness
R. Annealing	3. Hardening
S. Normalizing	4. Softening

Codes:

	P	Q	R	S
(a)	3	4	2	1
(b)	2	3	4	1
(c)	3	2	4	1
(d)	2	3	1	4

[ME 2014 : 1 Mark, Set-4]

12. Increase in carbon content in plain carbon steel raise its

(a) Ductility and UTS

(b) Tensile strength and malleability

(c) Tensile strength and hardness

(d) Ductility and melting temperature

[PI 1992 : 1 Mark]

Iron-Carbon diagram

13. The iron carbon diagram and the TTT curves are determined under

(a) Equilibrium and non-equilibrium conditions respectively

(b) Non-equilibrium and equilibrium conditions respectively

(c) Equilibrium conditions for both

(d) Non-equilibrium conditions for both

[ME 1996 : 1 Mark]

14. The percentage of carbon in grey cast iron is in the range of

(a) 0.25 to 0.75 percent

(b) 1.25 to 1.75 percent

(c) 3 to 4 percent

(d) 8 to 10 percent

[ME 2004 : 1 Mark]

15. If a particular Fe-C alloy contains less than 0.83% carbon, it is called

(a) high speed steel

(b) hypoeutectoid steel

(c) hypereutectoid steel

(d) cast iron

[ME 2007 : 1 Mark]

16. In a binary system of A and B a liquid of 20% A (80% B) is coexisting with a solid of 70% A(30%B). For an overall composition having 40% A the fraction of solid is

(a) 0.40

(b) 0.50

(c) 0.60

(d) 0.75

[ME 2016 : 2 Marks, Set-2]

17. In the phase diagram shown in the figure, four samples of the same composition are heated to temperatures marked by a, b, c and d.

Composition (Arbitrary Units)

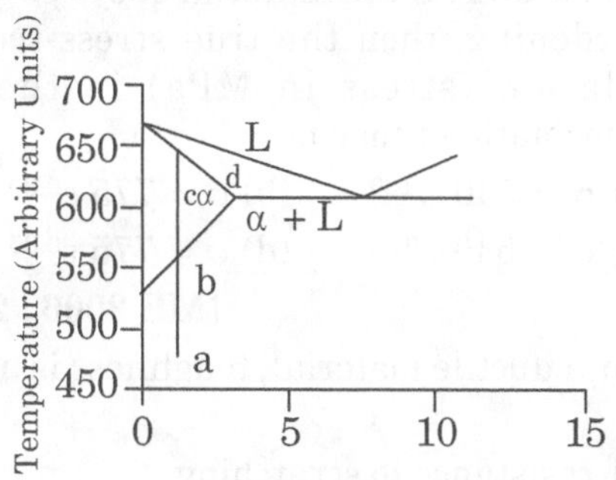

At which temperature will a sample get solutionized the fastest?

(a) a (b) b

(c) c (d) d

[ME 2016 : 1 Mark, Set-2]

18. Given the atomic weight of Fe is 56 and that of C is 12, the weight percentage of carbon in cementite (Fe_3C) is ______.

[ME 2017 : 1 Mark, Set-2]

19. Denoting L as liquid and M as solid in a phase-diagram with the subscripts representing different phases, a eutectoid reaction is described by

(a) $M \rightarrow M_2 + M_3$ (b) $L_1 \rightarrow M_1 + M_2$

(c) $L_1 \rightarrow M_1 + M_2$ (d) $M_1 + M_2 \rightarrow M_3$

[ME 2018 : 1 Marks, Set-2]

20. The binary phase diagram of metals P and Q is shown in the figure. An alloy X containing 60% P and 40% Q (by weight) is cooled from liquid to solid state. The fractions of solid and liquid (in weight percent) at 1250°C, respectively, will be

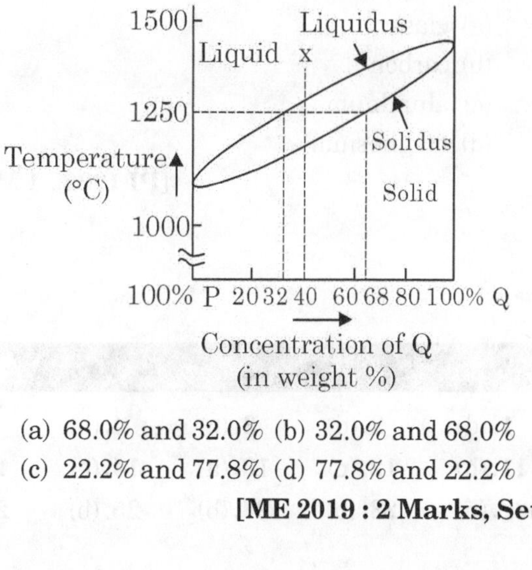

(a) 68.0% and 32.0% (b) 32.0% and 68.0%

(c) 22.2% and 77.8% (d) 77.8% and 22.2%

[ME 2019 : 2 Marks, Set-2]

Mechanical Properties

21. The ultimate tensile strength of a material is 400 MPa and the elongation up to maximum load is 35%. If the material obeys power law of hardening, then the true stress-true strain relation (stress in MPa) in the plastic deformation range is

(a) $\sigma = 540\ \varepsilon^{0.30}$ (b) $\sigma = 775\ \varepsilon^{0.30}$

(c) $\sigma = 540\ \varepsilon^{0.35}$ (d) $\sigma = 775\ \varepsilon^{0.35}$

[ME 2006 : 2 Marks]

22. For a ductile material, toughness is a measure of

(a) resistance to scratching

(b) ability to absorb energy up to fracture

(c) ability to absorb energy till elastic limit

(d) resistance to indentation

[ME 2013 : 1 Mark]

23. The "Jominy test" is used to find

(a) You ng's modulus

(b) hardenability

(c) yield strength

(d) thermal conductivity

[ME 2016 : 1 Mark, Set-1]

24. Hardenability of steel is a measure of

(a) the maximum hardness that can be obtained when it is austenitized and then quenched

(b) the depth to which required hardening is obtained when it is austenitized and then quenched

(c) the ability to harden when it is cold worked

(d) the ability to retain its hardness when it is heated to elevated temperatures

[ME 2019 : 1 Mark, Set-2]

25. The most widely used reinforcement in modern day FRP tennis racket is

(a) glass

(b) carbon

(c) aluminum

(d) magnesium

[PI 1992 : 1 Mark]

Crystal Structure

26. The effective number of lattice points in the unit cell of simple cubic, body centered cubic, and face centered cubic space lattices, respectively, are

(a) 1, 2, 2 (b) 1, 2, 4

(c) 2, 3, 4 (d) 2, 4, 4

[ME 2009 : 1 Mark]

27. Which of the following is the correct data structure for solid models?

(a) solid part → face → edges → vertices

(b) solid part → edges → faces → vertices

(c) vertices → edges → faces → solid parts

(d) vertices → faces → edges → solid parts

[ME 2009 : 1 Mark]

28. The material property which depends only on the basic crystal structure is

(a) fatigue strength (b) work hardening

(c) fracture strength (d) elastic constant

[ME 2010 : 1 Mark]

29. The crystal structure of austenite is

(a) body centered cubic

(b) face centered cubic

(c) hexagonal closed packed

(d) body centered tetragonal

[ME 2011 : 1 Mark]

30. The atomic packing factor for a material with body centered cubic structure is _____.

[ME 2015 : 1 Mark, Set-2]

31. The crystal structure of aluminium is

(a) body-centred cubic

(b) face-centred cubic

(c) close-packed hexagonal

(d) body-centred tetragonal

[ME 2017 : 1 Mark, Set-2]

32. The number of atoms per unit cell and the number of slip systems, respectively, for a face-centered cubic (FCC) crystal are

(a) 3, 3 (b) 3, 12

(c) 4, 12 (d) 4 ,48

[ME 2018 : 1 Marks, Set-1]

ANSWERS

1. (d)	2. (d)	3. (c)	4. (b)	5. (c)	6. (a)	7. (a)	8. (d)	9. (c)	10. (d)
11. (b)	12. (c)	13. (a)	14. (c)	15. (b)	16. (a)	17. (c)	19. (a)	20. (c)	21. (b)
22. (b)	23. (b)	24. (b)	25. (b)	26. (b)	27. (c)	28. (d)	29. (c)	31. (b)	32. (c)

EXPLANATIONS

1. The resulting structure will contain Austenite & cementite.

2. Spherodite, Coarse Pearlite, Fine Pearlite, Martensite

Structure	**Rockwell Hardness**
Coarse Pearlite	RC15
Fine Pearlite	RC25
Martensite	RC65

3. Cold working is a phenomenon in which material is deformed in recrystallization temperature.

4. Cyaniding: Carbon & Nitrogen diffuses to surface and rapidly quenched in case hardening.

5. In Tempering, both hardness and brittleness are reduced.

In Austempering, Austenite is converted into bainite.

In Martempering, Austenite gets converted into Martensite.

6. Strength decreases as dislocation movement becomes easier.

7. Improves machinability of high carbon steels.

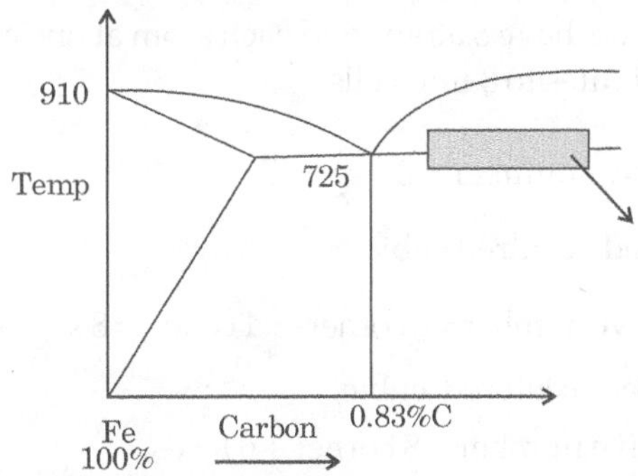

8. Charpy Test - Toughness

Knoop Test - Microhardness

Spiral Test - Fluidity

Cupping Test - Formability

9. LCT-Lower Critical Temperature

UCT-Upper Critical Temperature

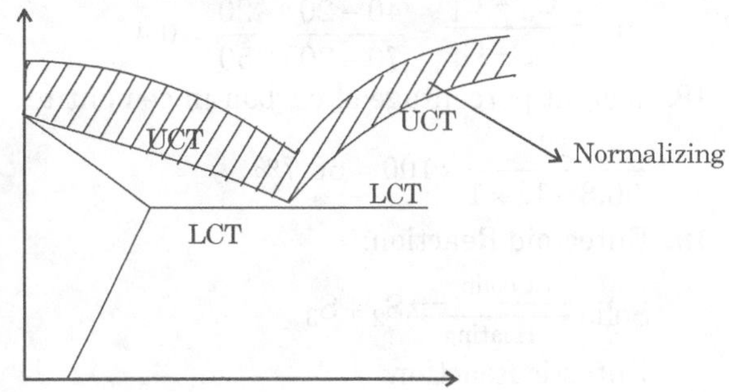

10.

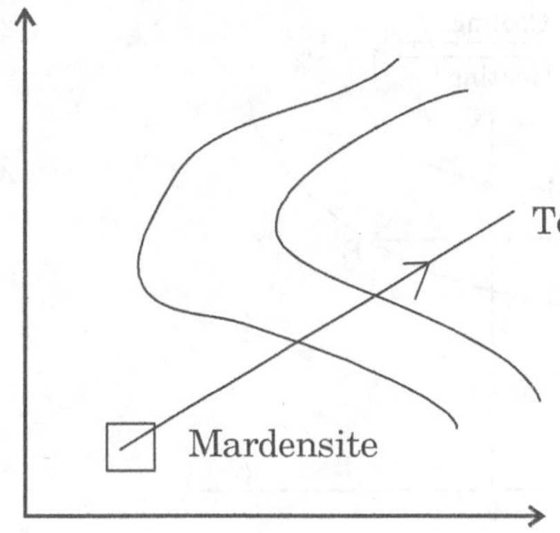

11. Tempering increases Toughness.

Quenching affects hardening.

Annealing affects softening.

Normalizing affects strengthening & Grain Refinement.

12. As carbon content increases, both tensile strength and hardness increases but ductility and toughness decreases.

13. Under TTT diagram cooling rate is variable.

14.

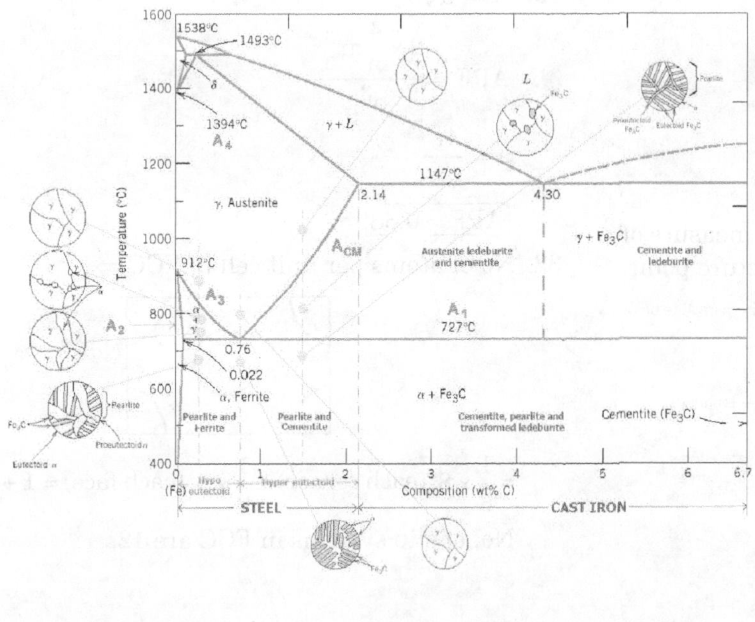

15. Refer above diagram

16. $C_o = 40\%$

$C_l = 20\%$

$C_s = 70\%$

$$m_s = \frac{C_o - C_l}{C_s - C_l} = \frac{40-20}{70-20} = \frac{20}{50} = 0.4$$

18. Weight percentage of carbon in cementite

$$\frac{12}{56.3 + 12 \times 1} \times 100 = 66.7\%$$

19. Eutectoid Reaction:

$$\text{Solid} \underset{\text{Heating}}{\overset{\text{Cooling}}{\rightleftharpoons}} S_2 + S_3$$

Eutectic Reaction:

$$\text{Liquid} \underset{\text{Heating}}{\overset{\text{Cooling}}{\rightleftharpoons}} S_1 + S_2$$

20.

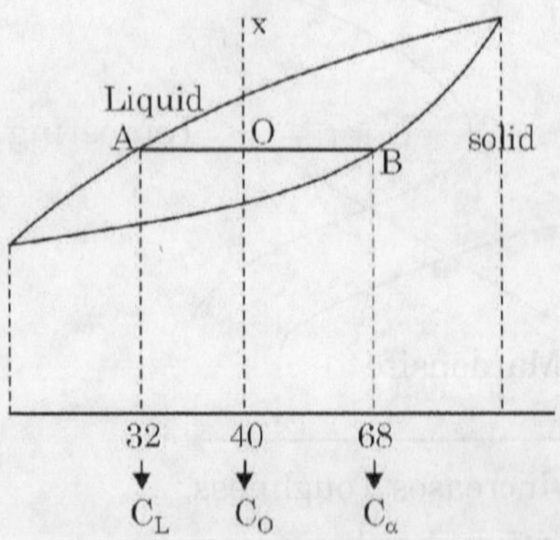

Mass fraction of liquid, $W_L = \dfrac{C_\alpha - C_0}{C_\alpha - C_L}$

$$W_L = \frac{68-40}{68-32} = 0.778$$

$\boxed{W_L = 77.8\%}$

$W_s = 1 - W_L = 1 - 77.8$

$\boxed{W_s = 22.2\%}$

21. $\in = n\,(1 + e) = 1n\,(1 + 0.35) = 0.3$

At UTS point: $n = \in$

$\sigma = \sigma_0(1 + e) = 400\,(1 + 0.35)$

$540 = k(0.3)^{0.3}$

$k = 774.9 \simeq 775;\ \sigma_f = 775 \in^{0.3}$

22. For a ductive material,toughness is a measure of ability to absorb energy up to the fracture point.

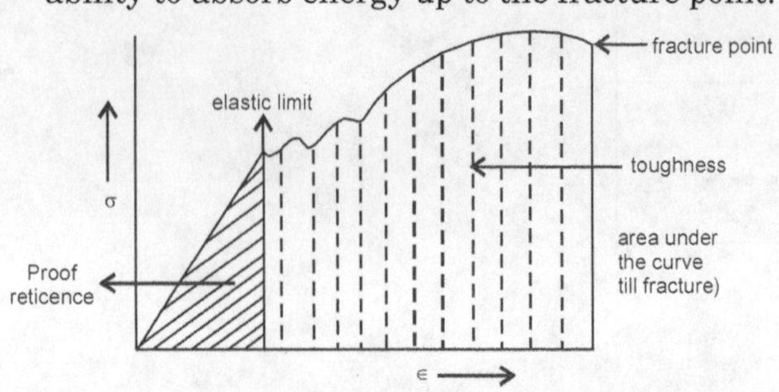

Verification:

(*a*) Resistance to scrateling is resistance to abrasion.

(*b*) Toughness is the ability to absorb energy upto fracture.

(*c*) Proof resilence is the ability to absorb energy till elastic limit.

(*d*) Hardness the resistance to indentation.

23. The Jominy End Quench Test ASTM A 255 measures Hardenability of steels. Hardenability is a measure of the capacity of a steel to be hardened in depth when quenched from its austenitizing temperature. Hardenability of a steel should not be confused with the hardness of a steel.

24. Hardenability is a measure of the capacity of a steel to be hardened in depth when quenched from its austenitizing temperature.

25. Carbon fibre because of light weight

26. For a simple cubic unit cell:

8 corners have 8 atoms and each atom at one corner contributes to 8 unit cells.

$$\text{Effective number} = 8 \times \frac{1}{8} = 1$$

For Body centred cubic,

$$\text{Effective number} = 8 \text{ corners} + 1 \text{ centre} = 8 \times \frac{1}{8} + 1 = 2$$

For Face centered cubic,

effective number = 8 corner + 6 faces

$$= 8 \times \frac{1}{8} + 6 \times \frac{1}{2} = 1 + 3 = 4$$

Therfore, answer is (1, 2, 4)

29. γ - Austenite has FCC Structure.

30. $4r = a\sqrt{3}$

$$APF = \frac{2 \times \frac{4}{3}\pi r^3}{a^3}$$

$$a = \frac{4r}{\sqrt{3}}$$

$\boxed{APF = 0.68}$

32. No of atoms per unit cell in FCC ⇒

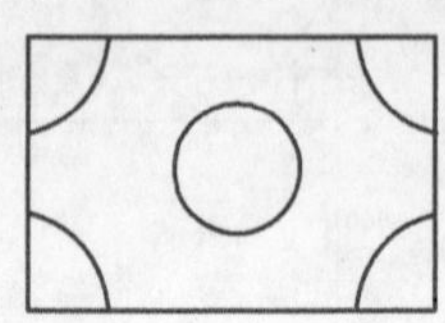

$$= \frac{1}{8} \times 8 \text{ (each corner)} + \frac{1}{2} \times 6 \text{ (each face)} = 1 + 3 = 4$$

No. of slip systems in FCC are 12.

2 CHAPTER

Metal Casting

Analysis of Previous GATE Papers

Topic	Year/Set → Marks Distribution ↓		2019 Set 1	2019 Set 2	2018 Set 1	2018 Set 2	2017 Set 1	2017 Set 2	2016 Set 1	2016 Set 2	2016 Set 3	2015 Set 1	2015 Set 2	2015 Set 3
Gating System	1 Mark	MCQ Type	1						1		1			
		Numerical Type												
	2 Marks	MCQ Type							1			1		
		Numerical Type		1									1	2
	Total		**1**	**2**					**3**		**1**	**2**	**2**	**4**
Casting System	1 Mark	MCQ Type												1
		Numerical Type												
	2 Marks	MCQ Type												
		Numerical Type				1	1			1				
	Total					**2**	**2**			**2**				**1**
Casting Defects	1 Mark	MCQ Type		1										
		Numerical Type												
	2 Marks	MCQ Type	2											
		Numerical Type												
	Total		**4**	**1**										

Gating System

1. Riser is designed so as to
   (a) freeze after the casting freezes
   (b) freeze before the casting freezes
   (c) freeze at the same time as the casting
   (d) minimize the time of pouring

[ME 1987 : 1 Mark]

2. The contraction allowance provided on the pattern and core boxes compensates for the following type of contraction
   (a) liquid contraction
   (b) solidification contraction
   (c) solid contraction
   (d) All of the above three types of contractions

[ME 1988 : 1 Mark]

3. Chills are used in moulds to
   (a) achieve directional solidification
   (b) reduce possibility of below holes
   (c) reduce the freezing time
   (d) smoothen the metal for reducing spatter

[ME 1989 : 1 Mark]

4. Two cubical castings of the same metal and size of 2 cm side and 4 cm side are moulded in green sand. If the smaller casting solidifies in 2 mins, the expected time of solidifications of large casting will be
   (a) 16 min (b) $2\sqrt{2}$ min
   (c) 8 min (d) 4 min

[ME 1989 : 2 Marks]

5. The pressure at the in-gate will be maximum with the gating system
   (a) 4 : 8 : 3 (b) 1 : 3 : 3
   (c) 1 : 2 : 4 (d) 1 : 2 : 1

[ME 1990 : 1 Mark]

6. When there is no room temperature change, the total shrinkage allowance on a pattern is independent of
   (a) pouring temperature of the liquid metal
   (b) freezing temperature of the liquid metal
   (c) the component size
   (d) coefficient of thermal contraction of solidified metal

[ME 1991 : 1 Mark]

7. Converging passage is used for feeding the liquid molten metal into the mould to
   (a) increase the rate of feeding
   (b) quickly break off the protruding portion of the casting
   (c) decrease wastage of cast metal
   (d) avoid aspiration of air

[ME 1991 : 1 Mark]

8. Light impurities in the molten metal are prevented from reaching the mould cavity by providing
   (a) Strainer (b) Bottom well
   (c) Skim bob (d) All of the above

[ME 1996 : 1 Mark]

9. Hardness of green sand mould increases with
   (a) increase in moisture content beyond 6 percent
   (b) increase in permeability
   (c) decrease in permeability
   (d) increases in both moisture content and permeability

[ME 2003 : 1 Mark]

10. With a solidification factor of 0.97 × 106 s/m^2, the solidification time (in seconds) for a spherical casting of 200 mm diameter is
   (a) 539 (b) 1078
   (c) 4311 (d) 3233

[ME 2003 : 2 Marks]

11. Gray cast iron blocks 200 x 100 x 10 mm are to be cast in sand moulds, Shrinkage allowance for pattern making is 1%. The ratio of the volume of pattern to that of the casting will be
   (a) 0.97 (b) 0.99
   (c) 1.61 (d) 1.03

[ME 2004 : 2 Marks]

12. A mould has downsprue whose length is 20 cm and the cross sectional area at the base of the downsprue is 1 cm^2. The downsprue feeds a horizontal runner leading into the mould cavity of volume 1000 cm^3. The time required to fill the mould cavity will be
   (a) 4.05 s (b) 5.05 s
   (c) 6.05 s (d) 7.25 s

[ME 2006 : 2 Marks]

13. In a sand casting operation, the total liquid head is maintained constant such that it is equal to the mould height. The time taken to fill the mould with a top gate is t_A. If the same mould is filled with a bottom gate, then the time taken is t_B. Ignore the time required to fill the runner and frictional effects, Assume Atmospheric pressure at the top molten metal surfaces. The relation between t_A and t_B is

(a) $t_B = \sqrt{2t_A}$ (b) $t_B = 2t_A$

(o) $t_B = \frac{t_A}{\sqrt{2}}$ (d) $t_B = 2\sqrt{2t_A}$

[ME 2006 : 2 Marks]

14. Volume of a cube of side l and volume of a sphere of radius r are equal. Both the cube and the sphere are solid and of same material. They are being east. The ratio of the solidification time of the cube to the same of sphere is

(a) $\left(\frac{2\pi}{6}\right)^3\left(\frac{r}{l}\right)^6$ (b) $\left(\frac{4\pi}{6}\right)\left(\frac{r}{l}\right)^2$

(c) $\left(\frac{4\pi}{6}\right)^2\left(\frac{r}{l}\right)^3$ (d) $\left(\frac{4\pi}{6}\right)^2\left(\frac{r}{l}\right)^4$

[ME 2007 : 1 Mark]

15. A 200 mm long down sprue has an area of cross-section of 650 mm² where the pouring basin meets the down sprue (i.e. at the beginning of the down sprue). A constant head of molten metal is maintained by the pouring basin. The molten metal flow rate is 6.5×10^5 mm³/s. Considering the end of down sprue to be open to atmosphere and acceleration due to gravity of 10^4 mm/s², the area of the down sprue in mm² at its end (avoiding aspiration effect) should be

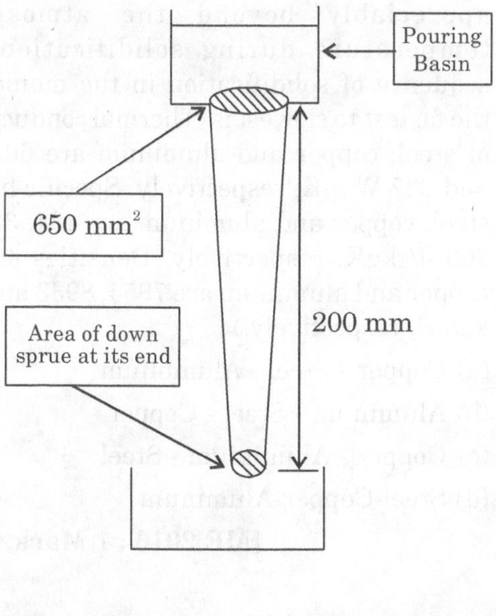

(a) 650.0 (c) 290.7

(b) 350.0 (d) 190.0

[ME 2007 : 2 Marks]

16. Match the items in Column I and Column II.

Column I	Column II
P. Metallic Chills	1. Support for the core
Q. Metallic Chaplets	2. Reservoir of the molten metal
R. Riser	3. Control cooling of critical sections
S. Exothermic Padding	4. Progressive solidification

(a) P–1, Q–3, R–2, S–4

(b) P–1, Q–4, R–2, S–3

(c) P–3, Q–4, R–2, S–1

(d) P–4, Q–1, R–2, S–3

[ME 2009 : 2 Marks]

17. In a gating system, the ratio 1:2:4 represents

(a) sprue base area : runner area : ingate area

(b) pouring basin area: ingate area: runner area

(c) sprue base area : ingate area : casting area

(d) runner area : ingate area : casting area

[ME 2010 : 1 Mark]

18. A cubic casting of 50 mm side undergoes volumetric solidification shrinkage and volumetric solid contraction of 4% and 6% respectively. No riser is used. Assume uniform cooling in all directions. The side of the cube after solidification and contraction is

(a) 48.32 mm (b) 49.90 mm

(c) 49.94 mm (d) 49.96 mm

[ME 2011 : 2 Marks]

19. A cube shaped casting solidifies in 5 min. The solidification time in min for a cube of the same material, which is 8 times heavier than the original casting, will be

(a) 10 (b) 20

(c) 24 (d) 40

[ME 2013 : 1 Mark]

20. A cylindrical blind riser with diameter d and height h, is placed on the top of the mold cavity of a closed type sand mould as shown in the figure. If the riser is of constant volume, then the rate of solidification in the riser is the least when the ratio h : d is

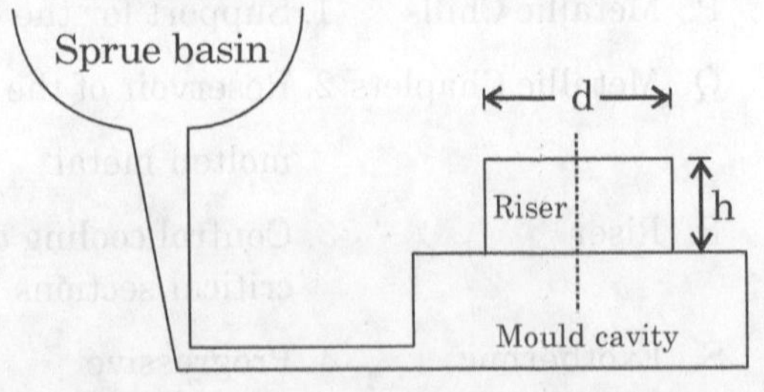

(a) 1:2 (b) 2:1
(c) 1:4 (d) 4:1

[ME 2014 : 2 Marks, Set-3]

21. A cylindrical riser of 6 cm diameter and 6 cm height has to be designed for a sand casting mould for producing a steel rectangular plate casting of 7 cm × 10 cm × 2 cm dimensions having the total solidification time of 1.36 minute. The total solidification time (in minute) of the riser is ______.

[ME 2014 : 2 Marks, Set-4]

22. The solidification time of a casting is proportional to $\left(\frac{V}{A}\right)^2$ where V is the volume of the casting and A is the total casting surface area losing heat. Two cubes of same material and size are cast using sand casting process. The top face of one of the cubes is completely insulated. The ratio of the solidification time for the cube with top face insulated to that of the other cube is

(a) $\frac{25}{36}$ (b) $\frac{36}{25}$
(c) 1 (d) $\frac{6}{5}$

[ME 2015 : 2 Marks, Set-1]

23. A cube and a sphere made of cast iron (each of volume 1000 cm^3) were cast under identical conditions. The time taken for solidifying the cube was 4 s. The solidification time (in s) for the for the sphere is ________.

[ME 2015 : 2 Marks, Set-2]

24. Ratio of solidification time of a cylindrical casting (height = radius) of that of a cubic casting of side two times the height of cylindrical casting is ______

[ME 2015 : 2 Marks, Set-3]

25. The dimensions of a cylindrical side riser (height = diameter) for a 25 cm × 15 cm × 5 cm steel casting are to be determined. For the tabulated shape factor values given below, diameter of the riser (in cm) is______.

Shape Factor	2	4	6	8	10	12
Riser volume/ Casting volume	1.0	0.70	0.55	0.50	0.40	0.35

[ME 2015 : 2 Marks, Set-3]

26. The part of a gating system which regulates the rate of pouring of molten metal is

(a) pouring basin (b) runner
(c) choke (d) ingate

[ME 2016 : 1 Mark, Set-1]

27. A cylindrical job with diameter of 200 mm and height of 100 mm is to be cast using modulus method of riser design. Assume that the bottom surface of cylindrical riser does not contribute as cooling surface. If the diameter of the riser is equal to its height, then the height of the riser (in mm) is

(a) 150 (b) 200
(c) 100 (d) 125

[ME 2016 : 2 Marks, Sat-1]

28. Equal amounts of a liquid metal at the same temperature are poured into three moulds made of steel, copper and aluminum. The shape of the cavity is a cylinder with 15 mm diameter. The size of the moulds are such that the outside temperature of the moulds do not increase appreciably beyond the atmospheric temperature during solidification. The sequence of solidification in the mould from the fastest to slowest is (Thermal conductivities of steel, copper and aluminum are 60.5, 401 and 237 W/mK, respectively. Specific heats of steel, copper and aluminum are 434, 385 and 903 J/ kgK, respectively. Densities of steel, copper and aluminum are 7854, 8933 and 2700 kg/m^3, respectively.)

(a) Copper - Steel - Aluminium
(b) Aluminum - Steel - Copper
(c) Copper - Aluminium - Steel
(d) Steel-Copper-Aluminum

[ME 2016 : 1 Mark, Set-3]

29. In a casting process, a vertical channel through which molten metal flows downward from pouring basin to runner for reaching the mould cavity is called

(a) riser (b) blister

(c) pin hole (d) sprue

[ME 2019 : 1 Mark, Set-1]

30. The figure shows a pouring arrangement for casting of a metal block. Frictional losses are negligible. The acceleration due to gravity is 9.81 m/s^2. The time (in s, round off to two decimal places) to fill up the mold cavity (of size 40 cm × 30 cm × 15 cm) is ____.

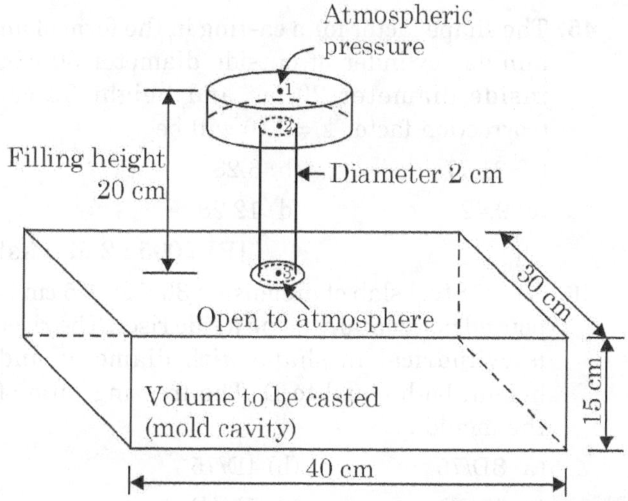

[ME 2019 : 2 Marks, Set-2]

31. What is the velocity of the steel at the bottom of the sprue if the sprue height is 30 cm? Assume the frictional and other losses to be 20%.

[PI 1989 : 2 Marks]

32. **Assertion (A):** Converging passage is used for feeding liquid metal into a mould.

Reason (R): Inhalation of air promotes blow holes in casting.

(a) Both (A) and (R) are true and (R) is the correct reason for (A)

(b) Both (A) and (R) are true but (R) is NOT the correct reason for (A)

(c) Both (A) and (R) are false

(d) (A) is false but (R) is true

[PI 1990 : 1 Mark]

33. The optimum pouring time for a casting depends on several factors. One important factor among them is

(a) location of riser

(b) porosity of sand mould

(c) fluidity of casting metal

(d) area of the pouring basin

[PI 1991 : 1 Mark]

34. The primary function of a riser is to

(a) feed molten metal to casting as it solidifies

(b) prevent atmospheric air from contaminating the metal in the mould

(c) allow gases to easily escape from mould cavity

(d) allow molten metal to rise above the mould cavity

[PI 1992 : 1 Mark]

35. Chaplets are placed between mould in order to

(a) promote directional solidification

(b) help alloying the metal

(c) facilitate easy removal of core from casting

(d) prevent core movement due to buoyancy

[PI 1992 : 1 Mark]

36. Application

A. Undercut in components

B. Large bells

C. Mass production of casting by machine moulding

D. Components with irregular parting lines

Type of pattern

1. Cope and drag
2. Follow board
3. Gated
4. Loose piece
5. Sweep

[PI 1992 : 2 Marks]

37. According to Chvorinov's rule, the solidification time of a casting is proportional to (Volume/surface area)^n, where 'n' equals to

(a) 0.5 (b) 1

(c) 2 (d) 4

[PI 1994 : 2 Marks]

38. The height of the down sprue is 175 mm and its CS are at the base is 200 mm^2. The CS area of the horizontal runner is also 200 mm^2. Assuming no losses, indicate the correct choice for-the time (second) required to fill a mould cavity of volume $10^6 mm^3$. (Use g = 10 m/s^2)

(a) 2.67 (b) 8.45

(c) 26.7 (d) 84.50

[PI 2002 : 2 Marks]

39. The permeability of moulding sand was determined using a standard AFS sample by passing 2000 cc of air at a gauge pressure of 10 g/cm^2. If the time taken for the air to escape was 1 min, the permeability number is

(a) 112.4 (b) 100.2

(c) 75.3 (d) 50.1

[PI 2002 : 2 Marks]

40. Proper gating design in metal casting

P. Influences the freezing range of the melt

Q. Compensates the loss of fluidity of the melt

R. Facilitates top feeding of the melt

S. Avoids misrun

(a) P, R (b) Q, S

(c) R, S (d) P, S

[PI 2002 : 2 Marks]

41. A 10 mm thick steel bar is to be horizontally cast with two correctly top risers of adequate feeding capacity. Assuming end effect without chill, what should be the theoretical length of the bar?

(a) 96 mm (b) 132 mm

(c) 192 mm (d) 156 mm

[PI 2002 : 2 Marks]

42. Gating ratio of 1 : 2 : 4 is used to design the gating system for magnesium alloy casting. This gating ratio refers to the cross-section areas of the various gating elements as given below:

1. Down sprue
2. Runner
3. Ingates

The sequence of the above elements in the ratio 1 : 2 : 4 is

(a) 1, 2 and 3 (b) 1, 3 and 2

(c) 2, 3 and 1 (d) 3, 1 and 2

[PI 2003 : 2 Marks]

43. A casting of size 100 mm × 100 mm × 50 mm is required. Assume volume shrinkage of casting as 2.6%. If the height of the riser is 80 mm and riser volume desired is 4 times the shrinkage in casting, what is the appropriate riser diameter in mm?

(a) 14.38 (b) 20.34

(c) 28.76 (d) 57.52

[PI 2003 : 2 Marks]

44. Wood flour is added to core sand to improve

(a) Collapsibility of core

(b) Dry strength of core

(c) Shear strength of core

(d) Tolerance on casting

[PI 2004 : 1 Mark]

45. The shape factor for a casting in the form of an annular cylinder of outside diameter 30 cm, inside diameter 20 cm and height 30 cm (correction factor k = 1.0) will be

(a) 21.77 (b) 6.28

(c) 9.42 (d) 12.28

[PI 2005 : 2 Marks]

46. A cast Steel slab of dimension 30 × 20 × 5 cm is poured horizontally using a side riser. The riser is cylindrical in shape with diameter and height, both equal to D. The freezing ratio of the mould is

(a) 8D/75 (b) 4D/75

(c) 75/8D (d) 75/4D

[PI 2005 : 2 Marks]

47. Cold shut is a defect in casting due to

(a) sand sliding from the cope surface

(b) internal voids or surface depression due to excessive gas trapped

(c) discontinuity resulting from hundred contraction

(d) two streams of material that are too cold to fuse properly

[PI 2006 : 1 Mark]

48. The volume flow rate (in mm^3/s) is

(a) 0.8×10^5 (b) 1.1×10^5

(c) 1.7×10^5 (d) 2.3×10^5

[PI 2007 : 2 Marks]

49. The mould filling time is seconds is

(a) 2.8 (b) 5.78

(c) 7.54 (d) 8.41

[PI 2007 : 2 Marks]

50. In sand casting of hollow part of lead, a cylindrical core of diameter 120 mm and height 180 mm is placed inside the mould cavity. The densities of core material and lead are 1600 kg/m^3 and 11300 kg/m^3 respectively. The net force (in N) that tends to lift the core during pouring of molten metal will be

(a) 19.7 (b) 64.5

(c) 193.7 (d) 257.6

[PI 2008 : 2 Marks]

51. A solid cylinder of diameter D and height equal to D, and a solid cube of side L are being sand cast by using the same material. Assuming there is no superheat in both the cases, the ratio of solidification times of the cylinder to the solidification time of the cube is

(a) $(L/D)^2$ (b) $(2L/D)^2$

(c) $(2D/L)^2$ (d) $(D/L)^2$

[PI 2009 : 2 Marks]

52. Solidification time of a metallic alloy casting is

(a) Directly proportional to its surface area

(b) Inversely proportional to the specific heat of the cast material

(c) Inversely proportional to the thermal diffusivity of the mould material

(d) Inversely proportional to the pouring temp

[PI 2010 : 1 Mark]

53. During the filling process of a given sand mould cavity by molten metal through a horizontal runner of circular C.S, the frictional head loss of the molten metal in the runner will increase with the

(a) Increase in runner diameter

(b) Decrease internal surface roughness of the runner

(c) Decrease in length of runner

(d) Increase in average velocity of molten metal

[PI 2010 : 2 Marks]

54. In sand casting process, a sphere and a cylinder of equal volumes are separately cast from the sand molten metal under identical conditions. The height and diameter of the cylinder are equal. The ratio of the solidification time of the sphere to that of the cylinder is

(a) 1.14 (b) 0.87

(c) 1.31 (d) 0.76

[PI 2011 : 2 Marks]

55. A mould having dimensions 100 mm × 90 mm × 20 mm is filled with molten metal through a gate with height h and C.S. area A, the mould filling time is t_1 The height is now quadrupled and the cross-sectional area is halved. The corresponding filling time is t_2. The ratio t_2/t_1 is

(a) $1\sqrt{2}$ (b) 1

(c) $\sqrt{2}$ (d) 2

[PI 2012 : 2 Marks]

Casting Method

56. Increase in water content in moulding sand causes

(a) flowability to go through a maxima

(b) permeability to go through a maxima

(c) compressive strength to go through a maxima

(d) strength to go through a maxima

[ME 1989 : 1 Mark]

57. In a green sand moulding process, uniform ramming leads to

(a) less chance of gas porosity

(b) uniform flow of molten into the mould cavity

(c) greater dimensional stability of the casting

(d) less sane expansion type of casting defect

[ME 1992 : 1 Mark]

58. Casting process

A. Slush casting

B. Shell moulding

C. Dry sand moulding

D. Centrifugal casting

Product

1. Turbine blade
2. Machine tool bed
3. Cylindrical block
4. Hollow castings like lamp shades
5. Rain water pipes
6. Cast iron shoe brake

[ME 1992 : 2 Marks]

59. Centrifugally casted products have

(a) large grain structure with high porosity

(b) fine grain structure with high density

(c) fine grain structure with low density

(d) segregation of slug toward the outer skin of the casting

[ME 1993 : 1 Mark]

60. Only 4 pairs can be matched

Casting process

A. Investment casting
B. Die casting
C. Centrifugal casting
D. Drop forging
E. Extrusion
F. Shell moulding

Product

1. Turbine rotors
2. Turbine blades
3. Connecting rods
4. Galvanized iron pipes
5. Cast iron pipes
6. Carburetor body

[ME 1994 : 2 Marks]

61. Match the following

A. Sand casting
B. Plaster mould casting
C. Shell mould casting
D. Investment casting

1. Symmetrical and circular shapes only
2. Parts have hardened skins and soft interior
3. Minimum post casting process
4. Parts have soft skin and hard interior
5. Parts have a tendency to warp
6. Suitable only for nonferrous metals

[ME 1998 : 2 Marks]

62. Which of the following materials requires the largest shrinkage allowance, while making a pattern for casting

(a) Aluminium (b) Brass
(c) Cast iron (d) Plain carbon steel

[ME 1999 : 1 Mark]

63. Disposable patterns are made of

(a) wood (b) rubber
(c) metal (d) polystyrene

[ME 2000 : 1 Mark]

64. Shrinkage allowance on pattern is provided to compensate for shrinkage when

(a) the temperature of liquid metal drops from pouring to freezing temp
(b) the metal changes from liquid to solid state at freezing temperature
(c) the temperature of solid phase drops from freezing to room temperature
(d) the temp of metal drops from pouring to room temperature

[ME 2001 : 1 Mark]

65. In centrifugal casting, the lighter impurities are

(a) uniformly distributed
(b) forced towards the outer surface
(c) trapped near the mean radius of the casting
(d) collected at the centre of the casting

[ME 2002 : 1 Mark]

66. The primary purpose of sprue in casting mould is to

(a) feed the casting at rate consistent with the rate of solidification
(b) Act as a reservoir for molten metal
(c) feed molten metal from the pouring basin to the gate
(d) help feed the casting until all solidification takes place

[ME 2002 : 1 Mark]

67. Match the items of List-I (Equipment) with the items of List-II (Process) the correct answer using the given codes.

Equipment

P. Hot chamber machine
Q. Muller
R. Dielectric baker
S. Sand blaster

Process

1, Cleaning
2. Core making
3, Die casting
4. Annealing and mixing

(a) P-2, Q-1, R-4, S-6
(b) P-4, Q-2, R-3, S-5
(c) P-4, Q-5, R-1, S-2
(d) P-3, Q-5, R-2, S-1

[ME 2006 : 2 Marks]

68. An expendable pattern is used in

(a) slush casting
(b) squeeze casting
(c) centrifugal casting
(d) Investment casting

[ME 2006 : 1 Mark]

69. Which of the following engineering materials is the most suitable candidate for hot chamber die casting?

(a) low carbon steel (b) titanium
(c) copper (d) tin

[ME 2007 : 1 Mark]

70. While cooling, a cubical casting of side 40mm undergoes 3%, 4% and 5% volume shrinkage during the liquid state, phase transition and solid state, respectively. The volume of metal compensated from the riser is

(a) 2% (b) 7%

(c) 8% (d) 9%

[ME 2008 : 2 Marks]

71. Green sand mould indicates that

(a) polymeric mould has been cured

(b) mould has been totally dried

(c) mould is green in colour

(d) mould contains moisture

[ME 2011 : 1 Mark]

72. During solidification of a pure molten metal, the grains in the casting near the mould wall area

(a) coarse and randomly oriented

(b) fine and randomly oriented

(c) fine and ordered

(d) coarse and ordered

[ME 2013 : 1 Marks, Set-2]

73. Match the casting defects (Group A) with the probable causes (Group B):

Group-A

P. Hot tears

Q. Shrinkage

R. Blow holes

S. Cold shunt

Group-B

1. Improper fusion of two streams of liquid metal
2. Low permeability of the sand mould
3. Volumetric contraction both in liquid and solid stage
4. Differential cooling rate

(a) P-1, Q-3, R-2, S-4

(b) P-4, Q-3, R-2, S-1

(c) P-3, Q-4, R-2, S-1

(d) P-1, Q-4, R-4, S-3

[ME 2014 : 1 Mark, Set-3]

74. In full mould (cavity-less) casting process, the pattern is made of

(a) expandable polystyrene

(b) wax

(c) epoxy

(d) plaster of Paris

[ME 2015 : 1 Mark, Set-3]

75. Gray cast iron blocks of size 100 mm × 50 mm × 10 mm with a central spherical cavity of diameter 4 mm are sand cast. The shrinkage allowance for the pattern is 3%. The ratio of the volume of the pattern to volume of the casting is

[ME 2016 : 2 Marks, Set-2]

76. A sprue in a sand mould has a top diameter of 20 mm and height of 200 mm. The velocity of the molten metal at the entry of the sprue is 0.5 m/s. Assume acceleration due to gravity as 9.8 m/s^2 and neglect all losses, If the mould well ventilated, the velocity (upto 3 decimal points accuracy) of the molten metal at the bottom of the sprue is ______ m/s.

[ME 2017 : 2 Marks, Set-1]

77. For sand-casting-a steel rectangular plate with dimensions 80 mm × 120 mm × 20 mm, a cylindrical riser has to be designed, The height of the riser is equal to its diameter. The total solidification time for the casting is 2 minutes. In Chvorinov's law for the estimation of the total solidification time, exponent is to be taken as 2. For a solidification time of 3 minutes in the riser, the diameter (in mm) of the riser is ______ (correct to two decimal places).

[ME 2018 : 2 Marks, Set-2]

78. Negative allowance is provided on the pattern to take care of

(a) The distortion allowance

(b) The draft allowance

(c) The machining allowance

(d) The shake allowance

[PI 1989 : 1 Mark]

79. The process by which shell like casting (like toys) are made is

(a) Shell moulding

(b) Centrifugal casting

(c) Slush casting

(d) Die casting

[PI 1989 : 1 Mark]

80. For the same material, powder metallurgy process in superior to casting for

(a) making large products

(b) better control over the density of product

(c) better strength of the finished product

(d) making parts with wide variations of thickness at different sections

[PI 1989 : 1 Mark]

81. Green strength in powder metallurgy refers to the strength of

(a) original material in the bulk form

(b) the powder before compaction

(c) the powder after compaction

(d) the powder after sintering and cooling

[PI 1989 : 1 Mark]

82. For casting of turbine blades made of high temperature and high strength alloys, the most suitable process is

(a) Die casting

(b) Investment casting

(c) Centrifugal casting

(d) Slush casting

[PI 1990 : 1 Mark]

83. The problem of shrinkage cavity is more likely to occur while casting ____ A ____ B ____ a chill block

A. metal/plastic B. with/without

[PI 1990 : 1 Mark]

84. Ceramics slurry is used in making the moulds for

(a) investment casting

(b) die casting

(c) shell mould casting

(d) centrifugal casting

[PI 1991 : 1 Mark]

85. Pressure die casting deals with casting metals of relatively ____ A ____ melting point and provides ____B. production rate > A. higher/ lower B. higher /lower

[PI 1991 : 1 Mark]

86. Investment casting uses ... A as pattern material and is preferred for small parts with .....

B ____designs.

A. wax/resin coated sand

B. simple/intricate

[PI 1992 : 1 Mark]

87. Electrical switches made out of thermoset materials are produced by

(a) Compression moulding

(b) Transfer moulding

(c) Injection moulding

(d) Vacuum moulding

[PI 1992 : 2 Marks]

88. Bentonite is commonly used bonding material for dry sand cores (T/F)

[PI 1994 : 1 Mark]

89. Shell moulding is an investment casting process (T/F)

[PI 1995 : 1 Mark]

90. Match the following

Process

P. Green sand moulding

Q. Shell moulding

R. Investment moulding

S. Ceramic moulding

Mould making technique

1. Pouring
2. Dipping
3. Compaction
4. Resin bonding

[PI 2002 : 2 Marks]

91. Which one of the following is not a property of a sand mould?

(a) Permeability (b)Collapsibility

(c) Strength (d)Fluidity

[PI 2003 : 1 Mark]

92. Match the following

Group-I

P. Sand casting

Q. Centrifugal casting

R. Investment casting

S. Die casting

Group-I I

1. Turbine blades
2. IC Engine
3. Large bells
4. Pulleys

	P	Q	R	S
(a)	4	1	3	2
(b)	2	4	3	1
(c)	3	4	1	2
(d)	3	2	1	4

[PI 2007 : 2 Marks]

93. In hollow cylindrical parts made by centrifugal castings, the density of the part is
(a) Maximum at the outer region
(b) Maximum at the inner region
(c) Maximum at the mid point between outer and inner surfaces
(d) Uniform throughout

[PI 2008 : 1 Mark]

94. Hot chamber die casting is not suited for
(a) Lead and its alloys
(b) Zinc and its alloys
(c) Tin and it's alloys
(d) Aluminum and its alloys

[PI 2009 : 1 Mark]

95. In sand casting fluidly of the molten metal increases with
(a) Increase in degree of superheat
(b) Decrease in pouring rate
(c) Increase in thermal conductivity of the mould
(d) Increase in sand grain size

[PI 2011 : 1 Mark]

96. Which of the following casting processes uses expandable pattern and expandable mould?
(a) Shell mould casting
(b) Investment casting
(c) Pressure die casting
(d) Centrifugal casting

[PI 2011 : 1 Mark]

Casting Defect

97. Misrun is a casting defect which occurs due to
(a) very high pouring temperature of the metal
(b) insufficient fluidity of the molten metal
(b) absorption of gases by the liquid metal
(d) improper alignment of the mould flasks

[ME 2004 : 1 Mark]

98. Two streams of liquid metal which are not hot enough to fuse properly result into a casting defect known -as
(a) cold shut (b) swell
(c) sand wash (d) scab

[ME 2009 : 1 Mark]

99. The hot tearing in a metal casting is due to
(a) high fluidity
(b) high melt temperature
(c) wide range of solidification temperature
(d) low coefficient of thermal expansion

[ME 2014 : 1 Mark, Set-3]

100. Match the following sand mold casting effects with their respective causes.

	Defect		Cause
P	Blow hole	1.	Poor Collapsibility
Q	Misrum	2.	Mold erosion
R	Hot tearing	3.	Poor permeability
S	Wash	4.	Insufficient fluidity

(a) P-3, Q-4, R-1, S-2 (b) P-3, Q-4, R-2, S-1
(c) P-4, Q-3, R-1, S-2 (d) P-2, Q-4, R-1, S-3

[ME 2019 : 2 Marks, Set-1]

101. The fluidity of molten metal of cast alloys (without any addition of fluxes) increases with increase in
(a) freezing range (b) viscosity
(c) degree of superheat (d) surface tension

[ME 2019 : 1 Mark, Set-2]

102. Poor collapsibility of the mould leads to distortion cracks type of casting defect in aluminium castings. (T/F)

[PI 1989 : 1 Mark]

103. Low permeability moulds and low pouring temp of the molten metal causes blow holes. Defect in castings.(T/F)

[PI 1991 : 1 Mark]

104. Manufacturing process
A. Hot tears
B. Porosity
C. Sand inclusions
D. Shrinkage cavity

Property
1. Mould restraint
2. Inadequate risering
3. High pouring temp
4. Loosely rammed sand mould
5. Gas entrapment

[PI 1993, 1995 : 2 Marks]

105. Cold shut is a defect in casting due to
(a) sand sliding from the cope surface
(b) internal voids or surface depression due to excessive gas trapped
(c) discontinuity resulting from hundred contraction
(d) two streams of material that are too cold to fuse properly

[PI 2006 : 1 Mark]

ANSWERS

1. (a)	2. (c)	3. (a)	4. (c)	5. (d)	6. (a)	7. (d)	8. (c)	9. (c)	10. (b)
11. (a)	12. (b)	13. (b)	14. (d)	15. (c)	16. (d)	17. (a)	18. (a)	19. (b)	20. (a)
22. (b)	26. (a)	27. (a)	28. (c)	33. (b)	34. (c)	35. (a)	36. (d)	38. (c)	39. (a)
40. (d)	41. (b)	42. (b)	43. (a)	44. (a)	45. (a)	46. (a)	47. (a)	48. (d)	49. (c)
50. (b)	51. (c)	52. (d)	53. (c)	54. (d)	55. (c)	56. (b)	57. (b)	58. (c)	59. (b)
60. (b)	63. (b)	64. (d)	65. (c)	66. (d)	67. (c)	68. (c)	69. (d)	70. (b)	71. (b)
72. (d)	73. (b)	74. (c)	75. (a)	78. (d)	79. (d)	80. (a)	81. (b)	82. (c)	83. (b)
84. (a)	85. (a)	87. (a)	88. (a)	91. (d)	92. (d)	93. (c)	94. (a)	95. (d)	96. (a)
97. (b)	98. (b)	99. (a)	100. (c)	101. (c)	102. (*c*)	105. (d)	106. (d)		

EXPLANATIONS

1. If riser will solidify before casting if cannot compensate the liquid shrinkage and phase Transition shrinkage .

2. Third stage of contraction solidification contraction compensated by pattern allowances.

3. Chills are used to promote uni-directional solidification.

4. Solidification time $= \left(\frac{V}{A}\right)^2$

$$\frac{T_1}{T_2} = \frac{\left(\frac{V}{A}\right)_1^2}{\left(\frac{V}{A}\right)_2^2} = \frac{\left(\frac{2^3}{6\times 2^2}\right)^2}{\left(\frac{4^3}{6\times 4^2}\right)^2} = \frac{\left(\frac{1}{3}\right)^2}{\left(\frac{2}{3}\right)^2}$$

$$\frac{T_1}{T_2} = \frac{1}{4}$$

$T_2 = 2\times 4 = 8$ min

5. 1 : 2 : 1 Ingate area equals to sprue Base.

6. Pattern allowances only taking care of solid shrinkage.

7. Pressure in runner is greater than atmospheric pressure so no aspiration takes place.

8. Skin bob

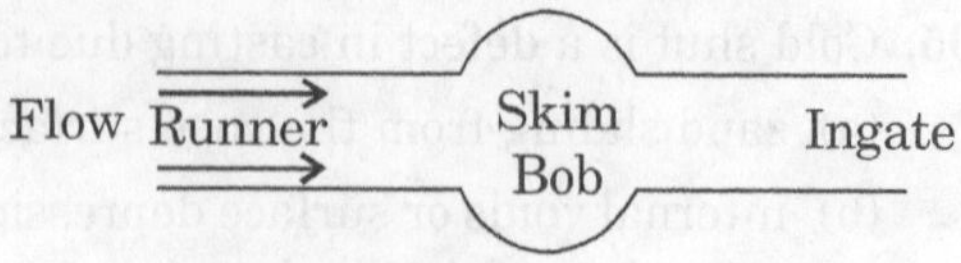

10. $t = k\left(\frac{V}{A}\right)^2$

D = 200 mm

$$t = 0.97\times 10^6\left(\frac{\frac{4}{3}\pi(0.1)^3}{4\pi(0.1)^2}\right)$$

$$t = 0.97\times 10^6\left(\frac{0.1}{3}\right)^2$$

t = 1077.8 sec

t = 1078 sec

11. 1 % shrinkage allowance

$\Rightarrow 200 \times 0.01 = 2$mm

$\Rightarrow 100 \times 0.01 = 1$ mm

$\Rightarrow 10 \times 0.01 = 0.1$ mm

Gray cast iron expands after solidification.

Vp = pattern volume = (200 – 2) × (100 – 1)

Vc = Casting volume = (200 × 100 × 10)

$$\frac{Vp}{Vc} = \frac{(200-2)\times(100-1)\times 999}{200\times 100\times 10}$$

$$\boxed{\frac{V_P}{Vc} = 0.97}$$

12. Height of sprue, h = 20 cm = 0.2m

Area of down sprue (Ingate) = 1m^2

= 1×10^{-4} m^2

Volume of casting = 1000 cm^3

(Mould cavity)

Velocity of sprue base

$V = \sqrt{2gh}$

$= \sqrt{2\times 9.81\times 0.2} = 1.98$m/s

10–3 = 10–4 × 1.98 × t

$$t = \frac{10^{-3} \times 10^4}{1.98}$$

$$\boxed{t = 5.05\,\text{sec}}$$

13. Top gate

$$t_{f1} = \frac{A \times H}{Ag\sqrt{2gh_m}}$$

$$h_m = H$$

$$t_{f1} = \frac{A \times H}{Ag\sqrt{2gh_m}} = \frac{A\sqrt{h_m}}{Ag\sqrt{2g}} \qquad .....(1)$$

Bottom gate

$$t_{f_2} = \frac{2A}{Ag\sqrt{2g}}\left(\sqrt{h_m} - \sqrt{h_m - H}\right)$$

$$t_{f_2} = \frac{2A}{Ag\sqrt{2g}}\sqrt{h_m} \qquad .....(2)$$

$$\frac{t_{f_2}}{t_{f_1}} = 2 \quad \boxed{t_{f_2} = 2t_{f_1}}$$

14. $V_{cube} = V_{sphere}$

$$\text{Solidification time} = \left(\frac{V}{A}\right)^2,$$

where, A = surface area

$$\text{Required ratio} = \left(\frac{\frac{V_{cube}}{A_{cube}}}{\frac{V_{sphere}}{A_{sphere}}}\right)^2 = \left(\frac{A_{sphere}}{A_{cube}}\right)^2$$

$$= \left(\frac{4\pi r^2}{6l^2}\right)^2 = \left(\frac{4\pi}{6}\right)^2\left(\frac{r}{l}\right)^4$$

15.

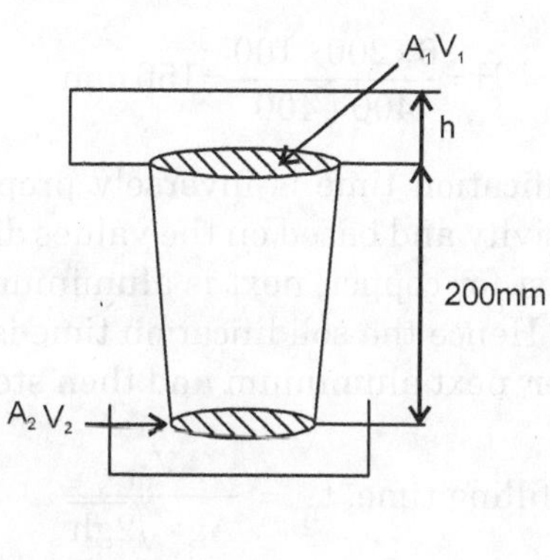

Here,

$$\text{Flow rate} = 6.5 \times 10^5\ \text{mm}^3/\text{s}$$

$$A_1V_1 = 6.5 \times 10^5$$

$$V_1 = 1000\ \text{mm/s} = \sqrt{2g \times h}$$

$$h = 50\ \text{mm}$$

At down, $V_2 = \sqrt{2g(200+50)}$

$$= 2236.06\ \text{mm/s}$$

$$A_2V_2 = 6.5 \times 10^5$$

$$\therefore \quad A_2 = 290.68\ \text{mm}^2$$

16. P – 4, Q – 1, R –2, S –3

17.

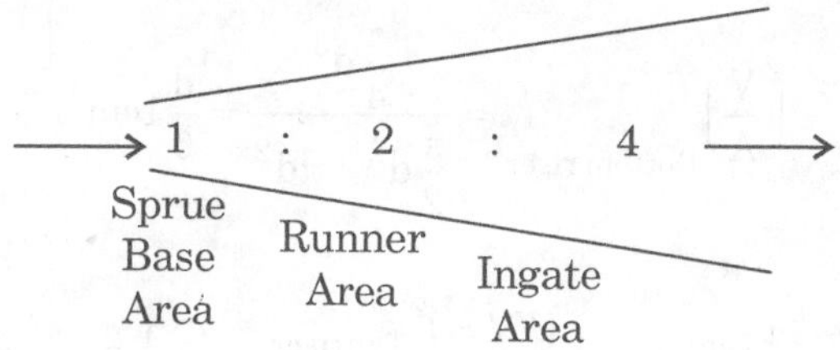

18. Side after

19. $t\,\alpha\,(V/A)^2$ and $m_2 = 8m_1$,

$$\rho V_2 = 8\rho V_1$$

$$\Rightarrow V_2 = 8V_1$$

$$a_2 = 2a_1 \Rightarrow t_2 = 4t_1$$

20.

$$v = \frac{\pi}{4}D^2h$$

$$h = \frac{4v}{\pi D^2} \qquad ...(1)$$

$$A_s = \pi Dh + \frac{\pi}{4}D^2$$

$$= \pi D\frac{4v}{\pi D^2} + \frac{\pi}{4}D^2 = \frac{4v}{D} + \frac{\pi}{4}D^2$$

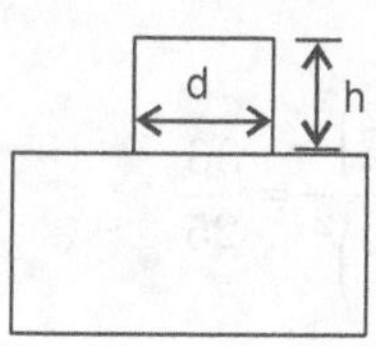

For min A_s,

$$\frac{dA_s}{dD} = 0$$

$$\frac{-4v}{D^2} + \frac{\pi}{4}2D = 0 \Rightarrow \frac{4v}{D^2} = \frac{d\pi}{2}$$

$$v = \frac{\pi D^3}{8} = \frac{\pi}{4}D^2h$$

$$D = 2h$$

$$\Rightarrow \frac{h}{D} = \frac{1}{2}$$

$$h : D = 1 : 2$$

21. $\left(\frac{V}{A}\right)_{casting} = \frac{7 \times 10 \times 2}{2(7 \times 10 + 7 \times 2 + 2 \times 10)}$

$= 0.673$ cm

$$\left(\frac{V}{A}\right)_{Top\,riser} = \frac{\frac{\pi}{4}d^3}{\frac{\pi}{4}d^2 + \pi d^2} = \frac{d}{S} = \frac{6}{5} = 1.25$$

$$\left(\frac{V}{A}\right)_{Bottom\,riser} = \frac{\frac{\pi}{4}d^3}{\frac{\pi}{2}d^2 + \pi d^2} = \frac{d}{6} 1\text{cm}$$

$$\frac{t_{Top\,riser}}{t_{casting}} = \frac{(V/A)^2_{Top\,riser}}{(V/A)^2_{casting}} = \left(\frac{1.2}{0.673}\right)^2$$

$$t_{top\,riser} = 4.133 \text{ min}$$

$$\frac{t_{Bottom\,riser}}{t_{casting}} = \frac{(V/A)_{Bottom\,riser}}{(V/A)_{casting}} = \left(\frac{1}{0.673}\right)^2$$

$$t_{Bottom\,riser} = 3 \text{ min}$$

22. $t = K\left(\frac{V}{A}\right)^2$

$$\frac{t_1}{t_2} = \frac{K\left(\frac{V_1}{A_1}\right)^2}{K\left(\frac{V_2}{A_2}\right)^2}$$

$$\frac{t_1}{t_2} = \frac{A_2^2}{A_1^2} = \frac{(6_a^2)^2}{(5_a^2)^2} = \frac{36}{25}$$

23. $t_s = k\left(\frac{V}{A}\right)^2$

$$t_{sphere} = t_{cube}\frac{A^2_{cube}}{A^2_{sphere}} \quad ...(i)$$

$$\frac{4}{3}\pi r^3 = 1000 \text{ cm}^3$$

$\therefore \quad r = 6.2035$ cm

$$A_{sphere} = 4\pi r^2 = 483.597 \text{ cm}^2$$

$$a^3 = 1000 \text{ cm}^3$$

$\therefore \quad a = 10$ cm

$$A_{cube} = 6a^2 = 600 \text{ cm}^2$$

From equation (*i*),

$$\therefore \quad t_{sphere} = 4 \times \frac{(600)^2}{(483.597)^2} = 6.157 \text{ sec}$$

25. $Z = \frac{L + W}{\text{Thickness}}$

$$= \frac{25 + 15}{5} = 8$$

$$\frac{\text{Riser volume}}{\text{Casting volume}} = 0.5$$

$$\frac{\frac{\pi}{4}d^2h}{25 \times 15 \times 5} = 0.5$$

but d = h

$$\frac{\pi d^3}{4} = 0.5 \times 25 \times 15 \times 5$$

$$\boxed{d = 10.60\text{cm}}$$

26. Rate of pouring of molten metal depends on pouring basin.

27. According to modulus method

$$M_R = 1.2\, M_C$$

$$\left[\frac{V}{As}\right]_R = 1.2\left[\frac{V}{As}\right]_C$$

If diameter of riser = height of riser for top riser D = H

$\Rightarrow D = 6\, M_C$

$$D = 6 \times \frac{\frac{\pi}{4} \times (200)^2 \times 100}{2 \times \frac{\pi}{4} \times (200)^2 + \pi \times 200 \times 100}$$

$$D = H = \frac{6 \times 200 \times 100}{400 + 400} = 150 \text{ mm}$$

28. Solidification time is inversely proportional to diffusivity and based on the values diffusivity is highest for copper, next is aluminium and then steel. Hence the solidification time is lowest for copper, next aluminium and then steel.

30. Mold filling time, $t_m = \frac{V_m}{Ag \cdot \sqrt{2gh}}$

$$= \frac{40 \times 30 \times 15}{\frac{\pi(2)^2}{4} \times \sqrt{2 \times 9.81 \times 100 \times 20}}$$

$$\boxed{t = 28.923 \text{ sec.}} \approx 28.92 s$$

31. $V = \sqrt{2gh \times 0.8}$

$= \sqrt{2 \times 9.81 \times 0.3 \times 0.8}$

$= 2.17$ m/s

33. Higher the fluidity, flow will be easier

34. Riser is used to compensate liquid shrinkage & phase transition shrinkage.

36. A – 4 , B – 5, C – 3, D – 1

37. $t = k\left(\frac{V}{A}\right)^2$

n = 2

38. $V_s = \sqrt{2 \times 10 \times 175 \times 10^{-3}}$

= 1870 mm/s

Pouring time

$$t = \frac{10^6}{200 \times 1870}$$

$$\boxed{t = 2.67 m/s}$$

39. $PN = \frac{VH}{pAT} = \frac{2000 \times 5.08}{10 \times \frac{\pi}{4}(5.08)^2 \times 1}$

$$\boxed{PN = 50.12}$$

42. Spruce base : Runner : Ingates

44. Wood flour get band out and provide easily removal of casting.

45. Shape factor $= \frac{L + W}{\text{Thickness}}$

$$W = \pi D_{mean} = \pi\left(\frac{30+20}{2}\right)$$

= 78.53 cm

$$t = \frac{30-20}{2} = 5cm$$

$$\text{Shape factor} = \frac{30 + 78.53}{5} = 21.72$$

47.

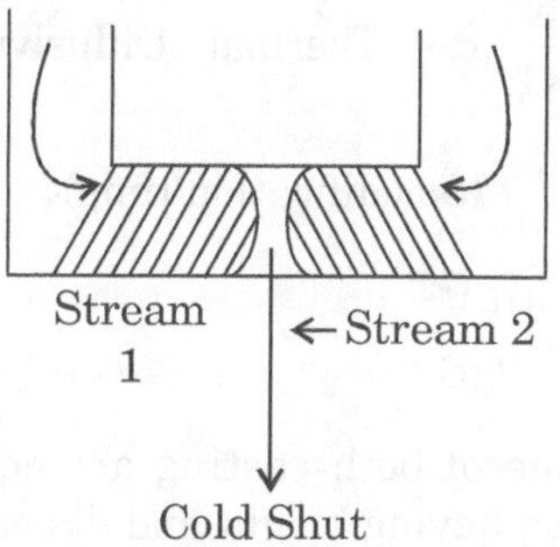

48. D = 10mm, h_s = 250 mm

$$V_s = \sqrt{2gh_s} = \sqrt{2 \times 9.81 \times 250 \times 10^3}$$

$$A_s = \frac{\pi}{4} \times 10^2$$

78.53 mm²

Flowrate = $V_s \cdot A_s = 1.39 \times 10^5$ mm³/s

49. Mould filling time $= \frac{\text{Volume}}{\text{Flow rate}}$

$$= \frac{100 \times 100 \times 100}{1.739 \times 10^5}$$

= 5.75 sec

50. Net Buoyancy force, F

$$F = (11300 - 1600) \times 9.81 \times \frac{\pi}{4} \times (120)^2 \times 180 \times 10^{-3}$$

= 193.71 N

51. $t = k\left(\frac{V}{A}\right)^2$

cube $V = L^3$ $A = 6L^2$

$$t_1 = k\left(\frac{L}{6}\right)^2$$

cylinder

$$V = \frac{\pi}{4}D^3$$

$$A = \frac{\pi}{2}D^2 + \pi D^2 = \frac{3\pi D^2}{2}$$

$$t_2 = k\left(\frac{P}{6}\right)^2$$

$$\frac{t_2}{t_1} = \left(\frac{P}{L}\right)^2$$

52. $\alpha = \frac{K}{\rho C_p}, \alpha \rightarrow$ Thermal Diffusivity

as $\alpha \uparrow$ t (Solodification time)$\downarrow$

53. $h_L = \frac{4fLV^2}{2gD}$

54. Volume of both casting are equal & cylinder casting having height and dia equal

$$\frac{\pi}{4}D_c^3 = \frac{\pi}{6} \times D_s^3$$

$$\boxed{Ds = 1.14Dc}$$

55. $T_1 = \frac{\text{Vol. of cavity}}{\text{flowrate}} = \frac{V}{A\sqrt{2gh}}$

$$T_2 = \frac{V}{\frac{A}{2} \times \sqrt{2g \times 4b}}$$

$$\boxed{\frac{T_1}{T_2} = 1}$$

56.

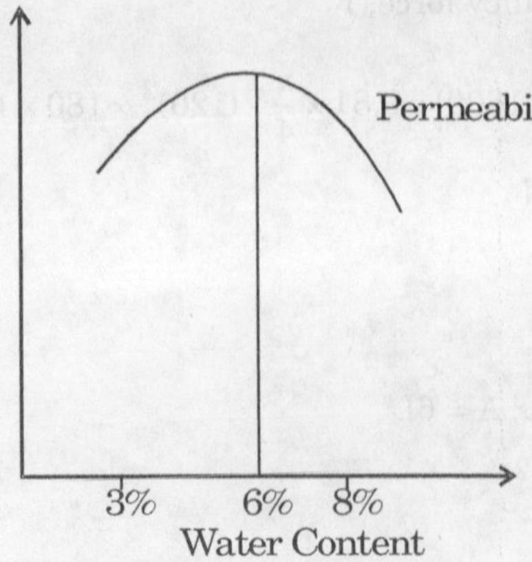

57. Sand will go to last corners of casting

58. A – 4, B – 3, C –2, D – 5

59. Due to faster cooling rate of casting

60. A – 2, B – 3, C – 2, D – 5

61. A – 5, B – 6, C – 2, D – 3

62. Brass

63. Thermosetting plastic

64. Liquid shrinkage is compensated by riser.

65. At the inner surface of the casting.

68. Expendable pattern is used in Investment casting.

69. For hot chamber die casting, titanium is used.

70. The riser can compensate for volume shrinkage only in the liquid stage and transition stage and not in the solid state.

Hence volume of metal that needs to be compensated from the riser = 3 + 4 = 7%

73. P – 3, Q – 4, R – 2, S – 1

74. The pattern is made of expanded polystyrene.

75. Ratio of volume of pattern to casting

$$\frac{[(0.97 \times 100)(0.97 \times 50)(0.97 \times 10)]}{(100 \times 50 \times 10)} = 0.913$$

Volume of cavity is assumed to be small and negligible.

76.

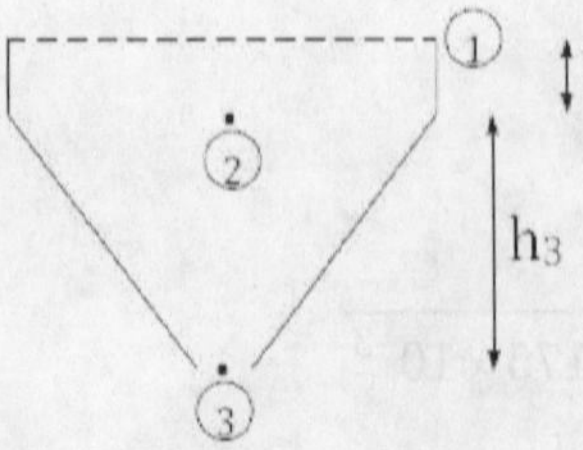

We know that,

$$V_2 = \sqrt{2gh_2} = \sqrt{2 \times 9.81 \times h_2} = 0.5$$

$$h_2 = 0.01274 \text{ m}$$

$$h_3 = h_2 + 0.2 = 0.2127 \text{ m}$$

$$V_3 = \sqrt{2gh_3} = 2.043 \text{ m/sec}$$

78. Allowance for ramming

79. Hollow sections made by this

80. Powder metallurgy reduce porosity

85. A – lower B – higher

86. A. wax B. intricate

88. Bentonite is an additive.

89. False

90. P – 3, a –4, R –2, S1

102. (T)

103. (T)

104. A – 1, B – 5, C – 4, D – 2

105.

3 CHAPTER

Metal Forming

Analysis of Previous GATE Papers														
Topic	**Year/Set → Marks Distribution ↓**		**2019 Set 1**	**2019 Set 2**	**2018 Set 1**	**2018 Set 2**	**2017 Set 1**	**2017 Set 2**	**2016 Set 1**	**2016 Set 2**	**2016 Set 3**	**2015 Set 1**	**2015 Set 2**	**2015 Set 3**
Rolling	**1 Mark**	MCQ Type		1										
		Numerical Type												
	2 Marks	MCQ Type												
		Numerical Type		1				1				1		1
	Total			**3**				**2**				**2**		**2**
Forging	**1 Mark**	MCQ Type			1	1		1			1	1		1
		Numerical Type												
	2 Marks	MCQ Type												
		Numerical Type					1						1	
	Total				**1**	**1**	**2**	**1**			**1**	**1**	**2**	
Extrusion and Wire Drawing	**1 Mark**	MCQ Type												
		Numerical Type			1									
	2 Marks	MCQ Type											1	
		Numerical Type			1	2								
	Total				**3**	**4**							**2**	

Rolling

1. Calculate the bite angle when rolling plates 12 mm thick using work rolls 600 mm diameter and reducing the thickness by 3 mm

[ME 1994 : 2 Marks]

2. A wire of 0.1 mm dia is drawn from a rod of 15 mm diameter dies giving reductions of 20%, 40% and 80% are available. For minimum error in the final size, the number of stages and reduction at each stage respectively would be

(a) 3 stages and 80% reduction for all three stages

(b) 4 stages and 80% reduction for first three stages followed by a finishing stage of 20% reduction

(c) 5 stage and reduction of 80%, 80%, 40%, 40%, 20% is sequence

(d) None of these

[ME 1996 : 2 Marks]

3. A strip with cross-sectional area 150 mm × 4.5 mm is being rolled with 20% reduction of area using 450 mm diameter rollers. The angle subtended by the deformation zone at the roll centre is (in radians)

(a) 0.01 (b) 0.02
(c) 0.03 (d) 0.06

[ME 1998 : 2 Marks]

4. Hot rolling of mild steel is carried out

(a) at recrystallization temperature

(b) between 100°C to 150°C

(c) below recrystallization temperature

(d) above recrystallization temperature

[ME 2002 : 1 Mark]

5. In a rolling process, sheet of 25 mm thickness is rolled to 20 mm thickness. Roll is of diameter 600 mm and it rotates at 100 rpm. The roll strip contact length will be

(a) 5 mm (b) 39 mm
(c) 78 mm (d) 120 mm

[ME 2004 : 2 Marks]

6. A 4 mm thick sheet is rolled with 300 mm diameter rolls to reduce thickness without any change in its width. The friction coefficient at the work-roll interface is 0.1. The minimum possible thickness of the sheet that can be produced in a- single pass is

(a) 1.0 mm (b) 1.5 mm
(c) 2.5 mm (d) 3.7 mm

[ME 2006 : 2 Marks]

7. The thickness of a metallic sheet is reduced from an initial value of 16 mm to a final value of 10 mm in one single pass rolling with a pair of cylindrical rollers each of diameter of 400 mm. The bite angle in degree will be

(a) 5.936 (b) 7.936
(c) 8.936 (d) 9.936

[ME 2007 : 2 Marks]

8. The maximum possible draft in cold rolling of sheet increases with the

(a) increase in coefficient of friction

(b) decrease in coefficient of friction

(c) decrease in roll radius

(d) increase in roll velocity

[ME 2011 : 1 Mark]

9. A solid cylinder of diameter 100 mm and height 50 mm is forged between two frictionless flat dies to a height of 25 mm. The percentage change in diameter is

(a) 0 (b) 2.07
(c) 20.7 (d) 41.4

[ME 2012 : 1 Mark]

10. In a single pass rolling process using 410 mm diameter steel rollers, a strip of width 140 mm and thickness 8 mm undergoes 10% reduction of thickness. The angle of bite in radians is

(a) 0.006 (b) 0.031
(c) 0.062 (d) 0.600

[ME 2012 : 2 Marks]

11. In a rolling process, the state of stress of the material undergoing deformation is

(a) pure compression

(b) pure shear

(c) compression and shear

(d) tension and shear **[ME 2013 : 1 Mark]**

12. With respect to metal working, match Group A with Group B:

Group A

P. Defect in extrusion
Q. Defect in rolling
R. Product of skew rolling
S. Product of rolling through cluster mill

Group B

I. alligatoring
II. scab
III. fish tail
IV. seamless tube
V. thin sheet with tight tolerance
VI. semi-finished balls of ball bearing

(a) P-II, Q-III, R-VI, S-V
(b) P-III, Q-I, R-VI, S-V
(c) P-III, Q-I, R-IV, S-VI
(d) P-I, Q-II, R-V, S-VI

[ME 2014 : 1 Mark, Set-2]

13. In a slab rolling operation, the maximum thickness reduction (Δh_{max}) is given by $\Delta h_{max} = \mu^2 R$, where R is the radius of the roll and μ is the coefficient of friction between the roll and the sheet. If μ = 0.1, the maximum angle subtended by the deformation zone at the centre of the roll (bite angle in degree) is _____.

[ME 2015 : 2 Marks, Set-1]

14. In a rolling operation using rolls of diameter 500 mm if a 25 mm thick plate cannot be reduced to less than 20 mm in one pass, the coefficient of friction between the roll and the plate is _____.

[ME 2015 : 2 Marks, Set-3]

15. A strip of 120 mm width and 8 mm thickness is rolled between two 300 mm diameter rolls to get a strip of 120 mm width and 7.2 mm thickness. The speed of the strip at the exit is 30 m/min. There is no front or back tension. Assuming uniform roll pressure of 200 MPa in the roll bite and 100% mechanical efficiency, the minimum total power (in kW) required to drive the two rolls is _______.

[ME 2017 : 2 Marks, Set-2]

16. The cold forming process in which a hardened tool is pressed against a workpiece (when there is relative motion between the tool and the workpiece) to produce a roughened surface with a regular pattern is

(a) Roll forming
(b) Knurling
(c) Chamfering
(d) Stip rolling

[ME 2019 : 1 Mark, Set-2]

17. The thickness of a sheet is reduced by rolling (without any change in width) using 600 mm diameter rolls. Neglect elastic deflection of the rolls and assume that the coefficient of friction at the roll-workpiece interface is 0.05. The sheet enters the rotating rolls unaided. If the initial sheet thickness is 2 mm, the minimum possible final thickness that can be produced by this process in a single pass is ____ mm (round off to two decimal places).

[ME 2019 : 2 Marks, Set-2]

18. The blank diameter used in thread rolling will be

(a) Equal to the major diameter of the thread
(b) Equal to the pitch diameter of the thread
(c) Little higher than the minor diameter of the thread
(d) Little higher than the pitch diameter of the thread

[PI 1989 : 1 Mark]

19. While rolling a strip the peripheral velocity of the roll is .... A .... than the entry velocity of the strip and is ... B... the exit velocity of the strip.

(a) less than/greater than
(b) greater than/less than/equal to

[PI 1990 : 1 Mark]

20. Thread rolling is restricted to

(a) ferrous materials
(b) ductile materials
(c) hard materials
(d) None of these

[PI 1992 : 1 Mark]

21. If the elongation factor during rolling of an ingot is 1.22. The minimum number of passes needed to produce a section 250 x 250 mm from an ingot of 750 × 750 mm are

(a) 8 (b) 9
(c) 10 (d) 12

[PI 1992 : 2 Marks]

22. In order to reduce roll pressure in strip rolling, back tension can be applied to strip. (T/F)

[PI 1994 : 1 Mark]

23. A copper strip of 200 mm width and 300 mm thickness is to be rolled to a thickness of 295 mm. The roll of radius 300 mm rotates at 100 rpm. The average shear strength of the work material is 180 MPa. The roll strip contact length and the roll force are

(a) 15.8 mm and 0.569 MN
(b) 18.97 mm 0.683 MN
(c) 38.73 mm and 1.395 MN
(d) 38.73 mm and 2.09 MN

[PI 2006 : 2 Marks]

24. The maximum possible percentage reduction in area per pass during wire drawing of an ideal plastic material without friction is of the order of

(a) 37 (b) 50
(c) 63 (d) 75

[PI 2007 : 2 Marks]

25. In a rolling process, thickness of a strip is reduced from 4 mm to 3 mm using 300 mm diameter rolls rotating at 100 rpm. The velocity of the strip (in m/sec) at the neutral point is

(a) 1.57 (b) 3.14

(c) 47.10 (d) 94.20

[PI 2008 : 2 Marks]

26. Anisotropy in rolled components is caused by

(a) Change in dimensions

(b) Scale formation

(c) Closure of defects

(d) Grain orientation

[PI 2009 : 1 Mark]

27. In a rolling process, the roll separating force can be decreased by

(a) Reducing the roller diameter

(b) Increasing friction between the rolls and the metal

(c) Reducing front tension to rolled material

(d) Providing back up rollers

[PI 2010 : 2 Marks]

28. The thickness of a plate is reduced from 30 mm to 10 mm by successive cold rolling passes using identical rolls of diameter 600 mm. Assume that there is no change in width. If the coefficient of friction between the rolls and the workpiece is 0.1, the minimum number of passes required is

(a) 3 (b) 4

(c) 6 (d) 7

[PI 2011 : 2 Marks]

29. In a single pass rolling process using 410 mm diameter steel rollers, a strip of width 140 mm and thickness 8 mm undergoes 10% reduction of thickness. The angle of bite in radians is

(a) 0.006

(b) 0.031

(c) 0.062

(d) 0.600

[PI 2012 : 2 Marks]

30. In a rolling process, the state of stress of the material undergoing deformation is

(a) pure compression

(b) pure shear

(c) compression and shear

(d) tension and shear

[PI 2013 : 1 Mark]

Forging

31. In forging operation the sticking friction condition occurs near the ... (centre/ends)

[ME 1987 : 1 Mark]

32. The true sliding strain for a low carbon Steel bar which is doubled in length by forging is

(a) 0.307 (b) 0.5

(c) 0.693 (d) 1.0

[ME 1992 : 1 Mark]

33. A test specimen is stressed slightly beyond the yield point and then unloaded. Its yield strength

(a) Decreases (b) Increases

(c) remains same (d) become equal to UTS

[ME 1995 : 1 Mark]

34. Match the following:

List-I	List-II
A. Rivets for aircraft body	1. Forging
B. Carburetor body	2. Cold heading
C. Crank shafts	3. Aluminum base alloy
D. Nails	4. Pressure die casting
	5. Investment casting

[ME 1996 : 2 Marks]

35. Ductility of material with work hardening

(a) Increases (b) Decreases

(c) Remains same (d) Unpredictable

[ME 2002 : 1 Mark]

36. Match the following

Product

P. Moulded luggage

Q. Packaging containers for liquid

R. Long structural shapes

S. Collapsible tubes

Process

1. Injection moulding
2. Hot rolling
3. Impact extrusion
4. Transfer moulding
5. Blow moulding
6. Coining

(a) P-1, Q-4, R-6, S-3

(b) P-4, Q-5, R-2, S-3

(c) P-1, Q-5, R-3, S-2

(d) P-5, Q-1, R-2, S-2

[ME 2004 : 2 Marks]

37. A 2 mm thick metal sheet is to be bent at an angle of one radian with a bend radius of 100 mm. If the stretch factor is 0.5, the bend allowance is

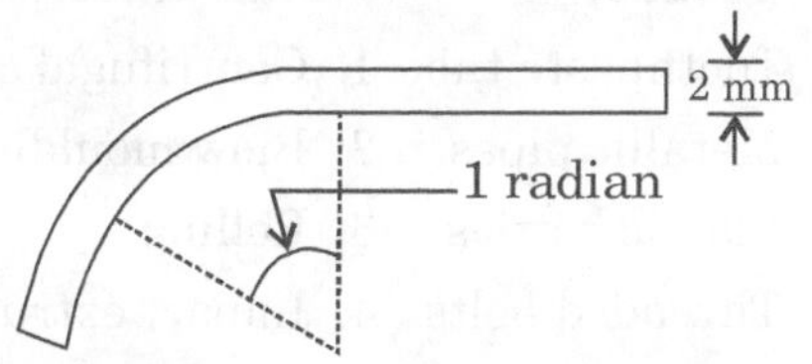

(a) 99 mm (b) 100 mm
(c) 101 mm (d) 102 mm

[ME 2005 : 2 Marks]

38. Match the items in columns I and II

Column-I

P. Wrinkling
Q. Orange peel
R. Stretcher strains
S. Earing

Column-II

1. Yield point elongation
2. Anisotropy
3. Large grain
4. Insufficient blank holding force
5. Fine grain size
6. Excessive blank holding force

(a) P-6, Q-3, R-1, S-2
(b) P-4, Q-5, R-6, S-1
(c) P-2, Q-5, R-3, S-1
(d) P-4, Q-3, R-1, S-2 **[ME 2006 : 2 Marks]**

39. In open-die forging, a disc of diameter 200 mm and height 60 mm is compressed without any barreling effect. The final diameter of the disc is 400 mm. The true strain is

(a) 1.986 (b) 1.686
(c) 1.386 (d) 0.602

[ME 2007 : 2 Marks]

40. Match the correct combination for following metal working processes.

Processes	Associated state of stress
P Blanking	1. Tension
Q. Stretch Forming	2. Compression
R. Coining	3. Shear
S. Deep Drawing	4. Tension and Compression
	5. Tension and shear

(a) P-2, Q-1, R-3, S-4
(b) P-3, Q-4, R-1, S-5
(c) P-5, Q-3, R-3, S-1
(d) P-3, Q-1, R-2, S-4 **[ME 2007 : 2 Marks]**

41. The force requirement in a blanking operation of low carbon steel sheet is 5.0 kN. The thickness of the sheet is 't' and diameter of the blanked part is 'd'. For the same work material, if the diameter of the blanked part is increased to 1.5d and thickness is reduced to 0.4t, the new blanking force in kN is

(a) 3.0 (b) 4.5
(c) 6.7 (d) 8.0

[ME 2007 : 2 Marks]

42. The operation in which oil is permeated into the pores of a powder metallurgy product is known as

(a) mixing (b) sintering
(c) impregnation (d) infiltration

[ME 2011 : 1 Mark]

43. Match the following metal forming processes with their associated stresses in the workpiece

Metal forming process

A. Coining
B. Wire Drawing
C. Blanking
D. Deep Drawing

Type of stress

1. Tensile
2. Shear
3. Tensile and compressive
4. Compressive Codes:

	A	B	C	D
(a)	4	1	2	3
(b)	4	1	3	2
(c)	1	2	4	3
(d)	1	3	2	4

[ME 2012 : 1 Mark]

44. The relationship between true strain (ε_T) and engineering strain (ε_E) in a uniaxial tension test is given as

(a) $\varepsilon_E = \ln(1 + \varepsilon_T)$ (b) $\varepsilon_E = \ln(1 - \varepsilon_T)$

(c) $\varepsilon_T = \ln(1 + \varepsilon_E)$ (d) $\varepsilon_T = \ln(1 - \varepsilon_E)$

[ME 2014 : 1 Mark, Set-2]

45. Match the following products with preferred manufacturing processes:

Product	Process
P. Rails	1. Blow molding
Q. Engine crankshaft	2. Extrusion
R. Aluminium channels	3. Forging
S. PET water bottles	4. Rolling

(a) P-4, Q-3, R-1, S-2
(b) P-4, Q-3, R-2, S-1
(c) P-2, Q-4, R-3, S-1
(d) P-3, Q-4, R-2, S-1

[ME 2015 : 1 Mark, Set-1]

46. The flow stress (in MPa) of a material is given by

$\sigma = 500\,\varepsilon^{0.1}$

where ε is true strain. The Young's modulus of elasticity of the material is 200 GPa. A block of thickness 100 mm made of this material is compressed to 95 mm thickness and then the load is removed. The final dimension of the block (in mm) is _____.

[ME 2015 : 2 Marks, Set-2]

47. The strain hardening exponent n of stainless steel SS 304 with distinct yield and UTS values undergoing plastic deformation is

(a) $n < 0$ (b) $n = 0$
(c) $0 < n < 1$ (d) $n = 1$

[ME 2015 : 1 Mark, Set-3]

48. Engineering strain of a mild steel sample is recorded as 0.100%. The true strain is

(a) 0.010% (b) 0.055%
(c) 0.099% (d) 0.101%

[ME 2016 : 1 Mark, Set-3]

49. A 10 mm deep cylindrical cup with diameter of 15 mm is drawn from a circular blank. Neglecting the variation in the sheet thickness, the diameter (upto 2 decimal points accuracy) of the blank is mm.

[ME 2017 : 2 Marks, Set-1]

50. It is desired to make a product having T-shaped cross-section from a rectangular aluminium block. Which one of the following processes is expected to provide the highest strength of the product?

(a) Welding (b) Casting
(c) Metal forming (d) Machining

[ME 2017 : 1 Mark, Set-2]

51. In a linearly hardening plastic material, the true stress beyond initial yielding

(a) increases linearly with the true strain
(b) decreases linearly with the true strain
(c) first increases linearly and then decreases linearly with the true strain
(d) remain constant

[ME 2018 : 1 Marks, Set-1]

52. Match the following products with the suitable manufacturing process

Product	Manufacturing Process
P. Toothpaste tube	1. Centrifugal casting
Q. Metallic pipes	2. Blow moulding
R. Plastic bottles	3. Rolling
S. Threaded bolts	4. Impact extrusion

(a) P-4,Q-3,R-1,S-2
(b) P-2, Q-1, R-3, S-4
(c) P-4, Q-1, R-2, S-3
(d) P-1, Q-3, R-4, S-2

[ME 2018 : 1 Marks, Set-2]

53. At the last hammer stroke the excess material from the finishing cavity of a forging die is pushed into _______.

[PI 1989 : 1 Mark]

54. At 1000°C the crystallographic structure of iron is _______.

[PI 1991 : 1 Mark]

55. To stress relieve a cold worked steel part, it is heated to a temperature close to .... A ... and grain obtained after cooling are ... B.

A. lower critical temperature/upper critical temp
B. fine/coarser

[PI 1992 : 2 Marks]

56. Production process application

List-I

A. Extrusion
B. Hot forging
C. Metal spinning
D. Explosive welding

List-II

1. Cladding of noble metal to base metal
2. Long continuous metal tubes
3. Connecting rod of IC engine
4. Long flanged pipes
5. House hold utensils.

[PI 1992 : 2 Marks]

57. Which one of the following manufacturing processes requires the provision of 'gutters'?

(a) closed die forging
(b) centrifugal casting
(c) investment casting
(d) impact extrusion

[PI 1994 : 1 Mark]

58. An annealed copper wire of 25 mm diameter is drawn into a wire of 5 mm diameter. The average yield stress in this operation if the curve of the material is given, $\sigma = 315\ \varepsilon^{0.54}$ MPa

(a) 592 MPa (b) 458 MPa
(c) 342 MPa (d) None of these

[PI 1994 : 2 Marks]

59. Cold working produces the following effects

1. Stresses are set up in the material
2. Grain structure get distorted
3. Strength and hardness of the metal are increased
4. Surface finish is reduced

(a) 1 and 2 (b) 1, 2 and 3
(c) 3 and 4 (d) 1 and 4

[PI 2003 : 2 Marks]

60. Consider the following steps in forging a connecting rod from the bar stock

1. blocking 2. trimming
3. finishing 4. edging

Select the correct sequence of these operations using the codes given below:

(a) 1-2-3-4 (b) 2-3-4-1
(c) 3-4-1-2 (d) 4-1-3-2

[PI 2003 : 2 Marks]

61. Cold shut (lap) may occur in products obtained by

(a) Casting (b) Forging
(c) Machining (d) Welding

[PI 2005 : 1 Mark]

62. Flash and gutter are provided in drop forging dies at the following stage:

(a) Blocking
(b) Preforming (Edging)
(c) Finishing
(d) Fullering

[PI 2005 : 1 Mark]

63. A lever having 90° bend is to be produced by drop forging using mild steel bar as raw material.

The various operations to be performed on it during forging are:

P-Cutting;
Q-Bending:
R-Fullering:
S-Blocking cum finishing;
T-Edging

The correct sequence for performing the operation is

(a) PQRS (b) RTQSP
(c) TRSQP (d) RQTSP

[PI 2005 : 2 Marks]

64. The true stress true curve is given by $\sigma = 1400\ \varepsilon^{0.33}$, where the stress σ is in MPa. The true stress maximum load (in MPa) is

(a) 971 (b) 750
(c) 698 (d) 350

[PI 2005 : 2 Marks]

65. Consider the following statements:

In comparison to hot working, in cold working,

1. higher forces are required
2. no heating is required
3. less ductility is required
4. better surface finish is obtained.

Which of the statements given above are correct?

(a) 1,2 and 3
(b) 1, 2 and 4
(c) 1 and 3
(d) 2, 3 and 4

[PI 2005 : 2 Marks]

66. Assertion (A): Cold working of metals results in increase of strength and hardness.

Reason (R): Cold working reduce the total number of dislocations per unit volume of the material

(a) Both A and R are true and R is the correct explanation of A
(b) Both A and R are true but R is not a correct explanation of A
(c) A is true but R is false
(d) A is false but R is true

[PI 2005 : 2 Marks]

67. A solid cylindrical stainless steel work piece of 200 mm diameter and 150 mm height. This component is reduced by 50% in height with flat die in open die forging. Assuming the flow stress of the material as 1000 MPa and the coefficient of friction to be 0.2 , the estimated forging force at the end of the stroke is

(a) 20.8 kN
(b) 31 kN
(c) 78.6 kN
(d) 78.6 MN

[PI 2006 : 2 Marks]

68. By application of tensile force, the cross-sectional area of bar P is first reduced by 30% and then by an additional 20%. Another bar Q of the same material is reduced in cross-sectional area by 50% in a single step by applying tensile force.

After deformation, the true strain in bar P and bar Q will, respectively be

(a) 0.5 and 0.5 (b) 0.58 and 0.69

(c) 0.69 and 0.69 (d) 0.78 and 1.00

[PI 2008 : 2 Marks]

69. Hot die steel, used for large solid dies in drop forging, should necessarily have

(a) high strength and high copper content

(b) high hardness and low hardenability

(c) high toughness and low thermal conductivity

(d) high hardness and high thermal conductivity

[PI 2010 : 1 Mark]

70. During open die forging process using two flat and parallel dies, a solid steel disc of initial radius (R_{IN}) 200 mm and initial height (H_{IN}) 50 mm attains a height (H_{FN}) of 30 mm and radius of R_{FN},

Along the die-disc interfaces,

(i) The coefficient of friction (μ) is:

$\mu = 0.35\,[1 + e^{-R_{IN}}/R_{FN}]$

(ii) In the region $R_{ss} \le r \le R_{FN}$ slIding friction prevails and

$P = \sqrt{3}\,K.e^{2\mu(R_{IN-r})/H_{FN}}$ and $\tau = \mu p$, sllding friction prevails and

Where p and τ are the normal and the shear stresses respectively; K is the shear yield strength of steel and r is the radial distance of any point.

(i) In the region $0 \le r \le R_{IN}$, sticking condition prevails.

The value of R_{ss} (in mm), where sticking condition changes to sliding friction is

(a) 241.76 (b) 254.55

(c) 265.45 (d) 278.20

[PI 2010 : 2 Marks]

Extrusion and Wire Drawing

71. For wire drawing operation, the work material should essentially be

(a) ductile (b) tough

(c) hard (d) malleable

[ME 1987 : 1 Mark]

72. The process of hot extrusion is used to produce

(a) Curtain rods made of aluminium

(b) Steel pipes of domestic water supply

(c) Stainless steel tubes used in furniture

(d) Large size pipes used in city water mains

[ME 1994 : 1 Mark]

73. A brass billet is to be extruded from its initial diameter of 100 mm to a final diameter of 50 mm. The working temperature of 700°C and the extrusion constant is 250 MPa. The force required for extrusion is

(a) 5.44 MN (b) 2.72 MN

(c) 1.36 MN (d) 0.36 MN

[ME 2003 : 2 Marks]

74. In a wire drawing operation, diameter of a Steel wire is reduced from 10 mm to 8 mm. The mean flow stress of the material is 400 MPa. The ideal force required for drawing (ignoring friction and redundant work) is

(a) 4.48 kN (b) 8.97 kN

(c) 20.11 kN (d) 31.41 kN

[ME 2006 : 2 Marks]

75. A metal rod of initial length L_0 is subjected to a drawing process. The length of the rod at any instant is given by the expression, $L(t) = L_0(1 + t^2)$, where t is the time in minutes. The true strain rate (in min^{-1}) at the end of one minute is ______.

[ME 2014 : 2 Marks, Set-1]

76. In a two-stage wire drawing operation, the fractional reduction (ratio of change in cross-sectional area to initial cross-sectional area) in the first stage is 0.4. The fractional reduction in the second stage is 0.3. The overall fractional reduction is

(a) 0.24 (b) 0.58

(c) 0.60 (d) 1.00

[ME 2015 : 2 Marks, Set-2]

77. A bar is compressed to half of its original length. The magnitude of true strain produced in the deformed bar is (correct to two decimal places).

[ME 2018 : 2 Marks, Set-1]

78. The maximum reduction in cross-sectional area per pass (R) of a cold wire drawing process is

$R = 1 - e^{-(n+1)}$

where n represents the strain hardening coefficient. For the case of a perfectly plastic material, R is

(a) 0.865 (b) 0.826

(c) 0.777 (d) 0.632

[ME 2018 : 2 Marks, Set-1]

79. The true stress (σ), true strain (ε) diagram of a strain hardening material is shown in figure. First, there is loading up to point A, i.e. up to stress of 500 MPa and strain of 0.5. then from point A, there is unloading up to point B, i.e. to stress of 100 MPa, Given that the Young's modulus E = 200 GPa, the natural strain at point B (ε_B) _____(correct to two decimal places).

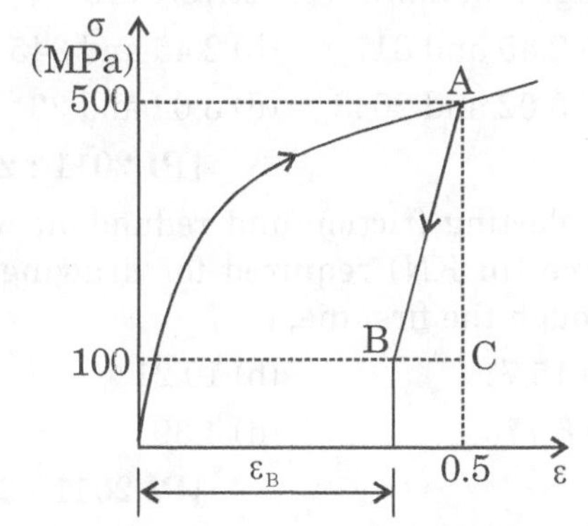

[ME 2018 : 2 Marks, Set-1]

80. A steel wire is drawn from an initial diameter (d_i) of 10 mm to a final diameter (d_f) of 7.5 mm. The half cone angle (α) of the die is 5° and the coefficient of friction (μ) between the die and the wire is 0.1. The average of the initial and final yield stress [$(\sigma_y)_{avg}$] is 350 MPa. The equation for drawing stress σ_f. (in MPa) is given as:

$$\sigma_f = (\sigma_Y)_{avg}\left\{1+\frac{1}{\mu\cot\alpha}\right\}\left[1-\left(\frac{d_f}{d_i}\right)^{2\mu\cot\alpha}\right]$$

The drawing stress (in MPa) required to carry out this operation is (correct to two decimal places).

[ME 2018 : 2 Marks, Set-2]

81. The true stress (in MPa) versus true strain relationship for a metal is given by

$\sigma = 1020\ \sigma^{0.4}$

The cross-sectional area at the start of a test (when the stress and strain values are equal to zero) is 100 mm^2. The cross-sectional area at the time of necking (in mm^2) is _____(correct to two decimal places).

[ME 2018 : 2 Marks, Set-2]

82. Collapsible tubes are made by

(a) Drawing (b) Spinning

(c) Extrusion (d) Rolling

[PI 1989 : 1 Mark]

83. Semi brittle materials can be extruded by

(a) impact extrusion

(b) closed cavity extrusion

(c) hydrostatic extrusion

(d) backward extrusion

[PI 1990 : 1 Mark

84. Seamless long steel tubes are manufactured by rolling, drawing and _________.

[PI 1991 : 1 Mark]

85. Forces in hot extrusion is a function of the strain hardening component of the billet material. (T/F)

[PI 1994 : 1 Mark]

86. A moving mandrel is used in

(a) wire drawing (b) tube drawing

(c) metal cutting (d) forging

[PI 1994 : 1 Mark]

87. The extrusion processes used for the production of toothpaste tubes is/are

1. tube extrusion
2. forward extrusion
3. impact extrusion

(a) 1 only (b) 1 and 2

(c) 2 and 3 (d) 3 only

[PI 2004 : 2 Marks]

88. A round billet made of brass is to be extruded (extrusion constant = 250 MPa) at 700°C. The billet diameter is 100 mm and the diameter of the extrusion is 50 mm. The extrusion force required (in MN) is

(a) 1.932 (b) 2.722

(c) 3.423 (d) 4.650

[PI 2005 : 2 Marks]

Linked Answer Questions 89 and 90

A 10 mm diameter annealed steel wire is drawn through a die at a speed of 0.5 m/s to reduce the diameter by 20%. The yield stress of the material is 800 MPa.

89. Neglecting friction and strain hardening, the stress required for drawing (in MPa) is

(a) 178.6

(b) 357.0

(c) 1287.5

(d) 2575.0

[PI 2008 : 2 Marks]

90. The power required for the drawing process (in kW) is

(a) 8.97

(b) 14.0

(c) 17.95

(d) 28.0

[PI 2008 : 2 Marks]

91. Using direct extrusion process, a round billet of 100 mm length and 50 mm diameter is extruded. Considering an ideal deformation process (no friction and no redundant work) extrusion ratio 4 and average flow stress of material 300 MPa, the pressure (MPa) on the ram will be

(a) 416

(b) 624

(c) 700

(d) 832

[PI 2009 : 2 Marks]

Linked Answer Questions 92 and 93

In a multi-pass operation, a round bar of 10 mm diameter and 100 mm length is reduced in cross-section by drawing it successively through a series of seven dies of decreasing exit diameter. During each of these drawing operations, the reduction in cross-section area is 35%. The yield strength of the material is 200 MPa. Ignore strain hardening.

92. The total true strain applied and the final length (in mm), respectively are

(a) 2.45 and 817 (b) 2.45 and 345

(c) 3.02 and 2043 (d) 3.02 and 33

[PI 2011 : 2 Marks]

93. Neglecting friction and redundant work, the force (in KN) required for drawing the bar though the first die, is

(a) 15.71 (b) 10.21

(c) 6.77 (d) 4.39

[PI 2011 : 2 Marks]

ANSWERS

2. (b) **3.** (d) **4.** (d) **5.** (b) **6.** (c) **7.** (b) **8.** (a) **9.** (d) **10.** (c) **11.** (c)
12. (b) **16.** (*b*) **18.** (d) **19.** (a) **20.** (b) **21.** (c) **23.** (c) **24.** (c) **25.** (a) **26.** (d)
27. (a) **28.** (d) **29.** (c) **30.** (c) **32.** (c) **33.** (b) **35.** (b) **36.** (b) **37.** (c) **38.** (d)
39. (c) **40.** (d) **41.** (a) **42.** (c) **43.** (a) **44.** (c) **45.** (b) **47.** (c) **48.** (c) **50.** (c)
51. (a) **52.** (c) **57.** (a) **58.** (a) **59.** (a) **60.** (d) **61.** (a, b) **62.** (c) **63.** (b) **64.** (a)
65. (b) **66.** (c) **67.** (d) **68.** (b) **69.** (c) **70.** (b) **71.** (a) **72.** (a) **73.** (b) **74.** (b)
76. (b) **78.** (d) **82.** (c) **83.** (c) **86.** (b) **87.** (d) **88.** (b) **89.** (b) **90.** (a) **91.** (a)
92. (c) **93.** (d)

EXPLANATIONS

1. D 600 mm, $R = \frac{D}{2} = 300\,\text{mm}$

$h_i = 12$ mm

$\Delta h = 3$mm.

$$\theta = \tan^{-1}\left(\frac{\sqrt{R\Delta h}}{R - \frac{\Delta h}{2}}\right)$$

$$\theta = \tan^{-1}\left(\frac{\sqrt{300 \times 3}}{300 - \frac{3}{2}}\right)$$

$\boxed{\theta = 5.73°}$

2. $\%\text{ reduction} = \frac{\text{Diameter reduced in draw}}{\text{Diameter before draw}}$

for option (a) $= \frac{d_0 - d_1}{d_0} = 0.8$

$d_1 = 0.2d0$

Given, $d_0 = 15$ mm, we get $d_1 = 3$ mm

Similarly, after 2nd pass, we get $d_2 = 0.6$ mm and after 3rd pass, we get $d_3 = 0.12$ mm

For option (b), 1st 3 stages are same as in option (a), in 4th pass we have

$$\frac{d_3 - d_4}{d_3} = 0.2$$

$\Rightarrow d_4 = 0.8d3$

$\Rightarrow d_4 = 0.8 \times 0.12 = 0.096$ mm

For option (c), first two stages are same as in option (a), in 3rd pass we have

$$\frac{d_2 - d_3}{d_3} = 0.4$$

$\Rightarrow d_3 = 6, d_2 = 0.6 \times 0.6 = 0.36$ mm

For 4th pass we have

$$\frac{d_3 - d_4}{d_3} = 0.4$$

$\Rightarrow d_4 = 0.6\, d_3 = 0.6 \times 0.36 = 0.216$ mm

For 5th stage, we have

$$\frac{d_5 - d_4}{d_4} = 0.2$$

$\Rightarrow d_5 = 0.8\, d_4 = 0.8 \times 0.216 = 0.1728$ mm

Ouptput in option (b) is closest to the desired diameter of 0.1 mm

3. Bite angle: $\theta = \tan^{-1}\left(\sqrt{\frac{R\Delta h}{R - \frac{\Delta h}{2}}}\right)$

From question, we have 20% reduction in cross-sectional area

$150 \times 4.5 \times 0.8 = 150 \times h$

$h = 3.6$ mm

$\Delta h = 4.5 - 3.6 = 0.9$ mm

Given, $R = \frac{450}{2} = 225$ mm

$$\theta = \tan^{-1}\left(\frac{\sqrt{225 \times 0.9}}{225 - \frac{0.9}{2}}\right) = 3.62° = 0.06 \text{ rad}$$

4. Hot working is carried out at above Recrystallization temperature.

Hot working > 0.6 T_m

and Recrystallization Temp = 0.5 T_m

T_m = melting point.

5. Given: $t_1 = 25$ mm,

$t_f = 20$ mm

R = 300 mm, N = 100 r.p.m

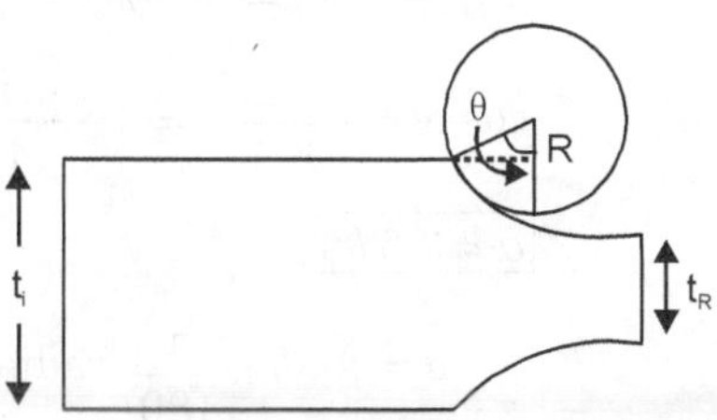

Angle subtended by the deformed zone at the roll center,

$$Q_i = \sqrt{\frac{t_i - t_R}{R}} = \sqrt{\frac{25 - 20}{300}} = 0.129 \text{ rad.}$$

$\therefore$ Roll strip contact length

$= \theta \cdot R = 0.129 \times 300 = 38.72$ mm

6. Given, Initial thickness of sheet,

$t_i = 4 \text{ mm} = 4 \times 10^{-3}$ m

Diameter of roll,

D = 300 mm = 0.3 m

Frictional coefficient at work-roll interface,

$\mu = 0.1$

Let t_f = Final thickness of sheet after rolling

For single pass without slipping minimum possible thickness,

$t_i - t_f = \mu^2 R$

$$\mu = \sqrt{\frac{t_1 - t_f}{R}}$$

$$0.1 = \sqrt{\frac{4 \times 10^{-3} - t_f}{0.15}}$$

or $t_f = 2.5$ mm

7. $\tan\alpha = \sqrt{\frac{t_1 - t_2}{R}} = \sqrt{\frac{16-10}{200}}$

$\tan\alpha = 0.173$

$\therefore$ Bite angle, $\alpha = \tan^{-1}(0.173) = 9.936°$

8. $\Delta h = \mu^2 R$, $\mu \uparrow \Delta h \uparrow$

9. From incompressibility,

$$\pi d_1^2 h_1 = \pi d_2^2 h_2$$

$$\Rightarrow \quad d_2 = d_1 \times \sqrt{\frac{h_1}{h_2}}$$

$$= 100 \times \sqrt{\frac{50}{25}} = 141.42$$

Percentage change in diameter

$$= \frac{d_2 - d_1}{d_1} \times 100 = 41.42\%$$

10. $$\Delta H = D(1 - \cos\alpha)$$

$$\Rightarrow \quad \cos\alpha = 1 - \frac{\Delta H}{D} = 1 - \frac{0.1 \times 8}{410}$$

$$\Rightarrow \quad \boxed{\alpha = 3.57°}$$

$$\therefore \quad \alpha = 3.57 \times \frac{\pi}{180} \text{ radians}$$

$$\boxed{\alpha = 0.062 \text{ radians}}$$

11. The material (metals) is subjected to high compressive stress as a resort of friction between the roll and the metal suface.

12. P – III, Q – I, R – VI, S – V

13. $\theta = \tan^{-1}\mu$, $\sqrt{\frac{\Delta h}{R}} = \tan^{-1}\mu$

$\Rightarrow$ 5.7106°

14. $$\mu = \sqrt{\frac{(\Delta h)_{max}}{R}}$$

$$(\Delta h)_{max} = 25 - 20 = 5\text{mm}$$

$$R = 250\text{mm}$$

$$\therefore \quad \mu = \sqrt{\frac{5}{250}} = 0.1414$$

15. Velocity of Neutral plane

By applying continuity Equation

$H_n = H_2 + 0.25\,\Delta H$

$H_n = 7.2 + 0.25 \times 0.8$

$= 7.4$ mm

$V_n H_n = V_2 H_2$

$V_2 \rightarrow$ Exit velocity

$H_2 \rightarrow$ Final Height

$V_n = 29.19$ m/min

$$V_n = \frac{\pi DN}{60}$$

$$N = \frac{29.19 \times 60}{\pi \times 300}$$

$N = 30.98$ rpn

Total power = $2T\omega$

$$= 2 \times 1439.9 \times 2\pi \times \frac{30.98}{60}$$

$$\boxed{= 9.34 \text{ kW}}$$

16. Knurling : It is a process of impressing a diamond shaped or straight line pattern on the surface by a workpiece by using specially shaped hardened metal wheels to improve its appearance and to provide a better gripping surface.

17. $(\Delta h)_{max} = \mu^2 . R$

$\Rightarrow h_i - h_f = \mu^2 . R$

$\Rightarrow 2 - h_f = (0.05)^2 \times 300$

$\Rightarrow h_f = 2 - 0.75$

$$\boxed{h_f = 1.25\text{mm}}$$

19. At entry, plate velocity is smaller than Roll velocity then it increases & becomes equal to roll velocity at Neutral plane after that it is higher than the roll velocity.

20. Intricate Deformation can take place only in Ductile material.

21. Elongation factor = $\frac{A_i}{A_f} = 1.22$

Let n $\rightarrow$ number of passes

$$\frac{750 \times 750}{(1.11)^n} = 250 \times 250$$

$(1 - 22)^n = 9$

n/n 1.22 = 9

$$\boxed{n = 11.04}$$

22. (T),Back Tension is applied. It decreases the amount of force in Rollers laterally.

23. $h_i = 300$ mm $h_f = 295$ mm

$\Delta h = h_i - h_f = 5$ mm

R = 300 mm, N = 100 rpm.

$\tau_{avg} = 180$ MPa

Roll strip contact length

$$L = \sqrt{\Delta h R}$$

$$= \sqrt{5 \times 300} = \sqrt{1500} = 38.73 \text{ mm}$$

Rolling force = $\tau_{avg}.L_f \times$ width = 1.394 MN

24. $\sigma_d = \sigma_0\left(\frac{A_i}{A_f}\right)$

$\frac{A_i}{A_f} = R \rightarrow$ Extrusion Ratio

If $\sigma_d = \sigma_0$ {Maximum possible Reduction}

$$1 = \ln\left(\frac{A_i}{A_f}\right)$$

$$e = \frac{A_i}{A_f}$$

$$\% \text{ Reduction} = \frac{A_i - A_f}{A_i} \times 100 = \left(1 - \frac{1}{e}\right) \times 100$$

% Reduction = 63%

25. $V = \frac{\pi DN}{60}$

At Neutral Point, Roll velocity and plate velocity are equal.

26. When a material having different value of property in different Direction, called as anisotropic material

27. As roll diameter decreases for the same reduction the contact length decreases. So less force will be required.

28. $h_i = 30$ mm $\mu = 0.1$ D = 60 mm

$h_f = 10$ mm R = 300 mm

$\Delta h = 20$ mm

Draft of single pass

$(\Delta h)_1 = \mu^2 R = (0.11)^2 \times 300$

$$\text{No = passes} = \frac{20}{3} = 6.67 \simeq 7 \text{ passes.}$$

29. $\cos\theta = 1 - \frac{\Delta h}{2R}$

$h_i = 8$ mm

$h_f = 8 \times 0.9 = 7.2$ mm

$$\cos\theta = 1 - \left(\frac{8 - 7.2}{2 \times 205}\right)$$

D = 410

R = 205

$\theta = 0.0624$ radians

30. Compression & Shear

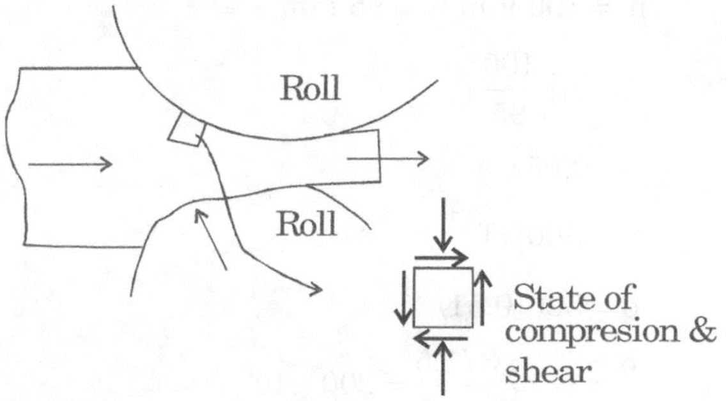

31. At the centre, because of maximum pressure at the centre.

32. Strain = $\in = \frac{2l - 1}{1} = 1$

strain = ln (1 + 1)

= ln 2 = 0.693

34. A-3, B-4, C – 1, D – 2

35. Because with Deformation, material becomes stronger.

36. P – 4, Q – 5 R – 2, S – 3

37. Bend Allowance = a (R + Kt)

R = Bend inside Radius

K = 0.5 because R > 2t

t = sheet thickness

Bend allowance = 1 × (100 + 0.5 × 2)

= 101 mm.

38. P – 4, Q – 3, R – 1, S – 2

39. Constancy of volume

$$\frac{\pi}{4} D_i^2 h_1 = \frac{\pi}{4} D_2^2 h_2$$

$(200)^2 \times 60 = 400^2 \times 4_2.$

Change is length = 45 mm

$$\text{True stain} = \ln\left(\frac{h_1}{h_2}\right) = \ln\left(\frac{60}{15}\right) = \ln 4$$

True strain = 1.386

40. P – 3, Q – 1, R–2, S –4

41. Let τ be the shear stress

$F = t \times d \times t = 5.0$ kN

$F_1 = r \times 1.5\, d \times 4 \times t = .6 \times 5 = 3$ kN

43. A – 4 B – 1 C – 2 D – 3

44. $\in_{True} = \int_{L_o}^{L_f} \frac{dL}{L} = \ln\left[1 + \frac{\Delta L}{L_o}\right]$

But $\frac{\Delta L}{L_o} = \in_E \Rightarrow \in_{True} = \ln(1 + \in_E)$

45. P–4, Q – 3, R –2, S – 1.

46. $\sigma = 500 \in^{0.1}$.

$h_i = 100$ mm $h_f = 95$ mm

$$\in = \ln\left(\frac{100}{95}\right)$$

$$\in = 0.051$$

$$\sigma = 500(\in)^{0.1}$$

$$\sigma = 500(0.51)^{0.1}$$

$$\frac{\sigma}{\in} = G \Rightarrow \frac{371.5}{\in} = 200 \times 10^5$$

$$\in = 1.85 \times 10^{-3} = \frac{dh}{95}$$

final height = 95.157mm

47.

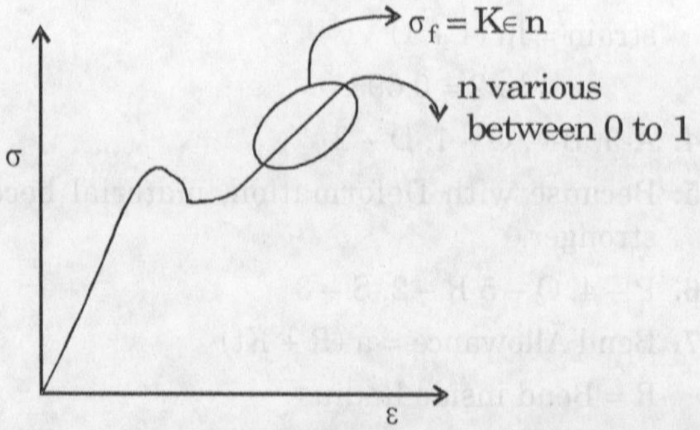

48. Engg. strain $\varepsilon = \frac{0.1}{100}$

True strain $\varepsilon_T = \ln(1 + \varepsilon)$

$$= \ln\left(1 + \frac{0.1}{100}\right)$$

$$= 0.0009995$$

ε_T in % = 0.0995%

49.

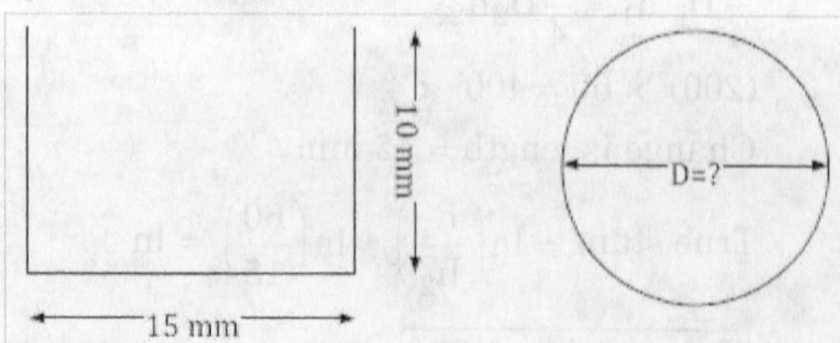

Here

d = 15 mm and h = 10 mm

Now, by formula

$$D = \sqrt{d^2 + 4dh}$$

$$= \sqrt{(15)^2 + 4 \times 15 \times 10}$$

$$= 28.72 \text{ mm}$$

50. Metal forming of Aluminimum is easy because there is negligible strain hardening in Aluminium.

51.

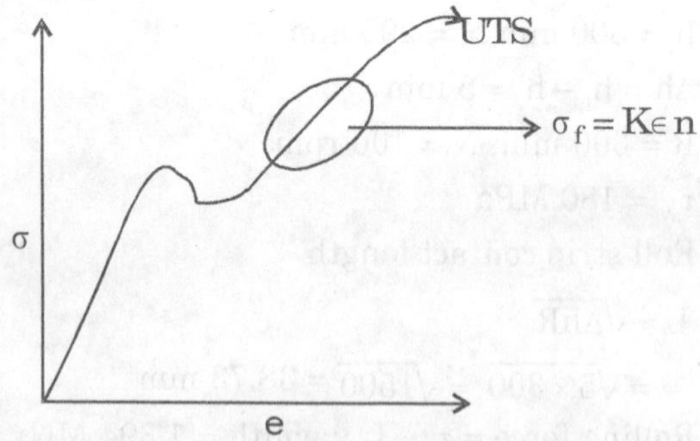

52. P – 4, Q – 1, R – 2, S – 3

53. Gulter is provided in finishing stroke.

54. FCC → due to Austenitic structure.

55. upper critical temp & causer grain structure.

56. A – 2 B 3 C – 5 D 1

57. Gutters are used to settle down excess material.

58. True strain $\in = \ln\frac{A_i}{A_f} = \ln\left(\frac{d_i}{d_f}\right)^2 = \ln\left(\frac{25}{5}\right)$

$$\in = 3.22$$

$$\sigma = 315 \in^{0.54} = 315\,(3.22)^{0.54}$$

$$\boxed{\sigma = 592 \text{ MPa}}$$

59. Stress are build up due to strain hardening & Grain structure distortion.

60. Edging – blocking – Finishing – Trimming.

61. Casting when molten metal is not able to reach in corners.

63. Fullering – Edging – Bending – Blocking cum finishing – Cutting (Trimming)

64. $\sigma = 1400\in^{0.33}$ MPa.

At UTS, n = 0.33

$$\sigma_f = 1400\,(0.33)^{0.33}$$

$$\boxed{\sigma = 971 \text{ MPa}}$$

68. True strain = $\in = \ln\left(\frac{A_i}{A_f}\right)$

$$\in_T = \ln\left(\frac{A_i}{0.8 \times 0.7 A_i}\right)$$

$$\in_T = \ln\left(\frac{1}{0.56}\right)$$

$$\boxed{\in_T = 0.58}$$

True strain in single step of 50%

$$\in_T = \ln\left(\frac{A_i}{A_f}\right)$$

$$= \ln\left(\frac{A_i}{0.5 A_i}\right)$$

$$\in_T = \ln(2) = 0.693$$

69. High toughness for hammering action and low thermal conductivity for retaining higher temperature.

70. At RSS,

$$\frac{\pi}{4} d_0^2 \times h_0 = \pi R_{FN}^2 \times h_1$$

R_{FN} = 258 m

Shear stress sticking (k)

= Shear stress in sliding (μP_{ss})

$$\therefore \text{RSS} = R_{FN} - \frac{h}{2\mu} \text{In}\left(\frac{1}{\sqrt{3}\mu}\right)$$

$$= 258.2 - \frac{30}{2 \times 0.51} \text{In}\left(\frac{1}{\sqrt{3} \times 0.51}\right)$$

= 254.55 mm

71. For wire drawing, work material should be ductile in nature.

72. Aluminimum is ductile at higher temperature.

73. $R = \frac{A_i}{A_f}$.

$$F = 250 \times 10^6 \times \frac{\pi}{4}(0.1)^2 \ln\left(\frac{0.1}{0.05}\right)^2$$

$\boxed{F = 2.72\text{MN}}$

74. D_i = 10 mm, D_f = 8mm

σ_0 = 400 MPa

$$\sigma_d = \sigma_0 \ln\left(\frac{D_i}{D_f}\right)^2$$

$$\sigma_d = 400 \ln\left(\frac{10}{8}\right)^2$$

σ_d = 178.51 MPa

$$\text{Ideal force} = 178.51 \times 10^6 \times \frac{\pi}{4}(0.008)^2$$

$\boxed{\text{I deal force} = 8.97 \text{ KN}}$

75. $L(t) = L_0(1 + t^2)$

$\Rightarrow dL = L_0 \times 2t\, dt$

$$\text{As } d\in_T = \frac{dL}{L}$$

$$d\in_T = \frac{L_0 2t\, dt}{L_0(1+t^2)}$$

$$\frac{d\in_T}{dt} = \frac{2t}{1+t^2}$$

at t = 1 min

$$\frac{d\in_T}{dt} = \frac{2 \times 1}{1+1^2} = 1 \text{ min}^{-1}$$

76. 40% Reduction = 1 × 0.4 = 0.4

30% Reduction of 0.6 = 0.6 × 0.3 = 0.18

Total Reduction = 0.58

77. $$\in_T = \ln\left(\frac{l_f}{l_i}\right)$$

$$= \ln\left(\frac{\left(\frac{Lo}{2}\right)}{(L_o)}\right) = \ln\left(\frac{1}{2}\right) = -0.6932$$

78. $\sigma_d = \sigma_0 l_n$.

For Maximum Reduction

$\sigma_d = \sigma_0$.

$$1 = \ln\left(\frac{A_i}{A_f}\right)$$

$\boxed{R = 0.63}$

79. Slope of AB line is E

$$E = \frac{AC}{BC}$$

$$200 \times 10^3 = \frac{500 - 100}{BC}$$

BC = 0.002

$\in_E = 0.5 - 0.002$

$\boxed{\in_E = 0.498}$

80. D_i = 10 mm, D_f = 7.5 mm

$\alpha = 5°$, $\mu = 0.1$

$(\sigma_y)_{avg}$ = 350 MPa

$$\sigma_f = (\sigma_y)_{avg} \times \left\{1 + \frac{1}{\mu \cot \alpha}\right\}$$

$$\left[1 - \left(\frac{d_f}{d_i}\right)^{2.4 \cot\alpha}\right]$$

$$= 350 \times \left[1 + \frac{1}{0.1 \cot 5}\right]\left[1 - \left(\frac{7.5}{10}\right)^{2\times 0.1\times \cot 5}\right]$$

$\boxed{\sigma_f = 316.25 \text{ MPa}}$

81. $\sigma = 1020 \in^{0.4}$.

at UTS $\Rightarrow \in = n = 0.4$

$$\in = 0.4 = \ln\left(\frac{A_i}{A_f}\right)$$

$$0.4 = \ln\left(\frac{100}{A_f}\right)$$

$\boxed{A_f = 67.032 \text{ mm}^2}$

82. Collapsible tubes are made by extrusion process.

83. Hydrostatic stresses increases the ductility of Brittle material.

84. Extrusion & Tube piercing.

85. Wrong or false

86. Tube drawing with moving mandrel.

87. Collapsible tube are made by impact extrusion ex. Toothpaste tube, Battery case.

88. $F = K \ln (R) A_i$.

$$R = \frac{A_i}{A_f} = 250 \times \frac{\pi}{4}(0.1)^2 \times \ln\left(\frac{(0.1)^2}{(0.05)^2}\right)$$

$$\boxed{F = 2.722 \text{ MN}}$$

89. Diameter of steel wire

$d_i = 10$ mm

$v = 0.5$ m/sec

Final diameter of wire

$d_f = 0.8 \times di = 8$ mm

(drawing stress)

$$\sigma_d = \sigma_{yt} \times \ln\left(\frac{d_i}{d_f}\right)^2$$

$$= 800 \times 2\ln\left(\frac{10}{8}\right) = 357 \text{ MPa}$$

90. $P = F_d \times V = 357 \times \frac{\pi}{4} \times (8)^2 \times 0.5$

$$\boxed{P = 8.972 \text{ Kw}}$$

91. A, $L_0 = 100$, $\sigma_y = 300$ MPa

$d_0 = 50$ mm

$$\text{E.R.} = \frac{A_0}{A_1} = 4, \ A_0 = 1962.5 \text{ mm}^2$$

$$A_1 = \frac{A_0}{4} = \frac{\frac{\pi}{4}50^2}{4} = 490 \text{ mm}^2$$

Extrusion pressure = 300 in (extrusion ratio)

= 415.8 MPa

92. $D_i = 10$ mm li = 100 mm n = 7

$$\text{True strain } \epsilon = \ln\left(\frac{d_i}{d_f}\right)^2$$

$$A_f = (0.67)^7 \times \frac{\pi}{4}(10)^2$$

$d_f = 2.21$ mm

$$\epsilon = 2\ln\left(\frac{10}{2.21}\right)$$

$$\boxed{\epsilon = 3.02}$$

Constancy of volume

$$\frac{\pi}{4}d_i^2 li = \frac{\pi}{4}d_f^2 l_f$$

$l_f = 2047.46$ mm

93. $F_d = K \ln(R) \sigma A_f$

$$\frac{\pi}{4} \times d_{f_1}^2 = 0.65\frac{\pi}{4} \times (10)^2$$

$$= 200 \times \frac{\pi}{4}(10)^2 \times 2\ln\left(\frac{10}{8.062}\right)$$

$$\boxed{F_d = 4.39 \text{ KN}}$$

4

CHAPTER

Sheet Metal Operations

Analysis of Previous GATE Papers														
Topic	**Year/Set → Marks Distribution ↓**		**2019 Set 1**	**2019 Set 2**	**2018 Set 1**	**2018 Set 2**	**2017 Set 1**	**2017 Set 2**	**2016 Set 1**	**2016 Set 2**	**2016 Set 3**	**2015 Set 1**	**2015 Set 2**	**2015 Set 3**
Shearing, Spinning and Stretch Forming	**1 Mark**	MCQ Type									1			
		Numerical Type												
	2 Marks	MCQ Type												
		Numerical Type			1						2			
	Total				**2**						**5**			
Drawing and Bending	**1 Mark**	MCQ Type		1						1				
		Numerical Type												
	2 Marks	MCQ Type												
		Numerical Type							1					
	Total			**1**					**2**	**1**				

Shearing, Spinning and Stretch Forming

1. For blanking and piercing operations, clearance is provided on the ... and the .. respectively.

[ME 1987 : 1 Mark]

2. The thickness of the blank needed to produce, by power spinning a missile cone of thickness 1.5 mm and half cone angle 30°

(a) 3.0 mm (b) 2.5 mm
(c) 2.0 mm (d) 1.5 mm

[ME 1992 : 2 Marks]

3. A 50 mm diameter disc is to be punched out from a carbon steel sheet 1.0 mm thick. The diameter of the die should be (Take clearance as 3% of sheet thickness) ______.

(a) 49.925 mm (b) 50.00 mm
(c) 50.075 mm (d) None of these

[ME 1996 : 2 Marks]

4. The cutting force in punching and blanking operations mainly depends on

(a) The modulus of elasticity of the material
(b) The shear strength of the material
(c) The bulk modulus of the material
(d) The yield strength of the material

[ME 2001 : 1 Mark]

5. In blanking operation the clearance is provided on

(a) The die
(b) The punch
(c) Both die and punch equally
(d) Neither the punch nor the die

[ME 2002 : 1 Mark]

6. A metal disc of 20 mm diameter is to be punched from a sheet of 2 mm thickness. The punch and the dia clearance is 3%. The required punch diameter is

(a) 19.88 mm (b) 19.94 mm
(c) 20.06 mm (d) 20.12 mm

[ME 2003 : 2 Marks]

7. 10 mm diameter holes are to be punched in a steel sheet of 3 mm thickness. Shear strength of the material is 400 N/mm^2 and penetration is 40%. Shear provided on the punch is 2 mm. The blanking force during the operation will be

(a) 22.6 kN (b) 37.7 kN
(c) 61.6 kN (d) 94.3 kN

[ME 2004 : 2 Marks]

Linked Answer Questions 8 and 9

In shear cutting operation, a sheet of 5 mm thickness is cut along a length of 200 mm. The cutting blade is 400 mm long (see figure) and zero-shear (S = 0) is provided on the edge. The ultimate shear strength of the sheet is 100 MPa and penetration to thickness ratio is 0.2. Neglect friction.

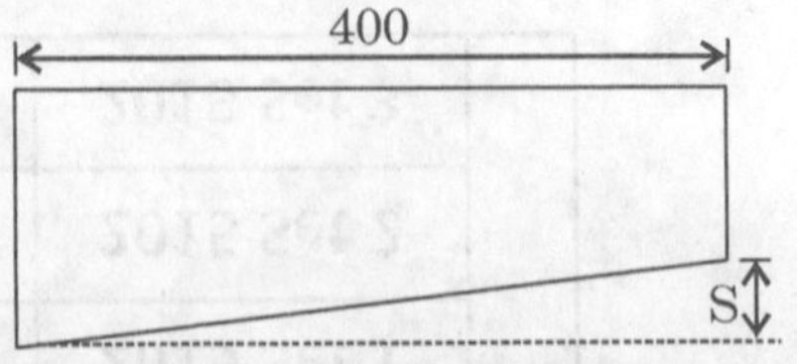

8. Assuming force vs displacement curve to be rectangular, the work done (in J) is

(a) 100 (b) 200
(c) 250 (d) 300

[ME 2010 : 2 Marks]

9. A shear of 20 mm (S = 20 mm) is now provided on the blade. Assuming force vs displacement curve to be trapezoidal, the maximum force (in kN) exerted is

(a) 5 (b) 10
(c) 20 (d) 40

[ME 2010 : 2 Marks]

10. The shear strength of a sheet metal is 300 MPa. The blanking force required to produce a blank of 100 mm diameter from a 1.5 mm thick sheet is close to

(a) 45 kN (b) 70 kN
(c) 141 kN (d) 3500 kN

[ME 2011 : 2 Marks]

11. Calculate the punch size in mm, for a circular blanking operation for which details are given below.

Size of the blank 25 mm
Thickness of the sheet 2 mm
Radial clearance between punch and die 0.06 mm
Die allowance 0.05 mm

(a) 24.83 (b) 24.89
(c) 25.01 (d) 25.17

[ME 2012 : 2 Marks]

12. A rectangular hole of size 100 mm × 50 mm is to be made on a 5 mm thick sheet of steel having ultimate tensile strength and shear strength of 500 MPa and 300 MPa, respectively. The hole is made by punching process. Neglecting the effect of clearance, the punching force (in kN) is

(a) 300 (b) 450
(c) 600 (d) 750

[ME 2014 : 2 Marks, Set-2]

13. Internal gears are manufactured by
(a) hobbing
(b) shaping with pinion cutter
(c) shaping with rack cutter
(d) milling

[ME 2016 : 1 Mark, Set-3]

14. In a sheet metal of 2 mm thickness a hole of 10 mm diameter needs to be punched. The yield strength in tension of the sheet material is 100 MPa and its ultimate shear strength is 80 MPa. The force required to punch the hole (in kN) is

[ME 2016 : 2 Marks, Set-3]

15. In a single point turning operation with cemented carbide tool and steel work piece, it is found that the Taylor's exponent is 0.25. If the cutting speed is reduced by 50% then the tool life changes by ______ times.

[ME 2016 : 2 Marks, Set-3]

16. The percentage scrap in a sheet metal blanking operation of a continuous strip of sheet metal as shown in the figure (correct to two decimal places) is ______.

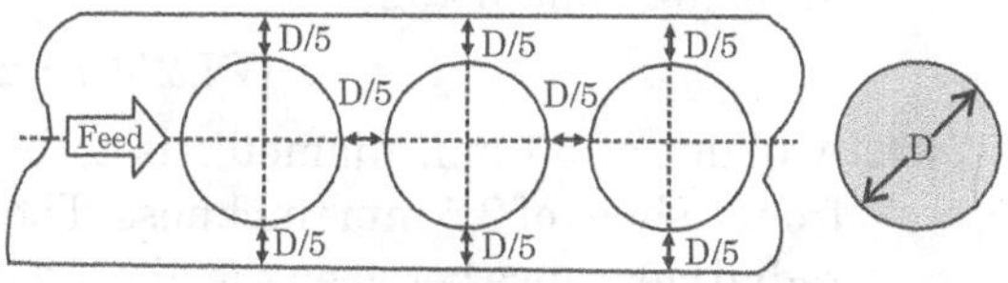

[ME 2018 : 2 Marks, Set-1]

17. Calculate the smallest punch diameter that can be designed for piercing sheet metal strip with the following data. Crushing strength of the punch material is 1500 MPa. Thickness of the sheet is 2 mm, factor of safety is 3, shear strength of the sheet material 500 MPa is
(a) 2 mm (b) 6 mm
(c) 8 mm (d) 24 mm

[PI 1990 : 2 Marks]

18. In ... A... operation the diameter of the desired hole will be smaller than the diameter of the ... B...
A. blanking/piercing
B. punch/die

[PI 1991 : 1 Mark]

19. In blanking operation, the best way to improve the smoothness and secureness of the edges is to
(a) have reduced gap between punch and die
(b) increase the ductility of the sheet
(c) decrease the speed of blanking
(d) provide shear on the punch

[PI 1994 ; 1 Mark]

20. In producing a punched hole of 20 mm diameter the dimensions of the punch is ______ given the radial die clearance of 0.1 mm.

[PI 1994 ; 1 Mark]

21. For 50% penetration of work material, a punch with single shear equal to thickness will
(a) reduce the punch load to half the value
(b) increase the punch load by half the value
(c) maintain the same punch load
(d) reduce the punch load to quarter load

[PI 2001 : 1 Mark]

22. Which one is not a method of reducing cutting forces to prevent the overloading of press?
(a) providing shear on die
(b) providing shear on punch
(c) increasing the clearance
(d) stepping punches

[PI 2003 : 1 Mark]

23. A hydraulic press is used to produce circular blanks of 10 mm diameter from a sheet of 2 mm thickness. If the shear strength of the sheet material is 400 N/mm^2, the force required for producing a circular blank is
(a) 8 kN (b) 25.13 kN
(c) 31.42kN (d) 125.66 kN

[PI 2003 : 2 Marks]

24. Consider the following statements related piercing and blanking
1. Shear on the punch reduces the maximum cutting force
2. Shear increases the capacity of the press needed
3. Shear increases the life of the punch
4. Total energy needed to make the cut remains unaltered due to provision of shear.

Which of the above statements are correct?
(a) 1 and 2 (b) 1 and 4
(c) 2 and 3 (d) 3 and 4

[PI 2003 : 2 Marks]

25. A cylindrical cup of 48.5 mm diameter and 52 mm height has a corner radius of 1.5 mm. Cold rolled steel sheet of thickness 1.5 mm is used to produce the cup. Assume trim allowance as 2 mm per 25 mm of cup diameter. What is the blank size in mm?
(a) 102.0 (b) 115.4
(c) 120.5 (d) 128.5

[PI 2004 : 2 Marks]

26. A hole of 40 mm diameter is pierced in a steel sheet of 4 mm thickness without shear on the tool. Shear strength of steel is 400 N/mm^2 and penetration is 25%. What is the expected percentage load reduction if a shear of 1 mm is provided on the punch?

(a) 25.00 (b) 33.33
(c) 50.00 (d) 66.67

[PI 2004 : 2 Marks]

27. Determine the load required to punch a 20 mm diameter hole in 2 mm thick sheet. The properties of the material of the sheet are:

Tensile strength – 580 MPa
Yield strength in tension – 410 MPa
Shear strength – 350 MPa
Yield strength in shear – 250 MPa
The load in kN is

(a) 31.40 (b) 43.98
(c) 64.40 (d) 91.10

[PI 2004 : 2 Marks]

28. A ϕ 25 mm hole is pierced in a t = 2.5 mm thick steel sheet having shear strength τ = 350 MPa.

If the diametral clearance is given by the expression $c = 0.0064\sqrt{\tau}$, the die diameter (in mm), punch diameter (in mm) and punch force in (kN) respectively are

(a) 25.0, 25.3, 171.8
(b) 25.0, 24.7, 68.75
(c) 24.7, 25.0, 171.8
(d) 25.3, 25.0, 68.75

[PI 2005 : 2 Marks]

29. It is require to punch a hole of diameter 10 mm on a sheet of thickness 3 mm. The shear strength of the work material is 420 MPa. The required punch force is

(a) 19.78 kN (b) 39.58 kN
(c) 98.9 kN (d) 395.6 kN

[PI 2006 : 2 Marks]

30. It is required to punch a hole of diameter 10 mm on a sheet of thickness 3 mm. The shear strength of the work material is 420 MPa. The required punch force is

(a) 19.78 kN
(b) 39.56 kN
(c) 98.9 kN
(d) 359.6 kN

[PI 2006 : 2 Marks]

Common Data Question **31** and **32**

A cup is to be drawn to a diameter of 70 mm with 35 mm depth from a 0.5 mm thick sheet metal. The cup is drawn in one operation. Assume σ_u = 430 MPa.

31. The required blank diameter is

(a) 86.7 mm (b) 119.5 mm
(c) 121.24 mm (d) 169 mm

[PI 2006 : 2 Marks]

32. The maximum drawing force is

(a) 0.047 MN (b) 0.082 MN
(c) 0.83 MN (d) 9.69 MN

[PI 2006 : 2 Marks]

33. Circular blanks of 35 mm dia are punched from a steel sheet of 2 mm thickness. If the clearance per side between the punch and die is to be kept as 40 microns. The sizes of punch and die should respectively be

(a) 35 and 35.040
(b) 34.92 and 35
(c) 35 and 35.080
(d) 35.040 and 34.92

[PI 2007 : 2 Marks]

34. A blank of 50 mm diameter is to be sheared from a sheet of 2.5 mm thickness. The required radial clearance between the die and the punch is 6% of sheet thickness. The punch and die diameter (mm) for this blanking operation respectively are

(a) 50.00 and 50.30 (b) 50.00 and 50.15
(c) 49.70 and 50.00 (d) 49.85 and 50.00

[PI 2008 : 2 Marks]

35. A disc of 200 mm diameter is blanked from a strip of an aluminum alloy of thickness 3.2 mm. The material shear strength to fracture is 150 MPa.

The blanking force (in kN) is

(a) 291 (b) 301
(c) 311 (d)421

[PI 2009 : 2 Marks]

36. Circular blanks of 10 mm diameter are punched from an aluminum sheet of 2 mm thickness. The shear strength of aluminum is 80 MPa. The minimum punching force required in kN is

(a) 2.57 (b) 3.29
(c) 5.03 (d) 6.33

[PI 2013 : 1 Mark]

Drawing and Bending

37. Wrinkling is a common defect found is
(a) Bent components
(b) Deep drawn components
(c) Embossed components
(d) Blanked component **[ME 1991 : 1 Mark]**

38. In deep drawing of sheets, the values of limiting drawing ratio depends on
(a) percentage elongation of the sheet
(b) yield strength of the sheet
(c) type of press used
(d) thickness of sheet **[ME 1994 : 1 Mark]**

39. Identify the stress strain in flange portion of a partially drawn cylindrical cup when deep drawing without a blank holder
(a) tensile in all three directions
(b) no stress in the flange at all, because there is no blank holder
(c) tensile stress in one direction and compressive stress in other direction
(d) tensile in two directions and compressive in third direction **[ME 1999 : 2 Marks]**

40. A shell of 100 mm diameter and 100 mm height with the corner radius of 0.4 mm is to be produced by cup drawing. The required blank diameter is
(a) 118 mm (b) 161 mm
(c) 224 mm (d) 312 mm
[ME 2003 : 2 Marks]

41. In the deep drawing of cups, blanks show a tendency to wrinkle up around the periphery (flange). The most likely cause and remedy of the phenomenon are, respectively.
(a) Buckling due to circumferential compression; Increase blank holder pressure.
(b) High blank holder pressure and high friction; Reduce blank holder pressure and apply lubricant.
(c) High temperature causing increase in circumferential length; Apply coolantto blank.
(d) Buckling due to circumferential compression; decrease blank holder pressure. **[ME 2008 : 2 Marks]**

42. A mild steel plate has to be rolled in one pass such that the final plate thickness is 2/3rd of the initial thickness, with the entrance speed of 10 m/min and roll diameter of 500 mm. If the plate widens by 2% during rolling, the exit velocity (in m/min) is______.
[ME 2014 : 2 Marks; Set-2]

43. In a rolling process, the maximum possible draft, defined as the difference between the initial and the final thickness of the metal sheet, mainly depends on which pair of the following parameters?
P: Strain
Q : Strength of the work material
R : Roll diameter
S: Roll velocity
T : Coefficient of friction between roll and work
(a) Q, S (b) R, T
(c) S, T (d) P, R
[ME 2014 : 1 Mark, Set-4]

44. A 300 mm thick slab is being cold rolled using roll of 600 mm diameter. If the coefficient of friction is 0.08, the maximum possible reduction (in mm) is______.
[ME 2016 : 2 Marks, Set-1]

45. The value of true strain produced in compressing a cylinder to half its original length is
(a) 0.69 (b) -0.69
(c) 0.5 (d) -0.5
[ME 2016 : 1 Mark, Set-2]

46. Consider a linear elastic rectangular thin sheet of metal, subjected to uniform uniaxial tensile stress of 100 MPa along the length direction. Assume plane stress condition in the plane normal to the thickness. The Young's modulus E = 200 MPa and Poisson's ratio μ = 0.3 are given. The principal strains in the plane of the sheet are
(a) (0.5, –0.15) (b)(0.5, 0.0)
(c) (0.35, –0.15) (d)(0.5, –0.5)
[ME 2019 : 1 Mark, Set-2]

47. In progressive die (for sheet metal), spring loaded stripper plate is used to clamp the stock until
(a) The punch penetrates twice the stock thickness
(b) It removes wrinkles on the product edges
(c) Automatic feeder plate releases it
(d) Punch is completely withdrawn from the stock
[PI 1991 : 1 Mark]

48. A cup of 10 cm height and 5 cm diameter is to be made from a sheet metal of 2 mm thickness.
The number of deductions necessary will be
(a) One (b) Two
(c) Three (d) Four
[PI 1991 : 1 Mark]

49. Wall thickness of drawn cup is controlled by

(a) Deep drawing (b) Reverse drawing

(c) Redrawing (d) Ironing

[PI 1992 : 1 Mark]

50. A hemispherical cup of radius R is formed in a press working operation. The radius of blank would be

(a) $\sqrt{2}R$ (b) $\sqrt{3}R$

(c) $\sqrt{\frac{2}{3}}R$ (d) $\sqrt{\frac{3}{2}}R$

[PI 1992 : 2 Marks]

51. In a bending operation, if the modulus of elasticity E is increased keeping all other parameters unehanged, the spring back will

(a) increase

(b) decrease

(c) remains unehanged

(d) be independent of E

[PI 1994 : 1 Mark]

52. The function of draw bead in the deep drawing operation is to

(a) produce a balance between the amount of stretching and drawing

(b) produce a circular groove on the flange

(c) reduce the drawing load

(d) reduce the wrinkles on the flange

[PI 1994 : 1 Mark]

53. In metal forming operation, the true strain in the X-direction is 0.3 and in the Y-direction −0.1. The true strain in the Z-direction is

(a) 0 (b) −0.2

(c) 0.2 (d) −0.03

[PI 1994 : 2 Marks]

54. The strain in the cup extrusion for the part shown in figure

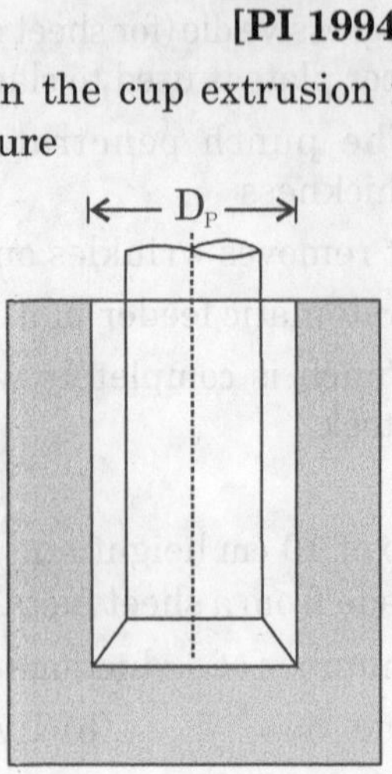

(a) ln $[D_b/D_p]$ (b) ln$[D_b/(D_b-D_p)]$

(c) ln$[D_p/(D_b-D_p)]$ (d) ln$[D_b^2/(D_b^2-D_p^2)]$

[PI 1995 : 2 Marks]

55. In sheet metal working, the spring back increase when

P. Ratio of bend radius to sheet thickness is small

Q. Young's modulus of the sheet is low

R. Yieid strength of the sheet is high

S. Tension is applied during bending

(a) P, Q, R (b) Q, R

(c) P, Q, S (d) P, R, S

[PI 2002 : 2 Marks]

56. Match the following

Product

P. Food cans

Q. Connecting rods

R. Metal foils

Manufacturing process

1. Forging
2. Rolling
3. Deep drawing
4. Extrusion

	P	Q	R
(a)	3	1	2
(b)	1	2	4
(c)	4	3	3
(d)	2	4	1

[PI 2003 : 1 Mark]

57. In deep drawing of sheet metal, spring loaded stripper clamps the work until:

(a) The punch penetrates twice the stock thickness

(b) It removes wrinkles on the product edges

(c) Shedder removes the work from the tools

(d) Punch is completely withdrawn from the work material

[PI 2004 : 2 Marks]

58. In a surface broaching process, a broach having 5 mm pitch and cut per teeth 0.05 mm is used for reducing the thickness by 6 mm. If the length of the workpiece is 50 mm and the cutting speed is 0.5 m/s, the time required/pass in second is

(a) 1.1 (b) 1.2

(c) 1.3 (d) 1200

[PI 2006 : 2 Marks]

59. Match the items in Group-I and II

Group-I

P. Yield point elongation
Q. Spring back
R. Wrinkling
S. Residual stresses

Group-II

1. Distortions of dimension
2. Leuder's bands
3. Plastic deformation
4. Compressive stress Codes:

	P	Q	R	S
(a)	3	4	2	1
(b)	4	3	1	2
(c)	2	3	4	1
(d)	2	1	4	3

[PI 2006 : 2 Marks]

60. Match the following:

Group-I	**Group-II**
P. Wrinkling	1. Upsetting
Q. Centre burst	2. Deep drawing
R. Barrelling	3. Extrusion
S. Cold shut	4. Closed die forging

	P	Q	R	S
(a)	2	3	4	1
(b)	3	4	1	2
(c)	2	3	1	4
(d)	2	4	3	1

[PI 2008 : 2 Marks]

61. Match the following metal forming processes with their associated stresses in the workpiece

Metal forming process

1. Coining
2. Wire drawing
3. Blanking
4. Deep drawing

Type of stress

P. Tensile
Q. Shear
R. Tensile and compressive
S. Compressive

	P	Q	R	S
(a)	2	3	4	1
(b)	1	2	3	4
(c)	2	4	3	1
(d)	1	3	2	4

[PI 2012 : 1 Mark]

ANSWERS

2. (a)	**3.** (b)	**4.** (b)	**5.** (b)	**6.** (a)	**7.** (a)	**8.** (a)	**9.** (d)	**10.** (c)	**11.** (a)
12. (b)	**13.** (b)	**17.** (c)	**18.** (a)	**19.** (a)	**21.** (a)	**22.** (a)	**23.** (b)	**24.** (b)	**25.** (b)
26. (c)	**27.** (b)	**28.** (d)	**29.** (b)	**30.** (b)	**31.** (c)	**32.** (b)	**33.** (b)	**34.** (c)	**35.** (b)
36. (c)	**37.** (b)	**38.** (d)	**39.** (c)	**40.** (c)	**41.** (a)	**43.** (b)	**45.** (b)	**46.** (a)	**47.** (d)
48. (c)	**49.** (d)	**50.** (a)	**51.** (b)	**52.** (d)	**53.** (b)	**54.** (d)	**55.** (a)	**56.** (b)	**57.** (b)
58. (c)	**59.** (c)	**60.** (c)	**61.** (a)						

EXPLANATIONS

1. Punch & Die
2. Thickness of blank

$$= \frac{\text{cone thickness}}{\cos(2\alpha)} = \frac{1.5}{\cos 60^\circ} = 3\text{mm}$$

3. Clearness should be given on punch in Blanking operation. Die should be exact.
4. Cutting force = $\tau \times A$

   τ = Shear strength
5. Blanking operation cleanness is provided on. Punch Die will be exact.
6. Punch diameter

$$= 20 - 2 \times 3\% \text{ of thickness}$$

$$= 20 - 2 \times \frac{3}{100} \times 2$$

$$= 20 - 0.1^2 = 19.88 \text{ mm}$$

7. Given, Shear strength,

$\tau = 400 \text{ N/mm}^2$, Sheet thickness,

$t = 3$ mm

Diameter of hole,

$d = 10$ mm,

Shear on punch,

$t_1 = 2$ mm

Penetration of punch on fraction,

$2p = 0.4$

Blanking force when shear is applied on the punch

$$= \pi dt\left(\frac{t_p}{t_1}\right)\tau = \pi \times 10 \times 3 \times 400 \times \left(\frac{3 \times 0.4}{2}\right)$$

$$= 22.61 \text{ kN}$$

8. Thickness = 5 mm
9. Length = 200 mm

   Blade length = 400 mm,

   Zero shear ($s = 0$)

   Ultimate shear strength of sheet = 100 MPa

$$\frac{\text{Penetration}}{\text{Thickness}} = 0.2 = k$$

$\therefore$ Shear area = (200 + 200) × 5 = 2000 mm^2

Now Work done = τ × shear area × k.t.

$$= 100 \times 2000 \times 5 \times 0.2 = 200 \text{ Joules}$$

$$\therefore \text{Maximum force} = \frac{\text{Work done}}{(kt + \text{shear})}$$

$$= \frac{200}{(0.2 \times 5 + 20)} = 9.5 \approx 10 \text{ kN}$$

10. Blanking force = $\tau \times A_s$

$$= 300 \times \pi dt$$

$$= 300 \times \pi \times 100 \times 1.5$$

$$= 141 \text{ kN}$$

11. Diameter of punch

    = Diameter of Blank – 2 × radial clearance – die allowance

    = 25 – 2 × 0.06 – 0.05 = 24.83 mm

12.

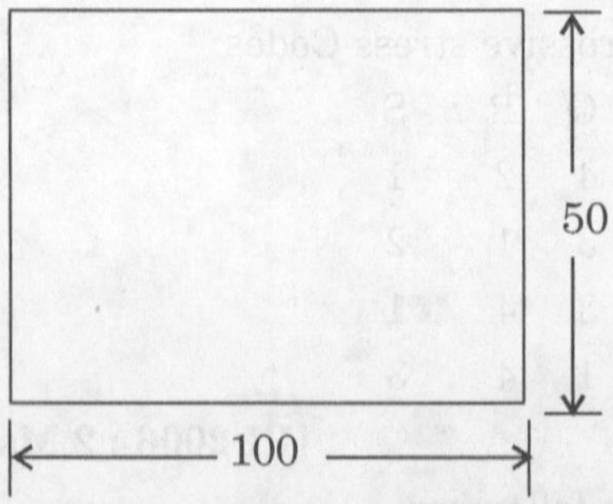

Perimeter = (50 + 100) × 2 = 300

Thickness = 5

so shown Area = 300 × 5 = 1500 mm^2.

Punching force = $\tau \times A_s$.

$$= 300 \times 1500 = 450 \text{ KN}$$

14. $F = \pi dt \times Ts = \pi \times 10 \times 2 \times 80 = 5026.5 \text{ N} = 5.026$ kN

15. $V_2 = 0.5V_1$,

$$\frac{V_1}{V_2} = \frac{V_1}{0.5V_1} = \frac{1}{0.5} = 2$$

$$V_1 T_1^n = V_2 T_2^n$$

$$\left(\frac{T_2}{T_1}\right)^n = \left(\frac{V_1}{V_2}\right)$$

$$\frac{T_2}{T_1} = \left(\frac{V_1}{V_2}\right)^{\frac{1}{n}} = \left(\frac{V_1}{V_2}\right)^{\frac{1}{0.5}} = \left(\frac{V_1}{V_2}\right)^4$$

$$= (2)^4 = 16$$

16. For blanking operation the piece obtained through hole is so, Scrap = hatched one area

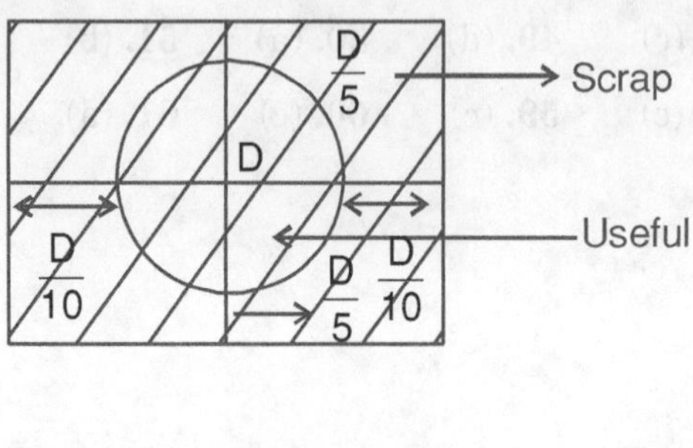

$$\% \text{ scrap} = \left(\frac{\text{scrap area}}{\text{Total Area}}\right) \times 100$$

$$= \left(\frac{\text{Total Area} - \text{hole area}}{\text{Total area}}\right) \times 100$$

$$= \left(1 - \frac{\text{hole area}}{\text{total area}}\right) \times 100$$

$$= \left(1 - \frac{\frac{\pi}{4}D^2}{\left(D + \frac{D}{10} + \frac{D}{10}\right) \times \left(D + \frac{D}{5} + \frac{D}{5}\right)}\right) \times 100$$

$= 53.25\%$

17. Punching force = $\tau = \pi dt$

= crushing strength [FOS = 3]

$$\text{Force} = \frac{1500}{3} \times \frac{\pi}{4} d^2$$

$$\frac{1500}{3} \times \frac{\pi}{4} d^2 = 500 \times \pi \times d \times 2$$

$\boxed{d = 8\text{mm}}$

18. Procuring (B) Die

20. Hole is inner profile so outer blade should be exact.

21. Punch load = $\pi D t \tau \left(\frac{P}{t_1}\right)$

t_1 = shear on punch.

P = 50%

$$\text{Punch load} = \pi D t \pi \times \frac{0.5t}{t}$$

$\Rightarrow 0.5\, \pi D t \tau$

22. Shear decrease the maximum force but Energy Requirement remains the same.

23. Punching force = $\pi D t \tau$

$= \pi \times 2 \times 10 \times 400 = 25.13$ KN

24. Over all Energy requirement Unaltered & Maximum Force decreased.

25. $D = \sqrt{d^2 + 4dh}$

$D = \sqrt{(48.5)^2 + 4 \times 48.5 \times 52}$

D = 111.53 mm

$$\text{Trim Allowance} = \frac{2}{25} \times 48.5 = 3.88 \text{ mm.}$$

Total blank size = 111.53 + 3.88

= 115.41 mm

26. $F = \frac{F_{max}\, Pt}{Pt + S} = \frac{F_{max} \times 0.25 \times 4}{0.25 \times 4 + 1}$

$$F = \frac{F_{max}}{2}$$

27. Punching force = $\pi D t \tau$

$= \pi \times 20 \times 2 \times 350 = 43.98$ KN.

28. t = 2.5

τ = 350 MPa

d_{hole} = 25 mm

Punch dia = 25 mm

Clearance = $= 0.0064 \times 2.5 \times \sqrt{350}$

$\simeq 0.3$mm

Die Dia = Punch Dia + Clearance

= 25 + 0.3 = 25.3

Punching force = π Dt τ = $\pi \times 25 \times 2.5 \times 350$

= 68.72 KN

30. Punching force = $\pi d t \tau$

$= \pi \times 10 \times 3 \times 420 = 39.58$kN

29. Punching force = $\pi D t \tau$

$= \pi \times 10 \times 3 \times 420 = 39.58$ KN

31. $D = \sqrt{d^2 + 4dh} = \sqrt{70^2 + (4 \times 70 \times 35)}$

D = 121.24 mm

32. Drawing force = $\pi D t \tau$

$= \pi \times 121.24 \times 0.5 \times 430 = 0.082$ MN

33. Dia of Die = 35 mm

Dia of Punch = 35 – 2 × 0.04 = 34.92mm

34. Radial clearance $= \frac{6}{100} \times t = 0.1 \leq 1.5$ mm

Total clearance = 2 × 0.15 = 0.3

Die Dia will be exact = 50 mm

Clearance will be provided on punch

= 50–0.30 = 49.70 mm

35. Blanking force = $\pi D t \tau$

$= \pi \times 200 \times 3.2 \times 150 = 301.59$ KN

36. Punching force = $\pi D t \tau$

$= \pi \times 10 \times 2 \times 80$

= 5.023 KN

37. Deep drawn components having defects of wrinkles due to less blank holding force

39. Tensile in Radial direction & compressive in circumferential direction.

40. $D = \sqrt{d^2 + 4hd}$

$D = \sqrt{100^2 + 4 \times 400 \times 100}$

$\boxed{D = 224\ mm}$

41. Due to circumferential compression buckling in circumference, by increasing blank holding force buckling decrease in lateral direction of load applied.

42. $b_1 h_1 V_1 = b_2 h_2 V_2.$

$\frac{h_2}{h_1} = \frac{2}{3}\frac{b_2}{b_1} = 1.02$

$V_2 = \frac{b_1}{b_2} \times \frac{h_i}{h_2} \times V_1 = \frac{1}{1.02} \times \frac{2}{3} \times 10$

$\boxed{V_2 = 14.7 m/min}$

43. $\Delta h = \mu^2 R$

44. Maximum possible reduction

$= \Delta H/pass = \mu^2 R$

$= 0.08^2 \times 300 = 1.92$ mm

45. Engg. strain $(\varepsilon) = \frac{\text{Change in Length}}{\text{Original Length}}$

$= \frac{\left(\frac{L}{2} - L\right)}{L} = -0.5$

True strain $= \ln(1 + \varepsilon) = \ln(1 - 0.5)$

$= -0.69$

46. Given

E = 200 MPa

$\sigma_x = 100$ MPa

$\sigma_y = 0$

Principal strain in x-direction

$\varepsilon_x = \frac{\sigma_x}{E} - \mu\frac{\sigma_y}{E} = \frac{100}{200} - 0 = 0.5$

$\varepsilon_x = 0.5$

Principal strain in y-direction,

$\varepsilon_y = \frac{\sigma_y}{E} - \mu\frac{\sigma_x}{E}$

$\Rightarrow 0 - 0.3.\frac{100}{200} = -0.15$

$(\varepsilon_x, \varepsilon_y) = (0,5, -0.15)$

48. $D = \sqrt{d^2 + 4hd} = \sqrt{5^2 + 4 \times 5 \times 10} = 15$ cm

first reduction 50%

Next reduction 30%

49. It is clearance between punch & die in subsequent reduction in single stroke.

50. Area of Blank = surface area of cup

$\pi R_b^2 = \frac{4\pi R^2}{2}$

$R_b = \sqrt{2}R$

52. Draw beads decreases circumferential stress or buckling in flange.

53. Total volumetric strain = 0

$\epsilon_x + \epsilon_y + \epsilon_z = 0$

$0.3 - 0.1 + \epsilon_z = 0$

$\boxed{\epsilon_z = -0.2}$

54. strain $= \ln\left(\frac{A_i}{A_f}\right)$

strain $= \ln\left(\frac{D_b^2}{D_b^2 - D_P^2}\right)$

55. P, Q, R

56. P – 1, Q – 2, R – 4

58. No. of teeth Required $= \frac{6}{0.05} = 120$

Cutting length = 120 × 5 = 600 mm

L = 50 + 600 mm = 650

$t = \frac{650}{0.5 \times 100} = 1.3$ seconds

59. P – 2 Q – 3 R – 4 S – 1

60. P – 2 Q – 3 R – 1 S – 4

61. P – 2 Q – 3 R – 4 S – 1

5 CHAPTER

Joining/Welding

Analysis of Previous GATE Papers

Topic	Year/Set → Marks Distribution ↓		2019 Set 1	2019 Set 2	2018 Set 1	2018 Set 2	2017 Set 1	2017 Set 2	2016 Set 1	2016 Set 2	2016 Set 3	2015 Set 1	2015 Set 2	2015 Set 3
Oxyacetylene Welding	1 Mark	MCQ Type												
		Numerical Type		1										
	2 Marks	MCQ Type												
		Numerical Type												
	Total			**1**										
Arc Welding	1 Mark	MCQ Type					1			1				
		Numerical Type										1		
	2 Marks	MCQ Type												
		Numerical Type		1						1			1	
	Total			**2**			**1**			**3**		**1**	**2**	
Resistance Welding	1 Mark	MCQ Type			1									
		Numerical Type												
	2 Marks	MCQ Type												
		Numerical Type									1			
	Total													
Solid and Advanced Welding, Brazing and Soldering	1 Mark	MCQ Type	1						1					1
		Numerical Type												
	2 Marks	MCQ Type												
		Numerical Type							1					
	Total		**1**						**3**					**1**

Oxyacetylene Gas Welding

1. For gas welding a particular job using a neutral oxyacetylene flame the acetylene consumption was 10 litres. The oxygen consumption from the cylinder in liters will be

(a) 5 (b) 10

(c) 15 (d) 20

[ME 1991 : 1 Mark]

2. The ratio of acetylene to oxygen is approximately for neutral flame used in gas welding

[ME 1994 : 1 Mark]

3. For butt welding 40 mm thick steel plates, when the expected quantity of such jobs is 5000 per month over a period of 10 years, choose the best suitable welding process out of the following available alternatives.

(a) SAW

(b) Oxyacetylene gas welding

(c) EBM

(d) MIG

[ME 1999 : 1 Mark]

4. The temperature of a carburizing flame in gas welding that of a neutral or an oxidizing flame

(a) Lower than (b) Higher than

(c) Equal to (d) Unrelated to

[ME 2002 : 1 Mark]

5. In Oxyacetylene gas welding, temperature at the inner cone of the flame is around

(a) 3500°C (b) 3200°C

(c) 2900°C (d) 2550°C

[ME 2003 : 1 Mark]

6. In an electrical discharge machining process, the breakdown voltage across inter electrode gap (IEG) is 200V and the capacitance of the RC circuit is 50 μF. The energy (in J) released per spark across the IEG is ____

[ME 2019 : 1 Mark, Set-2]

7. In welding brass with oxyacetylene flame, the type of flame used is ________ .

[PI 1989 : 1 Mark]

8. In welding a mild steel sheet using an oxyacetylene flame, the total amount of acetylene consumed was 8.0 litres, the oxygen consumption from the cylinder is

(a) 8 litres (b) 16 litres

(c) 20 litres (d) 24 litres

[PI 1989 : 2 Marks]

9. During gas welding of copper sheets a neutral flame was used consuming 8 litres of acetylene. But the weld was found to be defective. So it was rewelded using an oxidizing flame. The likely consumption of oxygen in litres will be

(a) 6 (b) 8

(c) 10 (d) 16

[PI 1993 : 2 Marks]

10. The composition of an oxyacetylene flame for cutting of steel contain more oxygen.

[PI 1994 : 1 Mark]

11. Which of the following powders should be fed for effective oxy-fuel cutting of stainless steel.

(a) Steel (b) Aluminum

(c) Copper (d) Ceramic

[PI 2001 : 1 Mark]

Arc Welding

12. Penetration is increased by

(a) Increasing welding current and welding speed

(b) Increasing welding current and decreasing welding speed

(c) Decreasing welding current and welding speed

(d) Decreasing welding current and increasing welding speed

[ME 1990 : 1 Mark]

13. At small variations of arc length at operating conditions, the manual metal arc welding transformer provides nearly

(a) Constant power

(b) Constant power factor

(c) Constant voltage

(d) Constant current

[ME 1990 : 1 Mark]

14. In DC welding, the straight polarity (electrode negative) results in

(a) Lower penetration

(b) Lower deposition rate

(c) Less heating of work piece

(d) Smaller weld pool

[ME 1993 : 1 Mark]

15. The electrodes used in arc welding are coated. This is not expected to

(a) Provide protective atmosphere to weld

(b) Stabilize the arc

(c) Add alloying elements

(d) Prevent electrode from contamination

[ME 1994 : 1 Mark]

16. Preheating before welding is done to
(a) Make the steel softer
(b) Burn away oil, grease etc. from the plate surfaces
(c) Prevent cold cracks
(d) Prevent plate distortion

[ME 1996 : 1 Mark]

17. Two plates of the same metal having equal thickness are to be butt welded with electric arc.
When the plate thickness changes, welding is achieved by
(a) Adjusting the current
(b) Adjusting the duration of the current
(c) Changing the electrode size
(d) Changing the electrode coating

[ME 2002 : 1 Mark]

18. Which of the following arc welding processes does not use consumable electrode
(a) GMAW (b) GTAW
(c) SAW (d) None

[ME 2002 : 1 Mark]

19. Two 1 mm thick steel sheets are to be spot welded at a current of 5000 A. Assuming effective resistance to be 200 micro-ohms and current flow time of 0.2 second, heat generated during the process will be
(a) 0.2 Joule (b) 1 Joule
(c) 5 Joule (d) 1000 Joule

[ME 2004 : 1 Mark]

20. In an arc welding process, the voltage and current are 25 V and 300 A respectively. The arc heat transfer efficiency is 0.85 and welding speed is 8 mm/s. The net heat input (in J/mm) is
(a) 64 (b) 797
(c) 1103 (d) 79700

[ME 2006 : 2 Marks]

21. A direct current welding machine with a linear power source characteristic provides open circuit voltage of 80 V and short circuit current of 800 A. During welding with the machine, the measured arc current is 500 A corresponding to an arc length of 5.0 mm and the measured arc current is 460 A corresponding to an arc length of 7.0 mm.
The linear voltage (E) - arc length (L) characteristic of the welding arc can be given as (where E is in Voit and L is in mm)
(a) E = 20 + 2L (b) E = 20 + 8L
(c) E = 80 + 2L (d) E = 80 + 8L

[ME 2007 : 2 Marks]

22. In arc welding of a butt joint, the welding speed is to be selected such that highest cooling rate is achieved. Melting efficiency and heat transfer efficiency are 0.5 and 0.7, respectively. The area of the weld cross section is 5 mm^2 and the unit energy required to melt the metal is 10 J/mm^3. If the welding power is 2 kW, the welding speed in mm/s is closest to
(a) 4 (b) 14
(c) 24 (d) 34

[ME 2008 : 2 Marks]

23. Which one among the following welding processes uses non-consumable electrode?
(a) Gas metal arc welding
(b) Submerged arc welding
(c) Gas tungsten arc welding
(d) Flux coated arc welding

[ME 2011 : 1 Mark]

24. In a DC arc welding operation, the voltage-arc length characteristic was obtained as $V_{arc} = 20 + 5l$ where the arc length l was varied between 5 mm and 7 mm. Here V_{arc} denotes the arc voltage in Volts. The arc current was varied from 400 A to 500 A. Assuming linear power source characteristic, the open circuit voltage and the short circuit current for the welding operation are
(a) 45 V, 450 A (b) 75 V, 750 A
(c) 95 V, 950 A (d) 150 V, 1500 A

[ME 2012 : 2 Marks]

25. In a linear arc welding process, the heat input per unit length is inversely proportional to
(a) welding current
(b) welding voltage
(c) welding speed
(d) duty cycle of the power source

[ME 2015 : 1 Mark, Set-1]

26. A DC welding power source has a linear voltage-current (V-I) characteristic with open circuit voltage of 80 V and a short circuit current of 300 A. For maximum arc power, the current (in Ampere) should be set as______.

[ME 2015 : 2 Marks, Set-1]

27. During a TIG welding process, the arc current and arc voltage were 50 A and 60 V, respectively, when the welding speed was 150 mm/min. In another process, the TIG welding is carried out at a welding speed of 120 mm/min at the same arc voltage and heat input to the material so that weld quality remains the same. The welding current (in A) for this process is ______
(a) 40.00 (b) 44.72
(c) 55.90 (d) 62.25

[ME 2015 : 2 Marks, Set-2]

28. The welding process which uses a blanket of fusible granular flux is

(a) tungsten inert gas welding

(b) submerged arc welding

(c) electroslag welding

(d) thermit welding

[ME 2016 : 1 Mark, Set-2]

29. The voltage-length characteristic of a direct arc in an arc welding process is V = (100 + 40l), where l is the length of the arc in mm and V is arc voltage in volts. During a welding operation, the arc length varies between 1 and 2 mm and the welding current is in the range 200-250 A. Assuming a linear power source, the short circuit current is ____ A.

[ME 2016 : 2 Marks, Set-2]

30. In an arc welding process, welding speed is doubled. Assuming all other process parameters to be constant, the cross sectional area of the weld bead will

(a) increase by 25%

(b) increase by 50%

(c) reduce by 25%

(d) reduce by 50%

[ME 2017 : 1 Mark, Set-1]

31. A welding operation is being performed with voltage = 30 V and current = 100 A. The cross-sectional area of the weld bead is 20 mm^2. The work-piece and filler are of titanium for which of the specific energy of melting is 14 J/mm^3.

Assuming a thermal efficiency of the welding process 70%, the welding speed (in mm/s) is (correct to two decimal places).

[ME 2018 : 2 Marks, Set-2]

32. A gas tungsten arc welding operation is performed using a current of 250 A and an arc voltage of 20 V at a welding speed of 5 mm/s. Assuming that the arc efficiency is 70%, the net heat input per unit length of the weld will be ____ kJ/mm (round off to one decimal place).

[ME 2019 : 2 Marks, Set-2]

33. The mode of metal transfer in CO_2-MIG welding at low temperature is

(a) Spray transfer

(b) Short circuiting type of transfer

(c) Globular transfer

(d) Droplet transfer

[PI 1990 : 1 Mark]

34. In TIG welding a .... A .... and .... B .... electrode is used

(a) non-consumable/consumable

(b) coated/bare

[PI 1990 : 1 Mark]

35. High alloy steel components are preheated before welding for reducing

(a) heat affected zone

(b) total energy consumption

(c) total time of welding

(d) welding stresses

[PI 1991 : 1 Mark]

36. The voltage arc length characteristics of a DC arc is given by V = 20 + 40L, where L = arc length in cm. The power source characteristics can be approximated by a straight line. Open circuit is 80 V and short circuit current 1000 Amps. The optimum arc length in mm

(a) 0.5 (b) 1

(c) 5 (d) 10

[PI 1991 : 2 Marks]

Common Data Question 37 & 38:

A weld is made using MIG welding process with the following welding parameters:

Current : 200 A : Voltage : 25 V; Welding speed: 18 cm/min; Wire diameter : 1.2 mm; Wire feed rate : 4 m/min. Thermal efficiency of the process: 65%

37. The heat input per unit length of the weld in kJ/cm is

(a) 0.18 (b) 0.28

(c) 10.83 (d) 16.66

[PI 2001 : 2 Marks]

38. The area of cross-section of weld bead in mm^2 is

(a) 16.3 (b) 25.1

(c) 30.3 (d) 38.6

[PI 2001 : 2 Marks]

39. Which of the following process can be used for welding of Aluminum alloys?

P. Submerged arc welding

Q. Gas metal arc welding

R. Electroslag welding

S. Gas tungsten arc welding

(a) P and Q (b) Q and S

(c) Q and R (d) R and S

[PI 2003 : 2 Marks]

40. **Group-I**

P. Arc welding
Q. Friction welding
R. Solid state Welding
S. Laser welding

Group-II

1. Diffusion
2. Polarity
3. Focusing
4. Kinetic energy

	P	Q	R	S
(a)	4	1	3	2
(b)	3	2	4	3
(c)	1	2	4	3
(d)	2	4	1	3

[PI 2003 : 2 Marks]

41. Match List-I (welding defects)with List-II (Causes) and select the correct answer using the codes given below the lists:

List-I

A. Spatter
B. Distortion
C. Slag inclusion
D. Porosity

List-II

1. Damp electrodes
2. Arc blow
3. Improper cleaning in multi pass welding
4. Poor joint selection

	A	B	C	D
(a)	4	2	3	1
(b)	4	2	1	3
(c)	2	4	1	3
(d)	2	4	3	1

[PI 2003 : 2 Marks]

42. The current in Amperes used in resistance spot welding of plain carbon steel sheets (1 to 3 mm thick) lies within the range

(a) 10-50 (b) 50-500
(c) 500-5000 (d) 5000-5000

[PI 2004 : 1 Mark]

43. Consider the following statements:
The size of the heat affected zone (HAZ) will increase with

1. Increased starting temperature
2. Increased welding speed
3. Increased thermal conductivity of the base metal
4. Increase in base metal thickness

Which of the statements given above are correct?

(a) 1, 2 and 3 (b) 1 and 3
(c) 1 and 4 (d) 2 and 3

[PI 2005 : 2 Marks]

44. The DC power source for are welding has the characteristic 3V + I = 240, where V = Voltage and I = Current in amp. For maximum are power at the electrode, voltage should be set at

(a) 20 V (b) 40 V
(c) 60 V (d) 80 V

[PI 2007 : 2 Marks]

45. Autogeneous gas tungsten arc welding of a steel plate is carried out with welding current of 500 A. Voltage of 20 V and weld speed of 20 mm/min. Consider the heat transfer efficiency from the arc to the weld pool as 90%. The heat input per unit length (in kJ/mm) is

(a) 27 (b) 35
(c) 45 (d) 55

[PI 2009 : 2 Marks]

46. During a steady gas metal arc welding with direct current electrode positive polarity, the welding current, voltage and weld speed are 150 A, 30 V and 6 m/min respectively. A metallic wire electrode of diameter 1.2 mm is being fed at a constant rate of 12 m/min. The density, specific heat and melting temp of the wire electrode are 7000 kg/m^3, 500 J/kg and 1530°C respectively.
Assume the ambient temp to be 30°C and neglect the latent heat of melting. Further consider that 2/3rd of the total electrical power is available for melting of the wire electrode. The melting efficiency (in percentage) of the wire electrode is

(a) 39.58 (b) 45.25
(c) 49.38 (d) 54.98

[PI 2010 : 2 Marks]

47. Which of the following welding process results in the smallest heat affected zone?

(a) Shielded metal arc welding
(b) Gas welding
(c) Laser beam welding
(d) Thermit welding

[PI 2011 : 1 Mark]

Resistance Welding

48. For resistance spot welding of 1.5 mm thick steel sheets, the current required is of the order

(a) 10 Amp (b) 100 Amp
(c) 1000 Amp (d) 10,000 Amp

[ME 1991 : 1 Mark]

49. Filler material is .... A ... in resistance welding and the heat generated in the process is directly proportional to .... B....

(a) used/not used
(b) square of the current/cube of the current

[ME 1992 : 1 Mark]

50. Resistance spot welding is performed on two plates of 1.5 mm thickness with 6 mm diameter electrode, using 15000 A current for a time duration of 0.25 sec. Assuming the interface resistance to be 0.0001 ohms. The heat generated to form the weld is

(a) 5625 W-sec (b) 8437 W-sec

(c) 22500 W-sec (d) 33750 W-sec

[ME 2001 : 2 Marks]

51. The strength of a brazed joint

(a) decreases with increase in gap between the two joining surfaces

(b) increases with increase in gap between the two joining surfaces

(c) decreases up to certain gap between the two joining surfaces beyond which it increases

(d) increases up to certain gap between the two joining surfaces beyond which it decreases

[ME 2005 : 1 Mark]

52. Spot welding of two 1 mm thick sheets of steel (density = 8000 kg/m^3) is carried out successfully by passing a certain amount of current for 0.1 second through the electrodes. The resultant weld nugget formed is 5 mm in diameter and 1.5 mm thick. If the latent heat of fusion of steel is 1400 kJ/kg and the effective resistance in the welding operation is 200 μ–ohms, the current passing through the electrodes is approximately

(a) 1480 A (b) 3300 A

(c) 4060 A (d) 9400 A

[ME 2005 : 2 Marks]

53. Two metallic sheets, each of 2.0 mm thickness, are welded in a lap joint configuration by resistance spot welding at a welding current of 10 kA and welding time of 10 millisecond. A spherical fusion zone extending up to the full thickness of each sheet is formed and properties of the metallic sheets are given as:

Ambient temperature = 293 K

Melting temperature = 1793 K

Density = 7000 kg/m^3

Latent heat of fusion = 300 kJ/kg

Specific heat = 800 J/kgK. Assume:

(i) contact resistance along sheet-interface is 500 micro-ohm and along electrode-sheet interface is zero;

(ii) no conductive heat loss through the bulk sheet materials; and

(iii) the complete weld fusion zone is at the melting temperature.

The melting efficiency (in %) of the process is

(a) 50.37 (b) 60.37

(c) 70.37 (d) 80.37

[ME 2007 : 2 Marks]

54. Two pipes of inner diameter 100 mm and outer diameter 110 mm each are joined by flash butt welding using 30 V power supply. At the interface, 1 mm of material melts from each pipe which has a resistance of 42.4 ohms. If the unit melt energy is 64.4 MJ/m^3, then time required for welding (in s) is

(a) 1 (b) 5

(c) 10 (d) 20

[ME 2010 : 2 Marks]

55. For spot welding of two steel sheets (base metal) each of 3 mm thickness, welding current of 10000 A is applied for 0.2 s. The heat dissipated to the base metal is 1000 J. Assuming that the heat required for melting 1 mm^3 volume of steel is 20 J and interfacial contact resistance between sheets is 0.0002 ohms, the volume (in mm3) of weld nugget is .

[ME 2014 : 2 Marks, Set-3]

56. Spot welding of two steel sheets each 2 mm thick is carried out successfully by passing 4 kA of current for 0.2 seconds through the electrodes. The resulting weld nugget formed between the sheets is 5 mm in diameter. Assuming cylindrical shape for the nugget, the thickness of the nugget is ________ mm.

Latent heat of fusion for steel	1400 kJ/kg
Effective resistance of the weld joint	200 micro-ohms
Density of steel	8000 kg/m3

[ME 2016 : 2 Marks, Set-3]

57. The type of weld represented by the shaded region in the figure is

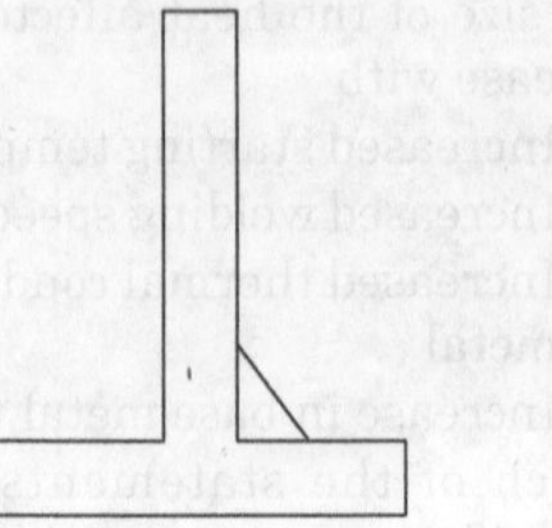

(a) groove (b) spot

(c) fillet (d) plug

[ME 2018 : 1 Marks, Set-1]

58. In spot pulsed laser welding of aluminum plates (density = 2700 kg/m^3, specific heat = 896 J/kg, melting temperature = 933 K, latent heat of melting = 398 kJ/kg) at a temp of 30°C, pulse with energy of 0.5 J is focused onto an area of 0.05 mm^2. If the entire energy is coupled into the material, what will be the depth of weld assuming the cross sectional area of the weld is circular and is uniform throughout its depth and only heat conduction in the direction of penetration.

(a) 5.34 mm (b) 2.15 mm
(c) 4.23 mm (d) 3.85 mm

[PI 2002 : 2 Marks]

59. In resistance welding, heat is generated due to the resistance between

(a) electrode and workpiece
(b) asperities between touching plates
(c) two dissimilar metals being in contact
(d) interatomic forces

[PI 2003 : 1 Mark]

60. A 900 mm long steel plate is welded by manual metal arc welding process using welding current of 150 A, arc voltage of 20 V and welding speed of 300 mm/min. If the process efficiency of is 0.8 and surface resistance is 36 micro-ohm, the heat input in J/mm

(a) 600 J/mm (b) 480 J/mm
(c) 146 J (d) 116.5 J

[PI 2003 : 2 Marks]

61. The metal powder used in Thermit welding of steel is

(a) AI (b) Cu
(c) Pb (d) W

[PI 2004 : 1 Mark]

62. In a fabrication shop, a T-joint is to be made between two plates using a chin intermittent, double fillet weld of 6 mm leg length. The intermittent welds are 40 mm long and spaced 80 mm apart, edge to edge. The welding symbol of the joint is

(a)
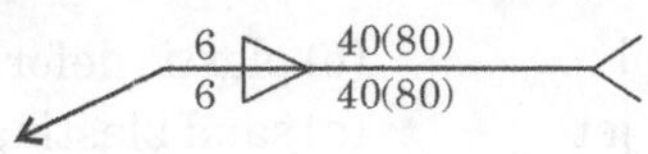

(b)
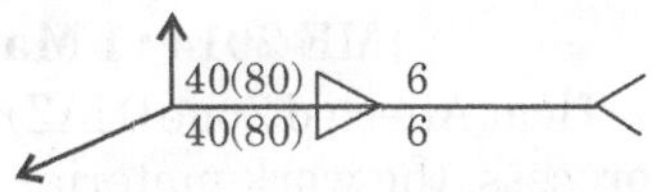

(c)
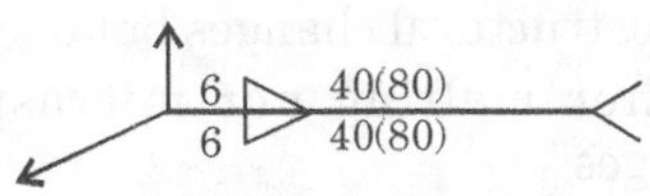

(d)
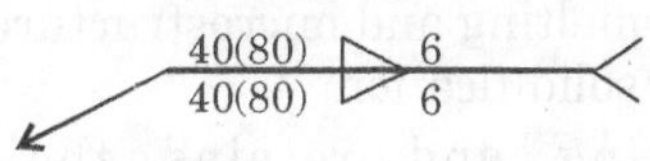

[PI 2005 : 2 Marks]

63. Two 8 mm thick steel plates are placed 5 mm apart and welded by a butt joint. Welding is carried out at 20 V and speed of 5 mm/s. Heat transfer efficiency is 0.80. If the heat required to melt steel is 10 J/mm^3 and melting efficiency is 0.625. The weld current (in amperes) will be

(a) 100 (b) 200
(c) 300 (d) 400

[PI 2005 : 2 Marks]

64. Consider the following statements:

The magnitude of residual stresses in welding depends upon

1. Metal melted/deposited
2. Design of weldment
3. Support and clamping of components
4. Welding process used

Which of the statements given above are correct?

(a) 1, 2 and 4 (b) 1, 2 and 3
(c) 1 and 3 (d) 2 and 3

[PI 2005 : 2 Marks]

65. Aluminum strips of 2 mm thickness are joined together by resistance spot welding process by applying an electric current 6000 A and 0.15 second. The heat required for melting aluminum is 2.9 J/mm^3. The diameter and thickness of the weld nugget are found to be 5 mm and 2.5 mm respectively. Assuming the electrical resistance to be 75 micro ohms, the percentage of total energy utilized in forming the weld nugget is

(a) 28 (b) 35
(c) 65 (d) 72

[PI 2008 : 2 Marks]

Linked Data Questions 66 & 67:

Resistance spot welding of two steel sheets is carried out in lap joint configuration by using a welding current of 3 kA and a weld time of 0.2 s. A molten weld nugget of volume 20 mm^3 is obtained. The effective contact resistance is 200 micro-ohms. The material properties of steel are given as (i) latent heat of melting 1400 kJ/kg, (ii) density 8000 kg/m^3, (iii) melting temp 1520°C (iv) specific heat 0.5 kJ/kg°C. The ambient temp is 20°C.

66. Heat (in Joules) used for production weld nugget will be (assuming 100% heat transfer efficiency)

(a) 324 (b) 334
(c) 344 (d) 354

[PI 2009 : 2 Marks]

67. Heat (Joules) dissipated to the base metal will be (neglecting all other heat losses)

(a) 10 (b) 16
(c) 22 (d) 32

[PI 2009 : 2 Marks]

68. In resistance seam welding, the electrode is in the form of a

(a) cylinder (b) flat plate
(c) coil of wire (d) circular disc

[PI 2011 : 1 Mark]

Solid and Advanced Welding, Brazing and Soldering

69. Welding process

A. Thermit welding
B. Seam welding
C. MIG welding
D. Friction welding

Heat source

1. Chemical reaction
2. Exothermic Chemical reaction
3. Electric spark
4. Ohmic resistance
5. Mechanical work
6. Electric arc

[ME 1991 : 2 Marks]

70. In an explosive welding process, the (maximum/ minimum) velocity of impact is fixed by the velocity of sound in the (flyer/ target) plate material.

[ME 1992 : 1 Mark]

71. Match the following

List-I

A. Welding of aluminum alloy
B. Ship building
C. Joining of HSS drill bit to shank
D. Deep penetration precision welds

List-II

1. Submerged arc welding
2. Electron beam welding
3. TIG welding
4. Friction welding
5. Gas welding

[ME 1996 : 2 Marks]

72. Match the following

Work material

P. Aluminium
Q. Die steel
R. Copper Wire
S. Titanium sheet

Type of joining

1. Submerged Arc Welding
2. Soldering
3. Thermit Welding
4. Atomic Hydrogen Welding
5. Gas Tungsten Arc Welding
6. Laser Beam Welding

(a) P-2, Q-5, R-1, S-3
(b) P-6, Q-3, R-4, S-4
(c) P-4, Q-1, R-6, S-2
(d) P-5, Q-4, R-2, S-6

[ME 2003 : 2 Marks]

73. Which one of the following is a solid State joining process?

(a) gas tungsten arc welding
(b) resistance spot welding
(c) friction welding
(d) submerged arc welding

[ME 2007 : 1 Mark]

74. Match the CORRECT pairs

Processes	Characteristics/ Applications
P. Friction Welding	1. Non-consumable electrode
Q. Gas Metal Arc welding	2. Joining of thick plates
R. Electroslag Welding	3. Consumable electrode wire
S. Tungsten Inert Gas welding	4. Joining of cylindrical dissimilar materials

(a) P-4, Q-3, R-2, S-1
(b) P-4, Q-2, R-3, S-1
(c) P-2, Q-3, R-4, S-1
(d) P-2, Q-4, R-1, S-3

[ME 2013 : 1 Mark]

75. The major difficulty during welding of aluminium is due to its

(a) high tendency of oxidation
(b) high thermal conductivity
(c) low melting point
(d) low density

[ME 2014 : 1 Mark, Set-1]

76. In solid-state welding, the contamination layers between the surfaces to be welded are removed by

(a) alcohol (b) plastic deformation
(c) water jet (d) sand blasting

[ME 2014 : 1 Mark, Set-1]

77. Within the Heat Affected Zone (HAZ) in a fusion welding process, the work material undergoes

(a) microstructural changes but does not melt
(b) neither melting nor microstructural changes
(c) both melting and microstructural changes after solidification
(d) melting and retains the original microstructure after solidification

[ME 2014 : 1 Mark, Set-4]

78. Which two of the following joining processes are autogenous?
i. Diffusion welding
ii. Electroslag welding
iii. Tungsten inert gas welding
iv. Friction welding
(a) i and iv (b) ii and iii
(c) ii and iv (d) i and iii
[ME 2015 : 1 Mark, Set-3]

79. Under optimal conditions of the process the temperatures experienced by a copper work piece in fusion welding, brazing and soldering are such that
(a) $T_{welding} > T_{soldering} > T_{brazing}$
(b) $T_{soldering} > T_{welding} > T_{brazing}$
(c) $T_{brazing} > T_{welding} > T_{soldering}$
(d) $T_{welding} > T_{brazing} > T_{soldering}$
[ME 2016 : 1 Mark, Set-1]

80. Heat is removed from a molten metal of mass 2 kg at a constant rate of 10 kW till it is completely solidified. The cooling curve is shown in the fig.

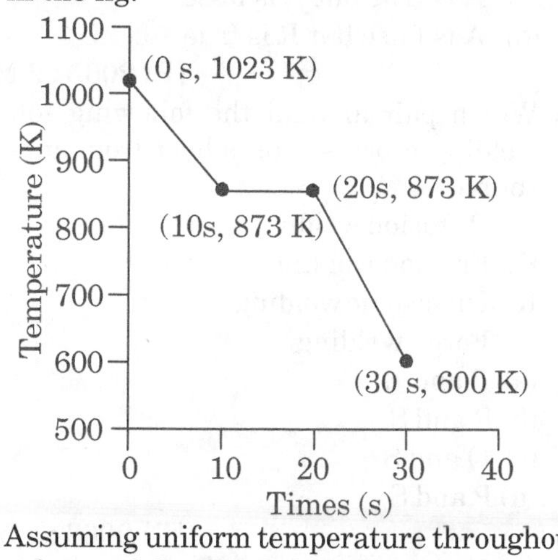

Assuming uniform temperature throughout the volume of the metal during solidifications, the latent heat of fusion of the metal (in kJ/kg) is _______
[ME 2016 : 2 Marks, Set-1]

81. Which one of the following welding methods provides the highest heat flux(W/mm²)?
(a) Tungsten inert gas welding
(b) Laser beam welding
(c) Oxyacetylene gas welding
(d) Plasma arc welding
[ME 2019 : 1 Mark, Set-1]

82. Match the following
List-I
A. Cellulose coating
B. Rutile coating
C. Iron powder coating
D. Basic coating
List-II
1. Alloying agent
2. Absorb gasses
3. Forming slag
4. Shielding gasses
[PI 1989 : 2 Marks]

83. With increasing Joint thickness, the tensile strength of a brazed Joint
(a) Continuously decreases
(b) First decreases and then increases
(c) Continuously increases
(d) First increases and then decreases
[PI 1990 : 1 Mark]

84. Match the following
Welding Process
A. Thermit welding
B. Projection welding
C. MIG welding
D. Friction welding
Heat Source
1. Electric arc
2. Mechanical work
3. Exothermic Chemical reaction
4. Ohmic resistance
[PI 1990 : 1 Mark]

85. Continuous rails for Indian railways are welded by ______ welding process.
[PI 1991 : 1 Mark]

86. Brazing Is the non-fusion method of joining dissimilar metals with metals whose melting temperature is greater than 400°C.
[PI 1991 : 1 Mark]

87. The joint configuration best suited for adhesive bonding is
(a) butt (b) fillet
(c) lap (d) spot
[PI 1992 : 1 Mark]

88. The type of coated electrode most widely used for welding low carbon steel is
(a) cellulose (b) acidic
(c) rutile (d) oxide
[PI 1992 : 1 Mark]

89. Welding application process
List-I
A. High reactive metals
B. Long straight butt joint in thick sections
C. Site weld of rails
D. MS rods to plate like components
List-II
1. Ultrasonic
2. Thermit
3. Electron beam
4. Submerged arc
5. Stud
[PI 1992 : 2 Marks]

90. High speed electron beam welding is focused on the weld spot using
(a) vacuum lens (b) inert gas lens
(c) optical lens (d) magnetic lens
[PI 2003 : 1 Mark]

91. The cold cracking susceptibility of the heat affected zone in an arc weld is influenced by
P. Entrapped hydrogen
Q. Residual stresses
R. Martensitically transformation
S. Slag inclusion
(a) P and S (b) P, Q and S
(c) P, Q and R (d) R and S
[PI 2003 : 2 Marks]

92. Match List-I (welding problems) with List-II (Causes) and select the correct answer using the codes given below the lists:

List-I
A. Cracking of weld metal
B. Cracking of base metal
C. Porosity
D. Inclusions

List-II
1. Excessive stresses
2. High joint rigidity
3. Failure to remove slag from previous deposit
4. Oxidation
5. Excessive H_2, O_2, N_2 in the welding atmosphere

Codes:

	A	B	C	D
(a)	2	1	5	3
(b)	3	4	2	1
(c)	2	4	5	3
(d)	3	1	4	2

[PI 2005 : 2 Marks]

93. Assertion (A): A sound welded joint should not only be strong enough but should also exhibits a good amount of ductility.
Reason (R): Welding process is used for fabricating mild Steel components only.
(a) Both A and R are true and R is the correct explanation of A
(b) Both A and R are true but R is not a correct explanation of A
(c) A is true but R is false
(d) A is false but R is true
[PI 2005 : 2 Marks]

94. Assertion (A): In electron beam welding process, vacuum is an essential process parameter.
Reason (R): Vacuum provides a highly efficient shield on weld zone.
(a) Both A and R are true and R is the correct explanation of A
(b) Both A and R are true but R is not a correct explanation of A
(c) A is true but R is false
(d) A is false but R is true
[PI 2005 : 2 Marks]

95. Which pair amount the following solid state welding processes uses heat from an external source
P. Diffusion welding
Q. Friction welding
R. Ultrasonic welding
S. Forge welding
(a) P and R
(b) R and S
(c) Q and S
(d) P and S
[PI 2008 : 2 Marks]

ANSWERS

1. (b)	**3.** (a)	**4.** (a)	**5.** (b)	**8.** (a)	**9.** (c)	**11.** (a)	**12.** (b)	**13.** (d)	**14.** (b)
15. (d)	**16.** (d)	**17.** (a)	**18.** (b)	**19.** (d)	**20.** (b)	**21.** (a)	**22.** (b)	**23.** (c)	**24.** (c)
25. (c)	**27.** (a)	**28.** (b)	**30.** (d)	**33.** (d)	**35.** (d)	**36.** (c)	**37.** (c)	**38.** (b)	**39.** (b)
40. (d)	**41.** (d)	**42.** (d)	**43.** (b)	**44.** (b)	**45.** (a)	**46.** (a)	**47.** (c)	**48.** (d)	**50.** (a)
51. (d)	**52.** (c)	**53.** (c)	**54.** (c)	**57.** (c)	**58.** (d)	**59.** (b)	**60.** (b)	**61.** (a)	**62.** (d)
63. (b)	**64.** (b)	**65.** (b)	**66.** (c)	**67.** (b)	**68.** (d)	**72.** (d)	**73.** (c)	**74.** (a)	**75.** (a)
76. (b)	**77.** (a)	**78.** (a)	**79.** (d)	**81.** (b)	**83.** (d)	**87.** (c)	**88.** (c)	**90.** (d)	**91.** (c)
92. (d)	**93.** (c)	**94.** (c)	**95.** (d)						

EXPLANATIONS

1. In neutral flame, volume of both oxygen and acetylene are equal.

2. 1 : 1 in terms of volume but not in weight.

3. No. of jobs is very high, therefore we need a welding process which is capable of welding in one pass. Therefore, SAW is chosen over others.

4. Due to insufficient acetylene, complete oxygen is not utilized that is why lower temperature than neutral or oxidising flame.

Carburising flame – 2900°C

Neutral flame – 3100°C

Oxidising flame – 3300°C

5.

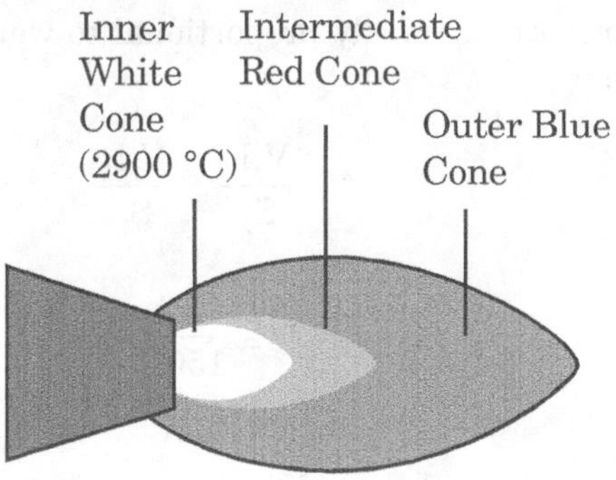

Carburizing Flame

6. Energy $= \frac{1}{2}CV^2$

$$\Rightarrow E = \frac{1}{2} \times 50 \times 10^{-6} \times (200)^2$$

$\boxed{E = 1J}$

7. Must use an oxidising flame, fuming is virtually eliminated.

9. 2 : 1 Ratio is maintained means 16 liters of oxygen will be used.

10. Yes. Oxyacetylene gas cutting is an oxidation process.

11. Powder of steel is used with oxygen such a flame. This powder burns producing large amount of heat facilitating cutting.

12. By increasing the current, large heat is generated and due to lesser speed heat is generated at particular area so that larger penetration can be carried out.

13. Constant current type of transformer is used in Manual Arc welding otherwise weld bead of irregular shape and penetration

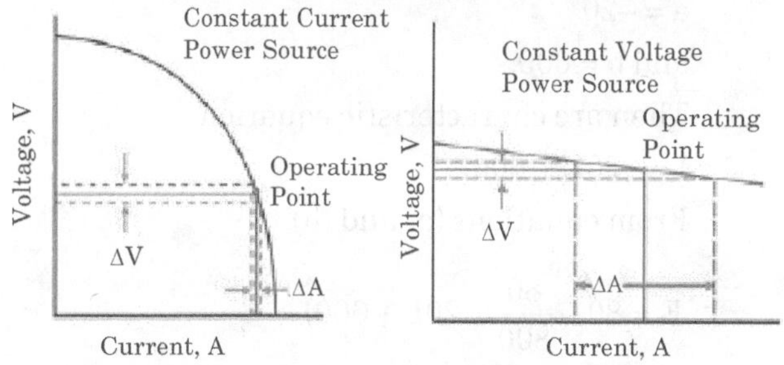

14. Lower penetration because of 2/3rd heat is generated at work piece so lower deposition rate of consumable electrode or filter material can be observed.

15. Coating is provided for first three options not for prevention from external contaminates.

16. Due to decreased Thermal conductivity, at higher temperature less heat transportation so less thermal stresses leads to less distortion.

17. Increasing the current setting increases the heat generation, so larger material Deposition takes place.

18. GTAW: Non consumable Tungsten Electrode is used & separate filler material is used.

19. Heat generated during the process = $1^2R\Delta t$

$= (5000)^2 \times (200 \times 10^{-6}) \times 0.2$

$= 1000$ Joules.

20. $V = 25$ Volt, $I = 300$ A, $\eta = 0.85$

Weld speed = 8 mm/s.

Power Generated at arc = V.I

$= 25 \times 300 = 7500$ J/S

Heat transferred $= 0.85 \times 7500$

$= 6375$ J/S

So net Heat Input = 6375/8 = 797 J/mm.

21. The power source characteristic can be written analytically as

$$E = 80 - \frac{80}{800}I \qquad (a)$$

The arc characteristic is given as

$I = aL + b$

where a and b are constant

Given,

When I = 500A

then L = 5.00 mm

$\therefore 500 = 5a + b \qquad ...(i)$

when I = 460A

then L = 7.00 mm

$\therefore 460 = 7a + b$ ...(ii)

Solving Eqs. (i) and (ii, we get

$a = -20$

and $b = 600$

Then arc characteristic equation

$I = -20L + 600$

From equations (a) and (b)

$$E = 80 - \frac{80}{800}(-20L + 600)$$

$= 80 - 0.1(-20L + 600)$

$= 80 + 2L - 60$

$= 20 + 2L$

22. Welding power applied = Heating power needed

$\Rightarrow P \times \eta_m \times \eta_{H.T.} = E_U \times (A) \times f$

$\Rightarrow 2 \times 10^3 \times 0.5 \times 0.7 = 10 \times 5 \times f$

$\Rightarrow f = 14$ mm/sec

23. Tungsten Electrode is used to create arc only. Reverse polarity is used to weld Aluminium metal.

24. $V_{ar} = 20 + 5l$

For $l = 5$ mm, $V_{arc} = 45$ V

For $l = 7$ mm, $V_{arc} = 55$V

$$V = V_0 - \frac{V_0}{I_S}I$$

where V_o is open circuit voltage and I_s is short circuit current.

$$\therefore \quad 45 = V_0 - \frac{V_0}{I_S} \times 500$$

and $$55 = V_0 - \frac{V_0}{I_S} \times 400$$

By solving, we get

$V_0 = 95$ Volts

$I_S = 950$ A

25. $$\text{Heat Input} = \frac{\text{Voltage} \times \text{current}}{\text{welding speed}}$$

$$\text{So}_{\text{Heat Input}} \propto \frac{1}{\text{welding speed}}$$

26. Constant Voltage characteristics

$$150A = v_{ocV} - \left(\frac{V_{ocV}}{1_{scc}}\right)I$$

$$V = 80 - \left(\frac{80}{300}\right)I$$

$P = VI$

$$= \left[80 - \left(\frac{80}{300}\right)I\right]I$$

For Maximum Power

$$\frac{dP}{dI} = 0$$

$80 - 2 \times 0.666\, I = 0$

$\boxed{I = 150A}$

27. Total heat input = VIt

Time t is inversely proportional to weld speed (S)

$$\therefore \quad \frac{V_2 I_2}{S_2} = \frac{V_1 I_1}{S_1}$$

$$\therefore \quad I_2 = \frac{120}{150} \times 50$$

$I_2 = 40$ A

28. In submerged arc welding, the arc is completely submerged inside the granular flux powder and thus forming as blanket.

29. Given, $V = 100 + 40\,l$

$l_1 = 1\text{mm} \Rightarrow V_1 = 140$ V

$I_1 = 250$ A

$l_2 = 2\text{ mm} \Rightarrow V_2 = 180$ V

$I_2 = 200$ A

$$\frac{V}{OCV} + \frac{I}{SSC} = 1$$

$$\frac{140}{OCV} + \frac{250}{SSC} = 1$$

$$\frac{180}{OCV} + \frac{200}{SSC} = 1$$

On solving we get

SSC = 424.6 A

30. If welding speed is doubled then original cross sectional area would become half if volume remains constant. Therefore, area reduces by 50%.

31. Voltage = 30 V, I = 100 A

Area = 30 mm^2

Specific Energy = 14 J/mm^3

$\eta_{th} = 70\%$

welding speed V = ?

Power = V × I = 30 × 100 = 300 J/s

Transferred Power = 0.7 × 3000 = 2100 J/S

$$\text{Specific Energy} = \frac{2100}{A \times V}$$

$$V = \frac{2100}{14 \times 20}$$

$$\boxed{V = 7.5\text{mm/s}}$$

32. Net heat input per unit length,

$$H_t = \frac{V.I}{v} \times \eta$$

V = voltage, I = current, A

v = welding speed

$$H_t = \frac{20 \times 250}{5} \times 0.70$$

$$H_t = 0.7\text{kJ/mm}$$

33. Droplet transfer because of higher surface tension force.

34. Non consumable base electrode is used.

35. Due to preheating, thermal conductivity will decrease so less thermal stressese and hence less distortion will take place.

36. Constant voltage Characteristics

V = 20 + 40 L ...(i)

Vocv = 80 V, I_{scc} = 1000 Amp.

$$V = V_{ocv} - \frac{V_{ocv}}{I_{scc}}.I$$

$$V = 80 - \frac{80}{1000}I \quad \text{...(ii)}$$

$$V_{transfermer} = V_{arc}$$

$$20 + 40L = 80 - \frac{80}{1000}L$$

$$I = \frac{1000}{80}(60 - 40L)$$

Power = V.I.

$$\text{For optimum Arc length } \frac{dP}{dL} = 0$$

$$= (20 + 40L)(60 - 40L)\frac{1000}{80}.$$

Hence L = 0.5 cm = 5mm

37. Current = 200A

Voltage = 25 V

d_w = 1.2 mm

Fr = 4 m/min

= 4000 mm/min

V = 18 cm/min

= 180 mm/min.

η_{Th} = 65%

Heat Input per unit length

$$= \frac{V \times I \times \eta}{\text{speed}} = \frac{200 \times 25 \times 0.65}{\frac{18}{60}} = 10.833 \text{ kJ/cm}$$

38. Volume of weld

$$= \frac{\pi}{4}d^2 \times V$$

$$= \frac{\pi}{4}(1.2)^2 \times 4000$$

= 4521.6 mm³/min

$$\text{Area of weld} = \frac{4521.6}{\text{speed}} = \frac{4521.6}{18 \times 10} = 25.12\text{mm}^2$$

39. To remove the ceramic layer of aluminium oxide, TIG is used.

40. Arc welding - Polarity

Friction welding - Kinetic energy

Solid state welding - Diffusion

Laser welding - Focusing

41. Spotter - Arc blow

Distortion - Poor joint selection

Slag Inclusion-Improper cleaning in multipass welding

Porosity - Damp electrodes

42. Very high current is used in air gap.

43. H A Z increases due to higher temperature & higher heat dissipation due to higher thermal conductivity

44. Constant current characteristics:

3 V + I = 240

I = 240 – 3 V

Drooping characteristics:

P = V I = V × (240 – 3 V)

$$\frac{dP}{dV} = 0$$

V = 40 V

45. Current = 500 A

Voltage = 20 V

V = 20 mm/min

η_{HT}= 0.9

Heat Input/length = $\eta_{HT} \dfrac{V \times I}{N}$ =

$$= \frac{0.9 \times 20 \times 500}{\frac{20}{60}}$$

= 27 kJ/mm

46. I = 150 A, V = 30 V

Speed V = 6m/min.

d_w = 1.2 mm Fr = 12 m/min

$\rho_{electrode}$ = 7000 kg/m³

C = 500 kJ/kg

T_m = 1530°C T_A = 30°C

Heat Required for welding

$$= \frac{\pi}{4} d_w^2 \times Fr \times \delta_{electrode} \times C\Delta T$$

$$= \frac{\pi}{4}(1.2 \times 10^{-3})^2 \times \frac{12}{60} \times 7000 \times 500 \times [1530 - 30]$$

= 1187.52 Joules

Power developed = P = VI = 30 × 150 = 4500 J

Heat Available for melting $= \frac{2}{3} \times 4500 = 3000$ j

Melting efficiency $= \dfrac{1157.52}{3000} = 34.58\%$

47. Laser beam welding will have smallest HAZ due to the fact that laser beam is very fine in nature.

48.

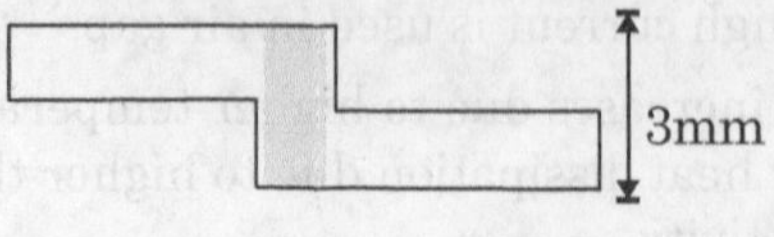

Air Resistance = 3300 ohm

Current required = 3300 × (– 1.5 + 1.5)

Current required = 9900 Amperes

= 10000 Amperes

49. Not used & square of the current

H = I² Rt.

50. t = 1.5mm, d = 6mm,I = 15000A, t = 0.25 sec

R = 0.000 1 Ω

H = I² R t

= 15000 × 0.0001 × 0.25 = 5625 W-sec

51. Strength increases up to some certain gap and beyond this strength decreases.

52. ρ = 8000 kJ/m³

t = 0.1 sec d_n = 5 mm

h_n = 1.5 mm

LH = 1400 kJ/kg R = 200 μ Ω

I = ?

$$\text{mass} = 8000 \times 10^{-9} \times \frac{\pi}{4} 5^2 \times 1.5$$

= 2.356 × 10⁻⁴kg.

Heat = 1400 × 8000 × 2.356 × 10⁻⁴

= 329.8 Joules

Heat = I² × 200 × 10⁻⁶ × 0.1

I = 4060 Amp.

53. Heat generated due to current,

H_1 = I²Rt= (10 × 10³)² × 500 × 10⁻⁶ × 10 × 10⁻³

= 500 J

Heat required for melting,

H_2 = 300 × 10³ × 7000 × volume + 800 × (1793 – 293) × 7000 × volume

Since fusion zone is spherical is shape, therefore

volume = $\frac{4}{3}\pi R^3$

where, *r* = *t* = thickness of sheet = 2 mm

Melting efficiency = 70.35%

54. Energy Required for melting/volume

E= 64.4 MJ/m³

V= 30 volts

R = 42.4 Ω

di = 100 mm

do = 110 mm

Melting Energy = $= \frac{\pi}{4}(d_0^2 - d_i^2) \times 644.4 \times 10^6$ j

= 212.32 Joules

$$H = I^2RT = \frac{V^2}{R} T$$

$\boxed{V = IR}$

$$T = \frac{H}{V^2} \times R = \frac{212.32 \times 42.4}{30^2}$$

$\boxed{T = 10 \text{sec}}$

55. Heat required = 20 J/mm^3

$I = 10000$ A

$R = 0.0002\ \Omega$

$t = 0.2$ sec

$H = I^2 R t$

$= (10000)^2 \times 0.0002 \times 0.2$

$= 4000$ Joules

Heat Required to melt the nugget

$= 4000 - 1000 = 3000$ Joules

Volume of Nugget

$$= \frac{3000}{20} = 150\text{mm}^3$$

56. $t = 2$ mm, $I = 4$ KA $= 4000$ A, $t = 0.25$ s

$d_n = 5$ mm, $h_n = ?$

Heat Generated = $I^2 R t$ ...(1)

Heat utilized = Volume × δ × LH ...(2)

(1) = (2)

$$4000^2 \times 200 \times 10^{-6} \times 0.2$$

$$= \frac{\pi}{4} 5^2 \times 8000 \times 1400$$

$$\boxed{h_n = 2.91\text{mm}}$$

57.

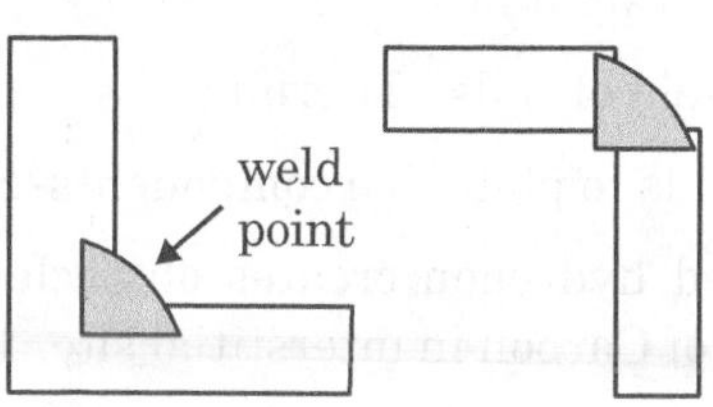

Fillet weld on corner joint

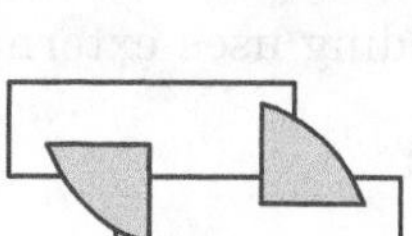

Fillet weld on lap joint

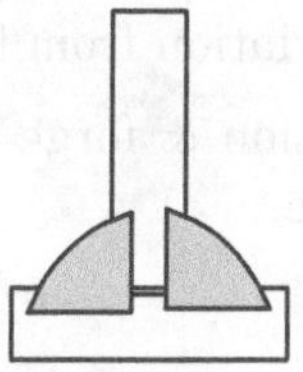

Fillet weld on T-joint

58. $\rho_{Al} = 2700$ kg/m^3

$C = 896$ J/kg

$T_m = 933$ k. $T_A = 30°C = 303$ k

LH = 398 kJ/kg

Heat supplied = 0.5 J

$A = 0.05$ mm^2

$$0.5 = 0.0.5x \times \frac{2700}{109}$$

$$\left[896(933 + 303) + 398 \times 10^3\right]$$

$$\boxed{x \simeq 3.85\text{mm}}$$

59. Because of Air Resistance is very high.

60. Heat Input to weld

$$= \eta\frac{VI}{V} = \frac{0.8 \times 20 \times 150}{\frac{300}{60}}$$

$= 480$ J/mm.

61. Exothermic Reaction

$3Fe_3O_4 + 8Al \rightarrow 9Fe + 4Al_2O_3 + \text{Heat}$

63. $t = 8$ mm, Gap = 5mm

$V = 20$ volts, $v = 5$ mm/s

$\eta_{Ht} = 0.8$, $H = 10$J/mm, $\eta_m = 0.625$

$I = ?$

Volume of weld pool = $8 \times 5 \times x$

$= 40\,x$ mm^3

Heat Required = $10 \times 40\,x = 400\,x$ J

$$\text{Heat Generated} = \eta_{HT}\eta_m \frac{VIx}{v}$$

$$400 = \frac{0.625 \times 20 \times I \times x \times 0.8}{5}$$

$$\boxed{I = 200\text{A}}$$

64. Residual stress depends on welding parameters such as current , voltage, speed, current.

65. $h_n = 2.5$ mm, $I = 6000$ A

$R = 75\ \mu\Omega$, $t = 0.15$sec

$d_n = 5$ mm Energy/Volume = 2.6 J/mm^3

Energy generated = $I^2 R t$

$= 6000^2 \times 75 \times 10^{-6} \times 0.15 = 405$ J

Energy Required = 2.9 × volume of Nugget

$$= 2.9 \times \frac{\pi}{4} d_n^2 h_n = 2.9 \times \frac{\pi}{4} 5^2 2.5 = 142.35 \text{ joules}$$

$$\text{Efficiency} = \frac{142.35}{405} \times 100$$

$\simeq 35\%$

66. I = 3000 A

t = 0.2 sec R = 200 μΩ

V = 200 mm³

LH = 1400 kJ/kg

ρ = 8000 kg/m³

C = 0.5 kJ/kg °C

Heat Generation = $I^2 RT$

$= 3000 \times 200 \times 10^{-6} \times 0.2$

= 360 Joules.

Heat utilised for production of weld Nugget

$= m\,[LH + C\Delta T]$

$800 \times 20 \times 10^{-9}[1400 + (1500 \times 0.5)] \times 1000$

= 344 Joules

67. Heat dissipated = 360 – 344 = 16 J

69. A-2, B-4, C-6, D-5

70. Maximum & Flyer

71. A-3, B-1, C-4, D-2

72. P-5, Q-4, R-2, S-6

73. Friction welding is a solid state welding.

75. Aluminum oxide layer which is ceramic in nature has higher melting point which decreases the quality of welding.

77. Temperature is above the Recrystalization temperature due to which microstructural changes takes place

78. Autogeneous → No filter material is used so Diffusion welding & friction welding

79. Temp of Brazing > 450°C

Temp of soldering < 450°C

80. Heat Removed Rate = 10 KJ/s = 10 kW

Time for Removal = (20 – 10) = 10 sec

Latent Heat of fusion = 10 × 10

$$= 100 \frac{\text{kJ}}{\text{Sec}} \times \text{Sec} = 100 \text{ kJ}$$

but mass = 2 kg

$$80 = \frac{100}{2} = 50 \text{ kJ/kg}$$

81. In laser beam welding, heat flux is highest upto 1 mW/mm²

82. A. Cellulose coating - Shielding gases

B. Rutile coating - Foaming slag

C. Iron Powder - Alloying Agent coating

D. Basic coating - Absorb Gases

84. Thermit welding - Exothermic chemical Reaction

Projection welding - Ohmic Resistance

MIG welding - Mechanical work

85. Continuous Rails for Indian Railways are welded by <u>Thermit welding</u> process. Basically, it is a casting process.

86. True. Yes, Brazing is used for high melting point dissimilar metal & soldering used for less 400°C.

87. Contact area is more and can be increased in case of lap joint. Therefore suitable for adhesive bonding.

88. Rutile: High proportion of titanium oxide create a shielding gas like co_2

This flux also contains cellulose.

89. A. High Reactive metals - EBM

B. Long straight butt joint - SAW

C. Site weld of rails - Thermit

D. MS Rods to plate like components- Stud

91. Entrapped hydrogen creates obstacle in the diffusion of Carbon in interstitial site.

92. A – 2 B – 1 C – 5 D – 3

93. Welding can be used for variety of material even it can be used for dissimilar materials.

94. EBW is carried out in vacuum to avoid dispersion or deviation from the focus.

95. Diffusion & forge welding uses external heat source.

CHAPTER 6

Machining and Machine Tool Operation

Analysis of Previous GATE Papers														
Topic	**Year/Set → Marks Distribution ↓**		**2019 Set 1**	**2019 Set 2**	**2018 Set 1**	**2018 Set 2**	**2017 Set 1**	**2017 Set 2**	**2016 Set 1**	**2016 Set 2**	**2016 Set 3**	**2015 Set 1**	**2015 Set 2**	**2015 Set 3**
Mechanism of Metal Cutting	**1 Mark**	MCQ Type												
		Numerical Type												
	2 Marks	MCQ Type	1											
		Numerical Type		1										
	Total		**2**	**2**										
Cutting tool geometry and tool life	**1 Mark**	MCQ Type				1								
		Numerical Type												
	2 Marks	MCQ Type				1								
		Numerical Type	1		1									
	Total		**2**		**2**	**3**								
Machining Process	**1 Mark**	MCQ Type	1											
		Numerical Type												
	2 Marks	MCQ Type												
		Numerical Type		1										
	Total		**1**	**2**										
Finishing Process	**1 Mark**	MCQ Type			2									1
		Numerical Type							1	1		1		
	2 Marks	MCQ Type						1	1		1			1
		Numerical Type					2	1		1	1	1	1	
	Total				**4**		**4**	**4**	**3**	**3**	**4**	**3**	**2**	**3**

Mechanism of Metal Cutting

1. The size of BUE in metal cutting increases with

(a) very high speed

(b) large uncut chip thickness

(c) use of cutting fluid

(d) increase in positive rake angle

[ME 1989 : 1 Mark]

2. Crater wear always starts at some distance from the tool tip because at that point

(a) cutting fluid does not penetrate

(b) chip tool interface temperature is maximum

(c) normal stress on rake face is maximum

(d) tool strength is minimum

[ME 1989 : 1 Mark]

3. Most of the metal cutting heat goes into the

(a) Moving chip (b) Cutting tool

(c) Work material (d) Machine tool

[ME 1990 : 1 Mark]

4. In comparison to HSS, super HS contains higher percentage of

(a) Tungsten (b) Carbon

(c) Vanadium (d) Cobalt

[ME 1990 : 1 Mark]

5. In a typical metal cutting operation, using a cutting tool a positive rake $\alpha = 10°$, it is as observed that the shear angle was 20°. The friction angle is

(a) 45° (b) 30°

(c) 60° (d) 40°

[ME 1997 : 2 Marks]

6. A cutting tool has a nose radius of 1.8 mm the feed rate for a theoretical surface roughness of Rt = 5 microns is

(a) 0.268 mm/rev (b) 0.187 mm/rev

(c) 0.036 mm/rev (d) 0.0187 mm/rev

[ME 1997 : 2 Marks]

7. In orthogonal machining operation, the chip thickness and the uncut chip thickness are equal to 0.45 mm. If the tool rake and is 0 deg. The shear plane angle is

(a) 45°

(b) 30°

(c) 18°

(d) 60°

[ME 1998 : 2 Marks]

8. What is the approximate % change in the life, f, of the tool with zero rake angle used in orthogonal cutting when its clearance angle, α, is changed from 10 to 7 deg?

(Hint : Flank wear rate is proportional to cot α)

(a) 30% increase (b) 30% decrease

(c) 70% increase (d) 70% decrease

[ME 1999 : 2 Marks]

9. A built up edge is formed while machining

(a) ductile material at high speed

(b) ductile materials at low speed

(c) brittle materials at high speed

(d) brittle materials at low speed

[ME 2000 : 1 Mark]

10. During orthogonal cutting of MS with a 10 deg rake angle tool, the chip thickness ratio was obtained as 0.4. The shear angle (in degrees) evaluated from this data is

(a) 6.53 (b) 20.22

(c) 22.94 (d) 50.00

[ME 20017 : 2 Marks]

11. Friction at the tool-chip interface can be reduced by

(a) decreasing the rake angle

(b) increasing the depth of cut

(c) decreasing the cutting speed

(d) increasing the cutting speed

[ME 2009 : 1 Mark]

12. Minimum shear-strain in orthogonal turning with a cutting tool of zero rake angle is

(a) 0.0 (b) 0.5

(c) 1.0 (d) 2.0

[ME 2009 : 2 Marks]

Common Data Q.13 and 14:

In Orthogonal turning of a bar of 100 mm diameter with a feed of 0.25 mm/rev, depth of cut of 4 mm and cutting velocity of 90m/min. it is observed that the main (tangential) cutting force is perpendicular to the friction force acting at the chip-tool interface. The main(tangential) cutting force is 1500 N.

13. The orthogonal rake angle of the cutting tool in degree is

(a) zero (b) 3.58

(c) 5 (d) 7.16

[ME 2013 : 2 Marks]

14. The normal force acting the chip-tool interface in N is

(a) 1000 (b) 1500

(c) 2000 (d) 2500

[ME 2013 : 2 Marks]

15. The main cutting force acting on a tool during the turning (orthogonal cutting) operation of a metal is 400 N. The turning was performed using 2 mm depth of cut and 0.1 mm/rev feed rate.

The specific cutting pressure (in N/mm^2) is

(a) 1000 (b) 2000

(c) 3000 (d) 4000

[ME 2014 : 1 Mark, Set-1]

16. Match the Machine Tools (Group A) with the probable Operations (Group B):

Group A	Group B
P. Centre Lathe	1. Slotting
Q. Milling	2. Counter-boring
R. Grinding	3. Knurling
S. Drilling	4. Dressing

(a) P-1, Q-2, R-4, S-3

(b) P-2, Q-1, R-4, S-3

(c) P-3, Q-1, R-4, S-2

(d) P-3, Q-4, R-2, S-1

[ME 2014: 1 Mark, Set-2]

17. In orthogonal turning of a cylindrical tube of wall thickness 5 mm, the axial and the tangential cutting forces were measured at 1259 N and 1601 N, respectively. The measured chip thickness after machining was found to be 0.3 mm. The rake angle was 10° and the axial feed was 100 mm/min. The rotational speed of the spindle was 1000 rpm. Assuming the material to be perfectly plastic and merchant's first solution, the shear strength of the martial is closest to

(a) 875 MPa (b) 200 MPa

(c) 722 MPa (d) 920 MPa

[ME 2019 : 2 Marks, Set-1]

18. In an orthogonal machining with a single point cutting tool of rake angle 10°, the uncut chip thickness and the chip thickness are 0.125 mm and 0.22 mm, respectively. Using Merchant's first solution for the condition of minimum cutting force, the coefficient of friction at the chip-tool interface is _____ (round off to two decimal places).

[ME 2019 : 2 Marks, Set-2]

19. In HSS, the tungsten can be substituted by

(a) Chromium (b) Nickel

(c) Molybdenum (d) Cobalt

[PI 1989 : 1 Mark]

20. BUE formation .... A... the cutting force and ... B.. the surface finish

(a) decreases/increases

(b) spoils/improves

[PI 1990 : 1 Mark]

21. **Assertion (A):** A diamond tool is used for USM of glass workpiece.

Reason (R): Diamond is harder than glass

(a) Both A and R are true and R is the correct explanation of A

(b) Both A and R are true but R is not a correct explanation of A

(c) A is true but R is false

(d) A is false but R is true

[PI 1990 : 1 Mark]

22. The diameter and rotational speed of a job are 100 m and 500 rpm respectively. The high spots (chatter marks) are found at a spacing of 30 deg on the job surface. The chatter frequency is

(a) 5 Hz (b) 12 Hz

(c) 100 Hz (d) 500 Hz

[PI 1990 : 2 Marks]

23. A milling cutter having 8 teeth is rotating at 100 rpm. The workpiece feed is set at 40 mm/min the feed per tooth is

(a) 5 mm (b) 0.05 mm

(c) 0.4 mm (d) 0.2 mm

[PI 1990 : 2 Marks]

24. The base of a brass bracket has to be rough ground to remove the unevenness. The 4 wheels available in the store. The appropriate wheel is

(a) A 30 K 12 V (b) C 90 M 4 B

(c) C 30 Q 12 V (d) C 50 G 8 V

[PI 1990 : 2 Marks]

25. Increase in rake angle causes ... A... in chip thickness and ...B... in shear angle

(a) increase/reduction

(b) increase/decrease **[PI 1991 : 1 Mark]**

26. Thrust force in drilling can be reduced by... A., the point angle and by ... B... the helix angle of the drill

(a) increasing/reducing

(b) increasing/reducing **[PI 1991 : 1 Mark]**

27. Only negative rake angles are used with the following tool material

(a) HSS (b) Carbon tool steels

(c) Carbides (d) Diamonds

[PI 1994 : 1 Mark]

28. In metal cutting BUE formation results in

(a) improvement in surface finish

(b) reduction in cutting force

(c) deterioration of surface finish

(d) improvement in tool life

[PI 1994 : 1 Mark]

29. The heat generated in metal cutting is dissipated in different proportions into environment, tool, chip and workpiece. The correct order of this proportion in decreasing magnitude is (no cutting fluid is used)

(a) tool, work, chip, environment

(b) work, tool, chip, environment

(c) tool, chip, environment, work

(d) chip, tool, work, environment

[PI 1994 : 2 Marks]

30. Hot hardness in an essential property for

(a) Gear materials

(b) Shaft materials

(c) Welding electronics

(d) Tool materials **[PI 1995 : 1 Mark]**

31. In HSS tool materials the element tungsten can be completely replaced, without changing the material property by

(a) Molybdenum (b) Carbon

(c) Cobalt (d) Vanadium

[PI 1995 : 1 Mark]

32. When the depth of cut in increased, the specific cutting energy

(a) Increases

(b) Decreases

(c) remains same

(d) Reaches an optimum value

[PI 1995 : 1 Mark]

33. The material most commonly used for manufacturing of machine tool beds is.

(a) MS

(b) Grey CI

(c) White CI

(d) Galvanized iron

[PI 1995 : 1 Mark]

34. A 60 deg symmetrical V tool is used in shaping at workpiece with a depth of cut of 0.1 mm and feed of 0.1 mm/stroke. The theoretical peak to valley height, in mm, of the surface produced is

(a) 0.043 (b) 0.067

(c) 0.086 (d) None of these

[PI 1995 : 2 Marks]

35. The cutting length of the broach for the key way cutting in a 150 mm long east iron gear hub. Given that width of the key way is 9 mm depth of key way is 4.5 mm of finishing teeth = 4 and no. of semi finishing teeth = 8.

	Rise per tooth	Pitch
Roughing teeth	0.1 mm	22 mm
Semi finishing	0.0125 mm	20 mm
Finishing teeth	0 mm	20 mm

(a) 968 mm (b) 160 mm

(c) 1208 mm (d) None of these

[PI 1995 : 2 Marks]

36. A single point cutting tool with a nose radius of 0.4 mm was used to turn a component in a lathe employing feed rate of 0.3 mm/rev. If the feed rate is doubled, the ideal surface roughness (peak-to-valley height) produced on the components will increase by a factor of

(a) 2 (b) 4

(c) 8 (d) 16

[PI 2003 : 2 Marks]

37. Orthogonal machining of a steel workpiece is done with a HSS tool of zero rake angle. The ratio of the cutting force and the thrust force on the tool is 1 : 0.372. The length of cut chip is 4.71 mm while the uncut chip length is 10 mm. What is the shear plane angle ϕ and friction angle β in degree?

Use Merchants theory

(a) 32.49, 10.22 (b) 25.22, 20.41

(c) 64.78 20.41 (d) 25.22, 23.21

[PI 2003 : 2 Marks]

38. A job of the shape shown in the figure below is to be machined on a lathe. The tool best suited for machining of this job must have

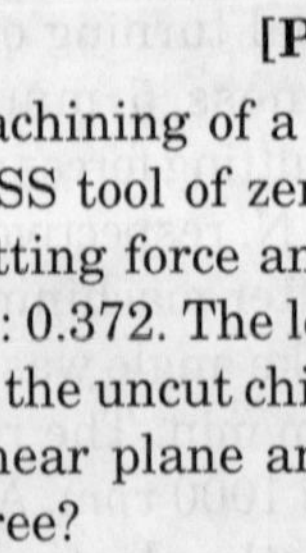

(a) positive side rake angle
(b) positive back rake angle
(c) positive cutting edge inclination angle
(d) negative cutting edge inclination angle

[PI 2004 : 2 Marks]

39. In a machining operation chip thickness ratio is 0.3 and the back rake angle of the tool is 10°. What is the value of the shear strain?

(a) 0.31 (b) 0.13
(c) 3.00 (d) 3.34

[PI 2005 : 2 Marks]

40. Consider the following statements:
During the third stage of tool wear, rapid deterioration of tool edge takes place because
1. Flank wear is only marginal
2. Flank wear is large
3. Temperature of the tool increases gradually
4. Temperature of the tool increases drastically

Which of the statements given above are correct?

(a) 1 and 3 (b) 2 and 4
(c) 1 and 4 (d) 2 and 3

[PI 2005 : 2 Marks]

41. Match List-I with List-II and select the correct answer using the codes given below:

List-I	List-II
A. Plane approach angle	1. Tool face
B. Rake angle	2. Tool flank
C. Clearances angle	3. Tool face and flank
D. Wedge angle	4. Cutting edge
	5. Tool nose

Codes:

	A	B	C	D
(a)	1	4	2	5
(b)	4	1	3	2
(c)	4	1	2	3
(d)	1	4	3	5

[PI 2005 : 2 Marks]

42. Match List-I (Cutting tool materials) with List-II (Manufacturing methods) and select the correct answer using the codes given below:

List-I	List-II
A. HSS	1. Casting
B. Satellite	2. Forging
C. Cemented carbide	3. Rolling
D. UCON	4. Extrusion
	5. Powder metallurgy

Codes:

	A	B	C	D
(a)	3	1	5	2
(b)	2	5	4	3
(c)	3	5	4	2
(d)	2	1	5	3

[PI 2005 : 2 Marks]

43. Assertion (A): The ratio of uncut chip thickness to actual chip thickness is always less than one and is termed as cutting ratio in orthogonal cutting.

Reason (R): The frictional force is very high due to the occurrence of sticking friction rather than sliding friction.

(a) Both A and R are true and R is the correct explanation of A
(b) Both A and R are true but R is not a correct explanation of A
(c) A is true but R is false
(d) A is false but R is true

[PI 2005 : 2 Marks]

44. Match the following materials with their most appropriate application.

Material	Application
1. Low carbon steel	P. Machine tool base
2. Stainless steel	Q. Aircraft parts
3. Gray cast iron	R. Kitchen utensils
4. Titanium alloys	S. Car body panels

Codes:

	P	Q	R	S
(a)	1	3	2	4
(b)	1	4	2	3
(c)	3	2	4	1
(d)	3	4	2	1

[PI 2011 : 2 Marks]

45. A steel bar 200 mm in a diameter is turned at a feed of 0.25 mm/rev. with a depth of cut of 4 mm. The rotational speed of the workpiece is 160 rpm. The material removal rate in mm^3/s is

(a) 160
(b) 167.6
(c) 1600
(d) 1675.5

[PI 2013 : 1 Mark]

46. Two cutting tools are being compared for a machining operation. The tool life equations are

Carbide tool $VT^{1.6} = 3000$

HSS tool $VT^{0.6} = 200$

where V is the cutting speed in m/min and T is the tool life in min. The carbide tool will provide higher tool life if the cutting speed in m/min exceeds

(a) 15.0 (b) 39.4

(c) 49.3 (d) 60.0

[PI 2013 : 2 Marks]

Common Data Q. 47 and 48:

A disc of 200 mm outer and 80 mm inner diameter is faced at a feed of 0.1 mm/rev with a depth of cut of 1 mm. The facing operation is undertaken at a constant cutting speed of 90 m/min in a CNC lathe. The main (tangential) cutting force is 200 N.

47. Assuming approach and over-travel of the cutting tool to be zero, the machining time in min is

(a) 2.93 (b) 5.86

(c) 6.66 (d) 13.33

[PI 2013 : 2 Marks]

48. Neglecting the contribution of the feed force towards cutting power, the specific cutting energy in J/mm^3 is

(a) 0.2 (b) 2

(c) 200 (d) 2000

[PI 2013 : 2 Marks]

Linked Data Q. 49 and 50:

In orthogonal turning of a bar of 100 mm diameter with a feed of 0.25 mm/rev, depth of cut of 4 mm and cutting velocity of 90 m/min, it is observed that the main (tangential) cutting force is perpendicular to the friction force acting at the chip-tool interface. The main (tangential) cutting force is 1500 N.

49. The orthogonal rake angle of the cutting tool in degree is

(a) Zero (b) 3.58

(c) 5 (d) 7.16

[PI 2013 : 2 Marks]

50. The normal force acting at the chip-tool interface in N is

(a) 1000 (b) 1500

(c) 2000 (d) 2500

[PI 2013 : 2 Marks]

51. A steel bar 200 mm in diameter is turned at a feed of 0.25 mm/rev with a depth of cut of 4 mm. The rotational speed of the workpiece is 160 rpm. The material removal rate in mm^3/s is

(a) 160 (b) 167.6

(c) 1600 (d) 1675.5

[PI 2013 : 1 Mark]

Cutting Tool Geometry and Tool Life

52. Cutting tools are provided with large positive rake angle mainly for

(a) Increasing the strength of the cutting edge

(b) Avoiding rubbing action with the finished surfaces

(c) Reducing the magnitude of the cutting force

(d) Better heat dissipation

[ME 1987 : 1 Mark]

53. In turning operation the feed rate could be doubled to increase the metal removal rate. To keep the same level of surface finish, the nose radius of the tool has to be

(a) Doubled

(b) Halved

(c) Multiplied by 4 times

(d) Kept unchanged **[ME 1989 : 2 Marks]**

54. The effect of rake angle on the mean friction angle in machining can be explained by

(a) Sliding model of friction

(b) Sticking and then sliding friction model

(c) Sticking friction model

(d) Sliding and then sticking friction model

[ME 1992 : 1 Mark]

55. Machine tool, structures are made ... for high process capability (tough/rigid/strong)

[ME 1995 : 2 Marks]

56. The rake angle in drill

(a) Increase from centre to periphery

(b) Decreases from centre to periphery

(c) Remains constant

(d) Is irrelevant to the drilling operation

[ME 1996 : 1 Mark]

57. Tool life of a 10 hours is obtained when cutting with a single point tool at 63 m/min. If Taylor's constant C = 257.35, tool life on doubling the velocity will be

(a) 5 hours (b) 25.7 min

(c) 38.3 min (d) unchanged

[ME 1996 : 2 Marks]

58. Helix angle of fast helix drill is normally

(a) 35 deg (b) 60 deg

(c) 90 deg (d) 5 deg

[ME 1997 : 1 Mark]

59. For turning NiCr alloy steel at cutting speeds of 64 m/min and 100 m/min, the respective tool lives are 15 min-and 12 min. The tool life for a cutting speed of 144 m/min is

(a) 8 min (b) 9 min

(c) 10 min (d) 11.5 min

[ME 2001 : 2 Marks]

60. Tool life testing on a lathe under dry cutting conditions gauge n and C of Taylor tool life equation as 0.12 and 130 m/min. respectively. When a coolant was used, C increased by 10%. The increased tool life with the use of coolant at a cutting speed of 90 m/min is

(a) 47.4 min (b) 37.4 min

(c) 27.4 min (d) 17.4 min

[ME 2001 : 2 Marks]

61. A batch of 10 cutting tools could produce 500 components while working at 50 rpm with a tool feed of 0.25 mm/rev and depth of cut of 1 mm. A similar batch of 10 tools of the same specification could produce 122 components while working at 80 rpm with a feed of 0.25 mm/rev and 1 mm depth of cut. How many components can be produced with one cutting tool at 60 rpm?

(a) 29 (b) 31

(c) 37 (d) 42

[ME 2003 : 2 Marks]

Common Data for Q. 62 and 63:

A cylinder is turned on a lathe with orthogonal machining principle. Spindle rotates at 200 rpm. The axial feed rate is 0.25 mm per revolution. Depth of cut is 0.4 mm. The rake angle is 10°. In the analysis it is found that the shear angle is 27.75°.

62. The thickness of the produced chip is

(a) 0.511 mm (b) 0.528 mm

(c) 0.818 mm (d) 0.846 mm

[ME 2003 : 2 Marks]

63. In the above problem, the coefficient of friction at the chip tool Interface obtained using Earnest and Merchant theory is

(a) 0.18 (b) 0.36

(c) 0.71 (d) 0.98

[ME 2003 : 2 Marks]

64. Through holes of 10 mm diameter are to be drilled in a steel plate of 20 mm thickness. Drill spindle speed is 300 rpm, feed 0.2 mm/rev and drill point angle is 120°. Assuming drill over travel of 2 mm, the time for producing a hole will be

(a) 4 seconds (b) 25 seconds

(c) 100 seconds (d) 110 seconds

[ME 2004 : 2 Marks]

65. In an orthogonal cutting test on mild Steel, the following data were obtained

Cutting speed	:	40 m/min
Depth of cut	:	0.3 mm
Tool rake angle	:	+5°
Chip thickness	:	1.5 mm
Cutting force	:	900 N
Thrust force	:	450 N

Using Merchants analysis, the friction angle during the machining will be

(a) 26.6° (b) 31.5°

(c) 45° (d) 63.4°

[ME 2004 : 2 Marks]

66. In a machining operation, doubling the cutting speed reduces the tool life to 1/8th of the original value. The exponent n in Taylor's tool life equation $VT^n = C$. is

(a) $\frac{1}{8}$ (b) $\frac{1}{4}$

(c) $\frac{1}{3}$ (d) $\frac{1}{2}$

[ME 2004 : 2 Marks]

67. The figure below shows a graph which qualitatively relates cutting speed and cost per piece produced.

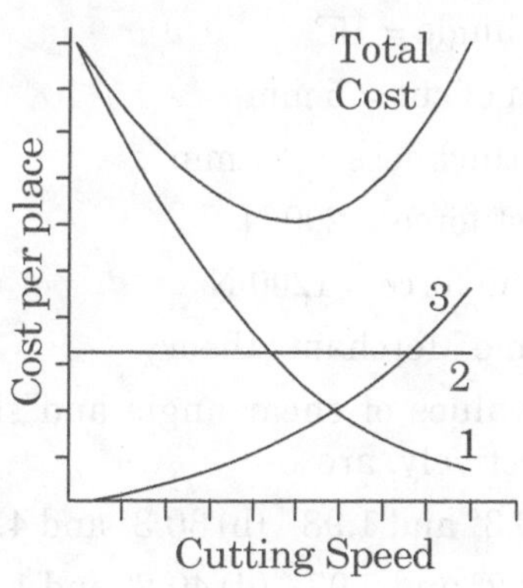

The three curves 1,2 and 3 respectively represent

(a) machining cost, non-productive cost, tool changing cost

(b) non-productive cost, machining cost, tool changing cost

(c) tool changing cost, machining cost, non-productive cost

(d) tool changing cost, non-productive. cost, machining cost

[ME 2005 : 1 Mark]

68. A 600 mm x 30 mm flat surface of a plate is to be finish machined on a shaper. The plate has been fixed with the 600 mm side along the tool travel direction. If the tool over-travel at each end of the plate is 20 mm, average cutting speed is 8 m/min, feed rate is 0.3 mm/stroke and the ratio of return time to cutting time of the tool is 1 : 2, the time required for machining will be

(a) 8 minutes (b) 12 minutes

(c) 16 minutes (d) 20 minutes

[ME 2005 : 2 Marks]

69. Two tools P and Q have signatures 5°-50-60-6°-8°-30°-0 and 5°-5°-7°-7°-8°-15°-0 (both ASA) respectively. They are used to turn components under the same machining conditions. If h_p and h_Q denote the peak-to-valley heights of surfaces produced by the tools P and Q, the ratio h_p/h_Q will be

(a) $\frac{\tan 8° + \cot 15°}{\tan 8° + \cot 30°}$ (b) $\frac{\tan 15° + \cot 8°}{\tan 30° + \cot 8°}$

(c) $\frac{\tan 15° + \cot 7°}{\tan 30° + \cot 7°}$ (d) $\frac{\tan 7° + \cot 15°}{\tan 7° + \cot 30°}$

[ME 2005 : 2 Marks]

Common Data for Q. 70 to 72:

In an orthogonal machining operation:

Uncut thickness = 0.5 mm

Cutting speed = 20 m/min

Rake angle = 15°

Width of cut = 5 mm

Chip thickness = 0.7 mm

Thrust force = 200 N

Cutting force = 1200 N

Assume Merchants theory.

70. The values of shear angle and shear strain, respectively, are

(a) 30.3° and 1.98 (b) 30.3° and 4.23

(c) 40.2° and 2.97 (d) 40.2° and 1.65

[ME 2006 : 2 Marks]

71. The coefficient of friction at the tool-chip interface is

(a) 0.23 (b) 0.46

(c) 0.85 (d) 0.95

[ME 2006 : 2 Marks]

72. The percentage of total energy dissipated due to friction at the tool-chip interface is

(a) 30% (b) 42%

(c) 58% (d) 70%

[ME 2006 : 2 Marks]

73. In orthogonal turning of a low carbon steel bar of diameter 150 mm with uncoated carbide tool, the cutting velocity is 90 m/min. The feed is 0.24 mm/rev and the depth of cut is 2 mm. The chip thickness obtained is 0.48 mm. If the orthogonal rake angle is zero and the principal cutting edge angle is 90°, the shear angle in degree is

(a) 20.56 (b) 26.56

(c) 30.56 (d) 36.56

[ME 2007 : 1 Mark]

74. In orthogonal turning of medium carbon steel, the specific machining energy is 2.0 J/mm^3. The cutting velocity, feed and depth of cut are 120 m/min, 0.2 mm/rev and 2 mm respectively. The main cutting force in N is

(a) 40 (b) 80

(c) 400 (d) 800

[ME 2007 : 2 Marks]

75. In orthogonal turning of low carbon steel pipe with principal cutting edge angle of 90°, the main cutting force is 1000 N and the feed force is 800 N. The shear angle is 25° and orthogonal rake angle is zero. Employing Merchants theory, the ratio of friction force to normal force acting on the cutting tool is

(a) 1.56 (b) 1.25

(c) 0.80 (d) 0.64

[ME 2007 : 2 Marks]

Linked Answer Q. 76 and 77:

A low carbon steel bar of 147 mm diameter with a length of 630 mm is being turned with uncoated carbide insert. The observed tool lives are 24 min and 12 min for cutting velocities of 90 m/min and 120 m/min respectively. The feed and depth of cut are 0.2 mm/rev and 2 mm respectively. Use the unmachined diameter to calculate the cutting velocity.

76. When tool life is 20 min, the cutting velocity in m/min is

(a) 87 (b) 97

(c) 107 (d) 114

[ME 2007 : 2 Marks]

77. Neglect over-travel or approach of the tool. When tool life is 20 min, the machining time in min for a single pass is

(a) 5 (b) 10
(c) 15 (d) 20

[ME 2007 : 2 Marks]

78. In a single point turning tool, the side rake angle and orthogonal rake angle are equal. ϕ is the principal cutting edge angle and its range is $0° \leq \phi \leq 90°$. The chip flows in the orthogonal plane. The value of ϕ is closest to

(a) 0° (b) 45°
(c) 60° (d) 90°

[ME 2008 : 2 Marks]

Linked Answer Q. 79 & 80:

Orthogonal turning is performed on a cylindrical workpiece with shear strength of 250 MPa. The following conditions are used: cutting velocity is 180 m/min, feed is 0.20 mm/rev, depth of cut is 3 mm, chip thickness ratio = 0.5. The orthogonal rake angle is 7°. Apply Merchants theory for analysis.

79. The shear plane angle (in degrees) and the shear force respectively are

(a) 52; 320 N (b) 52; 400 N
(c) 28; 400 N (d) 28; 320 N

[ME 2008 : 2 Marks]

80. The cutting and thrust forces, respectively, are

(a) 568 N; 387 N (b) 565 N; 381 N
(c) 440 N; 342 N (d) 480 N; 356 N

[ME 2008 : 2 Marks]

81. Friction at the tool-chip interface can be reduced by

(a) decreasing the rake angle
(b) increasing the depth of cut
(c) decreasing the cutting speed
(d) increasing the cutting speed

[ME 2009 : 1 Mark]

Linked Answer Q. 82 and 83:

In a machining experiment, tool life was found to vary with the cutting speed in the following manner:

Cutting speed (m/min)	Tool life (minutes)
60	81
90	36

82. The exponent (n) and constant (k) of the Taylor's tool life equation are

(a) n = 0.5. and k = 540
(b) n = 1 and k = 4860
(c) n = –1 and k = 0.74
(d) n = –0.5 and k = 1.155

[ME 2009 : 2 Marks]

83. What is the percentage increase in tool life when the cutting speed is halved?

(a) 50% (b) 200%
(c) 300 % (d) 400%

[ME 2009 : 2 Marks]

84. For tool A, Taylor's tool life exponent (n) is 0.45 and constant (K) is 90. Similarly for tool B, n = 0.3 and K = 60. The cutting speed (in 'm/min) above which tool A will have a higher tool life than tool B is

(a) 26.7 (b) 42.5
(c) 80.7 (d) 142.9

[ME 2010 : 2 Marks]

85. A single-point cutting tool with 12° rake angle is used to machine a steel work-piece. The depth of cut, i.e. uncut thickness is 0.81 mm. The chip thickness under orthogonal machining condition is 1.8 mm. The shear angle is approximately

(a) 22° (b) 26°
(c) 56° (d) 76°

[ME 2011 : 2 Marks]

86. In a single pass drilling operation, a through hole of 15 mm diameter is to be drilled in a steel plate of 50 mm thickness. Drill spindle speed is 500 rpm, feed is 0.2 mm/rev and drill point angle is 118°, Assuming 2 mm clearance at approach and exit, the total drill time (in seconds) is

(a) 35.1 (b) 32.4
(c) 31.2 (d) 30,1

[ME 2012 : 2 Marks]

87. Details pertaining to an orthogonal metal cutting process are given below:

Chip thickness ratio	0.4
Undeformed thickness	0.6 mm
Rake angle	+10°
Cutting speed	2.5 m/s
Mean thickness of primary shear zone	25 microns

The shear strain rate in s^{-1} during the process is

(a) 0.1781 × 105 (b) 0.7754 × 105
(c) 1.0104 × 105 (d) 4.397 × 105

[ME 2012 : 2 Marks]

88. If the Taylor's tool life exponent n is 0.2, and the tool changing time is 1.5 min, then the tool life (in min) for maximum production rate is___

[ME 2014 : 2 Marks, Set-1]

89. A hole of 20 mm diameter is to be drilled in a steel block of 40 mm thickness. The drilling is performed at rotational speed of 400 rpm and feed rate of 0.1 mm/rev. The required approach and over run of the drill together, is equal to the radius of drill. The drilling time (in minute) is

(a) 1.00 (b) 1.25
(c) 1.50 (d) 1.75

[ME 2014 : 2 Marks, Set-2]

90. An orthogonal cutting operations is being carried out in which uncut thickness is 0.010 mm, cutting speed is 130 m/min, rake angle is 15° and width of cut is 6 mm. It is observed that the chip thickness is 0.015 mm, the cutting force is 60 N and the thrust force is 25 N. The ratio of friction energy to total energy is ______. (correct to two decimal places)

[ME 2018 : 2 Marks, Set-1]

91. The preferred option for holding an odd-shaped workpiece in a centre lathe is

(a) live and dead centres
(b) three jaw chuck
(c) lathe dog
(d) four jaw chuck

[ME 2018 : 1 Marks, Set-2]

92. Feed rate in slab milling operation is equal to

(a) rotation per minute (rpm)
(b) product of rpm and number of teeth in the cutter
(c) product of rpm, feed per tooth and number of teeth in the cutter
(d) product of rpm, feed per tooth and number of teeth in contact

[ME 2018 : 1 Marks, Set-2]

93. Taylor's tool life equation is used to estimate the life of a batch of identical HSS twist drills by drilling through holes at constant feed in 20 mm thick mild steel plates. In test 1, a drill lasted 300 holes at 150 rpm while in test 2, another drill lasted 200 holes at 300 rpm.

The maximum number of holes that can be made by another drill from the above batch at 200 rpm is (correct to two decimal places).

[ME 2018 : 2 Marks, Set-2]

94. Taylor's tool life equation is given by $VT^n = C$, where V is in m/min and T is in min. In a turning operation, two tools X and Y are used. For tool X, n = 0.3 and C = 60 and for tool Y, n = 0.6 and C = 90. Both the tools will have the same tool life for the cutting speed (in m/min, round off one decimal place) of ______.

[ME 2019 : 2 Marks, Set-1]

95. If the number of double strokes per minute in a shaper is calculated by (0.643 × Cutting speed in mm/min)/length of the stroke in m. Then the return speed is faster than the cutting speed by

(a) 40% (b) 60%
(c) 80% (d) 100%

[PI 1989 : 2 Marks]

96. A single point cutting tool with 12° rake angle is used for orthogonal machining of a ductile material. The shear plane angle for the theoretically minimum possible shear strain to occur

(a) 51 (b) 45
(c) 30 (d) None of these

[PI 1990 : 2 Marks]

97. In a cutting test with 0.3 mm flank wear as tool failure criterion, a tool life of 10 min was obtained at a cutting velocity of 20 m/min. Taking tool life exponent as 0.25, the tool life in minutes at 40 m/min of cutting velocity will be

(a) 0.506 (b) 0.625
(c) 3.140 (d) 5.002

[PI 1993 : 2 Marks]

98. A workpiece of 2000 mm length and 300 mm width was machined by a planning operation with the feed set at 0.3 mm/stroke. If the machine tool executes 10 double strokes/min, the planning time for a single pass will be

(a) 50 min (b) 100 min
(c) 200 min (d) 220 min

[PI 1993 : 2 Marks]

99. In a single pass turning operation the cutting speed is the only variable based on the cutting time cost and the cutting edge cost. The tool life for minimum cost given that cost of 1 cutting edge is Rs. 5, operator wages including the machine tool cost is Rs. 75/hour and tool life equation is $VT^{0.1}$ is 100

(a) 10 (b) 17
(c) 29 (d) 36

[PI 1995 : 2 Marks]

100. Two different tools A and B having nose radius of 0.6 mm and 0.33 mm respectively are used to machine C-45 steel employing feed rate of 0.2 mm/rev and 0.1 mm/rev respectively. The tool that gives better finish and the value of ideal surface roughness are

(a) Tool A and 4.166 μm respectively
(b) Tool B and 3.78 μm respectively
(c) Tool A and 8.333 μm respectively
(d) Tool B and 8.333 μm respectively

[PI 2002 : 2 Marks]

101. A 31.8 MM HSS drill is used to drill a hole in cast iron block 100 mm thick at cutting speed of 20 m/min and feed 0.3 mm/rev. If the over travel of drill is 4 mm and approach 9 mm, the time required to drill the hole is

(a) 1 min 40 s (b) 1 min 44 s
(c) 1 min 49 s (d) 1 min 53 s

[PI 2002 : 2 Marks]

102. Tool life equations for two tools under consideration are as follows

HSS : $VT^{0.2} = 150$

Carbide : $VT^{0.3} = 250$

Where V is the cutting speed in m/min and T is the tool life in mm. The breakeven cutting speed above which the carbide tool will be beneficial is

(a) 54 m/min (b) 150 m/min
(c) 194 m/min (d) 250 m/min

[PI 2003 : 2 Marks]

103. A cylinder of 25 mm diameter and 100 mm length is turned with a tool, for which the relation $VT^{0.25} = 55$ is applicable. The cutting velocity is 22 m/min. For a tool feed of 0.046 mm/rev, the number of tool regrinds required to produce 425 cylinders is

(a) 12 (b) 22
(c) 43 (d) 85

[PI 2003 : 2 Marks]

104. Two identical cylindrical jobs are turned using

(a) a round nosed tool of nose radius 2 mm and

(b) a sharp corner tool having principal cutting edge angle = 45° and auxiliary cutting edge angle = 10°. If the operation is carried out of the feed of 0.08 mm/rev, the height at micro irregularities on the machined surfaces (in mm) in the two cases will be

(a) 0.0001, 0.024 (b) 0.0002, 0.012
(c) 0.0003, 0.024 (d) 0.0004, 0.012

[PI 2004 : 2 Marks]

105. The rake angle of a cutting tool is 15°, shear angle 45° and cutting velocity 35 m/min. What is the velocity of chip along the tool face?

(a) 28.5 m/min (b) 27.3 m/min
(c) 25.3 m/min (d) 23.5 m/min

[PI 2005 : 2 Marks]

Linked Data Q. 106 and 107:

A batch of 500 jobs of diameter 50 mm and length 100 mm is to be turned at 200 rev/min and feed 0.2 mm/rev.

106. Applying Taylor's equation $VT^{0.25} = 160$, the tool

(a) 20.36 (b) 22.43
(c) 674 (d) 28.20

[PI 2005 : 2 Marks]

107. The number of tool changes required to machine the whole batch is

(a) 138 (b) 270
(c) 62 (d) 61

[PI 2005 : 2 Marks]

Linked Data Q. 108 and 109:

A ϕ 40 mm job is subjected to orthogonal turning by a +10° rake angle tool at 500 rev/min. By direct measurement during the cutting operation, the shear angle was found equal to 25°.

108. The velocity (in m/min) with which the chip flows on the tool face is

(a) 32 (b) 63
(c) 22 (d) 27.5

[PI 2005 : 2 Marks]

109. If the friction angle at the tool chip interface is 58° 10' and the cutting force components measured by a dynamometer are 600 N and 200 N, the power loss due to friction (in kNm/min) is approximately

(a) 20 (b) 18
(c) 16 (d) 350

[PI 2005 : 2 Marks]

Linked Data Q. 110 and 111:

The following data relate to an orthogonal turning process: back rake angle = 15 deg. width of cut = 2 mm chip thickness = 0.4 mm, feed rate = 0.2 mm/rev

110. The shear angle is

(a) 14° (b) 25°
(c) 29° (d) 75°

[PI 2005 : 2 Marks]

111. If the cutting force and the thrust force are 900 N and 810 N, the mean strength in MPa

(a) 137.94 (b) 477.91
(c) 500.58 (d) 635.84

[PI 2005 : 2 Marks]

Common Data Q. 112 and 113:

In an orthogonal cutting test, the following observations were made

Cutting force = 1200 N

Thrust force = 500 N

Tool rake angle = zero

Cutting speed = 1 m/s

Depth of cut = 0.8 mm

Chip thickness = 1.5 mm

112. Friction angle during machining will be

(a) 22.6 (b) 32.6
(c) 57.1 (d) 67.4

[PI 2007 : 2 Marks]

113. Chip speed along the tool rake face will be

(a) 0.83 m/s (b) 0.53 m/s

(c) 1.2 m/s (d) 1.88 m/s

[PI 2007 : 2 Marks]

Linked Data Q. 114 and 115:

Blind hole 10 mm diameter, 50 mm deep are being drilled in steel block, drilling spindle speed is 600 rpm, feed 0.2 mm/rev, point angle of drill is 120.

114. Machining time (in min) per hole will be

(a) 0.08 (b) 0.31

(c) 0.44 (d) 0.86

[PI 2007 : 2 Marks]

115. During the above operation, the drill wears out after producing 200 holes. Taylor's tool life equation is of the form $VT^{0.3} = C$, where V = cutting speed in m/min and T= tool life in min. Taylor's constant C will be

(a) 15 (b) 72

(c) 93 (d) 490

[PI 2007 : 2 Marks]

116. In an orthogonal machining operation, the tool life obtained is 10 min at a cutting speed of 100 m/min, while at 75 m/min cutting speed, the tool life is 30 min. The value of index (n) in the Taylor's tool life equation is

(a) 0.262 (b) 0.323

(c) 0.423 (d) 0.521

[PI 2009 : 2 Marks]

Common Data Q. 117 and 118:

An orthogonal operation is carried out at 20 m/min cutting speed, using a cutting tool of rake angle 15 degree. The chip thickness is 0.4 mm and uncut chip thickness is 0.2 mm.

117. The shear angle (in degrees) is

(a) 26.8 (b) 27.8

(c) 28.8 (d) 29.8

[PI 2009 : 2 Marks]

118. The chip velocity (in m/min) is

(a) 8 (b) 10

(c) 12 (d) 14

[PI 2009 : 2 Marks]

Linked Data Q. 119 and 120:

During orthogonal machining of a mild steel specimen with a cutting tool of zero rake angle, the following data is obtained:

Uncut chip thickness = 0.25 mm

Chip thickness = 0.75 mm

Normal force = 950 N

Thrust force = 475 N

119. The shear angle and shear force, respectively, are

(a) 71.565°, 150.12 N

(b) 9.218°, 861.64 N

(c) 18.435°, 751.04 N

(d) 23.157°, 686.66 N

[PI 2011 : 2 Marks]

120. The ultimate shear stress (in N/mm²) of the work material is

(a) 235 (b) 139

(c) 564 (d) 380

[PI 2011 : 2 Marks]

121. Details pertaining to an orthogonal metal cutting process are given below:

Chip thickness ratio 0.4

Undeformed thickness 0.6 mm

Rake angle +10°

Cutting speed 2.5 m/s

Mean thickness of primary shear zone 25 microns

The shear strain rate in s-1 during the process is

(a) 0.1781×10^5 (b) 0.7754×10^5

(c) 1.0104×10^5 (d) 4.397×10^5

[PI 2012 : 2 Marks]

122. In a shaping process, the number of double strokes per minute is 30 and the quick return ratio is 0.6. If the length of the stroke is 250 mm, the average cutting velocity in mm/min.

(a) 3.0 (b) 4.5

(c) 7.5 (d) 12.0

[PI 2012 : 2 Marks]

Machining Process

123. The standard specification of a grinding wheel is A-46-M-6-V-21. It means a wheel of

(a) Aluminum oxide of mesh size 6

(b) Boron carbide of mesh size 46

(c) Aluminum oxide of mesh size 46

(d) Silicon carbide of mesh size 6

[ME 1987 : 1 Mark]

124. If in a turning operation both the feed rate and the nose radius are doubled the surface finish value will be

(a) Decreases by 50%

(b) Increases by 300%

(c) Increases by 100%

(d) Remains unaffected

[ME 1987 : 1 Mark]

125. The ideal cutting fluid for low speed machining of metals should be one which

(a) removes the heat faster from the cutting zone

(b) forms a coating on the cutting tools by Chemical reaction

(c) forms a low shear strength film of work material at the tool chip interface

(d) serves as a dielectric, minimizing thereby reactions due to EMF at the interface

[ME 1988 : 1 Mark]

126. In twist drills.... (small/large) point angle and ...(small/large) helix angle are provided for drilling soft, low strength steel.

[ME 1988 : 1 Mark]

127. Cutting speed in grinding is set to a high value to

(a) Reduce the cutting time

(b) Increase the bond strength

(c) Improve cooling of job and wheel

(d) Reduce the wheel wear

[ME 1988 : 1 Mark]

128. Pure metal pose machinability problem in turning operations. The reason is the

(a) Increased length of contact due to the production of continuous chip

(b) Susceptibility to Chemical reactions

(c) Tendency to from intense adhesion Joint with the tool face

(d) Absence of inclusions which aids chip formation

[ME 1988 : 2 Marks]

129. If the longitudinal feed in centreless grinding is expressed by $V_f = \pi DN \sin\alpha$, D stands for

(a) Diameter of blank

(b) Diameter of finished workpiece

(c) Diameter of control wheel

(d) Diameter of grinding wheel

[ME 1990 : 1 Mark]

130. For cutting double start screw threads of pitch 1.0 mm on a lathe, the thread cutting tool should have a feed rate of

(a) 0.5 mm/rev

(b) 1.0 mm/rev

(c) 2.0 mm/rev

(d) 4.0 mm/rev

[ME 1991 : 2 Marks]

131. A milling cutter having 10 teeth is rotating at 100 rpm. The table feed is set at 50-mm per minute. The feed per tooth in mm is

(a) 5 (b) 0.5

(c) 0.2 (d) 0.05

[ME 1991 : 2 Marks]

132. In horizontal milling process (up/down) milling provides better surface finish and (up/down) milling provide longer tool life.

[ME 1992 : 1 Mark]

133. Component process

A. Square hole in a high strength alloy

B. Square hole in a ceramic component

C. Blind holes in a die

D. Turbine blade profile on high strength alloy

1. Milling

2. Drilling

3. ECM

4. Jig boring

5. EDM

6. USM

[ME 1992 : 2 Marks]

134. A milling cutter having 8 teeth is rotating at 150 rpm. If the feed per tooth is 0.1 mm, the speed in mm per minute is

(a) 120 (b) 187

(c) 125 (d) 70

[ME 1993 : 2 Marks]

135. A grinding wheel A 27 K 7 V is specified for finish grinding of a HSS cutting tool. What do you understand about the wheel from the above code.

Is this an appropriate choice?

(a) yes

(b) no, because abrasive is not correct

(c) no, grain size is not correct

(d) no, because grade is not correct choice

[ME 1994 : 2 Marks]

136. Cutting power consumption in turning can be significantly reduced by

(a) Increasing rake angle of the tool

(b) Increasing cutting angle of the tool

(c) Widening the nose radius of the tool

(d) Increasing the clearance angle

[ME 1995 : 1 Mark]

137. Among the conventional machining processes, maximum specific energy is consumed in

(a) Turning (b) Drilling

(c) Planning (d) Grinding

[ME 1995 : 1 Mark]

138. Plain milling of mild steel plates produces

(a) irregular shaped discontinuous chips

(b) regular shaped discontinuous chips

(c) continuous chips without BUE

(d) jointed chips

[ME 1995 : 1 Mark]

139. Diamond wheels should not be used for grinding steel components (T/F)

[ME 1995 : 1 Mark]

140. Abrasive material used in grinding wheel selected for grinding ferrous alloys is

(a) SiC (b) Diamond

(c) Al_2O_3 (d) Boron carbide

[ME 2000 : 1 Mark]

141. A conventional lathe and a CNC lathe are under consideration for machining a given part. The relevant data are shown below:

	Preparation Cost	Production time per part(min)	Machine and Operator Cost
Conventional Lathe	30	30	75
CNC Lathe	150	15	120

The machine preferred for producing 100 pieces is

(a) conventional lathe

(b) CNC lathe

(c) any of the above

(d) None of these **[ME 2000 : 2 Marks]**

142. A lead screw with half nuts in a lathe, free to rotate in both directions has

(a) V-A threads

(b) With worth threads

(c) Buttress threads

(d) Acme threads **[ME 2002 : 1 Mark]**

143. The hardness of a grinding wheel is determined by the

(a) hardness of abrasive grains

(b) ability of the bond to retain abrasives

(c) hardness of the bond

(d) ability of the grinding wheel to penetrate the workpiece **[ME 2002 : 1 Mark]**

144. The time taken to drill a hole through a 25 mm thick plate with the drill rotating at 300 rpm and moving at a feed rate of 0.25 mm/rev is

(a) 10 sec

(b) 20 sec

(c) 60 sec

(d) 100 sec

[ME 2002 : 2 Marks]

145. The figure shows an incomplete schematic of a conventional lathe to be used for cutting threads with different pitches. The speed gear box U_v is shown and the feed gear box U_s is to be placed. P, Q, R and S denote locations and have no other significance. Changes in U_v should NOT affect the pitch of the thread being cut and changes in U_s should NOT affect the cutting speed.

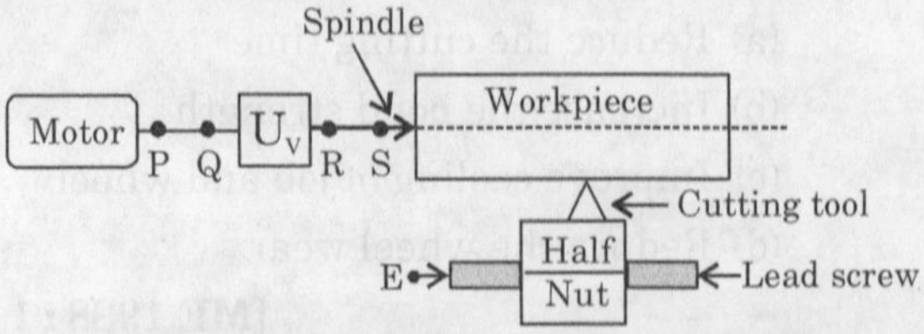

The correct connections and the correct placement of U_s are given by

(a) Q and E are connected. U_s is placed between P and Q.

(b) S and E are connected, U_s is placed between R and S.

(c) Q and E are connected. U_s is placed between Q and E.

(d) S and E are connected. U_s' is placed between S and E

[ME 2008 : 2 Marks]

146. During pure orthogonal turning operation of a hollow cylindrical pipe, it is found that the thickness of the chip produced is 0.5 mm. The feed given to the zero degree rake angle tool is 0.2 mm/rev. The shear strain produced during the operation is______.

[ME 2014 : 2 Marks, Set-1]

147. A straight turning operation is carried out using a single point cutting tool on an AIS 1020 steel rod. The feed is 0.2 mm/rev and the depth of cut is 0.5 mm. The tool has a side cutting edge angle of 60°. The uncut chip thickness (in mm) is _____.

[ME 2014 : 1 Mark, Set-3]

148. Cutting tool is much harder than the workpiece. Yet the tool wears out during the tool-work interaction, because

(a) extra hardness is imparted to the workpiece due to coolant used

(b) oxide layers on the workpiece surface impart extra hardness to it

(c) extra hardness is imparted to the workpiece due to severe rate of strain

(d) vibration of induced in the machine tool

[ME 2014 : 1 Mark, Set-3]

149. Two separate slab milling operations, 1 and 2, are performed with identical milling cutters. The depth of cut in operation 2 is twice that in operation 1. The other cutting parameters are identical. The ratio of maximum uncut chip thicknesses in operations 1 and 2 is

[ME 2014 : 1 Mark, Set-4]

150. Better surface finish is obtained with a large rake angle because

(a) the area of shear plane decreases resulting in the decrease in shear force and cutting force

(b) the tool becomes thinner and the cutting force is reduced

(c) less heat is accumulated in the cutting zone

(d) the friction between the chip and the tool is less

[ME 2014 : 1 Mark, Set-4]

151. The length, width and thickness of a steel sample are 400 mm, 410 mm, 40 mm and 20 mm, respectively. Its thickness needs to be uniformly reduced by 2 mm in a single pass by using horizontal slab miling. The miling cutter (diameter : 100 mm, width ; 50 mm) has 20 teeth and rotates at 1200 rpm. The feed per tooth is 0.05 mm. The feed direction is along the length of the sample. If the over-travel distance is the same as the approach distance, the approach distance and time taken to complete the required machining task are :

(a) 21 mm, 39.4 s (b) 14 mm, 18.4 s

(c) 14 mm, 21.4 s (d) 21 mm, 28.9 s

[ME 2019 : 1 Mark, Set-1]

152. A through hole is drilled in an aluminum alloy plate of 15 mm thickness with a drill bit of diameter 10 mm, at a feed of 0.25 mm/rev and a spindle speed of 1200 rpm. If the specific energy required for cutting this material is 0.7 N-m/mm^3, the power required for drilling is ___ W (round off to two decimal places).

[ME 2019 : 2 Marks, Set-2]

153. Assertion (A): Single start threads are used for fastening purposes.

Reason (R): Single start threads are easier to produce in a lathe.

(a) Both A and R are true and R is the correct explanation of A

(b) Both A and R are true but R is not a correct explanation of A

(c) A is true but R is false

(d) A is false but R is true

[PI 1990 : 1 Mark]

154. A drilling machine has to be designed with 8 spindle speeds ranging approximately between 120 to 1200 rpm. The 5th spindle speed is

(a) 445 (b) 480

(c) 620 (d) 865

[PI 1991 : 2 Marks]

155. In a grinding wheel marked with AA-48-L-7-V-25, L refers to

(a) abrasive type

(b) wheel structure

(c) wheel hardness

(d) manufactures code

[PI 1992 : 1 Mark]

156. Find the speed range ratio for the drilling machine spindle if the minimum and maximum diameters of drills used are 5 mm and 25 mm respectively and if the machinability indices for the work materials are 120 (brass) and 40 (alloy steel)

(a) 3/5 (b) 5/3

(c) 5 (d) 15

[PI 1992 : 2 Marks]

157. If the index crank of a dividing head is turned through one complete revolution and 10 holes in a 30 hole Circle plate, the workpiece turns through (in degrees)

(a) 6 (b) 12

(c) 240 (d) 480

[PI 1992 : 2 Marks]

158. A grinding wheel is said to be glazed if

(a) grains have become blunt, but do not fall out

(b) gap between the grains is filled by swarf

(c) it becomes black due to burning

(d) part of the wheel is chipped off

[PI 1995 : 1 Mark]

159. A side and face cutter 125 mm diameter has 10 teeth. If operates at a cutting speed of 14 m/min with a table traverse 100 mm/min the feed per tooth of the cutter is

(a) 10 mm (b) 2.86 mm

(c) 0.286 mm (d) 0.8 mm

[PI 2002 : 2 Marks]

160. A non standard thread of pitch 3.175 mm is to be cut on a lathe having lead screw of pitch 6 mm. A change gear set provided with the lathe has one gear each with the following number of teeth: 20, 30, 40, 50, 60, 70, 80, 90, 100, 110, 120, 127. The correct pairs of change gears (a/b x c/d) for machining the given thread are (assuming that the transmission ratio of the rest of the kinematic train between the lathe spindle and lead screw is equal to 1),

(a) $\frac{127}{30}\times\frac{20}{80}$ (b) $\frac{127}{40}\times\frac{20}{120}$

(c) $\frac{127}{20}\times\frac{30}{120}$ (d) $\frac{127}{50}\times\frac{30}{80}$

[PI 2004 : 2 Marks]

161. A ϕ–20 mm through hole is to be drilled in a 30 mm thick plate using a double fluted, 120° lip angle drill. The drill tip is at a distance of 3 mm from the plate surface when cutting is started and an over travel of 2 mm is recommended as a margin to ensure drilling through the full thickness of the plate. If the drill rotates at 500 rev/min and the feed per tooth is 0.01 mm, the machining time of the operation (in min) will be

(a) 4.5 (b) 7.5

(c) 9 (d) 10.5

[PI 2004 : 2 Marks]

162. Consider the following statements:
The helical flute in a twist drill provides the necessary

1. Clearance angle for the cutting edge.
2. Rake angle for the cutting edge.
3. Space for the chip to come out during drilling.
4. Guidance for the drill to enter into the workpiece.

Which of the statements given above are correct?

(a) 1 and 2 (b) 2 and 3

(c) 3 and 4 (d) 1 and 4

[PI 2004 : 2 Marks]

163. During turning of a low carbon steel bar with TiN coated carbide insert, one needs to improve surface finish without sacrificing material removal rate. The achieve improved surface finish one should

(a) Decrease nose radius of the cutting tool and increase depth of cut

(b) Increase nose radius of the cutting tool

(c) Increase feed and reduce nose radius of the cutting tool

(d) Increase depth of cut and increase feed

[PI 2010 : 2 Marks]

164. The binding material used in cemented carbide cutting tools is

(a) Graphite (b) Tungsten

(c) Nickel (d) Cobalt

[PI 2011 : 1 Mark]

165. Grinding ratio is defined as Volume of wheel wear

(a) $\frac{\text{Volume of wheel wear}}{\text{Volume of work material removed}}$

(b) $\frac{\text{Volume of work material removed}}{\text{Volume of wheel wear}}$

(c) $\frac{\text{Cutting speed}}{\text{feed}}$

(d) $\frac{\text{Longitudinal feed}}{\text{Transverse feed}}$

[PI 2011 : 1 Mark]

Finishing Process

166. Gear hobbing produces more accurate gears than milling because in hobbing

(a) there is a continuous indexing operation

(b) pressure angle is larger than in milling

(c) hob and work piece both are rotating

(d) a special multi-tooth cutter (hob) is used

[ME 1989 : 1 Mark]

167. Teeth of internal spur gears can be accurately cut in a

(a) Milling machine

(b) Gear shaping machine

(c) Slotting machine

(d) Hobbing machine

[ME 1989 : 1 Mark]

168. Minimum dimensional and form accuracy can be obtained in the cylinder bores of automobile engines if the bores are finished by

(a) Lapping (b) Reaming

(c) Internal grinding (d) Honing

[ME 1989 : 1 Mark]

169. In small lot production for machining T-slots on machine tables, it is expected to use

(a) Shaping machine

(b) Broaching machine

(c) Vertical milling machine

(d) Horizontal milling machine

[ME 1990 : 1 Mark]

170. To get good surface finish on a turned job, one should use a sharp tool with ... feed rate and ... speed of rotation of the job.

[ME 1994 : 1 Mark]

171. Match the following

Manufacturing process

A. Finish turning

B. Forming

C. Thread cutting

D. Down milling

Conditions

1. Backlash eliminator

2. Zero rake

3. Nose radius

4. Low speed

[ME 1995 : 2 Marks]

172. Determine the transformation ratio (TR) of the gear train connecting the spindle and the lead screw of 4TPI required to cut 1.0 mm pitch screw threads in a centre lathe.

[ME 1995 : 2 Marks]

173. Trepanning is performed for

(a) finishing a drilled hole

(b) producing a large hole without drilling

(c) truing a hole for alignment

(d) enlarging a drilled hole

[ME 2003 : 1 Mark]

174. Strong screw threads are produced by

(a) thread milling

(b) thread chasing

(c) thread cutting with single point tool

(d) thread rolling **[ME 2003 : 1 Mark]**

175. Internal gear cutting operation can be performed by

(a) milling

(b) shaping with rack cutter

(c) shaping with pinion cutter

(d) hobbing

[ME 2008 : 1 Mark]

176. Which pair of following statements is correct for orthogonal cutting using a single-point cutting tool?

P. Reduction in friction angle increases cutting force

Q. Reduction in friction angle decreases cutting force

R. Reduction in friction angle increases chip thickness

S. Reduction in friction angle decreases chip thickness

(a) P and R (b) P and S

(c) Q and R (d) Q and S

[ME 2014 : 2 Marks, Set-3]

177. A cast iron block of 200 mm length is being shaped in a shaping machine with a depth of cut of 4 mm, feed of 0.25 mm/stroke and the tool principal cutting edge angle of 30°.

Number of cutting strokes per minute is 60.

Using specific energy for cutting as 1.49 J/mm^3, the average power consumption (in watt) is ________ .

[ME 2014 : 2 Marks, Set-4]

178. Under certain cutting conditions, doubling the cutting speed reduces the tool life to $\left(\frac{1}{16}\right)^{th}$ of the original. Taylor's tool life index(n) for this tool- workpiece combination will be ________.

[ME 2015 : 1 Mark, Set-1]

179. An orthogonal turning operation is carried out under the following conditions; rake angle = 5°, spindle rotational speed = 400 rpm; axial feed = 0.4 m/min and radial depth of cut = 5 mm. The chip thickness t_c, is found to be 3 mm. The shear angle (in degree) in this turning process is______.

[ME 2015 : 2 Marks, Set-1]

180. A single point cutting tool with 0° rake angle is used in an orthogonal machining process. At a cutting speed of 180 m/min, the thrust force is 490 N. If the coefficient of friction between the tool and the chip is 0.7, then the power consumption (in kW) for the machining operation is______.

[ME 2015 : 2 Marks, Set-2]

181. In a machining operation, if the generatrix and directrix both are straight lines, the surface obtained is

(a) cylindrical (b) helical

(c) plane (d) surface of revolution

[ME 2015 : 1 Mark, Set-3]

182. Orthogonal turning of a mild steel tube with a tool of rake angle 10° carried out at a feed of 0.14 mm/rev. If the thickness of the chip produced is 0.28 mm, the values of shear angle and shear strain will be respectively

(a) 28°20'and 2.19 (b) 22°20' and 3.53

(c) 24°30' and 3.53 (d) 37°20' and 5.19

[ME 2015 : 2 Marks, Set-3]

183. A shaft of length 90 mm has a tapered portion of length 55 mm. The diameter of the taper is 80 mm at one end and 65 mm at the other. If the taper is made by tailstock set over method, the taper angle and the set over respectively are

(a) 15°32' and 12.16 mm

(b) 15°32' and 15.66 mm

(c) 11°22' and 10.26 mm

(d) 10°32' and 14.46 mm

[ME 2015 : 2 Marks, Set-3]

184. In an orthogonal cutting process the tool used has rake angle of zero degree. The measured cutting force and thrust force are 500 N and 250 N, respectively. The coefficient of friction between the tool and the chip is

[ME 2016 : 1 Mark, Set-1]

185. The tool life equation for HSS tool is $VT^{0.14}f^{0.7}d^{0.4}$ = Constant. The tool life (T) of 30 min is obtained using the following cutting conditions: V= 45 m/min, f = 0.35 mm, d =2.0mm. If speed (V), feed (f) and depth of cut (d) are increased individually by 25%, the tool life (in min) is

(a) 0.15 (b) 1.06

(c) 22.50 (d) 30.0

[ME 2016 : 2 Marks, Set-1]

186. The following data is applicable for a turning operation. The length of job is 900 mm, diameter of job is 200 mm, feed rate is 0.25 mm/rev and optimum cutting speed is 300 m/min. The machining time (in min) is_______.

[ME 2016 : 1 Mark, Set-2]

187. For a certain job, the cost of metal cutting is Rs. 180 C/V and the cost of tooling is Rs. 270C/(TV), where C is a constant, V is the cutting speed in m/min and T is the tool life in minutes. The Taylor's tool life equation is $VT^{0.25}$ = 150. The cutting speed (in m/min) for the minimum total cost is________. **[ME 2016 : 2 Marks, Set-2]**

188. For an orthogonal cutting operation, tool material is HSS, rake angle is 22°, chip thickness is 0.8 mm, speed is 48 m/min and feed is 0.4 mm/rev. The shear plane angle (in degrees) is

(a) 19.24 (b) 29.70

(c) 56.00 (d) 68.75

[ME 2016 : 2 Marks, Set-3]

189. A firm uses a turning center, a milling center and a grinding machine to produce two parts. The table below provides the machining time required for each part and the maximum machining time available on each machine. The profit per unit on parts I and II are Rs. 40 and Rs. 100, respectively. The maximum profit per week of the firm is Rs. ________

Type of machine	Machining time required for the machine part(minutes)		Maximum machining time available per week (minutes)
	I	II	
Turning Center	12	6	6000
Milling Center	4	10	4000
Grinding Machine	2	3	1800

[ME 2016 : 2 Marks, Set-3]

190. Two cutting tools with tool life equations given below are being compared:

Tool 1 : $VT^{0.1}$ = 150

Tool 2: $VT^{0.3}$ = 300

where V is cutting speed in m/minute and T is tool life in minutes. The breakeven cutting speed beyond which Tool 2 will have a higher tool life is ________ m/minute.

[ME 2017 : 2 Marks, Set-1]

191. A block of length 200 mm is machined by a slab milling cutter 34 mm in diameter. The depth of cut and table feed are set at 2 mm and 18 mm/minute, respectively. Considering the approach and the over travel of the cutter to be same, the minimum estimated machining time per pass is minutes.

[ME 2017 : 2 Marks, Set-1]

192. In an orthogonal machining with a tool of 9° orthogonal rake angle, the uncut chip thickness is 0.2 mm. The chip thickness fluctuates between 0.25 mm and 0.4 mm. The ratio of the maximum shear angle to the minimum shear angle during machining is_______.

[ME 2017 : 2 Marks, Set-2]

193. During the turning of a 20 mm diameter steel bar at a spindle speed of 400 rpm, a tool life of 20 minute is obtained. When the same bar is turned at 200 rpm, the tool life becomes 60 minute. Assume that Taylor's tool life equation is valid. When the bar is turned at 300 rpm, the tool life (in minute) is approximately ____.

(a) 25 (b) 32

(c) 40 (d) 50

[ME 2017 : 2 Marks, Set-2]

194. Using the Taylor's tool life equation with exponent n = 0.5, if the cutting speed is reduced by 50%, the ratio of new tool life to original tool life is

(a) 4 (b) 2

(c) 1 (d) 0.5

[ME 2018 : 1 Marks, Set-1]

195. A grinding ratio of 200 implies that the

(a) grinding wheel wears 200 times the volume of the material removed.

(b) grinding wheel wears 0.005 times the volume of the material removed

(c) aspect ratio of abrasive particles used in the grinding wheel is 200

(d) ratio of volume of abrasive particle to that of grinding wheel is 200

[ME 2018 : 1 Marks, Set-1]

196. Abrasives are not used in

(a) Buffering process

(b) Burnishing process

(c) Polishing process

(d) Super finishing process

[PI 1989 : 1 Mark]

197. In turning operation the surface finish can be improved by decreasing

(a) cutting speed (b) feed per revolution

(c) rake angle (d) Nose radius

[PI 1989 : 1 Mark]

198. Match the following

List-I	List-II
A. Grinding	1. Compounded oil
B. Tapping	2. EP additive oil
C. Automatics	3. Chemical fluids
D. Broaching	4. Straight fatty oils

[PI 1989 : 2 Marks]

199. Reamer is designed to have even number of flutes to

(a) Balance the cutting forces

(b) Conform to shop floor standard

(c) Enable measurement of the reamer diameter

(d) Help in regrinding of reamer

[PI 1990 : 1 Mark]

200. The teeth of both external and internal spur gears can be cut of

(a) milling machine

(b) gear shaping machine

(c) gear hogging machine

(d) gear shaving machine

[PI 1991 : 1 Mark]

201. A blind hole is better tapped with a tap having

(a) right hand flutes (b) left hand flutes

(c) straight flutes (d) no flutes

[PI 1992 : 1 Mark]

202. In metal cutting with a carbide tool, at the maximum recommended speed, the largest % of heat generated goes to be

(a) tool (b) chip

(c) work (d) tool post

[PI 1992 : 1 Mark]

203. Assertion (A): Grinding needs higher specific cutting energy than milling.

Reason (R): Milling cutter material is harder than abrasive grains.

(a) Both A and R are true and R is the correct explanation of A

(b) Both A and R are true but R is not a correct explanation of A

(c) A Is true but R is false

(d) A is false but R is true

[PI 1993 : 2 Marks]

204. Machinability of steels is improved by the addition of

(a) Sulphur (b) Silicon

(c) Phosphorous (d) Ali of the above

[PI 1994 : 1 Mark]

205. Reamers have usually even number of teeth so that the cutting forces on teeth are properly balanced. (T/F)

[PI 1995 : 1 Mark]

206. In whitworth quick return mechanism (used in shapers) the velocity of the ram is maximum at

(a) middle of the forward stroke

(b) beginning of return stroke

(c) end of return stroke

(d) middle of the return stroke

[PI 2001 : 1 Mark]

207. Match the following

Group I	Group II
R. Dressing	1. Blunting of grinding wheels
Q. Loading	2. Shaping of grinding wheels
R. Glazing	3. Sharpening of grinding wheels
S. Trueing	4. Clogging of grinding wheels

	P	Q	R	S
(a)	2	1	4	3
(b)	3	1	4	2
(c)	3	4	1	2
(d)	4	3	1	2

[PI 2002 : 2 Marks]

Common Data for Questions 208 and 209:

Data for a plain milling operation are given below.

Length of workpiece	200 mm
Cutter diameter	100 mm
No. of teeth	4
Cutter speed	100 rpm
Feed	200 mm/min
Depth of cut	2 mm
Total clearance (entry and exit)	5 mm

208. Mean up-deformed chip thickness (in microns) is

(a) 142 (b) 100
(c) 71 (d) 50

[PI 2002 : 2 Marks]

209. Machining time for one single pass (in seconds) is

(a) 60 (b) 66
(c) 126 (d) 150

[PI 2002 : 2 Marks]

210. Which one of the following is not a synthetic abrasive material?

(a) Silicon carbide (b) Aluminum oxide
(c) Titanium nitride (d) Cubic boron nitride

[PI 2003 : 1 Mark]

211. The drawing shows a machined shaft. The purpose of providing rounded grooves at M and N is to

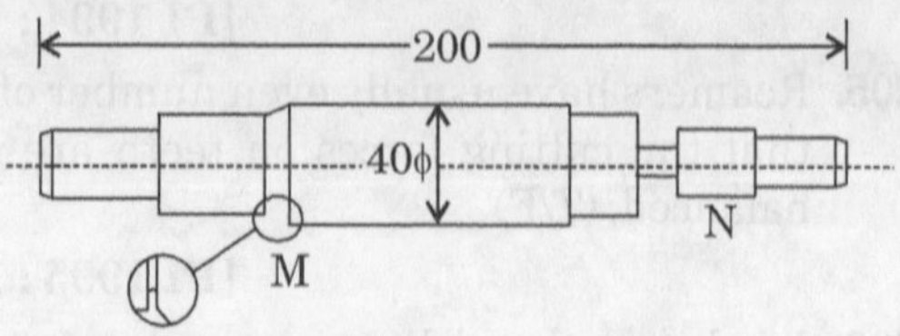

(a) Supply lubricating oil for bearing mounting
(b) Facilitate facing operation
(c) Facilitate grinding of bearing mounting surface
(d) Reduce stress concentration

[PI 2003 : 1 Mark]

212. The flanks of teeth of rack-type gear cutter used for cutting involute gear profiles are

(a) Cycloidal (b) Circular
(c) Involute (d) Straight

[PI 2003 : 1 Mark]

213. Match List-I (Machine tool) with List-II (Machine tool part)

List-I	List-II
A. Lathe	1. Lead screw
B. Milling machine	2. Rocker arm
C. Shaper	3. Universal indexing
D. Drilling machine	4. Flute

	A	B	C	D
(a)	4	2	3	1
(b)	1	3	2	4
(c)	4	3	2	1
(d)	1	2	3	4

[PI 2003 : 2 Marks]

214. The purpose of helical grooves in a twist drill is to

1. Improve the stiffness
2. Save a tool material
3. Provide space for chip removal
4. Provide rake angle for the cutting edge

Select the correct answer suing the codes given below:

(a) 1 and 2 (b) 2 and 3
(c) 3 and 4 (d) 1 and 4

[PI 2003 : 2 Marks]

215.

Mechanism	Machine
P. Pantograph	1. Automobile
Q. Quick return	2. Lathe
R. Ackermann	3. Engraver
S. Toggle	4. Shaper
	5. Press

	P	Q	R	S
(a)	3	1	2	4
(b)	3	4	1	5
(c)	2	4	5	3
(d)	2	3	4	5

[PI 2003 : 2 Marks]

216. The cutting portion of a drill is to be welded to its shank. The process best suited for this is

(a) ultrasonic welding
(b) electron beam welding
(c) friction welding
(d) laser welding

[PI 2004 : 1 Mark]

217. In which of the choices given below, the cutting tool materials are placed in the ascending order of permissible cutting speed for machining of steel?

(a) K group cemented carbide - P group cemented carbide - Ceramic - Cubic boron nitride
(b) Cubic boron nitride - K group cemented carbide - Cubic boron nitride - Ceramic
(c) P group cemented carbide - K group cemented carbide - Cubic boron nitride - Ceramic
(d) Ceramic - Cubic boron nitride - K group cemented carbide - P group cemented carbide

[PI 2004 : 2 Marks]

218. Consider the following statements with respect to the relief angle of cutting tool:

1. This affects the direction of chip flow.
2. This reduces excessive friction between the tool and workpiece
3. This affects tool life
4. This allows better access of coolant at the tool workpiece interface

Which of the statements given above are correct?

(a) 1 and 2 (b) 2 and 3
(c) 2 and 4 (d) 3 and 4

[PI 2004 : 2 Marks]

219. The effects of setting a boring tool above centre height leads to a/an

(a) increase in the effective rake angle and a decrease in the effective clearance angle
(b) increase in both effective rake angle and effective clearance angle
(c) decrease in effective rake angle and an increase in the effective clearance angle
(d) decrease in both effective rake angle and effective clearance angle

[PI 2005 : 1 Mark]

220. If the abrasive grain is viewed as a cutting tool, then which of the following represents the cutting parameters in common grinding operations?

(a) Large negative rake angle, low shear angle and high cutting speed
(b) Large positive rake angle, low shear angle and high cutting speed
(c) Large negative rake angle, high shear angle and low cutting speed
(d) Zero negative rake angle, high shear angle and high cutting speed **[PI 2005 : 1 Mark]**

221. Tool life in the case of a grinding wheel is the ti me

(a) between two successive regrinds of the wheel
(b) taken for the wheel to be balanced
(c) taken between two successive wheel dressings
(d) taken for a wear of 1 mm on its diameter

[PI 2005 : 1 Mark]

222. In a shaper machine, the mechanism for tool feed is

(a) Geneva mechanism
(b) Whitworth mechanism
(c) Ratchet and Pawl mechanism
(d) Ward-Leonard system

[PI 2005 : 1 Mark]

223. Consider the following statements:

1. As the cutting speed increase, the cost of production initially reduces, then after an optimum cutting speed it increases.
2. As the cutting speed increase the cost of production also increases and after a critical value it reduces.
3. Higher feed rate for the same cutting speed reduce cost of production.
4. Higher feed rate for the same cutting speed increases the cost of production.

Which of the statements given above are correct?

(a) 1 and 3 (b) 2 and 3
(c) 1 and 4 (d) 3 only

[PI 2005 : 2 Marks]

224. One brand of milling machining has the following two index plates supplied along with the indexing head.

Plate 1 : 15, 16, 17, 18, 19, 20 hole circles.

Plate 2 : 21, 23, 27, 29, 31, 33 hole circles.

It is proposed to mill a spur gear of 28 teeth using simple indexing method. Which one of the following combinations of index plate and number of revolutions is correct?

(a) Plate 1 : 1 revolution and 9 holes in 18 hole circles
(b) Plate 2 :1 revolution and 9 holes in 21 hole circles
(c) Plate 2 :1 revolution and 9 holes in 23 hole circles
(d) Plate 1 :1 revolution and 9 holes in 15 hole circles

[PI 2005 : 2 Marks]

225. Match List-I (Cutting tool) with List-II (Features) and select the correct answer using the codes given below the lists:

List-I	List-II
A. Turning tool	1. Chisel edge
B. Reamer	2. Flutes
C. Milling cutter	3. Axial relief
	4. Side relief

Codes:

	A	B	C
(a)	1	2	3
(b)	4	3	2
(c)	4	2	3
(d)	1	3	2

[PI 2005 : 2 Marks]

226. Match List-I (milling problem) with List-II (Probable causes) and select the correct answer using the codes given below the lists:

List-I

A. Chatter

B. Poor surface finish

C. Cutter burrs

D. Built-up edge formation

List-II

1. Too high feed
2. Lack of rigidity in machine, fixtures, bar or workpiece
3. High cutting load
4. Radial relief too great
5. Not enough lubricant

Codes:

	A	B	C	D
(a)	2	1	5	3
(b)	2	1	3	5
(c)	4	5	2	3
(d)	4	2	3	5

[PI 2005 : 2 Marks]

227. The cutting tool material normally used for turning steel of very high hardness is

(a) HSS (b) Tungsten carbide

(c) CBN (d) Diamond

[PI 2006 : 1 Mark]

228. The final finishing process for the surface plate made of cast iron which is used as a reference surface is

(a) Buffing (b) Grinding

(c) Hand scraping (d) Honing

[PI 2006 : 1 Mark]

229. In the manufacture of twist drills, the shank is joined to the body of the drill using

(a) spot welding

(b) TIG welding

(c) projection welding

(d) friction welding **[PI 2006 : 1 Mark]**

230. A workpiece of 100 cm length and 30 cm width is machined by planning operation with a feed of 0.3 mm/stroke. The machine tool executes 10 double strokes per minute. The planning time for a single pass is

(a) 50 min (b) 100 min

(c) 166.66 min (d) 333.33 min

[PI 2006 : 2 Marks]

231. A gear with 84 teeth is to be machined using a milling process with indexing. The index plate has the following four hole circles: 36, 38, 42, 48. The change gear ratio required is

(a) 8/21 (b) 10/21

(c) 21/10 (d) 21/8

[PI 2006 : 2 Marks]

232. Reaming is primarily used for achieving

(a) Higher MRR

(b) Improved dimensional tolerance

(c) Fine surface finish

(d) Improved positional tolerance

[PI 2007 : 1 Mark]

233. Which of following gear manufacturing processes is based on generation principle.

(a) Gear hobbing (b) Gear shaping

(c) Gear milling (d) Gear shaving

[PI 2007 : 1 Mark]

234. A tool with side cutting edge angle of 30° and end cutting edge angle 10° is used for fine turning with a feed of 1 mm/rev. Neglecting nose radius of the tool, the maximum (peak to valley) height of the surface roughness produced will be

(a) 0.16 mm (b) 0.26 mm

(c) 0.32 mm (d) 0.48 mm

[PI 2007 : 2 Marks]

235. Diamond cutting tools are not recommended for machining of ferrous metals to

(a) High tool hardness

(b) High thermal conductivity of work material

(c) Poor tool toughness

(d) Chemical affinity of tool material with iron

[PI 2008 : 1 Mark]

Linked Data Q. 236 & 237:

In an orthogonal cutting experiment, an HSS tool having the following tool signature in the orthogonal reference system (ORS) has been used : 0-10-7-7-10-75-1

Given: Width of cut = 3.6 mm, Shear strength of workpiece material = 460 N/mm^2; Depth of cut = 0.25 mm; Coefficient of friction at chip tool interface = 0.7.

236. Shear plane angle (in degrees) for minimum cutting force is

(a) 20.5 (b) 28.6

(c) 24.5 (d) 32.5

[PI 2008 : 2 Marks]

237. Minimum power requirement (in kW) at a cutting speed of 150 m/min is

(a) 2.5 (b) 3.25

(c) 3.35 (d) 3.45

[PI 2008 : 2 Marks]

238. During machining, the wear land (h) has been plotted against machining time (7) as given in the following figure

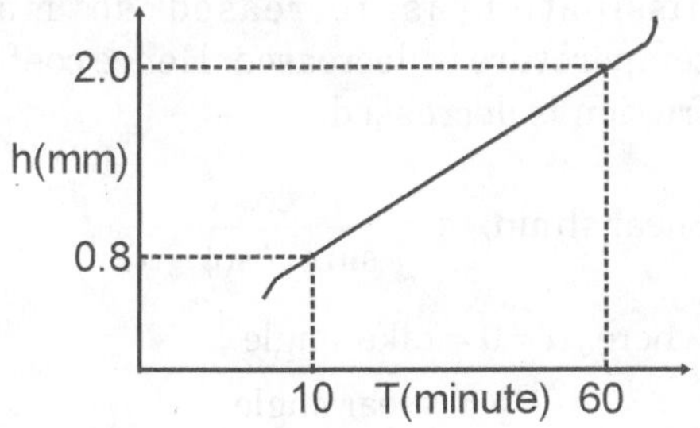

For a critical wear land of 1.8 mm, the cutting tool life (in min) is

(a) 52.00 (b) 51.67

(c) 51.50 (d) 50.00

[PI 2008 : 2 Marks]

239. Brittle materials are machined with tools having zero or negative rake angles because it

(a) Results in lower cutting force

(b) Improves surface finish

(c) Provides adequate strength to cutting forces

(d) Results in more accurate dimensions

[PI 2008 : 2 Marks]

240. The quick return mechanism used in shaper has rocker arm drive of length 200 mm. If the crank radius is 50 mm and the offset between crank centre and rocket arm pivot is 20 mm, length of the stroke (in meters) is

(a) 0.5 (b) 1.0

(c) 1.5 (d) 2.0

[PI 2008 : 2 Marks]

241. Which of the following powder production methods produces spongy and porous particles?

(a) Atomization

(b) Reduction of metal oxides

(c) Electrolytic deposition

(d) Pulverization

[PI 2010 : 1 Mark]

ANSWERS

1. (b)	**2.** (b)	**3.** (a)	**4.** (c)	**5.** (c)	**6.** (a)	**7.** (a)	**8.** (b)	**9.** (b)	**10.** (c)
11. (d)	**12.** (d)	**13.** (a)	**14.** (b)	**15.** (b)	**16.** (c)	**17.** (c)	**19.** (c)	**21.** (d)	**22.** (c)
23. (c)	**24.** (a)	**27.** (c)	**28.** (c)	**29.** (d)	**30.** (d)	**31.** (a)	**32.** (b)	**33.** (b)	**34.** (c)
35. (c)	**36.** (b)	**37.** (b)	**38.** (a)	**39.** (d)	**40.** (b)	**41.** (c)	**42.** (d)	**43.** (c)	**44.** (d)
45. (d)	**46.** (b)	**47.** (a)	**48.** (b)	**49.** (a)	**50.** (b)	**51.** (d)	**52.** (c)	**53.** (c)	**54.** (b)
56. (d)	**57.** (b)	**58.** (a)	**59.** (c)	**60.** (a)	**61.** (a)	**62.** (a)	**63.** (d)	**64.** (b)	**65.** (b)
66. (c)	**67.** (a)	**68.** (b)	**69.** (b)	**70.** (d)	**71.** (b)	**72.** (a)	**73.** (b)	**74.** (d)	**75.** (c)
76. (b)	**77.** (c)	**78.** (d)	**79.** (d)	**80.** (b)	**81.** (d)	**82.** (a)	**83.** (c)	**84.** (a)	**85.** (b)
86. (a)	**87.** (c)	**89.** (b)	**91.** (d)	**92.** (c)	**95.** (a)	**96.** (a)	**97.** (b)	**98.** (b)	**99.** (d)
100. (b)	**101.** (d)	**102.** (a)	**103.** (d)	**104.** (d)	**105.** (a)	**106.** (c)	**107.** (b)	**108.** (d)	**109.** (d)
110. (c)	**111.** (b)	**112.** (a)	**113.** (b)	**114.** (c)	**115.** (b)	**116.** (a)	**117.** (c)	**118.** (b)	**119.** (c)
120. (d)	**121.** (c)	**122.** (d)	**123.** (c)	**124.** (a)	**125.** (a)	**127.** (a)	**128.** (a)	**129.** (c)	**130.** (c)
131. (d)	**134.** (a)	**135.** (c)	**136.** (a)	**137.** (d)	**138.** (b)	**140** (c)	**141.** (b)	**142.** (d)	**143.** (b)
144. (b)	**145.** (c)	**148.** (c)	**150.** (a)	**151.** (c)	**153.** (b)	**154.** (a)	**155.** (c)	**156.** (b)	**157.** (b)
158. (a)	**159.** (c)	**160.** (b)	**161.** (a)	**162.** (b)	**163.** (b)	**164.** (d)	**165.** (b)	**166.** (d)	**167.** (b)
168. (d)	**169.** (c)	**173.** (b)	**174.** (d)	**175.** (c)	**176.** (d)	**181.** (c)	**182.** (a)	**183.** (a)	**185.** (b)
188. (b)	**193.** (b)	**194.** (a)	**195.** (b)	**196.** (b)	**197.** (b)	**199.** (a)	**200.** (b)	**201.** (a)	**202.** (b)
203. (c)	**204.** (a, c)	**206.** (d)	**207.** (c)	**208.** (c)	**209.** (b)	**210.** (c)	**211.** (a)	**212.** (c)	**213.** (b)
214. (c)	**215.** (b)	**216.** (c)	**217.** (a)	**218.** (b)	**219.** (c)	**220.** (a)	**221.** (c)	**222.** (b)	**223.** (a)
224. (b)	**225.** (c)	**226.** (b)	**227.** (b)	**228.** (a)	**229.** (d)	**230.** (b)	**231.** (b)	**232.** (b)	**233.** (a)
234. (a)	**235.** (d)	**236.** (d)	**237.** (b)	**238.** (b)	**239.** (c)	**240.** (b)	**241.** (b)		

EXPLANATIONS

2. Maximum temperature at the chip tool interface is 2 – 3 mm away from cutting edge because shank provides rapid heat dissipation just below the cutting edge.

3. Chip takes away heat from the working zone otherwise tool life and work piece surface finish would get spoiled.

4. Vanadium increases the wear resistance of High speed steel.

5. $\alpha = 10°, \phi = 20°$

Ernest & Merchants' theory

$$2\phi + \beta - \alpha = 90°$$
$$2 \times 20 + \beta - 10 = 90°$$
$$40 + \beta - 10 = 90°$$
$$\beta = 60°$$

6. Nose Radius = 1 – 8 mm, f = ?

R_t = 5 microns

(Peak to valley height.)

$$0.005 = \frac{f^2}{8R_t}$$

$$f = \sqrt{0005 \times 8 \times 1.8}$$

f = 0.268 mm/rev

7. $$r = \frac{\text{Uncut chip thickness}}{\text{Chip thickness after the metal is cut}} = \frac{0.45}{0.45} = 1$$

If ϕ = shear plane angle

and α = rake angle = 0°

$$\text{then } \tan\phi = \frac{r\cos\alpha}{1 - r\sin\alpha} = \frac{1\times\cos 0}{1 - 1\sin 0} = 1.$$

or $\phi = 45°$

8. Wear rate is proportional to $\tan\gamma$, i.e.

Wear rate $\propto \tan\gamma$

Where, γ = Clearance angle

When clearance angle is decreased from 10° to 7°, then

$$\text{percentage change in } f = \frac{\tan 10° - \tan 7°}{\tan 10°} \times 100$$

= 30.37 % decrease.

9. BUE is formed while machining ductile materials at low speed.

10. $$\tan\phi = \frac{r\cos\alpha}{1 - r\sin\alpha}$$

where, r = 0.4

and $\alpha = 10°$

$$\therefore \tan\phi = \frac{0.4\cos 10°}{1 - 0.4\sin 10°}$$

or ϕ = 22.94°

11. By increasing the cutting speed, heat dissipation is increased so maximum temperature is decreased. Hence coefficient of friction is decreased.

12. Shear strain, $$\tau = \frac{\cos\alpha}{\sin\phi\cos(\phi - \alpha)}$$

where, $\alpha = 0$ = rake angle

ϕ = shear angle

$$\therefore \tau = \frac{1}{\sin\theta\cos\theta}$$

$$\therefore T = \sin\phi\cos\phi$$

$$\frac{dt}{d\phi} = \cos^2\phi - \sin^2\phi$$

or $\phi = 45°$

$$\therefore \tau_{min} = \frac{1}{\sin\frac{\pi}{4}\cos\frac{\pi}{4}} = \frac{1}{\frac{1}{\sqrt{2}} \times \frac{1}{\sqrt{2}}} = 2$$

13. D = 100 mm feed, f = 0.25 mm/rev

d = 4 mm, V = 90m/min, Fc = 1500 N

$\boxed{\alpha = 0}$

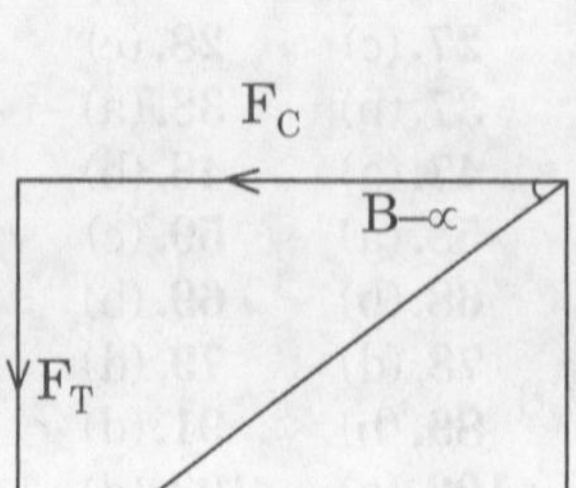

14. $N = F_c.\cos\alpha - F_T.\sin\alpha$

$\because \alpha = 0$

$\boxed{N = F_c = 1500\text{N}}$

15. Specific cutting pressure

$$\therefore \frac{F_c}{W.t} = \frac{400}{2 \times 0.1} = 2000 \text{ N/mm}^2$$

16. Centre lathe - knurling
Milling – Slotting
Grinding – Dressing
Drilling – Center Boring

17. For, orthogonal turning,

$\lambda = 90°$, $F_c = 1601$ N

Chip thickness 't_c' = 0.3,

$F_t = 1259$ N.

N = 1000 rpm, $\alpha = 10$ m

$$f = 100 \text{ mm/min} = \frac{100 \text{ mm}}{1000 \text{ rev}}$$

$$f = 0.1 \frac{\text{mm}}{\text{rev}} = t$$

$$r = \frac{t}{t_c} = \frac{0.1}{0.3} = \frac{1}{3}$$

$$F_a = \tau_s \times \frac{bt}{\sin\phi}$$

$$\Rightarrow \tau_s = \frac{F_s \sin\phi}{6 \times t}$$

$$\tan\phi = \frac{r\cos\alpha}{1 - r\sin\alpha}$$

$$\Rightarrow \phi = 19.19°$$

$$F_s = F_c\cos\phi - F_t\sin\phi$$

$$= 1601 \times \cos 19.19° - 1259 \sin 19.19$$

$$= 1098.20 \text{ N}$$

$$\tau_s = \frac{1098.20 \times \sin 19.19°}{5 \times 0.1} = 721.96$$

≈ 722 MPa

18. $t = 0.125$ mm

$t_c = 0.22$ mm

rake angle, $\alpha = 10°$

Chip thickness ratio, $r = \frac{t}{t_c}$

$$r = \frac{0.125}{0.22} = 0.568$$

$$\tan\phi = \frac{r\cos\alpha}{1 - r\sin\alpha} = \frac{0.568\cos 10°}{1 - 0.568\sin 10°}$$

Shear angle, $\phi = 31.83°$

$$2\phi + \beta - \alpha = \frac{\pi}{2} = 90°$$

$$\Rightarrow 2 \times 31.83 + \beta - 10 = 90°$$

Friction angle $\beta = 36.33°$

$\because \tan\beta = \mu$

$\Rightarrow$ Coefficient of friction,

$\mu = \tan 36.33 = 0.7355°$

$\boxed{\mu = 0.7355} \approx 0.74$

19. Tungsten is available in T-series & M- series.

20. BUE increases the cutting force due to decrease in apparent rake angle and spoils the surface finish.

21. Soft tool material must be used in order to avoid tool wear.

22. Chatter in every 30° .

So 12 chatter mark in one revolution.

$$\rightarrow 500 \text{ rpm} = \frac{500}{60} = 8.33 \text{ rps.}$$

$\Rightarrow$ Frequency = 12 × 8.33 = 100 Hz (cycle/sec)

$\Rightarrow$ Frequency = 12×8.33 = 100 Hz (cycle/sec)

23. Z = teeth table feed = 40 mm / min

N = 100 rpm

$$f_t = \frac{F}{ZN} = \frac{40}{8 \times 100}$$

$f_t = 0.05$ mm

24. A 30 (K) 12 V

(K) → Soft Wheel must be used for Hard & strong Material

25. Reduction in chip thickness and increase in shear angle. Because F_T, thrust force is decreasing in nature.

26. Reducing & Increasing.

Point angle in drill plays the role of side cutting edge angle as in single point turning tool.

27. Carbide tool are ceramics, which are having low toughness. So negative rakes angles are used.

28. Detoriation of surface finish

29. 10 : 20 : 10 : 5–10 Maximum heat generated must goes with chip and remaining goes to workplace & tool least goes to the Environment

30. Hot Hardness is the property of tool material to withstand at higher temperature with hardness.

31. Molybdenum and Tungsten both increases hot hardness & abrasion Resistance.

32. $\text{Specific cutting Energy} = \dfrac{Fc}{W.t_1}$

$t \uparrow SCE \downarrow$ $t_1 \rightarrow$ uncut chip thickness & depth of cut in orthogonal model

33. Gray Cast Iron because of its damping nature and characteristic to absorb shock.

34.

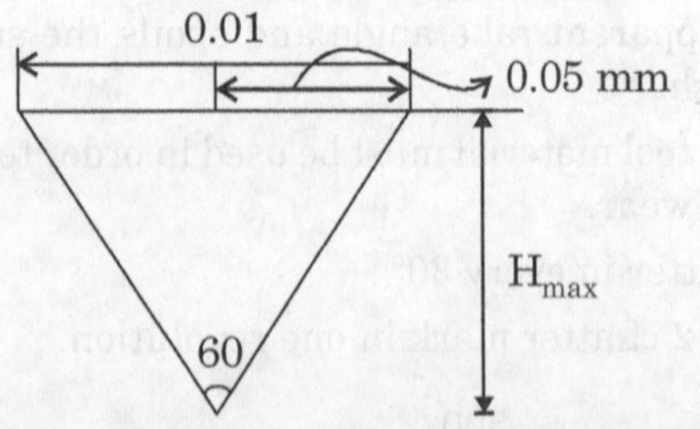

$$\tan 30° = \frac{0.05}{H_{max}}$$

$$H_{max} = 86 \text{ microns} = 0.086 \text{ mm}$$

36. (b) $H_{max_1} = \dfrac{f^2}{8R}$

$$H_{max_2} = \frac{2f^2}{8R} = 4H_{max_1}$$

37. $\alpha = 0, \dfrac{Fc}{F_T} = \dfrac{1}{0.372}, L_c = 4.71, L_{uc} = 10mm$

$$\frac{L_C}{L_{UC}} = \frac{t_1}{t_2} = \frac{4.71}{10} = r$$

r = 0.471 chip reduction coefficient

$$\tan\beta = \frac{F_T}{F_C}, \text{where } \alpha = 0$$

$\beta = \tan^{-1}(0.372)$

$\beta = 2071°$

$$\tan\phi = \frac{r\cos\alpha}{1 - r\sin\alpha} = \frac{0.471\cos 0}{1 - 0.471\sin 0}$$

$= \tan^{-1}(0.471)$

$\phi = 25.22°$

38. Should be cut by negative SCEA.

39. $r = 0.3 = \dfrac{t_1}{t_2}, \alpha = 10°$

$$\tan\phi = \frac{r\cos\alpha}{1 - r\sin\alpha} = \frac{0.3\cos 10°}{1 - 0.3\sin 10°}$$

$\phi = 17.3°$

Shear strain = $\tan(\phi - \alpha) + \cot\phi$

$= \tan(17.3 - 10) + \cot 17.3$

Shear strain = 3.34

40.

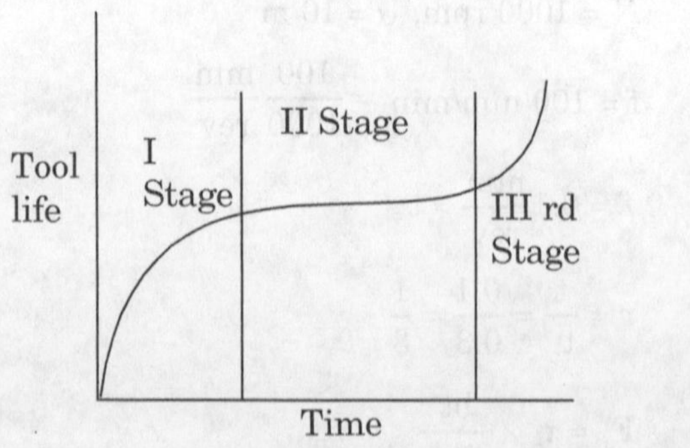

41.

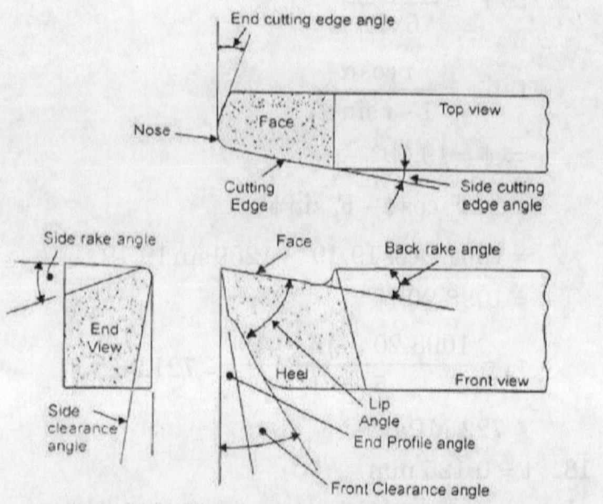

Tool Geometry of a Single point Cutting Tool

42. HSS – Forging

Stellite – Casting

Cementite – Powder metallurgy

UCON – Rolling

43. Sliding friction is very higher

45. Material removal rate (MRR) in mm^3/s

MRR = f.d × V

$$MRR = 0.25 \times 4 \times \frac{\pi \times 200 \times 160}{60}$$

$MRR = 1675.5 \text{ mm}^3/\text{s}$

46. Carbide tool: $VT^{1.6} = 3000$

HSS tool: $VT^{0.6} = 200$

$$T = \left(\frac{3000}{V}\right)^{1/1.6}$$

$$T = \left(\frac{200}{V}\right)^{1/0.6}$$

$$\left(\frac{3000}{V}\right)^{1/1.6} = \left(\frac{200}{V}\right)^{1/0.6}$$

V = 39.389 m/min

47. $$t = \frac{\pi\left(d_0^2 - d_1^2\right)}{4f_r.V} = \frac{\pi\left[(0.2)^2 - (0.08)^2\right]}{4 \times 0.1 \times 10^{-3} \times 90}$$

time, t= 2.93 minutes

48. Specific cutting energy(SCE)

$$SCE = \frac{F_C}{d \times f} = \frac{200}{0.1 \times 1 \times 1000}$$

SCE = 2 J/mm³.

51. $MRR = \pi DNd_f$

52. As rake angle α increases, shear plane angle ϕ decreases. So less force is required for cutting.

53. $$H_{max} = \frac{f^2}{8R} = \frac{2^2 f^2}{8R \times 4} = \frac{F^2}{8R}$$

55. Rigid

57. V_1=63m/min T_1 = 60 hours = 600 min

C = 257.35

$V_1 T_1^n = C$

$63 \times (100)^n = 257.35$

n = 0.22

$V_2 T_2^n = 257.35$

$2 V_1 T_2^n = 257.35$

$2 \times 63 (T_2)0.22 = 257.35$

$\boxed{T2 = 25.6\,min}$

58. α = 35°

60. n = 0.12, c = 130 m /min

$C_1 = 130 \times 1.1 = 143$ m/min

$V_1 = 90$ m /min

$V_1 T_1^n = C_1$

$90 (T_1)^n = 143$

$$T_1 = \left(\frac{143}{90}\right)^{1/0.12}$$

$\boxed{T_1 = 47.4\,min}$

61. Velocity of first cutting tool,

$v_1 = 50 \times 0.25 = 12.5$ mm/min

Velocity of second cutting tool,

$v_2 = 0.25 \times 80 = 20$ mm/min

Since tool life = number of component produce × tool constant

∴ Tool life for first tool = $T_1 = 500 \times k$

Took life for second tool = $T_{21} = 122 \times k$

Now, $v_1 T_1^n = v_2 T_2^n$

$\Rightarrow 12.5(500\,k)^n = 20 \times (22\,k)^n$

$$\Rightarrow \left(\frac{500k}{122k}\right)^n = \left(\frac{20}{12.5}\right)$$

$$\Rightarrow \left(\frac{500}{122}\right)^n = 1.6$$

$$\Rightarrow n \ln\left(\frac{500}{122}\right) = \ln(1.6)$$

$\Rightarrow n = 0.3332$

Hence number of components produced by one cutting tool at 60 rpm is:

$$\Rightarrow 12.5 \times \left(\frac{500k}{10}\right)^{0.3332} = 60 \times 0.25 \times (nk)^{0.332}$$

$$\Rightarrow \frac{12.5}{60 \times 0.25} = \left(\frac{nk}{50k}\right)^{0.332}$$

$$\Rightarrow \frac{n}{50} = 0.5786$$

∴ $n = 0.5786 \times 50 = 28.92 \approx 29$.

62. Given, rake angle, α = 10°

shear angle, ϕ = 27.75

depth of cut, $t_1 = 0.4$ mm

Let t_2 = thickness of chip produced

$$\text{Now, } \frac{t_1}{t_2} = \frac{\sin\phi}{\cos(\phi - \alpha)}$$

$$\Rightarrow t_2 = \frac{t_1 \cos(\phi - \alpha)}{\sin\phi}$$

$$= \frac{0.4\cos(27.75 - 10)}{\sin(27.75)}$$

$$= \frac{0.4\cos 17.75}{\sin 27.75} = 0.818 \text{ mm}$$

63. From Earnest and Merchants' theory

$2\phi + \lambda - \alpha = 90$

where μ = co-efficient of friction = $\tan \lambda$

$\therefore \lambda = 90 + 10 - 2 \times 27.75$

$\therefore = 44.5^\circ$

$\mu = \tan(44.5) = 0.98$

64. Given: Diameter of hole,

$d = 10$ mm

Thickness of steel plate,

$t = 20$ mm

$$\alpha = \frac{120}{2} = 60^\circ$$

Break through distance,

$$A = \frac{d}{2\tan\alpha} = \frac{10}{2\tan 60^\circ} = 2.8867 \text{ mm}$$

Total length of tool travel,

$l = t + A + 2$

$= 20 + 2.8867 + 2 = 24.88$

Time for drilling

$$= \frac{l}{fN} = \frac{24.88}{0.2 \times 300} = 0.4147 \text{ min}$$

$= 24.88$ sec ≈ 25 sec.

65. $$\mu = \frac{F}{N} = \frac{F_C \sin\alpha + F_T \cos\alpha}{F_C \cos\alpha - F_T \sin\alpha}$$

$$= \frac{900\sin 5^\circ + 450\cos 5^\circ}{900\cos 5^\circ - 450\sin 5^\circ} = \frac{526.72}{857.35} = 0.614$$

$\therefore \lambda = \tan^{-1}(\mu) = 31.5^\circ$

Alternative method

From Merchants' Circle

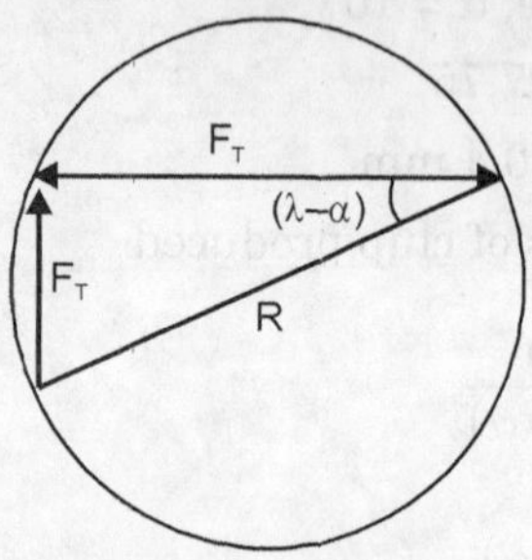

Given: $F_C = 900$ N

$F_T = 450$ N

Now

$$\tan(\lambda - \alpha) = \frac{F_T}{F_C} = \frac{450}{900}$$

$$\therefore (\lambda - \alpha) = \tan^{-1}\left(\frac{1}{2}\right)$$

[where, α = Rake angle]

Friction angle,

$\lambda = 26.5 + 5 = 31.50$

66. Taylor's tool life equation,

$VT^n = C$ ...(*i*)

Where, V = cutting speed and T = tool life

When cutting speed is doubled and tool life reduces to $\frac{1}{8}$, then

$$2V\left(\frac{1}{8}T\right)^n = C \quad ...(ii)$$

Dividing equation (*i*) by equation (*ii*), we get

$$\frac{VT^n}{2V\left(\frac{T}{8}\right)^n} = 1$$

or $VT^n = 2V\left[\frac{T}{8}\right]^n = 2VT^n 2^{-3n}$

or $1 = 2 \cdot 2^{-3n}$

or $1 = 2^{-3n+1}$

or $2^{-3n+1} = 2^0$

or $-3n + 1 = 0$

or $3n = 1$

or $n = \frac{1}{3}$.

67.

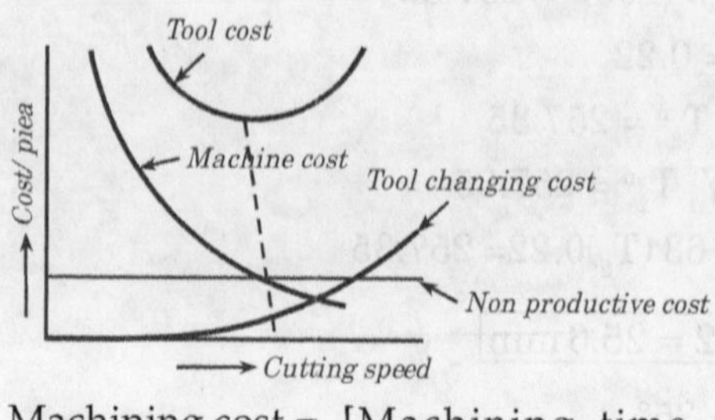

Machining cost = [Machining time × Direct Labour Cost]

So as cutting speed increases, machining time decreases and therefore machining cost decreases.

68. Length travelled in forwarded stroke = 640 mm

Number of strokes = 100

Time for cutting = 8 min

Return time = 4 min

Total time = 12 min.

69. $H_{max} = \dfrac{f}{\tan(\psi)+\cot(\psi)}$ $\quad \psi$ = SCEA, ψ_1 = ECEA

$$\frac{h_P}{h_Q} = \frac{\tan 15^\circ + \text{Cot}8^\circ}{\tan 30^\circ + \cot 8^\circ}$$

70. Here, r = cutting ratio,

$$n = \frac{t_1}{t_2} = \frac{0.5}{0.7} = 0.714$$

$$\therefore \ \tan\phi = \frac{r\cos\alpha}{1 - r\sin\alpha}$$

where, ϕ = shear angle

and α = rake angle

$\therefore \ \phi = \ 40.2^\circ$

Shear strain = $\cot\phi + \tan(\phi - \alpha) = 1.65$

71. From Merchant's theory,

$$2\phi + \lambda - \alpha = \frac{\pi}{2},$$

where, λ =friction angle

$\lambda = 24.6^\circ$

$\therefore \ \mu = \tan\lambda = 0.457$

72. Total heat generated, $Q_1 = F_c \times V$

= 24000 J/min

Heat dissipated due to friction,

$Q_2 = F \times r \times v$

Frictional force,

$F = F_c \sin\alpha + F_f \cos\alpha$ = 503.76 N

Q_2 =7193.69 J/min.

$$\therefore \frac{Q_2}{Q_2} = 0.299;\ 30\%$$

73. Chip thickness,

$$r = \frac{\text{feed}}{\text{chip thickness}} = \frac{0.24}{0.48} = .5$$

rake angle = 0

$$\tan\phi = \frac{r\cos\alpha}{1 - r\sin\alpha} = .5$$

$\therefore \ \phi = \ 26.56$

74. F_C = Cutting force;

V = Cutting velocity

Specific machining energy = 2.0 J/mm^3

$\therefore \ F_C \times V = 2 \times \text{MRR}$

$\Rightarrow F_C = 2 \times .2 \times 10^{-3} \times 2 \times 10^3 = 800$ N

75. Here, ϕ = shear angle = 25°

λ = Friction angle,

α = rake angle = 0°

From Merchant's theory, $2\phi + \lambda - \alpha = 90^\circ$

$\therefore \lambda = 90^\circ - 50^\circ = 40^\circ$

$$\therefore \ \mu = \frac{F}{N} = \frac{\text{Friction force}}{\text{Normal force}} = \tan\lambda = .83$$

76. Here, D = 147 mm, L = 630 mm,

f = .2 mm/rev,

d = 2 mm

Tool life equation,

VT^n = constant

Now, $V_1T_1^n = V_2T_2^n$

Here, n =.415

$\therefore \ VT^n = V_1T_1^n$

$\Rightarrow$ V =97.07 m/min

77. $V = \pi DN$

N =210.2 rpm

$\therefore$ Machining time,

$$t_m = \frac{L}{N} = .15\,\text{min}.$$

78. Side rake angle is equal to orthogonal rake angle when principle cutting edge angle become 90° and corresponding approach angle SCEA = 0°

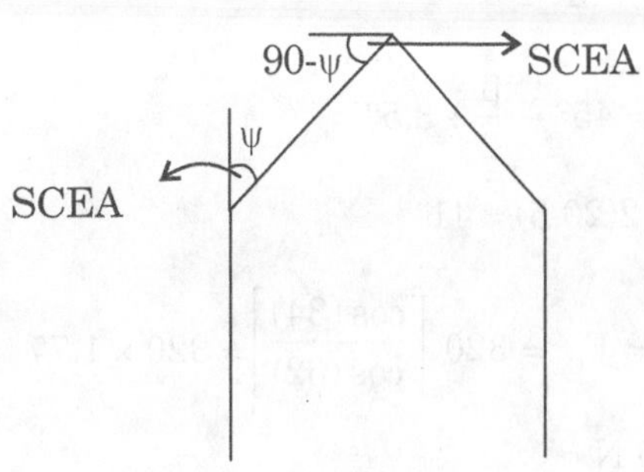

79. $$\tan\phi = \frac{r\cos\alpha}{1 - r\sin\alpha} = \frac{0.5\times\cos 7^\circ}{1 - 0.5\sin 7^\circ} = 0.503$$

$\phi = 28^\circ$

Now, shear force = shear stress × area

F = 250 × AB width of cut

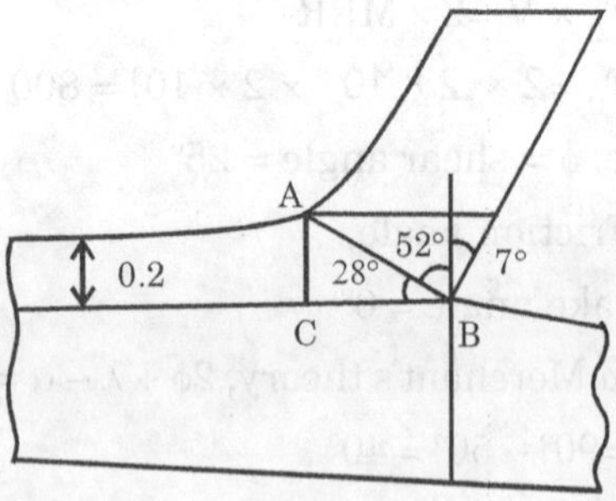

$$\therefore AB = \frac{AC}{\sin 28°} = \frac{0.2}{\sin 28°} = 0.43$$

where AC is incut chip thickness

Now width≅ depth of cut = 3 mm

F = 250 × 0.43 ×3

≅ 320 N

80. $F_H = F_S \left[\frac{\cos(\quad - \alpha)}{\cos(\quad\quad \alpha)}\right]$

where β is friction L.

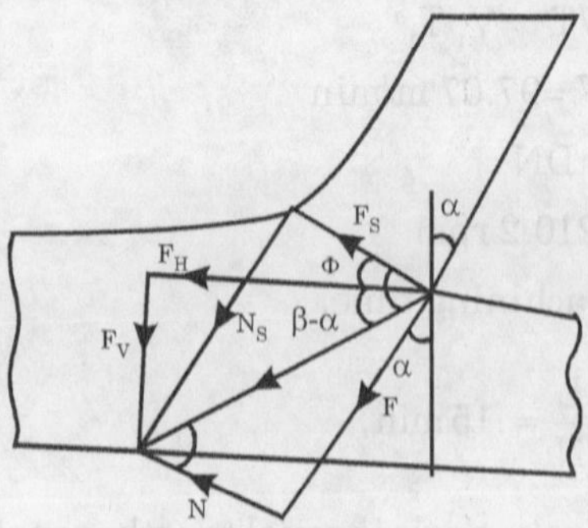

Now by Merchant's theory

$$\phi = \frac{\pi}{4} - \frac{1}{2}(\beta \quad - \alpha)$$

$$\therefore 28 = 45° - \frac{\beta}{2} + 3.5°$$

$$\therefore \beta = 2(20.5) = 41°$$

$$\therefore F_H = F_C = 320 \left[\frac{\cos(34)}{\cos(62)}\right] = 320 \times 1.77$$

= 565 N

Now $\frac{F_V}{F_H} = \tan(\beta - \alpha)$

$\therefore F_V = 565 \times \tan 34 = 381$ N

Frictional force F from the figure can be written as,

$F = F_H \sin + F_V \cos \alpha$

= 68 + 381 = 447N

Hence none of the options seems to be correct.

81. By increasing the cutting speed. Heat dissipation is increased hence there is lower temperature & lower friction coefficient.

82. According to Taylor's equation

$VT^n = C$

where, V = cutting speed in m/s

T = tool life (min)

C = constant (Taylors constant)

$\therefore V_1T_1^n = V_2T_2^n = C$

or $60 \times (81)^n = 90 \times (36)^n$

or $\frac{60}{90} = \left(\frac{36}{81}\right)^n$

or $\frac{2}{3} = \left(\frac{4}{9}\right)^n$

Taking log of both sides

$\log\frac{2}{3} = n \log\frac{4}{9}$

$$\therefore n = \frac{\log\frac{2}{3}}{\log\frac{4}{9}} = 0.5$$

$\therefore$ Taylor's constant $C = V_1T_1^n$

$= 60 \times (81)^{0.5}$

$= 60\sqrt{81} = 60 \times 9 = 540$

83. $VT^n = C$

$$\therefore T^n = \frac{C}{V}$$

$$\text{or } T = \left(\frac{C}{V}\right)^{1/n} = \left(\frac{C}{V}\right)^2$$

$$\text{or } T \propto \frac{1}{V^2}$$

$$\therefore \frac{T_0}{T} = \frac{V^2}{V_0^2}$$

$$\text{or } \frac{T_0}{T} = \frac{\left(\frac{V_0}{2}\right)^2}{V_0^2}$$

$$\text{or } \frac{T}{T_0} - 1 = 4 - 1$$

$$\text{or} \quad \frac{T - T_0}{T_0} = 3$$

Hence percent increase = 300%

Alternately:

When cutting speed is halved

$$60 \times (8.1)^{0.5} = \frac{60}{2} (T_2)^{0.5}$$

$$\text{or} \quad \left(\frac{T_2}{81}\right)^{0.5} = \frac{60 \times 2}{60}$$

$$\text{or} \quad \frac{T_2}{81} = (2)^2$$

or $T_2 = 4 \times 81$

$$\% \text{ change in tool life} = \frac{T_2 - T_1}{T_1}$$

$$= \frac{(4 \times 81) - 81}{81} \times 100$$

$$= 3 \times \frac{81}{81} \times 100 = \textbf{\textit{300\%}}$$

84. Taylor's tool life equation is

VT^n= Constant

$\therefore k_1 = v_1 T_1^{n_1}$ and $k_2 = v_2 T_2^{n_2}$

Given conditions for both tools :

Tool A : constant, $k_1 = 90$;

exponential constant, $n_1 = 0.45$

Tool B : Constant, $k_2 = 60$;

exponential constant, $x_2 = 0.3$

$$\therefore \left(\frac{k_1}{k_2}\right) = \left(\frac{v_1}{v_2}\right)\frac{T_1^{n_1}}{T_2^{n_2}}$$

$$\Rightarrow \left(\frac{90}{60}\right) = \left(\frac{v_1}{v_2}\right)\frac{T_1^{0.45}}{T_2^{0.3}}$$

At point of intersection

$v_1 = v_2$, and $T_1 = T_2 = T$

$$\therefore \frac{3}{2} = \frac{T^{0.45}}{T^{0.3}} = T^{0.15}$$

$$\Rightarrow T = (3/2)^{\frac{1}{0.15}} = 14.92 \text{ min}$$

$\therefore \ 90 = v_1 (14.92)^{0.45}$

$\Rightarrow v_1 = 26.7$ m/min

Above $v_1 = 26.7$ m/min, tool A will have a higher tool life them too B.

85. Relation between shear angle (ϕ), chip thickness ratio (r) and rake angle (α) is given by

$$\tan\phi = \frac{r\cos\alpha}{1 - r\sin\alpha}$$

$$\text{where,} \quad r = \frac{0.81}{1.8} = 0.45$$

$$\tan\phi = \frac{0.45\cos 12}{1 - 0.45\sin 12}$$

$\Rightarrow \phi = 26°$

86. $$T_c = \frac{(L_h + A + O + C)}{Nf}$$

Here, $L_h = 50$ mm; A = O = 2 mm;

$$C = \frac{D}{2}\cot\left(\frac{\beta}{2}\right) = 7.5 \times \cot(59) = 4.5 \text{ mm}$$

N = 500 rpm; f = 0.2 mm/rev

$T_c = 0.585$ min or 35.1 seconds

87. *Given :* r= 0.4

$t_1 = 0.6$ mm

$\alpha = 10°$

$V_C = 2.5$ m/s

$t_m = 25$ m

$$\tan\phi = \frac{r\cos\alpha}{1 - r\sin\alpha} = 0.4233$$

$\Rightarrow \phi = 22.9°$

$$\text{Shear strain rate} = \frac{V_C \cos\alpha}{\cos(\phi - \alpha) \times t_m}$$

$$= \frac{2.5 \times \cos 10}{\cos 12.9 \times 25 \times 10^{-6}}$$

$= 1.010^4 \times 10^5$/s

88. $$T_{opt} = \left[\frac{1-n}{n} \times T_C\right] = \frac{1 - 0.2}{0.2} \times 1.5 = 6 \text{ min.}$$

89. $$T = \frac{L}{f_N}$$

$L = t + Ap_1$

$Ap_1 = 0.5$ D (holes diameter)

$= 10$ mm

$t = 40$ mm

$$T = \frac{50}{0.1 \times 400} = 1.25 \text{ min.}$$

90. $t = 0.010$ mm

$v = 130$ m/min

$\alpha = 15°$

$b = 6$ mm

$t_c = 0.015$ mm

$F_c = 60$ N

$F_t = 25$ N

$F = F_c \sin\alpha + F_t \cos\alpha$

$= 60 \sin 15 + 25 \cos 15$

$= 39.6773$ N

Ratio of frictional energy to total energy

$$= \frac{F}{F_c} \cdot \frac{V_c}{V} = \frac{F}{F_c}\left(\frac{t}{t_c}\right)$$

$$\left[\because \frac{t}{t_c} = \frac{V_c}{V} = r\right]$$

$$= \frac{39.6773}{60} \times \frac{0.010}{0.015} = 0.4408$$

91. 4 jaw Chuck

92. Table feed or Feed rate

$$= f_t \quad Z \quad N$$

f_t → Feed/tooth; Z → Number of teeth; N → rpm

93. Number of holes can be drilled = 253.53

94. Taylor's tool life equation,

$VT^n = c$

for Tool X, for Tool y

$VT^{0.3} = 60$ $VT^{0.6} = 90$

Velocity when both tools have same tool life.

$$\frac{T^{0.3}}{T^{0.6}} = \frac{60}{90}$$

$$(T)^{0.6-0.6} = \frac{6}{9}$$

$$T^{0.3} = 1.5$$

$$T = (1.5)^{1/0.3} \text{ min}$$

$$V = \frac{60}{(1.5)^{0.3/0.3}} = 40 \text{ m/min}$$

96. From Ernest & Merchant's Theory for minimum power consumption:

$$\phi = \frac{90 + \alpha}{2} = \frac{90 + 12}{2} = 51°$$

$$\boxed{\phi = 51°}$$

97. $n = 0.25$ $V_1 = 20$m /min, $T_1 = 10$ min

$V_2 = 40$m/min , $T_2 = ?$

$$V_1 T_1^n = V_2 T_2^n = C$$

$$20\,(T_0)^{0.25} = 40\,(T_2)^{0.25}$$

$$\boxed{T_2 = 0.625 \text{ min}}$$

98. l = 2000 mm

w = 300 mm

f = 0.3 mm/ stroke

No. of stroke = 2 × 10

= 20 stroke / min {double stroke}

$$\text{Total Machining time} = \frac{2000}{20} = 100 \text{ min}$$

99. $T_{opt} = \left(\frac{1}{n} - 1\right) T_c$

$T_{c1} = 5$

$$n = \frac{5}{\frac{75}{60}} = \frac{5}{1.25}$$

n = 4 rs/piece

$$Topt = \frac{1 - 0.1}{0.1} \times 4$$

$$\boxed{Topt = 36 \text{ min}}$$

100. $R_A = 0.6$ mm. $R_B = 0.33$ mm

$F_A = 0.2$ mm/rev $F_B = 0.1$ mm/rev

$$(H_{max})_A = \frac{F_A^2}{8R_A} = \frac{0.2^2}{8 \times 0.6} = 0.000833\text{mm}$$

$= 8.33$um

$$(H_{max})_B = \frac{F_B^2}{8R_B} = \frac{0.1^2}{8 \times 0.33} = 3.78\text{um}$$

Tool B have higher surface finish them A.

101. D = 31.8mm, t = 100 mm, V = 20m /min
f = 0.3 mm/rev,
Approach length = 4 mm
Over Travel = 9 mm

$$\text{Time / hole} = \frac{100+9+9}{0.3 \times 200} = 1 \text{ min } 53 \text{ sec.}$$

102. HSS:
$VT^{0.2} = 150$
$VT^{0.3} = 250$

$$\left(\frac{150}{V}\right)^{1/0.2} = \left(\frac{250}{V}\right)^{1/0.3}$$

$$\boxed{V = 54\text{m/min}}$$

103. D = 25 mm,
h = 100 mm
$VT^{0.25} = 55$
V = 22 m/min, f = 0.046 mm/ rev

$$V = \frac{\pi DN}{60}, \quad N = \frac{1000 \times 22}{\pi \times 25}$$

$$\boxed{N = 280\text{r/mm}}$$

$$\text{time/piece} = \frac{L}{fN} = \frac{100}{0.046 \times 280} = 7.76 \text{ min}$$

425 piece time = 425 × 7.76 = 3299.7 min

$$\text{one tool life } T = \left(\frac{55}{V}\right)1/0.25 = \left(\frac{55}{22}\right)9 = 39$$

$$\text{No. of tool Regrinding} = \frac{3299.7}{39} = 85$$

104. ASA Signature
ψ= 45°, ψ=10°

$$\text{Hmax} = \frac{f^2}{8R} = \frac{0.08^2}{8 \times 2}$$

ORS signature = 0.0004 mm

$$H_{max} = \frac{f}{\tan\psi + \cot\psi} = \frac{0.08}{\tan 4548° + \cot 10°}$$

Hmax = 0.012 mm

105. α = 15°
ϕ = 45°
V = 35 m/min

$$\frac{V}{\cos(\phi-\alpha)} = \frac{Vc}{\sin\phi} = \frac{Vs}{\cos\alpha}$$

$$V_C = \frac{V.\sin\phi}{\cos(\phi\text{-}\alpha)} = \frac{35\sin 45}{\cos(45-10)}$$

$$\boxed{V_c = 25.58m/\text{min}}$$

106. No. of jobs = 500
D = 50 mm, L = 100 mm
N = 200 rpm, f = 0.2 mm/rev

$$V = \frac{\pi DN}{1000} = 31.4\text{m/min}$$

$$T = \left(\frac{160}{314}\right)^{1/0.25}$$

T = 6 min

107. $$\text{Time / piece} = \frac{L}{fN} = \frac{100}{0.2 \times 200} = 2.5\text{min}$$

$$\text{No of Tool Changes} = \frac{\text{cutting time}}{\text{Tool life}}$$

$$= \frac{674}{2.5} = 269.6 = 270$$

108. D = 40 mm
α = 10°
N = 500 pm
ϕ = 25°

$$V = \frac{\pi DN}{1000} = \frac{\pi \times 40 \times 500}{1000}$$

V = 62.8 m/min

$$\frac{V_c}{V_s} = \frac{\sin\phi}{\cos(\phi\text{-}\alpha)}$$

$$V_c = \frac{62.8 \times \sin 25}{\cos(25\text{-}10)}$$

$$\boxed{V_c = 27.5 \text{ m/min}}$$

109. $$r = \frac{t_1}{t_2} = \frac{V_c}{V} = \frac{27.5}{62.8} = 0.437$$

$$\frac{F}{Fc} = \frac{R\sin\beta}{R\cos(\beta-\alpha)}, \beta = 58°10'$$

F = 764 N
loss of energy due to friction = F × Vc

$= 764 \times 27.5 = 350.3$ Watts

110. $\alpha = 15°$

W = 2 mm

$t_2 = 0.4$mm

fr = 0.2 mm/rev = t_1

$$r = \frac{t_1}{t_2} = \frac{0.2}{0.4} = \frac{1}{2}$$

r = 0.5

$$\phi = \tan^{-1}\left(\frac{r\cos\propto}{1 - r\sin\propto}\right)$$

$$\phi = \tan^{-1}\left(\frac{0.5\cos 15}{1 - 0 - 5\sin 15}\right)$$

$\phi = 29°$

111. $F_c = 900$ N,

$F_t = 810$ N

$$\tan\beta = \frac{\frac{F_T}{Fc} + \tan\alpha}{1 - \frac{F_T}{Fc}\tan\alpha}$$

$$\tan\beta = \frac{\frac{810}{900} + \tan 15°}{1 - \frac{810}{900}\tan 15°}$$

$\boxed{\beta = 56.98°}$

$$\frac{F_S}{F_C} = \frac{R\text{Cos}(\phi + \beta - \propto)}{R\text{Cos}(\beta - \propto)}$$

$$Fs = \frac{900\,\text{Cos}(70.98)}{\text{Cos}(56.98 - 15)}$$

Fs = 394.56 N

$$Cs = \frac{Fs}{Wt_1} \times \text{Sin}\phi = \frac{394.56}{2 \times 0.2} \times \sin 29$$

$\boxed{Cs = 478\text{mpa}}$

112. Fc = 1200 N F_T = 500 N $\propto = 0$ VC = 1 m/s d = t_1 = 0.8 mm t_2 = 1.5 mm

$$\text{tab } \beta = \frac{\frac{F_T}{F_C} + \tan\propto}{1 - \frac{F_T}{F_C}\tan\propto}$$

$$\text{tab } \beta = \frac{\frac{500}{1200} + \tan 0}{1 - \frac{500}{1200}\tan 0}$$

$\beta = 22.62°$

113. $r = \frac{t_1}{t_2} = \frac{0.8}{1.5} = 0.533$

{chip Reduction coefficient}

$$\frac{t_1}{t_2} = \frac{Vc}{V}$$

$Vc = 0.533 \times 1$

$\boxed{Vc = 0.533 \text{m/s}}$

114. D = 10 mm Plate thickness = 50 mm N= 600 rpm feed = 0.2 mm/rev point Angle $= 2\propto = 120°$

Approach = 0.3D = 0.3 × 10 = 3mm

L = 3 + 50 = 53 mm

$$\text{Time/hole} = \frac{L}{fN} = \frac{53}{0.2 \times 600} = 0.442 \text{min}$$

115. VT0.3 =C

No of holes = 200

$$V = \frac{\pi DN}{60} = 18.84 \text{ min}$$

= Tool life = 200×0.442 = 88.33 min

$18.84 \times 88.33^{0.3} = 6 = 72.26.$

116. Ti = 10 min V_1 = 100 m/min

T_2 = 75 min V_2 = 30 m/min

$$n = \frac{1n\left(\frac{V_2}{V_1}\right)}{1n\left(\frac{T_1}{T_2}\right)} = \frac{1n\frac{75}{100}}{1n\left(\frac{10}{30}\right)}$$

$$n = \frac{-0.2876}{-1.09}$$

$\boxed{n = 0.262}$

117. V = 20m/min $\propto$ 15° t_2 = 0.4 mm t_1 = 0.2 mm

$$\tan\phi = \frac{r\cos\propto}{1 - r\sin\propto}$$

$$\tan\phi = \frac{\frac{0.2}{0.4}\cos 15}{1 - \frac{0.2}{0.4}\sin 15°}$$

$\phi = 29°$

118. $\frac{Vc}{V} = \frac{t_1}{t_2} = r$

$Vc = 0.5 \times 20 = 10$ m/min

119. $\propto = 0$; t_1-0.25 mm t_2= 0.75 mm
w = 2.5 mm Fc = 950 N, F_T = 475 N

$$\frac{t_1}{t_2} = r = \frac{0.25}{0.75} = \frac{1}{3} = 0.333$$

$$\tan\phi = r \Rightarrow \phi = \tan^{-1}(0.333)$$

$$\phi = 18.435°$$

$$\tan\beta = \frac{F_T}{F\tau} = \frac{475}{950}$$

$$\beta = \tan^{-1}(0.5)$$

$$b = 26.56°$$

$$Fs = \frac{F_C R\cos(\phi + \beta - \propto)}{R\mathrm{Cos}(\beta - \propto)}$$

$$Fs = \frac{950\mathrm{Cos}(45°)}{\mathrm{Cos}(26.56°)}$$

$$\boxed{Fs = 75\mathrm{In}}$$

120. $\tau s = \frac{Fs}{Wt_1}\mathrm{Sin}\phi$

$$= \frac{751}{2.5 \times 0.25}\sin 18.43°$$

$$\boxed{\tau = 380\mathrm{N/mm}^2}$$

121. r = 0.4 t_1 = 0.6 $\propto$ 10° V = 2.5 m/s
tm = 0.025

$$\phi = \tan^{-1}\left(\frac{r\cos\propto}{1 - r\sin\propto}\right); \phi = \tan^{-1}\left(\frac{0.9\cos 10}{1 - 0.4\sin 10}\right)$$

$$\phi = 22.94$$

$$= \frac{\cos\propto}{\cos(\phi - \propto)} \cdot \frac{V}{tm}$$

$$= \frac{\cos 10}{\cos(22.94 - 10)} \times \frac{2500}{0.025} = 1.0104 \times 10^5/s$$

122. Number of Double stroke = 30
Quick Return Ratio = 0.6 = M
Length of Strike = 250 mm = L
Average cutting velocity = L.N (1 + 19)
= 30 × 250 (1 + 0.6) = 12 m /min

123.

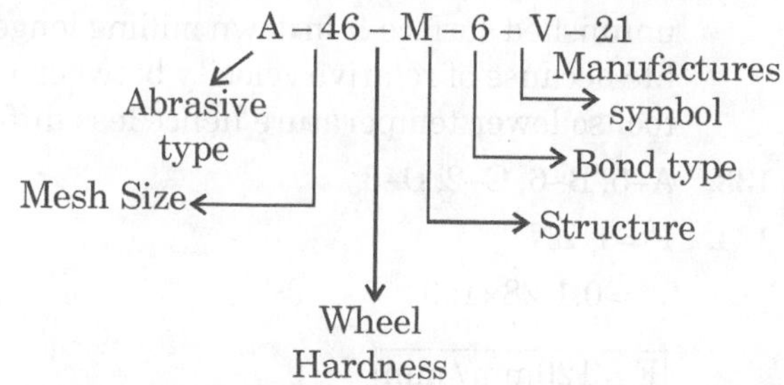

124. $H_{max_1} = \frac{f2}{8R}$

$$H_{max_2} = \frac{(2f)^2}{8 \times (2R)} = \frac{2f^2}{8R}$$

High of surface roughness be doubled means surface finished hill be decreased by 50%

125. Function of cutting fluid
(1) Cooling
(2) Lubrication
(3) Cleaning
(4) Save from external contamination.

126. Small point angle large Helix angle
Point angle similar to SCEA by Decreasing point angle chips become thinner and order.
Helix angle related to back rake angle. Higher back rake angle gain for soft & ductile material

128. Pure metals are ductile, produces long & continuous chip

129.

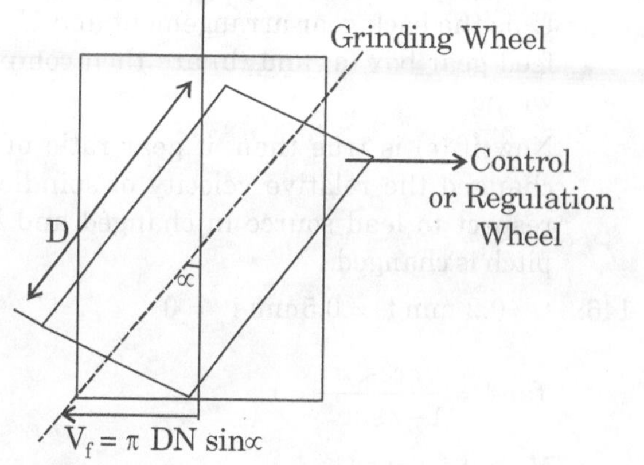

130. Pitch = 1mm
Start = Double Start
Feed rate = 2 × 1 = 2 mm/res.

131. $ft = \frac{f}{ZN} = \frac{50}{10 \times 100}$

ft = 0.05 mm

132. Down milling process provide better surface

finish because it start cutting process with unfinished surface & in down milling longer tool life because of relative velocity between chip & tool so lower temperature hence less diffusion.

133. A–5, B–6, C–2, D–3

134. $F = f_t ZN$

$= 0.1 \times 8 \times 150$

$\boxed{F = 120 \text{mm/min}}$

135. Grain size is given for rough cut.

27 → rough cut

136. By increasing the back rake angle the drag between chip & tool decreased hence cutting power consumption decreased.

137. Grinding because of smaller chips, large cutting area.

139. True yes because carbon and steel having higher affinity to carbon (diamond) can diffuse in to the steel

140. Less chance of wheel wear. Other wise all other to ferrous material.

142. Acme threads

143. Bond strength gives the hardness of wheel for soft material strong bond material is used for hard material soft bond material is used

144. $T = \frac{L}{fN} = \frac{25100}{0.25 \times 300} = \frac{1}{3} \text{min}$

T = 20 sec

145. Taking the engineering of a conventional lathe, U_v is the back gear arrangement and U_s is the feed gear box (a) and (b) are then completely wrong.

Now if (c) is true then, if gear ratio of U_v is changed the relative velocity of spindle with respect to lead source in changed and hence pitch is changed.

146. $t_1 = 0.2$ mm $t_2 = 0.5$mm i = 0

$$\tan\phi = \frac{r\cos\alpha}{1 - r\sin\alpha} = r$$

$V = \cot\phi + \tan(\phi - \alpha)$

$= \cot\phi + \tan\phi$

$$= \frac{1 + \tan^2\phi}{\tan} = \frac{1 + (0.4)^2}{0.4}$$

$$r = \frac{0.2}{0.5}$$

$\boxed{V = 2.9}$

147. $t_1 = f\cos\theta$

$= 0.2 \quad \cos 60$

$t_1 = 0.1$ mm

where t_1 =uncut chip thickness

148. Metal cutting takes place in plastic deformation zone where work hardening takes place as chip get strained.

149. Ratio of 1 and 2 is = $\frac{\sqrt{d_1}}{\sqrt{d_2}}$

$$\therefore \; d_2 = 2d_1 = \frac{\sqrt{d_1}}{\sqrt{2d_2}} = \sqrt{\frac{1}{2}} = 0.707.$$

150. As rake angle increase shear plane decreases so smaller force one required so smooth machining operation

$\tan\phi = \frac{r\cos\alpha}{1 - r\sin\alpha}$ Favourable machining condition $\phi\uparrow\alpha\uparrow r\uparrow$

151.

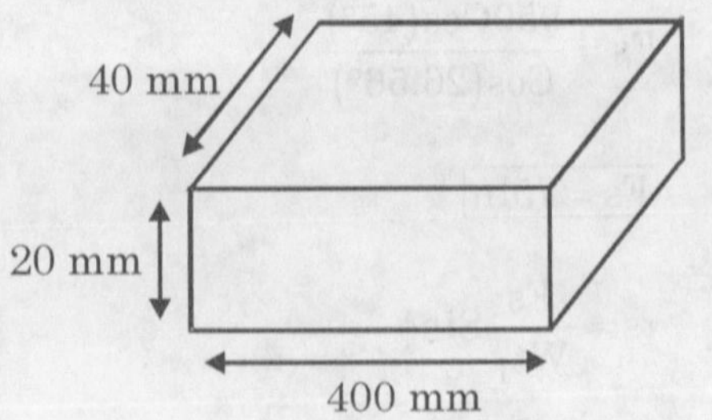

Depth of cut ⇒ (d = 2 mm) Milling cutter,

D = 100 mm

W = 50 mm

Z = 20 tooth,

N = 1200 rpm

$f_t = 0.05$ mm/tooth rev

$f = z \times f_t$

$= 20 \times 0.05$ mm/rev

= 1 mm/rev

Over travel = Approach (A)

$= \sqrt{d(D - d)}$

$= \sqrt{2 \times (100 - 2)}$

$= \sqrt{196} = 14$ mm

Total length covered by tool

$L + 2A = 400 + 2 \times 14 = 428$ mm

Time for one pass,

$$t = \frac{L}{fN} = \frac{428}{1 \times \left(\frac{1200}{60}\right)} = \frac{428}{20}$$

= 21.4 second

152. Power required, P = S × Q

Where S = specific power = 0.7 N-m/mm^3

$$Q = MRR = \frac{\pi D^2}{4} \times f \times N$$

$$= \frac{\pi \times (10)^2}{4} \times 0.25 \times \frac{1200}{60} = 125\pi\, mm^3/sec$$

Power, P = 0.7 × 125π

$\boxed{P = 274.889\ W} \approx 274.89\ W$

153. Single start thread on be made by single point cutting tool.

154. $N_{min} = 120$ $N_{max} = 1200$

N = 8

$$r = n-1\sqrt{\frac{N_{max}}{N_{min}}}$$

$$= 8-1\sqrt{\frac{1200}{120}}$$

$$= \sqrt[7]{10} = 1.384$$

$N_5 = N_1 r^4 = 120 \times 1.384^4$

$\boxed{N_5 = 447 rpm}$

155. 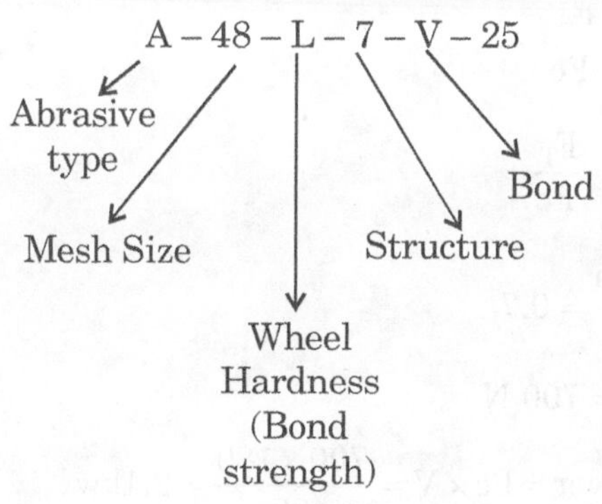

157. By one complete revolution of indexing crank the work piece turned through $= \frac{360}{40} = 9$

and by 1/3 Rotation $= \frac{9}{3} = 3$

total rotation or turn of workpiece = 9 + 3 =12°

158. Glazing is when hard material is grounded by hard wheel.

159. D = 125 mm Z = 10 teeth V = 14 m/min

F = 100 m/min

$$N = \frac{1000 \times 14}{\pi \times 125} = 35.67$$

$$f_t = \frac{F}{ZN} = \frac{100}{10 \times 35.67}$$

ft = 0.25 mm/tooth

160. $$\text{Geen Ratio} = \frac{\text{Pitch of job threads}}{\text{Pitch of lead screothreads}}$$

$$\Rightarrow \frac{3.175 \times 40}{6 \times 40} = \frac{127}{240} \rightarrow \text{Not possible}$$

$$\Rightarrow \frac{127}{40} \times \frac{1 \times 20}{6 \times 20} = \frac{127}{40} \times \frac{20}{120}\ \text{Possible.}$$

161. D = 20 mm

Plate thickness = 30 mm

Double fluted

AP = 3 mm

OR = 2 mm

N = 500 rpm

Feed/tooth = 0.01 mm

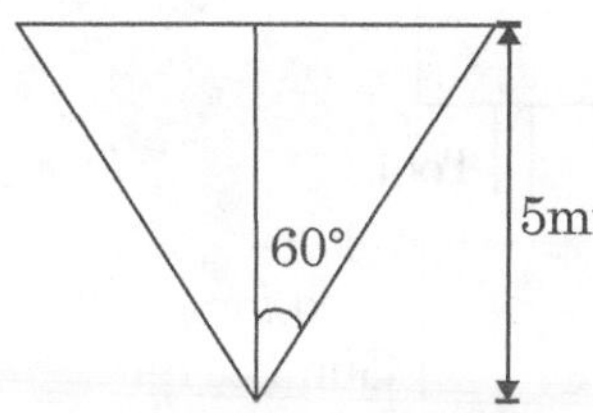

Total travel distance = 30 + 3+ 2 + 0.25 × 20

= 45 mm

Double fluted = so feed = 0.01×2 = 0.02mm

$$\text{Machining time} = \frac{L}{fN} = \frac{45}{0.02 \times 500} = 4.5\,min$$

162. Helical provide rake angle which is zero at centre and increase to helical angle at the periphery form which space chips are coming out.

163. By increasing nose radius, surface finish increases.

$$H_{max} = \frac{f^2}{8R}$$

$H max \downarrow R \uparrow$

164. Cobalt used as a binding agent in cemented carbide tool. which increases the grain size transverse rapture strength increases & hot hardness decreases.

166. Gear hobbing is generating process and gear milling is gear cutting process.

167. Gear hobbing produces most accurate external gear but cannot generate internal gear.

168. Honing is an abrasive process per formed by a set of banded abrasive sticks. surface finish 0.12 um.

169. Vertical milling machine

170. Small feed rate $H\max = \dfrac{f^2}{8R}$

$H\max \uparrow f \uparrow$

Higher speed for higher heat dissipation.

171. A–3, B–4, C–2, D–1.

172. Transformation ratio = train value

= pitch of job threads/pitch of lead screw threads

= 1.0 mm/(1/4 inch) = (1.0 × 4)/(127/5)

= 1 × 4 × 5/127 = 20/127. (The them 127/5 is used for converting inches into mm).

173. Trepanning

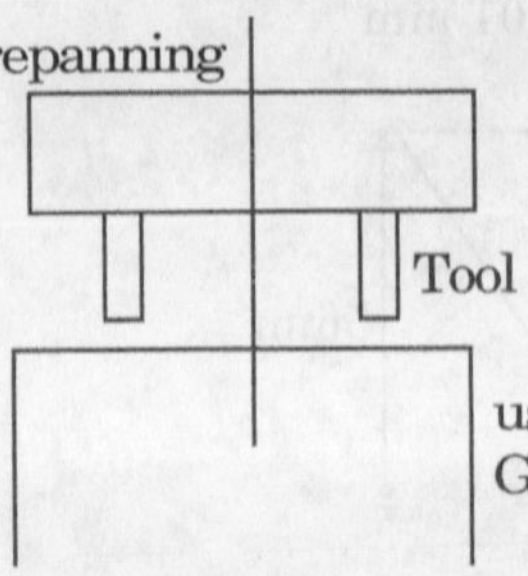

used for Gun Barrel

174. Thread rolling is forming process so work hardening takes place.

175. Internal gear cutting is performed by shaping with a pinion cutter.

176. $2\phi + \beta - \alpha = 90°$

as $\beta \downarrow \phi \uparrow$ so less force is required for cutting.

177. Specific cutting energy $= \dfrac{F_C}{b \times t_1} = 1.49\ \text{J/mm}^3$

$$= \frac{F_C}{4 \times 0.25}$$

F_C = 1.49 J/mm. stroke

$$F_C = 1.49\ \text{J/mm} \times \frac{60}{\text{min}}$$

F_C = 89.4 J/mm.min

Power = F_C.1 = 89.4 × 200 mm J/mm.min

= 17880 J/min

Power = 298 J/S or (W).

178. $VT^n = C$

$$V_1 T_1^n = 2V_1 \times \left(T_1/16\right)^2 \quad \text{on solving we get}$$

$n = 0.25$

179. μ= 5° N = 400 rpm f = 0.4m/min

$$f = \frac{0.4 \times 1000}{400} = 1\text{mm/res} = t_1$$

t_2 = 3mm

$$\tan\phi = \frac{r\cos\alpha}{1 - t\sin\alpha}$$

$$\tan\phi = \frac{\frac{1}{3}\cos 5°}{1 - \frac{1}{3}\sin 5°}$$

$$r = \frac{t_1}{t_2} = \frac{1}{3}$$

$\tan\phi = 0.342$

$\boxed{\phi = 18.88}$

180. α = 0.

so

$F = F_T$

N = Fc

$$\frac{F}{N} = \frac{F_T}{Fc}$$

$$\frac{490}{Fc} = 0.7$$

Fc = 700 N

$$\text{Power} = Fc \times V = \frac{700 \times 180}{60} = 2.1\text{kw}$$

181. The surface obtained is plane.

182. $r = \dfrac{0.14}{0.28} = 0.5$

$$\tan\phi = \frac{r\cos\alpha}{1 - r\sin\alpha}$$

$\therefore \phi = 28.3345°$

or 28° 20′

183. Rate of taper,

$$T = \frac{80-65}{55} = 0.27$$

$$\text{Set over} = \frac{T \times L}{2} = \frac{0.27 \times 90}{2} = 12.15$$

184. Because the rake angle is zero,
F = Friction force = F_c = Cutting force = 500N,
N = normal to friction force = F_t = thrust force = 250 N,
Coefficient of friction = F/N = 250/500 = 0.5

185. $VT^{0.14} f^{0.7} d^{0.4} = C$

$\Rightarrow T_1 = 30$ min,
$V_1 = 45$ m/min,
$f_1 = 0.35$ min,
$d_1 = 2.0$ mm,

$\Rightarrow C = V_1(T_1)^{0.14}(f_1)^{0.7}(d_1)^{0.4}$

$C = 45(30)^{0.14}(0.35)^{0.7}(2)^{0.4}$

$C = 45.8425$

$V_2(T_2)^{0.14}(f_2)^{0.7}(d_2)^{0.4} = 45.8425$

$(1.25 \times 45)(T_2)^{0.14}(1.25 \times 0.35)^{0.7}(1.25 \times 2)^{0.4}$
$= 45.8425$

$\Rightarrow T_2 = 1.06$ min

186. L = 900 mm
d = 200 mm
f = 0.25 mm/rev
v = 300 m/min
t = ?

$v = \pi$ DN m/min

$$N = \frac{v}{\pi D} = \frac{300}{\pi \times 0.2}$$

= 478 rpm

$$t = \frac{L}{f.N} = \frac{900}{0.25 \times 478}$$

≈ 7.539 min

187. $L_m = \frac{18C}{V}$, $C_g = \frac{270C}{TV}$, $VT^{0.25} = 150$

$$V_{opt} = V = C\left[\frac{n}{1-n}\frac{L_m}{C_g}\right]^n$$

$$= 150\left[\frac{0.25}{1-0.25} \times \frac{\frac{18C}{V}}{\frac{270C}{TV}}\right]^{0.25}$$

$$= 150\left[\frac{0.25}{1-0.25} \times \frac{18}{270}T\right]^{0.25}$$

$$= 150\left[\frac{1}{3} \times \frac{18}{270} \times \frac{150^4}{V^4}\right]^{0.25}$$

$$= 150(57.914) \times \left(\frac{1}{V^4}\right)^{0.25}$$

$$V = 150 \times 57.914 \times \frac{1}{V}$$

$V^2 = 150 \times 57.914$

V = 93.2 m/min

188. $\alpha = 22°$
Chip thickness (t_2) = 0.8 mm
feed = 0.4 mm/rev
speed = 48 m/mm
For orthogonal cutting
feed= thickness of uncut chip (t_1)

$$r = \frac{t_2}{t_1} = \frac{0.8}{0.4} = 2$$

$$\tan\theta = \frac{\cos\alpha}{r - \sin\alpha} = \frac{\cos 22}{2 - \sin 22} = 0.57$$

$\theta = \tan^{-1}(0.57)$
$\theta = 29.7$
Taper angle = $\tan^{-1}(0.27) = 15.10$

189.

	M/C I	M/C II	
T C	12	6	6000
M C	4	10	4000
G M	2	3	1800
Profit/unit	40 x	100 y	

$Z_{max} = 40x + 100y$
S.t
$12x + 6y \le 6000,\ 4x + 10y \le 4000$
$2x + 3y \le 1800\ x,\ y \ge 0$

$$\frac{x}{500}+\frac{y}{1000}\le 1 \quad \frac{x}{1000}+\frac{y}{400}\le 1$$

$$\frac{x}{900}+\frac{y}{600}\le 1$$

$12x+6y=6000$

$4x+10y=4000$

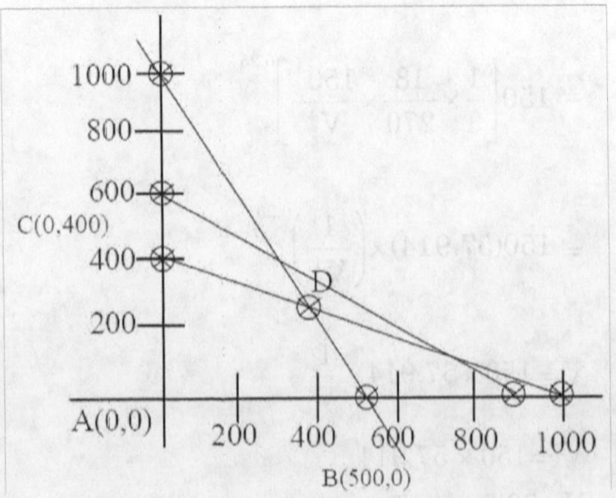

$12x+6y=6000$

$12x+30y=12000$

$-24y=-6000$

$$y=\frac{6000}{24}=250$$

$4x+10y=4000$

$4x+250\times 10=4000$

$4x=1500$

$$x=\frac{1500}{4}=375$$

$Z_{max}=40x+100y$

$=40\times 375+100\times 250$

$=15000+25000$

$Z_D=$ 40,000/-

$Z_A=0$

$Z_B=40\times 500+100\times 0$

$=20,000$

$Z_C=40\times 0+100\times 400=40,000$

Optional at (D) & (C).

190. $VT^{0.1}=150$

$VT^{0.3}=300$

For breakeven cutting speed,

$3\ln 150-3\ln V-\ln 300-\ln V$

$V=106.66m/\text{min}$

191.

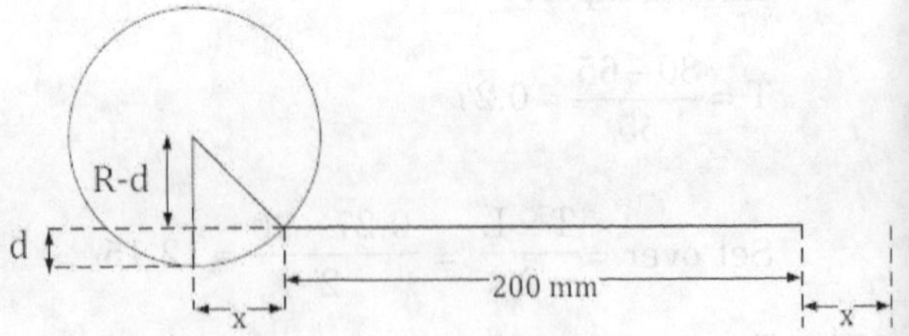

$L=200+2x$

$x=\sqrt{17^2-15^2}=8$

$\therefore$ L= 216 mm

$$\text{Time}=\frac{L}{fN}=\frac{216}{18}=12\text{ min}$$

192. Given $\alpha=9°$

t = 0.2 mm

$t_{c\,max}$ = 0.4 mm

t_{cmin} = 0.25 mm

$$\tan\phi_{max}=\frac{r_{max}\cos\alpha}{1-r_{max}\sin\alpha}=\frac{\frac{0.2}{0.25}\cos 9°}{1-\frac{0.2}{0.25}\sin 9°}$$

or $\phi_{max}=42.09°$

$$\tan\phi_{min}=\frac{r_{min}\cos\alpha}{1-r_{min}\sin\alpha}=\frac{\frac{0.2}{0.4}\cos 9°}{1-\frac{0.2}{0.4}\sin 9°}$$

or $\phi_{sin}=28.18°$

$$\text{Ratio}=\frac{\phi_{max}}{\phi_{min}}=\frac{42.09}{28.18}=1.494$$

194. $V_2=\frac{V_1}{2}$

$V_1T_1^{0.5}=V_2T_2^{0.5}$

$V_1T_1^{0.5}=\frac{V_1}{2}T_2^{0.5}$

$$\therefore\ \left(\frac{T_2}{T_1}\right)^{0.5}=2$$

$$\therefore\ \frac{T_2}{T_1}=2^{\frac{1}{0.5}}=4$$

195. Grinding ratio is defined as

$$= \frac{\text{Volume of work material removed } (V_m)}{\text{Volume of wheel wear } (V_w)}$$

$$200 = \frac{\text{Volume of work material removed } (V_m)}{\text{Volume of wheel wear } (V_w)}$$

Grinding wheel wears 0.005 times the volume of material removed.

196. It is a rubbing process not a cutting process by abrasive particles

197. $H\max = \frac{f^2}{8R}$

$H\max \downarrow f \downarrow$

198. A–3, B–4, C–1, D–2

199. (a) Reamer always have even number of flutes to balance the cutting forces.

202. (b) Chip takes away maximum heat otherwise surface finish of work piece would get spoiled and tool life decreases

205. True

207. P–3 Q–1 R–4 S–2

208. $t_{avg} = \frac{t_{max} + t_{min}}{2}$

$$= \frac{t_{max}}{2} = \frac{t}{Nz}\sqrt{\frac{d}{D}} = \frac{200}{4 \times 100}\sqrt{\frac{2}{100}}$$

$= 0.0707$ mm $= 71$ mm

209. $AP = \frac{1}{2}\left(D - \sqrt{D^2 - d^2}\right)$

$$= \sqrt{d(D-d)}$$

$$= \sqrt{2(100-2)} = 14\text{mm}$$

$$\text{Time/cut} = \frac{L}{f_m} = \frac{200 + 5 + 14}{200} \times 60$$

$= 65.7$ s ≈ 66 s

210. Titanium Nitride.

212. Involute

213. A–1 B–3 C–2 D–4

214. Flutes gives the chip removal space & also provide rake angle which is minimum at centre & equal to helix angle at periphery

215. P–3, Q–4, R–1, S–5

216. Friction welding solid state welding softness due to rubbing action.

217. K– Group having higher cobalt content to higher transverse rapture strength hence higher speed.

218. Provides flank inclination to relief from elastic recovery so less rubbing action, hence low flank wear.

220. Large negative rake angle=

$\alpha = (-40°) \rightarrow$ average value

hence low shear angle & higher cutting speed.

221. Dressing is the operation to remove logged chip between grits.

223.

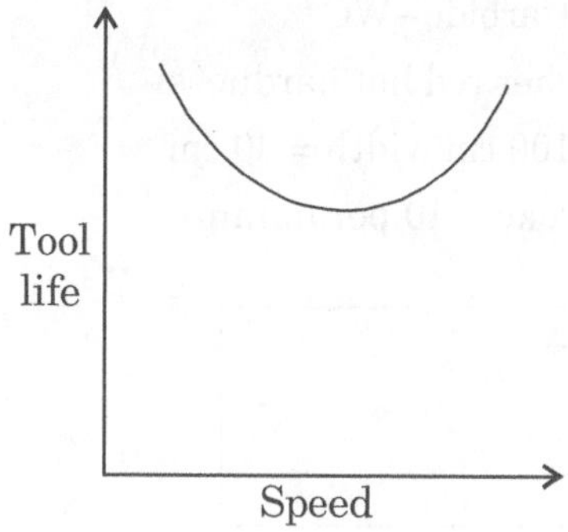

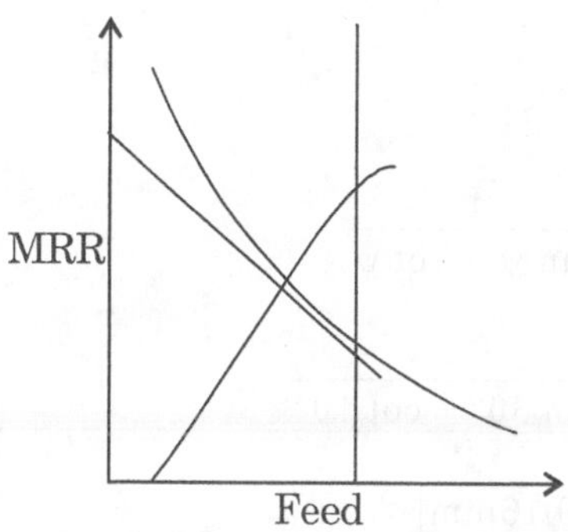

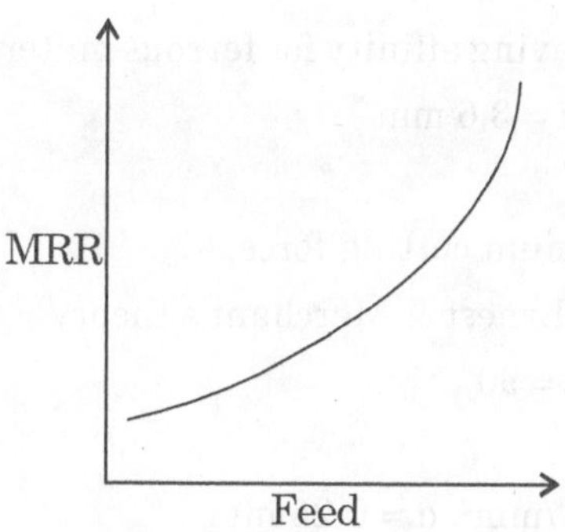

224. $\frac{40}{N} = \frac{40}{28} = 1 + \frac{12}{28}$

$$= 1 + \frac{6}{14} = 1 + \frac{9}{21}$$

In 1 revolution ,9 hole of 21 hole circle

225. Turning tool → Relief Angle
Reamer → Flutes
Milling cutter → Axial relief

226. A–2 B–1 C–3 D–5

A → End cutting Edge Engle
B → $H_{max} = \frac{f^2}{8R}$
C → Excersine Clip Defermation
D → Less Heat Dissipation

227. Said "Normally"
Tungsten Carbide –WC
having higher red hot hardness.

230. Length = 100 cm width = 30 cm
Double stroke = 10 per /minute

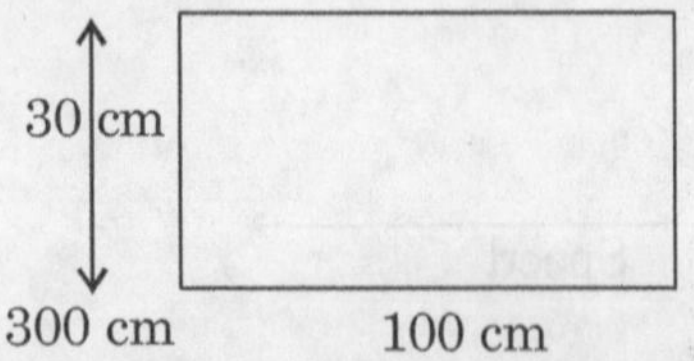

231. $\frac{40}{N} = \frac{40}{84} = \frac{10}{21}$

234. $H_{max} = \frac{f}{\tan\psi + \cot\psi}$

$H_{max} = \frac{f}{\tan 30° + \cot 10°}$

$\boxed{H_{max} = 0.16\text{mm}}$

235. Carbon having affinity for ferrous material.

236. $\alpha = 10°$, w = 3.6 mm
$\mu = 0.7$
For minimum cutting force,
By using Ernest & Merchant's theory
$2\phi + \beta - \alpha = 90°$
$\phi = 32.50°$
$\tau_s = 460$ N/mm², d = 0.25 mm
$f_t = f\cos\psi$
$\psi = 90 - 75$
$\psi = 15°$
$f_t = t_1 = \cos 15° \times 0.2$

237. $F_s = \frac{\tau_s \times w \times t_1}{\sin\phi}$

$= \frac{\tau_s \times 3.6 \times 0.2 \times \cos 15°}{\sin 32.5}$

$F_s = 770\text{N}$

$\frac{F_c}{F_s} = \frac{R\cos(\beta - \alpha)}{R(\cos(\phi + \beta - \alpha)}$

$F_c = 1299$ N

Power = $F_c \times V$

$= \frac{1299 \times 150}{60}$

Power = 3.25 kW

238. Apply line equation

slope $= \frac{2.0 - 0.8}{60 - 10} = 0.024$

If wear land = 1.8 mm²

Tool life $= \frac{10 + 1.8 - 0.8}{0.024}$

= 51.67 min

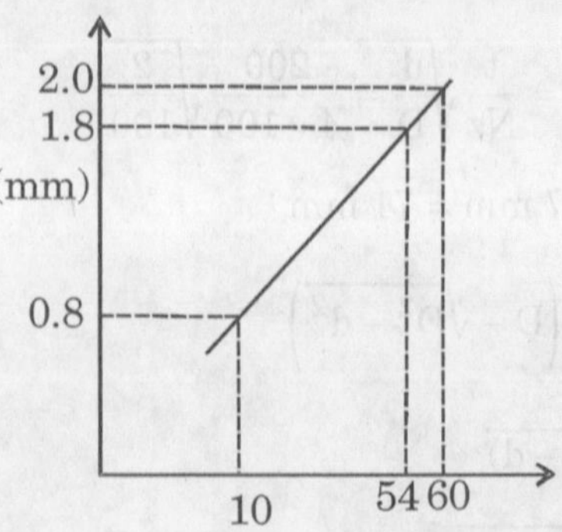

239. As back rake angle decreases, lip angle increases so the higher strength to tool.

7 CHAPTER

Metrology and Inspection

Analysis of Previous GATE Papers

Topic	Year/Set → Marks Distribution ↓		2019 Set 1	2019 Set 2	2018 Set 1	2018 Set 2	2017 Set 1	2017 Set 2	2016 Set 1	2016 Set 2	2016 Set 3	2015 Set 1	2015 Set 2	2015 Set 3
Limit, Tolerance and Fits & Gauge Design	1 Mark	MCQ Type		1								1	1	
		Numerical Type			1									
	2 Marks	MCQ Type	1									1		
		Numerical Type						1						
		Total	2	2	1			2				3	1	
Linear and Angular measurement	1 Mark	MCQ Type	1						1					
		Numerical Type					1							
	2 Marks	MCQ Type					1							
		Numerical Type									1			
		Total	1				3		1		2			

Limit, Tolerance and Fits and Gauge Design

1. The hole $40^{+0.020}_{+0.000}$ and shaft, $40^{+0.010}_{+0.000}$ when assembled will result in
(a) clearance fit (b) interference fit
(c) transition fit (d) drive fit
[ME 1987 : 1 Mark]

2. In an engineering drawing one finds the designation of 20G7f8, the position of tolerance of the hole is indicated by
(a) Letter G (b) Letter f
(c) Number 7 (d) Number 8
[ME 1988 : 1 Mark]

3. A bush was turned after mounting the same on a mandrel. The mandrel diameter in millimeters is $40^{+0.00}_{-0.05}$ and bore diameter of bush is $40^{+0.06}_{-0.010}$. The maximum eccentricity of the bush in mm will be
(a) 0.01 (b) 0.055
(c) 0.1 (d) 0.11
[ME 1991 : 2 Marks]

4. Two shafts A and B have their diameters specified as 100 ± 0.1 mm and 0.1 ± 0.0001 mm respectively. Which of the following statements is/are true
(a) Tolerance in the dimension is greater in shaft A
(b) The relative error in the dimension is greater in shaft A
(c) Tolerance in the dimension is greater in shaft B
(d) The relative error in the dimension is same for shaft A and shaft B
[ME 1992 : 2 Marks]

5. A shaft of diameter $20^{+0.06}_{-0.15}$ mm and a hole of diameter $20^{+0.20}_{+0.10}$ mm when assembled would yield
(a) Transition fit (b) Interference fit
(c) Clearance fit (d) None of these
[ME 1993 : 2 Marks]

6. Checking the diameter of a hole using GO and NOGO gauges in an example of inspection by Variable/attributes)
[ME 1995 : 1 Mark]

7. The fit on a hole-shaft pair system is specified as H7-h6, the type of fit is
(a) Clearance fit
(b) Running fit (Sliding fit)
(c) Push fit (transition fit)
(d) Force fit (Interference fit)
[ME 1996 : 1 Mark]

8. 3 blocks of B_1, B_2 and B_3 are to be inserted in a channel of width S maintaining a minimum gap of width T = 0.125 mm, as shown in figure. For P = 18.75 ± 0.08, Q = 25.00 ± 0.12, R = 28.125 ± 0.1 and S = 72.35 + X. The tolerance X is

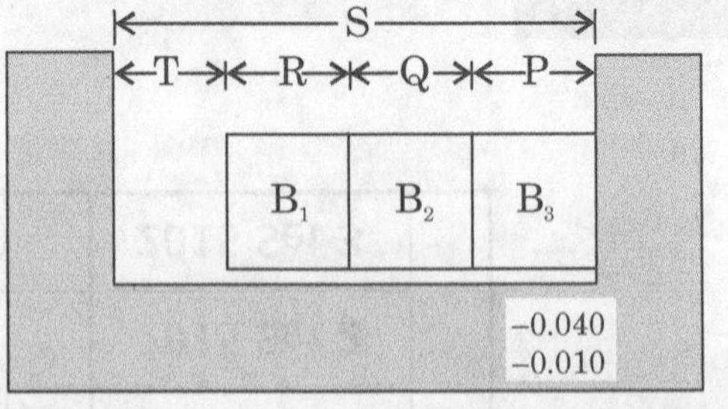

[ME 1997 : 2 Marks]

9. Allowance in limits and fits refers to
(a) Maximum clearance between the shaft and hole
(b) Minimum clearance between the shaft and hole
(c) Difference between the maximum and minimum size of hole
(d) Difference between the maximum and minimum size of shaft
[ME 2001 : 1 Mark]

10. The dimensional limits on a shaft of 25h7are
(a) 25.000, 25.021 mm
(b) 25.000, 24.979 mm
(c) 25.000, 25.007 mm
(d) 25.000, 24.993 mm
[ME 2003 : 1 Mark]

11. A part shown in the figure is machined to the sizes given below
P = 35.00 ± 0.08 mm
Q = 12.00 + 0.02 mm
R = $13.00^{+0.04}_{-0.02}$ mm

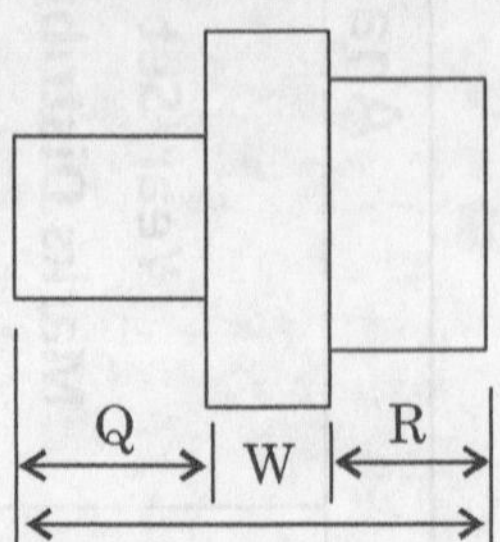

With 100% confidence, the resultant dimension W will have the specification
(a) 9.99 ± 0.03 mm
(b) 9.99 + 0.13 mm
(c) 10.00 ± 0.03 mm
(d) 10.00 ± 0.13 mm
[ME 2003 : 2 Marks]

12. In an interchangeable assembly, shafts of size $-25.000^{+0.040}_{-0.0100}$ mm mate with holes of size $25.000^{+0.020}_{-0.000}$ mm. The maximum possible clearance in the assembly will be

(a) 10microns (b) 20microns

(c) 30 microns (d) 60 microns

[ME 2004 : 1 Mark]

13. GO and NO-GO plug gauges are to be designed for a hole $20.000^{+0.050}_{+0.010}$ mm. Gauge tolerances can be taken as 10% of the hole tolerance. Following ISO system of gauge design, sizes of GO and NO-GO gauge will be respectively.

(a) 20.010 mm and 20.050 mm

(b) 20.014 mm and 20.046 mm

(c) 20.006 mm and 20.054 mm

(d) 20.014 mm and 20.054 mm

[ME 2004 : 2 Mark]

14. In order to have interference fit, it is essential that the lower limit of the shaft should be

(a) greater than the upper limit of the hole

(b) lesser than the upper limit of the hole

(c) greater than the lower limit of the hole

(d) lesser than the lower limit of the hole

[ME 2005 : 1 Mark]

15. A hole is specified as $40^{0.050}_{0.000}$ mm. The mating shaft has a clearance fit with minimum clearance of 0.01 mm. The tolerance on the shaft is 0.04 mm. The maximum clearance in mm between the hole and the shaft is

(a) 0.04 (b) 0.05

(c) 0.10 (d) 0.11

[ME 2007 : 2 Marks]

16. What are the upper and lower limits of the shaft represented by 60 f_8?

Use the following data :

Diameter 60 lies in the diameter step of 50-80 mm. Fundamental tolerance unit, i, in μm = 0.45 $D^{1/3}$ + 0.001 D, where D is the representative size in mm; Tolerance value for IT8 = 25i. Fundamental deviation for f shaft = –5.5 D 0.41

(a) lower lim. = 59.924 mm, Upper lim. = 59.970 mm

(b) lower lim. = 59.954 mm, Upper lim. = 60.000 mm

(c) lower lim. = 59.970 mm, Upper lim. = 60.016 mm

(d) lower limit = 60.000 mm, Upper limit = 60.046 mm

[ME 2009 : 2 Marks]

17. A shaft has a dimension, $\phi 35^{-0.009}_{-0.025}$ respective values of fundamental deviation and tolerance are

(a) –0.025, ± 0.008 (b) –0.025, 0.016

(c) –0.009, ± 0.008 (d) –0.009, 0.016

[ME 2010 : 1 Mark]

18. A hole is of dimension $\phi 9^{+0.015}_{+0}$ mm. The corresponding shaft is of dimension $\phi 9^{+0.010}_{+0.001}$

The resulting assembly has

(a) loose running fit (b) close running fit

(c) transition fit (d) interference fit

[ME 2011 : 1 Mark]

19. In an interchangeable assembly, shafts of size $25.000^{+0.040}_{-0.010}$ mm mate with holes of size $25.000^{+0.030}_{-0.020}$ mm. The maximum interference (in microns) in the assembly is

(a) 40 (b) 30

(c) 20 (d) 10

[ME 2012 : 1 Mark]

20. Cylindrical pins of $25^{+0.020}_{+0.010}$ mm diameter are electroplated in a shop. Thickness of the plating is $30^{\pm 2.0}$ micron. Neglecting gauge tolerance, the sized of the GO gage in mm to inspect the plated component is

(a) 25.042 (b) 25.052

(c) 25.074 (d) 25.084

[ME 2013 : 2 Marks]

21. For the given assembly: 25 H78g8, match Group A with Group B

Group-A	**Group B**
P. H	I. Shaft Type
Q. IT8	II. Hole Type
R. IT7	III. Hole tolerance Grade
S. 9	IV. Shaft Tolerance Grade

(a) P–I, Q–III, R–IV, S–II

(b) P–I, Q–IV, R–III, S–II

(c) P–II, Q–III, R–IV, S–I

(d) P–II, Q–IV, R–III, S–I

[ME 2014 : 2 Marks, Set-1]

22. The diameter of a recessed ring was measured by using two spherical balls of diameter $d_1 = 60$ mm and $d_2 = 40$ mm as shown in the figure.

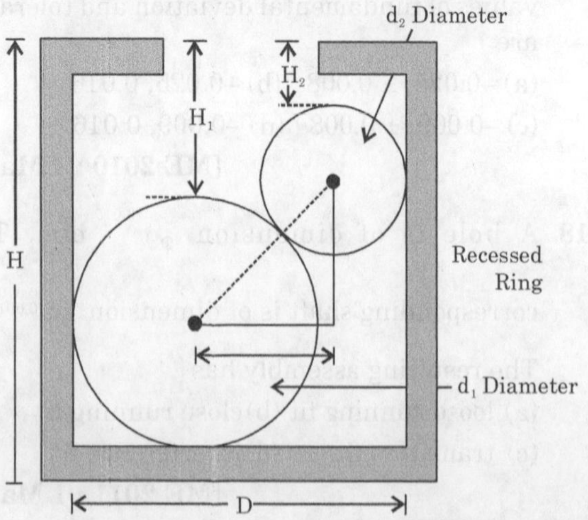

The distance $H_1 = 35.55$ mm and $H_2 = 20.55$ mm. The diameter (D, in mm) of the ring gauge is ______

[ME 2014 : 2 Marks, Set-3]

23. A Go-NO Go plug gauge is to be designed for measuring a hole of nominal diameter 25 mm with a hole tolerance of ± 0.015 mm. considering 10% of work tolerance to be the gauge tolerance and no wear condition, the dimension (in mm) of the GO plug gauge as per the unilateral tolerance system is

(a) $24.985^{+0.003}_{-0.003}$ (b) $25.015^{+0.000}_{-0.006}$

(c) $24.985^{+0.03}_{-0.03}$ (d) $24.985^{+0.000}_{-0.003}$

[ME 2014 : 2 Marks, Set-4]

24. Holes of diameter $25^{+0.040}_{+0.020}$ mm are assembled interchangeably with the pins of diameter $25^{+0.005}_{-0.008}$ mm. The minimum clearance in the assembly will be

(a) 0.048 mm (b) 0.015 mm

(c) 0.005 mm (d) 0.008 mm

[ME 2015 : 1 Mark, Set-1]

25. In the assembly shown below, the part dimensions are:

$L_1 = 22.0 \pm 0.01$ mm

$L_2 = L_3 = 10.0 \pm 0.005$ mm

Assuming the normal distribution of part dimensions, the dimension L_4 in mm for assembly condition would be:

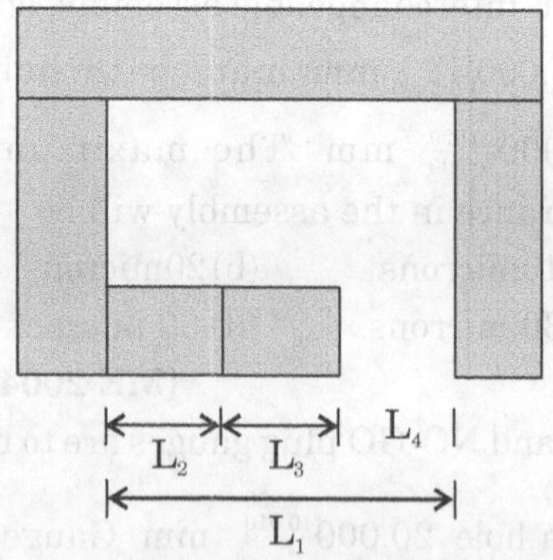

(a) 2.0 ± 0.008 (b) 2.0 ± 0.012

(c) 2.0 ± 0.016 (d) 2.0 ± 0.020

[ME 2015 : 2 Marks, Set-1]

26. Which one of the following statements is TRUE?

(a) The 'GO' gauge controls the upper limit of a hole

(b) The 'NO GO' gauge controls the lower limit of a shaft

(c) The 'GO' gauge controls the lower limit of a hole

(d) The 'NO GO' gauge control the lower limit of a hole

[ME 2015 : 1 Mark, Set-2]

27. A cylindrical pin of $25^{+0.020}_{+0.010}$ mm diameter is electroplated, plating thickness is 2.0 ± 0.005 mm. Neglecting the gauge tolerance, the diameter (in mm, up to 3 decimal points accuracy) of the GO ring gauge to inspect the plated pin is ____.

[ME 2017 : 2 Marks, Set-2]

28. The height (in mm) for a 125 mm sine bar to 'measure a taper of 27'32' on a flat work piece is ________(correct to three decimal places).

[ME 2018 : 1 Marks, Set-1]

29. A circular shaft having diameter $65.00^{+0.01}_{-0.05}$ mm is manufactured by turning process. A 50 μm thick coating of TiN is deposited on the shaft Allowed variation in TiN film thickness is ±5μm. The minimum hole diameter (in mm) to just provide clearance fit is

(a) 65.01 (b) 64.95

(c) 65.12 (d) 65.10

[2019 : 2 Marks, Set-1]

30. The most common limit gauge used for inspecting the hole diameter is
(a) Plug gauge (b) Master gauge
(c) Snap gauge (d) Ring gauge
[ME 2019 : 1 Mark, Set-2]

31. Outside diameter of bush is turned using a mandrel. The mandrel diameter is maintained as $30.000^{0.000}_{-0.050}$ mm and the bore diameter is $30.000^{+0.050}_{-0.000}$ mm. The maximum value in mm of the eccentricity on the bush due to the locating mandrel
(a) 0.050 (b) 0.100
(c) 0.020 (d) 0.025
[PI 1989 : 2 Marks]

32. The combination of slip gauges to obtain a dimension of 10.35 mm will be
(a) 10.00 + 0.30 + 0.05
(b) 8.00 + 1.30 + 1.05
(c) 10.00 + 0.35
(d) 5.00 + 4.00 + 1.00 + 0.35
[PI 1991, 1994 : 1 Mark]

33. In an interference microscope, a groove produces a band distortion of 4 band spacing. If the wave length of the monochromatic light source is 0.5 microns, the groove depth is .... mm
[PI 1992 : 1 Mark]

34. A shaft of diameter $20^{+0.20}_{-0.10}$ mm and a hole of diameter $20^{+0.20}_{+0.10}$ mm when assembled would yield.
(a) Transition fit (b) Interference fit
(c) Clearance fit (d) None of these
[PI 1992 : 1 Mark]

35. A 30 mm hole shaft assembly results in minimum and maximum clearances of 0.03 mm and 0.30 mm respectively. The hole has a unilateral tolerance with zero fundamental deviation. If the tolerance on the shaft is 0.08 mm, the maximum size of the hole is
(a) 230.11 (b) 30.16
(c) 30.19 (d) 30.27
[PI 1992 : 2 Marks]

36. To obtain dimension of 61.18 mm using slip gauges, the most appropriate combination is
(a) 1.18 + 50.00 + 10.00
(b) 0.08 + 1.10 + 60.00
(c) 1.08 + 0.10 + 50.00 + 10.00
(d) 1.08 + 1.10 + 50.00 + 9.00
[PI 1992 : 2 Marks]

37. The maximum interference in mm after assembly between a bush of size $30.00^{+0.06}_{+0.03}$ and a shaft of size $30.00^{+0.04}_{+0.02}$ is
(a) 0.07 (b) 0.05
(c) 0.02 (d) 0.01
[PI 1993 : 2 Marks]

38. Holes 1, 2, 3, 4 drilled in sequence, with angular positions and tolerances are shown in figure. Evaluate closing angle X with its tolerance band.

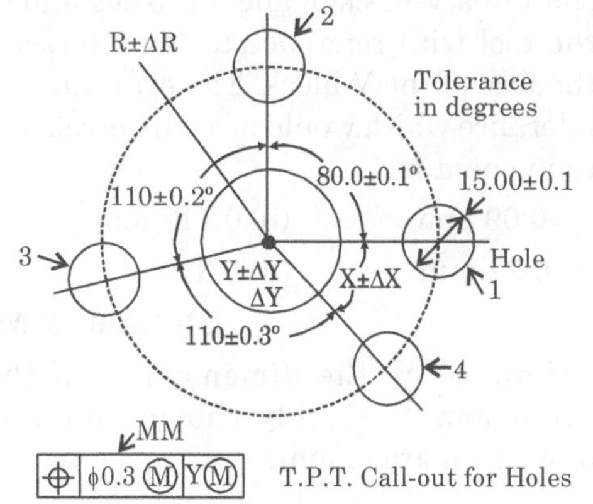

39. A hole is specified as ϕ 30± 0.04 mm with – ϕ0.01 mm. The virtual diameter of the hole (i.e., the maximum diameter of the pin that can enter the hole) is _______.
[PI 1994 : 2 Marks]

40. Abbe's principle of alignment is used to be followed in
(a) vernier calipers
(b) depth vernier
(c) internal caliper micrometer
(d) height vernier
[PI 1995 : 1 Mark]

41. The diameter of a hole is given as $50^{+0.15}_{+0.00}$ mm. The upper limit on the dimensions in mm, of the shaft for achieving maximum interference of 50 microns is _______.
[PI 1995 : 2 Marks]

42. In an interchangeable assembly, a shaft of $20.00^{+0.015}_{-0.005}$ mm diameter and a hole of $20.00^{+0.025}_{-0.010}$ mm diameter form a mating pair. In the worst assembly conditions, clearance between them will be
(a) 40 microns (b) 30 microns
(c) 25 microns (d) 15 microns
[PI 2004 : 2 Marks]

43. A cylindrical pin of diameter $1.996^{+0.0015}_{-0.0015}$ mm is assembled into a hole of diameter $2.000^{+0.0015}_{-0.0015}$ mm. The allowance (in mm) provides for this assembly is

(a) 0.001 (b) 0.015

(c) 0.025 (d) 0.035

[PI 2005 : 2 Marks]

44. A keyway of 5 mm depth is to be milled in a shaft of diameter 60 ± 0.1 by positioning the shaft in a V-block of angle 120 deg and setting the tool with reference to the intersection of the face of the V-block. The error due to shaft tolerance which would be transferred on to the component is

(a) 0.09 mm (b) 0.119 mm

(c) 0.22 mm (d) 0.3 mm

[PI 2006 : 2 Marks]

45. Tolerance on the dimension x in the two component assembly shown below in (all dimension are in mm)

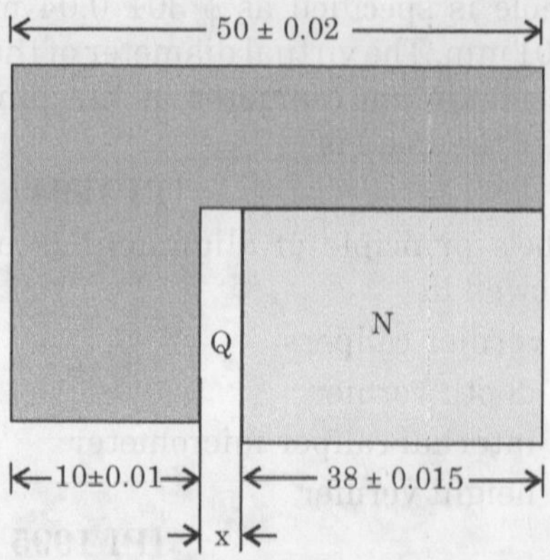

(a) ±0.025 (b) ±0.030

(c) ±0.040 (d) ±0.045

[PI 2007 : 2 Marks]

46. Diameter of a hole after plating needs to be controlled between $30^{+0.050}_{-0.010}$ mm. If the plating thickness varies between 10-15 microns, diameter of the hole before plating should be

(a) $30^{+0.07}_{+0.030}$ mm (b) $30^{+0.65}_{+0.020}$

(c) $30^{+0.08}_{+0.030}$ mm (d) $30^{+0.70}_{+0.040}$

[PI 2007 : 2 Marks]

47. The following data are given for calculating limits of dimension and tolerances for a hole: Tolerance unit i (microns) = 0.45 ($D^{1/3}$) + 0.001 D. The unit of D is mm. Diameter step is 18 – 30. If the fundamental deviation for hole H is zero and IT8 = 26i, the maximum and minimum limits of dimension for a 25H8 hole (in mm) are

(a) 24.984 and 24.967

(b) 25.017 and 24.984

(c) 25.033 and 25.00

(d) 25.00, 24.967

[PI 2008 : 2 Marks]

48. A small bore is designated 25H7. The lower (minimum) and upper (maximum) limits of the bore are 25.000 mm and 25.021 respectively. When the bore is designated as 25H8, then the upper limit is 25.033 mm. When the bore is designated as 25H6, then the upper limit of the bore in mm is

(a) 25.001 (b) 25.005

(c) 25.009 (d) 25.013

[PI 2010 : 2 Marks]

49. To measure the effective diameter of an external metric thread (included angle is 60°) of 3.5 mm pitch, a cylindrical standard of 30.5 mm diameter and two wires of 2 mm diameter each are used. The micrometer reading over the standard and over the wires are 16.532 mm and 15.398 mm respectively. The effective diameter (in mm) of the thread is

(a) 33.336 (b) 30.397

(c) 29.366 (d) 26.397

[PI 2011 : 2 Marks]

50. Observation of a slip gauge on flatness interferometer produced fringe counts numbering 10 and 14 for two readings. The second reading is taken by rotating the setup by 180°. Assume that both faces of the slip gauge are flat and the wavelength of the radiation is 0.5086 µm. The parallelism error (in µm) between the two faces of the slip gauge is

(a) 0.2543 (b) 1.172

(c) 0.5086 (d) 0.800

[PI 2011 : 2 Marks]

51. In an interchangeable assembly, shafts of size $25.000^{+0.040}_{-0.010}$ mm mate with holes of size $25.000^{+0.030}_{+0.020}$ mm. The maximum interference (in microns) in the assembly is

(a) 40 (b) 30

(c) 20 (d) 10

[PI 2012 : 1 Mark]

Linear and Angular Measurement

52. Appropriate instrument to check the flatness of slip gauge is
(a) dial indicator
(b) pneumatic comparator
(c) optical interferometer
(d) tool makers microscope with projection facility

[ME 1989 : 1 Mark]

53. Slip gauges are calibrated by outside micrometer (T/F)

[ME 1995 : 1 Mark]

54. A threaded nut of M 16, ISO metric type, having 2 mm pitch with a pitch diameter of 14.701 mm is to be checked for its pitch diameter using two or three number of balls or rollers of the following sizes
(a) Rollers of 2 mm ϕ
(b) Rollers of 1.155 mm ϕ
(c) Balls of 2 mm ϕ
(d) Balls of 1.155 mm v

[ME 2003 : 2 Marks]

55. Two slip gauges of 10 mm width measuring 1.000 mm and 1.002 mm are kept side by side in contact with each other lengthwise. An optical flat is kept resting on the slip gauges as shown in the figure. Mono-chromatic light of wavelength 0.0058928 mm is used in the inspection. The total number of straight fringes that can be observed on both slip gauges is

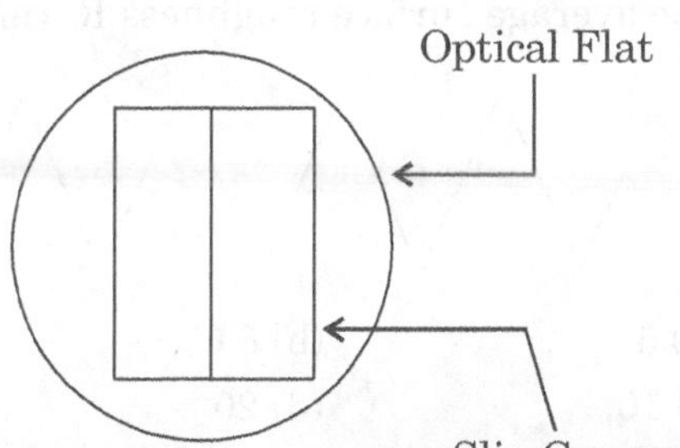

(a) 2 (b) 8
(c) 6 (d) 13

[ME 2003 : 2 Marks]

56. Match the following

Feature to be inspected
P. Pitch and Angle errors of screw thread
Q. Flatness error of a surface
R. Alignment error of a machine sideway
S. Profile of a cam

Instrument
1. Auto Collimator
2. Optical interferometer
3. Dividing Head and Dial Gauge
4. Spirit Level
5. Sine bar
6. Tool maker's Microscope

(a) P-6, Q-2, R-4, S-6
(b) P-5, Q-2, R-1, S-6
(c) P-6, CM, R-1, S-3
(d) P-1, Q-4, R-4, S-2

[ME 2004 : 2 Marks]

57. A ring gauge is used to measure
(a) outside diameter but not roundness
(b) roundness but not outside diameter
(c) both outside diameter and roundness
(d) only external threads

[ME 2006 : 1 Mark]

58. A displacement sensor (a dial indicator) measures the lateral displacement of a mandrel mounted on the taper hole inside a drill spindle. The mandrel axis is an extension of the drill spindle taper hole axis and the protruding portion of the mandrel surface is perfectly cylindrical. Measurements are taken with the sensor placed at two positions P and Q as shown in the figure.
The readings are recorded as R_x = maximum deflection minus minimum deflection, corresponding to sensor position at X, over one rotation.

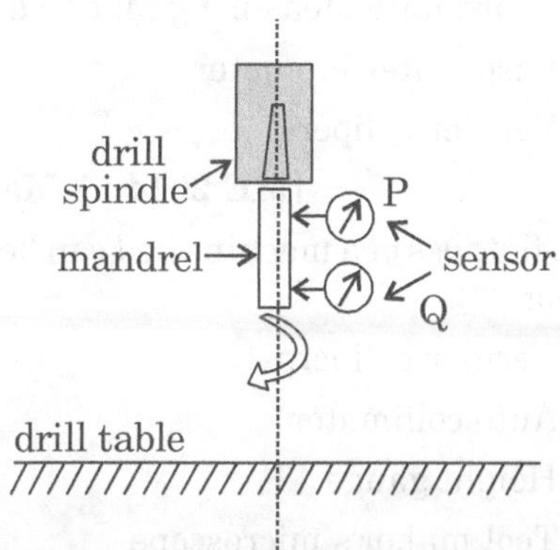

If $R_p = R_Q > o$, which one the following would be consistent with the observation?
(a) The drill spindle rotational axis is coincident with the drill spindle taper hole axis.
(b) The drill spindle rotational axis intersects the drill spindle taper hole axis at point P.
(c) The drill spindle rotational axis is parallel to the drill spindle taper hole axis.
(d) The drill spindle rotational axis intersects the drill spindle taper hole axis at point Q.

[ME 2008 : 2 Marks]

59. A taper hole is inspected using a CMM, with a probe of 2 mm diameter. At a height, Z=10 mm from the bottom, 5 points are touched and a diameter of circle (not compensated for probe size) is obtained as 20 mm. Similarly, a 40 mm diameter is obtained at a height Z= 40 mm. The smaller diameter (in mm) of hole at Z = 0 is

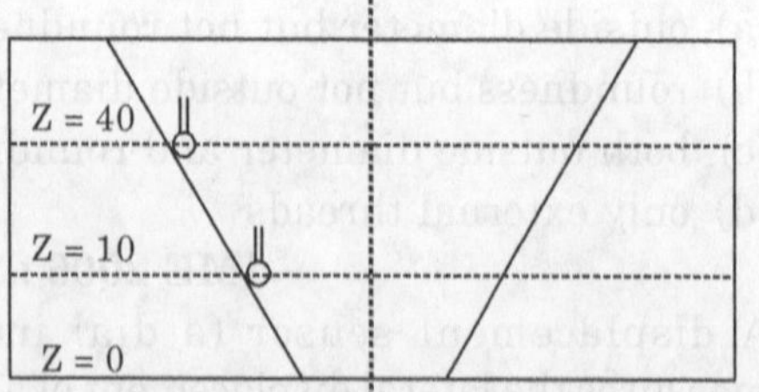

(a) 13.334 (b) 15.334

(c) 15.442 (d) 15.542

[ME 2010 : 2 Marks]

60. Which one of the following instruments is widely used to check and calibrate geometric features of machine tools during their assembly?

(a) Ultrasonic probe

(b) Coordinate Measuring Machine (CMM)

(c) Laser interferometer

(d) Vernier calipers

[ME 2014 : 1 Mark, Set-1]

61. The flatness of a machine bed can be measured using

(a) Vernier calipers

(b) Auto collimator

(c) Height gauge

(d) Tool maker's microscope

[ME 2014 : 1 Mark, Set-2]

62. Match the following :

P. Feeler gauge

Q. Fillet gauge

R. Snap gauge

S. Cylindrical plug gauge

(i) Radius of an object

(ii) Diameter within limits by comparison

(iii) Clearance or gap between components

(iv) Inside diameter of straight hole

(a) P-(iii), Q-(i), R-(ii), S-(iv)

(b) P-(iii), Q-(ii), R-(i), S-(iv)

(c) P-(iv), Q-(ii), R-(i), S-(iii)

(d) P-(iv), Q-(i), R-(ii), S-(iii)

[ME 2016 : 1 Mark, Set-1]

63. Two optically flat plates of glass are kept at a small angle θ as shown in the figure. Monochromatic light is incident vertically.

Monochromatic Light

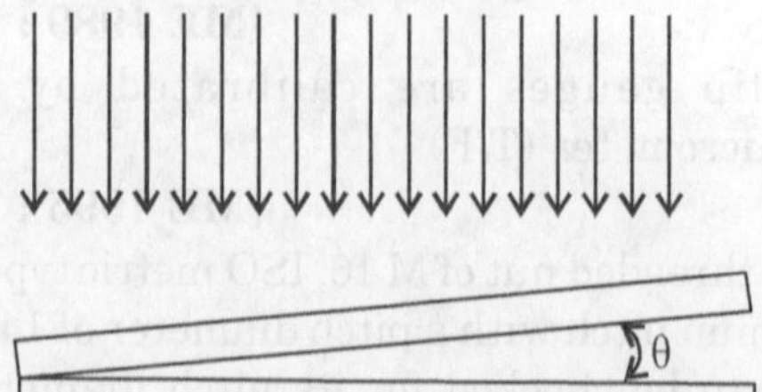

If the wavelength of light used to get a fringe spacing of 1 mm is 450 mm, the wavelength of light (in mm) to get a fringe spacing of 1.5 mm is __________

[ME 2016 : 2 Marks, Set-3]

64. Metric thread of 0.8 mm pitch is to be cut on a lathe. Pitch of the lead screw is 1.5 mm. If the spindte rotates at 1500 rpm, the speed of rotation of the lead screw (rpm) will be __________.

[ME 2017 : 1 Mark, Set-1]

65. Assume that the surface roughness profile is triangular as shown schematically in the figure. If the peak to valley height is 20 μm, the central line average surface roughness R_0 (in μm) is

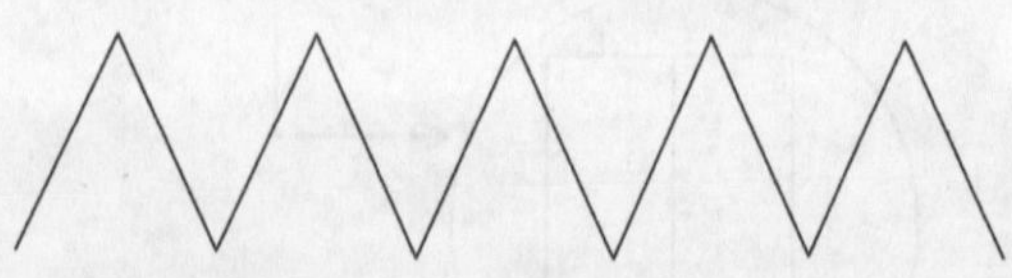

(a) 5 (b) 6.67

(c) 10 (d) 20

[ME 2017 : 2 Marks, Set 1

66. In a casting process, a vertical channel through which molten metal flows downward from pouring basin to runner for reaching the mould cavity is called

(a) riser

(b) blister

(c) pin hole

(d) sprue

[ME 2019 : 1 Mark, Set-1]

67. Flatness of slip gauges is checked with
(a) Interferometer
(b) Optical flat
(c) Electronic comparator
(d) Linear bar and bevel protractor
[PI 1989 : 1 Mark]

68. Assertion (A): Basic hole system is advocated by Indian Standards for fit design.
Explanation (E): Holes can be produced to any size
(a) Both A and E are correct and E explains A
(b) Both A and E are true but E does not explains A
(c) A is true but E is false
(d) A is false but E is true
[PI 1990 : 2 Marks]

69. For angle measurement in metrology, the following pair can be used in conjunction with each other
(a) sine bar and vernier calipers
(b) bevel protractor and slip gauges
(c) slip gauges and sine bar
(d) sine bar and bevel protractor
[PI 1991 : 1 Mark]

70. Outside micrometer is calibrated with the help of
(a) inside micrometer
(b) depth micrometer
(c) ring gauges
(d) slip gauges
[PI 1991 : 1 Mark]

71. Abbe's principle of alignment states that the scale axis must coincide with line.
[PI 1991 : 1 Mark]

72. In mass manufacturing, two holes in a plate type of component are ideally located with the help of
(a) one solid pin in the larger hole and one diamond pin in the smaller hole
(b) two solid pin in either holes
(c) two diamond pins in either holes
(d) one solid pin in the smaller hole and one demand pin in the larger hole
[PI 1992 : 1 Mark]

73. The skid in a stylus type roughness measuring instrument is used to provide ... A .... and define .... B.... geometrical profile of the surface.
A. support/datum B. micro/macro
[PI 1992 : 1 Mark]

74. The diameter of the best wire to measure the effective diameter of a metric screw threads of 0.5 mm pitch is .... mm
[PI 1992 : 1 Mark]

75. A straight edge is inspected on a coordinate measuring machine. The readings obtained at intervals of 100 mm along the top edge are 100.1, 100.5 and 100.4 mm. The straightness error, making the end points as reference is ...
[PI 1992 : 1 Mark]

76. Assertion (A): End standards are used for precision component inspection.
Reason (R): In end standards line of measurement coincides with line of indication.
(a) Both A and R are correct and R explains A
(b) Both A and R are true but R not explains A
(c) A is true but R is false
(d) A is false but R is true
[PI 1993 : 2 Marks]

77. Abbe's principle of alignment is used in measurements made in ______ .
[PI 1994 : 1 Mark]

78. The least count of a 200 mm long spirit level is given as 10 sec per division. The gradient of the spirit level expressed as mm over one meter is
[PI 1995 : 2 Marks]

79. Instrument
P. Dial indicator
Q. Pneumatic gauge
R. GO/NOGO gauge
Principle of inspection
1. Non contact
2. Limit of size
3. Comparator

	P	Q	R
(a)	2	3	1
(b)	3	1	2
(c)	1	2	3
(d)	2	1	3

[PI 2003 : 1 Mark]

80. The surface finish of the component produced by a grinding process was measured. Based on roughness parameter obtained, which one of the following is true.
(a) $R_a < R_s < R_z < R_t$
(b) $R_s < R_a < R_z < R_t$
(c) $R_z < R_a < R_s < R_t$
(d) $R_a < R_z < R_a < R_t$
[PI 2006 : 2 Marks]

Linked Data Question 81 & 82:

In a three wire inspection of external thread of M16 × 2, the measurement over the wires was found to be 16.455 mm

81. The best wire size is

(a) 1.155 mm (b) 2 mm

(c) 2.309 mm (d) 4 mm

[PI 2006 : 2 Marks]

82. The effective pitch diameter of the thread is

(a) 14.723 (b) 16.705

(c) 17.122 (d) 17.455

[PI 2006 : 2 Marks]

83. The geometric tolerance that does not need a datum for its specification is

(a) Concentricity (b) Run out

(c) Perpendicularity (d) Flatness

[PI 2007 : 1 Mark]

84. Which one of the instrument is a comparator

(a) Tool makers microscope

(b) GO/NOGOgauge

(c) Optical interferometer

(d) Dial gauge

[PI 2007 : 1 Mark]

85. An experimental setup is planned to determine the taper of workpiece as shown in the figure. If the two precision rollers have radii 8 mm and 5 mm and the total thickness of slip gauges inserted between the rollers is 15.54 mm, the taper angle "θ" is

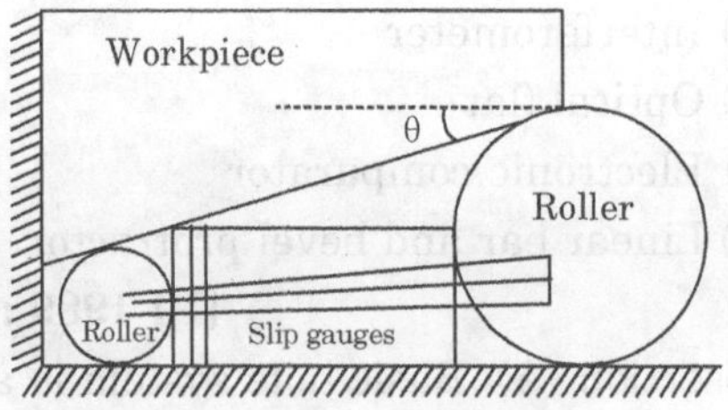

(a) 6 deg (b) 10 deg

(c) 11 deg (d) 12 deg

[PI 2008 : 2 Marks]

86. An autocollimator is used to

(a) Measure small angular displacements on flat surfaces

(b) Compare known and unknown dimensions

(c) Measure the flatness error

(d) Measure roundness error between enters

[PI 2009 : 2 Marks]

87. A sine bar has a length of 250 mm, each roller has a diameter of 20 mm, during taper angle measurement of a component, the height from the surface plate to the centre of a roller is 100 mm. The calculated taper angle (in degrees) is

(a) 21.1 (b) 22.8

(c) 23.6 (d) 68.9

[PI 2012 : 2 Marks]

88. Which one of the following is modeled based on adaptation capabilities of biological systems?

(a) Relational database

(b) Fuzzy system

(c) Simulated annealing algorithm

(d) Genetic algorithm

[PI 2013 : 1 Mark]

ANSWERS

1. (c)	**2.** (a)	**3.** (b)	**4.** (a)	**5.** (c)	**7.** (a)	**9.** (b)	**10.** (b)	**11.** (b)	**12.** (d)
13. (d)	**14.** (a)	**15.** (c)	**16.** (a)	**17.** (d)	**18.** (c)	**19.** (c)	**20.** (d)	**21.** (d)	**23.** (d)
24. (b)	**25.** (d)	**26.** (c)	**29.** (c)	30. (a)	**31.** (a)	**32.** (b)	**34.** (c)	**35.** (c)	**36.** (a)
37. (d)	**40.** (c)	**42.** (b)	**43.** (a)	**44.** (c)	**45.** (d)	**46.** (d)	**47.** (c)	**48.** (d)	**49.** (b)
50. (c)	**51.** (c)	**52.** (c)	**54.** (b)	**55.** (a)	**56.** (c)	**57.** (c)	**58.** (c)	**59.** (a)	**60.** (c)
61. (b)	**62.** (a)	**65.** (a)	**66.** (d)	**67.** (a)	**68.** (a)	**69.** (c)	**70.** (d)	**72.** (a)	**76.** (c)
79. (b)	**80.** (c)	**81.** (a)	**82.** (a)	**83.** (d)	**84.** (d)	**85.** (d)	**86.** (c)	**87.** (a)	**88.** (d)

EXPLANATIONS

1. 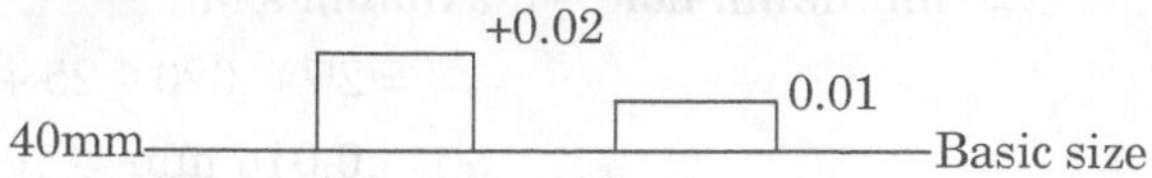

2. Capital letter for Hole $\rightarrow$ G
Small letter for shaft $\rightarrow$ f
7 for $\rightarrow$ IT7
8 for $\rightarrow$ IT8
IT $\rightarrow$ Tolerance zone width

3. 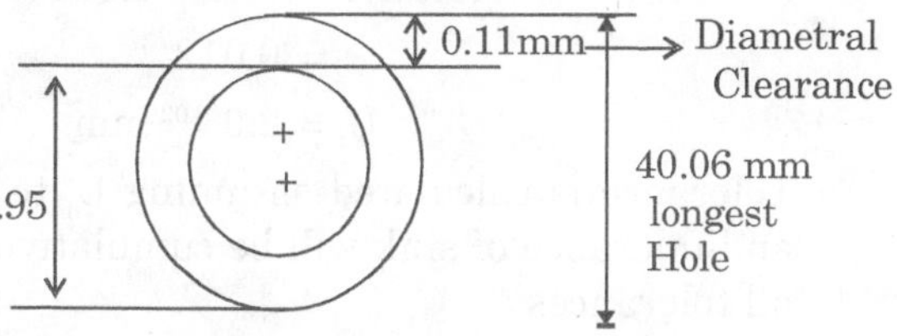

Eccentricity in terms of Radius always

$$= \frac{0.11}{2} = 0.055 \text{ mm.}$$

4. Tolerance of shaft A
= 100.1 – 99.9 = 0.2
Tolerance of shaft B
= 0.1001 – 0.0999 = 0.002m
$Tol_A > Tol_B$

5. 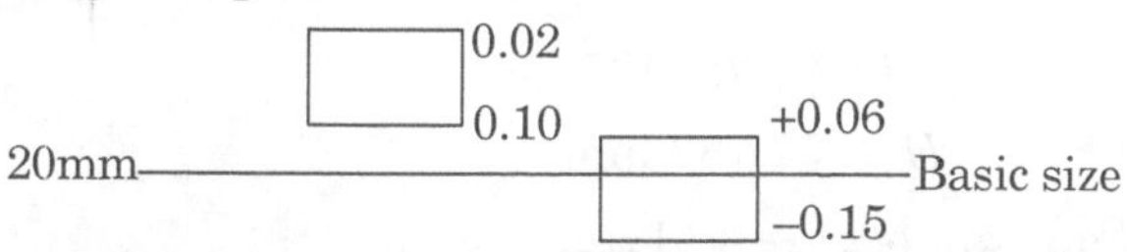

6. Attributes.

8. Consider, $P + Q + R + T_{(max.)} = 18.83 + 25.12 + 28.225 + 0.125 = 72.3 = s$
Given s = 72.35 + X
$\therefore\ 72.35 + X = 72.3$
or $X = \boldsymbol{-0.05}$

9. Allowances is always the difference between Maximum material limits.

10. Shaft Basic system

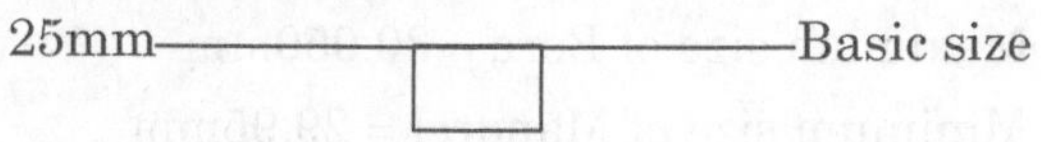

$$D = \sqrt{18 \times 30} = 23.24$$

$$i = 0.45\sqrt[3]{D} + 0.001(D)$$

$$i = 0.45\sqrt{2324} + 0.001(23.24)$$

i = 0.31 microns
IT7 = 16i = 16 × 1.31= 20.96 microns.
shaft Tolerance = 25 – (20.96 × 10^{-3})
= 24.979mm.

11. P = 35.00 ± 0.08 mm
Q = 12.00 ± 0.02 mm
$R = 13.00^{+0.04}_{-0.03}$ mm
P = Q + W + R

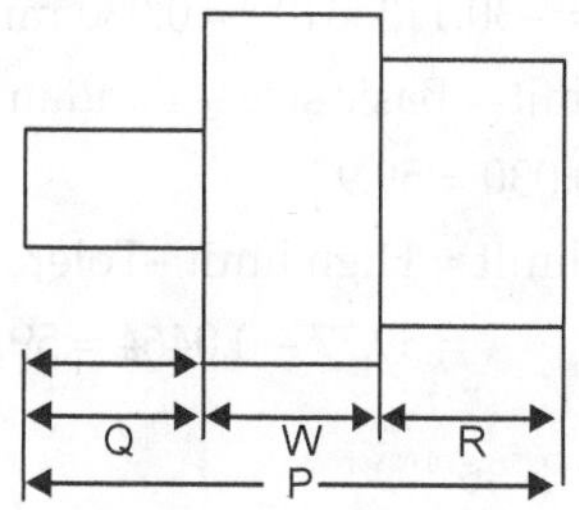

Now

$$35.00 \pm 0.08 = 12.00 \pm 0.02 + W + 13.00^{+0.04}_{-0.03}$$

$$\Rightarrow\ W = 35.00 \pm 0.08 - \left[12 \pm 0.02 + 13.00^{+0.04}_{-0.03}\right]$$

$$= (35 - 12 - 13)^{+0.08-0.02-0.04}_{-0.08+0.02+0.03}$$

$$= 10.00^{+0.02}_{-0.03} = 9.99 \pm 0.03 \text{ mm}$$

12. Minimum shaft size = 25.000 – 0.04 = 24.96 mm
Maximum hole size = (25.000 + 0.02) = 25.02 mm

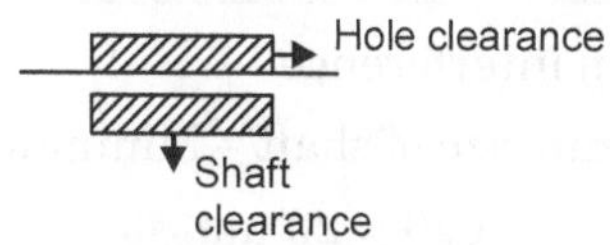

$\therefore$ Maximum possible clearance
= Maximum size of hole – minimum size of shaft
= 25.02 – 24.96 = 0.060 mm = 60 micron.

13. Dimension of hole = $20.000^{+0.050}_{+0.010}$
$\therefore$ Tolerance = 20.05 – 20.01 = 0.04 mm
Gauge tolerance = $0.04 \times \frac{10}{100}$ = 0.004 mm
Size of NO – GO gauge
= Maximum limit – gauge tolerance
= 20.05 – 0.004 = 20.046

14. For making interference fit, seeing the figure it is essential that lower limit of shaft should be greater than upper limit of hole.

15. Minimum clearance = Minimum hole size – Maximum shaft size
$\therefore$ Maximum shaft size
= 40.00 – .01 = 39.99 mm
Tolerance on shaft =.04 mm
Minimum shaft size = 39.99 – .04 = 39.95
Maximum clearance = Maximum hole size – Minimum shaft size
= 40.05 – 39.95 = .10

16. $D = \sqrt{50 \times 80} = 63.25$ mm

$i = 0.45(D)^{1/3} + .001\, D\ \mu m = 1.856 \times 10^{-6}$ m

Tolerance $= 25i = 46.4 \times 10^{-6}$ m

Fundamental deviation

$\delta = -(5.5)\, D^{0.41}$

$= -30.113 \times 10^{-6} = 0.030$ mm (absolute)

High limit = Basic size – Fundametal deivation

= 60 – 0.030 = 59.97

Lower limit = High limit –Tolerance

= 59.77 – 0.0464 = 59.924mm

17. Given : $\phi\, 35^{-0.009}_{-0.025}$

Tolerance = upper deviation – lower deviation

= 0.025 – 0.009 = 0.016

Fundamental deviation for basic shaft is lower deviation = – 0.009.)

18. Transition Fit.

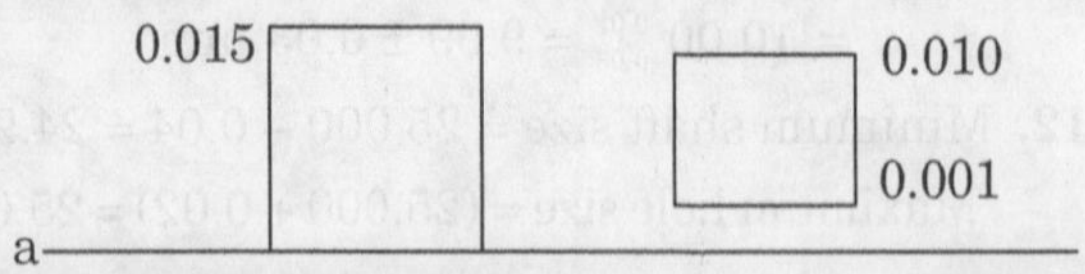

19. Maximum interference

= Maximum size of shaft – Minimum size of hole

= 25.040 – 25.020 = 20 μm

20. Size of Go Gauge

$= (25 + 0.20) + [(30+2) \times 10^{-3}] \times 2 = 25.084$ mm.

21. H7 is for hole where 7 indicates its tolerance grade

g8 is for shaft where 8 indicates its tolerance grade

22. $D_{ring} = d_1 \sec\frac{\theta}{2} + 2(h_1 + r_1)\tan\frac{\theta}{2}$

$\therefore \theta = 60^{\circ} = 4 \times \sec 30 + 2\,(20.55 + 20) \tan 30^{\circ}$

$D_{ring} = 93$ mm.

23.

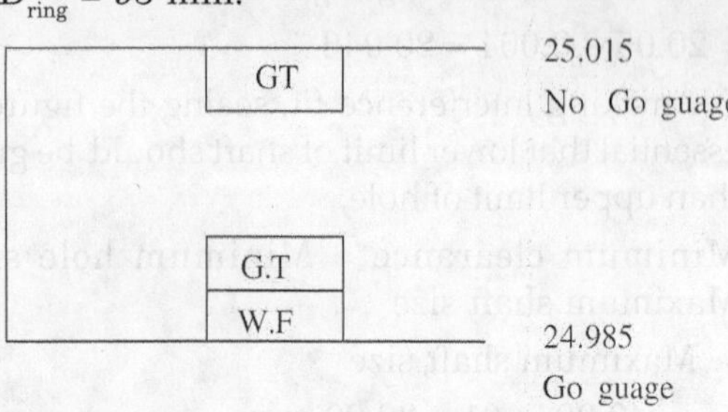

25 ± 0.015

GO-Gauge

U.L = 24.985

L.L = 24.985

$24.985^{+0.003}_{+0.000}$

24. Minimum clearance

⇒ minimum hole – maximum shaft

= 25 + .020 – 25 + .005

= 0.015 mm

25. Since all dimensions have bilateral tolerances

$L_4 = L_1 - L_2 - L_3$

= 22 – 10 – 10

$L_4 = 2$ mm

Tolerance = 0.01 + 0.0015 + 0.0005

= 0.012

$\therefore \quad L_4 = 2.0^{\pm 0.02}$ mm

Tolerance is calculated assuming L_4 to be sink and tolerance of sink will be cumulative sum of all tolerances

26. Flatness of machine bed can be measured from Auto collimator.

27. D = 25.02 + 2 × 2.005 = 29.030 mm.

28. $\theta = 27^{\circ}32' = 27 + \left(\frac{32}{60}\right)^{\circ} = 27.533^{\circ}$

In sine bar, $\sin\theta = \frac{H}{L}$

$\sin 27.533^{\circ} = \frac{H}{125}$

$H = 57.782427$

$H \approx 57.782$ mm

29. Shaft $= 65^{+0.01}_{-0.05}$ mm

Coating thickness 't' = (50 ± 5)μm

To provide clearance fit, given condition should be satisfied,

$(D_H)\min \geq ((D)\text{shaft})_{max}$

$\geq (D_s)_{max} + 2t_{max}$

$= 65.01 + \frac{2 \times 55}{1000} = 65.12$mm

30. Plug gauge is a cylindrical type of gauge, used to check the accuracy of holes.

31. Maximum size of Bore = 30.050mm

Minimum size of Mandrel = 29.95mm

Diametral Gap = 0.1mm

Eccentricity $= \frac{0.1}{2} = 0.05$mm

32. Normally, slip gauge box set does not have slip of less than 1 mm thickness. Always use minimum number of slipgauges in combination & less than 1mm thickness is not in set.

33. Groove depth $= \frac{h}{2}.\frac{\lambda}{2} = \frac{4}{2} \times \frac{0.5}{2}\ \mu m = 0.5\mu m$

35. Min. clearance

$= Hole_{min} - Shaft_{max} = 0.03$ ... (i)

Max clearance

$= Hole_{min} - Shaft_{max} = 0.30$ ... (ii)

$Shaft_{max} - Shaft_{min} = 0.08$ ... (iii)

Since, hole has unilateral tolerance,

$H_{min} = 30$ mm

From eq. (i), we get

$Shaft_{max} = 30 - 0.03 = 29.97$ min

From eq. (iii), we get

$Shaft_{min} = 29.89$ mm

From eq. (ii), we get

$Hole_{max} = 30.19$ mm

36. Normally, slip gauge box set does not have slip of less than 1 mm thickness.

37.

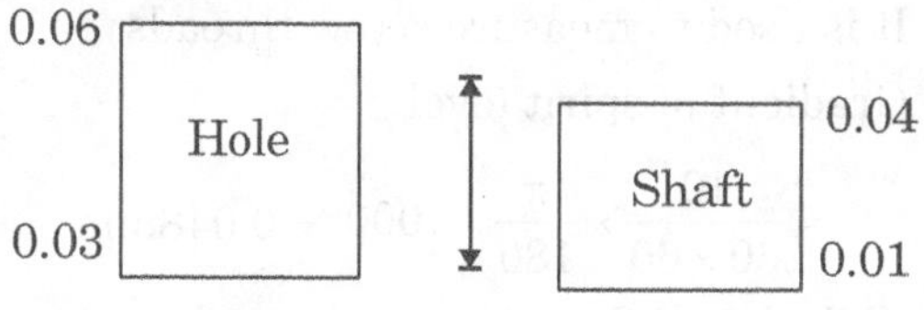

= 30.04 – 30.03 = 0.01mm.

38. $X_{max} = 360 - [79.9 + 10.88 + 109.7] = 60.6$

$X_{min} = 360 - [80.1 + 110.2 + 110.3] = 59.4.$

$Clearance = X_{max} - X_{min} = 1.2$

$$X_{mean} = \frac{X_{max} + X_{min}}{2} = \frac{60.6 + 59.4}{2}$$

$X_{mean} = 60.$

$\boxed{X = 60 \pm 0.6}$

when dia is minimum = 59.9mm

$$(OC)_2 = \frac{\frac{59.9}{2}}{\sin 60°} = 34.58\text{mm}.$$

Error in depth $= 2[(OC) - (OC)_2] = 0.22$ mm.

39. In maximum material condition

$Dia_{hole} = 30 - 0.04 - 0.01 = 29.95$ mm

This is virtual dia of the hole.

41. Maximum interference = Maximum shaft dia – Minimum hole diameter

$0.05 = Shaft_{max} - 50$

$Shaft_{max} = 50.05$ mm

42. In worst assembly condition, clearance will maximum

Clearance (max) = Max hole dia – Minimum shaft diameter

= 20.025 – 19.995 = 0.030 = 30 microns

43. Allowance = minimum hole dia – maximum shaft dia = 1.9985 – 1.9975 =- 0.001 mm

45. $X_{max} = 50.02 - (38 - 0.015) - (10 - 0.01)$

= 2.045mm

$X_{min} = 49.98 - (38 + 0.015) - (10.01) = 1.955$mm

$\boxed{X = 2 \pm 0.045}$

46. Remember flating comes from both side

So $D_{max} - 2 \times 0.01 = 30.05$

$D_{max} = 30.07$mm

$D_{min} - 2 \times 0.015 = 30.01$

$D_{min} = 30.04$

47. $D = \sqrt{18 \times 30} = 23.24$mm

$i = 0.45\sqrt[3]{D} + 0.001D$

$= 0.45\sqrt[3]{23.24} + 0.001(23.24) = 1.307$mm

IT8 = 26i = 26 × 1.307 = 33.98 ≈ 34μm = 0.034mm

25H8 = 250.00 + 0.034mm.

48. $i = 0.45\sqrt[3]{D} + 0.0011)$

$= 0.45\sqrt[3]{23.24} + 0.001(23.24)$

i = 1.31mm

IT7 = 16i = 16 × 1.31 = 21 μm

IT6 = 10i = 10 × 1.31 = 13.1μm

upper limit for Hole $25H_6 = 250.013$mm

50. Parallelism

$$= \left(\frac{n_2 - n_1}{2}\right) \times \frac{\lambda}{2} = \left(\frac{14 - 10}{2}\right) \times \frac{0.5086}{2}$$

= 0.5086 μm.

51. Maximum Interference = Allowances = Difference Between Maximum Material limit.

$\Rightarrow D_{max}$ for shaft = 25.040mm.

D_{min} for Hole = 25.020mm.

Max. Interference = 25.040 – 25.020 = 20μm

53. False

54. Metric Thread M16

Pitch = 2mm

Pitch Dia = 14.701

θ = 30°

$$d_w = \frac{P}{2}\sec 30° = \frac{2}{2}\sec 30° = 1.155\text{mm}.$$

55. $\Delta h = \frac{N}{2} \times \frac{\lambda}{2.}$

Δh = 1.002 – 1.00 = 0.002mm

λ = 0.0058928mm

$$N = \frac{4 \times \Delta h}{\lambda} = \frac{4 \times 0.002}{0.0058928}$$

$N = 1.357 \simeq 2$

56. P-6, Q-4, R-1, S-3

57. Ring gauge is used to measure both outside diameter and roundness.

58. Equal deflection in both sensors means eccentricity parallel to the axis.

59.

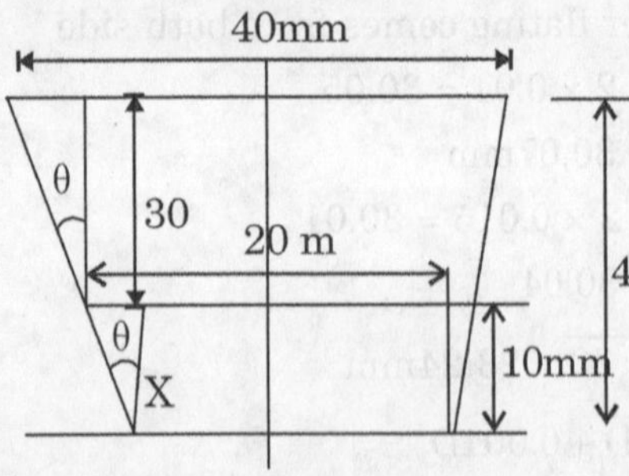

$$\tan\theta = \frac{x}{10}$$

$$\frac{1}{3} = \frac{x}{10}$$

x = 3.33

(d) = 20 – 2x = 20 – 2 × 3.33 = 13.334 mm

62. P(iii), Q(i), R(ii), S(iv)

63. 1mm = 450 mm spacing = 1.5mm

$$n = \frac{2}{\lambda}$$

$$1.5 = n \times \frac{\lambda_2}{2}$$

$$\lambda_2 = 1.5 \times 450$$

$$\boxed{\lambda_2 = 675\text{mm}}$$

64. Pitch = 0.8mm

N_t = 1500rpm

P_s = 1.5mm

$N_s \times P_3 \times Z_z = N_t \times P_t \times Z_t$

$[Z_s = Z_t = 1]$

$$N_s = \frac{1500 \times 0.8 \times 1}{1.5 \times 1}$$

$$\boxed{N_s = 800\text{rpm}}$$

65. $R_a = \frac{h_{max}}{4} = \frac{20}{4} = 5\mu m.$

71. False

Abbe's principle of alignment states that the scale axis must be coincide with the axis of work piece.

73. A. Datum

B. Macro

74. P = 0.5

$$d_w = \frac{0.5}{2} \times \sec 30°$$

$$\boxed{d_w = 0.28\text{mm}}$$

75.

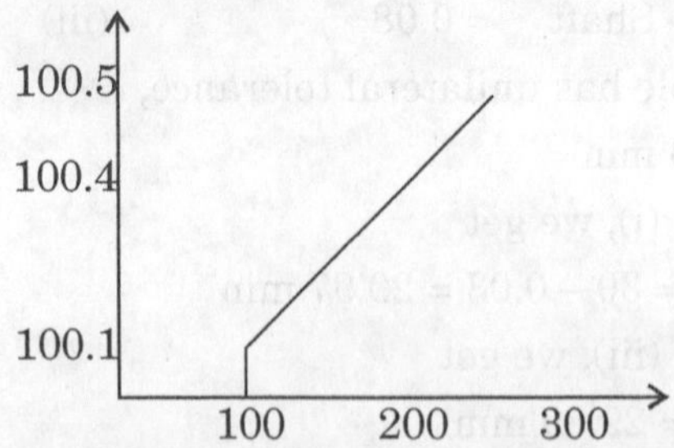

$$(y - 100.1) = \frac{0.3}{200}(x - 100)$$

at x = 200 y = 100.25mm

Error = y – 100.5 = 0.25mm

77. It is used to measure screw threads

78. Gradient of spirit level

$$= \frac{10}{60 \times 60} \times \frac{\pi}{180} \times 1000 = 0.048\text{mm}$$

79. P-3, Q-1, R-2

81. $d_w = \frac{2}{2}\sec 30° = 1.155\text{mm}$

83. Flat doesn't Required initial Datum because it is departure from the three point considered plane.

85. $\tan\theta = \frac{6}{5 + 8 + 15.54}$

$$\boxed{\theta = 12°}$$

87.

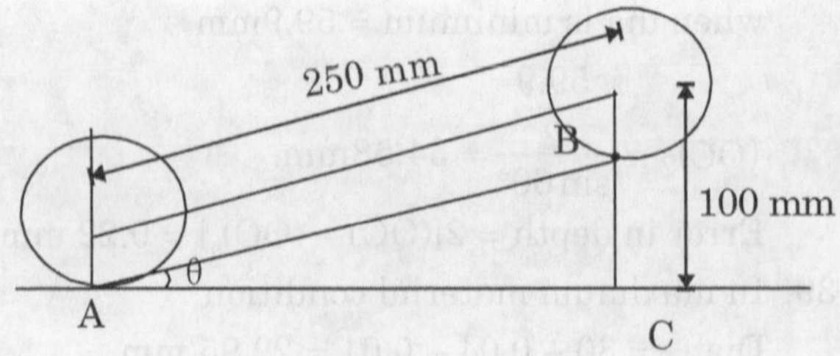

$$\sin\theta = \frac{BC}{AB} = \frac{100 - \frac{20}{2}}{250}$$

$$\theta = \frac{90}{250}$$

θ = 21.1°

Chapter 8
CIMS

Topic	Year/Set → Marks Distribution ↓		2019 Set 1	2019 Set 2	2018 Set 1	2018 Set 2	2017 Set 1	2017 Set 2	2016 Set 1	2016 Set 2	2016 Set 3	2015 Set 1	2015 Set 2
CAD and CAM	1 Mark	MCQ Type		1									
		Numerical Type											
	2 Marks	MCQ Type							1	1	1	1	
		Numerical Type											
	Total			**1**					**2**	**2**	**2**	**2**	
Numeric Control	1 Mark	MCQ Type			1						1	1	
		Numerical Type											
	2 Marks	MCQ Type					1						
		Numerical Type											
	Total				**1**		**2**				**1**	**1**	
Group Technology	1 Mark	MCQ Type											
		Numerical Type											
	2 Marks	MCQ Type											
		Numerical Type											
	Total												

Analysis of Previous GATE Papers

CAD and CAM

1. In computer aided drafting practice, an arc is defined by
(a) two end points only
(b) centre and radius
(c) radius and one end point
(d) two end points and centre

[ME 2000 : 1 Mark]

2. In a 2–D CAD package, clockwise circular arc of radius 5, specified from P_1 (15, 10) to P_2 (10, 15) will have its center at
(a) (10, 10) (b) (15, 10)
(c) (15, 15) (d) (10, 15)

[ME 2004 : 2 Marks]

3. For generating a Coon's surface we require
(a) a set of grid points on the surface
(b) a set of grid control points
(c) four bounding-curves defining the surface
(d) two bounding curves and a set of grid control points

[ME 2008 : 1 Mark]

4. A robot arm PQ with end coordinates P(0,0) and Q(2,5) rotates counter clockwise about P in the XY plane by 90°. The new coordinate pair of the end point Q is
(a) (–2,5) (b) (–5,2)
(c) (–.5,-2) (d) (2, –5)

[ME 2014 : 1 Mark, Set-2]

5. A triangular face in a CAD model has vertices: P1(0, 0, 0); P2(1, 1, 0) and P3(1,1, 1).The area of the facet is
(a) 0.500 (b) 0.707
(c) 1.414 (d) 1.732

[ME 2015 : 2 Marks, Set-1]

6. The figure below represents a triangle PQR with initial coordinates of the vertices as P(1, 3), Q(4, 5) and ft(5.3.5). The triangle is rotated in X-Y plane about the vertex P by angle θ in clockwise direction. If sinθ = 0.6 and cosθ = 0.8, the new coordinates of the vertex Q are

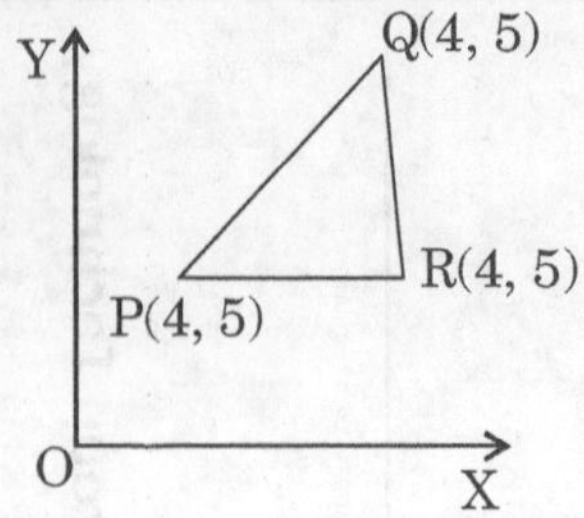

(a) (4.6, 2.8) (b) (3.2, 4.6)
(c) (7.9, 5.5) (d) (5.5, 7.9)

[ME 2016 : 2 Marks, Set-1]

7. For the situation shown in the figure below the expression for H in terms of r, R and D is

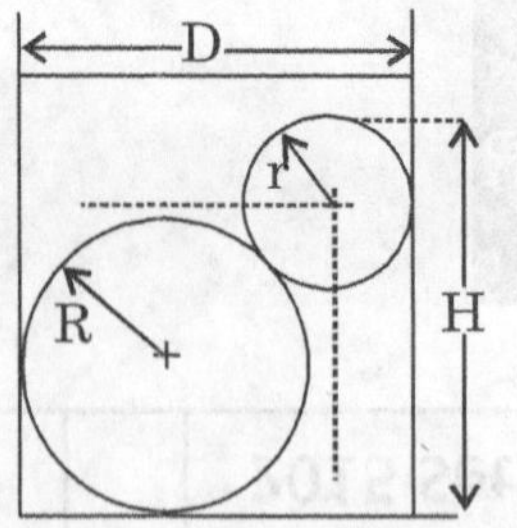

(a) $H = D + \sqrt{r^2 + R^2}$
(b) $H = (R + r) + (D + r)$
(c) $H = (R + r) + \sqrt{D^2 + R^2}$
(d) $H = (R + r) + \sqrt{2D(R+r) - D^2}$

[ME 2016 : 2 Marks, Set-2]

8. A point P(1, 3, -5) is translated by $2\hat{i}+3\hat{k}+4\hat{j}$ and then rotated counter clockwise by 90° about the z-axis. The new position of the point is:
(a) (–6, 3, –9) (b) (–6, –3, –9)
(c) (6, 3, –9) (d) (6, 3, 9)

[ME 2016 : 2 Marks, Set-3]

9. The transformation matrix for mirroring a point in x-y plane about the line y = x is given by

(a) $\begin{bmatrix} 0 & -1 \\ -1 & 0 \end{bmatrix}$ (b) $\begin{bmatrix} 1 & 0 \\ 0 & -1 \end{bmatrix}$

(c) $\begin{bmatrix} -1 & 0 \\ 0 & 1 \end{bmatrix}$ (d) $\begin{bmatrix} 0 & 1 \\ 1 & 0 \end{bmatrix}$

[ME 2019 : 1 Mark, Set-2]

10. Use of robot for spray painting is primarily justified for improved.... A.... and reduced ... B...
A. productivity/consistency
B. cost/hazard

[PI 1990 : 1 Mark]

11. The component development and manufacturing activities when assisted by a computer area: computer aided graphics (CAG), CAD, CAM, CAPP and computer aided quality control (CAQC).
The most appropriate sequence of these activities is
(a) CAD-CAG-CAPP-CAM-CAQC
(b) CAG-CAD-CAM-CAPP-CAQC
(c) CAD-CAG-CAPP-CAQC-CAM
(d) CAD-CAPP-CAG-CAM-CAQC

[PI 1992 : 2 Marks]

12. Given that the notations of robot joints are L-linear, O-orthogonal, R-rotation, T-twisting, and V- revolving, the anatomy of a polar configuration robot is described by the alphabetical string

(a) TOO (b) TLL
(c) TRL (d) TVO

[PI 2004 : 2 Marks]

13. Complete the manual part program for drilling a hole located at point (20, 20) from the target point. The depth of the hole to be drilled is 15 mm FROM/TARG

GOTO/20, 20, 0

(a) GODLTA/0, 0,-15
GODLTA/0, 0, 15
GOTO/SETPT

(b) GODLTA/0, 0, 15
GODLTA/0, 0, -15
GOTO/SETPT

(c) GOTO/SETPT
GODLTA/0, 0, -15
GODLTA/0, 0, 15

(d) GOTO/SETPT
GODLTA/0, 0, 15
GODLTA/0, 0, -15

[PI 2006 : 2 Marks]

14. In a CAD package, a point P(6,3,2) is projected along a vector $\vec{V}$ (-2,1,-1). The projection of this point on X-Y plane will be

(a) (4, 4, 0) (b) (8, 2, 0)
(c) (7, 4, 0) (d) (2, 5, 0)

[PI 2007 : 2 Marks]

15. The geometric transformation specified by

$$[x\ y\ 1] = [x\ y\ 1]\begin{bmatrix} 0.5 & 0 & 0 \\ 0 & 0.25 & 0 \\ 1 & 2 & 1 \end{bmatrix} \text{ in a 2D}$$

In a 2D CAD system represents

(a) Scaling and translation
(b) Scaling and rotation
(c) Rotation and translation
(d) Rotation

[PI 2007 : 2 Marks]

16. In a CAD package, mirror image of a 2D point P(5, 10) is to be obtained about a line which passes through the origin and makes an angle of 45° counterclockwise with the X-axis coordinates of the transformed point will be

(a) (7.5,5) (b) (10,5)
(c) (7.5,-5) (d) (10,-5)

[PI 2013 : 2 Marks]

Numerical control

17. In NC part programming spindle speed of 730 rpm will be coded by the magic -3 rule as

(a) S673 (b) S730
(c) S630 (d) S037

[ME 1989 : 1 Mark]

18. Numerically controlled machine tools are better suited for... (batch/mass) production, because their setup time is.... (larger/smaller) in comparison to special purpose machine tools

[ME 1991 : 1 Mark]

19. In PTP type of NC system

(a) Control of position and velocity of the tool is essential
(b) Control of only position of the tool is sufficient
(c) Control of only velocity of the tool is sufficient
(d) Neither position nor velocity need be controlled

[ME 1992 : 1 Mark]

20. With reference to NC machine, which of the following statement is wrong?

(a) Both closed loop and open loop control systems are used
(b) Paper tapes, floppy drives and cassettes are used for data Storage
(c) Digitisers may be used as interactive input devices
(d) Post-processor is an item of hardware

[ME 1993 : 2 Marks]

21. CNC machines are more accurate than conventional machines because they have a high resolution encoder and digital readouts for positioning. (T/F)

[ME 1994 : 1 Mark]

22. CNC machines are more economical to use even for simpie turning jobs. (T/F)

[ME 1994 : 1 Mark]

23. In a point to point control NC machine, the slides positioned by an integrally mounted stepper motor drive. If the specification of the motor is 1 deg/pulse, and the pitch Of the lead screw is 3.6 mm.
What is the expected positioning accuracy?
(a) 1 micron (b) 10 micron
(c) 50 micron (d) 100 micron

[ME 1997 : 2 Marks]

24. A cylinder of 155 mm diameter is to be reduced to 150 mm diameter In one turning cut with a feed of 0.15 mm/rev and a cutting speed of 150 m/min on a NC lathe. What are the programmed feed rate and the material removal rate.

[ME 1998 : 2 Marks]

25. In finish machining of an island on a casting with CNC milling machine, an end mill with 10 mm diameter is employed, The corner points of the Island are represented by (0, 0), (0, 30), (60, 30) and (50, 0). By applying cutter radius compensation, the trajectory of the cutter will be
(a) (-5, 0), (-5, 35), (55, 35), (55, -5), (-5, 5)
(b) (0, -5), (55, -5). (55, 35), (-5, 35), (-5, -5)
(c) (5,5), (5, 25), (45,25),(45, 5), (5, 5)
(d) (5, 5), (45, 5), (45, 25), (5, 25), (5, 5)

[ME 2000 : 2 Marks]

26. Machining of complex shapes on CNC machine requires
(a) simultaneous control of x, y and z axes
(b) simultaneous control of x and y axes
(c) independent control of x and y axes
(d) independent control of x, y and z axes

[ME 2002 : 1 Mark]

27. During the execution of a CNC part program block
NO20 GO2 X45.0 Y25.0 R5.0
the type of tool motion will be
(a) circular Interpolation - clockwise
(b) circular Interpolation - counterclockwise
(c) linear Interpolation
(d) rapid feed

[ME 2004 : 1 Mark]

28. Which among the NC operations given below are continuous path operations?
Arc Welding (AW)
Drilling (D)
Laser Cutting of Sheet (LC)
Milling (M)
Punching in Sheet Metal (P)
Spot Welding (SW)
(a) AW, LC and M
(b) AW, D, LC and M
(c) D, LC, P and SW
(d) D, LC, and SW

[ME 2005 : 1 Mark]

29. The tool of an NC machine has to move along a circular arc from (5,5) to (10,10) while performing an operation. The centre of the arc is at (10,5). Which one of the following NC tool path commands performs the above mentioned operation?
(a) NO10 G02 X10 Y10 X5 Y5 R5
(b) NO10 G03 X10 Y10 X5 Y5 R5
(c) NO10 GO1 X5 Y5 X10 Y10 R5
(d) NO10 G02 X5 Y5 X10 Y10 AS

[ME 2005 : 2 Marks]

30. NC contouring is an example of
(a) continuous path positioning
(b) point-to-point positioning
(c) absolute positioning
(d) incremental positioning

[ME 2006 : 1 Mark]

31. Which type of motor is NOT used in axis or spindle drives of CNC machine tools?
(a) induction motor (b) dc servo motor
(c) stepper motor (d) linear servo motor

[ME 2007 : 1 Mark]

Linked Answer Questions 31 and 32:

In the feed drive of a Point-to-Point open loop CNC drive, a stepper motor rotating at 200 steps/rev drives a table through a gear box and lead screw-nut mechanism (pitch = 4 mm, number of starts =1). The gear ratio

$$\left(= \frac{\text{Outpur rotational speed}}{\text{Input rotational speed}}\right) \text{ is given by}$$

$U = \frac{1}{4}$. The 1 stepper motor (driven by voltage pulses from a pulse. The frequency of the pulse train from the pulse generator is f = 10,000 pulses per minute.

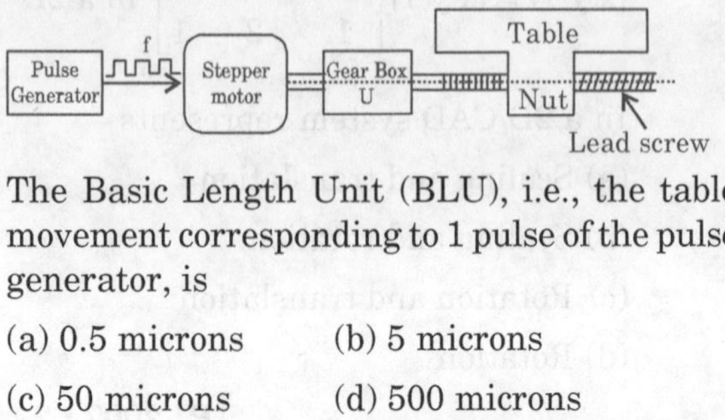

32. The Basic Length Unit (BLU), i.e., the table movement corresponding to 1 pulse of the pulse generator, is
(a) 0.5 microns (b) 5 microns
(c) 50 microns (d) 500 microns

[ME 2008 : 2 Marks]

33. A customer insists on a modification to change the BLU of the CNC drive to 10 microns without changing the table speed. The modification can be accomplished by

(a) changing U to $\frac{1}{2}$ and reducing f to $\frac{f}{2}$

(b) changing U to $\frac{1}{8}$ and increasing f to 2f

(c) changing U to $\frac{1}{2}$ and keeping f unchanged

(d) keeping U unchanged and increasing f to 2f.

[ME 2008 : 2 Marks]

34. Match the following:

NC Code	Definition
P. M05	1. Absolute coordinate system
Q. Q01	2. Dwell
R Q04	3. Spindle stop
S. G90	4. Linear interpolation

(a) P–2, Q–3, R–4, S–1
(b) P–3, Q–4, R–1, S–2
(c) P–3, Q–4, R–2, S–1
(d) P–4, Q–3, R–2, S–1

[ME 2009 : 2 Marks]

35. In a CNC program block, N002 G02 G91 X40 Z40..., G02 and G91 refer to

(a) circular interpolation in counterclockwise direction and incremental dimension
(b) circular interpolation in counterclockwise direction and absolute dimension
(c) circular interpolation in clockwise direction and incremental dimension
(d) circular interpolation in clockwise direction and absolute dimension

[ME 2010 : 1 Mark]

36. A CNC vertical milling machine has to cut a straight slot of 10 mm width and 2 mm depth by a cutter of 10 mm diameter between points (0, 0) and (100, 100) on-the XY plane (dimensions in mm). The feed rate used for milling is 50 mm/min. Milling time for the slot (in seconds) is

(a) 120 (b) 170
(c) 180 (d) 240

[ME 2012 : 1 Mark]

37. For machining a rectangular island represented by coordinates P(0,0), Q(100,0), fl(100,50) and S(0,50) on a casting using CNC milling machine, an end mill with a diameter of 16 mm is used. The trajectory of the cutter centre to machine the island PQRS is

(a) (- 8, - 8), (108, - 8), (108,58), (- 8,58), (- 8, -8)
(b) (8,8), (94,8), (94,44), (8,44), (8,8)
(c) (- 8,8), (94,0), (94,44), (8,44), (- 8,8)
(d) (0,0), (100,0), (100,50), (50,0), (0,0)

[ME 2014 : 1 Mark, Set-1]

38. For the CNC part programming, match Group A with Group B :

Group A

P. Circular interpolation, counter clock wise
Q. dwell
R. circular interpolation, clock wise
S. point to point countering

Group B

I. G02
II. G03
III. G04
IV. G00

(a) P-II, Q-III, R-I, S-IV
(b) p -I, Q-III, R-II, S-IV
(c) P-I, Q-IV, R-II, S-III
(d) p-II, Q-I, R-III, S-IV

[ME 2014 : 2 Marks, Set-2]

39. The function of interpolator in a CNC machine controller is to

(a) control spindle speed
(b) coordinate feed rates of axes
(c) control tool rapid approach speed
(d) perform Miscellaneous (M) functions (tool change, coolant control etc.)

[ME 2015 : 1 Mark, Set-1]

40. A drill is positioned at point P and its has to proceed to point Q. The coordinates of point Q in the incremental system of defining position of a point in CNC part program will be

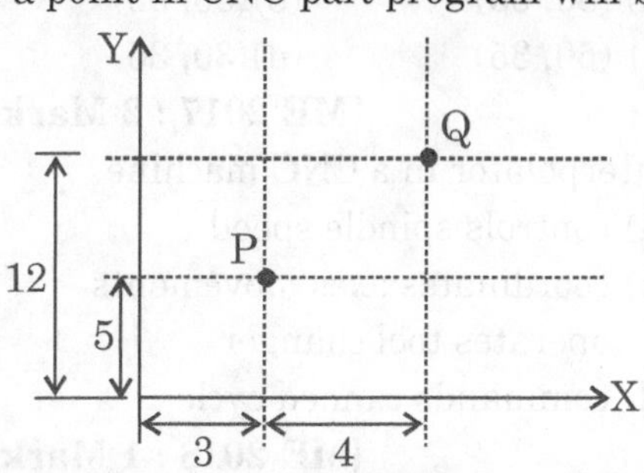

(a) (3,12) (b) (5,7)
(c) (7,12) (d) (4,7)

[ME 2015 : 1 Mark, Set-3]

41. In a CNC milling operation, the tool has to machine the circular arc from point (20, 20) to (10, 10) at sequence number 5 of the CNC part program. If the center of the arc is at (20,10) and the machine has incremental mode of defining position coordinates, the correct tool path command is
(a) N05 G90 G01 X-10 Y-10 R10
(b) N05 G91 G03 X-10 Y-10 R10
(c) N05 G90 G03 X20 Y20 R10
(d) N05 G91 G02 X20 Y20 R10
[ME 2015 : 2 Marks, Set-3]

42. Match the following part programming codes with their respective functions
Part Programming Codes
P. G01 Q. G03
R. M03 S. M05
Functions
I. Spindle stop
II. Spindle rotation, clockwise
III. Circular interpolation, anticlockwise
IV. Linear interpolation
(a) P-II, Q-I, R-IV, S-III
(b) P-IV, Q-II, R-III, S-I
(c) P-IV, O-III, R-II, S-I
(d) P-III, Q-IV, R-II, S-I
[ME 2016 : 1 Mark, Set-3]

43. Circular arc on a part profile is being machined on a vertical CNC milling machine, CNC part program using metric units with absolute dimensions is listed below:
....
N60 G01 X 30 Y 55 Z 5 F50
N70 G02 X 50 Y 35 R 20
N80 G01 Z 5
....
The coordinates of the centre of the circular arc are:
(a) (30, 55) (b)(50, 55)
(c) (50, 35) (d)(30, 35)
[ME 2017 : 2 Marks, Set-1]

44. Interpolator in a CNC machine
(a) controls spindle speed
(b) coordinates axes movements
(c) operates tool changer
(d) commands canned cycle
[ME 2018 : 1 Marks, Set-1]

45. If the z-axis and x-axis of CNC lathe are provided with straight line controls, it is possible to carry out
(a) Turning and facing only
(b) Turning, facing and taper turning
(c) turning, facing and thread cutting
(d) Turning, facing and drilling
[PI 1989 : 1 Mark]

46. The most common interpolation methods in continuous path NC machining are
(a) linear and circular
(b) linear and parabolic
(c) circular and parabolic
(d) circular and elliptic
[PI 1990 : 1 Mark]

47. A $2\frac{1}{2}$ axis control in a typical CNC machine provides
(a) simultaneous tool movements along 2 axes
(b) programmable movement along 3 axes but allow simultaneous movement along 2 axes
(c) simultaneous programmable movement along 3 axes in both directions
(d) simultaneous programmable movement along 3 axes but only in one direction
[PI 1990 : 2 Marks]

48. Feed motion can be provided with stepper motors in CNC
(a) lathe (b) drilling machine
(c) milling machine (d) grinding machine
[PI 1991 : 1 Mark]

49. Control of group of NC machine by a supervisory computer is known as ... control.
[PI 1992 : 1 Mark]

50. A DC servomotor is directly driving an NC table. The pitch of the lead screw of the table is 5 mm. The motor rotates at 100 rpm for an applied voltage of 10 V. If the voltage speed characteristic of the motor is linear, the applied voltage for a table of 3 m/min is equal to
(a) 30 V (b) 60 V
(c) 33 V (d) 50 V
[PI 1992 : 2 Marks]

51. A. CNC 1. Zero inventory
B. MRP 2. Rout sheet
C. JIT 3. DNC
D. FMS 4. Cutter path
5. Lumped demand
[PI 1993 : 2 Marks]

52. In DNC each machine tool is controlled by a dedicated computer.
[PI 1994 : 1 Mark]

53. Match the following

Group-I

P. G09

Q. G41

R. G01

S. G03

Group-II

1. Linear interpolation
2. Retardation
3. Circular interpolation
4. Cutter radius compensation

	P	Q	R	S
(a)	3	1	4	2
(b)	4	1	3	2
(c)	3	4	1	2
(d)	2	4	1	3

[PI 1994 : 2 Marks]

54. For machining the circular arc shown in the figure below on a NC machining, the interpolation parameters (I, K) and the incremental movements in the direction of the X and Z axes are programmed as

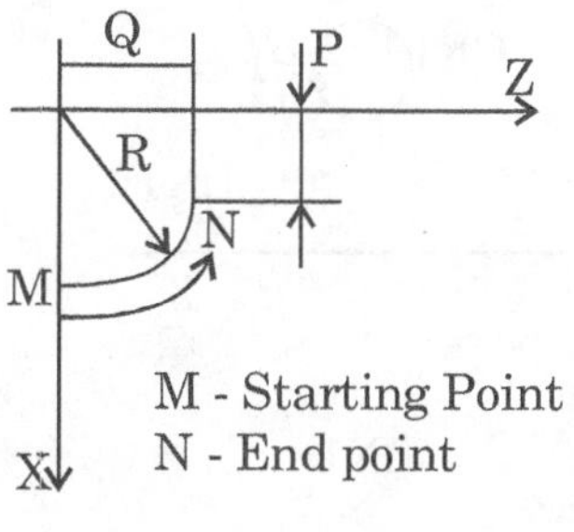

(a) I = 0, K = R, X = -(R–P), Z = –0

(b) I = R, K = 0, X = -(R–P), Z = –Q

(c) I = P, K= Q, X= R, Z= R–Q

(d) I = Q, K = P, X = (R-P), Z= R

[PI 2004 : 2 Marks]

55. Match the following:

Group-I

P. G08

Q. G41

R. G01

S. G02

Group-II

1. Linear interpolation
2. Acceleration
3. Circular interpolation
4. Cutter radius compensation

	P	Q	R	S
(a)	3	1	4	2
(b)	4	1	3	2
(c)	3	4	1	2
(d)	2	4	1	3

[PI 2006 : 2 Marks]

56. The interpolator in a CNC machine controls

(a) Spindle speed

(b) Coolant flow

(c) Feed rate

(d) Tool change

[PI 2007 : 1 Mark]

57. In a CNC feed drive, a stepper motor with step angle of 1.8 degree drives a lead screw with pitch of 2 mm. The basic length unit (BLU) for this drive is

(a) 10microns (b) 20 microns

(c) 40 microns (d) 100 microns

[PI 2008 : 2 Marks]

58. For a 3-axes CNC table, the slide along the vertical axis of the table driven by a DC servo motor via lead screw nut mechanism. The lead screw has pitch of 5 mm. This lead screw is fitted with a relative (incremental) circular encoder. The basic length unit (BLU) of the slide along the vertical axis of the table is 0.005 mm. When the table moves along the vertical axis by 9 mm. The corresponding number of pulses generated by the encoder is

(a) 1400 (b) 1800

(c) 4200 (d) 9000

[PI 2010 : 2 Marks]

Group technology

59. Cellular manufacturing is suitable for

(a) a single production in large volumes

(b) one-off production of several varieties

(c) production with similar features made in batches

(d) large variety of products in large volumes

[ME 2000 : 1 Mark]

60. A minimal spanning tree in network flow models involves

(a) all the nodes with cycle/loop allowed

(b) all the nodes with cycle/loop allowed

(c) shortest path between start and end nodes

(d) all the nodes with directed arcs

[ME 2014 : 1 Mark, Set-3]

61. FMS is well suited to mass scale manufacturing. (T/F)

[PI 1994 : 1 Mark]

62. FAPT is part programming language. (T/F)

[PI 1994 : 1 Mark]

63. Automation strategy type of equipment

Type of equipment

P. Stand-alone programmable manufacturing automation

Q. Programmable automation with automatic tool changing

R. Dedicated automation with automated material transfer

S. Programmable automation with programmable material handling and transfer

Automation strategy

1. Transfer line
2. CNC
3. FMS
4. Machining centre

	P	Q	R	S
(a)	2	3	4	1
(b)	4	2	3	1
(c)	1	2	3	4
(d)	2	4	1	3

[PI 2005 : 2 Marks]

64. Consider the following characteristics:

1. Single machine tool
2. Manual material handling system
3. Computer control
4. Random sequencing of part of machines

Which of the above characteristics are associated with flexible manufacturing system?

(a) 1,2 and 3 (b) 1 and 2

(c) 3 and 4 (d) 2, 3 and 4

[PI 2005 : 2 Marks]

65. A stepper motor has 150 steps. The output shaft of the motor is directly coupled to a lead screw of pitch 4 mm, which drives a table. If the frequency of pulse supply to the motor is 200 Hz, the speed of the table (in mm/min) is

(a) 400 (b) 320

(c) 300 (d) 280

[PI 2008 : 2 Marks]

66. Suppose point P1 and APT (Automatically Programmed Tool) programming is coded by statement

P1 = POINT/XSMALL, INTOF, LN1, CR1

The coded geometric situation without causing error is

(a)

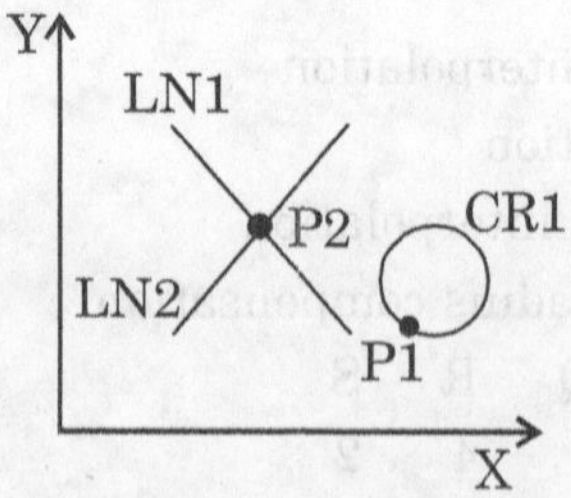

(b)

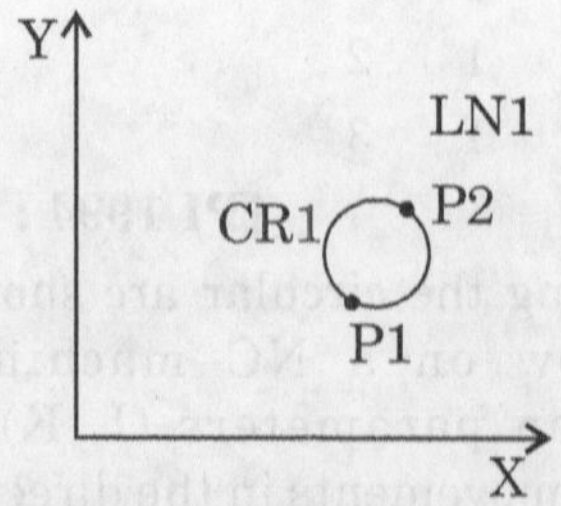

(c)

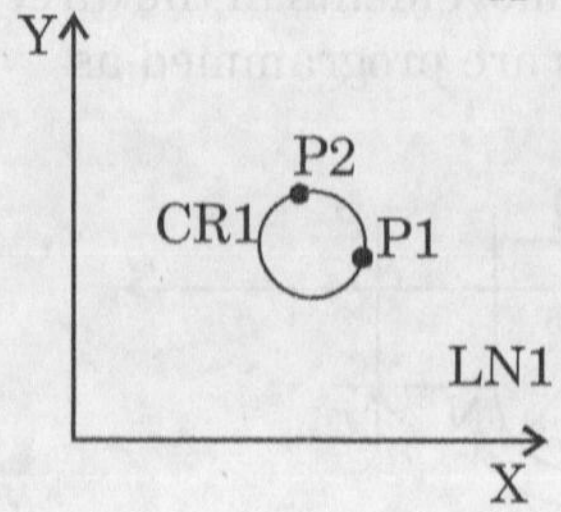

(d)

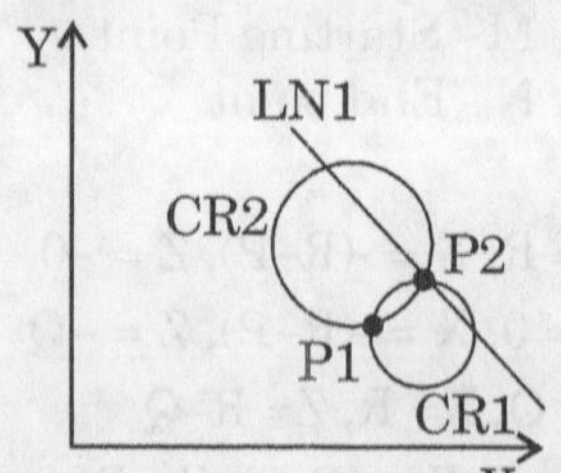

67. The total angular movement (in degrees) of a lead screw with a pitch of 5.0 mm to drive the work table by a distance of 200 mm in a NC machine is

(a) 14400 (b) 28800

(c) 57600 (d) 72000

[PI 2009 : 2 Marks]

ANSWERS

1. (d) **2.** (c) **3.** (c) **4.** (b) **5.** (b) **6.** (a) **7.** (d) **8.** (a) **9.** (d) **11.** (a)
12. (c) **13.** (a) **14.** (d) **15.** (c) **16.** (b) **17.** (b) **19.** (b) **20.** (d) **23.** (b) **25.** (b)
26. (a) **27.** (a) **28.** (a) **29.** (d) **30.** (a) **31.** (b) **32.** (b) **33.** (a) **34.** (a) **35.** (c)
36. (b) **37.** (a) **38.** (a) **39.** (b) **40.** (d) **41.** (b) **42.** (c) **43.** (d) **44.** (b) **45.** (b)
46. (a) **47.** (b) **48.** (b) **50.** (b) **53.** (d) **54.** (b) **55.** (d) **56.** (c) **57.** (a) **58.** (b)
59. (c) **60.** (b) **63.** (d) **64.** (c) **65.** (b) **66.** (b) **67.** (a)

EXPLANATIONS

2. Given: $P_1(15, 10)$
$P_2(10, 15)$

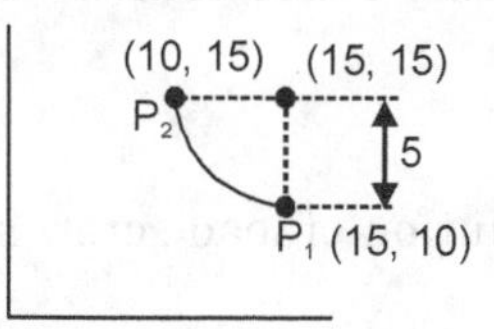

Clear from figure,
Centre of arc having radius = 5 is (15, 15)

4. For the new coordinate pair

$$\begin{bmatrix} x' \\ y' \end{bmatrix} = \begin{bmatrix} \cos\theta & -\sin\theta \\ \sin\theta & \cos\theta \end{bmatrix} \begin{bmatrix} x \\ y \end{bmatrix}$$

where (x', y') is coordinate at point θ.

by putting θ = 90°

$$\begin{bmatrix} x' \\ y' \end{bmatrix} = \begin{bmatrix} 0 & -1 \\ 1 & 0 \end{bmatrix} \begin{bmatrix} x \\ y \end{bmatrix} = \begin{bmatrix} 0 & -1 \\ 1 & 0 \end{bmatrix} \begin{bmatrix} 2 \\ 5 \end{bmatrix}$$

$$\begin{bmatrix} x' \\ y' \end{bmatrix} = \begin{bmatrix} -5 \\ 2 \end{bmatrix}$$

5.

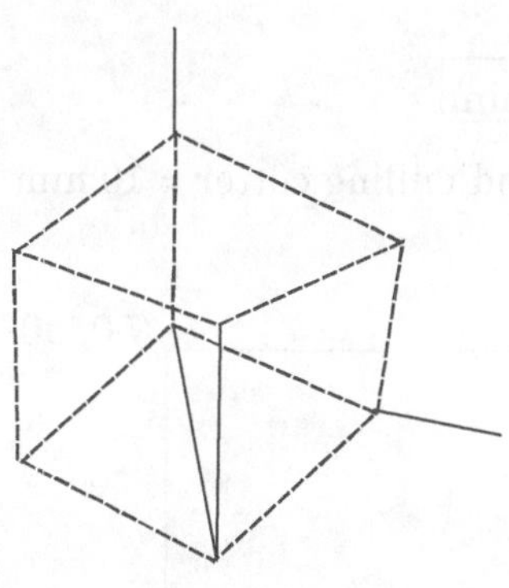

$$\text{Area} = \frac{1}{2} \times \sqrt{2} \times 1 = \frac{\sqrt{2}}{2} = 0.707$$

6. If the Δ PQR rotator by θ Clockwise then abscissa value of Q must increase whereas ordinate value of Q must decreases and only satisfying option is (4.6, 2.8) i.e. A.

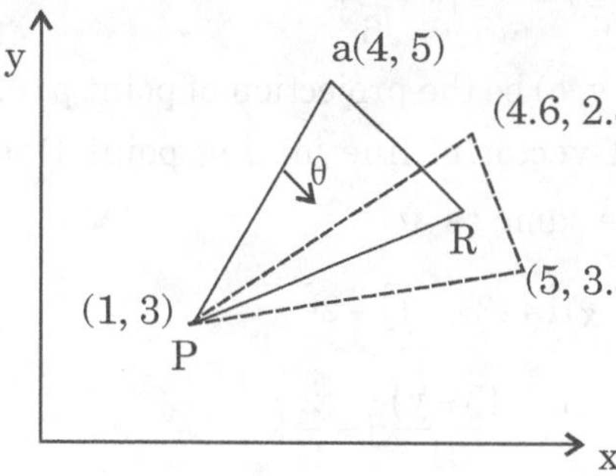

7. $AB = \sqrt{(R + r)^2 - (D - R - r)^2}$

$H = R + AB + r$

$H = R + r + \sqrt{(R + r + D - R - r)(R + r - D + R + r)}$

$H = R + r + \sqrt{D[2(R + r) - D]}$

$H = R + r + \sqrt{2D(R + r - D^2}$

8. P = (1, 3, – 5)

First translated by $2\hat{i} + 3\hat{j} - 4\hat{k}$

and them rotate about z – axis by 90°

Homogeneous Translation Matrix

$$\begin{bmatrix} 1 & 0 & 0 & 2 \\ 0 & 1 & 0 & 3 \\ 0 & 0 & 1 & -4 \\ 0 & 0 & 0 & 1 \end{bmatrix} \begin{bmatrix} 1 \\ 3 \\ -5 \\ 1 \end{bmatrix} = \begin{bmatrix} 3 \\ 6 \\ -9 \end{bmatrix}$$

Then rotated about Z – axis.

$$\begin{bmatrix} \cos\theta & -\sin\theta & 0 \\ \sin\theta & \cos\theta & 0 \\ 0 & 0 & 1 \end{bmatrix} \begin{bmatrix} 3 \\ 6 \\ 9 \end{bmatrix}$$

$$= \begin{bmatrix} 0 & -1 & 0 \\ 1 & 0 & 0 \\ 0 & 0 & 1 \end{bmatrix} \begin{bmatrix} 3 \\ 6 \\ 9 \end{bmatrix} = \begin{bmatrix} -6 \\ 3 \\ -9 \end{bmatrix} = [-6, 3, -9]$$

9. For a reflection in the line y = x

$$\begin{bmatrix} 0 & 1 \\ 1 & 0 \end{bmatrix}$$

10. (a) Consistency
(b) Hazard
use of robots improves consistency and reduce hazards for human life.

13. After executing drilling process through command (– 154 mm), drill bit were, return for O level and 90 to set point.

14. $\overline{V} = -2\hat{i} + \hat{j} - \hat{k}$

$$\Rightarrow \hat{V} = \frac{-2}{\sqrt{6}}\hat{i} + \frac{1}{\sqrt{6}}\hat{j} - \frac{1}{\sqrt{6}}\hat{k}$$

Let X (x, y, o) be the projection of point p (6, 3, 2)
The unit vector of line joining point P and X should be same as $\hat{V}$

$$\overrightarrow{P_x} = (6-x)\hat{i} + (3-y)\hat{j} + 2\hat{k}$$

$$\overrightarrow{P_x} = \frac{(6-x)}{0}\hat{i} + \frac{(3-y)}{D}\hat{j} + \frac{2}{D}\hat{k}$$

where $D = \sqrt{(6-x)^2 + (3-y)^2 + 2^2}$.

on equating, $\overrightarrow{P_x}$ and $\overrightarrow{V}$, we get

$$\frac{6-x}{D}, = \frac{-2}{\sqrt{6}}, \frac{3-y}{D} = \frac{1}{\sqrt{6}}, \frac{2}{D} = \frac{-1}{\sqrt{6}}$$

So $D = -2\sqrt{6}$

$$j = 3 - \frac{(-2\sqrt{6})}{\sqrt{6}} = 5$$

$$x = 6 - \frac{(-2 \times -2\sqrt{6})}{\sqrt{6}} = 2.$$

$$z = 0$$

15. Transformation matrix has components of both rotation and translation.

16. Transformation matrix for mirror image about y = x line

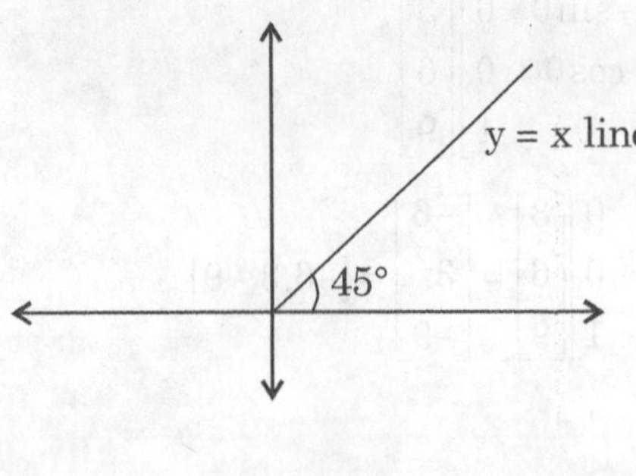

$$\begin{bmatrix} x' \\ y' \end{bmatrix} = \begin{bmatrix} 0 & 1 \\ 1 & 0 \end{bmatrix}\begin{bmatrix} x \\ y \end{bmatrix}$$

$$= \begin{bmatrix} 0 & 1 \\ 1 & 0 \end{bmatrix}\begin{bmatrix} 5 \\ 10 \end{bmatrix} = \begin{bmatrix} 10 \\ 5 \end{bmatrix}.$$

Mirror image will be at (10, 5).

17. G code for spindle speed is denoted by S.

18. Set-up time for NC machines is smaller in comparison to special purpose machines hence, used for batch Production

19. In a point to point type of NC system, only end position of tool is controlled. Neither path nor velocity of the tool is important.

20. Post-Processor is not used in NC machine.

21. True

22. False

23. For 1 revolution of lead screw side will move by 3.6 mm

1 revolution = 360° = 360 pulses

360 pulses will cause motion of 3.6 mm. 1 pulse will result in $\frac{3.6}{360}$ mm of motion

i.e. 0.01 mm or 10 μm.

24. $V = \omega r = 2\pi Nr$.

$= \pi dN \{2r = d, \omega = 2\pi N\}$

$$\text{So, } N = \frac{V}{\pi d} \Rightarrow N = \frac{150 \times 10^3}{\pi \times 155} = 308 \text{ rpm}.$$

Programmed feed rate

$= f \times N = 0.15 \times 308 = 46.2$ mm/rpm

$$= f \times V \times \frac{(D_2 - D_1)}{2} = 0.15 \times 150 \times 10^3 \times 2.5$$

$$\boxed{= 56250 \text{mm}^3/\text{min.}}$$

25. Given = Dia of end milling cutter = 10 mm

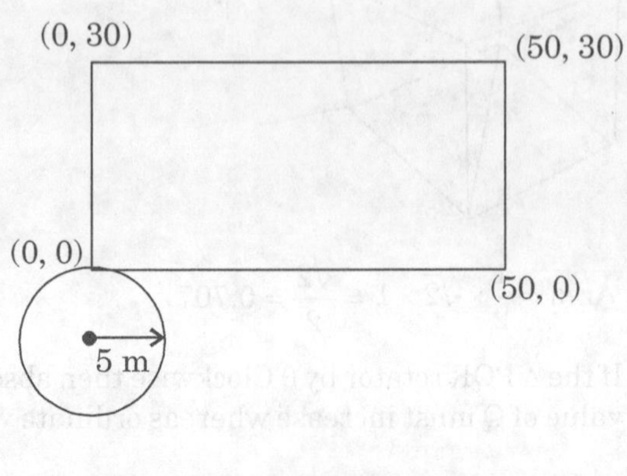

31. Step angle $(\alpha) = \frac{360}{200} = 1.8°$

$$(\alpha) = \frac{360}{200} = 1.8°$$

$$x = \frac{PA}{360}, A = \alpha n_p$$

n_p is number of pulse.

n_p = 1 for BLU

$$x = \frac{4 \times 1.8}{360} \times \text{Gear Ratio}$$

$$x = \frac{4 \times 1.8}{360} \times \frac{1}{4}$$

x = 5 microns

40.

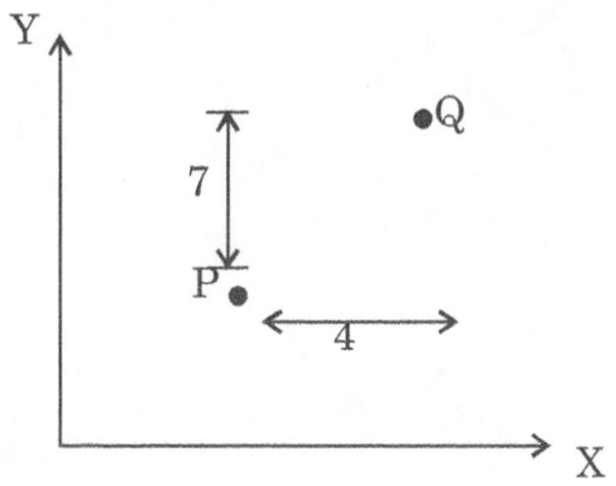

41.

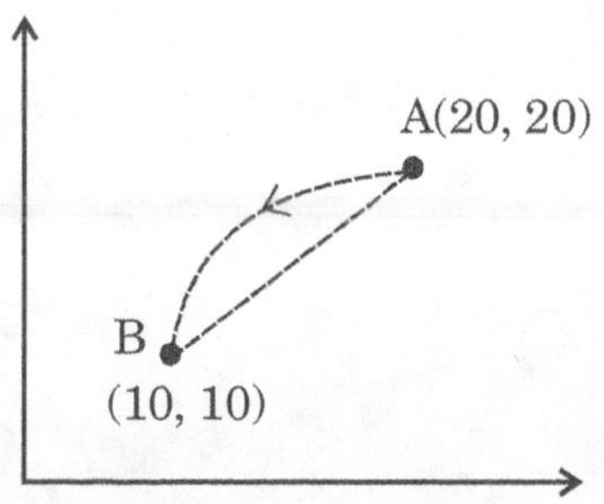

Radius of circle = 10

Incremental system is used

42. P-IV, Q-III, R-II, S-I

43.

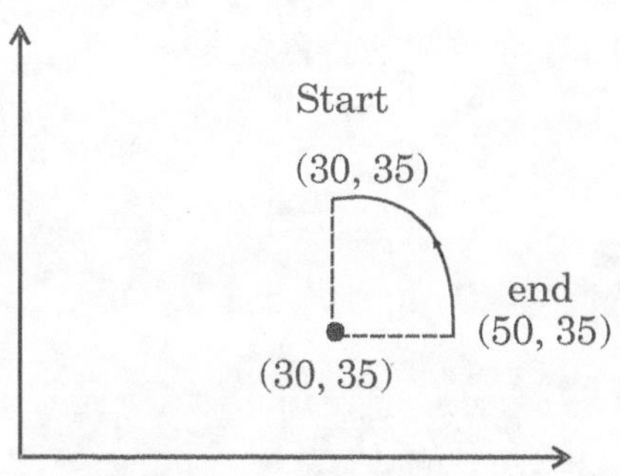

48. In PTP type motion like in drilling, stepper motors could be used.

49. DNC (Distributed numerical control)

50. Motor speed = 10 rpm.

Distance = travelled on

Lead screw = 5 mm.

Line for one revolution $= \frac{1}{100}$ min.

Speed of motion on lead serew

$$= \frac{5}{\frac{1}{100}} = 500\text{mm} / \text{min}.$$

Considering line characteristic of 0.5m/min to get travel speed of 3m/min, we have to increase the voltage by

$$= \frac{3}{0.5} = 6 \text{ times}$$

i.e.

$10 \times 6 = \boxed{60V}$

51. A – 4, B – 5, C – 1, D–3.

CNC - Cutter path.

MRP - Lumped demand.

JIT - Zero inventory

FMS - DNC.

52. True

IN DNC, Several machine tools connected to a supervisory computer, which controles all their functions.

57. Step Angle = 1.8°/pulse.

pitch = 2 mm.

Distance travelled in one revolution (360°)

= 2 m.

No of pulse in 360° $= \frac{360}{1.8} = 200$ pulse.

Distance travelled in one pulse $= \frac{2}{200}$ mm.

= 0.01 mm = 10 mm

58. p = 5 mm.

BLU = 0.005 mm.

Now

1 Rev. → x × BLU

x = 1000 pulse/Rev

y = 9 mm.

So $n = \frac{Y}{P}$ rev.

or no. of pulses

$$= \frac{Y}{P} \times x = \frac{9}{5} \times 1000$$

No. of pulses = 1800

59. Production of similar features products in Batches

60. All the nodes with cycle/loop allowed.

61. F

62. T

64. FMS is a highly automated GT Machine cell, consisting of a group of processing stations (usually CNC machines Tools), Interconnected by an automated material handling and storage system and controlled by an integrated computer system.

65. $V_t = \frac{60 \times fp}{n_s} \times P$

$$= \frac{60 \times 200}{150} \times 4$$

$f_p = 200$ Hz

p = 4mm (Pitch)

= 320 mm/min.

66. Point P is at the intersection of Line LN with Circle.

67. pitch = 5 mm

No. of Revolution required for 200 mm travel distance

$$= \frac{200}{5} = 40\text{rev.}$$

$\therefore 40 \times 360 = 1400°$

9 CHAPTER

Nonconventional Machining (EDM, High Energy Rate Forming)

Analysis of Previous GATE Papers

Topic	Year/Set → Marks Distribution ↓		2019 Set 1	2019 Set 2	2018 Set 1	2018 Set 2	2017 Set 1	2017 Set 2	2016 Set 1	2016 Set 2	2016 Set 3	2015 Set 1	2015 Set 2	2015 Set 3
Nonconventional Machining	1 Mark	MCQ Type				1	1	1	1	1	1		1	
		Numerical Type												
	2 Marks	MCQ Type											1	
		Numerical Type			1	1								
		Total			2	3	1	1	1	1	1		3	
High Energy rate Forming	1 Mark	MCQ Type												
		Numerical Type												
	2 Marks	MCQ Type												
		Numerical Type												
		Total												

Nonconventional Machining (ECM, EDM, ASM, USM)

1. In ultrasonic machining the tool... at very high frequency with the help of ... transducers.

[ME 1987 : 1 Mark]

2. ECM... (can/cannot) be used for all such materials for which ultrasonic machining is possible, while EDM... (can/cannot) be used for all such materials for which ECM is possible.

[ME 1987 : 1 Mark]

3. In ECM, the material removal rate will be higher for metal with

(a) large density

(b) larger valency

(c) larger Chemical absorption tendency

(d) large chemical weight

[ME 1989 : 2 Marks]

4. In USM the metal removal rate would ... with increasing mean grain diameter of the abrasive material

(a) increase

(b) decrease

(c) increase and then decrease

(d) decrease and then increase

[ME 1992 : 1 Mark]

5. The two main criteria for selecting the electrolyte in ECM is that the electrolyte should

(a) chemically stable

(b) not allow dissolution of cathode material

(c) not allow dissolution of anode material

(d) have high electrical conductivity

[ME 1992 : 1 Mark]

6. In ultrasonic machining process, the material removal rate will be higher for material with

(a) higher toughness

(b) higher ductility

(c) lower toughness

(d) higher fracture strain

[ME 1993 : 1 Mark]

7. EDM imposes larger forces on tool than ECM (T/F)

[ME 1994 : 1 Mark]

8. EDM is more efficient process than ECM for producing large non-circular holes (T/F)

[ME 1994 : 1 Mark]

9. USM is about the best process for making holes in glass which are comparable in size with thickness of the sheet.

[ME 1994 : 1 Mark]

10. Selection of electrolyte for ECM is as follows

(a) non-passivating electrolyte for stock removal and passivating electrolyte for finish control

(b) passivating electrolyte for stock removal and non-passivating electrolyte for finish control

(c) selection of electrolyte is dependent on current density

(d) electrolyte selection is based on tool work electrodes

[ME 1997 :1 Mark]

11. Inter electrode gap in ECG is controlled by

(a) controlling the pressure of electrolyte flow

(b) controlling the applied static load

(c) controlling the size of diamond particle in the wheel

(d) controlling the texture of the work

[ME 1997 : 1 Mark]

12. Match the following lists

List-I

A. ECM

B. EDM

C. USM

D. LBM

List-II

1. Plastic shear
2. Erosion/brittle fracture
3. Corrosive reaction
4. Melting vaporization
5. lon displaeement
6. Plastic shear and ion displacement

[ME 1998 : 2 Marks]

13. Match the following lists

List-I

A. Aluminium brake shoe

B. Plastic water bottles

C. Stainless steel cups

D. Soft drink can (aluminum)

List-II

1. Deep drawing
2. Blow moulding
3. Centrifugal casting
4. Impact extrusion
5. Sand casting
6. Upset forging

[ME 1998 : 2 Marks]

14. In EDM, the tool is made of

(a) Copper (b) HSS

(c) Cast iron (d) Plain carbon steel

[ME 1999 : 1 Mark]

15. Choose the correct statement

(a) Fixture is to guide the tool as well as to locate and clamp the work piece

(b) Jig is to guide the tool as well as to locate and clamp the work piece

(c) Jigs are used on CNC machines to locate and clamp the workpiece and guide the tool

(d) No arrangement to guide the tool is provided in jig

[ME 1999 : 1 Mark]

16. Deep hole drilling of small diameter, say 0.2 mm is done with EDM by selecting the tool material as

(a) Copper wire (b) Tungsten wire

(c) Brass wire (d) Tungsten carbide

[ME 2000 : 1 Mark]

17. In ECM the MRR is due to

(a) Corrosion (b) Erosion

(c) Fusion (d) Ion displacement

[ME 2001 : 1 Mark]

18. Estimate the MRR (in cc/sec) of an alloy containing 18% cobalt, 62% nickel and 20% chromium during ECM with current of 500 ampere. The density of the alloy is 8.28 g/cc. The following data is available.

Metal	Gram atomic weight	Valency
Cobalt	58.93	2
Nickel	58.71	2
Chromium	51.99	6

Assume Faraday's constant is 96500 coulombs/mole.

[ME 2002 : 2 Marks]

19. As tool and work are not in contact in EDM process

(a) no relative motion occurs between them

(b) no wear of tool occurs

(c) no power is consumed during metal cutting

(d) no force between tool and work occurs

[ME 2003 : 1 Mark]

20. The mechanism of material removal in EDM process is

(a) melting and evaporation

(b) melting and corrosion

(c) erosion and cavitation

(d) cavitation and evaporation

[ME 2004 : 1 Mark]

21. Typical machining operations are to be performed on hard-to-machine materials by using the processes listed below. Choose the best set of Operation- Process combinations

Operation

P. Deburring (internal surface)

Q. Die sinking

R. Fine hole drilling in thin sheets

S. Tool sharpening

Process

1. Plasma Arc Machining
2. Abrasive Flow Machining
3. Electric Discharge Machining
4. Ultrasonic Machining
5. Laser beam Machining
6. Electrochemical Grinding

(a) P-1, Q-5, R-3, S-4

(b) P-1, Q-4, R-1, 5-2

(c) P-5, Q-1, R-2, S-6

(d) P-2, Q-3, R-5, S-6

[ME 2004 : 2 Marks]

22. A zigzag cavity in a block of high strength alloy is to be finish machined. This can be carried out by using

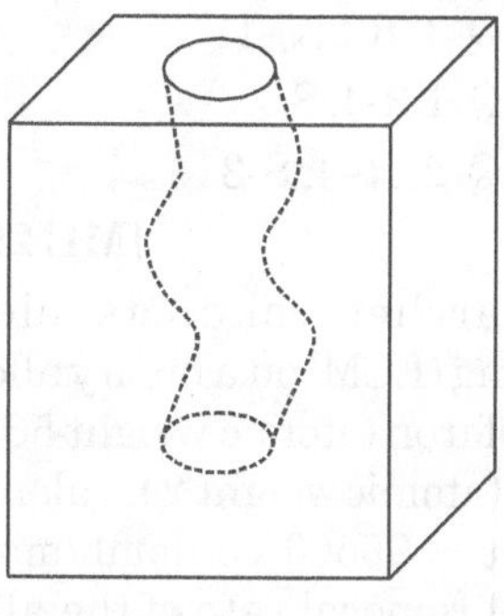

(a) electric discharge machining

(b) electro-chemical machining

(c) laser beam machining

(d) abrasive flow machining

[ME 2005 : 1 Mark]

23. Arrange the processes in the increasing order of their maximum material removal rate.

Electrochemical Machining (ECM)

Ultrasonic Machining (USM)

Electron Beam Machining (EBM)

Laser Beam Machining (LBM)

Electric Discharge Machining (EDM)

(a) USM, LBM, EBM, EDM, ECM

(b) EBM, LBM, USM, ECM, EDM

(c) LBM, EBM, USM, ECM, EDM

(d) LBM, EBM, USM, EDM, ECM

[ME 2006 : 2 Marks]

24. In Electrodischarge Machining (EDM), if the thermal conductivity of tools is high and the specific heat of work piece is low, then the tool wear rate and material removal rate are expected to be respectively

(a) high and high (b) low and low
(c) high and low (d) low and high

[ME 2007 : 2 Marks]

25. Match the most suitable manufacturing processes for the following parts,

Parts

P. Computer chip
Q. Metal forming dies and moulds
R. Turbine blade
S. Glass

Manufacturing Processes

1. Electrochemical Machining
2. Ultrasonic Machining
3. Electrodischarge Machining
4. Photochemical Machining

(a) P-4, Q-3, R-1, S-2
(b) P-4, Q-3, R-2, S-1
(c) P-3, Q-1 R-4, S-2
(d) P-1, Q-2, R~4, S-3

[ME 2007 : 2 Marks]

26. A researcher conducts electrochemical machining (ECM) on a binary alloy (density 6000 kg/m^3) of iron (atomic weight 56, valency 2) and metal P (atomic weight 24, valency 4). Faraday's constant = 96500 coulomb/mole. Volumetric material removal rate of the alloy is 50 mm^3/s at a current of 2000 A. The percentage of the metal P in the alloy is closest to

(a) 40 (b) 25
(c) 15 (d) 79

[ME 2008 : 2 Marks]

27. Electrochemical machining is performed to remove material from an iron surface of 20 mm × 20 mm under the following conditions:

Inter electrode gap = 0.2 mm
Supply voltage (DC) = 12 V
Specific resistance of electrolyte = 2 ohm- cm
Atomic weight of iron = 55.85
Valency of Iron = 2
Faraday's constant = 96540 Coulombs
The material removal rate (in g/s) is

(a) 0.3471 (b) 3.471
(c) 34.71 (d) 347.1

[ME 2009 : 2 Marks]

28. In abrasive jet machining, as the distance between the nozzle tip and the work surface increases, the material removal rate

(a) increases continuously
(b) decreases continuously
(c) decreases, becomes stable and then increases
(d) increases, becomes stable and then decreases

[ME 2012 : 1 Mark]

29. Match the following non-traditional machining processes with the corresponding material removal mechanisms:

Machining process

P. Chemical machining
Q. Electro-chemical machining
R. Electro-discharge machining
S. Ultrasonic machining

Mechanism of material removal

1. Erosion
2. Corrosive reaction
3. Ion displacement
4. Fusion and vaporization

(a) P-2, Q-3, R-4, S-1
(b) P-2, Q-4, R-3, S-1
(c) P-3, Q-2, R-4, S-1
(d) P-2, Q-3, R-1, S-4

[ME 2011 : 2 Marks]

30. During the electrochemical machining (ECM) of iron (atomic weight = 56, valency = 2) at current of 1000 A with 90% current efficiency, the material removal rate was observed to be 0.26 cc/s. If Titanium (atomic weight = 48, valency = 3) is machined by the ECM process at the current of 2000 A with 90% current efficiency, the expected material removal rate in gm/s will be

(a) 0.11 (b) 0.23
(c) 0.30 (d) 0.52

[ME 2013 : 2 Marks]

31. The following four unconventional machining processes are available in a shop floor. The most appropriate one to drill a hole of square cross section of 6 mm × 6 mm and 25 mm deep is

(a) Abrasive Jet Machining
(b) Plasma Arc Machining
(c) Laser Beam Machining
(d) Electro Discharge Machining

[ME 2014 : 1 Mark, Set-2]

32. The process utilizing mainly thermal energy for removing material is

(a) Ultrasonic Machining

(b) Electrochemical Machining

(c) Abrasive Jet Machining

(d) Laser Beam Machining

[ME 2014 : 1 Mark, Set-3]

33. The principle of material removal in Electrochemical machining is

(a) Fick's law (b) Faraday's law

(c) Kirchhoffs law (d) Ohm's law

[ME 2014 : 1 Mark, Set-4]

34. The primary mechanism of material removal in electrochemical machining (ECM) is

(a) Chemical corrosion

(b) Etching

(c) Ionic dissolution

(d) Spark erosion

[ME 2015 : 1 Mark, Set-2]

35. A resistance-capacitance relaxation circuit is used in an electrical discharge machining process. The discharge voltage is 100 V. At a spark cycle time of 25 μs, the average power input required is 1 kW. The capacitance (in μF) in the circuit is

(a) 2.5 (b) 5.0

(c) 7.5 (d) 10.0

[ME 2015 : 2 Marks, Set-2]

36. In a wire-cut EDM process the necessary conditions that have to be met for making a successful cut are that

(a) wire and sample are electrically non-conducting

(b) wire and sample are electrically conducting

(c) wire is electrically conducting and sample is electrically non-conducting

(d) sample is electrically conducting and wire is electrically non-conducting

[ME 2016 : 1 Mark, Set-3]

37. In an ultrasonic machining (USM) process, the material removal rate (MRR) is plotted as a function of the feed force of the USM tool. With increasing feed force, the MRR exhibits the following behavior:

(a) increases linearly

(b) decreases linearly

(c) does not change

(d) first increases and then decreases

[ME 2016 : 1 Mark, Set-2]

38. The non-traditional machining process that essentially requires vacuum is

(a) electron beam machining

(b) electro Chemical machining

(c) electro Chemical discharge machining

(d) electro discharge machining

[ME 2016 : 1 Mark, Set-1]

39. Match the processes with their characteristics.

Process	**Characteristics**
P. Electrical Discharge Machining	1. No residual stress
Q. Ultrasonic machining	2. Machining of electrically conductive materials
R. Chemical machining	3. Machining of glass
S. Ion Beam Machining	4. Nano-machining

(a) P-2, Q-3, R-1, S-4

(b) P-3, Q-2, R-1, S-4

(c) P-3, Q-2, R-4, S-1

(d) P-2, Q-4, R-3, S-1

[ME 2017 : 1 Mark, Set-1]

40. Which one of the following statements is TRUE for the ultrasonic machining (USM) process?

(a) In USM, the tool vibrates at subsonic frequency.

(b) USM does not employ magnetostrictive transducer.

(c) USM is an excellent process for machining ductile materials.

(d) USM often uses a slurry comprising abrasive-particles and water.

[ME 2017 : 1 Mark, Set-2]

41. An electrochemical machining (ECM) is to be used to cut a through hole into a 12 mm thick aluminium plate. The hole has a rectangular cross- section, 10 mm × 30 mm. The ECM operation will be accomplished in 2 minutes, with efficiency of 90%. Assuming specific removal rate for aluminium as 3.44×10^{-2} mm^3/(A s), the current (in A) required is _____ (correct to two decimal places).

[ME 2018 : 2 Marks, Set-1]

42. Metal removal in electric discharge machining takes place through

(a) ion displacement

(b) melting and vaporizatron

(c) corrosive reaction

(d) plastic shear

[ME 2018 : 1 Marks, Set-2]

43. A circular hole of 25 mm diameter and depth of 20 mm is machined by EDM process. The material removal rate (in mm³/min) is expressed as $4 \times 10^4 IT^{-1.23}$ where I = 300 A and the melting point of the material, T = 1600° C. The time (in minutes) for machining this hole is (correct to two decimal places).

[ME 2018 : 2 Marks, Set-2]

44. The maximum depth to width ratio in EBM welding is

(a) 1 : 1 (b) 2 : 1
(c) 10 : 1 (d) 25 : 1

[PI 1989 : 1 Mark]

45. Match the following

List-I
A. ECM
B. USM
C. EDM
D. LBM

List-II
1. Erosion
2. Fusion and vaporization
3. Vaporization and ablation
4. Ion displacement

[PI 1989 : 2 Marks]

46. Machining process job

List-I
A. LBM
B. EDM
C. Chemical milling
D. USM

List-II
1. Making a printed circuit board
2. Drilling a hole in glass plate
3. Making a die out of a die steel block
4. Cutting a polyester sheet

[PI 1990 : 2 Marks]

47. Match the following

Process	**Defect**
A. Casting	1. Lack of penetration
B. Welding	2. Penetration
C. Extrusion	3. Flange wrinkles
D. Deepdrawing	4. Centre burst

[PI 1990 : 2 Marks]

48. Match the following

Products
A. Porous bearings
B. Fly wheels
C. Double end spanners
D. Plastic bottles

Manufacturing Process
1. Powder metallurgy
2. Casting
3. Closed die forging
4. Blow moulding

[PI 1990 : 2 Marks]

49. A 10 mm square hole can be cut in a 2 mm thick glass plate by

(a) USM (b) EDM
(c) LBM (d) ECM

[PI 1991 : 1 Mark]

50. In ECM, the metal removal rate is directly proportional to

(a) Density of work material
(b) Gram atomic weight of the work material
(c) Concentration of the electrolyte used
(d) Thermal conductivity of the work material

[PI 1991 : 2 Marks]

51. Match the following

Process	**Properties of work material**
A. Casting	1. Ductility
B. Forging	2. Malleability
C. Wire drawing	3. Machinability
D. Turning	4. Fluidity

[PI 1991 : 2 Marks]

52. Match the following

Machine Components
A. Machine tool bed
B. Machine tool spindle
C. Cutting tool
D. Worm wheel

Material used
1. Forged used
2. HSS
3. Bronze
4. Cast iron

[PI 1991 : 2 Marks]

53. Match the following

Process	Media used
A. EDM	1. Electrolyte
B. ECM	2. Dielectric
C. USM	3. Abrasive slurry
D. EBM	4. Vacuum

[PI 1991 : 2 Marks]

54. Match the following

Machining Process

A. Machining
B. Casting
C. Welding
D. Drawing

Associated Problem

1. Wrinkling
2. Heat affected zone
3. Hot tear
4. Built up edge

[PI 1991 : 2 Marks]

55. Electrochemical grinding removes material by... A ... And ... B... processes.

A. electrochemical/chemical
B. mechanical/chemical

[PI 1992 : 1 Mark]

56. Match the following

Manufacturing process

A. Cold rolling
B. Milling
C. Brazing
D. Powder metallurgy

Property

1. Machinability
2. Weldability
3. Ductility
4. Tensile strength
5. Compactability

[PI 1993 : 2 Marks]

57. Match the following

A. Polyester resin
B. Methyl methacrylate
C. Polyurethane
D. Polyvinyl chloride

Product

1. Moulded luggage
2. Refrigerator insulation
3. FRP
4. Contact lenses
5. Floor tiles

[PI 1993 : 2 Marks]

58. Match the following

A. Abrasive jet machining
B. PAM
C. Wire EDM
D. Chemical machining

1. Masking
2. Stamping die
3. Engraving
4. Stainless steel profiling
5. Drilling fine holes

[PI 1993 : 2 Marks]

59. Match the following

A. Stamping die plate
B. Gas turbine blades
C. Twist drill
D. Carbide tool inserts
E. Hole in a ceramic plate

1. EDM
2. LBM
3. ECM
4. EBM
5. ECG
6. Friction welding
7. Wire EDM
8. USM
9. Ultrasonic welding

[PI 1993 : 2 Marks]

60. Match the following

A. Abrasive jet machining	1. Horn
B. EDM	2. Wire as cutting tool
C. LBM	3. Light
D. USM	4. Vacuum
	5. Nozzle
	6. Cladding

[PI 1995 : 2 Marks]

61. Match the following

Process	Mechanism of material removal
P. EDM	1. Erosion
Q. ECM	2. Thermal evaporation
R. AJM	3. Anodic dissolution
	4. Etching

	P	Q	R
(a)	2	3	1
(b)	2	4	1
(c)	1	3	4
(d)	4	3	1

[PI 2003 : 2 Marks]

62. Match List-I (Material) with List-II (Machining) and select the correct answer using the code given below the lists;

List-I	**List-II**
A. Machining of conductive material	1. ECM
B. Ruby rod	2. EDM
C. Electrolyte	3. USM
D. Abrasive slurry	4. LBM

	A	B	C	D
(a)	4	2	1	3
(b)	4	2	3	1
(c)	2	4	1	3
(d)	2	4	3	1

[PI 2004 : 2 Marks]

63. Assertion (A): In electron beam welding process, vacuum is an essential process parameter.

Reason (R): Vacuum provides a highly efficient shield on weld zone.

(a) Both A and R are true and R is the correct explanation of A

(b) Both A and R are true but R is not a correct explanation of A

(c) A is true but R is false

(d) A is false but R is true

[PI 2005 ; 1 Mark]

64. Mechanism of material removal process R Abrasive grain throwing and hammering Q. Erosion due to vaporization R. Chipping and erosion S. Melting and partial vaporization T. Electrolysis

1. ECM
2. USM
3. PAM
4. EDM
5. AJM

	P	Q	R	S	T
(a)	3	2	5	1	4
(b)	4	3	2	1	5
(c)	2	4	5	3	1
(d)	2	5	3	1	4

[PI 2005 : 2 Marks]

65. Match List-I (Machining processes) with List-II (Operating Media) and select the correct answer using the codes given below the lists:

List-I

A. Abrasive jet machining

B. Electron beam machining

C. Electrochemical machining

D. Electro-discharge machining

List-II

1. Dielectric
2. Electrolyte
3. Abrasive slurry
4. Vacuum
5. Air

	A	B	C	D
(a)	5	4	2	1
(b)	4	5	2	1
(c)	4	2	3	5
(d)	2	5	3	4

[PI 2005 : 2 Marks]

66. Match the following

Group-I

P. Permaloy

Q. Dielectric

R. Accumulator

S. Anodic erosion

Group-II

1. EDM
2. USM
3. De-ionized water
4. Water jet machining

	P	Q	R	S
(a)	3	4	1	2
(b)	2	3	4	1
(c)	4	1	2	3
(d)	2	3	1	4

[PI 2006 : 2 Marks]

Linked Data Question 67 and 68:

In an EDM process using RC relaxation circuit, a 12 mm diameter, through hole is made in a steel plate of 50 mm thickness using a graphite tool and kerosene as dielectric. Assume discharge time to be negligible. Machining is carried out under the following conditions.

Resistance = 40 ohms

Capacitance = 20 μF

Supply voltage = 220 V

Discharge voltage = 110 V

67. The time for one cycle, in milliseconds, is

(a) 0.55 (b) 0.32
(c) 0.89 (d) 0.24

[PI 2006 : 2 Marks]

68. Average power input (in kW) is

(a) 0.373 (b) 0.137
(c) 0.218 (d) 0.5

[PI 2006 : 2 Marks]

69. Which one of the following process conditions leads to higher MRR in ECM?

(a) Higher current, large atomic weight
(b) Higher valency, lower current
(c) Lower atomic weight, lower valency
(d) Higher valency, lower atomic weight

[PI 2007 : 1 Mark]

70. In an AJM process, if Q = flow rate of abrasives and d = mean diameter of the abrasives, then MRR is proportional to

(a) Q/d^2 (b) Qd
(c) Qd^2 (d) Qd^3

[PI 2007 : 1 Mark]

71. In a electrochemical machining (ECM) operation, a square hole of dimensions 5 mm × mm is drilled in a block of copper. The current used is 5000 A. Atomic weight of copper is 63 and valence of dissolution is 1. Faradays constant is 96500 coulomb. The material removal rate (g/s) is

(a) 0.326 (b) 3.26
(c) 3.15×10^3 (d) 3.15×10^5

[PI 2008 : 2 Marks]

72. Match the following:

Device	Function
P. Jig	1. Helps to place workpiece in the same position cycle after cycle
Q. Fixture	2. Holds the workpiece only
R. Clamp	3. Holds and position the workpiece
S. Locator	4. Holds and position the workpiece and guides the cutting tool during a machining operation

	P	Q	R	S
(a)	4	3	1	2
(b)	1	2	3	4
(c)	1	4	3	2
(d)	4	3	2	1

[PI 2009 : 2 Marks]

73. A titanium sheet of 5.0 mm thickness is cut by wire cut EDM process using a wire of 1.0 mm diameter. A uniform spark gap of 0.5 mm on both sides of the wire is maintained during cutting operation. If the feed rate of the wire into the sheet is 20 mm/min, the material removal rate (in mm^3/min) will be

(a) 150 (b) 200
(c) 300 (d) 400

[PI 2009 : 2 Marks]

74. Ultrasonic machines used in material removal process, requires ultrasonic transducers. The transducers work on different working principles.

One of the working principles of such ultrasonic transducers is based on

(a) Eddy current effect
(b) Seeback effect
(c) Piezoresistive effect
(d) Piezoelectric effect

[PI 1010 : 1 Mark]

75. Keeping all other parameters unchanged, the tool wear in electrical discharge machining (EDM) would be less if the tool material has

(a) High thermal conductivity and high specific heat
(b) High thermal conductivity and low specific heat
(c) Low thermal conductivity and low specific heat
(d) Low thermal conductivity and high specific heat

[PI 2010 : 1 Mark]

76. Match List-I (Type of material) with List-II (Name of material) and select the correct answer using the codes given below the lists;

List-I	List-II
1. Thermoplastic	P. SiALoN
2. Thermosets	Q. Polyvinylchloride
3. Elastomers	R. Epoxy
4. Ceramics	S. Latex

	P	Q	R	S
(a)	4	1	2	3
(b)	4	2	1	3
(c)	4	3	2	1
(d)	3	2	1	4

[PI 2011 : 2 Marks]

77. While removing material from iron (atomic weight = 56, valency = 2 and density = 7.8 g/cc by electrochemical machining, a metal removal rate of 2 cc/min is desired. The current (in A) required for achieving this material removal rate is

(a) 896.07 (b) 14.93

(c) 448.03 (d) 53764.29

[PI 2011 : 2 Marks]

High Energy rate forming

78. Generally cylindrical parts produced by powder metallurgy should not have non-uniform cross section and length to diameter ratio exceeding.

[ME 1994 : 1 Mark]

79. 3-2-1 method of location in jig or fixture would collectively restrict the workpiece in 'n' degree of freedom, where the value of 'n' is

(a) 6 (b) 8

(c) 9 (d) 12

[ME 2001 : 2 Marks]

80. Fixtures are used in batch production for

(a) Clamping the job

(b) Supporting the clamping the job

(c) Supporting and clamping the job and guiding the tool

(d) Supporting, locating and clamping the job

[PI 1990 : 1 Mark]

81. The use of fixture reduces

(a) Only operation time

(b) Tooling cost

(c) Only setting time

(d) Both setting time and operation

[PI 1991 : 1 Mark]

82. In powder manufacturing process of manufacturing maximum temperature is associated with

(a) Briquetting (b) Sintering

(c) Pre-sintering (d) Blending

[PI 1991 : 1 Mark]

83. The process of shot peening increases fatigue life of steel springs mainly because it results in

(a) Hardening of surface

(b) Increased stiffness of material

(c) Compressive residual stresses

(d) Microstructural changes in material

[PI 1991 : 1 Mark]

84. Match the following

Process

A. Explosive welding

B. Brazing

C. Thermit welding

D. Manual metal arc welding

Associated Mechanism

1. Liquid state
2. Solid state
3. Solid liquid state
4. Liquid state Chemical

[PI 1991 : 2 Marks]

85. Powder metallurgical components have

(a) same density as that of cast products

(b) higher density than cast products

(c) lower density than cast products

(d) higher density than forged products

[PI 1992 : 1 Mark]

86. Match the following

A. Rail sections	1. Powder metallurgy
B. Plastic sheets	2. Rolling
C. Allen bolts	3. Cold forming
D. Porous bearing	4. Transfer moulding
	5. Calendering
	6. Injection moulding

[PI 1995 : 2 Marks]

87. Requirement in the design of jigs and fixtures

P. Heavy clamping force on the workpiece

Q. Clamping on rough surfaces of workpiece

R. Four workpieces in a line to be machined in one loading

S. Drilling and reaming of workpiece in one loading

Recommended device

1. Clamp wlth a floating pad
2. Slip renewable bush
3. Indexing bush
4. Equalizing clamp
5. Strap clamp

	P	Q	R	S
(a)	5	1	4	2
(b)	1	4	5	3
(c)	5	1	2	3
(d)	4	5	3	2

[PI 2004 : 2 Marks]

88. The surface irregularities of electrodes used in an electrochemical machining (ECM) process are 3 µm and 6 µm as shown in the figure. If the work-piece is of pure iron and 12V DC is applied between the electrodes, the largest feed rate is mm/min.

Conductivity of the electrolyte	$0.02\ ohm^{-1}\ mm^{-1}$
Over-potential voltage	1.5 V
Density of iron	$7860\ kg/m^3$
Atomic weight of iron	55.85 gm

Assume the iron to be dissolved as Fe^{+2} and the Faraday constant to be 96500 Coulomb.

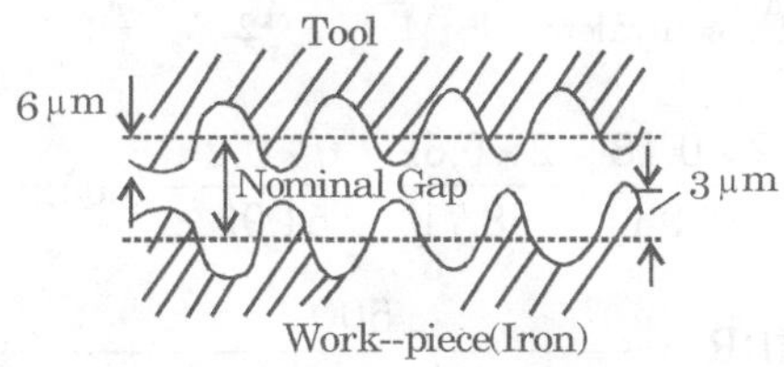

[2016 : 2 Marks, Set-2]

ANSWERS

3. (d)	**4.** (d)	**5.** (b,d)	**6.** (c)	**10.** (b)	**11.** (c)	**14.** (a)	**15.** (b,c)	**16.** (b)	**17.** (d)
19. (d)	**20.** (a)	**21.** (d)	**22.** (a)	**23.** (d)	**24.** (d)	**25.** (a)	**26.** (b)	**27.** (a)	**28.** (d)
29. (a)	**30.** (c)	**31.** (d)	**32.** (d)	**33.** (b)	**34.** (c)	**35.** (b)	**36.** (b)	**37.** (d)	**38.** (a)
39. (a)	**40.** (d)	**42.** (b)	**44.** (d)	**49.** (a)	**50.** (b)	**61.** (a)	**62.** (c)	**63.** (b)	**64.** (c)
65. (a)	**66.** (b)	**67.** (a)	**68.** (c)	**69.** (a)	**70.** (d)	**71.** (b)	**72.** (d)	**73.** (b)	**74.** (d)
75. (a)	**76.** (a)	**77.** (a)	**79.** (c)	**80.** (d)	**81.** (c)	**82.** (b)	**83.** (c)	**85.** (c)	**87.** (b)

EXPLANATIONS

1. Vibrate, magnetostrictive. Tool vibrates at very high frequency with the help of magnetostrictive transducer.

2. Cannot, can. Ultrasonic machining can be done on any material but ECM and EDM can be done for conductive material.

3. For ECM, Material removel rate (MRR) is dependent on atomic weight of material (MRR $\propto$ A.W.M.)

5. Does not allow dissolution of cathode material and have high electrical conductivity.

6. In USM, brittle fracture is responsible for removing material, so for removing material it's toughness should be lower.

7. (False) EDM is non–contact type machining while in ECM material removal takes place through ion displacement.

8. (False) MRR : ECM > EDM > USM hence ECM is more efficient.

9. (True) USM is used for machining, non-conducting brittle material like glass, ceramics and it takes place through brittle fracture.

10. For stock removal, conductive electrolyte called passivating electrolyte is used.

12. ECM – Ion displacement.
EDM – Corrosive reaction.
USM – Erosion / brittle fracture.
LBM – Melting vaporization

13. a. Aluminium brake shoe – sand casting
b. Plastic water bottles. – Blow moulding
c. Stainless steel cups – Deep drawing
d. Soft drink can – Impact extrusion.

14. Tool required for EDM should be made of conductive material like cu, graphite etc.

15. In jigs provision for guiding tool is there along with locating and clamping arrangement for workpiece while fixture are for locating and clamping workpiece only.

16. Tungsten wire has very high melting point so it is more suitable.

18. Given data:
Current (I) = 500 cump.
Density (f) = 8.28 g/cc.

$$MRR = \frac{AI}{\rho ZF}$$

where,
A = atomic weight
Z = Valcncy

F = Faraday's constant= 96500

$$\left(\frac{Z}{A}\right)_{equivalent} = \frac{Z_1 x_4}{A_1} + \frac{Z_2 x_2}{A_2} + \frac{Z_3 x_3}{A_3}$$

$$= \frac{2 \times 0.18}{58.93} + \frac{2 \times 0.62}{58.71} + \frac{6 \times 0.2}{51.99} = 5.03 \times 10^{-2}$$

$$MRR = \frac{500}{8.28 \times 5.03 \times 10^{-2} \times 96500}$$

$= 1.24 \times 10^{-2}$ CC/sec.

20. Melting and evaporation.

21. P–2, Q–3, R–5, 5–6

22. For making complex shape for higher strength alloy, electro-chemical machining is used because in it no mechanical force is exerted as well as tool left its impression over machined surfaced.

23.

		MRR
EDM	–	5×10^3 mm³/min
LBM	–	5 mm³/min
EBM	–	10 mm³ / min
USM	–	180 mm³ /min
ECM	–	10×10^3 mm³ /min

24. Material removal rate ∝ Thermal conductivity of tool

∴ Tool wear ∝ material removal rate

25. P–4, Q–3, R–1, S–2

26. Given data :

$\rho = 6000$ kg /m³ = 6 gm/cc.

MRR (Q) = 50 mm³ /s

50×10^{-3} CC/S

$$MRR = \frac{eI}{\rho ZF}$$

$$50 \times 10^{-3} = \frac{e \times 2000}{96500 \times 6}$$

$$\boxed{e = 14.475}$$

and for Iron Atomic weight

$A_{FE} = 56$, Valency : $r_{Fe} = 2$

For metal P, at weight

$(A_t)_p = 24$, Valency $r_p = 4$

Now, let % of P in alloy = x.

$$\frac{1}{e} = \frac{x}{100} \times \frac{Vp}{Ap} + \frac{(100-x)}{100} \times \frac{V_{Ft}}{A_{Fet}}$$

$$\frac{1}{14.75} = \frac{x}{100} \times \frac{4}{27} + \frac{2}{56}\left(\frac{100-x}{100}\right)$$

$x = 24.48 \approx 25$

27. We know, $R = \frac{\rho l}{A}$

where,

ρ = resistivity

= special resistance of electrolyte

= 2Ω

R = resistance of electrolyte between electrodes

l = interelectrode gap = 0.2mm

A= electrodes cross-sectional area = 20 × 20 mm²

$$\therefore R = (2 \times 10) \times \frac{0.2}{20 \times 20} = 0.01\Omega$$

$$\therefore \text{ Current, } I = \frac{V}{R} = \frac{12}{0.01} = 1200 \text{ A}$$

Faraday of electricity, $E = \frac{M}{v}$

where, E = equivalent weight

M = molecular weight

v = valency

$$\text{M.R.R (gm/sec)} = \frac{I}{F} \times \frac{M}{v}$$

$$= \frac{1200}{96540} \times \frac{55.85}{2} = 0.3471$$

28. In abrasive jet machining the variation is as shown

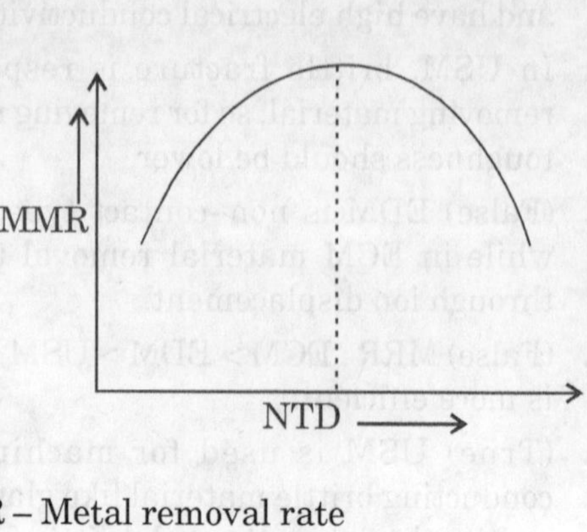

MRR – Metal removal rate

NTD – Nozzle tip distance.

29. P – 2, Q – 3, R – 4, S – 1

30. $MRR = \frac{AI}{\rho ZF}$

$$0.26 = \frac{56 \times 1000 \times 0.9}{\rho \times 2 \times 96500}$$

$\rho = 0.9956.$

$$\text{Now, Expected MRR} = \frac{48 \times 2000 \times 0.9}{0.9956 \times 3 \times 96500}$$

= 0.30 gm/s

31. EDM is most suitable process of square holes, intricate triangular holes etc.

32. Machining by laser beam is achieved through thermo – optic interaction between the beam and work material.

35. $\text{Power} = \frac{\text{work}}{\text{time}} = \frac{1}{2}\frac{CV^2}{t}$

$$C = \frac{2t \times P}{V^2}$$

$$= \frac{2 \times 25 \times 10^{-6} \times 1000}{(100)^2}$$

Where:–

V = Voltage

t = time

36. In EDM or wirecut EDM, the work and tool must be electrically conductive otherwise the current passage will not takes place.

37.

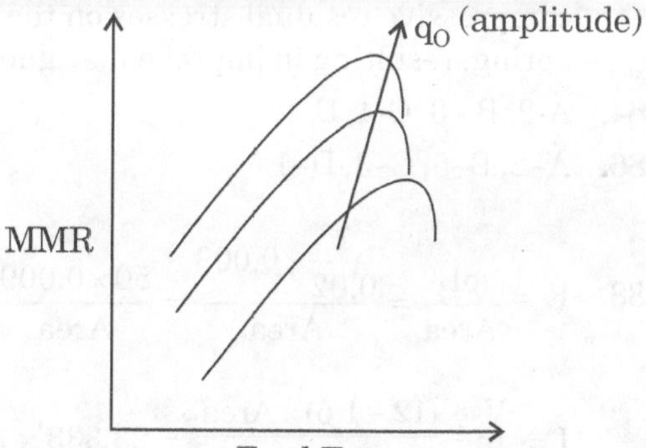

38. EBM requires vaccum chamber and therefore collision of accelerated electron is avoided with the air molecules to obtain fine machining.

40. USM uses an abrasive slurry which transfers the impact load provided by the vibrating tool to the work piece.

41. Volume of metal to be removed.

= (10 × 30) × 12 mm^3 = 3600 mm^3

Ideal energy required.

$$= \frac{3600 \quad mm^3}{3.44 \times 10^{-2} mm^3 / as.} = 104651.163 \text{ As.}$$

$$\text{Actual energy required} = \frac{104651.163}{0.9} \text{As.}$$

= 116279.07 As.

Time is 2 min = 120 s

$$\therefore \text{ Current} = \frac{116279.07}{120} = 968.992 \text{ A.}$$

43. MRR = 4 × 10^4 $IT^{-1.23}$ mm^3/min

Volume required to remove.

$$= \frac{\pi}{4}D^2 \times L = \frac{\pi}{4} \times 25^2 \times 20 = 9817.477 \text{ mm}^3$$

MRR = 4 × 10^4 × I × $T^{-1.23}$

= 4 × 10^4 × 300 × $(1600)^{-1.23}$

= 1374.40 mm^3/min

$$\text{Time required} = \frac{9817.77}{1374.4} = 7.1431 \text{ min}$$

45. A–4, B–1, C–2, D–3

Erosion	–	USM
Fusion and vapoirization	–	EDM
Vaporization and ablation	–	LBM
Ion Displacement	–	ECM

46. A – 4, B – 3, C – 1, D – 2

LBM – cutting a polyester sheet

EDM – Making a die out of a die steel block.

Chemical milling – Making a printed circuit board.

USM – Drilling a hole in glass plate

47. A– 1, B–2, C–4, D–3

48. A–1, B–2, C–3, D–4

49. USM is used for machining brittle material like glass.

51. A–4, B– 2, C–1, D–3,

52. A–4, B–1, C–2, D–3

53. A–2, B–1, C–3, D–4

EDM – Dielectric

ECM – Electrolyte

USM – Abrasive slurry

EBM – Vaccum.

54. A–4, B–3, C–2, D–1

55. A. (Electrochemical) and B. (Mechanical)

56. A–3, B–1, C–4, D–5.

57. A–1, B–2, B–3, C–4

58. A–5, B–3, C–2, D–1

59. A–7, B–1, C–6, D–5, E–8

60. A–5, B–2, C–3, D–1

61. P–2, Q–3, R–1

62. A–2, B–4, C–1, D–3

64. P–2, Q–4, R–5, S–3, T–1

65. A–5, B– 4, C–2, D–1

66. P–2, Q–3, R–4, S–1

67. Cycle time:

$$T_c = R.C.\ln\left(\frac{V_S}{V_s\text{-}V_d}\right)$$

$$= 40 \times 20 \times 10^{-6} \times \ln\left(\frac{220}{220-110}\right)$$

Tc = .55 miliseconds.

68. Total Energy consumed

$$E = \frac{1}{2}CV^2 = 0.121 \text{ J/cycle.}$$

Average Power,

$$P = \frac{0.121}{0.55 \times 10^{-3}} \text{J/s.}$$

= 220 W = 0.22 kW.

69. In ECM, high current improve material removal rate, also material removal rate increase with atomic weight of material.

70. In abrasive jet machining.MRR $\propto QD^3\sqrt{V^3}$

71. $MRR = \frac{AI}{ZF}$ kg / s.

given:

A = 63

I = 5000 A

F = 96500 couloms.

Z = 1

So,

$$MRR = \frac{63 \times 5000}{96500 \times 1} = 3.26 \text{ gm/s}$$

72. P – 4, Q – 3, R – 2, S – 1

73. Cross – Section Covered by wire

= (1 + 0.5 × 2) × 5 mm²

Feed rate = 20 mm / min.

MRR rate = 20 × 5 × 2 = 200 × mm³ / min

75. With the high thermal conductivity, heat dissipation rate is more.

76. P–4, Q–1, R–2, S–3

77. $MRR = \frac{AI}{\rho ZF}$

Given:

$$MRR = 2 \text{ cc / min} = \frac{2}{60} \text{cc / sec.}$$

A = 56,

Z = 2,

F = 96500

ρ= 7.8 g/cc

So,

$$\frac{2}{60} = \frac{56 \times I}{7.8 \times 2 \times 96500}$$

I = 869.07 A.

78. Generally cylindrical parts produced by powder metallurgy should have length to diameter ratio less than or equal to 2.5

79. Out of 12 DOF, 3–2–1method restrict 9 DOF leaving it with 3 degress of freedom for mounting and dismounting workpiece.

80. Guide for tool is not present in fixtures.

81. Use of fixtures reduces setting time for job.

82. In sintering, compacted job is heated to a high temperature for proper diffusion of metal powder.

83. Shot peening with the help of wire cuts inflicts compressive residual stresses on the surface of spring, resulting in improved fatigue life.

84. A-2, B - 3, C-4, D-4

86. A–2, B–5, C–3, D–1

88. $$R = \frac{\rho L}{\text{Area}} = \frac{\frac{1}{0.02} \times 0.009}{\text{Area}} = \frac{50 \times 0.009}{\text{Area}}$$

$$I = \frac{V}{R} = \frac{(12-1.5) \times \text{Area}}{50 \times 0.009} = 23.333 \times \text{Area}$$

L = 3 + 6 = 9 μm = 0.009

$$MRR = \frac{AI}{\rho ZF} = \frac{55.85 \times 23.333 \times \text{Area}}{7860 \times 10^{-6} \times 2 \times 96500}$$

= 0.98189 × Area

$$\frac{MRR}{\text{Area}} = 0.8590 \text{ mm/sec} = 0.8590 \times 60 \text{ mm/min}$$

= 51.542 mm/min

Unit - XI

Industrial Engineering

Syllabus

Production Planning and Control: Forecasting models, aggregate production planning, scheduling, materials requirement planning. Inventory Control: Deterministic models; safety stock inventory control systems. Operations Research: Linear programming, simplex method, transportation, assignment, network flow models, simple queuing models, PERT and CPM.

Contents

1 CHAPTER Forecasting

Analysis of Previous GATE Papers

Topic	Year/Set → Marks Distribution ↓		2019 Set 1	2019 Set 2	2018 Set 1	2018 Set 2	2017 Set 1	2017 Set 2	2016 Set 1	2016 Set 2	2016 Set 3	2015 Set 1	2015 Set 2	2015 Set 3
Forecasting	1 Mark	MCQ Type	1		1								1	
		Numerical Type												
	2 Marks	MCQ Type									1			
		Numerical Type										1		
	Total		1		1						2	2	1	

Forecasting

1. Which one of the following forecasting techniques is not suited for making forecasts for planning
Production schedules in the short range?
(a) Moving average
(b) Exponential moving average
(c) Regression analysis
(d) Delphi

[1988 : 1 Mark]

2. Which of the following is a technique for forecasting?
(a) Exponential smoothing
(b) PERT/CPM
(c) Gantt chart technique
(d) Control charts

[1989 : 2 Marks]

3. In a forecasting model, at the end of period 13, the forecasted value for period 14 is 75. Actual value in the periods 14 to 16 are constant at 100.
If the assumed simple exponential smoothing parameter is 0.5, then the MSE at the end of period 16 is
(a) 820.31 (b) 273.44
(c) 43.75 (d) 14.58

[1997 : 2 Marks]

4. The most commonly used criteria for measuring forecast error is
(a) Mean absolute deviation
(b) Mean absolute percentage error
(c) Mean standard error
(d) Mean square error

[1997 : 1 Mark]

5. In a time series forecasting model, the demand for five time periods was 10,13,15,18 and 22. A linear regression fit resulted in an equation F = 6.9 + 2.9 where F is the forecast for period f. The sum of absolute deviations for the five data is
(a) 2.2 (b) 0.2
(c) –1.2 (d) 24.3

[2000 : 2 Marks]

6. When using a simple moving average to forecast demand, one would
(a) given equal weight to all demand data
(b) assign more weight to the recent demand data
(c) include new demand data in the average without discarding the earlier data
(d) include new demand data in the average after discarding some of the earlier demand data

[2001 : 1 Mark]

7. A regression model is used to express a variable V as a function of another variable X, this implies that
(a) There is a causal relationship between Y and X
(b) A value of X may be used to estimate a value of Y
(c) Values of X exactly determine values of Y
(d) There is no causal relationship between Y and X

[2002 : 1 Mark]

8. The sale of cycles in a shop in four consecutive months are given as 70, 68, 82, 95. Exponentially smoothing average method with a smoothing factor of 0.4 is used in forecasting. The expected number of sales in the next month is
(a) 59 (b) 72
(c) 86 (d) 136

[2003 : 2 Marks]

9. For a product, the forecast and the actual sales for December 2002 were 25 and 20 respectively. If the exponential smoothing constant (α) is taken as 0.2, the forecast sales for January 2003 would be
(a) 21 (b) 23
(c) 24 (d) 27

[2004 : 1 Mark]

10. The sales of a product during the last four years were 860, 880,870 and 890 units. The forecast for the fourth year was 876 units. If the forecast for the fifth year, using simple exponential smoothing, is equal to the forecast using a three period moving average the value of the exponential smoothing constant α is
(a) 1/7 (b) 1.5
(c) 2/7 (d) 2/5

[2005 : 1 Mark]

11. In an MRP system, component demand is
(a) forecasted
(b) established by the master production schedule
(c) calculated by the MRP system from the master production schedule
(d) ignored

[2006 : 1 Mark]

12. A moving average system is used for forecasting weekly demand. $F_1(t)$ and $F_2(t)$ are sequences of forecasts with parameters m_1 and m_2, respectively, where m_1 and m_2 $(m_1 > m_2)$ denote the numbers of weeks over which the moving averages are taken. The actual demand shows a step increase from d_1 to d_2 at a certain time. Subsequently,

(a) neither $F_1(t)$ and $F_2(t)$ will catch up with the value d_2

(b) both sequences $F_1(t)$ and $F_2(t)$ will reach d_2 in the same period

(c) $F_1(t)$ will attain the value d_2 before $F_2(t)$

(d) $F_2(t)$ will attain the value d_2 before $F_1(t)$

[2008 : 2 Marks]

13. Which of the following forecasting methods takes a fraction of forecast error into account for the next period forecast?

(a) simple average method

(b) moving average method

(c) weighted moving average method

(d) exponential smoothening method

[2009 : 1 Mark]

14. The demand and forecast for February are 12000 and 10275, respectively. Using single exponential smoothening method (smoothening coefficient = 0.25), forecast for the month of March is

(a) 431 (b) 9587

(c) 10706 (d) 11000

[2010 :1 Mark]

15. In simple exponential smoothing forecasting, to give higher weightage to recent demand information, the smoothing constant must be close to

(a) – 1 (b) zero

(c) 0.5 (d) 1.0

[2013 : 1 Mark]

16. In exponential smoothening method, which one of the following is true?

(a) $0 \leq \alpha \leq 1$ and high value of α is used for stable demand

(b) $0 \leq \alpha \leq 1$ and high value of α is used for unstable demand

(c) $\alpha \geq 1$ and high value of α is used for stable demand

(d) $\alpha \leq O$ and high value of α is used for unstable demand.

[2014 : 1 Mark, Set-1]

17. The actual sales of a product in different months of a particular year are given below:

September	October	November	December	January	February
180	280	250	190	240	?

The forecast of the sales, using the 4 months moving average method, for the month of February is_____.

[2014 : 1 Mark, Set-3]

18. Sales data of a product is given in the following table:

Month	January	February	March	April	May
Number of units sold	10	11	16	19	25

Regarding forecast for the month of June, which one of the following statements is TRUE?

(a) Moving average will forecast a higher value compared to regression

(b) Higher the value of order N, the greater will be the forecast value by moving average.

(c) Exponential smoothing will forecast a higher value compared to regression.

(d) Regression will forecast a higher value compared to moving average.

[2015 : 1 Mark, Set-2]

19. For a canteen, the actual demand for disposable cups was 500 units in January and 600 units in February. The forecast for the month of January was 400 units. The forecast for the month of March considering smoothing coefficient as 0.75 is.

[2015 : 2 Marks, Set-1]

20. The demand for a two-wheeler was 900 units and 1030 units in April 2015 and May 2015, respectively.

The forecast for the month of April 2015 was 850 units. Considering a smoothing constant of 0.6, the forecast for the month of June 2015 is

(a) 850 units (b) 927 units

(c) 965 units (d) 970 units

[2016 : 2 Marks, Set-3]

21. The time series forecasting method that gives equal weightage to each of the most recent observation is

(a) Moving average method
(b) Exponential smoothing with linear trend
(c) Triple Exponential smoothing
(d) Kalman Filter

[2018 : 1 Marks, Set-1]

22. The table presents the demand of a product. By simple three-months moving average method, the demand-forecast of the product for the month of September is

Month	Demand
January	450
February	440
March	460
April	510
May	520
June	495
July	475
August	560

(a) 510 (b) 530
(c) 536.67 (d) 490

[2019 : 1 Mark, Set-1]

ANSWERS

1. (d)	**2.** (a)	**3.** (b)	**4.** (a)	**5.** (a)	**6.** (d)	**7.** (a)	**8.** (c)	**9.** (c)	**10.** (c)
11. (c)	**12.** (d)	**13.** (d)	**14.** (c)	**15.** (d)	**16.** (b)	**18.** (d)	**20.** (d)	**21.** (a)	**22.** (a)

EXPLANATIONS

3. $F_{14} = 75, D_{14} = 100$

$F_{15} = F_{14} + \alpha (D_{14} - F_{14})$

$= 75 + 0.5\ (100.75)$

$F_{15} = 87.5$

$D_{15} = 100$

$F_{16} = 87.5 + 0.5\ (100 - 87.5)$

$F_{16} = 93.75$

$$MSE = \frac{\sum e_i^2}{n}$$

Where $e_i = D_i - F_i$

$$MSE = \frac{(100-75)^2 + (100-87.5)^2 + (100-93.75)^2}{3}$$

$MSE = 273.44$

5. $F = 6.9 + 2.9\,t$

$F_1 = 6.9 + 2.9 \times 1 \Rightarrow 9.8$

$F_2 = 6.9 + 2.9 \times 2 \Rightarrow 12.7$

$F_3 = 6.9 + 2.9 \times 3 \Rightarrow 15.6$

$F_4 = 6.9 + 2.9 \times 4 \Rightarrow 18.5$

$F_5 = 6.9 + 2.9 \times 5 = 21.4$

$D_1 = 10\ D_2 = 13\ 0_3 = 15$

$D_4 = 16\ D_5 = 22$

$SAD = \sum |e_i|$

$SAD = |10 - 9.8| + |13 - 12.7| + |15 - 15.6| + |18 - 18.5| + |22 - 21.4|$

Absolute Deviation = 2.2

8. $F_{t+1} = D_t + \alpha(1-\alpha)\,D_{t-1} + \alpha(1-\alpha)^2\,D_{t-2}$

$F_{t+1} = 0.4\ (95) + 0.4\ (1 - 0.4)^2 + \alpha(1-\alpha)^3\,D_{t-3}$

$+ (0.4)\,(1 - 0.4)^2 \times 68 + (0.4)\,(1 - 0.4)^3 \times 70$

$F_{t-1} = 73.52$ units

9. $F_{t+1} = F_t + \alpha\,(D_t - F_t)$

$F_{Jan\,(2003)} = F_{Dec} + \alpha\,(D_{dec} - F_{dec})$

$D_{dec} \Rightarrow 20$ units

$F_{dec} = 25$ units

$F_{Jan} = 25 + 0.2\ (20 - 25)$

$= 25 + 0.2\ (-5) = 25 - 1$

$F_{jan} \Rightarrow 24$ units

10. $F_4 = 876$ units

S.No	Demand	Forecast
1	860	
2	860	
3	870	
4	890	
5	–	880

$F_5 = F_3$

$$F_3 = \frac{890 + 870 + 880}{2}$$

$F_5 = F_4 + \alpha (D_4 - F_4)$

$880 = 876 + \alpha (890 - 876)$

$$\left[\alpha = \frac{2}{7}\right]$$

12. Given that, at certain the demand increased from d_1 to d_2

The weightage of the latest demand should be therefeace $f_2(t)$ become $M_2 > M_2$.

14. $F_{march} = F_{Feb} + \alpha (F_{Feb} - F_{Feb})$

$F_{Feb} = 10275$

$D_{Feb} = 12000$

$\alpha = 0.25$

$F_{March} = 10275 + 0.25 (12000 - 10275) = 10706$

15. We know that

$f_t = f_{t-1} + \alpha(D_t - f_{t-1})$

$= f_{t-1} (1 - \alpha) + D_t$

For, $f_t = D_t$

$1 - \alpha = 0$

$\Rightarrow \alpha = 1$

Conclusion:

It is showing the limit of responsiveness

16. $0 \le \alpha \le 2$

high value of 'α' means more weightage for immediate forecast.

Less value of 'α' means relatively less weightage for immediate forecast, or almost equal weightage for all previous forecast.

Hence high value of forecast is only chosen when nature of demand is not reliable rather unstable.

17.

Month	Demand
Sept	180
Oct	280
Nov	250
Dec	190
Jan	240
Feb	

$$F_{Feb} = \frac{F_{Jan} + F_{Dec} + F_{Nov} + F_{Oct}}{4}$$

$$F_{Feb} = \frac{240 + 190 + 250 + 280}{4}$$

$F_{Feb} = 240$

19. $\alpha = 0.75$

$D_{Jan} = 500$ $D_{Feb} = 600$

$F_{Jan} = 400$ units

$F_{Feb} = F_{Jan} + \alpha (D_{Jan} - F_{Jan})$

$F_{Feb} = 400 + 0.75 (500–400)$

$F_{Feb} = 475$ units

$F_{March} = F_{Feb} + \alpha (D_{Feb} - F_{Feb})$

$F_{March} = 475 + 0.75(600 - 475)$

$= 568.75$

$F_{March} = 569$

i.e. 569 cups

20.

	Demand	Forecast
April	800	850
May	1030	
June		

$x = 0.6$

$F_{May} = F_{April} - (D_{April} - F_{April})$

$\Rightarrow 850 - 0.6 (800 - 850)$

$F_{May} = 850$

$F_{June} = F_{May} - (D_{May} - F_{May})$

$= 850 - 0.6 (1030 - 850)$

$\Rightarrow 850 - 90$

$\Rightarrow$ 970 units

21. (a) Simple moving average method gives equal weight to each of the most recent observation

(b) Weighted moving average method gives more weight to the recent values.

(c) Triple exponential smoothing is a rule of thumb technique for smoothing time series data using the exponential window function.

(d) Kalman filter is also used for time series data using adaptive models.

22. Three month moving average.

Forecast of the product for the month of September is given by,

$$F = \frac{495 + 475 + 560}{3} = \frac{1530}{3} = 510$$

2

CHAPTER

Inventory Control

Analysis of Previous GATE Papers

Topic	Marks Distribution		2019 Set 1	2019 Set 2	2018 Set 1	2018 Set 2	2017 Set 1	2017 Set 2	2016 Set 1	2016 Set 2	2016 Set 3	2015 Set 1	2015 Set 2	2015 Set 3
Basics of Inventory	1 Mark	MCQ Type												
		Numerical Type												
	2 Marks	MCQ Type												
		Numerical Type												
	Total													
EOQ Model	1 Mark	MCQ Type				1								
		Numerical Type											1	
	2 Marks	MCQ Type												
		Numerical Type	1						1	1			1	1
	Total		2			1			2	2			3	2

Basics of Inventory

1. Setup costs do not include

(a) Labour cost of setting up machines

(b) Ordering cost or raw material

(c) Maintenance cost of the machines

(d) Cost of processing the work piece

[1997 : 1 Mark]

2. One of the following statements about PRS (Periodic Reordering System) is not true identify.

(a) PRS requires continuous monitoring of inventory levels

(b) PRS is useful in control of perishable items

(c) PRS provides basis for adjustments to account for variations in demand

(d) In PRS, inventory holding costs are higher than in Fixed Reorder Quantity System

[1998 : 1 Mark]

EOQ Model

3. In an ideal inventory control system, the economic lot size for a part is 1000. If the annual demand for the part is doubled, the new economic lot size required will be

(a) 500 (b) 2000

(c) $1000/\sqrt{2}$ (d) $1000\sqrt{2}$

[1989 : 2 Marks]

4. When the annual demand of a product is 24000 units, the EOQ (Economic Order Quantity) is 2000 units. If the annual demand is 48000 units the most appropriate EOQ will be

(a) 1000 units (b) 2000 units

(c) 2800 units (d) 4000 units

[1991 : 2 Marks]

5. If the demand for an item is doubled and the ordering cost halved, the economic order quantity

(a) remains unchanged

(b) increases by factor of $\sqrt{2}$

(c) is doubled

(d) is halved

[1995 : 1 Mark]

6. In inventory planning, extra inventory is unnecessarily carried to the end of the planning period when using one of the following lot size decision policies:

(a) Lot for lot production

(b) Economic Order Quantity (EOQ) lot size

(c) Period Order Quantity (POQ) lot size

(d) Part period total cost balancing

[1998 : 1 Mark]

7. In computing Wilson's economic lot size for an item, by mistake the demand rate estimate used was 40% higher than the true demand rate. Due to this error in the lot size computation, the total cost of setup plus inventory holding per unit time. Would rise above the true optimum by approximately

(a) 1.4% (b) 3.3%

(c) 18.3% (d) 8.7%

[1999 : 2 Marks]

8. Market demand for springs is 8,00,000 per annum.

A company purchases these spring in lots and sells them. The cost of making a purchase order is Rs. 1,200. The cost of Storage of springs is Rs. 120 per stored piece per annum. The economic order quantity is

(a) 400 (b) 2,828

(c) 4,000 (d) 8,000

[2003 : 2 Marks]

9. There are two products P and Q with the following characteristics

Product	Demand (units)	Order cost (Rs / order)	Holding Cost (Rs.unit/year)
P	100	50	4
Q	400	50	1

The Economic Order Quantity (EOQ) of products P and Q will be in the ratio

(a) 1 : 1 (b) 1: 2

(c) 1 : 4 (d) 1 : 8

[2004 : 1 Mark]

10. A company has an annual demand of 1000 units, ordering cost of Rs. 100/ order and carrying cost of Rs. 100/unit-year. If the stock-out costs are estimated to be nearly Rs. 400 each time the company runs out-of-stock, the safety stock justified by the carrying cost will be

(a) 4 (b) 20

(c) 40 (d) 100

[2004 : 2 Marks]

11. Consider the following data for an item. Annual demand : 2500 units per year Ordering cost: Rs. 100 per order, Inventory holding rate: 25% of unit price

Price quoted by a supplier

Order quantity (units)	Unit price (Rs.)
<500	10
>500	9

The optimum order quantity (in units) is

(a) 447 (b) 471

(c) 500 (d) $\geq$600

[2006 : 2 Marks]

12. A stockist wishes to optimize the number of perishable items he needs to stock in any month in his store. The demand distribution for this perishable item is

Demand (in units)	2	3	4	5
Probability	0.10	0.35	0.35	0.20

The stockist pays Rs. 70 for each item and he sells each at Rs. 90. If the stock is left unsold in any month, he can sell the item at Rs. 50 each. There is no penalty for unfulfilled demand. To maximize the expected profit, the optimal stock level is

(a) 5 units (b) 4 units

(c) 3 units (d) 2 units

[2006 : 2 Marks]

13. The maximum level of inventory of an item is 100 and it is achieved with infinite replenishment rate. The inventory becomes zero over one and half month due to consumption at a uniform rate. This cycle continues through out the year. Ordering cost is Rs. 100 per order and inventory carrying cost is Rs. 10 per item per month. Annual cost (in Rs.) of the plan, neglecting material cost, is

(a) 800 (b) 2800

(c) 4800 (d) 6800

[2007 : 2 Marks]

14. In machine shop, pins of 15 mm diameter are produced at a rate of 1000 per month and the same is consumed at a rate of 500 per month. The production and consumption continue simultaneously till the maximum inventory is reached. Then inventory is allowed to reduce to zero due to consumption. The lot size of production is 1000. If backlog is not allowed, the maximum inventory level is

(a) 400 (b) 500

(c) 600 (d) 700

[2007 : 2 Marks]

15. The net requirements of an item over 5 consecutive weeks are 50-0-15-20-20. The inventory carrying cost and ordering cost are Rs. 1 per item per week and Rs. 100 per order respectively. Starting inventory is zero. Use "Least Unit Cost Technique" for developing the plan. The cost of the plan (in Rs.) is

(a) 200 (b) 250

(c) 225 (d) 260

[2007 : 2 Marks]

16. A company uses 2555 units of an item annually. Delivery lead time is 8 days. The reorder point (in number of units) to achieve optimum inventory is

(a) 7 (b) 8

(c) 56 (d) 60

[2009 : 2 Marks]

17. Annual demand for window frames is 10000. Each frame costs Rs. 200 and ordering cost is Rs. 300 per order. Inventory holding cost is Rs. 40 per frame per year. The supplier is willing to offer 2% discount if the order quantity is 1000 or more, and 4% if order quantity is 2000 or more. If the total cost is to be minimized, the retailer should

(a) order 200 frames every time

(b) accept 2% discount

(c) accept 4% discount

(d) order Economic Order Quantity

[2010 : 2 Marks]

18. A component can be produced by any of the four processes I, II, III and IV. The fixed cost and the variable cost for each of the processes are listed below. The most economical process for producing a batch of 100 pieces is

Process	Fixed cost (in Rs.)	Variable cost per piece (in Rs.)
I	20	3
II	50	1
III	40	2
IV	10	4

(a) I (b) II

(c) III (d) IV

[2014 : 1 Mark, Set-2]

19. Consider the following data with reference to elementary deterministic economic order quantity model

Annual demand of an item	100000
Unit price of the item (in Rs.)	10
Inventory carrying cost per unit per year (in Rs.)	1.5
Unit order cost (in Rs.)	30

The total number of economic orders per year to meet the annual demand is ______.

[2014 : 2 Marks, Set-2]

20. A manufacturer can produce 12000 bearings per day. The manufacturer received an order of 8000 bearings per day from a customer. The cost of holding a bearing in stock is Rs. 0.20 per month. Setup cost per production run is Rs. 500. Assuming 300 working days in a year, the frequency of production run should be

(a) 4.5 days (b) 4.5 months

(c) 6.8 days (d) 6.8 months

[2014 : 2 Marks, Set-3]

21. Annual demand of a product is 50000 units and the ordering cost is Rs. 7000 per order. Considering the basic economic order quantity model, the economic order quantity is 10000 units. When the annual inventory cost is minimized, the annual inventory holding cost (in Rs.) is_____.

[2015 : 1 Mark, Set-2]

22. A manufacturer has the following data regarding a product:

Fixed cost per month = Rs. 50000

Variable cost per unit = Rs. 200

Selling price per unit = Rs. 300

Production capacity = 1500 units per month .If the production is carried out at 80% of the rated capacity, then the monthly profit (in Rs.) is___.

[2015 : 2 Marks, Set-2]

23. The annual requirement of rivets at a ship manufacturing company is 2000 unit. The rivets are supplied in units of 1 kg costing Rs. 25 each. The costs Rs. 100 to place an order and the annual cost of carrying one unit is 9% of its purchase cost, the cycle length of the order (in days) will be ______.

[2015 : 2 Marks, Set-3]

24. The annual demand for an item is 10,000 units. The unit cost is Rs. 100 and inventory carrying charges are 14.4% of the unit cost per annum. The cost of one procurement is Rs. 2000. The time between two consecutive orders to meet the above demand is ______, month(s).

[2016 : 2 Marks, Set-1]

25. A food processing company uses 25,000 kg of corn flour every year. The quantity-discount price of corn flour is provided in the table below.

Quantity (kg)	Unit Price(Rs/kg)
1-749	70
750-1499	65
1500 and above	60

The order processing charges are Rs. 500/ order. The handling plus carry-over charge on an annual basis is 20% of the purchase price of the corn flour per kg. The optimal order quantity (in kg) is ______.

[2016 : 2 Marks, Set-2]

26. A local tyre distributor expects to sell approximately 9600 steel belted radial tyres next year. Annual carrying cost is Rs. 16 per tyre and ordering cost is Rs. 75. The economic order quantity of the tyres is

(a) 64 (b) 212

(c) 300 (d) 1200

[2018 : 1 Marks, Set-2]

27. The annual demand of valves per year in a company is 10,000 units. The current order quantity is 400 valves per order. The holding cost is Rs. 24 per valve per year and the ordering cost is Rs. 400 per order. If the current order quantity is changed to Economic Order Quantity, then the saving in the total cost of inventory per year will be Rs. _____(round off to two decimal places).

[2019 : 2 Marks, Set-2]

ANSWERS

1. (d)	**2.** (c)	**3.** (d)	**4.** (c)	**5.** (a)	**6.** (b)	**7.** (c)	**8.** (c)	**9.** (c)	**10.** (c)
11. (c)	**12.** (c)	**13.** (c)	**14.** (b)	**15.** (b)	**16.** (c)	**17.** (c)	**18.** (b)	**20.** (c)	**26.** (c)

EXPLANATIONS

3. $EOQ \propto \sqrt{D}$

$EOQ = 1000$

$$\frac{EOQ^1}{\sqrt{2D}} = \frac{EOQ}{\sqrt{D}}$$

$$EOQ^1 = 1000 \times \sqrt{\frac{2D}{D}}$$

$\Rightarrow 1000\sqrt{2}$

4. $EOQ \propto \sqrt{D}$

$$\frac{EOQ_2}{\sqrt{D_2}} = \frac{EOQ_1}{\sqrt{D_1}}$$

$$EOQ_2 = EOQ_1\sqrt{\frac{D_2}{D_1}} = 2000\sqrt{\frac{48000}{24000}}$$

$EOQ_2 \Rightarrow 2828$

5. $EOQ \propto \sqrt{DC_o}$

$$\frac{EOQ'}{\sqrt{D'C_o'}} = \frac{EOQ}{\sqrt{DCo}}$$

$D' = 2\,D$

$$C_o' = \frac{1}{2}Co$$

$$\therefore \frac{EOQ'}{\sqrt{2D \times \frac{1}{2}Co}} = \frac{EOQ}{\sqrt{DCo}}$$

$\Rightarrow EOQ' = EOQ$

7. Annual cost of setup & inventory holding for economic lot size

$TAC_1 = \sqrt{2DC_o C_h}$ → Actual Cost

$TAC_2 = \sqrt{2(1.4D)C_o C_h}$ → Computed Cost

$$\text{Erros} = \frac{(TAC)_2 - (TAC)_1}{(TAC)_1} \times 100$$

$$\Rightarrow \left(\frac{1.183 - 1}{1}\right) \times 100 = 18.3\%$$

8. $D = 8{,}00{,}000$ units/year

$C_o = 1200$/Order

$C_h = 120$/unit / year

$$Q^* = \sqrt{\frac{2DC_o}{C_h}}$$

$$\Rightarrow \sqrt{\frac{2 \times 800000 \times 1200}{120}}$$

$Q^* = 4000$ units

9.
$$\frac{EOQ_P}{EOQ_Q} = \frac{\sqrt{\left(\frac{2DC_o}{C_h}\right)_P}}{\sqrt{\left(\frac{2DC_o}{C_h}\right)_Q}} = \frac{\sqrt{\left(\frac{2 \times 100 \times 50}{4}\right)}}{\sqrt{\left(\frac{2 \times 400 \times 50}{1}\right)}}$$

$$= \frac{EOQ_P}{EOQ_Q} = \frac{1}{1}$$

10. $D = 1000$ Units

$C_o = 100$ Rs/order

$C_h = 100$ / unit/order

$C_s = 400$ each time

$$Q^* = \sqrt{\frac{2DC_O}{C_h}\left(\frac{C_h + C_s}{C_s}\right)}$$

$$Q^* = \sqrt{\frac{2(1000)(100)}{100}\left(\frac{100 + 400}{400}\right)}$$

$\Rightarrow 50$

$$S^* = Q^*\left(\frac{C_h}{C_S + C_h}\right) = 50\left(\frac{100}{500}\right)$$

$S^* \Rightarrow 10$

Softly Stock Justified by the carrying cost = Maximum Inventory level

$M = Q^* - S^* = 40$

11. D = 2500 units/years

C_O = Rs. 100/order

C_h = (0.25) Unit + Price

$$Q^* = \sqrt{\frac{2DCo}{C_h}} = \sqrt{\frac{2 \times 2500 \times 100}{0.25 \times 10}}$$

$Q^* \Rightarrow 447.2 \pm$

$$TAC(O^*) = D \times C + \sqrt{2DC_OC_h}$$

$$\Rightarrow 2500 \times 10 + \sqrt{2 \times 2500 \times 10 \times 2.5}$$

$$\Rightarrow 26118.033$$

$$\text{TAC at 500} = D \times C + \frac{D}{Q}Co + \frac{Q}{2}C_h$$

$$\Rightarrow 2500 \times 9 + \frac{2500}{500} \times 100 = \frac{500}{2}(23.25 \times 9)$$

$$\Rightarrow 23562.5$$

Since Total cost minimum at Q = 500 units

∴ the optimal order quality is Q = 500 units

13. Maximum Inventory = 100

Inventory Cycle = 1.5 months

Number of cycle required per year

$$= \frac{12}{1.5} = 8\,\text{cycle}$$

$C_o = 8 \times 100 = 800$/year

$C_h = 8 \times 10 = 80$/year/unit

$$TAC = \frac{Q}{2}C_h + \frac{D}{Q}Co$$

$$\Rightarrow \frac{100}{2}(80) + \frac{100}{100}(800) \Rightarrow 4800$$

14.

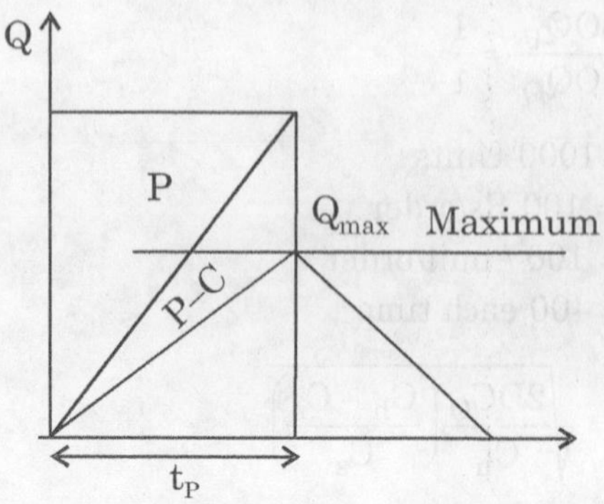

$$Q_{max} = t_p(P - C) = \frac{Q}{P}(P - C)$$

$$\Rightarrow Q\left(1 - \frac{C}{P}\right)$$

$$Q_{max} = 1000\left(1 - \frac{500}{1000}\right)$$

$$\Rightarrow 1000\left(\frac{1}{2}\right) = 500$$

So the maximum inventory level is 500

15. C_h = 1 per unit per week

C_O = 100/order

16. Reorder level = $L_T \times d$ + SS

↑ If guessed

Q ⇒ 2555 units

L_s = 8 days

Cycle Time = 365 days

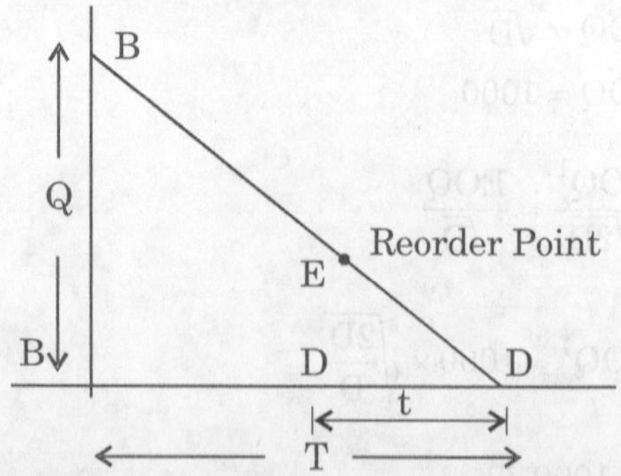

$$\frac{Q}{T} = \frac{n}{t}$$

$$n = \frac{Q}{T}t$$

$$t = \frac{2555 \times 8}{365} = 56 \text{ units}$$

17. D = 10000 units/year

C = 200/unit

C_o = Rs 300/order

C_h = 40 Rs/unit-year

$$Q^* = \sqrt{\frac{2DC_o}{C_h}} = \sqrt{\frac{2 \times 10000 \times 300}{40}}$$

Q^* = 387.29 units

$Q^* \simeq 388$ units

$$TAC(Q^*) = D \times C + \sqrt{2DC_oC_h}$$

$$\Rightarrow 10000 \times 200 + \sqrt{2 \times 1000 \times 300 \times 40}$$

$$\Rightarrow \text{Rs } 2015492$$

$$TAC(Q = 1000) = D \times C + \frac{D}{Q}C_o + \frac{Q}{2}C_h$$

$$\Rightarrow 10000 \times 196 + \frac{10000}{1000} \times 300 + \frac{1000}{2} \times 40$$

$$\Rightarrow \text{Rs.}1983000$$

$$TAC(Q = 2000) = 10000 \times 192 + \frac{10000}{100} \times 300 + \frac{2000}{2} \times 40$$

$$\Rightarrow \text{Rs.}196150$$

Since total cost is minimum at Q = 2000

Therefore 4% discount can be accepted.

18. TC = Fixed Cost + Variable cost

$TC_I = 20 + 100 \times 3 = Rs.320$

$TC_{II} = 50 + 100 \times 1 = Rs.150$

$TC_{III} = 40 + 100 \times 2 = Rs.240$

$TC_{IV} = 10 + 100 \times 4 = Rs.410$

Since total cost for the Process II is minimum

So II is cost for economical point of view

19. D = 100000

C = 10 Rs

$C_h = 1.5$

$C_o = 30$/order

$$EOQ(Q^*) = \sqrt{\frac{2 \times D \times C_o}{C_h}}$$

$$= \sqrt{\frac{2 \times 100000 \times 30}{1.5}} = 2000 \text{ units}$$

No of orders $N^* = \frac{D}{Q^*}$

$$N^* = \frac{100000}{2000}$$

$\Rightarrow 50$

20. D = 8000 × 300 bearings/year

P = 12000 bearings/day

C = 8000 bearings/day

$C_h = 0.12 \times 12$/year

$C_o = 500$/Production

$$Q^* = \sqrt{\frac{2DCo}{C_h}\left(\frac{P}{P-C}\right)}$$

$$\Rightarrow \sqrt{\frac{2 \times 240000 \times 500}{0.20 \times 12} \times \left(\frac{12000}{12000-8000}\right)}$$

$\Rightarrow$ 54772.526

$$T = \frac{Q^*}{C}$$

$T \Rightarrow 6.84$ days

21. $Q = \sqrt{\frac{2DC_O}{C_h}}$

$$Q^2 = \frac{2DC_O}{C_h}$$

$$C_h = \frac{2 \times 50000 \times 7000}{(10000)^2}$$

$C_h = 7$

Annual Carrying cost

$$= \frac{d}{2} \times C_h$$

$$\Rightarrow \frac{10000}{2} \times 7 = 35000$$

22. Fixed Cost (Fl) = Rs 50000

Variable Cost = Rs 200

Selling price/unit = Rs 300

Production Capacity = 1500 unit per month

At 80% Rated capacity the production is

= 0.80 × 1500 = 1200 unit/month

Total cost = FC + Variable cost

= 50000 + 1200 × 200 ⇒ 290000

Profit = Sales – Total Cost

= 3,60,000 – 2,90,000

⇒ Rs 70,000

23. D = 2000 unit

C = Rs 25

C_o = Rs 100

$$C_h = \frac{9}{100} \times 25 = 2.25$$

$$Q^* = \sqrt{\frac{2DC_o}{C_h}}$$

$$\Rightarrow \sqrt{\frac{2 \times 2000 \times 100}{2.25}}$$

= 421.637

No of orders per year

$N = \frac{P}{Q^*}$

$= \frac{2000}{421.637}$

$\Rightarrow 4.7434$

Cycle length of order

$\Rightarrow \frac{\text{No of days in a Year}}{N}$

$\Rightarrow \frac{365}{4.474}$

$\Rightarrow 76.9457$ days

24. D = 10000

C = Rs 100

$C_h = 14.4 \times C$

$\Rightarrow \frac{14.4}{100} \times 100$

$\Rightarrow 14.4$

C_o = Rs 2000 / order

$\text{EOQ } (Q^*) = \sqrt{\frac{2DC_O}{C_h}}$

$\Rightarrow \sqrt{\frac{2 \times 1000 \times 2000}{14.4}} = 1666.67$

Time between two consecutive order

$T = \frac{Q^*}{D} = \frac{1}{6}$ years

$\Rightarrow \frac{12}{6} = 2$ weeks

25. D = 25000 kg

C_O = Rs 500/order

C_h = 20% of unit purchase price

$Q^* = \sqrt{\frac{2DCo}{C_h}}$

$\Rightarrow \sqrt{\frac{2 \times 25000 \times 500}{0.20 \times 60}}$

$Q^* = 1443.376$

Not a feasible solution, since price of Rs 60 is only for purchases of 1500 & above.

$Q^* \sqrt{\frac{2DCo}{Ch_2}} = \sqrt{\frac{2 \times 25000 \times 500}{0.20 \times 65}}$

$\Rightarrow 13.86.75$

$TA(Q^*) = DC + \sqrt{2DCoCh}$

$\Rightarrow 65 \times 2500 + \sqrt{2 \times 2500 \times 500 \times 0.20 \times 65}$

$\Rightarrow 163027.76$

$TAC(Q = 1500) = DC + \frac{D}{Q^*}Co + \frac{Q}{2}Ch$

$\Rightarrow 2500 \times 60 + \frac{2500}{1500} \times 500 + \frac{1500}{2} \times 0.20 \times 60$

$\Rightarrow 159833.33$

So minimum cost is for 1500 quantity

Hence EOQ = 1500

26. D = 9600

$C_O = 75$

C_h = 16 Rs/type

$Q = \sqrt{\frac{2DC_O}{C_h}} \Rightarrow 300$

27. D = 10,000 units

Q = 400 units/order

C_h = 24 Rs/value/year

C_o = 400 Rs/order

$TIC = \frac{D}{Q}C_o + \frac{Q}{2}C_h$

$= \frac{10,000}{400} \times 400 + \frac{400}{2} \times 24$

= 10000 + 4800 = Rs. 14800

$(TIC)^* = \sqrt{2DC_oC_h}$

$= \sqrt{2} \times 10000 \times 400 \times 24$ = Rs. 13856.41

Saving = 14,800 – 13856.41

Saving = Rs. 943.59

3 CHAPTER

Linear Programming

Analysis of Previous GATE Papers

Topic	Year/Set → Marks Distribution ↓		2019 Set 1	2019 Set 2	2018 Set 1	2018 Set 2	2017 Set 1	2017 Set 2	2016 Set 1	2016 Set 2	2016 Set 3	2015 Set 1	2015 Set 2	2015 Set 3
Graphical Method	1 Mark	MCQ Type												
		Numerical Type												
	2 Marks	MCQ Type				1			1		1			
		Numerical Type			1		1							
	Total				2	2	2		2		2			
Simplex Method	1 Mark	MCQ Type												
		Numerical Type												
	2 Marks	MCQ Type												
		Numerical Type												
	Total													

Graphical Method

1. In an assembly line for assembling toys, five workers are assigned tasks which take times of 10, 8, 6, 9 and 10 minutes respectively. The balance delay for line is

(a) 43.3% (b) 14.8%
(c) 14.0% (d) 16.3%

[1996 : 2 Marks]

2. If at the optimum in a linear programming problem, a dual variable corresponding to a particular primal constraint is zero, then it means that

(a) Right hand side of the primal constraint can altered without affecting the optimum solution
(b) Changing the right hand side of the primal constraint will disturb the optimum solution
(c) The objectives function is unbounded
(d) The problem is degenerate

[1996 : 1 Mark]

3. A manufacturer produces two types of products, 1 and 2, at production levels of x_1 and x_2 respectively. The profit is given is $2x_1 + 5x_2$. The production constraints are:

$x_1 + 3x_2 < 40$

$3x_1 + x_2 < 24$

$x_1 + x_2 < 10$

$x_1 > 0$

$x_2 > 0$

The maximum profit which can meet the constraints is

(a) 29 (b) 38
(c) 44 (d) 75

[2003 : 2 Marks]

4. A company produces two types of toys : P and Q. Production time of Q is twice that of P and the company has a maximum of 2000 time units per day. The supply of raw material is just sufficient to produce 1500 toys (of any type) per day. Toy type Q requires an electric switch which is available @ 600 pieces per day only. The company makes a profit of Rs. 3 and Rs. 5 on type P and Q respectively. For maximization of profits, the daily production quantities of P and Q toys should respectively be

(a) 1000, 500 (b) 800, 600
(c) 500, 1000 (d) 1000, 1000

[2004 : 2 Marks]

5. A component can be produced by any of the four processes, I, II, III and IV. Process I has fixed cost of Rs. 20 and variable cost of Rs. 3 per piece. Process II has a fixed cost of Rs. 50 and variable cost of Rs. 1 per piece. Process III has a fixed cost of Rs. 40.00 and variable cost of Rs. 2 per piece. Process IV has fixed cost of Rs. 10 and Variable cost Rs. 4 per piece. If company wishes to produce 100 pieces of the component, from economic point of view it should choose

(a) Process I (b) Process II
(c) Process III (d) Process IV

[2005 : 2 Marks]

6. A company has two factories S_1, S_2 and two warehouses D_1, D_2. The supplies from S_1 and S_2 are 50 and 40 units respectively. Warehouse D_1 requires a minimum of 20 units and a maximum of 40 units. Warehouse D_2 requires a minimum of 20 units and, over and above, it can take as much as can be supplied. A balanced transportation problem is to be formulated for the above situation. The number of supply points, the number of demand points, and the total supply (or total demand) in the balanced transportation problem respectively are

(a) 2, 4, 90 (b) 2, 4, 110
(c) 3, 4, 90 (d) 3, 4,110

[2005 : 2 Marks]

Linked Answer Questions 7 and 8:

Consider a linear programming problem with two variable and two constraints. The objective function is maximize $x_1 + x_2$. The corner points of the feasible region are (0, 0), (0,2) (2, 0) and (4/3, 4/3)

7. If an additional constraint $X_1 + X_2 \leq 5$ is added, the optimal solution is

(a) (5/3, 5/3) (b) (4/3,4/3)
(c) (5/2, 5/2) (d) (5, 0)

[2005 : 2 Marks]

8. Let Y_1 and Y_2 be the decision variables of the dual and v_1 and v_2 be the slack variables of the dual of the given linear programming problem. The optimum dual variables are

(a) Y_1 and Y_2 (b) Y_1 and v_1
(c) Y_1 and Y_2 (d) v_1 and v_2

[2005 : 2 Marks]

9. A firm is required to procure three items (P, Q, and R). The prices quoted for these items (in Rs.) by suppliers S_1, S_2 and S_3 are given in table. The management policy requires that each item has to be supplied by only one supplier and one supplier supply only one item. The minimum total cost (in Rs.) of procurement to the firm is

Item	Supplier		
	S_1	S_2	S_3
P	110	120	130
Q	115	140	140
R	125	145	165

(a) 350 (b) 360
(b) 385 (d) 395

[2006 : 2 Marks]

10. Consider the following Linear Programming Problem (LPP):

Maximize $z = 3x_1 + 2x_2$,
Subject to $x_1 \leq 4$
$x_2 \leq = 6$
$3x_1 + 2x_2 < 18$
$x_1 \geq 0, x_2 \geq 0$

(a) The LPP has a unique optimal solution
(b) The LPP is infeasible
(c) The LPP is unbounded
(d) The LPP has multiple optimal solution

[2009 : 2 Marks]

Common Data for Questions 11 and 12:

One unit of product P_1 requires 3 kg of resource R_1 and 1 kg resource R_2. One unit of product P_2 requires 2 kg of resource R_1 and 2 kg of resource R_2. The profits per unit by selling product P_1 and P_2 and Rs. 12000 and Rs 3000 respectively. The manufacturer has 90 kg of resource R_1, and 100 kg of resource R_2.

11. The unit worth of resource R_2, i.e. dual price of resource R_2 in Rs per kg is

(a) 0 (b) 1360
(c) 1500 (d) 2000

[2011 : 2 Marks]

12. The manufacturer can make a maximum profit of Rs.

(a) 60000 (b) 135000
(c) 150000 (d) 200000

[2011 : 2 Marks]

13. A linear programming problem is shown below:

Maximise $3x + 7y$
Subjeot to $3x + 7y \leq 10$
$4x + 6y \leq 8$
$x, y \geq 0$

(a) an unbounded objective function
(b) exactly one optimal solution
(c) exactly two optimal solution
(d) infinitely many optimal solution

[2013 : 2 Marks]

14. Consider an objective function $Z(x_1, x_2) = 3x_1 + 9x_2$ and the constraints

$x_1 + x_2 \leq 8$
$x_1 + 2x_2 \leq 4$
$x_1 \geq 0, x_2 \geq 0$

The maximum value of the objective function is __________.

[2014 : 2 Marks, Set-3]

15. Maximize $z = 15x_1 + 20x_2$

Subject to

$12x_1 + 4x_2 \geq 36$
$12x_1 + 6x_2 \geq 24$
$x_1, x_2 \geq 0$

The above linear programming problem has

(a) infeasible solution
(b) unbounded solution
(c) alternative optimum solutions
(d) degenerate solution

[2016 : 2 Marks, Set-1]

16. For the linear programming problem:

Maximize $z = 3x_1 + 2x_2$

Subject to
$-2x_1 + 3x_2 \leq 9$
$x_1 - 5x_2 \geq -20$
$x_1, x_2 \geq 0$

The above problem has

(a) unbounded solution
(b) infeasible solution
(c) alternative optimum solution
(d) degenerate solution

[2016 : 2 Marks, Set-3]

17. Two models, P and Q of a product earn profits of Rs. 100 and Rs. 80 per piece, respectively. Production times for P and Q are 5 hours and 3 hours, respectively, while the total production time available is 150 hours. For a total batch size of 40, to maximize profit, the member of units of P to be produced is ______.

[2017 : 2 Marks, Set-1]

18. The minimum value of $3x + 5y$

such that:

$3x + 5y \leq 15$

$4x + 9y \leq 8$

$13x + 2y \leq 2$

$x \geq 0, y \geq 0$

is ________.

[2018 : 2 Marks, Set-1]

19. The problem of maximizing $Z = x_1 - x_2$ subject to constrains $x_1 + x_2 \leq 10$, $x_1 \geq 0$, $x_2 \geq 0$ and $x_2 \leq 5$ has

(a) no solution

(b) one solution

(c) two solutions

(d) more than two solutions

[2018 : 2 Marks, Set-2]

Simplex Method

20. For the standard transportation linear program with m sources and n destinations and total supply equaling total demand, an optimal solution (lowest cost) with the smallest number of non-zero x_{ij} values (amounts from source i to destination j) is desired. The best upper bound for this number is

(a) m n (b) 2(m + n)

(c) m + n . (d) m + n – 1

[2008 : 2 Marks]

Common Data for Questions 21 and 22:

Consider the linear Programme (LP)

Max $4x + 6y$

Subject to

$3x + 2y < 6$

$2x + 3y < 6$

$x, y > 0$

21. After introducing slack variables s and t, the initial basic feasible solution is represented by the table below (basic variables are s = 6 and t = 6, and the objective function value is 0).

	–4	–6	0	0	0
s	3	2	1	0	6
t	2	3	0	1	6
	x	y	s	t	RHS

After some simplex iterations, the following table is obtained

	0	0	0	2	12
S	5/3	0	1	–1/3	2
y	2/3	1	0	1/3	2
	x	y	s	t	RHS

From this, one can conclude that

(a) the LP has a unique optimal solution

(b) the LP has an optimal solution that is not unique

(c) the LP is infeasible

(d) the LP is unbounded

[2008 : 2 Marks]

22. The dual for the LP in Q. 21 is

(a) Min 6u+6v
subject to
$3u + 2v \geq 4$
$2u + 3v \geq 6$
$u, v \geq 0$

(b) Max 6u + 6v
subject to
$3u + 2v < 4$
$2u + 3v < 6$
$u, v \geq 0$

(c) Max 4u + 6v
subject to
$3u + 2v \geq 6$
$2u + 3v \geq 6$
$u, v \geq 0$

(d) Min 4u + 6v
subject to
$3u + 2v \leq 6$
$2u + 3v \leq 6$
$u, v \geq 0$

[2008 : 2 Marks]

23. Simplex method of solving linear programming problem uses

(a) all the points in the feasible region

(b) only the corner points of the feasible region

(c) intermediate points within the infeasible region

(d) only the interior points in the feasible region

[2010 : 1 Mark]

ANSWERS

1. (c)	**2.** (c)	**3.** (c)	**4.** (a)	**5.** (b)	**6.** (c)	**7.** (b)	**8.** (d)	**9.(d)**	**10.** (d)
11. (a)	**12.** (b)	**13.** (b)	**15.** (b)	**16.** (c)	**19.** (b)	**20.** (c)	**21.** (b)	**22.** (a)	**23.** (b)

EXPLANATIONS

1. Assuming cycle time = 10 + 8 + 6 + 9 + 10 = 43

Balance delay $= 1 - \frac{\sum t_i}{n \times t_c} = 1 - \frac{43}{5 \times 10} = 0.14$

= 14%

3.

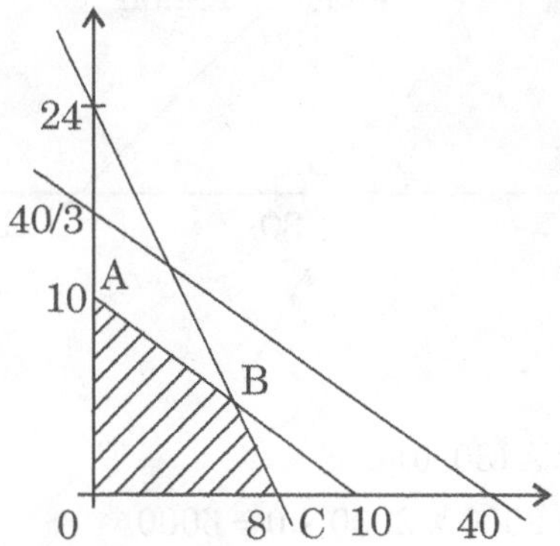

Feasible Region (0– A – B – C– 0)
At point (A) 0, 10)
Z = 2 (0) + 5 (10) = 50
At point B (7,3)
Z = 2(7) + 5(3) = 2y
At point C (8, 0)
Z = 2(16) + 5(0) = (16) 2
⇒ 32
Hence maximum profit is
Z_{max} = 50
No correct option is given.

4. $Z_{max} = 3P + 5Q$
subject to
$P + 2Q \leq 2000$
$P + Q \leq 1500$
$Q \leq 600$
$P, Q \geq 0$

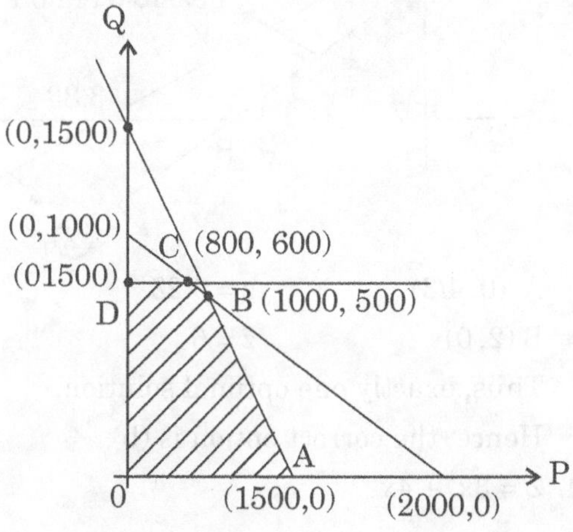

Feasible solution (O A B C D)
Since
At point A (1500,0)
Z = 3 × 1500 + 5 × 0 = 4500
At Point B (1000, 500)
Z = 3 × 1000 + 5 × 500 = 5500
At point C (800, 600)
Z = 3 × 800 + 5 × 600 = 5400
At Point O (0, 600)
Z = 3 × 0 + 5 × 600 = 3000
Hence Z in maximum at B (1000,500)
P = 1000 units
Q = 500 units

5.

	Fixed Cost (FC)	Variable Cost per piece (VC)
I	20	3
II	50	1
III	40	2
IV	10	4

Total cost = Fixed Cost (FC) + Number of piece (n) × Variable Cost per piece (VC)
TC = FC + (n) x V.C
For I
TC_I = 20+ (100) 3 = 320
For II
TC_{II} = 50 + (100) 1 = 150
For III
TC_{III} = 40 + 100 (2) = 240
For IV
TC_{IV} = 10 + 100 (4) = 410
So from economical point of view, one should chose process II

6.

	D_1	D_2	Supply
S_1	x_{11}	x_{12}	$a_1=50$
S_2	x_{21}	x_{22}	$a_2=40$
Demand	$20 \leq b_1 \leq 40$	$b_2 \geq 20$	90 / 90

Total no of supply point
⇒ m + n – 1
⇒ 2 + 2 – 1
⇒ 3

Total no of Demand point

$= 4\ (x_{11}, x_{12}, x_{21}, x_{22})$

Total supply = Total Demand

$\Rightarrow$ 90 units

7. Max $z = x_1 + x_2$

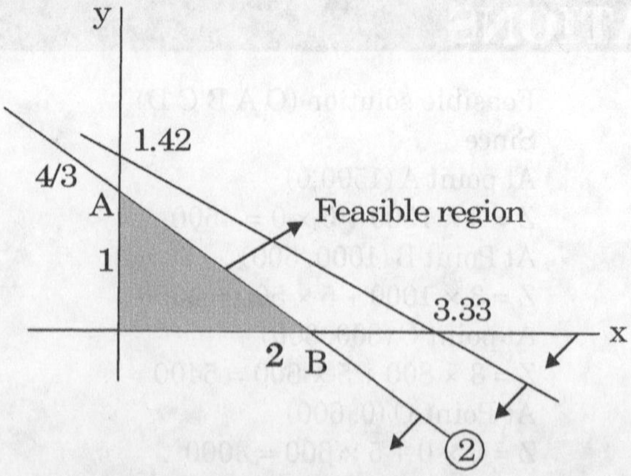

As feasibly region remains the same solution remains the same (4/3, 4/3).

Hence, the correct option is (b).

8. The optimal dual variables are V_1 & V_2.

10. $z= 3 \times 1 + 2 \times 2$

$x_1 \le 4$

$x_2 \le 6$

$3x_1 + 2x_2 \le 18$

$x_1, x_2 \ge 0$

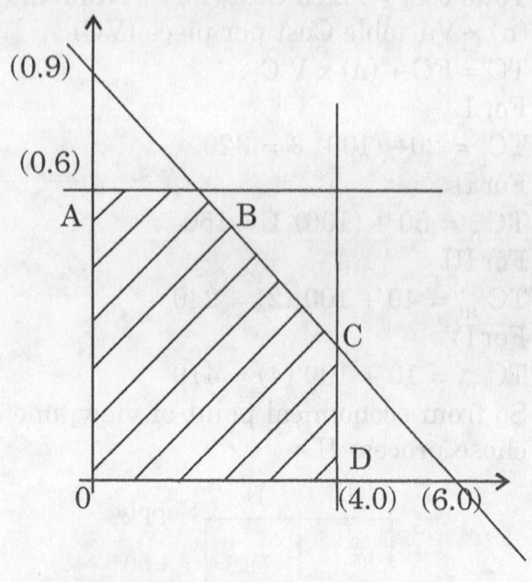

The objective function is parallel to constraint. Therefore multiple solution can be obtained.

11. Because second constraint is redundant in nature. Therefore, resource R_2 has no effect on the feasible solution.

12. $z = 2000\ P_1 + 3000\ P_2$

Subjected to

$3P_1 + 2P_2 \le 90$

$P_1 + 2P_2 \le 100$

$P_1, P_2 \ge 0$

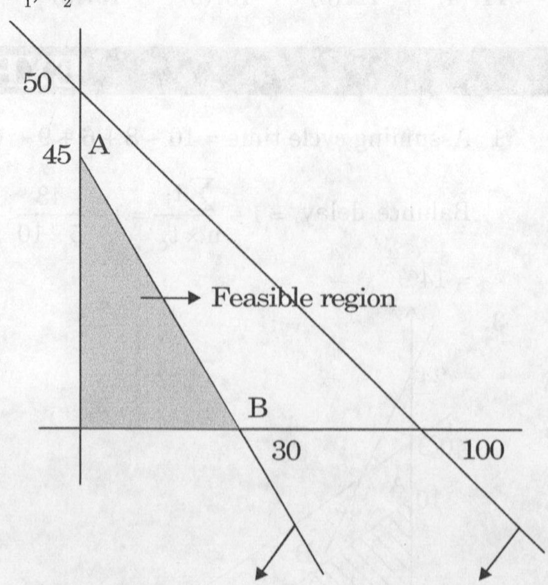

At point A (30, 0)

$Z = 30 \times 2000 \times 3000 \times 0 = 6000$

At point B (0, 45)

$Z = 2000\times 0 + 3000 \times 45 = 135000$

Hence Maximum Profit

$[Z_{max} = 135000]$

13. $z = 3x + 7y$

Constraints $3x + 7y \le 10$

$4x + 6y < 8$; $x, y \ge 0$

Corresponding equations

$3x + 7y = 10$; $4x + 6y = 8$

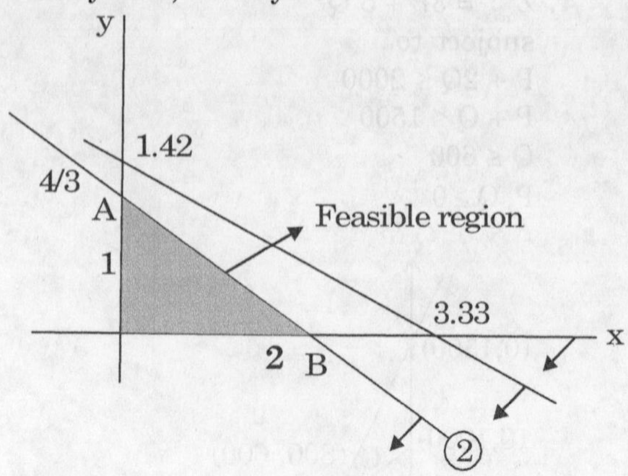

A (0, 4/3) $z = 9.23$

B (2, 0) $z = 6$

Thus, exactly one optimal solution.

Hence, the correct option is (b).

14. $Z = 3x_1 + 4x_2$

$x_1 + x_2 \leq 8$

$x_1 + 2x_2 \leq 4$

$x_1 , x_2 \geq 0$

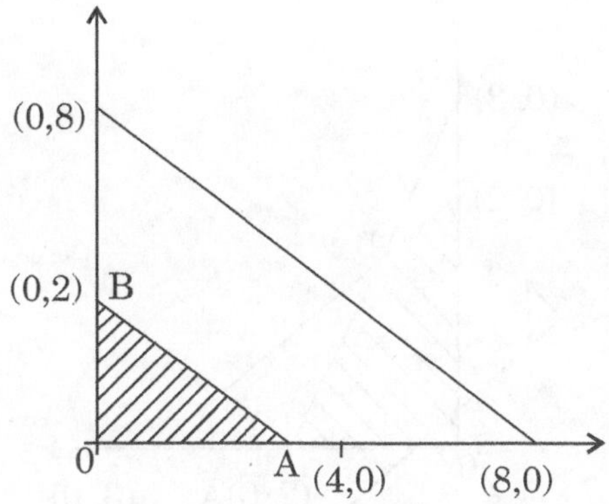

At point A:

z= 3 (4) +9 (0) = 12

At point B:

z = 3 (0) + 9 (2) =18

15. $z_{max} = 16x_1 + 20x_2$

$12x_1 + 4x_2 \geq 36$

$12x_1 + 6x_2 \geq 24$

$x_1\ x_2 \geq 0$

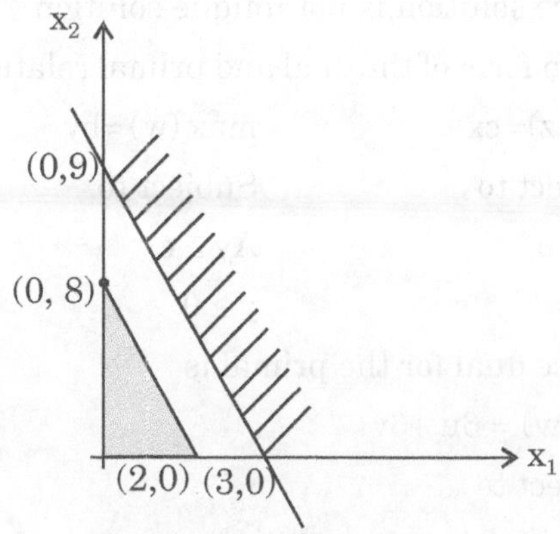

16. Maximize $Z = 3X_1 + 2X_2$

Subject to

$$-2X_1 + 3X_2 \leq 9$$

$$X_1 - 5X_2 \geq 20$$

$$X_1, X_2 \geq 0$$

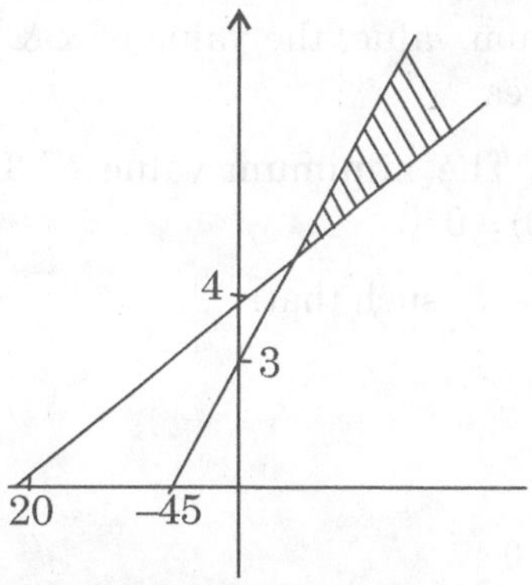

The LPP is unbounded.

17. z = 100P + 80Q

5P + 3Q ≤ 150

P + Q = 40

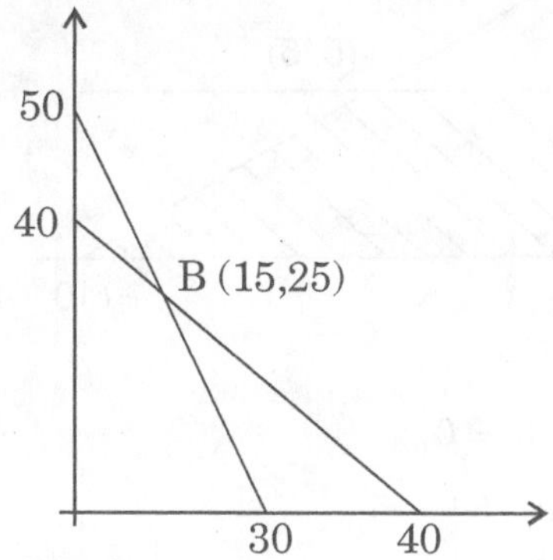

z at (0, 0) = 0

z at (0,40) = 3200

z at (15, 25) = 3500

z at (30, 0) = 300

$z_{max} = 3500$

∴ P=15

18. $3x+5y \leq 15 \rightarrow \frac{x}{b}+\frac{y}{3} \leq 1$

$4x+9y \leq 8 \rightarrow \frac{x}{2}+\frac{y}{(8/9)} \leq 1$

$13x+2y \leq 2 \rightarrow \frac{x}{(2/13)}+\frac{y}{1} \leq 1$

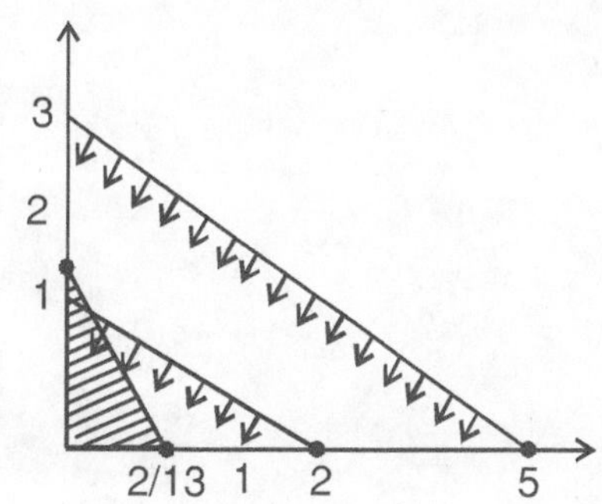

For minimum value, the value of x & y is in common area.

Therefore, The minimum value of 3x + 5y = 3(0) + 5(0) = 0

19. max $z = x_1 - x_2$, such that:

$x_1 + x_2 \le 10$

$x_2 \le 5$

$x_1 \ge 0, x_2 \ge 0$

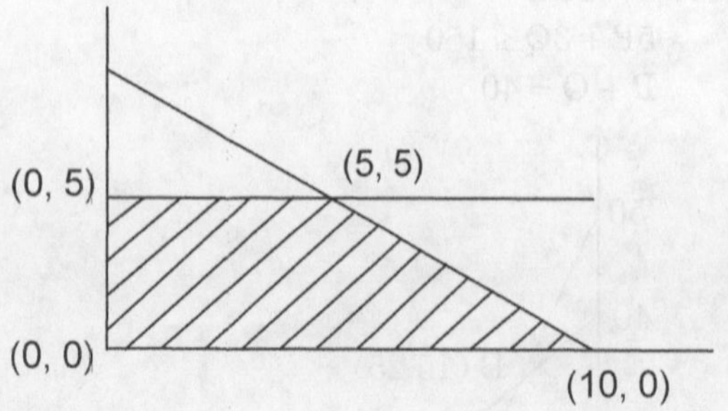

Now, z (0, 0) = 0

z (0, 5) = 0 – 5 = – 5

z (10, 0) = 10 – 0 = 10

z (5, 5) = 5 – 5 = 0

Thus, max z occurs at only one point (10, 0) and the maximum value of z is 10 which is unique optimal solution.

20. In such an L.P.P, m × n variables are there and m + n equations/constraints are there (satisfying the demand-supply requirements). But one constraint is removed as total supply equals total demand. The best upper bound x_{ij} values is (m + n – 1).

21. z = 4x + 6y

$3x + 2y \le 6$

$2x + 3y \le 6$

$x, y > 0$

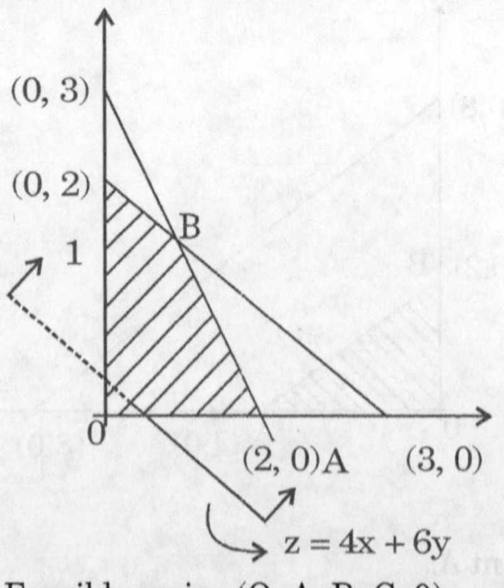

Feasible region (O–A–B–C–0)

Since, Slope of objective function is equal to the slope of constraint Hence LPP has multiple optimal solution

At B (6/5, 6/5)

Z = 12

At C (0,2)

Z = 12

Hence solution is not unique solution.

22. Given form of the dual and primal relationship

max(z)= cx	max (w) =by
Subject to	Subject to
$Ax \le b$	$Ay \ge c$
$x \ge 0$	$y \ge 0$

Hence dual for the primal is

min (w) = 6u +6v

Subject to

$3u + 2v \ge 4$

$2u + 3v \ge 6$

$u, v \ge 0$

4 CHAPTER

PERT and CPM

Analysis of Previous GATE Papers

Topic	Year/Set → Marks Distribution ↓		2019 Set 1	2019 Set 2	2018 Set 1	2018 Set 2	2017 Set 1	2017 Set 2	2016 Set 1	2016 Set 2	2016 Set 3	2015 Set 1	2015 Set 2	2015 Set 3
PERT	1 Mark	MCQ Type									1			
		Numerical Type												
	2 Marks	MCQ Type												
		Numerical Type	1											
	Total		2								1			
CPM	1 Mark	MCQ Type												
		Numerical Type												
	2 Marks	MCQ Type		1								1		
		Numerical Type				1		1		1				
	Total			2		2		2		2		1		
Sequencing	1 Mark	MCQ Type												
		Numerical Type												
	2 Marks	MCQ Type												
		Numerical Type	1											
	Total		2											

PERT

1. In PERT, the distribution of activity times is assumed to be

(a) Normal (b) Gamma

(c) Beta (d) Exponential

[1995 : 1 Mark]

2. A dummy activity is used in PERT network to describe

(a) Precedence relationship

(b) Necessary time delay

(c) Resource restriction

(d) Resource idleness

[1997 : 1 Mark]

3. In PERT analysis a critical activity has

(a) maximum float (b) zero float

(c) maximum cost (d) minimum cost

[2004 : 1 Mark]

Linked Answer Questions 4 and 5:

Consider a PERT network for a project involving six tasks (a to f)

Task	Predecessor	Expected task time (in days)	Variance of the task time (in days^2)
a	–	30	25
b	a	40	64
c	a	60	81
d	b	25	g
e	b,c	45	36
f	d,e	20	9

4. The expected completion time of the project is

(a) 238 days (b) 224 days

(c) 171 days (d) 155 days

[2006 : 2 Marks]

5. The standard deviation of the critical path of the project is

(a) $\sqrt{151}$ days (b) $\sqrt{155}$ days

(c) $\sqrt{200}$ days (d) $\sqrt{238}$ days

[2006 : 2 Marks]

6. The expected time (t_e) of a PERT activity in terms of optimistic time (t_0), pessimistic time (t_p) and most likely time (t_l) is given by

(a) $t_e = \frac{t_0 + 4t_l + t_p}{6}$ (b) $t_e = \frac{t_0 + 4t_p + t_l}{6}$

(c) $t_e = \frac{t_0 + 4t_l + t_p}{3}$ (d) $t_e = \frac{t_0 + 4t_p + t_l}{3}$

[2009 : 1 Mark]

Common data for Questions 7 and 8:

Consider the following PERT network:

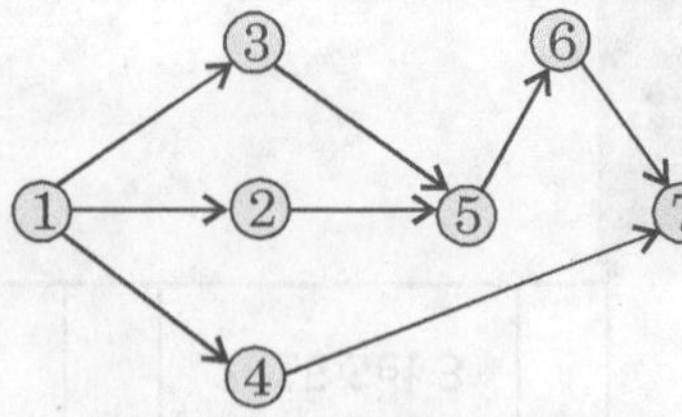

The optimistic time, most likely time and pessimistic time of all the activities are given in the table below:

Activity	Optimistic (days)	Most likely time (days)	Pessimistic time (days)
1–2	1	2	3
1–3	5	6	7
1–4	3	5	7
2–5	5	7	9
3–5	2	4	6
5–6	4	5	6
4–7	4	6	8
6–7	2	3	4

7. The critical path duration of the network (in days) is

(a) 11 (b) 14

(c) 17 (d) 18

[2009 : 2 Marks]

8. The standard deviation of the critical path is

(a) 0.33 (b) 0.55

(c) 0.88 (d) 1.66

[2009 : 2 Marks]

9. In PERT chart, the activity time distribution is

(a) Normal (b) Binomial

(c) Poisson (d) Beta

[2016 : 1 Mark, Set-3]

10. A project consists of six activities. The immediate predecessor of each activity and the estimated duration is also provided in the table below:

Activity	Immediate predecessor	Estimated duration (weeks)
P	-	5
Q	-	1
R	Q	2
S	P, R	4
T	P	6
U	S, T	3

If all activities other than S take the estimated amount of time, the maximum duration (in weeks) of the activity S without delaying the completion of the project is ______.

[2019 : 2 Marks, Set-1]

CPM

11. In the construction of networks, dummy activities are introduced in order to

(a) Compute the slack on all events

(b) Transfer resources, if necessary, during monitoring

(c) Clearly designate a precedence relationship

(d) Simplify the crashing plan

[1990 : 2 Marks]

12. The symbol used for Transport in work study is

(a) ⇒ (b) T

(c) ▭ (d) ▽

[2003 : 1 Mark]

13. A project consists of activities A to M shown in the net in the following figure with the duration of the activities marked in days

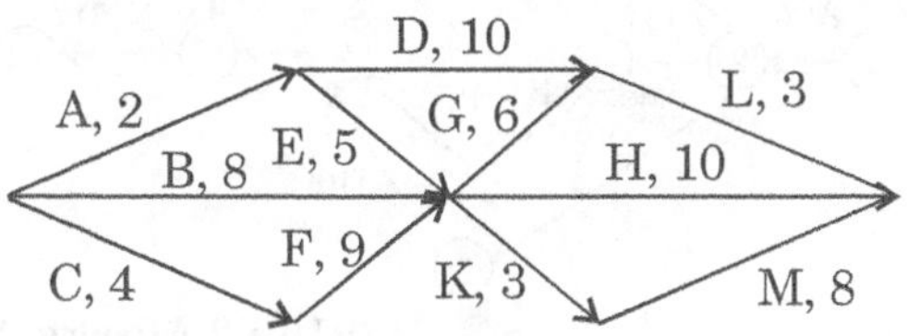

The project can be completed

(a) between 18,19 days

(b) between 20, 22 days

(c) between 24, 26 days

(d) between 60,70 days

[2003 : 2 Marks]

14. An assembly activity is represented on an Operation Process Chart by the symbol

(a) ▭ (b) A

(c) D (d) O

[2005 : 1 Mark]

15. A project has six activities (A to F) with respective activity durations 7,5,6,6,8,4 days. The network has three path A-B, C-D and E-F. All the activities can be crashed with the same crash cost per day.

The number of activities that need to be crashed to reduce the project duration by 1 day is

(a) 1 (b) 2

(c) 3 (d) 6

[2005 : 2 Marks]

16. For the network below, the objective is to find the length of the shortest path from node P to node G. Let d_{ij} be the length of directed arc from node i to node j.

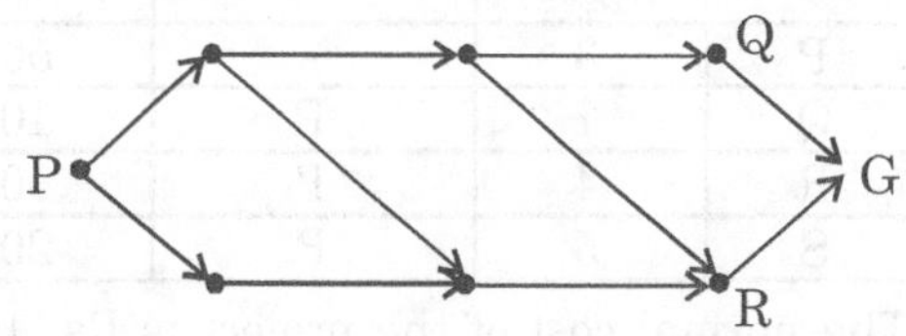

Let s_j, be the length of the shortest path from P to node j. Which of the following equations can be used to find SG?

(a) $S_G = \text{Min} \{S_Q, S_R\}$

(b) $S_G = \text{Mn} \{S_Q – d_{QG}, S_R – d_{RG}\}$

(c) $S_G = \text{Min} [SQ + d_{QG}, S_R + d_{RG}]$

(d) $S_G = \text{Mn} [S_{QG} – d_{RG}]$

[2008 : 1 Mark]

17. The project activities, precedence relationships and durations are described in the table. The critical path of the project is

Activity	Precedence	Duration (in days)
P	-	3
O	-	4
R	P	5
S	O	5
T	R, S	7
U	R, S	5
V	T	2

(a) P–R–T–V (b) Q–S–T–V

(c) P–R–U–W (d) Q–S–U–W

[2010 : 2 Marks]

18. The critical path for the project is

(a) a-b-e-g-h (b) a-c-g-h

(c) a-d-f-h (d) a-b-c-f-h

[2012 : 2 Marks]

19. If duration of activity f is changed to 10 days, then the critical path for the project is

(a) Critical path remains the same and the total duration to complete the project changes to 19 days.

(b) Critical path and the total duration to complete the project remains the same.

(c) Critical path changes but the total duration to complete the project remains the same.

(d) Critical path changes and the total duration to complete the project changes to 17 days.

[2012 : 2 Marks]

20. A project has four activities P, O, H and S as shown below.

Activity	Normal duration (days)	Predecessor	Cost slope (Rs/day)
P	3	-	500
Q	7	P	100
R	4	P	400
S	5	R	200

The normal cost of the project is Rs. 10,000/- and the overhead cost is Rs 200/- per day. If the project duration has to be crashed down to 9 days, the total cost (in Rupees) of the project is_________ .

[2014 : 2 Marks, Set-2]

21. Consider the given project network, where numbers along various activities represent normal time. The free float on activity 4-6 and the project duration, respectively, are

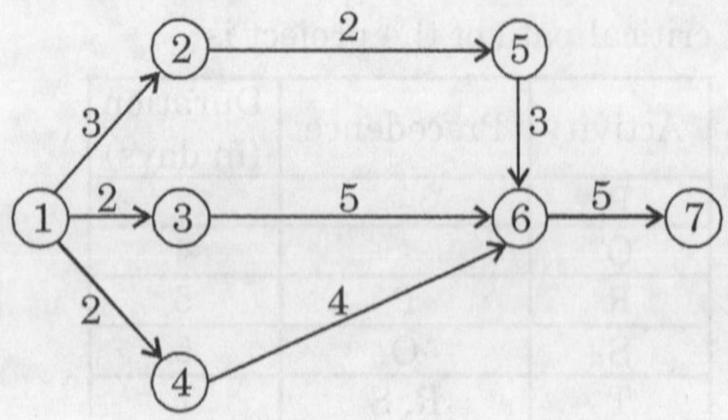

(a) 2 and 13 (b) 0 and 12

(c) –2 and 13 (d) 2 and 12

[2014 : 2 Marks, Set-3]

22. The precedence relations and duration (in days) of activities of a project network are given in the table. The total float (in days) of activities e and f, respectively, are

Activity	Predecessors	Duration (days)
a	-	2
b	-	4
c	a	2
d	b	3
e	c	2
f	c	4
g	d, e	5

(a) 0 and 4 (b) 1 and 4

(c) 2 and 3 (d) 3 and 1

[2014 : 2 Marks, Set-4]

23. Following data refers to the activities of a project, where, node 1 refers to the start and node 5 refers to the end of the project

Activity	Duration (days)
1-2	2
2-3	1
4-3	3
1-4	3
2-5	3
3-5	2
4-5	4

The critical path (CP) in the network is

(a) 1-2-3-5 (b) 1-4-3-5

(c) 1-2-3-4-5 (d) 1-4-5

[2015 : 2 Marks, Set-1]

24. A project consists of 14 activities, A to N. The duration of these activities (in days) are shown in brackets on the network diagram. The latest finish time (in days) for node 10 is_______.

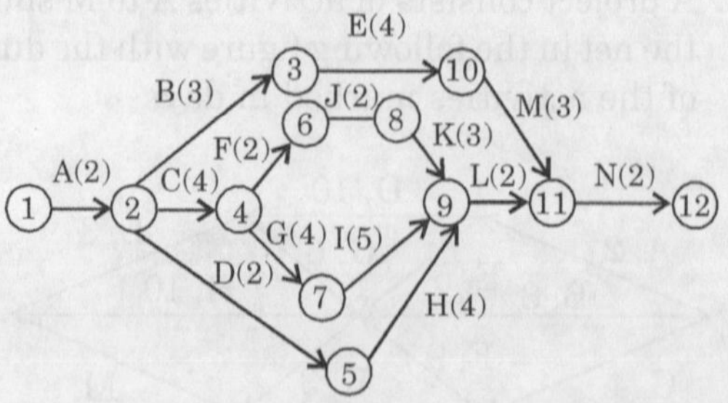

[2016 : 2 Marks, Set-2]

25. A project starts with activity A and ends with activity F. The precedence relation and durations of the activities are as per the following table:

Activity	Immediate predecessor	Duration (days)
A	-	4
B	A	3
C	A	7
D	B	14
E	C	4
F	D, E	9

The minimum project completion time (in days) is _______ .

[2017 : 2 Marks, Set-2]

26. The arc lengths of a directed graph of a project are as shown in the figure. The shortest path length from node 1 to node 6 is________.

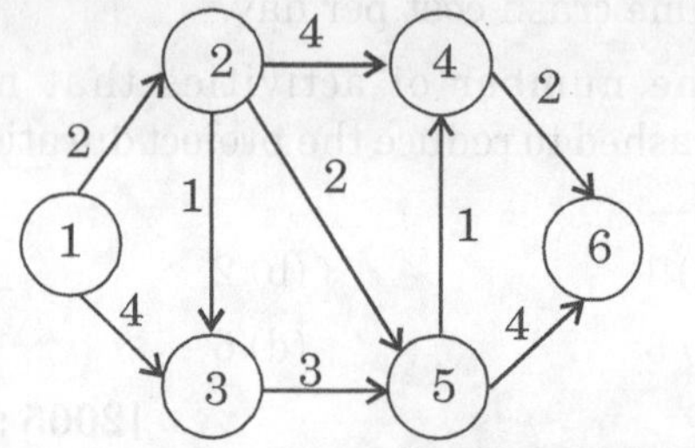

[2018 : 2 Mark, Set-2]

27. The activities of a project, their duration and the precedence relationships are given in the table. For example, in a precedence relationship "X < Y, Z" means that X is predecessor of activities Y and Z. The time to complete the activities along the critical path is ____ weeks.

Activity	Duration (weeks)	Precedence relationship
A	5	A < B, C, D
B	7	B < E, F, G
C	10	C < I
D	6	D < G
E	3	E < H
F	9	F < I
G	7	G < I
H	4	H < I
I	2	–

(a) 25 (b) 17

(c) 21 (d) 23

[2019 : 2 Marks, Set-2]

Sequencing

28. A set of 5 jobs is to be processed on a single machine. The processing time (in days) is given in the table below. The holding cost for each job is Rs K per day.

Job	Processing time
P	5
Q	2
R	3
S	2
T	1

(a) T-S-Q-R-P (b) P-R-S-Q-T

(c) T-R-S-Q- P (d) P-Q-R-S-T

[2008 : 1 Mark]

29. Six jobs arrived in a sequence as given below:

Jobs	Processing Time (days)
I	4
II	9
III	5
IV	10
V	6
VI	8

Average flow time (in days) for the above jobs using Shortest Processing Time rule is

(a) 20.83 (b) 23.16

(c) 125.00 (d) 139.00

[2009 : 2 Marks]

30. If the Earliest Due Date (EDD) rule is used to sequence the jobs, the number of jobs delayed is

(a) 1 (b) 2

(c) 3 (d) 4

[2010 : 2 Marks]

31. Using the Shortest Processing Time (SPT) rule, totel tardiness is

(a) 0 (b) 2

(c) 6 (d) 8

[2010 : 2 Marks]

32. Five jobs (J1, J2, J3, J4 and J5) need to be processed in a factory. Each job can be assigned to any of the five different machines (M1, M2, M3, M4 and M5). The time duration taken (in minutes) by the machines for each of the jobs, are given in the table. However, each job is assigned to a specific machine in such a way that the total processing time is minimum. The total processing time is _____ minutes.

	M1	M2	M3	M4	M5
J1	40	30	50	50	58
J2	26	38	60	26	38
J3	40	34	28	24	30
J4	28	40	40	32	48
J5	28	32	38	22	44

[2019 : 2 Marks, Set-1]

ANSWERS

1. (c) **2.** (a) **3.** (b) **4.** (d) **5.** (a) **6.** (a) **7.** (d) **8.** (c) **9.** (d) **11.** (c)

12. (a) **13.** (c) **14.** (d) **15.** (c) **16.** (c) **17.** (d) **18.** (c) **19.** (a) **21.** (a) **22.** (b)

23. (b) **27.** (d) **28.** (a) **29.** (a) **30.** (c) **31.** (c)

EXPLANATIONS

2. Dummy activities often have a zero completion time & are used to represents precendence relationship.

3. $\text{Earliest Start} = \dfrac{\text{Least Start}}{\text{Float}} = 0$

4.

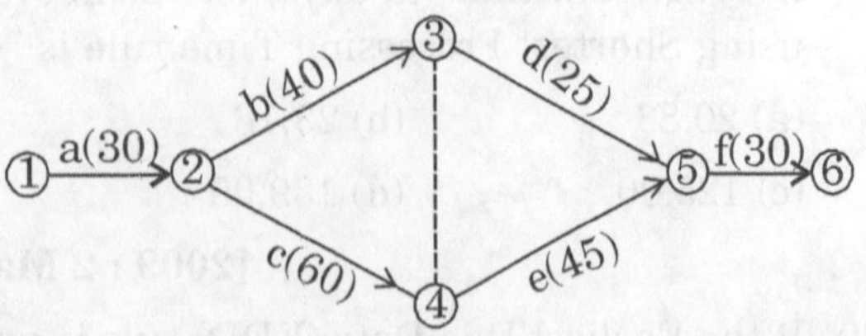

$\Rightarrow a - c - e - f$

$\Rightarrow 30 + 60 + 45 + 20 = 155$ days

5. $\sigma = \sqrt{25+81+36+9} = \sqrt{151}\text{days}$.

7.

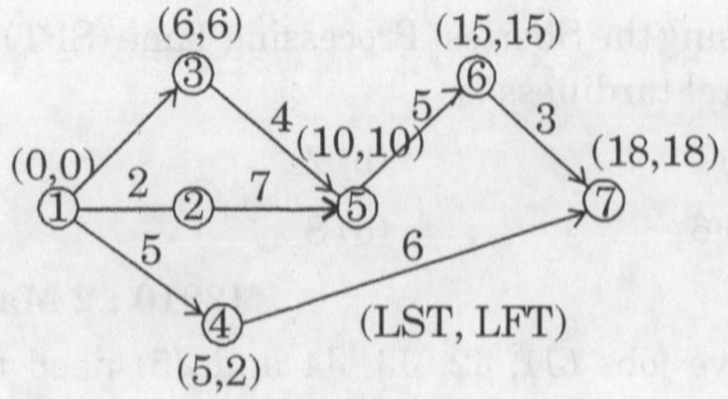

EST = Earliest start time

LFT = Latest Finish Time

$\therefore$ Path along which EST & LFT is equal are called critical path

Critical path duration is 18 days.

8. $\text{Variance} = \left(\dfrac{t_p - t_O}{6}\right)^2$

Critical path	Variance
1 – 3	$\left(\frac{1}{3}\right)^2$
3 – 5	$\left(\frac{2}{3}\right)^2$
5 – 6	$\left(\frac{1}{3}\right)^2$
6 – 7	$\left(\frac{1}{3}\right)^2$

$\text{Total variance} = \left(\frac{1}{3}\right)^2 + \left(\frac{2}{3}\right)^2 + \left(\frac{1}{3}\right)^2 + \left(\frac{1}{3}\right)^2$

$= \dfrac{7}{9}$

$\text{Standard deviation} = \sqrt{\text{variance}}$

$\Rightarrow \sqrt{\dfrac{7}{9}} = 0.88$

9. Beta distribution extended to the Maximum & Minimum & gives strict definition for the mean & variance

10.

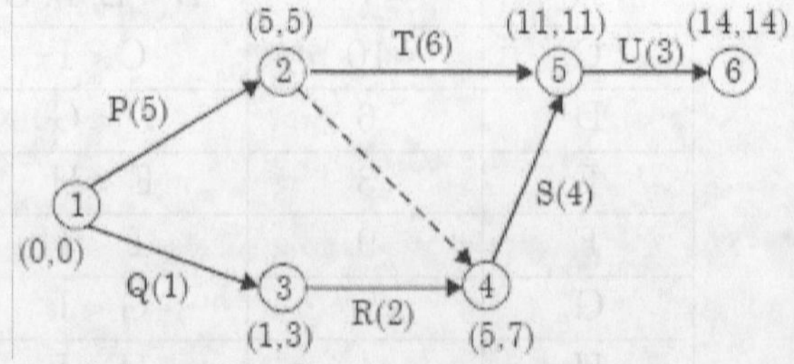

$\text{Total float} = L_j - E_i - t_{ij} = 11 - 5 - 4 = 2$

13.

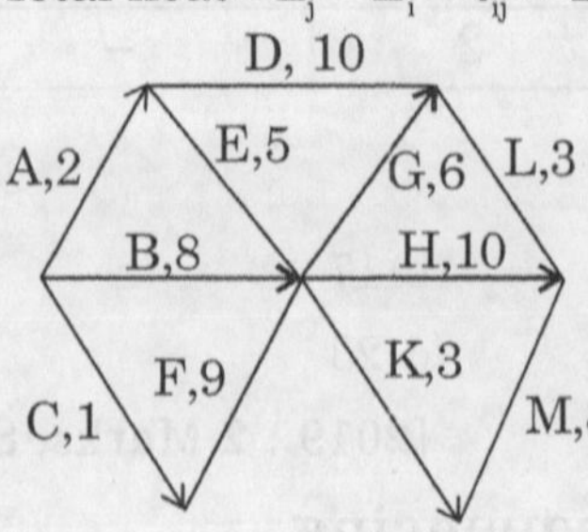

Project completed = Activity C + Activity F + Activity K + Activity M

$\Rightarrow 4 + 9 + 3 + 8 = 24$

15.

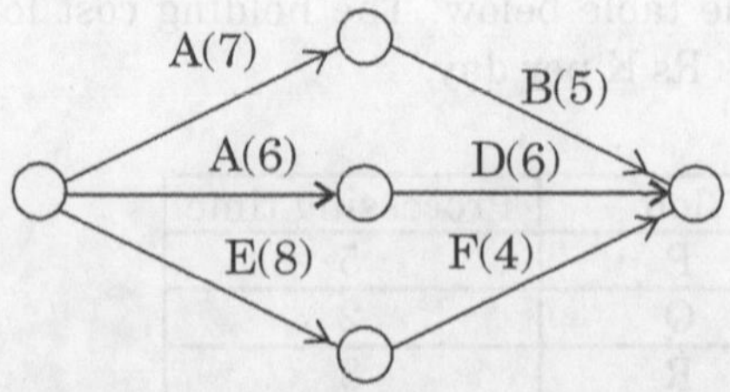

Minimum number of activity to be crashed = 3

$\therefore$ All three are critical path.

16. The only way to reach G is then by reading Q and then QG or reaching R and then RG.

$\therefore SG = \text{Min}\left[SQ + d_{QG}, SR + d_{RG}\right]$

17.

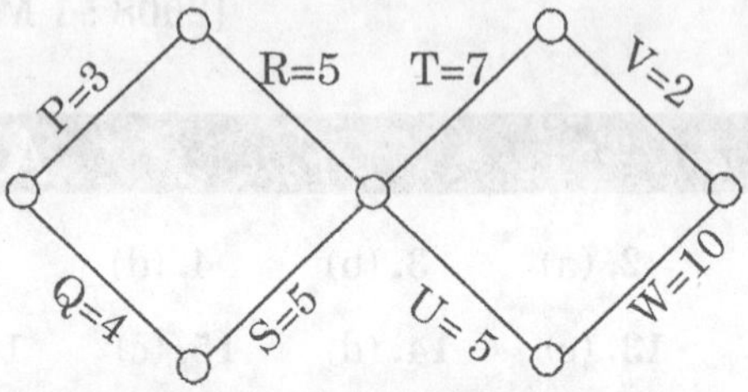

Path QSUW = Q + S + U + W

= 4 + 5 + 5 + 10 = 24 days.

18.

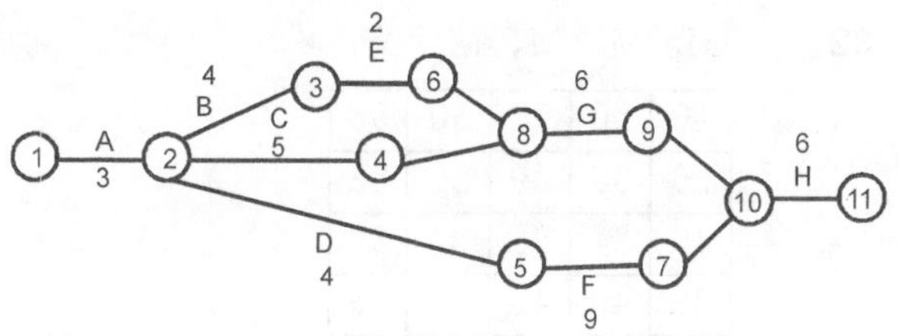

Various path are

A B E G H – 17 days

A C G H – 16 days

A D F H – 18 days;

Hence A D F H is critical path

19. If duration of activity F has changed to 10 days, critical path remains the same and project duration will increase to 19 days.

20.

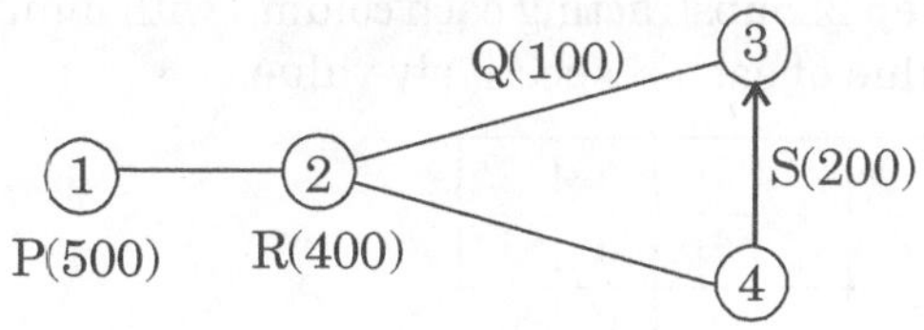

Critical path ⇒1 – 2 – 3 – 4

⇒ 5 + 4 + 1 + 2 = 12 days

For 12 days :-

10,000 + 12 (200) = 12400

For 9 days :-

= 10,000 + 9 (200) + (400 + 300)

= 12,500

21.

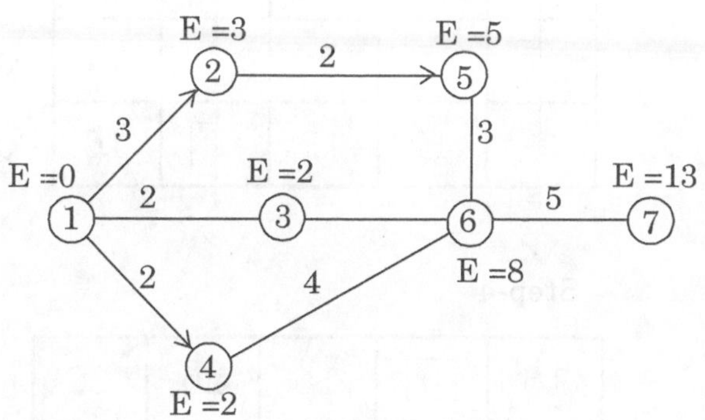

For 4 – 6

Free float = $(E_j - E_i) - d_{ij}$

= (8 – 2) – 4

= 6 – 4 = 2

Project duration , E = 13

∴ (2, 13)

22.

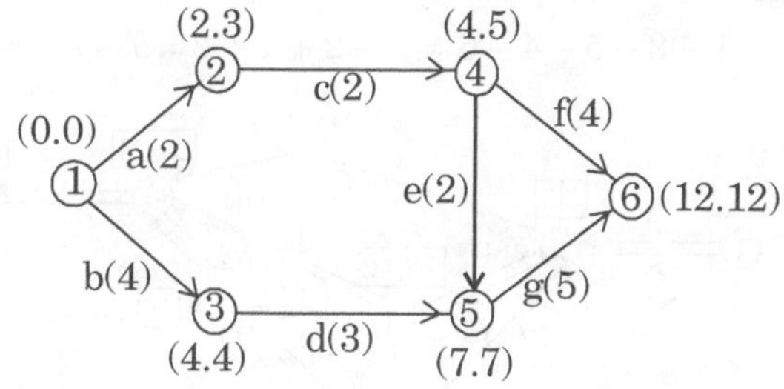

= (E, L)

TF = LFT – EFT = LST – EST

$(TF)_F \Rightarrow 7 - 4 - 2 = 1$

$(TF)_F \Rightarrow 12 - 4 - 4 = 4$

23.

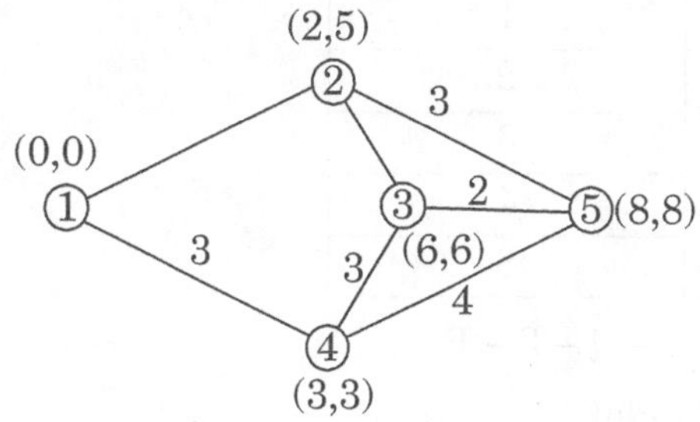

Critical path ⇒ 1 – 4 – 3 – 5

Time takes = 8 days

24.

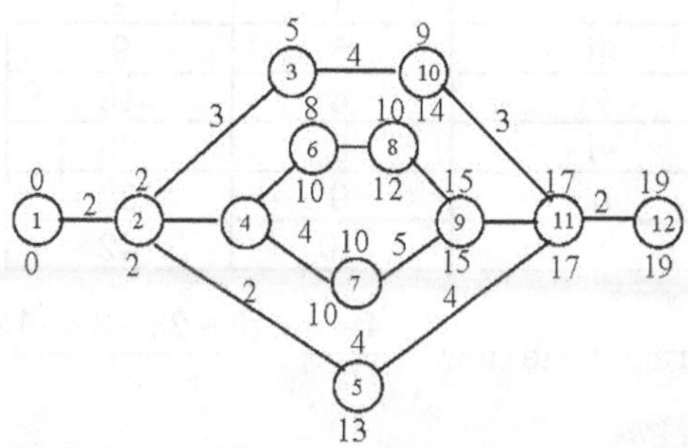

(LFT) for node 10 = 14 days

25.

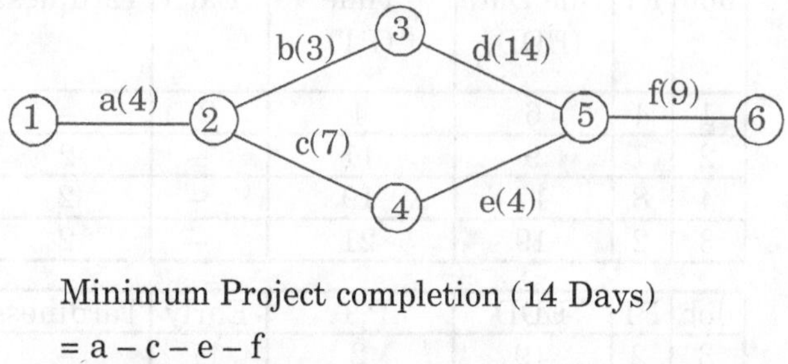

Minimum Project completion (14 Days)

= a – c – e – f

= 4 + 7 + 4 + 9

= 24 days

26. = 1 – 2 – 5 – 4 – 6 = 2 + 2 + 1 + 2 = 7

27.

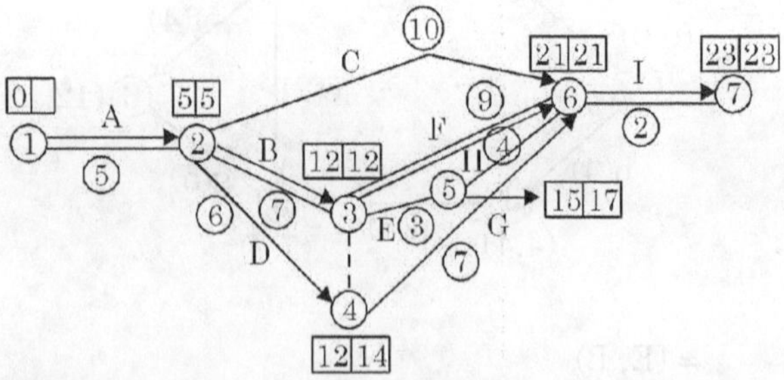

Critical path : A → B → F → I

Time to complete the activities along the critical path = 23 weeks

28.

Job	Processing Time
T	1
S	2
Q	3
R	4
P	S

= T – S – Q – R – P

29. Jobs using

Jobs using shortest possible time rule	Processing time(days)	camulative time (days)
i	4	4
iii	5	9
v	6	15
vi	8	23
ii	9	32
iv	10	42

$$\text{Average time flow} = \frac{4+9+15+23+32+12}{6}$$

= 20.83

31.

Job	PT	Eearliest Due Date (EDD)	Processing Time (PT)	Early	Tardiness
1	4	6	4	2	–
2	7	9	11	–	2
4	8	17	19	–	2
3	2	19	21	–	2

Job	PT	EDD	P.T.	Early	Tardiness
3	2	19	2	17	–
1	4	6	6	–	–
2	7	9	13	–	4
4	8	17	21	–	4

Total = 8

32.

	M_1	M_2	M_3	M_4	M_5
J_1	40	30	50	50	58
J_2	26	38	60	26	38
J_3	40	34	28	24	30
J_4	28	40	40	32	48
J_5	28	32	38	22	54

Step-1 Substracting each row with minimum value of each respective row.

10	0	20	20	28
0	12	34	0	12
16	10	4	0	6
0	12	12	4	20
6	10	16	0	22

Step 2: Substracting each column with minimum value of each respectively value.

10	0	16	20	22
0	12	30	0	6
16	10	0	0	0
0	12	8	4	14
6	10	12	0	16

Step-3 By assigning row and column

10	[0]	16	20	22	
~~0~~	12	30	~~0~~	6	✓
16	10	[0]	~~0~~	~~0~~	
[0]	12	8	4	14	✓
6	10	12	[0]	16	✓
✓			✓		

Step-4

16	[0]	16	26	22
~~0~~	6	24	~~0~~	[0]
22	10	[0]	6	~~0~~
[0]	6	2	4	8
6	4	6	[0]	10

Step-5 Minimum time = 28 + 30 + 28 + 22 + 38

= 146 min

5 CHAPTER

Probability and Statistics

Analysis of Previous GATE Papers

Topic	Year/Set → Marks Distribution ↓		2019 Set 1	2019 Set 2	2018 Set 1	2018 Set 2	2017 Set 1	2017 Set 2	2016 Set 1	2016 Set 2	2016 Set 3	2015 Set 1	2015 Set 2	2015 Set 3
Probability	1 Mark	MCQ Type												
		Numerical Type												
	2 Marks	MCQ Type												
		Numerical Type												
	Total													
Statistics	1 Mark	MCQ Type	1					1				1		
		Numerical Type						1						
	2 Marks	MCQ Type												
		Numerical Type												
	Total		1					2				1		

Probability

1. A residential school stipulates the study hours as 8.00 pm to 10.30 pm. Warden makes random checks on a certain student 11 occasions a day during the study hours over a period of 10 days and observes that he is studying on 71 occasions. Using 95% confidence interval, the estimated minimum hours of his study during that 10 day period is

(a) 8.5 hours (b) 13.9 hours
(c) 16.1 hours (d) 18.4 hours

[2003 : 2 Marks]

2. The distribution of lead time demand for an item is as follows:

Lead time demand	Probability
80	0.20
100	0.25
120	0.30
140	0.25

The reorder level is 1.25 times the expected value of the lead time demand. The service level is

(a) 25% (b) 50%
(c) 75% (d) 100%

[2005 : 2 Marks]

Statistics

3. The jobs arrive at a facility, for service, in a random manner. The probability distribution of number of arrivals of jobs in a fixed time interval is

(a) Normal (b) Poisson
(c) Erlang (d) Beta

[2014 : 1 Mark, Set-1]

4. Demand during lead time with associated probabilities is shown below:

Demand	50	70	75	80	85
Probability	0.15	0.14	0.21	0.20	0.30

Expected demand during lead time is

[2014 : 1 Mark, Set-4]

5. Jobs arrive at a facility at an average rate of 5 in an 8 hour shift. The arrival of the jobs follows Poisson distribution. The average service time of a job on the facility is 40 minutes. The service time follows exponential distribution. Idle time (in hours) at the facility per shift will be

(a) $\frac{5}{7}$ (b) $\frac{14}{3}$

(c) $\frac{7}{3}$ (d) $\frac{10}{3}$

[2014 : 2 Marks, Set-1]

6. Among the four normal distributions with probability density functions as shown below, which one has the lowest variance?

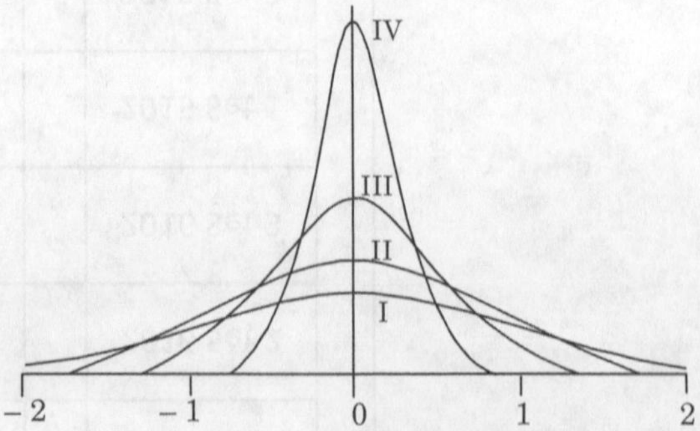

(a) I (b) II
(c) III (d) IV

[2 015 : 1 Mark, Set-1]

7. The standard deviation of linear dimensions P and O are 3 μm and 4 μm, respectively. When assembled, the standard deviation (in μm) of the resulting linear dimension (P+ Q) is____.

[2017 : 1 Mark, Set-2]

8. A sample of 15 data is as follows: 17, 18, 17, 17, 13, 18, 5, 5, 6, 7, 8, 9, 20, 17, 3. The mode of the data is

(a) 4 (b) 13
(c) 17 (d) 20

[2017 : 1 Mark, Set-2]

9. The lengths of a large stock of titanium rods follow a normal distribution with a mean (μ) of 440 mm and a standard deviation (σ) of 1 mm. What is the percentage of rods whose lengths lie between 438 mm and 441 mm ?

(a) 68.4% (b) 99.75%
(c) 81.85% (d) 86.64%

[2019 : 1 Mark, Set-1]

ANSWERS

1. (c) **2.** (d) **3.** (b) **5.** (b) **6.** (d) **8.** (c) **9.** (c)

EXPLANATIONS

1. Number of total observation in 10 days
$= 11 \times 10 = 110$
No. of observation when studying = 71

$$\therefore \underset{\downarrow}{P} = \frac{71}{110} = 0.6455$$

(Probability of studying)
Total studying hour in 10 days = (2.5 hr) × 10
Hence, Minimum no. of hours of studying in 10 days = (25 hr) × p = 25 × 0.6455 = 16.13 hrs.

2. The expected value of lead time demand
$\Rightarrow 80 \times 0.20 + 100 \times 0.25 + 120 \times 0.30 + 140 \times 0.25 = 112$
Reorder level is 1.25 times the lead time demand.
So, Reorder value = 1.25 × 112 = 140
∴ Here, we can see that both the maximum demand and reorder value are equal, service level = 100%

3.

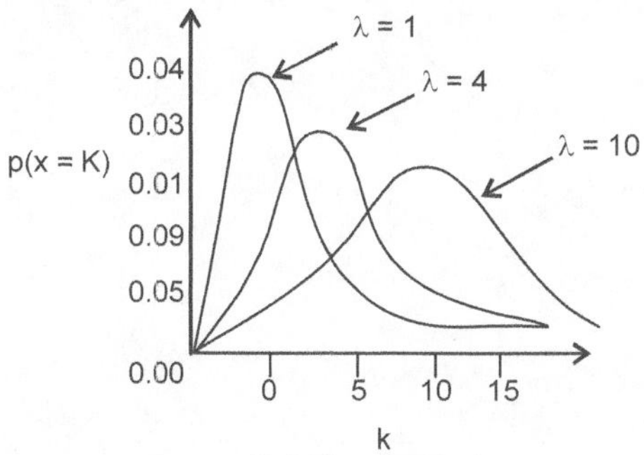

Poission distribution

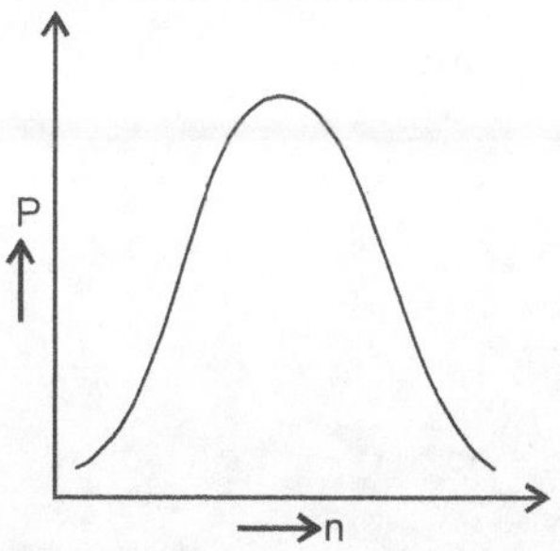

Since arrival rates depends upon the time factor, so accordingly graph can be chosen from Poisson distribution, but normal distribution expresses same result throughout.

4. Probability

x_i	f_i	$x_i f_i$
50	0.15	7.5
70	0.14	9.8
75	0.21	15.75
80	0.2	16
85	0.3	25.5

Expected demand during lead time
$\Rightarrow 7.5 + 9.8 + 15.75 + 16 + 25.5 = 74.55$

5. Arrival rate = 5 jobs in 8 hr
service time = 40 min/job
∴ Total service time = 40 × 5 = 200 min

$$= \frac{200}{60} = \frac{10}{3}\text{hrs}$$

$$\therefore \text{ Idle time 1 shift } = \frac{24-10}{3} = \frac{14}{3}\text{hrs}$$

6. We have Probability distribution function of Normal Distribution

$$f(x) = \frac{1}{\sigma\sqrt{x}} e^{-\frac{(x-\mu)^2}{2\sigma^2}} \quad ...(1)$$

Variance = σ^2 is lowest
$\Rightarrow$ σ also lowest
$\Rightarrow$ If σ decreases $\Rightarrow f(x)$ increases(∵ from (1))
$\Rightarrow$ Curve will have highest peak

7. P = 3 μ m (Standard deviate)
Variance of P = 9 μm
Standard deviate of Q = 4 μm
Variance of Q is 16 μm
∴ Variance of P & Q = 9 + 16 = 25
Standard deviation of P & Q

$$\Rightarrow \sqrt{\text{variance}} = \sqrt{25} = 5$$

8. The data which is repeated for maximum number of times is 17.

9. Given, mean, (μ) = 440 mm
Standard deviation, σ = 1 mm

$$Z = \frac{X - u}{\sigma}$$

lower limit,

$$Z(x = 438) = \frac{438 - 440}{1} = -2$$

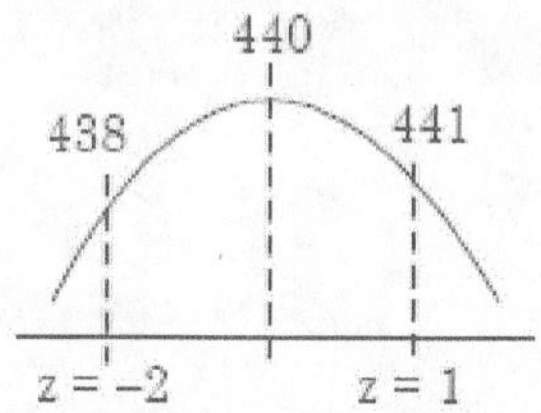

$$Z(x = 441\text{ mm}) = \frac{441 - 440}{1} = 1$$

Percentage of rods whose lengths lie between 438 mm and 441 mm.
= 0.3413 + (0.5 – 0.0228)
= 0.81854 = 81.854%

EXPLANATIONS

1. Number of total observation in 10 days

$= 11 \times 10 = 110$

No. of observation when studying = 71

$$P = \frac{71}{110} = 0.6455$$

(Probability of studying)

Total studying hour in 10 days = (2.5 hr) × 10

Hence, Minimum no. of hours of studying in 10 days = (25 hr) × p = 25 × 0.6455 = 16.13 hrs

2. The expected value of lead time demand

⇒ 30 × 0.20 + 100 × 0.25 + 120 × 0.30 + 140 × 0.25 = 112

Reorder level is 1.25 times the lead time demand.

So, Reorder value = 1.25 × 112 = 140

∴ Here, we can see that both the maximum demand and reorder value are equal, service level = 100%

3.

Poisson distribution

Since arrival rates depends upon the time factor, so accordingly graph can be chosen from Poisson distribution, but normal distribution expresses same result throughout.

4. Probability

X	P	XP
50	0.15	7.5
70	0.14	9.8
75	0.21	15.75
80	0.2	16
85	0.3	25.5

Expected demand during lead time

⇒ 7.5 + 9.8 + 15.75 + 16 + 25.5 = 74.55

5. Arrival rate = 5 jobs in 8 hr

service time = 40 min/job

Total service time = 40 × 5 = 200 min

$$= \frac{200}{60} = \frac{10}{3} \text{ hrs}$$

∴ Idle time 1 shift $= \frac{24 - 10}{3} = \frac{14}{3}$ hrs

6. We have Probability distribution function of Normal Distribution

$$f(x) = \frac{1}{\sigma\sqrt{2\pi}} e^{-\frac{(x-\mu)^2}{2\sigma^2}} \quad ...(i)$$

⇒ Variance = σ^2 is lowest

⇒ σ also lowest

⇒ If σ decreases ⇒ f(x) increases (from (i))

⇒ Curve will have highest peak

7. P = 3 µm (Standard deviate)

Variance of P = 9 µm

Standard deviate of Q = 4 µm

Variance of Q is 16 µm

Variance of P & Q = 9 + 16 = 25

Standard deviation of P & Q

$= \sqrt{\text{variance}} = \sqrt{25} = 5$

8. The data which is repeated for maximum number of times is 17.

9. Given, mean, µ = 140 mm

Standard deviation, σ = 1 mm

$$Z = \frac{X - \mu}{\sigma}$$

∴ lower limit

$$Z_{X = 138} = \frac{138 - 140}{1} = -2$$

$$Z_{X = 141\ \text{mm}} = \frac{141 - 140}{1} = 1$$

Percentage of rods whose lengths lie between 138 mm and 141 mm

= 0.3413 + 0.5 − 0.0228

= 0.8185 = 81.85%

6 CHAPTER

Production, Planning and Control

Analysis of Previous GATE Papers

Topic	Year/Set → Marks Distribution ↓		2019 Set 1	2019 Set 2	2018 Set 1	2018 Set 2	2017 Set 1	2017 Set 2	2016 Set 1	2016 Set 2	2016 Set 3	2015 Set 1	2015 Set 2	2015 Set 3
Production Planning and Control	1 Mark	MCQ Type					1						1	
		Numerical Type												
	2 Marks	MCQ Type												
		Numerical Type			2									
	Total				4		1						1	

Production, Planning and Control

1. For planning the procurement or production of dependent demand items, the technique most suited is

[1995 : 1 Mark]

2. The manufacturing area of a plant is divided into four quadrants. Four machines have to located one in each quadrant. The total number of possible layouts is

(a) 4 (b) 8

(c) 16 (d) 24

[1995 : 1 Mark]

3. Match the following

List-I (Problem areas)	**List-II** (Techniques)
A. JIT	1. CRAFT
B. Computer assisted layout	2. PERT
C. Scheduling	3. Johnson's rule
D. Simulation	4. Kanbans
	5. EOQ rule
	6. Monte Carl

[1995 : 2 Marks]

4. Production flow analysis (PFA) is a method of identifying part families that uses data from

(a) engineering drawings

(b) production schedule

(c) bill of materials

(d) route sheets

[2001 : 1 Mark]

5. Two machines of the same production rate are available for use. On machine 1, the fixed cost is Rs. 100 and the variable cost is Rs. 2 per piece produced. The corresponding numbers for the machine are Rs. 200 and Rs. 1 respectively. For certain strategic reasons both the machines are to be used concurrently. The sale price of the first 800 units is Rs. 3.50 per unit & and subsequently it is only Rs. 3.00. The breakeven production rate for each machine is

(a) 75 (b) 100

(c) 150 (d) 600

[2003 : 2 Marks]

6. A standard machine tool and an automatic machine tool are being compared for the production of a component. Following data refers to the two machines.

	Standard Machine tool	Automatic Machine tool
Setup time	30 min.	2 hours
Machining time per piece	22 min.	5 min
Machine rate	Rs. 200 per hour	Rs. 800 per hour

The breakeven production batch size above which the automatic machine tool will be economical to use, will be

(a) 4 (b) 5

(c) 24 (d) 225

[2004 : 2 Marks]

7. A soldering operation was work-sampled over two days (16 hours) during which an employee soldered 108 joints, actual working time was 90% of the total time and the performance rating was estimated to be 120 percent. If the contract provides allowance of 20 percent of the total time available, the standard time for the operation would be

(a) 8 min. (b) 8.9 min

(c) 10 min (d) 12 min

[2004 : 2 Marks]

8. An electronic equipment manufacturer has decided to add a component subassembly operation that can produce 80 units during a regular 8-hour shift.

This operation consists of three activities as below

Activity	Standard time (min)
M. Mechanical assembly	12
E. Electric wiring	16
T. Test	3

For line balancing the number of work stations required for the activities M, E and T would respectively be

(a) 2, 3, 1 (b) 3, 2, 1

(c) 2, 4, 2 (d) 2, 1,3

[2004 : 2 Marks]

9. A welding operation is time-studied during which an operator was pace-rated as 120%. The operator took, on an average, 8 minutes for producing the weld-joint. If a total of 10% allowances are at lowed for this operation. The expected standard production rate of the weld-joint (in units per 8 hour day) is

(a) 45 (b) 50

(c) 55 (d) 60

[2005 : 2 Marks]

10. A manufacturing shop processes sheet metal jobs, wherein each job must pass through two machines (M1 and M2, in that order). The processing time (in hours) for these jobs is

Machine	Jobs					
	P	Q	R	S	T	U
M1	15	32	8	27	11	16
M2	6	19	13	20	14	7

The optimal make-span (in hours) of the shop is

(a) 120 (b) 115
(c) 109 (d) 79

[2006 : 2 Marks]

11. The table gives details of an assembly line.

Work station	I	II	III	IV	V	VI
Total task time at the workstation (in minutes)	7	9	7	10	9	6

What is the line efficiency of the assembly line?

(a) 70% (b) 75%
(c) 80 % (d) 85%

[2006 : 2 Marks]

12. Capacities of production of an item over 3 consecutive months in regular time are 100,100 and 80 and in overtime are 20, 20 and 40. The demands over those 3 months are 90, 130 and 110. The cost of production in regular time and overtime are respectively Rs. 20 per item and Rs. 24 per item. Inventory carrying cost is Rs. 2 per item month. The levels of starting and final inventory are nil. Backorder is not permitted. For minimum cost of plan, the level of planned production in overtime in the third month is

(a) 40 (b) 30
(c) 20 (d) 0

[2007 : 2 Marks]

13. The Product structure of an assembly P is shown in the figure.

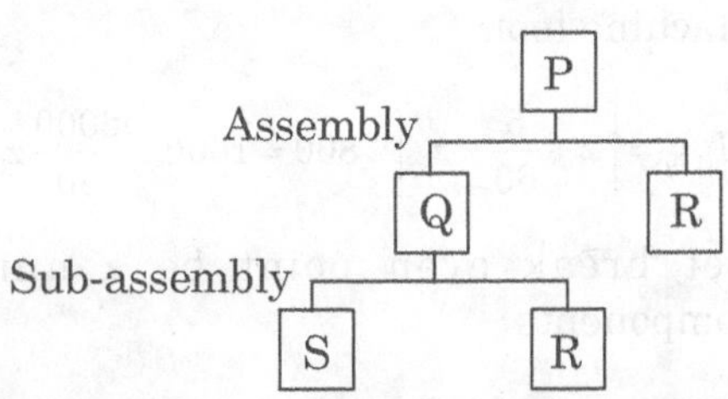

Estimated demand for end product P is as follows:

Week	1	2	3	4	5	6
Demand	1000	1000	1000	1000	1200	1200

Ignore lead times for assembly and subassembly. Production capacity (per week) for component R is the bottleneck operation. Starting with zero inventory, the smallest capacity that will ensure a feasible production plan up to week 6 is

(a) 1000 (b) 1200
(c) 2200 (d) 2400

[2008 : 2 Marks]

14. Vehicle manufacturing assembly line is an example of

(a) product layout (b) process layout
(c) manual layout (d) fixed layout

[2010 : 1 Mark]

15. The word kanban is most appropriately associated with

(a) economic order quantity
(b) just-in-time production
(c) capacity planning
(d) product design

[2011 : 1 Mark]

16. Which one of the following is NOT a decision taken during the aggregate production planning stage?

(a) Scheduling of machines
(b) Amount of labour to be committed
(c) Rate at which production should happen
(d) Inventory to be carried forward

[2012 : 1 Mark]

17. During the development of a product, an entirely new process plan is made based on design logic, examination of geometry and tolerance information. This type of process planning is known as

(a) retrieval (b) generative
(c) variant (d) group technology based

[2015 : 1 Mark, Set-2]

18. Cylindrical pins of diameter $15^{\pm 0.020}$ mm are being produced on a machine. Statistical quality control tests show a mean of 14.995 mm and standard deviation of 0.004 mm. The process capability index Cp is

(a) 0.833 (b) 1.667
(c) 3.333 (d) 3.750

[2017 : 1 Mark, Set-1]

19. Following data refers to the jobs (P, Q, R, S) which have arrived at a machine for scheduling. The shortest possible average flow time is ____ days.

Job	Processing Time (days)
P	15
Q	9
R	22
S	12

[2018 : 2 Marks, Set-1]

20. Processing times (including step times) and due dates for six jobs waiting to be processed at a work centre are given in the table. The average tardiness (in days) using shortest processing time rule is (correct to two decimal places).

Job	Processing time (days)	Due date (days)
A	3	8
B	7	16
C	4	4
D	9	18
E	5	17
F	13	19

[2018 : 2 Marks, Set-1]

ANSWERS

2. (d) **4.** (d) **5.** (a) **6.** (d) **7.** (d) **8.** (a) **9.** (a) **10.** (b) **11.** (c) **12.** (b)
13. (c) **14.** (a) **15.** (b) **16.** (b) **17.** (b) **18.** (b)

EXPLANATIONS

1. MRP

2. Way to locate machine at four plans

$\underline{|4} = 24$

3. (a) JIT – (4) Kanbans

(b) Computer –(1) CRAFT
Assisted
Layout

(c) Scheduling –(3) John son's Rule

(d) Simulations – (6) Monte carl

$\Rightarrow$ A –4, B–1, C–3, D–6.

4. Production flow analysis (PFA) is a comprehensive method for material analysis part family formations, design of manufacturing cell & facility layout design. These operation are taken from the Route sheet.

5. For Machine M1:

Fixed cost = Rs 100

Variable cost = Rs 2 per piece

For Machine M2:

Fixed cost = Rs 200

Variable cost = 1 Rs per piece

Total cost = Fixed cost + Variable cost × Number of units.

Total cost of production on machine M1 & M2 is

$\Rightarrow$ 100 + 2n + 200 + n = 300 + 2n

For the first 300 units, selling price is 3.50 Rs per unit.

So the break even point

300 + 3n = 3.50 (n+n)

300 + 3n = 3.50 (2 n)

7n = 300 + 3n = n = 75

6. Given data

	Standard Machine Tool	Automatic Machine Tool
Set up time	30 min	2 hours
Machining time per piece	22 min	5 min
Machine rate	200 per hour	Rs. 800 per hour

Total cost of z_1 component by using standard machine tool,

$$(T_C)_1 = \left[\frac{30}{60} + \frac{22 \times z_1}{60}\right] \times 200 = 100 + \frac{2200}{30} z_1$$

Total cost of z_2 component by using Automatic Machine tool,

$$(T_C)_2 = \left[2 + \frac{5}{60} \times z_2\right] \times 800 = 1600 + \frac{2000}{30} z_2$$

Let break even point be z number of components

$$\therefore\ 100 + \frac{2200}{30} z = 1600 + \frac{2000z}{30}$$

or $\frac{200}{30}z = 1500$

or $z = \frac{1500 \times 30}{200} = 225$

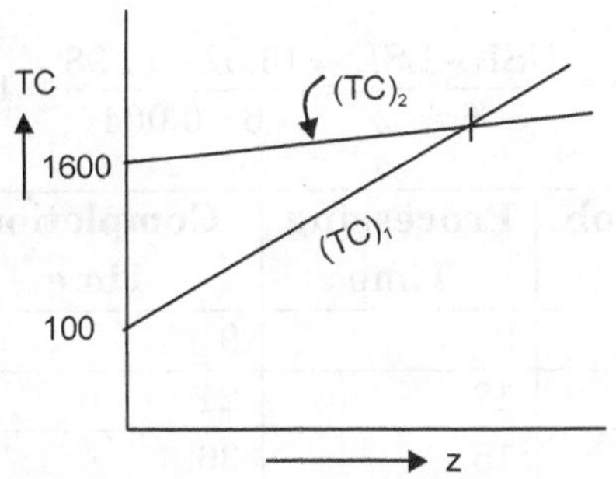

Alternately:

Let N be the Break even number

At Break even point

$$\left(\frac{1}{2}+\frac{22}{60}\cdot N\right) = \left(2+\frac{5N}{60}\right)\times\frac{800}{200}$$

$$\Rightarrow \quad \frac{1}{2}+\frac{22}{60}N = \left(2+\frac{5N}{60}\right)\times\frac{800}{200}$$

$$\Rightarrow \quad \frac{1}{2}+\frac{22}{60}N = 8+\frac{20N}{60}$$

$$\Rightarrow \quad \frac{22}{60}N - \frac{20}{60}N = -\frac{1}{2}$$

$$\Rightarrow \quad \left(\frac{22-20}{60}\right)N = \frac{16-1}{2}$$

$$\Rightarrow \quad N = \frac{15 \times 60}{2 \times 2} = 225$$

7. Total time = 16 hours

T = 16 × 60 = 960 min

$$\text{Actual time} = \frac{90}{100} \times 960$$

= 864 minute

Normal time T normal = 120% of 864

$$\Rightarrow \frac{120}{100} \times 864 = 1036.8 \text{ min}$$

Total Standard time: $T_{standard}$

$$T_{standard} = \frac{T_{normal}}{1-\frac{20}{100}}$$

$$= \frac{1036.8}{1-0.2} = \frac{1036.8}{0.8} = 1296 \text{ min}$$

No. of Joints soldered(N)

N = 108.

Standard time for operation

$$\Rightarrow \frac{1296}{108} = 12 \text{ min}$$

8. Number of units produce in a day = so units working hour in a day = 8 hours

$$\text{Time taken to produce 1 unit} = \frac{8}{80} \times 60 = 6 \text{ min}$$

T = 6 min

Activity	Standard Time (min)	No. of Work Station S.T/T
M. Mechanical Assembly	12	1216 = 2
E. Electrical Wiring	16	1616 = 2.666 = 3
T. Test	3	316 = 0.5 = 1

Number of work station are the whole number more the function.

So no of work station required for the activities, M, E & T would be 2, 3, 4, 1

9. Normal time = Actual Time × Rating factor

$$T_{normal} = 8 \times \frac{120}{100} = 9.6 \text{ min}$$

$$\text{Standard time} = T_{standard} = \frac{T_{normal}}{1-\frac{10}{100}} = \frac{9.6}{0.9}$$

= 10.67 min

$$\text{Standard Production rate} = \frac{8 \times 60}{10.67} = 45 \text{ units}$$

10. Using Johnson's algorithm

	M1		M2	
	IN	OUT	IN	OUT
R	0	8	8	21
T	8	19	21	35
S	19	46	35	55
Q	46	78	78	97
U	78	94	97	104
P	94	109	109	115

11. Line efficiency

$$= \frac{\text{Total time used}}{\text{Number of work stations} \times \text{Cycle time}} \times 100$$

$$= \frac{48}{6 \times 10} \times 100 = 80\%$$

12. For minimum total cost the entire regular time production capacity has to be used.

Hence, total over time capacity required =

(90 + 130 + 110) – (100+ 100 + 80) = 50

20 Overtime units will be used up in first 2 weeks

∴ Hence 30 overtime units are needed to fulfill the 3rd month demand.

13. The M.R.P. for R is given by

Week	1	2	3	4	5	6
Demand	2000	2000	2000	2000	2400	2400
Inventory Level I =P	P – 2000	2P – 4000	3P – 6000	4P – 8000	5P – 10400	6P-12800

If 6P – 12800 ≥ 0 then the capacity is feasible

∴ Minimization, 6P = 12800

⇒ P ≅ 2200

18. Diameter of cylindrical Pin = 15 ±.020

Upper Specification limit = 15 + 0.02

= 15.02 mm

Lower Specification limit = 15 – 0.02

= 14.98 mm

Standard deviation (σ) = 0.004 mm

Process capability Index

$$Cp = \frac{USL - LSL}{6\sigma} = \frac{15.02 - 14.98}{6 \times 0.004} = 1.667.$$

19.

Job	Processing Time	Completion Time
Q	9	9
S	12	21
P	15	36
R	22	58
		Σci = 124

$$\text{Near flow time} = \frac{\sum c_i}{4} = \frac{124}{4} = 31$$

20.

Job	Processing Time	Completion Time	Due Date	Lateness	Tardinese
A	3	3	8	-5	0
C	4	7	4	3	3
E	5	12	17	-5	0
B	7	19	16	3	3
D	9	28	18	10	10
F	13	41	19	22	22

$$\text{Average tardiness} = \frac{\text{Total tardiness}}{\text{No. of Jobs}}$$

$$= \frac{3 + 3 + 10 + 22}{6} = \frac{38}{6} = 6.33 \text{ days}$$

7
CHAPTER

Queuing Theory and Transportation

Analysis of Previous GATE Papers

Topic	Year/Set → Marks Distribution ↓		2019 Set 1	2019 Set 2	2018 Set 1	2018 Set 2	2017 Set 1	2017 Set 2	2016 Set 1	2016 Set 2	2016 Set 3	2015 Set 1	2015 Set 2	2015 Set 3
Queuing theory	1 Mark	MCQ Type						1						1
		Numerical Type				1				1				
	2 Marks	MCQ Type												
		Numerical Type												
	Total					1		1		1				1
Transportation	1 Mark	MCQ Type												
		Numerical Type												
	2 Marks	MCQ Type						1						
		Numerical Type						1						
	Total							4						

Queuing Theory

1. On an average 100 customers arrive at a place each hour, and on the average the server can process 120 customers per hour. What is the proportion of time the server is idle? .

[1995 : 2 Marks]

2. The cost of providing service in a queuing system increases with

(a) Increased mean time in the queue

(b) Increased arrival rate

(c) Decreased mean time in the queue

(d) Decreased arrival rate **[1997 : 1 Mark]**

3. At a production machine, parts arrive according to a Poisson process at the rate of 0.35 parts per minute. Processing time for parts have exponential distribution with mean of 2 minutes. What is the probability that a random part arrival finds that there are already 8 parts in the system (in machine + in queue)?

(a) 0.0247 (b) 0.0576

(c) 0.0173 (d) 0.082

[1999 : 2 Marks]

4. In a single server infinite population queuing model. Arrivals follow a Poisson distribution with mean $\lambda = 4$ per hour. The service times are exponential with mean service time equal to 12 minutes. The expected length of the queue will be

(a) 4 (b) 3.2

(c) 1.25 (d) 24.3

[2000 : 2 Marks]

5. A maintenance service facility has Poisson arrival rates, negative exponential service time and operates on a 'first come first served' queue discipline. Break-downs occur on an average of 3 per day with a range of zero to eight. The maintenance crew can service an average of 6 machines per day with a range of zero to seven. The mean waiting time for an item to be serviced would be

(a) 1/6 day (b) 1/3 day

(c) 1 day (d) 3 day

[2004 : 2 Marks]

6. Consider a single server queuing model with Poisson arrivals (λ = 4/hour) and exponential service (λ = 4/hour). The number in the system is restricted to a maximum of 10. The probability that a person who comes in leaves without joining the queue is

(a) 1/11 (b) 1/10

(c) 1/9 (d) 1/2

[2005 : 1 Mark]

7. The number of customers arriving at a railway reservation counter is Poisson distributed with an arrival rate of eight customers per hour. The reservation clerk at this counter takes six minutes per customer on an average with an exponentially distributed service time. The average number of the customers in the queue will be

(a) 3 (b) 3.2

(c) 4 (d) 4.2 **[2006 : 1 Mark]**

8. In an M/M/1 queuing system, the number of arrivals in an interval of length T is a Poisson random variable i.e. the probability of there being n arrivals in an interval of length T is $\frac{e^{\lambda T}(\lambda T)^n}{n!}$ The probability density function f(t) of the inter- arrival time is given by

(a) $\lambda^2(e^{-\lambda 2t})$ (b) $\frac{e^{-\lambda^2 t}}{\lambda^2}$

(c) $\lambda e^{-\lambda t}$ (d) $\frac{e^{-\lambda t}}{\lambda}$ **[2008 : 1 Mark]**

9. Little's law is a relationship between

(a) stock level and lead time in an inventory system

(b) waiting time and length of the queue in a queuing system

(c) number of machines and job due dates in a scheduling problem

(d) uncertainty in the activity time and project completion time **[2010 : 1 Mark]**

10. Cars arrive at a service station according to Poisson's distribution with a mean rate of 5 per hour. The service time per car is exponential with a mean of 10 minutes. At steady state, the average waiting time in the queue is

(a) 10 minutes (b) 20 minutes

(c) 25 minutes (d) 50 minutes

[2011 : 1 Mark]

11. Customers arrive at a ticket counter at a rate of 50 per hr and tickets are issued in the order of their arrival. The average time taken for issuing a ticket is 1 min. Assuming that customer arrivals form a Poisson process and service times are exponentially distributed, the average waiting time in queue in min is

(a) 3 (b) 4

(c) 5 (d) 6 **[2013 : 1 Mark]**

12. At a work station, 5 jobs arrive every minute. The mean time spent on each job in the work station is 1/8 minute. The mean steady state number of jobs in the system is

[2014 : 2 Marks, Set-4]

13. In the notation (a/b/c) : {d/e/f) for summarizing the characteristics of queuing situation, the letters b and c stand respectively for

(a) service time distribution & queue discipline
(b) number of servers & size of calling source
(c) number of servers & queue discipline
(d) service time distribution and maximum number allowed in system

[2015 : 1 Mark, Set-3]

14. In a single-channel queuing model, the customer arrival rate is 12 per hour and the serving rate is 24 per hour. The expected time that a customer is in queue is minutes.

[2016 : 1 Mark, Set-2]

15. For a single server with Poisson arrival and exponential service time, the arrival rate is 12 per hour. Which one of the following service rates will provide a steady state finite queue length?

(a) 6 per hour (b) 10 per hour
(c) 12 per hour (d) 24 per hour

[2017 : 1 Mark, Set-2]

16. The arrival of customers over fixed time intervals in a bank follow a Poisson distribution with an average of 30 customers/hour. The probability that the time between successive customer arrival is between 1 and 3 minutes is (correct to two decimal places).

[2018 : 1 Marks, Set-2]

Transportation

17. The supply at three sources is 50, 40 and 60 units respectively while the demand at the four destinations is 20, 30,10 and 50 units. In solving this transportation problem.

(a) a dummy source of capacity 40 units is needed
(b) a dummy destination of capacity 40 units is needed
(c) no solution exists as the problem is infeasible
(d) none solution exists as the problem is degenerate

[2002 : 2 Marks]

18. If there are m sources and n destinations in a transportation matrix, the total number of basic variables in a basic feasible solution is

(a) m + n (b) m + n + 1
(c) m + n – 1 (d) m

[2014 :1 Mark, Set-2]

19. The total number of decision variables in the objective function of an assignment problem of size n × n{n jobs and n machines) is

(a) n^2 (b) 2n
(c) 2n – 1 (d) n

[2014 : 1 Mark, Set-4]

20. Maximize $Z = 5x_1 + 3x_2$, subject to

$x_1 + 2x_2 \leq 10$

$x_1 - x_2 \leq 8$

$x_2 , x_2 \geq 0$

In the starting simplex tableau, x_1 and x_2 are non- basic variables and the value of Z is zero. The value of Z in the next simplex tableau is.

[2017 : 2 Marks, Set-2]

21. A product made in two factories, P and Q, is transported to two destinations, Rand S. The per unit costs of transportation (in Rupees) from factories to destinations are as per the following matrix:

Factory \ Destination	R	S
P	10	7
Q	3	4

Factory P produces 7 units and factory Q produces 9 units of the product. Each destination requires 8 units. If the north-west corner method provides the total transportation cost X(in Rupees) and the optimized (the minimum) total transportation cost is Y(in Rupees), then(X-Y), in Rupees, is

(a) 0 (b) 15
(c) 35 (d) 105

[2017 : 2 Marks, Set-2]

ANSWERS

2. (c) **3.** (c) **4.** (b) **5.** (a) **6.** (a) **7.** (b) **8.** (c) **9.** (b) **10.** (d) **11.** (c)
13. (a) **15.** (d) **17.** (b) **18.** (c) **19.** (a) **21.** (c)

EXPLANATIONS

3. Given, $\lambda = 0.35/\text{min.},$

$$\mu = \frac{1}{2}/\text{min} = 0.5 \text{ min.}$$

$$\rho = \frac{\lambda}{\mu} = \frac{0.35}{0.5} = 0.7$$

Probability of '*n*' numbers in the system,

$$p_n(t) = \rho^n(1-\rho)$$

$$= (0.7)^8(1-0.7) = 0.0173.$$

4. $\lambda = \frac{4}{60}, \ \mu = \frac{1}{12}$

$$\therefore \ \rho = \frac{\lambda}{\mu} = \frac{48}{60}$$

$$W_q = \text{queue length} = \frac{\rho^2}{1-\rho} = \frac{(0.8)^2}{(1-0.8)} = 3.2$$

5. Given: $\lambda = 3$ per day (arrival rate)

$\mu = 6$ per day (service rate)

Mean waiting time for an item to serviced

$$= \frac{\lambda}{\mu(\mu-\lambda)} = \frac{3}{6(6-3)} = \frac{1}{2} \times \frac{1}{3} = \frac{1}{6} \text{ day.}$$

6. When $\rho = 1,$

$$p_n = \frac{1}{N+1}$$

7. Average number of customer in queue

$$= \frac{\lambda^2}{\mu(\mu-A)} = \frac{8\times8}{10\times(10-8)} = 3.2$$

8. For any Poisson distribution, the probability density function f(t) is given by $\lambda e^{-\lambda t}$.

10. $W_q = \frac{\rho}{\mu-\lambda}$

where $\lambda = 5$/hr

$\mu = 6$/hr

$$\rho = \frac{\lambda}{\mu} = \frac{5}{6}$$

$$\therefore \ W_q = \frac{\frac{5}{6}}{6-5} = \frac{5}{6} \text{ hr} = 50 \text{ min}$$

11. $\lambda = 50, \quad = 60$

$$t = \frac{\lambda}{\mu(\mu-\lambda)} = .0833\text{hr}$$

$$= 5 \text{ min.}$$

13. b : Service time distribution (usually represented by 'm')

d : Queuing discipline (usually represented by 'GD')

14. $\lambda = 12$ /hr, $\mu = 24$ /hr

$$\rho = \frac{\lambda}{\mu} = \frac{12}{24} = 0.5$$

Expected time that the customer in the queue

$$W_q = \frac{L_q}{\lambda} = \frac{L_s - \rho}{\lambda}$$

$$= \frac{\frac{\rho}{1-\rho} - \rho}{\lambda} = \frac{\frac{0.5}{1-0.5} - 0.5}{12}$$

$$= \frac{0.5}{12} \text{hr} = \frac{0.5}{12} \times 60 \text{ min}$$

$$= 2.5 \text{ min}$$

15. For steady state, $\mu > \lambda$

as $\lambda = 12$ customers/hour, we need to select option (d)

18.

ORIGIN \ DESTINATION	1	2	-- Σ --	n	TOTAL
1	X_{11}	X_{12}	----		a_1
2	X_{21}	X_{22}	----		a_2
⋮	----	----	----	----	----
m	X_{m_1}	X_{m_2}		X_{m_n}	a_m
TOTAL	b_1	b_2	----	b_n	$\sum_{i=1} a = \sum_{J=1} b_J$

$$\sum_{J=1}^{n} X_{iJ} = a_i, \ (i = 1, 2, \ldots m) \quad \ldots(1)$$

$$\sum_{I=1}^{m} X_{iJ} = b_J, \ (J = 1, 2, \ldots n) \quad \ldots(2)$$

$$X_{iJ} \geq 0 \quad \ldots(3)$$

$$\sum_{i=1}^{m} \sum_{J=1}^{m} C_{iJ} X_{iJ} = mn \quad \ldots(4)$$

It is evident from the above diagram, that m + n equations (1) and (2) constitute m+n–1 independent equations in *mn* unknowns.

19. A, $n \times n$ assignment problem, if it is solved as a LPP it will have n^2 variables.

Unit - XII

General Aptitude

Syllabus

Verbal Ability: English grammar, sentence completion, verbal analogies, word groups, instructions, critical reasoning and verbal deduction.

Numerical Ability: Numerical computation, numerical estimation, numerical reasoning and data interpretation.

Contents

1
CHAPTER

Verbal Ability

Analysis of Previous GATE Papers

Topic	Year/Set → Marks Distribution ↓		2019 Set 1	2019 Set 2	2018 Set 1	2018 Set 2	2017 Set 1	2017 Set 2	2016 Set 1	2016 Set 2	2016 Set 3	2015 Set 1	2015 Set 2	2015 Set 3
Verbal Ability	1 Mark	MCQ Type	3	3	2	2	2	2	3	3	3	2	3	4
		Numerical Type												
	2 Marks	MCQ Type	3	2	1		2	2	2	1	1	3	3	2
		Numerical Type												
	Total		9	7	4	2	6	6	7	5	5	8	9	8

1. Which of the following options is the closest in meaning to the word below?
Circuitous
(a) Cyclic (b) Indirect
(c) Confusing (d) Crooked
[2010 : 1 Mark]

2. The question below consists of a pair of related words followed by four pairs of words. Select the pair that best expresses the relation in the original pair.
Unemployed: Worker
(a) Fallow: Land (b) Unaware: Sleeper
(c) Wit: Jester (d) Renovated: House
[2010 : 1 Mark]

3. Choose the most appropriate word from the options given below to complete the following sentence:
If we manage to .... our natural resources, we would leave a better planet for our children.
(a) uphold (b) restrain
(c) cherish (d) conserve
[2010 : 1 Mark]

4. Choose the most appropriate word from the options given below to complete the following sentence:
His rather casual remarks on politics ...... his lack of seriousness about the subject.
(a) masked (b) belied
(c) betrayed (d) suppressed
[2010 : 1 Mark]

5. Modern warfare has changed from large scale clashes of armies to suppression of civilian populations. Chemical agents that do their work silently appear to be suited to such warfare; and regretfully, there exist people in military establishments who think that Chemical agents are useful fools for their cause. Which of the following statements best sums up the meaning of the above passage?
(a) Modern warfare has resulted in civil strife.
(b) Chemical agents are useful in modern warfare.
(c) Use of Chemical agents in warfare would be undesirable.
(d) People in military establishments like to use Chemical agents in war.
[2010 : 2 Marks]

6. Choose the word from the options given below that is most nearly opposite in meaning to the given word:
Amalgamate
(a) merge (b) split
(c) collect (d) separate
[2011 : 1 Mark]

7. Which of the following options is the closest in the meaning to the word below:
Inexplicable
(a) Incomprehensible
(b) Indelible
(c) Inextricable
(d) Infallible
[2011 : 1 Mark]

8. Choose the most appropriate word(s) from the options given below to complete the following sentence.
I contemplated .... Singapore for my vacation but decided against it.
(a) to visit (b) having to visit
(c) visiting (d) for a visit
[2011 : 1 Mark]

9. Choose the most appropriate word from the options given below to complete the following sentence.
If you are trying to make a strong impression on your audience, you cannot do so by being understated, tentative or
(a) hyperbolic (b) restrained
(c) argumentative (d) indifferent
[2011 : 1 Mark]

10. **Few school curricula include a unit on how to deal with bereavement and grief, and yet all students at some point in their lives suffer from losses through death and parting.**
Based on the above passage which topic would not be included in a unit on bereavement?
(a) how to write a letter of condolence
(b) what emotional stages are passed through in the healing process
(c) what the leading causes of death are
(d) how to give support to a grieving friend
[2011 : 2 Marks]

11. Choose the most appropriate alternative from the options given below to complete the following sentence.
Suresh's dog is the one ....was hurt in the stampede.
(a) that (b) which
(c) who (d) whom
[2012 : 1 Mark]

12. Choose the most appropriate alternative from the options given below to complete the following sentence:
Despite several .... the mission succeeded in its attempt to resolve the conflict.
(a) attempts (b) setbacks
(c) meetings (d) delegations
[2012 : 1 Mark]

13. Which one of the following options is the closest in meaning to the word given below?

Mitigate

(a) Diminish (b) Divulge

(c) Dedicate (d) Denote

[2012 : 1 Mark]

14. Choose the grammatically INCORRECT sentence:

(a) They gave us the money back less the service charges to Three Hundred rupees.

(b) This country's expenditure is not less than that of Bangladesh.

(c) The committee initially asked for a funding of Fifty Lakh rupees, but later settled for a lesser sum.

(d) This country's expenditure on educational reforms is very less. **[2012 : 1 Mark]**

15. Wanted Temporary, Part-time persons for the post of Field Interviewer to conduct personal interviews to collect and collate economic data. Requirements: High School-pass, must be available for Day, Evening and Saturday work. Transportation paid, expenses reimbursed.

Which one of the following is the best inference from the above advertisement?

(a) Gender-discriminatory

(b) Xenophobic

(c) Not designed to make the post attractive

(d) Not gender-discriminatory

[2012 : 2 Marks]

16. Complete the sentence:

Universalism is to particularism as diffuseness is to _____.

(a) specificity (b) neutrality

(c) generality (d) adaptation

[2013 : 1 Mark]

17. Were you a bird, you _____ in the sky.

(a) would fly (b) shall fly

(c) should fly (d) shall have flown

[2013 : 1 Mark]

18. Which one of the following options is the closest in meaning to the word given below?

Nadir

(a) Highest (b) Lowest

(c) Medium (d) Integration

[2013 : 1 Mark]

19. Choose the grammatically INCORRECT sentence:

(a) He is of Asian origin.

(b) They belonged to Africa

(c) She is an European.

(d) They migrated from India to Australia.

[2013 : 1 Mark]

20. After several defeats in wars, Robert Bruce went in exile and wanted to commit suicide. Just before committing suicide, he came across a spider attempting tirelessly to have its net. Time and again, the spider failed but the did not deter it to refrain from making attempts. Such attempts by the spider made Bruce curious. Thus, Bruce started observing the near-impossible goal of the spider to have the net. Ultimately, the spider succeeded in having its net despite several failures. Such act of the spider encouraged Bruce not to commit suicide. And then, Bruce went back again and won many a battle, and the rest is history.

Which one of the following assertions is best supported by the above information?

(a) Failure is the pillar of success

(b) Honesty is the best policy

(c) Life begins and ends with adventures

(d) No adversity justifies giving up hope

[2013 : 2 Marks]

21. Choose the most appropriate phrase from the options given below to complete the following sentence. The aircraft _____ take off as soon as its flight plan was filed.

(a) is allowed to (b) will be allowed to

(c) was allowed to (d) has been allowed to

[2014 : 1 Mark, Set-1]

22. Read the statements:

All women are entrepreneurs.

Some women are doctors.

Which of the following conclusions can be logically inferred from the above statements?

(a) All women are doctors

(b) All doctors are entrepreneurs

(c) All entrepreneurs are women

(d) Some entrepreneurs are doctors

[2014 : 1 Mark, Set-1]

23. Choose the most appropriate word from the options given below to complete the following sentence.

Many ancient cultures attributed disease to supernatural causes. However, modern science has largely helped ______ such notions.

(a) impel (b) dispel

(c) propel (d) repel

[2014 : 1 Mark, Set-1]

24. Choose the most appropriate word from the options given below to complete the following sentence.

Communication and interpersonal skills are ______ important in their own ways.

(a) each (b) both

(c) all (d) either

[2014 : 1 Mark, Set-2]

25. Which of the options given below best completes the following sentence?
She will feel much better if she ________.
(a) will get some rest
(b) gets some rest
(c) will be getting some rest
(d) is getting some rest **[2014 : 1 Mark, Set-2]**

26. Choose the most appropriate pair of words from the options given below to complete the following sentence.
She could not ______, the thought of ______ election to her bitter rival.
(a) bear, loosing (b) bare, loosing
(c) bear, losing (d) bare, losing
[2014 : 1 Mark, Set-2]

27. "India is a country of rich heritage and cultural diversity". Which one of the following facts best supports the claim made in the above sentence?
(a) India is a union of 28 states and 7 union territories.
(b) India has a population of over 1.1 billion.
(c) India is home to 22 official languages and thousands of dialects.
(d) The indian cricket team draws players from over ten states.
[2014 : 1 Mark, Set-3]

28. The value of one U.S. dollar is 65 Indian Rupees today, compared to 60 last year. The Indian Rupee has______.
(a) depressed (c) appreciated
(b) depreciated (d) stabilized
[2014 : 1 Mark, Set-3]

29. 'Advice' is ________.
(a) a verb (b) a noun
(c) an adjective (d) both a verb and a noun
[2014 : 1 Mark, Set-3]

30. Which of the following options is the closest in meaning to the word underlined in the sentence below?
In a democracy, everybody has the freedom to <u>disagree</u> with the government.
(a) dissent (b) descent
(c) decent (d) decadent
[2014 : 1 Mark, Set-4]

31. After the discussion, Tom said to me, 'Please revert!' He expects me to ________.
(a) retract (b) get back to him
(c) move in reverse (d) retreat
[2014 : 1 Mark, Set-4]

32. While receiving the award, the Scientist said, "I feel vindicated". Which of the following is closest in meaning to the word 'vindicated'?
(a) punished (b) substantiated
(c) appreciated (d) chastened
[2014 : 1 Mark, Set-4]

33. Lights of four colors (red, blue, green, yellow) are hung on a ladder. On every step of the ladder there are two lights. If one of the lights is red, the other light on that step will always be blue. If one of the lights on a step is green, the other light on that step will always be yellow. Which of the following statements is not necessarily correct?
(a) The number of red lights is equal to the number of blue lights
(b) The number of green lights is equal to the number of yellow lights
(c) The sum of the red and green lights is equal to the sum of the yellow and blue lights
(d) The sum of the red and blue lights is equal to the sum of the green and yellow lights
[2014 : 2 Marks, Set-2]

34. What is the adverb for the given word below?
Misogynous
(a) Misogynousness (b) Misogynity
(c) Misogynously (d) Misogynous
[2015 : 1 Mark, Set-1]

35. Choose the appropriate word/phrase, out of the four options given below, to complete the following sentence:
Dhoni, as well as the other team members of Indian team, ________ present on the occasion.
(a) were (b) was
(c) has (d) have
[2015 : 1 Mark, Set-1]

36. Choose the word most similar in meaning to the given word:
Awkward
(a) Inept (b) Graceful
(c) Suitable (d) Dreadful
[2015 : 1 Mark, Set-1]

37. Choose the correct verb to fill in the blank below:
Let us ______.
(a) introvert (b) alternate
(c) atheist (d) altruist
[2015 : 1 Mark, Set-2]

38. Choose the most appropriate word from the options given below to complete the following sentence.
If the athlete had wanted to come first in the race, he ________ several hours every day.
(a) should practise
(b) should have practised
(c) practised
(d) should be practising **[2015 : 1 Mark, Set-2]**

39. Choose the most suitable one word substitute for the following expression. Connotation of a road or way

(a) Pertinacious (b) Viaticum
(c) Clandestine (d) Ravenous

[2015 : 1 Mark, Set-2]

40. Fill in the blank with the correct idiom/phrase. That boy from the town was a ____ in the sleepy village.

(a) dog out of herd (b) sheep from the heap
(c) fish out of water (d) bird from the flock

[2015 : 1 Mark, Set-3]

41. Choose the appropriate word/phase, out of the four options given below, to complete the following sentence:

Apparent lifelessness ________ dormant life.

(a) harbours (b) lead to
(c) supports (d) affects

[2015 : 1 Mark, Set-3]

42. Tanya is older than Eric.

Cliff is older than Tanya

Eric is older than Cliff.

If the first two statements are true, then the third statement is:

(a) True (b) False
(c) Uncertain (d) Data insufficient

[2015 : 1 Mark, Set-3]

43. Choose the statement where underlined word is used correctly.

(a) When the teacher eludes to different authors, he is being <u>elusive</u>
(b) When the thief keeps eluding the police, he is being <u>elusive</u>
(c) Matters that are difficult to understand, identify or remember are <u>allusive</u>
(d) Mirages can be <u>allusive</u>. but a better way to express them is illusory

[2015 : 1 Mark, Set-3]

44. Lamenting the gradual sidelining of the arts in school curricula, a group of prominent artists wrote to the Chief Minister last year, asking him to allocate more funds to support arts education in schools. However, no such increase has been announced in this year's Budget. The artists expressed their deep anguish at their request not being approved, but many of them remain optimistic about funding in the future. Which of the statement(s) below is/are logically valid and can be inferred from the above statements?

i. The artists expected funding for the arts to increase this year
ii. The Chief Minister was receptive to the idea of increasing funding for the arts.
iii. The Chief Minister is a prominent artist.
iv Schools are giving less importance to arts education nowadays.

(a) iii and iv (b) i and iv
(c) i, ii and iv (d) i and iii

[2015 : 2 Marks, Set-1]

45. Given below are two statements followed by two conclusions. Assuming these statements to be true, decide which one logically follows.

Statement:

i. All film stars are playback singers.
ii. All film directors are film stars.

Conclusions:

i. All film directors are playback singers.
ii. Some film stars are film directors.

(a) Only conclusion I follows
(b) Only conclusion II follows
(c) Neither conclusion I nor II follows
(d) Both conclusions I and II follow

[2015 : 2 Marks, Set-1]

46. In the following sentence certain parts are underlined and marked P, Qand R. One of the parts may contain certain error or may not be acceptable in standard written Communication. Select the part containing an error. Choose D as your answer if there is no error.

The student corrected <u>all the errors</u> (P) that <u>the instructor marked</u> (Q) on the <u>answer book</u> (R)

(a) P (b) Q
(c) R (d) No Error

[2015 : 2 Marks, Set-1]

47. Ms. X will be in Bagdogra from 01/05/2014 to 20/05/2014 and from 22/05/2014to 31/05/2014.

On the morning of 21/05/2014, she will reach Kochi via Mumbai.

Which one of the statement below is logically valid and can be inferred from the above sentences?

(a) Ms. X will be in Kochi for one day, only in May.
(b) Ms. X will be in Kochi for only one day in May.
(c) Ms. X will be only in Kochi for one day in May.
(d) Only Ms. X will be in Kochi for one day in May.

[2015 : 2 Marks, Set-2]

48. In the following questions, the first and the last sentence of the passage are in order and numbered 1 and 6. The rest of the passage is split into 4 parts and numbered as 2, 3, 4 and 5. These 4 parts are not arranged in proper order. Read the sentences and arrange them in a logical sequence to make a passage and choose the correct sequence from the given options.
1. On Diwali, the family rises early in the morning.
2. The whole family, including the young and the old enjoy doing this.
3. Children let off fireworks later in the night with their friends.
4. At sunset, the lamps are lit and the family performs various rituals.
5. Father, mother, and children visit relatives and exchanges gifts and sweets.
6. Houses look so pretty with lighted lamps are around.

(a) 2, 5, 3, 4 (b) 5, 2, 4, 3
(c) 3, 5, 4, 2 (d) 4, 5, 2, 3

[2015 : 2 Marks, Set-2]

49. Ram and Shyam shared a secret and promised to each other that it would remain between them.
Ram expressed himself in one of the following ways as given in the choices below. Identify the correct way as per standard English.
(a) It would remain between you and me
(b) It would remain between I and you
(c) It would remain between you and I
(d) It would remain with me

[2015 : 2 Marks, Set-2]

50. Select the appropriate option in place of underlined part of the sentence.
<u>Increased productivity necessary</u> reflects greater efforts made by the employees.
(a) Increase in productivity necessary
(b) Increase productivity is necessary
(c) Increase in productivity necessarily
(d) No improvement required

[2015 : 2 Marks, Set-3]

51. Given below are two statements followed by two conclusions. Assuming these statements to be true, decide which one logically follows:
Statements:
I. No manager is a leader.
II. All leaders are executive.
Conclusions:
I. No manager is a executive.
II. All executive is a manager.
(a) Only conclusion I follows.
(b) Only conclusion II follows.
(c) Neither conclusion I nor II follows.
(d) Both conclusion I and II follow.

[2015 : 2 Marks, Set-3]

52. Which of the following is CORRECT with respect to grammar and usage?
Mount Everest is ________.
(a) the highest peak in the world
(b) highest peak in the world
(c) one of highest peak in the world
(d) one of the highest peak in the world

[2016 : 1 Mark, Set-1]

53. The policeman asked the victim of a theft, "What did you ______?"
(a) loose (b) lose
(c) loss (d) louse

[2016 : 1 Mark, Set-1]

54. Despite the new medicine's ______ in treating diabetes, it is not ______ widely.
(a) effectiveness __ prescribed
(b) availability __ used
(c) prescription __ available
(d) acceptance __ prescribed

[2016 : 1 Mark, Set-1]

55. In a world filled with uncertainty, he was glad to have many good friends. He had always assisted them in times of need and was confident that they would reciprocate. However, the events of the last week proved him wrong.
Which of the following inference(s) is/are logically valid and can be inferred from the above passage?
(i) His friends were always asking him to help them.
(ii) He felt that when in need of help, his friends would let him down.
(iii) He was sure that his friends would help him when in need.
(iv) His friends did not help him last week.
(a) (i) and (ii) (b) (iii) and (iv)
(c) (iii) only (d) (iv)only

[2016 : 2 Marks, Set-1]

56. Leela is older than her cousin Pavithra. Pavithra's brother Shiva is older than Leela. When Pavithra and Shiva are visiting Leela, all three like to play chess. Pavithra wins more often than Leela does.
Which one of the following statements must be **TRUE** based on the above?
(a) When Shiva plays chess with Leela and Pavithra, he often loses.
(b) Leela is the oldest of the three
(c) Shiva is a better chess player than Pavithra
(d) Pavithra is the youngest of the three

[2016 : 2 Marks, Set-1]

57. The unruly crowd demanded that the accused be ________ without trial.
(a) hanged (b) hanging
(e) hankering (d) hung
[2016 : 1 Mark, Set-2]

58. Choose the statement(s) where the underlined word is used correctly:
(i) A prone is a dried plum.
(ii) He was lying prone on the floor.
(iii) People who eat a lot of fat are prone to heart disease.
(a) (i) and (iii) only (b) (iii)only
(c) (i) and (ii) only (d) (ii) and (iii) only
[2016 : 1 Mark, Set-2]

59. Fact: If it rains, then the field is wet.
Read the following statements:
(i) It rains (ii) The field is not wet
(iii) The field is wet (iv) It did not räin
Which one of the options given below is NOT logically possible, based on the given fact?
(a) If (iii), then (iv), (b) If (i), then (iii).
(c) If (i), then (ii), (d) If (ii), then (iv).
[2016 : 1 Mark, Set-2]

60. A smart city integrates all modes of transport, uses clean energy and promotes sustainable use of resources. It also uses technology to ensure safety and security of the city, something which critics argue, will lead to a surveillance state. Which of the following can be logically inferred from the above paragraph?
(i) All smart cities encourage the formation of surveillance states.
(ii) Surveillance is an integral part of a smart city.
(iii) Sustainability and surveillance go hand in hand in a smart city.
(iv) There is a perception that smart cities promote surveillance.
(a) (i) and (iv) only (b) (ii) and (iii) only
(c) (iv) only (d) (i)only
[2016 : 2 Marks, Set-2]

61. Based on the given statements, select the appropriate option with respect to grammar and usage. Statements
(i) The height of Mr. X is 6 feet.
(ii) The height of Mr. Y is 5 feet.
(a) Mr. X is longer than Mr. Y.
(b) Mr. X is more elongated than Mr. Y.
(c) Mr. X is taller than Mr. Y.
(d) Mr. X is lengthier than Mr. Y.
[2016 : 1 Mark, Set-3]

62. The students ______ the teachers' on teacher's day for twenty years of dedicated teaching.
(a) facilitated (b) felicitated
(c) fantasized (d) facilitated
[2016 : 1 Mark, Set-3]

63. After India's cricket world cup victory in 1985, Shrotria who was playing both tennis and cricket till then, decided to concentrate only on cricket. And the rest is history.
What does the underlined phrase mean in this context?
(a) history will rest in peace
(b) rest is recorded in history books
(c) rest is well known
(d) rest is archaic
[2016 : 1 Mark, Set-3]

64. Social science disciplines were in an amorphous form until the colonial period when they were institutionalized. In varying degree, they were intended to further the colonial interest. In the time of globalization and the economic rise of post-colonial countries like India, conventional ways of knowledge production have become obsolete.
Which of the following can be logically inferred from the above statements?
(i) Social science disciplines have become obsolete
(ii) Social science disciplines had a pre-colonial origin
(iii) Social science disciplines always promote colonialism.
(iv) Social science must maintain disciplinary boundaries.
(a) (ii) only (b) (i) and (iii) only
(c) (ii) and (iv) only (d) (iii) and (iv) only
[2016 : 2 Marks, Set-3]

65. He was one of my best ___ and I felt his loss ___ .
(a) friend, keenly (b) friends, keen
(c) friend, keener (d) friends, keenly
[2017 : 1 Mark, Set-1]

66. As the two speakers became increasingly agitated, the debate became ________ .
(a) lukewarm (b) poetic
(c) forgiving (d) heated
[2017 : 1 Mark, Set-1]

67. Two very famous sportsmen Mark and Steve happened to be brothers, and played for country K. Mark teased James, an opponent from country E. There is no way you are good enough to play for your country. James replied, 'May be not, but at least I am the best player in my own family'.
Which one of the following can be inferred from this conversation.
(a) Mark was known to play better than James
(b) Steve was known to play better than Mark
(c) James and Steve were good friends
(d) James played better than Steve
[2017 : 2 Marks, Set-1]

68. Here, throughout the early 1820s, Stuart continued to fight his losing battle to allow his sepoys to wear their caste-marks and their own choice of facial hair on parade, being again reprimanded by the commander-in-chief. His retort that a stronger instance than this of European prejudice with relation to this country has never come under my observations had no effect on his superiors.

According to this paragraph, which of the statements below is most accurate?

(a) Stuart's commander-in-chief was moved by this demonstration of his prejudice.
(b) The Europeans were accommodating of the sepoys desire to were their caste-marks
(c) Stuart's losing battle refers to his inability to succeed in enabling sepoys to wear caste-marks.
(d) The commander-in-chief was exempt from the European prejudice that dictated how the sepoys were to dress

[2017 : 2 Marks, Set-1]

69. The ways in which this game can be played _______ potentially infinite.

(a) is (b) is being
(c) are (d) are being

[2017 : 1 Mark, Set-2]

70. If you choose plan P, you will have to ____ , plan Q, as these two are mutually_______.

(a) forgo, exclusive
(b) forget, inclusive
(c) accept, exhaustive
(d) adopt, intrusive **[2017 : 1 Mark, Set-2]**

71. All people in a certain island are either 'Knights' or 'Knaves' and each person knows every other person's identity. Knights NEVER lie, and knaves ALWAYS lie.

P says 'Both of us are knight'. Q says 'None of us are knaves'

Which one of the following can be logically inferred from the above?

(a) Both P and Q are knights
(b) P is a knight; Q is a knave
(c) Both P and Q are knaves
(d) The identities of P, Q cannot be determined

[2017 : 2 Marks, Set-2]

72. If you are looking for a history of India, or for an account of the rise and fall of the British Raj, or for the reason of the cleaving of the subcontinent into two mutually antagonistic parts and the effects this mutilation will have in the respective sections, and ultimately on Asia, you will not find it in these pages; for though I have spent a lifetime in the country, I lived too near the seat of events, and was too intimately associated with the actors, to get the perspective needed for the impartial recording of these matters."

Which of the following is closest in meaning to 'cleaving'?

(a) deteriorating (b) arguing
(c) departing (d) splitting

[2017 : 2 Marks, Set-2]

73. "Her _______ should not be confused with miserliness; she is ever willing to assist those in need."

The word that best fills the blank in the above sentence is:

(a) cleanliness (b) punctuality
(c) frugality (d) greatness

[2018 : 1 Mark, Set-1]

74. Going by the _____that many hands make light work, the school _______ involved all the students in the task."

(a) principle, principal (b) principal, principle
(c) principle, principle (d) principal, principal

[2018 : 1 Mark, Set-1]

75. Consider the following three statements:

(i) Some roses are red
(ii) All red flowers fade quickly.
(iii) Some roses fade quickly

Which of the following statements can be logically inferred from the above statements?

(a) If (i) is true and (ii) is false, then (iii) is false.
(b) If (i) is true and (ii) is false, then (iii) is true.
(c) If (i) and (ii) are true, then (iii) is true.
(d) If (i) and (ii) are false, then (iii) is false.

[2018 : 2 Marks, Set-1]

76. "The dress _______ her so well that they all immediately _______ her on her appearance."

The words that best fill the blanks in the above sentence are

(a) complemented, complemented
(b) complimented, complemented
(c) complimented, complimented
(d) complemented, complimented

[2018 : 1 Mark, Set-2]

77. "The judge's standing in the legal community, though shaken by false allegations of wrongdoing, remained ______."

The word that best fill the blanks in the above sentence is

(a) undiminished (b) damaged
(c) illegal (d) uncertain

[2018 : 1 Mark, Set-2]

78. John Thomas, an __ writer, passed away in 2018.

(a) imminent (b) dominant
(c) eminent (d) prominent

[2019 : 1 Mark, Set-1]

79. The minister avoided any mention of the issue of women's reservation in the private sector. He was accused of _____ the issue.
(a) collaring (b) belting
(c) tying (d) skirting
[2019 : 1 Mark, Set-1]

80. _____ I permitted him to leave, I wouldn't have had any problem with him being absent _____ I?
(a) Have, would (b) Have, wouldn't
(c) Had, would (d) Had, wouldn't
[2019 : 1 Mark, Set-1]

81. Under a certain legal system, prisoners are allowed to make one statement. If their statement turns out to be true then they are hanged. If the statement turns out to be false then they are shot. One prisoner made a statement and the judge had no option but to set him free. Which one of the following could be that statement ?
(a) I committed the crime
(b) You committed the crime
(c) I will be shot
(d) I did not commit the crime
[2019 : 2 Marks, Set-1]

82. M and N had four children P, Q, R and S. Of them, only P and R were married. They had children X and Y respectively. If Y is a legitimate child of W, which one of the following statements is necessarily FALSE ?
(a) R is the father of Y
(b) W is the wife of R
(c) M is the grandmother of Y
(d) W is the wife of P
[2019 : 2 Marks, Set-1]

83. Congo was named by Europeans. Congo's dictator Mobuto later changed the name of the country and the river to Zaire with the objective of Africanising names of persons and spaces. However, the name Zaire was a Portuguese alteration of Nzadi o Nzere, a local African term meaning 'River that swallows Rivers'. Zaire was the Portuguese name for the Congo river in the 16th and 17 centuries.
Which one of the following statements can be inferred from the paragraph above ?
(a) The term Nzadi o Nzere was of Portuguese origin
(b) As a dictator Mobuto ordered the Portuguese to alter the name of the river to Zaire
(c) Mobuto's desire to Africanise names was prevented by the Portuguese
(d) Mobuto was not entirely successful in Africanising the name of his country
[2019 : 2 Marks, Set-1]

84. A final examination is the _____ of a series of evaluations that a student has to go through.
(a) culmination (b) consultation
(c) desperation (d) insinuation
[2019 : 1 Mark, Set-2]

85. Once the team of analysts identify the problem, we ____ in a better position to comment on the issue. Which one of the following choices CANNOT fill the given blank?
(a) are going to be (b) might be
(c) were to be (d) will be
[2019 : 1 Mark, Set-2]

86. Are there enough seats here? There are _____ people here than I expected.
(a) most (b) least
(c) many (d) more
[2019 : 1 Mark, Set-2]

87. While teaching a creative writing class in India, I was surprised at receiving stories from the students that were all set in distant places: in the American West with cowboys and in Manhattan penthouses with clinking ice cubes. This was, till an eminent Caribbean writer gave the writers in the once-colonised countries the confidence to see the shabby lives around them as worthy of being "told".
The writer of this passage is surprised by the creative writing assignments of his students because________.
(a) None of the students had written about ice cubes and cowboys
(b) Some of the students had written about ice cubes and cowboys
(c) None of the students had written stories set in India
(d) Some of the students had written stories set in foreign places
[2019 : 2 Marks, Set-2]

88. X is an online media provider. By offering unlimited and exclusive online content at attractive prices for a loyalty membership, X is almost forcing its customers towards its loyalty membership. If its loyalty membership continues to grow at its current rate, within the next eight years more housholds will be watching X than cable television.
Which one of the following statements can be inferred from the above paragraph?
(a) Most households that subscribe to X's loyalty membership discontinue watching cable television
(b) Non-members prefer to watch cable television
(c) The X is cancelling accounts of non-members
(d) Cable television operators don't subscribe to X's loyalty membership
[2019 : 2 Marks, Set-2]

ANSWERS

1. (b)	2. (a)	3. (d)	4. (c)	5. (c)	6. (b)	7. (a)	8. (c)	9. (b)	10. (c)
11. (a)	12. (b)	13. (a)	14. (a)	15. (d)	16. (a)	17. (a)	18. (b)	19. (c)	20. (d)
21. (c)	22. (d)	23. (b)	24. (b)	25. (b)	26. (c)	27. (c)	28. (b)	29. (b)	30. (a)
31. (b)	32. (b)	33. (d)	34. (c)	35. (b)	36. (a)	37. (b)	38. (b)	39. (b)	40. (c)
41. (a)	42. (b)	43. (b)	44. (b)	45. (d)	46. (b)	47. (b)	48. (b)	49. (a)	50. (c)
51. (c)	52. (a)	53. (b)	54. (a)	55. (b)	56. (d)	57. (a)	58. (d)	59. (c)	60. (c)
61. (c)	62. (b)	63. (c)	64. (a)	65. (d)	66. (d)	67. (b)	68. (c)	69. (c)	70. (a)
71. (d)	72. (d)	73. (c)	74. (a)	75. (c)	76. (d)	77. (a)	78. (c)	79. (d)	80. (c)
81. (c)	82. (d)	83. (a)	84. (a)	85. (c)	86. (d)	87. (d)	88. (a)		

EXPLANATIONS

1. The word Circuitous is used for routes or journey and it means longer than the most direct way. The closest meaning would be indirect.
2. Unemployed and Worker reflects contradiction. Among the given option only (a) Fallow : Land reflects such a relation.
   Fallow is used for a farmland i.e. ploughed and harrowed but left for a period without being sown in order to restore its fertility or to avoid surplus production.
3. The word Conserve is used when we talk about protecting natural resources.
4. The word betray also means to unintentionally reveal or be the evidence of something.
5. The word "regretfully" used in the passage supports statement (c) i.e. use of chemical agents being undesirable.
   Option (b) will be the second best option becaue of the line "appear to be suited to such warfare" which means that it seems to be useful but can't be right always.
6. Amalgamate means combine or unite to form one organization or structure. So best option here is split. Separate on the other hand, although a close synonym, it is too general to be the best antonym in the given question while Merge is the synonym; Collect is not related.
7. Inexplicable means not explicable; that cannot be explained, understood, or accounted for. So the best synonym here is incomprehensible.
8. Contemplate is a transitive verb and hence is followed by a gerund. Hence the correct usage of contemplate is verb + ing form.
9. Tone of the sentence clearly indicates a word that is similar to understated is needed for the blank. Alternatively, word should be antonym of strong (fail to make strong impression). Therefore, best choice is restrained which means controlled/reserved/timid.
10. Given passage clearly deals with how to deal with bereavement and grief and so after the tragedy occurs and not about precautions. Therefore, irrespective of the causes of death, a school student rarely gets into details of causes—which is beyond the scope of the context. Rest all are important in dealing with grief.
11. 'That' introduces what is called an essential clause (also known as a restrictive or defining clause).
    In the given statement 'hurt in the stampede' is the main clause.
12. Despite is used to define contrasting situations. Hence, 'setbacks' will be the right answer.
13. Mitigate means to make (something bad) less severe, serious, or painful,
    Hence, diminish would be the best option.
14. **Very less is wrong usage. The correct use is 'very little'.**
15. Gender is not mentioned in the advertisement and (b) clearly eliminated
16. Universalism is to particularism ← global consideration to a particular one is moving towards oposite
    ⇓
    difference is to—←opposite the idea of difference which pertains to reaching out to surrounding environment opposite idea to differences is specify.
17. Were and Would are used together in conditional sentences.
18. Zenith is used for the highest point and Nadir for the lowest point.

19. The incorrect sentence is "She is an European." 'an' applies to those words which get pronounced with a, e, i, o, u.
Here pronunciation is "Yuropean" (E-silent; has a "Y" sound) , similar to ' a University'.
20. Options (a), (b) and (c) are out of scope.
Only option (d) i.e. "No adversity justifies giving up hope" is related to the given information.
21. "was allowed" will be the correct option as the "plan was filed" which is the past tense.
22. Option (d) would be the most logical answer. Rest are the possibilities.
23. The word 'dispel' means to make (a doubt, feeling, or belief) disappear.
Rest of the three words describes a force or drift of some kind.
24. Both would be the most appropriate word.
25. In conditional sentences, we use Present Indefinite clauses with Future Indefinite clauses.
26. Bear, losing would be the right option. Loosing means to weaken the grip.
27. Diversity is shown in terms of difference languages.
28. The word depreciated is used when currency loses its value.
29. Advice is a noun while its verb form is Advise.
30. The word disagree means to dissent.
31. The word revert means to get back to someone.
32. The word vindicated means to clear (someone) of blame or suspicion.
Substantiated would be the closest meaning which means to provide evidence to prove something as truth.
34. Misogyny means dislike of, contempt for, or ingrained prejudice against women.
It's adverb form is misogynously.
35. The subject is plural so the correct word would be were.
36. The word awkward means causing difficulty; hard to do or deal with.
It's closest meaning would be inept which means incompetent.
37. The blank will take verb. Among the given options only alternate is a verb.
38. For condition regarding something which already happened, should have practiced is the correct choice.
39. Viaticum is an archaic word which means a supply of provisions or an official allowance of money for a journey.
Among the given options this is the only word which has a connotation related to road.
40. From the statement, it appears that boy found it tough to adapt to a very different situation.
41. Apparent: looks like
dormant: hidden
Harbour: give shelter
Effect (verb): results in
42. If the first two statements are right then it means: Cliff > Tanya > Eric.
Hence, the third statement will be false.
43. Elude means escape from or avoid (a danger, enemy, or pursuer), typically in a skilful or cunning way.
Elusive means difficult to find, catch, or achieve.
Hence, options (b) is the best answer.
44. Statements ii and iii are out of scope and statement i is the most logical argument but it is not alone in any option. Hence, i and iv will be the right option.
45. Both the conclusions i.e. i and ii follow.
46. The is not required in 'Q'.
47. Second sentence says that Ms. X reaches Kochi on 21/05/2014. Also she has to be in Bagdogora on 22/05/2014.
∴ She stays in Kochi for only one day in may.
48. 2 will follow 5 as both the sentences talk about the entire family.
Hence, the correct answer will be opiton (b).
49. **The objective pronoun of I is me. Hence, option (a) would be the correct answer.**
50. "Increase in productivity necessarily" would be the correct phrase.
51. S + 1: S + 2:

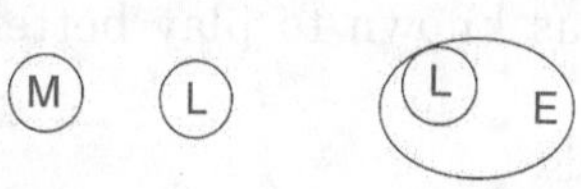

Therefore concluding diagram can be

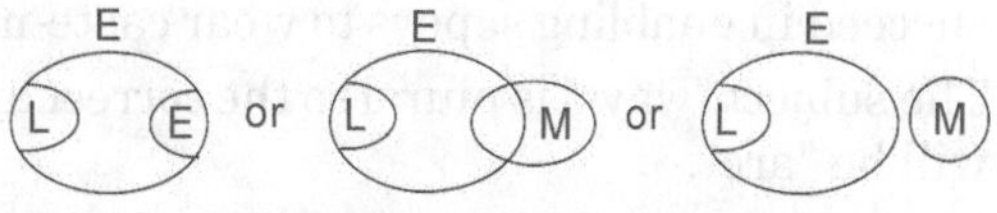

52. Before superlative article 'the' has to be used. "One of" the expression should take plural noun and so option 'C' and 'D' can't be the answer.
53. 'lose' is verb; to complete that sentence a verb is required so 'B' is correct.

54. 'Effectiveness' is noun and 'prescribed' is verb. These words are apt and befitting with the word 'medicine'.

55. The words 'was confident that they would reciprocate' and 'last week proved him wrong' lead to statements iii and iv as logically valid inferences.

56. From given data, the following arrangement is possible

Shiva

Leela

Pavithra

Among four alternatives, option D is TRUE.

57. 'hanged' means death by hanging 'hung' is used only with things and not with people.

58. 'lying prone' means lie down flat. 'Prone to' means vulnerable to.

59. Statements i and ii are not logically possible based on the given fact.

61. In degrees of comparison Mr. X is taller than Mr. Y is apt.

Positive degree – tall

Comparative degree – taller

Superlative degree – tallest

62. Felicitate means honour.

63. 'Rest is history' is an idiomatic expression which means 'rest is well known.'

64. Until the colonial period means pre-colonial origin. Other options can't be inferred.

65. If something is 'one' out of something else, then that something else has to be in plural numbers in order to part with one of itself.

Thus, one of my best friends will be the correct answer.

Further, the word keenly will be used because it is an adverb.

Hence, option (d) will be the right answer.

66. The word heated is used to denote intense debates.

67. The statement "but at least I am the best player in my own family' is a witty reply by James which implies Steve was known to play better than Mark.

68. The correct option is:

Stuart's losing battle refers to his inability to succeed in enabling sepoys to wear caste-marks.

69. The subject "ways" is plural so the correct answer will be "are".

70. If two things are mutually exclusive, they are separate and very different from each other, so that it is impossible for them to exist or happen together.

Forgo means to refrain or avoid.

Hence, these two words will be used together.

71. Both P and Q can either be knights or knaives i.e. either truth tellers or liers.

Hence, we cannot determine the right answer.

72. The closest meaning is splitting.

73. Frugality would be the correct and most appropriate option which means the quality of being economical with money or food.

74. Principle means fundamental truth and principal means head of the school. Hence, option (a) would be the right answer.

75.

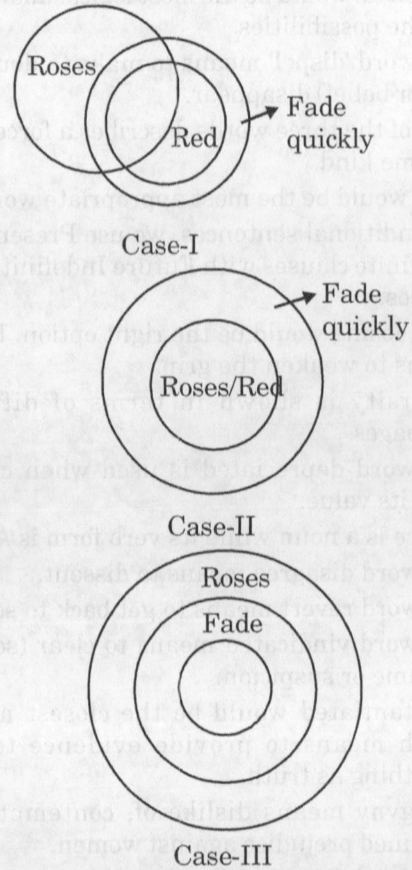

76. Complement means to support or contribute while compliment means to praise.

77. The word though is used to connect contrasting statements. Hence, the correct word will be undiminished.

81. If prisoner make the statement "I will be shot" then only judge have no option and prisoner will be set to free.

82.

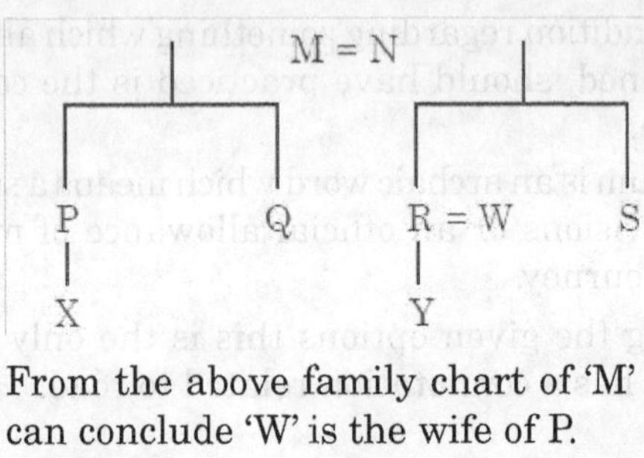

From the above family chart of 'M' and 'N', we can conclude 'W' is the wife of P.

2

CHAPTER

Numerical Ability

Analysis of Previous GATE Papers

Topic	Year/Set →Marks Distribution ↓		2019 Set 1	2019 Set 2	2018 Set 1	2018 Set 2	2017 Set 1	2017 Set 2	2016 Set 1	2016 Set 2	2016 Set 3	2015 Set 1	2015 Set 2	2015 Set 3
Numerical Ability	1 Mark	MCQ Type	1	2	3	1	3	3	2	2	2	2	2	1
		Numerical Type	1											
	2 Marks	MCQ Type	2	2	5	6	3	3	3	4	4	2	2	3
		Numerical Type		1										
	Total		**6**	**8**	**13**	**13**	**9**	**9**	**8**	**10**	**10**	**6**	**6**	**7**

1. 25 persons are in a room. 15 of them play hockey, 17 of them play football and 10 of them play both hockey and football. Then the number of persons playing neither hockey nor football is

(a) 2 (b) 17

(c) 13 (d) 3

[2010 : 1 Mark]

2. If 137+ 276 = 435 how much is 731 + 672?

(a) 534 (c) 1623

(b) 1403 (d) 1531

[2010 : 2 Marks]

3. 5 skilled workers can build a wall in 20 days; 8 semiskilled workers can build a wall in 25 days; 10 unskilled workers can build a wall in 30 days.

If a team has 2 skilled, 6 semiskilled and 5 unskilled workers, how long will it take to build the wall?

(a) 20 days (b) 18 days

(c) 16 days (d) 15 days

[2010 : 2 Marks]

4. Given digits 2, 2, 3, 3, 3, 4, 4, 4, 4 how many distinct 4 digit numbers greater than 3000 can be formed?

(a) 50 (b) 51

(c) 52 (d) 54

[2010 : 2 Marks]

5. Hari (H), Gita (G), Irfan (I) and Saira (S) are siblings (i.e. brothers and Sisters). All were born on 1st January. The age difference between any two successive siblings (that is born one after another) is less than 3 years. Given the following facts:

1. Hari's age + Gita's age > lrfan's age + Saira's age.
2. The age difference between Gita and Saira is 1 year. However, Gita is not the oldest and Saira is not the youngest.
3. There are no twins.

In what order were they born (oldest first)?

(a) HSIG (b) SGHI

(c) IGSH (d) IHSG

[2010 : 2 Marks]

6. If Log(P) = (1/2)Log(Q) = (1/3)Log(P), then which of the following options is TRUE?

(a) $P^2 = Q^3R^2$ (b) $Q^2 = PR$

(c) $Q^2 = R^3P$ (d) $R = P^2Q^2$

[2011 : 1 Mark]

7. A container originally contains 10 litres of pure spirit. From this container 1 litre of spirit is replaced with 1 litre of water. Subsequently, 1 litre of the mixture is again replaced with 1 litre of water and this processes is repeated one more time. How much spirit is now left in the container?

(a) 7.58 litres (b) 7.84 litres

(c) 7 litres (d) 7.29 litres

[2011 : 2 Marks]

8. The variable cost (V) of manufacturing a product varies according to the equation V= 4q, where q is the quantity produced. The fixed cost (f) of production of same product reduces with q according to the equation F= 100/q. How many units should be produced to minimize the total cost(V + F)?

(a) 5 (b) 4

(c) 7 (d) 6

[2011 : 2 Marks]

9. P, Q, R and S are four types of dangerous microbes recently found in a human habitat. The area of each circle with its diameter printed in brackets represents the growth of a single microbe surviving human immunity system within 24 hours of entering the body. The danger to human beings varies proportionately with the toxicity, potency and growth attributed to a microbe shown in the figure below:

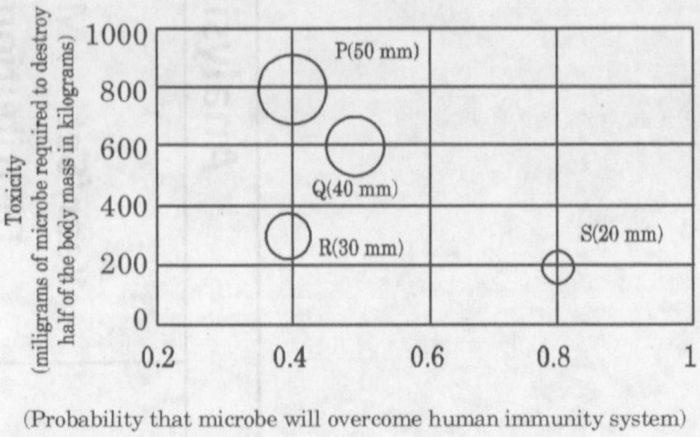

A pharmaceutical company is contemplating the development of a vaccine against the most dangerous microbe. Which microbe should the company target in its first attempt?

(a) P (b) Q

(c) R (d) S

[2011 : 2 Marks]

10. A transporter receives the same number of orders each day. Currently, he has some pending orders (backlog) to be shipped. If he uses 7 trucks, then at the end of the 4th day he can clear all the orders. Alternatively, if he uses only 3 trucks, then all the orders are cleared at the end of the 10th day. What is the minimum number of trucks required so that there will be no pending order at the end of the 5th day?

(a) 4 (b) 5
(c) 6 (d) 7

[2011 : 2 Marks]

11. The cost function for a product in a firm is given by $5q^2$, where q is the amount of production. The firm can sell the product at a market price of Rs 50 per unit. The number of units to be produced by the firm such that the profit is maximized is

(a) 5 (b) 10
(c) 15 (d)25

[2012: 1 Mark]

12. Given the sequence of terms, AD, CG, FK, JP, the next term is

(a) OV (b) OW
(c) PV (d) PW

[2012 : 2 Marks]

13. A political party orders an arch for the entrance to the ground in which the annual conventions is being held. The profile of the arch follows the equation $y = 2x - 0.1x^2$ where y is the height of the arch in meters. The maximum possible height of the arch is

(a) 8 meters (b) 10 meters
(c) 12 meters (d) 14 meters

[2012 : 2 Marks]

14. An automobile plant contracted to buy shock absorbers from two suppliers X and Y. X supplies 60% and V supplies 40% of the shock absorbers. All shock absorbers are subjected to a quality test. The ones that pass the quality test are considered reliable. Of X's shock absorbers, 96% are reliable. Of Y's shock absorbers, 72% are reliable.

The probability that a randomly chosen shock absorber, which is found to be reliable, is made by Y is

(a) 0.288 (c) 0.667
(b) 0.334 (d) 0.720

[2012 : 2 Marks]

15. Which of the following assertions are CORRECT?

P: Adding 7 to each entry in a list adds 7 to the mean of the list

Q: Adding 7 to each entry in a list adds 7 to the standard deviation of the list

R: Doubling each entry in a list doubles the mean of the list

S: Doubling each entry in a list leaves the standard deviation of the list unchanged

(a) P, Q (b) Q, R
(c) P, R (d) R, S

[2012 : 2 Marks]

16. What will be the maximum sum of 44, 42, 40, 2, 0?

(a) 502 (b) 504
(c) 506 (d) 500

[2013 : 1 Mark]

17. Out of all the 2-digit integers between 1 and 100, a 2-digit number has to be selected at random.

What is the probability that the selected number is not divisible by 7?

(a) 13/90 (c) 78/90
(b) 12/90 (d) 77/90

[2013 : 2 Marks]

18. A tourist covers half of this journey by train at 60 km/h, half of the remainder by bus at 30 km/h and the rest by cycle at 10 km/h. The average speed of the tourist in km/h during his entire journey

(a) 36 (c) 24
(b) 30 (d) 18

[2013 : 2 Marks]

19. Find the sum of the expression

$$\frac{1}{\sqrt{1}+\sqrt{2}}+\frac{1}{\sqrt{2}+\sqrt{3}}+\frac{1}{\sqrt{3}+\sqrt{4}}\cdots\cdots\frac{1}{\sqrt{80}+\sqrt{1}}$$

(a) 7 (b) 8
(c) 9 (d) 10

[2013 : 2 Marks]

20. The current erection cost of a structure is Rs.13,200. If the labour wages per day increase by 1/5 of the current wages and the working hours decrease by 1/24 of the current period, then the new cost of erection in Rs. is

(a) 16,500 (b) 15,180
(c) 11,000 (d) 10,120

[2013 : 2 Marks]

21. The statistics of runs scored in a series by four batsmen are provided in the following table. Who is the most consistent batsman of these four?

Batsman	Average	Standard Deviation
K	31.2	5.21
L	46.0	6.35
M	54.4	6.22
N	17.9	5.90

(a) K (b) M

(c) L (d) N

[2014 : 1 Mark, Set-1]

22. What is the next number in the series?

12 35 81 173 357 ______.

[2014 : 1 Mark, Set-1]

23. For submitting tax returns, all resident males with annual income below Rs 10 lakh should fill up Form P and all resident females with income below Rs 8 lakh should fill up Form Q. All people with incomes above Rs 10 lakh should fill up Form R, except non residents with income above Rs 15 lakhs, who should fill up Form S. All others should fill Form T. An example of a person who should fill Form T is

(a) a resident male with annual income Rs. 9 lakh

(b) a resident female with annual income Rs. 9 lakh

(c) a non-resident male with annual income Rs. 16 lakh

(d) a non-resident female with annual income Rs 16 lakh

[2014: 2 Marks, Set-1]

24. You are given three coins: one has heads on both faces, the second has tails on both faces, and the third has a head on one face and a tail on the other. You choose a coin at random and toss it, and it comes up heads. The probability that the other face is tails is

(a) 1/4 (b) 1/3

(c) 1/2 (d) 2/3

[2014 : 2 Marks, Set-1]

25. A train that is 280 metres long, travelling at a uniform speed, crosses a platform in 60 seconds and passes a man standing on the platform in 20 seconds. What is the length of the platform in metres?

[2014 : 2 Marks, Set-1]

26. The exports and imports (in crores of Rs) of a country from 2000 to 2007 are given in the following bar chart. If the trade deficit is defined as excess of imports over exports, in which year is the trade deficit 1/5th of the exports?

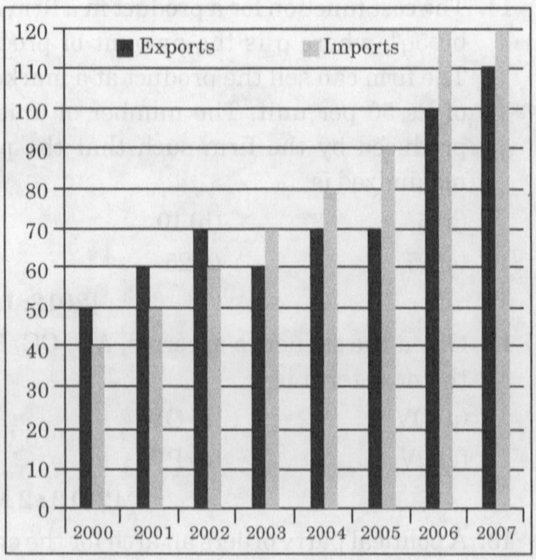

(a) 2005

(b) 2004

(c) 2007

(d) 2006

[2014: 2 Marks, Set-1]

27. A regular die has six sides with numbers 1 to 6 marked on its sides. If a very large number of throws show the following frequencies of occurrence:

$1 \rightarrow 0,167$; $2 \rightarrow 0,167$; $3 \rightarrow 0,152$; $4 \rightarrow 0,166$; $5 \rightarrow 0,168$; $6 \rightarrow 0,180$. We call this die

(a) irregular (b) biased

(c) Gaussian (d) insufficient

[2014 : 1 Mark, Set-2]

28. Fill in the missing number in the series.

2 3 6 15 _?_ 157.5 630

[2014 : 1 Mark, Set-2]

29. The sum of eight consecutive odd numbers is 656. The average of four consecutive even numbers is 87. What is the sum of the smallest odd number and second largest even number?

[2014 : 2 Marks, Set-2]

30. The total exports and revenues from the exports of a country are given in the two charts shown below. The pie chart for exports shows the quantity of each item exported as a percentage of the total quantity of exports. The pie chart for the revenues shows the percentage of the total revenue generated through export of each item. The total quantity of exports of all the items is 500 thousand tonnes and the total revenues are 250 crore rupees. Which item among the following has generated the maximum revenue per kg?

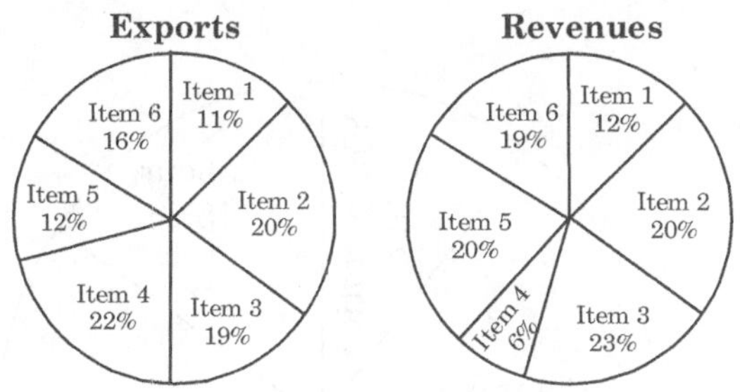

(a) Item 2 (b) Item 3

(c) Item 6 (d) Item 5

[2014 : 2 Marks, Set-2]

31. It takes 30 minutes to empty a half-full tank by draining it at a constant rate, It is decided to simultaneously pump water into the half-full tank while draining it. What is the rate at which water has to be pumped in so that it gets fully filled in 10 minutes?

(a) 4 times the draining rate

(b) 3 times the draining rate

(c) 2.5 times the draining rate

(d) 2 times the draining rate

[2014 : 2 Marks, Set-2]

32. The next term in the series 81, 54, 36, 24, ...is _______ .

[2014 : 1 Mark, Set-3]

33. Find the next term in the sequence:

7 G, 11 K, 13M, _____.

(a) 15Q (b) 17Q

(c) 15P (d) 17P

[2014 : 2 Marks, Set-3]

34. The multi-level hierarchical pie chart shows the population of animals in a reserve forest. The correct conclusions from this information are:

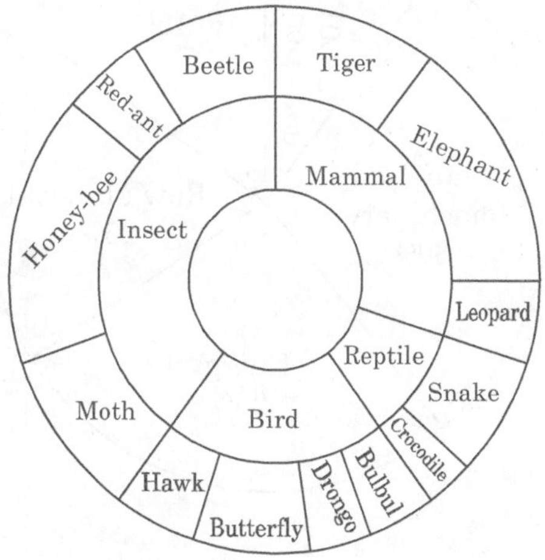

(i) Butterflies are birds

(ii) There are more tigers in this forest than red ants

(iii) All reptiles in this forest are either snakes or crocodiles

(iv) Elephants are the largest mammals in this forest

(a) (i) and (ii) only

(b) (i),(ii), (iii) and (iv)

(c) (i), (iii) & (iv) only

(d) (i),(ii) and (iii) only

[2014 : 2 Marks, Set-3]

35. A man can row at 8 km per hour in still water. If it takes him thrice as long to row upstream, as to row downstream, then find the stream velocity in km per hour.

[2014 : 2 Marks, Set-3]

36. A firm producing air purifiers sold 200 units in 2012. The following pie chart presents the share of raw material, labour, energy, plant & machinery, and transportation costs in the total manufacturing cost of the firm in 2012. The expenditure on labour in 2012 is Rs. 4,50,000. In 2013, the raw material expenses increased by 30% and all other expenses increased by 20%. If the company registered a profit of Rs. 10 lakhs in 2012, at what price (in Rs.) was each air purifier sold?

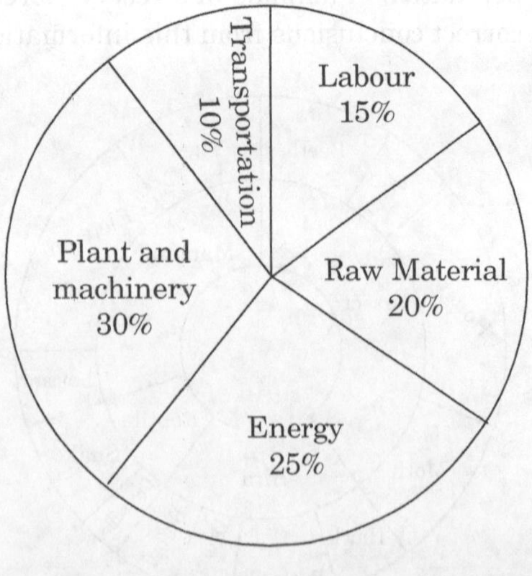

[2014 : 2 Marks, Set-3]

37. A batch of one hundred bulbs is inspected by testing four randomly chosen bulbs. The batch is rejected if even one of the bulbs is defective. A batch typically has five defective bulbs. The probability that the current batch is accepted is ________.

[2014 : 2 Marks, Set-3]

38. Let f[x, y) = x^n y^m = P. If x is doubled and y is halved, the new value of f is

(a) 2^{n-m}P (b) 2^{m-n}P

(c) 2(n-m)P (d) 2(m–n)P

[2014: 1 Mark; Set-4]

39. In a sequence of 12 consecutive odd numbers, the sum of the first 5 numbers is 425. What is the sum of the last 5 numbers in the sequence?

[2014 : 1 Mark, Set-4]

40. Find the next term in the sequence:

13M, 17O, 19S,

(a) 21 W (b) 21V

(c) 23 W (d) 23I

[2014 : 2 Marks, Set-4]

41. If 'KCLFTSB' stands for 'best of luck' and 'SHSVVDG' stands for 'good wishes', which of the following indicates 'ace the exam'?

(a) MCHTX (b) MXHTC

(c) XMHCT (d) XMHTC

[2014 : 2 Marks, Set-4]

42. Industrial consumption of power doubled from 2000-2001 to 2010-2011. Find the annual rate of increase in percent assuming it to be uniform over the years.

(a) 5.6 (b) 7.2

(c) 10.0 (d) 12.2

[2014 : 2 Marks, Set-4]

43. A firm producing air purifiers sold 200 units in 2012. The following pie chart presents the share of raw material, labour, energy, plant & machinery, and transportation costs in the total manufacturing cost of the firm in 2012. The expenditure on labour in 2012 is Rs. 4,50,000. In 2013, the raw material expenses increased by 30% and all other expenses increased by 20%. What is the percentage increase in total cost for the company in 2013?

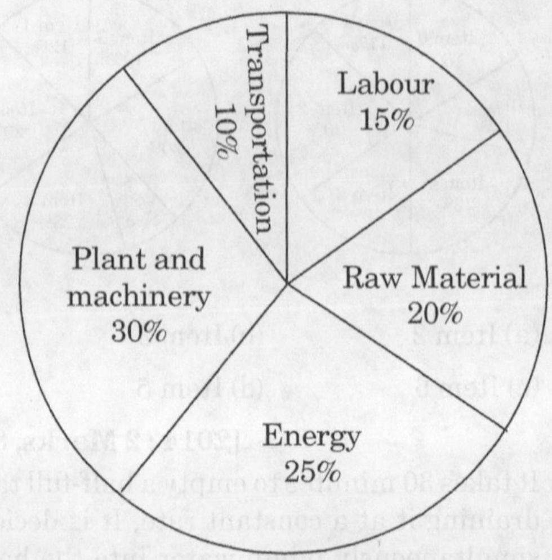

[2014 : 2 Marks, Set-4]

44. A five digit number is formed using the digits 1,3,5,7 and 9 without repeating any of them. What is the sum of all such possible five digit numbers?

(a) 6666660 (b) 6666600

(c) 6666666 (d) 6666606

[2014 : 2 Marks, Set-4]

45. An electric bus has onboard instruments that report the total electricity consumed since the start of the trip as well as the total distance covered. During a single day of operation, the bus travels on stretches M, N, O and P, in that order. The cumulative distances travelled and the corresponding electricity consumption are shown in the table below.

Stretch	Cumulative distance (km)	Electricity used (kWh)
M	20	12
N	45	25
O	75	45
P	100	57

The stretch where the electricity consumption per km is minimum is

(a) M (b) N

(c) O (d) P

[2015 : 1 Mark, Set-1]

46. Ram and Ramesh appeared in an interview for two vacancies in the same department. The probability of Ram's selection is 1/6 and that of Ramesh is 1/8. What is the probability that only one of them will be selected?

(a) $\frac{47}{48}$ (b) $\frac{1}{4}$

(c) $\frac{13}{48}$ (d) $\frac{35}{48}$

[2015 : 1 Mark, Set-1]

47. If x > y > 1, which of the following must be true?

i. In x > In y ii. $e^x > e^y$

ii. $y^x > x^y$ iv. cos x > cosy

(a) i and ii (b) i and iii

(c) iii and iv (d) ii and iv

[2015 : 1 Mark, Set-2]

48. Find the missing sequence in the letter series below:

A, CD, GHI, ?, UVVVXY

(a) LMN (b) MNO

(c) MNOP (d) NOPQ

[2015 : 1 Mark, Set-2]

49. Five teams have to compete in a league, with every team playing every other team exactly once, before going to the next round. How may matches will have to be held to complete the league round of matches?

(a) 20 (b) 10

(c) 8 (d) 5

[2015 : 1 Mark, Set-3]

50. A tiger is 50 leap of its own behind a deer. The tiger takes 5 leaps per minute to the deer's 4. If the tiger and the deer cover 8 metre and 5 metre per leap respectively, what distance in meters will the tiger have a run before it catches the deer?

[2015 : 2 Marks, Set-1]

51. If $a^2 + b^2 + c^2 = 1$, then ab + bc+ ac lies in the interval

(a) $\left[1, \frac{2}{3}\right]$ (b) $\left[\frac{-1}{2}, 1\right]$

(c) $\left[-1, \frac{-1}{2}\right]$ (d) [2, – 4]

[2015 : 2 Marks, Set-1]

52. log tan 1° + log tan 2° + _____ + log tan 89° is 1

(a) 1 (b) $\frac{1}{\sqrt{2}}$

(c) 0 (d) – 1

[2015 : 2 Marks, Set-2]

53. From a circular sheet of paper of radius 30 cm, a sector of 10% area is removed. If the remaining part is used to make a conical surface, then the ratio of the radius and height of the cone is___

[2015 : 2 Marks, Set-2]

54. A coin is tossed thrice. Let X be the event that head occurs in each of the first two tosses. Let Y be the event that a tail occurs on the third toss. Let Z be the event that two tails occurs in three tosses. Based on the above information, which one of the following statements is TRUE?

(a) Xand Y are not independent

(b) Yand Z are dependent

(c) Yand Z are independent

(d) Xand Z independent

[2015 : 2 Marks, Set-3]

55. Right triangle PQR is to be constructed in the xy– plane so that the right angle is at P and line PR is parallel to the-axis. The x and y coordinates of P, Q, and R are to be integers that satisfy the inequalities:

$-4 \leq, x \leq 5$ and $6 \leq y \leq 16$. How many different triangles could be constructed with these properties?

(a) 110 (b) 1,100

(c) 9,900 (d) 10,000

[2015 : 2 Marks, Set-3]

56. In the given figure angle Q is a right angle, PS:QS = 3 : 1, RT : QT = 5 : 2 and PU : UR = 1 : 1. If area of triangle QTS is 20 cm, then the area of triangle PQR in cm² is _______ .

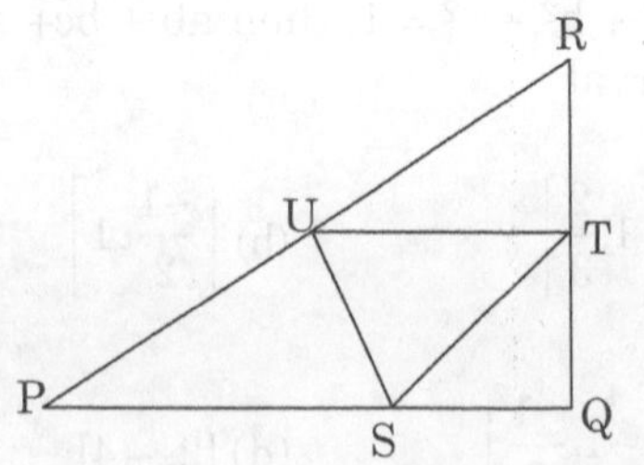

[2015 : 2 Marks, Set-3]

57. In a huge pile of apples and oranges, both ripe and unripe mixed together, 15% are unripe fruits. Of the unripe fruits, 45% are apples. Of the ripe ones, 66% are oranges. If the pile contains a total of 5692000 fruits, how many of them are apples?

(a) 2029198 (b) 2467482
(c) 2789080 (d) 3577422

[2016 : 1 Mark, Set-1]

58. Michael lives 10 km away from where I live. Ahmed lives 5 km away and Susan lives 7 km away from where I live. Arun is farther away than Ahmed but closer than Susan from where I live. From the information provided here, what is one possible distance (in km) at which I live from Arun's place?

(a) 3.00 (b) 4.99
(c) 6.02 (d) 7.01

[2016 : 1 Mark, Set-1]

59. A person moving through a tuberculosis prone zone has a 50% probability of becoming infected. However, only 30% of infected people develop the disease. What percentage of people moving through a tuberculosis prone zone remains infected but does not show symptoms of disease?

(a) 15 (b) 33
(c) 35 (d) 37

[2016 : 2 Marks, Set-1]

60. If $q^{-a} = \frac{1}{r}$ and $r-b = \frac{1}{s}$ and $s-c = \frac{1}{q}$ of abc is. the value of abc is ______.

(a) $(rps)^{-1}$ (b) 0
(c) 1 (d) r + q + s

[2016 : 2 Marks, Set-1]

61. P, Q, R and S are working on a project. Q can finish the task in 25 days, working alone for 12 hours a day. R can finish the task in 50 days, working alone for 12 hours per day. Q worked 12 hours a day but took sick leave in the beginning for two days, R worked 18 hours a day on all days. What is the ratio of work done by Q and R after 7 days from the start of the project?

(a) 10:11 (b) 11 : 10
(c) 20 : 21 (d) 21 : 20

[2016 : 2 Marks, Set-1]

62. The volume of a sphere of diameter 1 unit is ______ than the volume of a cube of side 1 unit.

(a) least (b) less
(c) lesser (d) low

[2016 : 1 Mark, Set-2]

63. A window is made up of a square portion and an equilateral triangle portion above it. The base of the triangular portion coincides with the upper side of the square. If the perimeter of the window is 6 m, the area of the window in m^2 is ______.

(a) 1.43 (b) 2.06
(c) 2.68 (d) 2.88

[2016 : 1 Mark, Set-2]

64. Students taking an exam are divided into two groups, P and Q such that each group has the same number of students. The performance of each of the students in a test was evaluated out of 200 marks. It was observed that the mean of group P was 105, while that of group Q was 85. The standard deviation of group P was 25, while that of group Q was 5. Assuming that the marks were distributed on a normal distribution, which of the following statements will have the highest probability of being TRUE?

(a) No student in group Q scored less marks than any student in group P

(b) No student in group P scored less marks than any student in group Q.

(c) Most students of group Q scored marks in a narrower range than students in group P.

(d) The median of the marks of group Pis 100.

[2016 : 2 Marks, Set-2]

65. Find the missing sequence in the letter series.
B, FH, LNP, ______.

(a) SUWY (b) TUVW
(c) TVXZ (d) TVVXZ

[2016 : 2 Marks, Set-2]

66. The binary operation – is defined as a–b = ab + (a + b), where a and b are any two real numbers. The value of the identity element of this operation, defined as the numberx such that a–x = a, for any a, is ______.

(a) 0 (b) 1
(c) 2 (d) 10

[2016 : 2 Marks, Set-2]

67. Which of the following curves represents the function $y = \left(\left|e^{\left[\sin C(|x|)\right]}\right|\right)$ for $|x| < 2\pi$?

Here, x represents the abscissa and y repreents the ordinate.

(a) 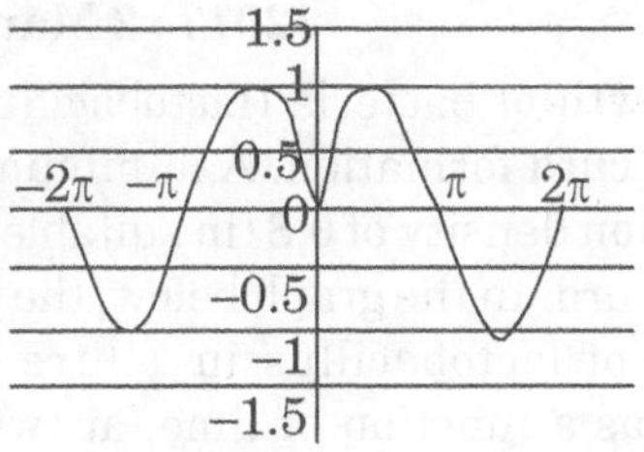

(b) 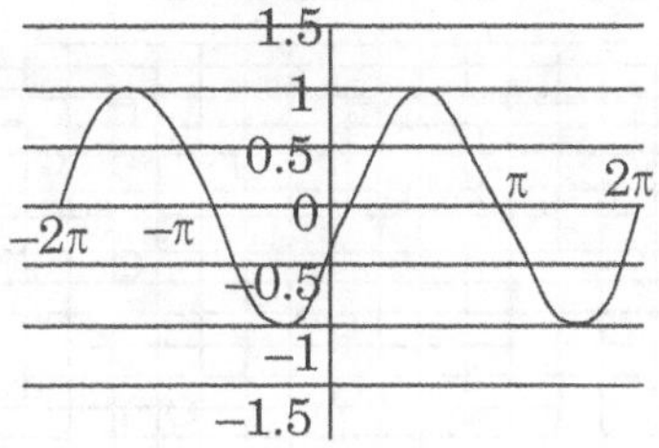

(c) 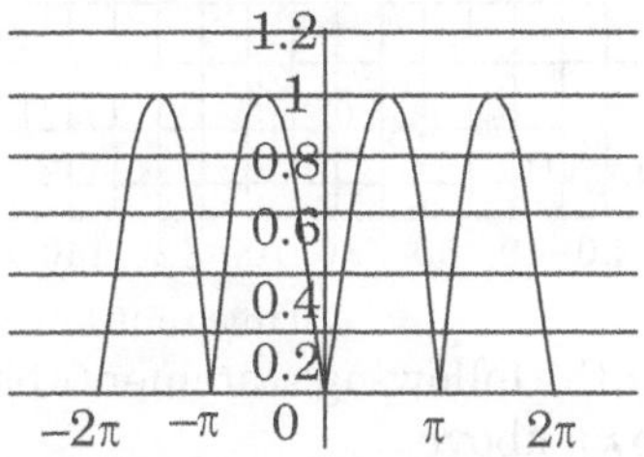

(d) 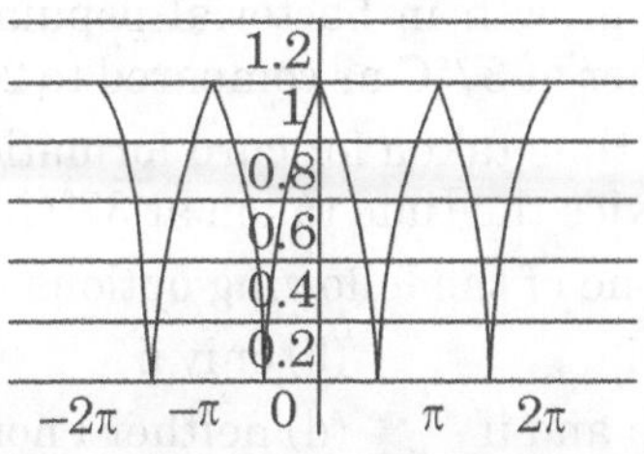

[2016 : 2 Marks, Set-2]

68. Given (9 inches)$^{1/2}$ = (0.25 yards)$^{1/2}$, which one of the following statements is TRUE?

(a) 3 inches - 0.5 yards

(b) 9 inches = 1.5 yards

(c) 9 inches = 0.25 yards

(d) 81 inches = 0.0625 yards

[2016 : 1 Mark, Set-3]

69. S, M, E and F are working in shifts in a team to finish a project. M works with twice the efficiency of others but for half as many days as E worked, S and M have 6 hour shifts in a day, whereas E and F have 12 hours shifts. What is the ratio of contribution of M to contribution of E in the project?

(a) 1 : 1 (b) 1 : 2

(c) 1 : 4 (d) 2 : 1

[2016 : 1 Mark, Set-3]

70. The Venn diagram shows the preference of the student population for leisure activities.

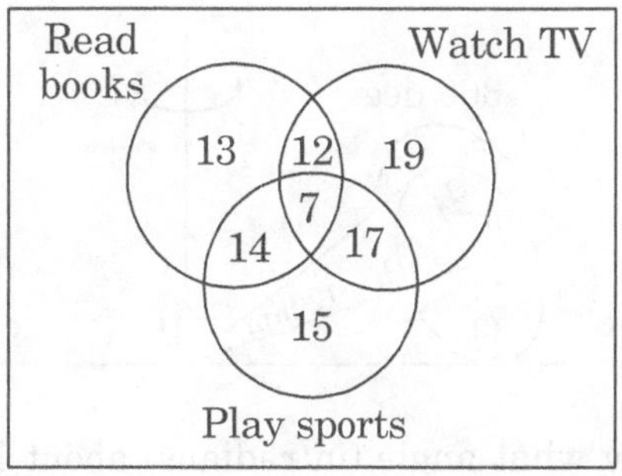

From the data given, the number of students who like to read books or play sports is

(a) 44 (b) 51

(c) 79 (d) 108

[2016 : 2 Marks, Set-3]

71. Two and a quarter hours back, when seen in a mirror, the reflection of a wall clock without number markings seemed to show 1:30. What is the actual current time shown by the clock?

(a) 8:15 (b) 11 :15

(c) 12:15 (d) 12:45

[2016 : 2 Marks, Set-3]

72. M and N start from the same location. M travels 10 km East and then 10 km North-East. N travels 5 km South and then 4 km South-East. What is the shortest distance (in km) between M and N at the end of their travel.

(a) 18.60 (b) 22.50

(c) 20.61 (d) 25.00

[2016 : 2 Marks, Set-3]

73. A wire of length 340 mm is to be cut into two parts. One of the parts is to be made into a square and the other into a rectangle where sides are in the ratio of 1 :2. What is the length of the side of the square (in mm) such that the combined area of the square and the rectangle is a **MINIMUM**?

(a) 30 (b) 40

(c) 120 (d) 180

[2016 : 2 Marks, Set-3]

74. A right-angled cone (with base radius 5 cm and height 12 cm), as shown in fte figure below, is rolled on the ground keeping the point P fixed until the point Q (at the base of the cone, as shown) touches the ground again.

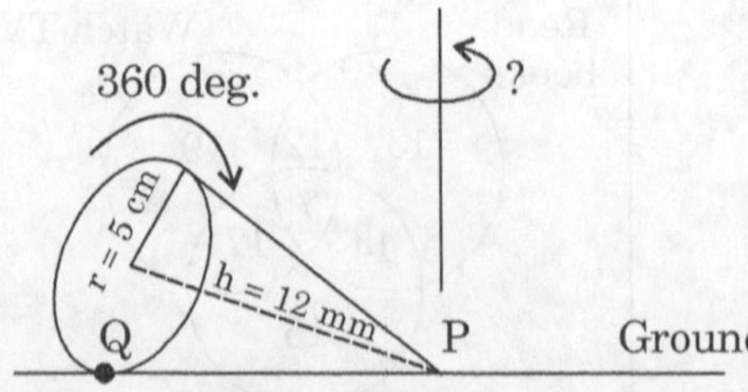

By what angle (in radians) about P does the cone travel ?

(a) $\frac{5\pi}{12}$ (b) $\frac{5\pi}{24}$

(c) $\frac{24\pi}{5}$ (d) $\frac{10\pi}{13}$

[2017 : 1 Mark, Set-1]

75. P, Q, and R talk about S's car collection, P states that S has at least 3 cars. Q believes that S has less than 3 cars, R says that to his knowledge, S has at least one car. Only one of P, Q and R is right. The number of cars owned by S is

(a) 0 (b) 1

(c) 3 (d) Cannot be determined

[2017 : 1 Mark, Set-1]

76. In a company with 100 employees, 45 earn Rs. 20000 per month, 25 earn Rs. 30000,20 earn Rs. 40000,8 earn Rs. 60000, and 2 earn Rs. 150000. The median of the salaries is

(a) Rs. 20000 (b) Rs. 30000

(c) Rs. 32300 (d) Rs. 40000

[2017 : 1 Mark, Set-1]

77. What is the sum of the missing digits in the subtraction problem below?

5..........
−48_89
1111

(a) 8 (b) 10

(c) 11 (d) Cannot be determined

[2017 : 2 Marks, Set-1]

78. Let S_1 be the plane figure consisting of the points (x, y) given by the inequalities $|x-1| \leq 2$ and $|y+2| \leq 3$. Let S_2 be the plane figure given by the inequalities $x - y \geq -2$, $y \geq 1$, and $x \leq 3$. Let S be the union of S_1 and S_2. The area of S is

(a) 26 (b) 28

(c) 32 (d) 34

[2017 : 2 Marks, Set-1]

79. The growth of bacteria (lactobacillus) in milk leads to curd formation. A minimum bacterial population density of 0.8 (in suitable) is needed to form curd. In the graph below, the population density of lactobacillus in 1 litre of milk is plotted as a function of time, at two different temperature, 25°C and 37°C.

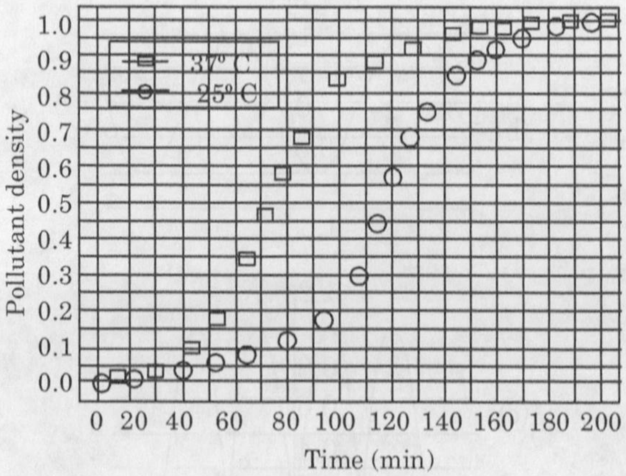

Consider the following statements based on the data shown above:

i. The growth in bacterial population stops earlier at 37°C as compared to 25°C

ii. The time taken for curd formation at 25°C is twice the time taken at 37°C

Which one of the following options is correct?

(a) only i (b) oniy ii

(c) both i and ii (d) neither i nor ii

[2017 : 2 Marks, Set-1]

80. If a and b are integers and a – b is even, which of the following must always be even?

(a) ab (b) $a^2 + b^2 + 1$

(c) $a^2 + b + 1$ (d) $ab - b^2$

[2017 : 1 Mark, Set-2]

81. P looks at Q while Q looks at R, P is married, R is not. The number of pairs of people in which a married person is looking at an unmarried person is

(a) 0

(b) 1

(c) 2

(d) cannot be determined

[2017 : 1 Mark, Set-2]

82. A couple has 2 children. The probability that both children are boys if the older one is a boy is

(a) 1/4 (b) 1/3

(c) 1/2 (d) 1

[2017 : 1 Mark, Set-2]

83. X bullocks and Y tractors take 8 days to plough a field. If we halve the number of bullocks and double the number of tractors, it takes 5 days to plough the same field. How many days will it take X bullocks alone to plough the field?

(a) 30 (b) 35

(c) 40 (d) 45

[2017 : 2 Marks, Set-2]

84. There are 4 women P, Q, R, S and 5 men V, W, X, Y, Z in a group. We are required to form pairs each consisting of one woman and one man. P is not to be paired with Z, and Y must necessarily be paired with someone. In how many ways can 4 such pairs be formed ?

(a) 74

(b) 76

(c) 78

(d) 80

[2017 : 2 Marks, Set-2]

85. In the graph below, the concentration of a particular pollutant in a lake is plotted over (alternate) days of a month in Winter (average temperature 10°C) and a month in summer (average temperature 30°C)

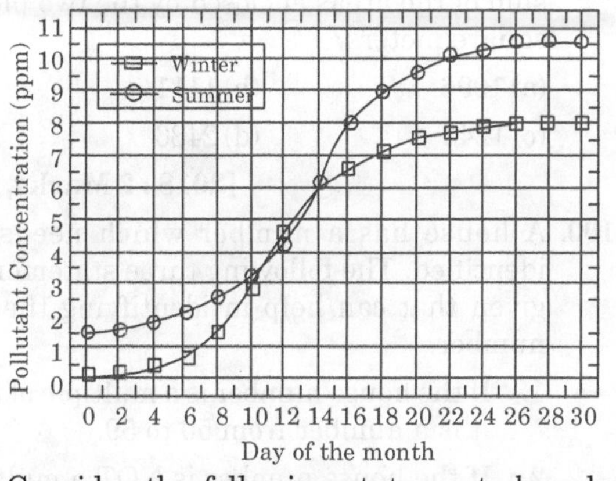

Consider the following statements based on the data shown above:

i. Over the given month, the difference between the maximum and the minimum pollutant concentrations is the same in both Winter and summer.

ii. There are at least four days in the summer month such that the pollutant concentrations on those days are within 1 ppm of the pollutant concentrations on the corresponding days in the Winter month.

Which one of the following options is correct?

(a) only i (b) only ii

(c) both i and ii (d) Neither i nor ii

[2017 : 2 Marks, Set-2]

86. A rectangle becomes a square when its length and breadth are reduced by 10 m and 5 m, respectively. during this process, the rectangle loses 650 m^2 of area. What is the area of the original rectangle in square meters?

(a) 1125 (b) 2250

(c) 2924 (d) 4500

[2018 : 1 Mark, Set-1]

87. A number consists of two digits. The sum of the digits is 9. If 45 is subtracted from the number, its digits are interchanged. What is the number?

(a) 63 (b) 72

(c) 81 (d) 90

[2018 : 1 Mark, Set-1]

88. Seven machines take 7 minutes to make 7 identical toys. At the same rate, how many minutes would it take for 100 machines to make 100 toys?

(a) 1

(b) 7

(c) 100

(d) 700

[2018 : 1 Mark, Set-1]

89. Which of the following functions describe the graph shown in the below figure.

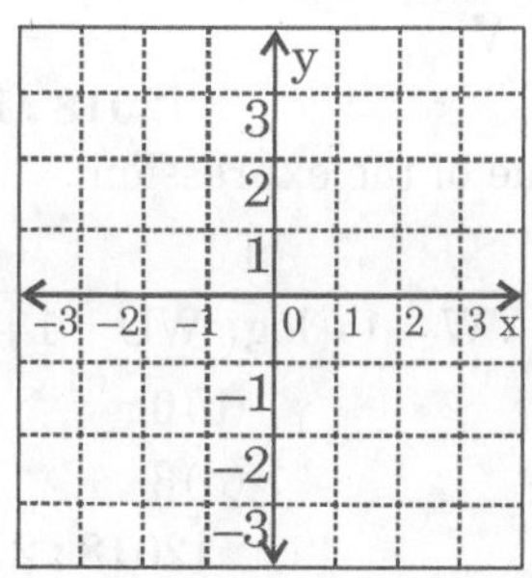

(a) $y = ||x|+1|-2$ (b) $y = ||x|-1|-1$

(c) $y = ||x|+1|-1$ (d) $y = ||x-1|-1|$

[2018 : 2 Marks, Set-1]

90. For integers a, b and c, what would be the minimum and maximum values respectively of $a + b + c$ if $\log|a| + \log|b| + \log|c| = 0$

(a) –3 and 3 (b) –1 and 1

(c) –1 and 3 (d) 1 and 3

[2018 : 2 Marks, Set-1]

91. From the time the front of a train enters a platform, it takes 25 seconds for the back of the train to leave the platform, while travelling at a constant speed of 54 km/h. At the same speed, it takes 14 seconds to pass a man running at 9 km/h in the same direction as the train. What is the length of the train and that of the platform in meters, respectively?

(a) 210 and 140 (b) 162.5 and 187.5

(c) 245 and 130 (d) 175 and 200

[2018 : 2 Marks, Set-1]

92. The perimeters of a circle, a square and an equilateral triangle are equal. Which one of the following statements is true?

(a) The circle has the largest area

(b) The square has the largest area

(c) The equilateral triangle has the largest area

(d) All the three shapes have the same area

[2018 : 2 Marks, Set-1]

93. Given that a and b are integers and $a + a^2b^3$ is odd, which one of the following statements is correct?

(a) a and b are both odd

(b) a and b are both even

(c) a is even and b is odd

(d) a is odd and b is even

[2018 : 2 Marks, Set-1]

94. Find the missing group of letters in the following series:

BC, FGH, LMNO

(a) UVWXY

(b) TUVWX

(c) STUVW

(d) RSTUV

[2018 : 1 Mark, Set-2]

95. The value of the expression

$$\frac{1}{1+\log_u VW}+\frac{1}{1+\log_V WU}+\frac{1}{1+\log_W UV} \text{ is —}$$

(a) –1 (b) 0

(c) 1 (d) 3

[2018 : 2 Marks, Set-2]

96. An unbiased coin is tossed six times in a row and four different such trials are conducted. One trial implies six tosses of the coin. If H stands for head and T stands for tail, the following are the observations from the four trials:

(1) HTHTHT (2) TTHHHT

(3) HTTHHT (4) HHHT ______.

Which statement describing the last two coin tosses of the fourth trial has the highest probability of being correct?

(a) Two T will occur

(b) One H and one T will occur

(c) Two H will occur

(d) One H will be followed by one T

[2018 : 2 Marks, Set-2]

97. A contract is to be completed in 52 days and 125 identical robots were employed, each operational for 7 hours a day. After 39 days, five-seventh of the work was completed. How many additional robots would be required to complete the work on time, if each robot is now operational for 8 hours a day?

(a) 50 (b) 89

(c) 146 (d) 175

[2018 : 2 Marks, Set-2]

98. Forty students watched films, A, B and C over a week. Each student watched either only one film or all three. Thirteen students watched film A, sixteen students watched film B and nineteen students watched film C. How many students watched all three films?

(a) 0 (b) 2

(c) 4 (d) 8

[2018 : 2 Marks, Set-2]

99. A wire would enclose an area of 1936 m^2, if it is bent into a square. The wire is cut into two pieces. The longer piece is thrice as long as the shorter piece. The long and the short pieces are bent into a square and a circle, respectively. Which of the following choices is closest to the sum of the areas enclosed by the two pieces in square meters?

(a) 1096 (b) 1111

(c) 1243 (d) 2486

[2018 : 2 Marks, Set-2]

100. A house has a number which needs to be identified. The following three statements are given that can help in identifying the house number.

1. If the house number is a multiple of 3, then it is a number from 50 to 59.
2. If the house number is NOT a multiple of 4, then it is a number from 60 to 69.
3. If the house number is NOT a multiple of 6, then it is a number from 70 to 79.

Which is the house number?

(a) 54 (b) 65

(c) 66 (d) 76

[2018 : 2 Marks, Set-2]

101. The sum and product of two integers are 26 and 165 respectively. The difference between these two integers is _____.

(a) 4 (b) 2

(c) 6 (d) 3

[2019 : 1 Mark, Set-1]

102. A worker noticed that the hour hand on the factory clock had moved by 225 degrees during her stay at the factory. For how long did she stay in the factory ?

(a) 8.5 hours

(b) 3.75 hours

(c) 4 hours and 15 minutes

(d) 7.5 hours

[2019 : 1 Mark, Set-1]

103. A person divided an amount of Rs. 100,000 into two parts and invested in two different schemes. In one he got 10% profit and in the other he got 12%. If the profit percentages are interchanged with these investments he would have got Rs. 120 less. Find the ratio between his investments in the two schemes.

(a) 9 : 16 (b) 47 : 53

(c) 11 : 14 (d) 37 : 63

[2019 : 2 Marks, Set-1]

104. A firm hires employees at five different skill levels P, Q, R, S, T. The shares of employment at these skill levels of total employment in 2010 is given in the pie chart as shown. There were a total of 600 employees in 2010 and the total employment increased by 15% from 2010 to 2016. The total employment at skill levels P, Q and R remained unchanged during this period. If the exmployment at skill level S increased by 40% from 2010 to 2016, how many employees were there at skill level T in 2016 ?

Percentage share of skills in 2010

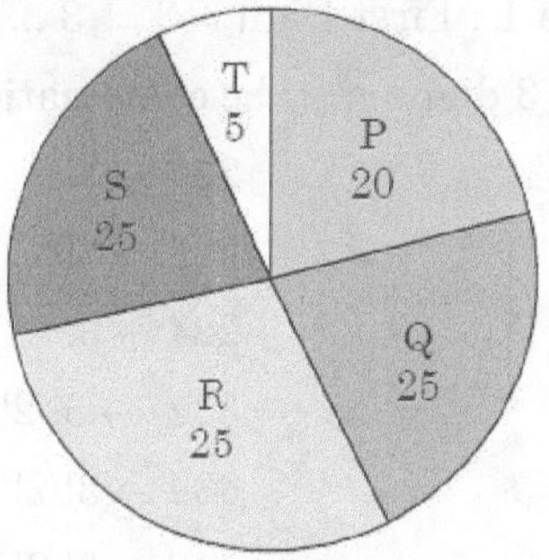

(a) 72 (b) 60

(c) 30 (d) 35

[2019 : 2 Marks, Set-1]

105. If IMHO = JNIP; IDK = JEL; and SO = TP, then IDC =

(a) JDC (b) JED

(c) JCD (d) JDE

[2019 : 1 Mark, Set-2]

106. The product of three integers X, Y and Z is 192. Z is equal to 4 and P is equal to the average of X and Y. What is the minimum possible value of P?

(a) 6 (b) 9.5

(c) 7 (d) 8

[2019 : 1 Mark, Set-2]

107. Fiscal deficit was 4% of the GDP in 2015 and that increased to 5% in 2016. If the GDP increased by 10% from 2015 to 2016, the percentage increase in the actual fiscal deficit is

(a) 25.00 (b) 37.50

(c) 35.70 (d) 10.00

[2019 : 2 Marks, Set-2]

108. Two pipes P and Q can fill a tank in 6 hours and 9 hours respectively, while a third pipe R can empty the tank in 12 hours. Initially, P and R are open for 4 hours. Then P is closed. The total time taken to fill the tank (in hours) is _____.

(a) 13.50 (b) 16.50

(c) 14.50 (d) 15.50

[2019 : 2 Marks, Set-2]

109. Mola is a digital platform for taxis in a city. It offers three types of rides – Pool, Mini and Prime. The table below presents the number of rides for the past four months. The platform earns one US dollar per ride. What is the percentage share of revenue contributed by Prime to the total revenues of Mola, for the entire duration?

	Month			
Type	January	February	March	April
Pool	170	320	215	190
Mini	110	220	180	70
Prime	75	180	120	90

(a) 23.97 (b) 16.24

(c) 38.74 (d) 25.86

[2019 : 2 Marks, Set-2]

ANSWERS

1. (d) **2.** (c) **3.** (d) **4.** (b) **5.** (b) **6.** (b) **7.** (d) **8.** (a) **9.** (d) **10.** (c)
11. (a) **12.** (a) **13.** (b) **14.** (b) **15.** (c) **16.** (c) **17.** (d) **18.** (c) **19.** (b) **20.** (b)
21. (b) **23.** (b) **24.** (b) **26.** (d) **27.** (b) **30.** (d) **31.** (a) **33.** (b) **34.** (d) **38.** (a)
40. (c) **41.** (b) **42.** (b) **44.** (b) **45.** (d) **46.** (b) **47.** (a) **48.** (c) **49.** (b) **51.** (b)
52. (c) **54.** (b) **55.** (c) **57.** (a) **58.** (c) **59.** (c) **60.** (c) **61.** (c) **62.** (b) **63.** (b)
64. (c) **65.** (c) **66.** (a) **67.** (c) **68.** (c) **69.** (b) **70.** (d) **71.** (d) **72.** (c) **73.** (b)
74. (d) **75.** (a) **76.** (b) **77.** (d) **78.** (c) **79.** (a) **80.** (d) **81.** (b) **82.** (c) **83.** (a)
84. (c) **85.** (b) **86.** (b) **87.** (b) **88.** (b) **89.** (b) **90.** (a) **91.** (d) **92.** (a) **93.** (d)
94. (b) **95.** (c) **96.** (b) **97.** (d) **98.** (c) **99.** (c) **100.** (d) **101.** (a) **102.** (d) **103.** (b)
104. (b) **105.** (*b*) **106.** (*c*) **107.** (*b*) **108.** (*c*) **109.** (*a*)

EXPLANATIONS

1.

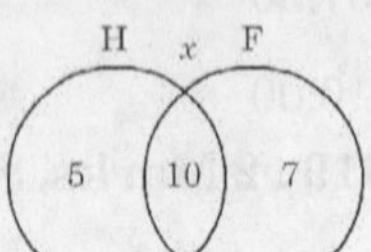

$\therefore \quad 5 + 10 + 7 + x = 25$

$\Rightarrow \quad x = 3$

2. By observation of summation, it can be seen that numbers are not in decimal. Let they have base n. Then converting all number in decimal

$$(137)_n = \left[1 \times n^2 + 3n + 7n^0\right]_{10}$$

$$(276)_n = \left[2n^2 + 7n + 6n^0\right]_{10}$$

$$(435)_n = \left[4n^2 + 3n + 5n^0\right]_{10}$$

$$(731)_n = \left[7n^2 + 3n + 1n^0\right]_{10}$$

$$(672)_n = \left[6n^2 + 7n + 2n^0\right]_{10}$$

$\therefore \quad (137)_n + (276)_n = (435)_n$

$\Rightarrow \quad n^2 + 3n + 7 + 2n^2 + 7n + 6n = 4n^2 + 3n + 5$

$\Rightarrow \quad n^2 - 7n - 8 = 0$

$\Rightarrow \quad n = -1, 8$

Possible base $n = 8$ (+ ve)

$(435)_n = (435)_8 = (4 \times 8^2 + 3 \times 8 + 5)_{10} = (285)_{10}$

$(731)_n = (731)_8 = (7 \times 8^2 + 3 \times 8 + 1)_{10} = (473)_{10}$

$(672)_n = (672)_8 = (6 \times 8^2 + 7 \times 8 + 2)_{10} = (442)_{10}$

So $(731)_8 + (672)_8 = (473)_{10} + (442)_{10} = (915)_{10}$

$(915)_{10} = (1623)_8$

3. 5 skilled workers build wall in 20 days

$\therefore$ 1 skilled worker build wall in 20×5 days

Hence in 1 day, part of work done by skilled work $= \dfrac{1}{100}$

Similarly in 1 day part of work done by semi-skilled workers

$$= \frac{1}{25 \times 8}$$

and in 1 day part of work done by un-skilled worker

$$= \frac{1}{30 \times 10}$$

So part of work done in 1 day by 2 skilled, 6 semi-skilled and 5 unskilled

$$= \frac{2}{100} + \frac{6}{200} + \frac{5}{300} = \frac{1}{15}$$

So work done by given workers in days = 15

4. Case 1 : First Digit = 3 ⇒ 3 ...

Rest 3 digits may be combination of

$234 \rightarrow 3!$

$223 \rightarrow 3!/2!$

$224 \rightarrow 3!/2!$

$332 \rightarrow 3!/2!$

$334 \rightarrow 3!/2!$

$442 \rightarrow 3!/2!$

$443 \rightarrow 3!/2!$

$444 \rightarrow 1$

$\therefore$ Total combinations = 25

Case 2 : First digit = 4 ⇒ 4 ....

Rest 3 digits may be combination of

$234 \rightarrow 6$

$223 \rightarrow 3$

$224 \rightarrow 3$

$332 \rightarrow 3$

$334 \rightarrow 3$

$442 \rightarrow 3$

$443 \rightarrow 3$

$444 \rightarrow 1$

$333 \rightarrow 1$

∴ Total combinations = 26

∴ Required answer = 25 + 26 = 51

5. (*i*) $H + G > I + S$

$\Rightarrow \quad I - H < G - S \qquad ...(i)$

(*ii*) $G - S = 1 \qquad ...(ii)$

From equations (*i*) and (*ii*), we get

$I - H < 1$

$\Rightarrow \quad H > 1$

Hence order is SGHIbecause G is not oldest and S is not youngest.

6. $$\log P = \frac{1}{2} \log Q$$

$$= \frac{1}{3} \log (R) = K$$

$$\therefore \quad P = b^k, Q = b^{2k}, R = b^{3k}$$

Now, $Q^2 = b^{4k} = b^{3k} b^k = PR$

7. $$10\left(\frac{10-1}{10}\right)^3 = 10\left(\frac{9}{10}\right)^3$$

$$= \frac{729}{1000}$$

$$\therefore \quad \frac{729}{1000} \times 1 = 7.29 \text{ litres}$$

8. Checking with all options in formula ($4q+100/q$) i.e. (V + F).

Option (*a*) gives minimum cost.

9. By observation of the table, we can say S

	P	Q	R	S
Requirement	800	600	300	200
Potency	0.4	0.5	0.4	0.8

10. Let each truck carry 100 units.

$2800 = 4n + e$

where, n = normal

$3000 = 10n + e$

where, e = excess/pending

$$\therefore \quad n = \frac{100}{3}$$

$$e = \frac{8000}{3}$$

$$\Rightarrow \quad 5 \text{ days} \rightarrow 500x = \frac{5 \times 100}{3} + \frac{8000}{3}$$

$$\Rightarrow \quad 500x = \frac{8500}{3}$$

$$\Rightarrow \quad x = \frac{17}{3}$$

$$\Rightarrow \quad x > 5$$

Minimum possible = 6

11. $$p = 50q - 5q^2$$

$$\therefore \quad \frac{dp}{dq} = 50 - 10q$$

$$\frac{d^2p}{dq^2} < 0$$

Hence p is maximum at $50 - 10q = 0$ or $q = 5$

Alternately check with options.

12. 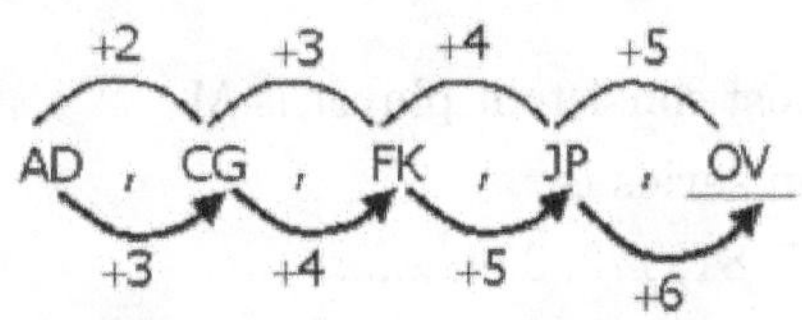

13. $$y = 2x - 0.1x^2$$

$$\frac{dy}{dx} = 2 - 0.2x$$

$$\frac{d^2y}{dx^2} < 0$$

Hence maximises at $2 - 0.2x = 0$

$$\Rightarrow \quad x = 10$$

$$\therefore \quad y = 20 - 10 = 10 \text{ m}$$

14.

	X	Y
Supply	60%	40%
Reliable	96%	72%
Overall	0.576	0.288

$$\therefore \quad P(X) = \frac{0.288}{0.576 + 0.288} = 0.334$$

15. P and R always holds true

Else consider a sample set {1, 2, 3, 4} and check accordingly.

16. $$S_n = \frac{n}{2}[2a + (n-1)d]$$

44, 42, 40, ------ 2, 0

n = 22 terms

17. $$P(A) = 1 - P(\bar{A})$$
$$= 1 - \frac{13}{90} = \frac{77}{90}$$

18. $$V_{avg} = \frac{\text{Total distance}}{\text{Total time}}$$

19. Multiply each term and divide each term by its conjugate.

20. Cost $\propto$ (wages $\times$ hours)
$$c \propto wh$$

21. Most consistent player is the one with least dispersion in the scores. For this, Coefficient of Variance is used to measure relative measure of dispersion between two sets.

Coefficient of Variance(CV) =

StandardDeviation/Mean

Calculating it for all the four players.

K's CV = 0.167

L's CV = 0.138

M's CV = 0.114

N's CV = 0.329

Larger the value of CV, less consistent is the player.

So the most consistent player is M.

22. The given series is

12, 35, 81, 173, 357, ...

The given series follows the following pattern

$12 \times 2 + 11 = 35$

$35 \times 2 + 11 = 81$

$81 \times 2 + 11 = 173$

$173 \times 2 + 11 = 357$

$357 \times 2 + 11 = 725$

23. By reading the instructions, it is clear that person who fills form T is a resident female with annual income of ₹ 9 Lakh.

24. The probability that the other face is tail is $\frac{1}{3}$.

25. Let the length of platform = x.m

and train length = 280 m (given)

According to Question

$$\frac{x + 280}{60} = \frac{280}{20}$$
$$\therefore \quad x = 560 \text{ m.}$$

26. **For 2005,**

trade deficit = (90 – 70) crores = 20 crores

Now $\frac{1}{5}^{th}$ of export = $\frac{1}{5}$(70) crores

= 14 crores $\neq$ 20 crores.

Hence option (*a*) is wrong.

For 2004,

Trade deficit = (80 – 70) crores

= 10 crores

Now $\frac{1}{5}^{th}$ of export = $\frac{1}{5}$(70) crores

= 14 crores $\neq$ 20 crores.

Hence option (*b*) is wrong.

For 2007

Trade deficit = (120 – 110) crores

= 10 crores

Now $\frac{1}{5}^{th}$ of export = $\frac{1}{5}$(110) crores

= 22 crores $\neq$ 10 crores.

Hence option (*c*) is wrong.

For 2006,

Trade deficit = (120 – 100) crores

= 20 crores

Now $\frac{1}{5}^{th}$ of export = $\frac{1}{5}$(100) crores

= 20 crores

= Trade deficts

Hence option (*d*) is correct.

27. For a very large number of throws, the frequency should be same for unbiased throw. As it not same, then the die is baised.

28. 2   3   6   15   [45]   157.5   630

   1.5   2   2.5   3   3.5   4

$\frac{\text{2nd number}}{\text{1st number}}$ is in increasing order as shown above

29. Eight consecutive odd number = 656

$a - 6, a - 1, a - 2, a, a + 2, a + 4, a + 6$

$a + 8 = 656$

$a = 81$

Smallest $m = 75$ ...(1)

Average consecutive even numbers

$$\Rightarrow \frac{a - 2 + a + a + 2 + a + 4}{4} = 87$$
$$\Rightarrow \quad a = 86$$

Second largest number = 88

$1 + 2 = 163$

30. Item : 2

$$\frac{\frac{20}{100}\times 250\times 10^7}{\frac{20}{100}\times 500\times 10^3}$$

$0.5\times 10^4 = 5\times 10^3 \boxed{1} = \text{Item 2}$

Item : 3

$$\frac{23\times 250\times 10^7}{19\times 500\times 10^3}$$

1.2 = Item 3

Item : 6

$\frac{19}{16} = 1.18 = \text{Item 6}$

Item : 5

$\frac{20}{12} = \frac{5}{3} = 1.6 \Rightarrow \boxed{1.6 = \text{Item 5}}$

31.

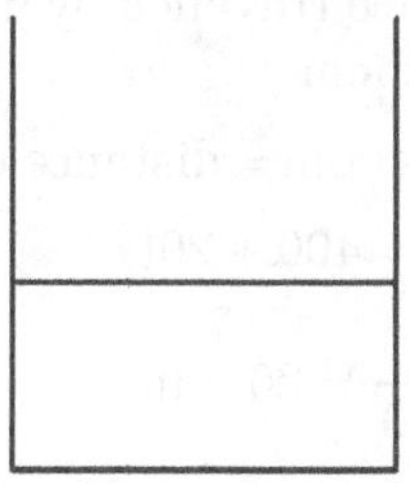

$V_{half} = 30(s)$ drawing rate = s

Total volume = 60 S tank

$(s^1)(10) - (s)10 = 30s$

$s^1(s) - s = 3s$

$s1 = 4s$

$s^1 = 4$ drawing rate

32. $81 - 54 = 27; 27 \times \frac{2}{3} = 18$

$54 - 36 = 18; 18 \times \frac{2}{3} = 12$

$36 - 24 = 12; 12 \times \frac{2}{3} = 8$

$\therefore\ 24 - 8 = 16$

33. In 7G - G is the 7th alphabet

11K - K is the 11th alphabet

13M - M is the 13th alphabet

Also, 7, 11, 13 are prime numbers, so next prime number will be 17 with correct answer as 17Q.

34. It is not mentioned that elephant is the largest animal.

35. Speed of man = 8;

Left distance = d

Time taken = $\frac{d}{8}$

Upstream: Speed of stream = s

$\Rightarrow$ speed upstream = S' = (8 – s)

$$t' = \left(\frac{d}{8-s}\right)$$

Downstream:

Given speed downstream = $t'' = \frac{d}{8+s}$

$\Rightarrow\ 3t' = t''$

$\Rightarrow\ \frac{3d}{8-s} = \frac{d}{8+s}$

$\Rightarrow\ \frac{3d}{8-s} = \frac{d}{8+s}$

$\Rightarrow$ s = 4 km/hr

36. Total expenditure = $\frac{15}{100}x = 4{,}50{,}000$

$x = 3 \times 10^6$

Profit = 10 lakhs

So, total selling price = 40,00,000 ...(1)

Total purifies = 200 ...(2)

S.P. of each purifier = (1)/(2) = 20,000

38. $P' = 2^n X^n \left(\frac{1}{2}\right)^m y^m$

$= 2^{n-m} X^n Y^m = 2^{n-m} P$

39. 8th observation is $7 \times 2 = 14$ more than 1st observation

9th observation is 14 more than 2nd observation

10th observation is 14 more than 3rd observation

11th observation is 14 more than 4th observation

12th observation is 14 more than 5th observation

Total $14 \times 5 = 70$

Sum of the first five numbers = 425

Sum of last five numbers = 495

40. In 13M - M is the 13th alphabet

17O - O is the 17th alphabet

19S - S is the 19th alphabet

Also, 13, 17, 19 are prime numbers, so next prime number will be 23 with correct answer as 23W.

41. KCLFTSB: BST-Best, F-Of, LCK-Luck (Reverse order)

SHSWDG: GD-Good, WSHS-Wishes (Reverse order)

Similarly "ace the Exam'- C-Ace, T-The, XM-Exam

42. 10 or more than 10 can't be the right answer because of compounding. Hence, 7.2 will be the right answer.

43.

	2012	2013
Transport (10%)	300,000	360,000
Labour (15%)	450,000	540,000
Raw material (20%)	750,000	780,000
Energy (25%)	750,000	900,000
Plant and Machinery (30%)	900,000	1,080,000
Total	**3,000,000**	**3,660,000**

Percentage increase in total cost = 22%

44. The digit in unit place is selected in 4! Ways

The digit in tens place is selected in 4! Ways

The digit in hundreds place is selected in 4! Ways

The digit in thousands place is selected in 4! Ways

The digit in ten thousands place is selected in 4! Ways

Sum of all values for 1

$4! \times 1 \times (10^0 + 10^1 + 10^2 + 10^3 + 10^4)$

$= 4! \times 11111 \times 1$

Similarly for '3' $4! \times (11111) \times 3$

Similarly for '5' $4! \times (11111) \times 5$

Similarly for '7' $4! \times (11111) \times 7$

Similarly for '9' $4! \times (11111) \times 9$

$\therefore$ sum of all such numbers

$4! \times (11111) \times (1 + 3 + 5 + 7 + 9)$

$= 24 \times (11111) \times 25$

$= 6666600$

45. For $M \Rightarrow \frac{12}{20} = 0.6$

$N \Rightarrow \frac{13}{25} = 0.52$

$O \Rightarrow \frac{20}{30} = 0.66$

$P \Rightarrow \frac{12}{25} = 0.48$

46. P (Ram) = 1/6; p(Ramesh) = 1/8

p(only at) = p (Ram) × p (not ramesh)

$+ p(\text{Ramesh}) \times p(n_0 \times R_{am})$

$= \frac{1}{6} + \frac{7}{8} + \frac{1}{8} \times \frac{5}{6}$

$\Rightarrow \frac{12}{40} = \frac{1}{4}$

47. For whole numbers, greater the value greater will be its log.

Same logic for power of *e*.

48. 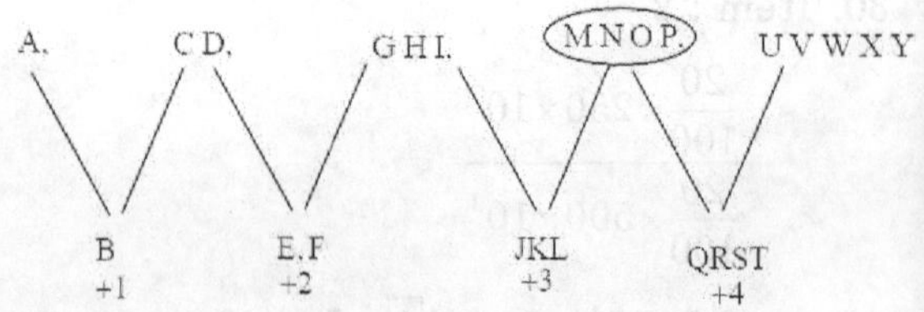

49. For a match to be played, we need 2 teams

No. of matches = No. of ways of selections 2 teams out of 5

$= 5_{c2} = 10$

50. Tiger – 1 leap ⇒ 8 meter

Speed = 5 leap/hr = 40m/min

Deer → 1 leap = 5 meter

speed = 4hr = 20m/min

Let at time '*t*' the tiger catches the deer.

∴ Distance travelled by deer + initial distance between them

50 × 8 ⇒ 400m = distance covered by tiger.

$\Rightarrow 40 \times t = 400 + 20t$

$\Rightarrow t = \frac{400}{200} = 20 \text{ min}$

$\Rightarrow$ total distance $\Rightarrow 400 + 20 \times t = 800$ ms

51. $(a + b + c)^2 >= 0$

$\Rightarrow a^2 + b^2 + c^2 + 2(ab + bc + ca) >= 0$

$\Rightarrow 1 + 2(ab + bc + ca) >= 0$

Thus, $(ab + bc + ca) >= -1/2$

Similarly,using:

$(a - b)^2 + (b - c)^2 + (c - a)^2 >= 1$

We will get $ab + bc + ca <= 1$.

Hence, [–1/2, 1] will be the right answer.

52. $\log \tan 1° + \log \tan 89° = \log(\tan 1° \times \tan 89°)$

$= \log(\tan 1° \times \cot 1°)$

$= \log 1 = 0$

Using the same logic total sum is '0'.

53. 90% of area of sheet = Cross sectional area of cone

$\Rightarrow 0.9 \times \pi \times 30 \times 30 = \pi \times r_1 \times 30$

$\Rightarrow 27 \text{ cm} = r_1$

$\therefore$ height of the cone $= \sqrt{30^2 - 27^2} = 13.08$ cm

54. x = {HHT, HHH}

y depends on x

z = {TTH, TTT}

∴ '*d*' is the correct choice.

55. We have the rectangle with dimensions 10 ×11 (10 horizontal dots and 11 vertical). PQ is parallel to y-axis and PR is parallel to x-axis.

Choose the (x, y) coordinates for vertex P (right angle): ${}^{10}C_1 \times {}^{11}C_1$

Choose the x coordinate for vertex R (as y coordinate is fixed by A): ${}^{9}C_1$, (10 – 1 = 9 as 1 horizontal dot is already occupied by A)

Choose the y coordinate for vertex Q (as x coordinate is fixed by A): ${}^{10}C_1$, (11 – 1 = 10 as 1 vertical dot is already occupied by A).

Hence, required number of triangles will be ${}^{10}C_1 \times {}^{11}C_1 \times {}^{9}C_1 \times {}^{10}C_1 = 9900$.

56. Let area of triangle PQR be 'A'

$$\frac{SQ}{PQ} = \frac{1}{1+3} = \frac{1}{4}$$

$$\frac{QT}{QR} = \frac{2}{2+5} = \frac{2}{7}$$

$\therefore$ Area of Δ^{le} QTS $= \frac{1}{2} \times SQ \times QT$

$$= \frac{1}{2} \times \left(\frac{1}{4} PQ\right) \times \left(\frac{2}{7} QR\right)$$

$$= \frac{1}{4} \times \frac{2}{7} \times \left(\frac{1}{2} \times PQ \times QR\right)$$

$$= \frac{1}{14} \times \text{Area of } \Delta^{le} \text{ PQR}$$

Given $20 \text{ cm}^2 = \frac{1}{14} \times A$

$\therefore$ $A = 14 \times 20 = 280 \text{ cm}^2$

57. Total no. of fruits = 5692000

Unripe type of apples = 45% of 15% of 5692000

$$= \frac{45}{100} \times \frac{15}{100} \times 5692000$$

= 384210

Ripe type of apples $= \frac{34}{100} \times \frac{85}{100} \times 5692000$

= 1644988

$\therefore$ Total number of apples

= 384210 + 1644988

= 2029198

58.

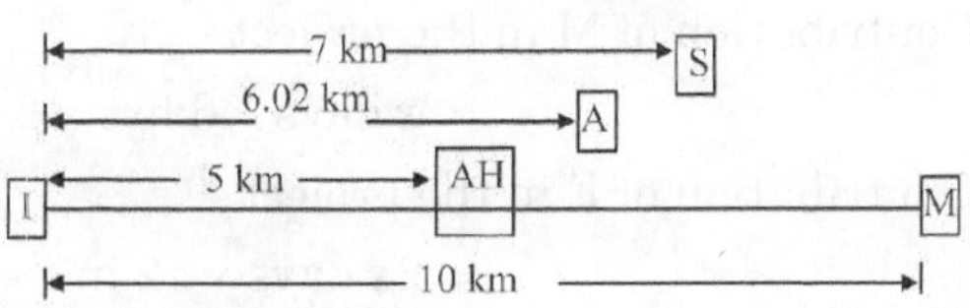

I = I live

AH = Ahmed lives

M = Michel lives

S = Susan lives

A = Arun lives

59.

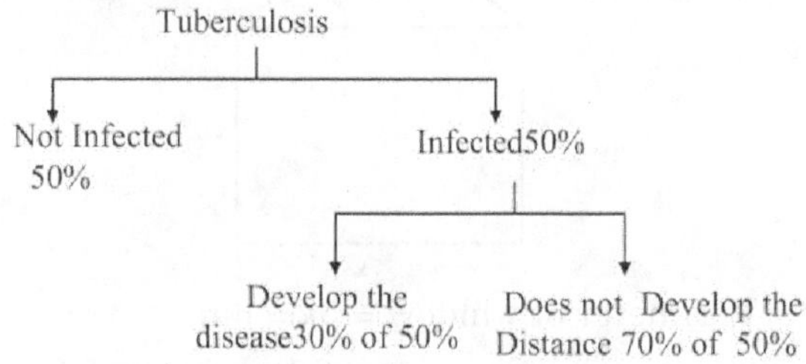

$$\frac{70}{100} \times \frac{50}{100} \Rightarrow \frac{35}{100} = 35\%$$

$\therefore$ Option (*c*) is correct

60. $q^{-a} = \frac{1}{r} \Rightarrow \frac{1}{q^a} = \frac{1}{r} \Rightarrow q^a = r$

$r^{-b} = \frac{1}{s} \Rightarrow \frac{1}{r^b} = \frac{1}{s} \Rightarrow s = r^b$

$s^{-c} = \frac{1}{q} \Rightarrow \frac{1}{s^c} = \frac{1}{q} \Rightarrow s^c = q$

$q^a = r \Rightarrow \left(s^c\right)^a = r \Rightarrow s^{ac} = r$

$\left(s^{ac}\right)^b = s \Rightarrow s^{abc} = s^1$

$\therefore$ $abc = 1$

$\therefore$ Option (*c*) is correct

61. Q can finish the task = 25 days, 12 hrs/day = 300 hrs

$\Rightarrow$ 1 hr $= \frac{1}{300}$th

R can finish the task = 50 days, 12 hrs/day

= 50 × 12 = 600 hrs

$\Rightarrow$ 1 hr $= \frac{1}{600}$th

Q working hours $\Rightarrow (7-2) \times 12 = 60$ hrs

R working hours $\Rightarrow 7 \times 18 = 126$ hrs

After 7 days, the ratio of work done by Q and R

Q	: R
$\frac{60}{300}$	$\frac{126}{600}$
20	: 21

62. 'Lesser than' is apt because the sentence should be in comparative degree.

63. The possible window will be as follows :

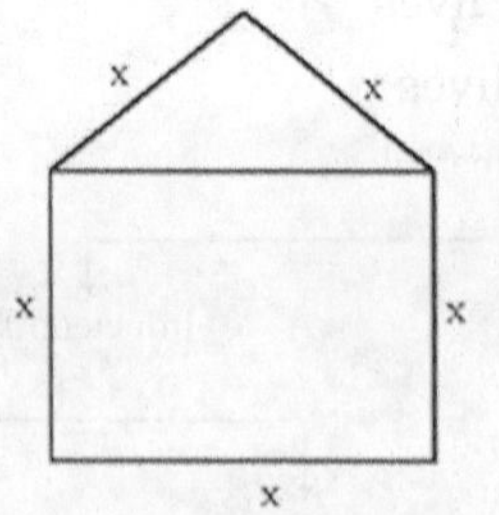

Perimeter of window = $5x = 6$m

$x = 1.2$ m

So area of window = Area of triangle + Area of square

$$= \frac{\sqrt{3}}{4}x^2 + x^2$$

$$= \frac{\sqrt{3}}{4}(1.2)^2 + (1.2)^2$$

$$= 2.06 \text{ m}^2$$

64. Group 'P'

Mean (μ) = 105

Standard deviation (σ_1) = 25

Pr ($\mu - \sigma \leq x \leq \mu + \sigma$) $\approx$ 0.6827

$\therefore$ 68% within one standard deviation

$$\mu_1 - \sigma_1 = 105 - 25 = 80$$

$$\mu_1 + \sigma_1 = 105 + 25 = 130$$

$\therefore$ range = 80 to 130

Distribution of P:

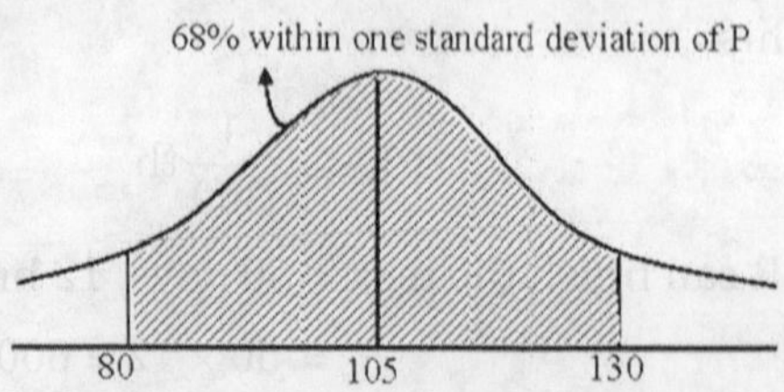

Group 'Q'

Mean (μ_2) = 85

Standard deviation (σ_2) = 5

Pr ($\mu - \sigma \leq x \leq \mu + \sigma$) $\approx$ 0.6827

$\therefore$ 68% within one standard deviation

$$\mu_2 - \sigma_2 = 85 - 5 = 80$$

$$\mu_2 + \sigma_2 = 85 + 5 = 90$$

$\therefore$ Range of Q in one standard deviation is 80 to 90

Distribution of 'Q'

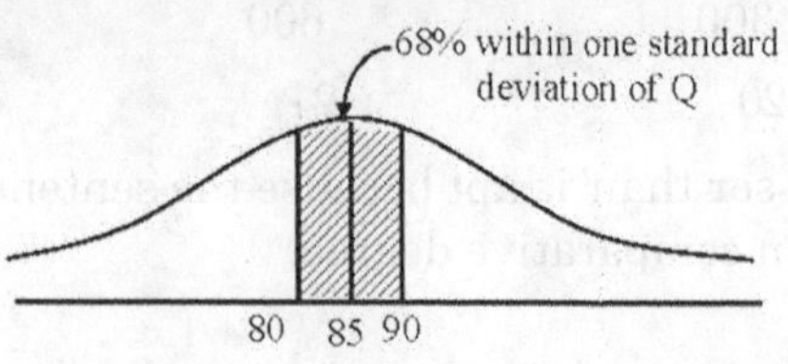

68% within one standard deviation of Q is narrower

$\therefore$ 68% within one standard deviation of Q means most students of group Q.

$\therefore$ Most students of group 'Q' scored marks in a narrower range than students in group 'P'

65. The following letter series is in the order of even letters series

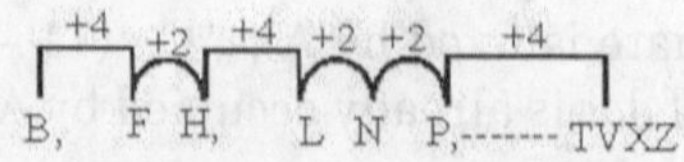

66. The binary operation – is defined

$\Rightarrow a - b = ab + (a + b)$

$a - x = a$

$\therefore$ From the equation 'b' is the variable

Option A: $x = 0$

$a - o = a \times 0 + (a + 0) = 0 + a = a$

Option B: $x = 1$

$a - 1 = a \times 1 + (a + 1) = a + a + 1 = 2a + 1$

Option C: $x = 2$

$a - 2 \Rightarrow a \times 2 + (a + 2) = 2a + a + 2 = 3a + 2$

Option D: $x = 10$

$a - 10 \Rightarrow a \times 10 + (a + 10) = 10a + a + 10$

$= 11a + 10$

$\therefore$ Option 'A' only True.

67. $f(x) = \ln(|e^{|\sin(|x|)|}|)$

1. $|x| \Rightarrow f(x)$ is Even. Hence, option (b) is not possible

2. $m = |\sin(|x|)| >= 0 \Rightarrow e^m >= 1$

3. $\Rightarrow f(x) = \ln(|e^m|) = \ln(e^m) >= 0$ {from 2}

Hence, option (a) is not possible

4. and $f(x)_{x=0} = 0$

Hence, option (d) is not possible

Therefore, correct option is (c).

68. 9 inches = 0.25 yards will be the right answer. Option (a) and (d) are wrong because we need to do the square root and square of not just the numbers but the units as well i.e. inches and yards can't remain the same.

69. M efficiency = 2 [efficiency of S, E and F]

Contribution of M in the project

$= x$ days $\times$ 6 hrs $\times$ 2

Contribution of E in the project

$= 2x$ days $\times$ 12 hrs $\times$ 1

Contribution of M : Contribution of E

$x \times 6 \times 2 : 2x \times 12 \times 1$

1 : 2

70.

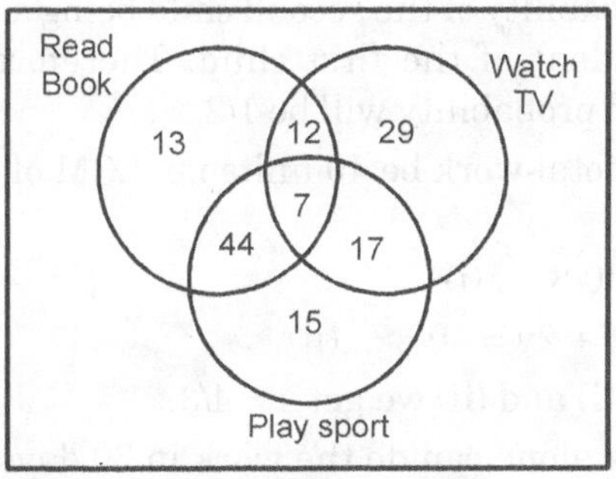

No. of person reading Books = 13 + 44 + 12

$n(A) + 7 = 76$

No. of person playing = 15 + 44 + 7 + 17

$n(B) = 83$

No. of person watch TV = 29 + 12 + 7 + 17

$n(C) = 65$

$n(A \cap B) = 44 + 7 = 51$

$n(A \cup B) = n(A) + n(B) - n(A \cap B)$

$= 76 + 83 - 51$

$= 108$

71.

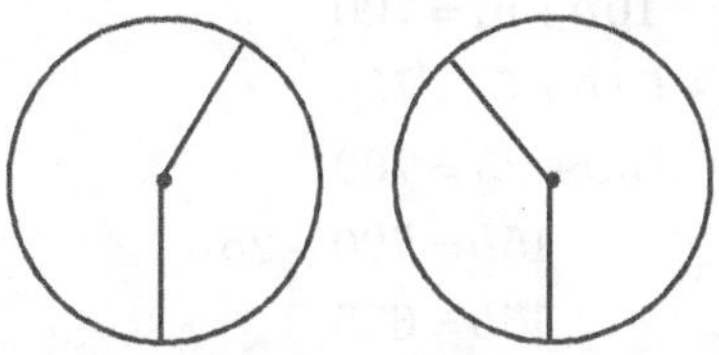

Time on mirror Image = 1:30 | Time on Actual Wall clock = 10:30

So present time = 10:30 + 2:15

= 12:45

72. From the given data, the following diagram is possible

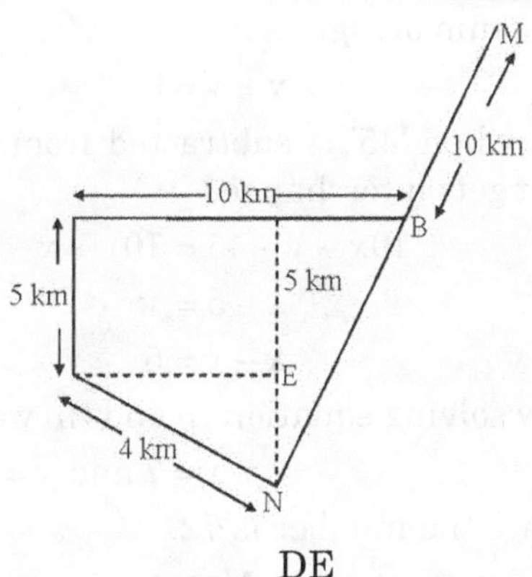

$$\cos 45° = \frac{DE}{4}$$

$$DE = \cos 45° \times 4 = 2.828 \text{ km}$$

$$\sin 45° = \frac{EN}{4}$$

$$EN = \sin 45° \times 4 = 2.828 \text{ km}$$

$$\sin 45° = \frac{EN}{4}$$

$$EN = \sin 45° \times 4 = 2.828 \text{ km}$$

$$CN = NE + CE = 2.828 + 5 = 7.828 \text{ km}$$

$$CB = AB - AC = 10 - 2.828 = 7.171 \text{ km}$$

$$(NB)^2 = (NC)^2 + (BC)^2 = (7.828)^2 + (7.171)^2$$

$$\therefore \quad NB = \sqrt{(7.828)^2 + (7.171)^2} = 10.616 \text{ km}$$

$$\therefore \quad NM = NB + BM = 10.616 + 10 = 20.61 \text{ km}$$

73.

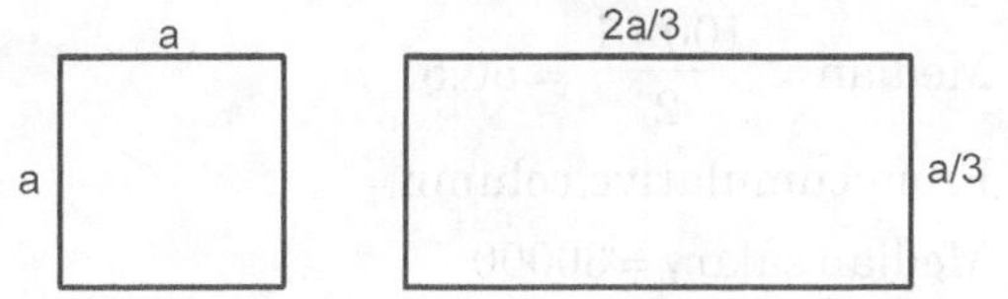

$a + b = 340$

$$\text{Perimeter of rectangle} = 2\left[\frac{a}{3} + \frac{2a}{3}\right] = 2a$$

$$\text{Length of square} = \frac{340 - 2a}{4}$$

$$\text{Total area} = \left(\frac{340 - 2a}{4}\right)^2 + \frac{2a^e}{4} = f'(a)$$

Condition for minimum area

$$f'(a) = 0$$

$$2\left(\frac{340 - 2a}{4}\right)\left(\frac{-a}{2}\right) + \frac{4}{9}a = 0$$

$$\frac{4}{9}a = \frac{1}{4}(340 - 20)$$

$$a = 90$$

$$\text{Length of the square} = \frac{340 - 2a}{4} = 40\text{mm}$$

74. Base radius = 5cm, Height = 12 cm. So Slant Height = 13cm.

If, we open it now then it appears as an arc with its radius = Slant Height of the cone = 13 cm.

Length of the arc = Circumfernce of the cone = 2×pie×5 = 10 pie cm

Now, when, length of arc is 2×p×13, angle subtended is 2pie.

So, angle subtended when length of arc is 10pie = 2pie/26pie × 10pie = 10pie/13.

75. According to P, S >= 3

According to Q, S < 3

According to R, S >= 1

From this we get that P's and Q's statements contradict each other while R supports the statement of P.

However, only one among them is right so it has to be Q when S has zero cars.

76.

Salary	No. of employees	Cumulative
2000	45	45
30000	25	70
40000	20	90
60000	8	98
150000	2	100

$$\text{Median} = \frac{100+1}{2} = 50.5$$

From cumulative column,

Median salary = 30000

77.

$$\begin{array}{r} 50000 \\ 48889 \\ \hline 1111 \end{array}$$

$sum = 0 + 0 + 0 + 0 + 8 = 8$

78.

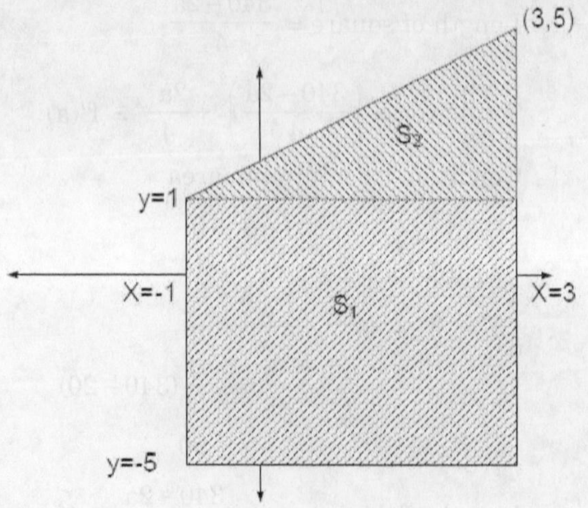

$$S = S_1 + S_2 = 6 \times 4 + \frac{1}{2} \times 4 \times 4$$

$$= 24 + 8 = 32$$

79. I is true as very obvious from the given graph.

II is false as the given graph shown only bacterial growth in milk but curd formation in milk may have other factors too like the purity of milk which might affect the speed of curd formation so II is false.

Hence, (a) is the correct answer.

80. As a – b is even so a and b will be both even or both odd. Hence, option (d) will be the right answer.

82. The probability of the second child being a boy is independent of the first child. Therefore, the required probability will be 1/2.

83. Let the total work be 40 units i.e. LCM of 8 and 5 days.

x + y = 40/8 .... (i)

and (x/2) + 2y = 40/5 ....(ii)

Solving (i) and (ii) we get x = 4/3.

Hence, x alone can do the work in 30 days

86. Let the initial length of rectangle is l and breath is b. Now if length is reduced by 10 m and breath is reduced by 5 m then it becomes square.

$\therefore \quad l - 10 = b - 5$

[∵ In square length = breadth]

$\therefore \quad l - b = 5$

Given, initial, Area (A) = $l \times b$

Now according to question,

$$(l - 10)(b - 5) = A - 650$$

$$\therefore \quad lb - 10b - 5l + 50 = A - 650$$

$$\therefore \quad A - 10b - 5l = A - 700$$

$$\therefore \quad 10b + 5l = 700$$

$$\therefore \quad 10b + 5(b + 5) = 700$$

$$\therefore \quad 15b + 25 = 700$$

$$\therefore \quad 15b = 700 - 25$$

$$\therefore \quad 15b = 675$$

$$\Rightarrow \quad b = 45 \text{ m}$$

$$\therefore \quad l = 45 + 5 = 50 \text{ m}$$

$\therefore$ Area of original rectangle (A) = $(45 \times 50)\text{m}^2$

$= 2250 \text{ m}^2$.

87. Let two digit number be xy.

Now sum of digit is 9

$\therefore \quad x + y = 9 \quad$ ....(*i*)

and when '45' is subtracted from number then digit gets interchanged

$$\therefore \quad 10x + y - 45 = 10y + x$$

$$\therefore \quad x - 5 = y$$

$x - y = 5 \quad$ ....(*ii*)

Now solving equation (*i*) and (*ii*) we get

$x = 7$ and $y = 2$

Hence the number is 72.

88. $$\frac{\text{Machine} \times \text{time}}{\text{Number of toys}} = \frac{M_1 \times t_1}{T_1} = \frac{M_2 \times t_2}{T_2}$$

$$\therefore \quad \frac{7 \times 7}{7} = \frac{100 \times t_2}{100}$$

$$\therefore \quad \boxed{t_2 = 7 \text{ min}}$$

89. Here from graph for x = 1, y = – 1

Checking the same from options.

From option (*a*)

for $x = 1$

$$y = ||1| + 1| - 2$$
$$= 2 - 2 = 0$$

∴ Option (*a*) is wrong

From option (*c*)

for $x = 1$

$$y = ||x| + 1| - 1$$
$$= |1 + 1| - 1 = 1$$

∴ Option (*c*) is wrong.

90. log |*a*| + log |*b*| + log |*c*| = 0

It is possible only,

when |*a*|, |*b*| and |*c*| all are equal to 1.

∴ *a, b, c* may be ±1, ±1, ±1 respectively.

Now for minimum value of all three will be negative.

∴ minimum value = - 3

and maximum value of all three will be positive.

∴ maximum value = + 3

91. Train speed (S_T) = 54 km/h

Time = 25 sec for travelling length of train and length of platform

Man speed (S_M) = 9 km/h

Speed of train to man = 45 km/h

Time = 14 sec

So, length of train = time × speed

$$= 14 \times 45 \times \frac{5}{18}$$

Length of train (L_T) = (35 × 5 m) = 175 m

Length of platform (L) + length of train (L_T) = speed × time

$$= 54 \times \frac{5}{18} \times 25 = 15 \times 25 = 375\text{m}$$

∴ Length of platform (L) = 375 - 175 = 200 m

92. The most symmetric figure will have the largest area. Hence, circle will be the right answer.

97. 125 × 7 × 39 = 5x/7

∴ x or total work = 25 × 39 × 49

Hence, [125 + y(additional robot)] × 8 × 13 = 25 × 39 × 49 × 2/7

Thus, the given data is insufficient to answer the question.

98. 13 + 16 + 19 = 48 > 40

Hence, 4 students would have watched all three due to which there is a double counting of 4 × 2 = 8.

99. Side of the square = 44 m

Long piece = 44 × 3 = 132 m and short piece = 44 m

Hence, equating these values with the perimeter of square and circle respectively we will get the required answer as 1243 m^2.

100. Either the number is a multiple of 3 or it is not. If it is then it is between 50 and 59, if it is not, then it is not a multiple of 6 either, and thus it is between 70 and 79.

Therefore we know it is not between 60 and 69, and therefore it is a multiple of 4.

If it is between 50 and 59, and is a multiple of 4 it must be 52 or 56. But neither of these are multiples of 3. Therefore the number is not between 50 and 59, but between 70 and 79.

Of the numbers between 70 and 79, it can be 72 or 76. Since 72 is a multiple of 6 however, this leaves 76 as the only possible solution.

101. From prime factorisation of

$$165 = 3 \times 5 \times 11$$
$$165 = 15 \times 11$$

So the two numbers whose sum is 26 and multiplication is 165 is 15 and 11

Now, difference = 15 – 11 = 4

102. Since, we know that

$$360° = 12 \text{ hours}$$

$$\therefore \quad 10 = \frac{12}{360} \text{ hours}$$

$$\therefore \quad 225° = \frac{12 \times 225}{360} \text{ hours}$$
$$= 7.5 \text{ hours}$$

103. Let the first part = x

So second part = 100,000x

According to question,

$$[0.1x + 0.12(100{,}000 - x)] - [0.12x + 0.1(1000{,}000 - x)] = 120$$

$$\Rightarrow 0.2x - 0.24x + 1200 - 1000 = 120$$

$$\Rightarrow \quad -0.04x = -1880$$

$$x = 47000$$

$$\frac{x}{100{,}000 - x} = \frac{47000}{53000} = \frac{47}{53}$$

104. 2010 2016

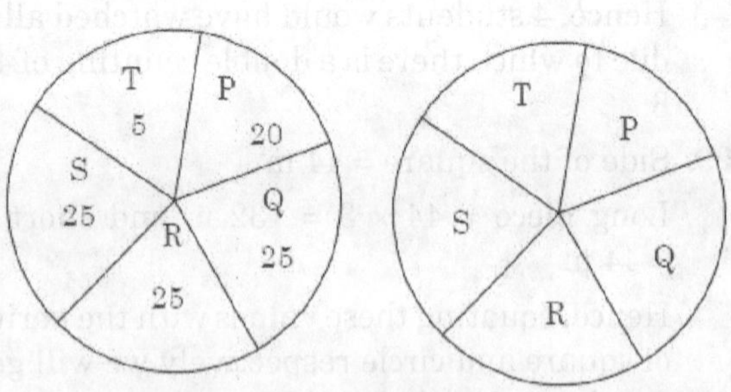

$$P = 20\% = 120\% \xrightarrow{\uparrow 15\%} 1.15 \times 600$$

$$P = 25\% = 150 \xrightarrow{\uparrow 0\%} 150$$

$$R = 25\% = 150 \xrightarrow{\uparrow 0\%} 150$$

$$S = 25\% = 150 \xrightarrow{\uparrow 40\%} 210$$

$$T = 5\% = 30 \longrightarrow ?$$

Now, employees at skill level T in 2016,

$= 690 - (120 + 150 + 150 + 210)$

$= 60$

105. For IMHO = JNIP

I + 1 = J

M + 1 = N

H + 1 = I

O + 1 = P

For

SO = TP

S + 1 = T

O + 1 = P

Similarly, for

IDC = JED

I + 1 = J

D + 1 = E

C + 1 = D

106. Given product of three integers,

X.Y.Z = 192

Given : Z = 4

$$\therefore X.Y = \frac{192}{4} = 48$$

$$\because\ P = \frac{X+Y}{2}$$

Now,

Possible value of (X, Y) are

(48, 1), (24, 2), (16, 3), (12, 4), (8, 6)

$$\text{So, } (P)_{min} = \frac{8+6}{2} = 7$$

So, min value of P is 7.

107. Let GDP in 2015 be x

Now, in 2016, GDP will be 1.1x

Fiscal deficit in 2015 = $0.04 \times x$

Fiscal deficit in 2016 = $0.05 \times 1.1\ x$

Here percentage increase in the actual fiscal deficit

$$= \left(\frac{0.05 \times 1.1x - 0.04 \times x}{0.04x}\right) \times 100 = 37.5\%$$

108. Here,

In 1 hr, part of tank filled by P alone is = $\frac{1}{6}$

In 1 hr, part of tank filled by Q alone is = $\frac{1}{9}$

In 1 hr, part of tank emptied by R is = $\frac{1}{12}$

Initially, P and R opened for 4 hr,

$$\text{So, part of tank filled} = 4\left(\frac{1}{6} - \frac{1}{12}\right) = \frac{4}{12} = \frac{1}{3}$$

After 4 hr, P is closed and for 6 more hr Q and R are opened.

$$\text{So, part of tank filled} = 6\left(\frac{1}{9} - \frac{1}{12}\right) = \frac{1}{6}$$

$$\text{So, total tank filled in 10 hr} = \frac{1}{6} + \frac{1}{3} = \frac{2+4}{12} = \frac{1}{2}$$

So, remaining 1/2 part of the tank is filled by Q alone

Now, Q will take 9/2 = 4.5 hr to fill remaining half of the tank.

Hence, total time taken = 4 + 6 + 4.5 = 14.5 hr

109. Here,

Revenue shared by pool = 170 + 320 + 215 + 190 = 895$

Similarly,

Revenue shared by mini = 110 + 220 + 180 + 70 = 580$

and,

Revenue shared by prime = 75 + 180 + 120 + 90 = 465$

Now, percentage share of revenue contributed by prime

$$= \frac{465}{895 + 580 + 465} = 23.97\%$$